Power Plant Engineering

Power Plant Engineering

Farshid Zabihian

CRC Press
Taylor & Francis Group
Boca Raton London New York

CRC Press is an imprint of the
Taylor & Francis Group, an **informa** business

First edition published 2021
by CRC Press
6000 Broken Sound Parkway NW, Suite 300, Boca Raton, FL 33487-2742

and by CRC Press
2 Park Square, Milton Park, Abingdon, Oxon, OX14 4RN

CRC Press is an imprint of Taylor & Francis Group, LLC

ISBN: 978-1-4987-0712-1 (hbk)
ISBN: 978-0-367-77396-0 (pbk)
ISBN: 978-0-429-06945-1 (ebk)

Typeset in Palatino
by codeMantra

Access the Support Material: https://www.routledge.com/9781498707121

To my wife and daughter, Niloufar and Diana, who

sacrificed so much so I can complete this book

and

To the memory of my father, who inspired most of what

I have done in my life, including this book.

Contents

Preface

Energy consumption in general and electricity consumption in particular are so intertwined with our daily life that often we do not even notice them. Electricity is perhaps the most important contributor to our standard of living and the economic development of the society we live in. Our daily life increasingly depends on electricity consumption. This dependency will continue in the foreseeable future. It is predicted that with the increase in the population and improvement in the quality of life on a global scale, global electricity consumption will continue to increase. While the decrease in global energy inequality is welcome news, inevitable environmental impacts should be concerning for all of us. The electricity generation industry is one of the most important contributors to global climate change as an existential threat to humanity's future on Earth. Addressing the environmental impacts of the power generation industry is a critical component of how we can mitigate climate change and other environmental problems that are threatening our future. While usually taken for granted, electricity generation, transmission, and distribution are incredibly complex and sophisticated. This complexity requires educating and training a new generation of fresh minds to address challenges we are currently facing. This book is written with this objective in mind.

While the transition from fossil fuels to renewable energy sources is currently underway, in the mid-term future, both of these energy sources are going to have a significant contribution to the energy mix in the U.S. and worldwide. As a consequence, typically individuals interested in a career in this industry are expected to be knowledgeable in both traditional and renewable power generation technologies. This text will provide the resources needed for this purpose.

This book is the result of teaching Power Plant Engineering courses in the past 10 years in various universities across North America. While there are very good books in the market that cover either conventional or renewable power generation systems, I had difficulty identifying a single book that covers all these topics. This textbook is an attempt to fill this gap.

Considering the wide variety of subject matters that have been covered in this book, authoring it required a wide range of expertise and knowledge in various fields. I have been fortunate that my industrial, research, and teaching experience on conventional power plants and renewable energy systems equipped me to cover both topics in this book. However, this diversity of topics may inherently lead to some inaccuracy in this book. I bear sole responsibility for the correctness, accuracy, and completeness of the selection of topics, their technical aspects, and my inevitable subjectivity toward their presentations. I look forward to receiving feedback from the readers to improve this book in future revisions.

Access the support material at https://www.routledge.com/9781498707121.

Acknowledgments

This book is an accumulation of over a quarter of a century of my involvement in the power generation industry in a wide range of capacities. Accordingly, many graduate and undergraduate students and colleagues from academia and industry have directly or indirectly contributed to what eventually manifested in this book. I am grateful to all of them!

It took me about 7 years to author this text and many individuals contributed to its completion. Here, I would like to name just a few with the most contribution. I would first like to thank Jonathan Plant, former Mechanical Engineering and Applied Mechanics Executive Editor at Taylor & Francis/CRC Press. He was the first one to see the value of what I envisioned for this text. He is the reason for the creation of this book. I have missed our annual rendezvous and enjoyable discussions at the American Society for Engineering Education (ASEE) annual conferences since he retired. I am also thankful for his understanding of what I was trying to achieve with this book and provided me with the flexibility to accomplish them.

My deepest gratitude to my friend of over 20 years, Mehdi Razzaghi. He has been my go-to source whenever I needed help in some topics. Our late-night phone conversations immensely contributed to this book. He also patiently proofread the entire manuscript and provided me with invaluable feedback.

My foremost appreciation is to Ideh Ghanbarpour, who painstakingly converted my chaotic drawings to the images you can see all over this book.

I am thankful to the entire team at Taylor & Francis and CRC Press, particularly Kyra Lindholm, Ed Curtis, Saranya PN, and Nishant Bhagat for making this book a reality.

Many thanks are due to the hundreds of students who have reviewed various chapters of this book in the past 10 years and provided me with their perspective and constructive criticisms, which helped me immeasurably to improve this book.

Finally, my ultimate gratitude to my best friend and wife Niloufar and my daughter Diana, who sacrificed so much for the creation of this book.

Author

Farshid Zabihian, PhD, has over 25 years of academic and industrial experience in the energy industry. He is currently an Associate Professor in the Department of Mechanical Engineering at California State University, Sacramento. He earned a MASc and BSc, both in mechanical engineering, at the Iran University of Science and Technology and Amirkabir University of Technology, Iran, respectively. He worked in the power generation industry for about a decade in various capacities before returning to academia. He earned a PhD in Mechanical Engineering in 2011 from Ryerson University, Canada. He was the recipient of the Governor General's Academic Gold Medal, the highest academic honor at the graduate level in Canada. He was previously an Assistant Professor at West Virginia University Institute of Technology between 2011 and 2016. He joined Sacramento State in 2016. His research focus is on sustainable electricity generation, including renewable energy resources and advanced fossil fuel power plants, through experimental and numerical approaches. He has published over 110 journal and conference papers, books, and book chapters. He also enjoys integrating sustainability into his teaching at the college level and beyond.

1

Introduction to Electricity

Abstract

In this first chapter, you will get familiar with the electrical infrastructure in general and the electricity generation industry in particular. We will learn the terminology that we will be using in later chapters. We will start with energy generation and consumption in general. We will discuss how energy consumption and economic development (and living standard) are related. We will also learn about energy-use sectors and primary and secondary energy sources. Then, we will briefly review the history of electricity. Next, we will take a look at how our life is intertwined with electricity. We will explore electricity consumption in several countries and energy inequality in the world. We will then learn about the components of electrical infrastructure, namely electricity generation, transmission, and distribution. We will introduce many parameters that can be used to characterize electricity generation and consumption and how they can be calculated. We will learn about electricity transmission lines and will conduct simple relevant calculations. We will discuss regulated and deregulated electricity markets and learn how to calculate the cost of electricity consumption. Finally, we will review some thermodynamic bases for power generation systems and their various categorizations.

LEARNING OBJECTIVES

After successful completion of this chapter, you will be able to:

- explain relationship between energy consumption and economic development (and living standard)
- identify energy-use sectors and primary and secondary energy sources
- explain a brief history of electricity
- describe relationship between electricity consumption, economic development, and energy inequality in the world
- describe components of electric infrastructure: electricity generation, transmission, and distribution
- define and calculate characteristics of electric load and power plants

- describe electricity transmission lines and conduct simple relevant calculations
- explain regulated and deregulated electricity markets
- calculate cost of electricity consumption
- explain some thermodynamic bases for power generation systems
- categorize various power generation technologies

1.1 Introduction to Energy and Society

The objective of every society is to sustain and improve the standard of living for its population, sustain and increase economic growth, and achieve the above two objectives sustainably. Energy in general and electrical energy in particular contribute significantly to all these objectives. Energy is intimately intertwined with every aspect of our life in a way that it is impossible to even imagine how our life could have been without it. We use energy to produce what we eat, to move around, to manufacture goods and transport them, to heat or cool our living environments to make them comfortable, to communicate with each other for pleasure or business, to operate equipment at home or work, and many other activities. In most societies there is a direct relationship between the standard of living and economic development and the level of energy consumption. We will start this chapter with a discussion on this relationship.

1.1.1 Economic Growth, Living Standard, and Energy Consumption

Considering the complexities of a modern society and its economic structure, in order to study an economy, we need to have a metric that can summarize all economic activities of the society in a single numerical parameter. **Gross domestic product (GDP)** is commonly used as a measure of economic development. GDP is a parameter that is determined based on the **market value of all products, including goods and services, produced in a country or a region in a particular period**, typically a year.

GDP per capita has been used as an indication of the quality of living in a society. However, some economists have criticized using GDP per capita as a measure of living standard because for example it does not take into account sustainability of the economy, environmental impacts of economic development, and inequality in the distribution of the generated wealth among the population.

The coupling between the living standard and energy consumption can be demonstrated by comparing these two parameters per capita basis. Figure 1.1 compares the annual GDP per capita and the annual primary energy consumption per capita in the U.S. between 1960 and 2018. As can be observed

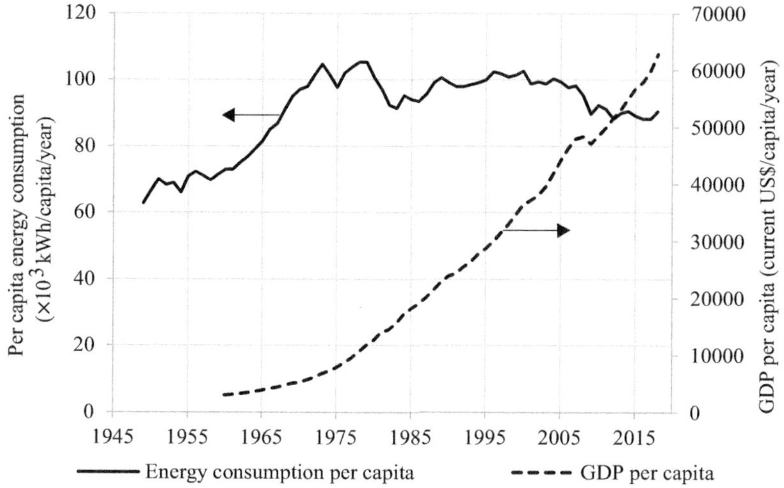

FIGURE 1.1
Annual GDP per capita and annual primary energy consumption per capita in the U.S. between 1960 and 2018. (Based on the data provided by the U.S. Energy Information Administration, EIA and the World Bank.)

from the diagrams, it seems there is no meaningful relationship between the two parameters. However, to clearly demonstrate the relationship, we need to compare the rate of change in these two parameters. Figure 1.2 shows the percentage of changes in both the annual energy consumption per capita and the annual GDP per capita compared to the prior year in the U.S. between 1960 and 2018. The diagrams clearly illustrate a relationship between the two parameters. Whenever one increases or decreases, the other follows the suit. Note that sometimes there are some delays and changes are not always in the same proportion. This dependency is undesirable, even dangerous, because we all want to maintain our living standard while we do not want to consume more energy with devastating impacts on the environment and our limited resources. This cycle is not sustainable.

A similar pattern can be observed in all countries. Figure 1.3 shows the per capita annual energy consumption as a function of the annual GDP per capita for many countries.[1] It can be seen that for most countries having a high annual GDP per capita means a high annual per capita energy consumption and vice versa. This is also evident by the fact that there is no country on the top left and bottom right corners of the diagram.

---

[1] The World Bank provides a great source of information and raw data. Its website contains data for hundreds of indicators (you can find these indicators at https://data.worldbank.org/indicator) for all countries for many years. You can choose any indicator and then download the data for all countries.

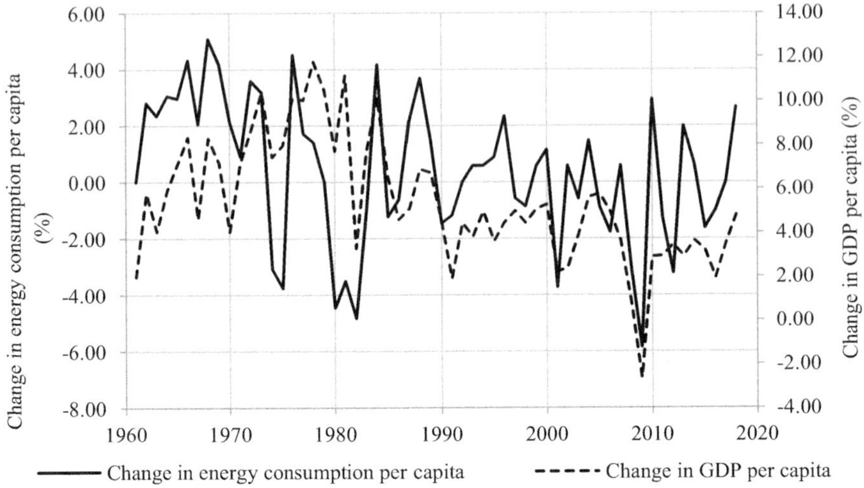

FIGURE 1.2
Percentage of changes in annual energy consumption per capita and annual GDP per capita in the U.S. between 1960 and 2018. (Based on the data provided by EIA and the World Bank.)

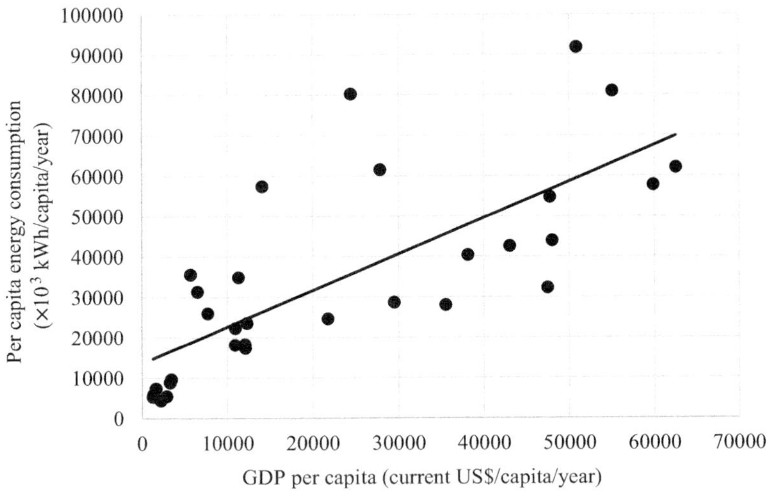

FIGURE 1.3
Relationship between annual GDP per capita and annual energy consumption per capita for various countries. (Based on the data provided by The World Bank.)

1.1.2 Energy Consumption

The energy we need in society to improve our living standard and maintain economic growth should be extracted or collected from natural resources of energy such as fossil fuels, nuclear energy, or renewable energy resources.

Furthermore, these natural energy resources are not typically in the form that we can use and are not located where we need them. We can use solar energy to heat our homes but cannot use it for cooling or for operating appliances. Fossil fuel resources and dams are typically far away from consumption centers. Also, their energy needs to be converted to suitable forms of energy such as electricity. Thus, we need an infrastructure to convert naturally occurring energy forms to more useful forms and to transport them. For example, crude oil is first extracted from oil wells and sent to refineries where gasoline (or petrol) and diesel fuels are produced. These fuels are then transported to gas stations for distribution. Finally, vehicles use these fuels to transport people and goods. In order to understand the energy industry, we need to look into three sectors:

- Energy use
- Energy conversion, transportation, and distribution
- Energy resources

1.1.3 Energy Consumption in Society

First, we need to take a look at energy usage in society. But before we discuss energy consumption, we need to understand the difference between primary and secondary energy sources to avoid the double counting of energy consumption. The **primary sources of energy** are forms of energy that can be directly found in nature. Fossil fuels, nuclear energy, hydro energy, biofuels, geothermal energy, solar energy, and wind energy are examples of primary energy sources. Sometimes these energy sources can be used without or with minimal energy conversion, for example when natural gas is used for heating purposes. However, in most cases, these energy sources need to be converted to useful energy forms. For example, coal, nuclear energy, or hydropower needs to be converted to electricity in power plants or crude oil needs to be processed into refined fuel oil. These converted or processed energy forms are called **secondary energy sources or energy carriers**. These energy sources cannot be directly found in nature and should be derived from primary energy sources. The distinction between primary and secondary energy sources is important in accurately reporting and analyzing energy consumption. For instance, one must be careful not to consider both fossil fuels used to produce electricity and the generated electricity in energy counting.

The global annual primary energy consumption between 1990 and 2017 has increased by about 60% since 1990 and continues to increase with no sign of meaningful reduction. It is predicted that the global annual primary energy consumption will increase at the rate of about 1.2% per year in the foreseeable future.[2]

[2] https://www.eia.gov/outlooks//ieo/tables_ref.php.

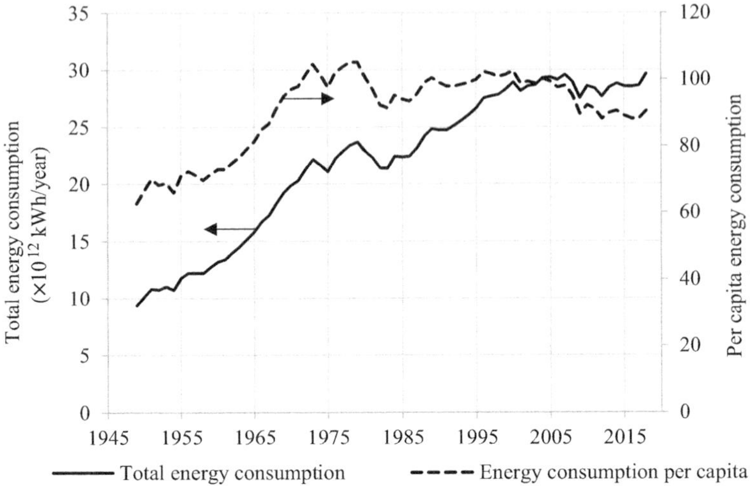

FIGURE 1.4
Total and per capita annual primary energy consumption in the U.S. between 1949 and 2018. (Based on the data provided by EIA.)

Figure 1.4 illustrates the annual primary energy consumption in the U.S. between 1949 and 2018 (solid line). The curve shows that the rate of increase in the energy consumption has decreased over time. In fact, the curve indicates that while the total energy consumption has increased significantly in the 20th century, in the past two decades the consumption has been fluctuating within a certain range without any net increase. This total energy consumption is directly proportional to the population. Thus, it can be very informative to study the per capita annual primary energy consumption. In Figure 1.4, the dashed-line curve presents the per capita annual primary energy consumption in the U.S. between 1949 and 2018. The curve shows that not only the net increase in per capita energy consumption in the U.S. stopped in the 1970s, but also the per capita energy consumption has markedly decreased in the new millennium.

1.1.4 Energy Consumption Sectors and Share of Each Sector

There are several ways to categorize energy consumption sectors. The most common categorization is to divide energy consumption into the following three major groups:

- **Building sector:** such as energy use for building heating and cooling, lighting, and operating appliances. This sector is typically divided into two subgroups:

    - Residential sector
    - Commercial sector
- **Industrial sector:** (manufacturing, agriculture, construction, mining, refining, etc.) which includes energy sources used as **input materials** such as crude oil used for plastic manufacturing or natural gas used for the production of fertilizers, and energy needed for **manufacturing processes**, such as energy used for heating materials or running machines.
- **Transportation sector:** to **move people and goods**.

Figure 1.5 illustrates the share of each of these sectors in the total primary energy consumption in the globe and in the U.S. in 2018. As the diagrams indicate, globally over half of the total energy is consumed in the industrial sector while in the U.S. the share of this sector is about one-third. In contrast, the share of the building sector is 40% in the U.S. compared to 20% in the world. Figure 1.6 shows the annual energy consumption in each sector in the U.S. between 1950 and 2018. The diagram shows that energy consumption in the industrial sector since the early 1970s and in other sectors since the early 2000s has not had a net growth and has been fluctuating within certain ranges. It also indicates that in the late 1990s, the energy consumption in the building sector exceeded that of the industrial sector.

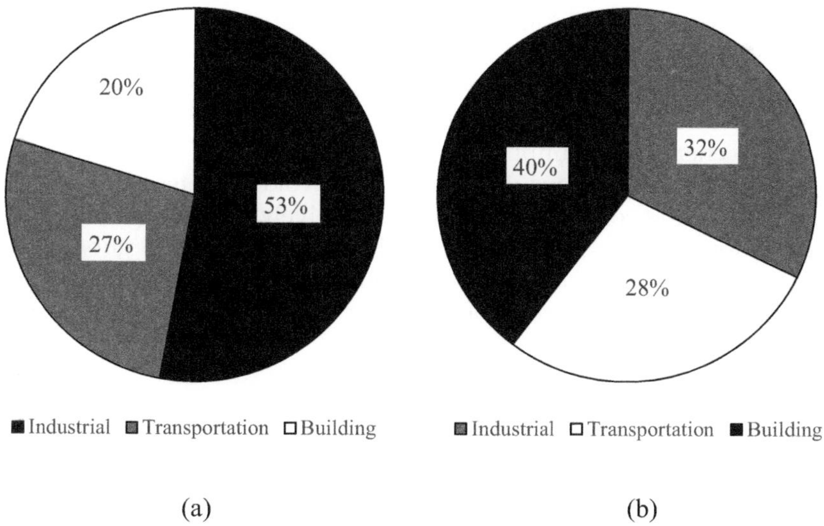

FIGURE 1.5
Share of various end-use energy consumption sectors (a) global in 2018 (b) the U.S. in 2019. (Based on the data provided by EIA (a), and International Energy Outlook 2019 (b).)

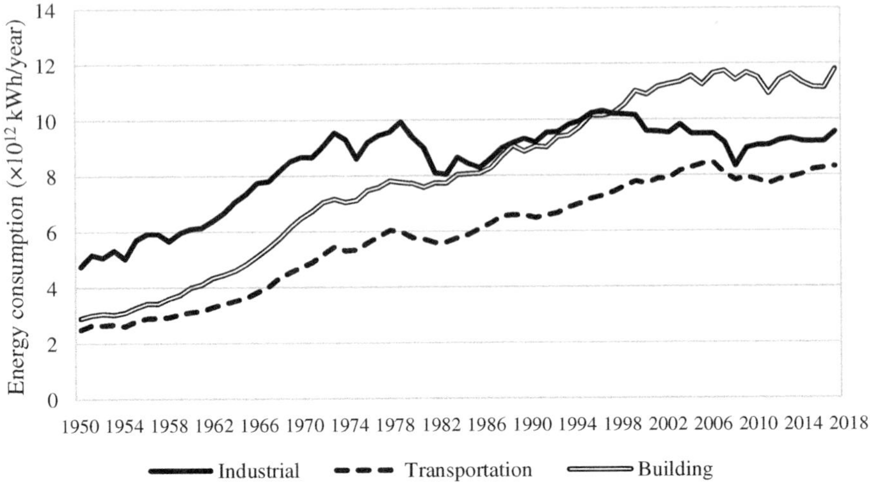

FIGURE 1.6
Annual energy consumption in the U.S. based on end-use sector between 1950 and 2018. (Based on the data provided by EIA.)

These variations in the amount of energy consumption and where energy is consumed in different countries and even different regions within a country primarily depend on the **level of economic development and living standard, economic structure, geography, and climate**. The relationship between energy consumption and economic development and the standard of living was discussed in the previous section. Economies focused on manufacturing tend to consume more energy in the industrial sector while in service-focused economies more energy is consumed by the building sector. Clearly, the energy consumption in the building sector strongly depends on the local climate. Regions with very hot or cold climates consume more energy on building cooling or heating, respectively. The energy used in the transportation sector is proportional to distances people and goods should be transported, which, in turn, depend on the size, the distribution of population, and the public transportation infrastructure and its popularity in the country or region of interest.

1.2 Primary Energy Sources

As noted earlier, a variety of primary energy sources are used to provide the energy demand of society. Figure 1.7 shows the contribution of fossil fuels, including oil, coal, and natural gas; biomass and waste; nuclear energy; hydropower; and renewable energy sources, including solar, wind, and geothermal

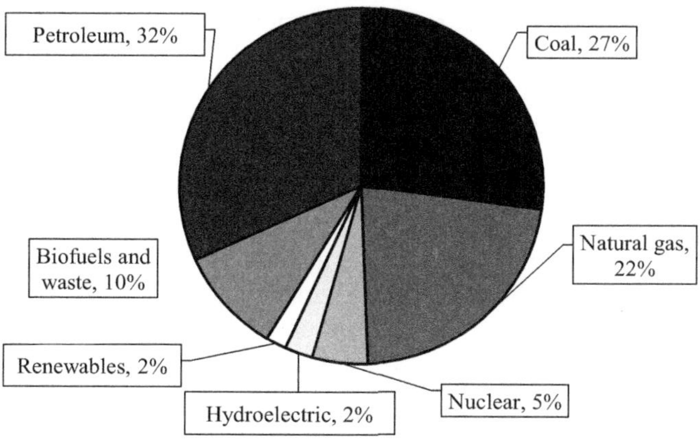

FIGURE 1.7
Global primary energy production sources in 2017. (Based on the data provided by IEA.)

energy, to the global primary energy consumption in 2017. The chart indicates that over 80% of the global energy consumption is provided by fossil fuels while all forms of renewable energy sources and nuclear energy are only responsible for about 15% and 5% of the global energy consumption, respectively.

Figure 1.8 shows the contribution of various primary energy sources to the global annual energy production between 1990 and 2017. The curves indicate that the consumption of oil and natural gas have been rapidly and continuously increasing since 1990. While the consumption of coal has slightly declined in recent years, in the first decade of the 21st century the coal consumption was

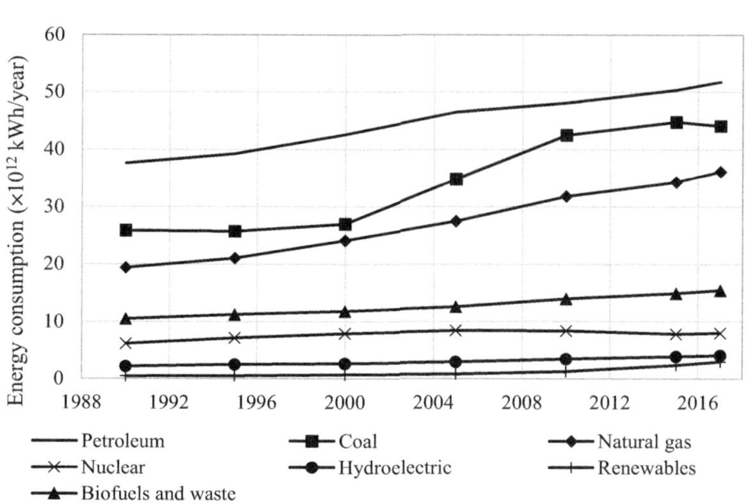

FIGURE 1.8
Global annual primary energy production sources between 1990 and 2017. (Based on the data provided by IEA.)

significantly increased. The share of other energy sources has been steady or has slowly increased. Also, it can be observed that while the contribution of renewable energy sources has increased in recent years, they are still responsible for a tiny percentage of energy consumption at the global level.

Figures 1.9 and 1.10 illustrate similar diagrams for the U.S. in 2018. The evaluation of these diagrams and the observation of variation in the contributions of various primary energy sources over many years as well as their comparison to the similar curves in the global scale are left as a practice for the readers. Also, Figure 1.11 shows the U.S. annual energy consumption by source and

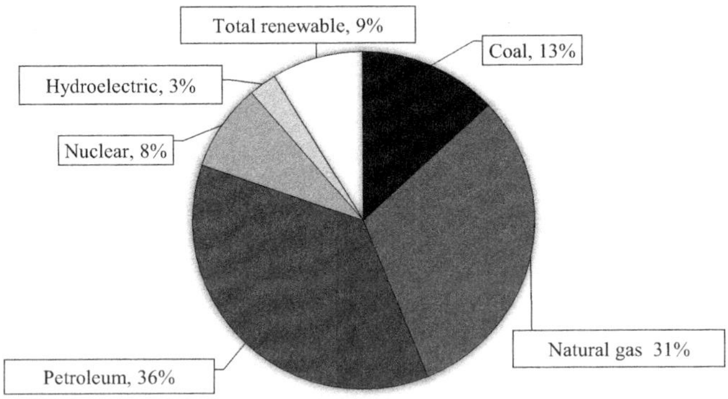

FIGURE 1.9
Share of various primary energy sources in energy consumption in the U.S. in 2018. (Based on the data provided by EIA.)

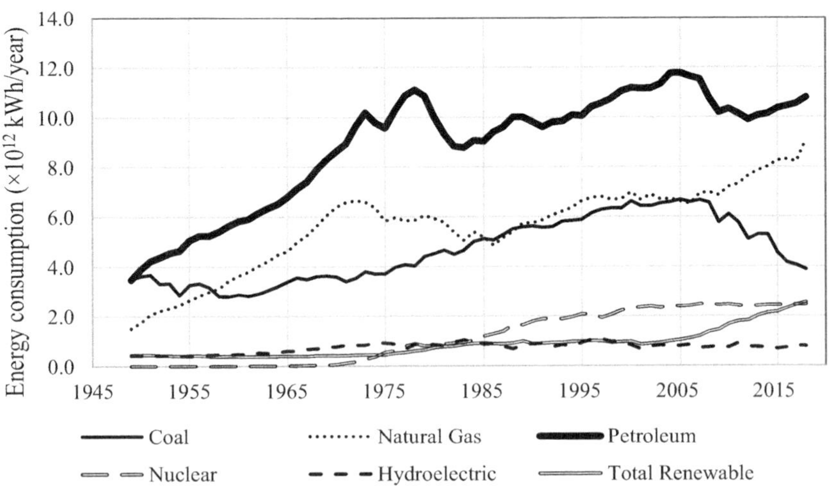

FIGURE 1.10
Annual primary energy consumption by sources between 1949 and 2018 in the U.S. (Based on the data provided by EIA.)

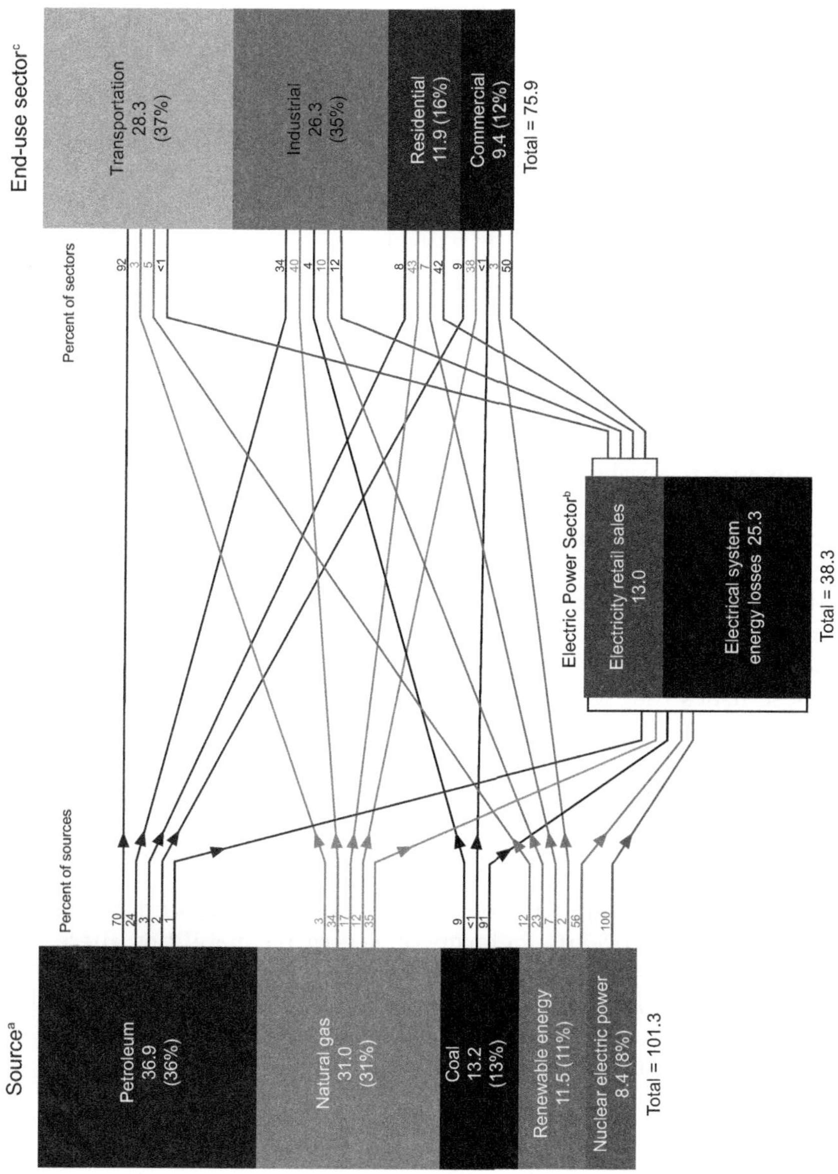

FIGURE 1.11

The U.S. energy consumption by source and end-use sector in 2018 (all values are in quadrillion Btu). (Source: The U.S. Energy Information Administration.)

end-use sector in 2018, provided by EIA.[3] This diagram indicates how each primary energy source (and electricity) contributes to various end-use sectors. The diagram provides a wealth of information for understanding the nature of the U.S. energy industry and infrastructure.

1.3 Introduction to Electricity

Electricity is indispensable to human beings comfort. Also, it is essential for productivity improvement and technological advancement without which growth is impossible. In the meantime, while the power generation industry has been responsible for a lot of damage to the environment, it has come a long way in the improvement of efficiency and the reduction of negative impacts on the environment.

1.3.1 Brief History of Electricity

Although this text is all about electricity generation, it does not deal with the field of electrical engineering itself. However, it is informative to know a little bit about the history of electricity. Probably the first time ancient people encountered a phenomenon related to electricity was when they accidentally touched electric fishes such as electric torpedo rays or electric catfishes (for a review on electric fishes and their amazing internal organs refer to FYI Box 1.1 and Moller (1995)[4]). Ancient texts described the effects of the electric shock of these fishes that was used to supposedly treat some diseases, an early form of electrotherapy.[5]

FYI BOX 1.1

ELECTRIC FISHES

Electric fishes can be found in the freshwaters of Africa and South America as well as in oceans. In terms of electrical capabilities, there are two types of fishes, electroreceptive and electrogenic fishes. Electroreceptive fishes, such as sharks, rays, and catfish, can only detect electric fields whereas electrogenic fishes, such as electric eels, electric catfishes, and electric rays, can generate electricity. About 1%

---

3 https://www.eia.gov/totalenergy/data/monthly/pdf/flow/css_2018_energy.pdf.

4 Moller, P. *Electric Fishes: History and Behavior*, Vol. 17. Berlin, Germany: Springer (1995).

5 Bullock, T.H., C.D. Hopkins, and R.R. Fay, eds. *Electroreception*, Vol. 21. Berlin, Germany: Springer Science & Business Media (2006).

of fish species are electrogenic. Electrogenic fishes use this capability for several purposes. In some species the voltage of produced electricity is high enough, from 10 to 600 V, to shock their prey or to be used for defensive purposes. These electrogenic fishes are classified as strongly electric fishes. Other electrogenic fishes produce very low-voltage electricity, <1 V, hence the name weakly electric fishes. They use this electricity to sense the environment, e.g., to navigate and detect objects, and to communicate with other fishes, e.g., to attract a mate.

How do these fishes generate electricity? We know that regular cells communicate with electrical impulses. In electric fishes, many modified nerve or muscle cells are interconnected in a dedicated electric organ, mostly in the tail. The way these cells are connected, in series or parallel, has evolved according to the salinity of water in their habitat, which determines the electrical resistance of water. In all cases, the result is the maximized delivery of electrical power to the target. Some electric fishes produce continuous and almost sinusoidal electricity called a wave-type discharge. Other fishes produce step waves that consist of brief pulses and long gaps called a pulse-type discharge. In order to protect their internal vital organs, these fishes have conductive tissues inside their body. The produced electricity goes through these low-resistance organs rather than high-resistance vital organs. Furthermore, to avoid jamming each other's electrical signals, they have a frequency control mechanism that adjusts the frequency of the produced electricity to avoid equal frequencies.[6]

Greeks knew about static electricity when they rubbed an amber rod (fossilized resin with a vegetable origin) with a cat's fur and realized that the rod can attract light objects. But mistakenly, they thought friction made amber magnetic. In fact, the word electricity was derived from the Greek word "elektron", which means amber. Until the 18th century, the accepted hypothesis about the origin of electricity was that this phenomenon had a mechanical nature.

Perhaps the oldest object that may be related to electricity generation is the Parthian Battery, also known as the Baghdad Battery, found in 1938, which belongs to the Persian dynasties of the Parthian or Sassanid period (the early centuries CE) (FYI Box 1.2).

[6] Nelson, M.E. "Electric fish." *Current Biology* 21(14) (2011): R528–R529.

FYI BOX 1.2

THE PARTHIAN BATTERY

The Parthian Battery is similar to a galvanic battery. It has been suggested that it may have been used for electroplating gold onto silver. Its reconstruction showed that it could generate around 1–2 V of electricity. Grape juice, lemon juice, or vinegar could have been used as an electrolyte. However, some scientists and archaeologists contradicted this theory and suggested that it is only a storage vessel for sacred scrolls.[7]

In the modern era, static electricity was studied for the first time in the 17th century. In the 18th century, Benjamin Franklin (1706–1790), one of the Founding Fathers of the U.S., conducted extensive research on electricity.[8] He conducted (and miraculously survived) his famous experiment to prove that the nature of lighting is electrical. In 1800, Alessandro Volta (Italian physicist, 1745–1827) invented the Volta battery or voltaic pile. Hans Christian Ørsted (Danish physicist and chemist, 1777–1851), André-Marie Ampère (French physicist and mathematician, 1775–1836), and James Clerk Maxwell (Scottish mathematical physicist, 1831–1879) showed the interrelation and eventually the unity of electric and magnetic phenomena. Based on this principle, Michael Faraday (English scientist, 1791–1867) invented electric motors in 1821, a concept that is still being used to convert shaft work to electrical energy and vice versa.

In the 19th century, electricity generation was evolved from a scientific field of research to the backbone of the Second Industrial Revolution, with the first one being the invention and utilization of steam engines. In the beginning, electricity was generated locally in the form of **direct current** (**DC**). Then, it was recognized that **alternating current** (**AC**) can be transported with lower cost and losses than DC. This recognition did not come easy. In the late 1880s, there was the so-called **War of Currents** between several giant historical figures of electricity: Thomas Edison (American inventor and businessman, 1847–1931) promoting DC and George Westinghouse (American entrepreneur and engineer, 1846–1914) along with Nikola Tesla (Serbian-American inventor and engineer, 1856–1943) supporting AC (everybody knows the outcome of the war!).[9] The domination of AC power made it possible to transmit electricity across great distances at high voltage, which resulted in large centralized power plants and a vast network of power distribution systems.

7 Frood, A. "Riddle of 'Baghdad's batteries'". *BBC News*, February 27, 2003. http://news.bbc.co.uk/2/hi/science/nature/2804257.stm.

8 It is interesting to know that he sold his personal possessions to fund his research.

9 Edison famously said "Fooling around with alternating currents is just a waste of time. Nobody will use it."

Since then electricity has been a fundamental building block of modern civilization. Until the 1960s, there were great interests and investments in the field of fossil fuel electricity production. Ever since the level of interest has depended upon the cost of fossil fuels. In the 1970s, the surge in oil prices resulted in huge investments in research and development (R&D) activities toward improving the efficiency of electricity generation and fostering renewable energy sources. However, these activities quickly declined when fuel costs were reduced.

The electricity generation industry, which was started from local generation and evolved to central power generation has gone through a complete circle. In recent years, the idea of local power generation, typically in the form of cogeneration, has gained popularity. This approach, also called **distributed generation (DG)**, is compatible with the move toward renewable energy sources.

Recently, there has been a new concern: the challenge of sustainability. Sustainable development, in general, means **meeting the needs of the current generation without destroying the future generation's ability to meet theirs**. Improving the living standard of a growing population can be only maintained by providing sufficient energy supplies. The consequences of the current approach to providing energy are evident not only in the declining of non-renewable energy resources but also in negative local impacts on air and water quality in addition to global climate change. In the U.S. and globally, the electricity generation industry is responsible for over a quarter of the total greenhouse gas (GHG) emission.[10]

While renewable energy sources will be the energy sources of the future, the short-term electricity needs should be provided mainly by conventional power generation units as sustainable as possible. This needs a two-frontal approach: reduce energy consumptions and better use of available energy sources. Energy conservation should be approached not only through advanced technologies but also through adjustment in our lifestyle. Better utilization of energy sources can be achieved partly by improving the efficiency of electricity generation. The electricity generation industry has experienced remarkable improvement in the efficiency of fossil fuel-based power plants from <15% in the early 20th century to above 50% (up to about 60%) in the 21st century. Furthermore, improved technologies for the extraction and processing of fossil fuels can extend their availability while reducing their negative environmental impacts.

1.3.2 Electricity and Society

In this section, the total and per capita electricity generation and consumption in the world, the U.S., and some other countries will be discussed and the relationship between this consumption and economic development and living standard will be evaluated. Figure 1.12 shows the global annual

[10] https://www.epa.gov/ghgemissions/inventory-us-greenhouse-gas-emissions-and-sinks.

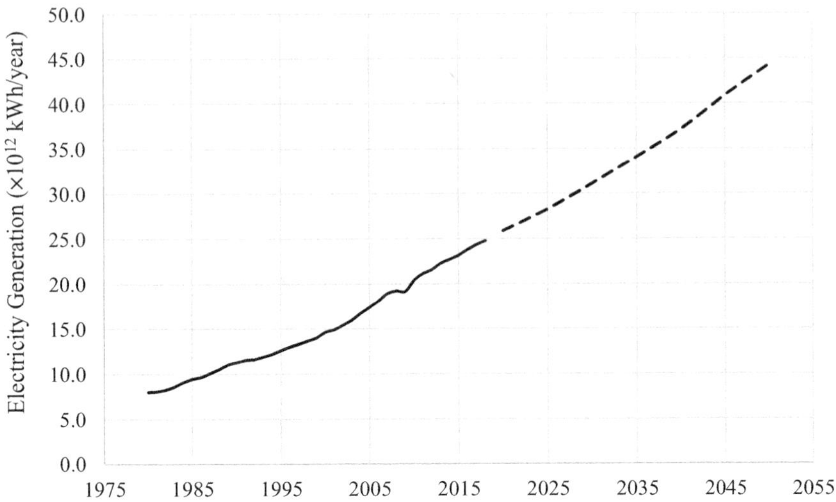

FIGURE 1.12
Global annual electricity generation between 1980 and 2050. (Based on the data provided by EIA, data before 2018 are from actual generation and beyond that are projection.)

electricity generation[11,12] between 1980 and 2050. The solid line is based on the actual electricity generation and the dashed line is based on the projected data. The curve indicates that in the past four decades the electricity generation has tripled. It is also expected to almost double by 2050 compared to the current consumption. It should be noted that while the electricity generation and consumption are slightly different, for our analyses (which is focused on trends rather than actual values) we can use them interchangeably because increase or decrease in electricity generation and consumption is always concurrent with a few percent difference.[13] In this discussion, energy generation is mostly used because it is what matters most from the utilization of resources as well as environmental impacts point of view.

[11] The electricity generated by both power plants and combined heat and power plants is included. Also, both utility and non-utility electricity generation are included.

[12] The reported electricity generation here is net generation (not gross generation). The difference between net and gross electricity generation is in whether we include the electricity consumed in power plants. The net electricity generation excludes the electricity consumed in the generating unit itself and only measures electricity at the outlet of the plant. The difference between these two depends on the type of the plant. It can be from <1% for hydropower plants to over 5% for fossil fuel power plants.

[13] Note that while in this text we present electricity generation or consumption interchangeably, they are not the same. Two factors distinguish electricity generation and consumption: electricity import and export and losses during electricity transmission and distribution as shown in the following equation:

Electric power consumption = net electricity generation (refer to the footnote above) + imported electricity − exported electricity − electricity losses during transmission and distribution.

Since electricity consumption is directly proportional to the population, it is much more meaningful to separately study the population and per capita electricity consumption (and generation). Figure 1.13 shows the world population between 1960 and 2018. The diagram indicates that the world population has been continuously increasing. The world population has increased from 3 to 7.6 billion people in this period, about 150% increase since 1960.

Figure 1.14 shows the global per capita electricity generation between 1980 and 2018. The diagram indicates that the global per capita electricity generation has increased by about 80% since 1980. The combined impact of increase

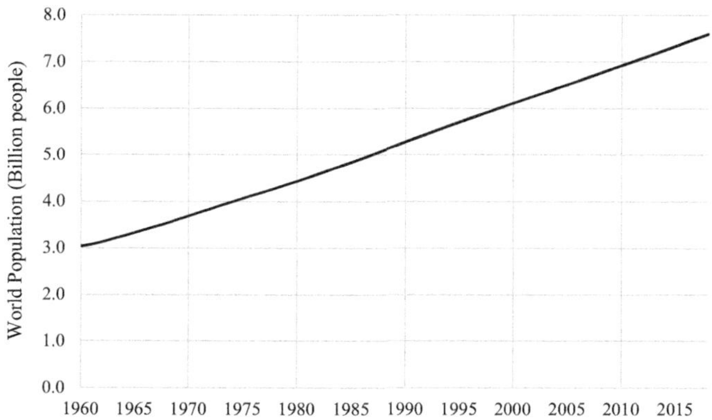

FIGURE 1.13
World population between 1960 and 2018 (Based on the data provided by The World Bank.)

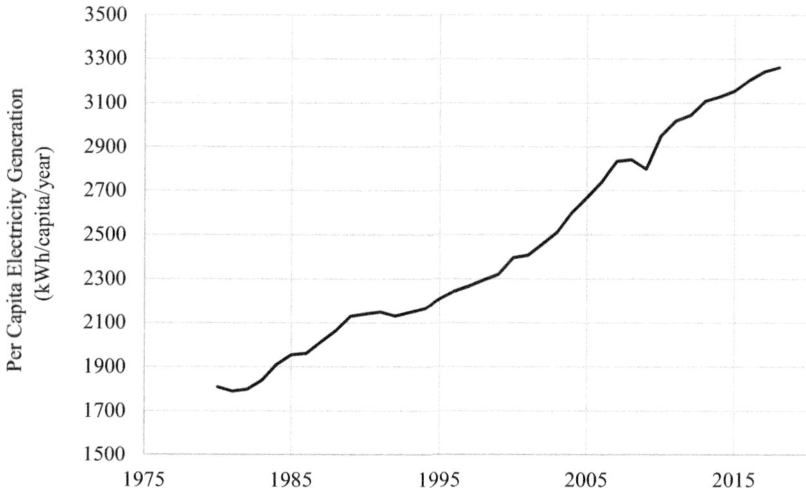

FIGURE 1.14
Global annual per capita electricity generation between 1980 and 2018 (Based on the data provided by EIA.)

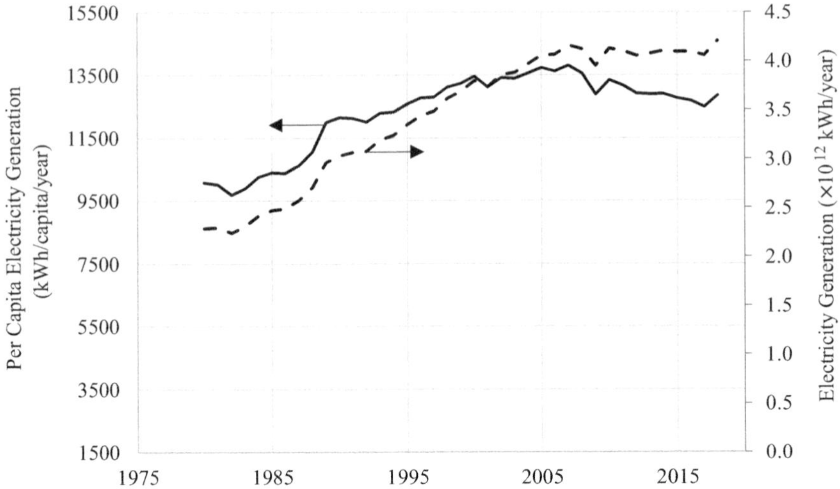

FIGURE 1.15
Annual total and per capita electricity generation in the U.S. between 1980 and 2018 (Based on the data provided by The World Bank.)

in electricity consumption by each person and increase in the world population has substantially increased the total global electricity consumption.

It is projected that both the world population and the per capita electricity consumption will continue to increase in the foreseeable future. The United Nation predicted that the world population will reach 9.8 and 11.2 billion by 2050 and 2100, respectively.[14] This means about 30% population increase by 2050. It is predicted that the per capita electricity generation will increase by 38% by 2050.[15] The combined impact will be about 78% increase in the global electricity generation (and consumption) by 2050.

Now let us take a look at the similar data for the U.S. and evaluate if we can observe similar patterns. Figure 1.15 illustrates the total and per capita annual electricity generation in the U.S. between 1980 and 2018. A close inspection of the curves shows a different pattern compared to those observed on the global scale. The curve indicates that the per capita electricity generation in the U.S. increased until 2000. Then, the growth stopped and the per capita electricity generation was flat at about 13,500 kWh/capita/year until 2007. Since then the annual per capita electricity generation has actually declined. However, the total electricity generation in the U.S. has been steady since 2005 and a similar decline cannot be seen for the total electricity production. This is due to the continued growth in the population as the U.S. population increased by about 6% between 2010 and 2018.

[14] https://www.un.org/development/desa/en/news/population/world-population-prospects-2017.html.

[15] https://www.eia.gov/international/data/world/electricity/electricity-generation.

If this diagram is created for all other industrially developed economies, the similar pattern can be observed. All of them **reached a steady annual per capita electricity generation albeit at different values and times**. For example, Figure 1.16 shows a similar diagram for Japan. A similar pattern can be seen but in this case, the annual per capita electricity generation growth stopped earlier, in the mid-1990s, and at a lower value, around 8000 kWh/capita/year. So the increase in the per capita and total global annual electricity generation observed in Figures 1.12 and 1.14 has not been driven by these economies.

Figures 1.17 and 1.18 illustrate similar diagrams for China and India (chosen for their immense population and rapid industrial development). The patterns observed for the U.S. and Japan in Figures 1.15 and 1.16, respectively, cannot be seen in these two cases. The per capita electricity generation in these countries is rapidly increasing. Note that regardless of this fast increase, still each person in China or India consumes far less than what an American or even a Japanese consumes (11,800 and 7500 kWh/capita/year for the U.S. and Japan, respectively, versus 5000 and 1100 kWh/capita for China and India, respectively).

This flattening of electricity consumption per capita in developed countries is both **a good news and a bad news**. It is a good news because it shows that there is a limit for electricity consumption per capita, and this consumption does not increase indefinitely. It is a bad news due to the fact that the majority of the world population lives in economies that have not reached to the level of economic development that results in flat or declining electricity consumption per capita. **For this underdeveloped population, the growth**

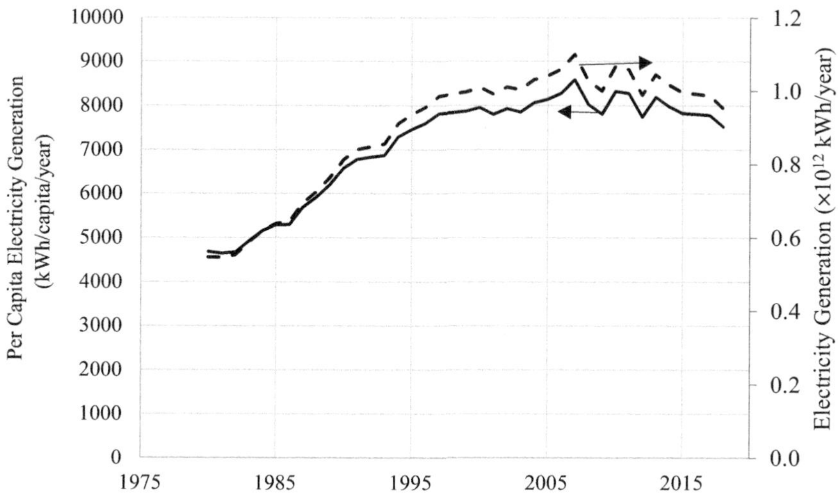

FIGURE 1.16
Annual total and per capita electricity generation in Japan between 1980 and 2018. (Based on the data provided by The World Bank.)

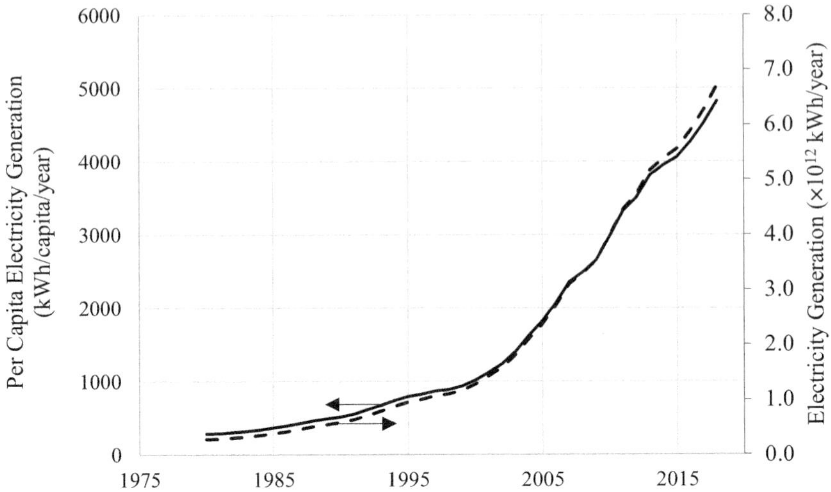

FIGURE 1.17
Annual total and per capita electricity generation in China between 1980 and 2018. (Based on the data provided by The World Bank.)

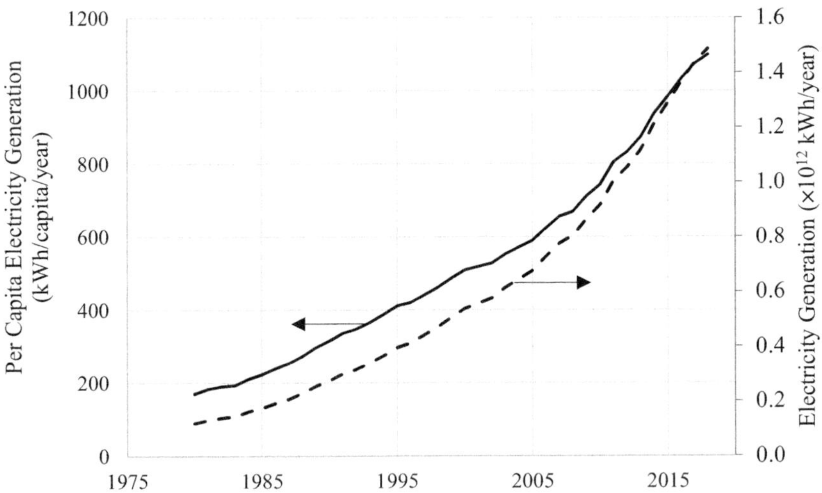

FIGURE 1.18
Annual total and per capita electricity generation in India between 1980 and 2018. (Based on the data provided by The World Bank.)

in electricity consumption will continue, along with improved living standard, until they are adequately developed. This means, as noted above, the total global electricity consumption will continue to grow in the foreseeable future. The impacts of this growth on the environment and energy sources are unpredictable but certainly not positive and potentially devastating.

We will discuss this issue further in the next section, which is about the relationship between electricity consumption, economic development, and energy inequality.

1.3.3 Electricity Consumption, Economic Development, and Energy Inequality

In Section 1.1.1 we observed that per capita energy consumption can be used as an indication of the quality of life in a society. However, electricity consumption per capita is a more accurate measure of living standard. Consider a society that burns a lot of wood for heating and cooking. While this society may consume a lot of energy, this consumption may not reflect a high living standard. Figure 1.19 shows the annual per capita GDP as a function of the annual per capita electricity generation for countries with a population over 20 million people in 2018. The figure illustrates a clear correlation between these two factors though not a one-to-one relationship. There is a significant variation in the diagram. For example, for the annual per capita GDP of 40,000 $/capita/year, the annual per capita electricity generation of <5000 kWh/capita to over 10,000 kWh/capita can be observed. However, the lack of any point in the top left and bottom right corners of the diagram indicates that the high living standard (manifested in a high GDP per capita) cannot be achieved without high electricity generation (and consumption) per capita.

To get some sense of electricity generation per capita in various countries, Figure 1.20 shows the annual per capita electricity generation for some countries in 2018. The diagram shows how immensely the annual per capita

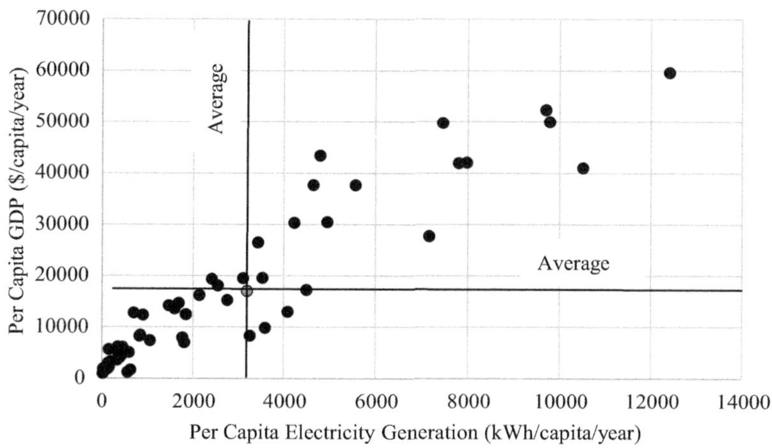

FIGURE 1.19
Annual per capita GDP as a function of per capita electricity generation for countries with population over 20 million people – 2018. (Based on the data provided by The World Bank.)

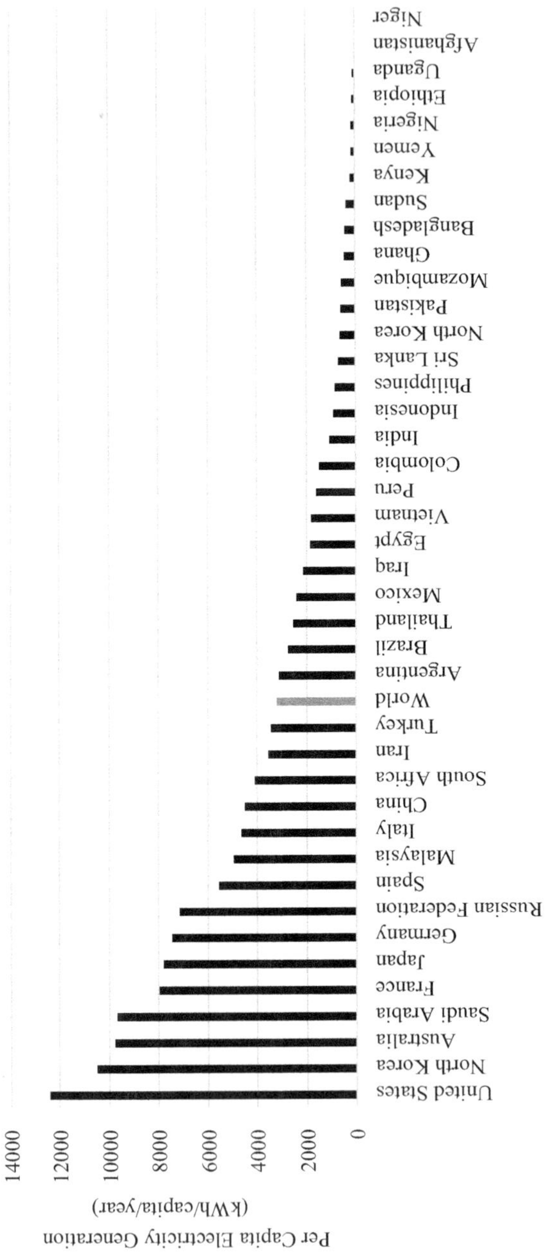

FIGURE 1.20
Annual per capita electricity generation for some countries in 2018. (Based on the data provided by The World Bank.)

electricity generation varies among developed, emerging, and developing economies.

To get a clearer picture, Table 1.1 lists the population and annual electricity generation, GDP, per capita electricity generation, and per capita GDP for some selected countries in 2018. The table indicates **a deep inequality in the per capita electricity consumption in the global level.** For example, each American on average consumes roughly as much as two Spaniards, three Chinese, 12 Indians, 63 Kenyans, 365 Afghans, and 827 Chadians. The good news is that this gap has been closing for most countries in the past few decades. For example, the ratios of annual per capita electricity consumption in the U.S. to that of China and India were about 50 and 77, respectively, in the early 1970s. These ratios reduced to about 14 and 35, respectively, in 2000[16] and about 3 and 12 in 2018. The bad news is that according to the World Bank, only about 16% of the world population lives in the high-income economies, and remaining 84% of the population lives in low- and middle-income countries (for example China

TABLE 1.1

Population and Annual Electricity Generation, GDP, per Capita Electricity Generation, and per Capita GDP for Some Selected Countries in 2018. (Based on the data provided by The World Bank.)

Country	Population (Million People)	Annual Electricity Generation (Billion kWh/year)	Annual GDP (Billion $/year)	Annual Electricity Generation per Capita (kWh/capita/year)	Annual GDP per Capita ($/capita/year)
Norway	5.3	146.5	376	27,575	70,672
Canada	37.1	640.5	1,743	17,284	47,041
Kuwait	4.1	68.4	289	16,542	69,925
United States	327.2	4,058.3	19,517	12,404	59,655
Germany	82.9	618.3	4,128	7,456	49,781
Spain	46.7	259.9	1,761	5,563	37,705
China	1,392.7	6,266.2	24,007	4,499	17,237
World	7,594.3	24,343.6	128,866	3,206	16,969
Mexico	126.2	305.1	2,439	2,418	19,325
India	1,352.6	1,438.2	10,025	1,063	7,412
Bangladesh	161.4	68.8	721	427	4,471
Kenya	51.4	10.1	169	197	3,285
Ethiopia	109.2	13.7	208	126	1,906
Haiti	11.1	1.0	20	86	1,767
Afghanistan	37.2	1.3	70	34	1,893
Chad	15.5	0.2	29	15	1,850

[16] Based on the data from the World Bank.

FIGURE 1.21
Night view of Earth created by NASA through compiling of over 400 satellite images. (Source: NASA, https://www.nasa.gov/sites/default/files/images/324350main_11_full.jpg.)

and India are upper and lower middle-income countries, respectively[17]). This means, as noted earlier, the rapid growth in electricity consumption will continue with unknown consequences.

Figure 1.21 illustrates a night view of Earth created by NASA through compiling of over 400 satellite images. The concentration of lighting shown in the figure can be used as an indication of electricity consumption in various parts of the world. As expected, North America, West Europe, and parts of East Asia are much brighter than Africa and most parts of Asia and South America.

1.4 Structure of Electrical Infrastructure

Modern electrical infrastructure is composed of three main components, which are integrated into a complex system: **power generation**, **transmission**, and **distribution**. Power generation units are facilities that convert various forms of energy, such as coal, natural gas, nuclear energy, solar energy, and wind energy, to electrical energy. A transmission system is a network of electricity transmission lines, either overhead or underground, which transmit high-voltage electricity produced in power plants to substations near consumption centers,

[17] https://data.worldbank.org/indicator/SP.POP.TOTL.

typically over a long distance. These substations receive high-voltage electricity from transmission lines and reduce the voltage through transformers. The low-voltage electricity is then distributed to end-users by distribution systems. A distribution system is mainly composed of power lines and transformers to safely and reliably deliver the required quantities of electricity at a specified voltage to consumers. Generally, power lines with voltages above 69 kV are considered transmission lines and those below 69 kV as distribution lines.

The focus of this text is on power generation systems. Transmission and distribution systems are out of the scope of this text. However, since transmission systems are the interface between power plants and the rest of the electrical infrastructure, a brief discussion on this topic will be presented here. Also, it is critical to understand load characteristics because it provides information about customer's needs.

1.4.1 Characteristics of Load Demand for Electricity

The most important characteristic of electricity, which distinguishes it from other commodities produced by human beings is the fact that in many cases **electricity cannot be economically stored in a meaningful quantity**. Note that there are several ways to store electrical energy, the most common being batteries, but these methods are neither efficient nor economically viable yet in many locations (the entire Chapter 16 is dedicated to this topic). This means that for the most part electricity should be generated whenever needed and transmitted and distributed to wherever it is needed. This unique characteristic has a profound impact on the selection of technology, capacity, configuration, and geographical location of power generation systems to meet demand loads. The load is the power consumed by end-users to produce work, heat, and light. So depending on the context, the term load may refer to the electrical power consumption of a TV or an entire nation. The electrical demand of consumers, and as a result, the total demand vary over a given period of time and involve significant uncertainty. Therefore, the analysis of the electrical demand is a complex process. This section deals with this process by introducing several basic concepts and how they affect the electricity generation industry.

To analyze the electrical demand, the pattern of variation of electricity consumption should be studied. The scope of this study could be regional or national, depending on the objective of the study. It starts with a load curve, also known as a load profile, which is a graph of the electrical load (power consumption) as a function of time. The graph is typically based on a 24-hour period and is prepared for every day for several years and then averaged to get general patterns. Figure 1.22 shows a sample of a load profile. The variation in this graph depends on the time of the day; customer type, e.g., residential, commercial, or industrial; the season of the year (summer or winter); ambient temperature; the day of the week (weekdays or weekends); and special occasions, such as holidays or sports events. In order to take these variations into account in a single graph, they can be

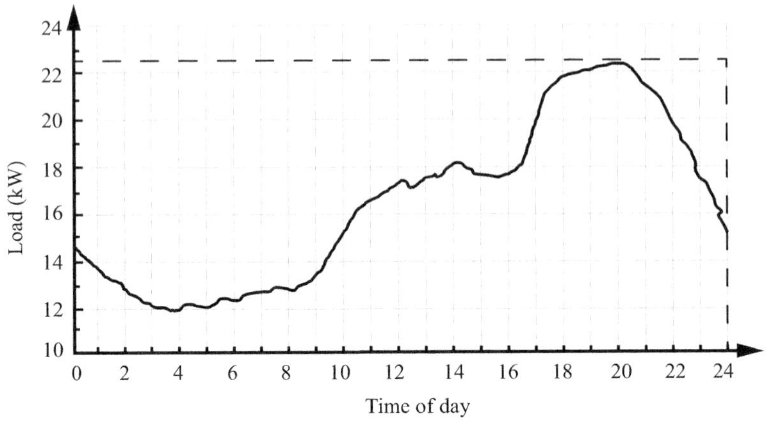

FIGURE 1.22
An example of a load profile of a typical day.

averaged for a certain period of time, e.g., month, season, or year. Typically the behavior of a single user can be unpredictable. But when many users are compounded, their overall load variation is to some degree predictable. The average load curve can help to express this predicted behavior. At this point, it is important to remind you of the difference between energy and power (FYI Box 1.3).

FYI BOX 1.3

ENERGY AND POWER

Although the terms energy and power are commonly used interchangeably in everyday discussions, they are different. Though interrelated, these terms should be properly used in engineering and scientific contexts. Broadly speaking energy is the capacity of a system to do work. It can be in the form of heat, work, kinetic, potential, electrical, chemical, nuclear, light, or sound energy among others. According to the first law of thermodynamics, these energy forms cannot be created or destroyed but can be converted from one form to another. Although useful energy can be wasted it cannot be destroyed. Power is the time rate of energy conversion or transmission. Power is associated with the size or capacity of the system whereas energy is related to both the size of the system and the duration of system operation. While power can be defined for a point in time, energy is only meaningful when expressed for a period of time. When you are buying an electric system, its power consumption is important. But when you are paying for your utility bill, it is for energy usage. The units of energy and power are:

Energy: N m=J (joule), ft.lbf, and Btu[18] (British thermal unit)[19]

Power: Nm/s=J/s=W, ft.lbf/s, Btu/h, and hp[20] (horse power)

Based on these definitions, the power consumption or generation of a system multiplied by the duration of system operation will give energy consumption or production in the system for that period. In some applications, especially related to electrical energy, the unit of Wh or kWh is used for energy.

$$\text{Power (demand)} = \frac{\text{Energy use or production}}{\text{Duration of energy use or production}} \tag{1.1}$$

A load curve can provide valuable information about the nature of electricity consumption in a specific region. The highest point of the curve is called the **maximum load demand**. The power generation systems under investigation should be able to provide this maximum load albeit for a short period of time to avoid power shortage and outage. The minimum load in the curve is called the **baseload demand**. This is the load that should be provided during the entire given period.

The area under the load curve is equal to the **total electrical energy** consumed in the given period (with the unit of 'power×time'). The total electrical energy usage, the area under the load curve, divided by the total time of the use is the **average load**.

$$\text{Average load} = \frac{\text{Total electrical energy usage}}{\text{Total time of the use}} \tag{1.2}$$

A **load factor** is the average load divided by the maximum load and is always ≤1.

$$\text{Load factor} = \frac{\text{Average load}}{\text{Maximum load}} \tag{1.3}$$

The load factor indicates how effective a power generation system is being utilized. A high load factor is desirable because it indicates that the power generation system operates close to its maximum capacity, which means more electricity to sell. Graphically speaking, in a load curve, the ratio of the area under the curve to the area of the rectangle containing the load curve (the rectangle in Figure 1.22) is equal to the load factor.

---

[18] Btu = 1055.1 J.

[19] 1 Btu is roughly equal to the energy released from burning a wooden match.

[20] hp = 550 ft lbf/s

The next two terms cannot be determined using solely the load curve. A **demand factor** is the ratio of the maximum load to the maximum possible load.

$$\text{Demand factor} = \frac{\text{Maximum load}}{\text{Maximum possible load}} \tag{1.4}$$

The maximum possible load, also known as the total connected load, is the load when all electrical devices (electricity consumers) are operating at the maximum capacity at the same time for a single consumer. For example, assume that in your house all electrical appliances operate at the same time. The power consumption in this case is the maximum possible load, say 1000 W. However, you generally do not operate all equipment at once, so the maximum load is usually lower than the maximum possible load, e.g., 500 W. In this case, the demand factor for your house is 0.5. The value of the demand factor is always less than or equal to unity.

A **coincidence factor** is the ratio of the maximum load to the sum of all maximum demands of individual customers in the region of interest.

$$\text{Coincidence factor} = \frac{\text{Maximum load}}{\text{Sum of all maximum demands of individual customers}} \tag{1.5}$$

Since typically the maximum demands of all consumers do not coincide (this is called diversity), the maximum demand will be less than the sum of individual demands. Therefore, the value of the coincidence factor is always ≤1. Note that the demand factor and coincidence factor are different. The former is based on all consumers connected, while the latter is based on the observed maximum demand of individual consumers, e.g., each home.

The demand and coincidence factors are the indication of the possibility of having a load greater than the maximum load. The greater demand factor suggests that a greater spare power generation capacity in the network is needed.

Example 1.1

In the load curve presented in Figure 1.22, determine the maximum and base demands, total electrical energy consumed, daily average load, load factor, and demand factor if the unit of the electrical load in the diagram is kW. Assume that the maximum possible demand is 30 kW.

Solution

Given: The load curve and maximum possible demand are given.

Find: The maximum and base demands, total electrical energy consumed, daily average load, load factor, and demand factor

Maximum demand = 24.6 kW
Base demand = 11.8 kW
Total electrical energy consumed = Area under the curve ≈ 75 × 2 + 10 × 24 = 390 kWh

$$\text{Daily average load} = \frac{390}{24} = 16.25 \text{ kW}$$

$$\text{Load factor} = \frac{16.25}{24.6} = 0.66 \text{ or } 66\%$$

$$\text{Demand factor} = \frac{24.6}{30} = 0.82 \text{ or } 82\%$$

Example 1.2

A neighborhood has ten identical houses. The maximum possible power consumption in each house is 7 kW (if all electrical appliances operate at the same time). However, in practice, the maximum power consumption in each house is 3 kW. If the maximum power consumption of the neighborhood is 20 kW, determine the demand factor for each house and the coincidence factor for the neighborhood.

$$\text{Demand factor} = \frac{\text{Maximum load}}{\text{Maximum possible load}} = \frac{3\,\text{kW}}{7\,\text{kW}} = 0.429$$

$$\text{Coincidence factor} = \frac{\text{Maximum load}}{\text{Sum of all maximum demands of individual customers}}$$

$$= \frac{20 \text{ kW}}{3 \times 10 \text{ kW}} = 0.667$$

1.4.2 Characteristic of Power Generation Systems

The above terms were defined for the demand (consumer) side of the system. The following terms are defined for the electricity generation side. A **capacity factor** is the ratio of the actual electrical energy produced by a single power generation unit or a system of power generation units to the maximum possible electrical energy that could have been produced, i.e., if the system was operated at the full installed capacity at all time, during a certain period.

Capacity factor

$$= \frac{\text{Actual electrical energy produced}}{\text{Maximum possible electrical energy that could have been produced}} \tag{1.6}$$

The following three factors contribute to the less than a unity capacity factor: (1) if the system operates at a reduced load or is out of service due to a failure in equipment, routine maintenance, or unplanned maintenance, (2) if the system operates at a reduced load or is shut down because electrical energy is not needed or it is not economical to produce electricity due to a low price of sale, and (3) if the system does not generate electricity due to unavailability of energy sources, for example in a gas turbine when natural gas from the pipeline is not available, in a wind turbine when wind does not blow, or in a hydroelectric plant when water behind the dam is not available. Furthermore, a hydroelectric plant may not be able to operate at the full capacity due to the limitations on the water flow rate or the limitations on the water level behind the dam. In many dams, too low water flow rates should be avoided to protect fish downstream. Also, too high or too low water level behind the dam should be avoided to enable the operation of recreational facilities, if any. The capacity factor also depends on the fuel type as well as plant design, age, and configuration. Typical capacity factors (or more accurately availability factors, discussed next) for various power generation technologies are:

- Steam power stations such as coal, geothermal, and nuclear power plants: between 50% and 95%.
- Gas turbines: from 80% to 99%.
- Wind farms: 25%–40%.
- Photovoltaic solar units: <25%.
- Hydroelectricity: 20%–99%

Note that for wind and solar power plants this factor depends on whether periods when the plant is ready but there is no wind or sunlight are counted as available, unavailable, or disregarded.

Example 1.3

Determine the capacity factor for a power plant with the installed capacity of 2000 MW that produces 15,000 GWh per year.

Solution

Actual electrical energy produced = 15,000,000 MWh

Maximum possible electrical energy = 2000 MW × 24 hour/day × 365.25 = 17,532,000 MWh

$$\text{Capacity factor} = \frac{15,000,000}{17,532,000} = 0.856 = 85.6\%$$

Since the second and third items above (when electrical energy is not needed or the primary energy source is not available) are out of control of the power plant, the availability factor is defined to provide a more accurate indication of plant operation. The **availability factor** is the ratio of the time that the plant or system is able to generate electricity, regardless of if it is actually operating or not, to the total time in a given period.

$$\text{Availability factor} = \frac{\text{Time that the plant or system is able to generate electricity}}{\text{Total time in a given period}} \tag{1.7}$$

The duration in which the system operates at part loads due to maintenance may or may not be considered in this definition and should be clarified for each case. The availability of a power generation system depends on fuel type, system design, system age, and duration of plant operation. For all other factors being equal, the newer plants and those that are operated less frequently have higher availability factors due to fewer maintenance requirements.

In contrast to the availability factor, a **utilization factor** or **use factor** is the ratio of the time that a power generation plant or system is in use to the total time that it is available to be used.

$$\text{Utilization factor or use factor} = \frac{\text{Time that a power generation unit or plant is in use}}{\text{Total time that it is available to be used}} \tag{1.8}$$

A low utilization factor and capacity factor along with a high availability factor for a plant indicates that the plant is maintained and operated properly, and the reason for the low utilization factor and capacity factor is the lack of demand or the shortage of primary energy.

Example 1.4

A power plant operates 4 hours a day, 40 weeks a year. If on average the plant is not available for 5 hours a week for scheduled and unscheduled maintenance, determine the plant's availability and utilization factors.

Solution

The system's actual operational period = 4 hours × 7 days × 40 weeks = 1120 hours

$$\text{The system's unavailable period} = 5 \times \left(\frac{365.25}{7}\right) = 260.9 \text{ hours}$$

$$\text{The system's available period} = 365.25 \times 24 - 260.9 = 8505.1 \text{ hours}$$

$$\text{Availability factor} = \frac{\textit{available period}}{\text{total period}} = \frac{8505.1}{(365.25 \times 24)}$$

$$= \frac{8505.1}{8766} = 0.97 \text{ or } 97\%$$

$$\text{Utilization factor} = \frac{\textit{operational period}}{\text{available period}} = \frac{1120}{8505.1} = 0.132 \text{ or } 132\%$$

This indicates that the plant is in an excellent shape but it is not properly utilized due to the lack of demand or primary energy source.

1.4.3 Selecting Capacity and Type of Electricity Generating Units

Selecting power plant technology and size is a complicated process that requires many parameters to be taken into account. These factors are not all technical and include economic, social, and environmental considerations to meet the interests of stakeholders, including owners, customers (consumers), local communities, and operators.

The number and capacity of power generation units should be chosen in a way that the produced power can approximately meet the load curves for various periods, i.e., daily, monthly, and annual curves, according to the projection of the load demand increases in the future. Many factors affect this projection, including the trend in population growth, planned new industries and expansion of existing ones, impacts of technological advances on electricity consumptions, and regulation changes in the region of interest. The urgency of the need for power affects the type and technology of the plant because the time needed to obtain required permissions as well as to design, construct, and install the plant depends on its technology. The selected units should be of various capacities to roughly fit the variations in the load curves with minimum partial-load operation. However, the number of small units should be limited to avoid the typically low efficiency of these units. The total capacity of the power generating system should be large enough to permit maintenance of units without affecting the capability of the system to provide the maximum demand load. Furthermore, the total capacity of the power generating system should include some extra capacity for future expansion and emergencies. This reserve margin is to provide sufficient reliability. In order for a power generation system to be able to meet required load demands, it should include a proper mixture of two types of power plants: baseload and peak load.

1.4.4 Baseload and Peak-Load Power Plants

Baseload power plants are plants that are dedicated to meeting the baseload demand. Since these power plants should produce the minimum required power for a given period, they should operate continuously at full capacity at all times throughout the period, except for scheduled or unscheduled maintenance downtimes. Typically if a power plant is operational for 6000 hours or more a year (out of 8760 hours in a year), it is considered to be a baseload plant. The power output of these units does not vary to match load demands. These units are typically able to generate electricity at a low cost, which means they should be highly efficient power plants. This typically requires high capital costs. They also usually have high plant capacity factors and low marginal costs. The following power plant technologies are commonly employed, or can be employed in the future, as baseload plants: steam power plants both fossil fuel and nuclear, hydroelectric, geothermal, solar thermal with storage, and ocean thermal energy conversion power plants. It should be noted that the available fuel is also important for the decision on the power plant type. If natural gas is the inexpensively available fuel, for example, a combined cycle power plant might be the best choice, while for coal, a pulverized coal steam power plant might be the best.

In contrast to baseload power stations, **peak-load power plants**, also known as peaking or peaker plants, operate to provide the difference between the baseload and the actual load demand, especially during a high demand period. Their period of operation may range from many days to a few hours a year with highly variable loads. Typically, if a power plant is operational for 2000 hours or less a year, it is considered to be a peaking power plant. Sometimes if a power plant operates between 2000 and 6000 hours annually, it is called an intermediate plant. Due to their limited operational period, it is not economical to use high-efficiency capital-intensive plants as peaking load power plants. Furthermore, in most baseload plants, load variations take a long time, which makes them unsuitable for variable loads. Therefore, peak-load power plants typically have low efficiency, low capital costs, fast load-following character, and low plant capacity factor with high marginal costs such as gas turbines, reciprocating engines, pumped storage and other storage technologies, and hydroelectric plants. Typically, to meet peak loads, power plants with higher efficiency are the ones operated first.

When planning for a new power plant, the projected load profiles can determine which type, baseload or peaking load, should be constructed. Note that regardless of the operating mode that the plant is designed for (i.e., baseload, intermediate, or peaking power plant), during later operation it may be used for an entirely different operation mode. Furthermore, constructing a new power plant may not be the best option. Retrofitting existing power plants to extend their useful life and/or upgrading operating plants to improve their efficiency, reliability, and capacity as well as reduce their environmental impacts should be studied before making any decision. This is particularly important if the existing plant's operating permit is still valid.

The determination of the location of a new power plant depends on many factors including availability of an appropriate site (for the current plant and the future possible expansions), availability of reliable inexpensive fuel sources (current and future projection), availability of cooling water sources, availability of transmission lines, environmental considerations, impacts on local communities, availability of labor during construction and operation, regional climate and potential hazardous events (such as high wind, flood, and earthquake), and regulations (environmental, taxes, etc.). In all selection processes, the best tool to choose the optimum solution is to conduct the life-cycle cost analysis for each scenario and compare them using this criterion. The decision on the type of plants and their capacity requires a detailed study of load curves along with fuel availability and environmental considerations.

In order to take advantage of baseload and peak-load power plants, power stations should be interconnected in a power grid. The concept of electrically connected power plants also improves the reliability of produced power, ensures economical operation of plants, and permits continued operation of older plants.

1.5 Electricity Transmission Lines

In the late 19th century, when commercial electricity was in its infancy, power generation units were located near consumption centers. This was mainly because generated power was direct current (DC), and it was difficult to change the voltage of DC power. This meant that the voltage of generated electricity and that of the consumed electricity must be equal. There were even different classes of electricity at different voltages with dedicated power generation and distribution systems. The low-voltage electricity meant that to transmit a significant amount of electric power, high current was required. Note that the power transmitted through a line can be determined by

$$P_T = IV \tag{1.9}$$

where P_T is transmitted electric power (W), I is current (A), and V is the voltage (V) of the transmission line. Also, the power loss in the transmission and distribution lines is directly proportional to the square of the current.

$$P_L = I^2R \tag{1.10}$$

where P_L is the loss of power during the transmission of electric power (W), I is the current (A), and R is the resistance (Ω) of the transmission line. The low-voltage and high-current electricity transmission meant a significant loss of power in long-distance transmissions (Example 1.5). Therefore, in those days electric power generation was mainly localized, which is now called distributed generation (DG). With the introduction and spread of alternating current (AC) power in the last decade of the 19th century and the early 20th century, the voltage of electricity transmission increased gradually up to 150 kV by the 1920s and the limitation in long-distance transmissions was eliminated. With this improvement, the locations of power plants were no longer constrained by proximity to demand centers. For example, hydropower plants could be built on sites that were commonly in remote locations or fossil fuel power plants could be built close to fuel sources. Also, central large power plants could be built to take advantage of the economy of scale. To reduce cost and increase reliability, power companies started to interconnect these power plants, which created local and regional networks. During World War I, to support wartime needs, governments took over and rapidly interconnected these regional power networks to create what is now called the **power grid, the national grid**, or simply **the grid**.

Example 1.5

A power plant produces 100 MW electric power, which is transmitted through a high-voltage line at 115 kV to a consumption center. If the resistance of the line is 10 Ω, determine the power loss in the transmission line. Determine the loss if the power was transmitted through a 34 kV line.

Solution

Given: The total power that needs to be transmitted, the voltage of the transmission line, and the resistance of the line are given.

Find: The power losses in two cases, when the voltage of transmission is 115 and 34 kV.

WHEN THE VOLTAGE IS 115 KV

The power transmitted through a line is $P_T = IV$.

$$P_T = Transmitted\ electric\ power = 100\ MW$$

$$V = Total\ voltage = 115\ kV$$

So

$$I = Current\ of\ transmitted\ electric\ power = \frac{P_T}{V} = \frac{100 \times 10^6}{115 \times 10^3} = 869.6\ A$$

The power loss in the line is $P_L = I^2R$.

$$P_L = 869.6^2 \times 10 = 7.56 \times 10^6\ W = 7.56\ MW$$

This means that out of 100 MW power, 92.44 MW is delivered to the destination and 7.56 MW is lost during the transmission.

WHEN THE VOLTAGE IS 34 KV

$$I = \frac{100 \times 10^6}{34 \times 10^3} = 2941.2\ A$$

$$P_L = 2941.2^2 \times 10 = 8.65 \times 10^7\ W = 86.5\ MW$$

So when the power is transmitted using the 115 kV line, only 7.56% of the power is lost while when the 34 kV line is used 86.5% of the power is lost.

Example 1.6

A power plant produces 100 MW electric power, which is transmitted through a high-voltage line. If the resistance of the line is 10 Ω, determine the transmission voltage that no power can be transmitted because all power will be lost.

Solution

In this case $P_L = P_T$.

$$I = \sqrt{\frac{P_L}{R}} = \sqrt{\frac{100 \times 10^6}{10}} = 3162.3\ A$$

$$V = IR = 3162.3 \times 10 = 31,623\ V$$

Note that in this case for the transmission voltage of lower than 31,623 V, no power can be transmitted because all power will be lost.

Example 1.7

The power loss in power lines can be determined using the equation $P_L = I^2R$. In this equation, if the current is replaced by voltage from the Ohm's law, $V = IR$, the power loss equation will be $P_L = \frac{V^2}{R}$. This new equation implies that increasing voltage not only does not reduce the power loss but increases it. Explain (use the numerical values for Example 1.5).

Solution

In the equation $P_L = \frac{V^2}{R}$, V is the voltage drop across the transmission line (the voltage difference between two ends of the line) and not the total voltage of the power line. For example, in Example 1-5 where the current and resistance of the line are 869.6 A and 10 Ω, the voltage drop across the line is

$$V_{Line} = IR = 869.9 \times 10 = 8696\ V$$

So using the new equation, the power loss is

$$P_L = \frac{V^2}{R} = \frac{8696^2}{10} = 7.56\ MW$$

This is equal to the value of the loss using the equation $P_L = I^2R$.

Nowadays in North America (the U.S. and Canada), there are **two major grids**, the Eastern and Western Interconnections (which are roughly separated by the Rocky Mountains), and **three minor grids**, the Texas, Quebec, and Alaska interconnections.[21] Sometimes the term a "wide area synchronous grid" is used instead of an interconnection. All power utilities in an interconnection are operated in harmony (or synchronized), i.e., they have the same voltage, frequency, and phase. Although these interconnections are independent and separated, they can exchange electricity to some degree during normal operation, except Alaska interconnection which is isolated. To avoid complications due to phase differences (they all have the same frequency and voltage), exchange lines typically operate at high-voltage direct current (HVDC). For most of continental Europe, there is only one major grid.

FYI BOX 1.4

TRANSMISSION LINES IN THE U.S.

According to the Department of Energy, there are 450,000 miles of high-voltage transmission lines across the U.S. (enough to connect Earth to the Moon and back!), the largest power grid in the world. These lines are able to interconnect over 1,000,000 MW power generation capacity.[22] The grid not only improves the reliability of power generation by interconnecting power plants but also provides several paths to deliver electricity to consumers, so problems in one line will not cause a blackout in a region. The negative effect of the grid is that a major disruption in one area can spread to other areas, which can eventually cause a major blackout.

[21] You can find interconnection maps by a simple internet search for "Interconnections in North America" or at https://www.energy.gov/oe/services/electricity-policy-coordination-and-implementation/transmission-planning/recovery-act-0#:~:text=North%20America%20is%20comprised%20of,(excluding%20most%20of%20Texas).

1.6 Regulated versus Deregulated Electricity Market

Before introducing the concept of an electricity market, we first need to define what a utility is. In general, an **electric utility** is an entity active in all or some aspects of electricity generation, transmission, and distribution businesses. But in this section, a utility is a company that owns and operates the entire power infrastructure in a region. A utility can be nationalized or owned either privately or publically. Each utility typically provides service to a particular region or territory.

An electricity market can be regulated or unregulated. Until the 1990s, almost all countries had a regulated electricity market. But since then many countries or states within countries have liberalized or deregulated their electric markets to varying degrees.

In a **regulated market**, based on the belief that certain markets are natural monopolies, due to the heavy required investment and the importance of the commodity to national security, the government controls the supply and demand process in the market. The list of regulated markets used to be relatively long, including telecommunication, airlines, water, natural gas, and electricity but now has reduced to a very few markets. Typically, in a regulated market, all three components of the electric power industry (generation, transmission, and distribution) are owned and managed by a single company called a utility (also known as a vertically integrated company). The price of electricity is determined and dictated to the utility by a regulatory body or a federal or state government entity by setting the price of electricity generation, transmission, and distribution (in the U.S. according to the Public Utility Holding Company Act of 1935). In exchange for this oversight, the utility's monopoly in the market is guaranteed. In this market, customers do not have any option to choose from for their power supplier. Due to the lack of competition, this type of market is usually inefficient and slow in responding to changes. However, they are reliable and stable.

In a **deregulated electricity market**, the electricity generation, transmission, and distribution businesses are separated. This means utilities do not have a monopoly on the market and independent power suppliers can enter the market and compete to sell their electricity. The utility still owns, operates, and maintains the power transmission and distribution sectors to deliver electricity to end-consumers but customers can buy power from the retailer of their choice. The deregulated natural gas market also works in the same way.

In the U.S., the deregulation was started with the Public Utilities Regulatory Policies Act in the 1970s to allow independent power producers to enter the

[22] https://www.eia.gov/electricity/capacity/.

market and was continued by the Energy Policy Act in 1992, which eliminated the limitations on wholesale electricity prices. Today some of the U.S. states are deregulated while others are regulated and some partially deregulated.

Many different models of deregulated markets have been adapted by different countries but their underlying principles are similar. A deregulated electric infrastructure is typically composed of two competitive markets: (1) a wholesale electricity market and a retail electricity market, and (2) a monopoly market of power transmission and distribution by the utility. In a wholesale electricity market, electricity generators, retailers, and sometimes large electricity consumers can trade electricity by bidding for or offering electrical energy. In this market, the pricing process is generally bid-based. In this a day-ahead mechanism, power generators present a list containing the amount of power they can generate and price per unit of power for certain time intervals of the day. The time interval can be an hour, half-hour, or 5-minute. The supply and demand for each interval and the cost of power generation for each generator are deciding factors for the cost of power in each interval. This market also may include real-time bids. Furthermore, there is a capacity or power market where some power plants are assigned as reserve units to provide power during unexpected peak periods. These reserve units can be spinning reserves or non-spinning reserves, the former being ready to go online at moment's notice.

In a retail electricity market, retailers re-price electricity bought in the wholesale market and offer it to end-consumers. Customers can choose their electricity provider from competing retailers (in the U.S. this is called "energy choice") and regardless of the retailer of choice, the utility transmits electricity to end-users. Customers may pay a flat rate or a variable rate according to the time of consumption. In the latter, the unit of electric energy is more expensive during the peak period (refer to FYI Box 1.5). Customers may be billed by the utility on behalf of the supplier or directly by the supplier. Note that the utility needs to be paid for the transmission and distribution of electricity. These expenses may be included in the electricity bill or customers may have an independent contract with the utility for this purpose.

The ultimate objective of the electricity market deregulation is to reduce the cost of electricity for consumers by introducing competition to the market. In this respect, the results of the introduction and implementation of the deregulated market have been mixed. Some markets have been able to reduce the cost for end-users, at least to some degree, while others have been disastrous and major failures have occurred. The most notable example of this failure was in California, particularly between 2000 and 2002, where as a result of the market switch, the wholesale prices increased drastically (in certain regions and periods over an order of magnitude) and retail prices for end-users increased significantly (sometimes over 50% increase). Also, there were power shortages and rolling blackouts (also referred to as rotating outages). As a consequence of this and similar experiences, the push for deregulation has subsided. Although most stakeholders and analysts believe that deregulated markets can work, how this can be achieved is an open question and there is an ongoing debate about it.

The design of a power plant is a complex trade-off between thermodynamics and economic considerations, which is bounded by regulations, particularly environmental regulations. The concept of base and peak load has a profound impact on the cost of electricity. In the following FYI Box, the procedure that is used to calculate the cost of electricity is explained.

FYI BOX 1.5

COST OF ELECTRICITY USAGE

There are two methods to calculate the cost of electrical energy consumed by customers. In the first method users only pay for the amount of consumed electrical energy (kWh) and a relatively small fixed (meter) charge. The calculation of the cost of electricity for many small users, including many residential applications is based on this procedure. It should be noted that in this method the cost of a unit of electrical energy may vary seasonally and/or based on the monthly energy usage.

In the second method, which is used for large consumers, the cost of electricity depends on both the amount of consumed electrical energy and the time that the energy is consumed. In this method, customers pay for the electrical energy usage and a fixed monthly meter charge, similar to the first method. They also have to pay for what is called a demand charge. The demand charge is the charge for the maximum power (time rate of energy) usage during the peak demand period. As the name implies, the peak demand period is the period of time when the grid demand is maximum. This period is predetermined for each region, for example in areas with a hot climate, it is typically during the hottest midday hours in summer.

Example 1.8

A large commercial unit in Arizona consumes 10 MWh in July with a maximum power consumption of 0.1 MW during the peak demand period. Determine the cost of electricity consumed by the facility for that month if the fixed monthly meter charge, the flat rate of energy consumption, and the demand charge are $90.00 per month, 8.0 cents per kWh, and $14.00 per kW for the month of July.

Solution

Given: The electrical energy consumption, the maximum power consumption during the peak demand period, the meter charge, the flat rate of energy consumption, and the demand charge are given for the month of July.

Find: The cost of electricity consumed by the facility for the month of July.

Breakdown of the charges:

Fixed monthly meter charge: \$90.00

Energy usage charge: $10\ MWh \times \left[\frac{1000\ kWh}{1\ MWh}\right] \times \$0.08/kWh = \$800.00$

Demand charge: $0.1\ MW \times \left[\frac{1000\ kW}{1\ MW}\right] \times \$14.00/kW = \$1,400.00$

Total electric utility bill for July: $\$90.00 + \$800.00 + \$1,400.00 = \$2,290.00$

Note that over half of the bill is for the demand charge. The cost of electricity depends on both the amount of electrical energy consumption and the time of consumption. If the unit can reduce its maximum power consumption during the peak demand period, it can save on the electricity bill without reducing the overall energy consumption. Also, note the different units for meter, energy, and demand charges.

1.7 Introduction to Conventional Power Generation Systems

The objective of many industrial systems is to convert energy from available forms, e.g., thermal or chemical energy, to desirable forms, e.g., work or electrical energy. In the following chapters, several power generation systems that convert thermal energy from fossil fuels, solar radiation, geothermal, etc., to work and electrical energy will be studied.

1.7.1 Methods to Categorize Electricity Generation

Electricity generation, or more precisely the direct conversion of various forms of energy to electricity, can be accomplished by the following seven methods:

- Electromagnetic induction (in electric generators)
- Electrochemistry (in batteries and fuel cells)
- Photoelectric effect (in solar-photovoltaic cells)

- Thermoelectric effect (in thermocouples)
- Static electricity (in lightning)
- Piezoelectric effect (in strain gauges)
- Nuclear transformation (in betavoltaics that generates electric current from nuclear radiation)

FYI BOX 1.6

BETAVOLTAIC BATTERIES

This type of battery can convert beta nuclear radiation directly into electrical energy. The battery is composed of layers of semiconductors with radioactive isotopes filled in between enclosed in a containment cover. The internal operation of the battery is similar to that of solar panels (Chapter 11) but beta particles (electrons) rather than photons from solar radiation are captured and converted to electricity in the semiconductors.

These batteries are suitable for applications with very low power requirements (in the order of nanowatt, 10^{-9} W, not nearly enough to power a cell phone). However, they can provide power for a very long time, from a few years to 100 years. The battery life is directly proportional to the half-life of the isotope. Due to their robustness, they are capable of operating and withstanding very harsh environments such as very low or high operating temperatures (–65°C to 150°C), high humidity, high-frequency vibration, and corrosive environments. These characteristics make betavoltaic batteries ideal for medical implants, deep space and deep ocean devices, military applications, and microelectromechanical systems (MEMS).

Early research on the idea started in the 1950s, and the first application was for heart pacemakers in the 1970s. In early systems, due to the low efficiency of semiconductors, a relatively large amount of radioactive isotopes were used, which resulted in expensive and potentially hazardous systems. Because of these, they lost the competition to lithium-ion batteries. With recent advancements in nanotechnologies, more efficient semiconductors are available, which means less radioactive isotopes can be used. Modern betavoltaic batteries use a hydrogen isotope called tritium. Beta particles from this isotope can be stopped by a dead layer of human skin. This isotope has a half-life of 12.3 years. By loading the battery twice as much isotope as needed at the beginning of the battery life, this battery can be operated for about 25 years without recharging. Similar principles are used for alphavoltaic batteries.

FYI BOX 1.7

TINIEST ELECTRICITY GENERATORS

The piezoelectric effect is the ability of some materials to generate electrical current when mechanical stress in the form of either compressing or stretching is applied. This phenomenon has been extensively used in strain gauges and pressure gauges. Recent developments in nanotechnologies resulted in new avenues for piezoelectric electricity generators. Some materials do not demonstrate any piezoelectric property in their bulk form while when reduced to an atomically thin film, their characteristics are entirely different. For example, while molybdenum disulfide (MoS_2) is a brittle opaque matter in its bulk form, when it is converted to a few atom film, it becomes transparent and bendable with very good piezoelectric character. This film can be integrated into textile and be used to make a wearable device to produce power from body motion or a flag-like device to produce electricity from wind-induced motions.

Electricity generation systems can be categorized based on the original source of energy as follows:

- Fossil fuels (oil, coal, or natural gas)
- Nuclear energy
- Renewable energies (originated from solar energy)
    - Hydropower
    - Solar energy
    - Wind energy
    - Wave energy
    - Ocean thermal energy
    - Biomass
- Renewable energies (non-solar energy-based)
    - Tidal energy
    - Geothermal
    - Osmotic energy
- Other resources
    - Solid waste
    - Heat recovery

In terms of machines that either convert some forms of energy to kinetic energy that eventually can be converted to electrical energy, i.e., prime mover, or directly convert different forms of energy to electrical energy, electricity generation systems can be grouped as:

- Steam power plants
- Gas turbine power plants
- Combined steam and gas turbine power plants (combined cycles)
- Hydroelectric power plants
- Wind turbines
- Reciprocating engines
- Photovoltaic solar cells
- Fuel cells
- Wave energy converters
- Osmotic power generators

1.7.2 Thermodynamics of Conventional Power Generation Systems

All thermal engines (i.e., thermal power plants) operate between a high-temperature heat source and a low-temperature heat sink and require four main processes to convert thermal energy to work:

- Pressurization (compression) of the working fluid
- Heat addition to the working fluid at a high temperature
- Expansion of the working fluid to produce work
- Heat rejection from the working fluid at a low temperature

It should be noted that these processes do not necessarily take place separately or independently and may have overlaps. In a steam power plant, water is pressurized in a pump. Heat is added in a steam generator (boiler) at a high temperature, which converts the inlet water to the outlet steam. Then, the steam expands in a steam turbine to produce work. Finally, the outlet flow of the turbine is condensed in a condenser by extracting heat at a low temperature.

In a gas turbine power plant, fresh air from the atmosphere is compressed in a compressor. Then, the temperature of the high pressure air is increased in a combustor typically by burning fuel. Next, the high temperature and pressure combustion products are expanded in a turbine to produce work. Last, the low temperature and pressure gas flow is exhausted to the atmosphere, rejecting the remaining thermal energy of the working fluid from the cycle. Similar processes occur in a cylinder and piston engine through the suction of fresh air, compression, combustion, expansion, and exhaust processes.

In order to understand why these four processes are needed for every conventional power generation cycle, first the work produced in a system should be reviewed from the thermodynamic point of view. Note that in this discussion, the working fluid of a system is the fluid that directly participates in thermodynamic processes to generate power.

1.7.2.1 Categorization of Power Cycles

Power generation systems can be categorized based on various characteristics of their thermodynamic cycles. Most of these classifications are based on the working fluid of the system.

1.7.2.1.1 *Open Power Systems versus Closed Power Systems*

In a closed system, the working fluid operates in a thermodynamic cycle, which means it undergoes a series of processes and finally returns to the initial state such as steam power plants. In an open system, the state of the working fluid does not return to its initial state, and it does not go through a thermodynamic cycle such as gas turbines and cylinder and piston engines. Note that in open systems, the engine usually operates in a mechanical cycle but not a thermodynamic cycle.

1.7.2.1.2 *Phase-Changing or Vapor (Working Fluid) Power Systems versus Gaseous (Working Fluid) Power Systems*

In a phase-changing or vapor power plant, the working fluid can be both in the gaseous phase and liquid phase such as steam power plants. In this case, the phase of the working fluid can change alternatively by vaporization and condensation. Conversely, in gaseous power systems, the working fluid remains in the gaseous form throughout the cycle (no phase change) such as gas turbines and cylinder and piston engines.

1.7.2.1.3 *Internal Combustion Engines versus External Combustion Engines*

While in internal combustion engines the working fluid participates in the combustion process and its composition alters throughout the cycle, in external combustion engines, heat is transferred from external sources, e.g., gaseous combustion products or nuclear reactions, to the working fluid and its composition remains unchanged. It should be noted that the term internal combustion engine typically refers to spark- or compression-ignition cylinder and piston power systems, although technically gas turbines are also internal combustion engines.

1.7.2.1.4 *Shaft Work versus Boundary-Moving Work*

Power generation systems can be either steady-state devices involving shaft work or cylinder and piston devices involving boundary-moving work. In the former, the continuous flow of the working fluid in a fixed control volume produces power, such as gas turbine, whereas in the latter a certain amount of the working fluid produces power by moving the boundary of the system, such as a spark- or compression-ignition cylinder and piston engine. While steady-state devices involving shaft work can have either a phase-changing or gaseous working fluid and can operate based on either open or closed cycles, cylinder and piston devices involving boundary-moving work typically operate in an open cycle with a gaseous-only working fluid.

1.7.2.2 Work in Moving Boundary Systems (Closed Systems)

Figure 1.23 illustrates gas or liquid enclosed in a piston and cylinder assembly that goes through a process. The particular case shown in the figure is a reversible expansion process from state 1 to state 2.

The work done on or by the system can be determined as follows:

$$\delta W = F dx \tag{1.11}$$

where F is the force exerted on the piston by the fluid (N).

$$F = PA \tag{1.12}$$

where A is cylinder/piston cross-sectional area (m^2) and P is the pressure of the fluid (Pa). So

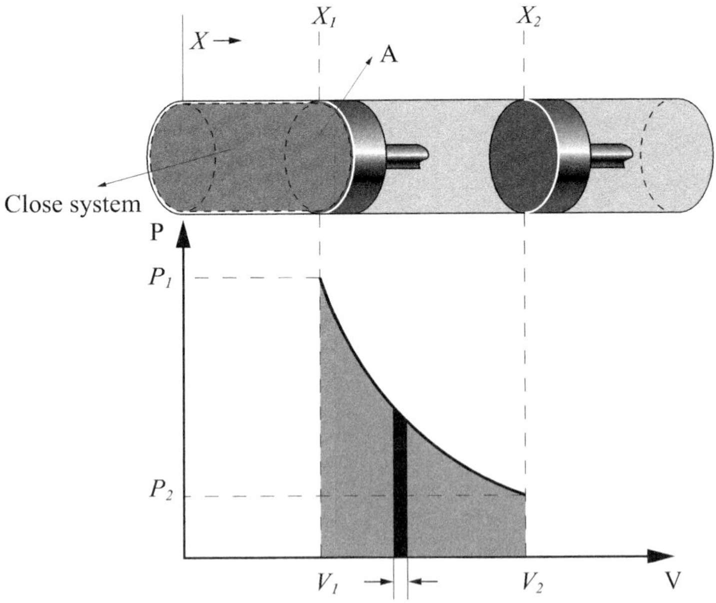

FIGURE 1.23
Work in a piston and cylinder assembly and graphical representation of the relationship between *P*, *V*, and work in a close system.

$$\delta W = PAdx \tag{1.13}$$

The change in the volume is $dV = Adx$. Thus,

$$\delta W = PdV \tag{1.14}$$

This equation can be applied to systems with any shape. In order to find work in the process 1–2 (W_{1-2}), the integral of this equation should be taken:

$$W_{1-2} = \int_1^2 W = \int_{V_1}^{V_2} PdV \tag{1.15}$$

When dV and thus W_{1-2} are positive (expansion), work is done **by** the system and when dV and thus W_{1-2} are negative (compression), work is done **on** the system. In order to find the above integral, the relationship between p and V should be known or determined, i.e., $P = P(V)$. This relationship can be presented either graphically or analytically. In the graphical approach, the work during this process is equal to the area under the P-V diagram (Figure 1.23). Note that in a constant volume process the work is zero. Thus, in a reversible process involving a boundary movement in a close system, work per unit of mass is:

$$w_{1-2} = \int_1^2 Pdv \tag{1.16}$$

1.7.2.3 Work in Control Volumes

In an internally reversible, steady-state, single-inlet, single-outlet control volume with negligible change in the kinetic and potential energy between the inlet and outlet flows, the first and second laws of thermodynamics can be simplified to the following forms (the state of the working fluid changes from 1 to 2):

$$w_{1-2} = q_{1-2} + (h_1 - h_2) \quad (\text{the 1st law of thermodynamics}) \tag{1.17}$$

$$q_{1-2} = \int_1^2 Tds \quad (\text{the 2nd law of thermodynamics}) \tag{1.18}$$

where w_{1-2} and q_{1-2} are the specific work and heat transfer (energy per unit of the mass flow rate through the control volume, kJ/kg). Substituting the second equation in the first one,

$$w_{1-2} = \int_1^2 T\,ds + (h_1 - h_2) \tag{1.19}$$

If the Gibbs equation ($Tds = dh - vdP$) is substituted in the above equation,

$$w_{1-2} = \int_1^2 (dh - vdP) + (h_1 - h_2) \tag{1.20}$$

$$w_{1-2} = -\int_1^2 v\,dP \tag{1.21}$$

The negative sign in Equation 1.21 is an indication of the direction of energy transfer, i.e., negative for compressors and pumps (work done on the system), and positive for turbines (work done by the system). However, since typically the direction of energy transfer is directly taken into account in the development of the energy balance equations, this negative sign is usually ignored. In a *P-v* diagram of a control volume with the aforementioned specifications, the specific work of the system is equal to the area to the left of the diagram (Figure 1.24). Thus, in a constant pressure process in a control volume, the work is zero.

Equation 1.21 indicates that for a fixed pressure change across a system, a higher specific volume results in more work involved (in an expansion process, more work production, and in a compression process, more work input). This means, if everything else is equal, the compression of liquids needs less work input than that of gases since the specific volume of liquids is much smaller than that of gases. Also, the expansion of high-temperature gases produces more work than that of low-temperature gases.

Since in ideal pumps, compressors, and turbines the control volume approach should be used to analyze the system, the work involved in the process can be determined using Equation 1.21. If there were no

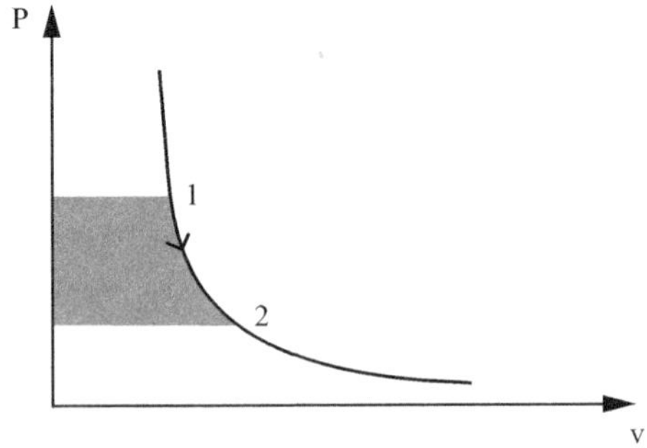

FIGURE 1.24
Graphical representation of the relationship between *p*, *v*, and specific work in a control volume.

heat-transfer processes in a power cycle, ignoring pressure losses in the components, the change in the pressure during the compression and expansion processes will be the same. This means all the work produced in the expansion process is consumed in the compression process (in reality to overcome losses, extra work input is required). For a power cycle to be able to produce net work, the specific volume of the working fluid during the compression process should be smaller than the specific volume of the working fluid during the expansion process. As a result, work produced in the expansion process is greater than the work consumption in the compression process. The larger difference between the specific volumes during these two processes increases the output power and improves the efficiency of the cycle. Understanding this concept is very important because many approaches to increase the efficiency and output of power cycles can be explained using this principle. In steam power plants, the difference between the specific volume of the working fluid during the compression and expansion processes is created by evaporating water in the steam generator and condensing steam in the condenser. In gas turbine cycles, this difference is created by taking fresh air and increasing its temperature in the combustor. Note that a similar discussion can be presented for cylinder and piston power cycles according to Equation 1.16. In this case, the volume change during the expansion and compression processes is equal and heat-transfer processes are used to increase the pressure of the working fluid in the expansion process compared to that of the compression process.

Example 1.9

Everything else being equal, compare the efficiency of cycles with a phase-changing working fluid and the efficiency of cycles with a gaseous working fluid.

Answer

The reason that power production cycles based on steady-state devices involving shaft work are able to produce net work is that during the compression process the specific volume of the working fluid is smaller than that during the expansion process. The greater this difference in specific volumes, the higher the efficiency will be. That is why generally the efficiency of cycles with phase-changing working fluids is higher than that of the gaseous working fluid cycles (remember that the specific volume of steam can be several orders of magnitude greater than liquid water). In fact, in steam cycles about 1%–2% of the work produced in the turbine is consumed in the pump while in gas turbines over 40% of the work produced in the turbine is consumed in the compressor.

1.7.2.4 Performance Evaluation Parameters

There are several parameters that can be used to measure the performance of a power generation unit. The **thermal efficiency** is the most common parameter. It is a quantity that indicates what percentage of the energy input to the working fluid (the cycle) is converted to the net work output (or power output) and is defined as

$$\eta_{Th} = \frac{w_{Net}}{q_{Input}} = \frac{\dot{W}_{Net}}{\dot{Q}_{Input}} \tag{1.22}$$

The higher efficiency means for a certain output power, less input fuel is needed, which reduces operating costs.The lower fuel flow rate also requires a smaller fuel preparation system, smaller heat exchangers, and less air flow rate. Moreover, a lower flow rate of the working fluid is needed. All these mean the higher efficiency also reduces the capital cost of the plant.

Similar to efficiency, **heat rate** (**HR**) is a quantity to express the performance of a power generation unit. But unlike the efficiency, it is defined as the thermal energy input required to produce a unit of work output. So heat rate is inversely proportional to efficiency. Typically, in calculating heat rate, since the work output is in the form of electricity, it is expressed in kWh and thermal energy input is expressed in Btu. Therefore, heat rate is not presented in percentage and it has the unit of Btu/kWh. Unlike efficiency, the lower values of heat rate are more desirable. The relationship between efficiency and heat rate is (note that 1 kWh=3412 Btu):

$$\eta\,[\%] = \frac{1}{HR\left[\dfrac{Btu}{kWh}\right]} \times \left[\frac{3412\ Btu}{kWh}\right] \times 100 = \frac{3412}{HR\left[\dfrac{Btu}{kWh}\right]} \times 100 \tag{1.23}$$

Another useful parameter to measure the performance of a power generation unit is the **back-work ratio** (**BWR**). It is defined as the ratio of the work consumed in the compression process (either a pump or a compressor) to the work produced in the expansion process (the turbine). The BWR indicates what portion of the power developed in the turbine is consumed internally in the compression process and does not contribute to net power output. The BWR is typically small for phase-changing working fluid power plants while it is significantly larger for gaseous working fluid power plants.

Both efficiency and heat rate can be defined based on either gross output energy (or power) or net output energy (or power). The gross output is the energy (power) at the outlet of the electric generator. The net outlet, however, is the energy (power) at the outlet of the unit or the plant (at the plant busbars[23]) after all internal energy consumptions for plant operation are tapped. Thus, the net output is the gross output minus the internal consumption of the plant, including the power needed to run pumps, fans, compressors, fuel preparation equipment, instruments and sensors, control system, and lighting as well as the heating and cooling of buildings.

[23] The plant busbars are where all generated electricity is accumulated to be transmitted to the grid.

1.7.2.5 Idealization of Power Generation Systems

Actual power systems are very complex and theoretically very difficult to analyze. Even the most sophisticated computer simulations need to make some assumptions to simplify the complicated reality of processes in a power cycle. It is very common and useful to analyze an ideal cycle with similar essential characteristics as in an actual cycle. Ideal cycles can be used to determine an upper limit for the performance of actual cycles. Also, the effects of certain design and operating parameters on the performance of the ideal cycle can be investigated. The influence of these parameters on the performance of actual cycles is qualitatively (but not quantitatively) similar to that in ideal cycles. These analyses can help to determine the right course of action to improve the performance of power systems. For example, it will be shown that increasing the temperature of the inlet flow to the turbine from the combustor in an ideal gas turbine cycle improves the efficiency of the ideal system. Increasing the temperature of the inlet flow to the turbine can be considered as the first step in optimizing the performance of an actual gas turbine, although several parameters limit the practical increase of the inlet temperature to the turbine in real gas turbines. The following ideal cycles have been universally accepted for common power generation systems (Table 1.2).

In the next few chapters, for each power generation technology, the ideal thermodynamic cycle for the power system of interest will be identified and analyzed. Then, the influence of certain design and operating parameters on the performance of the ideal cycle will be evaluated. Based on the results, the ideal cycle will be modified to optimize its performance. Finally, the deviation of actual systems from ideal ones will be identified, so the analyses of ideal cycles can be interrelated to the performance of actual systems.

TABLE 1.2

Ideal Cycles for Common Power Generation Systems.

Type of Power Plant	Ideal Thermodynamics Cycle	Working Fluid Phase	Energy Source	Renewable (R) or Non-renewable (N)
Steam power plant	Rankine	Vapor and liquid	Coal	N
			Natural gas	N
			Petroleum products	N
			Biomass	R
			Nuclear reaction	N
			Geothermal	R
			Solar thermal	R
			Ocean thermal	R
Gas turbine	Brayton	Gas	Natural gas	N
			Petroleum products	N

(Continued)

TABLE 1.2 (*Continued*)

Ideal Cycles for Common Power Generation Systems.

Type of Power Plant	Ideal Thermodynamics Cycle	Working Fluid Phase	Energy Source	Renewable (R) or Non-renewable (N)
Combined cycle	Rankine+Brayton	Vapor and liquid+gas	Natural gas	N
			Petroleum products	N
Integrated gasification combined cycle	Rankine+Brayton	Vapor and liquid+gas	Coal	N
			Biomass	R
Fuel cell	None	None	All gaseous and liquid fuels	N
Hydroelectric	None	None	Kinetic energy of water	R
Wind turbine	None	None	Kinetic energy of wind	R
Solar-photovoltaic	None	None	Solar radiation	R
Ocean energies (waves, tides, currents)	None	None	Ocean waves and tides	R
Spark-ignition piston and cylinder engine	Otto	Gas	Gasoline	N
Compression-ignition piston and cylinder engine	Diesel	Gas	Diesel	N

1.8 Questions and Problems

1.8.1 Energy and Society

1.1. We use energy to ....

a. produce what we eat
b. move around
c. manufacture goods and transport them
d. heat or cool our living environments
e. communicate with each other
f. operate appliances at home
g. all of the above

1.2. Gross domestic product (GDP) is a parameter that is determined based on ....

a. the market value of all products, including goods and services, exported by a country or a region in a particular period

b. the market value of all products, including goods and services, imported by a country or a region in a particular period

c. the market value of all products, including goods and services, imported or exported by a country or a region in a particular period

d. the market value of all products, including goods and services, produced in a country or a region in a particular period

1.3. In most societies, there is a direct relationship between the standard of living and the level of energy consumption.

a. True

b. False

1.4. The coupling between economic development and energy consumption can be demonstrated by directly comparing these two parameters per capita basis.

a. True

b. False

1.5. The following figure compares the annual GDP per capita and the annual primary energy consumption per capita in the U.S. between 1960 and 2018. Can this diagram be used to show the relationship between the annual GDP per capita and the annual primary energy consumption per capita? If so, explain the relationship. If not, explain how we can adjust the diagram so can be used for this purpose.

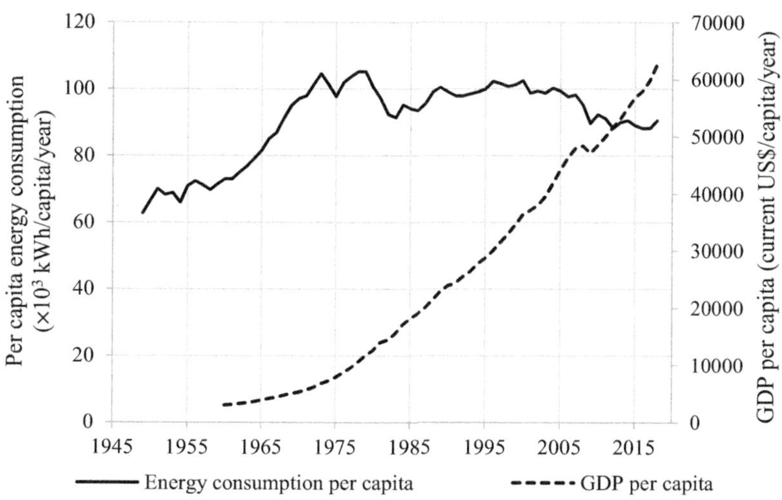

1.6. The following figure shows the per capita annual energy consumption as a function of the annual GDP per capita for many countries. Using the diagram explain the interdependency between the annual energy consumption per capita and the annual GDP per capita.

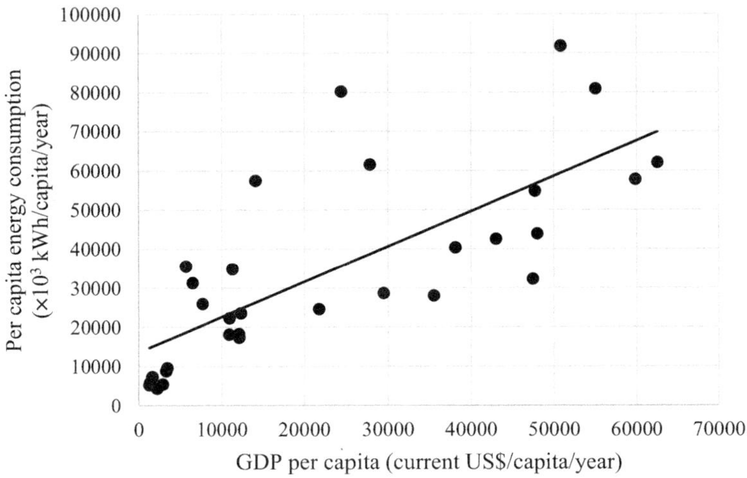

1.7. A high living standard (manifested in a high GDP per capita) can be typically achieved without high electricity generation (and consumption) per capita.

a. True

b. False

1.8. The energy we need in society to improve our living standard and maintain economic growth should be extracted or collected from natural resources of energy or produced from non-natural resources.

a. True

b. False

1.9. Usually we cannot use natural resources of energy directly without energy conversion because they are not typically in the form that we can use and are not located where we need them.

a. True

b. False

1.10. Forms of energy that can be directly found in nature are ……..

a. primary sources of energy

b. secondary energy sources

c. energy carriers

1.11. .... should be derived from ......
   a. Primary energy sources – primary energy sources
   b. Primary energy sources – secondary energy sources
   c. Secondary energy sources – primary energy sources
   d. Secondary energy sources – secondary energy sources

1.12. Why is the distinction between primary and secondary energy sources important in accurately reporting and analyzing energy consumption?

1.13. Which of the following energy sources are primary sources of energy? (Select all the correct ones)
   a. Crude oil
   b. Hydrogen
   c. Solar energy
   d. Biodiesel fuel
   e. Wind energy

1.14. Which of the following energy sources is not a primary source of energy?
   a. Crude oil
   b. Diesel fuel
   c. Solar energy
   d. Natural gas

1.15. Which of the following energy sources is a secondary source of energy?
   a. Nuclear energy
   b. Electricity
   c. Hydro energy
   d. Geothermal energy

1.16. Primary energy sources always need processing and/or energy conversion before they can be used.
   a. True
   b. False

1.17. Which ones are not a conversion of a primary energy source to a secondary energy source?
   a. Coal, nuclear energy, or hydropower converted to electricity in power plants.
   b. Crude oil processed to refined fuel oil in a refinery.
   c. Electricity is converted to mechanical work in an electric motor.

1.18. Why is the study of per capita annual primary energy consumption informative?

1.19. List the three main energy consumption sectors.

1.20. What are the two subcategories of energy consumption in the building sector?

1.21. Provide an example of when energy sources are used as input materials in the industrial sector.

1.22. The following figure illustrates the share of each energy end-use sector in the total primary energy consumption in the globe and in the U.S. in 2018. Compare and contrast the two charts and explain what they indicate about the U.S. economy and the global economy and why.

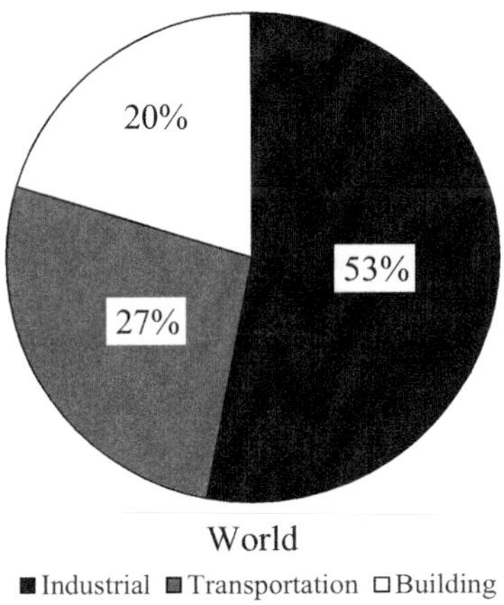

(a)

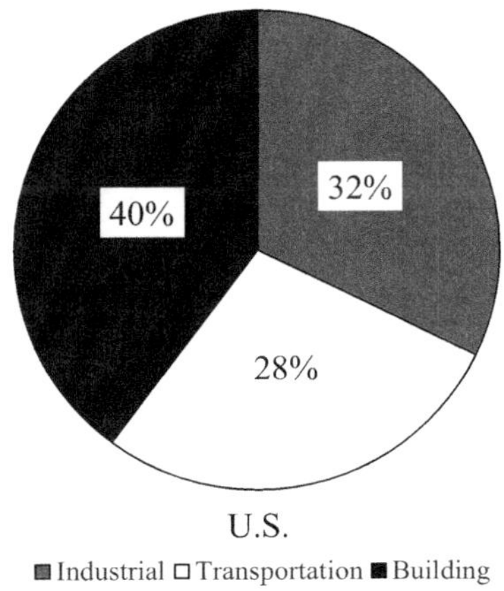

(b)

1.23. Rank the following three sectors according to their share of energy consumption in the U.S. (highest to lowest).

1. Transportation sector
2. Building sector
3. Industrial sector
   a. 1, 2, 3
   b. 2, 1, 3
   c. 3, 2, 1
   d. 2, 3, 1

1.24. Rank the following three sectors according to their share of energy consumption in the world (highest to lowest).

1. Building sector

2. Industrial sector
3. Transportation sector
    a. 1, 2, 3
    b. 2, 1, 3
    c. 3, 2, 1
    d. 2, 3, 1

1.25. List four factors that impact the amount of energy consumption and share of various energy consumption sectors for a particular region or country.

1.26. Explain how the economic structure of a particular region or country impacts the amount of energy consumption and the share of various energy consumption sectors.

1.27. Explain how the climate of a particular region or country impacts the amount of energy consumption and the share of various energy consumption sectors.

1.28. Which economy does consume more energy in the building sector?
    a. A manufacturing-focused economy
    b. A service-focused economy

1.29. What are the factors that impact energy consumption in the transportation sector?

1.8.2 Primary Energy Sources

1.30. The following figure shows the contribution of various primary energy sources to the global energy consumption in 2017. What does the chart indicate about the share of renewable and non-renewable energy sources?

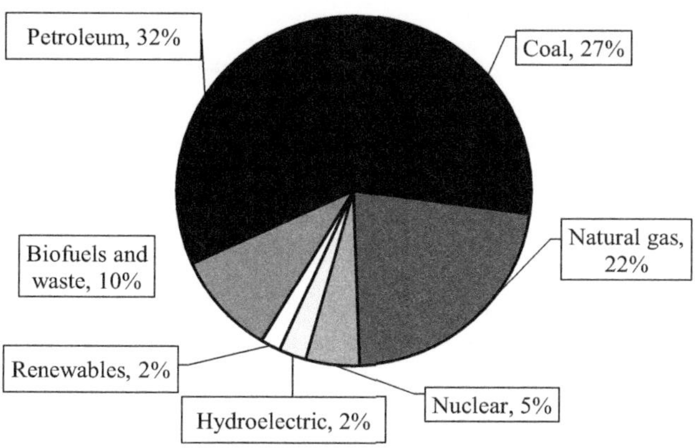

1.31. The following figure shows the contribution of various primary-energy sources to the U.S. primary energy consumption. Compare and contrast this chart with the one for the world in the previous question.

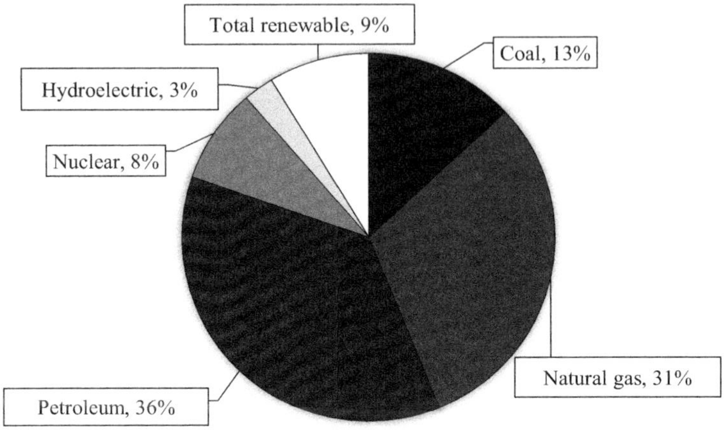

1.32. The following figure shows the contribution of various primary energy sources to the U.S. energy consumption between 1949 and 2018. Compare and contrast the share of coal and other energy sources over the years.

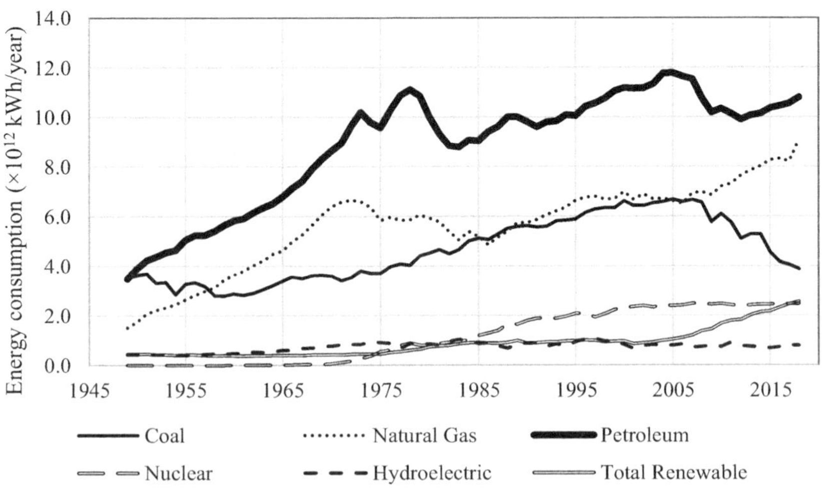

1.33. Using Figure 1.11 determine:
- where the great majority of coal is consumed?
- what source of energy provides most of the energy consumed in residential buildings?
- what source of energy provides most of the energy consumed in commercial buildings?
- what source of energy provides most of energy consumed in the industrial sector?
- what percentage of natural gas is consumed by the building sector?
- what percentage of energy used by the residential sector is provided by fossil fuels?
- what percentage of petroleum is collectively consumed by the non-transportation sectors?
- what percentage of renewable energy is collectively consumed by the non-building sectors?

1.34. Using Figure 1.11 rank natural gas consumers (descending order).

1.35. Using Figure 1.11 rank coal consumers (descending order).

1.36. Using Figure 1.11 rank petroleum consumers (descending order).

1.37. Using Figure 1.11 rank renewable energy consumers (descending order).

1.38. Rank the consumption of the following sources of primary energy in the U.S. in 2018 (in descending order).

1. Petroleum
2. Coal
3. Nuclear energy
4. Natural gas
5. Renewable energy sources
6. Hydroelectric

   a. 1, 4, 2, 5, 3, 6
   b. 1, 2, 3, 4, 5, 6
   c. 2, 1, 3, 6, 4, 5
   d. 4, 1, 5, 2, 6, 3

1.39. The consumption of coal has slightly declined in recent years in the U.S. and was primarily replaced by natural gas and renewable energy sources.

a. True
b. False

1.40. Globally, the majority of energy consumption is provided by ……..
while the contribution of ……. is relatively small.

a. fossil fuels – renewable energy sources
b. renewable energy sources – fossil fuels
c. nuclear energy – fossil fuels
d. renewable energy sources – nuclear energy

1.41. The global consumption of oil and natural gas have been rapidly and continuously increasing in the past three decades.

a. True
b. False

1.42. The contribution of renewable energy sources in the global and the U.S. energy mix has increased in recent years and they are responsible for a significant percentage of energy consumption.

a. True
b. False

1.8.3 Introduction to Electricity

1.43. How did ancient people encounter a phenomenon related to electricity for the first time?

1.44. Briefly explain how electric fishes generate electricity.

1.45. Explain how electricity generation and the industrial revolution were related.

1.46. Chronologically order the following milestones in the history of electricity generation.

a. The War of Currents
b. Local direct current (DC) power generation
c. Benjamin Franklin experiments
d. Alternating current (AC) and transmission of electricity across great distances
e. Volta battery
f. The Second Industrial Revolution
g. Invention of electric motors
h. Observation of natural static electricity

1.47. The domination of AC power made it possible to transmit electricity across great distances at high voltage, which resulted in large centralized power plants and a vast network of power distribution systems.

a. True
b. False

1.48. How did the oil crisis in the 1970s impact investments in R&D activities for improving the efficiency of electricity generation and fostering renewable energy?

1.49. Explain how the level of interest and investment in the field of electricity generation has depended upon the cost of fossil fuels.

1.50. What is distributed generation (DG)?

1.51. Define sustainability. Why does it matter?

1.52. In the U.S., which economic sector is responsible for over a quarter of the total greenhouse gas (GHG) emissions?

1.53. The following figure illustrates the global electricity generation between 1980 and 2050. Evaluate the global electricity generation in the past and future projection.

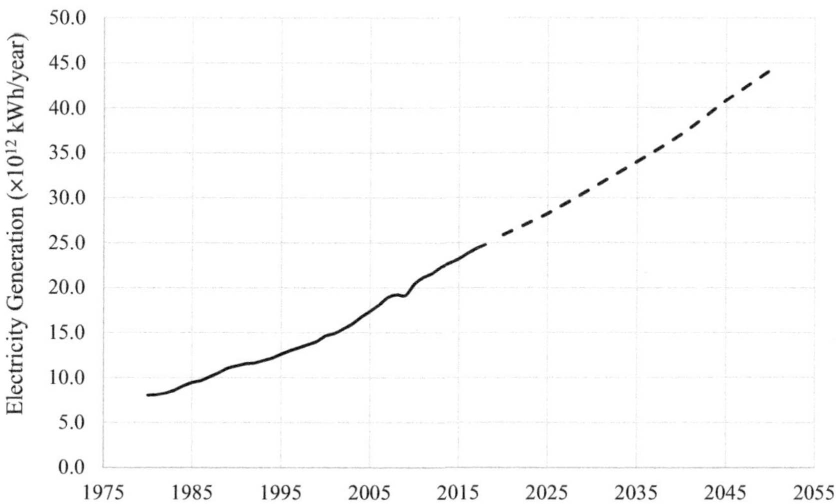

1.54. Explain the difference between net and gross electricity generation.

1.55. The difference between net and gross electricity generation depends on the type of a power plant.

a. True

b. False

1.56. List two factors that distinguish electricity generation and consumption.

1.57. Write the equation that relates electric power consumption to electricity generation, electricity import and export, and losses during electricity transmission and distribution.

1.58. Electric power consumption depends on .... (Check all correct items.)

a. Electricity generation

b. Electricity import

c. Electricity export

d. Losses during electricity transmission and distribution

1.59. It is expected that the global electricity generation will almost double by 2050 compared to the current consumption.

a. True

b. False

1.60. What can we say about the world population in the past seven decades and its future projection? Use the figure below to justify your answer.

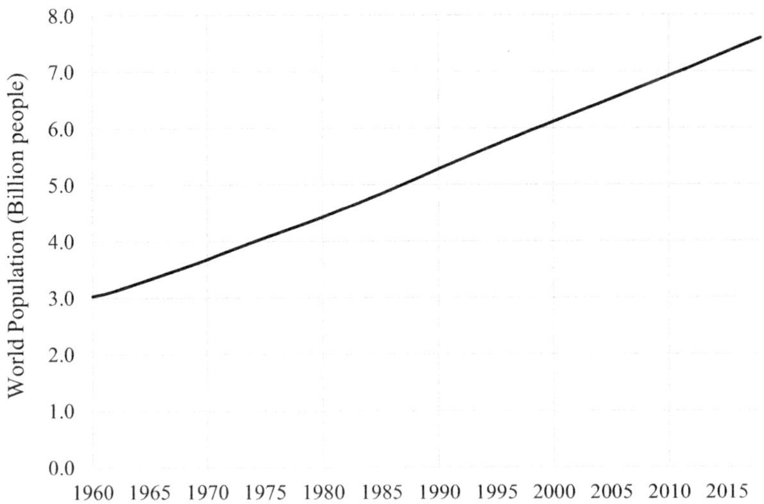

1.61. The world population has increased from 3 to 7.6 billion people between 1960 and 2018, about 150% increase.

a. True

b. False

1.62. The combined impact of increased electricity consumption by each person and increased in the world population has substantially increased the total global electricity consumption.

a. True

b. False

1.63. It is projected that both the world population and per capita electricity consumption will continue to increase in the foreseeable future.

a. True

b. False

1.64. It is predicted that the average global per capita electricity generation will increase by 38% by 2050.

a. True

b. False

1.65. The figure below illustrates the total and per capita electricity generation in the U.S. between 1980 and 2018. Explain what the diagram indicates.

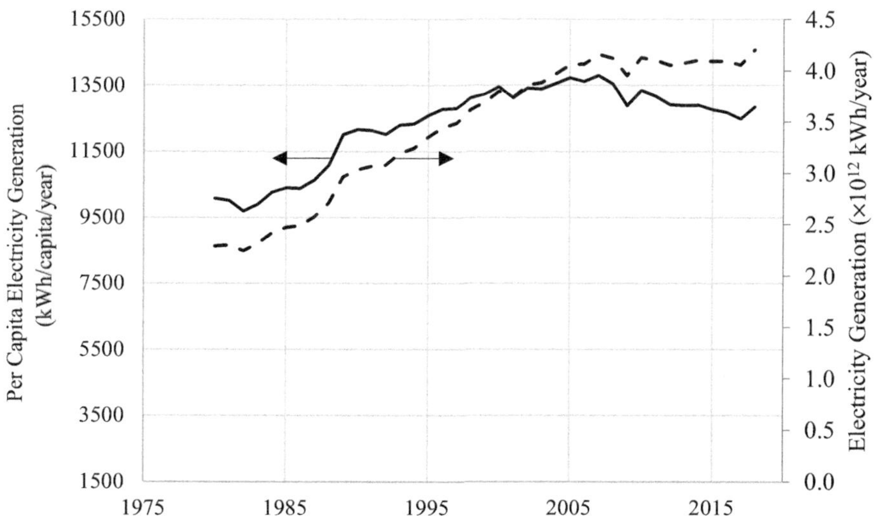

1.66. The patterns of variation in the total and per capita electricity generation in the U.S. and the world are similar.

a. True

b. False

1.67. The per capita electricity generation in the U.S. increased until 2000. Then, the growth stopped and the per capita generation was flat at about 13,500 kWh/capita until 2007.

a. True

b. False

1.68. The per capita electricity generation has been steady since 2007 in the U.S.

a. True

b. False

1.69. The total electricity generation in the U.S. has been steady since 2005.

a. True

b. False

1.70. While in recent years the per capita electricity generation has declined in the U.S., a similar decline cannot be seen for the total electricity production. Why?

1.71. For all industrially developed economies, the per capita electricity generation has declined in recent years. Why?

1.72. Which statement is correct?

a. All the industrially developed economies reached a steady flat per capita electricity generation at different values and times.

b. All the industrially developed economies reached a steady flat per capita electricity generation at different values but more or less at the same time.

c. All the industrially developed economies reached a steady flat per capita electricity generation at different times but more or less at the same values.

d. Not all industrially developed economies reached a steady flat per capita electricity generation.

1.73. All industrially developed economies reached a steady flat per capita electricity generation at different values but more or less at the same time.

a. True

b. False

1.74. The recent increase in the per capita and total global electricity generation has been driven by the increase in all economies.

a. True

b. False

1.75. The following two figures illustrate the total and per capita electricity generation in the U.S. and Japan between 1980 and 2018. Compare the two countries. What are the similarities and differences?

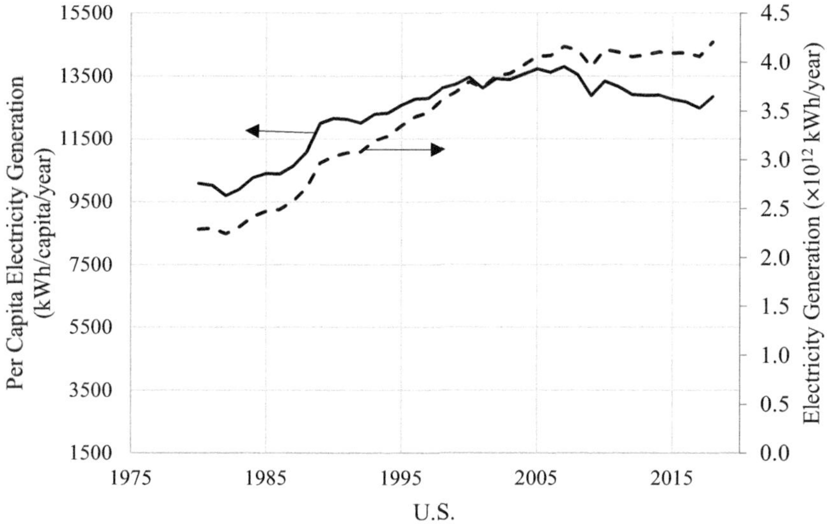

(a)

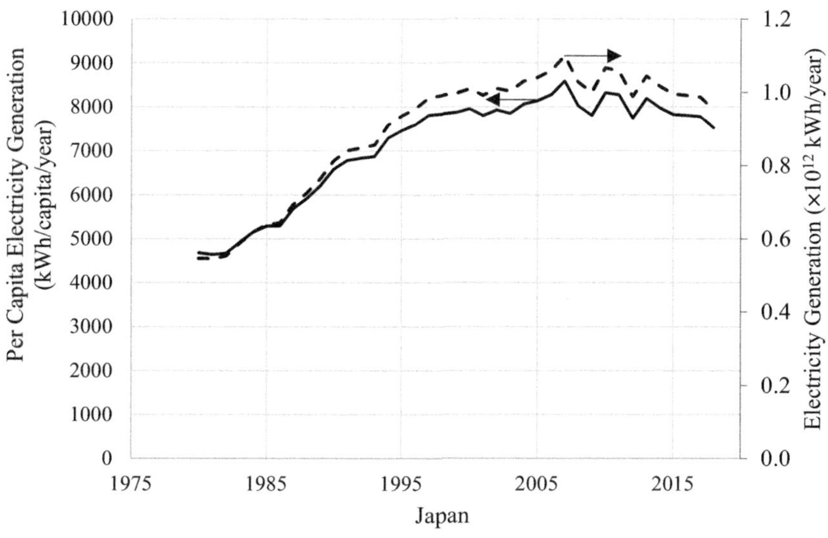

(b)

1.76. The following two figures illustrate the total and per capita electricity generation in the U.S. and China between 1980 and 2018. Compare the two countries. What are the similarities and differences?

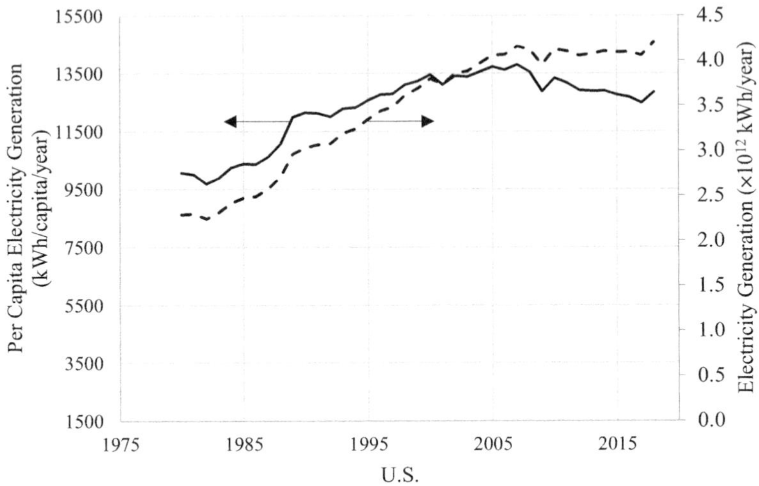

(a)

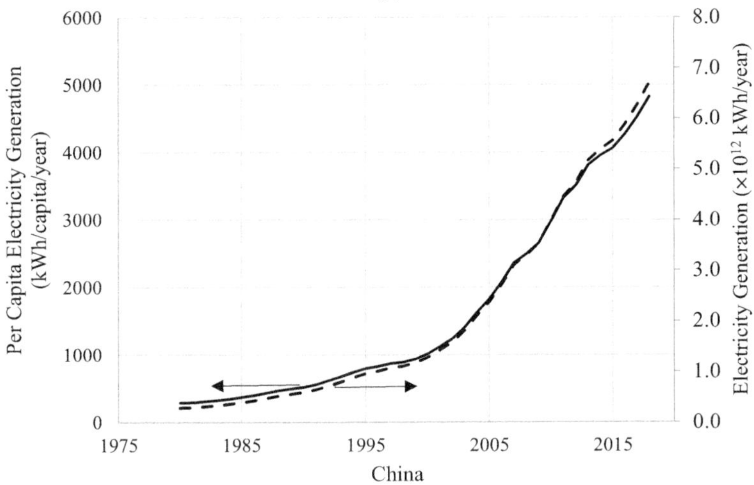

(b)

1.77. The per capita electricity generation in developing and emerging countries is rapidly increasing.

a. True

b. False

1.78. Why is the flattening of electricity consumption per capita in developed countries both a good news and a bad news?

1.79. Which statement is not correct?

a. There is a limit for electricity consumption per capita and this consumption does not increase indefinitely.

b. The majority of the world population lives in economies that have not reached the level of economic development that results in flat or declining electricity consumption per capita.

c. For the underdeveloped population, the growth in the electricity consumption will continue, along with improved living standard until they are adequately developed.

d. The total global electricity consumption will not continue to grow in near future.

1.80. There is no limit for electricity consumption per capita, and this consumption can increase indefinitely.

a. True

b. False

1.81. The majority of the world population lives in economies that have reached the level of economic development that results in flat or declining electricity consumption per capita.

a. True

b. False

1.82. The total global electricity consumption will continue to grow in near future with potentially devastating impacts.

a. True

b. False

1.83. Why is electricity consumption per capita a more accurate measure of living standard than per capita energy consumption?

1.84. Each American on average consumes roughly as much as ….. Chinese (use Table 1.1).

a. 2

b. 3

c. 12

d. 63

e. 368

f. 837

1.85. The per capita electricity consumption gap between developed and developing countries has been closing for most countries in the past few decades.

a. True

b. False

1.86. Explain how Figure 1.21 can be used as an indication of electricity consumption in various parts of the world.

1.8.4 Structure of Electrical Infrastructure

1.87. Name the main components of the modern electrical infrastructure.

1.88. How are transmission lines and distribution lines distinguished?

1.89. Which statement is correct?

a. There is no way to store electrical energy.

b. There is no way to economically store electrical energy.

1.90. What is the most important characteristic of electricity that distinguishes it from other commodities produced by human beings?

1.91. Electricity should be generated whenever it is needed and transmitted and distributed to wherever it is needed. Why?

a. True

b. False

1.92. Provide two units for power and two units for energy.

1.93. Typically the behavior of a single user can be unpredictable. But when many users are compounded, their overall load variation is to some degree predictable.

a. True

b. False

1.94. What is a load profile? Name three parameters that can be determined from it?

1.95. Why do we study a load curve or a load profile?

1.96. A load curve is typically based on a ......

a. a year

b. a season

c. a weak

d. 24-hour period

1.97. A load curve is typically based on a 24-hour period and is prepared for everyday for several years and then is averaged to get a general pattern.

a. True

b. False

1.98. Explain how the customer type, e.g. residential, commercial, or industrial; the season of year e.g. summer or winter; ambient temperature; the day of week, e.g., weekdays or weekends; and special occasions, e.g., holidays or sports events impact electricity consumption is residential applications.

1.99. The variation in a load curve (or a load profile) depends on:

a. ambient temperature
b. day of week, e.g., weekdays or weekends
c. special occasions, e.g., holidays or sport events
d. all of the above

1.100. Define each of the following parameters.

- Maximum load demand
- Baseload demand
- Average load
- Load factor (LF)
- Demand factor
- Coincidence factor

1.101. In a load curve, how you can graphically determine each of the following parameters.

- Maximum load demand
- Baseload demand
- Total electrical energy consumption
- Average load
- Load factor (LF)

1.102. Explain the difference between the demand factor and the coincidence factor.

1.103. Define the plant capacity factor (PCF). Why is the plant capacity factor important?

1.104. The capacity factor depends on the

a. Fuel type
b. Plant design
c. Age
d. Configuration
e. All of the above

1.105. Determine the capacity factor for a power plant with the installed capacity of 2000 MW that produces 150,000 MWh per year.

1.106. What are the three factors contributing to the lower than unity capacity factor? Which one(s) are out of the control of the power plant?

1.107. If a power generation system operates at a reduced load or is shut down because electrical energy is not needed or it is not economical to produce electricity due to a low price of sale, the capacity factor increases.

a. True

b. False

1.108. Explain the situations when the following power generation systems may not be available due to unavailability of energy sources.

- Gas turbine
- Wind turbine
- Solar photovolaic unit
- Hydroelectric plant

1.109. Define the availability factor of a power plant. What are the differences between the capacity factor and the availability factor in a power plant?

1.110. Define the utilization factor or use factor of a power plant.

1.111. A power plant operates 4 hours a day, 45 weeks a year. If on average the plant is not available for 5 hours a week for scheduled or unscheduled maintenance, determine the plant's availability and utilization factors.

1.112. In the load curve presented below, determine the maximum and base demands, total electrical energy consumed, daily average load, load factor, and demand factor. Assume that the maximum possible demand is 2 GW.

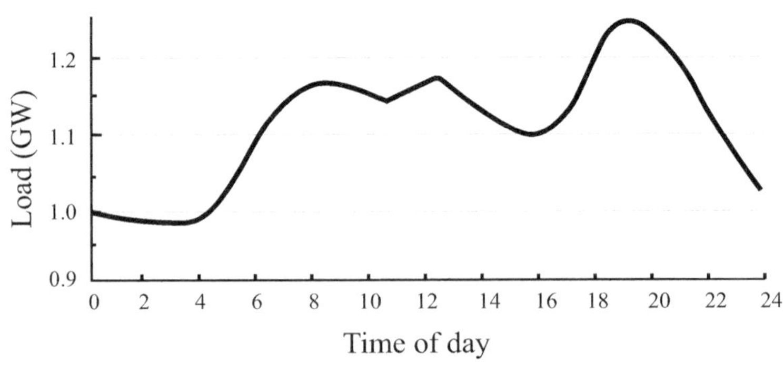

1.113. The factors impacting the selection of power plant technology and size are .... (check all the correct ones)

a. economic considerations

b. social considerations

c. environmental considerations
d. urgency of the need for power plant

1.114. The number and capacity of power generation units should be in a way that the produced power can approximately meet the load curves.
a. True
b. False

1.115. Many factors affect the projection of the load demand increase in the future, including ...... (check all the correct ones).
a. trend in population growth
b. planned new industries
c. expansion of the existing industries
d. impacts of technological advances on electricity consumptions
e. regulation changes in the region of interest

1.116. Why should power plants in a network be of various capacities?

1.117. The number of small units should be limited to avoid typically low efficiency of these units.
a. True
b. False

1.118. The total capacity of a power generating system should be large enough to permit maintenance of units without affecting the capability of the system to provide the maximum demand load.
a. True
b. False

1.119. Which statement is accurate?
a. Baseload power plants are devoted to the production of baseload supply.
b. Baseload power plants meet continuous energy demand.
c. Baseload power plants typically produce energy at a constant rate.
d. All of the above.

1.120. Select the characteristics of a baseload power plant (check all the correct ones).
a. Low-cost generation
b. High efficiency
c. Low plant capacity factor
d. High fixed costs
e. Very high marginal costs

1.121. List power plant technologies that are commonly employed or can be employed in future as baseload plants.

1.122. When are peak-load power plants, also known as peaking or peaker plants, used?

1.123. Which one is correct for peak-load power plants?

a. Usually not very efficient

b. Low fixed costs

c. High marginal costs

d. All of the above

1.124. Peak-load power plants can operate for many hours a day or as little as a few hours per year.

a. True

b. False

1.125. If natural gas is the inexpensively available fuel, a combined cycle power plant might be the best choice for the power plant technology.

a. True

b. False

1.126. To meet peak loads, power plants with .... efficiency are the ones operated first.

a. higher

b. lower

c. does not matter

1.127. Regardless of the operating mode that the plant is designed for (i.e., baseload, intermediate, or peaking power plant), during the later operation it may be used for an entirely different operation mode.

a. True

b. False

1.128. Retrofitting existing power plants is to ....

a. extend the useful life and/or upgrading operating plants

b. improve their efficiency, reliability, and capacity

c. reduce their environmental impacts

d. all of the above

1.129. The location of a power plant depends on many factors including ...... (check all the correct ones).

a. availability of appropriate site (for the current plant and for the future possible expansions)

b. availability of reliable inexpensive fuel sources (current and future projection)

c. availability of cooling water sources
d. availability of transmission lines
e. environmental considerations
f. impacts on local communities
g. availability of labor during construction and operation
h. regional climate and potential hazardous events (such as high wind, flood, and earthquake)
i. regulations (environmental, taxes, etc.)

1.130. What is the best tool in selecting the specifications of a new power plant? Why?

1.8.5 Electricity Transmission Lines

1.131. In order to take advantage of baseload and peak-load power plants, power stations should be interconnected by a power grid.
a. True
b. False

1.132. What are the benefits of the concept of electrically connected power plants?

1.133. Why was DC electricity only used for local power production and distribution?

1.134. The low-voltage electricity meant that to transmit a significant amount of electric power, low current was required. Why?
a. True
b. False

1.135. The power grid is created to .....
a. ensure economical operation
b. reduce reserve capacity
c. increase reliability of supply
d. exchange peak loads
e. continue using less efficient plants
f. all of the above

1.136. Write the equation to determine the power transmitted through a line?

1.137. The low-voltage and high-current electricity transmission meant a significant loss of power in long-distance transmissions.
a. True
b. False

1.138. Explain why nowadays hydropower plants can be built on sites that commonly are in remote locations or fossil fuel power plants can be built close to fuel sources.

1.139. Explain how the grid can help to take advantage of economy of scale.

1.140. Name two major and three minor grids or interconnections in North America (the U.S. and Canada).

1.141. All power utilities in an interconnection are operated in harmony, i.e. they are synchronized.

a. True

b. False

1.142. What does differentiate the electricity in two different interconnections?

1.143. How can interconnections exchange electricity?

1.144. What is the application of high-voltage direct current (HVDC) in a grid?

1.145. A power plant produces 50 MW electric power, which is transmitted through a high-voltage line at 115 kV to a consumption center. If the resistance of the line is 5 Ω, determine the power loss in the transmission line. Determine the lost if the power was transmitted through a 34 kV line.

1.146. A power plant produces 50 MW electric power, which is transmitted through a high-voltage line. If the resistance of the line is 5 Ω, determine the transmission voltage that no power can be transmitted because all power will be lost.

1.147. The power loss in power lines can be determined using the equation $P_L = I^2R$. In this equation, if the current is replaced by voltage from the Ohm's law, $V = IR$, the power loss equation will be $P_L = \frac{V^2}{R}$. This new equation implies that increasing voltage not only does not reduce the power loss but increases it. Explain (use the numerical values in Problem 1.145).

1.148. A power plant produces 100 MW electric power, which is transmitted through a high-voltage line at 115 kV to a consumption center. If the resistance of the line is 10 Ω, determine the power loss in the transmission line. Determine the loss if the power was transmitted through a 34 kV line.

1.149. A power plant produces 100 MW electric power, which is transmitted through a high-voltage line. If the resistance of the line is 10 Ω, determine the transmission voltage that no power can be transmitted because all power will be lost.

1.8.6 Regulated versus Deregulated Electricity Market

1.150. What is a utility?

1.151. Why some industry were/are regulated even in free-market economies?

1.152. What is a regulated electricity market? What are its advantages and disadvantages?

1.153. Due to the lack of competition, a regulated electricity market is usually inefficient and slow in responding to changes. However, they are reliable and stable.

a. True
b. False

1.154. In a region, you can choose your electricity provider from several providers competing in the market. What type of market is this?

a. Regulated electricity market
b. Deregulated electricity market
c. Combined electricity market

1.155. Describe the "California experience" in a deregulated electricity market.

1.156. Compare regulated and deregulated electricity markets.

1.157. Explain a-day-ahead-mechanism in a deregulated electricity market.

1.158. The ultimate objective of the electricity market deregulation is to reduce the cost of electricity for customers by introducing competition to the market.

a. True
b. False

1.159. A typical residential user only pays for the amount of consumed electrical energy (kWh) and a relatively small fixed (meter) charge.

a. True
b. False

1.160. The cost of the unit of electrical energy may vary seasonally and/or based on the monthly energy usage.

a. True
b. False

1.161. List the items that a large consumer pays for in the utility bill.

1.162. A manufacturing facility is located in a very hot region. On average, the facility consumes 200,000 kWh electricity per month. During the peak demand period of June to September, its average maximum power consumption is 1,200 kW. The utility company charges the facility the fixed monthly meter charge of $100/month, the flat electrical energy rate of 6 cents per kWh, and $14.00 per kW for maximum power consumption during peak periods.

a. Determine the average annual cost of electricity for the facility.

b. After an energy audit and an energy consumption optimization project, the facility reduces its electricity consumption by 15% and its maximum power consumption during the peak demand period by 25%. Determine the percentage of cost reduction.

1.163. A large commercial unit in Arizona consumes 10 MWh in July with a maximum power consumption of 0.1 MW during the peak demand period. Determine the cost of consumed electricity by the facility for that month if the fixed monthly meter charge, the flat rate of energy consumption, and the demand charge are $90.00 per month, 8.0 cents per kWh, and $14.00 per kW for the month of July.

1.164. A manufacturing facility is located in a hot region. On average, the facility consumes 300,000 kWh electricity per month. During the peak demand period of June to September, its average maximum power consumption is 1000 kW. The utility company charges the facility the fixed monthly meter charge of $100/month, the flat electrical energy rate of 5 cents per kWh, and $14.00 per kW for maximum power consumption during peak periods.

a. Determine the average annual cost of electricity for the facility.

b. After an energy audit and an energy consumption optimization project, the facility reduces its electricity consumption by 15% and its maximum power consumption during the peak demand period by 25%, determine the percentage of cost reduction.

1.8.7 Introduction to Conventional Power Generation Systems

1.165. Provide an example for each of the following electricity generation methods:

- Electromagnetic induction
- Electrochemistry
- Photoelectric effect
- Thermoelectric effect
- Piezoelectric effect
- Nuclear transformation

1.166. Electricity generation systems can be categorized based on the original source of energy. List these categories with examples.

1.167. List the four required main processes to convert thermal energy to work in a thermodynamic cycle.

1.168. Compare the following power systems:

a. Open power systems versus closed power systems

b. Vapor power systems versus gas power systems

c. Internal combustion engines versus external combustion engines

d. Device involving shaft work versus device involving boundary-moving work

1.169. Identify which of the following equations is used to determine work is for a closed system.

a. $w = \int_1^2 pdv$

b. $w_{1-2} = -\int_1^2 vdp$

1.170. Explain the importance of $w_{1-2} = -\int_1^2 vdp$ in power generation cycles.

1.171. Everything else being equal, compare the efficiency of cycles with a phase-changing working fluid and the efficiency of cycles with a gaseous working fluid.

1.172. Compare efficiency and heat rate of a power plant. Describe their similarities and differences.

1.173. In a gas turbine, the turbine, compressor, and generator are all on a single shaft. The turbine produces 300 MW power. The compressor consumes 120 MW. The efficiency of the generator is 98%. If the plant consumes 5% of the generator's output power internally for its operation, determine the BWR and the unit gross and net efficiency and heat rate. The input thermal energy to the unit is 500 MW.

1.174. Explain why we study idealized power generation systems rather than actual systems. Also explain what we learn from these idealized systems.

1.175. Complete the following table.

Type of Power Plant	Ideal Thermodynamic Cycle	Working Fluid Phase
Steam power plant		
Gas turbine		
Combined cycle		

1.8.8 Open-Ended Questions

1.176. Select three countries, one developed, one emerging, and one developing countries. Find their populations, GDP, and electricity consumption. Determine their per capita GDP and electricity consumption. Compare the three countries.

1.177. Select two geographical regions such as countries, continents, and states. Make sure that the required data are available for your region of choice. Find data you need to recreate the diagrams for energy consumption in your regions of interest. Choose at least two diagrams from this chapter. Alternatively, you can come up with other energy-related diagrams. Create the appropriate diagrams. Interpret what the diagrams are representing. Compare your two regions for each diagram. Do not forget to properly cite your sources. Make sure that you distinguish the difference between the primary and secondary energy sources.

1.178. For your state or country identify the share of various primary sources in the energy mix. What is the share of each end-use energy consumers?

1.179. For your state or country find the current electricity generation and consumption. What is the share of various primary sources for electricity generation? What are the possible future electricity demands? What are potential primary sources to meet future demand?

1.180. What is the peak demand period in your city? What is the peak electrical demand in your state, or country? Both in summer and winter.

1.181. There is a lot of truth in the statement "we can learn more from our mistakes than from our successes". Unfortunately, usually the mistakes are not properly documented and the lessons are sometimes lost. Find a disaster related to the energy industry. Then, explain the background, what happened, why happened, how could have been prevented, and most importantly the lessons and the actual impacts of the event. Report your findings. Prepare a 10-minute presentation based on your report.

1.182. Repeat the previous assignment for a failure in the electricity generation industry. Write a one-page executive summary about your findings.

1.183. Research what the differences between brownout and blackout are. Identify the latest and/or the largest blackout in your region. Explain the reason(s) for the blackout and how it can be prevented in the future.

1.184. List at least five electricity consuming devices in your home or workplace. Find out how much power they consume. What is the greatest source of electrical energy consumption in your home? What is the device with the highest electrical power? Use the specification label for devices and/or a wattmeter.

1.185. Prepare an oral presentation on the impacts, positive and negative, of energy on your life on both the individual level and societal level.

1.186. Identify at least five major future problems that the energy sector of your region of choice is facing. Select one issue and write a report about it.

1.187. Identify at least three major future problems that the U.S. (or your country's) electrical infrastructure is facing. Select one issue and write a report about it.

1.188. Prepare a poster on the impacts of energy on the environment on the local, national, or global scale.

1.189. Write an essay on if and how energy issues impacted election(s) in your country/region.

1.190. Locate three power plants in the region you live (the closer to where you live, the better). Identify what technology they use, what the sources of primary energy are, the size and age of the plants, whether they are peak-load or baseload power plants, and any reported safety incidents they had in the past or potentially can have in the future. What about environmental impacts?

1.191. Locate the largest power plants in your country or state. Identify what technology they use, what the sources of primary energy are, the size and age of the plants, and any reported safety incidents they had in the past or potentially can have in the future. What about environmental impacts?

1.192. Check your home's electric and natural gas bills. What is the unit of consumed energy? What is the cost of energy? How the total cost is calculated? Explain your answers. If you do not have access to these bills, just google one.

1.8.9 General

1.193. Determine the capacity factor for a power plant with the installed capacity of 1000 MW that produces 100,000 MWh per year.

1.194. The demand charge is the charge for the maximum power (time rate of energy) usage during the peak demand period.

a. True

b. False

1.195. Peak-load power plants typically have ....(chose all correct ones).

a. low efficiency

b. high capital costs

c. fast load-following character

d. low plant capacity factor

e. high marginal costs

1.196. Explain how betavoltaic batteries work.

1.197. The following figure shows the contribution of various primary energy sources to the global energy consumption in 2017. Explain what the chart signifies.

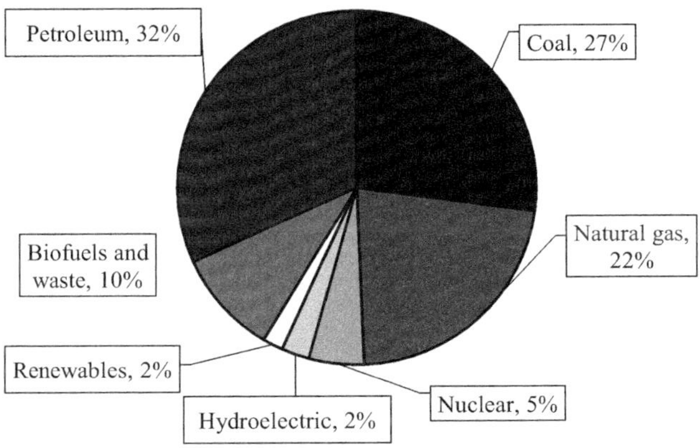

1.198. For the underdeveloped population, the growth in electricity consumption will continue until they improve their living standard.

a. True

b. False

1.199. In the above diagram explain why the share of biofuels is so high in the global primary energy consumption.

1.200. What is a regulated market?

1.201. What are the typical availability factors for the following power generation technologies?

Steam power stations

Gas turbines

Wind farms

Photovoltaic solar units

Hydroelectricity plants

1.202. What is the peak demand period?

1.203. The electrical demand of consumers and as a result the total demand vary over a given period of time and is totally unpredictable.

a. True

b. False

1.204. Explain how the type of electricity customer impacts electricity consumption in residential applications.

1.205. Constructing a new power plant is typically the best option to meet growing power demand.

a. True

b. False

1.206. What is the criterion to distinguish transmission lines and distribution lines?

1.207. For all industrially developed economies, the per capita electricity generation has remained steady or declined in recent years.

a. True

b. False

1.208. Explain how the time of the day impacts electricity consumption in commercial applications.

1.209. What is the availability factor?

1.210. The total electricity generation in the U.S. has declined since 2005.

a. True

b. False

The following figures illustrate the share of each of energy end-use sectors in the total primary energy consumption in the globe and the U.S. in 2018. Answer 1.211–1.216 using these diagrams.

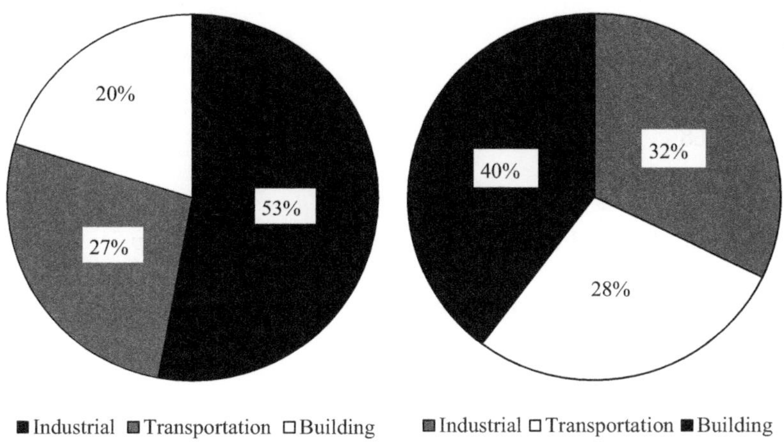

(a) (b)

1.211. Globally, over half of the total energy is consumed in the building sector while in the U.S. the share of this sector is about one-third.

a. True

b. False

1.212. The share of the building sector in energy consumption is 40% in the U.S. compared to 20% in the world.

a. True

b. False

1.213. The share of the building sector in energy consumption is 40% in the U.S. compared to 20% in the world. Why?

1.214. What does retrofitting an existing power plant mean?

1.215. Explain how piezoelectric electricity generators work.

1.216. In recent years the per capita electricity generation has declined in the U.S.

a. True

b. False

1.217. All industrially developed economies reached a steady flat per capita electricity generation at different values and times.

a. True

b. False

1.218. What are the advantages and disadvantages of a regulated electricity market?

1.219. Each American on average consumes roughly as much as ….. Indians.

a. 2

b. 3

c. 12

d. 63

e. 368

f. 837

1.220. Explain the application of this equation: $w_{1-2} = -\int_1^2 v dp$

1.221. The first encounter of human being with electricity was with …..

a. eel fish

b. rubbing an amber rod with cat's fur

c. lightening

d. the Parthian Battery

1.222. Electricity consumption per capita is a more accurate measure of living standard than per capita energy consumption.

a. True

b. False

1.223. Two factors distinguish electricity generation and consumption: electricity import and export and losses during electricity transmission and distribution. Write the equation that relates electric energy consumption to electricity generation.

1.224. When one rubs an amber rod on cat's fur, ... electricity is generated.

a. static

b. AC

c. DC

d. dynamic

1.225. Which one of the capacity factor or availability factor does provide a more accurate indication of how available the plant is in terms of maintenance?

1.226. Depending on the context, the term load may refer to the electrical power consumption in a TV or on an entire nation.

a. True

b. False

1.227. The following figure shows the contribution of various primary energy sources to the global energy production between 1990 and 2017. Compare and contrast the share of coal and other energy sources over the years.

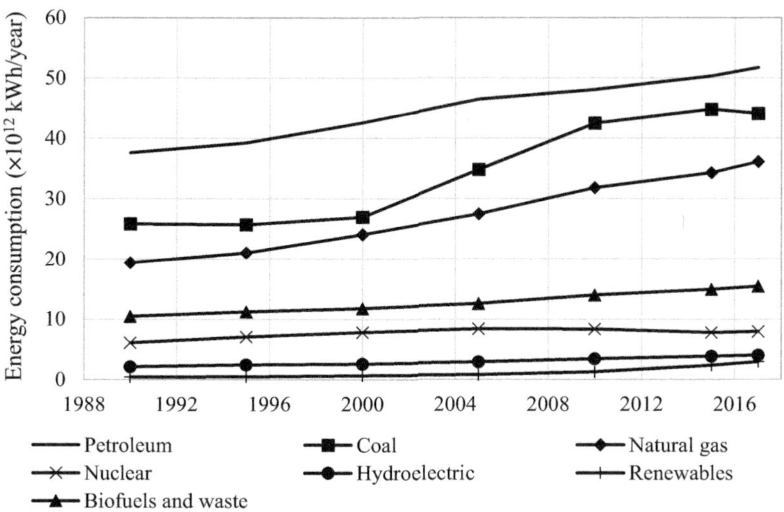

1.228. Define the demand factor and explain how it affects power generation planning.

1.229. The global annual primary energy consumption continues to increase with no sign of meaningful reduction.

a. True

b. False

1.230. Explain how the season, e.g., summer or winter, impacts electricity consumption.

1.231. List some of the important stakeholders in the selection of the power plant technology.

1.232. The capacity factor reduces if the system operates at a reduced load or is out of service due to a failure in equipment, or routine or unplanned maintenance.

a. True

b. False

1.233. What is the demand charge?

1.234. What is the relationship between efficiency and heat rate?

1.236. The number and capacity of power generation units should be in a way that the produced power can approximately meet the load curves for various periods of time, i.e., daily, monthly, and annual curves, according to the projection of the load demand increases in the future.

a. True

b. False

1.236. List three factors that demands for electricity depend on.

1.237. In a regulated electricity market, the utility still owns, operates, and maintains the power transmission and distribution sectors to deliver electricity to end-consumers but customers can buy power from the retailer of their choice.

a. True

b. False

1.238. Conversion of which one of the following forms of energy to electricity is most popular for large scale electricity generation?

a. Electromagnetic induction

b. Electrochemistry
c. Photoelectric effect
d. Thermoelectric effect
e. Static electricity
f. Piezoelectric effect
g. Nuclear transformation

1.239. Why do we always transmit electricity at high voltages?

1.240. When (time of the day) are the typical peak demand periods in winter and summer? Why?

1.241. The demand for electricity variation depends on:

a. season
b. day of the week
c. hour on the day
d. climate of a region
e. extent of air conditioning
f. presence of heavy industry
g. weather
h. all of the above

1.242. It is predicted that the global annual primary energy consumption will .... in the foreseeable future.

a. increase
b. decrease
c. remain unchanged

1.243. It is predicted that the global annual primary energy consumption will increase at the rate of about 1.2% per year in the foreseeable future.

a. True
b. False

1.244. The factors impacting the selection of the power plant technology and size are not all technical.

a. True
b. False

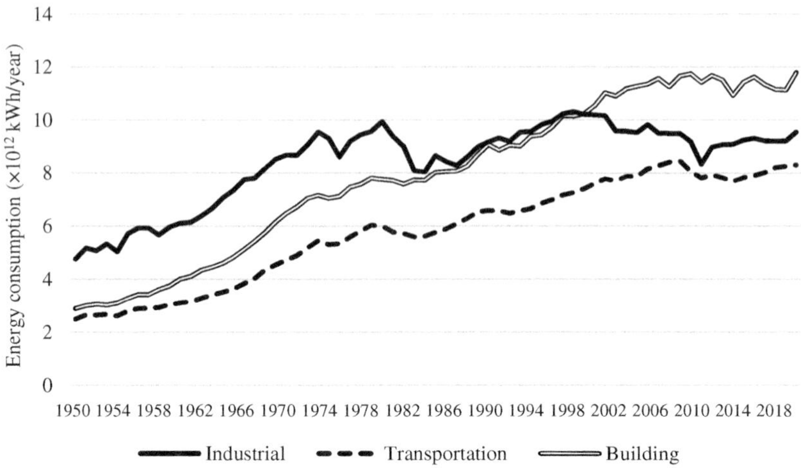

1.245. The following figure shows the energy consumption in each sector in the U.S. between 1950 and 2018. Explain the growth of energy consumption in each sector over years.

1.246. The patterns of variations in the total and per capita electricity generation in the U.S. and the world are totally different.

a. True

b. False

1.247. Which one is incorrect for peak-load power plants?

a. Their period of operation may range from many days to a few hours a year with highly variable loads.

b. Due to the limited operational period, it is not economical to use high-efficiency capital-intensive plants as peaking load power plants.

c. They are slow load-following plants.

1.248. A variable electric resistance is connected to a constant voltage. If we increase the resistance, what will happen to the current?

1.249. The results of the introduction and implementation of the deregulated electricity market policies have been mixed. Some markets have been able to reduce the cost for end-users, at least to some degree, while others have been a disaster and major failure

a. True

b. False

1.250. What are the two methods to calculate the cost of electrical energy consumed by customers?

1.251. Explain how the time of the day impacts electricity consumption in industrial applications.

1.252. Define the efficiency of a power cycle.

1.253. For each of the following cases identify if the energy source is primary or secondary and renewable or non-renewable.

Fossil fuels

Nuclear energy

Hydropower

Electricity

Solar energy

Wind energy

Wave energy

Hydrogen

Ocean thermal energy

Biomass

Tidal energy

Diesel fuel

Geothermal

Solid waste

Heat recovery

1.254. To meet the demand, the most efficient power plants are brought online first.

a. True

b. False

1.255. What are the three types of electricity customers?

1.256. The efficiency of a power plant is ….. the heat rate of the plant.

a. directly proportional to

b. inversely proportional to

c. equal to

d. independent of

1.257. The urgency of the need for s power plant affects the type and technology of the plant because the time needed to obtain required permissions as well as to design, construct, and install the plant depend on the technology.

a. True

b. False

1.258. What are the two items in the electricity bill of a typical residential user?

1.259. Which equation is for work in a control volume and which one is for work in a control mass?

a. $w = -\int_1^2 v\,dp$

b. $w = \int_1^2 p\,dv$

1.260. Do you prefer a higher or lower plant capacity factor?

1.261. In a control volume, process 1–2 is a constant pressure process. Determine the work involved in this process.

1.262. Each person in China or India consumes far less than what an American or a Japanese consumes

a. True

b. False

1.263. List three reasons that the capacity factor of a power plant is reduced.

1.264. How does the urgency of the need for a power plant affect the type and technology of a new power plant?

1.265. The following two figures illustrate the total and per capita electricity generation in India and Japan between 1980 and 2018. Compare the two countries. What are the similarities and differences?

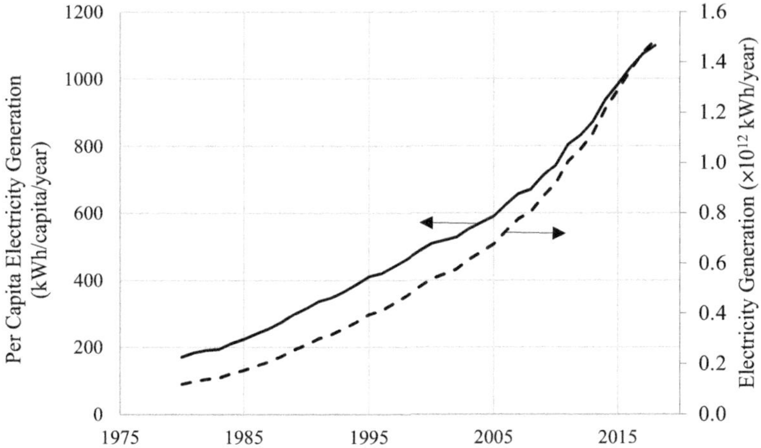

(a)

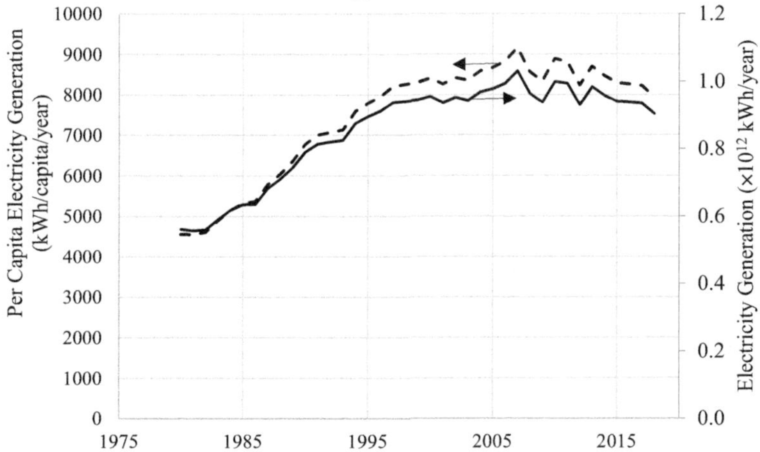

(b)

1.266. It is predicted that the world population will increase by about 30% by 2050.

a. True

b. False

1.267. Explain how the day of the week, e.g., weekdays or weekends, impacts the load profile.

1.268. What is a deregulated electricity market?

1.269. Large customers pay for the electrical energy usage, a fixed monthly meter charge, and a demand charge.

a. True

b. False

1.270. What are the ranges of the period of operation in a year for base, peak, and intermediate load power plants?

1.271. The variation in a load curve or a load profile depends on:

a. time of day

b. customer type, e.g., residential, commercial, or industrial

c. season of year, e.g., summer or winter

d. all of the above

1.272. Primary energy sources .... be directly found in nature.

a. cannot

b. can

1.273. The solid line in the following figure illustrates the annual primary energy consumption in the U.S. between 1949 and 2018. Explain your interpretation of the diagram.

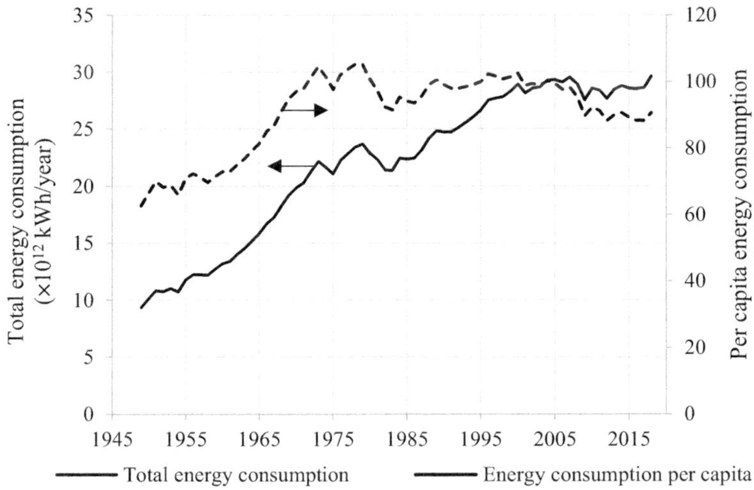

1.274. The dashed line in the above figure illustrates the per capita annual primary energy consumption in the U.S. between 1949 and 2018. Explain your interpretation of the diagram.

1.275. The per capita energy consumption in the U.S. has markedly decreased in the new millennium (use above diagram).

a. True

b. False

1.276. Explain how the time of the day impacts electricity consumption in residential applications.

1.277. What does the plant capacity factor indicate?

1.278. The variation in a load curve or a load profile depends on the time of day; customer type, e.g., residential, commercial, or industrial; season of year, e.g., summer or winter; ambient temperature; the day of week, e.g., weekdays or weekends; and special occasions, e.g., holidays or sports events.

a. True

b. False

1.279. The combined impact of the increase in the world population and the higher per capita electricity generation will increase the global electricity generation (and consumption) by about 78% by 2050.

a. True

b. False

1.280. The figure below illustrates the total and per capita electricity generation in China between 1980 and 2018. Explain what the diagram indicates.

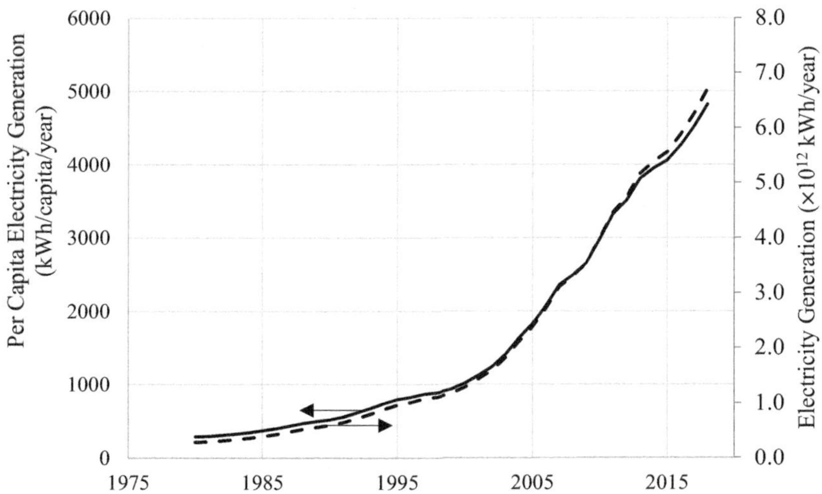

1.281. While the electricity consumption per capita in China has increased significantly in the 20th century, in the past two decades the consumption has been fluctuating within a certain range without any net increase (use the above diagram).

a. True

b. False

1.282. How is the plant capacity factor different from the load factor?

1.283. Peak-load power plants operate to provide the difference between the baseload and the actual load demand, especially during a high demand period.

a. True

b. False

1.284. What are some of the energy consumption examples in the building sectors (three examples)?

1.285. What is the ultimate objective of the electricity market deregulation?

1.286. All industrially developed economies reached a steady flat per capita electricity generation at different times but more or less at the same value.

a. True

b. False

1.287. Explain how special occasions, e.g., holidays or sports events, impact the load profile.

1.288. Which one is incorrect for baseload power plants?

a. The power output of these units does not vary to match load demands.

b. These units are typically able to generate electricity at a low cost which means they should be highly efficient power plants.

c. They typically require low capital costs.

d. They usually have high plant capacity factor and low marginal costs.

1.289. In the industrial sector, energy sources can be used ....

a. as input materials

b. for manufacturing processes

c. as input materials and for manufacturing processes

1.290. Explain the difference between power and energy.

1.291. Explain regulated and deregulated power markets very briefly (2 lines). Identify who owns the transmission and distribution systems in each case and how the price of electricity is determined.

1.292. Which one is correct for baseload power plants?

a. Baseload power plants are plants that are dedicated to meeting the baseload demand.

b. Since these power plants should produce the minimum required power for a given period, they should operate continuously at full capacity at all times throughout the period, including for scheduled or unscheduled maintenances.

c. Typically if a power plant is operational for 6000 hours or less a year (out of 8760 hours in a year), it is considered to be a baseload plant.

d. The power output of these units varies to match load demands.

1.293. Name two major grids or interconnections in North America (the U.S. and Canada).

1.294. GDP per capita is commonly used as an indication of the quality of living for a specific region.

a. True

b. False

1.295. The following two figures illustrate the total and per capita electricity generation in China and Japan between 1980 and 2018. Compare the two countries. What are the similarities and differences?

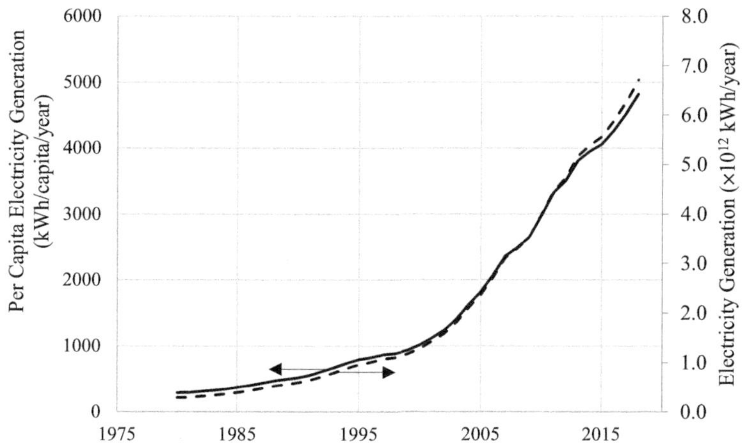

(a)

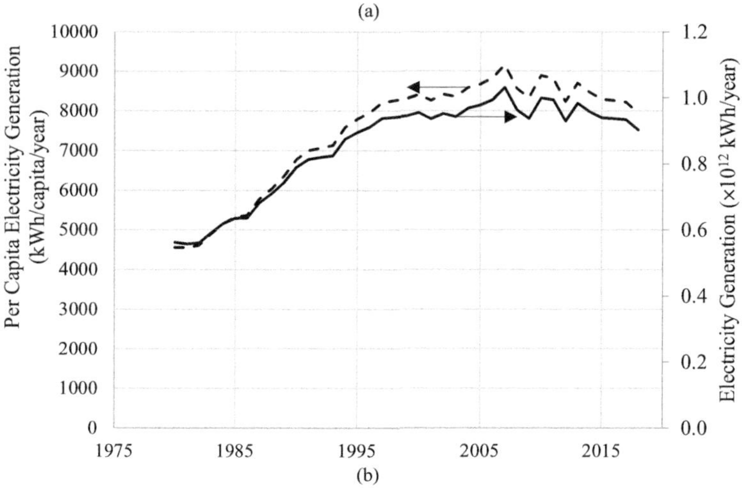

(b)

1.296. Define gross domestic product (GDP).

1.297. Why should a power generation system include a proper mixture of baseload and peak-load power plants?

1.298. According to the World Bank, only about 16% of the world population lives in high-income economies, and remain 84% of the population lives in low- and middle-income countries.

a. True
b. False

1.299. Is the demand load power or energy?

1.300. The capacity factor depends on the fuel type as well as plant design, age, and configuration.

a. True
b. False

1.301. Which one(s) cannot be used to provide the peak load?

a. Gas turbines
b. Steam power plants both fossil fuel and nuclear
c. Hydroelectric power plants
d. Reciprocating engines
e. Pumped storage units
f. Other storage technologies
g. Geothermal power plants

1.302. Explain how the grid helped to determine the locations of power plants by parameters other than just closeness to centers of demand.

1.303. If coal is the inexpensively available fuel, a combined cycle power plant might be the best choice for a power plant technology.

a. True
b. False

1.304. The total energy consumption of a society is ……. the population.

a. proportional to
b. independent of

1.305. Name at least three natural sources of energy.

1.306. If we reduce the current by half while the electrical resistance is constant, the voltage will …..

a. double
b. reduce by half
c. not change
d. depends on many factors

1.307. In order for a power generation system to be able to meet required load demands, it should include a proper mixture of two types of power plants: baseload and peak load.

a. True

b. False

1.308. The selected units in a power generation system should be of various capacities to roughly fit the variations in the load curve with minimum partial-load operation.

a. True

b. False

1.309. A power grid can ……..

a. improve reliability of produced power

b. ensure economical operation of plants

c. permit continued operation of older plants

d. all of the above

1.310. It is predicted that the world population will reach 9.8 and 11.2 billion by 2050 and 2100, respectively.

a. True

b. False

1.311. Explain how the day of week, e.g., weekdays or weekends, impacts electricity consumption.

1.312. List three main charges in the cost of electricity usage for a large commercial unit.

1.313. While in recent years the per capita electricity generation has declined in the U.S., a similar decline cannot be seen for the total electricity production. This is due to the continued growth in the population as the U.S. population increased by about 6% between 2010 and 2018.

a. True

b. False

1.314. The figure below shows the global electricity generation between 1980 and 2018. Explain what the diagram indicates about the past and the future projection.

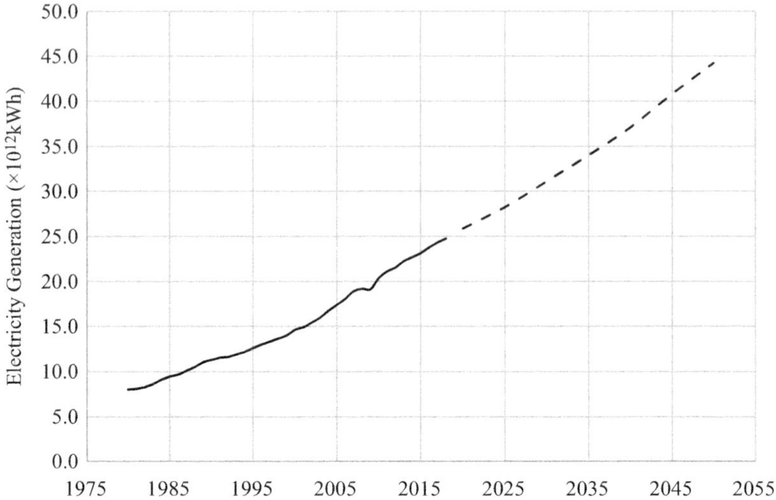

1.315. Gross domestic product (GDP) is a parameter that is determined based on ……..

a. the market value of all products, including goods and services, exported by a country or a region in a particular period
b. the market value of all products, including goods and services, imported by a country or a region in a particular period
c. the market value of all products, including goods and services, imported or exported by a country or a region in a particular period
d. none of the above

1.316. A too high or too low water level behind the dam should be avoided to enable the operation of recreational facilities if any.

a. True
b. False

1.317. Write the equation to determine the power loss through a line?

1.318. Name three sectors of the energy industry.

1.319. When selecting generating units' specifications the following considerations should be taken into account. Describe each item and its importance.

- Number and size of units
- Minimal partial-load operation
- Units to be of different capacities
- Spare generating capacity
- Avoid selecting smaller units

1.320. How can the voltage of AC power be changed? Can the same method be used for DC power?

1.321. Typically the behavior of a single user can be unpredictable. But when many users are compounded, their overall load variation is to some degree predictable.

a. True

b. False

1.322. Which one(s) cannot be used as a baseload power plant?

a. steam power plants both fossil fuel and nuclear

b. hydroelectric power plants

c. geothermal power plants

d. solar *PV* panels

e. solar thermal power plants with storage

f. ocean thermal energy conversion power plants

1.323. Each American on average consumes roughly as much as ….. Chadians.

a. 2

b. 3

c. 12

d. 63

e. 368

f. 837

1.324. Secondary energy sources are also called energy carriers. Why?

a. True

b. False

1.325. Name two types of electricity generation technologies that are suitable for baseload power generation and two technologies suitable for peak-load power generation.

1.326. If we double the voltage while the electrical resistance is constant, the current will …..

a. double

b. reduce by half

c. not change

d. depend on many factors

1.327. The power grid can be created by connecting several electricity generating stations together in parallel.

a. True

b. False

1.328. When planning for a new power plant, the projected load profiles can determine which type, baseload or peaking load, should be constructed.

a. True

b. False

1.329. Explain how the level of economic development of a particular region or country impacts the amount of energy consumption and share of various energy consumption sectors.

1.330. Complete the following table, which is listing actual power systems and corresponding ideal cycles.

Actual System	**Ideal Cycle**
Steam power plant	
Gas turbine	
	Otto cycle
Compression-ignition internal combustion engine	

1.331. The global per capita electricity generation has increased by about 80% since 1980.

a. True

b. False

1.332. What is the unit of voltage?

1.333. Which of the following energy sources are primary sources of energy? (Select all the correct ones)

a. nuclear energy

b. electricity

c. hydro energy

d. diesel fuel

e. geothermal energy

1.334. What is the relationship between voltage, current, and electrical resistance?

1.335. In the past four decades, the global electricity generation has tripled.

a. True

b. False

1.336. The figure below illustrates the total and per capita electricity generation in Japan between 1980 and 2018. Explain what the diagram indicates.

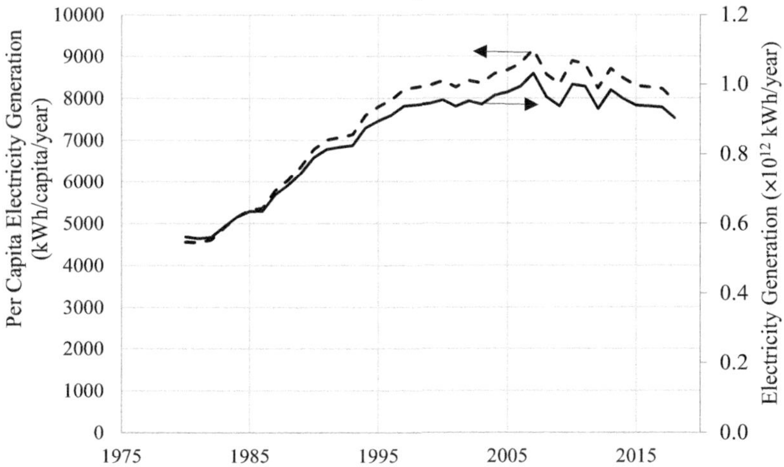

1.337. Explain how the living standard of a particular region or country impacts the amount of energy consumption and share of various energy consumption sectors.

1.338. What is the unit of electrical resistance?

1.339. Not all industrially developed economies reached a steady flat per capita electricity generation.

a. True

b. False

1.340. Which economy does consume more energy in the building sector: a manufacturing-focused economy or a service-focused economy? Why?

1.341. What is the equation for electricity transferred by an electric line?

1.342. We need an infrastructure to convert naturally occurring energy forms to more useful forms and to transport them.

a. True

b. False

1.343. Is there an interdependency between the annual energy consumption per capita and the annual GDP per capita? Why?

1.344. Why should the number of small units in an electric network be limited?

1.345. In most cases primary energy sources need to be converted to useful energy forms.

a. True

b. False

1.346. What are the factors that impact energy consumption in the industrial sector?

1.347. In order to understand the energy industry, we need to look into three sectors:

a. Energy use
b. Energy conversion, transportation, and distribution
c. Energy resources
d. All of the above

1.348. Explain how the geography of a particular region or country impacts the amount of energy consumption and share of various energy consumption sectors.

1.349. There is a deep inequality in per capita electricity consumption at the global level.

a. True
b. False

1.350. Explain how the customer type, e.g. residential, commercial, or industrial, impacts electricity consumption.

1.351. In many dams, too low water flow rates should be avoided to protect fish downstream.

a. True
b. False

1.352. Why should the total capacity of a power generating system include some extra capacity (over the maximum demand load)?

1.353. The per capita electricity generation in the U.S. increased until 2007 at the rate of about 13,500 kWh/capita/year.

a. True
b. False

1.354. Explain why a hydroelectric plant may not be able to operate at a full capacity due to limitations on downstream water flow rate or limitations on the water level behind the dam.

1.355. The energy we need in society to improve our living standard and maintain economic growth should be extracted or collected from natural resources of energy.

a. True
b. False

1.356. The figure below illustrates the total and per capita electricity generation in India between 1980 and 2018. Explain what the diagram indicates.

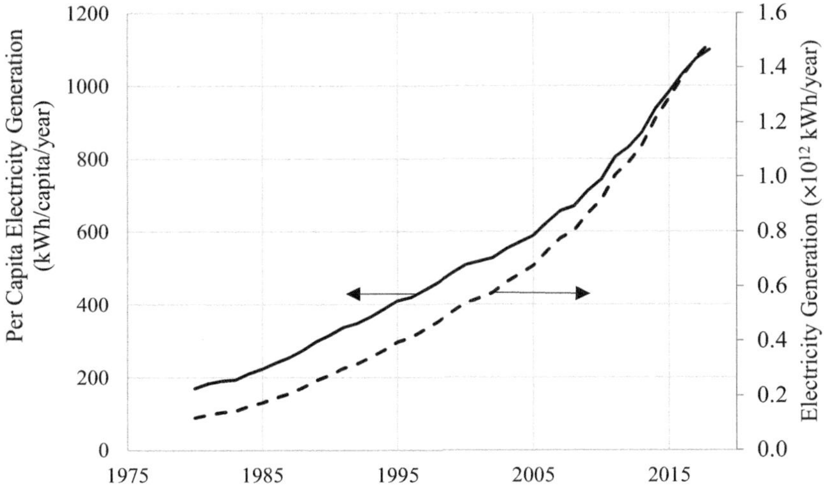

1.357. Provide two examples of when energy sources are used for manufacturing processes in the industrial sector.

1.358. For most countries, having a high annual GDP means a high per capita energy consumption and vice versa.

a. True

b. False

1.359. What is the difference between primary and secondary sources of energy?

1.360. Which of the following energy sources are secondary sources of energy? (Select all the correct ones)

a. Crude oil

b. Hydrogen

c. Solar energy

d. Biodiesel fuel

e. Wind energy

1.361. The recent increase in the per capita and total global electricity generation have been driven by an energy consumption increase in developed economies.

a. True

b. False

1.362. Explain how special occasions, e.g. holidays or sports events, impact electricity consumption.

1.363. We typically cannot directly use natural resources of energy. Explain the solution to this problem. Provide two examples.

1.364. The primary sources of energy are forms of energy that can be directly found in nature.

a. True

b. False

1.365. List three examples of energy consumption in the industry.

1.366. There is a clear correlation between per capita GDP and per capita electricity consumption though not a one–to-one relationship.

a. True

b. False

1.367. Can we use natural resources of energy directly without any energy conversion? Why? Provide two reasons with examples.

1.368. The electrical demand of individual consumers vary over a given period of time and involve significant uncertainty.

a. True

b. False

1.369. Explain what a load curve or a load profile is.

1.370. If a system does not generate electricity due to the unavailability of the energy source, the capacity factor increases.

a. True

b. False

1.371. The energy we need in society to improve our living standard and maintain economic growth should be extracted or collected from natural resources of energy such as .... (Choose all the correct items.)

a. fossil fuels

b. nuclear energy

c. renewable energy resources

d. electricity

1.372. Name at least three primary sources of energy.

1.373. A load curve or a load profile is a graph of electrical load (power consumption) as a function of time.

a. True

b. False

1.374. A power plant with the electricity generation capacity of 500 MW, operates 250 days a year. The plant cannot produce electricity 65 days of the year due to the required maintenance. The rest of the year plant does not operate because there is no demand. Determine the capacity and availability factors of the plant.

1.375. Explain why economies focused on manufacturing tend to consume more on the industrial sector while in service-focused economies more energy is consumed by the building sector.

1.376. The objective of every society is to ……..

a. sustain and improve the standard of living

b. sustain and increase growth

c. achieve the above two objectives sustainably

1.377. The recent increase in per capita and total global electricity generation has been driven by the increase in developing and emerging economies.

a. True

b. False

2

Introduction to Steam Power Plants

Abstract

Steam cycle power plants are the workhorse of the electricity generation industry. They are used in fossil fuel, nuclear, geothermal, and solar thermal energy as well as combined cycle power plants. We will first review the history of the development of steam power plants. Due to their major contributions to the industry, the rest of the chapter is entirely dedicated to the Rankine cycle, as the thermodynamic cycle that represents steam power plants. While we study this cycle, we will also review some fundamental thermodynamic principles relevant to topics. We will introduce the Carnot cycle first and learn why it cannot be used to express steam cycles. We will adjust the Carnot cycle and will discuss the specifications of the simple Rankine cycle. We will then evaluate the impacts of several operating parameters on the performance of the cycle. Then, we will take a look at how the performance of the cycle can be improved by modifying the simple Rankine cycle. Finally, we will explore how real steam cycles deviate from ideal Rankine cycles and how we can take them into account in our calculations. For each of these topics, we will have examples to evaluate their design implications.

LEARNING OBJECTIVES

After successful completion of this chapter, you will be able to:

- describe the Rankine cycle
- explain the impacts of operating conditions on the performance of the Rankine cycle
- discuss various modifications in the simple Rankine cycle to improve its performance
- identify the distinction between an ideal Rankine cycle and an actual steam power plant
- calculate the thermodynamic performance of various forms of Rankine cycles

2.1 Introduction

Steam generation and the production of work using steam marked the dawn of the industrial revolution. Despite the recent remarkable growth in renewable energy sources, electricity produced in steam power plants is still playing a major role in our modern life. When you use electricity in your house, there is a very high chance that the electricity you are consuming comes from a natural gas, coal, or nuclear steam power plant. Steam power plants have dominated conventional power generation units as nearly all coal and nuclear plants operate based on steam cycles. Before the introduction of very large capacity gas turbines, the majority of natural gas and petroleum product power plants were also operated based on steam cycles. Currently, combined cycle power plants (a combination of gas turbines and steam cycles, refer to Chapter 10) dominate natural gas-fueled power generation units. While in recent years many coal and nuclear steam power plants have retired and have not been replaced with similar technologies, many new natural gas-fueled combined cycle power plants have been installed. While in recent years renewable energy sources are steadily growing, this growth does not necessarily mean the decline of steam power plant cycles. In fact, some renewable energy sources use steam or other vapor cycles such as geothermal, solar thermal, biomass, and solid waste power plants.

Figure 2.1 shows the share of steam power plants in total utility-scale electricity generation in the U.S. between 2009 and 2018 (based on the data from the U.S. Energy Information Administration modified based on some common sense assumptions). As the diagram illustrates, while the share of electricity generated in steam power plants has decreased continuously in the

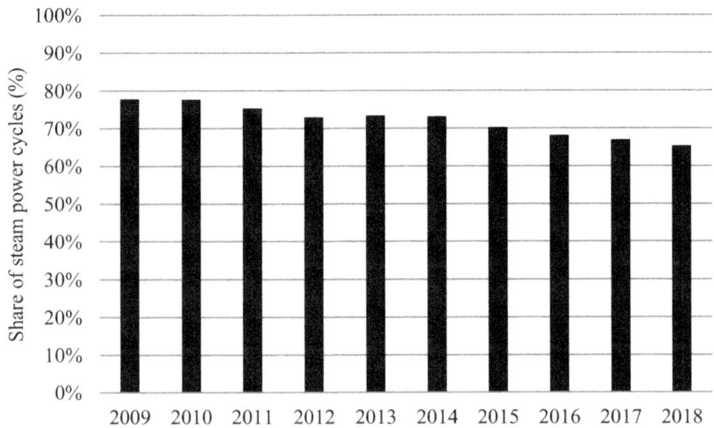

FIGURE 2.1
Share of steam power plants in total utility-scale electricity generation in the U.S. between 2009 and 2018. (Based on the data from the U.S. Energy Information Administration at eia.gov.)

past decade, about two-thirds of electricity in the U.S. is still produced by these power plants. A similar pattern is expected for the global electricity generation. This popularity is primarily due to the flexibility of thermal energy sources, relatively high efficiency, and low operating cost of these plants. These characteristics make these units ideal for baseload power generation although some steam power plants can be used as peak-load power plants too.

2.2 History of Steam Power Generation

The idea of using steam to generate work is an ancient one. The Hero of Alexandria in the 1st century AD was the first one who described the production of steam and the conversion of its energy to work through a reaction turbine (refer to Chapter 6, Figure 6.12) but he did not provide any application for the produced work.

In more recent history, the conversion of steam power to mechanical work started well before the discovery of electricity. Early publications regarding the idea appeared in the late 16th century. However, practical usage did not begin until the 17th century. Interestingly, early steam power generators were exclusively developed for pumping water out of coal mines. Until the 17th century, thermal energy required for the heating of buildings and cooking as well as industrial energy requirements was provided mainly by wood and coal. But forests were rapidly disappearing especially in England. In addition to population growth, thermal energy for smelting ores and minerals for metal extraction, particularly the utilization of coke for smelting iron, significantly increased the consumption of coal. So the coal mining industry was a booming one. The increase in coal mining resulted in digging deeper mines. Deeper mines faced the problem of water flooding, which required the pumping of water out of mines. Initially human- and animal-powered pumps were used for dewatering mines. But ever-increasing water volume to be pumped made this method practically insufficient; thus, alternative approaches to generating mechanical work were desperately needed.

FYI BOX 2.1

CAPITAL PUNISHMENT FOR BURNING COAL

Did you know that at one time burning coal for heating and cooking in London was a crime punishable by death penalty? By as early as the 12th century, wood in London was so rare and expensive that poor people could not afford it. So they started to burn coal. However, the effect of burning coal on air quality was so adverse that in the early 14th century Edward, the king of England, banned the practice, and it was punishable by a fine and destroying the furnace. Since the law did not work, he increased the punishment to death penalty. Reportedly one person was even executed for this crime (although some historians disputed this)! Of course this did not work as well because there was no alternative for the poor. The problem of air pollution and poor air quality in London remained unsolved until 1952 when ~4000 people died in London after 4 days of smog, now known as the Great Smog. After this disaster, the government took drastic actions that eventually resolved the problem for good (almost!).

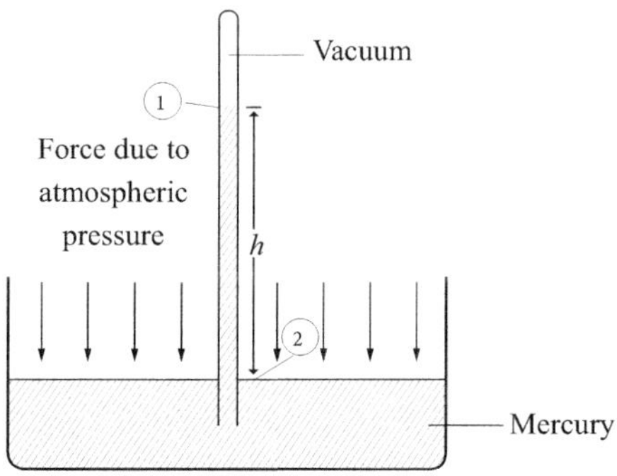

FIGURE 2.2
Schematic of operating principle of barometers (raising liquids to a higher elevation using the effect of the atmospheric pressure against vacuum).

The objective of the initial attempts for power generation was not to produce electricity or even mechanical work. As mentioned, the objective was to pump water out of coal mines. The principles that eventually led to early steam engines were around for a long time. Scientists knew that water can be elevated by the force of the atmospheric pressure if vacuum can be created. In 1643, Evangelista Torricelli (1608–1647, Italian physicist and mathematician who was an assistant to Galileo and succeeded him as the court mathematician after his death) used this principle to invent the first barometer where the atmospheric pressure pushed mercury up in a glass tube (Figure 2.2). The height of mercury in the tube indicated the atmospheric pressure.

The main question was how to create vacuum. Denis Papin (1647–1712, a French physicist and mathematician) suggested using the condensation of steam enclosed in a container to create vacuum. In 1690, he built a piston and cylinder assembly where steam was condensed inside the cylinder and created vacuum. The force of the atmospheric pressure against the created vacuum could generate work in the form of raising a weight. Although this was the first steam engine, it was never used for practical applications.

In 1698, Thomas Savery (1650–1715, an English inventor and engineer) patented the first successfully operating steam engine. His engine was

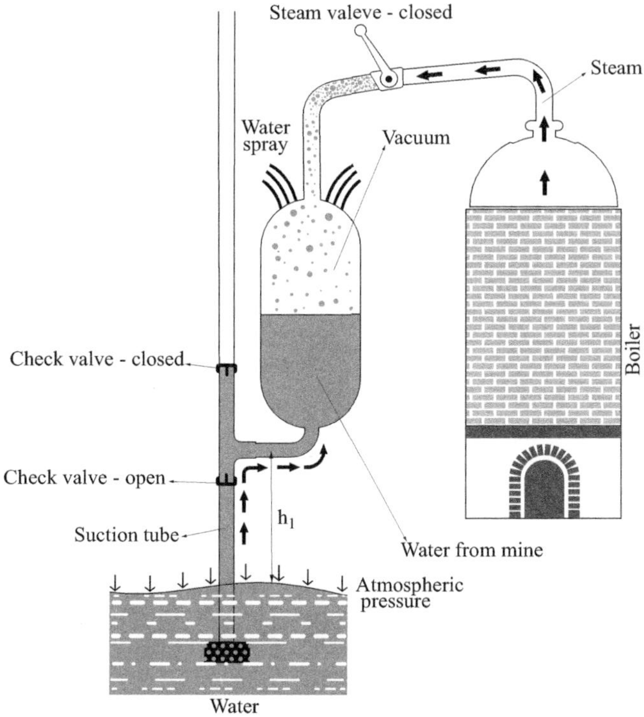

FIGURE 2.3
Savery's engine, water suction process.

composed of a boiler, a metal vessel, and a system of pipes and valves (Figures 2.3 and 2.4). First, steam was generated in the boiler and fed to the vessel. Then, the vessel was disconnected from the boiler. The steam enclosed in the vessel was then condensed by pouring cold water to the outside of the vessel (Figure 2.3). The vacuum created by the steam condensation sucked the water into the vessel from the mine. Finally, steam was again allowed to enter the vessel, and water was pumped to the surface due to the steam pressure (Figure 2.4). The cycle was repeated by condensing the steam in the vessel again.

The efficiency of the engine was extremely low mainly due to the heat loss from steam to water during their direct contact in the water pumping process. Furthermore, the height that water could be raised in the water suction and pumping processes was limited. The height that water could be raised in the water suction process (h_1 in Figure 2.3) was theoretically bounded by the maximum of 10.33 m (33.89 ft), which could be achieved if a perfect vacuum could be created (Example 2.1). Although there is no theoretical limit for the height that water could be pumped during the second process (h_2 in Figure 2.4), due to technological limitations of those days, the steam pressure was limited to slightly above the atmospheric pressure. These limitations meant that several stages of the engine were needed to be installed in

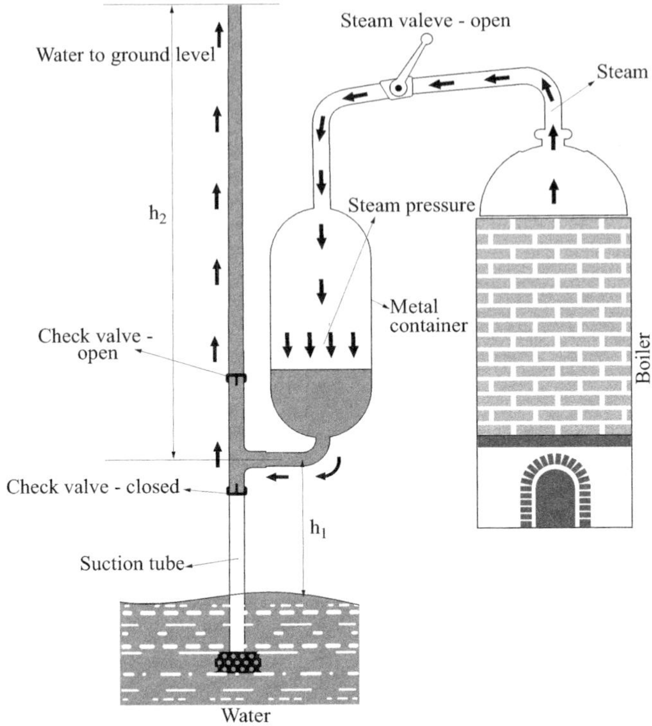

FIGURE 2.4
Savery's engine, water pumping process.

the mine shaft to pump water from deep in mines all the way to the surface. Moreover, it could not generate any external mechanical work. The engine was very slow and operating the valves manually was a laborious job. In practice, Savery used two vessels in his engine to accelerate the processes (Figure 2.5). His engines were called the "fire engine" or the "miner's friend". Although the Savery's engines were installed in several sites, its application never became widespread due to the aforementioned limitations. This design was later improved by adding an internal water jet for condensation (instead of spraying water to the outer shell) and adding a safety valve.

Example 2.1

Determine the maximum height that water can be raised using the effect of the atmospheric pressure against vacuum (h in Figure 2.2 and h_1 in Figure 2.3). What about mercury? Assume that the ambient pressure is equal to the standard atmospheric pressure.

In Figure 2.2, the pressure difference between points 1 and 2 is

$$P_2 - P_1 = \Delta P = \rho g h \tag{2.1}$$

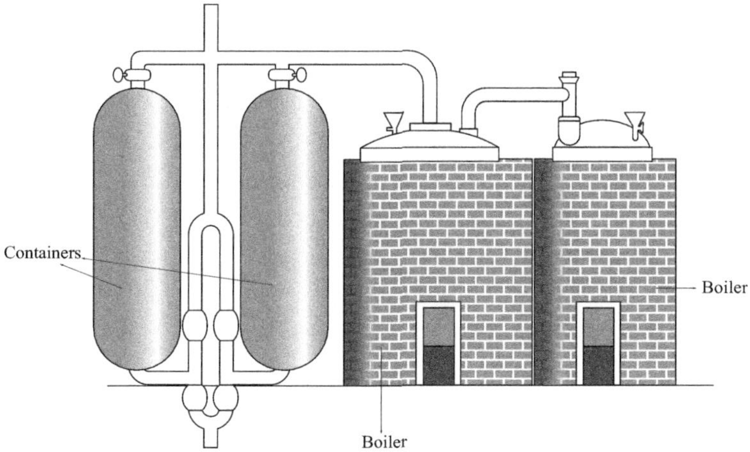

FIGURE 2.5
Schematic of an actual Savery's engine.

where ρ is the density of the liquid inside the tube, g is the gravity, and h is the height of the liquid inside the tube.

$$P_1 = 0 \text{ (vacuum)}$$

$$P_2 = P_{atm} = \text{Standard atmospheric pressure} = 101,325 \text{ Pa abs}$$

Since the densities of water and mercury are 1000 and 13,594 kg/m^3, respectively,

$$\text{For water}: h = \frac{101,325}{1000 \times 9.81} = 10.33 \text{ m} (33.89 \text{ ft})$$

$$\text{For mercury}: h = \frac{101,325}{13,594 \times 9.81} = 0.760 \text{ m} (2.50 \text{ ft})$$

Note that in Figure 2.2, a full vacuum can only be created if no liquid evaporates into the void above the surface of the column of the liquid. In this sense, using mercury in the tube comes close to satisfy this condition.

FYI BOX 2.2

THE FIRE ENGINE PATENT

The Savery's 1698 patent for his fire engine was very interesting. It had no illustration or even description; however, it granted him the right for not just his engine but every engine that raised water using fire for 14 years. Later, due to the importance of his invention and high hopes for its contribution to the mining industry, the parliament of England approved the Fire Engine Act to extend his patent for 21 more years.

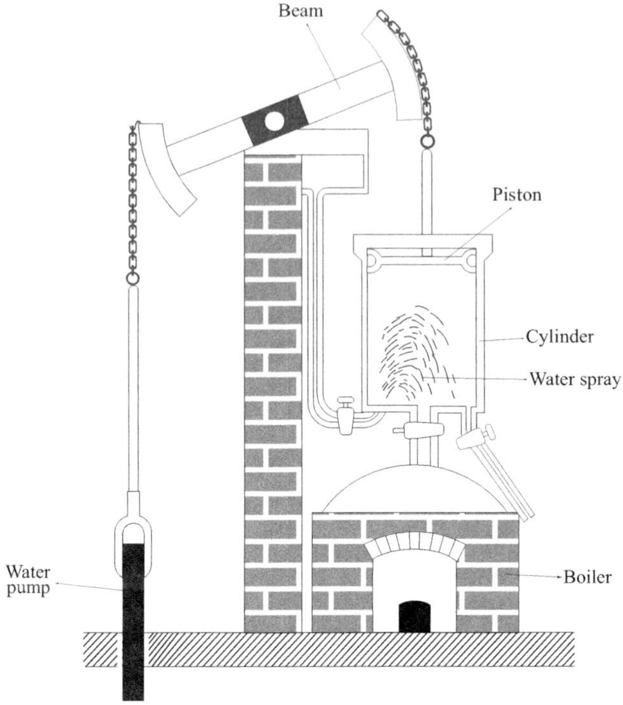

FIGURE 2.6
Schematic of Newcomen's atmospheric steam engine.

Around 1710 Thomas Newcomen (1664–1729, English inventor and ironmonger) ingeniously combined three already existing concepts: the Savery's engine, the Papin's piston and cylinder, and a conventional reciprocating water pump to develop his steam engine. Similar to the Savery's engine, it consisted of a boiler and a container, in this case a cylinder (Figure 2.6), but there was no direct interaction between steam and pumped water.

First, steam from the boiler entered the cylinder and the pressure of the steam raised the piston. The steam was then condensed by spraying cold water into the steam which created vacuum. As a result, the atmospheric pressure pushed the piston down. Since his engine always operated under the atmospheric pressure, it was called the Newcomen's atmospheric steam engine. The reciprocating motion of the piston was transferred through a beam pivoted in the center to the water pump. Since the diameter of the steam piston was greater than that of the pump piston, the water could be pumped to a higher elevation than that of previous engines, which was limited by the vacuum effect (Example 2.1).

In early engines, the operation of the valves was reportedly manual but later it was automated. Newcomen's atmospheric engines were simple, safe, and reliable and until 1775 hundreds of them were built with little modifications. Some of these engines were operational even in the early 20th century.

FYI BOX 2.3

INNOVATIVE COCK BOY

As stated, early Newcomen's engines were manually operated and controlled. This required the opening and shutting of the valves (then called plugs) in a precise manner. The operator's job (called a plug man or a cock boy) was very demanding and mistakes could be very costly and possibly fatal because of the lack of safety valves. There is a common legend that the first automated steam engine was developed by a young cock boy named Humphrey Potter. In 1713, he reportedly observed the regular patterns in the operation of the plugs and beans and used them to operate the plugs by a string mechanism called the "Potter cord".

Despite all its advantages, the efficiency of Newcomen's atmospheric engines was terribly low though much greater than that of Savery's engines. The cooling of the cylinder by water injection in one stroke and then reheating it with fresh steam in the next stroke wasted a lot of thermal energy. Due to the availability of fuels, this was not a major problem when engines were installed in coal mines. However, the applications of the engine quickly spread to other mines (other than coal mines) and other industries where in many installation sites coal was not readily available and was expensive.

The next big steps in the development of steam engines were introduced by James Watt (1736–1819, a Scottish mechanical engineer) (Figure 2.7). In order to reduce heat losses, he moved the condensation process out of the cylinder. The introduction of the separate chamber for steam condensation, known as the condenser, approximately doubled the efficiency of steam engines. He initially used water spray for steam condensation (called jet condensers, refer

FIGURE 2.7
An actual James Watt's steam engine. (By Chris Allen, CC BY-SA 2.0, https://commons.wikimedia.org/w/index.php?curid=2308677.)

to Chapter 7) but later used a heat exchanger (called surface condensers, refer to Chapter 7) for this purpose.

He made many other improvements too. To increase the power output of the engine, he used steam in the upper part of the piston to further push it downward during condensation, instead of just relying on the atmospheric pressure. This later led to the invention of double-acting engines in which both ends of the cylinder were connected to the boiler and the condenser. But before this could happen, he added a mechanism to convert the reciprocating motion of the piston to the rotational one. To avoid infringement of a patent on the crank mechanism, he developed the sun and planet gear mechanism.

This arrangement not only made double-acting engines possible but also expanded the utilization of these engines to many other industries because the rotational power output was more suitable for various industrial applications. Although Watt's engines used both vacuum and low-pressure steam to drive the piston, most of the pressure difference on the two sides of the piston came from vacuum as a result of steam condensation. These engines are also known as condensing engines in contrast to non-condensing engines that operate solely using steam pressure force with no steam condensation. Furthermore, to generate smooth and constant speed rotation, he added a flywheel and a centrifugal governor. The Watt's steam engine was one of the most important driving forces of the industrial revolution in Europe.

FYI BOX 2.4

JAMES WATT

He was born in Scotland. His father was a shipbuilder and his grandfather was a mathematics and surveying teacher. He started his career as an instrument maker and repairer at the University of Glasgow. The legend has that he was first inspired to work on steam engines when he observed that the power of steam in a boiling kettle was able to raise the kettle lid. When he started to work on the concept, the Newcomen's engine was in the market for about half a century without much change. He was tasked to fix a model of Newcomen's engine in the university during which he noticed the engine's problems, especially its extremely low efficiency. He correctly attributed this to repeated heating and cooling of the cylinder. He suggested and later patented the idea of a separate vessel for steam condensation (a condenser). While his ideas were brilliant, he did not have financial means to implement and commercialize them for about a decade. He then formed a partnership with Matthew Boulton called Boulton and Watt. This enabled him to finalize his engine and offer it to the market. The firm's financial model was very interesting. They charged their customers an annual fee based on the saving in fuel consumption compared to the equivalent Newcomen's engine (the actual charge was one-third of the saving). That meant the more efficient their engines became, the more income they had. In fact, later Watt's engines were about five times more efficient than Newcomen's engines. Regardless of his technical genius, he was not a businessman by nature and hated financial negotiations. He invented several other machines including a machine to copy letters and another to copy sculptures. He was a "hero of the industrial age" (as a plaque in The Science Museum, London put it). His remarkable ability to combine engineering theoretical knowledge and practical skills helped him immensely to contribute to the industrial revolution.

In the early 19th century, there was a great interest to increase the pressure of steam to build "strong engines" that were smaller, faster, and less expensive compared to Watt's engines. Note that those days steam at the pressure of few bars was considered high-pressure steam, which is extremely low pressure with today's standards.

In 1849, George Henry Corliss (an American mechanical engineer, 1817–1888) invented rotary valves and introduced them to steam engines. In Corliss steam engines, these fast-closing valves were able to reduce steam throttling losses and improve the engine efficiency. The next major improvement was the "uniflow steam engine" (Figure 2.8). The idea was first introduced in 1827 in Britain and was later patented in 1885. However, the first commercial uniflow engine was built by Johann Stumpf (a German engineer) in 1909. In these engines, steam entered the cylinder from two ends and exhausted in the middle. The thermal efficiency of the engine was higher than that of earlier engines. The reason was that steam always entered the hot ends of the cylinder and exhausted in the cooler midpoint, which reduced the energy loss due to the alternating heating and cooling of different parts of the cylinder in earlier engines.

In the 19th century, there were many other improvements on steam engines such as using two cylinders in series to expand steam on high-pressure and low-pressure stages (compound engines); using insulated boiler, engine, and pipes; using cylindrical boilers; using more efficient and accurate valves; and better speed control. In the late 19th century, large reciprocating steam engine units with the capacity of several MW (the largest being 5 MW) were built.

As discussed, early steam power generators were all reciprocating engines. The first commercial central power generation systems that were established in New York and London in the early 1880s were of this type. Later turbine-based steam power plants gradually took over because of their higher efficiency and lower cost of construction. The capacity of early steam turbine-based electricity generation plants was a few hundred kWs. In the late 1890s, the first megawatt size power plants were installed, and by the late 1910s units with capacity over 10 MW were installed. By the early 20th century,

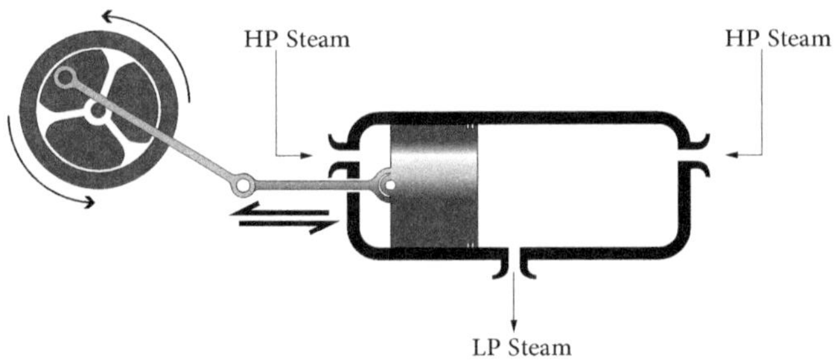

FIGURE 2.8
Schematic of a uniflow steam engine.

reciprocating steam engines were mostly replaced by steam turbines. The trend continued with increasingly larger units being installed. By the late 1920s and 1930s, units with capacities of about 200 and 500 MW, respectively, were installed. After World War II units with the capacity of over 1000 MW became standard. These improvements were achieved primarily due to more advanced technologies that allowed using superheated steam with increasingly higher pressures and temperatures.

2.3 Thermodynamics of Steam Power Plants

In this section, first the main components and operation principles of today's steam power plants are introduced. Then, the ideal thermodynamic cycle to model steam power plants is discussed followed by the effects of various operating parameters. The ways to improve the performance of steam power plants are then discussed. Finally, the deviations of actual steam power plants from ideal cycles are introduced.

Steam power plants are power generation systems that operate in a closed cycle during which the phase of the working fluid, usually water but not always, changes between the liquid and vapor phases alternatively. In the previous chapter, it was explained that conventional thermal power generation systems are able to produce net power output due to the difference between specific volumes of the working fluid during the compression and expansion processes. In phase-changing working fluid power cycles, since the working fluid is pressurized in the liquid form and expands in the vapor phase, a relatively large specific volume difference of the working fluid can be achieved (Example 2.2).

Example 2.2

Determine the ratio of the specific volume of saturated vapor to the specific volume of saturated liquid for water at the pressure of 0.1 and 150 bar.

From steam property tables (Table A.2(SI)),

$$v_{f\,@\,0.1\text{ bar}} = 0.001010\text{ m}^3/\text{kg}$$

$$v_{g\,@\,0.1\text{ bar}} = 14.67\text{ m}^3/\text{kg}$$

$$\frac{v_{g\,@\,0.1\text{ bar}}}{v_{f\,@\,0.1\text{ bar}}} = \frac{14.67}{0.001010} = 14525.8$$

$$\frac{v_{g\,@\,150\text{ bar}}}{v_{f\,@\,150\text{ bar}}} = \frac{0.01034}{0.001658} = 6.2$$

Like other power cycles, there are four main components in a steam power generation cycle (Figure 2.9): pump, steam generator (boiler), turbine, and condenser. The low-pressure water (liquid phase, flow 1) is pressurized in the pump (flow 2) and is sent to the steam generator where water is heated and converted to high-temperature steam (flow 3). The high-pressure and high-temperature steam then expands through the turbine producing work, which can be converted to electricity in the generator. The low temperature and pressure mixture of steam and water (flow 4) is then condensed in the condenser before being sent to the pump to complete the cycle. In order to condense the steam and reject heat to the surrounding, cooling water from a nearby body of water may be used. Alternatively, a wet or dry cooling tower may be employed. There are various sources of thermal energy for steam generators, including fossil fuels, such as coal, natural gas, or petroleum products; nuclear reactions; or renewable sources, such as geothermal, solar thermal energy, ocean thermal energy, or biomass. These thermal energy

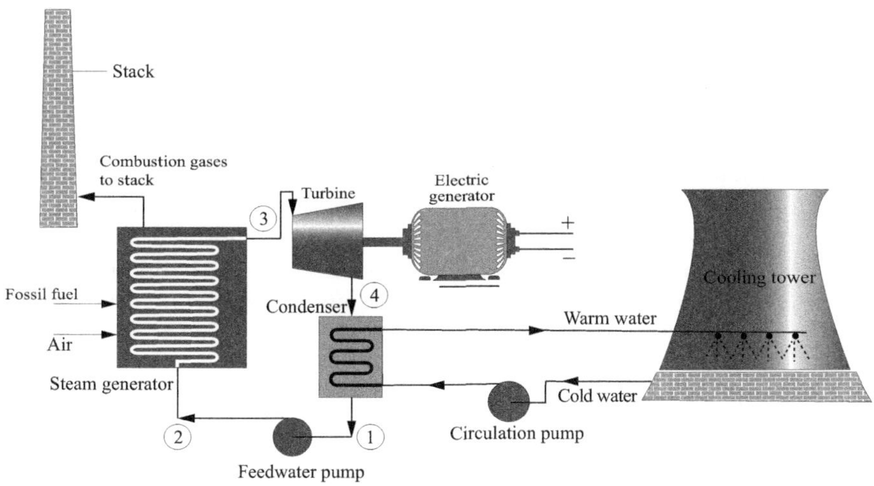

FIGURE 2.9
Schematic of a simple steam power generation cycle.

sources will be studied in details in later chapters; however, regardless of the thermal energy source, the fundamental operation of steam power cycles will remain unchanged.

REMINDER BOX 2.1

CARNOT CYCLES

The Carnot cycle (after Sadi Carnot, 1796–1832, a French engineer and physicist) is an ultimate ideal cycle for any thermal engine. All four processes in this cycle are reversible, so the cycle is reversible. The Carnot cycle is composed of two constant temperature processes (isothermal processes) and two adiabatic processes. Note that the latter reversible adiabatic processes are constant entropy or isentropic. It can be shown that it is impossible to construct an engine that operates between a low-temperature thermal energy reservoir and a high-temperature heat reservoir with an efficiency higher than that of the Carnot cycle. Also, it can be shown that all Carnot cycles operating between the same two thermal energy reservoirs have the same efficiency. Therefore, the Carnot cycle is an ideal cycle that theoretically can operate with any working fluid with the highest thermal efficiency possible for the given thermal energy reservoirs. The Carnot cycle efficiency is a function of the absolute temperatures of the thermal energy reservoirs that the cycle interacts with

$$\eta_{\text{Carnot}} = 1 - \frac{T_L}{T_H} \tag{2.2}$$

where η_{Carnot} is the thermal efficiency of the Carnot cycle, and T_L and T_H are the absolute temperatures of the low-temperature and high-temperature thermal energy reservoirs (in K or °R), respectively. Although this cycle is not practical in the real world, it provides an upper bound for the achievable efficiency of thermal engines and a benchmark to compare the efficiency of actual cycles with.

For most power cycles, the T-s and P-v diagrams are used to demonstrate the cycle. The former is typically able to provide the most useful information to study the cycle. Figures 2.10 and 2.11 illustrate the T-s and P-v diagrams for the Carnot cycle when the working fluid can have a phase change and is entirely located under the vapor dome and when the working fluid is always in the gaseous form, respectively. Accordingly, the efficiency of the cycle is given by the following equation in the T-s diagram:

$$\eta_{\text{Carnot}} = \frac{w_{\text{Net}}}{q_{\text{In}}} = \frac{\text{Area 1-2-3-4-1}}{\text{Area a-2-3-b-a}} \tag{2.3}$$

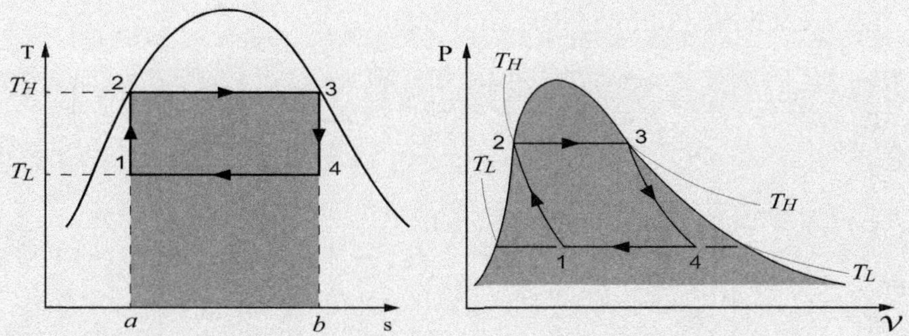

FIGURE 2.10
T-s and P-v diagrams of a Carnot cycle for a phase-changing working fluid.

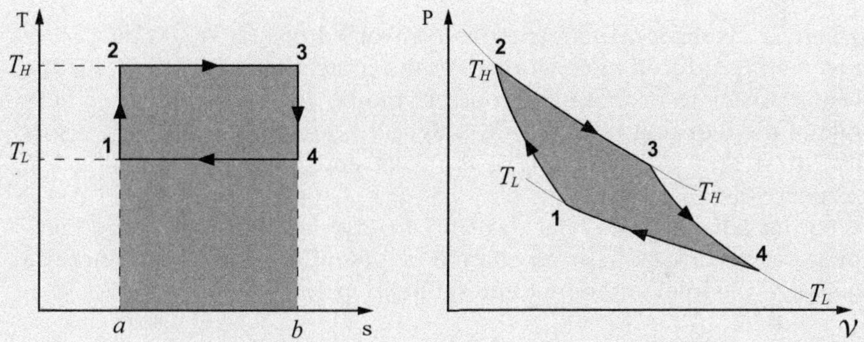

FIGURE 2.11
T-s and P-v diagrams of a Carnot cycle for a gaseous working fluid.

REMINDER BOX 2.2

THE FIRST LAW OF THERMODYNAMICS

The thermodynamic analysis of any cycle or any component of a cycle involves the conservation of mass and energy laws and the second law of thermodynamics as well as the thermodynamic data needed for these laws. The first law of thermodynamics for a control volume (CV) (the energy balance based on the rate of energy transfer) can be written as follows[1].

[1] Note that to distinguish volume and velocity, a vector sign is placed over V for the velocity.

For a CV with multiple inlets and outlets,

$$\frac{dE_{CV}}{dt} = \dot{Q}_{CV} - \dot{W}_{CV} + \sum_i \dot{m}_i\left(h_i + \frac{1}{2}\vec{V}_i^2 + g\,z_i\right) - \sum_e \dot{m}_e\left(h_e + \frac{1}{2}\vec{V}_e^2 + g\,z_e\right) \tag{2.4}$$

For a CV with a single inlet and outlet,

$$\frac{dE_{CV}}{dt} = \dot{Q}_{CV} - \dot{W}_{CV} + \dot{m}_i\left(h_i + \frac{1}{2}\vec{V}_i^2 + g\,z_i\right) - \dot{m}_e\left(h_e + \frac{1}{2}\vec{V}_e^2 + g\,z_e\right) \tag{2.5}$$

For a steady-state CV,

$$\frac{\dot{Q}_{CV}}{\dot{m}} - \frac{\dot{W}_{CV}}{\dot{m}} + (h_i - h_e) + \frac{1}{2}\left(\vec{V}_i^2 - \vec{V}_e^2\right) + g(z_i - z_e) = 0 \tag{2.6}$$

where E_{CV} is the total energy of the control volume (J); $\dot{W}_{CV}$ is the power involved (produced or consumed) in the control volume (W); $\dot{Q}_{CV}$ is the heat transfer to or from the control volume (W); $\dot{m}$ is the mass flow rate of the working fluid (kg/s); h, $\vec{V}$, and Z are enthalpy (J/kg), velocity (m/s), and elevation (m) of the inlet and outlet flows in the CV, respectively.

For an adiabatic (no heat transfer) steady-state expansion and compression processes with no change in potential and kinetic energies between the inlet and outlet flows, Equation 2.6 is reduced to

$$\frac{\dot{W}_{Turb}}{\dot{m}} = (h_i - h_e) \quad \text{or} \quad w_{Turb} = (h_i - h_e) \tag{2.7}$$

$$\frac{\dot{W}_{Pump}}{\dot{m}} = (h_e - h_i) \quad \text{or} \quad w_{Pump} = (h_e - h_i) \tag{2.8}$$

Similarly, the heat transfers during the heat addition and rejection processes are (with the same assumptions, i.e., steady state and no change in potential and kinetic energies)

$$\frac{\dot{Q}_{SG}}{\dot{m}} = (h_e - h_i) \quad \text{or} \quad q_{SG} = (h_e - h_i) \tag{2.9}$$

$$\frac{\dot{Q}_{Cond}}{\dot{m}} = (h_i - h_e) \quad \text{or} \quad q_{Cond} = (h_i - h_e) \tag{2.10}$$

where $\dot{W}$ and $\dot{Q}$ are the rate of energy transfer (kW), q and w are the specific energy transfers (kJ/kg), and $\dot{m}$ is the mass flow rate (kg/s).

Note that in these equations, the sign convention of heat and work transfers is not applicable. Typically, positive values correspond to the direction of arrows in schematic diagrams. The sign convention of heat and work transfer (positive for when work is done by the system and when heat is transferred to the system) should be only considered when writing energy balance equations (Equations 2.4 and 2.5).Since these equations are developed based upon the mass and energy conservation laws, they can be applied for ideal and non-ideal cycles (with or without irreversibilities). The efficiency of a cycle is given by

$$\eta_{Th} = \frac{w_{Net}}{q_{SG}} = \frac{w_{Turb} - w_{Pump}}{q_{SG}} \quad (2.11)$$

Another parameter that is being used to describe the performance of power generation cycles is the back work ratio (BWR),

$$\text{Back work ratio} = \text{BWR} = \frac{w_{Pump}}{w_{Turb}} \quad (2.12)$$

The BWR indicates what percentage of the work (power) produced in the turbine is consumed in the pump. The lower BWR is desirable because it means less work (power) is consumed in the pump and more work (power) is available as the net output. Alternatively, some texts used the work ratio (WR) for this purpose defined as

$$\text{Work ratio} = \text{WR} = \frac{w_{Net}}{w_{Turb}} \quad (2.13)$$

Example 2.3

In a Carnot cycle with the working fluid of water, the operating pressures of the steam generator and the condenser are 10,000 and 10 kPa, respectively. If the turbine inlet flow is saturated vapor and the pump outlet flow is saturated liquid, determine the specific heat transfers in the steam generator and the condenser (in kJ/kg), the specific work involved in the turbine and the pump (in kJ/kg), and the thermal efficiency and BWR of the cycle.

Solution

Given: A Carnot cycle with the working fluid of water operates at the given operating pressures of the steam generator and the condenser. The turbine inlet flow is saturated steam and the pump outlet flow is saturated liquid.

Find: The specific heat transfers in the steam generator and the condenser, the specific work involved in the turbine and the pump, and the thermal efficiency and BWR of the cycle.

Schematic diagrams: Figure 2.10
Assumptions:

1. The cycle is an ideal Carnot cycle; therefore, all processes are internally reversible. The heat-transfer processes are at constant temperatures. The expansion and compression processes are adiabatic and thus isentropic. The working fluid is water.
2. The steady-state control volume mass and energy balance analyses with no kinetic and potential energy changes are applied to each component.
3. There is no pressure drop due to friction and no heat loss to the surrounding in the components and piping systems.

Solution

First, the state and properties of the working fluid (water/steam) at the four main states in the cycle (states 1–4 in Figure 2.10) should be identified.

State 2 (the pump outlet flow):

At this state, the working fluid is saturated liquid at the pressure of 10,000 kPa, from Table A.2 (SI)[2]:

$$h_2 = h_{\text{sat liq@ 10,000 kPa}} = h_{f\,@\,10,000\text{ kPa}} = 1407.56\text{ kJ/kg}$$

$$v_2 = v_{f\,@\,10,000\text{ kPa}} = 0.001452\text{ m}^3/\text{kg}$$

$$s_2 = s_{f\,@\,10,000\text{ kPa}} = 3.3596\text{ kJ/kg K}$$

$$T_2 = T_{\text{sat@ 10,000 kPa}} = 311.06°\text{C}$$

State 1:

At this state, the working fluid is a mixture of saturated liquid and vapor at the pressure of 10 kPa and another independent property is needed to determine the state of the stream. Since the process 1-2 is isentropic,

$$s_1 = s_2 = 3.3596\text{ kJ/kg K}$$

$$s_1 = s_{f\,@\,10\text{ kPa}} + x_1 s_{fg\,@\,10\text{ kPa}} \rightarrow 3.3596 = 0.6493 + x_1(7.5009)$$

$$x_1 = 0.3613\text{ or }36.13$$

$$\%$$

$$h_1 = h_{f\,@\,10\text{ kPa}} + x_1 h_{fg\,@\,10\text{ kPa}} = 191.83 + 0.3613 \times 2392.8 = 1055.6\text{ kJ/kg}$$

$$T_1 = T_{\text{sat@10 kPa}} = 45.81°\text{C}$$

[2] You can find thermodynamics properties of water from the property tables at the appendix of this book or any other thermodynamic textbook. Alternatively, you can use online tools or smartphone applications, such as SteamProperty App. Note that depending on the source that you use to find properties, their values may slightly differ from the values used in this text.

State 3:

At State 3, the working fluid is at the saturated steam state at the pressure of 10,000 kPa from Table A.2 (SI),

$$h_3 = h_{\text{sat vap@10,000 kPa}} = h_{g\,@\,10,000\ \text{kPa}} = 2724.7\ \text{kJ/kg}$$

$$s_3 = s_{g\,@\,10,000\ \text{kPa}} = 5.6141\ \text{kJ/kg K}$$

State 4:

At this state, the working fluid is the mixture of saturated water and steam at the pressure of 10 kPa. Since process 3-4 is isentropic,

$$s_3 = s_4 = s_{f\,@\,10\ \text{kPa}} + x_4 s_{fg\,@\,10\ \text{kPa}} \rightarrow 5.6141 = 0.6493 + x_4\,(7.5009)$$

$$x_4 = 0.6619 \text{ or } 66.19\%$$

$$h_4 = h_{f\,@\,10\ \text{kPa}} + x_4 h_{fg\,@\,10\ \text{kPa}} = 191.83 + 0.6619 \times 2392.8 = 1775.6\ \text{kJ/kg}$$

Now, the specific heat transfers in the steam generator and the condenser, the specific work involved in the turbine and the pump, and the thermal efficiency and BWR of the cycle can be calculated as follows:

$$w_{\text{Pump}} = h_2 - h_1 = 1407.56 - 1055.6 = 352.0\ \text{kJ/kg}$$

$$q_{\text{SG}} = h_3 - h_2 = 2724.7 - 1407.56 = 1317.1\ \text{kJ/kg}$$

$$w_{\text{Turb}} = h_3 - h_4 = 2724.7 - 1775.6 = 949.1\ \text{kJ/kg}$$

$$q_{\text{Cond}} = h_4 - h_1 = 1775.6 - 1055.6 = 720.0\ \text{kJ/kg}$$

$$w_{\text{Net}} = w_{\text{Turb}} - w_{\text{Pump}} = 949.1 - 352.0 = 597.1\ \text{kJ/kg}$$

$$\eta_{\text{Carnot}} = \frac{w_{\text{Net}}}{q_{\text{SG}}} = \frac{597.1}{1317.1} = 0.4533 \text{ or } 45.33\%$$

$$BWR = \frac{w_{Pump}}{w_{Turb}} = \frac{352.0}{949.1} = 0.3709 \text{ or } 37.09\%$$

The correctness of the solution can be checked by applying the energy balance for the cycle, as follows:

$$w_{\text{Net}} = w_{\text{Turb}} - w_{\text{Pump}} = q_{\text{SG}} - q_{\text{Cond}}$$

$$(949.1 - 352.0) = (1317.1 - 720.0\ ) \rightarrow 597.1\ \text{kJ/kg} = 597.1\ \text{kJ/kg}$$

The efficiency of the Carnot cycle can also be found as follows:

$$\eta_{\text{Carnot}} = 1 - \frac{T_L}{T_H} = 1 - \frac{T_1}{T_2} = 1 - \frac{45.81 + 273.15}{311.06 + 273.15} = 0.4540 \text{ or } 45.40\%$$

which is almost equal to the efficiency found using steam properties. The reason for the different is the rounding error in the two methods of calculations.

2.3.1 Simple Rankine Cycles

Although the Carnot cycle is the most efficient ideal power cycle, it is **impractical** for steam power cycles. The first main reason is that a heat-transfer process at a constant temperature is not practical because in order to keep the temperature constant, the pressure of the working fluid should be continuously manipulated. In reality, heat-transfer processes in steam generators and condensers are constant pressure (if the pressure drop, which is typically small, is ignored). Therefore, the Carnot cycle should be modified to fit the conditions of steam power cycles.

The universally accepted ideal cycle for a contemporary steam power plant is called a Rankine cycle (named after William John Macquorn Rankine, 1820–1872, a Scottish engineer and physicist). The Rankine cycle is a closed cycle composed of four major reversible processes (Figure 2.12)

- **Process 1-2**: a reversible adiabatic (thus constant entropy or isentropic) compression process in a pump
- **Process 2-3**: a reversible constant-pressure process involving heat transfer at high temperature in a steam generator
- **Process 3-4**: a reversible adiabatic (thus constant entropy or isentropic) expansion process in a turbine
- **Process 4-1**: a reversible constant-pressure heat transfer from the working fluid to the surrounding at low temperature in a condenser

Note that if the Rankine cycle takes place entirely inside the two-phase liquid-vapor dome, it will be a Carnot cycle because in this region saturation pressure and temperature are dependent, and thus a constant-pressure process is also a constant temperature one (cycle 1′-2′-3-4-1′ in Figure 2.12). Though this new cycle can fundamentally express steam power cycles, it needs some adjustments to truly represent them.

First, pumping the mixture of vapor and liquid at state 1′ is very difficult and requires a lot of energy (in Example 2-3 about 37% of the work produced in the turbine is consumed in the pump). So it is much easier to condense the mixture of vapor and liquid at state 1′ to saturated liquid at state 1 and then pump it. Furthermore, in state 4 (the turbine exit) the quality of steam is relatively low, and there are water droplets in the stream. This is not good for

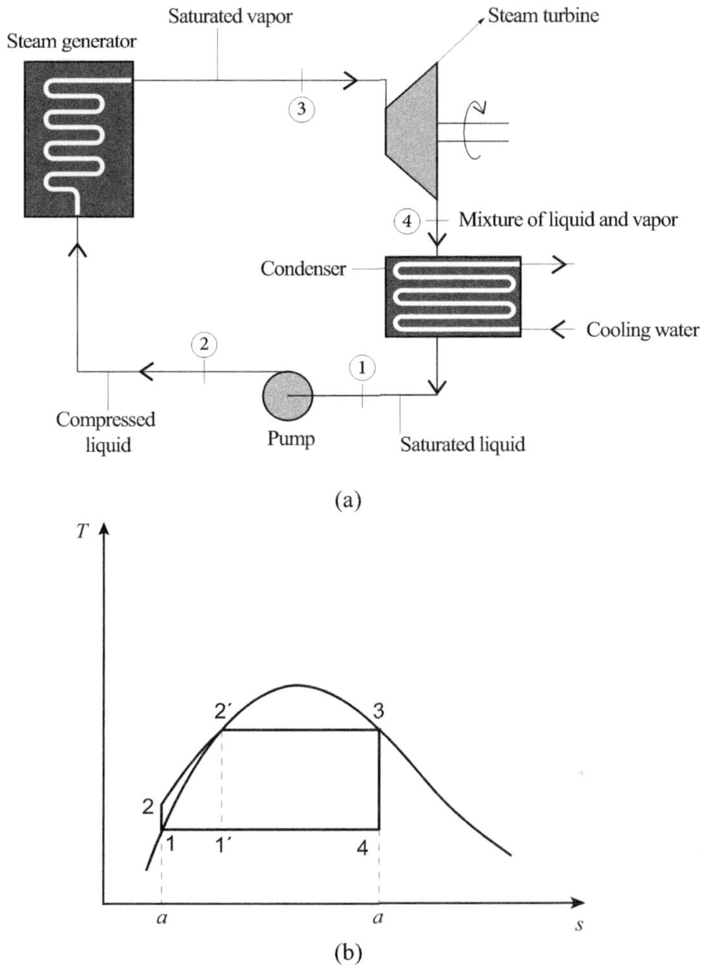

FIGURE 2.12
An ideal Rankine cycle, (a) the schematic of the cycle, (b) the T-s diagram of the Rankine cycle.

the low-pressure turbine blades because it reduces the turbine efficiency and causes erosion in the turbine blades. Generally, as a rule of thumb, the moisture content of the turbine outlet stream should be less than about 10%–15% (the quality of steam should be higher than about 85%–90%); otherwise, there will be a significant efficiency drop and serious blade erosion. Keeping this limitation in mind is important, and for every action to improve the efficiency of the cycle, the effect on the quality of steam at the outlet of the turbine should be investigated. Also, in the Carnot cycle, the temperature of the combustion products discharged to the surrounding is high (about T_3) which means the loss of high-grade thermal energy and thus a lower efficiency. Therefore, in practical Rankine cycles, the outlet of the condenser (state 1) is at the saturated liquid state, which is then pressurized in the pump to subcooled (compressed) state 2 and is heated in the steam generator to state 3. In the majority of modern steam power plants, the steam leaving the steam generator is superheated. Note that for clarity and instructive purposes, in all *T-s* diagrams, the difference between process 2-2′ and the saturated liquid line is unrealistically exaggerated.

For the sake of simplicity of analyses, inspired by the definition of the efficiency of the Carnot cycle, it can be shown that the efficiency of any cycle is directly proportional to the average temperature of the heat transfer to the cycle (at high temperature) and is inversely proportional to the average temperature of the heat transfer from the cycle (at low temperature). So **if the average temperature of the heat source is increased or if the average temperature of the heat sink is decreased, the efficiency of the cycle will increase**. For instance, it can be seen that the efficiency of a Rankine cycle is less than that of a Carnot cycle operating between the same heat source and sink because the average temperature of the heat transfer to the cycle in the steam generator in the Rankine cycle is lower than that of the Carnot cycle (due to the heating of liquid at the lower temperature, process 2-2′, at the left side of the T-s diagram in Figure 2.12). Also, note that in the T-s diagram of any reversible process, the area under the curve of the process is equal to the heat transfer during the process. Furthermore, the area enclosed by the cycle in the T-s diagram is equal to the net specific heat input as well as the net specific work output of the cycle.

The most common working fluid for Rankine cycle-based power generation systems is purified (demineralized) water due to its abundance, low cost, and desirable thermodynamic characteristics of water. Note that the working fluid of steam cycles sometimes is referred to as feedwater, especially water leaving the pump after the condenser. The latent heat of water (change in the specific enthalpy during the phase change from liquid to vapor) is high, which results in smaller components for the same power output. Similarly, the change in the specific volume of water at the typical operating pressure of steam power plants is high (Example 2.1), which is manifested in the much smaller power consumption in the pump compared to the power production in the turbine. Also, physical and chemical characteristics of water, including its chemical stability as well as its nontoxic and noncorrosive nature, are favorable. However, not all water properties are favorable for this application. The critical pressure of water is very high, which makes achieving supercritical steam power plants (will be discussed in Section 2.3.2.3) a challenge. Furthermore, the relatively high boiling temperature of water makes it unsuitable for power generation from low-temperature thermal energy resources, including geothermal energy, solar thermal energy, and waste thermal energy resources. To address the latter problem, alternative organic working fluids, including ammonia (and other substances that are ordinarily used as refrigerants), hydrocarbon mixtures, pentane, and silicon oil, all with low boiling temperatures have been used as the working fluid in some Rankine cycles. These cycles are referred to as Organic Rankine Cycles (ORC) and will be discussed in Chapter 15.

FYI BOX 2.5

WILLIAM JOHN MACQUORN RANKINE

William John Macquorn Rankine was a professor of civil engineering and mechanics at the University of Glasgow. He is known for his work in mechanical engineering, civil engineering, physics, mathematics, and engineering education. Along with Rudolf Clausius and Lord Kelvin (William Thomson), he is considered as one of the founding fathers of thermodynamics, which he called "the science of energetics". He conducted his theoretical studies with the objective of addressing practical applications. As a teacher, he promoted the same approach by integrating theories into applications. He actively and aggressively promoted the idea of engineering discipline independent from science. He believed that engineering subjects (defined as the utilization of science in practical applications) should be taught as an independent field in universities. He published hundreds of books, articles, and notes. Many of his textbooks were instantaneous classics and were taught for decades to come. He set the cornerstone for engineering education methods in Britain. He was also the first president of the Institution of Engineers in Scotland as well as the fellow of the Royal Society of Edinburgh, the Royal Society of London, the American Academy of Arts and Sciences, and the Royal Swedish Academy of Sciences.

As a child, he was interested in music and mathematics. He started to study natural philosophy (the term used for science at the time) and natural history at the University of Edinburgh. Although he was a very good student, he never finished his studies and left the university after 2 years. His first encounter with thermodynamics was when he tried to mathematically explain the heat phenomenon (which he rightfully believed was a branch of mechanics), which remained his fascination for the rest of his life. He developed a molecular theory to describe thermodynamic phenomena. He also developed the relationship between the saturation temperature and pressure of vapors; the relationship between temperature, pressure, and density of gases; the expressions for the latent heat of evaporation for liquids; the properties of steam and other gases; and the theory of shock waves. He also proposed an absolute temperature scale corresponding to the Fahrenheit scale, which is now known as the Rankine scale.

His work on thermal engines started with an attempt to determine their efficiency. He and a couple of other scientists showed that the efficiency of the Carnot cycle is a function of the temperature of the heat reservoirs that the cycle exchanges heat with. He introduced and developed the theory of the ideal cycle for a heat engine that used a condensable vapor as the working fluid. He determined the steam properties and used them to calculate the efficiency of the cycle numerically. The cycle now bears his name.

Some of his interests and achievements were related to railways, probably because his father was a superintendent for a railway company, including a method for laying out circular curves (known as the Rankine's method) and studies of metal fatigue. He also studied naval architecture and ship design, soil mechanics, hydrodynamics, the stability of walls, and the force distribution in frame structures, among others. He was an accomplished musician, singer, composer, and poet. He never married and remained a bachelor for his entire life. He opposed the metric system and wrote the following humorous poem called "The Three-Foot Rule" to express his opposition:

> "When I was bound apprentice, and learned to use my hands,
> Folk never talked of measures that came from foreign lands:
> Now I'm a British Workman, too old to go to school;
> So whether the chisel or file I hold, I'll stick to my three-foot rule.
> Some talk of millimetres, and some of kilogrammes,
> And some of decilitres, to measure beer and drams;
> But I'm a British Workman, too old to go to school,
> So by pounds I'll eat, and by quarts I'll drink, and I'll work by my three-foot rule.
> A party of astronomers went measuring the earth,
> And forty million metres they took to be its girth;
> Five hundred million inches, though, go through from pole to pole;
> So let's stick to inches, feet and yards, and the good old three-foot rule."

His other similarly humorous poem was called "The Mathematician in Love" (available at http://www.poemhunter.com/poem/the-mathematician-in-love/).

Example 2.4

In a steam power plant that operates based on a Rankine cycle, the operating pressures of the steam generator and the condenser are 10,000 and 10 kPa, respectively. If the turbine inlet stream is saturated vapor and the condenser outlet flow is saturated liquid, determine the specific heat transfers in the steam generator and the condenser, the specific work involved in the turbine and the pump, and the thermal efficiency and the BWR of the cycle. Also, if the power plant produces 250 MW power, determine the mass flow rate of the cycle's working fluid.

Solution

Given: An ideal Rankine cycle with the working fluid of water/steam operates at the given operating pressures of the steam generator and the condenser. The turbine inlet stream is saturated steam and the condenser outlet flow is saturated water. The power output is also given.

Find: The specific heat transfers in the steam generator and the condenser, the specific work involved in the turbine and the pump, the thermal efficiency and the BWR of the cycle, and the mass flow rate of the cycle's working fluid.

Schematic diagrams: Figure 2.12.

Assumptions:

1. The cycle is an ideal Rankine cycle; therefore, all processes are reversible. The heat-transfer processes are at constant pressures. The expansion and compression processes are adiabatic and thus isentropic. The working fluid is water/steam.
2. The steady-state control volume mass and energy balance analyses with no kinetic and potential energy changes are applied to each component.
3. There is no pressure drop due to friction and no heat loss to the surrounding in the components and piping systems.

Analysis:

First, the state and properties of the working fluid (water/steam) at the four main states in the cycle (states 1–4 in Figure 2.12) should be identified.

State 1:

At this state the working fluid is saturated liquid at the pressure of 10 kPa, from Table A.2 (SI)

$$h_1 = h_{\text{sat liq @ 10 kPa}} = h_{f\,@\,10\,\text{kPa}} = 191.83\,\text{kJ/kg}$$

$$v_1 = v_{f\,@\,10\,\text{kPa}} = 0.001010\,\text{m}^3/\text{kg}$$

$$s_1 = s_{f\,@\,10\,\text{kPa}} = 0.6493\,\text{kJ/kg K}$$

Note that the pressure in the condenser is well below the atmospheric pressure (partial vacuum).

State 2:

At this state, the water is at the compressed or subcooled liquid state at the pressure of 10,000 kPa. There are two ways to find the properties of the working fluid at this state.

Method 1: Using the compressed liquid water table (Table A.4 (SI))

Since the process 1-2 is isentropic,

$$s_1 = s_2 = 0.6493\,\text{kJ/kg K}$$

For the given pressure of 10,000 kPa and entropy of 0.6493 kJ/kg K, the enthalpy and temperature of the working fluid can be found by interpolation in Table A.4 (SI).

$$T_2 = 46.3°\text{C}$$

$$h_2 = 202.46 \text{ kJ/kg}$$

The problem with this method is that the range of the values given in the subcooled/compressed tables is limited and may not cover the required data. Note that the temperature increase of the working fluid during the pressurization process is very small, from 45.81°C in the pump inlet (the saturation temperature of water at 10 kPa) to 46.5°C in the pump outlet.

Method 2: Using the work required in the pump

The work input in the reversible compression (as indicated by the subscript "Rev") can be found by Equation 2-14.

$$\left(w_{Pump}\right)_{\text{Rev}} = \int_1^2 v dP \tag{2.14}$$

Since the working fluid in the pump is in the liquid phase, and the specific volume of the liquid can be assumed constant $(v_2 = v_1)$, the specific work in the pump can be estimated as follows:

$$w_{Pump} = v_1\left(P_2 - P_1\right) = 0.001010 \text{ m}^3/\text{kg} \times \left(10,000 - 10\right) \text{kPa}$$

$$\left[\frac{1000 \text{ N/m}^2}{1 \text{ kPa}}\right]\left[\frac{1 \text{ kJ}}{1000 \text{ Nm}}\right] = 10.09 \text{ kJ/kg}$$

$$w_{Pump} = h_2 - h_1 \rightarrow h_2 = w_{Pump} + h_1 = 10.09 + 191.83 = 201.92 \text{ kJ/kg}$$

Comparing the results of the two methods shows <0.3% error which is acceptable for most engineering applications.

State 3:

At state 3, since the working fluid is at the saturated steam state at the pressure of 10,000 kPa,

$$h_3 = h_{\text{sat vap@10,000 kPa}} = h_{g\text{@10,000 kPa}} = 2724.7 \text{ kJ/kg}$$

$$s_3 = s_{g\text{@10,000 kPa}} = 5.6141 \text{ kJ/kg K}$$

State 4:

At this state, the working fluid is the mixture of saturated water and steam at the pressure of 10 kPa and another independent property is needed to determine the state of the stream. Since process 3-4 is isentropic,

$$s_3 = s_4 = s_{f@10\,\text{kPa}} + x_4 s_{fg@10\,\text{kPa}} \rightarrow 5.6141 = 0.6493 + x_4(7.5009)$$

$$x_4 = 0.6619\,\text{or}\,66.19\%$$

$$h_4 = h_{f@10\,\text{kPa}} + x_4 h_{fg@10\,\text{kPa}} = 191.83 + 0.6619 \times 2392.8 = 1775.6\ \text{kJ/kg}$$

Now, the specific heat transfers in the steam generator and the condenser, the specific work involved in the turbine and the pump, and the thermal efficiency and the BWR of the cycle can be calculated as follows (note that for h_2 the value from the second method is used):

$$w_{\text{Pump}} = h_2 - h_1 = 201.92 - 191.83 = 10.09\ \text{kJ/kg}$$

$$q_{\text{SG}} = h_3 - h_2 = 2724.7 - 201.92 = 2522.8\ \text{kJ/kg}$$

$$w_{\text{Turb}} = h_3 - h_4 = 2724.7 - 1775.6 = 949.1\ \text{kJ/kg}$$

$$q_{\text{Cond}} = h_4 - h_1 = 1775.6 - 191.83 = 1583.8\ \text{kJ/kg}$$

$$w_{\text{Net}} = w_{\text{Turb}} - w_{\text{Pump}} = 949.1 - 10.09 = 939.0\ \text{kJ/kg}$$

$$\eta_{\text{Th}} = \frac{w_{\text{Net}}}{q_{\text{SG}}} = \frac{939.0}{2522.8} = 0.3722 \text{ or } 37.22\%$$

$$\text{BWR} = \frac{w_{\text{Pump}}}{w_{\text{Turb}}} = \frac{10.09}{949.1} = 0.0106 \text{ or } 1.06\%$$

The correctness of the solution can be checked by applying the energy balance for the cycle as follows:

$$w_{\text{Net}} = w_{\text{Turb}} - w_{\text{Pump}} = q_{\text{SG}} - q_{\text{Cond}}$$

$$(949.1 - 10.09) = (2522.8 - 1583.8) \rightarrow 939.0\,\text{kJ/kg} = 939.0\,\text{kJ/kg}$$

Note that the BWR is very small compared to the Carnot cycle in Example 2-3, which means a very small percentage of the work produced in the turbine is consumed in the pump. This is expected because while in the pump, liquid with a very low specific volume is compressed ($v_1 = 0.001\ \text{m}^3/\text{kg}$), in the turbine steam with a high specific volume is expanded ($v_3 = 0.018\ \text{m}^3/\text{kg}$ in the turbine inlet and $v_4 = 9.711\ \text{m}^3/\text{kg}$ in the turbine outlet).

The mass flow rate of the cycle's working fluid is

$$\dot{W}_{\text{Net}} = \dot{m} w_{\text{Net}}$$

$$\dot{m} = \frac{\dot{W}_{\text{Net}}}{w_{\text{Net}}} = \frac{250{,}000\ \text{kJ/s}}{939\ \text{kJ/kg}} = 2662\ \text{kg/s} = 9.58 \times 10^5\ \text{kg/h}$$

TABLE 2.1

Summary of the Results for Examples 2.3 and 2.4

Type of Cycle	$P_2=P_3$ (kPa)	$P_1=P_4$ (kPa)	x_4	q_{SG} (kJ/kg)	q_{Cond} (kJ/kg)	w_{Pump} (kJ/kg)	w_{Turb} (kJ/kg)	w_{Net} (kJ/kg)	η_{Th} (%)
Carnot cycle	10,000	10	0.662	1317.1	720.0	352.0	949.1	597.1	45.33
Simple Rankine –Turbine inlet: saturate steam	10,000	10	0.662	2522.8	1583.8	10.1	949.1	939.0	37.22

Question: What would be the efficiency of the cycle if the mass flow rate of the working fluid was 5×10^5 kg/h ?

Answer: The efficiency of the cycle is independent of the mass flow rate of the working fluid, so the efficiency remains 37.22%.

The summary of the results for the Rankine cycle in this example and those of the Carnot cycle in Example 2.3 is compared in Table 2.1:

Note that while the pump energy consumption is much higher in the Carnot cycle (352.0 kJ/kg versus 10.1 kJ/kg), the efficiency of the Carnot cycle is still higher.

Question: Determine the rate of heat addition in the steam generator and the rate of heat rejection to the environment in the condenser.

Answer: By multiplying the mass flow rate of the working fluid to the specific heat transfers and specific works of various components, the total heat transfer and work of each component can be determined.

$$\dot{Q}_{SG} = \dot{m}q_{SG} = 266.2\,[\text{kg/s}]\times 2522.8\,[\text{kJ/kg}] = 671,569\text{ kW}$$

$$\dot{Q}_{Cond} = \dot{m}q_{Cond} = 266.2\,[\text{kg/s}]\times 1583.8\,[\text{kJ/kg}] = 421,608\text{ kW}$$

For practice, determine the heat rate of the cycle in this example.

Example 2.5

In Example 2.4, if the cooling water enters the condenser at 40°C and leaves at 15°C, determine the cooling water mass flow rate through the condenser to provide sufficient cooling for 1 kg of the working fluid.

Solution

Given: In the ideal Rankine cycle with the given operating conditions in Example 2.4, the inlet and outlet temperatures of the cooling water in the condenser are specified.

Find: The cooling water mass flow rate through the condenser for 1 kg of the working fluid.

Schematic diagrams: Figure 2.12.

Assumptions: Similar to Example 2.4.

Analysis:

From Example 2.4,

$$q_{Cond} = 1583.8\,\text{kJ/kg}$$

q_{Cond} is the amount of heat that the cooling water should remove from 1 kg of the working fluid. If the energy balance equation is written for the control volume around the cooling water side of the condenser,

$$q_{\text{Cond}} = m_{\text{CW}}\left(h_{\text{CW, out}} - h_{\text{CW, in}}\right)$$

m_{CW} is the mass flow of the cooling water for 1 kg of the working fluid (kg) and h_{CW} is the enthalpy of the cooling water (kJ/kg). Since the property of compressed liquid at the atmospheric pressure is not covered in the compressed water table, the enthalpy of the cooling water can be approximated by the enthalpy of the saturated water at the same temperature.

$$h_{\text{CW,out}} = h_{f\,@\,40°\text{C}} = 167.57\ \text{kJ/kg}$$

$$h_{\text{CW, in}} = h_{f\,@\,15°\text{C}} = 62.99\ \text{kJ/kg}$$

$$m_{\text{CW}} = \frac{q_{\text{Cond}}}{\left(h_{\text{CW,out}} - h_{\text{CW,in}}\right)} = \frac{1583.8}{(167.57 - 62.99)} = 15.14\frac{\text{kg}_{\text{Cooling water}}}{\text{kg}_{\text{Working fluid}}}$$

This means for every 1 kg of the working fluid circulated in the cycle, 15.14 kg of the cooling water through the condenser is required to reject the low-temperature heat to the surrounding. The total cooling water mass flow rate is

$$\dot{m}_{\text{CW}} = \dot{m} \times m_{\text{CW}} = 266.2\left[\frac{\text{kg}_{\text{Working fluid}}}{\text{s}}\right] \times 15.14\left[\frac{\text{kg}_{\text{Cooling water}}}{\text{kg}_{\text{Working fluid}}}\right]$$

$$= 4030.3\frac{\text{kg}_{\text{Cooling water}}}{\text{s}}$$

2.3.2 Effects of Various Operating Parameters on Rankine Cycles

2.3.2.1 Effects of Superheating Steam in Steam Generator

The efficiency of the Rankine cycle will increase if the steam produced in the steam generator is superheated because of the increase in the average temperature of the heat addition to the cycle (Figure 2.13). Note that the temperature of process 3-3′ is higher than that of the other parts of the process 2-3′. This also reduces the moisture content of the turbine outlet flow from state 4 to 4′. So both the efficiency and the turbine performance improve (a win-win situation). This can be achieved by adding a superheater (a separate heat exchanger) to the steam generator.

Example 2.6

In Example 2.4, the outlet steam from the steam generator is superheated to the pressure and temperature of 10,000 kPa and 600°C, respectively. The condenser outlet flow is saturated liquid at the pressure of 10 kPa.

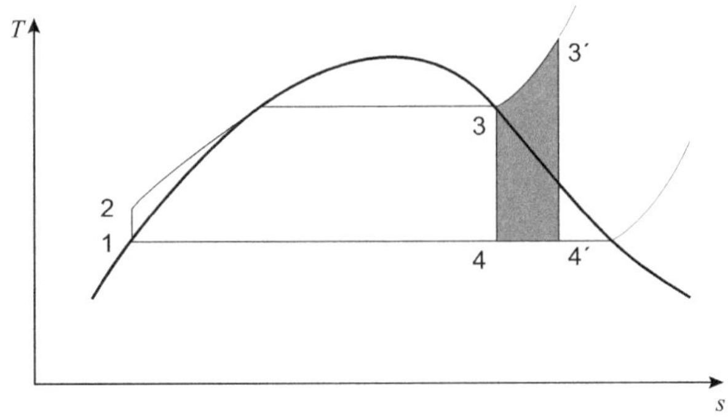

FIGURE 2.13
Effect of superheating steam in the steam generator on a Rankine cycle performance.

Determine the specific heat transfers in the steam generator and the condenser, the specific work involved in the turbine and the pump, and the thermal efficiency of the cycle. Also, if the power plant produces 250 MW (net power output), determine the mass flow rate of the cycle's working fluid.

Note: This steam power plant is expressed as 100/600 (100 bar/600°C, steam generator pressure/turbine inlet temperature).

Given: An ideal Rankine cycle with the working fluid of water/steam operates at the given operating pressures of the steam generator and the condenser. The turbine inlet stream is superheated steam and the condenser outlet flow is saturated liquid water.

Find: The specific heat transfers in the steam generator and the condenser, the specific work involved in the turbine and the pump, the thermal efficiency of the cycle, and the mass flow rate of the cycle's working fluid.

Schematic diagrams: Figure 2.13

Assumptions:

1. The cycle is an ideal Rankine cycle; therefore, all processes are reversible. The heat-transfer processes are at constant pressures. The expansion and compression processes are adiabatic and thus isentropic. The working fluid is water/steam.
2. The steady-state control volume mass and energy balance analyses with no kinetic and potential energy changes are applied to each component.
3. There is no pressure drop due to friction and no heat loss to the atmosphere in the components and piping systems.

Analysis:

First, the state and properties of the working fluid at the four main states in the cycle (states 1, 2, 3′, and 4′ in Figure 2.13) should be identified.

States 1 and 2:

The properties of these states are the same as Example 2.4.

$$h_1 = 191.83\ \text{kJ/kg}$$

$$h_2 = 201.92\ \text{kJ/kg}$$

State 3′:

Flow 3′ is superheated steam, so from Table A.3 (SI)

$$h_{3'} = 3625.3 \text{ kJ/kg}$$

$$s_{3'} = 6.9029 \text{ kJ/kg K}$$

State 4′:

Process 3′-4′ is isentropic; therefore,

$$s_{3'} = s_{4'} = s_{f@10\text{ kPa}} + x_{4'} s_{fg@10\text{ kPa}} \rightarrow 6.9029$$

$$= 0.6493 + x_{4'}(7.5009) \rightarrow x_{4'} = 0.8337 \text{ or } 83.37\%$$

As expected, the quality of the outlet flow of the turbine is considerably higher when the turbine is fed by superheated steam rather than by saturated steam (83.37% versus 66.19%, respectively).

$$h_{4'} = h_{f@10\text{ kPa}} + x_{4'} h_{fg@10\text{ kPa}} = 191.83 + 0.8337 \times 2392.8 = 2186.7 \text{ kJ/kg}$$

The energy transfers in different components are

$$w_{\text{pump}} = h_2 - h_1 = 201.92 - 191.83 = 10.09 \text{ kJ/kg}$$

$$q_{\text{SG}} = h_{3'} - h_2 = 3625.3 - 201.92 = 3423.4 \text{ kJ/kg}$$

$$w_{\text{Turb}} = h_{3'} - h_{4'} = 3625.3 - 2186.7 = 1438.6 \text{ kJ/kg}$$

$$q_{\text{Cond}} = h_{4'} - h_1 = 2186.7 - 191.83 = 1994.9 \text{ kJ/kg}$$

$$w_{\text{Net}} = w_{\text{Turb}} - w_{\text{Pump}} = 1438.6 - 10.09 = 1428.5 \text{ kJ/kg}$$

$$\eta_{\text{Th}} = \frac{w_{\text{Net}}}{q_{\text{SG}}} = \frac{1428.5}{3423.4} = 0.4173 \text{ or } 41.73\%$$

The result indicates that although both heat addition in the steam generator and net work production increase when the steam generator outlet is superheated steam, the rate of increase in the net work is faster than the increase in the rate of the heat addition, which results in the efficiency improvement (41.73% versus 37.22%).

The mass flow rate of the cycle's working fluid is

$$\dot{m} = \frac{\dot{W}_{\text{Net}}}{w_{\text{Net}}} = \frac{250{,}000 \text{ kJ/s}}{1428.5 \text{ kJ/kg}} = 175.0 \text{ kg/s} = 6.30 \times 10^5 \text{ kg/h}$$

Comparing the mass flow rate of the working fluid in the cycle with superheated steam entering the turbine (175.0 kg/s) with that of Example 2.4 with saturated steam entering the turbine (266.2 kg/s) shows that for the same power output, the mass flow rate of the working fluid is less in the former because of the higher efficiency. This lower mass flow rate of the working fluid means smaller and potentially less expensive components are required.

Let us repeat Example 2.6 for the superheated steam temperature between 400°C and 700°C with the increment of 100°C. Table 2.2 reflects the performance of the cycle at various superheated steam temperatures of the turbine inlet flow. Figure 2.14 illustrates the efficiency, specific net work, specific heat input, and quality of the flow at the outlet of the turbine as a function of the temperature of the superheated steam inlet to the turbine. The presented results indicate that the higher temperature of the superheated steam inlet to the turbine improves both the efficiency of the cycle and the quality of the outlet flow from the turbine as well as the specific net work. However, due to the material

TABLE 2.2

Effects of the Superheated Steam Temperature of the Turbine Inlet Flow on the Performance of the Cycle in Example 2.6

Type of Cycle	$P_2=P_3$ (kPa)	$P_1=P_4$ (kPa)	x_4	q_{SG} (kJ/kg)	q_{Cond} (kJ/kg)	w_{Pump} (kJ/kg)	w_{Turb} (kJ/kg)	w_{Net} (kJ/kg)	η_{Th} (%)
Simple Rankine – Turbine inlet: saturate steam (Example 2.4)	10,000	10	0.662	2522.8	1583.8	10.1	949.1	939.0	37.22
Superheated at 400°C	10,000	10	0.742	2894.6	1774.5	10.1	1130.2	1120.1	38.70
Superheated at 500°C	10,000	10	0.793	3171.3	1896.8	10.1	1284.7	1274.6	40.19
Superheated at 600°C (Example 2.6)	10,000	10	0.834	3423.4	1994.9	10.1	1438.6	1428.5	41.73
Superheated at 700°C	10,000	10	0.869	3668.6	2079.7	10.1	1599.0	1588.9	43.31

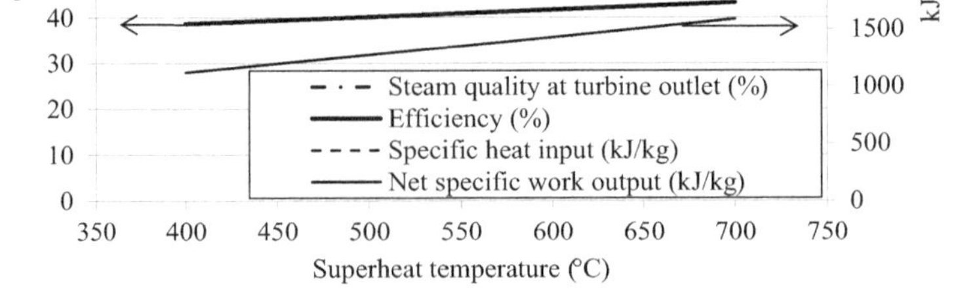

FIGURE 2.14
Efficiency, specific net work output, specific heat input, and quality of the flow at the outlet of the turbine as a function of the temperature of the superheated steam inlet to the turbine in Example 2.6.

limitations of the steam generator and the turbine, the temperature of the inlet stream to the turbine is limited. Note that your calculations results may slightly differ from values reported in this table or other tables in this chapter depending on the source of thermodynamic properties that you use.

2.3.2.2 Effects of Steam Generator Pressure

As Figure 2.15 indicates, increasing the pressure of the outlet flow from the pump (from state 2 to 2′), which means the higher operating pressure of the steam generator, increases the average temperature at which heat is supplied to the cycle. This increases the thermal efficiency of the cycle. However, the maximum temperature of the steam inlet to the steam turbine (and indeed the maximum temperature in the entire cycle) is limited by the material and design of both the steam generator and the early stages of the turbine blades. Therefore, to realistically study the effects of the operating pressure of the steam generation on the performance of the cycle, the maximum temperature of the inlet flow to the turbine should be kept constant. From Figure 2.15, it is evident that in this case (the limited inlet temperature to the turbine) also the efficiency of the cycle improves with an increase in the operating pressure of the steam generator while the moisture content of the turbine outlet flow increases.

Example 2.7

In a Rankine cycle, similar to the one in Examples 2.4 and 2.6, the inlet steam to the turbine is superheated steam at the temperature of 600°C and the pressure of 20,000 kPa. If the outlet stream of the condenser is saturated water at the pressure of 10 kPa, determine the specific heat transfers in the steam generator and the condenser, the specific work involved in the turbine and the pump, and the thermal efficiency of the cycle.

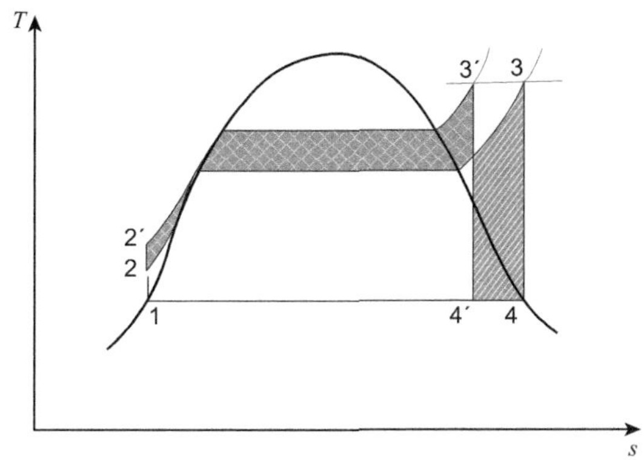

FIGURE 2.15 Effect of the steam generator operating pressure at the constant temperature of the steam turbine inlet flow.

Solution

Given: Similar to Example 2.6.

Find: Similar to Example 2.6.

Schematic diagrams: Figure 2.15

Assumptions: Similar to Example 2.6.

Analysis:

The working fluid (water) properties at the four main states in the cycle (states 1, 2′, 3′, and 4′ in Figure 2.15) are as follows:

State 1:

Similar to Example 2.4,

$$h_1 = h_{f\,@\,10\,\text{kPa}} = 191.83\ \text{kJ/kg}$$

$$v_1 = v_{f\,@10\,\text{kPa}} = 0.0010102\ \text{m}^3/\text{kg}$$

State 2′:

$$w_{Pump} = \int_1^{2'} v dP = v_1\left(P_{2'} - P_1\right)$$

$$= 0.001010\ \text{m}^3/\text{kg} \times (20{,}000 - 10)\,\text{kPa}\left[\frac{1000\,\dfrac{\text{N}}{\text{m}^2}}{1\,\text{kPa}}\right]\left[\frac{1\,\text{kJ}}{1000\,\text{Nm}}\right]$$

$$= 20.19\ \text{kJ/kg}$$

$$w_{Pump} = h_{2'} - h_1 \rightarrow h_{2'} = w_{Pump} + h_1 = 20.19 + 191.83 = 212.02\,\text{kJ/kg}$$

State 3′:

$$h_{3'} = h_{\text{super heat@20,000 kPa and 600°C}} = 3537.6\,\text{kJ/kg}$$

$$s_{3'} = s_{\text{superhea @20,000 kPa and 600°C}} = 6.5048\,\text{kJ/kg K}$$

State 4′:

Since process 3′-4′ is isentropic,

$$s_{3'} = s_{4'} = s_{f\,@10\,\text{kPa}} + x_{4'} s_{fg\,@10\,\text{kPa}} \rightarrow 6.5048 = 0.6493 + x_{4'}(7.5009)$$

$$x_{4'} = 0.7806 \text{ or } 78.06\%$$

As expected, the quality of the flow at the outlet of the turbine in this case is lower than that in the case of Example 2.6 with the steam generator pressure of 10,000 kPa (78.1% versus 83.4%, respectively)

$$h_{4'} = h_{f\,@10\,\text{kPa}} + x_{4'} h_{fg\,@10\,\text{kPa}} = 191.83 + 0.7806 \times 2392.8 = 2059.7\ \text{kJ/kg}$$

The specific heat transfers in the steam generator and the condenser, the specific work involved in the turbine and the pump, and the thermal efficiency of the cycle can be calculated as follows:

$$w_{\text{Pump}} = h_{2'} - h_1 = 212.02 - 191.83 = 20.19\ \text{kJ/kg}$$

TABLE 2.3

Effects of the Steam Generator Operating Pressure at the Constant Turbine Inlet Temperature on the Performance of the Cycle in Example 2.7

Type of Cycle: Simple Rankine–Turbine Inlet:	$P_2'=P_3'$ (kPa)	$P_1=P_4'$ (kPa)	x_4'	q_{SG} (kJ/kg)	q_{Cond} (kJ/kg)	w_{Pump} (kJ/kg)	w_{Turb} (kJ/kg)	w_{Net} (kJ/kg)	η_{Th} (%)
Superheated at 4000 kPa	4000	10	0.896	3478.5	2143.5	4.0	1339.0	1335.0	38.38
Superheated at 10,000 kPa (Example 2.6)	10,000	10	0.834	3423.4	1994.9	10.1	1438.6	1428.5	41.73
Superheated at 12,000 kPa	12,000	10	0.820	3404.4	1963.3	12.1	1453.2	1441.1	42.33
Superheated at 16,000 kPa	16,000	10	0.799	3365.5	1911.0	16.2	1470.7	1454.5	43.22
Superheated at 20,000 kPa (Example 2.7)	20,000	10	0.781	3325.6	1867.9	20.2	1477.9	1457.7	43.83
Superheated at 24,000 kPa	24,000	10	0.765	3284.6	1830.5	24.2	1478.4	1454.1	44.27

$$q_{SG} = h_{3'} - h_{2'} = 3537.6 - 212.02 = 3325.6 \text{ kJ/kg}$$

$$w_{Turb} = h_{3'} - h_{4'} = 3537.6 - 2059.7 = 1477.9 \text{ kJ/kg}$$

$$q_{Cond} = h_{4'} - h_1 = 2059.7 - 191.83 = 1867.9 \text{ kJ/kg}$$

$$w_{Net} = w_{Turb} - w_{Pump} = 1477.9 - 20.19 = 1457.7 \text{ kJ/kg}$$

$$\eta_{Th} = \frac{w_{net}}{q_{SG}} = \frac{1457.7}{3325.6} = 0.4383 \text{ or } 43.83\%$$

The efficiency of the cycle in this case is higher than that in the case of Example 2.6 with the steam generator pressure of 10,000 kPa (43.83% versus 41.73%, respectively). Table 2.3 shows the performance of the cycle at various steam generator pressures. The presented results indicate that the higher steam generator pressure improves the cycle efficiency but reduces the quality of the outlet flow from the turbine.

2.3.2.3 Supercritical Steam Power Plants

With improved materials used in the manufacturing of steam turbine blades and steam generators, the operating pressure of steam generators can exceed the critical pressure of water (221 bar or 3203.6 psia). Based on the operating pressure of the steam generator, steam power plants can be categorized into two groups: **subcritical** (with the steam generator operating pressure below 221 bar) and **supercritical** (with the steam generator operating pressure above 221 bar). The efficiency of the former is typically <40% while the latter can achieve the efficiency of as high as about 47%. In supercritical steam

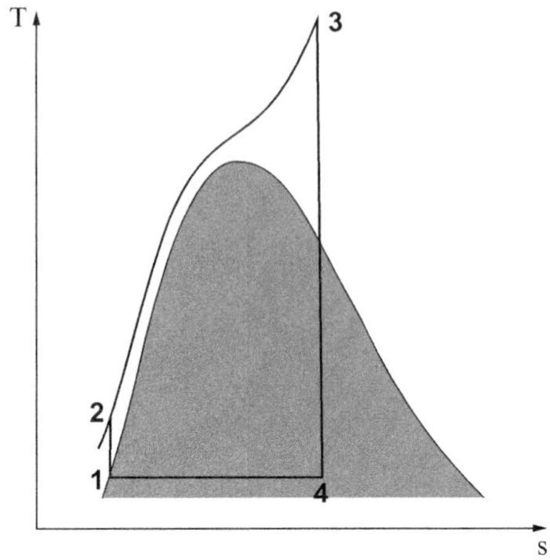

FIGURE 2.16
T-s diagram of a supercritical Rankine cycle.

generators, there is no boiling process, i.e., no two-phase mixture of liquid water and steam. Instead, liquid water is gradually converted to superheated steam (Figure 2.16). Modern supercritical steam power plants can operate at the turbine inlet pressure and temperature of as high as 300 bar (4350 psia) and 600°C (1110°F), respectively or higher. Note that in supercritical steam cycles, the moisture content of the turbine outlet flow is unacceptably high. This problem is addressed by adding reheaters (discussed later in this chapter).

Recently, with the advancement of materials used in plant components, a new category of steam power plants is introduced, **ultra-supercritical steam power plants** with the turbine inlet steam pressure and temperature of about 350 bar (5075 psia) and 700°C (1290°F), respectively. The efficiency of these power plants can exceed 50%.

2.3.2.4 Effects of Pressure and Temperature of Turbine Exhaust

If the pressure of the turbine exhaust, which is equal to the operating pressure and temperature of the condenser, is reduced from P_4 to P_4' (Figure 2.17), the efficiency of the cycle increases since the average temperature of the heat transfer from the cycle to the environment decreases. But the moisture content of the stream leaving the turbine increases, which can potentially limit the extent to which this approach can be used to improve the efficiency of the cycle. This problem can be addressed by adding a reheater (discussed later).

Example 2.8

Similar to Example 2.6, in a Rankine cycle, the inlet flow to the turbine is superheated steam at the pressure of 10,000 kPa and the temperature of 600°C. If the outlet stream of the condenser is saturated water and the operating

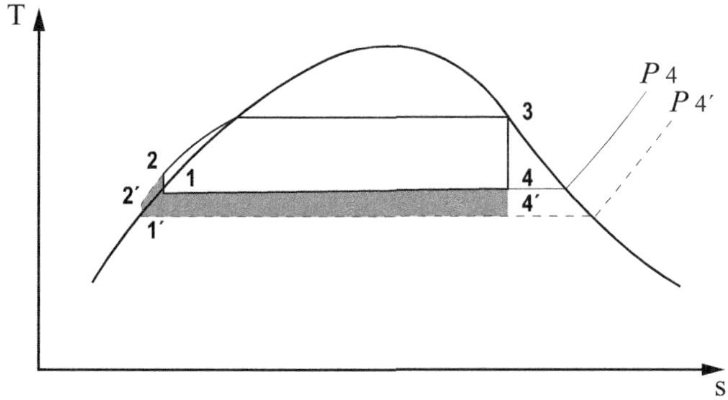

FIGURE 2.17
Effects of the operating temperature and pressure of the condenser on the performance of the cycle.

pressure of the condenser is 40 kPa, determine the specific heat transfers in the steam generator and the condenser, the specific work involved in the turbine and the pump, and the thermal efficiency of the cycle.

Solution

Given: Similar to Example 2.6.
Find: Similar to Example 2.6.
Schematic diagrams: Figure 2.17
Assumptions: Similar to Example 2.6.
Analysis:
State 1′:
At this state, the working fluid is saturated liquid water at the pressure of 40 kPa, from Table A.2 (SI):

$$h_{1'} = h_{\text{sat liq@ 40 kPa}} = h_{f\,@\,40\,\text{kPa}} = 317.58\ \text{kJ/kg}$$

$$v_{1'} = v_{f\,@\,40\,\text{kPa}} = 0.001027\ \text{m}^3/\text{kg}$$

State 2′:

$$w_{Pump} = \int_{1'}^{2'} v dP = v_{1'}\left(P_{2'} - P_{1'}\right)$$

$$= 0.001027\ \text{m}^3/\text{kg} \times (10{,}000 - 40)\,\text{kPa}\left[\frac{1000\ \text{N/m}^2}{1\ \text{kPa}}\right]\left[\frac{1\ \text{kJ}}{1000\ \text{Nm}}\right]$$

$$= 10.23\ \text{kJ/kg}$$

$$w_{Pump} = h_{2'} - h_{1'} \rightarrow h_{2'} = w_{Pump} + h_{1'} = 10.23 + 317.58 = 327.81\ \text{kJ/kg}$$

State 3:

$$h_3 = h_{\text{superheat@10,000 kPa and 600°C}} = 3625.3\ \text{kJ/kg}$$

$$s_3 = s_{\text{superheat@10,000 kPa and 600°C}} = 6.9029\ \text{kJ/kg K}$$

State 4′:

Since process 3-4′ is isentropic,

$$s_3 = s_{4'} = s_{f\,@\,40\text{ kPa}} + x_{4'} s_{fg\,@\,40\text{ kPa}} \rightarrow 6.9029 = 1.0259 + x_{4'}(6.6441)$$

$$x_{4'} = 0.8845 \text{ or } 88.45\%$$

This indicates that increasing the operating pressure of the condenser from 10 to 40 kPa improves the quality of the flow at the turbine outlet from 83.4% to 88.5%.

$$h_{4'} = h_{f\,@\,40\text{ kPa}} + x_{4'} h_{fg\,@\,40\text{ kPa}} = 317.58 + 0.8845 \times 2319.2 = 2368.9 \text{ kJ/kg}$$

The specific heat transfers in the steam generator and the condenser, the specific work involved in the turbine and the pump, and the thermal efficiency of the cycle can be calculated as follows:

$$w_{\text{Pump}} = h_{2'} - h_{1'} = 327.81 - 317.58 = 10.23 \text{ kJ/kg}$$

$$q_{\text{SG}} = h_3 - h_{2'} = 3625.3 - 327.81 = 3297.5 \text{ kJ/kg}$$

$$w_{\text{Turb}} = h_3 - h_{4'} = 3625.3 - 2368.9 = 1256.4 \text{ kJ/kg}$$

$$q_{\text{Cond}} = h_{4'} - h_{1'} = 2368.9 - 317.58 = 2051.3 \text{ kJ/kg}$$

$$w_{\text{Net}} = w_{\text{Turb}} - w_{\text{Pump}} = 1256.4 - 10.23 = 1246.2 \text{ kJ/kg}$$

$$\eta_{\text{Th}} = \frac{w_{\text{Net}}}{q_{\text{SG}}} = \frac{1246.2}{3297.5} = 0.3779 \text{ or } 37.79\%$$

This indicates that increasing the operating pressure of the condenser from 10 to 40 kPa reduces the efficiency of the cycle from 41.73% to 37.79%. Also, note that the net specific work of the cycle reduced from 1428.5 to 1246.2 kJ/kg.

Table 2.4 compares the performance of the cycle with different operating pressures of the condenser. The presented results indicate that the

TABLE 2.4

Effects of the Operating Pressure of the Condenser on the Performance of the Cycle in Example 2.8

Type of Cycle: Simple Rankine–Condenser Pressure:	$P_2' = P_3$ (kPa)	$P_1' = P_4'$ (kPa)	x_4'	q_{SG} (kJ/kg)	q_{Cond} (kJ/kg)	w_{Pump} (kJ/kg)	w_{Turb} (kJ/kg)	w_{Net} (kJ/kg)	η_{Th} (%)
5 kpa	10,000	5	0.811	3478.8	1966.0	10.05	1522.8	1512.8	43.49
10 kPa (Example 2.6)	10,000	10	0.834	3423.4	1994.9	10.09	1438.6	1428.5	41.73
20 kPa	10,000	20	0.858	3363.7	2023.2	10.15	1350.7	1340.6	39.85
40 kPa (Example 2.8)	10,000	40	0.885	3297.5	2051.3	10.23	1256.4	1246.2	37.79
80 kPa	10,000	80	0.914	3223.3	2079.1	10.30	1154.5	1144.2	35.50
100 kPa	10,000	100	0.925	3197.5	2087.8	10.33	1120.0	1109.7	34.71

lower operating pressure of the condenser improves the efficiency of the cycle but adversely affects the quality of the outlet flow from the turbine. The lower operating pressure of the condenser also increases the net specific work of the cycle, which means for the same power output, smaller and less expensive components are required.

Comparing the results presented in Tables 2.3 and 2.4 can lead us to a very interesting observation. Table 2.3 shows that increasing the operating pressure of the steam generator from 4000 to 24,000 kPa (20,000 kPa increase) can increase the efficiency of the cycle from 38.38% to 44.27% (5.89 point increase). Similar observation from Table 2.4 indicates that reducing the operating pressure of the condenser from 100 to 5 kPa (only 95 kPa reduction) can increase the efficiency of the cycle from 34.71% to 43.49% (8.78 point increase). This reemphasizes the importance of the condenser and its operating pressure. The effect of severaltens of kilo Pascal reductions in the operating pressure of the condenser can be greater than the effect of a few thousand kilo Pascal increase in the operating pressure of the steam generator. The typical condenser absolute pressure is about 7 kPa (1 psi).

To summarize this section, we can say that the improvement of the net specific output power and efficiency of the cycle can be achieved by:

- **Superheating steam in the steam generator**
- **Increasing the operating pressure of the steam generator**
- **Lowering the operating pressure of the condenser**

Increasing the quality of the flow leaving the turbine can be achieved by

- **Superheating steam in the steam generator**
- **Decreasing the operating pressure of the steam generator**
- **Increasing the operating pressure of the condenser**

Up to this point, the operating conditions of the simple Rankine cycle were optimized. In the following few sections, the additional components to the cycle will be explored to improve the system performance.

2.3.3 Reheat Rankine Cycles

It was noted (Section 2.3.2.2) that increasing the steam generator operating pressure is beneficial for the efficiency of the cycle but adversely affects the quality of steam in the turbine outlet flow. To take advantage of increasing the operating pressure of the steam generator at the fixed turbine inlet temperature without suffering from increasing the moisture content of the turbine outlet, the reheat Rankine cycle can be used. In this cycle, the steam from the steam generator is expanded in the turbine to an intermediate pressure (state 4 in Figure 2.18) and then is returned to the steam generator to be reheated. The reheated steam at state 5 is then further expanded in a lower pressure turbine to state 6. By this process, not only the efficiency of the cycle increases but the quality of the steam at the turbine outlet flow also improves. Note that if the material for turbine blades could stand the temperature of state 3′,

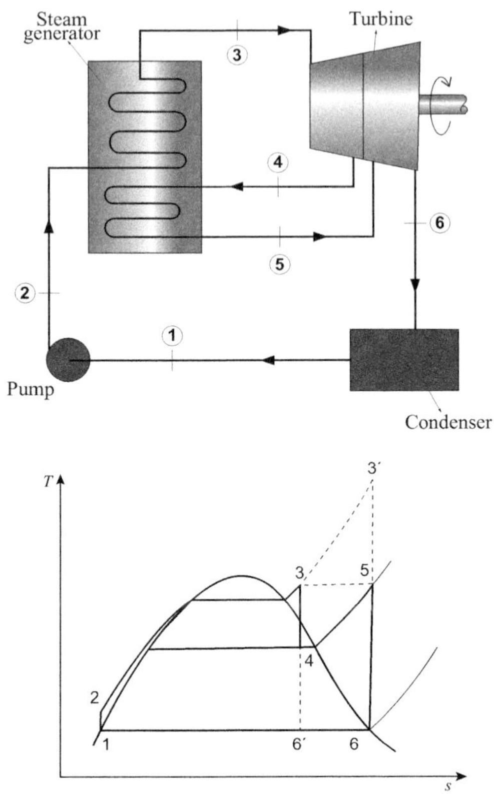

FIGURE 2.18
A reheat Rankine cycle.

the efficiency of the simple Rankine cycle would be higher. Modern large fossil fuel steam power plants typically employ at least one and sometimes two reheaters.

Example 2.9

Reconsider the Rankine cycle in Example 2.6, where the operating pressure of the steam generator and the condenser are 10,000 and 10 kPa, respectively, and the turbine inlet steam is superheated at the temperature of 600°C. If the expansion process occurs in two stages with a reheater with the pressure of 1000 kPa in between, determine the specific heat transfers in the steam generator and the condenser, the specific work involved in the turbine and the pump, and the thermal efficiency of the cycle. The outlet flows of the steam generator and the reheater are at the same temperature.

Note: This steam power plant is sometimes expressed as 100/600/600 (100 bar/600°C/600°C, steam generator pressure/high-pressure turbine inlet temperature/low-pressure turbine inlet temperature).

Given: An ideal reheat Rankine cycle with the working fluid of water operates at the given operating pressure of the steam generator and the condenser. The turbine inlet flows aresuperheated steam and the condenser outlet flow is saturated liquid. The processes in the turbine and pump are isentropic. The turbine has two stages with a reheater between two stages.

Find: The specific heat transfers in the steam generator and the condenser, the specific work involved in the turbine and the pump, and the thermal efficiency of the cycle.

Schematic diagrams: Figure 2.18

Assumptions:

1. The cycle is an ideal reheat Rankine cycle. The working fluid is water.
2. The steady-state control volume mass and energy balance analyses with no kinetic and potential energy changes are applied to each component.
3. The outlet flows of the steam generator (state 3) and the reheater (state 5) are at the same temperature.
4. There is no pressure drop due to friction and no heat loss to the surrounding in the components and piping systems, including in the reheater.

Analysis:

States 1, 2, and 3:

The properties of these states are same as that in Example 2.6.

$$h_1 = 191.83 \text{ kJ/kg}$$

$$h_2 = 201.92 \text{ kJ/kg}$$

$$h_3 = 3625.3 \text{ kJ/kg}$$

$$s_3 = 6.9029 \text{ kJ/kg K}$$

State 4:

$$s_3 = s_4 = 6.9029 \text{ kJ/kg K}$$

At state 4, the flow is still superheated steam, so from Table A.3 (SI) at the pressure of 1000 kPa (by interpolation)

$$h_4 = 2931.8 \text{ kJ/kg}$$

$$T_4 = 245.3°\text{C}$$

State 5:

At state 5, the working fluid is superheated steam at the pressure of 1000 kPa and the temperature of 600°C, from Table A.3 (SI),

$$h_5 = 3697.9 \text{ kJ/kg}$$

$$s_5 = 8.0290 \text{ kJ/kg K}$$

State 6:

$$s_6 = s_5 = s_{f@10\,\text{kPa}} + x_6 s_{fg@10\,\text{kPa}} \rightarrow 8.0290$$

$$= 0.6493 + x_6(7.5009) \rightarrow x_6 = 0.9838 \text{ or } 98.38\%$$

$$h_6 = h_{f@10\,\text{kPa}} + x_6 h_{fg@10\,\text{kPa}} = 191.83 + 0.9838 \times 2392.8 = 2545.9 \text{ kJ/kg}$$

Since there are two stages of expansion and heat addition,

$$w_{\text{Turb}} = (h_3 - h_4) + (h_5 - h_6) = (3625.3 - 2931.8) + (3697.9 - 2545.9) = 1845.5 \text{ kJ/kg}$$

$$q_{\text{SG}} = (h_3 - h_2) + (h_5 - h_4) = (3625.3 - 201.92) + (3697.9 - 2931.8) = 4189.5 \text{ kJ/kg}$$

$$w_{\text{Pump}} = h_2 - h_1 = 201.92 - 191.83 = 10.09 \text{ kJ/kg}$$

$$q_{\text{Cond}} = h_6 - h_1 = 2545.9 - 191.83 = 2354.1 \text{ kJ/kg}$$

$$w_{\text{Net}} = w_{\text{Turb}} - w_{\text{Pump}} = 1845.5 - 10.09 = 1835.4 \text{ kJ/kg}$$

$$\eta_{\text{Th}} = \frac{w_{\text{net}}}{q_{\text{SG}}} = \frac{1835.4}{4189.6} = 0.4381 \text{ or } 43.81\%$$

Table 2.5 compares Examples 2.6 and 2.9 for the similar cycles with and without a reheater.

Let us repeat Example 2.9 for when the steam leaves the high-pressure turbine at the pressure of 1500, 2000, 4000, and 8000 kPa and then is sent to the reheater. Table 2.6 shows the results.

The results in Table 2.6 indicate that lowering the reheat pressure from 8000 to 1500 kPa improves the overall cycle efficiency because more heat is added in the reheater, which increases the average temperature of the heat addition to the cycle. However, further reduction of the reheat pressure will result in the slight reduction in the efficiency. So there is an optimum reheater pressure that maximizes the efficiency of the cycle. Also, reducing the reheater pressure improves the quality of the turbine outlet flow.

TABLE 2.5

Effects of the Reheater on the Performance of the Cycle in Examples 2.6 and 2.9

Type of Cycle: Rankine Cycle	$P_2 = P_3$ (kPa)	$P_1 = P_6$ (kPa)	x_4 or x_6	q_{SG} (kJ/kg)	q_{Cond} (kJ/kg)	w_{Pump} (kJ/kg)	w_{Turb} (kJ/kg)	w_{Net} (kJ/kg)	η_{Th} (%)
Without reheater (Example 2.6)	10,000	10	0.834	3423.4	1994.9	10.1	1438.6	1428.5	41.73
With reheater (Example 2.9)	10,000	10	0.984	4189.5	2354.1	10.1	1845.5	1835.4	43.81

Note: A double-reheat steam power plant is expressed as: steam generator pressure/high-pressure turbine inlet temperature/intermediate pressure turbine inlet temperature/low-pressure turbine inlet temperature.

TABLE 2.6

Effects of the Reheater Pressure on the Performance of the Cycle

Reheater Pressure at Rankine Cycle (kPa)	$P_2=P_3$ (kPa)	$P_1=P_6$ (kPa)	x_6	q_{SG} (kJ/kg)	q_{Cond} (kJ/kg)	w_{Pump} (kJ/kg)	w_{Turb} (kJ/kg)	w_{Net} (kJ/kg)	η_{Th} (%)
1000 (Example 2.9)	10,000	10	0.984	4189.5	2354.1	10.1	1845.5	1835.4	43.81
1500	10,000	10	0.958	4087.6	2293.4	10.1	1804.3	1794.2	43.89
2000	10,000	10	0.940	4008.6	2250.0	10.1	1768.8	1758.7	43.87
4000	10,000	10	0.896	3791.5	2143.5	10.1	1658.1	1648.0	43.46
8000	10,000	10	0.849	3523.6	2032.5	10.1	1501.2	1491.1	42.32

2.3.4 Regenerative Rankine Cycles

In a steam generator, between states 2 and 2′ (Figure 2.19) the working fluid at the liquid phase is heated. This process reduces the average temperature at which heat is supplied to the cycle which, in turn, reduces the efficiency of the cycle. That is why the efficiency of the Rankine cycle is lower than the efficiency of the Carnot cycle operating between the same high- and low-temperature thermal energy reservoirs (Figure 2.19).

The idea of a regenerative cycle is to increase the temperature of the steam generator inlet flow from state 2 to as close to state 2′ as possible to increase the average temperature at which heat is supplied to the cycle. This also reduces external irreversibilities due to the temperature difference between the working fluid and the flue gas.

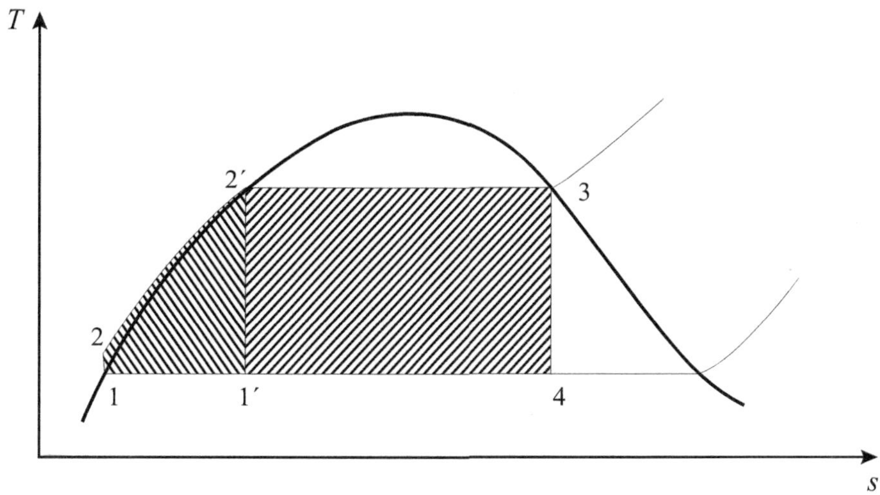

FIGURE 2.19
Comparison of the Carnot cycle and the Rankine cycle operating between the same high- and low-temperature thermal energy reservoirs.

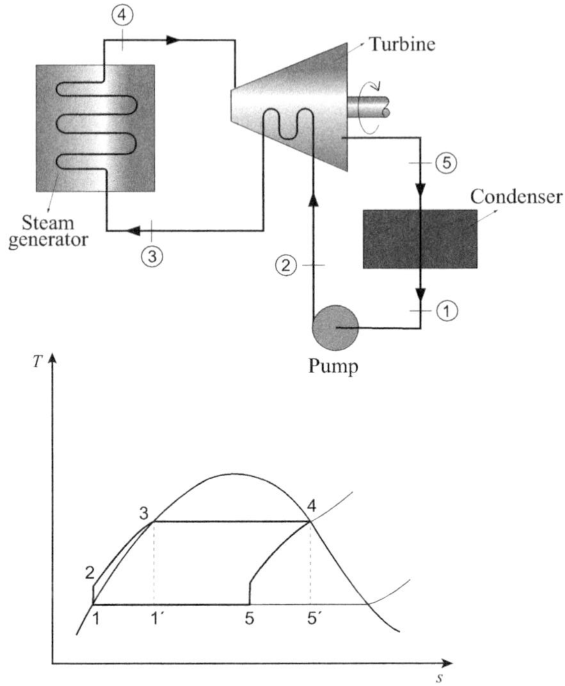

FIGURE 2.20
An ideal regenerative Rankine cycle.

For this purpose, some heat should be transferred from other parts of the cycle to the compressed liquid leaving the pump. In the ideal regenerative cycle (Figure 2.20), the liquid leaving the pump (state 2) circulates around the turbine casing (at the opposite direction of the steam in the turbine), so the subcooled water from the pump is heated, and its temperature increases. In this ideal case, the heat-transfer surface area should be large enough so that the inlet flow to the steam generator is at a saturated liquid phase (state 3). As a result of this reversible heat transfer from the steam expanding in the turbine, the outlet flow of the turbine would be at state 5 instead of state 5′.

In this ideal Rankine cycle (Figure 2.20), process 2-3 is an internal process and heat transfer from the external thermal energy source to the subcooled working fluid is eliminated. As a result, the average temperature at which heat is added to the cycle increases, and irreversibilities reduce. In Figure 2.20, lines 1-3 and 5-4 are parallel, so the enclosed areas by two cycles 1-2-3-4-5-1 and 1′-3-4-5′-1′ are exactly equal; therefore, the efficiency of the ideal regenerative Rankine cycle and the Carnot cycle is equal. But the idealized regenerative Rankine cycle is not practical because it is impossible to provide such a huge heat-transfer surface area around the turbine at a reasonable cost. Also, the quality of steam at the turbine outlet will drop to an unsafe zone.

In a real regenerative steam power generation cycle, some of the steam in the turbine is extracted (or bled) after partial expansion and used to heat the pump outlet stream (subcooled liquid) through a feedwater heater (or a feed-liquid heater for working fluids other than water). In real regeneration, since the continuous heating of the working fluid from state 2 to state 3 (Figure 2.20) is not possible, the heating takes place in several discrete steps. This will not entirely eliminate the process of heat transfer from the external heat source to subcooled feedwater but will significantly reduce its extent and magnitude. In an actual regenerative process, the steam generator inlet flow is not saturated water and is somewhere between states 2 and 3. Note that the extraction or bleeding of the steam from the turbine will reduce the power production in the turbine. However, the benefit gained in the steam generator will overcome this loss. There are two types of feedwater heaters: open or direct-contact and closed.

2.3.4.1 Open Feedwater Heaters

In this type of feedwater heaters, the extracted steam from the turbine (state 6 in Figure 2.21) is directly mixed with feedwater (high-pressure water after the pump) at state 2 in a pressure vessel to form saturated water at state 3. The amount of the extracted steam should be sufficient to produce saturated water at state 3. Depending on the pressure of the extraction point, the steam can be superheated steam or a two-phase mixture. The mass flow rate of the extracted flow from the turbine should be carefully selected. Too much extracted steam will produce two-phase mixture of saturated steam and water at the outlet flow of the open feedwater heater, which is difficult to pump. Too little extracted steam will reduce the effectiveness of the feedwater heating process. In this regenerative cycle, the heat addition to the working fluid in the steam generator takes place in process 4-5 (Figure 2.21) instead of process 2-3 (Figure 2.18) for the non-regenerative cycle. The former requires less thermal energy input in the steam generator.

It can be observed that the average temperature of heat addition in the regenerative cycle is higher than that of the non-regenerative cycle, which indicates a higher efficiency in the former. Also, it is evident that when some steam is bled from the turbine at state 6, there is less mass flow rate to expand from state 6 to state 7, which reduces the total work developed in the turbine. However, considering the effects of both less fuel consumption in the steam generator and less work production in the turbine, the overall efficiency of the cycle increases. Since the pressures of the inlet streams to the open feedwater heater (at states 2 and 6) **should be equal** (otherwise the liquid working fluid may enter the turbine), feedwater cannot be pressurized to the steam generator pressure in the first pump after the condenser, and a second pump after the open feedwater heater is required to pump feedwater to the steam generator pressure. In the first pump (process 1-2), commonly referred to as the **condensate pump**, saturated water (**condensate water**) at the outlet of the condenser is pumped from the condenser pressure (state 1) to the pressure that steam is extracted from the turbine (state 6). Then, the second pump, the **feedwater pump**, pressurizes saturated feedwater at the outlet of the open feedwater heater to the steam generator pressure (process 3-4).

In the analysis of an open feedwater heater, both mass and energy balance equations should be considered. For the CV around the open feedwater heater (Figure 2.21),

$$\text{Mass balance}: \dot{m}_2 + \dot{m}_6 = \dot{m}_3 \tag{2.15}$$

$$\text{Energy balance}: \dot{m}_2 h_2 + \dot{m}_6 h_6 = \dot{m}_3 h_3 \tag{2.16}$$

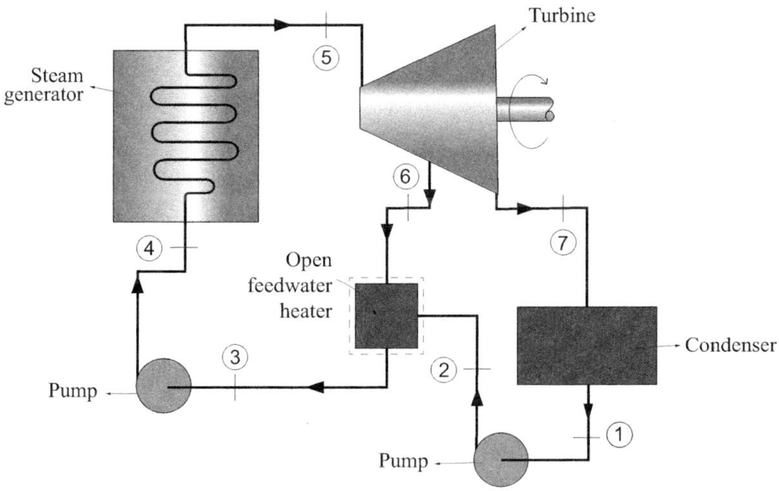

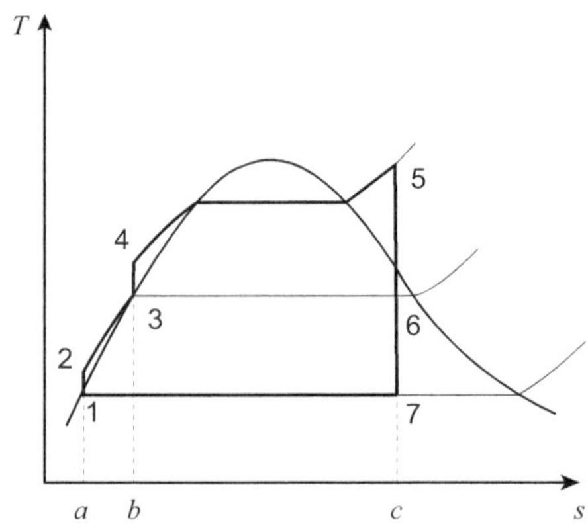

FIGURE 2.21
A regenerative Rankine cycle with an open feedwater heater.

Dividing Equations 2.15 and 2.16 by the total mass flow rate of the working fluid (which is equal to $\dot{m}_3$).

$$\frac{\dot{m}_2}{\dot{m}_3} + \frac{\dot{m}_6}{\dot{m}_3} = 1 \tag{2.17}$$

$$\frac{\dot{m}_2}{\dot{m}_3} h_2 + \frac{\dot{m}_6}{\dot{m}_3} h_6 = h_3 \tag{2.18}$$

There are two equations with two unknowns ($\dot{m}_2$ and $\dot{m}_6$). Note that if there are more feedwater heaters, for each heater one system of equations like this one can be written. It is convenient to show the ratio of the mass flow rate of the extracted steam to the total mass flow rate of the working fluid by λ, so

$$\frac{\dot{m}_6}{\dot{m}_3} = \frac{\dot{m}_{ext}}{\dot{m}_t} = \lambda \tag{2.19}$$

$$\frac{\dot{m}_2}{\dot{m}_3} = 1 - \lambda \tag{2.20}$$

Substituting into the energy balance equation (Equation 2.18),

$$(1-\lambda)h_2 + \lambda h_6 = h_3 \tag{2.21}$$

$$\lambda = \frac{h_3 - h_2}{h_6 - h_2} \tag{2.22}$$

Note that in Equation 2.22, λ is the ratio of the mass flow rate of the extracted steam from the turbine to the total mass flow rate of the working fluid and can be determined for the given states 2, 3, and 6. Since $h_6 - h_2$ is the enthalpy change during the condensation of extracted steam, which is a very large number compared to $h_3 - h_2$ which is the enthalpy change of liquid water, λ is a relatively small number. This means a small fraction of the expanding steam in the turbine is extracted, which will limit the negative impact of the steam extraction on the output power of the turbine. Due to the two-step expansion and compression processes, the power and work involved in the turbine and pump can be estimated as follows:

$$\dot{W}_{Turb} = \dot{m}_t(h_5 - h_6) + (\dot{m}_t - \dot{m}_6)(h_6 - h_7) \quad \text{or} \quad w_{Turb}$$

$$= \frac{\dot{W}_{Turb}}{\dot{m}_t} = (h_5 - h_6) + (1-\lambda)(h_6 - h_7) \tag{2.23}$$

$$\dot{W}_{\text{Pump}} = (\dot{m}_t - \dot{m}_6)(h_2 - h_1) + \dot{m}_t(h_4 - h_3) \quad \text{or} \quad w_{\text{Pump}}$$

$$= \frac{\dot{W}_{\text{Pump}}}{\dot{m}_t} = (1-\lambda)(h_2 - h_1) + (h_4 - h_3) \tag{2.24}$$

For the heat transfers in the steam generator and the condenser,

$$\dot{Q}_{\text{SG}} = \dot{m}_t(h_5 - h_4) \quad \text{or} \quad q_{\text{SG}} = \frac{\dot{Q}_{\text{SG}}}{\dot{m}_t} = h_5 - h_4 \tag{2.25}$$

$$\dot{Q}_{\text{Cond}} = (\dot{m}_t - \dot{m}_6)(h_7 - h_1) \quad \text{or} \quad q_{\text{Cond}} = (1-\lambda)(h_7 - h_1) \tag{2.26}$$

Example 2.10

In an ideal regenerative Rankine cycle, the operating pressures of the steam generator and the condenser are 10,000 and 10 kPa, respectively, and the turbine inlet steam is superheated at the temperature of 600°C (similar to Example 2.6). In the turbine, some steam is extracted at the pressure of 1000 kPa and diverted to an open feedwater heater. If the outlet of the open feedwater is saturated liquid water, determine the specific heat transfers in the steam generator and the condenser, the specific work involved in the turbine and the pumps, and the thermal efficiency of the cycle.

Given: An ideal regenerative Rankine cycle with the working fluid of water operates at the given operating conditions. The cycle also has an open feedwater system operating at the given pressure.

Find: The specific heat transfers in the steam generator and the condenser, the specific work involved in the turbine and the pumps, and the thermal efficiency of the cycle.

Schematic diagrams: Figure 2.21

Assumptions:

1. The cycle is an ideal regenerative Rankine cycle with the working fluid of water.
2. The steady-state control volume mass and energy balance analyses with no kinetic and potential energy changes are applied to each component.
3. There is no pressure drop due to friction and no heat loss to the surrounding in the components and piping systems, including in the open feedwater heater.

Analysis:

State 1:

$$h_1 = 191.83 \text{ kJ/kg}$$

$$v_1 = 0.001010 \text{ m}^3/\text{kg}$$

State 2:

$$h_2 = w_{Pump1} + h_1 = v_1(P_2 - P_1) + h_1 = 0.001010\ \text{m}^3/\text{kg} \times (1000 - 10)\ \text{kPa}$$

$$\left[\frac{1000\ \dfrac{\text{N}}{\text{m}^2}}{1\ \text{kPa}}\right]\left[\frac{1\ \text{kJ}}{1000\ \text{Nm}}\right] + 191.83\ \frac{\text{kJ}}{\text{kg}} = 192.83\ \text{kJ/kg}$$

State 3:

$$h_3 = h_{f\,@\,1000\ \text{kPa}} = 762.81\ \text{kJ/kg}$$

$$v_3 = v_{f\,@\,1000\ \text{kPa}} = 0.001127\ \text{m}^3/\text{kg}$$

State 4:

$$h_4 = w_{Pump2} + h_3 = v_3(P_4 - P_3) + h_3 = 0.001127\ \text{m}^3/\text{kg}(10{,}000 - 1000)\ \text{kPa}$$

$$\left[\frac{1000\ \text{N/m}^2}{1\ \text{kPa}}\right]\left[\frac{1\ \text{kJ}}{1000\ \text{N m}}\right] + 762.81\ \text{kJ/kg} = 772.95\ \text{kJ/kg}$$

State 5:

$$h_5 = 3625.3\ \text{kJ/kg}$$

$$s_5 = 6.9029\ \text{kJ/kg K}$$

State 6:

$$s_6 = s_5 = 6.9029\ \text{kJ/kg K}$$

At state 6, the flow is superheated steam, so from Table A.3 (SI) at the pressure of 1000 kPa (by interpolation)

$$h_6 = 2931.8\ \text{kJ/kg}$$

State 7:

$$s_7 = s_5 = 6.9029\ \text{kJ/kg K}$$

$$s_7 = s_{f\,@\,10\ \text{kPa}} + x_7 s_{fg\,@\,10\ \text{kPa}} \rightarrow 6.9029$$

$$= 0.6493 + x_7(7.5009) \rightarrow x_7 = 0.8337 \text{ or } 83.37\%$$

$$h_7 = h_{f\,@\,10\ \text{kPa}} + x_7 h_{fg\,@\,10\ \text{kPa}} = 191.83 + 0.8337 \times 2392.8 = 2186.7\ \text{kJ/kg}$$

Now that the properties of flows at states 1–7 are known, the ratio of the mass of the extracted steam from the turbine to the unit of mass of the working fluid can be determined as follows (Equation 2.22):

$$\lambda = \frac{h_3 - h_2}{h_6 - h_2} = \frac{762.81 - 192.83}{2931.8 - 192.83} = 0.2081$$

With the enthalpy of all states determined, the work involved in the turbine and pumps can be calculated from Equations 2.23 and 2.24.

$$w_{\text{Turb}} = (h_5 - h_6) + (1 - \lambda)(h_6 - h_7) = (3625.3 - 2931.8)$$
$$+ (1 - 0.2081)(2931.8 - 2186.7) = 1283.5 \text{ kJ/kg}$$

$$w_{\text{Pump}} = (1 - \lambda)(h_2 - h_1) + (h_4 - h_3)$$
$$= (1 - 0.2081)(192.83 - 191.83) + (772.95 - 762.81) = 10.93 \text{ kJ/kg}$$

For the heat transfers in the steam generator and the condenser,

$$q_{\text{SG}} = h_5 - h_4 = 3625.3 - 772.95 = 2852.3 \text{ kJ/kg}$$

$$q_{\text{Cond}} = (1 - \lambda)(h_7 - h_1) = (1 - 0.2081)(2186.7 - 191.83) = 1579.7 \text{ kJ/kg}$$

$$w_{\text{Net}} = w_{\text{Turb}} - w_{\text{Pump}} = 1283.5 - 10.93 = 1272.6 \text{ kJ/kg}$$

$$\eta_{\text{Th}} = \frac{w_{\text{Net}}}{q_{\text{SG}}} = \frac{1272.6}{2852.3} = 0.4462 \quad \text{or} \quad 44.62\%$$

Table 2.7 compares the results of Examples 2.10 and 2.6, the Rankine cycle with the open feedwater heater and the Rankine cycle without one, respectively. As discussed earlier, although the work developed in the turbine reduces, the thermal efficiency of the cycle improved considerably because the heat added in the steam generator also decreases. Also, note that while the pump work slightly increases due to the two-step pumping process, the rejected heat in the condenser reduced substantially.

TABLE 2.7

Comparison of the Performance of the Rankine Cycle with the Open Feedwater Heater and the Rankine Cycle without One in Examples 2.10 and 2.6

Type of Cycle	P_2 (kPa)	P_1 (kPa)	x_4 or x_7	q_{SG} (kJ/kg)	q_{Cond} (kJ/kg)	w_{Pump} (kJ/kg)	w_{Turb} (kJ/kg)	w_{Net} (kJ/kg)	η_{Th} (%)
Simple Rankine cycle without feedwater heater (Example 2.6)	10,000	10	0.834	3423.4	1994.9	10.1	1438.6	1428.5	41.73
Rankine cycle with open feedwater heater (Example 2.10)	10,000	10	0.834	2852.3	1579.7	10.9	1283.5	1272.6	44.62

2.3.4.2 Closed Feedwater Heaters

In this type of feedwater heaters, feedwater and extracted steam are not mixed. The heat transfer takes place in a shell and tube heat exchanger. The temperature of feedwater inside the tubes increases from state 2 to state 3 while extracted steam outside the tubes condenses from state 6 to state 6a (Figure 2.22).

Since feedwater and extracted steam are not mixed in this type, they can be at different pressures; thus, feedwater can be directly pumped from state 1 to state 2 in just one feedwater pump. However, extracted steam condensate downstream of the closed feedwater heater should be handled. There are two approaches for this purpose (Figure 2.22). The condensate

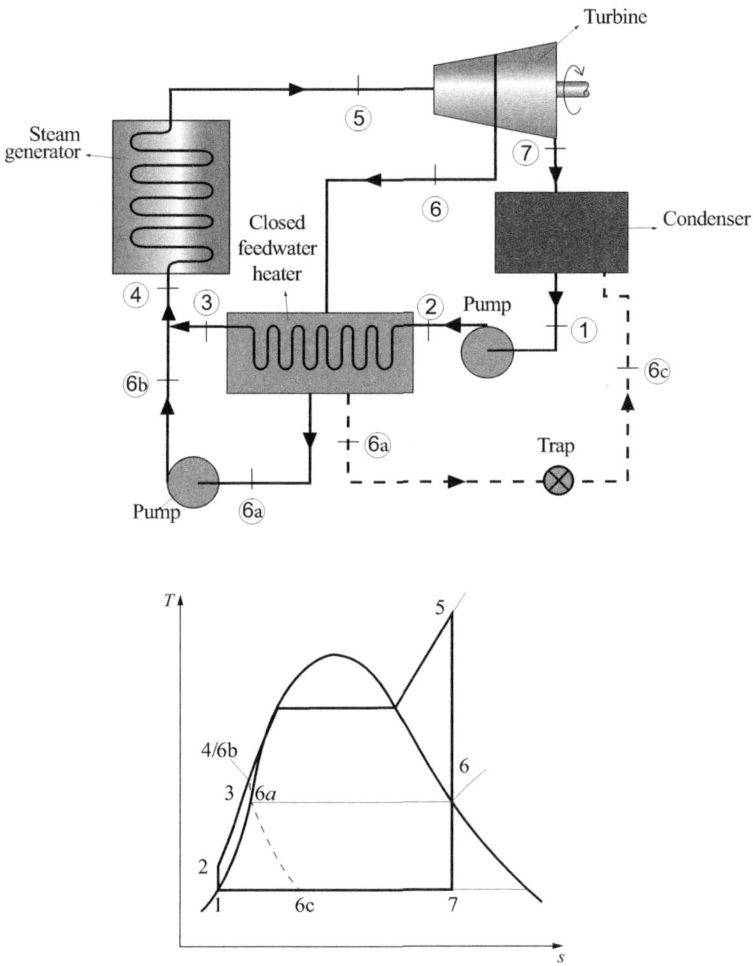

FIGURE 2.22
A regenerative Rankine cycle with a closed feedwater heater.

can be pumped forward to the pressure of the feedwater outlet from the heater (which is at a higher pressure than the extracted steam condensate) and then mixed with it (state 6a to 6b). This is called the **closed feedwater heater with drain pumped forward**. Alternatively, the extracted steam condensate can go through a trap to reduce its pressure and separate steam from the saturated mixture. Then, the water is returned to the next lower pressure closed feedwater heater, to the open feedwater heater, or to the condenser. This is called the **closed feedwater heater with drain cascaded backward**. The latter feedwater heater is shown in Figure 2.22 by a dashed line (state 6a to 6c). Note that in the T-s diagram in Figure 2.22, process 6a-6c (from the closed feedwater heater outlet to the condenser) is a constant enthalpy process (a throttling process). The closed feedwater heater with drain pumped forward is more efficient but more expensive because an extra pump, though a small one, is required. The closed feedwater heater with drain cascaded backward is inexpensive but less efficient because of the loss of availability in the trap.

In order to determine the mass flow rate of extracted steam from the turbine, the energy balance equation for the closed feedwater heater can be written for the forward and backward systems (Figure 2.22).

For a closed feedwater heater with drain pumped forward (note that in this case $\dot{m}_3 = \dot{m}_t - \dot{m}_6$),

$$\dot{m}_6 (h_6 - h_{6a}) = \dot{m}_3 (h_3 - h_2) \tag{2.27}$$

$$\dot{m}_6 (h_6 - h_{6a}) = (\dot{m}_3 - \dot{m}_6)(h_3 - h_2) \tag{2.28}$$

$$\frac{\dot{m}_t}{\dot{m}_6} - 1 = \frac{1}{\lambda} - 1 = \frac{h_6 - h_{6a}}{h_3 - h_2} \tag{2.29}$$

Note that λ is the ratio of the mass flow rate of extracted steam to the total mass flow rate of the working flow $\left(\frac{\dot{m}_6}{\dot{m}_t}\right)$. For a closed feedwater heater with drain cascaded backward (note that in this case $\dot{m}_3 = \dot{m}_t$),

$$\dot{m}_6 (h_6 - h_{6a}) = \dot{m}_3 (h_3 - h_2)$$

$$\lambda = \frac{\dot{m}_{\text{ext}}}{\dot{m}_t} = \frac{\dot{m}_6}{\dot{m}_3} = \frac{h_3 - h_2}{h_6 - h_{6a}} \tag{2.30}$$

Note that $h_{6a} = h_{6c}$ (the throttling process).

Example 2.11

In an ideal regenerative Rankine cycle, the operating pressures of the steam generator and the condenser are 10,000 and 10 kPa, respectively, and the turbine inlet flow is superheated steam at the temperature of 600°C (similar to Example 2.6). In the turbine, some steam is extracted at the

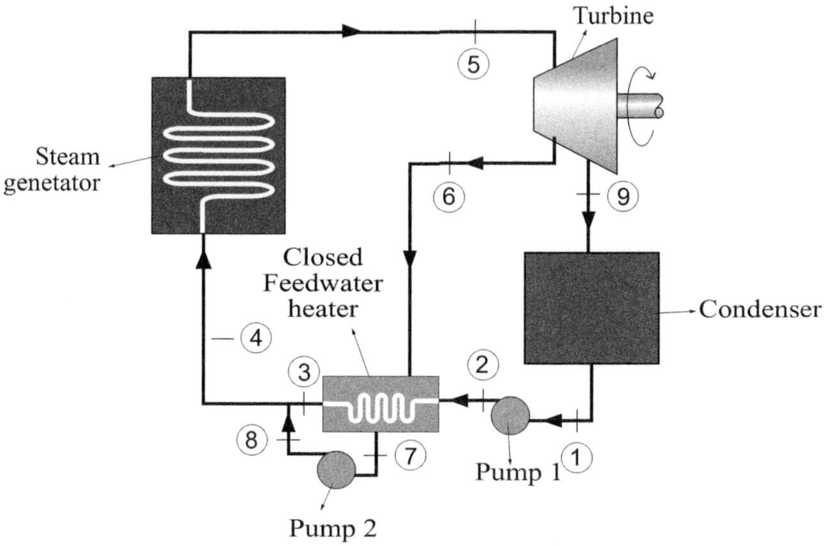

FIGURE 2.23
Schematic of the regenerative Rankine cycle for Example 2.11.

pressure of 1000 kPa and diverted to a closed feedwater heater. If the outlet of the closed feedwater heater is at the temperature of 150°C and is forwarded to the outlet of the heater by a small pump, determine the specific heat transfers in the steam generator and the condenser, the specific work involved in the turbine and the pumps (both the feedwater pump and the condensate extracted steam), and the thermal efficiency of the cycle.

Given: An ideal regenerative Rankine cycle with the working fluid of water operates at the given operating conditions. The cycle has a closed feedwater heater operating at the given conditions.

Find: The specific heat transfers in the steam generator and the condenser, the specific work involved in the turbine and both pumps, and the thermal efficiency of the cycle.

Schematic diagrams: Figure 2.23.

Assumptions:

1. The cycle is an ideal regenerative Rankine cycle with the working fluid of water.
2. The steady-state control volume mass and energy balance analyses with no kinetic and potential energy changes are applied to each component.
3. The outlet flow from the heater is at the temperature of 150°C and is forwarded to the outlet of the closed feedwater by a small pump.
4. There is no pressure drop due to friction and no heat loss to the surrounding in the components and piping systems, including the closed feedwater heater.

Analysis:
State 1:

$$h_1 = 191.83\ \text{kJ/kg}$$

$$v_1 = 0.0010102\ \text{m}^3/\text{kg}$$

State 2:

$$h_2 = w_{Pump1} + h_1 = v_1(P_2 - P_1) + h_1$$

$$= 0.001010\ \text{m}^3/\text{kg} \times (10{,}000 - 10)\,\text{kPa}\left[\frac{1000\ \text{N/m}^2}{1\ \text{kPa}}\right]\left[\frac{1\ \text{kJ}}{1000\ \text{Nm}}\right]$$

$$+ 191.83\ \text{kJ/kg} = 201.92\ \text{kJ/kg}$$

State 3:

In state 3, the pressure is 10,000 kPa (since there is no pressure drop in the closed feedwater heater) and the temperature is 150°C, which is subcooled liquid. The properties of the stream can be found in Table A.3 (SI).

$$h_3 = 638.25\ \text{kJ/kg}$$

Alternatively, the properties can be approximated by the properties of the saturated liquid at the same temperature from Table A.1 (SI).

$$h_3 = h_{f\,@\,150°\text{C}} = 632.20\ \text{kJ/kg}$$

State 5:

$$h_5 = 3625.3\ \text{kJ/kg}$$

$$s_5 = 6.9029\ \text{kJ/kg K}$$

State 6:

At this state, the steam is extracted from the turbine to be used in the closed feedwater heater

$$s_6 = s_5 = 6.9029\ \text{kJ/kg K}$$

$$P_6 = 1000\ \text{kPa}$$

At state 6, the flow is superheated steam, so from Table A.3 (SI) at the pressure of 1000 kPa (by interpolation)

$$h_6 = 2931.8\ \text{kJ/kg}$$

State 7:

The condensate extracted steam at the outlet of the closed feedwater heater is saturated liquid water at the pressure of 1000 kPa, so from Table A.2 (SI),

$$h_7 = 762.81\ \text{kJ/kg}$$

$$v_7 = 0.001127\ \text{m}^3/\text{kg}$$

State 8:

Pump 2 pressurizes the feedwater from state 7 to state 8, so

$$h_8 = w_{Pump2} + h_7 = v_7(P_8 - P_7) + h_7$$

$$= 0.001127\ \text{m}^3/\text{kg} \times (10{,}000 - 1000)\,\text{kPa}\left[\frac{1000\ \text{N/m}^2}{1\ \text{kPa}}\right]\left[\frac{1\ \text{kJ}}{1000\ \text{Nm}}\right]$$

$$+762.81\ \text{kJ/kg} = 772.95\ \text{kJ/kg}$$

State 9:

$$s_9 = s_5 = 6.9029\ \text{kJ/kg K}$$

$$s_9 = s_{f@10\,\text{kPa}} + x_9 s_{fg@10\,\text{kPa}} \rightarrow 6.9029$$

$$= 0.6493 + x_9(7.5009) \rightarrow x_9 = 0.8337 \text{ or } 83.37\%$$

$$h_9 = h_{f@10\,\text{kPa}} + x_9 h_{fg@10\,\text{kPa}} = 191.83 + 0.8337 \times 2392.8 = 2186.7\ \text{kJ/kg}$$

In order to determine the properties of the working fluid at state 4, first the mass flow rate of the extracted steam is needed. From Equation 2.29,

$$\frac{1}{\lambda} - 1 = \frac{h_6 - h_7}{h_3 - h_2}$$

$$\frac{1}{\lambda} - 1 = \frac{2931.8 - 762.81}{638.25 - 201.92}$$

$$\frac{1}{\lambda} - 1 = 4.97$$

$$\lambda = 0.167$$

By writing the energy balance equation for the point where the outlet stream from pump 2 (for pumping the condensate from the closed feedwater heater) at state 8 and the feedwater leaving the closed feedwater heater at state 3 are mixed, the enthalpy of state 4 can be found

$$h_4 = \lambda h_8 + (1 - \lambda)h_3$$

$h_4 = 0.167 \times 772.95 + (1 - 0.167) \times 638.25 = 660.74\ \text{kJ/kg}$ With the enthalpy of all stated determined, the work involved in the turbine and pumps can be calculated.

$$w_{\text{Turb}} = (h_5 - h_6) + (1 - \lambda)(h_6 - h_9)$$

$$= (3625.3 - 2931.8) + (1 - 0.167)(2931.8 - 2186.7) = 1314.2\ \text{kJ/kg}$$

$$w_{\text{Pump1}} = (1 - \lambda)(h_2 - h_1) = (1 - 0.167)(201.92 - 191.83) = 8.40\ \text{kJ/kg}$$

TABLE 2.8

Comparison of the Performance of the Rankine Cycle with the Closed Feedwater Heater, with the Open Feedwater Heater, and without a Feedwater Heater in Examples 2.6, 2.10, and 2.11

Type of Cycle	P_2 (kPa)	P_1 (kPa)	x_4, x_7, or x_9	q_{SG} (kJ/kg)	q_{Cond} (kJ/kg)	w_{Pumps} (kJ/kg)	w_{Turb} (kJ/kg)	w_{Net} (kJ/kg)	η_{Th} (%)
Simple Rankine cycle without feedwater heater (Example 2.6)	10,000	10	0.834	3423.4	1994.9	10.1	1438.6	1428.5	41.73
Rankine cycle with open feedwater heater (Example 2.10)	10,000	10	0.834	2852.3	1579.7	10.9	1283.5	1272.6	44.62
Rankine cycle with closed feedwater heater (Example 2.11)	10,000	10	0.834	2964.6	1661.7	10.1	1314.2	1304.1	43.99

$$w_{Pump2} = \lambda(h_8 - h_7) = 0.169(772.95 - 762.81) = 1.69 \text{ kJ/kg}$$

As expected, since just a portion of the total flow of the working fluid is pressurized in pump 2, the work input of this pump is much smaller than the feedwater pump. For the heat transfers in the steam generator and condenser,

$$q_{SG} = h_5 - h_4 = 3625.3 - 660.74 = 2964.6 \text{ kJ/kg}$$

$$q_{Cond} = (1-\lambda)(h_9 - h_1) = (1-0.167)(2186.7 - 191.83) = 1661.7 \text{ kJ/kg}$$

$$w_{Net} = w_{Turb} - w_{Pump} = 1314.2 - 8.40 - 1.69 = 1304.1 \text{ kJ/kg}$$

$$\eta_{Th} = \frac{w_{Net}}{q_{SG}} = \frac{1304.1}{2964.6} = 0.4399 \quad \text{or} \quad 43.99\%$$

Table 2.8 compares the results of Examples 2.11, 2.10, and 2.6, the Rankine cycles with the closed feedwater heater, with the open feedwater heater, and without a feedwater heater (the simple Rankine cycle). As discussed earlier, for cycles with open or closed feedwater heaters, although the work developed in the turbine is reduced, the thermal efficiency of the cycle is considerably improved.

The results show that although the net specific work developed in the Rankine cycle with the closed feedwater heater is slightly more than that of the cycle with the open feedwater heater, the efficiency of the latter is greater.

Comparing open and closed feedwater heaters indicates that open feedwater heaters are less expensive and have better heat-transfer efficiency. But one extra pump is required for each open feedwater heater to pump almost the entire feedwater. Note that these pumps are large expensive ones and significantly increase capital and operating costs, so only one open feedwater heater is typically used in modern plants. The open feedwater heater acts as the deaerator as well to separate air and other noncondensable gases from feedwater to protect the steam generator and other cycle components from corrosion. Open feedwater heaters are typically referred to as **deaerators** or **deaerator heaters (DA)**. On the other hand, closed feedwater heaters do not add to the required feedwater pumps. Instead, they need either a trap or a small pump to pressurize the condensate extracted steam but the pump is much smaller than the feedwater pump. However, closed feedwater heaters are less efficient and the heat exchanger itself is more expensive.

In actual steam power plants, several feedwater heaters are used. The more feedwater heaters in the cycle will result in the higher cycle efficiency because the cycle approaches the ideal regenerative cycle. But due to economic considerations, no more than seven or eight feedwater heaters are used, one of which is open and the rest are closed. While adding more feedwater heaters can slightly increase the efficiency, the increase in the costs and complexity of the cycle cannot be justified by this slight efficiency improvement. The number of feedwater heaters is a trade-off between cycle efficiency, capital costs, and system complexity.

In the next example, a Rankine cycle with a reheater, a closed feedwater heater, and an open feedwater heater will be analyzed. The operational conditions of the components are close to those in Examples 2.6, 2.9, 2.10, and 2.11.

Example 2.12

In an ideal reheat-regenerative Rankine cycle, the operating pressures of the steam generator and the condenser are 10,000 and 10 kPa, respectively, and the turbine inlet steam is superheated at the temperature of 600°C. The expansion process occurs in two stages with a reheater in between at the pressure of 1000 kPa. The outlet flow of the reheater is at the temperature of 600°C. Steam is extracted from the turbine at pressures of 2000 and 500 kPa, which is fed to the closed feedwater heater and the open feedwater heater, respectively. The outlet flow of the closed feedwater heater (high-pressure feedwater side) is at the temperature of 230°C and the outlet flow of the open feedwater heater is saturated liquid. The condensate from the closed feedwater heater, which is saturated liquid water, is discharged into the open feedwater heater after going through the trap. Determine the specific heat transfers in the steam generator and the condenser, the specific work involved in the turbines and the pumps, and the thermal efficiency of the cycle.

Given: An ideal reheat-regenerative Rankine cycle with the working fluid of water operates at the given operating conditions. The cycle has a closed feedwater heater, an open feedwater heater, and a reheater operating at the given conditions.

Find: The specific heat transfers in the steam generator and the condenser, the specific work involved in the turbines and the pumps, and the thermal efficiency of the cycle.

Schematic diagrams: Figure 2.24.

Assumptions:

1. The cycle is an ideal Rankine cycle with the working fluid of water.
2. The steady-state control volume mass and energy balance analyses with no kinetic and potential energy changes are applied to each component.
3. There is no heat loss in any equipment and no pressure drop due to friction in the cycle except in the trap, which is a throttling process.
4. The feedwater outlet flows from the condenser and open feedwater heater are saturated liquid. Also, the condensate outlet flow from the closed feedwater heater is saturated liquid.

Analysis:

State 1:

$$h_1 = 191.83 \text{ kJ/kg}$$

$$v_1 = 0.001010 \text{ m}^3/\text{kg}$$

State 2:

$$h_2 = w_{Pump1} + h_1 = v_1 (P_2 - P_1) + h_1$$

$$= 0.001010 \text{ m}^3/\text{kg} \times (500 - 10)\text{kPa} \left[\frac{1000 \text{ N/m}^2}{1 \text{ kPa}}\right]\left[\frac{1 \text{ kJ}}{1000 \text{ Nm}}\right]$$

$$+ 191.83 \text{ kJ/kg} = 192.32 \text{ kJ/kg}$$

State 3:

$$h_3 = h_{f\,@\,500\text{ kPa}} = 640.23 \text{ kJ/kg}$$

$$v_3 = v_{f\,@\,500\text{ kPa}} = 0.001093 \text{ m}^3/\text{kg}$$

State 4:

$$h_4 = w_{Pump2} + h_3 = v_3 (P_4 - P_3) + h_3$$

$$= 0.001093 \text{ m}^3/\text{kg} \times (10{,}000 - 500)\text{kPa} \left[\frac{1000 \text{ N/m}^2}{1 \text{ kPa}}\right]\left[\frac{1 \text{ kJ}}{1000 \text{ Nm}}\right]$$

$$+ 640.23 \text{ kJ/kg} = 650.61 \text{ kJ/kg}$$

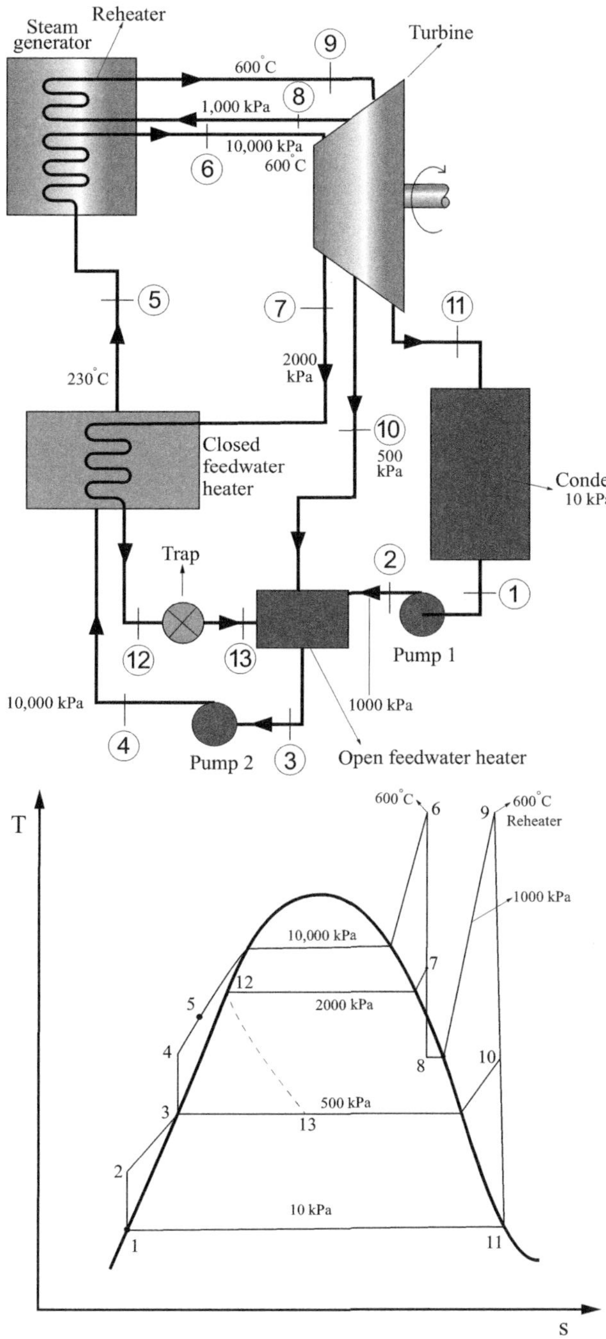

FIGURE 2.24
Schematic and T-s diagrams of the Rankine cycle in Example 2.12.

State 5:

In state 5, the working fluid is subcooled liquid at the pressure of 10,000 kPa (since there is no pressure drop in the closed feedwater heater) and the temperature of 230°C. Table A.4 (SI) can be used for the properties of compressed liquid water

$$h_5 = 990.9 \text{ kJ/kg}$$

Alternatively, the properties of the stream can be approximated by the properties of saturated liquid at the same temperature from Table A.1.

$$h_{f @ 230°\text{C}} = 990.12 \text{ kJ/kg}$$

State 6:

$$h_6 = 3625.3 \text{ kJ/kg}$$

$$s_6 = 6.9029 \text{ kJ/kg K}$$

State 7:

At this state, the steam is extracted from the turbine for the closed feedwater heater.

$$s_7 = s_6 = 6.9029 \text{ kJ/kg K}$$

$$P_7 = 2000 \text{ kPa}$$

At state 7, the flow is superheated steam, so from Table A.3 (SI) at the pressure of 2000 kPa (by interpolation),

$$h_7 = 3105.1 \text{ kJ/kg}$$

State 8:

At this state, the steam leaves the first stage of the turbine and is sent to the reheater.

$$s_8 = s_6 = 6.9029 \text{ kJ/kg K}$$

$$P_8 = 1000 \, kPa$$

At state 8, the flow is superheated steam, so from Table A.3 (SI) at the pressure of 1000 kPa (by interpolation),

$$h_8 = 2931.8 \text{ kJ/kg}$$

State 9:

At this state, the steam returns to the second stage of the turbine at the temperature of 600°C.

$$P_9 = 1000 \text{ kPa}$$

$$h_9 = 3697.9 \text{ kJ/kg}$$

$$s_9 = 8.0290 \text{ kJ/kg K}$$

State 10:

At this state, the steam is extracted from the low-pressure turbine for the open feedwater heater.

$$s_{10} = s_9 = 8.0290 \text{ kJ/kg K}$$

$$P_{10} = 500 \text{ kPa}$$

At state 10, the flow is superheated steam, so from Table A.3 (SI) at the pressure of 500 kPa (by interpolation),

$$h_{10} = 3436.0 \text{ kJ/kg}$$

State 11:

$$s_{11} = s_9 = 8.0290 \text{ kJ/kg K}$$

$$s_{11} = s_{f@10\text{ kPa}} + x_{11}s_{fg@10\text{ kPa}} \rightarrow 8.0290 = 0.6493 + x_{11}(8.1502 - 0.6493) \rightarrow x_{11} = 0.9838 \text{ or } 98.38\%$$

$$h_{11} = h_{f@10\text{ kPa}} + x_{11}h_{fg@10\text{ kPa}} = 191.83 + 0.9838 \times 2392.8 = 2546.0 \text{ kJ/kg}$$

State 12:

At this state, the condensate extracted steam from the turbine is saturated liquid.

$$P_{12} = 2000 \text{ kPa}$$

$$h_{12} = h_{f@p=2000} = 908.79 \text{ kJ/kg}$$

State 13:

Since in the trap the steam is throttled,

$$h_{13} = h_{12} = 908.79 \text{ kJ/kg}$$

To determine the mass flow rates of extracted steams, the energy and mass balance equations should be written for the closed and open feedwater heaters.

For the closed feedwater heater,

$$\lambda' = \frac{h_5 - h_4}{h_7 - h_{12}}$$

For the unit of total mass flow rate:

$$\lambda' = \frac{990.9 - 650.61}{3105.1 - 908.79}$$

$$\lambda' = 0.1549$$

For the open feedwater heater,

$$\lambda'' h_{10} + (1 - \lambda' - \lambda'')h_2 + \lambda' h_{13} = h_3$$

$$\lambda''(h_{10} - h_2) = h_3 - (1 - \lambda')h_2 - \lambda' h_{13}$$

$$\lambda'' = \frac{h_3 - (1 - \lambda')h_2 - \lambda' h_{13}}{(h_{10} - h_2)}$$

For the unit of total mass flow rate,

$$\lambda'' = \frac{640.23 - (1 - 0.1549)192.32 - 0.1549 \times 908.79}{3436.0 - 192.32} = 0.1039$$

With the enthalpy of all states determined, the work involved in the turbines and pumps can be calculated

$$w_{\text{Turb}} = (h_6 - h_7) + (1 - \lambda')(h_7 - h_8) + (1 - \lambda')(h_9 - h_{10}) + (1 - \lambda' - \lambda'')(h_{10} - h_{11})$$

$$= (3625.3 - 3105.1) + (1 - 0.1549)(3105.1 - 2931.8) + (1 - 0.1549)(3697.9 - 3436.0)$$

$$+ (1 - 0.1549 - 0.1039)(3436.0 - 2546.0) = 1547.66 \text{ kJ/kg}$$

$$w_{\text{Pump}} = (1 - \lambda' - \lambda'')(h_2 - h_1) + (h_4 - h_3)$$

$$= (1 - 0.1549 - 0.1039)(192.32 - 191.83) + (650.61 - 640.23) = 10.74 \text{ kJ/kg}$$

For heat transfers in the steam generator and the condenser,

$$q_{\text{SG}} = (h_6 - h_5) + (1 - \lambda')(h_9 - h_8)$$

$$= (3625.3 - 990.9) + (1 - 0.1549)(3697.9 - 2931.8) = 3281.8 \text{ kJ/kg}$$

$$q_{\text{Cond}} = (1 - \lambda' - \lambda'')(h_{11} - h_1) = (1 - 0.1549 - 0.1039)(2546.0 - 191.83)$$

$$= 1744.9 \text{ kJ/kg}$$

$$w_{\text{Net}} = w_{\text{Turb}} - w_{\text{Pump}} = 1547.66 - 10.74 = 1536.9 \text{ kJ/kg}$$

$$\eta_{\text{Th}} = \frac{w_{\text{Net}}}{q_{\text{SG}}} = \frac{1536.9}{3281.8} = 0.4683 \quad \text{or} \quad 46.83\%$$

Table 2.9 summarizes and compares the results of the major examples for the Rankine cycles with various configurations. They all operate at the steam generator and the condenser operating pressures of 10,000 and

TABLE 2.9

Comparison of the Performance of the Rankine Cycle with Various Configurations (Examples 2.4, 2.6, 2.9, 2.10, and 2.12)

Type of Cycle	P_2 (kPa)	P_1 (kPa)	x_4, x_7, x_9, or x_{11}	q_{SG} (kJ/kg)	q_{Cond} (kJ/kg)	w_{Pump} (kJ/kg)	w_{Turb} (kJ/kg)	w_{Net} (kJ/kg)	η_{Th} (%)
Simple Rankine –Turbine inlet: saturate steam (Example 2.4)	10,000	10	0.662	2522.8	1583.8	10.1	949.1	939.0	37.22
Simple Rankine –Turbine inlet: superheated at 600°C (Example 2.6)	10,000	10	0.834	3423.4	1994.9	10.1	1438.6	1428.5	41.73
Reheat Rankine cycle (Example 2.9)	10,000	10	0.984	4189.5	2354.1	10.1	1845.5	1835.4	43.81
Rankine cycle with open feedwater heater (Example 2.10)	10,000	10	0.834	2852.3	1579.7	10.9	1283.5	1272.6	44.62
Rankine cycle with closed feedwater heater (Example 2.11)	10,000	10	0.834	2964.6	1661.7	10.1	1314.2	1304.1	43.99
Reheat Rankine cycle with open and closed feedwater heaters (Example 2.12)	10,000	10	0.984	3281.8	1744.9	10.7	1547.66	1536.9	46.83

10 kPa, respectively. The turbine inlet is superheated steam at the temperature of 600°C for all cycles, except for the first one, which is saturated steam.

As Table 2.9 demonstrates, the efficiency of the cycle significantly improved from 41.7% in the simple superheated cycle to 48.88% in the cycle with the open and closed feedwater heaters and the reheater. Note that all these cycles are ideal, but similar trends should be expected in real-world cycles. Also, note that even a small increase in the efficiency may result in a significant reduction in fuel consumption and fuel cost for a large modern fossil fuel steam power plant.

For actual steam power plants where there are many more components with losses that should be taken into account, these manual calculations can be very cumbersome. Nowadays, design engineers are using modeling software to simulate and optimize steam power plants with different configurations and operating conditions. The final product of this process is a heat balance diagram, also known as a **flow diagram**. This diagram provides the mass flow rate, pressure, and temperature of major cycle flows. This diagram is the first step and backbone of the design of any steam power plant.

2.3.5 Deviation of Actual Steam Cycles from Ideal Rankine Cycles

2.3.5.1 Internal Irreversibilities

Actual steam power plants deviate from ideal Rankine cycles due to irreversibilities associated with processes. They can be categorized as internal irreversibilities, those that occur within the system, and external irreversibilities, those that take place on the boundaries of the system. Internal irreversibilities are due to friction (mostly fluid friction indicated by pressure drops), throttling, and mixing of flows. Most important losses in an actual steam power cycle take place in the steam generator, turbine, pump(s), pipes, and condenser.

2.3.5.1.1 Turbine Losses

Turbine losses are one of the most important losses in actual cycles. The primary cause of turbine losses is due to the flow of the working fluid through blades and passages. As a result of these losses, the expansion process is not isentropic, and the entropy increases in the process (in Figure 2.25, process 3-4 instead of state 3-4_s). In order to take these irreversibilities into account, the **isentropic efficiency** of the turbine, also known as the **adiabatic** or **polytropic efficiency**, can be defined as the ratio of the work (or power) produced in the actual turbine to the work (or power) produced in the isentropic expansion process in the ideal cycle. Note that the latter is always greater than the former for an actual turbine.

$$\eta_{\text{Turb}} = \frac{\dot{W}_{\text{Turb}}}{\left(\dot{W}_{\text{Turb}}\right)_s} = \frac{w_{\text{Turb}}}{\left(w_{\text{Turb}}\right)_s} = \frac{h_3 - h_4}{h_3 - h_{4s}} \tag{2.31}$$

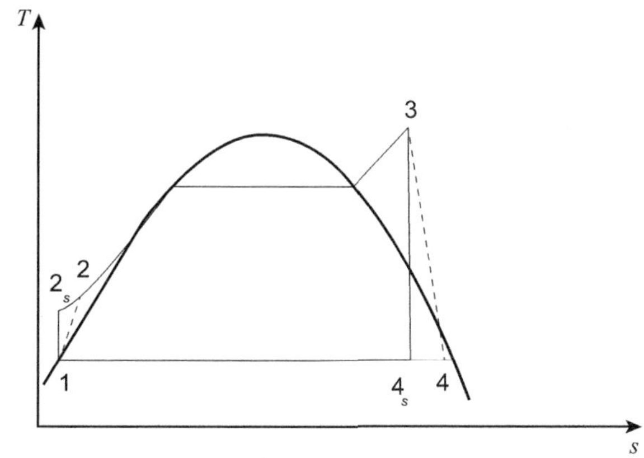

FIGURE 2.25
Losses associated with expansion and compression processes (in the turbine and the pump, respectively) in a Rankine cycle.

where subscript s represents the isentropic expansion process.

The isentropic efficiency is adversely affected by the existence of moisture in the low-pressure parts of the turbine. The larger turbines have higher efficiency, above 90% for well-designed turbines. Other losses of secondary importance in steam turbines are associated with heat transfer to the surrounding and turbine governing procedures. However, in most typical turbines, an adiabatic process is a reasonable assumption because the heat loss per unit mass of the working fluid is negligible (due to a very large mass flow rate of the working fluid). Similar justification can be presented for heat losses in other main components. Note that these less important losses will not be considered in the analyses in this text.

2.3.5.1.2 Pump Losses

Irreversibilities in pumps are similar to those in turbines in nature but much smaller (Figure 2.25). The losses in pumps due to the flow of the working fluid through blades and passages can be modeled by defining the **isentropic efficiency** of pumps (also known as the **adiabatic** or **polytropic efficiency**). Note that in pumps, the actual work required for the compression (the denominator in Equation 2.23) is always greater than the isentropic compression work (the numerator).

$$\eta_{\text{Pump}} = \frac{\left(\dot{W}_{\text{Pump}}\right)_s}{\dot{W}_{\text{Pump}}} = \frac{\left(w_{\text{Pump}}\right)_s}{w_{\text{Pump}}} = \frac{h_{2s} - h_1}{h_2 - h_1} \tag{2.32}$$

Figure 2.25 illustrates that while irreversibilities in the pump increase the temperature of the outlet flow compared to the isentropic compression process, irreversibilities in the turbine increase the quality of the outlet flow at a constant temperature. Note that in both cases, the entropy of the outlet flows increases compared to the isentropic processes.

2.3.5.1.3 Steam Generator

The greatest irreversibility in any steam power plant is associated with the combustion process in the fireside of the steam generator and the heat transfer from high-temperature combustion products to the working fluid. Note that since these losses take place in the boundary of the cycle, they are considered as external irreversibilities. Another loss in steam generators is pressure drop due to friction in tubes.

2.3.5.1.4 Condenser

The losses in the condenser, including fluid friction and irreversibilities in heat-transfer processes from the working fluid to the cooling water (the latter being external irreversibilities), are typically small. Since the working fluid goes through the condensation process and is at the saturated pressure, there is no pressure drop in the working fluid of the condenser. It should be noted that most of the heat added in the steam generator is rejected to the surrounding in the condenser. However, since the temperature at which this thermal energy is rejected is a little higher than the ambient temperature, availability losses (i.e., the value of energy losses) in the condenser is small.

Another loss that commonly occurs in a condenser in actual steam power plants is having subcooled water in the outlet of the condenser instead of saturated water, which requires more cooling in the condenser and more heat addition in the steam generator and thus resulting in lower efficiency.

2.3.5.1.5 Piping Losses

In every steam power plant, extensive piping systemsare required to transport the working fluid to and from components. Piping losses are mainly a result of pressure drops in pipes, valves, bends due to friction as well as heat losses. For instance, in the steam transfer from the steam generator to the turbine, in Figure 2.26, the process a-b represents the pressure drop and the heat loss is represented by process b-c. Figure 2.26 illustrates all the losses in the working fluid from the pump outlet to the turbine inlet due to pressure drops and heat losses in various heat-transfer surfaces within the steam generator, feedwater heaters, and the piping systems.

Example 2.13

Reconsider the Rankine cycle in Example 2.6, where the operating pressures of the steam generator and the condenser are 10,000 and 10 kPa, respectively, and the turbine inlet steam is superheated at the temperature of 600°C. If the isentropic efficiencies of the turbine and pump are both 90%, determine the specific heat transfers in the steam generator and the condenser, the specific work involved in the turbine

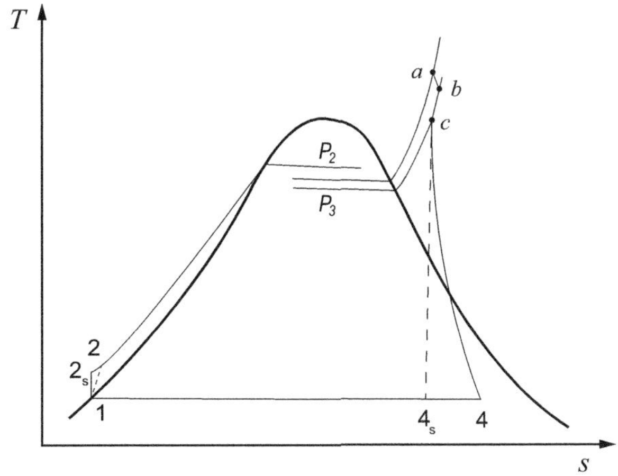

FIGURE 2.26
T-s diagram of deviation of actual steam power plants from Rankine cycles.

and the pump, and the thermal efficiency and the BWR of the cycle. Also, if the power plant produces 250 MW power, determine the mass flow rate of the cycle's working fluid.

Given: A non-ideal Rankine cycle with the working fluid of water operates at the given operating pressure of the steam generator and the condenser. The turbine inlet flow is superheated steam and the condenser outlet flow is saturated liquid. The power output is also given. The processes in the turbine and pump are not isentropic and the isentropic efficiencies of the turbine and the pump are given.

Find: The specific heat transfers in the steam generator and the condenser, the specific work involved in the turbine and the pump, the thermal efficiency and the BWR of the cycle, and the mass flow rate of the working fluid.

Schematic diagrams: Figure 2.26

Assumptions:

1. The cycle is a non-ideal Rankine steam cycle. The expansion and compression processes are adiabatic but not isentropic with the isentropic efficiencies of 90%.
2. The steady-state control volume mass and energy balance analyses with no kinetic and potential energy changes are applied to each component.
3. There is no pressure drop due to friction and no heat loss to the surrounding in the components and piping systems.

Analysis:

From Examples 2.4 and 2.6,

$$h_1 = 191.83 \text{ kJ/kg}$$

$$h_{2s} = 201.92 \text{ kJ/kg}$$
$$h_3 = 3625.3 \text{ kJ/kg}$$
$$h_{4s} = 2186.7 \text{ kJ/kg}$$

The enthalpy of the pump and turbine outlet flows can be determined using the isentropic efficiencies of the pump and turbine.

$$\eta_{\text{Pump}} = \frac{h_{2s} - h_1}{h_2 - h_1}$$

$$0.9 = \frac{201.92 - 191.83}{h_2 - 191.83}$$

$$h_2 = 203.04 \text{ kJ/kg}$$

Note that if Method 1 (using Table A.4 (SI)) in Example 2.4 is used to find the enthalpy of state 2,

$$h_2 = 203.64 \text{ kJ/kg}$$

$$\eta_{\text{Turb}} = \frac{h_3 - h_4}{h_3 - h_{4s}}$$

$$0.9 = \frac{3625.3 - h_4}{3625.3 - 2186.7}$$

$$h_4 = 2330.56 \text{ kJ/kg}$$

The quality of the turbine outlet flow can be calculated as follows:

$$h_4 = h_{f\,@\,10\text{ kPa}} + x_4 h_{fg\,@\,10\text{ kPa}}$$

$$2330.56 = 191.83 + x_4 \times 2392.8$$

$$x_4 = 0.8938 \text{ or } 89.38\%$$

Comparing this quality with that of the ideal Rankine cycle in Example 2.6 (83.37%) indicates that the irreversibilities in the turbine actually help to increase the quality of the mixture at the turbine outlet, which is desirable. The energy transfers in different components are

$$w_{\text{Pump}} = h_2 - h_1 = 203.04 - 191.83 = 11.21 \text{ kJ/kg}$$
$$q_{\text{SG}} = h_3 - h_2 = 3625.3 - 203.04 = 3422.3 \text{ kJ/kg}$$
$$w_{\text{Turb}} = h_3 - h_4 = 3625.3 - 2330.56 = 1294.7 \text{ kJ/kg}$$
$$q_{\text{Cond}} = h_4 - h_1 = 2330.56 - 191.83 = 2138.7 \text{ kJ/kg}$$
$$w_{\text{Net}} = w_{\text{Turb}} - w_{\text{Pump}} = 1294.7 - 11.21 = 1283.5 \text{ kJ/kg}$$

$$\eta_{Th} = \frac{w_{net}}{q_{SG}} = \frac{1283.5}{3422.3} = 0.3750 \text{ or } 37.50\%$$

$$BWR = \frac{w_{Pump}}{w_{Turb}} = \frac{11.21}{1294.7} = 0.0087 \text{ or } 0.87\%$$

The mass flow rate of the working fluid is

$$\dot{W}_{Net} = \dot{m}w_{Net} \quad \text{so} \quad \dot{m} = \frac{\dot{W}_{Net}}{w_{Net}} = \frac{250{,}000 \text{ kJ/s}}{1283.5 \text{ kJ/kg}} = 194.78 \text{ kg/s} = 7.01 \times 10^5 \text{ kg/h}$$

The results of Examples 2.6 and 2.13 are compared in Table 2.10. While the specific work developed in the turbine in the non-ideal cycle is less than that of the ideal one, the specific work consumed in the pump is higher in the former. These two effects reduce the net work output. As mentioned, the irreversibilities in the turbine are beneficial to improve the quality of the steam at the turbine outlet. Similarly, the irreversibilities in the pump slightly reduce the amount of heat addition in the steam generator. However, the rejected heat in the condenser and the work required in the pump are increased, and the work developed in the turbine is reduced. Overall, as a result of irreversibilities in the turbine and pump, the efficiency of the cycle considerably reduces.

Table 2.11 shows the performance characteristics of the non-ideal Rankine cycle with various turbine and pump isotropic efficiencies in the range of 60%–100% with an increment of 10%.

Table 2.11 shows that reducing the isentropic efficiencies of the turbine and pump significantly reduces the thermal efficiency of the cycle while the quality of the steam leaving the turbine increases. At the isentropic efficiency of 60% and 70%, the quality of steam at the exhaust of the turbine is greater than 1 (shown in parenthesis). This means in these cases the turbine outlet flow is superheated steam. This situation is undesirable because the superheated steam leaves the turbine with unused thermal energy and the load of the condenser unnecessarily increases.

TABLE 2.10

The Effects of the Turbine and Pump Irreversibilities on the Performance Characteristics of the Rankine Cycles in Examples 2.6 and 2.13

Type of Cycle: Simple Rankine	$P_2=P_3$ **(kPa)**	$P_1=P_4$ **(kPa)**	x_4	q_{SG} **(kJ/kg)**	q_{Cond} **(kJ/kg)**	w_{Pump} **(kJ/kg)**	w_{Turb} **(kJ/kg)**	w_{Net} **(kJ/kg)**	η_{Th} **(%)**
Ideal turbine and pump (Example 2.6)	10,000	10	0.834	3423.4	1994.9	10.1	1438.6	1428.5	41.73
Non-ideal turbine and pump (Example 2.13)	10,000	10	0.894	3422.3	2138.7	11.2	1294.7	1283.5	37.50

TABLE 2.11

Effects of the Turbine and Pump Isentropic Efficiencies on the Performance of the Rankine Cycle in Example 2.13

Isentropic Efficiency (%)	$P_2=P_3$ (kPa)	$P_1=P_4$ (kPa)	x_4	q_{SG} (kJ/kg)	q_{Cond} (kJ/kg)	w_{Pump} (kJ/kg)	w_{Turb} (kJ/kg)	w_{Net} (kJ/kg)	η_{Th} (%)
60	10,000	10	(1.07)	3416.7	2570.1	16.8	863.3	846.5	24.78
70	10,000	10	(1.01)	3419.1	2426.2	14.4	1007.2	992.8	29.04
80	10,000	10	0.95	3420.9	2282.4	12.6	1151.1	1138.5	33.28
90	10,000	10	0.89	3422.3	2138.7	11.2	1294.7	1283.5	37.50
100	10,000	10	0.83	3423.4	1994.9	10.1	1438.6	1428.5	41.73

2.3.5.2 External Irreversibilities

The concept of the pinch point has critical importance in this discussion. In a heat exchanger, the point with the minimum temperature difference between the cooling flow and the heating flow is called the pinch point. Figure 2.27 illustrates the heat-transfer processes in parallel flow (a) and counterflow (b) heat exchangers. Clearly, the latter has more favorable operating conditions. Note that if the temperature of one flow is constant (as in the condenser), the performance of the two types will be the same.

For heat exchangers with similar configurations, a smaller pinch-point temperature difference reduces the overall temperature difference during the heat-transfer process which, in turn, reduces the irreversibilities and improves the efficiency. However, the small pinch-point temperature difference reduces the heat-transfer rate, which means for the same heat transfer a larger heat exchanger is required. Therefore, the pinch-point temperature difference is determined by a trade-off between the system size (thus capital costs) and the system efficiency (thus fuel and operational costs).

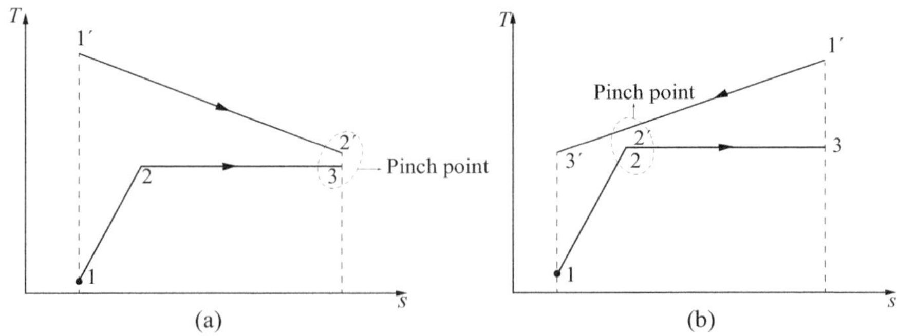

FIGURE 2.27
Heat-transfer processes in heat exchangers (a) parallel flow heat exchanger (b) counterflow heat exchanger.

External irreversibilities occur on the boundaries where the system interacts with the environment. In a Rankine cycle, these irreversibilities are in the steam generator and the condenser due to the heat-transfer processes at a finite temperature difference. In the combustion-based steam generator in Figure 2.28, the temperature difference is between the working fluid in process 1-3 and the combustion products in the process 1′-3′ with the pitch point 2/2′. Note that the changes in the entropy of the two flows in the heat-transfer process are not equal. According to the second law of thermodynamics, the entropy of the system should increase; therefore, $S_3 - S_1 > S_{1'} - S_{3'}$.

In the condenser in Figure 2.28, the temperature difference is between the working fluid in process 4-5 and the condenser cooling fluid in the process 1″-2″ with the pinch point 2″/4 . Note that the changes in the entropy of the two flows in the condenser are not equal, too. According to the second law of thermodynamics, the entropy of the system should increase; therefore, $S_{2''} - S_{1''} > S_4 - S_5$. There is not much we can do to reduce external irreversibilities in the condenser other than design optimization. Also, the magnitude of the loss is small due to the low temperature of the working fluid in the condenser.

As Figure 2.29 indicates, superheating steam in the steam generator (a), adding the reheater (b), and increasing the pressure to supercritical pressures can reduce the temperature difference between the working fluid and the combustion products, which reduces the external irreversibilities in the steam generator.

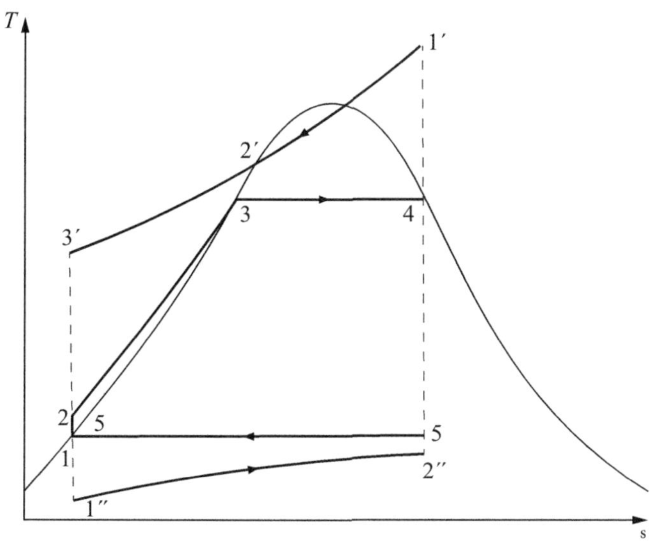

FIGURE 2.28
External irreversibilities in the steam generator and the condenser.

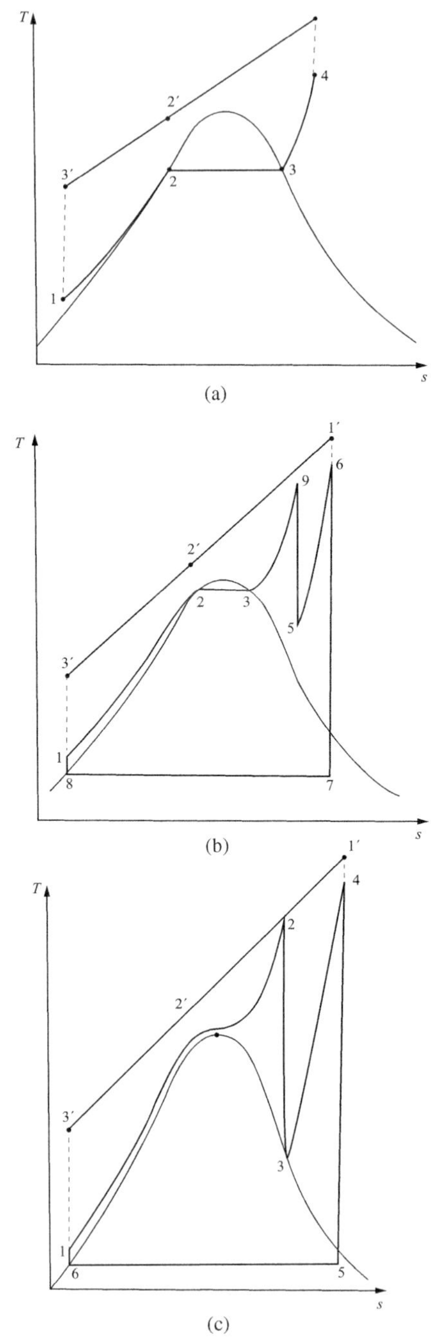

FIGURE 2.29
Impacts of superheating stream (a), adding the reheater (b), and increasing the pressure to supercritical pressures (c) on external irreversibilities.

2.4 Questions and Problems

2.4.1 Introduction

2.1. About ….. of total utility-scale electricity generation in the U.S. is still produced by steam power plants.

a. 25%
b. 50%
c. 65%
d. 90%

2.2. Explain why the share of electricity produced in steam power plants has reduced in the past decade.

2.3. In recent years many coal and nuclear steam power plants have retired and have not been replaced with similar technologies.

a. True
b. False

2.4. Steam cycles are only used in conventional fossil fuel burning power generation units.

a. True
b. False

2.5. What are the main advantages of steam power cycles? (choose all correct items.)

a. Flexibility of thermal energy sources
b. Relatively high efficiency
c. Low operating cost
d. Suitability for peak-load power generation

2.4.2 History of Steam Power Generation

2.6. Early steam engines were ……….

a. to generate electricity
b. to generate mechanical work
c. to pump water from mines
d. to run industrial machines

2.7. Explain how the invention of steam engines and invention of barometer were related.

2.8. Explain how steam engines contributed to the first and second industrial revolution.

2.9. Explain why the application of Savery's engines did not become widespread.

2.10. Why was the height that water could be raised in Savery's engines limited?

2.11. Explain why the extremely low efficiency of Newcomen's atmospheric engines in their early applications in coal mines was not a major problem.

2.12. Identify which of the following schematics is Newcomen's engine and which one is Savery's engine.

(a) (b)

2.13. Using the above schematic diagrams, explain why the efficiency of the Savery's engine was extremely low and how the Newcomen's engine improved the efficiency.

2.14. Explain the improvements in steam engines that were introduced by James Watt.

2.15. List some of the improvements on steam engines introduced in the 19th century.

2.16. Why did turbine-based steam power plants eventually replace steam engines?

2.17. Which one is incorrect?

a. Steam power plants are power generation systems that operate in a open cycle.

b. In steam power plants, the phase of the working fluid changes between the liquid and vapor phases alternatively.

c. In Rankine power plants the working fluid is usually but not always water.

d. Rankine cycle power plants are able to produce net power output due to the difference between pressure during the compression and expansion processes.

2.4.3 Thermodynamics of Steam Power Plants

2.18. In phase-changing working fluid power cycles, since the working fluid is pressurized in the liquid form and expands in the vapor phase, the maximum specific volume difference is provided.

a. True

b. False

2.19. Determine the ratio of the specific volume of saturated vapor to the specific volume of saturated liquid for water at the pressure of 1.5 and 10 bar.

2.20. Sketch a steam power plant showing the four main components. Explain the processes in each of the four components.

2.21. Name various sources of thermal energy for steam generators.

2.22. Draw the T-s and P-v diagrams for a phase-changing working fluid Carnot cycle.

2.23. Define the back work ratio (BWR) and work ratio (WR). Explain what they signify.

2.24. Which statements are correct for Carnot cycles? (Choose all correct ones.)

a. The Carnot cycle is composed of two constant temperature processes (isothermal processes) and two adiabatic processes.

b. The compression and expansion processes are constant entropy or isentropic.

c. It is impossible to construct an engine that operates between a low-temperature thermal energy reservoir and a high-temperature heat reservoir with an efficiency higher than that of the Carnot cycle.

d. All Carnot cycles operating between the same two thermal energy reservoirs have the same efficiency.

e. The Carnot cycle is an ideal cycle that theoretically can operate with only phase-changing working fluid with the highest thermal efficiency possible for a given thermal energy reservoirs.

2.25. Write the equation for the Carnot cycle efficiency.

2.26. Which one of the following assumptions is applicable to a typical turbine?

1. $\dot{Q}=0$
2. $\dot{W}=0$
3. hi = he
4. kei = kee
5. Pei = Pee
6. Steady-state process
   a. 1, 2, 3, and 6
   b. 3, 4, 5, and 6
   c. 1, 4, 5, and 6
   d. 2, 3, 4, and 6

2.27. Simplify the first law of thermodynamics for a typical turbine considering the appropriate assumptions from the previous question?

2.28. When can we make the following assumptions?

$$\frac{dm_{CV}}{dt}=0 \quad \text{and} \quad \frac{dE_{CV}}{dt}=0$$

2.29. Heat transfer from a low-temperature source to a high-temperature source …. the first law of thermodynamics and …. the second law of thermodynamics.
   a. violates, does not violate
   b. does not violate, violates
   c. violates, violates
   d. does not violate, does not violate

2.30. If we reduce the temperature difference of a heat-transfer process, the process will be more irreversible or less irreversible. Why?

2.31. A coal-fired power plant burns 100,000 kg/h coal to produce 250 MW. Determine efficiency of the power plant. (Burning 1 kg of coal releases 25,000 kJ).

2.32. Draw the T-s diagram for an ideal superheat Rankine cycle with double-reheat. Explain two main sources of irreversibilities in this cycle and show them in the diagram.

2.33. Steam at 4.0 MPa and 400°C enters an insulated turbine operating at steady state and expands to saturated vapor at 0.1 MPa. Kinetic and potential energy effects can be neglected. Determine the specific work developed (kJ/kg).

2.34. Which one(s) improve the net output power and efficiency of the Rankine cycle?
   a. Superheating the steam in the boiler
   b. Lowering pressure of the condenser
   c. Increasing pressure of the boiler

2.35. Explain why the Carnot cycle is impractical for steam power cycles.

2.36. Which process is not included in a Rankine cycle?

a. a reversible adiabatic (thus constant entropy or isentropic) compression process in a pump
b. a reversible constant-pressure process involving heat transfer at high temperature in a steam generator
c. a reversible adiabatic (thus constant entropy or isentropic) expansion process in a turbine
d. a reversible adiabatic constant-pressure heat transfer from the working fluid to the surrounding at low temperature in a condenser

2.37. A heat-transfer process at a constant temperature is not practical because in order to keep temperature constant, the pressure of the working fluid should be manipulated continuously.

a. True
b. False

2.38. If a Rankine cycle takes place entirely inside the two-phase liquid-vapor dome, it will be a Carnot cycle.

a. True
b. False

2.39. Explain why the moisture content of the turbine outlet stream should be limited.

2.40. In a majority of modern steam power plants, the steam leaving the steam generator is ………

a. superheated
b. saturated steam
c. saturated mixture of steam and liquid
d. all of the above

2.41. The efficiency of a Rankine cycle is generally ….. than that of a Carnot cycle operating between the same heat source and sink.

a. less
b. more
c. equal or less

2.42. Why is the most common working fluid for Rankine cycle-based power generation systems purified (demineralized) water?

2.43. Explain why high critical pressure and high boiling temperature of water are not desirable characteristics of water as the working fluid for Rankine cycles?

2.44. When are organic working fluids used in Rankine cycles?

2.45. What is the main application of Organic Rankine Cycles (ORC)?

2.46. The efficiency of the Rankine cycle .... and the moisture content of the turbine outlet flow ..... if the steam produced in the steam generator is superheated.

a. reduces – increases

b. increases – reduces

c. increases – increases

d. reduces – reduces

2.47. Explain the effects of increasing the operating pressure of the condenser on the efficiency of a Rankine cycle. Justify your answer.

2.48. Explain how a reheat Rankine cycle works.

2.49. Explain why an ideal regenerative Rankine cycle is not practical (two reasons).

2.50. List two types of feedwater heaters and explain their pros and cons.

2.51. Explain how the outlet flow of a closed feedwater heater can be handled. Which method is more efficient? Which method is less expensive?

2.52. The steam required in feedwater heaters extracted from ....

a. the turbine

b. the condenser

c. the steam generator

d. the reheater

2.53. When using an open feedwater heater, why cannot feedwater be pressurized to the steam generator pressure in the first pump after the condenser?

2.54. Explain why just a small fraction of the expanding steam in the turbine is extracted for feedwater heaters.

2.55. Explain why feedwater can be pumped directly in one feedwater pump in a Rankine cycle with a closed feedwater heater.

2.56. Explain why only one open feedwater heater is used in modern steam power plants.

2.57. What is the function of a deaerator?

2.58. How many open and closed feedwater heaters are used in modern large steam power plants?

2.59. In which type of feedwater heaters are not feedwater and extracted steam mixed?

2.60. What is the definition of the isentropic efficiency of a pump?

2.61. List the sources of irreversibilities in a steam generator.

2.62. What is a pinch point in a heat exchanger?

2.63. Identify the pinch points in the following two diagrams for two heat exchangers.

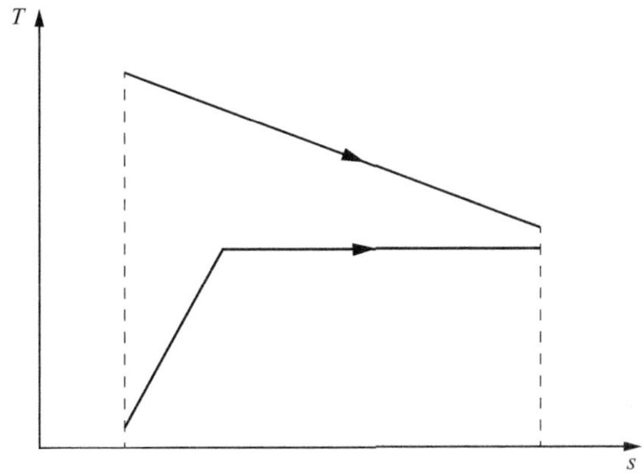

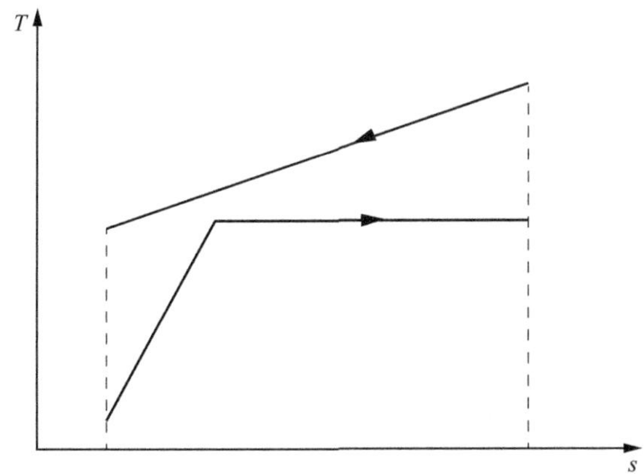

2.64. Explain why a smaller pinch-point temperature difference is desirable.

2.65. What factor does limit how small the pinch-point temperature difference in a heat exchanger can be?

2.66. Which one of the following actions can reduce the temperature difference between the working fluid and the combustion products?

a. Superheating stream in the steam generator
b. Adding a reheater
c. Increasing the operating pressure of the steam generator to supercritical pressures
d. All of the above

2.67. The power produced in a steam turbine is 250 MW and the power consumed in the pump is 4 MW. Determine the BWR and the WR of the cycle.

2.68. In a steam power cycle, the input thermal energy is 150 MW and the turbine power generation is 60 MW. Determine the efficiency of the cycle.

2.69. In a steam power cycle, the input thermal energy is 1500 MW and the turbine power generation is 725 MW. What is the rate of thermal energy removal in the condenser?

2.70. In a steam power plant that operates based on a Rankine cycle, the inlet and output pressures of the pump are 0.15 and 50 bar, respectively. If the steam generator outlet stream is saturated vapor and the inlet stream of the pump is saturated liquid, determine the specific heat transfers in the steam generator and the condenser, the specific work involved in the turbine and the pump, and the thermal efficiency of the cycle.

2.71. Repeat Example 2.6 for the superheated steam temperature between 400°C and 700°C with the increment of 100°C. List the results in a table and draw the curves of the main performance parameters (the efficiency, specific net work, specific heat input, and quality of the flow at the outlet of the turbine) as a function of the superheated steam temperature (results presented in Table 2.2).

2.72. Repeat Example 2.7 for the steam generator pressure of 4000, 10,000, 12,000, 16,000, 20,000, and 24,000 kPa (results presented in Table 2.3).

2.73. In a supercritical Rankine cycle, the inlet steam to the turbine is at the pressure and temperature of 32,000 kPa and 600°C, respectively. If the outlet stream of the condenser is saturated water at the pressure of 10 kPa, determine the specific heat transfers in the steam generator and the condenser, the specific work involved in the turbine and the pump, and the thermal efficiency of the cycle.

2.74. Repeat Example 2.8 for the condenser pressure of 5, 10, 20, 40, 80, and 100 kPa (results presented in Table 2.4).

2.75. Repeat Example 2.9 when the steam leaves the high-pressure turbine at the pressure of 1500, 2000, 4000, and 8000 kPa and then is sent the reheater (results presented in Table 2.6).

2.76. Repeat Example 2.10 if the isentropic efficiency of the turbine is 0.90.

2.77. Repeat Example 2.12 by exchanging the location of the open and closed feedwater heaters.

2.78. Repeat Example 2.13 for the turbine and pump isentropic efficiencies varying between 60% and 100% with the increment of 10 (results presented in Table 2.11).

2.79. Repeat Examples 2.10, 2.11, and 2.12 when the isentropic efficiency of 90% for the turbine(s).

2.80. In a Carnot cycle with the working fluid of water, the operating pressure of the steam generator and the condenser are 15,000 and 20 kPa, respectively. If the turbine inlet flow is saturated vapor and the pump outlet flow is saturated liquid, determine the specific heat transfers in the steam generator and the condenser (in kJ/kg), the specific work involved in the turbine and the pump (in kJ/kg), and the thermal efficiency and BWR of the cycle.

2.81. In a steam power plant that operates based on a Rankine cycle, the operating pressure of the steam generator and the condenser are 15,000 and 20 kPa, respectively. If the turbine inlet stream is saturated vapor and the condenser outlet flow is saturated liquid, determine the specific heat transfers in the steam generator and the condenser, the specific work involved in the turbine and the pump, the thermal efficiency and the BWR of the cycle. Also, if the power plant produces 500 MW power, determine the mass flow rate of the cycle's working fluid. If the cooling water enters the condenser at 35°C and leaves at 15°C, determine the cooling water mass flow rate through the condenser to provide sufficient cooling for 1 kg of the working fluid.

2.82. In a simple Rankine cycle, the steam generator produces saturated steam at 320°C and the condenser operating temperature is 42°C. Determine the cycle efficiency.

2.83. In a steam power plant, the outlet steam from the steam generator is superheated to the pressure and temperature of 20,000 kPa and 550°C, respectively. The condenser outlet flow is saturated liquid at the pressure of 25 kPa. Determine the specific heat transfers in the steam generator and the condenser, the specific work involved in the turbine and the pump, and the thermal efficiency of the cycle. Also, if the power plant produces 750 MW (net power output), determine the mass flow rate of the cycle's working fluid. Do the problem for two cases, when the processes in the turbine and pump are isentropic and when their isentropic efficiency is 85%.

2.84. In a Rankine cycle, the operating pressure of the steam generator and the condenser are 18,000 and 15 kPa, respectively, and the turbine inlet steam is superheated at the temperature of 550°C. If the expansion process occurs in two stages with a reheater with the pressure of 1500 kPa in between, determine the specific heat transfers in the steam generator and the condenser, the specific work involved in the turbine and the pump, and the thermal efficiency of the cycle. The outlet flows of the steam generator and the reheater are at the same temperature.

2.85. In an ideal regenerative Rankine cycle, the operating pressure of the steam generator and the condenser are 17,000 and 20 kPa, respectively, and the turbine inlet steam is superheated at the temperature of 500°C. In the turbine, some steam is extracted at the pressure of 2000 kPa and diverted to an open feedwater heater. If the outlet of the open feedwater is saturated water, determine the specific heat transfers in the steam generator and the condenser, the specific work involved in the turbine and the pump, and the thermal efficiency of the cycle. Repeat the problem when the steam extraction from the turbine is at 1500 kPa and compare the results.

2.86. In a simple Rankine cycle, the steam generator produces saturated steam at 320°C and the condenser operating temperature is 42°C. The isentropic efficiency of the turbine and pump are 90% and 85%, respectively. Determine the cycle efficiency. Repeat this problem for the condenser temperature of 32°C and compare the results of the two cases.

2.87. In an ideal reheat-regenerative Rankine cycle, the operating pressure of the steam generator and the condenser are 15,000 and 15 kPa, respectively, and the turbine inlet steam is superheated at the temperature of 600°C. The expansion process occurs in two stages with a reheater in between at the pressure of 2000 kPa. The outlet flow of the reheater is at the temperature of 600°C. Steam is extracted from the turbine at the pressure of 3000 and 800 kPa which is fed to the closed feedwater heater and the open feedwater heater, respectively. The outlet flow of the closed feedwater heater (high-pressure feedwater side) is at the temperature of 250°C and the outlet flow of the open feedwater heater is saturated liquid. The condensate from the closed feedwater heater, which is saturated liquid water, is discharged into the open feedwater heater after going through the trap. Determine the specific heat transfers in the steam generator and the condenser, the specific work involved in the turbines and the pumps, and the thermal efficiency of the cycle.

2.4.4 Open-Ended Questions

2.88. Find a steam power plant in your region (city, county, state, or country). Try to find as much information as you can about the configuration of the plant and the operating temperature and pressure of the major components. Based on the information you collected, determine the efficiency of the cycle. Use the available online resources and/or visiting the plant in your search for information.

2.89. Try to find examples of steam power plants in your region that use each of the following thermal energy sources. For cases that you could not find any plant, think of a reason.

- Coal
- Natural gas
- Petroleum products
- Nuclear reactions
- Geothermal
- Solar thermal energy
- Ocean hot water
- Biomass

2.4.5 General

2.90. If a Rankine cycle takes place entirely inside the two-phase liquid-vapor dome, it will be a Carnot cycle because ....

2.91. What are the pumps after the condenser and after the open feedwater heater called?

2.92. Which ones of the following power plants work based on a steam power cycles? (Choose all correct types).

a. Coal power plants
b. Nuclear power plants
c. Geothermal power plants
d. Natural gas-fueled gas turbines
e. Solar thermal power plants
f. Solar photovoltaic plants

2.93. Explain the motivation for early steam engines.

2.94. List the drawbacks of Savery's engines.

2.95. In recent years, many natural gas-fueled combined cycle power plants have retired and have not been replaced with similar technologies.

a. True
b. False

2.96. Briefly explain how uniflow steam engines worked.

2.97. Which component is the site of the greatest irreversibilities in any steam power plant?

2.98. What is the function of the condenser in a steam power plant?

2.99. What is the function of a trap after a closed feedwater heater?

2.100. There are two flows in a heat exchanger. What do we need to have an actual heat transfer?

2.101. In a throttle process, enthalpy .....

a. remains constant

b. increases

c. decreases

d. can increase, decrease, or remain constant

2.102. What is the function of a pump?

2.103. Compare the efficiency and quality of steam in the outlet flow of the turbine in a reheat Rankine cycle with those of a simple Rankine cycle.

2.104. A Rankine cycle consists of four processes. Name these four processes.

2.105. Why are open feedwater heaters typically known as deaerators or deaerator heaters (DA)?

2.106. Explain how an ideal regenerative Rankine cycle works.

2.107. Considering the effects of both less fuel consumption in the steam generator and less work production in the turbine, the overall efficiency of the Rankine cycle increases if feedwater heaters are added to a steam power plant.

a. True

b. False

2.108. What is the definition of the isentropic efficiency of a steam turbine?

2.109. Why is a high change in the specific volume of the working fluid desirable?

2.110. List the sources of irreversibilities in a condenser.

2.111. It is impossible to construct a heat engine that operates between two heat reservoirs with an efficiency higher than that of the Carnot cycle operating between the same two heat reservoirs.

a. True

b. False

2.112. Draw the T-s diagram of a Rankin cycle.

2.113. Two heat engines operates between the same two heat reservoirs and both receives the same Q_H. One engine is reversible and the other one is not. Which one has larger Q_L? Why?

2.114. Pumping the mixture of vapor and liquid is very difficult and requires a lot of energy.

a. True

b. False

2.115. Why is a high latent heat of the working fluid desirable?

2.116. The outlet flow of an open feedwater heater is .....

a. saturated liquid

b. saturated steam

 c. superheated steam
 d. compressed liquid

2.117. In modern steam power plants only one closed feedwater heater is used.
 a. True
 b. False

2.118. Why is having subcooled water instead of saturated water in the outlet of a condenser undesirable?

2.119. Explain the effects of increasing steam generator pressure on the performance of a steam cycle with or without limitation on the inlet temperature of the steam turbine.

2.120. In supercritical cycles, the moisture content of the turbine outlet is unacceptably .... (high or low). How the problem can be addressed?

2.121. Draw the T-s diagram of a reheat Rankine cycle.

2.122. Select all correct statements.
 a. Pumping the mixture of vapor and liquid is very difficult and requires a lot of energy.
 b. In a Rankine cycle, the mixture of vapor and liquid is condensed to saturated liquid in the contender and then saturated liquid is pumped.
 c. A heat-transfer process at a constant temperature is not practical.
 d. Heat-transfer processes in steam generators and condensers are constant temperature.

2.123. The most common working fluid for Rankine cycle-based power generation systems is purified (demineralized) water due to its ......
 a. abundance
 b. low cost
 c. desirable thermodynamic characteristics of water
 d. all of the above

2.124. Explain the effects of reducing condenser pressure on the performance of a steam cycle. What limits the extent that the condenser pressure can be reduced?

2.125. Which one is desirable for improving the efficiency of a steam power plant?
 a. A high-pressure condenser
 b. A low-pressure condenser

2.126. What is the typical range for isentropic efficiency of large steam turbines?

2.127. Select the correct statements.
 a. Too much extracted steam will produce two-phase mixture of saturated steam and water at the outlet flow of an open feedwater heater which is difficult to pump.

b. Too little extracted steam will reduce the effectiveness of the feedwater heating process.

c The average temperature of heat addition in a regenerative Rankine cycle is higher than that of the non-regenerative cycle Rankine.

d When some steam is bled from the turbine for feedwater heaters, there is less mass flow rate to expand which reduces the total work developed in the turbine.

2.128. Why should the pressure of inlet flows to an open feedwater heater be equal? What will happen if the flows are not at the same pressure?

2.129. A steam generator's operating pressure is 180 bar. This steam generator is supercritical.

a. True

b. False

2.130. Which one of the following actions can reduce external irreversibilities in the steam generator?

a. Superheating stream in the steam generator

b. Adding a reheater

c. Increasing the operating pressure of the steam generator to supercritical pressures

d. All of the above

2.131. The pinch-point temperature difference is determined by a trade-off between the system size (thus the capital cost) and the system efficiency (thus the fuel and operational cost).

a. True

b. False

2.132. Which parameter is used to take irreversibilities in a steam turbine into account?

2.133. In most typical turbines, the adiabatic process is a reasonable assumption because the heat loss per unit mass of the working fluid is negligible.

a. True

b. False

2.134. List the sources of irreversibilities in a piping system.

2.135. Which is not correct?

a. Almost all coal plants operate based on steam cycles.

b. Almost all nuclear plants operate based on steam cycles.

c. Almost all natural gas plants operate based on steam cycles.

d. Almost all solar thermal plants operate based on steam cycles.

2.136. Explain briefly the principal of operation of Newcomen's engine and how it was different from Savery's engine.

2.137. The efficiency of a power cycle is directly proportional to the average temperature of the heat transfer to the cycle (at high temperature) and is inversely proportional to the average temperature of the heat transfer from the cycle (at low temperature).

a. True

b. False

2.138. Using a reheater in a steam power plant .... the efficiency of the cycle and ..... the quality of the steam at the turbine outlet.

a. reduces – increases

b. increases – reduces

c. increases – increases

d. reduces – reduces

2.139. List factors that distinguish actual steam power plants from an ideal Rankine cycle.

2.140. Briefly explain the impact of increasing the pressure of the feedwater pump outlet to supercritical steam pressure on the design of the steam generator.

2.141. The idea behind early steam engines was that water can be elevated by the force of the atmospheric pressure if vacuum can be created.

a. True

b. False

2.142. Why in a regenerative Rankine cycle are there two pumps?

2.143. What was the main contribution of George Henry Corliss to development of steam engines?

2.144. How the efficiency of a Carnot cycle can be improved?

2.145. The extraction or bleeding of steam from the turbine for regeneration will increase both the power production in the turbine and the efficiency of the cycle.

a. True

b. False

2.146. What are the correct temperature unit for the Carnot cycle efficiency equation? (Choose all correct ones.)

a. °C

b. K

c. °F

d. °R

2.147. How can we improve the efficiency of a heat engine working based on a Carnot cycle?

2.148. Which one(s) increases the quality of steam leaving the turbine?

a. Superheating the steam in the boiler
b. Increasing pressure of the condenser
c. Decreasing pressure of the boiler

2.149. Explain the effects of increasing the operating pressure of a steam generator on the efficiency of a Rankine cycle. Justify your answer.

2.150. Does h_{fg} (i.e. $h_g - h_f$) change with increase in the pressure? Explain.

2.151. The heat released during the condensation of 1 kg water at a given pressure is equal to the heat needed for the evaporation of 1 kg water at the same pressure.

a. True
b. False

2.152. What are some alternative organic working fluids used in Rankine cycles?

2.153. The share of steam power plants in total utility-scale electricity generation in the US ….. between 2009 and 2018.

a. increased
b. decreased
c. remained unchanged

2.154. Explain how the Savery's steam engine worked.

2.155. Currently, combined cycle power plants dominate natural gas-fueled power generation units.

a. True
b. False

2.156. List the problems of Newcomen's engine.

2.157. Draw the T-s diagram for a Carnot cycle and show the efficiency of the cycle using the areas in the diagram.

2.158. Which one is desirable for reducing the moisture content of the stream leaving the turbine of a steam power plant?

a. A high-pressure condenser
b. A low-pressure condenser

2.159. In a Rankine cycle, why is the power consumed in the pump much smaller than the power generated in the turbine?

2.160. Which ones are favorable physical and chemical characteristics of water as the working fluid for Rankine cycles?

a. chemical stability
b. nontoxic nature

c. noncorrosive nature
d. high critical pressure
e. high boiling temperature

2.161. Why the Carnot cycle cannot be used as ideal cycle for steam power plants?

2.162. For a steam power cycle, the net power production is 50 MW and the pump power consumption is 0.9 MW. What is the BWR?

2.163. In a steam power cycle, the input and output thermal energy are 700 and 450 MW, respectively. Determine the efficiency of the cycle.

2.164. In a real turbine and pump, entropy ..... in the absence of heat transfer.

a. increases
b. decreases
c. remains constant
d. can increase or decrease

2.165. In a Carnot cycle with the working fluid of water, the operating pressure of the steam generator and the condenser are 20,000 and 25 kPa, respectively. If the turbine inlet flow is saturated vapor and the pump outlet flow is saturated liquid, determine the specific heat transfers in the steam generator and the condenser (in kJ/kg), the specific work involved in the turbine and the pump (in kJ/kg), and the thermal efficiency and BWR of the cycle.

2.166. In a steam power plant that operates based on a Rankine cycle, the operating pressure of the steam generator and the condenser are 20,000 and 25 kPa, respectively. If the turbine inlet stream is saturated vapor and the condenser outlet flow is saturated liquid, determine the specific heat transfers in the steam generator and the condenser, the specific work involved in the turbine and the pump, the thermal efficiency and the BWR of the cycle. Also, if the power plant produces 750 MW power, determine the mass flow rate of the cycle's working fluid. If the cooling water enters the condenser at 38°C and leaves at 22°C, determine the cooling water mass flow rate through the condenser to provide sufficient cooling for 1 kg of the working fluid.

2.167. In a steam power plant that operates based on a Rankine cycle, the inlet and output pressures of the pump are 0.10 and 70 bar, respectively. If the steam generator outlet stream is saturated vapor and the inlet stream of the pump is saturated liquid, determine the specific heat transfers in the steam generator and the condenser, the specific work involved in the turbine and the pump, and the thermal efficiency of the cycle. Also, if the power plant produces 150 MW power, determine the mass flow rate of the cycle's working fluid. If the cooling water enters the condenser at 42°C and leaves at 22°C, determine the cooling water mass flow rate through the condenser to provide sufficient cooling for 1 kg of the working fluid.

2.168. In a steam power plant that operates based on a Rankine cycle, the inlet and output pressures of the pump are 0.10 and 70 bar, respectively. The steam generator outlet stream is superheated steam and the inlet stream of the pump is saturated liquid. Determine the specific heat transfers in the steam generator and the condenser, the specific work involved in the turbine and the pump, and the thermal efficiency of the cycle for two cases, when the superheated steam leaving the steam generator is at the temperature of 500 and 700°C. Compare the results of the two cases.

2.169. In an ideal regenerative Rankine cycle, the operating pressure of the steam generator and the condenser are 17,000 and 20 kPa, respectively, and the turbine inlet flow is superheated steam at the temperature of 500°C. In the turbine, some steam is extracted at the pressure of 2000 kPa and diverted to a closed feedwater heater. If the outlet of the closed feedwater heater is at the temperature of 170°C and is forwarded to the outlet of the heater by a small pump, determine the specific heat transfers in the steam generator and the condenser, the specific work involved in the turbine and the pumps (both the feedwater pump and the condensate extracted steam), and the thermal efficiency of the cycle. Repeat the problem when the steam extraction from the turbine is at 1500 kPa and compare the results.

2.170. In an ideal reheat-regenerative Rankine cycle, the operating pressure of the steam generator and the condenser are 18,000 and 10 kPa, respectively, and the turbine inlet steam is superheated at the temperature of 650°C. The expansion process occurs in two stages with a reheater in between at the pressure of 2000 kPa. The outlet flow of the reheater is at the temperature of 600°C. Steam is extracted from the turbine at the pressure of 3000 and 1000 kPa which is fed to the closed feedwater heater and the open feedwater heater, respectively. The outlet flow of the closed feedwater heater (high-pressure feedwater side) is at the temperature of 260°C and the outlet flow of the open feedwater heater is saturated liquid. The condensate from the closed feedwater heater, which is saturated liquid water, is discharged into the open feedwater heater after going through the trap. Determine the specific heat transfers in the steam generator and the condenser, the specific work involved in the turbines and the pumps, and the thermal efficiency of the cycle.

3

Steam Generators: Water-Steam Side

Abstract

This chapter deals with the water-steam side of steam generators. In this chapter, we will see what the function of a steam generator is, and why utility steam generators are considered one the most complicated industrial systems. We will review the historical development of steam generators and their components in the past three centuries. We will then learn about modern drum-type and once-through steam generators. Finally, we will extensively discuss about various components of modern steam generators and their operational principles, including waterwalls, drums, superheaters, reheaters, economizers, and control components.

LEARNING OUTCOMES

After successful completion of this chapter, you will be able to:

- describe the function of steam generators
- list components of a typical steam generator
- briefly explain historical development of steam generators and their components
- explain operation of modern drum-type and once-through steam generators
- describe functions of various components of modern steam generators and their operational principles, including waterwalls, drums, superheaters, reheaters, economizers, and control components

3.1 Introduction

The phase change of the working fluid from liquid to vapor is one of the four main processes in any power generation unit operating based on a Rankine cycle. In this process, thermal energy is transferred from a thermal energy source to the working fluid of the cycle. The source of thermal energy can be:

- fossil fuel combustion
- nuclear reactions
- biomass combustion
- renewable energy sources, such as solar thermal energy, geothermal energy, or ocean thermal energy
- municipal or industrial waste combustion
- waste heat
- electricity (for small industrial or residential applications where power generation is not the objective)

Due to its availability, nontoxicity, and favorable thermodynamic properties (including a relatively high specific heat and latent heat), water is the working fluid in most power plants operating based on the Rankine cycle (refer to Section 2.3.1). However, recently there has been a great deal of interests in Organic Rankine Cycles (ORC) where the working fluid is an organic low-temperature boiling point substance such as isobutane, pentane, and propane. These cycles will be covered in Chapter 15.

In steam power plants, the steam generator is where thermal energy is used to convert liquid water to steam at certain conditions to be sent to a steam turbine for power generation (much more on steam turbines in Chapter 6). Steam generators are considered as one of the most complex engineering systems. In this chapter and the next chapter, combustion-based steam generators will be discussed where the source of thermal energy can be the combustion of fossil fuels, biomass, or waste materials. When the thermal energy source for steam generation is provided by a-would-be-otherwise-wasted-heat from other processes, the steam generator is called a heat recovery steam generator (HRSG) and will be covered in Chapter 10. Renewable energy-based steam generators will be covered in Chapters 12 and 15. It should be noted that generated steam in a steam generator can also be used for non-utility applications such as industrial processes and district heating. Every combustion-based steam generator can be divided into two sides: water-steam side and fire side. The latter side is where the chemical energy of fuel is converted to heat by means of the combustion process and will be covered in the next chapter. The water-steam side is where the thermal energy generated by burning fuel is used to convert water to steam at rated conditions and will be covered in this chapter.

3.1.1 Introduction to Water-Steam Side of Steam Generators

The water-steam side of steam generators consists of various combinations of an economizer, evaporator (boiler), drum, superheater, and reheater, as well as many supporting components. The overall objective of any steam generator is to integrate all these components **to use thermal energy from fuel combustion to raise steam at desired conditions with maximum efficiency and reliability in the most economical manner (considering both capital and operating costs) with minimal environmental impacts.** It should be noted that often the entire steam generator is called the boiler. Technically speaking, while this term is very popular in the industry, it is not accurate because, as it will be explained later, the boiling process is only one of the processes in a steam generator.

Steam generators come in a wide variety of configurations, operating conditions, capacities, shapes, and complexities. Accordingly, they can be categorized based on many criteria. They can be classified based on the heat source, application, configuration, operating pressure, capacity, fire and water location, and water circulation mechanism.

The source of thermal energy in combustion-based steam generators can be the combustion of traditional fossil fuels, such as coal, natural gas, or petroleum products; alternative fossil fuels, such as peat; biomass, such as wood, corn husks, or straw; or municipal and industrial wastes. In some steam generators, steam generation is the secondary objective and the main objective is to get rid of materials used as fuel, such as domestic and clinical solid wastes.

Steam generators can generate steam either for power generation in steam power plants or for industrial and residential applications. The former steam generators are generally referred to as utility steam generators and the latter steam generators are called industrial steam generators. There are also units that produce steam for both electricity generation and industrial applications known as cogeneration or combined heat and power (CHP) units, which will be covered in Chapter 10. Industrial steam generators may be used in petroleum refineries, food processing, and paper manufacturing, as well as many other applications. They usually produce either saturated steam (not superheated steam) or hot water with relatively low capacities at typically low pressures.

Utility steam generators can be either subcritical at the operating pressure of below 180 bar (about 2600 psia) or supercritical at the pressure of higher than 240 bar (about 3500 psia), above the water critical pressure of 220.6 bar (3200 psia), up to 340 bar (about 5000 psia). The range of steam generation capacity in utility steam generators can be 125–1250 kg/s (about 1–10 million lbm/h) at the temperature of up to 750°C (about 1400°F). They can be either drum-type or once-through type. Regarding locations of fire and water-steam, steam generators can be shell-type, fire-tube, or water-tube units. The water circulation mechanism can be either natural circulation or controlled (forced) circulation. Small steam generators can be packaged factory-assembled but utility steam generators are typically field-erected.[1]

[1] Do not panic if these terms are unfamiliar! We will extensively discuss all of them throughout this chapter.

Our discussion on steam generators will begin with the thermodynamics of steam generation followed by the history of their development from early steam generators to modern ones.

3.1.2 Thermodynamics of Steam Generation for Steam Power Cycles

Thermodynamically speaking, a steam generator converts compressed liquid water to superheated steam at a constant pressure process. In Chapter 2, we learned that increasing the average temperature at which heat is supplied to a Rankine cycle improves the efficiency of the cycle. This can be achieved by increasing the temperature and pressure of the generated steam in the steam generator. The cycle efficiency can also benefit from superheating the outlet steam. However, due to material limitations of turbine blades, the maximum temperature of the outlet steam flow from the steam generator is limited. We also learned that reheating steam after the initial expansion process in the turbine can significantly improve the cycle efficiency.[2]

3.2 Historical Development of Steam Generators

Humanity has boiled water for cooking since 40,000 BCE. However, the early steam generators for power generation started to appear in the late 17th and early 18th centuries. Today's modern steam generators are the result of over 300 years of experience in the industry.

3.2.1 Early Steam Generators

The early steam generators were just like large kettles, hence the name **shell-type or kettle boilers**. In these boilers, fire directly heated the outer surface of the shell which, in turn, heated the water inside the vessel. Since water and steam were both present in the same vessel and were in equilibrium, these boilers could **only generate saturated steam**. The boiler that Newcomen used in his engine (Section 2.2) was of this type and was called a haycock boiler. The boiler of Watt's engine was similar but with the shape of a covered wagon, hence the name the wagon boiler. Until the late 18th century the only metal available to manufacture boilers was (riveted) copper plates. In the early 19th century, iron plates were also used for boiler manufacturing, which allowed a substantial increase in steam pressure (with standards of those days). Later boilers were iron-casted in cylindrical shapes with dished ends. In shell-type boilers, due to limited heat-transfer areas and short contact time between combustion products and shell surfaces, most of thermal

[2] For more detailed discussions on the thermodynamics of steam generators, please review Chapter 2.

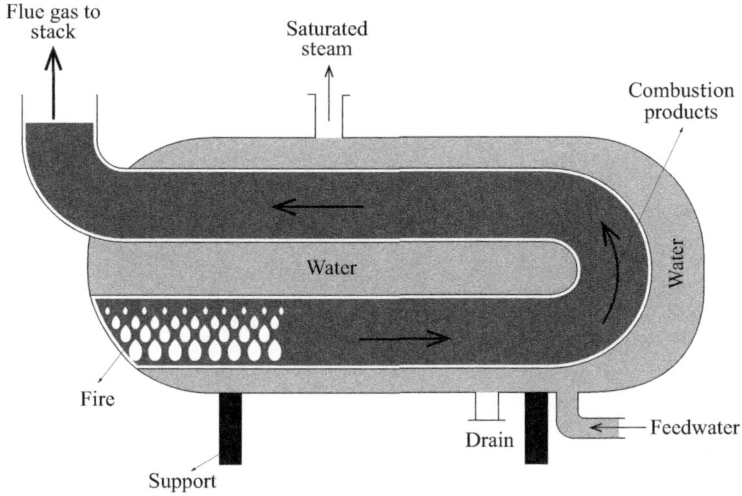

FIGURE 3.1
Schematic of Trevithick fire-tube boiler.

energy of fire was lost. To increase heat-transfer surfaces and improve the efficiency of these boilers, thermal energy was internally transferred to water through tubes inside the vessel while the water surrounded the tubes inside the shell (Figure 3.1). These boilers were the first generation of **fire-tube boilers** in which fire was inside the tubes and water was on the shell side. Richard Trevithick (a British inventor and mining engineer and the inventor of the first steam locomotive, 1771–1833) was a pioneer in these developments.

In later boilers, to further increase heat-transfer surfaces and steam generation capacity, the number of tubes and paths was increased (Figures 3.2 and 3.3). Regarding the location of the combustion process, fire-tube boilers could be categorized as **fire-box** and **Scotch marine** steam generators. In the former, combustion occurred in the fire-box outside or inside the vessel, and combustion products were directly fed to fire tubes (Figure 3.2). In Scotch marine boilers, combustion took place in the cylinder(s) located inside the shell at the bottom (Figure 3.3). In this type, combustion products were fed to fire tubes parallel to the combustion tube. Scotch marine boilers were commonly used to propel ships and were usually fueled with gaseous or liquid fuels. In terms of the orientation of tubes, fire-tube boilers could be horizontal, vertical, or inclined.

Since in fire-tube boilers the entire body of water should be heated to raise steam, these units were slow in steam generation and **slow in responding to demand changes**. Furthermore, in these boilers, the scale deposition of water outside tubes was difficult to remove.

As time passed, demand for electricity increased, which required more steam generation capacity. This meant larger steam generators. The large diameter of pressure vessels used in these steam generators resulted in high stresses on the shell which, in turn, required thick shells (refer to Reminder

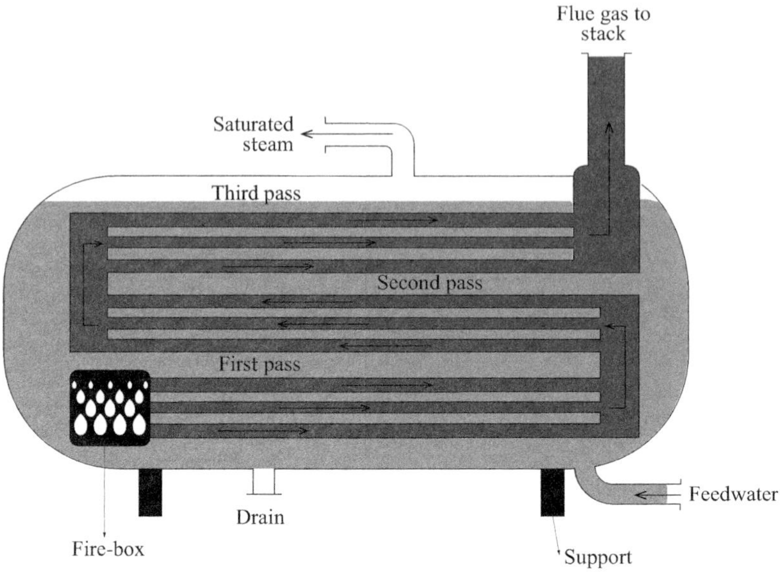

FIGURE 3.2
Schematic of horizontal fire-tube fire-box boiler.

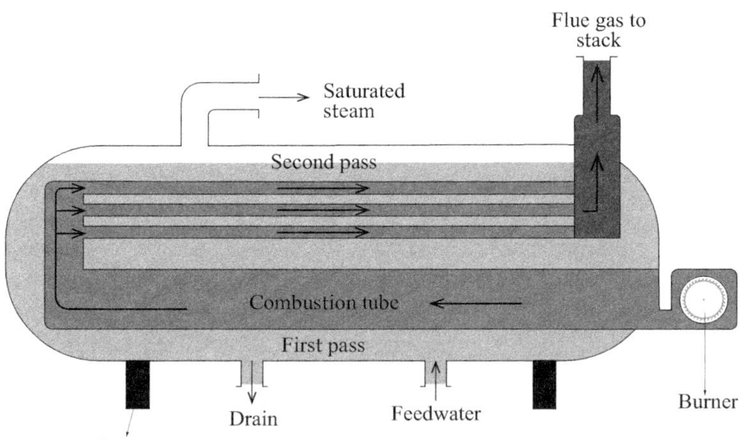

FIGURE 3.3
Schematic of horizontal fire-tube Scotch marine boiler.

Box 3.1). Moreover, as discussed in Section 2.3.2.2, higher steam pressures improve the efficiency of power cycles. Therefore, the pressure of steam generators was continuously increased. As a consequence of high steam pressures and larger boilers, there were **many disastrous boiler explosions** (refer to FYI Box 3.1).

REMINDER BOX 3.1

CIRCUMFERENTIAL STRESS IN CYLINDRICAL CONTAINERS

Let us consider a simple cylinder with the diameter of D (and the radius of R) and the thickness of t exposed to the internal pressure of P (Figure 3.4). The circumferential force induced on the shell is equal to PDL. Using the "thin wall" assumption (which means the radius of the cylinder is much greater than its thickness), the circumferential stress can be determined as follows:

$$\text{Circumferential stress} = \frac{PDL}{2tL} = \frac{PD}{2t} = \frac{PR}{t} \tag{3.1}$$

If the allowable stress of the material is S_{min}, the thickness of the shell can be calculated as

$$t = \frac{PR}{S_{min}} \tag{3.2}$$

This equation indicates that for a vessel made of metal with given allowable stress, the thickness of the wall is directly proportional to the radius of the cylinder and the internal pressure of the vessel.

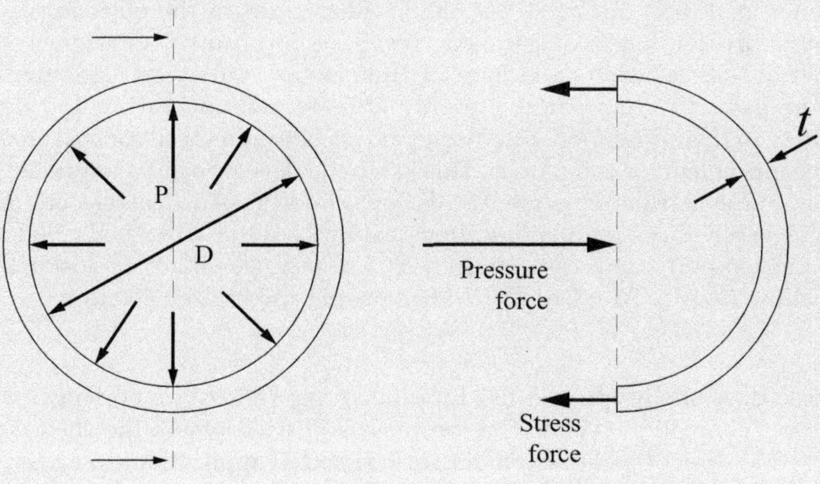

FIGURE 3.4
Schematic of circumferential stress in walls of cylindrical containers.

FYI BOX 3.1

BOILER EXPLOSIONS

Fatal boiler explosions with many casualties were common phenomena in the 19th and early 20th centuries. The problem was so intense that the progress in the industrial revolution depended on addressing this problem. In the 1880s, there were thousands of boiler explosions in the U.S. alone. Among them, the explosions of boilers in steamboats were particularly catastrophic.

In 1858, the boiler explosion at the steamboat Pennsylvania resulted in more than 250 dead. In 1865, at SS Sultana, a Mississippi River side-wheel steamboat, three of the ship's four boilers exploded killing 1800 of its 2400 passengers (more than the casualties of Titanic sinking). This was the greatest maritime disaster in the U.S. history.

The American Society of Mechanical Engineers (ASME) was founded in 1880 mainly for regulating boiler manufacturing and inspection. However, the efforts failed mainly due to the unwillingness of the stakeholders to reveal their trade secrets. Meanwhile, explosion incidents and casualties were mounting to hundreds of people every year. The Grover Shoe Factory disaster was a major turning point for mandatory safety regulations and the role of ASME in setting these regulations and enforcing them. In 1905, the boiler of the Grover Shoe Factory in Massachusetts exploded killing 58 people and injuring 150 more. This disaster and another fatal Massachusetts shoe factory boiler explosion outraged the nation. Regardless of the objections of manufacturers on this so-called "needless government interference", first in Massachusetts and later in other states, state-wide regulations were passed to improve the quality of boiler manufacturing and the safety of their operation. But discrepancies between these codes were a major problem for companies. This led to the development and publication of the first nation-wide ASME Boiler and Pressure Vessel Code in 1915. While these regulations almost eliminated those incidents, boiler explosions still cause casualties. In 2001, a steam generator explosion in Medina County Fair, Ohio killed five people and injured 40 more.

These days shell-type and fire-tube steam generators are no longer used in power generation cycles. The former is still used where the shell is not exposed to direct fire. For example, for residential applications, they are utilized in electric boilers or water heaters where heating electrical elements are submerged in water. In industrial applications, they are used in accumulators where heat is supplied by an external hot flow that goes through tubes inside the shell. Fire-tube steam generators are still used for some low pressure and low steam capacity industrial steam generation.

3.2.2 Water-Tube Steam Generators

As mentioned in the previous section, in fire-tube steam generators, the large diameter of the shell created unacceptably large stress on the vessel walls. The idea of water-tube boilers was to place water inside the tubes while flue gas surrounded the tubes in the shell. The configuration of early models was similar to fire-tube steam generators. Later more advanced designs of these steam generators had two main components: water tubes and a drum. Water was circulated inside the tubes and was heated by combustion product gases flowing outside the tubes. The separation of steam and water took place in a vessel, typically a cylindrical one, called the **drum**. Also, the drum sent saturated water into the tubes and received the mixture of water and steam from the tubes. Since a limited amount of water flowed inside the tubes, this type of boilers could be quickly heated to raise steam and could relatively rapidly respond to the variation in demand. Also, due to the relatively small diameter of the tubes and the drum, the stress level on the tube and drum walls was significantly lower compared to that of fire-tube steam generators.

The idea of water-tube steam generators was first introduced in the 1760s. During the next 100 years, many configurations were proposed and built. But the most important breakthrough was introduced by George Herman Babcock (an American inventor, 1832–1893) and Stephen Wilcox (an American inventor, 1830–1893). They invented inclined water-tube boilers (Figure 3.5). They called their steam generators "nonexplosive boilers" and founded the Babcock & Wilcox Company to manufacture their design.

Early water-tube steam generators had straight tubes (Figures 3.5 and 3.6). In these steam generators, the drum received hot water from the last

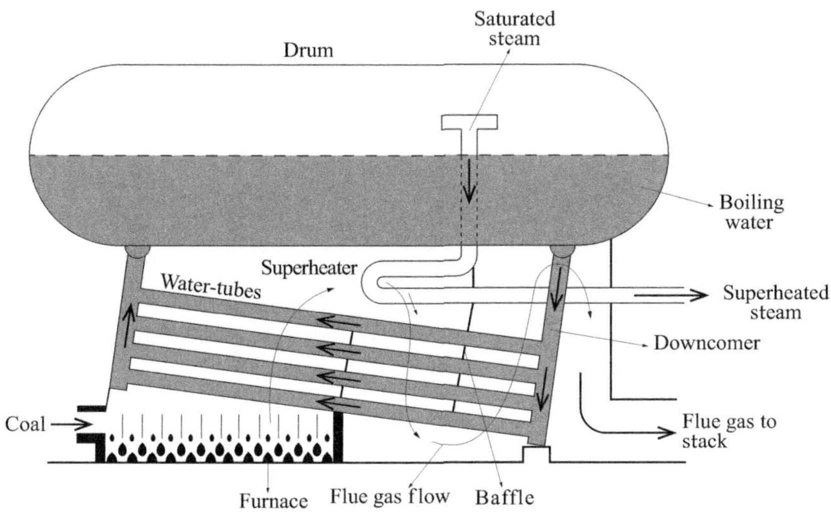

FIGURE 3.5
Schematic of a parallel-drum inclined water-tube steam generator with a superheater.

feedwater heater. The nearly saturated water was sent to the tubes through **downcomers** or **downtakes**. The water was then distributed among the tubes through the **headers**. Water inside the tubes received thermal energy from combustion product gases and was partially boiled. The mixture of water and steam was collected in another set of headers and was returned to the drum through the **risers** or **uptakes**. In the drum, water and steam were separated and steam was sent to the **superheater**. In addition to being safer, these steam generators increased heat-transfer surface areas and improved water circulation.

In early water-tube steam generators, the drum was arranged parallel to the tubes and was called the **parallel** or **longitudinal drum** (Figure 3.5). The number of tubes that could be connected to this configuration was limited and insufficient for high-capacity boilers. So sometimes more than one longitudinal drum was used. In order to overcome this problem, the **cross drum** arrangement was used where the drum was perpendicular to the tubes and could accept many more tubes (Figure 3.6).

The circulation of water in these steam generators was due to the **density difference** between flows in the downcomer and riser. The inclined tubes at about 15° enhanced water circulation. Since no external driving force, such as a pump, was used for water circulation in the system, they were called **natural circulation** steam generators. In order to increase the exposure of

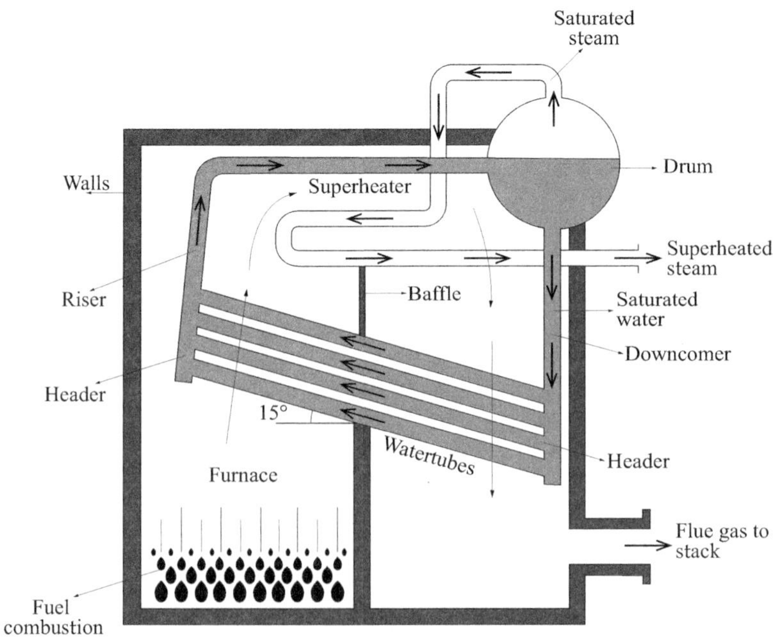

FIGURE 3.6
Schematic of a cross-drum inclined water-tube steam generator.

the tubes to combustion products (by increasing heat-transfer surfaces and the time of heat-transfer processes) and improve the efficiency of heat-transfer processes between combustion gases and tubes, baffles could be used to guide gases through up to three gas passages, which divided heat-transfer areas into several regions in series with respect to the flue gas flow (Figures 3.5 and 3.6).

3.2.3 Bent-Tube Steam Generators

Although cross-drum strait-tube steam generators allowed much more tubes to enter and leave the drum compared to their predecessors, they could not cope with increase in steam demand. Bent-tube steam generators could address this problem. The bent tubes in these steam generators could enter and leave drums radially. With this design, several rows of tubes could be connected to a single drum. This also allowed more than one drum to be used in bent-tube steam generators, top and bottom drums (Figure 3.7). Those days the quality of feedwater used in steam generators was poor. To collect and remove sediments from water, downcomers were connected to a bottom drum called the mud drum or water drum. Makeup feedwater also entered this drum.

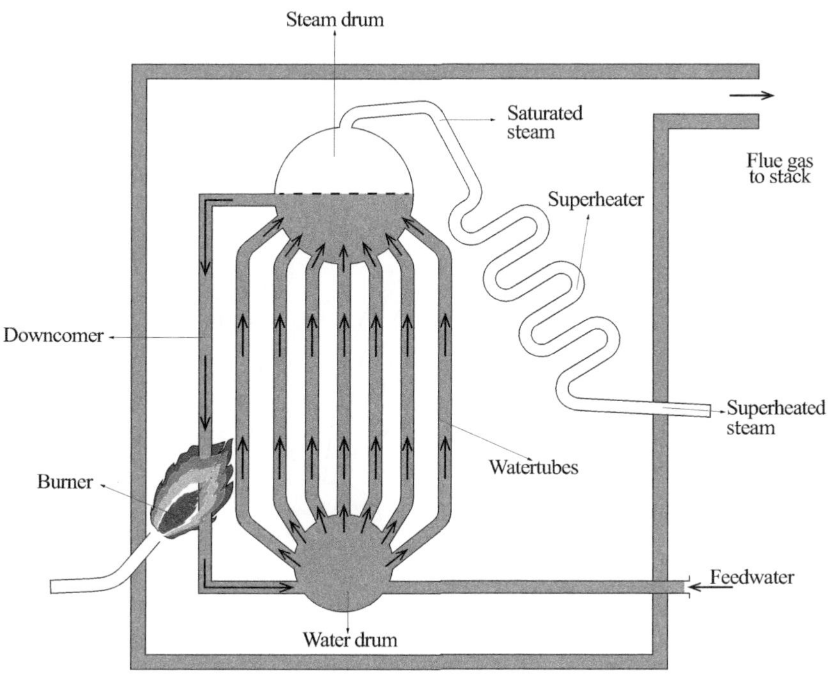

FIGURE 3.7
Schematic of a bent-tube water-tube steam generator.

In the early 20th century, **bent-tube Stirling** steam generators (named after its inventor Allan Stirling) were very popular for stationary and marine steam generation (Figure 3.8). They had two sets of drums: top drums (also known as steam drums) and bottom drums (also known as water drums or mud drums). In this type of steam generator, tube banks were directly connected to the drums (with no downcomers, risers, or distribution headers). These steam generators could have three, four, or five drums (Figure 3.8). The number of tube banks connecting these drums was one less than the number of drums, i.e., two, three, or four tube banks, respectively. In these steam generators, one or two tube banks acted as downcomers circulating water from the top drums to the bottom drums, and the rest of the tube banks acted as risers to return the mixture of water and steam to the top drums. Note that both water and steam halves of all top drums were connected (Figure 3.8). Similarly, the bottom drums were connected. These steam generators were able to operate with low-quality feedwater. Also, they could be fueled with various fuels while being able to follow steam demands.

Later simpler Milne boilers with two drums, one top drum and one bottom drum (Figure 3.9). There were also other types of steam generators, such as Yarrow, White-Forster, and Thornycroft among others. Some steam generators also used a pump to circulate water flow through tube banks (forced-circulation steam generators).

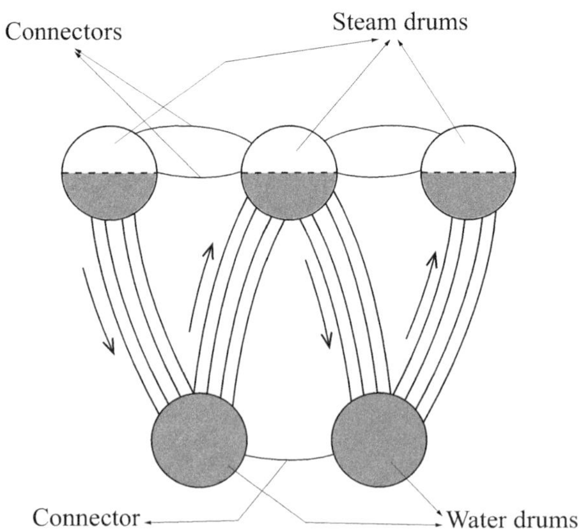

FIGURE 3.8
Schematic of a Stirling steam generator with five drums and four tube banks.

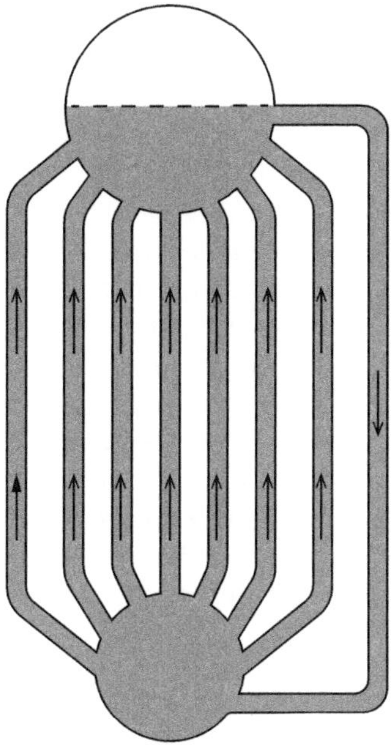

FIGURE 3.9
Schematic of a Milne boiler.

3.2.4 Waterwall Tubes

Until the 1920s, the increase in steam demand was met by employing several parallel steam generators. But this trend could not continue due to the complexity and costs of these systems; therefore, larger steam generators were needed. The high required thermal energy input of these larger units was met by furnaces with higher combustion rates. Note that in this discussion, the furnace refers to a large enclosed space where the combustion process takes place. But the higher temperature of these new furnaces required better furnace wall protection to reduce the temperature of combustion gases before they encountered the tube banks. The flue gases leaving the furnace could overheat the tubes downstream of the furnace if their temperatures were too high. Those days furnace walls were made of fire resistance blocks or bricks. They not only had to resist very high temperatures (sometimes as high as 1600°C, about 2900°F) but also needed to tolerate thermal stress due to temperature fluctuations, which could cause crack in the material.

The solution came in the form of furnaces with **water-cooled walls**, or in short **waterwalls**. Due to a very high heat-transfer rate during the boiling process, this process is ideal for cooling purposes. These water tubes in furnace walls not only reduced the temperature of furnace walls and combustion gases leaving the furnace but also helped to raise steam. This reduced the size of the boiler or evaporator (the section that converts water to saturated steam). The tubes used to cool down furnace walls gradually became an integral part of steam generators. The introduction of waterwalls eventually led to the integration of all components of steam generators to a single block.

The pressure of steam generators also increased over time. The latent heat of vaporization of water (h_{fg}) reduces significantly with increase in the pressure. The heat of vaporization is zero above the critical pressure of water because there is no two-phase mixture and water-like fluid is directly converted to steam-like fluid. As a result, at high pressures, less heat is required for the phase change from water to steam, and thus, the size of the evaporator of steam generators reduces. In fact, at pressures above 100 bar (about 1500 psia), **waterwall tubes can provide the entire heating surface required for the evaporator**. This frees more space in steam generators for more superheating and reheating surfaces, which can further improve the efficiency of steam power cycles.

3.2.5 Economizers

To increase the temperature of the inlet feedwater to the drum, **economizers** were introduced to steam generation units. In early units, since there were no feedwater heaters, the cold water from the condenser was directly fed to the economizer. This caused some problems including the condensation of vapor in combustion products on the outer surface of the economizer tubes (since the temperature of the tube's outer surface was below the dew point of the flue gas). This caused corrosion of the tubes due to the presence of sulfide oxides in the flue gas. Also, the fouling of the tubes due to the collection of ashes on the wet outer surfaces of the tubes required frequent cleaning. Furthermore, the heating of the cold feedwater in the economizer released oxygen dissolved in the feedwater, which caused corrosion in the inner surfaces of the tubes. To address these problems, feedwater heaters were added to increase the temperature of the inlet feedwater to the economizer to above the dew point of the combustion products (feedwater heaters will be discussed in Chapter 7). Moreover, the addition of an open feedwater heater (deaerator) to remove oxygen in the feedwater eliminated the oxygen corrosion problem.

3.2.6 Reheaters

As noted earlier in Section 2.3.3, **reheaters** are used to add more thermal energy to steam after a partial expansion in the turbine to increase the power output of the turbine and improve the quality of steam in the low-pressure sections of the turbine. While reheaters were used in a few steam power plants before World War II, they became popular and later a standard component of utility steam generators in the late 1940s and the 1950s.

3.2.7 Pressure and Temperature of Stream Generators

As stated earlier, a higher steam generator operating temperature and pressure result in a higher overall efficiency of steam power plants. Early boilers generated saturated steam at a few bar (a few tens of psia). By the early 1900s, the pressure of steam generators increased to over 10 bar (about 150 psia) and steam was superheated by 40°C (about 70°F) above saturation temperature. The steam pressures and temperatures increased to 30–40 bar (about 400–600 psia) and 250°C–300°C (about 500°F–600°F), respectively, by the mid-1920s. By the late 1920s, the steam pressure and temperature further increased to 80 bar (about 1200 psia) and 370°C (about 700°F), respectively.

Since a drum was the largest under-pressure vessel in steam generators, the operating condition of the drum were the most severe among steam generator components. Since the early days of the steam generation industry, drums were made of metal plates that were formed and riveted to build the vessel. To seal the joints, they were caulked. While metal used for these plates was improved over time, the overall manufacturing method did not change until the late 1920s. The first welded drum was operational in the early 1930s. This allowed a significant increase in the operating pressure of steam generators to about 125 bar (about 1800 psia), very impressive by the standards of those days. Further improvements in the metal properties led to the increase in the steam temperature up to 500°C (about 925°F) by the late 1930s. By the late 1940s, this temperature exceeded 540°C (about 1000°F). By the mid-1950s, the operating temperatures and pressures of 650°C (about 1200°F) and 365 bar (about 5300 psia) were achieved in a supercritical unit, which is still one of the highest operating temperature and pressures that have been ever attempted. However, due to some problems, such as ash deposition and corrosion, standard temperatures and pressures were settled at more moderate values.

Currently, a typical **subcritical** steam generator operates at about 180 bar (about 2600 psia) with a superheater and reheater temperature of about 560°C (1040°F). This arrangement is typically shown as 180 bar/560°C/560°C (about 2600 psia/1040°F/1040°F). **Supercritical** steam generators typically operate at similar temperatures but at much higher pressures. A typical arrangement can be 250 bar/570°C/570°C (about 3600 psia/1060°F/1060°F). The highest temperatures and pressures that are commercially available are **ultra-supercritical** steam power plants with the typical arrangement of 275 bar/600°C/620°C (about 4000 psia/1110°F/1150°F). Even higher temperatures and pressures are gradually becoming available in so-called **advanced ultra-supercritical** steam power plants operating at 330 bar/650°C/670°C (about 4800 psia/1200°F/1240°F).[3]

---

[3] https://www.ge.com/power/steam/steam-power-plants/advanced-ultra-supercritical-usc-ausc.

Note that with improved efficiency of power plants not only the fuel consumption rate reduces but also the size of equipment reduces. In steam generators, this means smaller heat-transfer areas. Also,increasing the operating pressure of steam cycles impacts the relative size of various components of steam generators. With increasing the operating pressure, the size of the evaporator reduces while the size of the superheater and reheater increases.

FYI BOX 3.2

RATING OF STEAM GENERATORS

One of the units used to rate steam generators has been boiler horsepower (BHP). Initially, one BHP was defined as 10 ft^2 of the steam generator's heat-transfer surface area. But obviously, this was not a proper definition because the amount of heat transfer depends on the temperature difference between the flows and heat-transfer mechanisms (and not just heat-transfer area). Clearly, the rating should refer to the heat-transfer capacity (the time rate of heat transfer) of the steam generator, and the heat-transfer surface area is only one of the parameters affecting the rate of heat transfer. Currently, one boiler horsepower is equal to 33,480 Btu/h.

3.3 Arrangement of Modern Steam Generators

The configuration of utility steam generators greatly varies depending on the fuel, combustion system, and required steam flow rate and its conditions, along with many other parameters. However, all modern steam generators have a furnace enclosed with waterwall tubes where the combustion process takes place. The combustion products then pass through many heat-transfer surfaces with various functions such as superheaters, reheaters, and economizers. In modern steam generators, all components to operate and control the system are integrated into a single unit. Figure 3.10 illustrates the schematic diagram of a typical modern drum-type steam generator. In Section 3.5, we will discuss another type of modern steam generators, i.e., once-through steam generators.

Now let us take a closer look at the processes within a drum-type steam generator. Figures 3.11 and 3.12 illustrate the schematic diagrams of an evaporator loop and a typical modern drum-type steam generator, respectively. First, hot feedwater from the last feedwater heater enters the economizer (Flow 1 in Figure 3.12) and leaves as saturated water or a low-quality

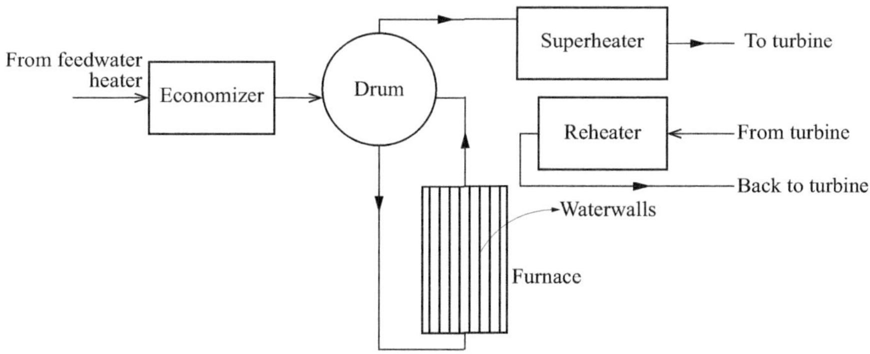

FIGURE 3.10
Schematic diagram of a typical modern drum-type steam generator.

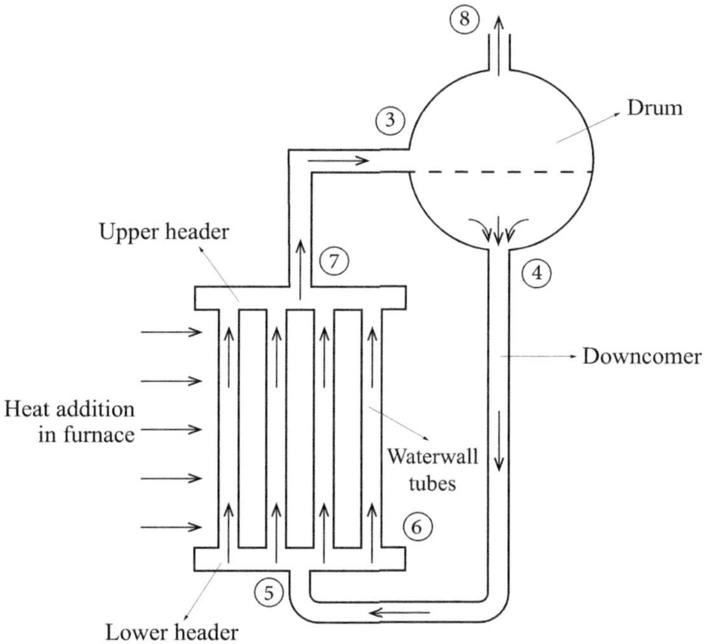

FIGURE 3.11
Schematic diagrams of an evaporator loop in a drum-type steam generator.

mixture of steam and water (Flow 2). The outlet of the economizer enters the midpoint of the drum (Flow 3 in Figures 3.11 and 3.12). Due to the natural circulation effect, the water circulates the waterwall tubes through the vertical downcomers and the bottom waterwall header system, which are typically located outside the furnace and well-insulated (Flows 4, 5, and 6). The headers, also known as manifolds, are to distribute or collect the working

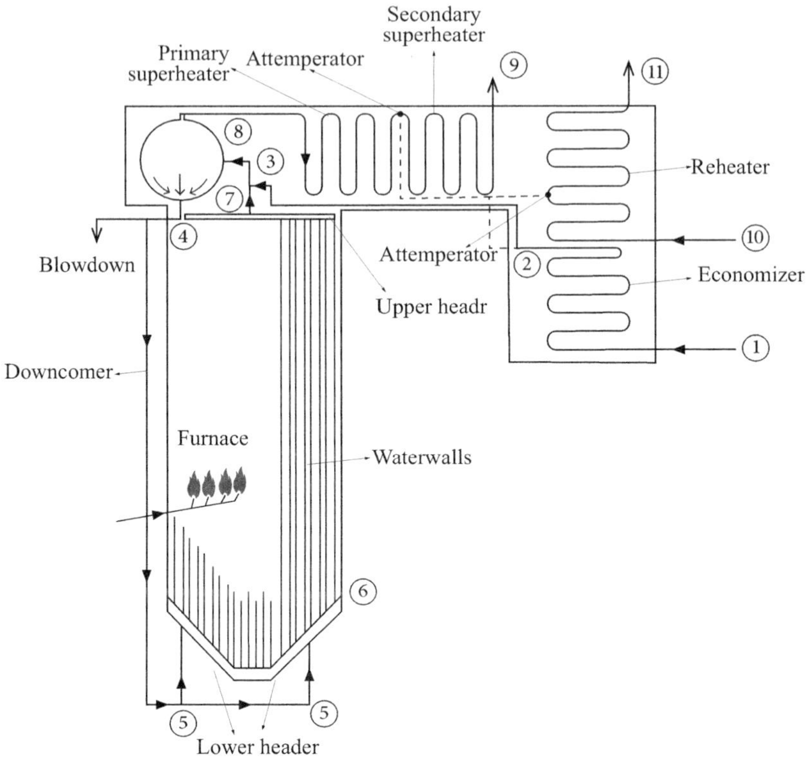

FIGURE 3.12
Schematic of a modern steam generator.

fluid to or from the waterwalls. The heat transfer from combustion gases in the furnace to the waterwall tubes boils the water (Figure 3.11). The boiling water returns to the drum through the upper headers and risers (Flow 7), sometimes mixed with the flow from the economizer. In the drum, the steam and water are separated and the steam is sent to the primary and secondary superheaters (Flow 8).

The superheated steam is then sent to the high-pressure turbine where it expands and produces shaft work (Flow 9). The intermediate pressure steam from the turbine (Flow 10) is returned to the steam generator where its temperature increases in the reheater. All modern utility steam generators have at least one stage, and often up to a maximum of two stages of reheating. The steam exiting the reheater(s) is returned to the lower pressure turbines (Flow 11).

While not common in utility steam generators, most industrial steam generators have a dedicated evaporator, called the boiler bank, to provide sufficient heat-transfer surface areas for the boiling process. They may be also equipped with a few rows of heat-transfer surfaces between the furnace and the superheater. This section, called the screen, is to shield and protect the superheater from intense radiation of the furnace (not shown in Figure 3.12). Note that there are a few parts that will be discussed later.

3.3.1 Radiant and Convection Sections of Steam Generators

The heat transfer to tubes in steam generators can be by radiation or convection (for more information on these heat-transfer mechanisms please refer to Reminder Box 3.2). In all furnaces, combustion gases emit radiant thermal energy but the amount of this energy and its contribution to the total heat transfer depend on many factors, including the fuel type and the combustion technology. Depending on the dominant heat-transfer mechanism, a certain heat-transfer surface may be characterized as a radiant or convection type. The radiant part of steam generators should have a direct view of the combustion in the furnace. The heat transfer in waterwall tubes is commonly radiation, which is why they are also called **radiant evaporators** (boilers). Depending on the configuration of a steam generator, superheaters can be either **convective** or **radiant superheaters**. The part of the steam generator where convection is the dominant heat-transfer mode is called the **convection pass**, which can include reheaters, economizers, and a portion of superheaters.

REMINDER BOX 3.2

HEAT-TRANSFER MECHANISMS

There are three modes of heat transfer: conduction, convection, and radiation. Conduction is heat transfer due to the vibration of molecules of a heat-transfer medium, which can be determined by Fourier's law:

$$\dot{Q} = kA\frac{dT}{dx} \tag{3.3}$$

where $\dot{Q}$ is the time rate of heat transfer (W), k is the conductivity (W/mK) which depends on the properties of the medium, A is the heat-transfer area (m^2), and $\frac{dT}{dx}$ is the temperature gradient along the medium (K/m).

Convection is heat transfer due to the flow (bulk motion) of a heat-transfer medium, which can be determined by Newton's law:

$$\dot{Q} = hA\Delta T \tag{3.4}$$

where h is the heat-transfer coefficient $\left(\frac{\text{W}}{\text{m}^2\text{K}}\right)$ which depends on the characteristics of the flow especially its velocity and ΔT is the temperature difference between the heat-transfer surface and the flow (K).

Radiation is heat transfer by electromagnetic waves, which can be determined by the following equation:

$$\dot{Q} = \varepsilon \sigma A \left(T_h^4 - T_c^4 \right) \tag{3.5}$$

where σ is the Stefan-Boltzmann constant $\left(5.67 \times 10^{-8} \frac{\text{W}}{\text{m}^2\text{K}^4} \right)$, ε is the emissivity (unitless), and T_h and T_c are the absolute temperatures of hot and cold bodies (K). The emissivity depends on the properties of the emitting surface. In heat transfer from combustion products to water or steam in the tubes within a steam generator, all three heat-transfer mechanisms are involved. This heat-transfer process can be divided into three distinct regions (Figure 3.13). The first region is the heat transfer to the outer surface of the tubes. This can be either through radiation directly from the flame, fuel bed, other tubes, or furnace walls to the tubes; through convection from the moving high-temperature flue gas to the outer surface of the tubes; or a combination of the two mechanisms. Second, the heat is transferred through the tube walls by conduction. Depending on the conditions of the tube wall, this may involve conduction through a layer of soot and a gas film on the outer surface of the tubes and a layer of scale and a water or steam film in the inner surface of the tubes. Finally, the heat is transferred to the water or steam inside the tubes through convection. Among these three heat-transfer modes, the greatest resistance is typically in the heat transfer from the flue gas to the outer surface of the tubes.

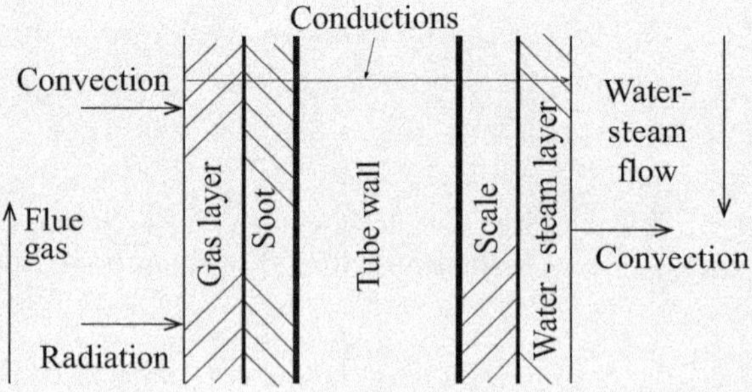

FIGURE 3.13
Heat transfer from flue gas flow to water or steam flow inside a tube.

3.4 Components of Modern Steam Generators

Modern steam generators are composed of all or some of the following components on the water-steam side: the boiler tube banks (usually in industrial steam generators), waterwalls, steam drum, superheater, reheater, economizer, and steam control equipment. All these components and combustion side components (which will be covered in Chapter 4) are integrated into a single unit. In the past, these components were designed and manufactured by various manufacturers and then were assembled to form a single coordinated system. However, in the modern steam generator manufacturing practice, many major components are designed, manufactured, and assembled by a single company in an integrated design. In this section, we will discuss these components, their functions, and their designs in more detail.

3.4.1 Waterwall Tubes

In modern utility steam generators, waterwall tubes form the entire walls, roof, and bottom of the steam generator, and **no other physical structure is needed to contain combustion gases** in the furnace and other components. They provide sufficient heat-transfer surface area for the evaporation process, and no dedicated evaporator is needed. In low-pressure steam generators, usually in industrial ones, waterwalls may not provide a sufficient heat-transfer surface area, and an extra **boiler bank** may be required. These boiler banks typically have steam and water drums connected with bent tubes (similar to the steam generator in Figure 3.9).

Waterwall tubes may be arranged in several configurations, including **bare tubes, embedded in the refractory, studded tubes, or membrane design** (Figure 3.14a–d, respectively). In all these configurations, waterwall tubes are welded to create pressure-tight walls, so no other casing is required to contain steam generator components. Membrane waterwalls are the most common design in modern steam generators. In this design the rows of parallel tubes with small spacing are connected by the welded bars (metal strips) between them. They are typically manufactured in the form of membrane panels in the factory, up to about 3 m (about 10 ft) wide and 30 m (about 100 ft) long, and then shipped and assembled in the field. The material of these tubes is relatively inexpensive carbon steel or low alloys because the boiling process limits the temperature of the tube walls, so expensive high-temperature-resistant alloys are not required. Note that the roof and floor of the furnace are also made of water-cooled tubes. In these areas, the tubes are inclined to ease water circulation and avoid steam trapping and potential tube overheating. In fact, the entire steam generator, including superheaters, reheaters, and economizer, are enclosed by gas-tight tubes. These tubes should be able to stand exposure to high temperature in the furnace

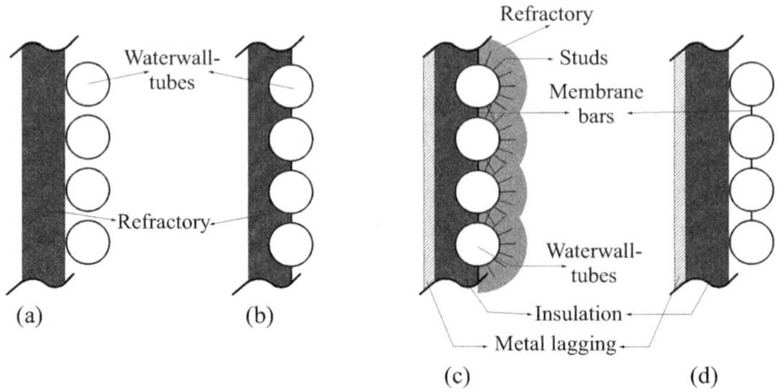

FIGURE 3.14
Different types of waterwall tubes: (a) bare tubes, (b) embedded in the refractory, (c) studded tubes, (d) membrane design.

(up to 2000°C), thermal expansion, ash and molten ash in the flue gas flow, and environmental elements, such as wind and earthquake. These tubes are backed by appropriate insulations and sheet metals to reduce heat losses and protect personnel from otherwise high-temperature walls. Before we discuss the internal shapes of waterwall tubes, we need to know different water boiling processes(Reminder Box 3.3).

REMINDER BOX 3.3

BOILING PROCESS

The function of the evaporator (boiler section) in a steam generator is to remove the maximum amount of heat from combustion gases and transfer it to the boiling water inside the tubes. Therefore, it is critical to understand the nature of the water boiling process. Consider a tube with water flowing inside the tube and the heat flux (heat transfer per unit area) is applied to the outer surface of the tube. Since the amount of heat transfer to the flow depends on the temperature difference between the bulk of the water flow and the tube inner surface, this difference will be used as an indication of the heat flux to the water inside the tube. At low-temperature differences, the heat transfer increases the water temperature but not the saturation temperature. With increase in the temperature difference, the temperature of the water adjacent to the wall will eventually reach saturation temperature and will start to boil locally. However, since the temperature of the bulk of the water has not reached saturation temperature yet, as soon as steam bubbles are separated from the surface, they will cool down and disappear. This process is called **local boiling, subcooled boiling, or subcooled-nucleate**. With further increase in temperature difference, the temperature of the bulk of the water will increase and eventually will reach saturation temperature. So when steam bubbles are formed on the tube surface and are separated from the wall, they will not condense and the two-phase flow will be formed. This water boiling mechanism, which has a very high heat-transfer coefficient, is called **nucleate boiling**. The higher temperature difference will result in larger bubbles at a higher generation rate. With even higher temperature difference, the rate of the steam bubble formation will eventually reach the point that the entire heat-transfer surface will be covered with superheated steam bubbles which will prevent the direct contact of the water and the tube walls. This boiling mechanism is called **film boiling** and the point that it starts is called **departure from nucleate boiling (DNB)** or **critical heat flux (CHF)**. This boiling mechanism is characterized by a low heat-transfer coefficient and should be avoided. The tube-wall temperature rapidly increases in this situation, which may cause tube failure due to either tube overheat or thermal fatigue due to temperature fluctuations. Also note that for a similar reason, the full evaporation of water in the tubes (100% quality of steam) is undesirable.

As noted above, the heat-transfer flux to waterwall tubes should be limited to avoid DNB in the evaporator. In modern steam generators, in order to delay DNB and thus increase the heat-transfer flux and steam generation rate, **internally grooved, corrugated,** or **ribbed (rifled) tubes** are utilized as waterwall tubes (Figure 3.15a–c, respectively). These tubes induce rotation in the water flow inside tubes. The centrifugal force produced by this rotation

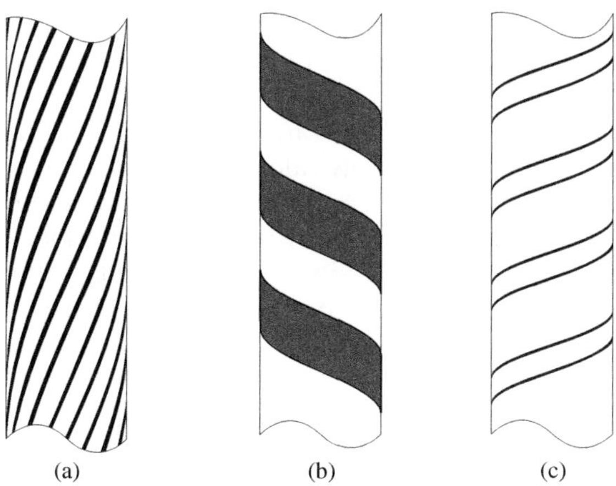

FIGURE 3.15
Schematics of interior of water tubes (a) internally grooved tube (b) internally corrugated tube (c) internally ribbed (rifled) tube (note that schematics show the inner surface of the tubes).

pushes heavier water droplets toward the tube walls while lighter steam moves to the center of the tubes which significantly delays DNB. Tubes with internal helical ribs are the most common design.

3.4.2 Drum

As mentioned before, some steam generators, such as Stirling steam generators, have both top (or steam) drums and bottom (water or mud) drums. However, most modern steam generators, except once-through steam generators which do not have any drum, have only one steam drum and in this section the drum refers to this drum. Although the main objective of a drum is to **separate steam from boiling water**, it also has other functions. The drum provides saturated water that is recirculated through downcomers to waterwall tubes. Then, the drum receives the mixture of water and steam back. The return tubes from the waterwalls to the drum are spaced evenly along the side of the drum to keep the distribution of the saturated mixture of steam and water as uniform and consistent as possible throughout the drum. Also, it receives feedwater from the economizer to makeup evaporated water and mixes it with saturated water already in the drum. Furthermore, the **chemical treatment** of the boiler water is done in the drum. Boiler blowdown to control water quality is also taken from the water in the drum. Finally, the drum functions as the accommodator of feedwater to control water flow variations due to small load changes. In fact, the volume of the drum is typically determined by this function and the space requirements of the internal components of the drum, particularly separators (refer to the next section).

Drums are enormous cylindrical vessels placed horizontally and about half full of feedwater. Their length, diameter, and weight can be perhaps about 30 m (about 100 ft), 1.8 m (about 6 ft), and hundreds of tons, respectively. They may have many steam outlet nozzles (to superheater headers) and many inlet nozzles from waterwall headers (containing boiling water) and a few downcomer nozzles (to lower waterwall headers). The number of steam outlet nozzles and boiling water inlet nozzles is determined to ensure the uniform distribution of these flows along the drum. Note that in pressure vessels, a nozzle refers to any inlet or outlet through the shell. Also a header is a container to distribute or collect flow among many tubes.

Drums are made of longitudinally welded rolled carbon steel plates, referred to as courses, connected by circumferential welds and capped at two ends with two typically hemispherical heads. The manufacturing process of the drum typically includes many steps of preheating, welding, and heat treatment processes. All welded joints are examined by radiography (X-ray) tests or ultrasonic tests to ensure their integrity. The latter is usually used for hard-to-reach joints. Moreover, all under pressure components are hydraulically tested at 1.5 times the design pressure.

One of the challenges in the design and manufacturing of drums is handling **thermal stress** on the walls due to the temperature gradient in the

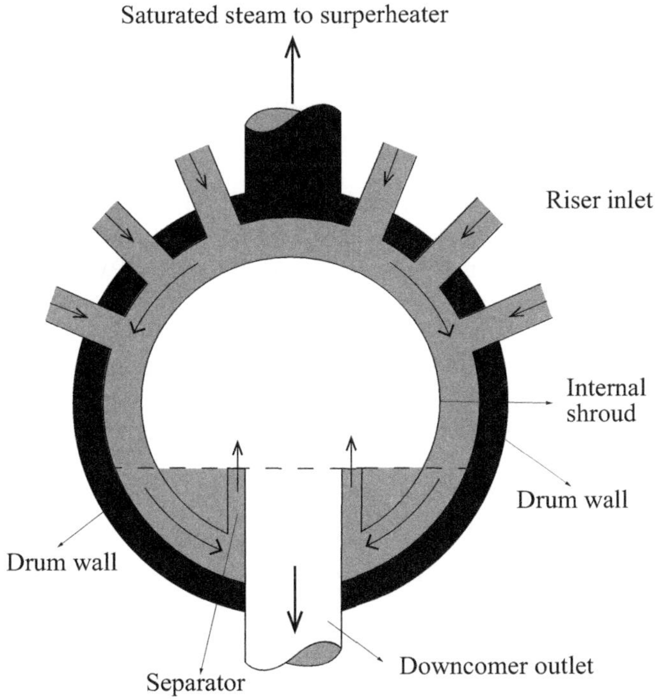

FIGURE 3.16
Schematic of a steam drum (internal components are not shown).

shell. As noted, the bottom half of the drum is full of saturated water and the top half contains saturated steam. This creates a temperature gradient between the two sections of the drum. One way to reduce this effect is using a **shrouded drum**, where the mixture of water and steam returning from the waterwalls is directed to the internal wall of the drum, so the entire wall along the drum is exposed to the same flow (Figure 3.16). It should be noted that this introduces a significant pressure drop to the flow and is typically used in forced-circulation steam generators (refer to Section 3.4.3).

As noted drum water is continuously or intermittently blew down to control dissolved solid concentrations in water. Reducing dissolved solids in drum water also helps to control foaming on the surface of water in the drun. A heat exchanger may be used to recover some of the thermal energy of the blow-down flow before discharging it from the system. The magnitude of the blow-down flow is measured relative to the steam flow rate leaving the drum (in %).

3.4.2.1 Calculation of Wall Thickness of Pressure Vessels

In Reminder Box 3.1 (circumferential stress in walls of cylindrical containers), the simple calculation for the required wall thickness of cylindrical pressure vessels was presented. In practice, this calculation is performed according to codes. In the U.S., Canada, and many other countries the ASME codes are used for this purpose. The European Union (EU), Japan, and some other countries have also developed similar codes. These codes are enforced by local, state, or national laws. Often, they are integrated into the contractual agreements.

As noted in FYI Box 3.1, the ASME was established to regulate the steam generator manufacturing industry. The **ASME Boiler & Pressure Vessel Code (BPVC)** is the result of this effort. The ASME BPVC is an extensive series of documents to ensure the high quality of all aspects of steam generator and pressure vessel design and manufacturing. They are developed to ensure the safe and reliable operation of steam generators during their expected life cycle. The BPVC consists of 12 sections, some with several divisions, sections, and appendices. The list of these sections is as follows:

- ASME BPVC Section I: Rules for Construction of Power Boilers
- ASME BPVC Section II: Materials
- ASME BPVC Section III: Rules for Construction of Nuclear Facility Components
- ASME BPVC Section IV: Rules for Construction of Heating Boilers
- ASME BPVC Section V: Nondestructive Examination
- ASME BPVC Section VI: Recommended Rules for the Care and Operation of Heating Boilers
- ASME BPVC Section VII: Recommended Guidelines for the Care of Power Boilers
- ASME BPVC Section VIII: Rules for Construction of Pressure Vessels
- ASME BPVC Section IX: Welding and Brazing Qualifications
- ASME BPVC Section X: Fiber-Reinforced Plastic Pressure Vessels
- ASME BPVC Section XI: Rules for Inservice Inspection of Nuclear Power Plant Components
- ASME BPVC Section XII: Rules for the Construction & Continued Service of Transport Tanks

While these codes have a mostly humble and relatively simple origin (the entire first Boiler and Pressure Vessel Code published in 1915 was 147 pages), over time they become very complicated technical-legal documents with many cross references. These codes are reviewed and revised regularly (for the BPVC every 3 years) and always the latest revision must be consulted. For example, the ASME BPVC Section VIII, which is typically used for pressure vessel design and manufacturing (including drums), have three divisions and Division 1 alone has 777 pages in its 2017 edition. The objective of Section VIII is "to enhance public safety and technological advancement to meet the needs of a changing world".[4] The BPVC Section VIII, Division 1 provides guidelines for the design, construction, and testing of fired and unfired pressure vessels when they are exposed to the operating pressure of over 15 psig (about 1 bar), either internally or externally.

For calculations of the shell thickness of pressure vessels, several codes can be utilized, including Section IV; Section VIII, Division 1; and Section VIII, Division 2. Each of these codes uses varying level of complexity in their analyses with different safety factors. A more detailed calculation requires a smaller safety factor. The safety factor is directly applied to the allowable stress used in the calculation. The equation used for circumferential stress in Section VIII, Division 1 is as follows (a similar equation exists for longitudinal stress):

$$t = \frac{PR}{SE - 0.6P} \tag{3.6}$$

where t is the required shell thickness (in), P is the design pressure (psig), R is the shell internal radius (in), S is the allowable stress at the design temperature (psi), and E is the efficiency of the welded joints. In this equation, the design pressure and the shell radius are dictated from thermodynamic and process requirements, the allowable stress can be determined from the material property tables in the ASME BPVC Section II – Materials, and E depends on the level of testing and inspection on welded joints. For example, if all joints are fully X-rayed, E will be equal to 1. As you can see, the equation does not have a safety factor within the equation. The safety factor is already taken into account when the allowable stress was determined and listed in the BPVC Section II. That is why the allowable stress used in the equation is not equal to any physical property of materials, such as the ultimate tensile stress, but it is proportional to them. This large safety factor, between 3 and 5, is to cover all unknowns, uncertainties, and imperfections such as nonuniform loadings, localized stresses, and material property variations.

Example 3.1

Determine the minimum wall thickness required for a drum with an internal radius of 3 ft operating at an internal pressure of 2500 psig. The drum is made of carbon steel SA-516 Grade 55 and all welded joints are radiographed. Use the ASME BPVC Section VIII, Division 1 for the calculation. Also, determine the safety factor of the design if the minimum tensile stress of SA-516 Grade 55 is 55,000 psi.

[4] https://www.asme.org/products/codes-standards/bpvcviii1-2017-bpvc-section-viiirules.

Solution

Givens:

$P=2500$ psig

$R=3$ ft$=36$ in

Fabrication material=Carbon steel SA-516 Grade 55

All welded joints are radiographed.

Minimum tensile stress of SA-516 Grade 55=55,000 psi

Find:

Determine the minimum wall thickness required for the drum.

Analysis:

In order to use the given equation for the shell thickness in the ASME BPVC Section VIII, Division 1, we need to determine S and E. As noted, the allowable stress can be determined from the material property tables in the ASME BPVC Section II – Materials, but we need to have the operating temperature to find the allowable stress. Since the water in the drum is boiling (water and steam are at the saturation condition), the operating temperature is roughly equal to the saturation temperature at the operating pressure of the drum (it is assumed that the drum is unheated and is located out of the pass of hot gases). Referring to Table A.2, the saturated temperature of water at the pressure of 2515 psia (note that the pressure in thermodynamic property tables is absolute pressure) is about 669°F. The allowable stress for carbon steel SA-516 Grade 55 at the operating temperature of 650°F and 700°F is 14,800 and 14,300 psi, respectively (you can find these values from the ASME BPVC Section II code or a simple internet search, for example at http://www.metalspiping.com/applicability-allowable-stress-of-astm-a516-gr-55.html). A linear interpolation between these two values gives an allowable stress of 14,610 psi. Note that the allowable stress reduces with increase in the operating temperature. Since all welded joints are fully X-rayed, $E=1$. So the thickness of the drum can be determined as follows:

$$t = \frac{PR}{SE - 0.6P} = \frac{2500 \times 36}{14{,}610 \times 1 - 0.6 \times 2500} = 6.86\,\text{in}$$

You can verify your answer using some online tools such as https://checalc.com/calc/vesselThick.html

In order to determine the safety factor of the design, Equation 3.2 can be used.

$$t_{\min} = \frac{PR}{S_{\min}} = \frac{2500 \times 36}{55{,}000} = 1.64\,\text{in}$$

$$\text{Safety factor} = \frac{t}{t_{\min}} = \frac{6.86}{1.64} = 4.2$$

If a more detailed analysis is used for the calculation, for example based on the ASME BPVC Section VIII, Division 2, the safety factor will be smaller.

3.4.2.2 Steam Separation

The presence of water droplets in saturated water leaving a drum toward superheaters is undesirable or even dangerous for the performance and durability of steam generator components as is the presence of steam in water outlet from the drum to waterwall tubes. The **carryover of water droplets to steam** and the **carryunder of steam to saturated water** can negatively impact the performance of a steam generator. The carryover of water containing solid particles and solutions to the superheaters can result in the deposition of these solids on the superheater tubes, which may locally increase the temperature of the tubes and result in their burn-out. These deposits, particularly silica deposits, may even reach to the turbine. The silica deposits on the turbine blades are not removable with water washing and may seriously damage the blades. The carryunder of steam to waterwall tubes reduces the density of the flow in the downcomers, which reduces the driving force for the natural circulation of the flow (refer to Section 3.4.3).

Several mechanisms are used to help to separate steam from boiling water in the drum. The simplest one, which was used in early steam generators, is **gravity separation**. The driving force in gravity separation is the density difference between water and steam. For low-capacity and low-pressure steam generators, this mechanism is effective. However, with increase in the operating pressure of steam generators, the difference in densities reduces significantly. Also, with increase in the capacity of steam generators, the velocity of steam leaving the water surface significantly increases, which can result in a higher likelihood of the carryover of water droplets. These two factors reduce the effectiveness of gravity separation. The latter problem can be partly overcome by a larger drum which increases the surface area and reduces the velocity of the steam leaving the water surface and allows more time for the separation process. For modern steam generators however, this will result in an unacceptably large drum. In modern utility steam generators, gravity separation alone cannot separate steam from boiling water, and other means of separation, such as mechanical separation methods, are required to supplement it.

Modern drums are commonly equipped with two series of steam separators, **primary separators** and **secondary separators** (Figure 3.17). The primary separation has two main functions to remove almost all steam from circulating water to prevent steam from being sucked into the downcomers and waterwall tubes and remove most of water droplets from steam.

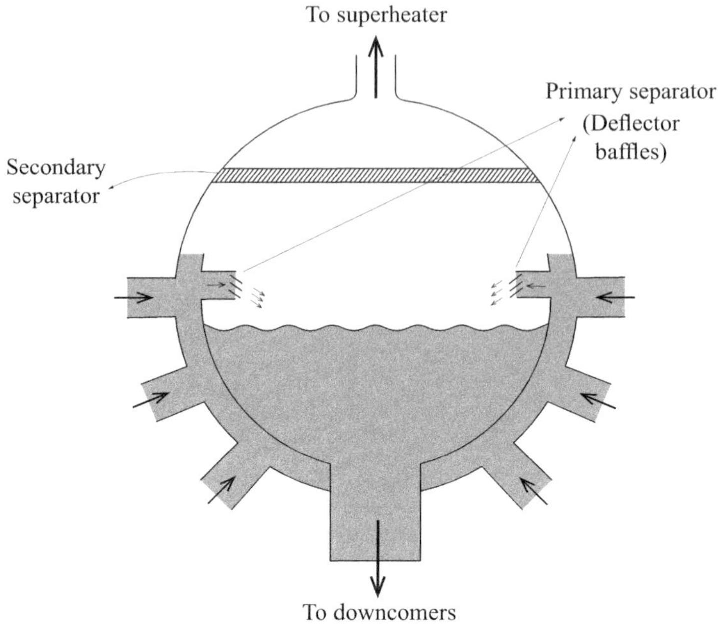

FIGURE 3.17
Schematic of a primary separator (deflector baffles) and secondary separator.

The simplest form of these separators is **deflector baffles** which separate steam from water by changing the direction of the flow (Figure 3.17). Other primary separators utilize **corrugated plates**, **screens** (wire meshes), or **cyclone** steam separators. In corrugated plates and screens, the large surface area accumulates water droplets and allows steam to pass. Cyclone steam separators operate based on the difference in the density of steam and water, similar to gravity separation. However, in this case, centrifugal force created by the tangentially admitted flow to the separator is much greater than gravity force. This force separates the downward moving water flow on the wall of the cyclone and the upward-moving steam flow in the center. This type of separator is suitable for high-pressure steam generators where the difference between the density of steam and water is small. In many designs, corrugated plates or screens are installed on top of cyclone steam separators to supplement the separation process.

The secondary separation process is to remove small water droplets that escape the primary separator and remove vaporized or solid contaminations that may be carried by saturated steam. The corrugated plates, perforated plates, or screens (discussed above) can be used for mechanical secondary separators. Sometimes contaminants may not be removed mechanically and water washing using relatively pure water from the economizer might be required. This process is also called steam scrubbing or steam drying. In the design and selection of all separators, the pressure drop through the device is an important factor that should be taken into account in circulating water calculations (refer to Section 3.4.3).

3.4.3 Circulation of Water

In order to have a flow of saturated water from a drum to waterwall tubes, a driving force is required to overcome the pressure drop in the tubes and connections. Steam generators can have either natural, thermosyphonic, or thermal circulation or forced, controlled, or pumped circulation. In the former, the density difference between saturated water in the downcomers and the mixture of water and steam in the risers and waterwall tubes provides a driving force to compensate for the pressure losses in the tubes. The mass flow rate of circulating water depends on the difference between the average density of the unheated leg of the loop and that of the heated one. This driving force depends on the operating pressure, rate of heat addition to circulating water, the height of the drum from the lowest point of the water circulation path, and configuration of the boiler. It can be determined by the following equation:

$$\Delta p = gh\left(\bar{\rho}_{\text{liq}} - \bar{\rho}_{\text{mix st \& liq}}\right) \tag{3.7}$$

where Δp is the pressure difference generated by the density difference between unheated water and heated mixture of steam and liquid water (kPa), g is the gravitational acceleration (m/s^2), h is the height of the water outlet of the drum from the lowest point of the water circulation path (m), $\bar{\rho}_{\text{water}}$ is the average density of unheated saturated water in the downcomers (kg/m^3), $\bar{\rho}_{\text{mix st \& w}}$ is the average density of heated mixture of water and steam in the risers and waterwall tubes (kg/m^3). In the equilibrium steady-state operation of a steam generator, the driving pressure (Δp) is equal to the pressure drop created by flow friction in the tubes, local pressure drops, and fluid acceleration. Also, note that the pressure difference is to overcome pressure losses in the circuit, so all efforts should be made to reduce the number of fittings and other sources of pressure loss, including using large-diameter tubes.

In order to increase the driving pressure and the flow rate of circulating water, the height of the drum and the density difference should increase. To increase h, the **drum is typically installed at the highest elevation possible in the boiler room**. Note that at high operating pressures, the density difference reduces, which, in turn, reduces the driving pressure and mass flow rate. When the combustion rate and heat-transfer rate increase, the quality of the water and steam mixture in the waterwall tubes increases. This, in turn, increases the density difference, and hence the driving pressure and mass flow rate are increased. Therefore, natural circulation steam generators have an inherent self-control mechanism. This phenomenon may not be necessarily desirable. The waterwall tubes are parallel with varying lengths, shapes, and heat absorption. These differences create uneven driving forces and pressure drops that can result in nonuniform flow rates in different tubes. Designers should calculate the flow rate in each tube at all possible operating conditions to make sure that all tubes receive needed water flow to prevent tube overheating and an unacceptable temperature gradient between adjacent tubes. These calculations involve the determination of heat absorption in each tube at all possible loads and transient conditions considering furnace configuration, all fuels, varying thickness of tubes with internal and external deposits (scale deposit and ash coverage, respectively) depending on the cleanliness of the tubes, and varying tube internal diameters (due to manufacturing tolerances). Any significant mistake in these calculations can be impossible to fix and may result in the permanent loss of steam generation capacity and/or reduced operating pressure. Another problem with natural circulation steam generators is their long start-up period. Since water circulation in the waterwall tubes depends on heat absorption by the tubes, during the start-up the firing rate should be gradually increased until there are sufficient water flows in all tubes for the full combustion rate.

In many large modern steam generators, the natural circulation force is sufficient to circulate water from the drum to the waterwall tubes, so they are called **natural, thermosyphonic or thermal circulation** steam generators (Figure 3.18). Some steam generators need a supplementary external driving force via a circulating pump between the drum and the waterwalls to pressurize saturated liquid water in the downcomer pass. In this case, the combined driving forces of the pump and density difference should be equal to the pressure drop in the circulating water-steam loop (Figure 3.18). Although the head (required pressure) of these pumps is relatively small, due to the high pressure of the feedwater and the water being saturated, their operating

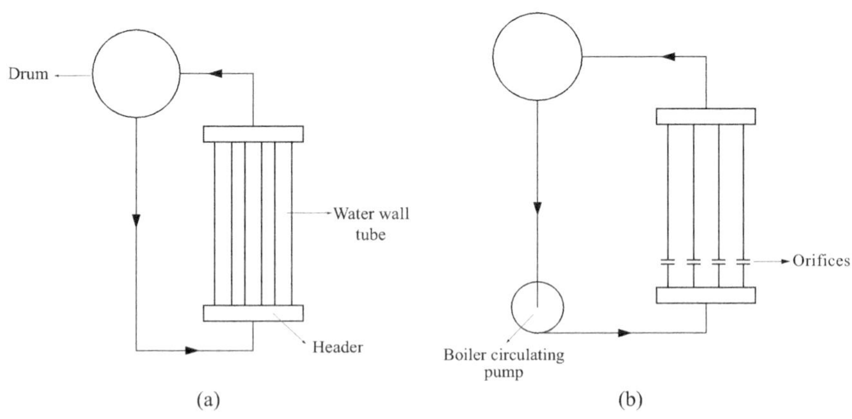

FIGURE 3.18
Schematics of (a) natural circulation and (b) forced-circulation steam generators.

condition is severe, and their design requires special attention. These units are called **forced, pumped, or controlled circulation** steam generators and were introduced in the early 1940s for the first time. This type is particularly useful for high-pressure steam generators (above 200 bar, about 2900 psia) because at high pressures, the difference between the density of saturated liquid water and saturated steam reduces, which reduces the natural driving force. If a sufficient driving pressure for circulation is not provided, the flow rate of feedwater in all or some tubes reduces, which may result in DNB and tube burnout.

As noted above, ensuring appropriate flow rates in waterwall tubes is a challenging task for the design of natural circulation steam generators. In forced-circulation steam generators, due to the availability of pressure head, this problem is addressed by adding flow distributing orifices in the inlets of waterwall tubes (Figure 3.18). These orifices are simply plates with a hole in the center that impose restriction (i.e. pressure drop) to the flow. The pressure drop depends on the diameter of the hole relative to the internal diameter of the tube. This pressure drop, which is greater than the pressure drop in the rest of the system, helps designers to balance the flow rate in the tubes regardless of many varying parameters (mentioned above), particularly the heat absorption of each tube.

While the addition of a boiler circulation pump adds to capital costs and slightly increases operational costs, it provided many other advantages. In forced-circulation steam generators, smaller diameter tubes can be used for waterwalls tubes, which reduces the weight and stress of the tubes. Also, unlike natural circulation steam generators, in these steam generators, the water flow in the waterwall tubes is independent of tube heating, which helps rapid start-up and load change. The fact that these steam generators can be equipped with a shrouded drum improves this attribute. This rapid load-flowing characteristic is a very important factor for power plants that are under cycling loads.

3.4.4 Superheaters and Reheaters

Superheaters receive saturated steam from the drum and send superheated steam to the high-pressure section of the turbine. They are typically configured in the primary and secondary sections to ease the steam temperature control process (this will be discussed later in this chapter). Reheaters receive steam from the outlet of the high-pressure steam turbine and return superheated steam to the low or intermediate pressure sections of the turbine after reheating the steam flow.

Due to the high temperature of steam inside tubes in superheaters and reheaters, typically between 550°C and 650°C (about 1000°F–1200°F) which is the highest metal temperature in a steam generator, they are subjected to severe operating conditions and require special materials. These tubes are exposed to high thermal stress due to the lower thermal conductance of the steam flow. The material of tubes should have high resistance to oxidation, fatigue, creep, erosion, and corrosion. The material and design of superheaters and reheaters also heavily depend on the ash properties of the fuel. Depending on operating temperatures, the tubes can be made of carbon steel (for low operating temperatures), high alloys, or stainless steel (for high operating temperatures). Reheater operating temperatures are similar to that of superheaters but at a much lower operating pressure, which results in lower stress in the tubes. This means the reheater tubes can be made of more common and less expensive metals.

For both superheaters and reheaters, typically bare simple tubes, both internally and externally are used. This means the tubes are neither internally ribbed nor externally finned unlike waterwall tubes for example where water boils to form a two-phase flow, and ribbed tubes are commonly used. The selection of the tube diameters is a trade-off between stress and pressure drop. The greater diameter of the tubes results in a lower pressure drop in the steam flow but greater stress in tube walls. A similar inverse relationship exists between the spacing of the tubes and the pressure drop in the flue gas. The tube spacing is also determined to reduce ash and slag buildup and to ease their removal.

As mentioned before, a superheater can be a convection or radiation superheater. The responses of these two types of superheaters to load variations are opposite. With an increase in steam load, the temperature of the outlet steam flow increases in convection superheaters while it decreases in radiation superheaters. To explain these characteristics, it should be noted that with an increase in the steam flow rate, the rate of combustion in the furnace increases. This means the mass flow rate of the combustion products increases but the temperature of the flame is not strongly affected. As explained in Reminder Box 3.3 (heat-transfer mechanisms), convection heat transfer depends on the velocity of the flue gas flow and increases at higher velocities. However, for a given surface, the heat transfer by radiation only depends on the temperatures of the flame and tube surfaces. In convective superheaters, the higher steam load increases the mass flow rates of both steam and combustion products which increases the total convection coefficient and the total heat transfer rate. The rate of the increase in the heat transfer is greater than the rate of the increase in the steam flow rate, which results in a higher temperature of the outlet superheated steam. In radiation superheaters, the rate of increase in the radiation heat transfer is smaller than the rate of increase in the steam flow, which reduces the temperature of the outlet superheated steam. Some modern steam generators use radiation and convection superheaters in series to reduce the dependency of the superheated steam temperature to the load.

The superheater and reheater units may be constructed in a pendant, inverted, or horizontal configuration (Figure 3.19). In the pendant configuration, the tube bundles hang vertically from the support structure, which results in mechanically robust units (in terms of bending stresses and

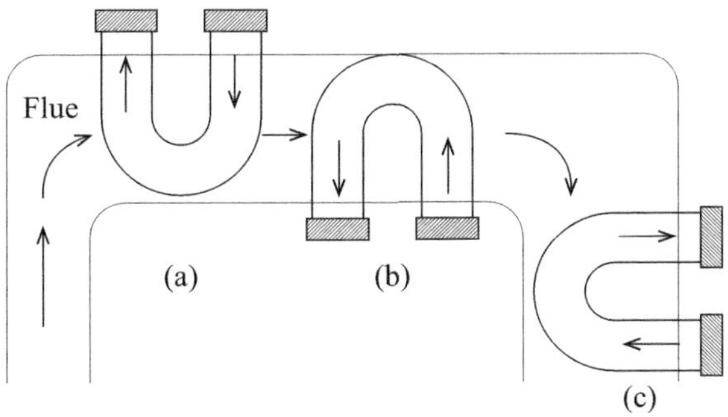

FIGURE 3.19
Schematics of superheater and reheater support systems (a) pendant (b) inverted (c) horizontal.

accommodation of thermal expansion, i.e., free expansion of tubes). But their start-up process is longer because of the trapped condensate at the bottom of the U-shape units. The inverted configuration is supported from the bottom and enjoys faster start-up; however, due to the lack of structural robustness, they are not common. The horizontal configuration is robust and has a fast start-up because the tubes can be fully drained but it is typically for convection superheaters and reheaters. The height of the pendant configuration is typically lower than that of the horizontal configuration, which can reduce the overall cost of the unit. Some superheaters have both pendant and horizontal sections. The entire heat-transfer areas of superheaters and reheaters are enclosed by water or steam cooled walls similar to waterwalls in a furnace.

3.4.5 Economizers

An economizer receives water from the last feedwater heater and sends near saturated water to the drum. This reduces the thermal load of the evaporator. The economizer is located at the low-temperature section of the steam generator after the reheater and before the air preheater (the latter will be covered in Chapter 4). It reduces the temperature of the outlet flue gas to the lowest practically possible temperature. It also reduces the possibility of thermal shock from cold feedwater entering the drum. While internally-bare tubes are commonly used in the economizer design, to improve the heat-transfer rate, they may be made of finned-tubes. Since in the economizer the temperature difference between the flue gas flow and the water flow is small, a large surface area is required, which necessitates a tight tube configuration (depending on the ash level of combustion products and ash properties). When high-ash fuels are used, the economizer may be equipped with water, steam, or air cleaning equipment to remove collected ash and soot.

3.4.6 Relative Heat Absorption in the Economizer, Evaporator, and Superheater

To understand the nature of heat-transfer processes in a steam generator, it is critical to have some knowledge about the relative heat absorption in the economizer, evaporators, and superheaters. This can be better shown via an example. While the design of heat-transfer surfaces is beyond the scope of this text, a very brief review of the process is presented in Reminder Box 3.4 (heat-transfer surface design).

Example 3.2

In a subcritical industrial steam generator, feedwater enters the economizer at the temperature and pressure of 140°C and 40 bar. The superheated steam leaves the superheater at the temperature of 450°C. Determine the relative heat absorption in the economizer, evaporator, and superheater. Assume the pressure losses in the steam generator components are negligible.

Analysis

The enthalpy of the inlet feedwater is (note that we are using saturated liquid water properties from Table A-1 (SI) instead of the compressed water properties)

$$h_{\text{Econ, in}} = h_{\text{sat water @}T=140°\text{C}} = 589.13\,\text{kJ/kg}$$

The enthalpy of the flow entering the evaporator, assuming the flow leaving the economizer is saturated liquid water, is (from Table A-2 (SI))

$$h_{\text{Econ, out}} = h_{\text{Evap, in}} = h_{\text{sat water@}P=40\text{ bar}} = 1087.31\,\text{kJ/kg}$$

The enthalpy of saturated steam leaving the evaporator is (from Table A-2 (SI))

$$h_{\text{Evap, out}} = h_{SH,\text{in}} = h_{\text{sat vapor@}P=40\text{ bar}} = 2801.4\text{ kJ/kg}$$

$$T_{\text{sat vapor@}P=40\text{bar}} = 250.40°\text{C}$$

Finally, the enthalpy of superheated steam leaving the superheater is (from Table A-3 (SI))

$$h_{SH,\text{out}} = h_{@T=450°\text{C and }P=40\text{ bar}} = 3330.3\text{ kJ/kg}$$

The heat absorption per kg and the relative heat absorption in the economizer, evaporator, and superheater can be calculated as follows and are presented in Table 3.1.

TABLE 3.1

Relative Heat Absorption in the Economizer, Evaporator, and Superheater for the Subcritical Industrial Steam Generator in Example 3.2

Heat-Transfer Surface	Economizer	Evaporator	Superheater	Total
Heat absorption per kg (kJ/kg)	498.18	1714.09	528.9	2741.17
Relative heat absorption (%)	18.2	62.5	19.3	100

TABLE 3.2

Relative Heat Absorption in the Economizer, Evaporator, and Superheater for the Subcritical Utility Steam Generator

Heat Surface	Economizer	Evaporator	Superheater	Total
Heat absorption per kg (kJ/kg)	1100	777.8	908.1	2785.9
Relative heat absorption (%)	39.5	27.9	32.6	100

$$\Delta h_{\text{Econ}} = h_{\text{Econ,out}} - h_{\text{Econ,in}} = 1087.31 - 589.13 = 498.18\,\text{kJ/kg}$$

$$\Delta h_{\text{Evap}} = h_{\text{Evap, out}} - h_{\text{Evap, in}} = 2801.4 - 1087.31 = 1714.09\text{kJ/kg}$$

$$\Delta h_{SH} = h_{SH,\text{out}} - h_{SH,\text{in}} = 3330.3 - 2801.4 = 528.9\,\text{kJ/kg}$$

Table 3.1 shows that most of the heat is absorbed in the evaporator (boiler) and the rest is almost equally divided between the superheater and the economizer. This was expected because at low-pressure steam generators, such as the one in this example, due to the large latent heat of water, most of the thermal energy is absorbed in the evaporator (boiler). This was explained in Section 3.2.4.

Table 3.2 presents similar results for a subcritical utility steam generator with feedwater entering the economizer at the temperature and pressure of 150°C and 180 bar and superheated steam leaving the superheater at the temperature of 550°C. The table shows that while the total heat absorption did not change much compared to Example 3.1, the heat absorption in the evaporator is dramatically decreased and the superheater and economizer have much larger share of heat absorption.

REMINDER BOX 3.4

HEAT-TRANSFER SURFACE DESIGN

In terms of the direction of the hot and cold flows and temperature variations within a heat exchanger, heat exchange surfaces can be parallel flow, counterflow, or crossflow. Note that in steam generators, the hot flow (or hot side) is always to the side where hot flue gas flows and the cold flow (or cold side) is where water, mixture of water and steam, or superheated steam flows. Superheaters, reheaters, and economizers can be either parallel or counterflow, while air heaters are typically crossflow. Also, note that the temperature of the cold flow remains constant when water boils in waterwalls or boiler banks. The calculation can be performed for one of the following three parameters as the unknown: temperatures, heat-transfer surface areas, or surface cleanliness. The calculation is always an iterative process. First, an initial value for the unknown is assumed based on past experiences. The calculation is then conducted to determine the unknown. If the assumed and calculated values are not close, the calculation is repeated with using the calculated values as the new assumed values. The iterative process is continued until the assumed and calculated values converge. For example, if the unknown is the outlet flow temperature, the calculation can be considered converged when the difference between the assumed and calculated temperature is <3°C (about <5°F).

3.5 Once-Through Steam Generators

In once-through steam generators, there is no designated component to separate saturated steam and boiling water. In this type of steam generators, heat is continuously added to water while passing through tubes. At some point, the water starts to boil and the boiling process inside the tubes continues until the entire flow is converted to steam. The heating process further continues until superheated steam at the desired temperature is generated. In these steam generators, there is neither a drum nor the circulation of water in the evaporator (boiler) section (at least in normal full-load operation of these steam generators, refer to the below discussion on start-up and low-load operation of these units), hence the name **once-through steam generators**. These steam generators are also called **universal-pressure** (**UP**) steam generators because they can operate at both subcritical and supercritical pressures, unlike drum-type steam generators in which the operating pressure is limited to about 200 bar (2900 psia). The introduction of these steam generators eliminated the limitation on the operating pressures of steam power plants.

The first once-through steam generator was built by Jacob Perkins (American mechanical engineer and physicist, 1766–1849) in 1822. However, the first commercial utility once-through supercritical steam generator was installed in 1957. This steam generator operated at the pressure of 310 bar (about 4500 psia) and an outlet steam temperature of 620°C (about 1150°F) and was installed in a 120 MW power plant. This followed by a unit at an operating pressure of 365 bar (about 5300 psia) and a temperature of 650°C (about 1200°F) in 1959.

Since in supercritical-pressure steam generators there is no boiling process, and the phase change process occurs gradually with increase in water temperature, no drum is required. Therefore, **all supercritical steam generators are a once-through type** but **subcritical steam generators can be of either a drum-type or a once-through type**.

Except the drum, once-through steam generators have the same components as drum-type steam generators. To improve the quality of the turbine outlet steam, all supercritical once-through steam generators are equipped with at least one reheater, flow double reheaters. One challenge of operating once-through steam generators is when the required steam flow rate is low, for example during low-load operation, start-up, or shutdown. In these situations, the water flow rate in waterwall tubes may not be sufficient to keep them within the acceptable wall temperature. In early designs, enough water was fed to the steam generator to protect the waterwall tubes; then the excess superheated steam bypassed the turbine and was directly discharged to the condenser. This solution was obviously wasteful and reduced the plant efficiency. In current designs the steam generators are equipped with a small drum (sometimes a spherical vessel) and a circulating pump to circulate water and provide sufficient cooling in the tubes. In these operational modes, a supercritical steam generator may be even operated at subcritical pressures.

In once-through steam generators, since there is no drum to separate suspended solids from superheated steam, in order to avoid deposition of solids in tubes, feedwater should go through an extensive treatment and have very high quality (Chapter 7).

In the design of once-through steam generators, special attention should be paid to the design of waterwall tubes. As noted in Section 3.4.3, depending on the heating pattern in the furnace, each tube in waterwalls may be exposed to a varying heat flux. In drum-type steam generators, this means that the returning water to the drum contains varying steam contents (quality), which is not a problem as long as there is adequate water to protect the tubes from overheating. But in once-through steam generators, this varying heat flux results in different temperatures of superheated steam at the outlet of each tube. A large number of tubes each with a low water flow rate can intensify the problem. So in once-through steam generators, it is desirable to have less tubes with a larger diameter and a greater water flow rate in each. But this means a larger spacing (or pitch) between tubes, which can result in overheating of metal strips between tubes. To address this problem, in many once-through steam generators, helical or spiral (**spiral wound**) waterwall tubes are used. In this design, waterwall tubes are placed on a typically 10°–25° angle from horizon (instead of vertical tubes), so they spiral around the furnace. Note that this configuration reduces the number of required tubes

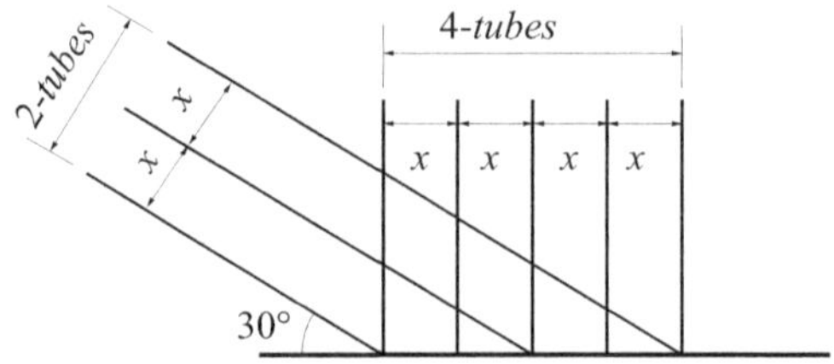

FIGURE 3.20
Using spiral wound waterwall design reduces the number of required tubes by half for the same area.

to create the wall surrounding the furnace with the same spacing between them. For example, at an angle of 30°, the number of required tubes is reduced by half (Figure 3.20). The design also reduces the variation of heat flux among tubes because all tubes rotate around the furnace and receive heat from all walls. This design also reduces the possibility of tube overheating due to a higher water mass flow rate in each tube.

This design however has the problem of higher pressure drops in waterwall tubes due to the longer length of tubes (which overweigh the effect of lager diameter tubes). Also, the wall support system is more challenging. To reduce these problems, typically only in the bottom portion of the furnace, where the heat flux is the highest, the spiral design is used. In the upper section of the furnace, where the heat flux is low, the vertical tubes are used (Figure 3.21).

3.6 Control of Steam Generators

The objective of any utility steam generator is to produce superheated steam at the desired flow rate, temperature, and pressure. A control system along with its sensors and instruments is to ensure safe, reliable, and economic operation of a steam generator. As an extremely complicated system, the safe operation of a steam generator requires an extensive control system that should be coordinated with other components of the cycle. The control system of a steam generator is essentially made of two major groups of devices. Sensors and instruments collect data from equipment. It is vital that all instrumentations and seniors are initially calibrated, and the calibrations are renewed before their expiration date. The collected data are then processed to generate control signals necessary to operate the system. The data are also recorded and critical performance information are reported to operators. Note that if required, operators can manually control the operation of a steam generator. While there is a central control system that monitors the entire power plant,

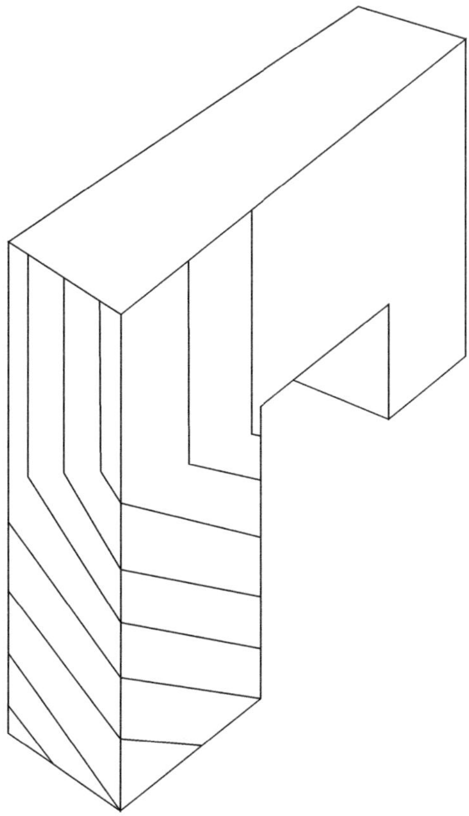

FIGURE 3.21
Typical spiral wound steam generator where only bottom half of waterwall tubes are angled.

some subsystems may have a local control system. Critical parts of a control system should have backup systems, so in the case of failures, a steam generator can continue to operate safely or can be safely shutdown.

The most important function of a steam generator control system is to control the rate of combustion (firing rate). For the steam generator to operate at steady-state condition, the rate of heat addition to the steam generator should be equal to the rate of heat withdrawal from the steam generator by produced steam. The objective is to provide sufficient thermal energy to raise required steam at needed conditions. Clearly, the required rate of combustion depends on the load. Understanding how the load impacts the performance of the steam generator requires a good understanding of the steam turbine performance. Since the control of any steam generator is tightly interconnected to the fire side of the steam generator and the turbine, the overall control of the system operation will be covered in Chapter 6. However, in this section, some of the most important independent control systems in steam generators will be discussed. But before doing so it might be informative to briefly review the basics of control systems first.

As noted above, in steam generators, the outlet steam flow rate, pressure, and temperature are the most important controlled variables. The first two of these parameters, steam flow rate and pressure, must be controlled and coordinated with the turbine-generator unit. So the control of these variables will be discussed after we covered the basic principles of steam turbines in Chapter 6. Also, note that there are many control systems for the fire side of steam generators that will be covered in the next chapter. In the following

REMINDER BOX 3.5

BASICS OF CONTROL SYSTEMS

In any control system, there are parameters that need to be kept at some desired values or ranges. These parameters are called **controlled variables** (also called output variables or simply outputs).the desired values for these parameters are called **set points** (or input demand signals or simply input signals). The independent variables that can be changed to maintain the controlled variables at their set points are called **manipulated variables**. For example, in a steam generator the flow rate, temperature, and pressure of the generated steam are the controlled variables, the required outlet steam flow rate, temperature, and pressure are the set points, and the fuel, air, and feedwater flow rates are the manipulated variables. Whenever an internal or external change, called a disturbance input, is imposed on the system that affects the controlled variables, the control system changes the manipulated variables to return the controlled variables to the set points. For example, when the load of a steam generator changes, the outlet steam condition changes, so the control system manipulates the fuel, air, and feedwater flow rates to return the steam conditions to the set point values. These corrective actions may be initiated by operators (manually) or automatically by a computerized system.

A control system can be an open-loop or a closed-loop system. In an **open-loop** control system (also known as a **non-feedback** or **feedforward system**), the controlled variables do not have any impact on the control actions of the manipulated variables. For example, in a steam generator when load increases, the control system increases the fuel flow rate by a predetermined value without checking if the controlled variables actually return to the set points. The predetermined values are typically provided by designers in the form of calibration curves. The advantages of open control systems are their simplicity, very fast response, and stability. However, the problem with this control strategy is difficulty in ensuring the accuracy of the calibration curves and control equipment considering the varying nature of component performance. For example, short-term variations of equipment cleanliness or long-term equipment aging can impact the accuracy of calibration curves.

In a **closed-loop** (also known as **feedback**) control system, the controlled variables are measured and compared to the set points. Any discrepancy (called an error signal) initiates corrective actions on the manipulated variables until the values of the controlled variables are equal to the set points (think of the cruise control in your car!). In this control system, no corrective action is taken until the disturbance influences the controlled variables.

There are several types of closed-loop control systems. In the simplest form, called on-off control, the error signal initiate an on or off action, e.g., a valve is opened or closed, until the set point is achieved. Due to the noise of valves, this strategy is sometimes called bang-bang control (think of an old-fashioned house HVAC system controlled by a thermostat). While very simple, this control system is not stable and can increase maintenance costs.

To address the stability problem, in proportional control systems the magnitude of the response to a disturbance, i.e., the increase or decrease of the manipulated variables (direct and reverse actions), is determined by and is proportional to the magnitude of the error signal. This means a larger difference between a controlled variable and its set point triggers a more aggressive corrective action. The problem with the proportional control strategy is that it always introduces an error (referred to as the offset) between the actual value of the controlled variable and its set point. The magnitude of the offset can be controlled by the sensitivity of the manipulated variable (called the proportional gain). However, there is an inverse relationship between the magnitude of the offset and the stability of the control system. To eliminate the offset in a proportional control system, an integral or reset control can be added (called a proportional plus integral control). However, this reduces the stability of the control system and increases its response time. Adding a derivative or rate control can increase the stability of the control system and reduce its response time (called a proportional plus integral plus derivative control). Finally, the open- and closed-loop systems can be combined so the open-loop control can provide fast response according to the calibration curve and the closed-loop control can correct the inaccuracies of the calibration curve.

sections, the temperature measurement, the superheated steam temperature control system, the drum water level control system, and safety valves will be discussed.

3.6.1 Temperature Measurement

Temperature is one of the most important parameters, if not the most important parameter, that is measured in a variety of power generation systems. Temperature measurement is vital for operation, control, maintenance, protection, and safety of any thermal power plant. A wide variety of physical phenomena have been used to measure temperature. While infrared thermometers (also known as non-contact thermometers, temperature guns, or laser thermometers) might be used for some special applications, a great majority of temperature measurement devices are either thermocouples or resistance temperature detectors (RTD).

3.6.1.1 Thermocouples

Thermocouples are temperature measurement devices that operate based on the thermoelectric or Seebeck effect (after Thomas Johann Seebeck, a German physicist, 1770–1831). If we connect two dissimilar wires at one end, and expose the connected and open ends to two different temperatures, a very small voltage will be created. The magnitude of the generated voltage depends on the metals that the wires are made of and the temperature gradient they are exposed to. For a given setup, if we measure the generated voltage, the temperature difference between the connected and open ends of the wires (called junctions) can be determined. So what a thermocouple measures is the temperature difference between the junctions. In order to measure the target temperature, we need to know the temperature of one of the two junctions. In all thermocouples, the end where two wires are connected is exposed to the temperature that needs to be measured and is referred to as the hot junction. The open end of the wires needs to have a known temperature, so it is referred to as the reference or cold junction. Generally speaking, the temperature of the reference junction can be determined either by immersing it to a mixture of distilled water and ice at the atmospheric pressure (called the ice bath) or by using a different thermometer. You may ask if we have another thermometer to measure the temperature of the reference junction, why not use it to measure the unknown temperature. The answer is that while this is possible, thermocouples are usually used to measure temperatures at harsh environments that the reference thermometer cannot tolerate. Note that the cold junction and its thermometer, which measures ambient temperature, are typically integrated to the electronic circuit that measures voltage and displays the temperature, so they are neither exposed to extreme conditions of thermocouples nor visible. Silicon (or generally semiconductor) thermometers are commonly used to measure the temperature of the reference junction. Thermocouples are categorized and named according to the combination of metals used to form the junctions. Each type is optimized for certain applications. The selection of an appropriate type of thermocouple for an application depends on the range of variation in the measured temperatures, the required accuracy, and the conditions of the environment that the thermocouple is exposed to. Some of the commonly used thermocouples in power plant applications are J (iron–constantan), N (chromel–alumel), K (Nicrosil–Nisil), R (platinum/rhodium – platinum), and S (platinum/Rhodium – platinum)[5] types. Their lower temperature range is between –270°C and 0°C and their higher temperature range is 750°C–1600°C. In the above list, thermocouples types are listed in the ascending order of the highest temperature they can measure. The following attributes contribute to the popularity of thermocouples:

- They can measure wide range of temperatures.
- They are very robust and have high durability when they are exposed to harsh environments such as high vibration.
- They do not need a power source to operate. This can make their circuit simpler compared to other temperature measurement technologies.
- They are relatively inexpensive.
- Generally they can rapidly respond to variations in temperature.

Because of these advantages, they are the most common temperature measurement devices in power plants. Their main problem is their limited accuracy (typically 1°C). While this level of precision is sufficient for many applications, it is unsuitable for applications where precise temperature measurements are important.

The hot junction of a thermocouple can be installed on a surface or can be placed in a shield, along with the immediate wires, to form a probe for protection from harsh environmental conditions. For extremely challenging environments, they may be installed into a thermowell for further protection by providing a barrier between the sensor and the hazardous environment.

The electronic circuit to process the produced voltage can be located in the closest location where a direct connection to the plant's control system can be made. But this can be problematic, particularly if the voltage has to be transferred for a long distance because extension wires must be made of same metals as the thermocouple itself, which can be expensive. Also, signals can be interfered by a variety of environmental parameters that may introduce noises to the signals. Therefore, in power plants, commonly a transmitter is used to convert the voltage signal from one or more thermocouples to control signals that can be transferred for long distances without any of

[5] Note that R and S types use a different ratio of platinum to rhodium.

the abovementioned problems. Recently, wireless (smart) transmitters with advanced control, calibration, safety, and diagnostic features are becoming popular.

3.6.1.2 Resistance Temperature Detectors

Resistance temperature detectors (RTD), also called resistance thermometers, measure the electrical resistance (in the form of a small voltage signal) of a piece of wire made from platinum, nickel, copper, or other metals that is exposed to an unknown temperature. Since the measured resistance is a function of temperature, it can be used to determine the temperature that the probe has been exposed to via a calibrated signal processing circuit. While more accurate than thermocouples (accuracy of about 0.1°C), the highest temperatures that they can measure is limited to about 600°C. They are also more expensive and more fragile than thermocouples. Moreover, they need an external excitation current for the probe to be able to measure the resistance. While all these disadvantages limit their use in many applications in power plants, they are commonly used for relatively low temperature and non-hazardous environments where precision, reliability, and repeatability are important. Similar to thermocouples, they can be installed on a surface, as a probe, or into a thermowell. Also, they can be connected to the control system either directly or via transmitters. Note that for critical systems and processes, temperature measurements with double or even triple redundancy can be used. Similar to any other measurement instruments, a regular calibration is required for temperature measurement devices. Most temperature measurement devices can be tested according to the ASME PTC 19.3 standard.

3.6.2 Superheated Steam Temperature Control

Many parameters influence the steam temperature in the outlet of a superheater, including load, combustion conditions (such as fuel type, burner combustion conditions, and excess air), cleanliness of heat-transfer surfaces, percentage of blowdown, and utilization of steam for auxiliary applications. The impacts of some of these parameters are more predictable than others. For example, as discussed earlier in this chapter, the impact of load on the steam temperature depends on whether the superheaters are convective or radiative with opposing impact in these two types. However, other parameters such as the cleanliness of heat-transfer surfaces or change in the fuel type are less predictable. Therefore, in order to protect the turbine, superheaters, and reheater from overheating and excessive thermal stress, an effective mechanism to regulate the temperature of superheated steam within an acceptable range is critical for the safe and reliable operation of any steam generator. Note that these control mechanisms are generally used to regulate the steam temperature during transient conditions. During steady-state operations, the control of the fuel flow rate (the firing rate) and the flow rate of feedwater are used to maintain the superheated steam temperature.

There are two general approaches for steam temperature control, controlling the amount of heat absorption by steam in superheater surfaces and regulating the steam temperature after heat is transferred to the steam. There are several mechanisms for the former, including recirculating a portion of combustion products from the economizer or the air preheater to the furnace through gas circulation fans, bypassing a portion of the hot combustion products from superheaters, and superheaters with separate firing. But due to technical complications and economic reasons none of these mechanisms are common. There are other more common methods, such as adjusting the flame position in the furnace, controlling the amount of excess air, and the burner selection that will be discussed in the next chapter. For controlling the steam temperature after heat is transferred to the steam, a surface method or direct contact method can be used. In the surface method, when the temperature of the steam exceeds the set point, a heat exchanger is used to remove heat from the steam and to reduce its temperature. This method is only used in low-capacity industrial steam generators.

The most common method used for steam temperature control in the majority of utility steam generators, both in drum-type and once-through steam generators, is the direct contact method, commonly known as **attemperator** or **desuperheater**. In this approach, the temperature of the outlet superheated steam from superheaters and reheaters is controlled by spraying water or injecting low-temperature steam into the superheated steam flow. Spray attemperators are the most common type. They have a rapid response and can provide a sensitive temperature control. From purely control point of view, the outlet of superheaters to the turbine might be the most suitable point to inject cooling water to provide a direct temperature control with no time daily. In fact in single superheater units, attemperators are located at the outlet of the superheater. However, in this design water droplets that are not evaporated can be carried to the turbine. Also, the temperature of steam within superheaters cannot be controlled, which may result in the overheating of the superheater tubes. To prevent these problems, the steam superheating process is typically divided between two stages of superheaters: the primary superheater and the secondary superheater. The attemperators are commonly located between these two superheaters. In this case, there is sufficient time for the creation of a superheated steam flow with a uniform temperature distribution before it reaches the turbine.

In a typical attemperator, a Venturi section is added to the piping between the primary and secondary superheaters to create a high-speed steam flow. The relatively cold water flow (at about 150°C) is sprayed into the steam flow through a nozzle installed at the throat of the Venturi to take advantage of the mixing effect of the high-velocity steam flow (Figure 3.22). The sprayed water evaporates and the latent heat of evaporation is extracted from the superheated steam flow, which, in turn, reduces the steam temperature. The

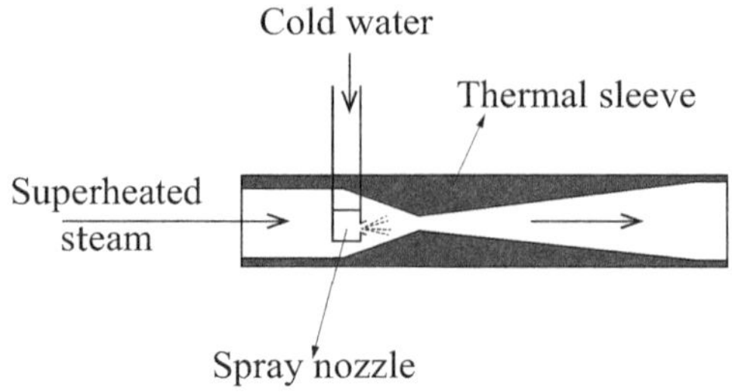

FIGURE 3.22
Schematic of an attemperator.

magnitude of the temperature reduction is determined by the mass flow rate of the injected water. A similar system is typically used at the upstream of the reheater inlet to control the reheated steam temperature. The sprayed water must have a high quality; otherwise solid impurities can be deposited on the superheater tubes or turbine blades. In many units, feedwater from the high-pressure feedwater heater (the inlet flow to the economizer) is sufficient. In some units, water is taken from the outlet of the feedwater pump. Attemperators can be single or multiple stage. The body of attemperators is vulnerable to thermal fatigue because of cold water droplets that continuously strike its hot surface. To reduce this effect, attemperators are equipped with a shock resist thermal sleeve. Sometimes steam and water attemperators are used in series for more effective control of the temperature of superheated steam particularly at low-load operations.

Example 3.3

Assume a steam generator that is designed to generate 50 kg/s superheated steam at 450°C and 40 bar. The steam generator at an off-design condition generates 20 kg/s superheated steam at 500°C and 40 bar. A water-spray attemperator is used to reduce the temperature of the steam to the design temperature. Determine the flow rate of the cold water (in kg/s) needed if the water at the inlet of the attemperator is at 140°C.

Given:

Superheated steam inlet to the attemprator: 500°C and 40 bar
Desired superheated steam outlet of the attemprator: 450°C and 40 bar
Superheated steam mass flow rate: 20 kg/s
Feedwater inlet to the attemperator: 140°C

Analysis:

Using the conservation of mass and energy laws:

$$\dot{m}_{st,in} h_{st,\,in} + \dot{m}_w h_w = \left(\dot{m}_{st,\,in} + \dot{m}_w\right) h_{st,\,out} \tag{3.8}$$

where $\dot{m}_{st,\,in}$ and $\dot{m}_w$ are the mass flow rates of the inlet steam and water, respectively, $h_{st,\,in}$ and $h_{st,\,out}$ are the enthalpy of the inlet and outlet steam flows, respectively, and h_w is the enthalpy of the inlet water. Note that while the pressure of the sprayed water is not given, we know that the pressure of the water is slightly higher than the pressure of the steam. Typically, the sprayed water is taken from the feedwater heaters and is directly transferred to the attemperator. This water is at a higher pressure than that of superheated steam due to the pressure losses of the steam flow in the steam generator tubes. If this pressure difference is not sufficient, a small pump can be used to pressurize the water flow. In this problem, however since we are using the saturated water property instead of the compressed water property, the water pressure will not have any impact on the solution.

$$\dot{m}_{st,\,in} = 20 \text{ kg/s}$$

$$h_{st,\,in} = h_{@T=500°\text{C and } P=40 \text{ bar}} = 3445.3 \text{ kJ/kg}$$

$$h_{st,\,out} = h_{@T=450°\text{C and } P=40 \text{ bar}} = 3330.3 \text{ kJ/kg}$$

$$h_w = h_{sat\ liq\ @T=140°\text{C}} = 589.13 \text{ kJ/kg}$$

Rearranging Equation 3.8,

$$\dot{m}_w = \frac{\dot{m}_{st,\,in}\left(h_{st,\,in} - h_{st,\,out}\right)}{\left(h_{st,\,out} - h_w\right)} = \frac{20(3445.3 - 3330.3)}{(3330.5 - 589.13)} = 0.84 \text{ kg/s}$$

3.6.3 Drum Water Level Control

The control of the drum water level is another critical factor for the safe operation of steam generators. A high level of water in the drum can impair the performance of the separators, which can result in the presence of water droplets in the superheaters and even turbine blades. A low water level in the drum can interfere with the water circulation in the waterwall tubes, which may result in tube overheating and failure. To regulate the drum water level (the controlled variable), the feedwater flow rate is controlled (the manipulated variable). In low-capacity steam generators or when large steam generators operate at low loads, this objective can be achieved simply by a level transmitter. This transmittercompares the actual water level and the set point and sends a signal to the inlet feedwater control valve to take an appropriate action to compensate any discrepancy. This simple control system however may not be able to control the drum water level in large steam generators. Due to the high evaporation rate in utility steam generators, a relatively large volume of bubbles is created over the water surface in the drum, which makes water level reading by the transmitter to be inaccurate. This problem is intensified during the load change when bubbles can swell or shrink due to pressure change. In order to address this problem, the outlet flows of steam, blowdown, and other outlets are also measured and used to balance the inlet feedwater flow rate with all outlet flows. In practice, an integrated open- and closed-loop control system is used to maximize the accuracy of water level control in the drum. Also, the drum is always equipped with a water level glass gauge for visual verification.

3.6.4 Safety and Relief Valves

Safety and relief valves are critical for the safe operation of high-pressure components of steam generators by protecting them from operating at pressures higher than safe operating pressures. They must pop when pressure exceeds a preset value and close when pressure is below the set point. They must be installed in a way that they can carry the sum of all steam flows. It should be impossible to isolate them from the component that they are protecting. Typically safety valves are referring to the valves for gas and vapor applications, while relief valves are referring to the valves for liquid applications. Safety relief valves can be used for both applications. The drum is always equipped with safety valves. The design pressure of the under-pressure equipment must be higher than the set point of safety valves.

3.7 Operation, Start-up and Shutdown, Manufacturing, and Maintenance of Steam Generators

The efficiency of a steam generator is strongly dependent on its design and the specifications of its components. However, a properly built, appropriately operated, and adequately maintained system can maintain and even improve the efficiency as well as reduce maintenance and improve safety, reliability, and availability of the entire plant. In this section, we will take look at these aspects of steam generators. Some aspects of what we cover in this section are applicable to other parts of steam power plants as well.

3.7.1 Operation of Steam Generators

While the proper operation of a steam generator requires complying with equipment-specific procedures, it involves much more than just that. The proper operation is all about coordinating, balancing, and optimizing the performance of all equipment at all time for safe operation of the system, for both personnel and equipment. The overall objective is to maximize the long-term profitability of the unit while meeting all emission and other environmental requirements. Regardless of all automations in the operation and control of modern systems, a major part of proper system operation depends on operator trainings. All system operators should be knowledgeable about all components and their relationships to each other, as well as their specifications, designs, and their performance limitations. The operators must be familiar and trained for all possible operational circumstances. Since some situations cannot be recreated during the safe operation of the system, advanced training simulators are commonly used during the training process. The best time for training is during the initial commissioning period of the system. The operation of steam generators depends on many factors and is always in coordination with other cycle components. Therefore, a more comprehensive discussion on the operation of steam generators will be covered in the coming chapters.

3.7.2 Start-up and Shutdown of Steam Generators

The start-up and shutdown of steam generators are considered as a part of steam generator's operational procedures. Rapid and safe start-up, shutdown, and load change are critical for the operation of power plants, especially in current markets where fast load-following characteristics are very important for the profitability of power plants (refer to Chapter 1). The start-up procedure depends on many factors, including the steam generator type (drum-type or once-through), fuel type, and combustion system, which will be covered in the following chapter. But regardless of these factors, the following considerations should be always taken into account when developing these procedures.

For drum-type steam generators, in early stages of start-up, since no steam is raised yet, there is no flow of steam through the superheater and reheater tubes. Similarly, there is no feedwater addition, so no flow through the economizer tubes. The absence of the cooling effects of these flows through the tubes necessitates the protection of these heat-transfer surfaces until adequate flows can be maintained. The economizer tubes are protected by the flow of feedwater through them without feeding the outlet flow to the drum. The water or steam at the outlet of the economizer is either vented or circulated until the rate of steam generation is high enough that the flow of replacement feedwater can provide sufficient tube cooling. The superheater and reheater tubes are protected against overheating by limiting the temperature of the flue gas flowing outside the tubes until an adequate flow rate of steam can be provided. In fact, this is one of the limiting factors for the rapid start-up of drum-type steam generators. Another equally important limiting parameter is the thermal stress on the drum's thick shell. To measure and control this thermal stress, the temperature differences between the drum top and bottom halves are measured by thermocouples. These temperature differences must be kept within an acceptable range determined by designers.

In once-through steam generators, it is vital to protect waterwall tubes and other high heat absorbing surfaces before firing starts by assuring the presence of a minimum feedwater flow through the tubes. As discussed before, this can be achieved by either discharging the outlet flow or temporarily circulating feedwater through a relatively small drum. Another important consideration for both types of steam generators is to bypass the turbine until appropriate superheated steam with acceptable temperature and pressure is generated in the steam generator (more on this topic in Chapter 6).

The steam generator shutdown procedure is less complicated than the start-up procedure. Depending on circumstances and urgency, a shutdown-can be either controlled shutdown or emergency shutdown.

The initial start-ups for a new steam generation system or after major redesign or repair are very important for a long-term healthy operation of the system. These start-ups may include visual inspection; water, steam, and chemical cleaning of tubes; equipment preparations and individual start-ups; hydrostatic testing (leak tests with water as filling liquid at the pressure of 1.5 times of the design pressure); pre-calibration of instruments and sensors; safety valves' testing and setting, and initial operation for procedure adjustments.

3.7.3 Manufacturing of Steam Generators

It is clear from the range of topics in this chapter and the following chapters that the manufacturing of a power plant in general and a steam generator in particular is very complex and requires many experts and a wide variety of tools and machines. There are three general approaches to the manufacturing process of a steam power plant. The first one is the turnkey approach where one contractor is in charge of the system design and supplying materials, construction, and installation of the power plant. Another approach is the system island approach where each major system, referred to as an island such as the steam generator system or the turbine system, is completed by an individual contractor. The last approach is the equipment package approach where each component is provided by a dedicated supplier.

3.7.4 Maintenance of Steam Generators

As noted several times already, modern steam generators are expected to operate safely and reliably for many years (sometimes up to 60 years). To achieve this level of service, rigorous scheduled and preventive maintenance programs are essential. These programs are particularly important for high-temperature sections where the majority of problems occur.

After a new power plant is commissioned, there is a short period with typically high maintenance requirements and low plant availability. This is because of both operators' unfamiliarity with the system and its limits and also immaturity of the equipment. After this initial period, a power plant can operate with low maintenance needs for a long while. This period can be between a few years to a few decades primarily depending upon the fuel type. After this period the plant starts to age and may need high maintenance. The condition assessment is a vital tool to keep the plant available in this period. In the condition assessment, the history of the system, nondestructive tests, and numerical analyses are used to determine the remaining life of components and required maintenance or replacements before components actually fail during operation. Nondestructive examinations (NDE) are procedures that are used to determine the properties of components without damaging them. There are a wide variety of these tests, including visual, liquid penetrant, ultrasonic, radiography, magnetic particles, acoustics, and eddy current tests.

3.8 Design of Steam Generators

The objective of any steam generator is to produce steam at the desired flow rate, temperature, pressure, and purity with the maximum recovery of thermal energy from combustion products. However, like design of any complex system, in order to design a steam generator, many input parameters are required. Acquiring these inputs starts through comprehensive feasibility studies well ahead of starting the design process. Some of these parameters are as follows:

- Plant's power generation capacity and possible future expansion
- Type of power plant, i.e., peak load, baseload, as well as its life expectancy
- Current and future availability of fuel(s)
- Current and future availability of water as the working fluid and for condenser cooling (much more on the latter in Chapter 8)
- Emission limitations
- Solid and liquid waste disposal requirements
- Land availability to build the plant considering all requirements during the power plant construction and operation (and possibly decommissioning)
- Possible utilization of system byproducts, including steam needs for processes (i.e., cogeneration, Chapter 10)
- Project-specific requirements such as project scope, project schedule, required performance guarantees, etc.

In collecting these information, the current and foreseeable future regulations and laws, particularly environmental regulations, should be taken into account. The information can be used to determine the power production capacity of the power plant, its technology, and the site location. This phase may be followed by acquiring construction and environmental permits. Specifically for steam generators, the customer may define performance requirements, including steam output conditions (pressure, temperature, and flow rate), feedwater conditions, fuel and ash analyses, load range, capacity, and efficiency. The above criteria and all other requirements should be communicated between interested parties, from owners to designers and suppliers, via specifications. These specifications establish the bases for the design of all major components. The objective of the steam generator design is to meet these specifications with a minimum overall cost.

The design of a steam generator is a multidisciplinary undertaking that involves not only many engineering fields, such as thermodynamics, heat transfer, fluid mechanics, material sciences, solid mechanics, control and instrumentations, process, structure, construction, and manufacturing engineering but also non-engineering fields, such as chemistry, economics, environmental science, and many others. However, the application of fundamentals of science and engineering is not sufficient for the design of a steam generator due to the complexity of the system. Many empirical data and practical experiences are required for a successful design of steam generators. The design should take into account steam requirements, efficiency of the system, fuel characteristics, environmental and other regulatory restrictions, operation and maintenance requirements, and local weather and geographical considerations. The design engineers should trade off these sometimes conflicting parameters considering capital costs and operating costs to create a design that balances them to achieve an optimum system with maximum profitability. Considering the long life expectancy of steam generators, an entire life cycle of the system should be integrated into the design process.

Generally speaking, in the design of a steam generator first the heat and material balance diagram is created to determine the requirements of major flows, i.e., air, fuel, feedwater, flue gas, and major components. When these requirements are established, heat-transfer surfaces can be calculated and all equipment can be sized. The design process of a steam generator typically follows the direction of flue gas from the furnace to the stack.

Considering the complexity of steam generators, it is natural to take advantage of computerized numerical and analytical tools in their design. These tools can be used for the evaluation and verification of a new design, modification of an existing system, prediction of the system performance, or addressing a localized problem.

Similar to any analysis, in a numerical analysis, we solve mathematical expressions of processes in a system, i.e., the governing equations, such as mass, energy, momentum, and species conservation laws, using boundary conditions through numerical methods (versus analytical ones). This typically involves solving a coupled system of ordinary and partial differential equations. A typical case is using numerical tools to provide 2- or 3-dimensional analyses of a complex system. It should be remembered that a numerical tool is as good as its assumptions and input data. It is imperative that the outputs of these tools are evaluated critically by means of their validation.[6]

3.9 Questions and Problems

3.9.1 Introduction

3.1. List some sources of thermal energy that can be used to raise steam in a steam generator.

3.2. When can electricity be used in a steam generator or water heater?

3.3. Why is water the working fluid in most power plants working based on a Rankine cycle?

3.4. What are the working fluids in Organic Rankine Cycles (ORC)?

3.5. When the thermal energy source for steam generation is provided by a-would-be-otherwise-wasted-heat from other processes, the steam generator is called the ....

3.6. What are some examples of non-utility applications of steam generators?

3.7. Every combustion-based steam generator can be divided into two sides. What are these sides? What are their functions?

3.8. Name the parts that the water-steam side of steam generators typical consist of.

3.9. What is the overall objective of a steam generator?

3.10. Which item is a desirable attribute of a steam generator?

a. To produce thermal energy from fuel combustion
b. To raise steam at desired conditions
c. To achieve maximum efficiency
d. To maximize reliability

[6] Always remember the old saying "garbage in, garbage out".

e. To minimize capital and operating costs
f. To minimize environmental impacts
g. All of the above

3.11. List some conventional and alternative fuels that can be combusted in steam generators.

3.12. Provide an example where steam generation is the secondary objective of a steam generator.

3.13. What are the differences between utility steam generators and industrial steam generators?

3.14. Units that produce steam for both electricity generation and industrial applications are referred to as …

3.15. List four ways to improve the efficiency of a steam power plant that are directly related to the thermodynamics of its steam generator.

3.9.2 Historical Development of Early Steam Generators

3.16. Why could shell-type or kettle boilers only generate saturated steam?

3.17. Why were fire-tube boilers more efficient than shell-type boilers?

3.18. Explain the difference between fire-box steam generators and Scotch marine steam generators.

3.19. Fire-tube boilers were slow in responding to load changes and hard to clean. Why?

3.20. Why did fire-tube boilers caused so much casualties and financial damages in the turn of the 20th century?

3.21. Two fire-tube boilers are made of the same metal operating at the same temperature. The diameter and operating pressure of one boiler is half of those for the other boiler. Determine the ratio of the thickness the larger boiler to that of the smaller one.

3.22. Why are shell-type and fire-tube steam generators no longer used in power generation plants?

3.23. Explain why shell-type boilers can still be safely used for some applications.

3.24. What is the main function of a drum?

3.25. What are the advantages of water-tube steam generators over fire-tube steam generators?

3.26. Identify if the following steam generators are fire tube or water tube. Explain main problems with fire-tube steam generators.

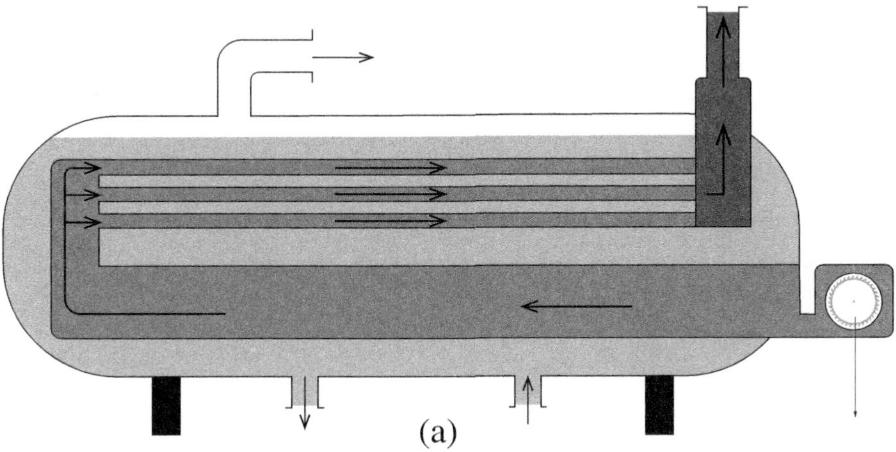

(a)

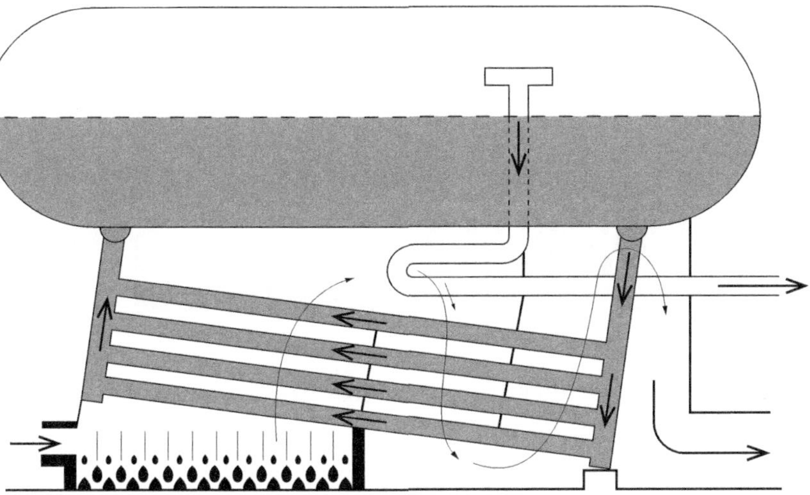

(b)

3.27. Why were Babcock & Wilcox steam generators called "nonexplosive boilers"?

3.28. In the following parallel-drum inclined water-tube steam generator (a Babcock & Wilcox steam generator), show where the drum, water tubes, headers, downcomer, riser, furnace, and superheater are.

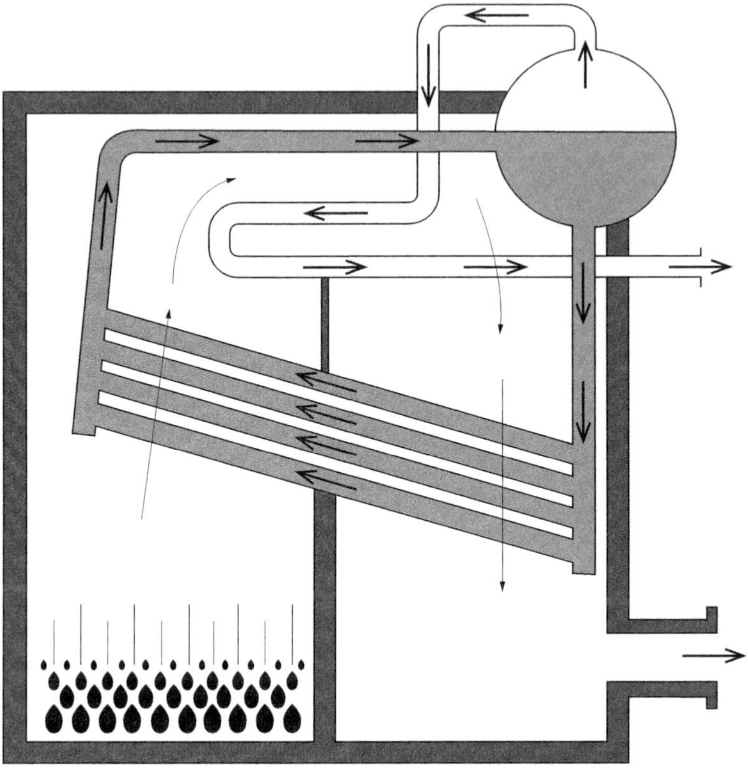

3.29. Compare a parallel drum and cross drum. What is the advantage of the latter over the former?

3.30. How does the inclination of water tubes assist water circulation?

3.31. What is the driving force for water circulation in natural circulation steam generators?

3.32. What is the function of baffles in the flue gas stream?

3.33. Why did bent-tube steam generators replace straight-tube steam generators?

3.34. What were the functions of mud drum(s)?

3.35. Explain how bent-tube Stirling steam generators worked.

3.36. A bent-tube Stirling steam generators with three steam drums and two water drums typically has .... tube banks.

a. 2

b. 3

c. 4

d. 5

3.37. Draw a simple schematic of a Milne boiler.

3.38. Why could the high rate of combustion cause problems in old steam generators?

3.39. In conventional steam generators, what are the main functions of waterwalls in a furnace?

3.40. Explain why at high-pressure steam generators waterwall tubes can provide the entire heating surface required for the evaporator.

3.41. Why is the heat of vaporization zero above the critical pressure of water?

3.42. Which component of a steam generator is used to increase the temperature of the inlet feedwater to the drum?

a. Reheater

b. Superheater

c. Economizers

d. Evaporator

3.43. How does the lack of feedwater heaters impact the performance of economizers?

3.44. Why did reheaters become a standard component of utility steam generators?

3.45. Why was the drum manufacturing a bottleneck for increase in the operating pressure of steam generators in the early 20th century?

3.46. What have been the highest operating temperature and pressure that ever attempted? Explain why the standard temperatures and pressures were settled at more moderate values.

3.47. Name four types of modern steam power plants (and steam generators). What are the typically operating pressure and temperature for each type?

3.9.3 Arrangement of Modern Steam Generators

3.48. Name the major components of a typical modern drum-type steam generator in the following schematic diagram.

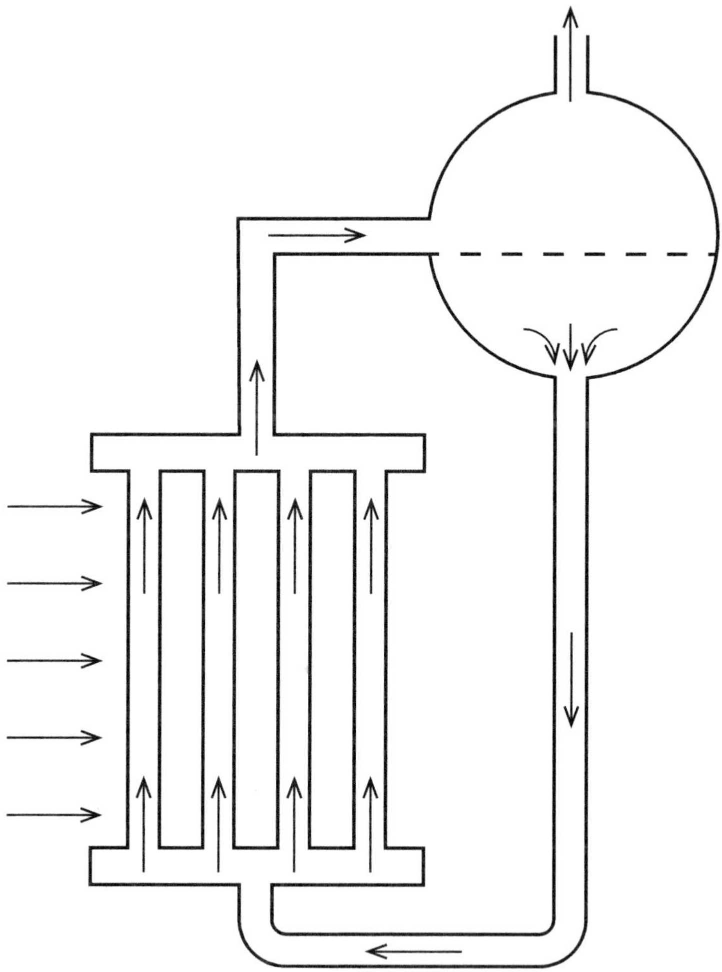

3.49. Describe an evaporator loop of a drum-type steam generator using the following schematic diagram.

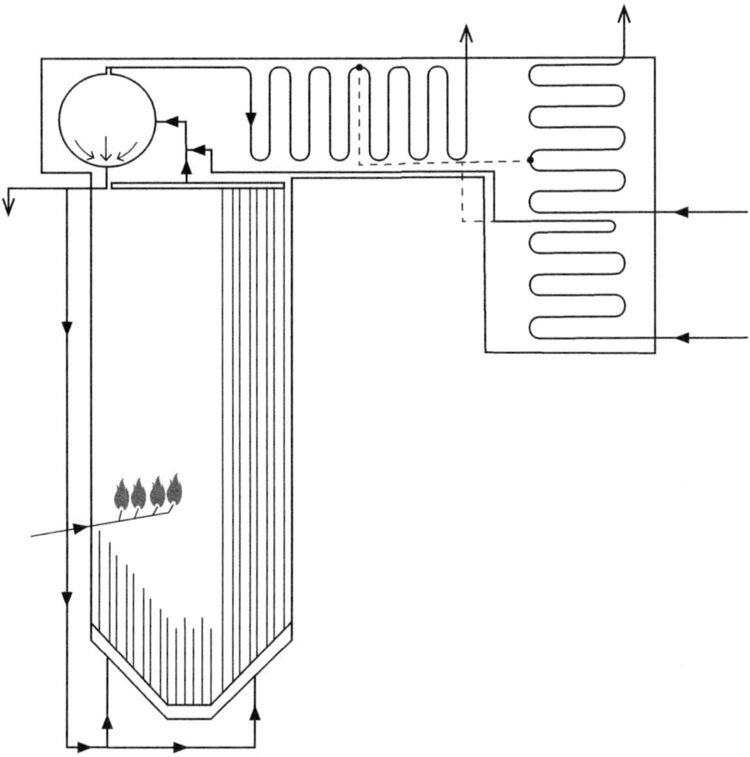

3.50. Use the following schematic diagram of a drum-type steam generator to explain how they operate. Label all major components in the diagram and show the direction on each flow.

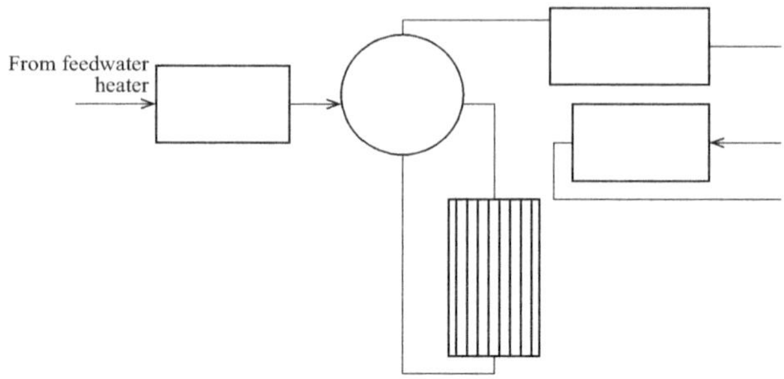

3.51. What are the differences between radiant and convection sections of steam generators?

3.52. On the schematic diagram of a steam generator for Question 3.50, mark where radiant and convection heat-transfer surfaces are.

3.53. A radiant section can be physically located in any part of a steam generator.

a. True

b. False

3.54. Determine if the following components of a steam generator can be a radiant, convection, or combination of two.

- Waterwalls:
- Reheater:
- Economizer:
- Screen:
- Superheater:

3.55. In heat transfer from combustion products to water or steam inside tubes within a steam generator, what heat-transfer mechanisms are involved?

3.56. In the following schematic of a tube, show the heat-transfer mechanism(s) for each part.

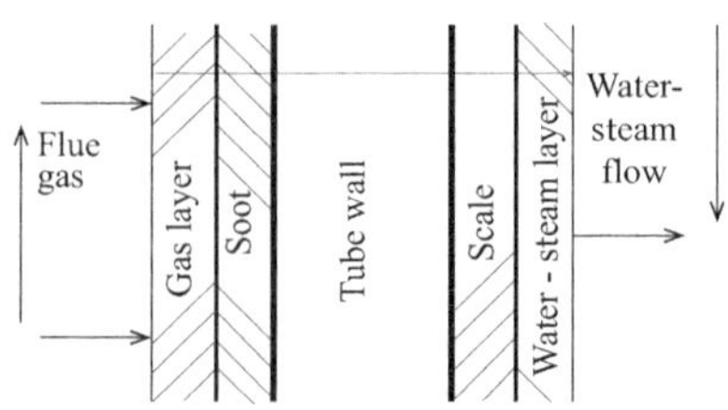

3.9.4 Components of Modern Steam Generators

3.57. In modern steam generators, all components of water side and combustion side are integrated into a single unit.

a. True

b. False

3.58. In modern steam generators, components are designed and manufactured by various manufacturers and then are assembled to form a single coordinated system.

a. True

b. False

3.59. Describe why in some steam generators, usually in industrial ones, waterwalls may not provide sufficient heat-transfer surface areas and an extra boiler bank may be required.

3.60. List different types of waterwall tube arrangement. Identify each in the following figure. Which one is the most common?

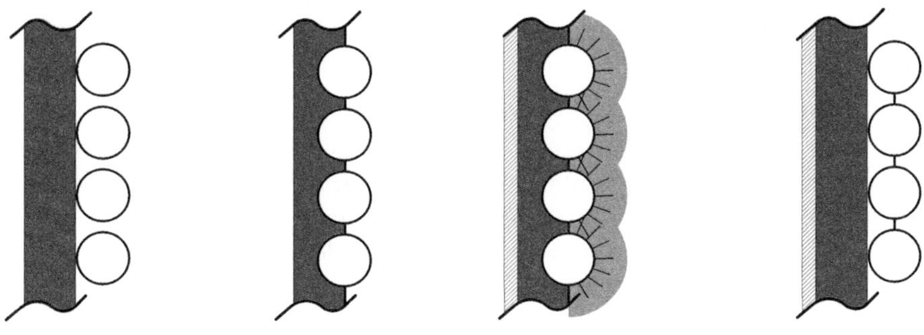

3.61. Why waterwall tubes can be made of relatively inexpensive carbon steel or low alloys?

3.62. The entire steam generator, including superheaters, reheaters, and economizer are enclosed by gas-tight tubes and no other structure is needed.

a. True

b. False

3.63. List three types of water boiling mechanisms. Describe each mechanisms. Which one is the most desirable and which one is the least desirable for water wall tubes? Why?

3.64. Explain why film boiling is characterized with a low heat-transfer coefficient while it occurs when there is a high-temperature gradient.

3.65. Why is the full evaporation of water in the tubes (100% quality of steam) undesirable?

3.66. In modern steam generators, how is the DNB delayed in waterwalls? Explain the mechanism.

3.67. All steam generators have one and only one drum.

a. True

b. False

3.68. List at least five different functions of a drum.

3.69. Describe how the quality of feedwater is maintained in a drum.

3.70. Explain how a drum functions as the accommodator of feedwater.

3.71. Briefly explain the manufacturing processes of drums.

3.72. List three tests conducted on drums (in that matter on any important pressure vessels) during manufacturing to ensure their integrity.

3.73. What is the boiler blowdown? What is its function?

3.74. Why is thermal stress a major challenge for the design and manufacturing of drums?

3.75. What is the function of internal shroud in a drum?

3.76. Which code is used for the design and manufacturing of steam generators and pressure vessels in the U.S., Canada, and many other countries?

3.77. What is the ASME Boiler & Pressure Vessel Code (BPVC)?

3.78. Determine the minimum wall thickness required for a pressure vessel with an internal radius of 5 ft operating at an internal pressure of 1000 psig at the operating temperature of 650°F. The vessel is made of carbon steel SA-516 Grade 70 and all welded joint are radiographed. Use the ASME BPVC Section VIII, Division 1, for the calculation. Also, determine the safety factor of the design if the minimum tensile stress of SA-516 Grade 70 is 70000 psi. Hint: The allowable stress of carbon steel SA-516 Grade 70 at the temperature of 650°F is 18,800 psi.

3.79. Determine the minimum wall thickness (in cm) required for a pressure vessel with an internal radius of 2.5 m operating at an internal pressure of 180 bar at the operating temperature of 400°C. The vessel is made of carbon steel SA-516 Grade 55 and all welded joint are radiographed. Use the ASME BPVC Section VIII, Division 1 for the calculation. Also, determine the safety factor of the design if the minimum tensile stress of SA-516 Grade 55 is 55,000 psi. Hint: The allowable stress of carbon steel SA-516 Grade 55 at the temperature of 400°C is 12,912 psi.

3.80. Why is the presence of water droplets in the saturated water leaving the drum toward superheaters undesirable or even dangerous?

3.81. Why is the presence of steam in the water outlet from the drum to the waterwall tubes undesirable?

3.82. What are the mechanisms to separate steam from boiling water in the drum?

3.83. What is the driving force in the gravity separation of steam in a drum?

3.84. Why is gravity separation alone ineffective in modern utility steam generators?

3.85. The modern drums are commonly equipped with two series of steam separators, primary separators and secondary separators.

a. True

b. False

3.86. What are the two main functions of primary steam separators?

3.87. List common primary steam separators in drums. Describe the basic operation principle of each.

3.88. What are the two main functions of secondary steam separators?

3.89. Why is a driving pressure needed to circulate saturated water through waterwall tubes?

3.90. What are the two types of water circulation mechanisms?

3.91. How is the driving pressure created in natural circulation steam generators? What factors impact this driving pressure?

3.92. What are the sources of pressure drop in a water circulation loop of drum-types steam generator?

3.93. Why is it critical to reduce pressure drops in natural circulation steam generators while it is not that important for forced-circulation steam generators?

3.94. Natural circulation steam generators have an inherent self-control mechanism. Why? Is this characteristic always desirable?

3.95. Why do different tubes in waterwalls have differing driving pressures? What parameters impact these driving pressures?

3.96. What happen if there are nonuniform flow rates in different tubes of waterwalls?

3.97. Explain how water is circulated in forced-circulation steam generators.

3.98. Why do forced-circulation steam generators can have more rapid start-up and load change compared to those of natural circulation steam generators?

3.99. Which steam generator is more suitable for a power plant that is under cycling loads?

a. Forced-circulation steam generators

b. Natural circulation steam generators

3.100. What is the function of orifices in the inlet to waterwall tubes?

3.101. Explain why superheaters and reheaters require special materials.

3.102. Why reheaters can be made of less expensive materials compared to superheaters?

3.103. Superheater and reheater tubes are ……. Why?

a. internally ribbed

b. externally finned

c. both internally and externally bare

3.104. Describe why the selection of superheater and reheater tube diameters is a trade-off between stress and pressure drop.

3.105. Explain how and why the responses of convection or radiation superheaters to load variations are different. What are the implications of this difference?

3.106. What are the three superheater and reheater support systems? What are the pros and cons of each?

3.107. Identify the pendant, inverted, and horizontal support system for superheaters and reheaters in the following figure.

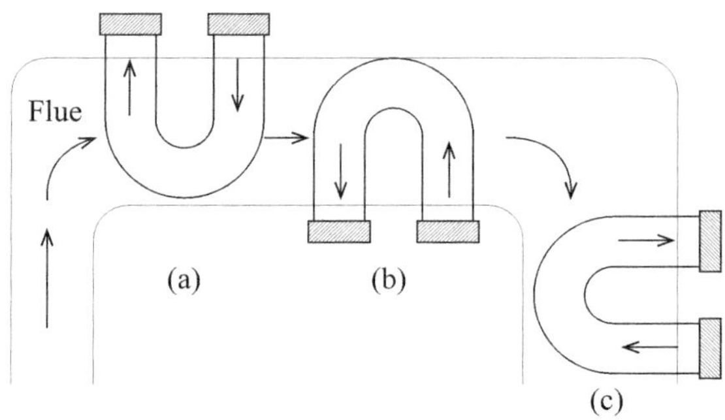

3.108. In a subcritical utility steam generator, the feedwater from the last high-pressure feedwater heater enters the economizer at the temperature and pressure of 150°C and 180 bar. The superheated steam leaves the superheater at the temperature of 550°C. Determine the relative heat absorption in the economizer, evaporator, and superheater. Assume the pressure losses in the steam generator components are negligible.

3.109. Economizer tubes can be ……. Why?

a. internally ribbed

b. externally finned

c. both internally and externally bare

3.110. Explain why a large surface area is required in an economizer.

3.9.5 Once-Through Steam Generators

3.111. What is the main difference between a once-through steam generator and a drum-type steam generator?

3.112. Why are once-through steam generators referred to as universal-pressure (UP) steam generators?

3.113. Which type of steam generators can operate at subcritical operating pressures?

a. Drum-type steam generators

b. Once-through steam generators

c. Both

d. None

3.114. Which type of steam generators can operate at supercritical operating pressures?

a. Drum-type steam generators

b. Once-through steam generators

c. Both

d. None

3.115. Why is operating once-through steam generators during low-load operation, start-up, or shutdown challenging? How can the problem be addressed?

3.116. Why do once-through steam generators require purer feedwater than drum-type steam generators?

3.117. Explain why once-through steam generators typically are equipped with spiral wound waterwall tubes. Also, explain why typically only the bottom half of waterwall tubes are angled.

3.9.6 Control of Steam Generators

3.118. Explain the basics of closed-loop and open-loop control systems. What are the pros and cons of each system?

3.119. Is the cruise control in your car an open or closed control system? Briefly explain how it works.

3.120. What type of closed-loop control system is used in an old-fashioned house HVAC system controlled by a thermostat? Briefly explain how it works.

3.121. Consider the lighting level in your office as the controlled variable. Compare using a light switch and a dimmer switch as your control strategy. Which one uses on-off control? What about proportional control?

3.122. Consider an object hung to a vertical spring as a simple proportional control. Explain this control mechanism if the disturbance is adding to the mass of the object. Explain the concept of the offset in this proportional control system.

3.123. The fly-ball governor is one of the oldest examples of proportional control systems. Identify the controlled and manipulated variables in this system and briefly explain how the system works.

3.124. What are the two most common temperature measurement devices that are used in steam power plants? Describe the operational principles of each device. What are their advantages and limitations?

3.125. Briefly describe the two general approaches used to control the temperature of superheated steam leaving superheaters and reheaters.

3.126. Explain the operational principles of attemperator or desuperheater. Draw schematic of the system. Where are they typically installed? Why?

3.127. Explain why it is dangerous if the water level in the drum is too high or too low.

3.128. Describe how the water level in the drum is controlled.

3.129. What is the function of safety and relief valves? What is their difference?

3.130. A supercritical steam generator is designed to generate 400 kg/s superheated steam at 550°C and 250 bar. The steam generator at an off-design conditions generates 300 kg/s superheated steam at 600°C and 250 bar. A water-spray attemperator is used to reduce the temperature of the steam to the design temperature. Determine the flow rate of the cold water (in kg/s) needed if the water at the inlet of the attemperator is at 150°C.

3.9.7 Operation, Start-up and Shutdown, Manufacturing, and Maintenance of Steam Generators

3.131. How and why the drum is a limiting factor in the start-up of a steam generator?

3.132. How are the economizer, superheater, and reheater tubes protected during early stages of start-up process in a drum-type steam generator?

3.133. How are the heat absorbing tubes protected during early stages of start-up process in a once-through steam generator?

3.134. Explain three stages of steam generators' life in terms of required maintenance.

3.135. List some parameters that are needed before starting the design of a steam generator.

3.136. What are the three contractual approaches to constructing a steam power plant (in that matter any plant)?

3.137. Why should the results of numerical tools be treated cautiously?

3.9.8 General

3.138. In evaporator tubes, what is the departure from nucleate boiling (DNB) and how can it be eliminated or delayed?

3.139. What is the function blowdown in a drum?

3.140. Compare the following terms: [6]

- Utility steam generators versus industrial steam generators
- Subcritical steam generators versus supercritical steam generators
- Water-tube drum-type steam generators versus once-through steam generators

3.141. Explain why grooved, ribbed, and corrugated tubes are utilized in the boiler/water walls.

3.142. Determine the minimum wall thickness required for a drum with an internal radius of 4 ft operating at an internal pressure of 2000 psig. The drum is made of carbon steel SA-516 Grade 70 and all welded joint are radiographed. Use the ASME BPVC Section VIII, Division, 1 for the calculation. Also, determine the safety factor of the design if the minimum tensile stress of SA-516 Grade 55 is 70,000 psi.

3.143. Explain the most common superheated steam temperature control system in steam generators (the location of the control system and its main operating mechanism).

3.144. Which one operates at more severe operating conditions (both temperature and pressure): superheater or reheater? Why?

3.145. Explain various steam separation mechanisms in a drum.

3.146. All modern steam power plants have superheater.

a. True

b. False

3.147. Sketch the schematic of a modern steam generation system showing all major components and flows.

3.148. What are the advantages and disadvantages of natural circulation and forced-circulation steam generators?

3.149. Why bent-tube steam generators are better than straight-tube ones?

3.150. What is the difference between a cross drum steam generator and a longitudinal drum steam generator?

3.151. Explain why fire-tube steam generators are no longer used.

3.152. Briefly explain the origins of the ASME.

3.153. What is a Scotch marine steam generator?

3.154. Explain the functions of waterwalls and economizers.

3.155. Describe centrifugal separators in the drum (function and operation mechanism).

3.156. List the major components of the water-steam side of a steam generator.

3.157. Once-through steam generators can be used for … power plants.

a. subcritical

b. supercritical

c. both subcritical and supercritical

3.158. The departure from nucleate boiling (DNB) in a boiler is ….

a. desirable

b. undesirable

3.159. In the following steam generator, show where steam and water are separated.

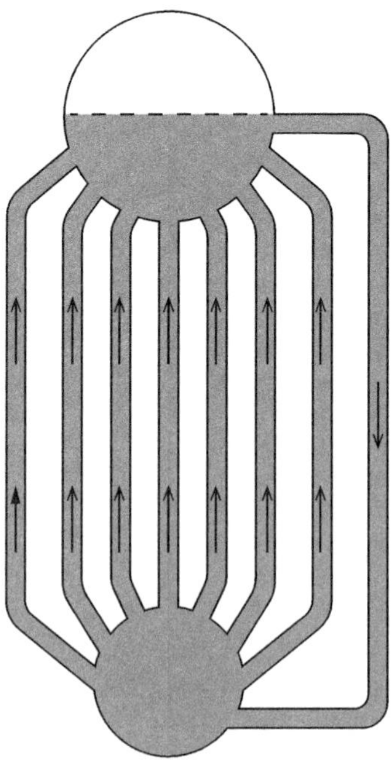

3.160. In the above steam generator (previous question), show where the following type of pipe is used. Why?

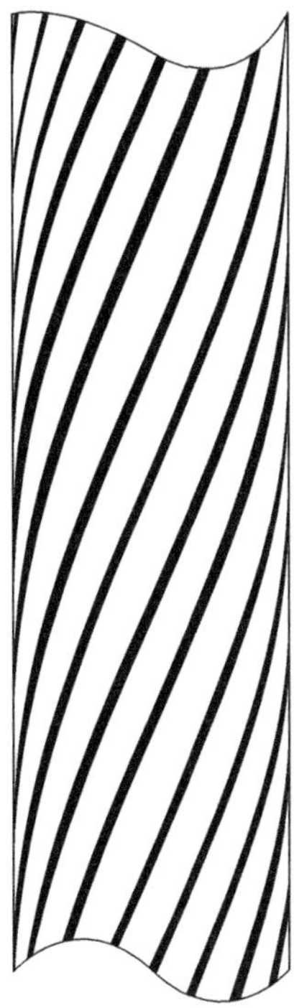

4

Steam Generators: Fire Side

Abstract

In this chapter, we deal with the fire side of steam generators. We will start with the thermodynamics of the combustion process. Then, various fuels types, such as coal, biomass, solid wastes, natural gas, and hydrocarbon liquid fuels and their properties will be discussed. Their extraction, transportation, preparation, and storage processes will be extensively discussed. Although the share of coal in power generation is declining in many countries, we will extensively discuss this fuel due to its historical significance. Also, most technologies that are being used in the combustion of other fuels have been originated in technologies that were initially developed for coal. Finally, coal is still one of the most important fuels, if not the most important one, on the global scale. Various coal combustion technologies, including mechanical stoker, pulverized coal, cyclone-furnace firing, and fluidized-bed combustion will be covered. It will be also explained how similar technologies are adapted for other fuel types. All components and their functions for each technology will be described. Some emission reduction technologies that are integrated into combustion systems will be evaluated. Many auxiliary components related to combustion systems and their operational principles will be also discussed.

LEARNING OUTCOMES

After successful completion of this chapter, you will be able to:

- conduct thermodynamic calculations of combustion reactions
- discuss various types of solid, liquid, and gaseous fuels and their properties
- describe extraction, transportation, preparation, and storage processes of various fuels
- identify various combustion technologies for different fuel types
- describe components for each technology and their functions
- describe many auxiliary components related to combustion systems and their operational principles

4.1 Introduction

In Chapter 3, it was explained that in modern utility steam generators all components required to generate steam are integrated into a single unit. The components of the water-steam side of a steam generator, including waterwalls, drum, superheaters, reheater, and economizer were also discussed in Chapter 3. In this chapter, we will focus on the fire side of steam generators. A great deal of this chapter will be spent on the combustion process and fuels. We will then look at how these fuels can be combusted in steam generators.

While the share of coal in electricity generation is declining in many countries, particularly in industrialized countries, it is still one of the most important fuels, if not the most important one, on a global scale. Also, most technologies that are currently being used in the combustion of other fuels one way or another have been originated in technologies that were originally developed for coal combustion. Moreover, since coal has contributed so much to our society and the human civilization that we share today (positively and negatively), it deserves our attention. Finally, our future and civilization may depend on how we get rid of our "addiction" to this cheap source of energy. For these reasons, we will extensively cover coal.

The combustion process (fire) converts the chemical energy of fuels to energy in the form of heat and light. Historical evidence show that our ancestors were probably able to control fire by observing naturally occurring fire as early as 1.9 million years ago,[1] certainly since 400,000 years ago, and widely since 100,000 years ago. To better control fire and to improve output heat, hearths with air-intake channels were in use since more than 40,000 years ago. This capability resulted in better diet due to cooked food; ability to move and live in colder climates; better protection against enemies; ability to preserve food by drying and smoking; ability to transform materials to make products such as glass, pottery, and bricks as well as metal smelting; extended active life to night by providing light; and providing social focus point; among other benefits. "Harnessing the power of fire was an important part of the development of civilization and had influenced every part of human life".[2] Fire provided human beings with a tool to control, alter, and shape nature and eventually conquer the world. However, there was no understanding of the nature of the combustion process and its properties until the late 19th century.

---

1 Current Anthropology Volume 40, Number 5, December 1999. The Raw and the Stolen Cooking and the Ecology of Human Origins, by Richard W. Wrangham, James Holland Jones, Greg Laden, David Pilbeam, and Nancy Lou Conklin-Brittain.

2 Science, Adam Hart-Davis (editor), 2009, DK publishing.

4.2 Thermodynamics of Ideal Gases

If there is a chemical reaction in a thermodynamic process, such as the combustion process as the most well-known one, the principles used to thermodynamically analyze the process using the first and second laws of thermodynamics do not change. However, some new terms and concepts should be considered to adjust our approaches. In this section, the concepts of thermodynamic properties, the mass conservation law, and the first and second laws of thermodynamics will be extended to include processes involving chemical reactions. For this purpose, some modifications will be introduced to the determination of internal energy, enthalpy, and entropy. While the presented materials will be about the combustion process, they can be similarly applied to other chemical reactions. Since in the combustion process always a mixture of several gases is involved, it is appropriate to review the ideal gas law and the properties of a mixture of ideal gases here.

REMINDER BOX 4.1

Ideal gas lawIn the ideal or perfect gas model, it can be assumed that gas molecules behave independently and are not affected by each other. The ideal gas equation of state is developed experimentally and can be expressed in several forms including

$$PV = n\bar{R}T \tag{4.1}$$

where P is pressure (kPa), V is volume (m^3), T is temperature (K), n is the number of moles of the gas, and $\bar{R}$ is the universal gas constant for all gases:

$$\bar{R} = 8.314\,\text{kJ} / \text{kmol}.K = 1.986\,\text{Btu} / \text{lbmol}.R = 1545\,\text{ft lbf} / \text{lbmol}.R$$

The gas constant for a particular gas (R) is defined as

$$R = \frac{\bar{R}}{M} \tag{4.2}$$

where M is the atomic or molecular weight of the gas (kg/kmol or lb/lbmol). The gas constant for air is

$$R_{\text{air}} = 0.2870\,\text{kJ/kg}\,K = 0.06855\,\text{Btu/lb}\,R$$

The relationship between the number of moles and the mass of the gas is

$$n = \frac{m}{M} \tag{4.3}$$

where m is the mass of the gas (kg or lb). Considering Equation 4.3, the ideal gas equation (Equation 4.1) can be rearranged as

$$PV = mRT \tag{4.4}$$

$$Pv = RT \tag{4.5}$$

where v is specific volume of gas (m^3/kg or ft^3/lb). Note that by considering the compressibility factor, similar equations can be developed and used for non-ideal gases. Please refer to your thermodynamic textbook for further information.

4.2.1 Ideal Gas Mixture

In many applications, the matter of interest is not a pure substance; instead, it is a mixture of pure matters. Unlike pure substances, due to unlimited combinations of substances that can be created by mixing pure matters, it is not possible to present properties of mixtures in the form of property tables, except for certain mixtures such as air. The objective of this review is to determine the properties of a mixture by using the properties of the pure substances that the mixture is made of and measurable properties of the mixture such as pressure, temperature, volume, mass, and composition.

In order to identify a mixture, its composition is needed that can be presented by either mass or moles of each component in the mixture. However, it is more convenient to present the mixture composition relative to the total mass or total number of moles so the composition is independent of the total amount of the mixture. For this purpose, the mass fraction and molar fraction can be defined.

For a mixture of N components of pure substances, the total mass and the total number of moles are

$$m_t = m_1 + m_2 + \cdots + m_N = \sum_{j=1}^{j=N} m_j \tag{4.6}$$

$$n_t = n_1 + n_2 + \cdots + n_N = \sum_{j=1}^{j=N} n_j \tag{4.7}$$

where m_j and n_j represent the mass and the number of moles of the jth component of the mixture.

The mass fraction (c) and mole fraction (y) of each component are defined as follows:

$$c_i = \frac{m_i}{m_t} \tag{4.8}$$

$$y_i = \frac{n_i}{n_t} \tag{4.9}$$

Dividing Equations 4.6 and 4.7 to the total mass and the total number of moles of the mixture, respectively,

$$\sum_{j=1}^{j=N} c_j = 1 \tag{4.10}$$

$$\sum_{j=1}^{j=N} y_j = 1 \tag{4.11}$$

Using the mass- and mole-based representations of the composition of the mixture is also called the gravimetric analysis and molar (or volumetric) analysis, respectively. It is important to be able to convert mass and molar fractions to each other. The relationship between mass and the number of moles of a component of the mixture is

$$n_i = \frac{m_i}{M_i} \tag{4.12}$$

where M_i is the molecular weight of component i of the mixture in kg/kmol or lb/lbmol. For the mass fraction of component I,

$$c_i = \frac{m_i}{m_t} = \frac{m_i}{\sum_{j=1}^{j=N} m_j} = \frac{n_i M_i}{\sum_{j=1}^{j=N}\left[n_j M_j\right]} = \frac{n_i M_i / n_t}{\sum_{j=1}^{j=N}\left[n_j M_j / n_t\right]} = \frac{y_i M_i}{\sum_{j=1}^{j=N}\left[y_j M_j\right]} \tag{4.13}$$

Similarly, for the mole fraction of component I,

$$y_i = \frac{n_i}{n_t} = \frac{n_i}{\sum_{j=1}^{j=N} n_j} = \frac{m_i / M_i}{\sum_{j=1}^{j=N}\left[m_j / M_j\right]} = \frac{m_i / (M_i m_t)}{\sum_{j=1}^{j=N}\left[m_j / (M_j m_t)\right]} = \frac{c_i / M_i}{\sum_{j=1}^{j=N}\left[c_j / M_j\right]} \tag{4.14}$$

The average or apparent molecular weight of the mixture can be found as a function of the molar fraction and molecular weight of the components as follows:

$$M_{\text{mix}} = \frac{m_t}{n_t} = \frac{\sum_{j=1}^{j=N} m_j}{n_t} = \frac{\sum_{j=1}^{j=N}\left[n_j M_j\right]}{n_t} = \sum_{j=1}^{j=N}\left[n_j M_j / n_t\right] = \sum_{j=1}^{j=N}\left[y_j M_j\right] \tag{4.15}$$

Similarly, the average molecular weight of the mixture can be found as a function of the mass fraction and molecular weight of the components as follows:

$$M_{\text{mix}} = \frac{m_t}{n_t} = \frac{m_t}{\sum_{j=1}^{j=N} n_j} = \frac{m_t}{\sum_{j=1}^{j=N}\left[{m_j}/{M_j}\right]} = \frac{1}{\sum_{j=1}^{j=N}\left[{m_j}/{m_t M_j}\right]} = \frac{1}{\sum_{j=1}^{j=N}\left[{c_j}/{M_j}\right]} \tag{4.16}$$

4.2.2 Dalton Model

In an ideal gas, the intermolecular forces and the volume of gas molecules are negligible. These properties of ideal gases lead to the Dalton model. In an ideal gas mixture, according to the Dalton model, each component of the mixture behaves as if it occupies the entire volume by itself and it exists separately and independently at the temperature and volume of the mixture. If an ideal gas mixture is composed of two gases, *A* and *B*, according to the Dalton model

$$T_A = T_B = T_{\text{Mix}} \text{ and } V_A = V_B = V_{\text{Mix}} \tag{4.17}$$

$$P_A V_A = n_A \bar{R} T_A,\ P_B V_B = n_B \bar{R} T_B,\ P_{\text{Mix}} V_{\text{Mix}} = n_{\text{Mix}} \bar{R} T_{\text{Mix}} \tag{4.18}$$

Note that P_A represents the partial pressure of gas *A*. Dividing two sides of Equation (4.18)

$$\frac{P_A V_A}{P_{\text{Mix}} V_{\text{Mix}}} = \frac{n_A \bar{R} T_A}{n_{\text{Mix}} \bar{R} T_{\text{Mix}}} \tag{4.19}$$

$$\frac{P_A}{P_{\text{Mix}}} = \frac{n_A}{n_{\text{Mix}}} = y_A \tag{4.20}$$

In general, the partial pressure of each component of the mixture is equal to the total pressure of the mixture multiplied by the mole fraction of the component.

$$P_i = y_i\ P_{\text{Mix}} \tag{4.21}$$

Also, it can be shown that the total pressure of the mixture is equal to the sum of the partial pressure of its components.

$$n_{\text{Mix}} = n_A + n_B \tag{4.22}$$

$$\frac{P_{\text{Mix}} V_{\text{Mix}}}{\bar{R} T_{\text{Mix}}} = \frac{P_A V_A}{\bar{R} T_A} + \frac{P_B V_B}{\bar{R} T_B} \tag{4.23}$$

$$P_{\text{Mix}} = P_A + P_B = y_A P_{\text{Mix}} + y_B P_{\text{Mix}} \tag{4.24}$$

In general, for an ideal mixture of N gases,

$$P_{\text{Mix}} = \sum_{j=1}^{j=N} P_j = \sum_{j=1}^{j=N} y_j P_{\text{Mix}} \tag{4.25}$$

Example 4.1

The temperature and pressure of air in a $2\,\text{m}^3$ container are 350 K and 1 bar. If the nitrogen mole fraction of the air is 79%, determine the temperature, pressure, volume, specific volume, and mass of the nitrogen in the air.

Solution

$$T_{N_2} = 350\,\text{K}$$

$$V_{N_2} = 2\,\text{m}^3$$

$$v_{N_2} = \frac{\overline{R}T}{MP} = \frac{8.314\left[\dfrac{\text{kJ}}{\text{kmol}\cdot\text{K}}\right]\times 350\ [\text{K}]}{28.01\left[\dfrac{\text{kg}}{\text{kmol}}\right]\times 79\ [\text{kpa}]} = 1.315\ \frac{\text{m}^3}{\text{kg}}$$

$$m_{N_2} = \frac{V}{v_{N_2}} = \frac{2}{1.315} = 1.521\,\text{kg}$$

4.2.3 Internal Energy, Enthalpy, and Entropy of Ideal Gas Mixture

Other properties of an ideal gas mixture can be determined using the respective properties of the components at the condition at which the component exists in the mixture.

$$U_{\text{Mix}} = U_1 + U_2 + \cdots + U_N = \sum_{j=1}^{j=N} U_j \tag{4.26}$$

$$H_{\text{Mix}} = H_1 + H_2 + \cdots + H_N = \sum_{j=1}^{j=N} H_j \tag{4.27}$$

$$S_{\text{Mix}} = S_1 + S_2 + \cdots + S_N = \sum_{j=1}^{j=N} S_j \tag{4.28}$$

Dividing Equations 4.26–4.28 to the total mass of the mixture

$$u_{\text{Mix}} = \frac{m_1}{m_{\text{Mix}}} u_1 + \frac{m_2}{m_{\text{Mix}}} u_2 + \cdots + \frac{m_N}{m_{\text{Mix}}} u_N = \sum_{j=1}^{j=N} c_j u_j \tag{4.29}$$

$$h_{\text{Mix}} = \frac{m_1}{m_{\text{Mix}}} h_1 + \frac{m_2}{m_{\text{Mix}}} h_2 + \cdots + \frac{m_N}{m_{\text{Mix}}} h_N = \sum_{j=1}^{j=N} c_j h_j \tag{4.30}$$

$$s_{\text{Mix}} = \sum_{j=1}^{j=N} c_j s_j \tag{4.31}$$

Similar equations can be developed for the properties of a mixture based on the mole fraction.

$$\bar{u}_{\text{Mix}} = \sum_{j=1}^{j=N} y_j \bar{u}_j \tag{4.32}$$

$$\bar{h}_{\text{Mix}} = \sum_{j=1}^{j=N} y_j \bar{h}_j \tag{4.33}$$

$$\bar{s}_{\text{Mix}} = \sum_{j=1}^{j=N} y_j \bar{s}_j \tag{4.34}$$

Using a similar approach for other properties,

$$c_{v,\text{Mix}} = \sum_{j=1}^{j=N} c_j c_{v,j} \tag{4.35}$$

$$c_{P,\text{Mix}} = \sum_{j=1}^{j=N} c_j c_{P,j} \tag{4.36}$$

$$R_{\text{Mix}} = \sum_{j=1}^{j=N} c_j R_j \tag{4.37}$$

$$k = \frac{c_{P,\text{Mix}}}{c_{v,\text{Mix}}} \tag{4.38}$$

and

$$\bar{c}_{v,\text{Mix}} = \sum_{j=1}^{j=N} y_j \bar{c}_{v,j} \tag{4.39}$$

$$\bar{c}_{P,\text{Mix}} = \sum_{j=1}^{j=N} y_j \bar{c}_{P,j} \tag{4.40}$$

$$\bar{R}_{\text{Mix}} = \sum_{j=1}^{j=N} y_j \bar{R}_j \tag{4.41}$$

Instead of all of the above equations, the following general equation can be written, where x can represent any property:

$$\bar{x} = Mx \text{ and } \bar{x}_i = M_i x_i \tag{4.42}$$

Note that the properties of all the components are evaluated at the temperature of the mixture and the partial pressure of the component.

4.3 Thermodynamics of Combustion Process

4.3.1 Combustion Reactions

Combustion is rapid oxidation of matters, notably carbon, hydrogen, sulfur, and their compounds.

$$\text{reactants (fuel + oxidizer)} \rightarrow \text{products} \tag{4.43}$$

In a complete combustion process, all fuel molecules are oxidized, e.g., all the carbon, hydrogen, and sulfur in the fuel are converted to carbon dioxide, water, and sulfur dioxide in the combustion products. If these conditions are not fulfilled, combustion is incomplete. The reactions for the complete combustion of carbon, hydrogen, and sulfur are

$$\text{C(fuel)} + \text{O}_2\,\text{(oxidizer)} \rightarrow \text{CO}_2\,\text{(product)} \tag{4.44}$$

$$H_2 + \frac{1}{2}O_2 \rightarrow H_2O \tag{4.45}$$

$$S + O_2 \rightarrow SO_2 \tag{4.46}$$

Similar to any other process, in combustion, the mass is conserved and the mass of the elements before and after the reaction remains the same. This means the number of atoms of each element in the reactants and products should be the same even though they may be within different molecules. For combustion of carbon,

$$1\,\text{mole}\,C + 1\,\text{mole}\,O_2 \rightarrow 1\,\text{mole}\,CO_2 \tag{4.47}$$

$$12\,\text{kg}\,C + 32\,\text{kg}\,O_2 \rightarrow 44\,\text{kg}\,CO_2 \tag{4.48}$$

or

$$1\,\text{kg}\,C + 2.67\,\text{kg}\,O_2 \rightarrow 3.67\,\text{kg}\,C \tag{4.49}$$

As is evident in these reactions, the number of moles of reactants and products is not conserved but the number of moles of each element is conserved. In combustion reactions, the numerical coefficients of reactants and products are called the **stoichiometric coefficients**. Note that the number of molecules associated with the number of moles according to Avogadro law.

Similarly, for the combustion of methane (CH_4)

$$CH_4 + 2O_2 \rightarrow CO_2 + 2H_2O \tag{4.50}$$

In this reaction, H_2O in the combustion products can be in the form of steam, liquid water, or even ice depending on the temperature and pressure of the products.

So far pure oxygen is considered as the oxidant for combustion, but in the real world, for the majority of applications, air is the oxidant and not pure oxygen. The typical composition of the dry atmospheric air in the molar or volume basis is

O_2: 21%

N_2: 78%

Ar+other gases*: 1%

* such as carbon dioxide, neon, helium, hydrogen, ozone, etc.

Usually argon and other gases in air are neglected and considered as a part of nitrogen, which is then referred to as **atmospheric nitrogen**. In this case, the composition of air is (refer to Example 4.2)

O_2: 21%
N_2: 79%
Ar (argon)+other gases: 0%

Note that for the humid air (or the wet atmospheric air) the vapor content of the air should be also taken into account. For the air at the temperature of 25°C and humidity of 60%, the vapor content is 1.4% per kg of dry air.

Example 4.2

There are several ways to describe the composition of air and take it into account in thermodynamic calculations. The atmospheric air is composed of several gases, e.g., N_2, O_2, Ar, and, as well as water vapor, dust, pollutions, and other contaminations. When water vapor and all contaminations are removed, it is called the standard air composition or dry air (Table 4.1). Typically, the composition of air is assumed N_2, 78%; O_2, 21%; and Ar 1%. Commonly argon is neglected in air and is considered as a part of nitrogen, which is called atmospheric nitrogen. Then, the air composition will be N_2, 79% and O_2, 21%. Determine the mass fraction of each component and the average molecular weight of air in each case. In Table 4.1, assume all other components (last row of the table) as helium.

TABLE 4.1
Standard Air Composition

Component	Mole Fraction (%)
Nitrogen	78.09
Oxygen	20.95
Argon	0.92
Carbon dioxide	0.03
Helium and others	0.01

Given:
The composition of air is given based on the molar analysis.
Find:
The mass fraction of each component and the average molecular weight of the mixture.
Analysis:
Table 4.2 summarizes the calculation for the atmospheric air. The values for the molecular weight of each component (third column) can be found from Table A.5. The fourth column is the result of the second column multiplied by the third column. The sum of the values in the fourth column is equal to the average molecular weight of the mixture. The fifth column shows the mass fraction of each component.

For air with the composition of O_2, 21%; N_2, 78%; and Argon 1%., the mass fraction and the average molecular weight of the mixture can be calculated as follows:

TABLE 4.2

Results for Example 4.2

Component	y_i (kmol)	M_i (kg/kmol)	$m_i = y_i M_i$ (kg)	$c_i = \frac{y_i M_i}{\sum_{j=1}^{j=N}[y_j M_j]} \times 100$ (%)
Nitrogen	0.7809	28.01	21.87	75.533
Oxygen	0.2095	32.00	6.70	23.151
Argon	0.0092	39.95	0.37	1.269
Carbon dioxide	0.0003	44.01	0.01	0.046
Helium and others	0.0001	4.003	0.0004	0.001
Sum	1		$\sum_{j=1}^{j=5}[n_j M_j] = 28.958$	100

$$\sum_{j=1}^{j=3}[n_j M_j] = 0.78 \times 28.01 + 0.21 \times 32.00 + 0.01 \times 39.94 = 28.9672 \text{ kg/kmol}$$

$$c_{N_2} = \frac{y_{N_2} M_{N_2}}{\sum_{j=1}^{j=3}[y_j M_j]} = \frac{0.78 \times 28.01}{28.9672} = 0.754225$$

$$c_{O_2} = \frac{y_{O_2} M_{O_2}}{\sum_{j=1}^{j=3}[y_j M_j]} = \frac{0.21 \times 32.00}{28.9672} = 0.231987$$

$$c_{Ar} = \frac{y_{Ar} M_{Ar}}{\sum_{j=1}^{j=3}[y_j M_j]} = \frac{0.01 \times 39.94}{28.9672} = 0.013788$$

The simplified composition of air introduces 1.27% error to the calculated average molecular weight of the mixture.

For air with atmospheric nitrogen (N_2, 79% and O_2, 21%),

$$\sum_{j=1}^{j=2}[n_j M_j] = 0.79 \times 28.01 + 0.21 \times 32.00 = 28.8479 \text{ kg/kmol}$$

$$c_{N_2} = \frac{y_{N_2} M_{N_2}}{\sum_{j=1}^{j=2}[y_j M_j]} = \frac{0.79 \times 28.01}{28.9672} = 0.763895$$

$$c_{O_2} = \frac{y_{O_2} M_{O_2}}{\sum_{j=1}^{j=2}[y_j M_j]} = \frac{0.21 \times 32.00}{28.9672} = 0.231987$$

Using air with atmospheric nitrogen introduces 0.9% error to the calculated average molecular weight of the mixture.

Question:

Compare the average molecular weight of pure nitrogen and atmospheric nitrogen.

$$\sum_{j=1}^{j=2}\left[n_j M_j\right]=\frac{0.78}{0.79}\times 28.01+\frac{0.01}{0.79}\times 39.94=28.16\ \text{kg/kmol}$$

This indicates 0.54% error.

The molecular weight of the pure nitrogen is 28.013, while the molecular weight of the atmospheric nitrogen, taking argon into account as a part of nitrogen, is 28.16. In this text, nitrogen always refers to the atmospheric nitrogen (unless specified otherwise). With this assumption, the molar ratio of nitrogen to oxygen in the air is 3.76 $\left(=\frac{0.79}{0.21}\right)$, which means for each mole of oxygen, there are 3.76 moles of nitrogen in the air.

Another assumption is that nitrogen, argon, and other gases in the air do not participate in the combustion process. They enter the combustion chamber at the inlet air temperature and leave at the temperature of the combustion products. This means although these gases do not participate in the reactions and act as inert gases, they still affect the overall process by absorbing thermal energy to change their temperature. Note that at high-temperature combustion processes, nitrogen and oxygen can react and produce oxides of nitrogen, which are major pollutants (refer to Chapter 5). But the rate of that reaction is negligible and is typically ignored in combustion calculations.

The complete combustion of methane in air is

$$CH_4+2\left(O_2+3.76N_2\right)\rightarrow CO_2+2H_2O+7.52N_2 \tag{4.51}$$

In a combustion process, the **theoretical air** is the minimum amount of air to provide enough oxygen for the complete combustion of all combustible atoms, e.g., H_2 and C, so there is neither free oxygen nor unburned fuel in the combustion products. The mixture of fuel and theoretical air is called the **stoichiometric mixture**. The combustion of methane in the above reaction is with theoretical air and the reaction is stoichiometric combustion. In general, the complete combustion of a hydrocarbon fuel with the theoretical air can be summarized in the following reaction:

$$C_xH_y+\left(x+\frac{y}{4}\right)\left(O_2+3.76N_2\right)\rightarrow xCO_2+\frac{y}{2}H_2O+3.76\left(x+\frac{y}{4}\right)N_2 \tag{4.52}$$

It means the combustion of 1 mole of the fuel C_xH_y requires $4.76\left(x+\frac{y}{4}\right)$ moles of air.

Usually more than the theoretical air is needed for a complete combustion process. This is referred to as **excess air**. The amount of the excess air depends on the fuel type and the combustion technology and may vary with the steam generator load. The actual amount of the air supplied to a combustion process is presented by either the percentage of the theoretical air or the percentage of the excess air. The percentage of the theoretical air is the fraction of the number of moles of the actual air to the number of moles of the theoretical air. The percentage of excess air is defined similarly by the ratio of the number of moles of the excess air (the actual air minus the theoretical air) to the number of moles of the theoretical air. For example, 200% theoretical air and 100% excess air are the same and it means the actual amount of air is twice the theoretical air. The complete combustion of methane with 200% theoretical air or 100% excess air can be written as

$$CH_4 + 2 \times 2\left(O_2 + 3.76N_2\right) \rightarrow CO_2 + 2H_2O + 2O_2 + 15.04N_2 \tag{4.53}$$

Using the same terminology, the theoretical air is sometimes referred to as 100% theoretical air. Similarly, 90% of the theoretical air or 10% deficiency of the air means that the actual amount of the air is less than the theoretical air.

Another method to express the relative amount of fuel and air in the combustion process is by the air-to-fuel ratio (AF) or the fuel-to-air ratio (FA). They both can be mass or mole basis and are defined as follows:

$$AF_{mass} = \frac{m_{air}}{m_{fuel}} \tag{4.54}$$

$$AF_{mole} = \frac{n_{air}}{n_{fuel}} \tag{4.55}$$

$$AF_{mass} = \frac{m_{air}}{m_{fuel}} = \frac{n_{air}M_{air}}{n_{fuel}M_{fuel}} = AF_{mole}\frac{M_{air}}{M_{fuel}} \tag{4.56}$$

$$M_{air} = 28.97$$

where n represents the number of moles and M represents the molecular weight. Also, the **equivalence ratio** can be used to express the relative amount of the actual air to the theoretical air. It is defined as the ratio of the fuel-to-air ratio of the actual combustion process to the fuel-to-air ratio of the complete combustion with the theoretical air and can be mass or mole basis.

$$\Phi = \frac{FA}{FA_s} = \frac{AF_s}{AF} \tag{4.57}$$

where *s* represents the combustion with the theoretical or stoichiometric air. Note that for a combustion process, all the following expressions are equal:

- 200% theoretical air
- 2 times stoichiometric air
- equivalence ratio of $\frac{1}{2}$
- 100% excess air

In all methods of presenting the relative amount of air and fuel, if the amount of air in the actual combustion is less than that in the stoichiometric combustion, the fuel and air mixture is a **rich mixture**. If the amount of air in the actual combustion is more than that in the stoichiometric combustion, the fuel and air mixture is a **lean mixture**.

Example 4.3

Ethylene (C_2H_4) fuel is completely combusted in air. Write the combustion reaction when the combustion is with the theoretical air and with 100% excess air. Determine the air-to-fuel ratio (both mole-based and mass-based), and the dew temperature of the combustion products for both reactions. Also, determine the equivalence ratio in the second reaction. The pressure of the combustion products is 1 bar.

Analysis

Given: The fuel is ethylene (C_2H_4). There are two combustion reactions, with the theoretical air and with 100% excess air.

Find:

- Write both combustion reactions.
- Determine the air-to-fuel ratio (both mole-based and mass-based) when the combustion is with the theoretical air and with 100% excess air.
- Determine the dew point temperature of the combustion products when the combustion is with the theoretical air and with 100% excess air.
- Determine the equivalence ratio when the combustion is with 100% excess air.

Assumptions:

1. The combustion process is complete.
2. Air is 21% oxygen and 79% nitrogen (mole fraction).
3. Nitrogen in the air is the atmospheric nitrogen.
4. Nitrogen does not participate in the combustion reaction.

Analysis:

With theoretical air:

First, the general reaction with the theoretical air is written as follows with the unknown stoichiometric coefficients:

$$C_2H_4 + a(O_2 + 3.76N_2) \rightarrow bCO_2 + cH_2O + dN_2$$

According to the conservation of mass law, the elements on the two sides of the reaction must be balanced:

C balance: $b = 2$
H balance: $2c = 4$
O balance: $2b + c = 2a$
N balance: $d = 3.76a$

Solving the equations, $b = 2$, $c = 2$, $a = 3$, and $d = 11.28$. So the combustion reaction will be

$$C_2H_4 + 3(O_2 + 3.76N_2) \rightarrow 2CO_2 + 2H_2O + 11.28\,N_2$$

This means for combustion of one mole of ethylene three moles of oxygen is needed that can be provided by 14.28 moles of air. The molecular weight of ethylene is 28.05. The air-to-fuel ratio can be found as follows:

$$AF_{mole} = \frac{n_{air}}{n_{fuel}} = \frac{14.28}{1} = 14.28$$

$$AF_{mass} = AF_{mole}\frac{M_{air}}{M_{fuel}} = 14.28\,\frac{28.97}{28.05} = 14.75$$

For the dew point, first the mole fraction of vapor in the combustion products should be determined.

$$\text{Mole fraction of vapor in the combustion products} = \frac{p_{vap}}{p_{tot}} = \frac{2}{2+2+11.28} = 0.131$$

This mole fraction of vapor in the combustion products is equal to the ratio of the partial pressure of the vapor to the total pressure of the combustion products.

$$\frac{p_{vap}}{p_{tot}} = 0.131$$

$$p_{vap} = 0.131 \times 1 = 0.131\,\text{bar}$$

The vapor in the combustion products is saturated steam at the determined partial pressure. Thus, the dew point temperature can be found from Table A.1 (SI) with interpolation.

$$T_{\text{Dew}} = 50.23°\text{C}$$

100% excess air:

In this case, the amount of the air in the combustion reaction doubles.

$$C_2H_4 + 2\times3(O_2 + 3.76N_2) \rightarrow bCO_2 + cH_2O + dN_2 + fO_2$$

Balancing two sides of the reaction:

C balance: $b = 2$
H balance: $2c = 4$
O balance: $2b + c + 2f = 12$
N balance: $d = 22.56$

Solving the equations, $b = 2$, $c = 2$, $f = 3$, and $d = 22.56$. So the combustion reaction will be

$$C_2H_4 + 6\ (O_2 + 3.76N_2) \rightarrow 2CO_2 + 2H_2O + 22.56N_2 + 3O_2$$

This means for combustion of one mole of ethylene fuel, 6 moles of oxygen are needed that can be provided by 28.56 $(=6+22.56=6\times4.76)$ moles of air. The air-to-fuel ratio can be found as follows:

$$AF_{\text{mole}} = \frac{n_{\text{air}}}{n_{\text{fuel}}} = \frac{28.56}{1} = 28.56$$

$$AF_{\text{mass}} = AF_{\text{mole}} \frac{M_{\text{air}}}{M_{\text{fuel}}} = 28.56\ \frac{28.97}{28.05} = 29.50$$

Note that AF_{mole} and AF_{mass} in the second case are twice those in the first case.

The equivalence ratio of the mixture in the second reaction is

$$\Phi = \frac{AF_s}{AF} = \frac{14.28}{28.56} = 0.5$$

The mixture of fuel and air is a lean mixture.

For the dew point,

$$\text{Mole fraction of vapor in the combustion products} = \frac{p_{\text{vap}}}{p_{\text{tot}}} = \frac{2}{2+2+22.56+3}$$
$$= 0.0677$$

$$p_{\text{vap}} = 0.0677 \times 1 = 0.061\ \text{bar}$$

From Table A.1 (SI) with interpolation

$$T_{\text{Dew}} = 38.22°\text{C}$$

The combustion calculation method presented so far sometimes is referred to as the mole method. The second combustion calculation method is called the million BTU method. In this method, the mass of the required fuel and air as well as the mass of the combustion products to generate one million BTU thermal energy are determined.

It should be noted that actual combustion is much more complicated than what is described above. There are many intermediate products produced in the rapid chemical reactions of combustion. In this section, only the initial reactants and the final products are considered. But for a detailed analysis of combustion, these intermediate products are very important.

In an actual combustion process, many parameters affect the reactions and products, such as the temperature and pressure of the reactants and products, the amount of oxygen (air) in the mixture and the degree of their mixing, the degree of turbulence during combustion, the time allocated for combustion, and many other factors. If incomplete combustion occurs, which can be with less than or more than theoretical air, the latter in the case of insufficient mixing of air and fuel, one of the following two cases may take place:

1. Slight shortage of oxygen, which may lead to the appearance of some CO instead of CO_2 in the combustion products.
2. Considerable shortage of oxygen, which may lead to the appearance of some CO and some unburned hydrocarbon in the products.

An accurate determination of the combustion products and their relative amount in an actual combustion process is typically possible through measurements using instrumentations, such as a gas chromatograph, flame ionization detector, infrared analyzer, and Orsat analyzer. The composition of the combustion products is typically reported on a dry basis where the amount of water vapor is ignored.

It should be noted that the vapor content of the combustion products is an important factor in designing any system involving combustion. This is particularly important when the combustion products are cooled down, such as in steam generators where the combustion products are used in the economizer and air preheater. If the temperature of the combustion products is brought down close to the dew point temperature, some vapor may condense and cause corrosion in metal components.

4.3.2 Conservation of Energy for Systems Involving Combustion

So far all thermodynamic analyses we looked at were about systems with fixed chemical compositions. The fundamental principle of energy conservation can be applied for processes with combustion, but the evaluation of enthalpy, internal energy, and entropy is different. One complication is about the arbitrary reference points where these properties are set zero. When internal energy and enthalpy are tabulated, for steam, for instance, the internal energy of saturated liquid at 0.01°C is assumed zero. It is fine for systems with fixed compositions to choose the references for individual matters independently because we always deal with changes in parameters and the value of the reference point is canceled in calculations. But when there is a chemical reaction, we need a common reference point for the properties of all substances. This point for enthalpy and internal energy is called the standard reference state and is set at the temperature of 25°C (298.15 K) and pressure of 1 atm. The enthalpy of all stable elements is zero at the standard reference state. The term "elements" here refers to molecules made of a single type of atom, e.g., O_2, H_2, Ar, and He (and not compound molecules), and "stable" means chemically stable elements, for example, while H_2 is stable, H (or H^+) is not.

4.3.2.1 Enthalpy of Formation

The enthalpy of compounds at the standard reference state is called the enthalpy of formation. The thermal energy released or absorbed by the reaction of the formation of a compound from its elements at the temperature of 25°C and pressure of 1 atm is the enthalpy of formation shown by $\bar{h}_f^\circ$. For instance, in the complete combustion of carbon ($C+O_2 \rightarrow CO_2$), if the reactants and the product are at the temperature of 25°C and pressure of 1 atm, 393,522 kJ/kmol heat is released, so the enthalpy of formation of carbon dioxide is equal to –393,520 kJ/kmol.

The enthalpies of formation for other substances can be found from any thermodynamic textbook. Note that for some substances there are two forms, such as oxygen (O and O_2) where the value of the enthalpy of formation for O_2 is zero.

The enthalpy of formation can be positive or negative. In the case of the negative enthalpy of formation, heat is released during the formation of the compound. This means the enthalpy of the compound is lower than that for the reacting elements, and since the enthalpies of the reacting elements are zero, the enthalpy of the compound should be negative. The positive enthalpy of formation means for forming these compounds from their elements, some heat should be added (the process is endothermic).

At any state other than the reference standard state, the enthalpy can be found by adding the difference between that state and the standard reference state to the enthalpy of formation.

$$\bar{h}(T, p) = \bar{h}_f^\circ + \left[\bar{h}(T, p) - \bar{h}(T_{\text{ref}}, p_{\text{ref}})\right] = \bar{h}_f^\circ + \Delta\bar{h} \tag{4.58}$$

The values of $\bar{h}(T, p)$ and $\bar{h}(T_{\text{ref}}, p_{\text{ref}})$ can be found from the thermodynamic property table for each compound (e.g., steam tables or ideal gas tables). The value of enthalpy can be negative because the enthalpy at the reference state is zero.

Example 4.4

Determine the enthalpy of CO_2, H_2O (vapor form), and N_2 at 1500 K and C_2H_4, O_2, and N_2 at the temperature and pressure of 298 K and 1 atm.

Analysis:

The enthalpy of formation for each substance and the change of specific enthalpy from the standard reference state to the temperature of 1500 K can be found from appropriate thermodynamic property tables for CO_2, H_2O, and N_2.

CO_2, H_2O, and N_2 at 1500 K

$$\left(\bar{h}_f^\circ + \Delta\bar{h}\right)_{CO_2} = -393,520 + (71,078 - 9,364) = -331,806 \text{ kJ/kmol}$$

$$\left(\bar{h}_f^\circ + \Delta\bar{h}\right)_{H_2O,\,g} = -241,820 + (57,999 - 9,904) = -193,725 \text{ kJ/kmol}$$

$$\left(\bar{h}_f^\circ + \Delta\bar{h}\right)_{N_2} = 0 + (47,073 - 8,669) = 38,404 \text{ kJ/kmol}$$

C_2H_4, O_2, and N_2 at the temperature of 298 K

$$\left(\bar{h}_f^\circ + \Delta\bar{h}\right)_{C_2H_4} = 52,280 + 0 = 52,280 \text{ kJ/kmol}$$

$$\left(\bar{h}_f^\circ + \Delta\bar{h}\right)_{O_2} = 0 + 0 = 0$$

$$\left(\bar{h}_f^\circ + \Delta\bar{h}\right)_{N_2} = 0 + 0 = 0$$

The first law of thermodynamics for a control volume with combustion can be written as follows:

$$\frac{\dot{Q}_{CV}}{\dot{n}_F} - \frac{\dot{W}_{CV}}{\dot{n}_F} = H_P - H_R = \sum_P n_i\bar{h}_i - \sum_R n_e\bar{h}_e = \sum_P n_i\left(\bar{h}_f^\circ + \Delta\bar{h}\right)_i - \sum_R n_e\left(\bar{h}_f^\circ + \Delta\bar{h}\right)_e \tag{4.59}$$

where *R* and *P* represent reactants and products, respectively.

Example 4.5

Ethylene (C_2H_4) with flow rate of 1 kg/s is completely combusted in theoretical air. The inlet air and fuel flows are at the temperature and pressure of 25°C and 1 atm, respectively. The combustion products are at the temperature of 1500 K. Assuming no work production in the process, determine thermal energy released in the process.

Analysis

Given:

Ethylene is completely combusted in theoretical air at the given inlet and outlet conditions.

Find:

Assumptions:

1. The combustion process is complete.
2. Air is 21% oxygen and 79% nitrogen.
3. Nitrogen is atmospheric nitrogen.
4. Nitrogen does not participate in the combustion reaction.
5. No work or change in kinetic and potential energies is involved in the process.

Analysis:

From Example 4.3, the reaction for complete combustion of ethylene is

$$C_2H_4 + 3(O_2 + 3.76N_2) \rightarrow 2CO_2 + 2H_2O + 11.28N_2$$

The energy balance for the process is

$$\frac{\dot{Q}_{CV}}{\dot{n}_F} - \frac{\dot{W}_{CV}}{\dot{n}_F} = \sum_P n_i \left(\bar{h}_f^\circ + \Delta\bar{h}\right)_i - \sum_R n_e \left(\bar{h}_f^\circ + \Delta\bar{h}\right)_e$$

$$\frac{\dot{Q}_{CV}}{\dot{n}_F} = 2\left(\bar{h}_f^\circ + \Delta\bar{h}\right)_{CO_2} + 2\left(\bar{h}_f^\circ + \Delta\bar{h}\right)_{H_2O} + 11.28\left(\bar{h}_f^\circ + \Delta\bar{h}\right)_{N_2}$$

$$-\left(\bar{h}_f^\circ + \Delta\bar{h}\right)_{C_2H_4} - 3\left(\bar{h}_f^\circ + \Delta\bar{h}\right)_{O_2} - 11.28\left(\bar{h}_f^\circ + \Delta\bar{h}\right)_{N_2}$$

From Example 4.4, the values of enthalpies can be substituted

$$\frac{\dot{Q}_{CV}}{\dot{n}_F} = 2\left[-393,520 + (71,078 - 9,364)\right] + 2\left[-241,820 + (57,999 - 9,904)\right]$$

$$+ 11.28\left[0 + (47,073 - 8,669)\right] - [52,280 + 0] - 3[0 + 0] - 11.28[0 + 0]$$

$$= 2 \times (-331,806) + 2 \times (-193,725) + 11.28 \times 38,404 - 52,280$$

$$= -670,145 \text{ kJ/kmol}$$

This is the heat generated for a mole of fuel. Now the flow rate of the total number of fuel moles can be estimated as follows:

$$\dot{n}_F = \frac{\text{Fuel mass flow rate}}{\text{Molecular weight of fuel}} = \frac{1}{28.05} = 0.03565 \ \frac{\text{kmol}}{\text{s}}$$

$$\dot{Q}_{CV} = \dot{n}_F \left[\sum_P n_i \left(\bar{h}_f^\circ + \Delta\bar{h} \right)_i - \sum_R n_e \left(\bar{h}_f^\circ + \Delta\bar{h} \right)_e \right]$$

$$= 0.03565 \times (-670,145) = -23890.7\ \frac{\text{kJ}}{\text{s}}$$

4.3.2.2 Enthalpy of Combustion and Heating Values

Since the composition of many fuels can vary significantly depending on mines and processes used to produce these fuels, the enthalpy of formation of these fuels cannot be tabulated. For these fuels, the enthalpy of combustion, which can be determined experimentally, is more convenient to use than the enthalpy of formation. The enthalpy of combustion, h_{RP}, is the difference between the enthalpy of the products and that of the reactants during complete combustion taking place at a given temperature and pressure (per kg or mole).

$$\bar{h}_{RP} = \sum_P n_e \bar{h}_e - \sum_R n_i \bar{h}_i \tag{4.60}$$

where n is the coefficient of the reactants and products in the reaction equation for a mole of fuel. When the enthalpy of combustion is evaluated at the standard state, it is shown as $\bar{h}_{RP}^\circ$.

$$\bar{h}_{RP}^\circ = \sum_P n_e \bar{h}_e^\circ - \sum_R n_i \bar{h}_i^\circ = \sum_P n_e \left(\bar{h}_f^\circ \right)_e - \sum_R n_i \left(\bar{h}_f^\circ \right)_i \tag{4.61}$$

From Equation 4.59

$$\frac{\dot{Q}_{CV}}{\dot{n}_F} - \frac{\dot{W}_{CV}}{\dot{n}_F} = \sum_P n_i \left(\bar{h}_f^\circ + \Delta\bar{h} \right)_i - \sum_R n_e \left(\bar{h}_f^\circ + \Delta\bar{h} \right)_e = \sum_P n_e \left(\bar{h}_f^\circ \right)_e - \sum_R n_i \left(\bar{h}_f^\circ \right)_i$$
$$+ \sum_P n_e \left(\Delta\bar{h} \right)_e - \sum_R n_i \left(\Delta\bar{h} \right)_i = \bar{h}_{RP}^\circ + \sum_P n_e \left(\Delta\bar{h} \right)_e - \sum_R n_i \left(\Delta\bar{h} \right)_i$$

(4.62 and 4.63)

where $\bar{h}_{RP}^\circ$ is the enthalpy of combustion and $\sum_P n_e \left(\Delta\bar{h} \right)_e$ and $\sum_R n_i \left(\Delta\bar{h} \right)_i$ can be determined from any thermodynamic textbook.

The **heating value** of a fuel is the thermal energy generated during combustion at a fixed temperature and pressure of the reactants and products. The heating value is equal to the negative of the enthalpy of combustion. There are two types of heating values, **higher heating value (HHV)** or **lower heating value (LHV)**. In the higher heating value, all water vapor in the combustion products is condensed to liquid form. The lower heating value is defined when the water in the combustion products is in vapor form. The HHV and LHV are also referred to as the gross and net heating values, respectively. Clearly for a given fuel, the HHV is greater than the LHV. To measure the heating value of a fuel, a device called a bomb calorimeter is used.

Example 4.6

Determine the higher and lower heating values of Ethylene in Example 4.5. Compare your results with the published values.

Analysis

From Example 4.3,

$$C_2H_4 + 3(O_2 + 3.76N_2) \rightarrow 2CO_2 + 2H_2O + 11.28N_2$$

Since the heating values are defined at the standard state, for the higher heating value we can write,

$$\bar{h}^\circ_{RP} = \sum_P n_e \left(\bar{h}^\circ_f\right)_e - \sum_R n_i \left(\bar{h}^\circ_f\right)_i = 2\left(\bar{h}^\circ_f\right)_{CO_2} + 2\left(\bar{h}^\circ_f\right)_{H_2O\,(l)} + 11.28\left(\bar{h}^\circ_f\right)_{N_2}$$

$$-\left(\bar{h}^\circ_f\right)_{C_2H_4} - 3\left(\bar{h}^\circ_f\right)_{O_2} - 11.28\left(\bar{h}^\circ_f\right)_{N_2}$$

$$= 2\times(-393,520) + 2\times(-285,830) + 0 - 52,280 - 0 - 0$$

$$= -1,410,980 \frac{\text{kJ}}{\text{kmol of fuel}}$$

$$h^\circ_{PR} = \frac{-1,410,980 \dfrac{\text{kJ}}{\text{kmol of fuel}}}{28.05 \dfrac{\text{kg}}{\text{kmol}}} = -50,302.3 \frac{\text{kJ}}{\text{kg}}$$

$$\text{HHV} = -h^\circ_{PR} = 50,302.3 \frac{\text{kJ}}{\text{kg}}$$

For the lower heating value,

$$\bar{h}^\circ_{RP} = \sum_P n_e \left(\bar{h}^\circ_f\right)_e - \sum_R n_i \left(\bar{h}^\circ_f\right)_i = 2\left(\bar{h}^\circ_f\right)_{CO_2} + 2\left(\bar{h}^\circ_f\right)_{H_2O\,(g)} + 11.28\left(\bar{h}^\circ_f\right)_{N_2}$$

$$-\left(\bar{h}^\circ_f\right)_{C_2H_4} - 3\left(\bar{h}^\circ_f\right)_{O_2} - 11.28\left(\bar{h}^\circ_f\right)_{N_2}$$

$$= 2\times(-393,520) + 2\times(-241,820) + 0 - 52,280 - 0 - 0$$

$$= -1,322,960 \frac{\text{kJ}}{\text{kmol of fuel}}$$

$$\overset{\circ}{h}_{PR} = \frac{-1,322,960 \dfrac{\text{kJ}}{\text{kmol of fuel}}}{28.05 \dfrac{\text{kg}}{\text{kmol}}} = -47,164.3 \ \frac{\text{kJ}}{\text{kg}}$$

$$\text{LHV} = -\overset{\circ}{h}_{PR} = 47,164.3 \ \frac{\text{kJ}}{\text{kg}}$$

The values of the HHV and LHV for ethylene are about 50,300 and 47,160 kJ/kg, respectively, which are very close to values from the calculations.

4.3.2.3 Adiabatic Flame Temperature

If the combustion process is adiabatic with no work or change in kinetic or potential energy, the temperature of the products is called the **adiabatic flame temperature** or the **adiabatic combustion temperature**. The maximum temperature that can be achieved for a specific combustion reaction is when there is complete combustion with theoretical air. This temperature cannot be achieved in real combustion because of the need for excess air for better combustion, incomplete combustion, heat losses, and dissociation of products at high temperatures. In most applications, reaching this temperature is not even desirable, such as gas turbines where the turbine inlet temperature is limited and the combustion temperature is controlled by the amount of excess air. The determination of the adiabatic flame temperature typically requires an iterative calculation process.

4.4 Fuels

Every substance that can be rapidly oxidized to produce heat can be considered as a fuel. Most commercial fuels are composed of three combustible elements, i.e., carbon, hydrogen, and sulfur. The contribution of the latter element to energy production is not significant, but it is a major source of pollution (refer to Chapter 5).

Fuels may be fossil fuels or non-fossil fuels (such as biomass). Fuels used for power generation purposes can be categorized based on several criteria. They can be grouped according to their origin, as follows:

- Natural fuels such as fossil fuels and biomass
- Synthetic fuels (synfuels) such as biofuels and fuels derived from coal, oil shale, and tar sands
- Industrial and residential wastes

Fuels may also be grouped based on their physical phase, which is useful for the determination of fuel handling processes.

- Solid such as coal, domestic waste, and biomass
- Liquid such as petroleum-derived fuels, biodiesel, liquefied coal, and fuels derived from oil shale and tar sand
- Gas such as natural gas and syngas

In any combustion process, reactants, i.e., a fuel and an oxidizer, react to produce the combustion products and some thermal energy. During the process, the molecular bonds among reactant molecules are rapidly broken and recreated to form the products.

The selection of the fuel for a steam generator depends on many factors including the cost of the fuel at the plant site (which includes both the fuel cost and transportation cost), cost of fuel preparation, operating cost (including maintenance cost and cost of power for the fuel preparation system), and current and future emission control regulations.

Figure 4.1 illustrates the variations of the share of different energy sources in the total electricity production in the U.S. between 2000 and 2019. The details of this diagram will be discussed in the section dedicated to each fuel. We will start this section with coal and then will proceed to other fuels.

While the consumption of coal for electricity generation is declining in many countries, it is still important to study coal steam generation systems for several reasons. From historical perspective, coal has been a dominant fuel

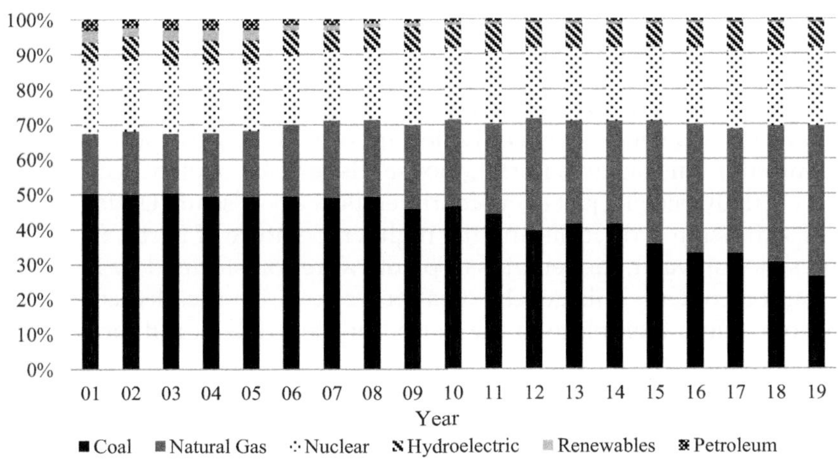

FIGURE 4.1
The share of various energy sources in the total electricity production (%) in the USA. (Based on the data provided by the U.S. EIA.)

for a very long time, and it is impossible to properly understand steam generators without knowing coal steam generators. The majority of advancements and breakthroughs were first introduced to coal steam generators. Second, in many countries, including some of the largest emerging economies, coal is still the most important fuel for power generation systems. Finally, the same technologies that initially were developed for coal steam generators are now being used for other solid fuels such as biomass and solid waste and even for natural gas steam generators, particularly those developed for pulverized coal. Fortunately, most concepts that we will discuss on coal can be directly or indirectly applied to other fuels too.

4.5 Coal

Historically speaking, coal is probably the most important substance in the development of the industrial civilization as we know. Coal not only plays a major role in the electricity production industry, but the majority of steel is also produced using coke processed from coal.

4.5.1 Coal Formation

Plants and trees utilize solar energy to convert water and carbon dioxide to carbohydrates, e.g., sugar, through the photosynthesis process. This process releases oxygen into the atmosphere and carbon is stored in the parts of plants. At the end of their life, remains of vegetation typically decay, and carbon returns to the atmosphere in the form of carbon dioxide or methane. However, in ancient times in some environments, such as swamps and wet regions, the remains of plants and trees were not exposed to atmospheric oxygen and thus were not completely decayed. They were buried under sedimentary deposits before they could decompose. Many generations of plants were buried in this fashion. In fact, thousands of years of plant accumulation may eventually become just a few centimeters of a coal seam. Under the pressure of sedimentary deposits and the high temperature of Earth, typically in the presence of water, the remains of plants were transformed by increase in their carbon content. This geochemical process, which is called coalification or carbonization, took place in the period of hundreds of millions of years. The first step in the process is the conversion of buried plants to peat. This follows by the formation of lignite then subbituminous, bituminous, and finally anthracite. In each step, the heating value of coal and its carbon content increase while the moisture and oxygen levels reduce. Coal can consist of carbon, hydrogen, oxygen, sulfur, and nitrogen, as well as ash-forming elements. It also contains a trace of mercury, arsenic, and selenium (in the order of just a few parts per million, ppm) which can be very harmful for human health and the environment when released through combustion products. Depending on the type of the original vegetation, the level of coalification, and the amount of impurities, coal may have a wide variety of compositions and properties.

4.5.2 Coal Consumption

According to the International Energy Agency (IEA), the share of coal in the global energy mix was 29% in 2013 which had experienced a significant increase compared to 23% in 2000.[3] This share peaked in 2014 but reduced to about 27% in recent years.[4] Furthermore, over 40% of the world electricity was produced using coal in 2013.[5] In 2018, this value reduced to 38.5%. Globally, about 80% of coal is used for steam generation. According to the U.S. Energy Information Administration (EIA), in the U.S., this percentage is higher, about 92% of coal is used for electricity generation, 8% for industrial usage and residential and commercial spacing heating (in 2018).[6] There were 491 coal power plants in the U.S. in 2014, which was significantly lower compared to 625 coal power plants in 2004. In 2018, this number further reduced to 336 coal power plants.[7]

According to the IEA, the annual global coal production was 3 Gt (billion metric tons) in 1972, and for a decade after that 1 Gt was added to the production. By the late 1980s, the production reached about 4.5 Gt and remained almost steady until the early 2000s. Between 2003 and 2013, however, about every three years the global coal production increased by 1 Gt to the peak of about 8 Gt in 2014. This meant the coal production experienced about 4.5% average annual growth. The coal production was reduced by 2016 to about 7.3 Gt. However, since then the production has increased again to about 8 Gt.[8] Since 1985 China has been the leading producer and consumer of coal by producing about half of the global coal production. India and the U.S. followed by about 9% of the global coal production each. In fact, about 81% of the global coal production is being produced by six countries, China, India, the U.S., Australia, Indonesia, and Russia. The per capita coal consumption in the U.S. is about twice the global average per capita coal consumption. Figure 4.2 illustrates the variations of the annual coal consumption in the world, the U.S., and China. The variation in the U.S. coal consumption, qualitatively speaking, represents the trend in many industrial countries. A similar statement can be expressed for China as representatives of many emerging economies. Note that the solid lines are representing the overall

[3] World Energy Outlook 2015, International Energy Agency (IEA).
[4] https://www.iea.org/reports/coal-2018.
[5] IEA Electricity Information (2015 edition), International Energy Agency (IEA).
[6] https://www.eia.gov/energyexplained/coal/use-of-coal.php.
[7] https://www.eia.gov/electricity/annual/html/epa_04_01.html.
[8] https://www.iea.org/data-and-statistics/charts/world-total-coal-production-1971-2019-provisional.

coal consumptions (the combination of coal consumptions for electricity production and any other applications), and the dashed-lines are representing the percentage of electricity production from coal resources relative to the total electricity production. The curves show that coal consumption was flat in the 1990s and 2000s but in the past few years the consumption has generally declined (though it has slightly increased since 2017). However, the ratio of the electricity production from coal to the total electricity production has been declining since the mid-1980s, because new electricity production plants were mostly non-coal units. Figures 4.1 and 4.2 indicate that until 2008 about half of the U.S. electricity was produced using coal but that share reduced to below 40% in 2014. This share has further reduced to about 32% in 2016, about 27% in 2018, and just below 24% in 2019.[9] In the global level, except for the past couple of years, the coal consumption and the share of coal in electricity generation have steadily increased. Currently about 38% of the world's electricity is generated from coal.

Regardless of the recent decline in the coal consumption for electricity production, according to the IEA, everyday 200 MW of new coal power generation capacity is being installed globally.[10] Currently, about **65% of electricity in China, 55% in India, about 24% in the U.S., and 38% of global electricity are produced in coal power plants**. In the U.S., as of 2018, there were 679 coal mines producing about 6.9×10^{11} kg (about 7.6×10^8 short tons) coal.[11] Note that 1 short tons = 2000 lb = 907.2 kg. The average price of coal in the U.S. in 2018 was about $40/metric ton ($36/short ton).

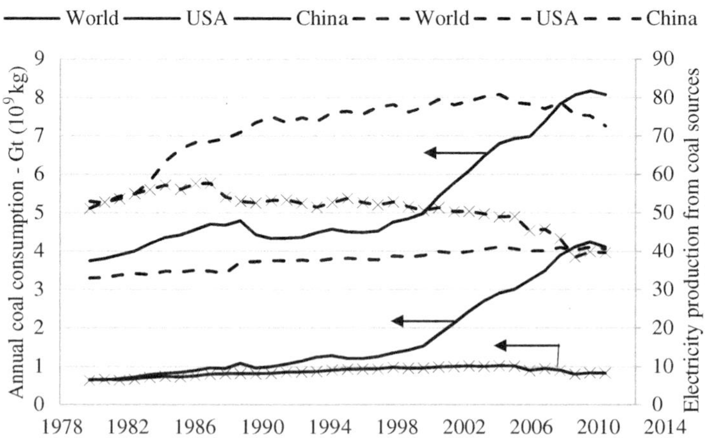

FIGURE 4.2
Annual coal consumption and electricity production from coal sources (% of total) in the world, the U.S., and China. (Based on the data from eia.gov and data.worldbank.org)

[9] https://www.eia.gov/tools/faqs/faq.php?id=427&t=3.
[10] http://www.iea.org/topics/coal/.
[11] Annual Coal Report 2015, EIA, https://www.eia.gov/coal/annual/pdf/acr.pdf.

Although coal has devastating impacts on the environment and human health at every level from extraction to preparation to combustion, its usage will continue, at least in the short- and mid-term future. So it is important to understand the design and operation of coal power plants and try to minimize the negative impacts of their operation.

4.5.3 Coal Resources

In order to evaluate the quantity of coal (for that matter any fossil fuel) in a certain region, two terms are being used: coal resources and coal reserves. Coal resources are the quantity of coal that exists or may exist in a deposit regardless of if it can be economically recovered or not (though there might be some limits on the depth and minimum thickness of coal seams for them to be considered as resources). Depending on the degree of confidence in the results of the exploration of a deposit, coal resources can be proven (proved or measured resources), probable (or indicated resources), or inferred, in the order of descending confidence. On the other hand, coal reserves are the quantity of coal that can be mined with the current (or the near future) technological and economic conditions. Note that coal reserves may vary with economic conditions. For example, if the price of coal goes down, the coal reserves will reduce because less coal is economically mineable. Also, advances in mining-related technologies and coal exploration techniques continuously affect these estimations. Moreover, for a given coalfield, due to the limitations of the used mining method, only a portion of the coal seams can be extracted. For example, to avoid collapsing of mine roofs, some coal columns must be left intact to support the roof (i.e., in the room and pillar mining method for underground mining). The quantity of coal that is actually extractable is called recoverable reserves. Different countries use different definitions for coal resources and reserves, which has caused considerable discrepancy and confusion in various estimations.

The U.S. Energy Information Administration (EIA) uses the following three terms to express the coal reserves: recoverable reserves at producing mines (the quantity of coal that can be economically recovered from active mines), estimated recoverable reserves (the quantity of coal that can be economically and practically recovered and used, excluding coal unavailable due to mining methods), and demonstrated reserve (the quantity of coal that is identified as recoverable, including the coal unavailable due to mining methods). As it is clear, even these definitions are somewhat confusing. The estimates for the reserves in the U.S. are about 14, 230, and 430 billion metric tons (about 15, 250, and 470 billion short tons), respectively,[12] in over 33 states. Most of the demonstrated coal reserves are in Montana, Illinois, Wyoming, West Virginia, Kentucky, and Pennsylvania. In terms of coal production, the states

[12] Energy Information Administration (EIA), https://www.eia.gov/energyexplained/coal/how-much-coal-is-left.php.

with the highest production are (in 2015) Wyoming (40%), West Virginia (13%), Pennsylvania (7%), Illinois (7%), and Kentucky (5%).[13] The coal producing regions in the U.S. can be divided into three main regions: Appalachian region (Alabama, Eastern Kentucky, Maryland, Ohio, Pennsylvania, Tennessee, Virginia, and West Virginia), Interior region (Arkansas, Illinois, Indiana, Kansas, Louisiana, Mississippi, Missouri, Oklahoma, Texas, and Western Kentucky), and Western region (Alaska, Arizona, Colorado, Montana, New Mexico, North Dakota, Utah, Washington, and Wyoming). According to the EIA, these three regions are responsible for producing 26%, 18%, and 55% of the total coal production in the U.S., respectively, in 2018.[14]

Also, the world's total recoverable coal is estimated to be about 1000 billion tons (1100 billion short tons). This is enough to produce about six thousand billion (6×10^{12}) MWh electricity. To get an idea, this is equal to the electricity a 2000 MW power plant (a large coal or nuclear power plant) can produce over 350,000 years if it operates continuously. Considering the fact that the global annual electricity consumption was about 22,300 TWh in 2018,[15] this much coal can power the world for about 112 years (Example 4.7) with no need for other energy sources if the consumption remains at the current level (assuming the efficiency of power generation is 40%). However, due to the existence of other energy resources and other factors, the estimates for how long coal can last vary from 100 to over 300 years.

Example 4.7

Do a rough calculation to show that the current coal reserves are enough for 112 years. Identify major sources of uncertainty of your calculation.

Assumptions:

The average heating value of the coal consumed in the U.S. in 2019 = 19.26 million Btu/short tons[16, 17] = 22400 kJ/kg

The world total recoverable coal = 1100 billion short tons = 1×10^{15} kg

The average efficiency of power plants = 40%

The global annual electricity consumption in 2018 = 22,300 TWh/year

Analysis

Heat content of the world total recoverable coal = $1\times10^{15}\times22400 = 2.24\times10^{19}$ kJ

The electricity that can be produced from the world total recoverable coal at the efficiency of 40% = $2.24\times10^{19}\times0.4 = 8.96\times10^{18}$ kJ = 2.49×10^{15} kWh = 2.49×10^{6} TWh

---

[13] https://www.eia.gov/tools/faqs/faq.php?id=69&t=2.

[14] https://www.eia.gov/energyexplained/coal/where-our-coal-comes-from.php.

[15] https://www.iea.org/reports/electricity-information-overview.

[16] https://www.eia.gov/tools/faqs/faq.php?id=72&t=2.

[17] Although EIA does not directly indicate if this is net (lower) heat content or gross (higher) heat content, typically the gross heat content is used in the USA unless it is specified otherwise.

Number of years this much electricity can power the world = 2.49×10^6 TWh / 22,300 TWh/year = 112 year

For the following reasons, this number is not accurate and actually coal can last much longer, and it is the largest fossil fuel energy resource.

- Variations in electricity consumption
- Not all electricity comes from coal
- Share of coal changes in the energy mix
- Improvement of power generation efficiency over time
- Reserves change with economics and technology

Unlike other fossil fuels, coal reserves are more uniformly distributed in the world. The distribution of recoverable coal reserves in different continents is: 31% Europe, 36% Asia, 29% America, and 4% Africa.[18] The countries with the largest reserves are the U.S., Russia, China, and India, but significant coal reserves can be found in over 70 countries in all continents.[19]

4.5.4 Coal Analysis

Coal composition varies significantly with the location of a mine and even within the same mine site. Thus, the analysis of coal composition is very important. There are two methods to analyze coal composition (for that matter any fuel composition): **proximate analysis and ultimate analysis**. In the proximate analysis the relative amount of moisture, volatile matter, fixed carbon, and ash is presented on a mass basis. The ultimate analysis is also expressed on a mass basis, but it presents relative amount of carbon, sulfur, hydrogen, nitrogen, oxygen, and ash. The former is typically used for business purposes, while the latter is used for combustion calculations. Both analyses can be of two types, either **as-received basis or dry basis** (no moisture). The latter is important because coal can lose or gain moisture during storage or transportation. Besides these analyses, other coal properties such as grindability or ease of pulverization, ash properties, and size might be of interest for various applications. Here is a list of some other properties and the standard to determine them:

- **Coal grindability:** according to ASTM standard D409
- **Ash composition:** according to ASTM standard D3174
- **Ash fusibility:** according to ASTM standard D1857
- **Coal free-swelling index:** according to ASTM standard D720

[18] World Energy Council, https://www.worldenergy.org/assets/images/imported/2013/10/WER_2013_1_Coal.pdf.

[19] The coal resource: a comprehensive overview of coal, World Coal Institute, 2009 https://www.worldcoal.org/file_validate.php?file=coal_resource_overview_of_coal_report(03_06_2009).pdf.

Note that **ASTM**, which stood for the American Society for Testing and Materials until 2001, is the oldest organization for international standards. In the following sections, we will study coal properties in much more details.

4.5.4.1 Proximate Analysis

The **proximate analysis** can provide a broad approximation of the properties of coal which can help in the extraction, preparation, combustion, and trading of coal. The analysis is typically conducted according to ASTM standard D3172. This analysis is simpler than the ultimate analysis and can be used to determine the rank of coal. For commercial applications, i.e., purchasing and selling of coal, typically the proximate analysis is used. In the proximate analysis, the **moisture, volatile matters, ash, and fixed carbon** in the coal sample are determined based on the percentage of the mass fraction. The analysis starts with sampling of coal according to ASTM standard D388, D6609, D4596, D5192, D6883, D7430, or D2234 depending on where the sample is being taken. Then, the sample should be pulverized and prepared according to ASTM standard D2013. In order to determine the percentage of each compound, several experiments should be conducted on 1 g of the prepared sample.

4.5.4.1.1 Moister Content

There are two types of moisture in coal, **inherent and surface moisture contents**. The former, also known as equilibrium, seem, or bed moisture, is a natural part of coal structure that is trapped during coal formation and is present in the natural state of coal. The latter, also known as free moisture, is added during coal extraction, storage, or transportation.

The moister content of coal is determined according to ASTM standard D3173. The pulverized coal sample is heated in the drying oven with the specifications described in the standard with the flow of dry air at the temperature of between 104°C and 110°C for the duration of one hour. The difference between the mass of the sample before and after the heating process is the moisture content of the sample. The moisture content of the fuel is a burden on combustion because it has to be evaporated and superheated during the combustion process.

4.5.4.1.2 Volatile Matter

Volatile matters are the compounds (mainly hydrocarbons such as methane, ethane, ethylene, and acetylene as well as carbon monoxide) that evaporate when coal is heated in the absence of oxygen. In actual combustion in a steam generator, the volatile matter content is very important for the combustion of the carbon content and for a stable flame. The percentage of the volatile content in coal is an indication of how easily coal can be ignited and if a supplementary flame stabilizing fuel is needed. The volatile matter content can contribute up to 50% of the total heating content of the coal.

The percentage of the volatile matters in an experiment depends on the composition of the coal, the temperature, and the duration of the experiment.

The volatile matters of coal are determined according to ASTM standard D3175. In order to determine the volatile matters in a coal sample, it is heated to the temperature of 950±20°C (1742±68°F) for seven minutes in an environment without oxygen. The difference between the mass of the sample before and after the experiment (excluding moisture) is the mass of the volatile matters in the sample.

4.5.4.1.3 Ash

Ash is composed of inorganic noncombustible substances, such as the compounds of silicon, aluminum, iron, and calcium, and to a lesser extent the compounds of titanium, magnesium, potassium, and sodium present in coal. For most types of coal, majority of the ash consists of silica (SiO_2) and alumina (Al_2O_3). For some types of coal also there are high concentration of iron oxide (Fe_2O_3) and calcium oxide (CaO) in the ash.

These minerals and inorganic substances were either in the original vegetation, were introduced during coal formation (via impurities or substances dissolved in the water where the plants fall and then buried), or are introduced during mining. The combination of the impurities from original vegetation and those introduced during coal formation is called **inherent impurities** and the impurities introduced during mining is called **extraneous impurities**. The coal extracted in the U.S. can have the ash content of 6%–20%. The ash content of coal is important because it directly affects the ash collection and removal systems and indirectly influences the design of heat-transfer surfaces. The ash content of coal is determined according to ASTM standard D3174. The sample from the moisture experiment is combusted in a furnace described in the standard. First, the temperature of the sample is increased to 500±10°C (932±50°F) in an hour period. Then, the temperature is further increased to 750±15°C (1382±59°F) in an hour period. The heating process continues for 2 hours at this temperature. By measuring the mass of the ash residue, the ash content of the sample can be determined. It should be noted that the ash content determined in this method is not exactly the same as the inorganic compounds of the coal sample because some inorganic substances are reacted during the process and leave the sample in the form of combustion products. This ash is not even the same as the ash in an actual combustion process because the amount and composition of the ash depend on the combustion technology and the combustor operating conditions.

Note that there are many ash properties that have significant impacts on the design and operation of a steam generator. The most important factor is the content of the ash. Other properties include the ash-softening temperature, acid content, and ash friability (the ease of ash crumbling). The ash behavior at high temperatures is generally analyzed by the ash fusibility test (to determine ash melting temperature) according to the ASTM standard D1857. These parameters impact the tube spacing in steam generator's heat-transfer surfaces and the furnace size.

While the ash content is typically expressed in mass base, the heat content base might be more meaningful for some applications. For example, the mass-based ash content of bituminous, subbituminous, and lignite may all be 10%, but the heat content-based ashes are 3.3, 5, and 10 kg of ash per 1 GJ thermal energy, respectively, for the same coal types (Example 4.8).

Example 4.8

The samples of bituminous, subbituminous, and lignite coal have the higher heating values of about 30,000, 20,000, and 10,000 kJ/kg, respectively. The mass-based ash content of all three samples is 10%. Determine the heat content-based ash in kg of ash/GJ.

Analysis:

$$\text{Heat content} - \text{based ash} = \frac{\text{Mass-based ash content}}{\text{HHV}} \times 10^6$$

$$\text{For bituminous coal: Heat content} - \text{based ash} = \frac{0.1}{30,000} \times 10^6 = 3.33 \frac{\text{kg}_{\text{ash}}}{\text{GJ}}$$

$$\text{For subbituminous coal: Heat content} - \text{based ash} = \frac{0.1}{20,000} \times 10^6 = 5 \frac{\text{kg}_{\text{ash}}}{\text{GJ}}$$

$$\text{Lignite coal: Heat content} - \text{based ash} = \frac{0.1}{10,000} \times 10^6 = 10 \frac{\text{kg}_{\text{ash}}}{\text{GJ}}$$

Example 4.9

A 250 MW coal-burning power plant uses bituminous coal with a higher heating value of 29,500 kJ/kg as fuel. If the mass-based ash content of the coal is 15%, determine the coal consumption (in kg/s) and the ash production (in kg/day). The higher heating value-based efficiency of the power plant is 34%.

Analysis:

$$\eta = \frac{\dot{W}}{\dot{Q}_H} = \frac{\dot{W}}{\dot{m}_{\text{fuel}} \times HHV}$$

$$\dot{m}_{\text{fuel}} = \frac{\dot{W}}{\eta \times HHV} = \frac{250,000}{0.34 \times 29,500} = 24.9 \frac{\text{kg}}{\text{s}} = 89.7 \frac{\text{tonne}}{\text{hr}}$$

$$\dot{m}_{\text{ash}} = 24.9 \times 0.15 = 3.74 \frac{\text{kg}}{\text{s}} = 323.0 \frac{\text{tonne}}{\text{day}}$$

4.5.4.1.4 Fixed Carbon

Fixed carbon is the carbon content of coal that is not bounded in hydrocarbons in the form of volatile matters and remains after all hydrocarbons are evaporated. The mass percentage of fixed carbon is not measured directly but calculated by subtracting the percentage of the moister, ash, and volatile matters in the coal sample from 100.

4.5.4.2 Ultimate Analysis

The ultimate analysis is more scientific than the proximate analysis and is needed for the combustion calculation of coal. It is conducted according to ASTM standard D3176. In this analysis, the mass fraction percentage of all chemical elements of coal, i.e., C, H_2, N_2, O_2, S, and ash, is determined. Although the moisture percentage is not included in the ultimate analysis, it is measured so the analysis can be converted to other bases if needed (refer to the types of the sample below). The sampling and its preparation processes are similar to the proximate analysis described above.

The mass percentage of carbon, hydrogen, and nitrogen in the coal sample is determined according to ASTM standard D5373. To determine the C, H_2, and N_2 contents in the coal sample, the sample is combusted in oxygen at a certain temperature (between 900°C and 1050°C in one method and 1350°C in another). Then, CO_2, H_2O, and N_2 contents of the combustion products are measured.

The sulfur content of the coal sample is determined according to ASTM standard D4239. In this standard, the coal sample is combusted at the temperature of above 1350°C in oxygen in the presence of tungsten trioxide (WO_3). At this temperature, sulfur and its compounds burn to produce SO_2. The amount of sulfur oxide is measured using infrared (IR) absorption detectors, which then can be used to determine the mass fraction of sulfur in the coal sample. The sulfur content of coal can also be measured according to ASTM standard D2492 where sulfur and its compounds can be extracted by some chemicals such as hydrochloric acid or by combustion in the presence of iron.

It should be noted that coal with a lower sulfur content is desirable. The sulfur content of coal can burn and slightly contribute to the heating value of coal; however, the combustion of sulfur forms sulfur oxide which is an environmental pollutant causing acid rain (refer to Chapter 5). Furthermore, if sulfur oxide is exposed to water in the flue gas path, it forms acid sulfuric, which can cause corrosion especially in the stack.

Similar to the proximate analysis, the ash content of coal for the ultimate analysis is measured according to ASTM standard D3174. The oxygen content of the sample is not measured directly and is determined by subtracting the mass of all measured components from the original mass of the sample.

As noted earlier, the proximate analysis is typically used for business purposes, while the ultimate analysis is used for combustion calculations. Both analyses can be of two types, either as-received basis or dry basis (no moisture). The latter is important because coal can lose or gain moisture during storage or transportation.

The sample used for both the approximate and ultimate analyses can be in several forms:

- **As-received**: The coal contents are presented based on the total mass of the coal sample, including moisture, ash, and mineral matters as received by the power plant.
- **As-fired**: Similar to the above definition, but for when the coal is actually fired.
- **Dry or moisture-free basis** (db): The coal contents are presented on a moisture-free basis.
- **Dry, ash-free** (daf): The coal contents are presented on an ash- and moisture-free basis.
- **Dry mineral-matter-free** (dmmf): The mineral matters and moisture are not included in the coal content percentages. Note that the mineral-matter and ash are not exactly the same.
- **Moist, ash-free** (maf): In the presentation of coal contents, the moisture is included but ash is removed.
- **Moist, mineral-matter-free** (mmmf): In the presentation of coal contents, the moisture is included but mineral matters are removed.

In the dry-based analysis, both inherent and surface moisture are removed. In the case that only the surface moisture is removed, the analysis is called air-dried-based analysis. Some of these different bases can be converted to each other according to ASTM standard D3180.

Finally, the higher heating value (HHV) or gross calorific value of coal can be determined using a calorimeter according to ASTM standard D5865 and D2015. In a calorimeter, the coal sample is burned in oxygen and the released heat is determined by measuring the temperature increase in the unit components and the water surrounding the unit.

4.5.4.3 Classification of Coal

As noted in the previous sections, coal properties and compositions can be very wide from one seam to another or even within one seam at different locations. Therefore, a coal classification system is needed to identify coal products with similar characteristics. Coal possesses many physical and chemical properties such as the carbon content, content of volatile matters, sulfur content, percentage of impurities, heating value, combustibility, hardness (mechanical strength), grindability, and ash melting temperature, and accordingly, it can be categorized based on these parameters. These properties can differ depending on the location of the coal mine. It is important to be able to identify the properties of coal and to be able to categorize coal based on these properties because the coal preparation and combustion processes heavily depend on these properties.

Typically, coal may be categorized to different **ranks, types, and grades** according to the level of carbonization (coalification), the type of old vegetation it is made of, and the degree of impurities in it, respectively.[20] The **rank of coal** depends on the length of time for the coalification process (i.e., the age of coal), the pressure and temperature that the vegetable remaining were exposed to, and the nature of the environment where the old vegetation were buried. This is also called the **organic maturity of coal**. Coal with a higher rank has a higher content of carbon and a lower content of volatile matters and moisture.

According to ASTM standard D388, various types of coal are classified into four ranks: **lignite, subbituminous, bituminous, and anthracite**. This categorization is based on the degree of their coalification which is identified by either the percentages of fixed carbon or the higher heating value (HHV), also known as the gross calorific value of coal.

Peat is the first stage in the coalification process. It is not actually a type of coal (at least not officially according to ASTM standards!). The decomposing vegetation still can be seen in it. Due to a very high percentage of moisture, peat is rarely used for steam generation and its global production is negligible compared to the other types of coal.

Lignite (means wood in Latin), also commonly known as **brown coal**, is the youngest and lowest state of coal with a very high percentage of moisture (between 20% and 70%) and volatile matters. Because of having high volatile matters, it ignites easily. The percent of the ash content can be from a few percent to mid-thirty. Typically, it is not economical to transport this type of coal far from the mine site, and it is commonly burned in the mine site or in a close by site. According to the World Coal Institute, about 17% of world coal reserves are lignite,[21] and according to the IEA about 10% of the global coal production has been lignite in recent years.[22] In the U.S., about 8% of the total coal mined annually is of this type. Lignite is mostly mined in Texas and North Dakota (over 90% of the lignite coal mined in the U.S.), but in other states such as Montana and some Gulf states also lignite coal is mined.[23]

The next step in the coalification process is **subbituminous** coal with a greater carbon content and heating value compared to those of lignite coal and a lower moisture content. This type of coal typically has a low sulfur content and low mining cost and is used for steam generators in the pulverized form. According to the World Coal Institute, about 30% of world reserves are subbituminous. In the U.S., it is mostly found in surface mines in Great

---

[20] Coal energy systems, Bruce G. Miller, ELSEVIER Academic Press, 2005.
[21] The coal resource: a comprehensive overview of coal, World Coal Institute, 2005.
[22] International Energy Agency (IEA) Coal Information (2015 edition).
[23] Annual Coal Report 2015, EIA, https://www.eia.gov/coal/annual/pdf/acr.pdf.

Plains, Western states, and Alaska[11]. About a half of the total coal production (mass-based) in the U.S. is of this type. Most of subbituminous coal in the U.S. is produced in Wyoming (about 90% of the total subbituminous coal produced annually).[24]

In terms of the degree of metamorphism (the level of advancement in the coalification process), **bituminous** coal is right after subbituminous with a greater carbon content and heating value compared to those of subbituminous coal and a lower moisture content. Bituminous coal is suitable for combustion in steam generators, particularly in the pulverized form, and for coke production for the steel industry[25] but typically has a high sulfur content. Bituminous coal resources are the largest in the world and in the U.S. Over half of the world reserves are bituminous[2]. In the U.S., this type of coal is about 45% of the total coal production. It is mostly produced in the Appalachia (primarily in West Virginia, Kentucky, and Pennsylvania) with about 54% of the U.S. bituminous coal production and in the interior region (primarily in Illinois and Indiana) with about 31%, and to a lesser extent in the western region with about 15%.

Anthracite is the highest state of coal in its evolution and is characterized by the high percentage of fixed carbon and low volatile matters. Although anthracite coal has a high heating value, it is not suitable for burning in modern steam generators due to its slow burning (due to the low volatile content) and is only used in stoker burners (will be discussed later in this chapter). This type of coal is commonly used for residential and commercial heating. There are currently very limited resources and production of anthracite in the U.S. (less than 0.01% of the total coal production) entirely in Pennsylvania. However, anthracite coal was the first type that was used for heating and industrial applications in the U.S. particularly because of its clean burning compared to other coal types.

In the above categorization, lignite and subbituminous are considered low-rank coal types, and bituminous and anthracite are considered high-rank coal types. According to the World Coal Institute, roughly half of the world reserves is low-rank coal and the other half is high-rank coal[11].

In the U.S. in 2018, there were 585 bituminous mines producing about 3.3×10^{11} kg (about 3.6×10^{8} short tons) coal, 23 subbituminous mines producing about 3.1×10^{11} kg (about 3.4×10^{8} short tons) coal, 17 lignite mines producing about 5.2×10^{10} kg (about 5.7×10^{7} short tons) coal, and 55 anthracite mines producing about 1.7×10^{9} kg (about 1.9×10^{6} short tons) coal.[26]

According to ASTM standard D388, for the coal with the fixed carbon content of lower than 69%, the higher heating value is used to determine the rank of the coal. For the coal with the fixed carbon content of higher than 69%, the fixed carbon content is the determining parameter to identify the rank of the

[24] Annual Coal Report 2015, EIA, https://www.eia.gov/coal/annual/pdf/acr.pdf.
[25] Annual Coal Report 2015, EIA, https://www.eia.gov/coal/annual/pdf/acr.pdf.
[26] Annual Coal Report 2018, EIA, https://www.eia.gov/coal/annual/pdf/acr.pdf.

coal (regardless of the heating value). In this standard, the fixed carbon content is determined on a dry, mineral-matter-free basis but the heating value is determined on a moist, mineral-matter-free basis (only inherent moisture is included). In order to determine the rank of the coal in a mine, at least three samples should be taken from the mine. The average of the proximate analyses on the samples should be used for the purpose of the rank determination. It should be noted that the heating value of coal increases when the rank improves from lignite to bituminous but slightly decreases for anthracite. The reason is the fact that although the carbon content provides most of the heating value of coal, other factors such as volatile matters also influence the heating value. The characteristics of different types of coal and subcategories of each are summarized in Table 4.3.

Other than the rank, coal can also be categorized by type and grade. The coal type depends on the vegetation that coal originated from, which has significant impacts on the physical properties of coal including its hardness and thermoplastic properties[10]. The grade of coal depends on mineral matters in coal and influences the coal preparation process and the behavior of its ash.

There are other standards to classify coal, especially in Europe. In some of these standards, the volatile matter content is used for the categorization of coal; however, as Table 4.3 indicates, the fixed carbon content and the volatile matter content are inversely proportional so the results of using the fixed carbon or the volatile matter contents are similar. In some European standards, for low-rank coal, a two digit code, and for high-rank coal a 14-digit code are used for the identification of different coal types. Although this categorization is more accurate and provides more information, it is more difficult to determine and complex to understand and interpret.

Note that the optimum design of the coal preparation and combustion systems heavily depends on coal characteristics; therefore, a careful analysis of the possible available coal resources for a plant is critical. Typically, coal properties are determined and dictated by the "plant owner". The coal properties can be stated in the form of a range or low/average/high properties. In either case, although typically the entire fuel preparation and combustion system are designed for either the average properties or the most available coal (currently and/or in the future), they should be able to operate safely with all possible inlet fuels.

4.5.4.4 Derivative Fuels from Coal

The availability and inexpensiveness of coal make it a source to derive other forms of fuels from it. Some of these fuels are listed here.

Coke: Coke can be produced by heating coal in the absence of air, which results in releasing volatile and liquid hydrocarbons from coal. The remaining of the process is porous carbon, which is ideal for iron production in blast furnaces. Some coke not suitable for this purpose (e.g., the size is too small) is used in steam generators.

TABLE 4.3

Coal Categorization according to ASTM Standard D388

Coal Category		Fixed Carbon (%)	Volatile (%)	Moisture (%)	Heating Value	Physical Characteristics	Applications	Subcategory
High-rank coal	Anthracite	86–100	0–14	Very low	25,600–32,600 kJ/kg (11,000–14,000 Btu/lb)	Shiny black, dense, hard, and brittle	Residential and commercial space heating	Meta-anthracite Anthracite Semi-anthracite
	Bituminous	45–86	14–40	Low	24,418–32,600 kJ/kg (10,500–14,000 Btu/lb)	Grayish black and granular	Electricity generation, coke production for steel industry	Low-volatile Medium-volatile High-volatile A High-volatile B High-volatile C
Low rank coal	Subbituminous	35–45	High	15–30	19,300–26,743 kJ/kg (8300–11,500 Btu/lb)	Black and homogenous	Electricity generation	A B C
	Lignite (Brown coal)	25–35	High	<30	14,650–19,300 kJ/kg (6300–8300 Btu/lb)	Various shades of brown, quite soft, with wood fibers commonly recognizable, and laminar	Electricity generation	A B
-	Peat			<90		Yellowish to brownish black, decomposing plant parts typically visible	Limited industrial combustion-Other industrial applications such as oil and water absorber	**Not listed in the standard**

There are several gaseous fuels derived from coal.

Syngas: This gas is produced from the gasification process and will be discussed later.

Producer gas: This gaseous fuel is produced as a result of the incomplete combustion of coal (with less than stoichiometric air) in the presence of steam.

Water gas: Water gas is produced when steam and hot coke interact and is mainly composed of hydrogen and carbon monoxide.

Coke oven gas: The gas is produced during the coke production process and is mainly composed of hydrogen and methane.

Blast furnace gas: The gas is produced during steel production in blast furnaces and is mainly composed of carbon monoxide.

In order for coal to be used as the source of thermal energy in steam generation in power plants, there should be a continuous and reliable flow of coal to the power plant site ready to be processed and/or combusted. This requires effective coal mining, preparation, transportation, and storage. In the following two sections, these topics will be briefly covered.

4.5.5 Coal Extraction

The first step for coal utilization is to extract it from a coal mine. Coal mining can be either **underground** (deep mining) or **surface** (open-cast or strip mining). When coal seams are relatively close to surface, the strip mining method can be used where the earth above the seams (known as overburden) is removed, and then coal is extracted (Figure 4.3). Typically, draglines or excavation shovels are used to remove the earth. However, for very large mine operations gigantic bucket wheel excavators are used (Figure 4.4). These are some of the largest moving machines that have ever been built (Figure 4.5). In this method, the topsoil is systematically removed and stored to be used to reclaim the land after the mining operation is ceased. This method is fast and

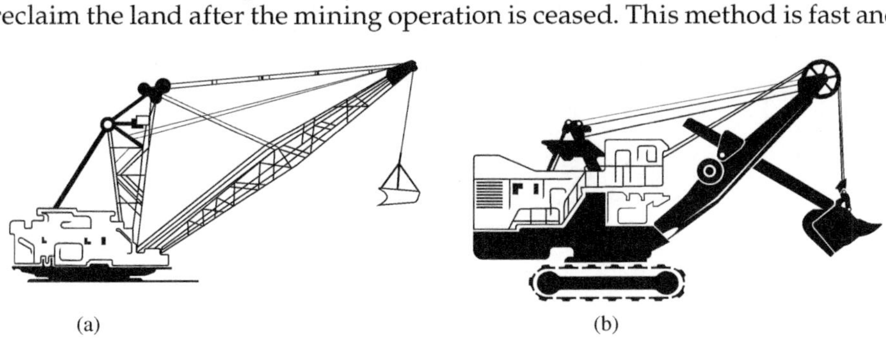

FIGURE 4.3
Schematics of a dragline (a) and an excavation shovel (b) used in strip mining.

FIGURE 4.4
A bucket wheel excavator in a brown coal mine (Note the front loader at the bottom for size comparison). (Image by Rolf Dobberstein from Pixabay, https://pixabay.com/photos/brown-coal-energy-garzweiler-283211/.)

FIGURE 4.5
A bucket wheel excavator (Note the front loader at the center for size comparison). (Image by Herbert Aust from Pixabay, https://pixabay.com/photos/garzweiler-brown-coal-2515764/.)

inexpensive, but reclaiming lands after mining operation is over, restoring vegetation, and handling acidic mine drainage are major challenges.

When coal seams are deep underground and are not available by surface mining, the underground mining method should be used. There are two main methods of underground coal mining: **room and pillar mining**; and **longwall mining**. In the former, only part of the coal seam is mined (about 50%) and the rest is left intact to support the roof. This remaining coal may be mined at the end of the mining operation, which results in the roof collapse while the operation retreats. In longwall mining, the entire coal seam is excavated using shearers or plows and the roof is supported by an artificial roof.

Most coal in the world and the U.S. are mined in surface mines. There were 430 surface coal mines in the U.S. producing 64% of the total coal, while the number of underground mines was 236 mines producing 36% of the coal in the country in 2018.[27]

4.5.6 Coal Preparation

After mining, coal needs to be processed. The extracted coal from mines, called **run-of-mine** (ROM), contains impurities that should be removed before it is used for combustion by end-users. The process can take place in a coal processing facility or rarely in the power plant site. The preparation process depends on the quality of coal and user requirements and may include the sizing, screening, crushing, and separation of impurities. This process can reduce the ash and sulfur content of the coal, which can significantly improve the combustion process and reduce the negative environmental impacts.

4.5.6.1 Coal Sizing

The mined coal, run-of-mine, is made of a wide range of coal pieces, from fine particles to very large pieces as large as or larger than 8 inches, and needs to be screened to the specified sizes. The coal sizing is mainly composed of screening and crushing processes. These processes typically take place in the coal preparation facility but sometimes some processes are left for the power plant site.

4.5.6.2 Screens

There are four types of screens commonly used in the coal preparation process:

- Gravity bars or grizzly
- Revolving type or trommel
- Vibrating type
- Shaker type

[27] Annual Coal Report 2018, EIA, https://www.eia.gov/coal/annual/pdf/acr.pdf.

Gravity bars or grizzlies are simply a series of inclined bars with the feed end at a higher elevation and the discharge end at a lower elevation. The size of non-passing materials depends on the gap between the bars, their slope, and their length. A trammel or revolving type screen is a cylinder made of wire screens or perforated plates with a small inclination from the inlet toward the outlet. To avoid coal breaking in the screen, this type of screen typically rotates slowly and is used for small size coal. This type of screen has a low screening capacity per unit of screening surface, which increases the capital cost of the system.

Shaker screens and vibrating screens (Figure 4.6) are both installed on a rectangular frame. Their difference is in the type of motion that they are exposed to. In shaker screens, the screen oscillates back and forth with a relatively low frequency and large magnitude, while in vibrating screens the motion is a high frequency and low magnitude. The motion is created by an unbalanced disk that is attached to an electric motor. The magnitude of the motion can be controlled by the location of the unbalanced weight on the disk, and the frequency is controlled by the rotational speed (rpm) of the motor. As a part of the screening process, magnetic separators can be also used to separate ferrous objects.

The screen (mesh) size is determined and expressed as the number of openings in a linear inch of the screen. A 100-mesh screen has 100 openings in every linear inch of the screen, which is equivalent to 10,000 openings in every square inch of the screen. The higher mesh screen means a smaller opening of the screen. The actual size of the openings depends on both the mesh size of the screen and the diameter of the wire that the screen is made of. The specifications of screens can be according to ASTM E11-01, W.S. Tyler, or ISO 565:1990 standards.

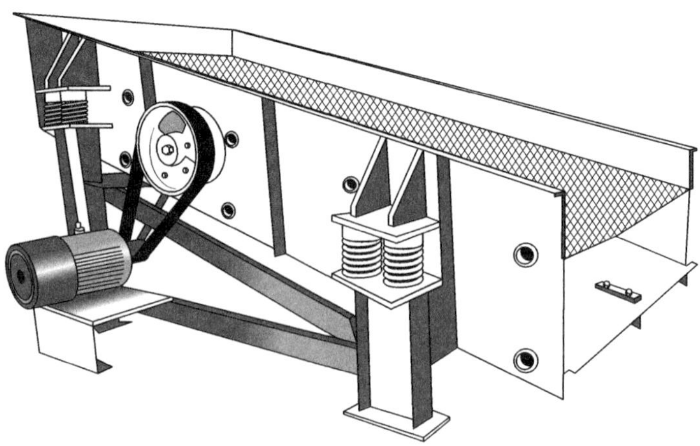

FIGURE 4.6
Schematic of a vibrating screen.

4.5.6.3 Coal Size Reduction

Typically the delivered coal to a power plant is clean and at the appropriate size to be burned in the stoker furnace or to be pulverized (these combustion systems will be discussed in later sections of this chapter). However, in some cases where the coal size is too large, it needs to be further crushed. In this case, crushers are part of the coal handling and preparation system. Note that crushing and screening are typically associated together. The most important parameter in the design and selection of the coal size reduction equipment is the reduction ratio, which is defined as the ratio of the feed particle size to the product particle size. For this purpose, the particle size for both the feed and the product is the screen size that 80% of the particles pass.

The breaking of coal can be achieved by one or a combination of the following mechanisms: impact, compression or crushing, and attrition. The particle size reduction by impact occurs when an object, such as a hammer, hits large pieces of coal at a high speed. In the compression (crushing) mechanism, coal pieces that are trapped between two surfaces break down because of the compressive forces of the surfaces. In the attrition-based grinding, the friction between coal particles breaks them when they move against each other under pressure. Note that according to Rittinger's law (after Peter von Rittinger, an Austrian pioneer in mineral processing, 1811–1872), the energy needed to reduce the size of minerals is proportional to the surface area ratio before and after size reduction. While this is not the only law to express the energy needed for material size reduction, probably it is the most commonly used one. Other similar laws include the Kick's law and Bond's law. They state that the energy needed to reduce the size of minerals is proportional to the volume reduction ratio and the square root of the average diameter reduction ratio, respectively. However, none of these laws are accurate, and typically empirical data are used in the design process.

Generally, crushers can be divided into two groups: **once-through crushers** and **retention crushers**. In once-through crushers, as the name implies, the feed coal goes through the crusher once and leaves the crusher regardless of the product size. But in the latter, the feed remains in the crusher until its size reaches a certain size so it can pass through a screen or a similar separation process that is integrated into the crusher. Generally speaking, the former types produce fewer fine particles but the product is more uniform in the latter types. Note that for some coal-burning methods, e.g., cyclone combustion (refer to Section 4.6.4.5), a high percentage of fine particles is not desirable. Among several types of crushers, the following types are commonly used for coal crushing:

- Bradford breakers
- Hammermills
- Ring crushers
- Roll crushers

4.5.6.3.1 Bradford Breaker

Bradford breakers, also known as rotating cylinders or rotary breakers, are retention crushers. They are made of a slowly rotating (at about 20 rpm) large cylinder made of perforated steel plates or screen plates (Figure 4.7). Inside the cylinder, there are lifting shelves that pick up the feed coal pieces and drop them from higher elevations. The impact of the falling coal pieces when they strike the screen or each other crushes them and when their size is small enough that they can pass through the screen, they leave the cylinder. The size of the product is determined by the mesh size of the screen. The feed enters in one end and rocks and other unwanted unbreakable materials are discharged in the other end. Since the gravitational force is relatively small, the fine particle production is limited. Other advantages of this type are a large capacity, ease of unwanted material rejection, low power consumption, and relatively uniform size of the product, which make this type the most effective crushing method for high-capacity applications.

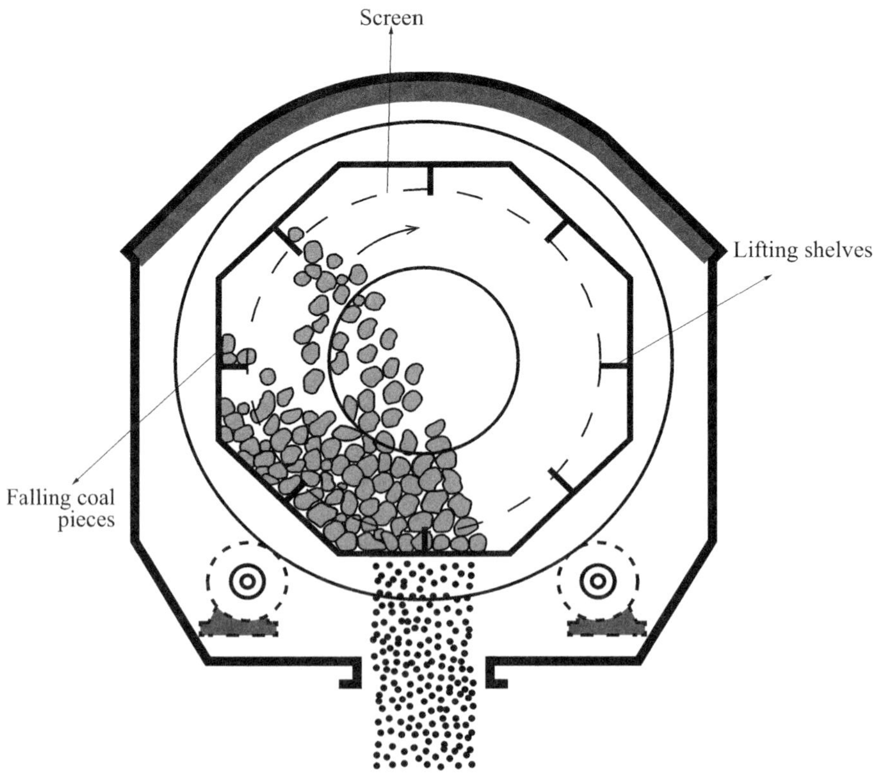

FIGURE 4.7
Schematic of a Bradford breaker.

4.5.6.3.2 *Hammermills*

Hammermills are made of a series of pivoting hammers attached to a rotor (Figure 4.8). The feed coal is crushed mainly by the impact force of the hammers when they strike the feed. The impact force depends on the hammers' size and shape and the rpm of the rotor. The product can leave the crusher through the grate at the bottom of the machine. The size of the product is determined by the gap between the bars in the grate. Like most other retention-type crushers, a considerable amount of fine particles is produced during the size reduction process. However, they are proven to be an economic and reliable crushing equipment for small-capacity applications.

4.5.6.3.3 *Ring Crusher or Granulator*

The operation of ring crushers is similar to hammermills (sometimes they are called the ring hammermills); except instead of swinging hammers, there are loose rings that can rotate off-center.

4.5.6.3.4 *Roll Crushers or Roller Mills*

In this type of once-through crushers, the feed coal is crushed by the compression and shear forces due to the rotation of rolls (Figure 4.9). There can be one or two rolls. The crushing action takes place between the roll and the adjustable wear plate in single-roll crushers and between two rolls rotating in the opposite directions in double-roll crushers. The size of coal pieces in the product depends on the size of the roll and the spacing between roll and plate or two rolls and can be adjusted. The most important problem with this type of crushers is the lack of size uniformity in the product.

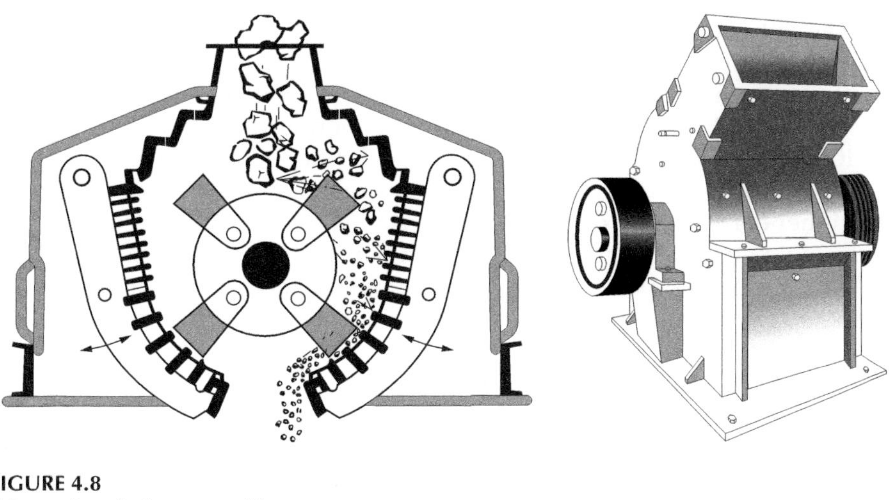

FIGURE 4.8
Schematics of a hammermill.

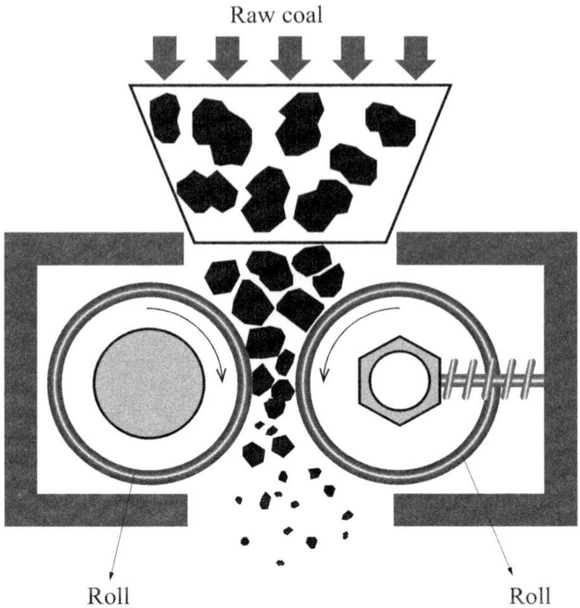

FIGURE 4.9
Schematic of a roll crusher or roller mill.

4.5.6.4 Coal Cleaning

In this section, the coal cleaning process to remove impurities to reduce ash and sulfur contents will be covered. The coal cleaning process and the removal of impurities can be either wet or dry. The application of the dry process is limited to small coal particles, and the majority of coal separation is achieved by the wet process. In the wet separation process, the difference between either **density** or **surface properties** of the coal and the impurities is used to separate them.

In the dense media separation method, the separation of larger size impurities from ROM can be achieved by the floatation process. In this process, a specific liquid with the density between the density of the coal and the density of impurities is used. The lower density coal pieces float over the liquid and higher density impurities sink.

In some methods, the different speed at which different pieces settle in a medium is used to separate them. In the jigging method, a continuous pulsation of water provides the suspension, settling, and separation processes. In the tabling method, this is achieved on an oscillating table. Note that a similar principle is used in dry separation systems too.

For the separation of finer coal particles, the mechanical separation or the froth flotation process may be used. In the former, the centrifugal force separates particles based on their mass. In the froth flotation process, fine coal particles and a specific liquid agent are mixed. The agent helps to create foam when air is blown into the mixture. The foam carrying the fine coal particles is concentrated on the surface of the mixture. The coal particles then can be recovered from the foam. In wet separation methods, the final product should be dewatered by either mechanical methods, such as vibrating screens and centrifuges, or by thermal dewatering methods, such as hot air driers.

4.5.6.5 Transportation of Coal to Power Plants

After the preparation of coal, it can be transported to users either continuously or by trucks, trains, barges, or ships. The cost of transportation can be a considerable portion of the coal cost depending on the distance between the coal source and the power plant site as well as the means of transportation. For a mine-mouth power plant, the cost of coal transportation is negligible. To minimize the cost and maintain the reliability of coal transportation, a combination of the above methods is commonly used. Typically a truck, rail car, and barge can carry about 23, 90, and 1400 metric tons of coal (about 25, 90, and 1500 short tons), respectively.

Alternatively, coal can be transported continuously by tramways or conveyors. Coal can be even transported in the form of coal slurry through a pipeline, though this is not common. In this method, coal is pulverized and then is mixed with water. The mixture of coal and water is pumped through the pipeline to the destination. At the destination, the water is separated and the coal is dried ready for combustion. These continuous methods are ideal for short distances.

According to EIA, on average about 40% of the total cost of coal is for transportation in the U.S.[28] As shown in Figure 4.10, about 70% of the coal consumed in the U.S. is transported by trains and the rest is roughly divided

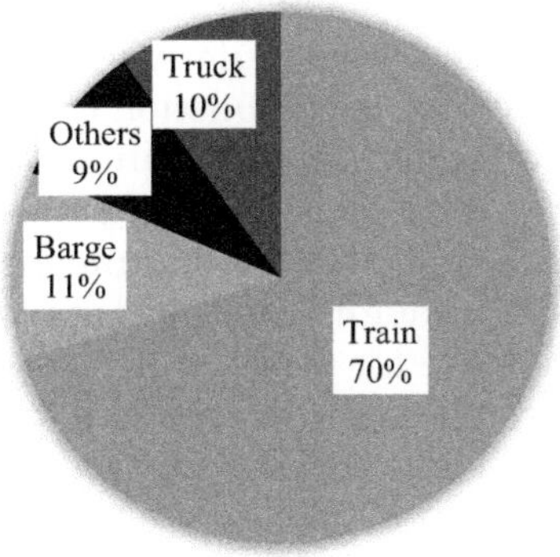

FIGURE 4.10
Share of different transportation modes in the total coal transportation in the U.S. (based on the data provided by EIA.) (Annual Coal Distribution Report 2015, EIA,)

[28] https://www.eia.gov/coal/transportationrates/.

equally between barges, trucks, and other methods (i.e., tramway, conveyor, and slurry pipeline). One reason for the popularity of railroad is the flexibility of the distant coal that can be economically transported, from a few ten miles to a few thousand miles. Note that the most economic mode of coal transportation is via rivers by barges, but it requires direct access to a river system at the plant location. In order to improve the reliability of coal transportation and to avoid delivery interruption, power plants try to make their coal sources as diverse as possible by receiving coal from several routs and modes simultaneously.

4.5.6.6 Storage of Coal in Power Plants

Upon arrival at the power plant, the delivered coal should be unloaded. The unloading method depends on the transportation mode. It can be self-unloading, such as bottom discharge or rotary dump rail cars. Other types of unloaders, such as clamshell buckets, can also be employed.

The unloaded coal can be directly delivered to bunkers for immediate or near future use or may be stored until the time arrives to burn it (long-term storage). The long-term storage is required to provide fuel for the plant when the coal delivery is interrupted or is not sufficient. The capacity of the storage depends on the capacity of the power plant. This storage is typically in the form of outdoor stockpiles where 30–90 days of coal supply for the plant are stored. However, for smaller power plants, or larger plants when the site experiences harsh winters or the land for outdoor storage is limited, silo storage may be used.

The pile is constructed in several layers and is compacted by a tractor to minimize the passage of air and the interchange of moisture with the atmosphere. To avoid these problems and to avoid the spreading of coal dust and coal freezing, sometimes a petroleum product is sprayed on the top and sides of the pile. The pile should be regularly inspected to identify any hot spot to avoid possible spontaneous combustion of coal. Whenever there is a need, the stored coal is reclaimed by either an under-pile system (through hoppers and feeders) or an over-pile system (by bucket wheels) and is sent to the bunkers. The most common method to transport coal from stockpiles to bunkers is conveyor belts, though sometimes screw and bucket conveyors are also used. The bunkers are for short-term storage of coal and typically can store 8 to 48 hours of coal supply for the plant, hence the other name for the bunkers, day-bins. These bunkers directly feed the coal to the preparation equipment, e.g., pulverizers, through a hopper and a feeder. The flow of coal pieces out of the bunkers can be challenging when coal, usually wet coal, adheres to the hopper walls. This can result either in so-called rat-hole where only coal directly above the opening can flow out of the bunker or coal arch (bridge), a total blockage. These problems can be addressed by manually ramming a long rod into the bunker, by vibrating the bin, or by sending strong air blasts.

Example 4.10

A 1000 MW coal-fired steam power plant with an efficiency of 30% needs to have 60 days of coal storage. If the plant is fueled with the average coal consumed in the U.S. (in terms of the heat content of coal) in 2019, determine the capacity of the coal storage in metric tons. The average heating value of the coal consumed in the U.S. in 2015 was 19.48 million Btu/short tons.[29, 30]

Solution

Given:

- A coal-fired steam power plant fueled with the average coal consumed in the U.S. (in term of the heat content of coal) in 2019
- Power plant capacity = 1000 MW
- Efficiency = 30%
- Number of days that the stored coal should last if the delivery of coal is interrupted = 60 days

Find:

- Determine the capacity of the coal storage

Assumptions:

- The average heating value of coal consumed in the U.S. in 2019 = 19.26 million Btu/short tons

Analysis:
The power produced in a fossil fuel steam power plant can be determined using the following equation.

$$P = \eta \, \dot{m}_f \, HV \tag{4.64}$$

Rearranging the equation, the mass flow rate of fuel can be found by this equation

$$\dot{m}_f = \frac{P}{\eta \, HV}$$

The unit of coal average heating value should be converted from million Btu/short ton to J/kg.

$$1 \frac{\text{million Btu}}{\text{short ton}} = 1 \frac{\text{million Btu}}{\text{short ton}} \times \frac{1055 \times 10^6 \text{ J}}{1 \text{ million Btu}} \times \frac{1 \text{ short ton}}{907 \text{ kg}} = 1.16 \times 10^6 \frac{\text{J}}{\text{kg}}$$

[29] https://www.eia.gov/tools/faqs/faq.php?id=72&t=2.

[30] Although EIA does not directly indicate if this is net (lower) heat content or gross (higher) heat content, typically the gross heat content is used in the U.S. unless it is specified otherwise.

$$\dot{m}_f = \frac{P}{\eta HV} = \frac{1000 \times 10^6}{0.3 \times 19.26 \times 1.16 \times 10^6} = 149.2\ \frac{\text{kg}}{\text{s}} = 1.29 \times 10^7\ \frac{\text{kg}}{\text{day}}$$

$$= 1.29 \times 10^4\ \frac{\text{metric tonnes}}{\text{day}}$$

$$\text{Capacity of the coal storage} = 1.29 \times 10^4\ \frac{\text{metric tonnes}}{\text{day}} \times 60\,\text{days}$$

$$= 7.73 \times 10^5\,\text{metric tonnes}$$

4.5.7 Other Solid Fuels

4.5.7.1 Biomass

Historically speaking human being's first fuel source to create fire was biomass. But due to its scarcity at the time of the industrial revolution, it was gradually replaced by fossil fuels, particularly coal. Biomass is defined as any living organisms that **are alive or were recently alive**. In this definition "recently" is needed to exclude fossil fuels since all fossil fuels have originated from living things millions of years ago. Based on this definition, both **biological** (**animal-based**) and **botanical** (**plant-based**) resources are considered biomass. The former is referred to as the primary biomass sources, while the latter is known as the secondary biomass sources. They can include crops and forest products and byproducts, animal fat, and food processing wastes. All biomass resources are directly or indirectly created by capturing CO_2 from air via the photosynthesis process. When biomass is burned, this CO_2 is returned to the atmosphere. However, it is not accurate to consider biomass resources as "**carbon neutral**" resources because the CO_2 released during their planting, cultivation, harvesting, processing, and transportation should be taken into account. Among various renewable sources of energy, only bioenergy can be directly converted to renewable liquid and gaseous fuels, particularly for transportation and materials for chemical synthesis to produce plastics and other similar products. On the negative side, significant variations in their chemical compositions and cell structures make the design and operation of biomass burning systems challenging. Furthermore, their growth requires large areas of farmland (which may be created by forest conversion), large amounts of water, and large quantities of fertilizers. Biofuels can be solid (such as wood and charcoal), liquid (such as ethanol, biodiesel, and various vegetable oils), or gaseous (such as syngas and methane). In the scope of this text, we are particularly interested in solid biomass resources, such as wood chips, sawdust, shipping pallets, lumber rejects, straw, vine clipping, leaves, bagasse (sugar cane remains after extraction of sugar), husks (from coconut), nutshells (from walnut, almond, etc.), peach pits, grasses, bamboo, coffee ground, rice hulls, bark, clarifier sludge, and many others.

The chemical composition of biomass can be expressed by proximate or ultimate analyses. Unsurprisingly the majority of technologies that are being used in biomass combustion systems have been originally developed for coal and then adjusted for biomass.

4.5.7.2 Municipal Solid Waste (MSW)

On average every American produces over 4.5 lb of garbage every day. This per capita solid waste generation pattern has been consistently unchanged since 1990.[31] Municipal solid waste (MSW) can be recycled, composted, landfilled, incinerated, and combusted with heat recovery in waste-to-energy plants. While increasingly more solid waste in the U.S. is being recycled or used for energy production, still over a half of the solid waste produced annually ends up in landfills. This is not only harmful to the environment but also increasingly more expensive. The suitable sites are becoming scarcer and farther away from population centers and more expensive to build because of more stringent regulations particularly to protect underground water. There is no single way to tackle this problem, and every country and region should come up with a combination of the following approaches: produce less waste, recycle and compost as much as possible, burn combustible remaining and recover the thermal energy for steam or hot water production, and finally landfill the ash from combustion.

In this section, we will briefly review the properties of MSW as fuel. Solid waste that is combusted is made of a varying mixture of paper, plastics, food scrap, wood and yard trimmings, metals, glass, textiles, and other wastes. The percentage of each component greatly changes from country to country and from region to region. More industrially advanced countries and regions have higher percentages of paper and plastic in their solid wastes because of the higher consumption of packaged goods. Also, the widespread usage of home garbage disposals in North America has lowered the percentage of food remains as the high moisture component of solid wastes. Both the higher paper and plastic contents and the lower moisture contents help to increase the heating value of the solid wastes. On the other hand, more recycling of paper and plastic reduces the heating value of the wastes and more recycling of glassware and aluminum cans and other metals increase the heating value. As a result of all these factors, the composition of solid wastes changes not only with geographical locations but also with time. The time variation can be long term or short term, even daily. The fuel composition can also be nonhomogeneous even in the same shipping. The higher heating value of solid waste can be in the range of 7000–14,000 kJ/kg (3000–6000 Btu/lb). One major advantage of solid wastes as fuel is the low concentration of nitrogen and sulfur in the fuel which simplify pollution control systems (Chapter 5).

[31] https://archive.epa.gov/epawaste/nonhaz/municipal/web/html/.

4.6 Combustion of Coal

Every fuel burning system for steam generators should be able to introduce and mix fuel and air for combustion, initially ignite the mixture, sustain the flame, handle ash, provide flue gas at appropriate conditions to heat-transfer surfaces downstream, and respond rapidly to load changes at all operating conditions. A steam generator should be able to perform these tasks with minimal excess air and pollutants in the combustion products (including unburned fuel, NO_x, and SO_2), high availability, and low maintenance while maintaining maximum efficiency, safety, and reliability and minimum cost.

Historically, there have been the following five different major coal combustion systems (ordered chronologically according to when they were invented):

1. Hand feeding
2. Mechanical stoker
3. Pulverized coal
4. Cyclone-furnace firing
5. Fluidized-bed combustion

4.6.1 Hand Feeding

We all have seen how coal was shoveled into locomotive boilers. While there have been many designs for hand-fed stationary steam generators, here we will only discuss one particular design called the Dutch oven (Figure 4.11). In Dutch ovens, the combustion process occurred in a cell adjacent and

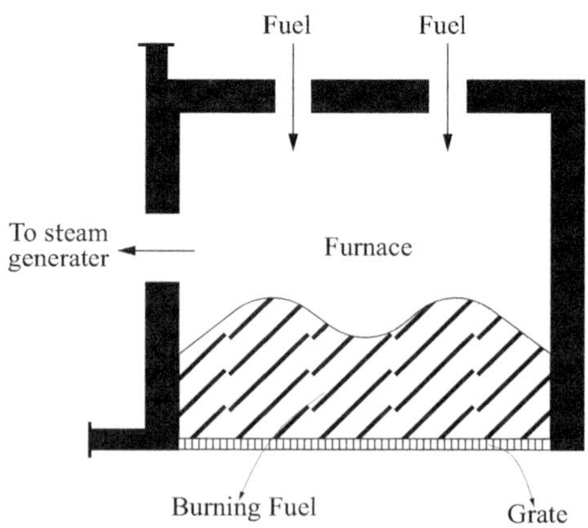

FIGURE 4.11
Schematic of a hand-fed Dutch oven.

connected to the steam generator. The walls were covered by refractory and sometimes water-cooled tubes. The solid fuels, typically wood or coal, were fed to the fire through openings on the roof and combusted in a pile. These ovens were suitable to burn high humidity fuels because the large burning bed could quickly dry the incoming fresh fuel. However, it needed to be shut down for ash removal and it could not adjust well to load variations and changes in the fuel properties.

4.6.2 Mechanical Stokers

In all stokers, the solid fuel is distributed over a grate to burn. A wide variety of solid fuels can be combusted in this system including the majority of coal types, biomass, and various solid wastes. They are responsive to rapidly changing steam demands. They do not require an extensive fuel preparation facility. In the case of coal, chunk coal is typically prepared in the coal preparation facility before being transported to the power plant site, and no cleaning or sizing processes is needed in the power plant location. However, their **efficiency and capacity are the lowest** among modern combustion systems. Because of these drawbacks, they are typically used for small steam generation units in industrial, utility, and cogeneration applications. The interest in this type of solid fuel firing system has been recently revived to some degrees for biomass and solid waste-fueled steam generators.

Stokers are composed of the **fuel feeding system**, **grate**, **air distribution system**, and **ash removal system**, which are all integrated into the bottom of a furnace. In all stoker combustion systems, first the fuel should be transferred from the hopper to the distributors through the gravity-based systems, reciprocating plates, or metering chain conveyors. Then, the fuel is distributed on the grate by one of the methods that will be discussed in this section (mass-feed, spreader, or underfeed systems). The fuel is combusted on the grate by the introduction of the combustion air through the grate. After complete combustion of the fuel, the remaining ash is discharged. The characteristics of mechanical stoker combustion systems are determined according to the grate type and the way the fuel is distributed over the grate. Accordingly, these burners are categorized based on these components.

4.6.2.1 Grates

The grate provides a surface for the solid fuel to burn. It can be **stationary or moving**. Early stokers had stationary grates with no continuous automatic ash removal system. In these systems, the ash was removed manually for example by raking after the furnace was turned off. Later, **dumping grates** allowed the ash removal from a section of the furnace while combustion continued on the other sections (Figure 4.12). In **inclined stationary grates**, the fuel moved along the grate by gravity (Figure 4.13). The stationary grates are no longer used in modern furnaces.

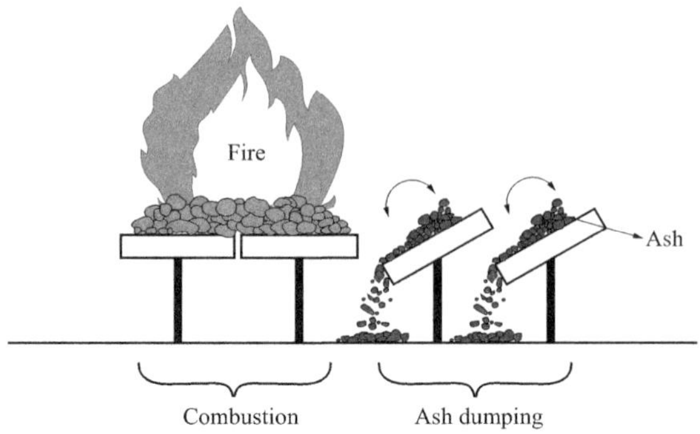

FIGURE 4.12
Schematic of a dumping grate.

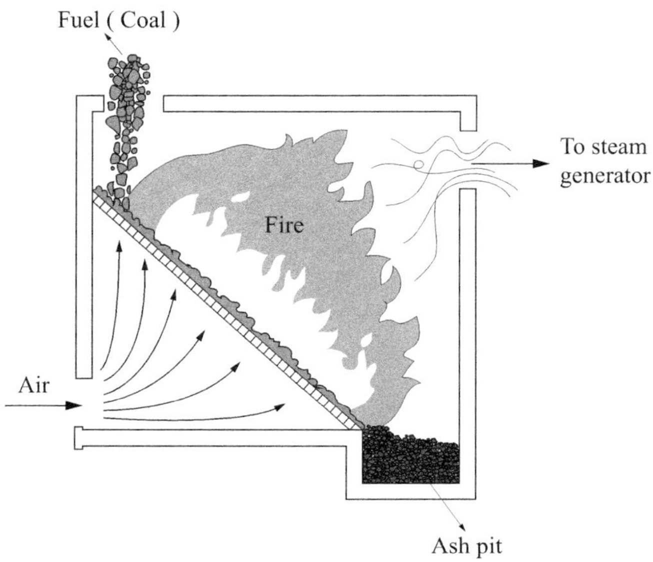

FIGURE 4.13
Schematic of an inclined stationary grate.

Traveling, moving, or chain grates are continuous (endless) conveyor belts where the fuel gradually burns and the ash is removed at the end of the belt where the grate turns around for the return run (Figure 4.14). The belt is driven by the drive drum or sprocket at one end and passes over the idle drum at the other end. This system allows a continuous feed of the fuel and a continuous discharge of the ash, which means the continuous operation of the burner is possible. This type of grate has many moving parts exposed to harsh operating conditions including high temperatures and high sliding contacts. This means the grate is subject to high wear, which in turn means

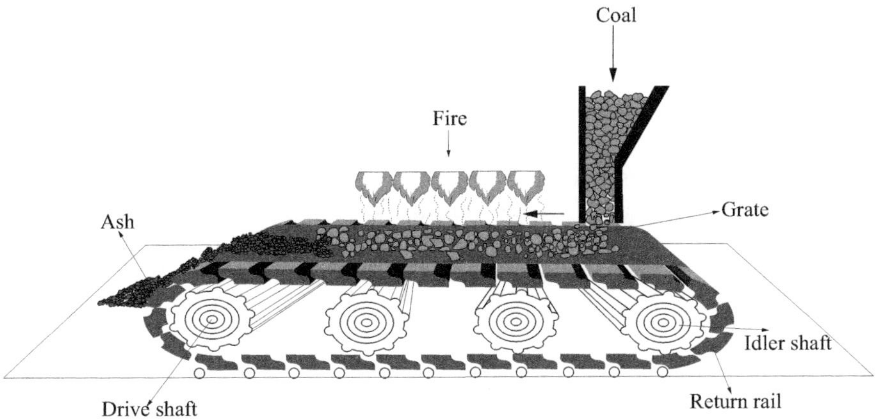

FIGURE 4.14
Schematic of a traveling grate stoker.

high maintenance costs. To reduce the wear and maintenance costs, the grate moves very slowly.

In **vibrating grates,** the fuel moves across the grate by intermittent vibration. The frequency and timing of the vibration can be adjusted to fit the fuel characteristics and its required residence time. Due to the simplicity and minimal moving parts, the costs of maintenance are lower in vibrating grates compared to traveling grates. Traveling and vibrating grates are the most common grate types used in mechanical stoker systems. They are typically made of cast iron bars with air holes in them. They can be either air-cooled or water-cooled. The latter is more effective, which in turn means the system can operate at higher operating temperatures.

4.6.2.2 Fuel Feeding Systems

There are two main ways to burn solid fuels: **thin-bed combustion** and **suspension firing**. In the former, the fuel is combusted while it forms a layer of solid fuel on a combustion surface. In the latter, the combustion process occurs while fuel is entrained in the airflow. In some combustion systems, a combination of two mechanisms is used.

Solid fuel in mechanical stokers can be distributed over the grate by several mechanisms. Depending on the position of the fuel feeding system relative to the grate, stokers can be either **overfeed** or **underfeed**. The former can be further divided into mass-feed and spreader systems.

In a **mass-feed system**, the fuel is continuously fed to one end of a horizontal grate, typically either a vibrating or a traveling grate. The height of the fuel bed is regulated before entering the furnace by an adjustable gate under which the fuel passes. The thickness of the bed controls the mass flow rate of the inlet fuel, which in turn controls the rate of combustion in the furnace. The fuel burns on the bed while traveling across the grate with the air flow passing through the grate. The thickness of the fuel bed reduces as it moves further in the furnace. The remaining ash is discharged to the ash pit at the end of the grate.

Unlike mass-feed systems, in **spreader systems,** the fuel pieces are projected or thrown into the furnace (Figure 4.15). The finer fuel particles are combusted in suspension, while the larger pieces fall over a burning bed and burn on the grate. Again, the grate is most commonly either a traveling grate or a vibrating grate. Depending on the size, up to a half of the inlet fuel may burn in suspension. While the entire system might be referred to as a stoker or spreader, the fuel distribution system is also sometimes confusingly called stoker, spreader, distributor, feeder, or spout. The fuel distributor should be able to spread the fuel as uniformly as possible over the grate. The fuel might be mechanically distributed by a rotor made of curved blades for uniform fuel distribution. Alternatively, the fuel might be pneumatically injected by the steam or air flow. Sometimes a combination of the two methods is used. To improve the uniformity of the bed, typically several distributors are used.

Note that the direction of the fuel motion on the grate depends on the type of the fuel distribution system always with the objective of maximizing the fuel residence time to provide more time for combustion. While in mass-feed systems, the fuel moves from the furnace front to its rear, away from the fuel feed end, in spreader systems, the direction is the opposite. Due to the nature of spreader systems, there is a potential for the high carryover of ash particulate and unburned fuel particles, depending on the fuel size. To reduce the negative impacts of this phenomenon, both on the environment and the

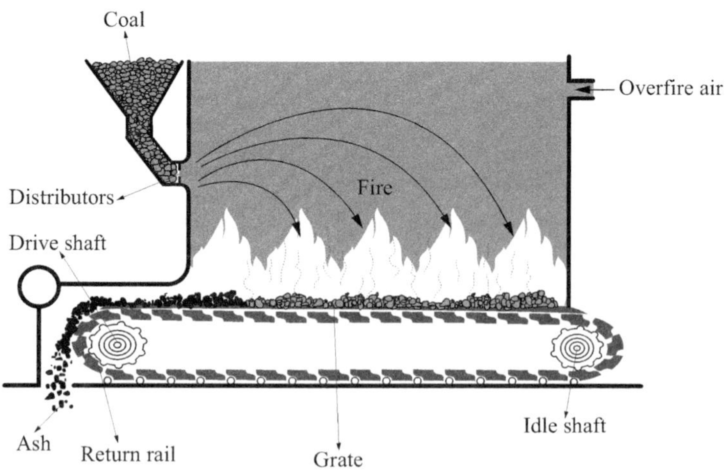

FIGURE 4.15
Schematic of a spreader stoker.

overall efficiency of the steam generator, dust collectors are used to capture and reinject unburnt particles to the furnace.

Another type of mechanical stoker system is underfed stokers where solid fuel is fed under the burning bed by a screw conveyor or a ram through one or several channels, called retorts. Among three different presented methods to feed the fuel to the grate, only spreader stokers are used in modern steam generators of small power plants. Other types can be only used for some biomass-fueled steam generators, for example, steam generators fueled with straw.

4.6.2.3 Air Systems

In mechanical stoker systems, the excess air can be as high as 50%. This air can be supplied as undergrate (underfire) air and/or overfire air (OFA). Also, some air may be used in the fuel distribution system in some spreader stokers. The amount of excess air and the distribution of air between undergrate and overfire air depend on the type of fuel and the type of the system.

The low-pressure undergrate air is to dry the fuel and provide combustion air. It also cools the grate in air-cooled grates. To uniformly distribute the undergrate air, it is important to have the fuel uniformly distributed over the grate, in terms of both the thickness and the density of the bed. Also, for proper distribution of the undergrate air, some systems compartmentalize air in several air plenums.

The overfire air is introduced above the fire (just above the fuel projectile in the case of spreaders). It is used for the staged air introduction to reduce NO_x production (refer to Section 4.6.4). It also provides turbulent mixing of the gases above the fire for better combustion. The pressure of OFA is much higher than the undergrate fire. Typically, the required air is provided by forced draft (FD) fans (refer to Section 4.9). While two smaller high- and low-pressure fans can be used to provide the undergrate and overfire air, one high-pressure large fan is more economical.

4.6.2.4 Design Considerations

The design of a stoker system depends on several parameters including the fuel type, fuel distribution type, grate type, desired steam production capacity, size limitations, and load characteristics. First, the required heat in the furnace can be determined based on the desired steam flow rate and its conditions. Then, the grate heat release rate can be used to determine the plan area of the grate. The grate heat release rate is the heat released from the combustion of the input fuel divided by the active or effective area of the grate. The value of the grate heat release rate depends on the type of fuel, fuel distributor, and grate type. Its values are provided by the manufacturers and can be roughly between 1 and 4 MW/m^2 (about 400–1100 Btu/h ft^2). The plan area of the grate and the furnace can be determined by dividing the required heat to the grate heat release rate. The depth (length) of the grate depends on the distance the distributer can propel the fuel in the spreader stoker and is limited by the grate manufacturers to typically less than 8 m (about 26 ft) to keep maintenance requirements at an acceptable level. It also depends on the type of the grate. Vibrating grates are typically shorter than traveling ones. The only remaining variable that can be manipulated to achieve the desired steam generation capacity is the width of the grate. The width of a single grate is limited to typically 6 m (about 20 ft). To overcome this limitation, two grates can be installed side by side in a furnace. As a result of all these limitations, the plan area of a grate is typically small. Due to the limitations on the physical size of the grate, the steam production capacity of stokers is limited depending on the fuel type, higher for biomass compared to coal.

4.6.3 Pulverized Coal Combustion Systems

The idea of pulverized coal combustion systems is that if coal or other solid fuels are broken down to a fine enough powder form with a high surface area, they can burn like a gaseous fuel. This is why **combustion systems in natural gas-fueled steam generators are very similar to pulverized coal combustion systems**. While pulverized coal steam generators are considered as one of the most important industrial and engineering achievements of the 20th century, this was a very old idea. Sadi Carnot (1796–1832) used powder coal as the fuel to describe his cycle. Rudolf Diesel (1858–1913, German inventor and mechanical engineer and the inverter of the diesel engine) tested his first internal combustion engine with powder coal. The first commercial application of pulverized coal was in the 1890s in the cement industry followed by the steal industry. In cement kilns, powder coal not only provides the heat required for the process but the ash content is also useful for the quality of the product. Thomas Edison tried to improve this process. Later, pulverized coal was used in steam generators for locomotives. In the 1920s, the previous experiences gained in the utilization of pulverized coal were used to manufacture the early pulverized coal steam generators for stationary power generation units with the first being operational in Milwaukee, Wisconsin. The introduction of pulverized coal steam generators removes many limitations on the size and capacity of steam generators imposed by mechanical stokers. In the decade following the introduction of these steam generators, their capacity increased ten folds, exceeding 125 kg/s (about 1 million lbm/h). The widespread utilization of pulverized coal power plants was a milestone for large, safe, reliable, and efficient power plants of modern era with relatively low emissions. These types of power plants have been the predominant coal combustion technology since the second half of the 20th century in many parts of the world. They are still the most common technology to combust coal for utility steam generation units.

Their virtually unlimited capacity is due to their high combustion rate. In the pulverized coal combustion process, fine coal powder provides a high surface area for combustion. During the combustion process, oxygen can diffuse to the combustion zone near the particle surface and the combustion product gases can diffuse out of the region. In pulverized coal, volatile matters can be combusted in a fraction of a second, while the entire particles burn within a couple of seconds, which is close to the combustion times of gaseous and liquid fuels. Also, pulverized coal steam generators can practically use all types of coal so their application is not limited to any particular region where a specific coal type is mined. They can be also fueled by liquid and gaseous fuels as well as other solid fuels, such as sawdust and sander dust. Furthermore, they require less maintenance since fuel is transported to the furnace by air. The fuel ignition and control is also simpler. They can rapidly respond to load changes. They require less excess air that improves the efficiency of the combustion process and reduces power consumption in the fans (Section 4.9). Finally, they reduce operational costs by reducing the labor cost and the maintenance cost.

Early systems were developed based on technical experiences and empirical data without much theoretical understanding. While theoretical understandings have significantly improved since then, still many design aspects are according to empirical data.

4.6.3.1 Overall Firing Systems

Any pulverized fuel combustion system should be capable of providing the required amount of thermal energy in the furnace whenever needed with high efficiency, high reliability, and low emissions. This requires a continuous flow of pulverized fuel to the burner with the rapid load-following capability. Early pulverizers were not reliable and their response to the load change was slow. In order to overcome this problem, early pulverized coal combustion systems had a storage system between the pulverizers and the burner. Understandably these systems are called **bin**, **storage**, or **indirect systems**. In these systems, pulverized coal after the pulverization process is separated from the air that carries the particles in a cyclone separator and bag filters (Figure 4.16). The separated pulverized coal is stored in storage bins. When the burner requires fuel, the pulverized coal is aerated again with new transport air and pneumatically conveyed to the burner for combustion. Due to the explosive nature of pulverized coal, the storage of the fuel makes these systems vulnerable to fire hazards. Also, the particle separation and then reaerating processes result in high power consumption in fans. The systems are large and dirty too. Regardless of all these problems, storage pulverized coal systems were widely used until more advanced pulverizers were developed. They are not being used in modern power plants.

The introduction of more reliable and faster-responding pulverizers opened the path to eliminate the storage process and the introduction of **direct firing systems**. In these systems, the pulverized coal from pulverizers is directly fed to the burner without any storage process (Figure 4.17).

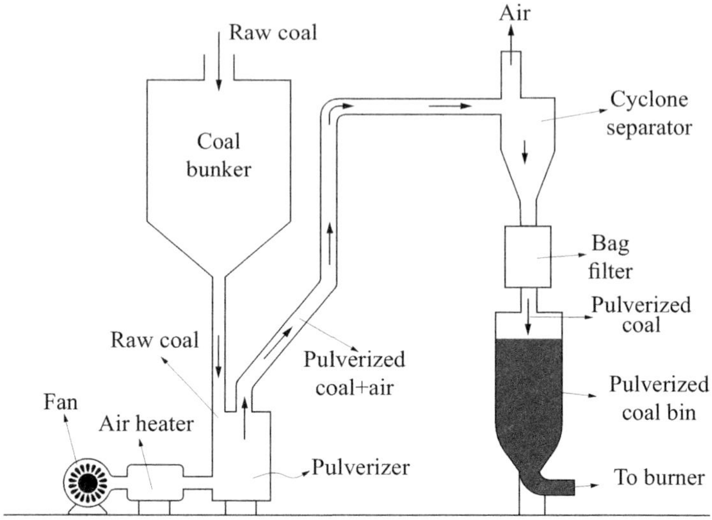

FIGURE 4.16
Schematic of an indirect firing system.

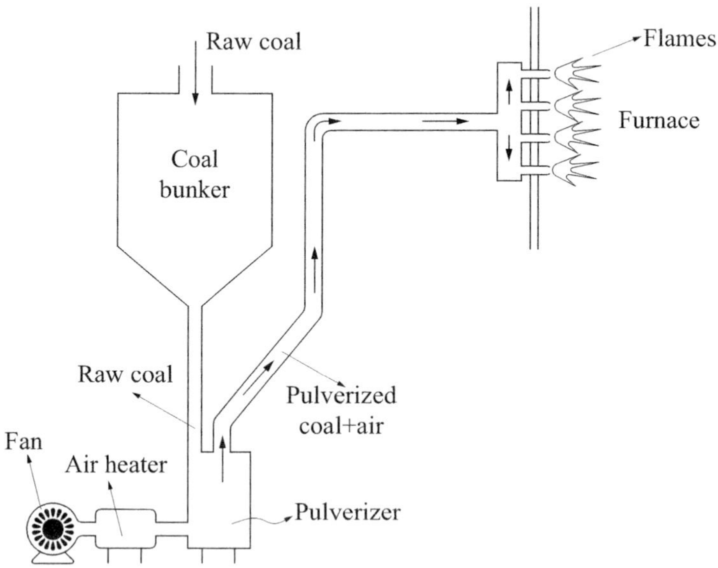

FIGURE 4.17
Schematic of a direct firing system.

The air used to dry and carry the pulverized coal is used for the combustion. The air that transports the pulverized coal to the burner is called the **primary air**. This air is typically a significant portion of the air required for combustion. This air is provided by the primary air fans, which can be placed after or before the pulverizers. If the fans are located after the pulverizer, the mixture of air and pulverized coal goes through the fan. But for the case where the fan is located before the pulverizer, the atmospheric air passes through the fan and then goes through the air preheater and the pulverizer.

The rest of the required air (the balance of air) is added to the burner and is called the **secondary** and the **tertiary air**. Since the fuel combusted in the burner is directly provided by the pulverizer, the control of the fuel flow rate (thus the steam flow rate) starts with the control of the inlet flow rate of coal into the pulverizers. This means the pulverizers must be able to respond to the load change rapidly, which was only possible with the advanced pulverizers introduced in the 1930s and 1940s. The continuous flow of the pulverized coal simplifies the system, which in turn reduces the capital costs. It also reduces the power consumption in the fans, since there is no need for reaeration, which reduces the operating costs. Moreover, the system is smaller (less floor space requirements), safer, and cleaner and requires less maintenance than storage systems.

Any pulverized fuel combustion system is composed of several major components: raw coal feeder, preheated primary air supply system, air fans, pulverizing system (pulverizer or mill), transporting piping system, burning system, and control systems. In the rest of this chapter, all these subsystems will be discussed in detail. The following few sections deal with the coal pulverization process. Note that typically the transported coal to the power plant is ready to be pulverized with no further preparation required.

4.6.3.2 Coal Feeder System

The pulverizer inlet feeder regulates the flow rate of the inlet coal to the mill. This is an important component, particularly for direct firing systems. Note that the steam demand and the required thermal energy in the furnace are the ultimate control objective. Since the pulverized coal is always entrained by the flow of hot air to control the coal inlet to the furnace both the coal mass flow rate and the air flow rate must be regulated. Typically, there are two ways to feed the coal and control its inlet flow rate: **belt feeders** and **overshot feeders**. A belt feeder is a conveyor belt where the speed of the belt regulates the fuel flow rate. They can be volumetric or gravimetric systems. In the former, the thickness of the coal bed on the belt is controlled to control the volumetric flow rate of the fuel. Alternatively, sensors might be used to directly measure and control the weight of the inlet coal. An overshot feeder is made of a rotor with redial long blades. The spaces between these blades are filled with coal. The coal carried by the overshot feeder depends on the rotational speed (rpm) of the rotor. A higher rpm means a higher fuel flow rate. Note that the inlet fuel weight is also needed to determine the efficiency of the power plant.

4.6.3.3 Pulverization-Related Coal Properties

The most important characteristic of coal for the pulverization process is its grindability. As the name implies, the grindability is an index that indicates how easy coal can be grinded or pulverized. The higher grindability of coal means that the coal is easier to be pulverized. Note that grindability is different from hardness. While lignite is soft and anthracite is hard, they are both hard to pulverize and have low grindability indexes. The grindability is expressed by the **Hardgrove Grindability Index (HGI)**. The HGI is measured by the level of grinding in a sample when it is grinded in a miniature grinder for a specific input work according to ASTM D409 standard. This method is not accurate because it does not take into account the moisture in the raw coal, which can be between 2% and 70% (measured according to ASTM D3302) and the degree of fineness of the pulverized coal. Also, it is difficult to directly relate this index to the operation of real pulverizers because the measurement method does not take into account the coal particle circulation, their continuous removal, and the much lower crushing force during the test. However, while more accurate methods have been developed recently, to use the vast data accumulated in the past decades, the HGI is still being used today.

The product sizing is very important for the operation of a pulverizing system. As noted earlier, the mesh of a screen is defined as the number of openings per linear inch of the screen. In order to have successful combustion of pulverized coal, it is critical to have right sizing of the fuel. This means to have a high percentage of very fine coal, say at least 80% passing through a 200-mesh (74 μm, 7.4×10^{-5}m) screen. But this alone is not sufficient. Even a small percentage of very large coal particles say more than 5% that is retained on a 50-mesh (297 μm, 2.97×10^{-4}m) (less than 95% passing through a 50-mesh screen), can result in the loss of unburned carbon, which in turn means lower efficiency and more air pollutions. Therefore, a pulverizer should be able to produce a very high percentage of fine particles and avoid even a small percentage of coarse particles. The fineness of produced pulverized coal is determined according to ASTM D197. Note that since the high-rank coal types have higher percentages of carbon, to have complete combustion in the pulverized form, they need to be finer than the low-rank coal types.

Another important coal characteristic for the design and operation of pulverizers is its abrasiveness. The abrasiveness of coal is related to its hardness and depends on the percentage of silica or pyrite (FeS_2) and quartz (SiO_4) in the coal which can be measured according to ASTM D2492 and ASTM D2799, respectively.

4.6.3.4 Pulverizers

The objective of pulverizers (mills) is to consistently prepare the pulverized (powder) coal with appropriate fineness and provide a uniform flow of the fuel for burners. The size reduction in pulverizers can be achieved by means of **compression** (crushing), **impact**, **attrition**, or a combination of them. There have been several types of pulverizers: ball-and-tube, ball-and-race, roll-and-race, and hammer mill. Since the dominant grinding mechanism in pulverizers is directly proportional to their rotational speed, they can be also categorized according to their rotational speed (rpm): low, medium, or high speed.

4.6.3.4.1 Drying and Separation of Coal

In early pulverizing systems, ambient air was used for the coal pulverization process with minimal drying capability. These units were only able to grind low moisture coal. Later systems were equipped with hot air external driers. In all modern units, the hot air drier is integrated into the pulverizer. The thermal energy to heat the hot air can be recovered from combustion products in the air preheater. Typically the same air, known as the primary air, that is used for drying, circulation, and separation of the pulverized coal is used for the transportation of the fuel to the burners and for the combustion process (in direct firing systems). Alternatively, the flue gas can be directly used for this purpose. This was common in the storage firing systems where the drier air was not the primary air for the combustion so the oxygen content of the flow was not a factor. This is not common in modern power plants.

The fans that provide the required air can be ahead of the mills or after them. In the former, the fans, also called the blowers, blow clean ambient air and the mills operate under pressure, hence the name **pressurized mills**. In the latter, the fans, called exhausters, suck the mixture of air, moisture, and pulverized coal out of the mill and the mill operates under partial vacuum (the **suction mills**). Typically, the combination of these fans is used in a **balanced system**. These fans should be able to overcome the pressure drops throughout the system in the pulverizer, classifier, pulverized coal transportation and distribution system, and burners. For more discussion on the fans refer to Section 4.9.

The control of the size of pulverized coal particles is critical for the proper operation of burners. There are two separation mechanisms for this purpose: **gravitational** and **centrifugal**. Separators can be either internally integrated separators or external separators. The centrifugal-based separation is typically the dominant mechanism in both of them. Generally, the former is the preliminary means of separating large particles from the pulverized coal in or near the grinding zone and the latter is the final means before the pulverized coal is sent to the burners. In the centrifugal-based separators, the centrifugal force is generated by inducing rotational motion by means of vanes or rotating classifiers. External classifiers are commonly made of a cyclone that the air flow can be controlled by adjustable inlet air vanes. The particle size distribution for each classifier is provided by the manufacturer.

It should be noted that in order to keep the fuel particles suspended at all loads, a minimum air flow rate is required regardless of the flow rate of the pulverized coal which depends on the steam generation load. This required minimum primary air can be as high as 70% of the primary air flow rate at the full load to keep a minimum speed of the air within the separators. For the rest of this section, we will take a look at some common pulverizers.

4.6.3.4.2 Ball-and-Tube Mills

Ball-and-tube mills were the dominant technology in early pulverized coal power plants. They were typically used with bin storage firing systems. They are made of a horizontal large cylinder internally lined with wear-resistance blocks (Figure 4.18). The cylinder is partly filled, to a quarter to one-third of its diameter, with forged steel or cast alloy balls. To avoid gaps between the balls, they have varying diameters between 25 and 100 mm (1–4 in.). The coal enters the mill from one end (in single end mills) or both ends (in double end mills) units along with hot air. The cylinder slowly rotates on its axis at the rotational speed of less than 40 rpm. The coal pieces are trapped between cascades of tumbling balls and are broken down. The larger pieces of coal break by impact, and the pulverization occur by attrition and crushing. When the coal pieces are reduced to small enough particles, they are carried by the flow of hot air from one or both ends (single or double end, respectively). The air flow also dries the grinded coal. The system is equipped with an external classifier to further separate the large particles and return them to the cylinder.

The ball-and-tube mills have a limited drying capacity and could not handle coal with a high moisture content. While they have high wear in the internal components, they are typically considered low maintenance machines due to their simplicity. They have a long coal residence time which results in the production of more than required fine particles. This means a lower capacity

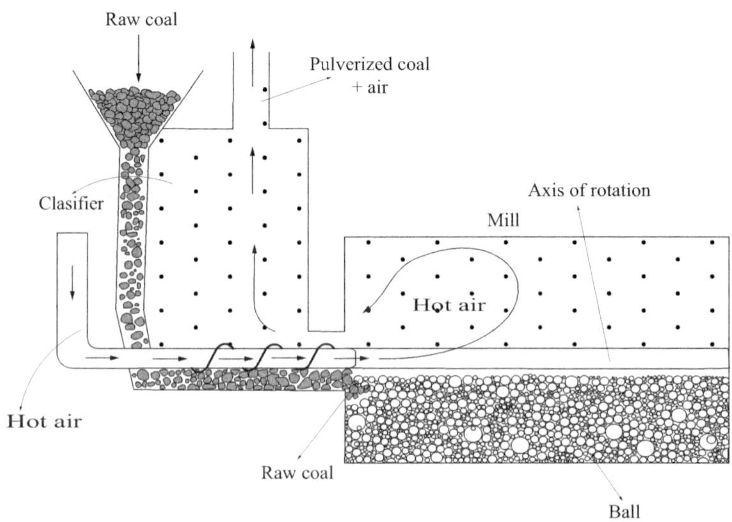

FIGURE 4.18
Schematic of a ball-and-tube mill.

for the same size and high capital costs as well as a higher specific power consumption (kW/kg of pulverized coal) and high operating costs. Furthermore, they are difficult to control and extremely noisy. Because of these problems, they are not typically used with direct firing systems. However, they are ideal for pulverizing hard to grind materials, including coal with low grindability.

4.6.3.4.3 Ball-and-Race and Roll-and-Race

As noted earlier, direct firing systems require more reliable and responsive pulverizers than what can be provided by ball-and-tube mills. In the 1930s, **ball-and-race** and then **roll-and-race** pulverizers were introduced to satisfy these requirements. These units are still the most common types of pulverizers in modern steam power plants. They are typically in the vertical configuration and hot air carries and classifies the pulverized particles. They are sometimes called vertical air-swept pulverizers. They can operate reliably with a variety of coal types, even very wet coal, with high availability. They can produce a uniform fineness even with aging components. They can also respond to the load change rapidly. They are compact and require small specific floor space, which is defined as the floor space per unit of pulverization mass capacity. Moreover, they have a low specific power consumption (kW/ton). They typically operate relatively quietly. Finally, they typically require low maintenance and their maintenance is easy.

In this type of pulverizer, coal enters the center of the unit at the top. The rolling elements roll over a surface, either race or ring, covered with coal that is being crushed (Figure 4.19). The rolling elements can be either balls or rolls. Depending on the design, either the rolling elements or the surface rotates, which in turn breaks coal pieces. The grinding mechanisms are mainly crushing and attrition. When coal particles are fine enough, they are suspended and carried away by the flow of hot air. The larger particles are then further separated by a classifier and returned for more grinding. So the separation process can occur in two stages: in the grinding zone and in an internal or external classifier. The internal classifier is much more common.

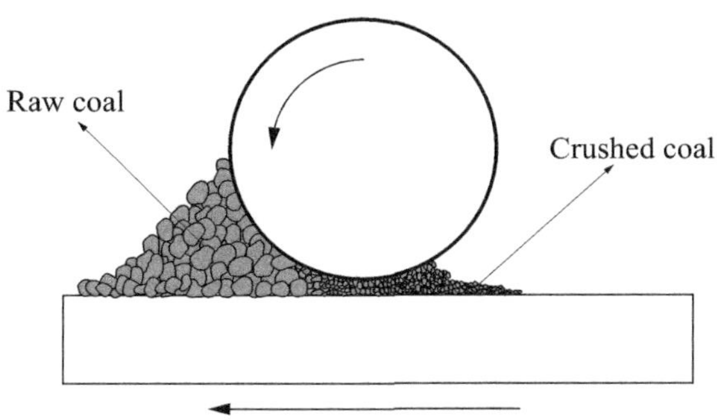

FIGURE 4.19
Pulverization mechanisms by a round object, crushing and attrition.

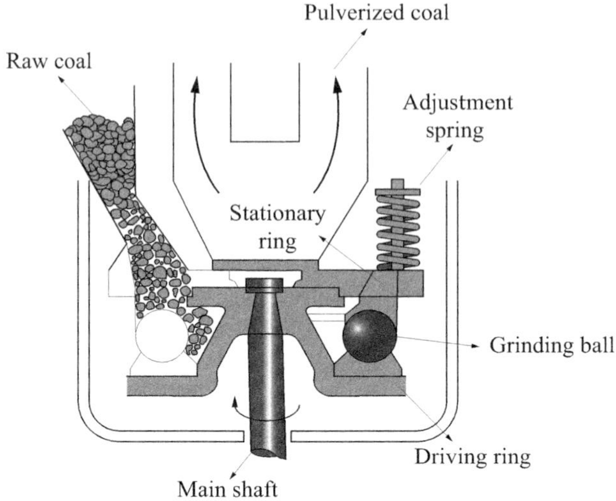

FIGURE 4.20
Schematic of a ball-and-race pulverizer.

In ball-and-race mills, the ball-shaped rolling elements are enclosed between two rings, Figure 4.20 (like a ball bearing). Typically the top ring is stationary and the bottom ring is driven at a medium rotational speed (up to about 250 rpm) which causes the rotation of the balls. The grinding pressure is applied to the upper ring by an external spring. This pressure can be adjusted to fit the requirements of the coal type and its grindability. The ball and mills are now typically used for industrial and small utility applications.

In roll-and-race (ring-roll) pulverizers, the grinding elements are several rollers, typically three, on a race or ring (Figure 4.21). Either the rollers are driven

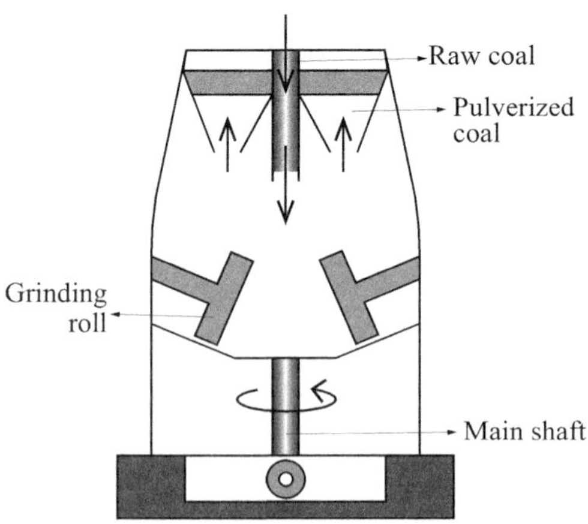

FIGURE 4.21
Schematic of a roll-and-race pulverizer.

and the ring is stationary or the rollers are stationary and the ring is driven. The rational speed of the system is much slower than ball-and-race mills.

Both ball-and-race and roll-and-race mills are sized according to the diameter of the race or ring which can be up to 2.5 m (about 100 in) for ball-and-race mills and up to 3.5 m (about 140 in) for roll-and-race mills. They can be over 10 m high.

4.6.3.4.4 *Other Types of Pulverizers*

The operation of impact or **hammer pulverizers** is similar to hammer mills discussed in Section 4.5.6.3.2 (Figure 4.8), but in this case, the pulverized coal is entrained by hot air. They are also equipped with an internal or external classifier. While these mills are suitable for small applications, they are not used in large modern power plants. One notable characteristic of these mills is their capability to grind rice hulls and grains that may open a new avenue for their applications.

A particular type of impact pulverizers is called the beater (wheel) mills that can handle very low-grade coal with very high humidity. To dry the coal, hot combustion products are used in this type of mill. These mills can be used in very large supercritical steam power plants (over 1000 MW) with an efficiency at low 40%. They are common in countries with large resources of lignite (brown) coal.

4.6.3.4.5 *Selection of Pulverizers*

The capacity of coal pulverizers depends on the grindability of the inlet coal and the required fineness of the pulverized coal. The actual grindability of the coal is a function of the coal Hardgrove Grindability Index (HGI) and its moisture level. There are empirical diagrams that relate the HGI to the percentage of moisture for various types of coal.

The required size of pulverized coal depends on the coal-burning characteristics and the specifications of the combustion system. The capacity of pulverizers, called the nominal capacity rating, is normalized and presented by manufacturers for the coal grindability of 50 HGI and pulverized products at 70% passing 200-mesh. The nominal capacity rating depends on the type, size, and manufacturer of the unit.

The required mass flow rate of pulverized coal is determined from the required steam flow rate and the combustion calculation. Generally, this mass flow rate of pulverized coal is not equal to the nominal capacity rating of the pulverizers because, as noted above, the nominal capacity rating is for the specific inlet coal HGI and the fineness of the products. The equivalent pulverizer capacity can be determined by the following equation:

$$\text{Equivalent pulverizer capacity} = \frac{\text{Required pulverized coal mass flow rate}}{\text{Capacity correction factor}} \tag{4.65}$$

The capacity correction factor depends on the HGI of the coal and the required fineness of the pulverized coal. The value of the factor is between 0.2 and 1.6. The higher HGI means the coal is easier to grind so a lower capacity pulverizer is needed. This is achieved by a higher capacity correction factor. In contrast, if finer pulverized coal is needed (i.e., higher percent passing through 200-mesh), a higher capacity pulverizer is needed, which in turn means a lower correction factor. In a design, all possible current and future inlet coal types should be considered and the worst-case scenario should be used for the design.

Now the number of the pulverizer units can be determined as follows:

$$\text{Number of the pulverizer units} = \frac{\text{Equivalent pulverizer capacity}}{\text{Nominal capacity rating of each pulverizer}} \tag{4.66}$$

The nominal capacity rating of a pulverizer is provided by the manufacturer. At this point, the designer can choose between fewer larger units or more but smaller units. The selection should be a compromise between capital costs, power consumption, and maintenance to minimize the overall cost while satisfying the operational requirements. It should be noted that the specific capital cost ($/kg of capacity) of smaller units is higher than that of larger units. It is common practice to add one extra pulverizer so the power plant can operate at its full load even with one pulverizer out of service.

Finally, the manufacturer should provide the pulverizer and burner coordination curves that interrelate the pulverized coal mass flow rates from each pulverizer to the steam mass flow rate, primary air flow rate, air-to-fuel ratio, and other factors. These parameters are critical for the design and selection of burners.

The performance of pulverizers is examined according to ASME Performance Test Code (PTC) 4.2. Due to the high combustibility of pulverized coal, pulverizers need to follow the suggestions presented at the National Fire Protection Association's NFPA 85 Standard chapter 9, "Pulverized Fuel Systems".

4.6.4 Burning of Pulverized Coal

A properly designed burner should receive fuel, primary air, and secondary air at a correct ratio; highly mix them; initiate combustion; create stable and safe flame; and achieve high combustion efficiency that approaches that of gas and oil burners while minimizing NO_x and other pollutant production.

As noted earlier, the main idea behind pulverized coal combustion systems is that if coal particles, for that matter generally any solid fuel particle, are reduced to small enough sizes, they will burn like gaseous and liquid fuels. The fuel residence time in pulverized coal burners is about 1–3 seconds which approaches that of natural gas and liquid fuels and far shorter than the resident time for other coal-burning technologies that can reach 60 seconds. As a result, pulverized coal burners are very similar to oil and natural gas burners, except they do not need atomizers for liquid fuels. They can also burn multiple fuels, including gaseous, liquid, or solid fuels.

In any burner, the fuel and air are mixed in an appropriate proportion and the mixture enters the furnace. In pulverized coal burners, the fuel at the rate dictated by the steam generation requirements is carried to the burner suspended in air. This air is called the primary air (PA) and is typically about 10%–30% of the total air. This is the same air that has gone through the pulverizers. The temperature and humidity of this air depend on the coal characteristics, coal mass flow rate, and pulverizer's technology. The temperature of PA is typically between 50°C and 100°C. Each pulverizer can typically feed 2–8 burners.

The rest of the combustion air, known as the secondary air, is added to the burner or the furnace for complete combustion. The secondary air is provided by the forced draft (FD) fans after going through the air preheater and is at the temperature of about 250°C–450°C (refer to Section 4.9.5). In order to avoid a complicated piping system to distribute air to the burners, the secondary air is fed to a duct system, called the windbox, outside of the furnace wall right behind the burners (Figure 4.22). The windbox can be open plenum or partitioned, the latter having a more uniform air distribution among burners. It can be located on opposite sides of the furnace or just on one side (as in Figure 4.22).

During the combustion process, the coal particles along with air enter the combustion zone. The heat transfer, via both radiation and convection, from the combustion already occurring in the furnace increases the temperature of the particles. Because of this increase in the temperature, the moisture and volatile matters evaporate. The gaseous volatile matters combust immediately, which further increases the temperature of the particles and helps to burn the remaining carbon content. Unlike volatile matters, the carbon content burns slowly. Eventually after burning the entire carbon content ash will remain.

It should be noted that the primary air must be able to carry coal particles in suspension and there is a minimum velocity of air that this can happen. Therefore, when the load of the steam generator and thus the mass flow rate of the fuel reduce (to as low as 25% of nominal load), the required reduction in the mass flow rate of air is mainly achieved by control of the secondary air mass flow rate while the flow of the primary air is mainly unchanged. The ratio of SA to PA is critical to achieve a stable flame. If the ratio goes below a certain level, the flame may become unstable.

The molecular interaction between fuel and air is critically important for the successful operation of pulverized coal combustion systems. There are two approaches to have an effective mixture of air and fuel: wall-fired systems and corner-fired or tangential combustion systems (single flame envelope).

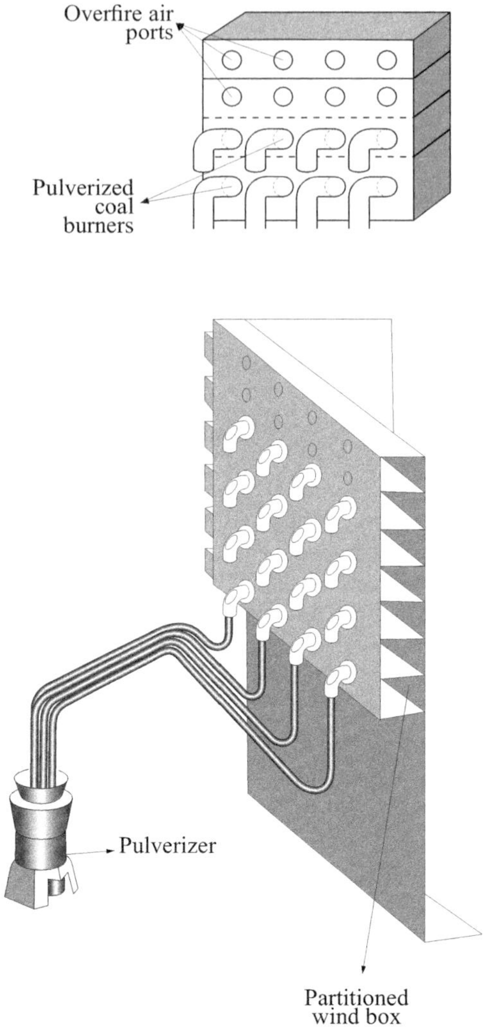

FIGURE 4.22
Schematic of a combustion system, including, furnace, windbox, and burners.

In wall-fired systems, many burners are installed on the walls of the furnace (Figure 4.22). Each burner creates its own individual and mostly independent flame envelop. This means while there are significant mixing and turbulence within each envelop, there is not much interaction between envelops and certainly no total unified turbulent flow pattern is created in the furnace. They are also called multiple flame envelope burners. Note that depending on the furnace design, the combustion products from various burners interact with each other after they leave the combustion zone. The wall-fired systems can be positioned horizontally or vertically. In horizontal wall-fired systems, the burners can be positioned on one wall (Figure 4.22) or two opposite walls. In the vertical firing systems, also known as arch firing or downshot, the flame enters the furnace vertically. This firing method is complicated and expensive and is not used except for very special fuels, such as hard-to-burn fuels. Tangential combustion systems will be covered in Section 4.6.4.4.

Pulverized coal burners can be grouped into the pre-NO_x restriction era designs and low NO_x designs. Since NO_x generation plays a very important role in the design of burners, a brief introduction to NO_x production will be presented here.

4.6.4.1 NO_x Production

NO_x is the generalized name for nitric oxide (NO) and nitrogen dioxide (NO_2), which are major air polluters (for more discussion on NO_x emissions and their environmental impacts refer to Chapter 5). They are generated from the chemical reaction of nitrogen and oxygen. Depending on the source of nitrogen, there are two types of NO_x: thermal NO_x and fuel NO_x. While air nitrogen and oxygen do not react at low temperatures, at high temperatures they can react. The produced nitrogen oxides in this case are referred to as **thermal NO_x**. Alternatively, if the fuel contains nitrogen in the compound forms, they can react with oxygen during the combustion process to generate NO_x, which is called **fuel NO_x**.

Coal typically contains 0.5%–2% nitrogen in both volatile matters and solid forms. The nitrogen in volatile matters is nitrogen compounds, such as hydrogen cyanide (HCN) and ammonia (NH_3), which are relatively unstable. In fuel-rich conditions, the nitrogen from these compounds can be reduced to nitrogen atoms (N_2) but they form NO_x under air-rich environments.

The portion of fuel NO_x that comes from nitrogen compounds in volatile matters is typically higher than that of the nitrogen in the solid phase due to the lower conversion rate of nitrogen to NO_x in char (carbon). Also, the combustion of char mostly occurs late in the combustion process where most oxygen in the air is consumed and less oxygen is available for NO_x formation.

The objective of any burner is to combust the fuel in the most economical manner. In the absence of environmental constraints, this means mixing the fuel and air as uniform as possible and then combusting them at the highest temperature possible. This design philosophy is to achieve the highest combustion rate for a given volume of the furnace. While this design can eliminate or minimize the production of CO and unburned carbon, the NO_x production level is typically unacceptable by modern standards and regulations.

While the details of pulverized coal burner design depend on coal properties, pulverized coal characteristics, and combustion conditions, the level of permissible NO_x production is critical in their design. So the design of these burners can be divided to pre- and post-NO_x restriction era designs (roughly before and after 1970, respectively).

4.6.4.2 Pre-NO_x Restriction Era Pulverized Coal Burners

In early burners, known as **circular burners**, the mixture of pulverized coal and primary air entered the burners axially at the center of the burner through a nozzle. To better mix the fuel and primary air before leaving the nozzle, some sort of swirl generator, such as an impeller (called coal impeller), was installed in the tip of the nozzle to induce mixing. The secondary air was admitted from the windbox through tangential register doors (vanes) (Figure 4.23). These interlinked adjustable vanes built into the windbox had two main functions, to control the secondary air flow rate and to induce rotation to the secondary air flow so it can properly mix with the coal and primary air mixture out of the nozzle. This induced swirl was a critical parameter for the stability of the flame and the shape of the flame envelop. Too much or too little swirl could hurt the combustion process. In this design, these two functions could not be controlled independently. It turned out that these two functions are not always complementary. To increase the air flow rate, the register door should further open, which reduced the imparted swirl. Despite these problems, this design was used for many decades with wide range of capacities, between 15 and 90 MW thermal load, not just for pulverized coal but for natural gas and liquid fuels too. In some designs, two or three of these burners integrated to a single assembly.

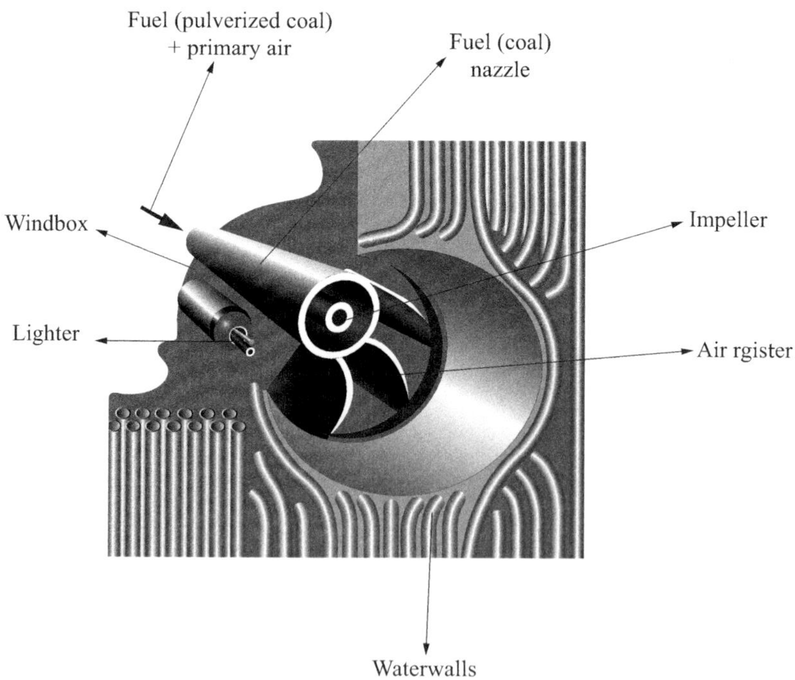

FIGURE 4.23
Schematic of a circular burner.

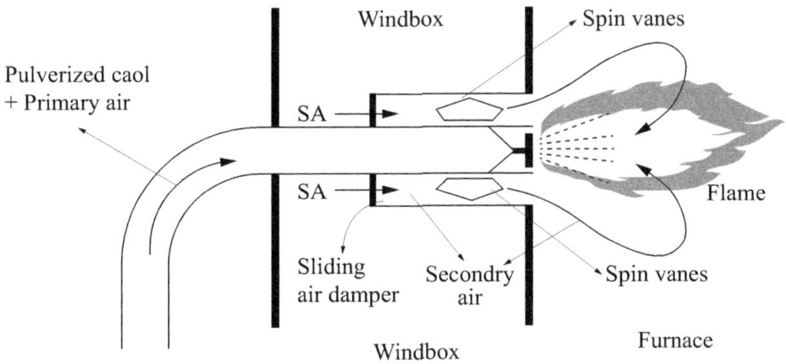

FIGURE 4.24
Schematic of S-type burners with sliding disks and adjustable spin vanes.

To address the problem of the dependency of the secondary air flow rate and its swirl, in later designs (called S-type burners), these two parameters were controlled by two different components. The air flow rate was controlled by **sliding disks** (air dampers) and the swirl was created and controlled by **adjustable spin vanes** (Figure 4.24).

The problem with all these designs was their high NO_x production. This level of NO_x emission could not meet the regulatory limits after the early 1970s. No power plant in the U.S. is using these burners nowadays.

4.6.4.3 Low NO_x Burners

There are several methods to reduce NO_x emissions in the combustion products, but avoiding their production during combustion is the least costly method. **Most of the NO_x produced during coal combustion comes from fuel** (fuel NO_x rather than thermal NO_x). For pulverized coal combustion, typically about three-quarters of the total NO_x production can be fuel NO_x. As noted earlier, most of fuel NO_x comes from nitrogen in volatile matters. It was also noted that while a fuel-rich environment promotes the formation of N_2, an air-rich environment promotes the formation of NO_x from nitrogen compounds in volatile matters. So if the temperature and available oxygen is controlled during the devolatilization process, the NO_x production can be significantly reduced. Note that this is only true for coal with high volatile matters. For low volatile matter coal, other NO_x reduction techniques should be employed.

There are three main methods for NO_x reduction in the combustion process. These three methods will be briefly introduced here and elaborated later in this section. In the first method, called **dual register burners**, while the entire air is introduced to the burner, the core of the flame where the majority of the devolatilization process takes place has less than the stoichiometric air. The rest of the air is added to the flame after the initial combustion stage at the center of the flame. The second method is called **air staging** where only some of the total air is provided in the burner and the rest of the air is introduced somewhere else in the furnace. The last method is called **fuel staging** or **reburning** where most of the fuel (but not all of it) is initially provided in the burner and is combusted in an air-rich environment and later the remaining of the fuel is added to create a fuel-rich environment where the previously produced NO_x is decomposed by striping oxygen from it.

4.6.4.3.1 Dual Register Burners

As noted earlier, in the pre-NO_x restriction era burners, the secondary air is introduced to the burner through a register built into the windbox. In those units, the fuel and air are premixed before leaving the burners for combustion (Figure 4.25). In dual register burners, however, there are **several layers in the flame in regard to the oxygen concentration** (Figure 4.25). The core of the flame is fuel-rich so the crucial devolatilization process takes place with less than the stoichiometric air and significantly reduces the initial NO_x production. When the fuel moves from the inner layers of the flame to the downstream, the rest of the air gradually diffuses to the combustion zone to complete the combustion process. The concentration of oxygen increases further away from the core of the flame. To achieve this distribution of oxygen, the air is admitted through two registers separately (Figure 4.26). The flow

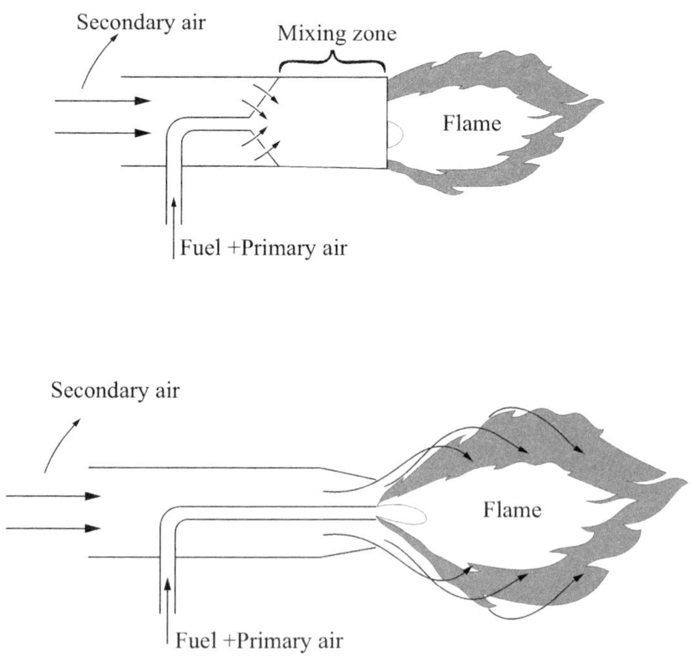

FIGURE 4.25
Premixed flame versus diffused flame.

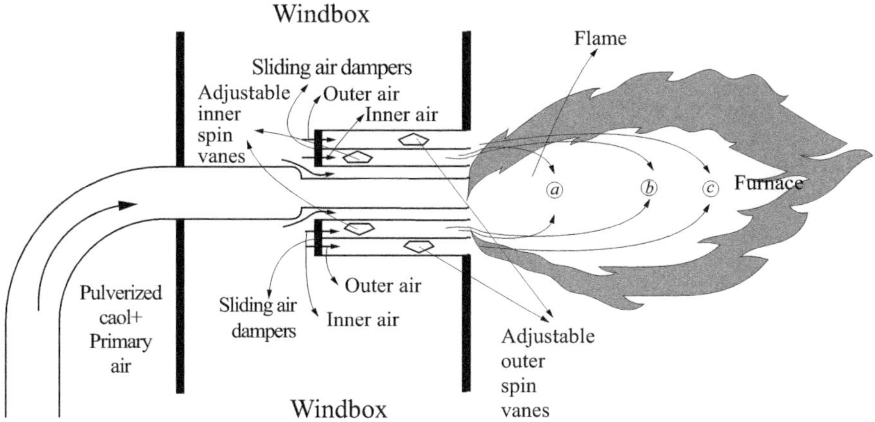

FIGURE 4.26
Schematic of a dual register burners. (a) Initial combustion (devolatilization with lean oxygen), (b) transition zone, (c) burnout zone.

of the inlet air in these registers are controlled independently. The register that provides the air for the fuel-rich core imparts high swirl to the air to create a stable flame in the core. The register that provides the air for the outer combustion zone imparts moderate swirl to allow sufficient time for the mixing of the air and completion of the combustion. These swirls are imparted by series of fixed, manually controlled, and automatically controlled vanes. They are to provide the minimum swirl needed, the permanent adjustment needed during the commissioning, and the variation needed during the operation, respectively.

In more advanced designs, a transition zone is also created between inner and outer combustion zones (Figure 4.26). These designs can reduce the NO_x production up to 80% compared to that of uncontrolled burners.

In this method, the **peak flame temperature also reduces**, which can help to reduce the thermal NO_x production. It should be noted that since in the core of the flame the entire air is not introduced, some fuel will remain unburned in the form of either CO or unburned carbon/hydrocarbon. This can cause an increase in CO and unburned carbon emissions if the system is not properly designed. One consequence of this design is a larger furnace requirement so sufficient time is provided to burnout the entire fuel. The larger furnace also increases the waterwall tube surfaces which can reduce the temperature of the combustion zone and can further reduce the thermal NO_x production. Typically, computational fluid dynamic (CFD) models are used to achieve an optimum design and minimum unburned fuel. Another approach to reducing the problem of unburned fuel is to have finer pulverized coal, which helps to burn the fuel faster and to reduce the required residence time. This method is also used in other NO_x reduction technologies.

In these burners, the components that are exposed to fire are typically manufactured from high-quality stainless steel. Also, critical components are equipped with thermocouples to monitor their operating temperatures. Finally, this burner design can be fueled with gaseous and liquid fuels as the primary fuel or in the dual fuel systems.

Note that in all these burners there should be a clearance between the burners, walls, and floor. This minimum spacing is to make sure the complete combustion process takes place and to minimize the interaction between flames from different burners. Also, the flame impingement on the wall tubes can increase the tube metal temperature to an unacceptable level and may result in tube fouling. These clearances are to avoid these negative impacts. Moreover, at part load operations, a selection of burners can be used to provide the required fire rate and to control the temperature of the produced superheated steam. Firing the burners located at higher elevations in the furnace provides more radiative heat transfer to the superheater, which in turn increases the temperature of the produced steam (more on this topic later).

4.6.4.3.2 Air Staging or Overfire Air (OFA)

In this method like the previous one, the entire air is not provided to the flame at the initial stages of the combustion process. But unlike the previous method where the entire air is provided to the burner, in this method, only **a portion of the total air is provided to the burner and the rest is added somewhere else in the furnace**. In both cases, the outcome is that initially, during the devolatilization process, less than required air is provided. The initial air should be sufficient so the combustion temperature can reach the temperature needed for volatile vaporization but not enough for NO_x generation. Later the balance of air is introduced to complete the combustion process and to maintain high efficiency. The reduction in NO_x production depends on how much of the air is diverted from the burners. The removed air from the burner could be a small percentage of the total air in which case the combustion in the burners could be with slight excess air or under slight air deficiency. In some designs, there is much deeper staging where a quarter or more of the entire air is removed from the burners. In all designs, the remaining air is added to the furnace in the **NO_x ports** as **overfire air (OFA)**. This air may be provided either through the windbox or by dedicated ducts. The NO_x ports provide OFA through the openings similar to the openings for the burners (Figure 4.22).

Similar to the previous method, the air staging may increase unburned fuel in the combustion products and this problem needs to be carefully considered in the furnace design. Some parameters that can impact this problem are the number, size, location, and distribution of NO_x ports. These parameters may have conflicting impacts on the NO_x and unburned carbon production. For instance, in terms of reduction in NO_x production, the farther the NO_x ports are from the main combustion zone the better, while for the unburned fuel problem, the NO_x ports closer to the combustion zone are more beneficial. Naturally, in these cases, there should be a compromised design to achieve optimum conditions. Also, as noted earlier, the finer pulverized coal can help to address the problem of unburned fuel.

In order to increase the effectiveness of NO_x reduction, the first and second methods may be combined in a single furnace. The air staging NO_x reduction method can be integrated into all types of coal combustion systems, including mechanical stokers, cyclones (refer to Section 4.6.4.5), wall-fired pulverized burners, and other methods that will be presented in the coming sections.

4.6.4.3.3 Fuel Staging or Reburning

In this method for NO_x reduction, NO_x is destroyed, i.e., decomposed, after it is formed in the main combustion zone. First, most of the fuel (but not all) is combusted in the burner with some excess air (similar to uncontrolled burners). Obviously the combustion products will contain a high concentration of NO_x. Then, the balance of the fuel with a small amount of air is introduced to the flow in a way that the combustion of the newly presented fuel is under less than stoichiometric air. In this reducing environment (because of the lack of oxygen), the fuel combustion process can **strip NO_x from its oxygen to leave N_2**. Finally, the balance of air is introduced to complete the combustion process. While the NO_x removal efficiency of this method is comparable to the previous two methods, it is not typically as cost effective.

4.6.4.4 Corner-Fired or Tangential Burners

Corner-fired or tangential burners are a coal combustion technology that uses the furnace to burn pulverized coal in a single flame envelope and has been utilized since the early days of pulverized coal combustion systems. In corner-fired burners, **a mixture of pulverized coal and primary air along with secondary air are injected horizontally from four corners of the furnace**. In some designs, there are four more burners in the middle of the furnace walls adding up to eight burners in total. The jets from burners are **slightly diverted off diagonals** (Figure 4.27a) to impart rotation in flows. The flames from the burners intertwine to create a single flame envelope that fills the furnace. The turbulence generated by the whirling effect creates intense combustion at a relatively uniform temperature since there is no concentrated combustion in the burners.

The secondary air is admitted to the burners from the windbox (Figure 4.27b). The burners are connected to windbox by dampers. The flow rate of the secondary air is controlled by these dampers. While the dampers at the same height of the furnace are controlled uniformly to avoid imbalance in the jets, great accuracy in the uniform distribution of fuel and air among the burners is not required. For the same reason, each pulverizer feeds all burners at a single height of the furnace.

The fuel nozzles provide the mixture of pulverized coal, primary air, and a portion of secondary air. Some designs may have multiple nozzles for various fuels (even including municipal solid wastes). The fuel nozzles and the secondary air (sometimes called the auxiliary air) are positioned alternatively in the corner burners, one after another (Figure 4.27b).

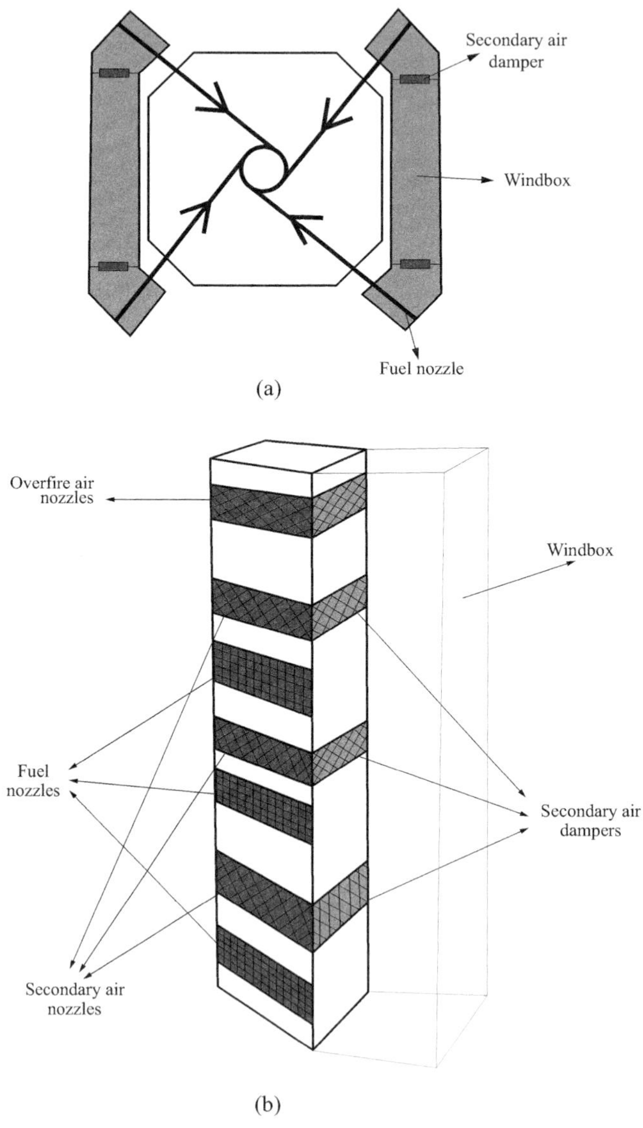

FIGURE 4.27
Schematic of a corner-fired or tangential burner. (a) Top view of a furnace, (b) a corner burner.

In the previous chapter, we discussed about the mechanisms to control the superheated steam temperature leaving the steam generator (after both superheaters and reheaters) toward the steam turbine. Here another way to achieve this objective will be explained. Since the early 1940s, the corner-firing systems have been equipped with **tilting air and fuel nozzles**. This capability can help to move the flame vertically along the height of the furnace (Figure 4.28). By **raising or lowering the flame, the heat-transfer pattern in the furnace changes,** which in turn changes the furnace exit-gas temperature. By controlling the furnace exit-gas temperature, which is the superheater inlet gas temperature, the temperature of the outlet superheated steam can be controlled. This control mechanism can be used to cope with the temperature variations due to load change, particularly at lower loads, and the temperature variations due to change in operational conditions. For example, when the slag and ash deposit on the walls of the furnace increases, the heat transfer to the waterwall tubes reduces. This will increase the furnace exit-gas temperature, which will increase the superheated steam temperature. To compensate for this effect, the nozzles can be tilted downward to allow more residence time for the combustion products in the furnace and more heat transfer to the waterwall tubes. Also, the lower portion of the furnace has more exposure to the flame. This overall increase of heat transfer in the furnace will balance the furnace exit-gas temperature and the temperature of the superheated steam. The opposite can happen after the furnace walls are cleaned by the sootblowers (Section 4.6.4.7). A similar analysis can be presented for the insulating effects of water deposits inside the waterwall tubes.

4.6.4.4.1 *Low NO_x Tangential Firing*

First, it should be noted that the nature of combustion in the corner-firing systems makes them an inherently low NO_x generating combustion technology. This is partly because while there is great global turbulence in the furnace, there is moderate local mixing between fuel and air. This slows down the combustion rate in the initial stages. For further reduction in the NO_x production, the air staging is the dominant technology. The air staging can

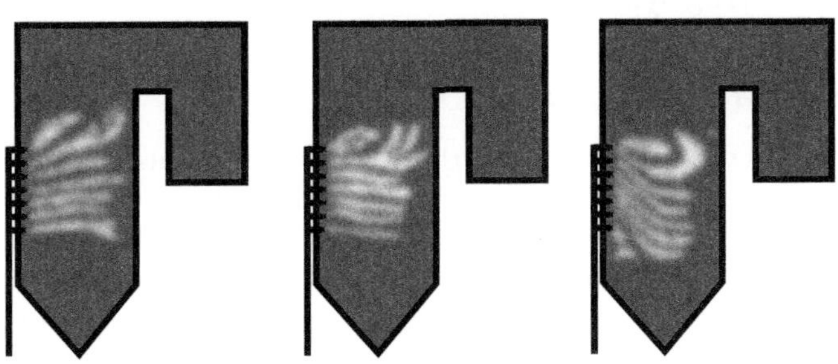

FIGURE 4.28
Schematic of a tilting corner-fired or tangential burner.

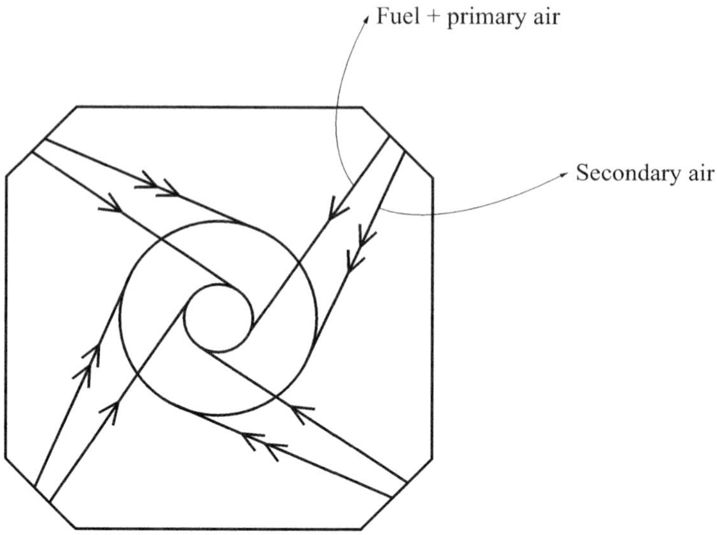

FIGURE 4.29
Schematic of an offset air system or concentric firing system to reduce NO_x production in corner-firing systems.

be done in both local and global levels. In the **local air staging** (sometimes called **closed-coupled OFA**), the air nozzles (NO_x ports) to provide the balance of air are right after one or several coal nozzles. In this case, the air for the OFA is provided through the windbox. In the **global air staging**, the NO_x ports are located downstream of the fuel and air nozzles, up to a few meters above the nozzles (Figure 4.27). In this case, the air can be provided by the windbox or a separate duct.

Another method to reduce NO_x production is to slightly offset the projection of the secondary air jets from that of the fuel jets (Figure 4.29). This delays the mixing of the secondary air and fuel and the introduction of air to the flame envelop, which can reduce the air during the crucial initial stages of the combustion process. This is called the **offset air system** or sometimes the **concentric firing system**.

4.6.4.5 Cyclone-Furnace Firing

Cyclone furnaces were developed in the 1940s **to accommodate low-grade coal** that could not be adequately burned in pulverized furnaces. The main problem with these coal types was their low ash melting temperature which resulted in molten fly ash that could cause slagging in heat-transfer surfaces. Cyclone furnaces actually take advantage of this low ash melting temperature to establish a layer of molten slag in the furnace which reduces the fly ash to about a quarter of that in pulverized coal furnaces. Since the size of coal particles are much larger than that in pulverized coal furnaces, the cost of the coal preparation process, both the capital and operational costs, reduces significantly.

A cyclone furnace is made of a cylinder with the diameter of typically between 2 and 3 m, Figure 4.30 (6–10 ft). One or several (up to few tens of units) of these furnaces are horizontally installed on the single side or opposite sides of the vertical walls of the main furnace of the steam generator. Their heat input capacity can be between 45 and 125 MW (150–425 × 10^6 Btu/h) per unit.

The inlet coal with the primary air is tangentially fed to the cylinder creating a swirling motion. The secondary air is also admitted tangentially at the top of the cylinder at the same rotational direction to create further swirling motion. This air induces strong turbulence to the combustion zone. A small quantity of tertiary air is admitted at the center of the cylinder to control the location of the main fire along the cylinder. The volatile gases and fine coal

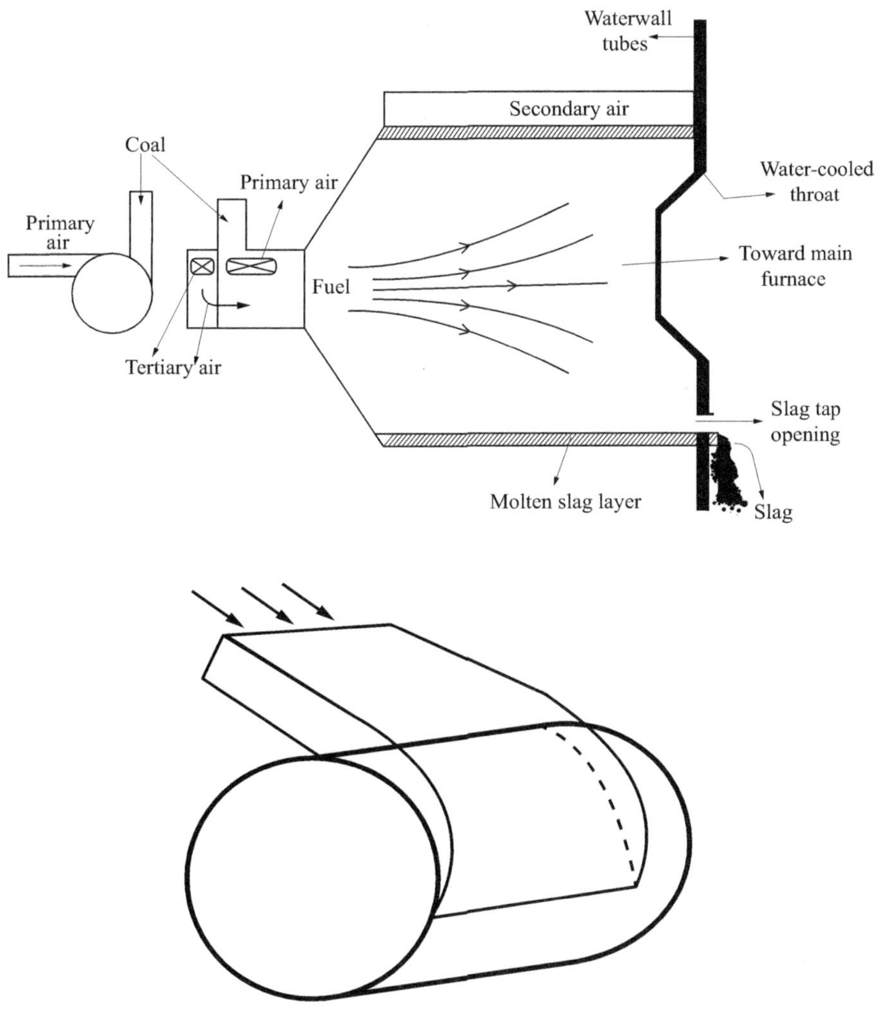

FIGURE 4.30
Schematic of a cyclone furnace.

particles are burned intensely with this turbulent air flow. The high temperature of the combustion melts the ash particles. The molten slag is accumulated on the inner wall of the cylinder due to the centrifugal force. The larger coal particles are drawn into this molten slag layer and are intensely burned with the flow of turbulent air over the layer. To sustain the layer, the coal must have a minimum percentage of ashes. The coal preparation process can be bin or direct-fired systems, with or without pre-drying.

The combustion product gases enter the main steam generator furnace through the throat to transfer heat to the waterwalls. The trapped slag flows out of the cyclone through the slag tap opening down the walls of the main furnace. The slag is quenched and solidified in a tank full of water under the floor of the main furnace. The accumulated solid slag is periodically broken and removed from the tank. This slag can be used for other applications, such as building and road construction materials.

In this method, about three-quarters of the ash is converted to slag and the rest leaves the system as fly ash. In cyclone furnaces, the main furnace size is smaller compared to that in other types of furnaces because all combustion takes place within the cylinder. Since there is less fly ash, there are less erosion and fouling in heat-transfer surfaces. For the same reason, the size of dust removal precipitators or filter houses significantly decreases.

The cylinder in cyclone furnaces is made of water-cooled tubes protected by refractory linings. Depending on where and how the crushed coal and primary air, secondary air, and tertiary air are admitted to the furnace, there are several types of cyclone furnaces, including scroll, vortex, and radial furnaces, the latter being the most common one.

This type of furnaces not only can burn a wide variety of coal types, but it can also be operated by various biomass sources, industrial and municipal wastes, and liquid or gaseous fossil fuels. However, the fuel should have a minimum amount of volatile and ash contents within certain ranges. Co-firing and fuel blending are also possible. The recommended size of coal particles depends on the type of coal.

Cyclone furnaces come with some disadvantages. The first and probably the most important problem, at least for older designs, is high NO_x emissions due to high operating temperatures. To address this problem, low NO_x cyclone furnaces have been developed using air staging technology. While the percentage of excess air in this type of furnace is relatively low (about 10%–15%), the pressure drop in the turbulent flow in the cylinder is high. This requires a high pressure forced draft fans, which in turn increases the power consumption of the unit. Also, while the furnace has a good performance at partial loads of up to about 50%, below this the fluidity of the molten slag cannot be maintained.

4.6.4.6 Furnaces

A furnace is a large unobstructed volume typically enclosed with water-cooled walls. The shape and volume of a furnace are determined to fulfill five main functions. The furnace volume should be large enough to provide a sufficient residence (retention) time for the mixture of fuel and air so that the complete combustion of fuel can take place. Otherwise, some combustion may occur among tubes downstream of the furnace (called secondary combustion).

Also, the maximum flame temperature in the furnace can reach over 1650°C (about 3000°F). The high **furnace exit-gas temperature (FEGT)** can create problems in the convection pass of a steam generator for two reasons. First, an unacceptably high FEGT can increase the tube metal temperature to a level that can overheat the tubes and can result in tube failure. This high temperature can also soften the ash, particularly for fuels with low ash-softening temperatures. This can result in fouling and slagging of heat-transfer surfaces. This will not only reduce the heat transfer to the flows inside the tubes but can also erode the components downstream of the furnace and increase the pressure drop of the flue gas. More substantial cleaning equipment is necessary to reduce these problems. Thus, the second function of the furnace is to cool the flue gas to **below the lowest ash-softening temperature** of all possible fuels before sending the gas to the convection pass. Moreover, to minimize the interaction between flames, there should be **sufficient spacing** between the burners (not applicable to stoker and tangential firing furnaces). Similarly to avoid excessive flame impingement on the furnace enclosing, which can cause tube overheating and fouling, a minimum clearance between burners and walls (and floors) should be provided. The furnace should be sufficiently large to accommodate all these required clearances. Finally, as discussed before, the furnace is typically enclosed with gas-tight welded membrane waterwall tubes. Inside these tubes boiling water (or supercritical pressure water with no boiling process) flows, in which heat is absorbed from the furnace via mainly radiation. Depending on the pressure of the water/steam, all or a significant portion of the evaporator surface area is provided by these waterwall tubes and the furnace should be large enough to accommodate all these tubes.

The minimum volume required to meet each of these functions is different for various steam generators and they may not necessarily match each other. In many steam generators, controlling the FEGT is the determining factor for differing reasons. While in coal and other solid fuel steam generators this is to reduce ash fouling, in natural gas steam generators it is to protect convective heat-transfer surfaces from overheating. However, the furnace volume set for the FEGT control can result in more than needed or less than needed evaporator surfaces depending on the pressure of the steam (refer to Example 3.2 where the percentage of heat transfer required in each component of a steam generator was determined compared to the total heat transfer).

The furnace size has a significant impact on the overall capital cost of a steam generator because it also determines the size of the boiler building and its metal structure. The amount of ash and its properties are also very important factors in determining the size of the furnace. Everything else being equal, the size of the furnace can be more than double from a favorable coal type to the worst coal type.

For the design of a furnace, the volumetric rate of heat input is not an important design parameter; rather, the heat input per plan area is the controlling design parameter. The heat input per plan area is defined as the rate of heat released in the combustion process (which is equal to the mass flow rate of fuel multiplied by the heating value of the fuel if the combustion process is assumed complete) divided by the horizontal cross-sectional area (plan area) of the furnace. The heat input per plan area is also an important factor for NO_x control. The lower heat input per plan area results in a larger furnace and lower NO_x production. The heat input per plan area can range from 3.5 to 6.5 MW/m^2 (about 1.2–2 × 10^6 Btu/h-ft^2).

4.6.4.7 Sootblowers

Sootblowers are devices that use a high-speed flow of steam (saturated or superheated), liquid water, or pressurized air to remove the ash accumulated on steam generator surfaces. These systems are used to clean furnace walls, convective heat-transfer surfaces in superheaters and reheaters, economizer, air preheater, and selective reduction systems (the latter two systems will be discussed later). Devices that use water and air as the cleaning medium are sometimes called waterlances and airlances. The most common cleaning medium is superheated steam due to its availability and minimal impacts on the heat-transfer surfaces. Many sootblowers are equipped with a retractable tube, called lance tube, with nozzles at the tip and/or side. They are installed on the steam generator walls. Whenever cleaning is needed, they are extended into the steam generator to direct jets of the cleaning medium on the target area. The lance tube may rotate or oscillate to enlarge the cleaning area. In some designs, water nozzles are installed on a steam generator wall to clean the opposite wall. There are systems that use sound waves at the frequency of about 75 Hz to dislodge ash deposits unavailable by other cleaning systems, such as within closely spaced tubes of the economizer.

Sootblowers are equipped with a control system to detect when cleaning is needed and to safely clean steam generator surfaces with minimum medium consumption. Various sensors are used for this purpose: heat flux sensors to detect heat-transfer reduction due to ash deposition, emissivity sensors to detect highly reflective ash accumulation, and temperature measurement sensors to monitor the flue gas temperature. The pressure drop in the heat-transfer banks and power consumption in the fans can be the indications of ash deposition but when they detect a problem, typically it is too late for sootblowers to be effective.

4.6.4.8 Ash Handling

The combustion of any type of fuel produces some ashes, some fuels more than others. The portion of the ash produced during the combustion that is carried by the flue gas can cause two major problems on downstream heat-transfer surfaces: slagging and erosion.

The molten ash can be deposited on heat-transfer surfaces, which can reduce the heat-transfer rate and can damage the surfaces. This ash behavior is called **slagging**. It can also block the gas flow and add extra weight and mechanical load to the system. To reduce slagging, the **furnace exit-gas temperature (FEGT)** is limited to an upper limit. There are several ways to regulate FEGT, including recirculating flue gas from the outlet of the air preheater to the furnace. The problem with this method is high power consumption and high maintenance of the fans needed for the recirculation.

The solid abrasive ash particles may erode heat-transfer surfaces. The magnitude of this problem depends on the amount of ash in the combustion products, the abrasiveness of the ash, and the velocity of the flue gas flow. Among these parameters, only the velocity of the flow can be controlled for a given type of coal. This velocity is limited and the maximum velocity depends on the amount and properties of the produced ash and can be roughly between 10 and 20m/s.

There are two types of coal ash collection and removal systems: dry-ash system and slag-tap system. The **dry-ash collection system**, also known as the **dry-bottom system**, is typically employed in pulverized coal-burning systems, particularly when coal with a high temperature melting point is burned. In this system, about one-quarter of the produced ash is collected at the bottom of the furnace while the rest is entrained by the flue gas as fly ash. The fly ash is collected in the hopper after the economizer or in the particulate control system (refer to Chapter 5). If the collected ash in the hopper or bottom of the furnace is in large pieces, they are crushed and then pneumatically transported to a silo where they are separated from air in a cyclone and/or bag (fabric) filters to be transported from the power plant site.

The **slag-tap system**, also known as the **wet-bottom** system, can be implemented in both pulverized coal and cyclone combustion systems. In this system, the high temperature of combustion melts the ash particles, which are then accumulated on the furnace surfaces and flows down to the bottom of the furnace where it is quenched in a water tank (hopper). When the tank is full, a gate at the bottom opens and the ash is emptied into a conveyor system. The slag is grinded in a clinker grinder and is sent to a dewatering tank where the ash particles are separated and removed from the site. This system is not popular in modern power plants because of its high NO_x production due to the high-temperature combustion.

The operating conditions of the combustion process can be adjusted to reduce ash production, including increasing the excess air and reducing the load. Also, the proper usage of sootblowers can reduce the impact of ash collection on heat-transfer surfaces.

The ash properties, particularly its potential for slogging, directly influence the size of the furnace (both the plan area and the height). The high potential for slagging requires a larger and taller furnace. The spacing of the convection passes should be sufficient to avoid slagging and bridging. The slagging potential in turn depends on the average inlet temperature of the flue gas to the pass. The higher the average inlet temperature of the flue gas, the more spacing should be provided in the heat-transfer surfaces, particularly in the superheater surfaces because they are exposed to the highest flue gas temperature.

The collected ash as solid waste byproducts may have commercial values. For example, they can be used to produce gypsum (used in the production of wallboards) or construction materials. These applications can reduce the cost associated with landfilling of wastes. The design and specifications of ash handling systems are heavily impacted by the requirements of these byproducts.

4.6.4.9 Excess Air

We know that sufficient excess air should be provided to have a complete combustion process. The excess air can be as low as 5% for gaseous and liquid fuel systems or as high as 60% for some stoker burning systems. **High excess air improves the combustion process, reduces unburned fuel, reduces CO formation, and reduces slagging**. However, **more than the bare minimum excess air will reduce the efficiency of the steam generator** because the portion of the thermal energy that is used to heat the excess air will be partly wasted in the stack without contributing to raising steam. Obviously less excess air than required will result in incomplete combustion and/or longer residence time. However, sometimes, particularly during part load operations of steam generators, the level of excess air is used as a means to control the temperature of the generated superheated steam.

4.6.4.10 Ignition Systems

We know that in order to start the combustion process, some initial thermal energy (ignition energy) is required. During the steady-state operation of a steam generator, this initial energy is provided by already existing fire in the furnace. However, during start-up when the combustion has not started and when the flame is unstable, for example, during shutdown and at very low loads, an external source of heat is required. The ignition system's functions are to initially light the fuel and stabilize the flame when it cannot sustain itself.

The operation of an ignition system typically involves two main steps. In the first step, high energy electrical spark is used to fire a limited flow of an easy-to-combust fuel. The flow rate of the initial fuel should be small due to the limited energy that the electrical spark can provide. When the initial flame is stable and can sustain itself, the produced thermal energy can be used to ignite the main fuel. If the main fuel is natural gas or most liquid fuels, a small portion of the main fuel is used as the initial fuel. In some cases, the entire process may take a few seconds. For hard-to-burn fuels, such as coal, biomass, or some heavy liquid fuels, an intermediate fuel other than the main fuel, such as natural gas, propane, or some light liquid fuels, is used for ignition. For these fuels, it may take a few hours before the main fuel combustion can sustain itself. The required energy of the initial fuel can be more than 10% of the main fuel energy at full load steady-state operation.

Ignition systems are typically equipped with a flame safety system to ensure the stability of the flame. In these systems, electronic flame scanners continuously monitor the flame in the furnace by detecting flame visible light to verify the presence and stability of the flame in the furnace.

4.6.4.11 Fluidized-Bed Combustion

In fluidized-bed systems, an upward gas flow passes through a bed of particles at a high enough velocity to separate particles and keep them suspended. At these conditions, the mixture of the gas and particles demonstrates **fluid-like behavior**, hence the name **fluidized bed**. The main characteristic of this process is the extreme turbulence induced in the mixture that results in the very high rate of mixing and reactions among the reactants in the bed. For the combustion process, the gas is air and the particles are solid fuel pieces and noncombustible particles.

The concept of fluidized-bed systems first appeared in the literature in the late 19th century. The chemical industry has used fluidized-bed systems in non-combustion applications to improve the reaction rate. The research and development on the application of fluidized-bed systems in utility steam generators started in the 1950s. The first fluidized-bed boiler was installed in a 30 MW unit in 1975 in West Virginia.

Fluidized-bed combustion systems come with many advantages. An inherent characteristic of the system results from intimate mixing between fuel and air and a very high rate of heat and mass transfers. This increases the rate of the combustion reaction and reduces the time required for the complete combustion, which in turn can reduce the size of the steam generator. These units are very flexible on the type of fuel they can burn. While they are more suitable for solid fuels, they can easily handle liquid and gaseous fuels as well. They can combust solid fuels that other combustion systems find difficult to burn. These systems also have an attribute that other combustion systems find hard to compete, particularly for fuels with high sulfur content. They can absorb sulfur oxides during the combustion process eliminating an after-combustion scrubbing system (these systems will be covered in Chapter 5). They also **operate at low temperatures, in the range of mid-700s to mid-900s °C** (about 1400°F and 1750°F), which reduces NO_x production (the reason for this low temperature operation will be discussed later). This combination of fuel flexibility and low emissions make these systems very attractive.

In some respects, fluidized-bed combustion systems are somewhere between stoker burners and pulverized coal combustion systems. While the combustion process takes place when the mass of fuel pieces form a bed (similar to stokers), the process occurs when the fuel particles are suspended on the air flow (similar to pulverized coal combustion systems). As a consequence, the fuel size and the air velocity in fluidized-bed combustion systems are somewhere between those of stokers and pulverized coal combustion systems. However, the combustion temperature is significantly lower in fluidized beds. These characteristics of fluidized-bed combustion systems provide them with unique attributes. The intimate contact between the fuel particles and the air flow as well as a long fuel residence time provide a high-efficiency combustion process at low temperatures. The presence of the bulk of fuel and bed materials at the combustion temperature creates a mass with high thermal inertia that enables the system to burn almost all fuels regardless of their composition and properties, including low-heating value fuels, fuels with very high ash and moisture contents, low volatile content fuels, waste fuels, and biomass. The low operating temperature (lower than the ash-softening temperature) helps the system to avoid the slagging problem and the fouling of the downstream heat-transfer surfaces by soft ash carryover. This independence from fuel ash properties further helps the fuel flexibility of the system. The low operating temperature also allows the construction of the system from inexpensive metals, which reduces the construction costs.

Fluidized-bed combustion systems are made of a container (the furnace) that encloses the bed. The walls of the furnace are covered with water-cooled tubes. The solid fuel pieces are combusted while they are held in suspension along with noncombustible particles by the high-velocity air flow. The noncombustible bed materials can be composed of fuel ash, sulfur sorbent, products of sulfur oxide removal processes (covered next), and/or inert material such as sand. A certain amount of bed materials is needed to provide a sufficient bed temperature for the system. If more than the required bed materials are accumulated in the bed, the excess material is drained. To reduce the sensible thermal energy loss accompanying the drained material, which can be significant for fuels with high ash contents, sometimes a heat exchanger is used to recover the thermal energy from the material before it is discharged. If not enough bed materials are available, for example for low ash fuels like biomass, inert external material, such as properly sized sand, is added to the bed.

The upward air flow is evenly distributed across the bed through a grid of air distributers. Due to the high velocity of air, a significant portion of ash and unburned fuel is carried away by the combustion products. These particles are collected in the separators after economizers and are returned to the bed. This recirculation improves the combustion efficiency by reducing the unburned fuel losses and by increasing the fuel residence time. It also helps to maintain the bed inventory. Note that the water side of fluidized-bed steam generators, similar to other steam generators, can be subcritical or supercritical, drum-type or once-through, and natural circulation or pumped circulation. The design considerations of the superheaters, reheaters, economizer, and air preheater are similar to other combustion systems (as discussed in the previous chapter).

4.6.4.11.1 Chemical Processes

As noted earlier, fluidized-bed combustion systems are able to operate with very low sulfur oxide and NO_x emissions. In this section, these aspects of fluidized-bed combustion systems will be discussed. Fluidized-bed combustion systems uniquely suit for sulfur oxide removal. While other combustion systems can only reduce sulfur oxide emissions by using low sulfur fuels, desulfurizing the fuel before combustion, or scrubbing them from combustion products, fluidized-bed combustion systems are capable of removing them within the furnace during the combustion process. For this purpose, when a high sulfur fuel is burned in the system, some sorbent typically **limestone** is added to the bed. Limestone is a sedimentary rock typically formed by the remains of marine organisms (more on this topic in Chapter 5). It is mainly composed of calcium carbonate ($CaCO_3$). Limestone can be delivered to the plant ready to be used or can be processed in the site. After limestone is added to the bed and heated by the bed materials, it goes through the following calcination process where lime (CaO) and carbon dioxide are produced:

$$CaCO_3 \rightarrow CaO + CO_2\,(\text{endothermic}) \tag{4.67}$$

Then, lime reacts with sulfur oxide and excess air to produce calcium sulfate through the following exothermic sulfation process:

$$SO_2 + CaO + 1/2O_2 \rightarrow CaSO_4\,(\text{exothermic}) \tag{4.68}$$

The produced calcium sulfate can be discarded or regenerated for further use. Theoretically, one mole of limestone is needed for each mole of sulfur to be removed. In practice, depending on the limestone composition and size, the sulfur content of the fuel, and the desired percentage of sulfur removal, more limestone is needed. Depending on the ratio of Ca to S, the combination of the above two reactions can provide net heat gain or loss to the bed. The **rates of the above reactions are optimum at the temperatures of mid-800°C** (mid-1500°F). At much lower temperatures, the first reaction cannot proceed to completion and at much higher temperatures the second reaction can be reversed. Both of these effects are undesirable. The former increases limestone consumption and the latter reduces the effectiveness of sulfur oxide removal. In practice, the fluidized-bed temperature is typically at the range of mid-700s°C to mid-900s°C (about 1400°F and 1750°F). This requires effective temperature control throughout the entire bed. The removal of a significant portion of sulfur oxides in the furnace and the low sulfur oxide concentration in the combustion products has the added benefit of markedly reducing the problem of acid corrosion in the downstream equipment. This enables designers to reduce the exhaust temperature of the flue gas, which improves the efficiency of the cycle. Limestone can also remove some other undesirable fuel contents such as some heavy metals.

The low operating temperature of the bed also reduces thermal NO_x production. However, it should be noted that limestone can act as a catalyst for nitrogen oxide production from the reaction of oxygen and nitrogen, so the consumption of limestone should be kept as low as possible. To reduce the fuel NO_x generation, air staging and using overfire air have been also employed.

4.6.4.11.2 Impact of Velocity

The most important characteristic of fluidized-bed combustion systems is burning the solid fuel particles while suspended in a high-velocity air flow. In this section, we will evaluate the interaction between the air flow and the fuel particles that will lead us to two very distinct types of fluidized combustion systems.

When the upward air flow passes through the bed, it exerts drag force on the fuel particles. The magnitude of this force depends on the air velocity and the properties of air, including air density and viscosity, and the size and shape of the particles. At low air velocities, this force is not sufficient to separate the solid fuel particles and they remain in contact. In this case, the air flow passes through voids between the fuel particles and the bed is called a **fixed bed** (Figure 4.31b). With the increase of the air velocity, the magnitude of the drag force increases. When the force exerted on a particle is equal to its weight, the particle will be suspended. This is called the minimum fluidization conditions and the air velocity is called the **minimum fluidization velocity** (Figure 4.31c). While the particles can move relatively freely in this condition, they have a limited movement space and there is a little change in the bed volume. With further increase in the air velocity, the bed will be more turbulent. Also, air bubbles will be created in the bed and the bed volume will increase (Figure 4.31d). This is called the **bubbling fluidized-bed (BFB)**. With even higher air velocity, the size of bubbles and the frequency

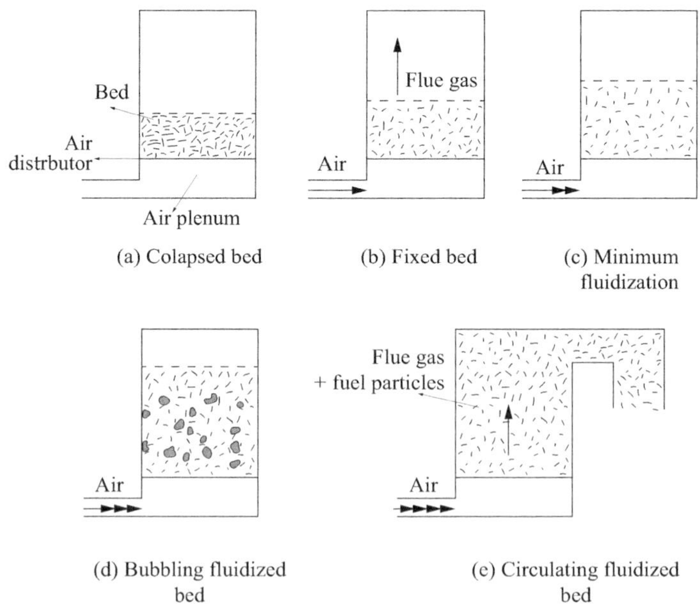

FIGURE 4.31
Various stages of fluidization with increase in air velocity.

of their creation increase. At very high air velocities, the air flow carries the particles out of the bed (Figure 4.31e). This is called the **circulating fluidized-bed (CFB)**. Accordingly, depending primarily on the air velocity and the fuel size, there are two types of fluidized-bed combustion systems: bubbling fluidized-bed and circulating fluidized-bed.

4.6.4.11.3 Bubbling Fluidized-Bed Combustion Systems

In bubbling fluidized-bed combustion systems, the fuel particles mostly remain at the bottom of the furnace at the typical temperature of 850°C–900°C (1550°F–1650°F). They operate with relatively **low air velocity** (lower than about 3 m/s or about 10 ft/s) and **relatively large fuel particles** (as large as 50 mm or 2 in for some systems). However, a small portion of the fuel is entrained by the air flow, which is mostly captured and returned to the bed by separation systems, typically cyclone separators after economizers, before the flue gas is discharged to the environment (Figure 4.32). There is also a clear boundary between the bed and the rest of the furnace with good mixing within the bed. The space above the bed in the furnace is called the freeboard. This space provides the needed volume for complete combustion. Also, it is where the NO_x control overfire air (OFA) is added to the flue gas leaving the bed.

The commercial power plants using the bubbling fluidized-bed combustion technology are typically **limited to smaller capacity utility plants** (less than 100 MW capacities) and industrial steam generators. They are particularly suitable for special fuels such as biomass and solid waste fuels. Per kW

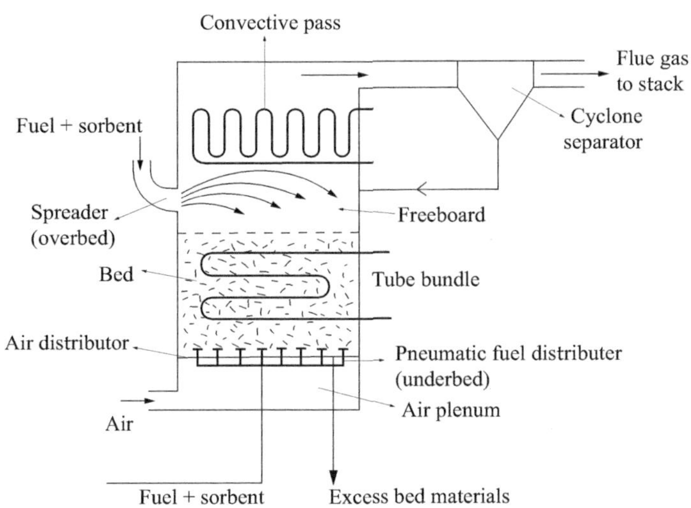

FIGURE 4.32
Schematic of a bubbling fluidized-bed combustion system (note that both underbed and overbed fuel and sorbent distribution systems are shown on a single bed).

basis, these systems are less expensive compared to circulating fluidized-bed systems.

The bed depth is typically about 1 m (3 ft) in the expanded condition and half of that in the collapsed condition (that is an expansion ratio of about 2). The crushed fuel and limestone (if needed, e.g., for high sulfur content fuels) can be injected into and spread across the bed either over the bed via fuel spreaders (like stoker spreaders) or in the bottom of the bed through air-conveyed fuels (Figure 4.32). The former is simpler and less sensitive to fuel composition and can accept much larger fuel pieces but less effective in sulfur oxide removal since most sulfur oxides bypass the limestone in the bed.

Typically heat-transfer tubes, either water- or steam-cooled, are submerged in the bed. These boiler tube bundles or superheaters not only improve the thermal performance of the combustion systems but also help to control the temperature of the bed within the proper limits for the optimum combustion and sulfur removal processes as well as the minimal NO_x production.

The air needed for fluidization and a portion of the combustion air is provided by a horizontal air distribution system. There are two types of distribution systems for this purpose: open bottom and flat floor air distribution systems. The open bottom system is made of a hopper and air ducts that are connected to many bubble caps (Figure 4.33). The large array of bubble caps act as the nozzles where air is uniformly distributed in the bed. Air distributers are designed to resist plugging and to have a minimal pressure drop. The actual fluidized bed is formed right above the level where air leaves the nozzles. Below that level, the hopper is filled with non-fluidized-bed materials. This type of the air distribution system is capable of easily removing large pieces of noncombustible objects in the fuel.

In flat floor air distribution systems, the air is also distributed via the bubble caps (Figure 4.34). The bottom of the furnace is made of horizontal waterwall tubes through which air is provided to the bubble caps from the windbox below the tubes. This system is suitable when the fuel contains no or few large noncombustible objects.

Bubbling bed combustion system designers must pay particular attention to the interaction between the air bubbles formed in the bed and the fuel particles. The air bubbles can cause turbulence in the bed, which increases

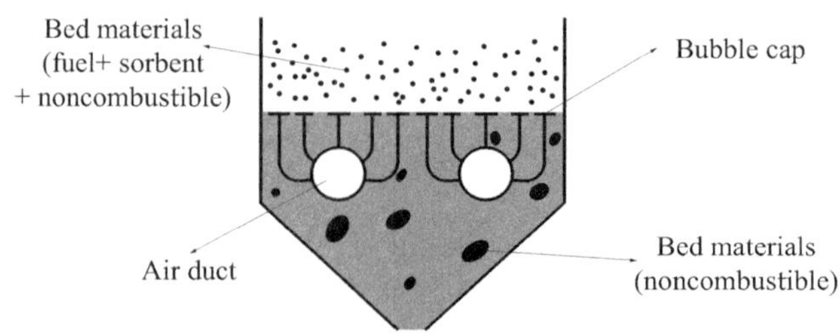

FIGURE 4.33
Schematic of an open bottom air distribution system.

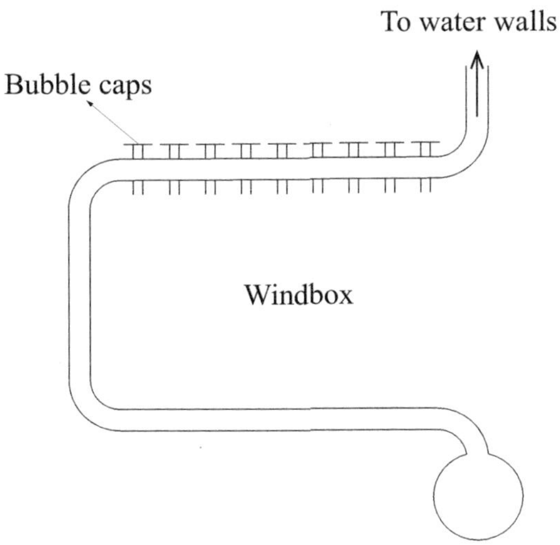

FIGURE 4.34
Schematic of a flat floor air distribution system.

the interaction between the air and the fuel particles. However, if not properly designed, they can provide a path for the air to go through the bed without participating in the combustion process. Similarly, they can limit the interaction between the combustion products and the limestone particles, reducing the effectiveness of sulfur oxide removal. This is a major challenge for a proper design of bubbling fluidized-bed combustion systems. This limitation is one of the reasons for the limited capacity (size) of bubbling fluidized-bed combustion systems. A successful design should combine the right air bubble and fuel particle properties to provide adequate mass transfer between the air inside bubbles and the gases within the bed while the bubbles pass through the bed.

4.6.4.11.4 Circulating Fluidized-Bed Combustion Systems

As noted earlier, in circulating fluidized-bed combustion systems, the fluidizing air velocity is much higher than that in the bubbling fluidized-bed systems. The air velocity of greater than 3 m/s (about 10 ft/s), typically about 5 m/s (about 16 ft/s), is high enough that not only can hold the fuel particles in suspension but can also entrain a significant portion of them into a close loop. Therefore, in this type of bed, there is no specific region of the furnace that can be referred to as the bed or the freeboard and the particles are present in the entire furnace with varying concentrations. Obviously, there is also no air bubbling effect. This bed-material circulation provides a very high mixing rate between the reactants of the reactions within the bed. These particles are captured by separators right after the furnace and returned to the furnace before the combustion products are sent to the convective heat-transfer surfaces. The recycled solid particles ensure that there is a sufficient bed inventory in the furnace. The bed materials and fuel particles are finer in these systems than in bubbling fluidized-bed systems. The circulating fluidized-bed systems are also more sensitive to the bed-material sizes and have more stringent bed particle size requirements. Many aspects of circulating and bubbling fluidized-bed combustion systems are similar, including the operating temperature range, low thermal NO_x production due to this low operating temperature, air staging via OFA nozzles to reduce fuel NO_x production, using limestone for sulfur oxide removal, fuel and limestone mechanical or pneumatic feeding systems, typically flat floor air distribution system from windbox, and heat-transfer surfaces submerged in the bed. The collection of entrained solid particles can be achieved by one or a combination of U-beams and cyclone separators. U-beam separators collect the particles from the flue gas flow by redirecting the gas and trapping the particles at the exit of the furnace. Cyclone separators can be installed downstream of the U-beam separators to remove the leftover particles or as the sole separation system. The U-beam separators can remove up to about 97% of the particles while the cyclones can collect above 99% of the particles.

Circulating fluidized-bed systems are the dominant technology for fluidized-bed combustion systems because of their capability to be built at large sizes and their capability to burn a wide variety of fuels. They have been built and installed in utility steam generator in power plants with a capacity of up to about 700 MW.

In all fluidized-bed combustion systems we discussed so far, the pressure of the furnace was equal or closed to the atmospheric pressure. If the furnace operates at elevated pressures, fluidized-bed combustion systems can be integrated into a combined gas turbine and steam cycle power plants. The pressurized fluidized-bed combustion systems will be briefly covered in Chapter 10.

4.6.4.12 Burning of Municipal Solid Wastes

While this section primarily deals with solid wastes, the presented material can be mostly applied to biomass resources as fuel as well. As discussed in Section 4.5.7.2, the composition and as a result, the heat content of municipal solid wastes (MSW) vary significantly with time. This variation in the heat content can make the design and operation of the **waste-to-energy plants** very challenging. The objectives of these steam generating plants are to combust a certain mass of MSWs while generating sufficient heat to raise a certain mass flow rate of steam. The parameter relating these two design objectives is the heating value of the fuel. If the fuel heating value drops below the average design heating value, the steam generator will burn the design mass flow rate of wastes but will not be able to raise the design steam flow rate. For heating values higher than the design waste heating value, the design steam mass flow rate will be met but the mass flow rate of MSWs will be less than the design value. The produced steam's design conditions depend on the application. Steam generators for utility power production operate at high temperatures and pressures. Steam generators for process and heating applications are typically at much lower temperatures and pressures.

The waste-to-energy plants are divided to three main categories depending on the level of preparation required for the fuel. They are in the order of popularity as follows: **mass burning**, **refuse-derived fuel**s, and **solid recovered fuel**s.

4.6.4.12.1 Mass Refuse Burning Systems

Mass burning units require the least fuel preparation process. In many cases, only large pieces like appliances are removed from delivered wastes. Solid wastes are transported to the plant by trucks, sometimes the same trucks that collected them (Figure 4.35). The arriving wastes are unloaded into a refuse pit. The pit is typically equipped with overhead cranes. The cranes use their grapples to remove large pieces of wastes. Also, the arriving wastes typically are not homogenous and the cranes mix them to make them as uniform as possible. The wastes are stored in the pit until it is needed for combustion. At that time the cranes move them to the feed hopper. Then, the fuel is mass-fed to the furnace by feeders, typically in the form of hydraulic rams. They also control the flow rate of the fuel to the furnace. The combustion system in the furnace is typically a vibrating (reciprocating) grate stoker system. The fuel burns on the bed over a typically inclined grate. The primary air, which is typically a little over half of the total air, is provided below the grate, which can also act as the cooling agent of the grate. However, some

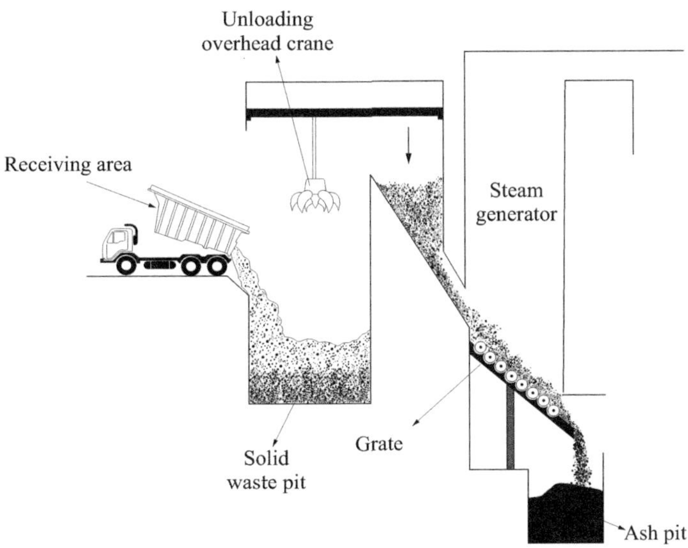

FIGURE 4.35
Schematic of a mass refuse burning system.

designs use water-cooled grates. The rest of the combustion air is injected to the furnace through the overfire air ports. When the fuel contains a high percentage of moisture, the primary air may be preheated to act as the fuel drying agent or an auxiliary burner may be used. The ash and noncombustible parts of the fuel are collected in the ash pit. They are removed by screw or chain conveyors. The combustion remains may be reused as construction materials or sent to landfills. The design of the furnace and the stoker burner are very similar to coal stokers discussed in Section 4.6.2. One major difference is the very corrosive nature of the combustion process in the furnace. This requires protection of the waterwall tubes in the lower portion of the furnace using refractory, ceramic tiles, and/or specially fabricated double-layer tubes. In the latter case, the inner layer is made of carbon steel and the outer layer is a corrosion-resistant metal. The combustion products also need substantial treatments to bring the exhaust gas in compliance with pollution regulations.

While the fuel preparation systems of mass burning units are simple and inexpensive, they require a more complicated and expensive combustion system, including the pollution control equipment and higher operation and maintenance costs. So their economic benefits should be carefully analyzed compared to other systems.

4.6.4.12.2 Refuse-Derived and Solid Recovered Fuel Burning Systems

Raw wastes from the source need to go through many preparation processes to be converted to **refuse-derived fuel (RDF)**. A **solid recovered fuel** requires even more preparation processes to increase its heating value so it can be used in industrial processes, such as cement kilns. Solid recovered fuel can technically replace conventional fuels with minimal adjustments to the burning system.

RDFs can be combusted as a main fuel in dedicated RDF-fired steam generators. They can be also combusted in dual- or multiple-fuel burners along with natural gas, biomass, and/or coal. These systems are suitable mainly when the produced electricity or steam is needed even when solid wastes are not available. Finally, RDFs can be used as the supplementary fuel in pulverized coal or cyclone combustion systems, replacing some of the main fuel. The latter two applications require much more preparation than the dedicated units. RDF-fired units can even replace old coal power plants in urban areas in retrofitting projects.

The fuel preparation process for solid recovered and refuse-derived fuels includes several crushing and shredding processes, several separation and classification processes to remove glass, dirt, and other noncombustible substances, ferrous metal removal process, aluminum and other no-ferrous metals removal process by hand pick up or eddy current separators, and tire shredding process (if needed). Some of the separated materials, such as ferrous and aluminum metallic pieces, are recycled and the rest is sent to landfill sites. The remaining mostly combustible substances are shredded and stored before being sent to the furnace for combustion. While some RDF-fired fluidized-bed units, both bubbling and circulating units, have been installed, majority of these systems have been spreader traveling grate stoker units.

RDF is metered and fed onto the furnace through feeders from the prepared fuel storage area. The fuel is distributed onto the bed using air-swept distributors with varying air pressure to make the spreading process as uniform as possible. Some fuel pieces combust while in suspension and the rest has complete combustion on the grate. While the preparation system in this type is much more complicated and expensive than mass-burning systems, the stoker system is less expensive and the ash handling system is smaller. This means the economic analysis can determine which system is more suitable for each application. Most aspects of both mass-burning and RDF-fired systems are similar to coal-fired systems. Another approach to use solid wastes and biomass as fuel for steam generation is the gasification process that will be covered later.

4.7 Liquid Fuels

As shown in Figure 4.1, liquid fuels are rarely used as a primary fuel in steam generators (for that matter any utility power generation). However, they are still sometimes used as a backup fuel and for the ignition process. In this section, these fuels and their combustion systems are briefly reviewed. Liquid fuels can be **hydrocarbon** or **alcohol**.

While petroleum-based fuels are not a major energy source for power generation, they are a vital player in other energy sectors, particularly for transportation. Their high volumetric energy content, easy transportation by piping systems on land and supertankers on sea, typically lower environmental impacts (at least compared to coal), and a wide variety of applications due to the diversity of fuel derivatives make them indispensable for human civilization at least in short- and mid-term future. In particular, the recent advancements in hydraulic fracturing (fracking) technology significantly improved the prospect of these fuels, especially in the U.S., regardless of adverse environmental impacts.

While petroleum contains oxygen, nitrogen, sulfur, and traces of some metals, it is primarily composed of carbon and hydrogen (hence the term hydrocarbons). Thus, in order to categorize the families of hydrocarbons, the following three concepts should be first understood. The structure of a hydrocarbon molecule can be either a ring or chain structure (Figure 4.36a and b). In a chain structure, each carbon atom bonds with only one other carbon atom. Also, a hydrocarbon structure can be either saturated or unsaturated. In saturated structures, all carbon atoms are joined by a single bond, while in unsaturated ones two or more carbon atoms have double or triple

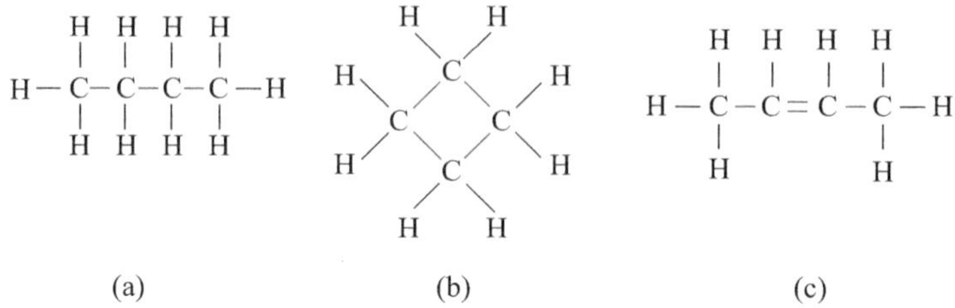

FIGURE 4.36
Examples of various hydrocarbon structures.

bonds (Figure 4.36a and c). Finally, two hydrocarbons can be an isomer if the same numbers of carbon and hydrogen atoms have different structures (Figure 4.36b and c).

According to these characteristics and the ratio of the number of carbon atoms to the number of hydrogen atoms, hydrocarbons are grouped by the following suffix identifications:

- Paraffins (or alkanes): C_nH_{2n+2} (chain, saturated) suffix in –ane, e.g., methane (n = 1), ethane (n = 2), propane (n = 3), and butane (n = 4)
- Olefins (or alkenes): C_nH_{2n} (chain, unsaturated) suffix in –ylene or –ene, e.g., ethene (n = 2), propene (n = 3), butane (n = 4), and pentene (n = 5)
- Diolefins (or alkadienes): C_nH_{2n-2} (chain, unsaturated) suffix in –diene, e.g., propadiene (n = 3) and butadiene (n = 4)
- Naphthenes (or cycloalkanes): C_nH_{2n} (similar formula as Olefin) (ring, saturated) prefix cyclo-, e.g., cyclobutane (n = 4) and cyclopentane (n = 7)
- Aromatic Benzene: C_nH_{2n-6} (ring, unsaturated), e.g., benzene (n = 6)
- Aromatic Naphthene: C_nH_{2n-12} (ring, unsaturated)

Petroleum-based fuels are typically a mixture of various hydrocarbons; for example, gasoline is primarily composed of 40 different hydrocarbons. The composition of liquid fuels is typically presented on mass basis. For simplicity and convenience, hydrocarbons are typically represented by an average single hydrocarbon. For example, octane (C_8H_{18}) is used for gasoline (gas) and dodecane ($C_{12}H_{26}$) is used for diesel.

Petroleum-based fuels are typically produced by the crude oil distillation process in a fractional distillation column. For this purpose, crude oil is first gasified (vaporized) and then enters the bottom of the column. Heavier hydrocarbons with higher boiling temperatures are condensed at lower levels in the column. Other hydrocarbons are gradually condensed and separated while the gasified oil moves upward at temperatures between 600°C and 20°C. They may also go through several refining processes such as cracking, reforming, and hydrogenation processes to produce products, such as gasoline, kerosene, diesel, residual fuel oil, and some gaseous fuels, at desired quantity and quality. Besides the ultimate analysis and the heating value as the most important characteristics of these fuels, viscosity and density (expressed in the form of specific gravity) are also critical for designing liquid fuel-fired steam generators. Residual fuel oil and in some special cases diesel are used in power production units.

Another group of fuels that are increasingly being used as fuel for various applications is alcoholic fuels. Alcohols are a family of hydrocarbons where one hydrogen atom is replaced by an OH radical. Some examples of these fuels are methanol (CH_3OH) and ethanol (C_2H_5OH). These fuels are gaining popularity as biofuels mostly for transportation applications. However, they do not have many applications in the utility power generation industry.

4.7.1 Transportation and Storage of Liquid Fuels

Petroleum products can be transported to a power plant site by pipelines, barges, rails, or trucks. The fuel should be then stored on the site in what is called a "tank farm". Several interconnected cylindrical steel containers are used for this purpose so if one is out of service, the plant can continue to operate. Since the fuel used in utility power plants is typically heavy fuel with low volatile matters, these containers do not require floating roofs. They can be located on surface or underground. If on the surface, each container is surrounded by protective walls to prevent spreading oil if there is a leakage. They are painted with light colors to avoid overheating under the sun. For very viscous fuels, such as residual fuel oil, containers and associated piping systems should be equipped with heaters to assist the flow of fuel. The source of thermal energy can be steam or electricity. The temperature of the heating process should be carefully set and controlled. Temperatures lower than the required temperature result in difficulty in fuel transfer, which may interrupt the flow and cause too much power consumption in the fuel pumps. However, a higher temperature may cause evaporation and cracking processes in the fuel. The fuel is pumped to the furnace when it is needed for burning. The fuel flow rate is also measured, which is used in the plant's efficiency calculation.

4.8 Gaseous Fuels

Gaseous fuels can be natural or manufactured gases. The former is mostly methane and ethane and the latter is commonly a mixture of CO and H_2. Similar to liquid fuels, gaseous fuels are typically made of a mixture of several hydrocarbons. The composition (ultimate analysis) of gaseous fuels is typically presented on molar (volumetric) basis. The heating value, density (in the form of specific gravity), sulfur content, and flammability are other important characteristics of gaseous fuels. Natural gas is the most common gaseous fuel used for steam generation.

4.8.1 Natural Gas

Similar to crude oil, natural gas can be extracted from underground reservoirs. Some natural gas reservoirs only contain natural gas with no or little crude oil. This type of natural gas is called **non-associated natural gas**. Alternatively, oil and natural gas can coexist in which case natural gas is referred to as **associated natural gas**. In this case, natural gas can be dissolved in oil (conveniently called dissolved gas) or can be accumulated on the void above oil (called free gas).

After extraction of natural gas, large solid impurities such as sand are scrubbed from the raw natural gas typically at or near the wellhead. The next step in the processing of natural gas is the separation of crude oil (if needed). This separation can typically happen by simply reducing the pressure of the extracted mixture with or without slight cooling. This can take place in a simple container also typically at or near the wellhead. Then, the raw natural gas is transported to the central processing facility by low-pressure gathering pipelines. In the processing plant, liquid water and steam are removed by condensation or a dehydrating medium. Next, the heavier hydrocarbons such as ethane, butane, and propane, known as **natural gas liquids (NGLs)**, are separated and liquefied. The separation can be achieved by their absorption by certain absorption agents and then their recovery by heating the agent. Alternatively, the separation can be achieved by condensation at very low temperature (at about –85°C or about –120°F). These NGLs are very valuable and have many industrial and household applications.

Finally, the extracted natural gas contains a high concentration of sulfur contents and some other gases, particularly hydrogen sulfide (H_2S) and carbon dioxide. The former can be recognized by its rotten smell. That is why untreated natural gas is sometimes referred to as **sour gas** and the removal process is called the **sweetening process**. The removal process typically involves the absorption of H_2S and CO_2 by an amine solution so the process is also called the amine process. The removed H_2S can be treated to produce sulfur, which is a commercially marketable byproduct.

4.8.2 Hydrogen

Nowadays hydrogen is considered as a gaseous fuel that can be produced from hydrocarbons or water electrolysis. Hydrogen is a flammable gas with very low ignition energy. It burns with a colorless flame. It can be used for high-efficiency power generation in fuel cells with water as the only product.

Hydrogen can be also mixed with natural gas to be burned in conventional power generation units. While the higher reactivity of hydrogen compared to that of methane can improve the combustion reaction and reduce unburned hydrocarbon and CO in the flue gas, it can increase the flame temperature and thermal NO_x production. It can also change the properties of the flame which need to be taken into account in the system design. The greater heating value of hydrogen can reduce the mass flow rate of the combustion products, which can be problematic for some power generation systems. In order to increase the energy intensity of hydrogen during storage and transportation, it can be liquefied at low temperatures and high pressures. This is known as cryogenic hydrogen storage. Alternatively, hydrogen can be stored in metal hydride systems.

There are other gaseous fuels such as blast furnace gas, coke oven gas, refinery gas (in refineries made of various hydrocarbons), regenerator offgas (in refineries made of mostly CO), landfill gas (mostly methane and CO_2), and many others. Some of these fuels are discussed in Section 4.5.4.4.

4.8.3 Transportation and Storage of Natural Gas

As noted earlier, processed natural gas can be transported to consumption sites by pipelines. There is an extended network of pipelines in many countries distributing natural gas to all major consumption centers. This is the most economical way to transport natural gas. The diameter of these pipes can be between 15 and 120 cm (about 6 to 48 in.). Larger pipes are typically welded carbon steel and are used for national interstate lines. The smaller ones can be seamless carbon steel or plastics and are used for local distribution lines. Pipes are buried at the depth of at least 70 cm (about 30 in.) and are typically protected by cathodic protection and outer coatings. Natural gas pressure in pipelines is increased to between 15 and 100 bar (200 to 1500 psi) to overcome the pressure drop of the gas flow in pipes. The high pressure also reduces the volume and increases the density of natural gas (in the order of several hundred folds) so for the same volumetric flow rate, a higher mass flow rate can be achieved. To maintain this high pressure and to compensate for the pressure loss, compression stations are located at a distance of up to 150 km (about 100 miles) between the stations. The compressors can be powered by natural gas-fueled gas turbines, gas-fueled internal combustion engines, or electric motors. Pipeline systems also include many valves, measuring stations, control systems, and distribution pipes. Natural gas can be also liquefied and transported by ships, rails, or trucks.

While natural gas can be economically transported by pipelines over land, it is not easy to transport it across oceans and where pipelines are not available. One solution for this problem is **liquefied natural gas (LNG)** because liquefying natural gas can reduce its volume by about 600 folds. This makes LNG transportation by ships and trucks economically feasible.

LNG is also very suitable for storing natural gas and for using it as a transportation fuel. Natural gas can be converted to liquid form at the temperature of about –160°C (about –260°F).[32] This process requires a significant amount of energy. Some power plants store natural gas as LNG on-site for periods of natural gas shortage. They may have their own small natural gas liquefaction plant on-site. The largest consumers of LNG are in Asia and Europe.

After delivery of LNG to its destination and before its consumption or injection into pipelines, it should be regasified and converted to gaseous fuel. The thermal energy needed for the regasification of LNG can be provided from ambient air. Also, it can be used to provide cooling for other applications such as air conditioning, cooling of inlet air to the compressor in gas turbine plants (refer to Chapter 9), or reducing the temperature of the condenser in steam power plants.

In order to provide fuel during high demand periods, some sort of energy storage should be provided for natural gas-fired power plants. This is particularly important for cold regions where in very cold days, a high residential consumption rate reduces the gas pressure in the pipeline. In these situations, the preference is always given to residential consumers and large industrial consumers must reduce their pipeline natural gas consumption. Also, natural gas pipelines are typically designed and operated at a fixed load for economic reasons. Therefore, the power plant should be equipped with a system to handle more or fewer gas consumptions than this fix design flow rate. There are three ways to handle these situations:

- to store some supplementary liquid fuel in tanks at or near the power plant site.
- to produce supplementary manufactured gas from other energy sources.
- to store natural gas on-site.

It should be noted that thousands of miles of large-diameter pipes can store a significant amount of natural gas that can provide fuel in an emergency. Natural gas can be stored in above-ground large steel tanks. It can be also liquefied and stored in above-ground pressurized insulated tanks. Moreover, it can be stored in underground pipe holders. If underground natural voids are available in a region, for example, depleted oil and/or gas wells, salt caverns, and aquifers, natural gas can be stored in them by pumping the gas into these voids. Natural gas can be also stored by being absorbed into a porous solid medium. The stored natural gas can then be released by passing a flow of hot gas through the solid substance.

[32] https://www.eia.gov/energyexplained/natural-gas/liquefied-natural-gas.php.

At the power plant site, the pressure of the natural gas flow is reduced from the pipeline pressure to the pressure required by the combustion system. This process can take place through a throttling device (constant enthalpy pressure reduction) or through an expansion turbine. Energy wise the latter is clearly preferred because some of the energy of expansion can be captured. The volumetric flow rate of the natural gas is measured for the efficiency calculation before the fuel is sent to burners.

4.8.4 Combustion Systems for Liquid and Gaseous Fuels

As noted in Section 4.6.3, the fundamental principle of pulverized combustion systems is that if fuel particles are sufficiently fine, their combustion will resemble the combustion of gaseous fuels. So it should not come as a surprise that many aspects of natural gas combustion systems are similar to pulverized coal combustion systems. In fact some systems can burn natural gas as well as pulverized coal, and/or liquid fuels. Some systems can even burn these fuels simultaneously and can switch from one fuel to another while the burners are operating. Some of the most important characteristics of liquid and gaseous fuel burning systems are their fast load-following capabilities and a wide range of combustion rate.

Regardless of the type of a combustion system, the fuel delivery system must transport the right amount of the fuel at the designed pressure to each burner. The uniform distribution of the fuel and air among burners according to the burner design is critically important. Natural gas can readily mix with air without any preparation device. There are two types of natural gas burners: wall-fired burners and corner-fired burners. These systems can be used for the combustion of natural gas, oil, or both in dual fuel burners.

4.8.4.1 Wall-Fired Burners

In wall-fired burners, several burners located on the furnace walls create individual flames without substantial interaction among them. The burners can be installed on one wall or opposing walls. Early wall-fired burners, which are used for both liquid and gaseous fuels, were circular burners. In this system, the fuel enters the burner at the center in the core of the flame, and air is introduced tangentially via an air register, which both controls the air flow and induces swirl in the air flow. But due to difficulty in independently controlling the air flow rate and the swirl induced in air, these burners are no longer used (please refer to Section 4.6.4 for more information about these systems).

In modern liquid and gaseous burners, the air flow control mechanism is separated from the air swirl mechanism. The air flow is controlled by sliding disks while adjustable vanes induce swirl. This allows more control of the air flow at a wider range. The air flow rate is also measured in each burner to ensure a uniform distribution of the combustion air among all burners in service.

Generally speaking, natural gas and lighter petroleum products have very low nitrogen content and as a result, thermal NO_x production is the main NO_x production mechanism. However, heavier hydrocarbons can contain a significant amount of nitrogen so both thermal and fuel NO_x contribute to the total NO_x production in heavy fuel oil-fired burners.

To reduce both thermal and fuel NO_x, modern liquid and gaseous fuel-fired burners are typically equipped with dual air registers. As the name implies, in these systems the combustion air is divided into two separate flows, which in turn creates two combustion zones in the flame. In the central core of the flame, there is air deficiency compared to the stoichiometric mixture. This slows the fuel NO_x production process due to the shortage of oxygen needed for the process and lowers the combustion temperature reducing thermal NO_x. The secondary air needed for the complete combustion is introduced to the outer combustion zone. To further reduce NO_x these burners can be also equipped with the air staging technology where some of the combustion air is introduced at overfire air ports. Similar to all modern steam generators, these combustion systems are also equipped with igniters, flame detection sensors, and other safety devices.

As noted, most aspects of liquid-fired burners and natural gas-fired burners are similar. However, in walled-fired burners, the fuel is introduced to the furnace through spuds for gaseous fuels and through atomizers for liquid fuels.

As you know by now, to increase the interaction between fuel and air, it is vital to **increase the ratio of the surface area to volume of fuel particles**. For coal, this was achieved by pulverized coal. With the same intent, liquid fuels must be reduced to fine mist-like flows before being mixed with air. This critical process to improve combustion efficiency is called the atomization process. The effectiveness of the atomization process is measured by the average size of fuel particles and their size distribution (similar to pulverized coal). Fuel atomizers can be divided into two types: mechanical atomizers or dual-fluid atomizers. The difference between the two methods is whether the atomizing medium, typically steam or air, is used during the atomization process.

In mechanical atomizers, the only atomizing mechanism is the fuel pressurization (hence the name). In these atomizers, the pressure of the fuel is increased to as high as 70 bar (about 1000 psi). Then, the high-pressure fuel is released to the air flow via the specially designed nozzles. Alternatively, the atomization process can take place in a chamber where the high-pressure fuel is tangentially introduced to the air flow. The mechanical atomizers are relatively simple and do not need a source of a secondary flow, but the quality of produced droplets is lower than the second method described below.

In dual-fluid atomization systems, the moderately pressurized fuel is atomized with the help of an atomizing agent, i.e., steam and air. These systems can provide a better quality atomization process. This method not only provides finer fuel particles which in turn has a higher combustion efficiency and lower emissions but also is safer. However, it has the disadvantage of requiring steam, which means the loss of latent and sensible thermal energy to the stack and the loss of feedwater. Both of these losses can be significant, especially where water is scarce.

In these systems, the fuel is pressurized to the pressure of up to 20 Bar (about 300 psi) and is mixed with steam or air at the pressure of up to 100 bar (about 1500 psi) higher than the pressure of the fuel. The mixture is then introduced directly to the furnace through nozzles or to a mixing chamber before entering the furnace. In the furnace, the temperature of the mixture rapidly increases. The quick expansion of the atomizing agent particles due to the rapid temperature increase atomizes the fuel to fine particles. The ratio of the steam mass flow rate to the fuel mass flow rate can typically be between 0.02 and 0.2 kg of steam/kg of fuel. The higher ratio is more desirable because it improves the quality of the atomization process. However, the benefit must balance the losses associated with using steam.

When the atomized fuel is injected to the furnace, the fuel particles burn in two stages. First, the volatile components of the fuel are evaporated and burned immediately. Then, the carbon-rich porous residues are burned, which require much longer time. Most other aspects of liquid-fired steam generators, including ash handling systems, are similar to pulverized coal steam generators.

The fuel delivery system in natural gas-fired burners includes a fuel manifold to distribute fuel uniformly among burners. The manifold is located outside the furnace either inside or outside of the windbox. Natural gas is typically delivered to the burners at the pressure of 0.5–1.5 bar (about 8–20 psi). The delivered fuel enters one or multiple gas elements (also known as spuds). These spuds are long enough to extend through the windbox. They enter to the furnace via the burner opening where they inject the fuel through specially designed caps.

4.8.4.2 Corner-Firing Systems

As the name implies, in these combustion systems burners are located on the corners of the furnace and each contributes to a single unified flame envelop. Both liquid and gaseous fuels can be burned in these burners. The burner is made of fuel and air compartments placed alternatively on a tower. The fuel is injected into the flow of air via liquid fuel atomizers or gaseous fuel nozzles in each fuel compartment. The combustion air from the windbox is typically divided equally between fuel and air compartments (as the primary and secondary air flows). Dampers are used to properly distribute the air among various compartments. To reduce NO_x production, these firing systems are also typically equipped with the air stage technology where some of the combustion air is diverted to overfire air ports on the top of the tower.

4.8.4.3 Igniters for Liquid and Gaseous-Fired Burners

Considering a wide range of characteristics that gaseous and liquid fuels can possess, there are a variety of igniters with varying specifications. They can be generally categorized to two groups: direct electrical igniters or electrically ignited liquid- or gas-fired igniters. In the latter case, the igniter fuel can be natural gas, liquefied petroleum gas (LPG), or light liquid hydrocarbons. In this case, the igniter itself needs an atomizer and even a heater.

Both types of igniters can be fixed or retractable. The latter is preferred because it does not need cooling air during the normal operation of the furnace. Igniters can be also used to heat the furnace during a cold start-up and complement or stabilize the main fire. The size and firing capacity of igniter systems depend on the requirements of these possible functions and can be as high as one-tenth of that of the burners in the full load operation. Note that the cold start-up of a furnace can take up to 6 hours, so the igniters may be operated for an extended period.

4.9 Combustion Air System

In steam generators, air is not only needed for the combustion of the fuel but may be required in the fuel preparation process, such as the coal pulverization process. In this section, we will discuss how this air can be provided and regulated. We will also cover air preheaters and stacks.

4.9.1 Providing Air for Combustion

The majority of the required air in a steam generator is for the combustion of the fuel. The fuel combustion air can be divided to stoichiometric air and excess air. The former is the minimum required air for the complete combustion process, which depends on the fuel composition. The latter is the extra air needed to make up for the imperfect nonuniform mixing of the fuel and air. The amount of excess air depends on the fuel composition, fuel state, and firing system. It could be between a few percent to a half of the stoichiometric air. While more excess air can facilitate a more complete combustion process, this air does not actually participate in the combustion reaction. The thermal energy to heat this air from the inlet air temperature to the flue gas temperature is considered a sensible heat loss. There is a balance point where the tradeoff between the maximum combustion efficiency and the minimum thermal losses provide the optimum operating conditions of combustion. Besides the combustion air, the provided air should cover other air uses, air leakage, and infiltration. The amount of this required air not only impacts the air handling system but also the design and volume of the furnace.

The combustion process in a furnace can be at a pressure above, below, or approximately equal to the atmospheric pressure. Similar to any other fluid flow, in order to air and flue gas to flow through a steam generator, the pressure drop of the flow should be overcome. The needed pressure can be provided **naturally** by the draft created in the stack or **mechanically** by fans. If the combustion products are drawn out of the furnace, the combustion process will be under partial vacuum (**suction firing**). These furnaces are typically referred to as induced-draft combustion systems. Draft here refers to the flow of air and combustion products due to the pressure difference between the inside of the furnace and the atmospheric pressure. The induced draft can be created by the stack, called **natural draft**, or by fans, called **induced-draft fans (ID fans)**. The main problem with this configuration is that the partial vacuum in the furnace can cause air infiltration into the furnace that reduces the efficiency of the combustion (similar to the effect of excess air). If the inlet air is forced into the furnace, the combustion process will be under pressure (**pressurized firing**). This type of furnace is typically referred to as forced draft and the fans are called **forced draft fans (FD fans)**. The main disadvantage of this configuration is that the higher pressure of the furnace compared to the atmospheric pressure can cause the leakage of the flue gas to the environment. This loss of hot combustion products reduces the efficiency of the steam generator and hurts air quality on the site. Often these two systems are combined to create a **balanced draft furnace**. In these systems, the total pressure drop of the system is distributed between FD fans and ID fans (or stack natural draft) in a way that the furnace operates at or near the atmospheric pressure. This system eliminates or significantly reduces the problems of air infiltration and flue gas leakage.

In all these systems, the stack natural draft, due to the pressure difference between the top and bottom of the stack, is always present. In small steam generators with relatively small gas flows and simple furnace configurations, the pressure drop is small enough that the natural draft provided by the stack can be sufficient to provide the entire needed draft. For large modern steam generators, the complexity of the steam generator, the presence of many components such as superheaters, reheaters, etc., and a high gas flow rate create such a large pressure drop in the flow that the stack alone cannot generate enough draft. In these cases, mechanical draft produced by fans is needed for the normal operation of the steam generator. So while the natural draft refers to a steam generator that the draft is purely provided by the stack, in forced and induced-draft steam generators, the stack natural draft is also present and supplements the fans in providing the total needed draft.

The fans can be located at the **furnace inlet** to pressurize the inlet air (FD fans) or at the **furnace outlet** to pressurize the outlet flue gas (ID fans). Depending on the configuration and location of these components, one of the following combinations can be utilized in a steam generator:

- Stack (induced draft)
- FD fans+stack (forced draft)
- ID fans+stack (induced draft, this combination is uncommon)
- FD fans+ID fans+stack (balanced draft)

FD fans receive the atmospheric air and provide a sufficient pressure head to the air. In forced draft steam generators, the pressure should be adequate to deliver the flue gas all the way to the stack by overcoming all the pressure losses. In balanced draft steam generators, the pressure should be sufficient to deliver the air to the furnace at about the atmospheric pressure. In pulverized coal steam generators, typically there are two sets of FD fans, primary air fans and secondary air fans.

ID fans are typically used in balanced draft steam generators to overcome pressure losses of the flue gas flow between the furnace and the stack.

While the functionality of FD and ID fans are similar, they pressurize very different flows. This provides FD fans with some major advantages over ID fans. An FD fan deals with the cold clean atmospheric air. The low temperature of the air flow significantly reduces the fan power consumption. Similarly, the high density of the cold air increases the mass flow rate of the air for the same air volumetric flow rate. Furthermore, unlike an ID fan that pressurizes the corrosive and erosive combustion products, an FD fan deals with the much less demanding atmospheric air flow.

While the ID and FD fans are the most important fans in any power plant, there are other fans in these plants. In the following section, a brief general discussion will be presented on fans.

4.9.2 Fans

Fans are devices that add energy to a gas flow to make up pressure losses on the flow path to sustain the flow. Various types of fans are commonly used in steam power plants for different purposes. Some of these fans are forced draft fans, induced-draft fans, primary air fans, overfire air fans, flue gas recirculating fans, ventilation fans, and fans in mechanical draft cooling towers (refer to Chapter 8).

The functions of the first four types of fans are similar to provide the air flow for the combustion process and/or remove the combustion products. The flue gas recirculating fans are used to control the temperature of the flue gas in the furnace. In terms of operating conditions, they are similar to ID fans but in even more severe conditions.

Like any other building, various parts of a power plant require ventilation. Depending on the requirements of a specific case, this can be achieved by creating pressure either above or below the atmospheric pressure in the target space. If the objective is to keep contaminations (for example, pulverized coal) in a space and to prevent them from spreading to other spaces, the pressure of the space is kept below the atmospheric pressure. If the objective is to keep contaminations out of a particular space, for example, the control room where highly sensitive equipment is located, the pressure is kept above the atmospheric pressure.

A typical fan has all or some of the following components:

- impeller
- housing
- driver
- inlet air screens and filters
- noise reduction devices
- gas flow control devices

The impeller, also known as the fan wheel, bladed rotor, or simply rotor, is the component that directly interacts with the gas flow and is where the energy is imparted in the flow. The flow leaving the impeller is collected by the housing and is directed toward the fan outlet. The driver provides the external energy needed to rotate the impeller.

In terms of the type of the impeller and the direction of the outlet flow with respect to the direction of the inlet flow, fans are divided into two categories: **centrifugal fans** or **axial fans**. Both centrifugal and axial fans are used as FD and ID fans in steam generators.

In centrifugal fans, the gas flow enters the impeller axially (Figure 4.37). As the gas flow goes through the impeller and energy is imparted to the flow to increase its velocity, the direction of the flow changes to a radial flow. The high-velocity flow is collected by the housing where the kinetic energy of the gas flow is converted to pressure. Generally speaking, centrifugal fans are suitable for applications where a relatively high pressure head and a low gas flow rate are required. In axial fans, both the inlet flow and the outlet flow are in the axial direction.

With regard to the orientation of fan blades with respect to the direction of the rotor rotation, centrifugal fans can be categorized into three groups: backward-inclined, forward-curved, and straight radial (Figure 4.38). As the name implies, in backward-inclined fans, the impellers are inclined away from the direction of the rotor rotation (Figure 4.38a). The blades can be airfoil, curved, or flat. The latter two types are single-thickness blades. These fans are generally very efficient and produce low noise. They can also operate at high rotational speeds, which results in high pressure heads. While some of these fans can operate with gases containing entrained solid particles, they are prone to particle accumulation behind the blades. The forward-curved fans can produce low pressure heads and are not commonly used in power plants (Figure 4.38b). The straight radial fans are not as efficient as backward-inclined fans and cannot produce high pressure heads (Figure 4.38c).

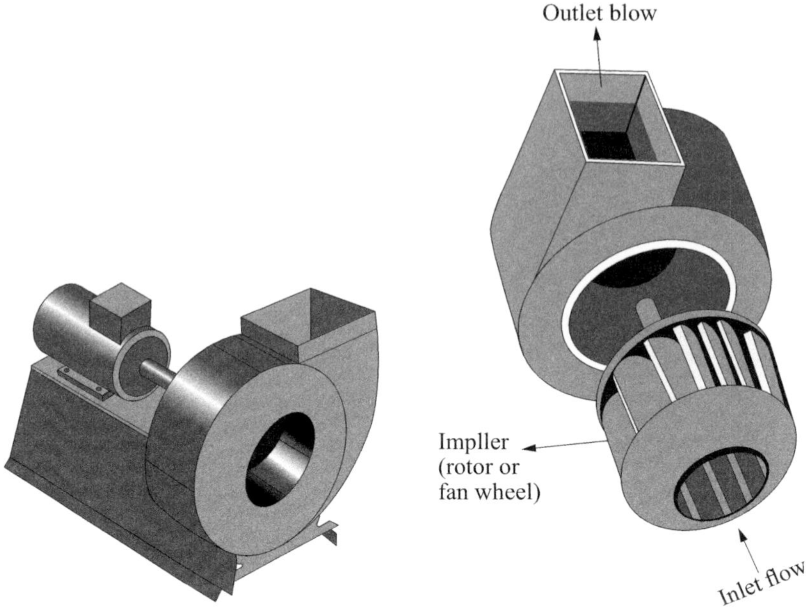

FIGURE 4.37
Schematic of a centrifugal fan.

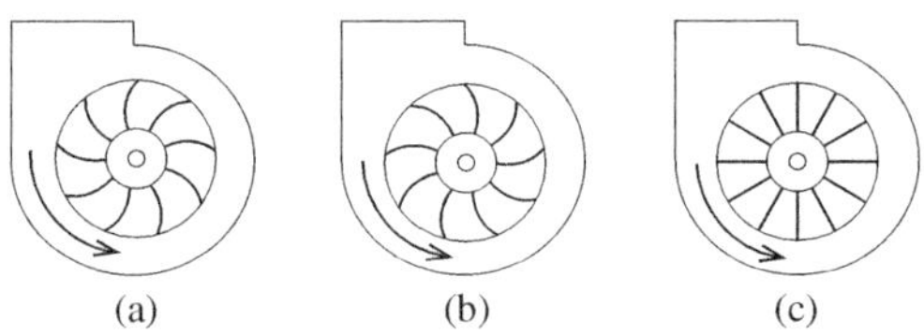

FIGURE 4.38
Orientation of fan blades. (a) Backward-inclined, (b) forward-curved, (c) straight radial.

However, they are inexpensive to manufacture and maintain. They are also suitable to operate with dust-laden flows with little particle accumulation behind the blades.

In axial fans, the gas flow enters the fan axially and it remains parallel to the axis of rotation throughout the fan while going through the blades and leaves the fan axially. Essentially they are similar to desk fans with an enclosed housing. These fans are typically cylindrical in shape. They can be installed horizontally or vertically. The attack angle of the blades (also pitch angle) in these fans can be fixed (called fixed-pitch blades), can be changed while the rotor is stopped (called adjustable-pitch blades), or can be changed while the rotor is rotating (called variable-pitch blades). The latter system can have a higher efficiency because the pitch angle can be altered to optimize the system's performance while it is operational. Generally speaking, axial fans are more suitable for applications with high gas flow rate and low pressure head requirements. One problem with axial fans is the possibility of stall in the fan. Stall can occur if the fan is pushed to operate at extreme operating conditions due to an inappropriate design and/or operation. In this condition, the flow is separated from the fan blades, which in turn results in the unstable operation and blade damage. We will discuss these topics in Chapter 9 when we cover compressors for gas turbine cycles.

4.9.3 Fan Selection or Sizing

The most important parameters in the selection of fans (also referred to as the **fan sizing**) are the required **gas volumetric flow rate** and the required **pressure head**. For ID and FD fans, these two factors are determined from steam generator calculations by including safety margins of as high as 30% for unforeseeable conditions. The volumetric flow rate is determined from the combustion air requirements plus the air leakage, infiltration, and other air utilizations. The pressure head is determined by the pressure drop throughout the steam generator including in the air ducts, windbox, burner or fuel bed, furnace, heat-transfer surfaces, and exhaust path. It should be noted that combustion air calculations are mass-based while fan calculations are volume-based. To convert the mass-based air flow rate to the volume-based flow rate, the density of the inlet air is needed. Since the density of the air depends on the ambient temperature and the local atmospheric pressure, the worst-case scenario should be considered for the determination of the air density. It should be borne in mind that for the given design and operation conditions, fans are constant volumetric flow rate devices and the volumetric flow rate is independent of the type and properties of the gas being pressurized.

In order to determine which type of fan is more suitable for a certain application, the parameter called specific speed can be used. The **specific speed** is defined as follows:

$$N_s = \frac{n\sqrt{Q}}{(H)^{3/4}} \tag{4.69}$$

where N_s is specific speed, n is rotational speed $\left(\frac{1}{s}\right)$, Q is volumetric flow rate (m^3), and H is pressure head (m). A higher specific speed indicates that axial fans are more suitable for the desired application. The fan performance is presented graphically in the fan performance curves (Figure 4.39). In these curves, the pressure head, fan efficiency, and required shaft input power are expressed as a function of the volumetric flow rate. Since for the same fan geometry at different sizes, with different pitch angles, and at various rotational speeds, the shape of the performance curves is similar but with different numerical values, a family of performance curves is typically presented in a single diagram.

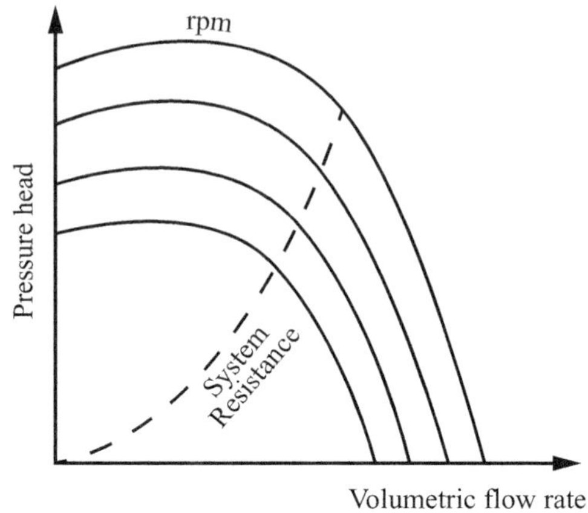

FIGURE 4.39
Performance curve of a fan for various rotational speeds.

For sensitive applications, often two or several parallel fans are used to provide some redundancy in case one of the fans requires maintenance. In the case of two fans, each fan can handle 50%–100% of the total load. In the parallel arrangement, both fans provide the same pressure head, but the total volumetric flow rate is equal to the sum of the flow rates of each fan. The volumetric flow rate of the fans may or may not be equal. The parallel arrangement adds to the complexity of the system operation.

4.9.4 Air Auxiliary Systems

As noted, the shaft power required to run a fan depends on the volumetric flow rate of the flow, pressure head, and efficiency of the system. This power can be provided by an electric motor or a steam turbine. Electric motors are the most common type of fan drives since they are inexpensive and efficient in a wide range of fan operation. They can be constant or variable speed, powered with AC or DC electricity. Steam turbine drives can be more economical for large fans where inexpensive steam is available.

Since the amount of the required air by combustion varies significantly depending on the plant load, an effective fan control strategy is critical for the proper operation of the system. **Dampers** are one way to regulate the gas flow rate by introducing resistance to the flow. They can be located on the air or flue gas flow path (for both ID and FD fans). They can be also used to isolate equipment when they are not in service or during maintenance. There are several types of dampers: louver, guillotine, round, and double blade. They can be controlled by electrical, hydraulic, or compressed air actuators. Dampers can also supplement variable speed motors (discussed below) to further regulate the flow. Dampers are simple and inexpensive to manufacture and maintain. They can also provide a continuous range of control.

However, their main problem is the waste of energy. The flow is pressurized in the fan and then throttled to reduce the pressure in the damper, which significantly increases the system power consumption.

An alternative for dampers to control a gas flow rate is **variable inlet vanes (VIV)**. Similar to dampers, they introduce resistance to the inlet flow to the fan. But while doing this they induce rotation to the flow, which reduces the power consumption in the fan. While their initial cost is higher than that of dampers, their lower operational cost can easily compensate that, which is why they are one of the most common and economical air flow control mechanisms for fans. They also enjoy all the advantages of dampers.

Since the gas flow rate of a fan is proportional to its rotational speed, **variable speed fans** can be used to control the gas flow rate. The lower volumetric flow rate can be achieved by reducing the rotational speed of the fan. The variable rotational speed of the fan can be created by **variable speed motors, two-speed motors, couplings** (magnetic or hydraulic), or **variable speed steam turbines**. While these systems have higher initial costs, their lower operational costs (due to lower power consumption) make them very attractive and they are commonly used in power plants.

Fans are one of the **noisiest components** in modern steam power plants (if not the noisiest). The noise is created because of the air flow interactions with solid surfaces and can be originated from various parts of the system. The noise can be reduced by the installation of silencers in ducts and acoustic laggings on gas flow passages.

For all components that are long enough to have a significant thermal expansion and contraction, a mechanism, called an **expansion joint**, should be included to absorb the size variation. These components for the air delivery system include the air and flue gas ducts and the windbox. For this purpose, expansion joints made of metals or non-metallic materials, such as elastomer and composite materials, can be used. The non-metallic expansion joints are capable of absorbing sheer movements as well as axial movements. Expansion joints can also absorb vibrations and prevent them from spreading to the entire system.

4.9.5 Air Preheater

At the outlet of the economizer, combustion products are at a temperature of about 300°C (about 600°F). This flow contains too much energy to be discharged to the surrounding. On the other hand, the combustion air flow after the FD fans is at slightly above the ambient temperature. The temperature difference between these two flows provides an opportunity for heat recovery in an air preheater to improve the efficiency of the steam generator and the cycle. Air preheating can reduce the temperature of the flue gas flow to about 150°C (about 300°F) and can increase the temperature of the combustion air flow up to about 300°C (about 600°F). Increasing the temperature of the combustion air flow not only eliminates the thermal energy that otherwise would have been required to heat the air in the furnace but also improves the combustion process itself. Heating the combustion air in pulverized coal steam generators is particularly important since it is responsible for drying and transporting pulverized fuel. It should be noted that while adding an air preheater improves the efficiency of the cycle and reduces the fuel cost, this should be balanced against the capital cost of the components and increase in the power requirement of the fans (to overcome the pressure drop of the air preheater). The performance of air preheaters is evaluated according to ASME Performance Test Code (PTC) 4.3.

As noted, an air preheater is located downstream of the economizer and forced draft fans. Since the temperature difference between the combustion products and the air is relatively small, an air preheater needs a large heat-transfer surface. Typically, air preheaters can be either **rotary regenerative** or **tubular recuperative**.

Rotary regenerative air preheaters are composed of a rotor containing the heat storage media as well as the air and flue gas ducts. The rotation of the rotor through the ducts results in heat transfer between the cold and hot flows by means of the heat storage media. There are two designs for regenerative preheaters. In the first design, the rotor rotates through stationary ducts. Alternatively, ducts can rotate to direct cold and hot streams to fixed heat storage media. This design is commonly referred to as the Rothemuhle design. The former design, the rotating heat storage media and stationary flow ducts, is the dominant type in large utility steam power plants. This design is commonly referred to as **Ljungström air preheaters** after its Swedish inventor Fredrik Ljungström (engineer, 1875–1964) who patented it in 1930.

As shown in Figure 4.40, in a Ljungström air preheater, the motor-driven rotor slowly rotates at the rotational speed of 1–3 rpm. As a result, the heat

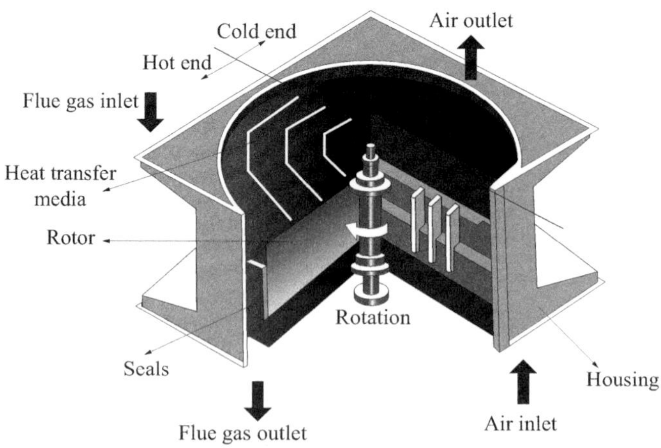

FIGURE 4.40
A rotary regenerative Ljungström air preheater.

storage media alternatively get in contact with the cold and hot flows. When exposed to the flue gas, the storage media absorb heat that increases its temperature. Then, when the media goes through the combustion air flow, the stored thermal energy is transferred to the air and the media cool down. This design of air preheaters is also referred to as the bisector heater. Some modern power plants use trisector or quadsector Ljungström air preheaters to independently preheat the primary and secondary combustion air flows.

The storage media, also known as heating elements, are assembled in a cylindrical rotor with a diameter up to 20 m. They are made of tightly spaced formed-steel plates, sometimes with coating, to allow the cold and hot gas flows to pass through. A maximum thermal energy storage capacity and minimum pressure drop in the flows are important factors in designing these units. The rotor is placed in a seal housing where one half of the rotor is always exposed to the flue gas from the economizer and the other half is exposed to the combustion air from the forced draft fans (in the bisector design). While both vertical- and horizontal-shaft Ljungström air preheaters are being used, the former is much more common due to its less cyclic stress. Note that in the vertical-shaft design the axis of rotation is vertical but the wheel itself rotates horizontally. The flows of the flue gas and combustion air are always counterflow, usually the former flowing downward to assist gravitational ash cleaning. To avoid leakage of the high-pressure combustion air flow to the low-pressure flue gas flow through the gaps between the rotor and the housing, a sealing mechanism is used to separate the two sides. However, there is always some mixing of the two flows due to several reasons, including leakage through the sealing, carryover of gases trapped and entrained in spaces between the energy storage media, and atmospheric air infiltration. The leakage problem can be intensified since two sides of the rotor are exposed to different temperatures that result in different thermal expansions and rotor distortion. Air preheaters are equipped with various air or superheated steam sootblowers and/or water-washing devices to remove ash deposits.

While Ljungström air preheaters are the most common air preheaters for utility applications, there are recuperative air preheaters that use stationary heat exchangers for the heat transfer between two flows. A major advantage of these preheaters is their little leakage between two flows. These heat exchangers can be tubular or plate types. In tubular counterflow air preheaters, the flue gas flows inside the tubes and the air flows outside. They are common for biomass-fueled steam generators because of the high contamination of the flue gas in these plants. Plate air preheaters are typically used in small steam generators.

4.9.6 Stack and Natural Draft

We have already discussed about natural circulation steam generators (refer to Section 3.4.3). The nature of the natural draft created in the stack in principle is exactly similar to natural circulation steam generators.

The temperature difference between the flue gas inside the stack and the atmospheric air surrounding the outside of the stack creates a pressure difference between the top and the bottom of the stack. This pressure difference generates a driving force that results in the natural flow of the combustion products through the stack. The same phenomenon in a home fire place causes the fresh air to be drawn into the fire place and forces the combustion products to be discharged through the chimney. That is why the phenomenon is sometimes referred to as the **stack** or **chimney effect**.

The driving force created in a stack can be determined by the following equation:

$$\Delta P_{\text{Stack}} = \left(\rho_{\text{amb}} - \rho_{\text{gas}}\right) gH \tag{4.70}$$

where ΔP_{Stack} is the driving force (pressure difference) created by the stack (kPa), ρ_{amb} is the density of the ambient air outside the stack (kg/m^3), ρ_{gas} is the average density of the combustion products inside the stack (kg/m^3), g is the gravitational acceleration (m/s^2), and H is the height of the stack (m). Assuming the linear changes of the flue gas temperature inside the stack, the ρ_{gas} is typically determined as the arithmetic average of the flue gas density in the inlet and outlet of the stack. Note that the temperature of the flue gas reduces while it flows up the stack due to the heat transfer to the surrounding and the air infiltration. In the above equation, the density of the flue gas is a function of the gas temperature in the stack inlet. The driving force increases with increase in the flue gas temperature. However, in steam generators, the flue gas temperature is reduced to the lowest possible to improve the overall efficiency of the steam generator and is only limited by the dew temperature of the flue gas to prevent condensation and corrosion from acid formation. The temperature of the flue gas in the inlet of the stack is generally about 150°C (about 300°F). The only parameter that designers can manipulate to increase the driving force created by the stack is the height of the stack. The taller the stack, the greater the driving force will be. That is why typically the **tallest structure in steam power plants is the stack**. Other factors that can significantly affect the stack effect are the atmospheric pressure and the ambient temperature. Furthermore, the pressure drop of the flue gas in the stack affects the net available pressure head. The main parameter in determining the flue gas pressure loss in the stack is the velocity of the flue gas, which in turn depends on the flue gas volumetric flow rate and the diameter of the stack. Another function of a tall stack is to disperse the exhaust gas in as large an area as possible. For this purpose, the temperature of the stack exhaust should be high enough to create sufficient buoyancy for rising the flue gas.

4.10 Questions and Problems

4.10.1 Introduction

4.1. How did the discovery of fire improve the life of ancient humans?

4.10.2 Thermodynamics of Ideal Gases

4.2. The first and second laws of thermodynamics cannot be extended to include processes involving chemical reactions.

a. True

b. False

4.3. Define the mass fraction and mole fraction of components of a mixture.

4.4. If an ideal gas mixture is composed of two gases, A and B, which one is not correct according to the Dalton model?

a. $T_A = T_B$

b. $V_A = V_B$

c. $P_A = P_B$

4.5. The temperature and pressure of a gas mixture in a 300 m^3 container are 500 K and 10 bar. If the mole fraction of gas A is 35%, determine the temperature, pressure, volume, specific volume, and mass of the gas A in the mixture. The molecular weight of the gas is 48 $\frac{\text{kg}}{\text{kmol}}$.

4.6. An ideal gas mixture is composed of three gases at the partial pressures of 2.1, 3.0, and 1.5 bar. What is the total pressure of the mixture?

4.7. Which equation is correct for the enthalpy of an ideal gas mixture?

a. $h_{\text{Mix}} = \sum_{j=1}^{j=N} h_j$

b. $h_{\text{Mix}} = \sum_{j=1}^{j=N} c_j h_j$

c. $h_{\text{Mix}} = \sum_{j=1}^{j=N} y_j h_j$

d. $h_{\text{Mix}} = \sum_{j=1}^{j=N} y_j c_j h_j$

4.8. What are the three most notable elements that participate in the combustion process? Write the combustion reaction for each.

4.9. In a combustion process, … should be balanced.

a. mass

b. mole

c. mass and mole

d. volume

4.10. Balance the following two mole- and mass-based reactions for carbon combustion by determining the stoichiometric coefficients.

$? \text{mole}\, C + ? \text{mole}\, O_2 \rightarrow ? \text{mole}\, CO_2$

$? \text{ kg } C + ? \text{ kg } O_2 \rightarrow ? \text{ kg } CO_2$

4.11. Write the complete combustion reaction of methane (CH_4) with pure oxygen (balance the reaction).

4.12. Water in the combustion products can be in the form of steam, liquid water, or even ice depending on the temperature and pressure of the products.

a. True

b. False

4.13. What is the difference between pure nitrogen and atmospheric nitrogen?

4.14. The molar-based composition of syngas (synthesis gas) produced in a biomass gasification process is presented in the table below (dry basis). Determine the mass fraction of each component and the average molecular weight of the mixture.

Syngas (synthesis gas) composition for Problem 4.14

Component	Molar fraction (%)
Nitrogen	48
Carbon monoxide	24
Carbon dioxide	10
Hydrogen	18

4.15. The molar-based composition of a gaseous fuel is presented in the table below (dry basis). Determine the mass fraction of each component and the average molecular weight of the mixture.

Gaseous fuel composition for Problem 4.15

Component	Molar fraction (%)
Nitrogen	30
Carbon monoxide	40
Carbon dioxide	15
Hydrogen	15

4.16. A designer takes the molar-based composition given in the previous problem as the mass-based composition of the gas. Determine the mole fraction of each component. Compare the average molecular weight of the mixture with that calculated in the previous problem.

4.10.3 Thermodynamics of Combustion Process

4.17. Write the complete combustion of methane in air.

4.18. Define the theoretical air of a combustion process and a stoichiometric mixture.

4.19. Determine the stoichiometric coefficients of the complete combustion of the following general hydrocarbon fuel with the theoretical air

$$C_xH_y + ?(O_2 + 3.76N_2) \rightarrow ?CO_2 + ?H_2O + ?3.76N_2$$

4.20. 100% excess air means that the actual amount of air is .... times of the theoretical air.

a. 1
b. 2
c. 3
d. 4

4.21. 90% of theoretical air or 10% deficiency of air means that the actual amount of air is more than the theoretical air.

a. True
b. False

4.22. Explain what a rich fuel and air mixture and a lean fuel and air mixture are.

4.23. Write the complete combustion of methane with 100% excess air.

4.24. Which one of the following expressions is different?

a. 200% theoretical air
b. 2 times stoichiometric air
c. equivalence ratio of 2
d. 100% excess air

4.25. Diesel fuel (with an average single hydrocarbon structure of $C_{12}H_{26}$) is completely combusted in air. Write the combustion reaction when the combustion is with theoretical air and with 100% excess air. Determine the air-to-fuel ratio (both mole base and mass base), and the dew point temperature of the combustion products for both reactions. Also, determine the equivalence ratio in the second reaction. The pressure of the combustion products is 1 bar.

4.26. In the previous problem, determine the mass of air and combustion products for 1 kg of fuel in both reactions, with theoretical air and with 100% excess air.

4.27. When there is a chemical reaction, why do we need a common reference point for the properties of all substances? Why is not this a problem for systems with fixed compositions?

4.28. The enthalpy of all stable elements is zero.

a. True
b. False

4.29. The enthalpy of all stable elements is zero at the standard reference state.

a. True
b. False

4.30. Define the enthalpy of formation for a compound.

4.31. Why is the enthalpy of formation for some matters positive and for others negative?

4.32. What is the difference between the higher heating value (HHV) and lower heating value (LHV) of a fuel?

4.33. Define the adiabatic flame temperature.

4.10.4 Fuels

4.34. Provide two examples of the following fuel categories:

a. Natural fuels
b. Synthetic fuels
c. Solid fuels
d. Liquid fuels
e. Gaseous fuels

4.10.5 Coal

4.35. The consumption of coal for electricity generation is declining in many countries. Why is it still important to study coal steam generators?

4.36. Briefly explain how coal resources have been formed.

4.37. Briefly discuss the coal production trends in the past few decades in the world, the U.S., and China.

4.38. Name the five countries with the highest coal production.

4.39. What percentage of electricity is currently generated from coal in the world, the U.S., China, and India?

4.40. What are the differences between the resources and reserves of fossil fuel?

4.41. What are the differences between proven, probable, and inferred coal resources?

4.42. What are the coal producing regions in the U.S.?

4.43. If the price of coal goes up, will coal reserves reduce or increase? Why?

4.44. Explain how improvement in mining-related technologies and coal exploration techniques affect estimations of coal reserves.

4.45. Determine the number of years that coal can last based on the following assumptions for the global electricity consumption of 2013. Compare your result with that of Example 4.7. Identify major sources of uncertainty of your calculation.

4.46. Name the two methods to analyze a fuel composition. What are the applications of each analysis? What factors are reported in each case?

4.47. What is the difference between as-received basis and dry basis of fuel compositions?

4.48. Briefly describe how the moisture, volatile matter, fixed carbon, and ash are measured for a proximate analysis according to ASTM standards.

4.49. What is the difference between inherent and surface moisture contents of coal?

4.50. How does the volatile matter content of coal affect the combustion process?

4.51. The volatile matter content can contribute up to 50% of the total heating content of the coal.

a. True

b. False

4.52. What are inherent impurities and extraneous impurities of coal?

4.53. Why is the ash content of coal important?

4.54. The samples of bituminous, subbituminous, and lignite coal have about 12,840, 8710, and 4505 Btu/lb of higher heating values, respectively. The mass-based ash content of three samples are 10.3, 11.0, and 10.7%, respectively. Determine the heat content-based ash in lb_{ash}/Btu.

4.55. A 250 MW coal-burning power plant uses subbituminous coal with the higher heating value of 19,850 kJ/kg as the fuel. If the mass-based ash content of the coal is 15%, determine the coal consumption (in kg/s) and the ash production (in kg/day). The higher heating value-based efficiency of the power plant is 34%. Repeat the problem for lignite coal with a higher heating value of 12,300 kJ/kg.

4.56. A 900 MW coal-burning power plant uses bituminous, subbituminous, and lignite coal with a higher heating value of 12,840, 8,710, and 4505 Btu/lb, respectively. If the mass-based ash contents of the coal are 10.3, 11.0, and 10.7%, respectively, determine the coal consumption rate (in lb/s) and the ash production rate (in lb_{ash}/day). The higher heating value-based efficiency of the power plant is 38%.

4.57. Explain how the fixed carbon content of a coal sample is determined for proximate analyses without directly measuring it.

4.58. Both proximate analysis and ultimate analysis of coal are ....

a. mass-based

b. mole-based

c. volume-based

4.59. How are the mass percentages of carbon, hydrogen, and nitrogen in a coal sample determined for ultimate analyses?

4.60. Explain how the oxygen content of a coal sample is determined for ultimate analyses without directly measuring it.

4.61. Describe the following bases used for both the approximate and ultimate analyses of coal:

- As-received
- As-fired
- Dry or moisture-free basis
- Dry mineral-matter-free
- Moist, ash-free

4.62. What are the applications of proximate analyses of coal?

4.63. The proximate analysis is ... than the ultimate analysis and can be used to determine the rank of coal.

a. simpler

b. more complicated

c. equally complicated

4.64. The coal rank depends on ...

a. the level of its carbonization (coalification)
b. the type of old vegetation it is made of
c. the degree of impurities
d. all of the above

4.65. The coal grade depends on …..
a. the level of its carbonization (coalification)
b. the type of old vegetation it is made of
c. the degree of impurities
d. all of the above

4.66. Which of the following ranks of coal are most suitable for combustion in power plants (select all correct ones)?
a. Peat
b. Lignite
c. Subbituminous
d. Bituminous
e. Anthracite

4.67. Which one of the following ranks of coal has the highest percentage of carbon?
a. Peat
b. Lignite
c. Subbituminous
d. Bituminous
e. Anthracite

4.68. Name some gaseous fuels derived from coal.

4.69. What are the two methods of coal extraction?

4.70. Name the following machines used in surface coal mines.

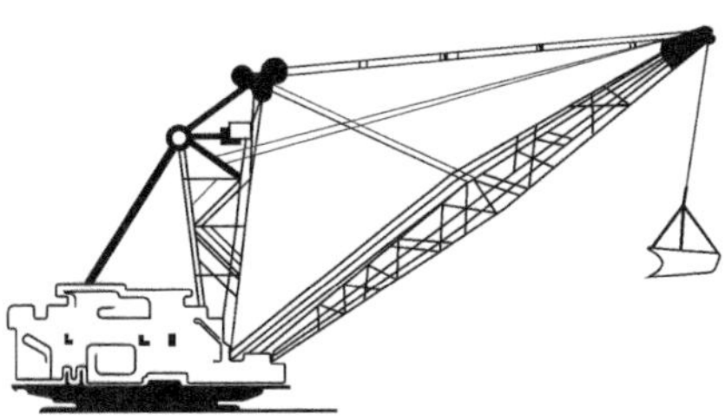

(a)

(b)

4.71. What are the two main methods of underground coal mining? Briefly describe them.

4.72. What is run-of-mine (ROM)?

4.73. What are the four processes that a coal preparation system may have?

4.74. What are the four screening systems typically used in coal preparation facilities?

4.75. Name the following coal sizing equipment (screens, revolving type, or shaker type).

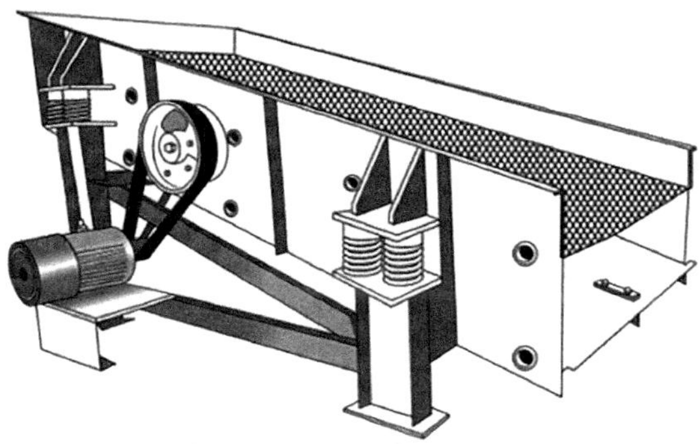

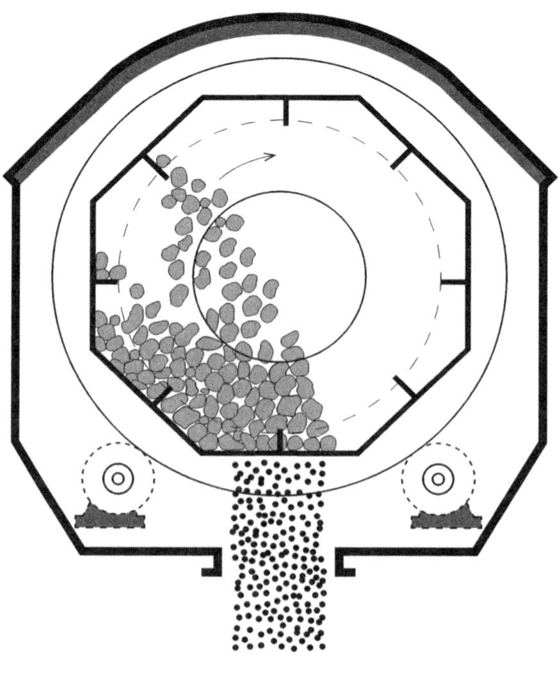

4.76. How many openings does a 10-mesh screen have in every linear inch and every square inch of the screen?

4.77. What are the differences between once-through crushers and retention crushers?

4.78. List the four types of coal crushing equipment. Briefly explain their operation principles and their advantages and disadvantages.

4.79. Name the following coal crushing equipment (Hammer mill, Ring crusher, or Bradford breaker).

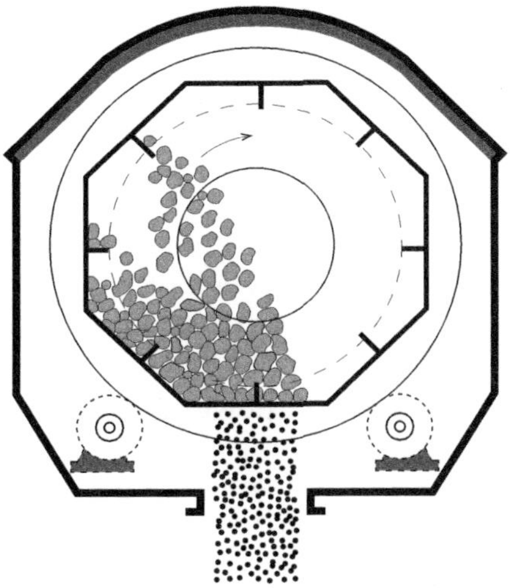

(a)

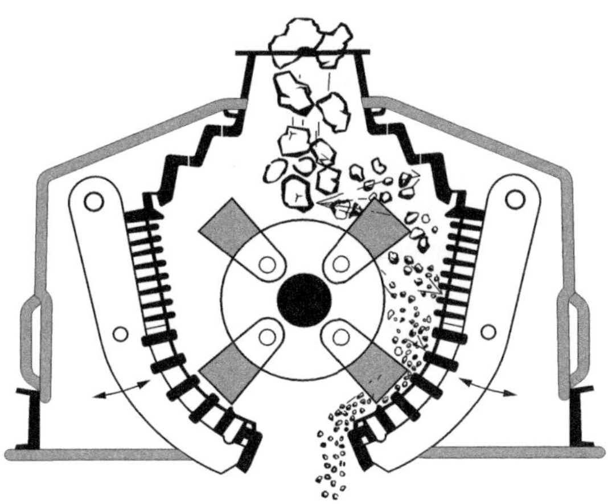

(b)

4.80. What are the objectives of the coal cleaning process?

4.81. In the wet separation process, the difference between either density or surface properties of the coal and the impurities is used to separate them.

a. True

b. False

4.82. After the preparation of coal, it can be transported to users by ... (name at least three methods). Which one is the most common mode of transportation for coal in the U.S.?

4.83. Why do power plants try to make their coal sources as diverse as possible by receiving coal from several routes and modes simultaneously?

4.84. Describe how coal is stored in power plants for long-term.

4.85. A 1000 MW coal-fired steam power plant with an efficiency of 30% needs to have 60 days of coal storage. If the plant is fueled with the average coal consumed in the U.S. (in terms of the heat content of coal) in 2015, determine the capacity of the coal storage in metric tons. The average heating value of the coal consumed in the U.S. in 2015 was 19.48 million Btu/short tons.

4.86. A 1500 MW coal-fired steam power plant with an efficiency of 33% needs to have 45 days of coal storage. The plant is fueled with coal with an average heating value of 19.88 million Btu/short tons. Determine the capacity of the coal storage in metric tons.

4.87. Define biomass fuel sources.

4.88. Explain why it is not accurate to consider biomass resources as "carbon neutral" resources.

4.89. What are some of the negative attributes of biomass energy sources?

4.90. List at least five biomass energy sources that can be combusted in a steam generator.

4.91. Why is it increasingly more expensive to landfill MSW?

4.92. Why does modern MSW, especially in North America, have increasingly higher energy content?

4.93. Explain the impacts of more paper and plastic, food remains, and glassware and aluminum cans in MSW on its heating value.

4.10.6 Combustion of Coal

4.94. Among modern coal combustion systems which system does have the lowest efficiency and capacity?

4.95. Which ones are the positive attributes of stoker combustion systems (select all correct ones)?

a. Capability to burn a wide variety of solid fuels
b. High efficiency
c. Rapid response to changes in steam demands
d. Do not require extensive fuel preparation facility
e. High capacity

4.96. Stoker combustion systems are typically used for small steam generation units in industrial, utility, and cogeneration applications.

a. True
b. False

4.97. Why has the interest in stoker combustion systems been recently revived?

4.98. List major components of stoker combustion systems.

4.99. What are the two types of grates in stoker combustion systems? How do they work?

4.100. Explain how dumping, inclined stationary, traveling, and vibrating grates work.

4.101. Which system does allow the continuous operation of a stoker burner by providing a continuous feed of fuel and a continuous discharge of ash (select all correct ones)?

a. dumping grates
b. inclined stationary grates
c. traveling grates
d. vibrating grates

4.102. Which system does have high maintenance costs for a stoker burner? Why? How can the wear and maintenance costs be reduced?

a. dumping grates
b. inclined stationary grates
c. traveling grates
d. vibrating grates

4.103. What are two main ways to burn solid fuels?

4.104. Describe how solid fuel in mechanical stokers can be distributed over the grate by fuel overfeeding or underfeeding.

4.105. How are the mass-feed fuel distribution systems different from the spreader systems in mechanical stokers?

4.106. Explain air distribution systems in mechanical stokers.

4.107. Why are combustion systems in natural gas-fueled and pulverized coal steam generators very similar?

4.108. Why are pulverized coal steam generators considered as one of the most important industrial and engineering achievements of the 20th century?

4.109. Explain the operational principles of indirect and direct pulverized coal combustion systems. What are the pros and cons of each? Which one is used in modern steam generators?

4.110. What are the primary air and secondary air?

4.111. Why do direct pulverized coal combustion systems require advanced reliable pulverizers?

4.112. Explain why the coal inlet flow rate to the combustion system is important. How is this flow controlled? Explain two systems for this purpose.

4.113. What is the grindability index of coal? Is it the same as coal hardness? How is the grindability index of coal measured?

4.114. A pulverizer should be able to produce a very high percentage of fine particles but a very small percentage of coarse particles is permitted.

a. True

b. False

4.115. List the three size reduction mechanisms in pulverizers.

4.116. Why is hot air used in all modern pulverizers rather than ambient air?

4.117. Describe the relationship between the location of fans and pressurized, suction, and balanced mills.

4.118. What are two separation mechanisms in pulverizers? Briefly explain their operational principles.

4.119. Name three types of pulverizers.

4.120. Briefly explain how ball-and-tube mills (pulverizers) work. List their pros and cons.

4.121. Briefly explain how ball-and-race and roll-and-race pulverizers work. List their pros and cons.

4.122. A properly designed burner should ... (select all appropriate ones).

a. receive fuel, primary air, and secondary air at the correct ratio

b. highly mix fuel and air

c. initiate combustion

d. create stable and safe flame

e. achieve high combustion efficiency

f. minimize pollutant production

g. handle ashes

4.123. What is the essential promise of coal pulverization for combustion?

4.124. What is the function of a windbox?

4.125. In the following schematic of a furnace, show where the windbox is.

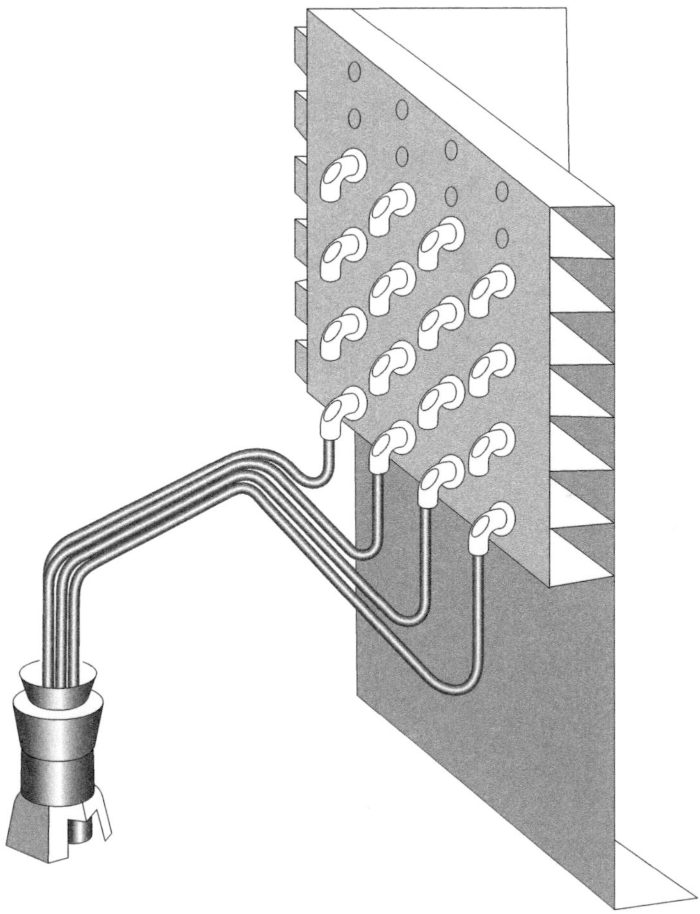

4.126. What are two approaches to have an effective mixture of air and fuel?

4.127. What are two mechanisms of NO_x production? Explain each mechanism.

4.128. Why is inducing swirl in the secondary air important for the performance of a burner? Why is the control of this swirl a challenge in circular burners? How is this problem addressed in S-type burners?

4.129. Name three techniques to reduce NO_x is pulverized coal burners.

4.130. How can dual register burners reduce NO_x production?

4.131. In dual register burners, there are several layers in the flame in regard to the oxygen concentration.

   a. True

   b. False

4.132. Which schematic shows a diffused flame and which is a premixed flame?

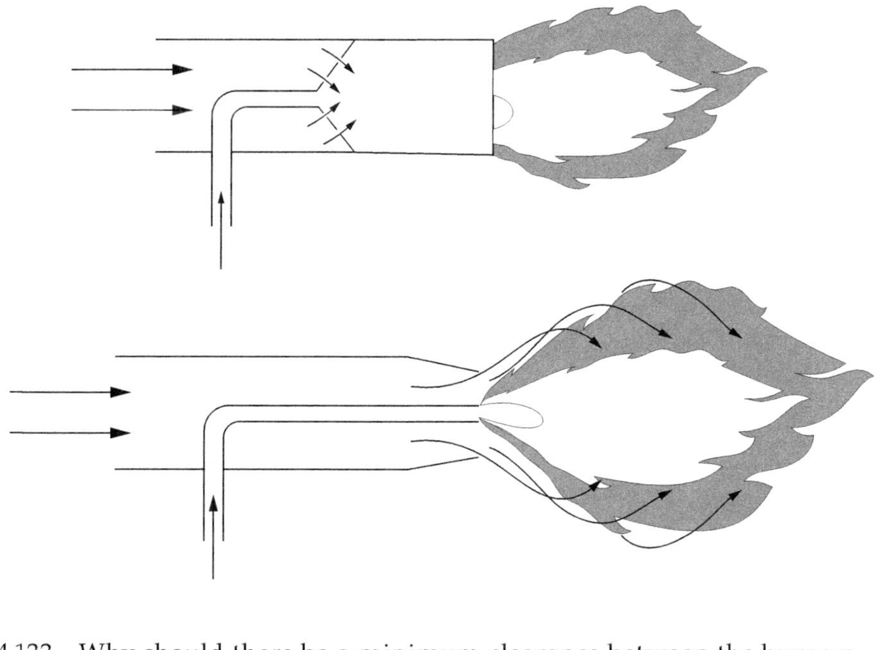

4.133. Why should there be a minimum clearance between the burners, walls, and floor?

4.134. Why is the amount of air during the devolatilization process important for NO_x reduction?

4.135. What is the common aspect of a dual register burner and air staging?

4.136. What is the overfire air (OFA)? What is the function of a NO_x port?

4.137. Why can both dual register burners and air staging increase the unburned fuel? How is the problem addressed in each case?

4.138. Which technology can be equipped with the air staging NO_x reduction technique (select all the correct ones)?

a. mechanical stokers

b. cyclones

c. wall-fired pulverized burners

d. corner-firing burners

e. fluidized bed

4.139. What are the similarities and differences between tangential burners and multiple flame envelope burners?

4.140. How do the flames from the corner-fired burners intertwine to create a single flame envelope?

4.141. What is the function of tilting air and fuel nozzles in corner-fired burners?

4.142. When the slag and ash deposit on the walls of the furnace increases, should the nozzles of corner-fired burners be tilted downward or upward? Explain.

4.143. Why are corner-firing systems inherently low NO_x generating combustion technologies?

4.144. Explain the difference between the local and global air staging in low NO_x corner-firing systems.

4.145. Explain how cyclone furnaces work? Why were they initially developed? What are their advantages and disadvantages?

4.146. What are five main functions that determine the shape and volume of a furnace?

4.147. What is secondary combustion? Is it desirable?

4.148. Why is a high furnace exit-gas temperature (FEGT) undesirable?

4.149. What problems can ash carried by the flue gas cause on downstream heat-transfer surfaces?

4.150. Explain what slagging is, how it is created, and how it can be reduced.

4.151. What are two types of coal ash collection and removal systems? Which combustion technologies are typically used?

4.152. Explain why too much or too little excess air is not desirable.

4.153. What are the functions of ignition systems?

4.154. Explain the general ignition process in steam generators.

4.155. What is the main characteristic of fluidized-bed combustion systems?

4.156. List the advantages of fluidized-bed steam generators over other steam generation technologies.

4.157. Explain why fluidized-bed combustion systems have high fuel flexibility.

4.158. What is the content of the bed in fluidized-bed combustion systems?

4.159. Why is a significant portion of ash and unburned fuel carried away by the combustion products in fluidized-bed combustion systems? How is this problem resolved?

4.160. How are sulfur oxide emissions removed from flue gases in fluidized-bed combustion systems? Why are the temperature of the bed and its control important for this purpose?

4.161. Explain different behaviors of fuel particles in a bed when the air velocity is increased from a low velocity to a very high velocity in a fluidized bed.

4.162. What is the difference between a bubbling fluidized-bed and a circulating fluidized-bed?

4.163. What are two methods to distribute fuel on a fluidized bed?

4.164. How is NO_x production reduced in fluidized-bed combustion systems?

4.165. What is the advantage of open bottom air distribution systems over flat floor air distribution systems in fluidized-bed steam generators?

4.166. Why is the varying heat content of municipal solid wastes a major challenge for steam generation units?

4.167. What are three types of municipal solid waste burning systems?

4.168. Explain the fuel preparation process for mass refuse burning systems?

4.169. What type of combustion system is typically used in mass refuse burning systems?

a. Mechanical stokers
b. Cyclones
c. Wall-fired pulverized burners
d. Corner-firing burners
e. Fluidized bed

4.170. Which type of solid waste burning strategy requires the least expensive fuel preparation process? Which one has the highest operation and maintenance costs?

a. Mass burning
b. Refuse-derived fuels
c. Solid recovered fuels

4.171. Explain the following terms for liquid fuels:
- Ring vs. chain structure
- Saturated vs. unsaturated
- Single, double, or triple carbon bonds
- Isomer hydrocarbons

4.172. Which one of the following hydrocarbons has chain structure?

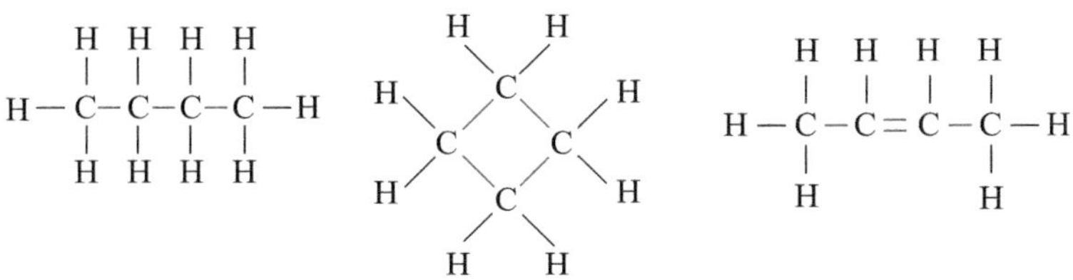

4.173. Liquid and gaseous hydrocarbon fuels are typically mixture of various hydrocarbons.

a. True

b. False

4.174. Explain the hydrocarbons production process from crude oil (the operating principle and the name of the system).

4.175. What is the difference between hydrocarbon fuels and alcoholic fuels in terms of their chemical structure?

4.176. The composition of gaseous fuels are typically presented on molar (volumetric) basis.

a. True

b. False

4.177. What is the difference between associated and non-associated natural gas?

4.178. Explain why untreated natural gas is sometimes referred to as sour gas.

4.179. What is the sweetening process in raw natural gas processing?

4.180. What is LNG? What are its applications?

4.181. Name two sources that can be used to produce hydrogen. Explain the production processes.

4.182. Hydrogen is a flammable gas with very high ignition energy burning with a colorless flame.

a. True

b. False

4.183. What are the advantages and disadvantages of mixing hydrogen with natural gas to be burned in conventional power generation units?

4.184. Name two ways to store hydrogen economically.

4.185. What are the benefits of pressurizing natural gas in pipelines?

4.186. Why do natural gas-fired power plants require some on-site fuel storage?

4.187. List some methods to store natural gas.

4.188. Describe how the combustion of liquid and gaseous fuels are similar to and different from pulverized coal combustion systems.

4.189. Regardless of the type of a combustion system, the fuel delivery system must transport the right amount of the fuel at the designed conditions to each burner.

a. True

b. False

4.190. What are the major differences between circular burners and dual register burners?

4.191. What is the function of an atomizer in a liquid fuel-fired combustion system?

4.192. Explain two types of atomizers and their operational principles. Discuss the pros and cons of each system.

4.193. What are two ignition systems for natural gas-fired combustion systems?

4.194. Explain the following two types of furnaces:

- Natural draft
- Mechanical draft

4.195. Identify if the following furnaces operate at a pressure above, below, or approximately equal to the atmospheric pressure.

a. Natural draft

b. Induced draft

c. Forced draft

d. Balanced draft

4.196. Explain how an induced-draft (ID) fan provides the required air for combustion.

4.197. Why cannot natural draft provide a sufficient driving force for the entire system in utility steam generators?

4.198. Explain how the gas flow in ID and FD fans is different. How do these differences impact the design and operational requirements?

4.199. List some fans used in a typical steam power plant.

4.200. Explain the difference between centrifugal fans or axial fans. Explain the function of each fan.

4.201. What are three types of centrifugal fans with regard to the orientation of the fan blades with respect to the direction of the rotor rotation?

4.202. Identify backward-inclined, forward-curved, and straight radial fan blades in the following figure.

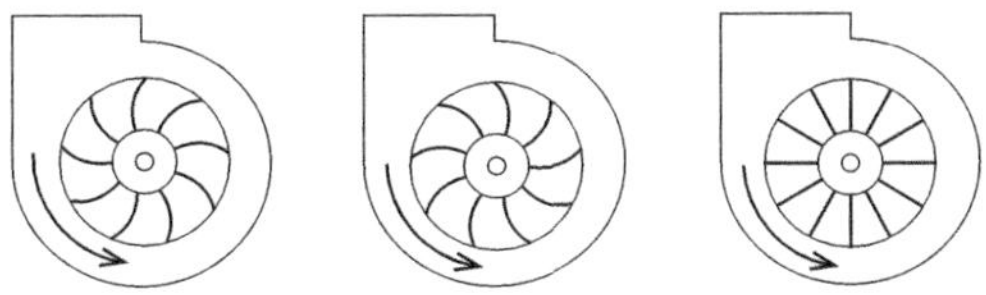

4.203. What are the most important parameters in the selection of fans? How are these parameters determined?

4.204. What is the specific speed of a fan? How is it used in fan sizing?

4.205. Name two ways a fan can be powered.

4.206. What are the similarities and differences between dampers and variable inlet vanes (VIV) to control gas flow rates in fans?

4.207. Why are fans the noisiest components in modern steam power plants?

4.208. What are the functions of expansion joints?

4.209. How can an air preheater improve the efficiency of a steam generator?

4.210. What are two types of air preheaters? Explain the operational principles of each type.

4.211. What are the functions of a stack in a steam generator?

4.212. Explain how the stack of a steam generator can produce natural draft. What factors impact the pressure difference created by natural draft?

4.10.7 General

4.213. Write the complete combustion of methane with 150% theoretical air.

4.214. Describe why a higher hydrogen content in fuel increases the difference between HHV and LHV.

4.215. How are the mass-based combustion air calculations converted to the volume-based fan calculations?

4.216. The as-fired mass-based analyses of a coal sample are given as follows:

Analysis A		**Analysis B**	
C	80%	Moisture	3%
H_2	5%	Volatiles	22%
S	1%	Fixed carbon	68%
O_2	3%	Ash	7%
N_2	1%		
H_2O	3%		
Ash	7%		

Which analysis (A or B) is proximate analysis and which one is ultimate analysis?

4.217. Compare an as-received coal analysis and a dry-based coal analysis.

4.218. What are the NO_x production reduction mechanisms in natural gas-fired steam generators?

4.219. Explain how the slagging problem and the furnace exit-gas temperature (FEGT) are related.

4.220. Explain why fans may be used in parallel. How does this arrangement impact the performance of the system?

4.221. How is the superheated steam temperature leaving the steam generator (after both superheaters and reheaters) controlled?

4.222. The ultimate analysis is more scientific and is needed for the combustion calculation of coal.

a. True

b. False

4.223. What are two designs for regenerative preheaters?

4.224. How are the moisture content, volatile matter, ash, and fixed carbon content measured for the proximate analysis?

4.225. Explain the tangential firing systems for natural gas and pulverized coal steam generators.

4.226. Explain the categorization of different fuels according to:

- the origin of the fuel
- the physical state of the fuel.

4.227. Provide examples of natural fuels and synthetic fuels (synfuels).

4.228. What is the main advantage of the fluidized-bed combustion technology that other conventional coal combustion methods cannot compete with?

4.229. How can variable rotational speeds be created in variable speed fans?

4.230. Which one of the following hydrocarbons has a ring structure?

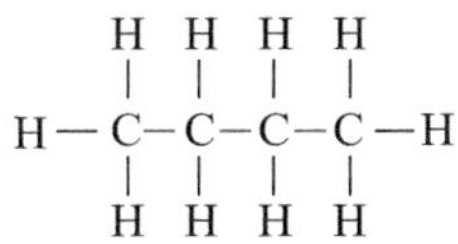
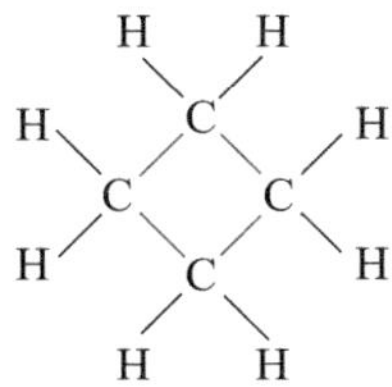
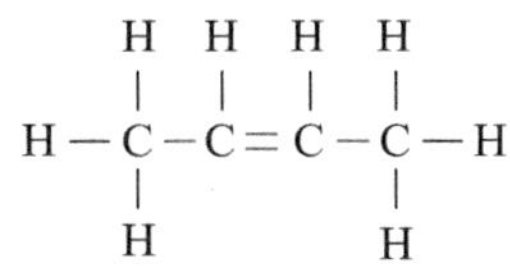

4.231. List different types of coal according to their age (oldest to youngest), carbon content, and volatile matters (highest to lowest).

4.232. Using the following two figures, explain bin storage (top) and direct firing (bottom) systems for burning pulverized coal.

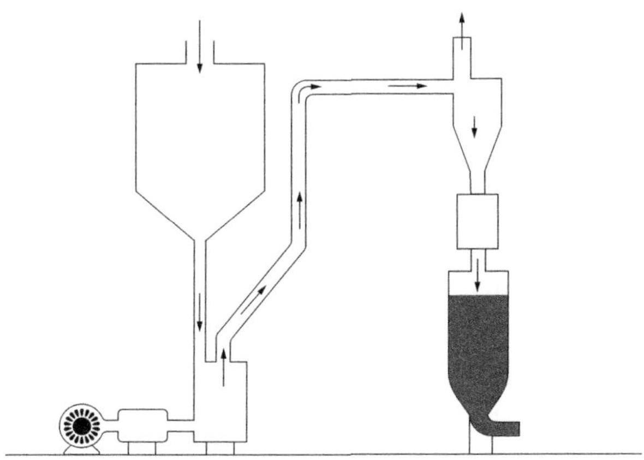

(a)

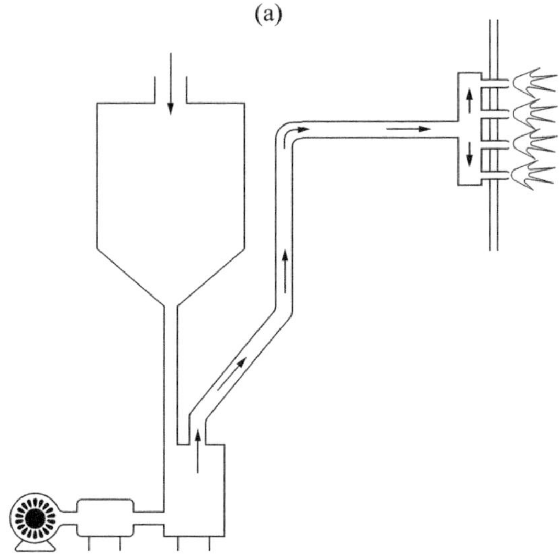

(b)

4.233. List the six biggest global coal producers.

4.234. Hydrogen can be used for high-efficiency power generation in fuel cells with water as a product.

a. True

b. False

4.235. Heavier hydrocarbons with higher boiling points are condensed at the lower levels in the distillation column.

a. True

b. False

4.236. Explain how Ljungström air preheaters work.

4.237. What is the difference between fuel resources and reserves?

4.238. Describe three types of fuel resources: proven (proved or measured), probable (or indicated), and inferred.

4.239. Which technology can be equipped with the dual register NO_x reduction technique (select all the correct ones)?

a. Mechanical stokers

b. Cyclones

c. Wall-fired pulverized burners

d. Corner-firing burners

e. Fluidized bed

4.240. If the price of coal goes down, coal reserves will … because …..

4.241. List major components of the fire side of a steam generator.

4.242. How do variable speed fans control the gas flow rate? How does the rotational speed of a fan vary?

4.243. Which one of the following hydrocarbons has saturated structure?

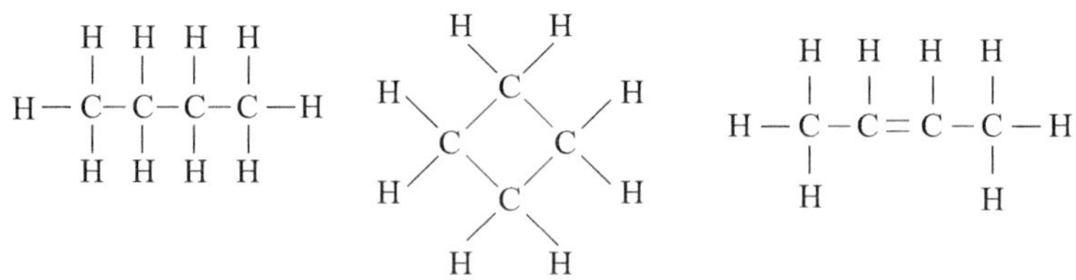

4.244. What type of coal ash collection and removal systems is typically used in pulverized coal-burning systems?

4.245. What is the difference between free gas and dissolved natural gas?

4.246. Explain each of the following methods to provide air for combustion and pros and cons of each:

a. Natural draft
b. Forced draft (FD) fan
c. Induced-draft (ID) fan
d. Balanced draft

4.247. Describe the advantages and disadvantages of dampers to regulate a gas flow rate.

4.248. Which one is a part of the fire side of a steam generator (check all applicable)?
a. Fuel preparation equipment
b. Stokers
c. Furnace
d. Drum
e. Burner
f. Fans
g. Superheater
h. Emission control equipment
i. Stack
j. Waterwalls
k. Ash handling equipment

4.249. Explain the titling of a burner in the tangential firing system as a steam temperature control system.

4.250. Which type of fuel needs a more extensive preparation process?
a. Solid
b. Liquid
c. Gas

4.251. Why is typically the tallest structure in a steam power plant the stack?

4.252. Identify the following coal-burning systems.

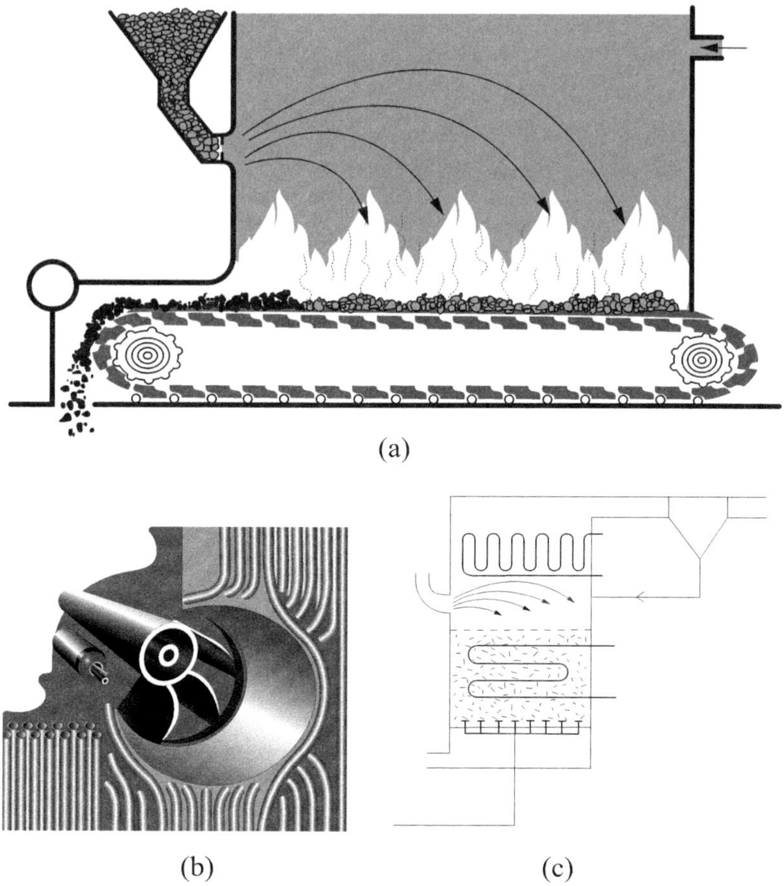

4.253. Describe the ultimate analysis of coal.

4.254. Briefly explain the operation principle of a cyclone furnace for coal-burning. Draw a schematic of the system. What are the advantages and disadvantages of the system?

4.255. Which one is not a measurable coal property?

a. Grindability
b. Ease of pulverization
c. Size
d. Combustibility

4.256. What are the three most common elements that can participate in the combustion process?

4.257. What are the products of the complete combustion of carbon, hydrogen, and sulfur? Write the combustion reactions.

4.258. Name two coal pulverizing mechanisms and briefly explain their operational principles.

4.259. Write the complete combustion reaction for methane and hydrogen with pure oxygen and then balance the reaction.

4.260. Briefly explain the operation principle of the cyclone-furnace firing and its advantages and disadvantages.

4.261. What is the theoretical air or stoichiometric air of combustion?

4.262. Describe the proximate analysis of coal.

4.263. In a combustion process, the theoretical air is the minimum amount of air to provide enough oxygen for the complete combustion of all combustible atoms.

a. True

b. False

4.264. In a combustion process with theoretical air, there is free oxygen in combustion products.

a. True

b. False

4.265. Complete the following combustion reaction:

$$C_xH_y + (x + y/4)(O_2 + 3.76N_2) \rightarrow$$

4.266. Use the following schematic to describe fluidized-bed combustion systems. What is the main advantage of the system?

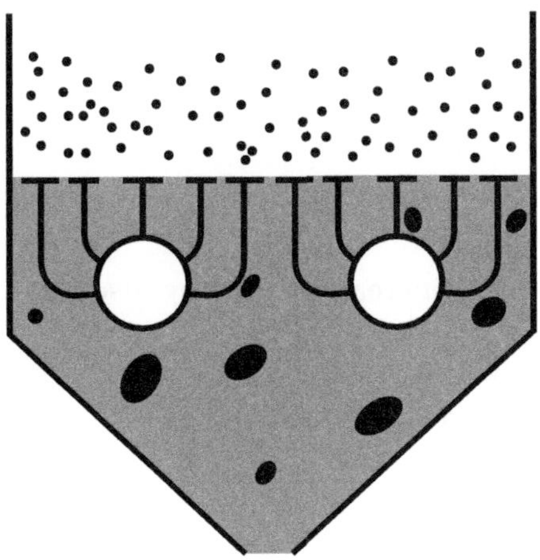

4.267. The coal composition analysis can be either .....

4.268. Which technology can be equipped with fuel staging or reburning?

4.269. How does the desulfurization of the coal combustion process take place in fluidized-bed combustion systems?

4.270. If the price of oil goes up, oil reserves will ....

a. increase

b. decrease

4.271. 200% of theoretical air means that the actual amount of air is .... times of the theoretical air.

a. 1

b. 2

c. 3

d. 4

4.272. A slight shortage of oxygen leads to the appearance of some CO instead of CO_2 in combustion products.

a. True

b. False

4.273. Why is the steam content of combustion products an important factor in designing any system involving combustion?

4.274. Write the complete combustion reaction for methane and hydrogen with oxygen from air and then balance the reaction.

4.275. What are the main components of coal?

4.276. Coal compositions and properties depend on:

a. the type of the original vegetation

b. the level of coalification

c. the amount of impurities

d. all of the above

4.277. Why alcohols are increasingly being used as fuel for various applications?

4.278. If the global consumption of coal remains at the current level, the global coal resources will be sufficient for ....

a. less than 50 years

b. between 100–300 years

c. about 500 years

d. more than 1000 years

4.279. Identify the following coal preparation systems.

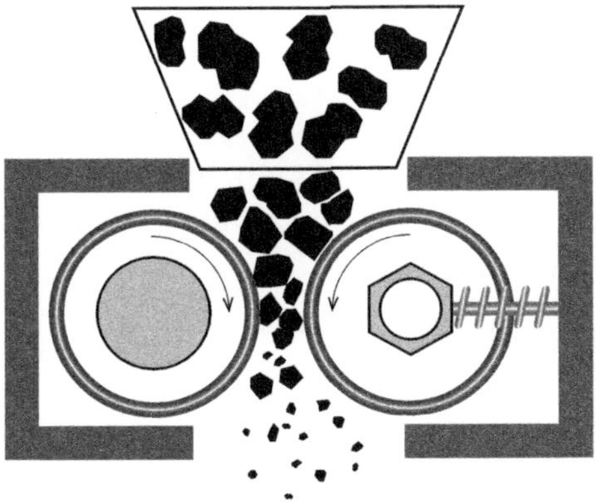

(a)

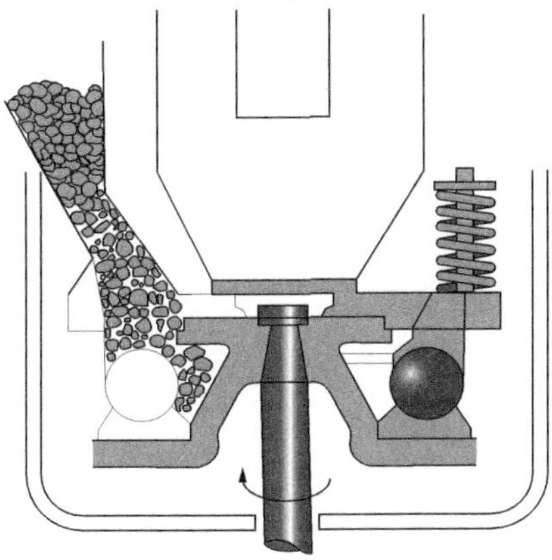

(b)

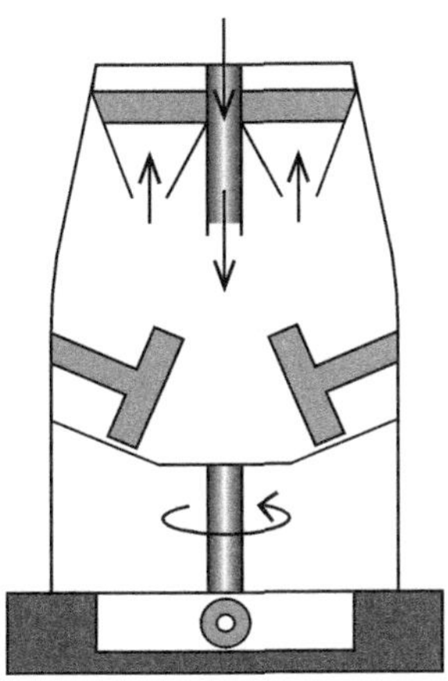

(c)

4.280. Describe the two combustion methods: suspension firing and thin-bed combustion.

4.281. List different types of mechanical stokers.

4.282. Which one is a requirement for any fuel burning system?

a. Introduce and mix fuel and air for combustion
b. Ignite the mixture
c. Remove combustion products
d. All the above

4.283. What are the products of the complete combustion of carbon, hydrogen, and sulfur? Write the combustion reactions.

4.284. Provide two examples of alcoholic fuels.

4.285. Identify different types of mechanical stokers (spreader stoker, inclined grate, traveling grate stoker, overfeed stoker, or underfed stoker)

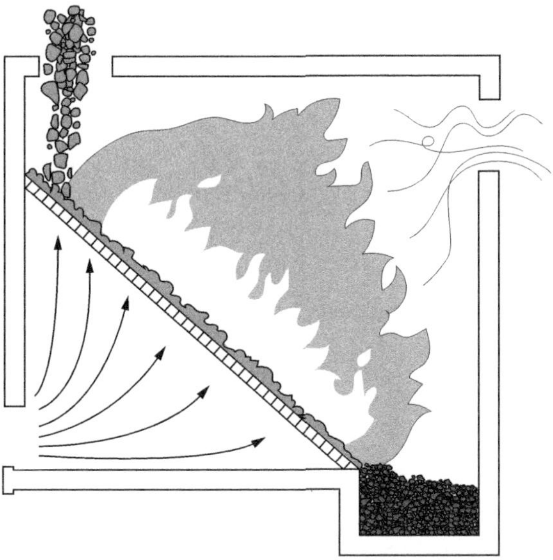

(a)

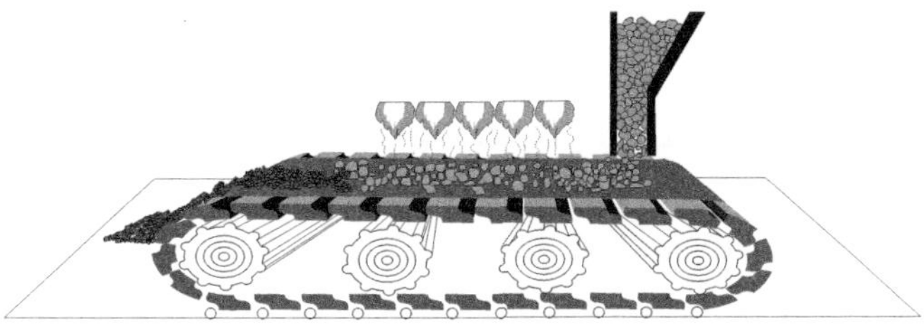

(b)

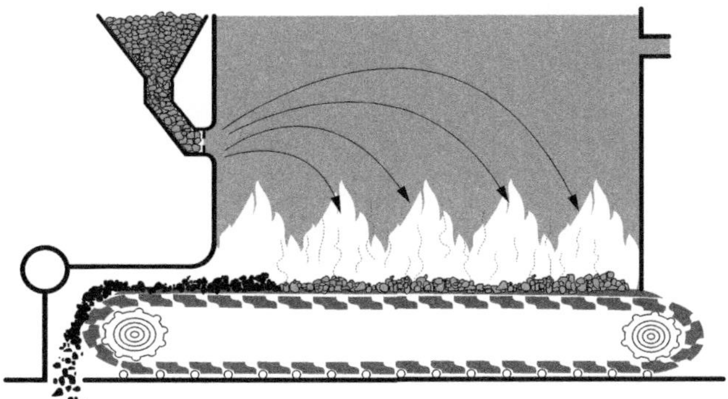

(c)

4.286. Name two types of coal pulverizer systems and briefly explain their operational concepts.

4.287. Explain the horizontal wall-fired burners for natural gas and pulverized coal.

4.288. Diesel is primarily composed of many hydrocarbons, which are represented by the average single hydrocarbon of octane (C_8H_{18}).

a. True

b. False

4.289. List three pollutants caused by the combustion of coal.

4.290. Why is not the vertical burner arrangement popular these days?

4.291. Gasoline is primarily composed of 40 hydrocarbons, which is represented by the average single hydrocarbon of octane (C_8H_{18}).

a. True

b. False

4.292. Explain the properties of anthracite coal.

4.293. Identify the characteristics of a good fuel burning system (select all appropriate ones).

a. Maximum excess air

b. Minimal unburned fuel in the combustion products

c. Minimal auxiliary ignition energy input

d. Minimal NO_x and SO_2

e. Effective handling of ashes

f. Wide and stable firing range

g. Fast response to the required firing rate change

h. High availability

i. Low maintenance

5

Emission Reduction in Steam Generators

Abstract

Our standard of living depends not only on our material possession but also on the environment that we live. The power generation industry plays a major role in emitting air polluters. In this chapter, we will take a look at the nature of these air polluters, their impacts on the environment, how they are produced during the combustion process, and most importantly how they can be mitigated. In each case, we will consider pre-combustion, during combustion, and post-combustion emission reduction strategies. We will start with one of the most notorious air polluters, nitrogen oxide emissions, also known as NO_x. It will be explained how NO_x production contributes to the formation of acid rain and particulate matter emissions, to the greenhouse effect, and to the ozone layer depletion. We will demonstrate how selective catalytic reduction can effectively reduce NO_x emissions. Next, we will take a look at sulfur oxide emissions and their contribution to acid rain and particulate matter pollutions. It will be described how wet flue gas desulfurization scrubbers are commonly used to reduce these emissions. Then, we will study particulate matter emissions and their removal by electrostatic precipitators and fabric filters. Finally, we will have an extensive discussion on global climate change and why an apparently simple concept of rising Earth's average temperature is so complex and can potentially change human civilization as we know. Then, various CO_2 capture systems including oxygen-firing systems, scrubbing post-combustion CO_2 capture systems, and CO_2 capture by microalgae will be discussed. Finally, we will conclude the chapter with a brief discussion on other emissions and environmental problems associated with fossil fuel-fired steam generators.

LEARNING OUTCOMES

After successful completion of this chapter, you will be able to:

- describe the contribution of the power generation industry to air pollution
- identify the sources of air pollution in steam generators

- explain the nature of nitrogen oxide emissions, their impacts on human health and the environment, their sources, and their mitigation strategies
- explain the nature of sulfur oxide emissions, their effects, their sources, and their mitigation techniques
- explain particulate matter emissions, their sources, and their removal methods
- describe global climate change, its causes, its feedback mechanisms, and its complexities
- explain various CO_2 capture systems

5.1 Introduction

So far in all our discussions, we repeatedly considered economic considerations as one of the most important factors, if not the most important factor, in the design and operation of power generation systems. However, we can all agree that our well-being is not limited to our material possession. The environment that we live in is a critical factor in determining our standard of living. In this chapter, we will look at some of the most important environmental impacts of fossil fuel power generation, and how they can be mitigated.

The combustion process, even in the most advanced and efficient combustion system, creates pollution. There are various methods to reduce or eliminate some of this pollution. The dilemma is that the construction of these emission control systems and providing resources for their operation require industrial activities and processes that lead to further pollution. A few years ago the author toured one of the nation's largest coal power plants where some of the most advanced air pollution reduction systems were installed. These systems consumed a lot of chemicals that the engineers working in the plant did not refer to the plant as a power plant; rather, they sarcastically called it a "chemical plant". Emission control systems may also reduce the plant efficiency, which, in turn, means more fuel should be combusted with further pollution to make up for the lost power. These conflicting factors require careful study and tradeoff between all the involved factors. Otherwise, we may end up with not only a higher cost of power production but also more negative environmental impacts. This requires life-cycle emission analyses of the entire system.

In the U.S., air pollution is regulated by the 1963 Clean Air Act (CAA). This act preceded the 1955 Air Pollution Control Act by providing limitations to the air pollution of stationary sources, including power generation units. The following Air Quality Act of 1967 added emission limitations for mobile pollutants such as vehicles. A major portion of these two acts were to develop the National Ambient Air Quality Standards (NAAQS) to limit the emissions of carbon monoxide, nitrogen dioxide, sulfur dioxide, particulate matters, hydrocarbons, and photochemical oxidants (as you may note carbon dioxide is noticeably missing in this list). The objective of these limitations is to mitigate air pollution-related health hazards such as acid rain and smog (again the greenhouse effect and climate change are missing). These limitations are categorized as those that are threat to human health, referred to as the primary standards, and those that are hazardous to the environment referred to as the secondary standards. Since then several amendments have been added to these acts. The collective consequences of these acts have been substantial improvements in air quality, particularly in large cities and urban areas.

Generally speaking, air pollutants are substances added to the atmosphere that can hurt human health or the environment. They can be in the form of gases, solid particles, or liquid droplets. In this section, we will review some of the most important air pollutants and how their impacts can be reduced. We will start with nitrogen oxide emissions, followed by sulfur oxide and particulate matter emissions. Then, we will discuss about climate change and greenhouse gas (GHG) emissions. We will conclude with a brief review of some other environmental problems associated with fossil fuel power generation. A schematic diagram of a steam plant with some of these air pollution reduction systems is presented in Figure 5.1. We will refer back to this figure throughout the chapter.

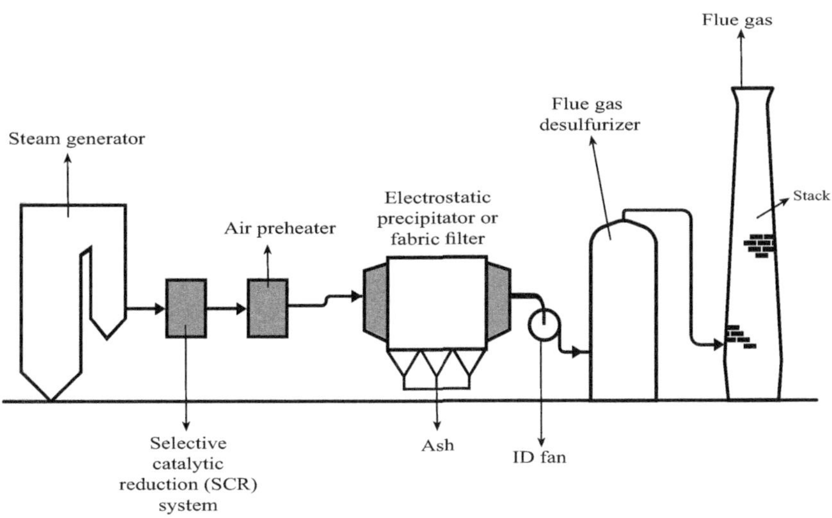

FIGURE 5.1
Schematic of a typical arrangement of air pollution control systems in a steam power plant.

5.2 Nitrogen Oxide Emissions

Nitric oxide (NO) and nitrogen oxide (NO_2), collectively referred to as the oxides of nitrogen or NO_x (rhymes like "fox"), are two major air pollutants. These should not be confused with nitrous oxide (N_2O), which is a greenhouse gas. A great majority of NO_x is NO with less than one-tenth being NO_2. The combustion of fossil fuels is the most important source of NO_x production. The mechanisms for NO_x formation were discussed in the previous chapter. The productions of NO_x and CO contradict each other. While the high concentration of oxygen during the combustion process reduces the production of CO, it can increase NO_x production. NO_x emissions can contribute to several environmental phenomena, including acid rain, smog, ground-level ozone, and particulate matter emissions (we will discuss these phenomena later in this chapter).

One of the major problems created by NO_x emissions is a phenomenon called the **photochemical smog** or just **smog**.[1] As the name implies, this event is assisted by solar radiation and is common in summer in much polluted urban areas. Perhaps the most famous smog in history was the 1952 Great Smog of London, which killed approximately 4000 people in just four days (refer to FYI box 2.1 on "Capital punishment for burning coal" in Chapter 2).

The processes that lead to this severe air quality problem are complicated and have several contributing factors. Here a rather oversimplified version will be briefly presented. For this phenomenon to happen, first solar radiation should decompose NO_2 according to the following reaction:

$$\text{Solar radiation} + NO_2 \rightarrow NO + O \tag{5.1}$$

The created free oxygen then reacts with an atmospheric oxygen atom forming ozone.

$$O + O_2 \rightarrow O_3 \tag{5.2}$$

Unlike ozone in the upper atmosphere (more on this later in the chapter), on the ground-level ozone is a major hazard for human health and the environment. Repeated or prolonged exposure to a high concentration of ozone can cause respiratory diseases. This ozone and other pollutants that are not directly released by emission sources are known as the **secondary pollutants**. The emissions that are directly emitted by sources are referred to as the **primary pollutants**.

Normally ozone can react with NO to reform NO_2 via the following reaction:

$$NO + O_3 \rightarrow O_2 + NO_2 \tag{5.3}$$

[1] The word smog is a combination of smoke and fog in reference to smoky fog associated with the phenomenon.

Depending on the atmospheric conditions, this reaction can reach equilibrium. While NO is colorless, NO_2 is brown and combines with other pollutants to give smog a reddish-brown color. NO_x emissions can also form nitric acid (HNO_3) through the following reaction and other acidic compounds that not only contribute to smog but can also contribute to acid rain (which will be discussed later in the chapter). However, the contribution of NO_x emissions to acid rain is less than that of SO_x.

$$NO_2 + OH \rightarrow HNO_3 \tag{5.4}$$

As if this is not bad enough, the presence of hydrocarbons from incomplete combustion makes the situation much worse. The reactions of hydrocarbons, ozone, and other primary pollution sources form organic radials such as peroxylacyl nitrates (commonly known as PANs). These nitrates can contribute to particulate matter pollution. They are not only harmful on their own but can also accelerate NO_2 production and slow the disassociation of ozone (reaction 5.3).

The combination of all these processes and some others creates an atmospheric condition that can negatively impact long-term human health and the environment. However, when they coincide with another atmospheric event called a temperature inversion or simply an inversion, they can cause immediate threats (for more on the temperature inversion phenomenon refer to FYI Box 5.1).

FYI BOX 5.1

TEMPERATURE INVERSION PHENOMENON

In normal atmospheric circumstances, solar radiations heat the ground. The hot ground then heats the air layer closest to the ground, which, in turn, heats air layers at higher elevations. This results in an air temperature profile where the temperature reduces with increase in elevation. Since the density of air increases with decrease in its temperature, this normal temperature profile creates a natural air circulation from the ground to higher elevations. This flow helps to remove the pollution over cities from the ground level. The pollution can eventually disperse by wind at higher elevations. If this air temperature profile is reversed for whatever reason a temperature inversion phenomenon, or simply an inversion, occurs. In this case, a layer of hot air blankets the cold air close to the ground level. This phenomenon stops the abovementioned air circulation and results in the accumulation of primary and secondary air pollution close to the ground. This can turn into smog where reddish-brown pollution can even block solar radiations during the day from reaching Earth's surface. This combination can be deadly, particularly for vulnerable population, such as elderly, children, and people with existing respiratory problems.

In summary, NO_x emissions **contribute to smog** as both a primary pollutant and a facilitator to form secondary pollutants. They can also contribute to the formation of **acid rain** and **particulate matter emissions**, the **greenhouse effect**, and **ozone layer depletion**. Therefore, they are the most notorious air pollutants with devastating impacts on humans and the environment. The fact that they can be transported for long distances from sources of emissions makes them hazardous even for locals living far from industrial facilities. As a consequence, they are heavily regulated, and there are strict limitations on how much NO_x industrial facilities can emit. In the rest of this section, we will briefly review how NO_x generation can be substantially reduced in power plants.

FYI BOX 5.2

OZONE LAYER DEPLETION

The impacts of ozone in the upper atmosphere and on the ground level are entirely different. While the former is essential to sustain life on Earth, the latter is quite harmful for human health and the environment. In the upper atmosphere, the ozone layer absorbs harmful ultraviolet (UV) solar radiation. It should be noted that while it is called the ozone layer, it is not a layer of pure ozone or even mostly ozone. In fact, the ozone layer is referred to as a portion of the upper atmosphere with a thickness of a few kilometers. This thickness varies at different locations and can be up to 20 km (about 12 miles). In this relatively thick layer, there is a high concentration of ozone. While this concentration is over an order of magnitude higher than the average ozone concentration in the entire upper atmosphere, it is in reality just several ppm. If all the ozone in the ozone layer is concentrated in a layer of pure ozone, it would be just a few millimeters thick. Any substance that can ascend to the level of the ozone layer and can release free radicals can react with ozone, and convert it to regular oxygen atom (the reverse of the reaction in Equation 5.2). This phenomenon is referred to as ozone depletion and can be accelerated by atomic chlorine (Cl), nitric oxide (NO), and some other substances. The most important contributors to ozone depletion are chlorofluorocarbons, which were preferred refrigerant for much of the 20th century. In the 1970s and 1980s, there were major concerns over the reduced concentration of ozone in the ozone layer, commonly but somewhat misleadingly referred to as "ozone holes". These concerns led to one of the most successful international collaborations ever known as the Montreal Protocol in 1987. The policy first banned the use of ozone-depleting substances in industrialized countries and then provided funds for developing countries to do so.

5.2.1 Reduction of Nitrogen Oxide Emissions

In Section 4.6.4.1 two mechanisms for NO_x production in the combustion of fossil fuels, particularly coal, were presented as thermal NO_x production and fuel NO_x production. The reduction of NO_x emissions can be achieved by **pre-combustion, during combustion**, and **post-combustion controls**. If possible, pre-combustion fuel switching is the simplest and most inexpensive NO_x reduction strategy for fuel NO_x production, particularly for coal. Generally, subbituminous coal, which is typically produced in the western states in the U.S., has a lower nitrogen content than that of bituminous coal, which is typically produced in the eastern states in the U.S. This significantly reduces fuel NO_x production through the combustion of these coal types. Furthermore, in some combustion methods, the high moisture content of subbituminous coal compared to that of bituminous coal can reduce the combustion temperature, which, in turn, can reduce thermal NO_x production.

Several techniques that can be utilized to reduce NO_x production during the combustion process, such as air or fuel staging, were extensively explained in Chapter 4. While these technologies are effective in lowering NO_x formation, they alone typically cannot meet regulatory limitations and post-combustion NO_x reduction techniques are needed to further reduce NO_x emissions. In this section, we will focus on post-combustion NO_x reduction technologies. These technologies can be divided into two categories: **selective catalytic reduction** (**SCR**) or **selective non-catalytic reduction** (**SNCR**). In both methods a reagent, such as ammonia (NH_3), is injected into the flue gas flow to reduce NO_x to nitrogen and water. For example, in the case of ammonia as the reagent, a combination of the following reactions takes place:

$$4NH_3 + 4NO + O_2 \rightarrow 4N_2 + 6H_2O \tag{5.5}$$

$$8NH_3 + 6NO_2 \rightarrow 7N_2 + 12H_2O \tag{5.6}$$

$$2NH_3 + NO + NO_2 \rightarrow 2N_2 + 3H_2O \tag{5.7}$$

Since a great majority of NO_x is NO, the first reaction is the dominant one. It should be noted that these reactions are exothermic but the amount of released heat is very small because of the small number of NO_x molecules in the flue gas flow. SCR systems are very effective in reducing the NO_x level by removing NO_x present in combustion products. While SNCR systems are less effective, they are less costly.

5.2.2 Selective Catalytic Reduction

SCR systems are currently the most commonly used NO_x reduction technology in power generation units (and a host of other industries). They can be applied to steam power plants, gas turbines (Chapter 9), combined cycle power plants (CCPP) (Chapter 10), and internal combustion engines. They are also compatible with almost every fuel type. In steam generators, they can be used for most combustion technologies.

In SCR systems, a reagent, primarily ammonia, is injected into combustion products via a network of injectors (Figure 5.2). The mixture then passes across a catalyst. The catalyst promotes the NO_x reduction process to produce nitrogen gas and steam. These reactions are favored at the temperature of about 250°C–400°C (about 500°F–750°F). In a utility steam generator, this temperature corresponds to the temperature of the flue gas after the economizer and before entering the air preheater (Figure 5.1). However, for high-ash fuels at this point the ash-laden flue gas may clog the catalysts. To avoid this problem, sometimes SCR systems are installed after the air preheater and dust collector. In these systems, the temperature of the flue gas should be increased by a heat exchanger or a burner before entering the SCR system.

In SCR systems, first ammonia should be prepared at an appropriate concentration. Ammonia can be delivered to the plant as nearly pure ammonia (since it contains no water, it is typically referred to as **anhydrous ammonia**), ammonia diluted in water (**aqueous ammonia**), or **urea** (in liquid or solid form). The latter is sometimes preferred due to the safety of its transportation and storage. In this case, ammonia is produced by the reaction (hydrolysis) of urea ($(NH_2)_2CO$) and water as follows:

$$(NH_2)_2CO + H_2O \rightarrow 2NH_3 + CO_2 \tag{5.8}$$

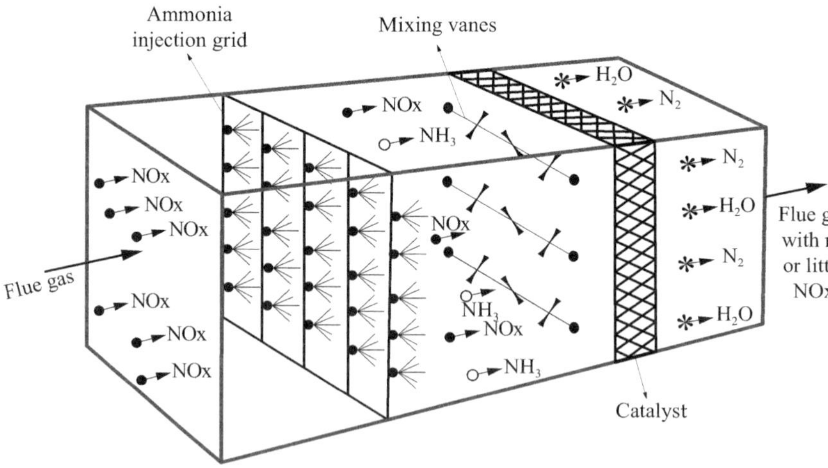

FIGURE 5.2
Schematic of a selective catalytic reduction (SCR) system.

The ammonia preparation subsystem is composed of the delivered raw material unloading, storage, and handling facility, as well as the ammonia preparation facility. The required components of the ammonia preparation facility depend on the nature of the delivered ammonia-containing substance and can include a water mixer, ammonia evaporator, air blower and mixer (to dilute evaporated ammonia), and flow controllers, as well as the required piping systems and valves. The reaction in Equation 5.5 (the predominant reaction for NO_x removal) indicates that in an ideal condition, i.e., a perfect mixture where all molecules of the reagent participate in the reaction, for each mole of NO, one mole of ammonia is needed. However, due to the nonuniform mixture of ammonia and flue gas, slightly more ammonia should be injected. This can result in excess ammonia, referred to as ammonia slip, in the outlet flow from the system, which is undesirable and should be minimized.

Whenever there is a need for NO_x reduction, the prepared ammonia is transferred to a grid of injectors where the mixture is injected into the flue gas flow. To create the maximum uniformity in the flow, static mixers (mixing vanes) can be installed after the injectors and before the catalyst. The system can be equipped with dampers to control and stop the flow if needed. To distribute the flow as uniformly as possible on the catalyst, flow distributors can also be used.

The catalyst is primarily made up of vanadium oxide or titanium dioxide. The catalyst can be made from these materials or they can be coated on a stainless steel plate or a glass fiber substrate in the shape of straight plates, corrugated plates, or honeycomb (Figure 5.3). Like any other catalyst, the catalytic materials are not consumed but accelerate the reactions, though they deteriorate with time. The main catalyst design parameters are its surface area and its concentration. To increase these factors and maximize the flue gas and catalyst contact, the catalyst is commonly made of a porous material. The catalyst can be equipped with a cleaning device, such as a sootblower, particularly if the flue gas is heavily ash-laden.

The entire system is enclosed in a large steel container, referred to as a reactor. The reactor should be large enough to provide the required residence time for the NO_x removal reaction to take place. Depending on the plant's

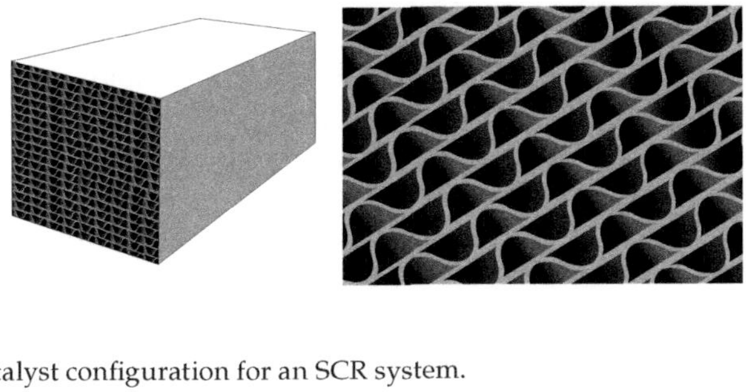

FIGURE 5.3
Schematic of catalyst configuration for an SCR system.

configuration and space availability, the flow of flue gas in the reactor can be vertical or horizontal.

5.2.3 Selective Non-catalytic Reduction

In SNCR systems, ammonia or urea is used to remove NO_x with no catalyst. This process takes place at a much higher temperature, typically 900°C–1100°C, than that in SCR units. Therefore, these units are typically installed inside the furnace at its upper section. The reactions of ammonia and NO_x are according to Equations 5.5–5.7. In the case of urea, the reaction is as follows:

$$2(NH_2)_2CO + 4NO + O_2 \rightarrow 4N_2 + 4H_2O + 2CO_2 \tag{5.9}$$

The effectiveness of SNCR systems is much less than that of SCR systems. They can typically remove between 30% and 50% of NO_x emissions from the flue gas flow.

5.3 Sulfur Oxide Emissions

In Chapter 4, we discussed that sulfur is one of the elements in fuels that can burn according to the following reaction and slightly contribute to its heat content:

$$S + O_2 \rightarrow SO_2 \tag{5.10}$$

A large majority of sulfur during the combustion process is converted to sulfur oxide (SO_2). However, a very small amount of sulfur oxide (typically less than 1%) is further oxidized to form sulfur trioxide through the following reaction:

$$2SO_2 + O_2 \rightarrow 2SO_3 \tag{5.11}$$

Sulfur oxides, also referred to as SO_x, are colorless gases with a strong smell (similar to the smell of a match right after being struck). When it rains in an area with a high concentration of SO_2 and SO_3 in the atmosphere, water droplets react with these gases and form diluted sulfuric acid according to the following reactions:

$$2SO_2 + 2H_2O + O_2 \rightarrow 2H_2SO_4 \tag{5.12}$$

$$SO_3 + H_2O \rightarrow H_2SO_4 \tag{5.13}$$

This acid along with nitric acid (formed by NO_x emissions according to the reaction in Equation 5.4) together forms **acid rain**. Acid rain can not only harm human health but also cause the defoliation of trees and plants as well as corrosion in buildings and other structures. An increase in the acidity of lakes, rivers, and streams can also severely damage their ecosystems. Sulfur oxides also contribute to particulate matter pollution by forming particulate sulfates (these sources of pollution will be discussed later in the chapter).

Reduction in sulfur oxide pollution can be achieved through one or a combination of the following strategies: **pre-combustion**, **during combustion**, or **post-combustion**. The pre-combustion sulfur oxide reduction approach primarily means switching from high sulfur fuels, such as bituminous coal, to low sulfur ones like subbituminous or lignite. In some plants, two types of coal are mixed with a varying ratio to achieve a desirable mixture that can meet regulatory limits while allowing the plant to operate efficiently and economically. Most of the sulfur oxide emission reduction in the U.S. is currently attributed to the abovementioned **fuel switching**. Furthermore, crystalized (pyritic) sulfur can be removed from coal by a coal washing process before combustion.

Some fuel combustion technologies allow the removal of simultaneous sulfur oxide and combustion processes. In the previous chapter, we discussed how fluidized-bed combustion systems have an inherent capability to remove the majority of sulfur oxide emissions during the combustion process. In the following section, we will briefly review post-combustion sulfur oxide removal systems.

5.3.1 Reduction of Sulfur Oxide Emissions

Since sulfur oxides are acidic gases, we naturally need an alkaline reagent to remove and neutralize them. While any alkaline substance, such as dolomite (calcium magnesium carbonate, $CaMg(CO_3)_2$), lime (CaO), caustic soda (sodium hydroxide, NaOH), or ammonia, can be used as a reagent in sulfur oxide reduction systems, the most common reagent is limestone ($CaCO_3$). These systems can be categorized as wet, semi-dry, or dry systems. A great majority of the currently used sulfur oxide reduction systems in power generation plants are wet systems. These wet systems are very effective in removing sulfur oxides. They can remove 85%–98% of SO_2 and SO_3 from combustion products. These systems along with several other systems that are used to remove other pollutants are collectively referred to as scrubbers.

5.3.2 Wet Flue Gas Desulfurization

As noted, the most common reagent used in wet flue gas desulfurization (WFGD) systems is limestone. In the most fundamental and simplest form, limestone (mostly $CaCO_3$) can react with sulfur oxide to form calcium sulfite ($CaSO_3$) and carbon dioxide through the following reaction:

$$SO_2 + CaCO_3 \rightarrow CaSO_3 + CO_2 \tag{5.14}$$

In wet scrubbers, limestone and water are mixed to form slurry. The heart of every scrubber is an **absorber tank or module** where the flue gas flow and limestone slurry can interact. Since the above reaction does not require high temperature, the absorber module is typically located right before the stack (Figure 5.1). Each steam generator may have its own dedicated scrubber unit or one scrubber unit can serve several steam generators. While there are several types of absorber modules, here we will focus on the most common design called the **spray tower**. A spray tower is typically a vertical vessel with a diameter of up to 25 m (about 80 ft) and a height of up to 50 m in which the flue gas flow (about 165 ft) enters approximately at the middle and moves upward (Figure 5.4). The bottom half of the absorber is filled with slurry. The slurry is pumped to the upper portion of the tank where it is sprayed into the ascending flue gas flow. To maximize the surface area of the interaction between the slurry and the flue gas, some nozzles are used to convert the flow of the slurry to fine droplets. The interaction between the flows moving at the counterflow direction results in the absorption of sulfur oxides by slurry droplets (Equation 5.14). To further increase the interaction between the flows and to increase the residence time, one or two perforated plates (trays) may be installed below the slurry nozzle grid. The perforated plates also assist in the more uniform distribution of the flue gas flow (in fact in this function of the plates can be more important than the one described in the previous sentence). To reduce the carryover of fine slurry droplets, a set of separators is installed above the nozzle grid. The mixture of the reacted and unreacted slurry drops back to the bottom of the vessel, which is then recycled back to the nozzles via a pump (Figure 5.4). To compensate for the reacted limestone, a fresh makeup slurry is continuously added to the absorber tank. The slurry accumulated at the lower half of the tank not only feeds the pump, collects the returning slurry droplets, and receives the makeup slurry, but also provides sufficient residence time for the completion of limestone and sulfur oxide reactions. For this reason, while the upper half of the tank is known as the **absorption zone**, the lower half is referred to as the **recirculation**, **reaction**, or **oxidation zone** (the reason for the latter name will be discussed below). It should be noted that wet scrubbers can also be helpful for the removal of particulate matters.

The clean flue gas exits at the top of the tower toward the stack. However, the interaction between the relatively hot flue gas flow and the slurry results in the evaporation of a significant amount of water into the flue gas, which, in turn, reduces the gas temperature. This presence of high humidity in the exhaust flow can cause condensation and corrosion in the stack and ducts leading to the stack. It also creates plumes after the exhaust flow leaves the

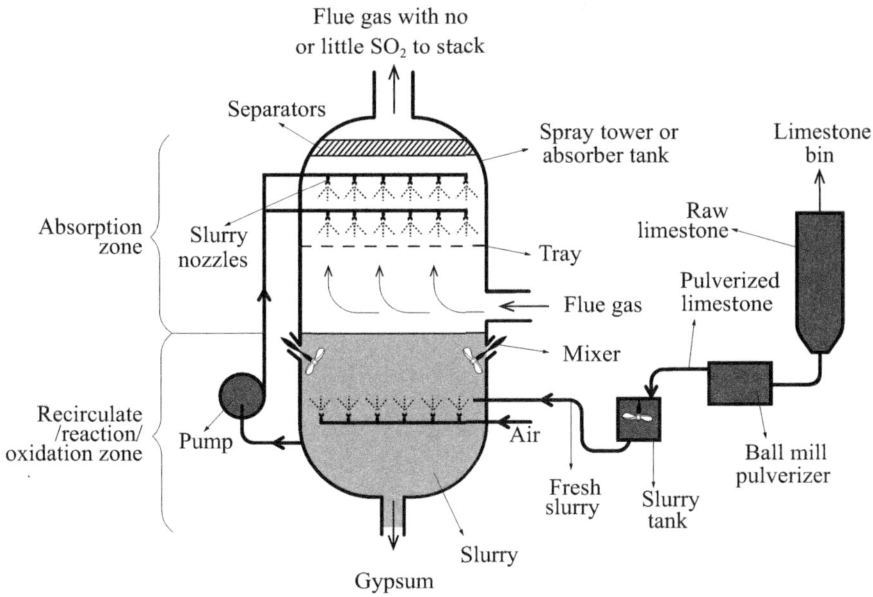

FIGURE 5.4
Schematic of a wet flue gas desulfurization system.

stack. Furthermore, the reduced temperature of the exhaust flow decreases its buoyancy and its ability to rise and properly intermix with the atmospheric air. To avoid the problem of corrosion, the ducts and stack can be made up of or lined with materials that can stand corrosion. However, this cannot address the other two problems. To eliminate all these problems, the clean humid flue gas is typically reheated after leaving the scrubber. For this purpose, the outlet flow of the scrubber can pass through a heat exchanger that is heated by the hot flue gas before entering the scrubber or steam from the power cycle. Alternatively, the outlet flow of the scrubber can be mixed with hot air from the air preheater or a portion of the flue gas that bypasses the scrubber.

The spent reagent either can leave the process by converting it to a byproduct or can be sent to a regeneration process where sulfur is extracted in the form of elemental sulfur (a marketable byproduct). In the latter approach, the regenerated reagent is recycled back to the scrubber. However, due to the complexity of the regeneration process and more importantly its high cost compared to purchasing fresh inexpensive reagent, the regeneration system is not commonly used.

In the case of limestone slurry as the reagent, the spent limestone (calcium sulfite, $CaSO_3$ from the reaction in Equation 5.14) is converted to $CaSO_4$ by injecting air into the lower portion of the scrubber while the mixture is stirred by mixers (Figure 5.4). That is why this part of the tank is sometimes referred to as the oxidation zone. After dewatering the byproduct ($CaSO_4$), it is industrially known as **gypsum**. Gypsum is a marketable byproduct that can be used to make drywall, plaster, cement, and fertilizers.

A sulfur oxide reduction facility also requires the reagent preparation unit. The components of this unit depend on the type of the reagent and the state in which it is delivered to the plant. The unit can include a dry-reagent storage tank, a reagent pulverizer, a reagent and water mixer, and a slurry storage system (Figure 5.4).

5.3.3 Dry and Semi-dry Flue Gas Desulfurization

Chemically speaking dry scrubbers are similar to wet ones but instead of wet slurry, a dry reagent is injected into the flue gas. In semi-dry systems, the reagent is in the form of thick slurry with little water. After injection, the water is immediately evaporated and a majority of the absorption reactions takes place while the reagent is solid. In these systems typically pulverized lime (CaO) with very fine particles is used. To collect reacted and unreacted reagent particles, dry scrubbers should be installed at the upstream of a dust collector. Either a filter collector or an electrostatic precipitator (ESP) can be used for this purpose (these collectors will be discussed later). Dry and semi-dry scrubbers are simpler and require lower capital costs compared to those of wet scrubbers. However, due to the higher cost of the reagent in these systems, they are commonly used in small units and for applications with a lower concentration of sulfur oxides in the flue gas.

5.3.4 SO_3 Reduction

The reaction presented in Equation 5.14 indicates that scrubbers can only absorb sulfur oxide (SO_2) and not sulfur trioxide (SO_3). As noted earlier, the concentration of sulfur trioxide in flue gas is very small compared to that of sulfur oxide. However, the presence of a catalyst in a SCR system can promote the conversion of sulfur oxide to sulfur trioxide. But even with this boost, the contribution of sulfur trioxide to sulfur oxide emissions is small, perhaps less than 4%. Nevertheless, even such a small amount of sulfur trioxide can be problematic. Sulfur trioxide can react with steam in the flue gas at the low temperature end of a steam generator where the gas temperature is well above the condensation temperature of steam. This reaction forms vapor sulfuric acid (H_2SO_4). This vapor can condense to very small droplets that cannot be removed with conventional filters. They not only cause corrosion of steam generator components but also create a visible plume, especially when a wet scrubber is used. Another problem with SO_3 is that in plants with SCR units, SO_3 reacts with ammonia slip to form solid particles that contribute to particulate emissions and the erosion of equipment.

Several methods have been used to reduce SO_3, including the injection of alkaline substances such as ammonia or calcium oxide in wet or dry forms, cooling the flue gas by water injection, and using wet precipitators. However, no predominant approach has universally emerged yet.

5.4 Particulate Matter Emissions

Most fuels produce a substantial amount of particulate matters when combusted. These emissions can be primary or secondary emissions. The primary emissions can be in the form of noncombustible matters (i.e., ash) or unburned carbon particles. For example, coal can contain up to 30% noncombustible components that are eventually converted to ash. As noted in Chapter 4, some of this ash is collected at the furnace of steam generators as the bottom ash, while the rest is carried from the furnace by the flue gas as fly ash. In previous sections, we discussed how nitrogen oxides and sulfur oxides can contribute to particulate matter emissions by forming secondary emissions. The amount, size, and shape of produced particulates during combustion depend on the combusted fuel and the firing system used to burn the fuel.

Several physical principles have been employed for the separation of solid particles from combustion product gases, namely the gravity, centrifugal effect, electrical fields, fabric filters, and scrubbers. In this section, two technologies that are most commonly used to remove these emissions, namely ESPs and fabric filters, will be discussed. Both these systems can be typically installed **before the sulfur oxide scrubber** or, if the plant is not equipped with a scrubber, right **before the stack** (Figure 5.1).

5.4.1 Electrostatic Precipitators

ESPs use electrical fields to remove solid particles from flue gas. They charge particulate matters with an electrical charge and then use an electrical field to collect the charged particles. With the capability of removing over 99% of solid particles from flue gas, these systems are very effective. Due to their mechanical simplicity, they require low maintenance and impose little pressure drop on the flow. Moreover, their performance is not impacted by conditions of the flue gas flow such as the size, shape, and amount of solid particles and the temperature and moisture content of the flow. They can be a dry type or wet type. We will first discuss the dry type.

Like any other electrical device working based on electromagnetic fields, ESPs need a positive electrode and a negative one. The **positive electrodes** are large parallel vertical plates, also referred to as **curtains** (Figure 5.5).

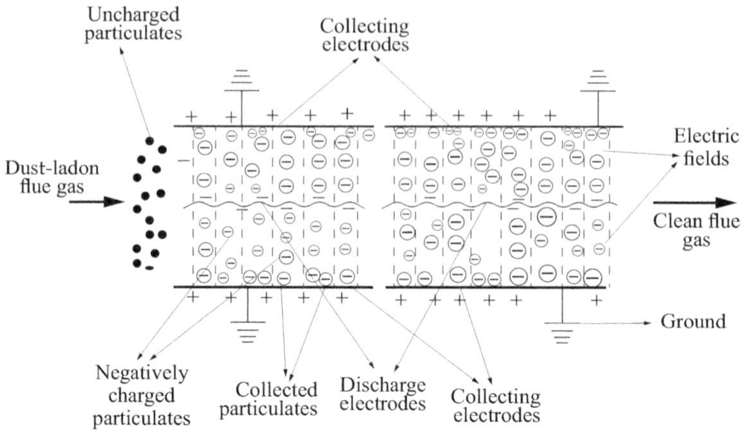

FIGURE 5.5
Schematic of a dry electrostatic precipitator.

These metallic electrodes are electrically grounded. The gap between two plates can be up to 0.5 m. The **negative electrode** is placed between the parallel plates. The solid particles in the flue gas are induced by negative charges from the electric field near the negative electrode. For this reason, the central negative electrode is sometimes called the **discharge electrode** (Figure 5.5). The negatively charged particles are then attracted to the positive electrodes where they adhere together to form a layer (a sheet). That is why the plates with positive charges are also known as **collecting electrodes**. When a certain amount of solid particles is collected, the collecting electrodes are mechanically struck by a weight, which separates sheets of collected particulates from electrodes. They fall into the hoppers below them and are removed from the system.

To ensure the sufficient removal of particles, a couple of these plates may be placed in series, the outlet flow of one being the inlet flow of the next (Figure 5.5). An ESP is composed of many of these units in parallel. The flow distributors are used to ensure a uniform distribution of the inlet flue gas among these units. The entire system is enclosed in a carbon steel casing. ESPs also require an electrical system where AC power is converted to DC power at the voltage of about 50–100 kV.

While dry ESPs are very effective in removing most solid particulates, they are not as effective in removing sulfuric acid mists (formed by the reaction of SO_3 and steam) and very fine particles. **Wet electrostatic precipitators (WESP)** can be used to remove these emissions as well as everything that an ESP can remove. In terms of the operational principles and physical components, WESPs are similar to ESPs but instead of the mechanical removal of collected particles, water spray or a water film on the collector surface is used to remove them.

5.4.2 Fabric Filters or Baghouses

Fabric filters or baghouses can be used to remove solid particles from flue gas. Like any other filter (such as face masks), in these systems flue gas passes through the porous fabric of filters where particles are trapped while gases pass through (more on this below). Similar to ESPs, they have a high particle removal efficiency; however, they are sensitive to the level of moisture in the flue gas flow. They can remove even very fine particles.

These units are typically composed of many small diameter long vertical bag filters commonly made of cotton cloth or glass fibers. Thousands of them can be arranged in several compartments. Each compartment can be isolated individually by dampers for cleaning[2] and replacement of the filters. The direction of the flue gas can be either from the inside of the filter to the outside or the opposite. In the former, particles are collected inside the filter and they are referred to as **inside-the-bag filters**. When the flue gas flows from the outside to inside of the filter, the particles are collected outside the filter and they are known as **outside-the-bag filters** (Figure 5.6).

As noted above in these filters, solid particles are trapped by the porous fabric of the filter. This is the main filtration mechanism when the filters are clean. When particulates are gradually accumulated, the flue gas has to pass through this layer of porous solid particles, which is referred to as a **dustcake**. The passing of the flue gas through the dustcake significantly contributes to

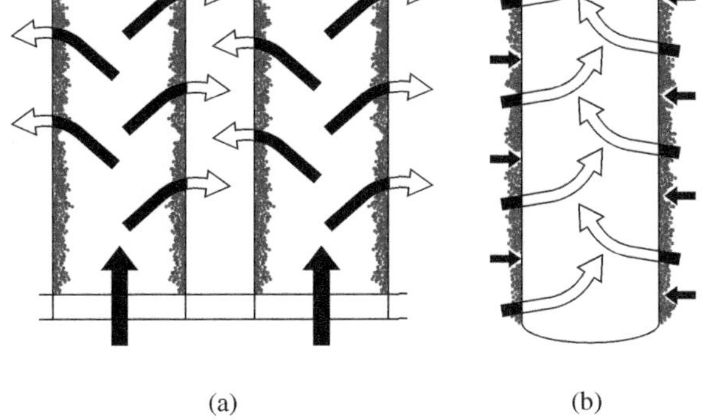

FIGURE 5.6
Schematic of (a) an inside-the-bag filter (b) an outside-the-bag filter.

[2] Some filters have a cleaning system that performs cleaning while the system is operating.

the filtration process. In fact, it may become the main separation process. While the presence of a dustcake can assist in the filtration process, too much particle accumulation can impose a high pressure drop on the flue gas flow.

To determine the appropriate time to clean filters, typically the **pressure drop** across them is used as the determining parameter, though some units use a predetermined number of hours of operation for this purpose. The filter cleaning mechanism depends on the type of filters. For inside-the-bag filters, either a **reverse gas flow** or a **mechanical shaker** is used (Figure 5.7a and b). In the former, the saturated filter compartment is isolated by isolation dampers. Then, a dedicated fan is used to send a reversed gas flow to remove the collected particles. Alternatively, the filters can be mechanically shaken to dislodge dustcakes. For outside-the-bag filters, the **pulse jet** cleaning technique is typically used where pulses of high pressure air are sent toward the filters to create shock waves to dislodge dustcakes (Figure 5.7c). This cleaning method can be implemented while the filters are either isolated or in service. In all cases, the removed dust from filters is collected by a hopper below the filters and transferred out of the system. In all these techniques the system is designed to leave some dust on the filters to assist the filtration process at the beginning of the next round.

While the particulate removal systems discussed above are the most commonly used systems in utility power plants, there are also other methods

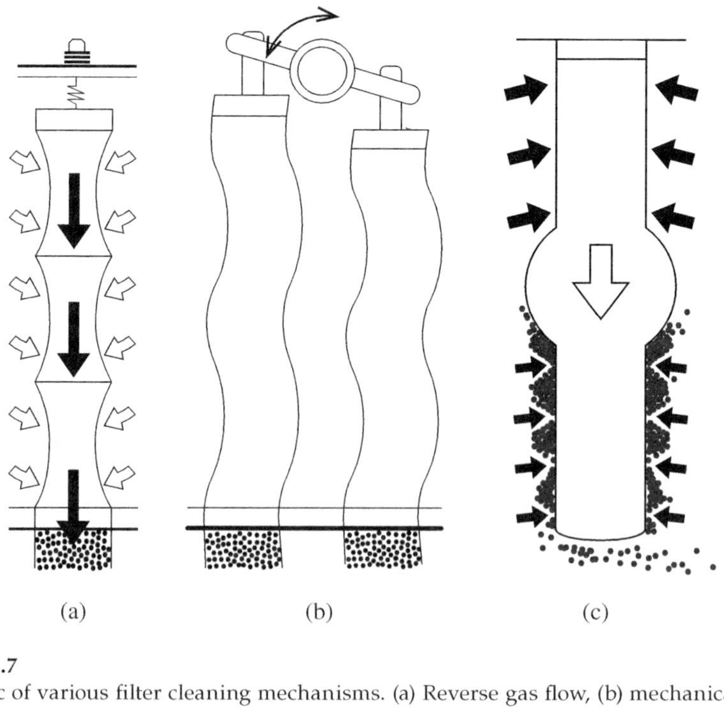

(a) (b) (c)

FIGURE 5.7
Schematic of various filter cleaning mechanisms. (a) Reverse gas flow, (b) mechanical shaker, (c) pulse jet.

for this purpose, including settling chambers, mechanical collectors such as cyclones, and wet particulate scrubbers. These systems are used in some special applications.

5.5 Global Climate Change and Greenhouse Gas (GHG) Emissions

Once, not long ago, a contested and perhaps unlikely doomsday scenario for the end of the 21st century or so, global climate change has become a new reality in our daily life. Hardly a day passes without new evidence for the phenomenon being presented. Likewise, every day there are big or small events that are being attributed to climate change. If unchecked, changes in global climate can cause disasters like of which have never been seen by the human race. These changes can be to a large degree irreversible.

Global climate change refers to a wide variety of changes on Earth as a result of the rising average temperature of Earth. It used to be (and still sometimes is) referred to as **global warming**. However, this term might be misleading since a **rise in the average temperature may not be reflected in a local temperature increase everywhere**. In this section, we will briefly review the causes of climate change, the contribution of the power generation industry, and the methods and technologies to address that.

5.5.1 Causes of the Climate Change

Earth receives solar radiation from the Sun and radiates energy back to space. Climate change is all about the imbalance of the received and emitted energy on Earth. Before we discuss this energy balance, we need to understand the relationship between the wavelength of the radiation emitted by an object, the temperature of the emitting object, and the greenhouse effect.

We know that any object at the temperature above the absolute zero, 0 K or –273.15°C (0°R, Rankine or –459.67°F), emits thermal energy in the form of electromagnetic waves. The distribution of this emitted energy is a function of the temperature of the body emitting the energy. According to Planck's radiation law (named after Max Planck, German physicist, 1858–1947)[3] increasing this temperature will increase the total radiated energy and will reduce the wavelength of the maximum emitted energy.[4] This means on average a body with a higher temperature will radiate at shorter wavelengths

[3] Planck's radiation law is considered as the birth point of the quantum theory for which he received the Nobel Prize in Physics in 1918.

[4] Planck's radiation law is for a black-body radiation but can be qualitatively applied to other objects too.

compared to that of a lower temperature object. This helps us to understand the nature or more precisely the wavelength of solar radiation and the energy emitted by Earth's surface. This indicates that considering the temperature of the Sun (at about 5500°C) compared to Earth's temperature, the solar radiations are at much shorter wavelengths compared to that of the energy radiated from Earth's surface.[5]

The last piece of the puzzle that we need to know is GHGs and their behaviors when they are exposed to electromagnetic waves. Most of Earth's atmosphere is composed of non-greenhouse gases such as nitrogen, oxygen, and argon (with about 78%, 21%, and 1% of the atmosphere, respectively). These gases are equally transparent to the entire spectrum of electromagnetic radiation. However, due to their molecular structure, GHGs such as water vapor and carbon dioxide are only transparent to short wavelength radiations. They absorb and emit radiation with long wavelengths. As will be observed next, while a tiny percentage of Earth's atmosphere is CO_2 (about 0.04%), this along with other GHGs, mainly water vapor, has a substantial impact on Earth.

Now let us take a look at what happens when solar energy is intercepted by Earth. Solar radiation first needs to pass through Earth's atmosphere (Figure 5.8). When solar energy reaches the atmosphere, some of its energy is reflected back into space by clouds. Due to the short wavelength of solar radiation, the rest of solar radiation reaches Earth's surface after passing through

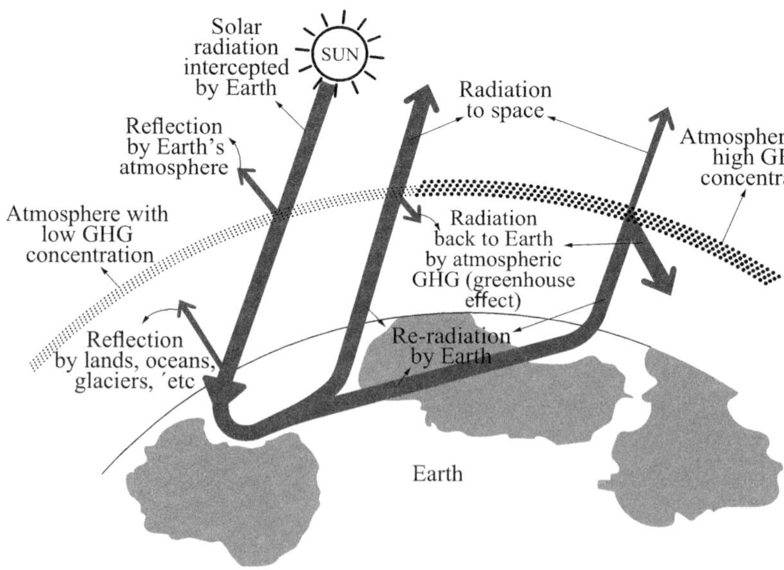

FIGURE 5.8
Balance of energy received and emitted by Earth.

[5] Solar radiation peaks at a wavelength that is about 20 times shorter than that for radiation from Earth.

the atmosphere without much absorption by GHGs in the atmosphere. Some of the solar energy that reaches Earth is reflected back into space, particularly by the areas covered by snow and ice. This reflected energy passes the atmosphere mostly intact (similar to when incoming solar radiation passes through the atmosphere). The remaining energy is absorbed by lands and oceans. This energy heats the ground and water on Earth's surface. They, in turn, re-radiate energy. Due to their low temperatures (on average about 15°C or about 60°F), this radiation is at high wavelengths (in the infrared region of the electromagnetic wave spectrum). Some of this emitted energy is absorbed by the GHGs in the atmosphere and the rest escapes to space. The absorbed energy heats the atmosphere, which again re-radiates energy in all directions some into space and some back to Earth. The downward portion of this radiation toward Earth is the **greenhouse effect** (Figure 5.8). The temperature of Earth's surface is determined by the balance of all these energy transfers. **If the sum of all energy flows that Earth receives is greater than the sum of all energy flows that are emitted by Earth, Earth's temperature will increase** and vice versa. So in order to understand the trend in Earth's temperature we need to evaluate the magnitude of the energy flows discussed above, at least qualitatively.

If Earth's atmosphere was free of any GHG, the entire solar radiation intercepted by Earth would have re-radiated back to space. This would have resulted in a freezing environment on the surface of Earth. In fact, the presence of the GHGs in the atmosphere is the pre-conditions for the existence of life on Earth. It is estimated that without the greenhouse effect the temperature of Earth could have been as low as –18°C (about 0°F) rather than its current temperature of about 15°C (about 60°F).[6] An example of how Earth would have looked like in the absence of the GHGs is the planet Mars. While over 95% of the Martian atmosphere is CO_2,[7] due to the lack of other GHGs, primarily water vapor, it has a weak greenhouse effect. As a result, the temperature of the Mars' surface is about –60°C (–76°F).[8] On the flip side too much GHGs on the atmosphere can trap so much solar radiation that make the temperature of Earth uninhabitable. An example of a planet in such a condition is Venus where over 96% of the atmosphere is CO_2. The presence of such a high percentage of carbon dioxide along with other GHGs, particularly water vapor, has created such a strong greenhouse effect that the temperature of the surface of Venus is about 500°C (about 930°F).[9]

While the above concept may not be complicated, there are many consequences of the increase in the average global temperature that make this

---

[6] https://www.giss.nasa.gov/research/briefs/ma_01/.

[7] Franz, H. B., et al. "Initial SAM calibration gas experiments on Mars: Quadrupole mass spectrometer results and implications." *Planetary and Space Science* 138 (2017): 44–54.

[8] Catling, D. C., and J. F. Kasting. *Atmospheric Evolution on Inhabited and Lifeless Worlds*. Cambridge: Cambridge University Press (2017).

[9] Basilevsky, A. T., and J. W. Head. "The surface of Venus." *Reports on Progress in Physics* 66.10 (2003): 1699.

phenomenon very complex. These factors are collectively referred to as climate change feedbacks. They can be reinforcing or balancing feedbacks. Here we just name a few of these sometimes conflicting feedbacks.

We noted above that the amount of energy radiated by an object depends on its temperature. This means increase in the average global temperature will increase re-radiation back to space, which, in turn, can decelerate the rate of increase in the temperature. Also, we know that the surfaces of ice and snow reflect most of solar radiation that strikes them while the surfaces of land and deep water absorb most of solar radiation. The increase in the global temperature accelerates melting of glaciers and ice-caps, particularly in Earth's North and South Poles, and replaces them with land or deep water. This not only increases the rate of temperature increase but also releases a tremendous amount of methane, a gas with strong GHG effects (below), trapped in them into the atmosphere. Furthermore, as was elaborated above, water vapor and clouds can reflect some solar radiation back to space before entering Earth's atmosphere while also being a strong GHG. The increase in the global temperature increases the atmospheric air's capability to hold vapor, which, in turn, can promote both of these opposing effects. Other conflicting impacts are related to tree and plant growth. Plants and plant-like ocean living things, such as microalgae, are responsible for the absorption of a significant amount of carbon dioxide from the atmosphere. While the warmer climate can improve their growth, climate change-induced droughts can hurt them.

There are a lot of scientists who do not know about these phenomena and there is a lot of uncertainty on quantifying what we know. This is why there are some discrepancies, sometimes substantial ones, between different models. Many scientists are working hard to improve our understanding of a wide variety of phenomena in the atmosphere, land, and ocean and their interactions with climate.

Clearly, this brief discussion by no means is sufficient for us to predict what can happen because of climate change. But we can make two certain conclusions. First, Earth's climate is extremely complicated and the **current climate is an intricate balance among many factors**. The anthropogenic GHG emissions can at least potentially tip this balance. If this happens, **Earth will certainly reach a new balance but there is no guarantee that the new climate will be hospitable to us**, to put it mildly. Climate change has happened many times in the past hundreds of millions of years and generally speaking they have not been kind to their contemporary species, especially complex species like us. The second important factor is that almost every metric that we can actually measure indicates that **we are not moving in the right direction**. The eight warmest years globally on the record have been between 2010 and 2020.[10] The level of ocean water increases at a higher rate (at about 3 mm/

[10] https://www.ncei.noaa.gov/news/projected-ranks#:~:text=Global%20temperature%20data%20document%20a,)%2C%20and%201998%2C%20respectively.

year) compared to that in the past 100 years. The area covered by ice sheets is decreasing at an alarming rate.[11] In summary, **climate change might be the greatest challenge that we have ever encountered and it should be treated as such**! The enormity of the problem and its scale require a universal approach that most likely will impact many aspects of our life.

Some GHGs are water vapor, carbon dioxide, methane, nitrous oxide (this is N_2O and should not be mistaken with NO_x), chlorofluorocarbons (CFCs), hydrochlorofluorocarbons (HCFCs), hydrofluorocarbons (HFCs), ozone, perfluorocarbons (PFCs), sulfur hexafluoride (SF_6), nitrogen trifluoride (NF_3), and aerosols (fine solid particles, such as carbon and sulfates, suspended in air). HFCs, PFCs, SF_6, and NF_3 are sometimes collectively referred to as fluorinated gases or F-gases. While water vapor is the most important naturally occurring GHG, carbon dioxide is the main gas driving the current climate change. In the U.S. over 80% of the GHG emissions by human activities is CO_2, about 10% is methane, about 3% is nitrous oxide, and the rest is a combination of other pollutants.[12] It should be noted that **CO_2 is naturally released to the atmosphere and absorbed from it in significant quantities** by plants, animals, volcanic activities, oceans, and Earth's surface soil. However, they collectively form a natural balance called the carbon cycle. What jeopardizes this balance is human activities, particularly combustion of fossil fuels.

According to the U.S. Environmental Protection Agency (EPA), transportation, electricity generation, and industry are responsible for most of GHG emissions in the U.S. by emitting 28%, 27%, and 22% of the emissions, respectively. They are followed by commercial-residential and agricultural sources by 12% and 10%, respectively.[13] China, the U.S., the European Union, India, Russia, and Japan are together responsible for about 70% of the global CO_2 emissions.[14] However, if we consider the per capita CO_2 emissions, the picture is totally different where this value for China and the European Union is less than half of that in the U.S. In the case of India, the per capita CO_2 emissions is about a tenth of that in the U.S.

Unlike other emissions where there is some universal consensus that they are harmful and should be regulated, unfortunately there is no such consensus on GHG emissions (at least in the U.S.). For this reason, there is a wide range of policies regarding these emissions from no regulation at all to policies that impose economic costs for emissions. The two most common such policies are **carbon tax** and **cap-and-trade** policies. In the former, industries are taxed according to their carbon emission intensity. The carbon emission intensity is defined as the emission per unit of economic activities or product production. For example, gram (or kilogram) of carbon (or CO_2)

---

[11] There are debates on if the total volume of ice is also decreasing. You can read about this at https://www.nasa.gov/feature/goddard/nasa-study-mass-gains-of-antarctic-ice-sheet-greater-than-losses.

[12] https://www.eia.gov/environment/emissions/ghg_report/ghg_overview.php.

[13] https://www.epa.gov/ghgemissions/sources-greenhouse-gas-emissions.

[14] https://www.epa.gov/ghgemissions/global-greenhouse-gas-emissions-data.

per kWh is the carbon emission intensity unit for electricity generation. In the cap-and-trade (CAT) policy, industries are given permission to emit a certain amount of emissions. A carbon trade market is created, so industries can sell their excess credits or purchase extra credits. As noted, the CO_2 emission of power generation units is measured by a parameter called the CO_2 emission intensity or the carbon emission intensity. The CO_2 emission intensity is defined as the mass of CO_2 emitted by a unit to generate 1 kWh of electrical energy.

5.5.2 Reduction of Greenhouse Gas Emissions

The most effective and affordable means to reduce GHG emissions in the energy sector is reducing energy consumptions by improving the efficiency of energy systems and minimizing losses. This includes using more efficient equipment and appliances, reducing building energy losses (to reduce heating and cooling loads), and **adapting a less energy-intensive lifestyle** by ordinary citizens. In the power generation industry, this is reflected in using higher efficiency power plants, for example by increasing the temperature and pressure of the inlet steam to the turbine. Also improving the efficiency of power plant components reduces the internal consumption of the plant, which, in turn, increases the net output power. Similarly using CCPP, which is a combination of a gas turbine and a steam cycle (much more on them in Chapter 10), can substantially increase the efficiency of power plants and reduce their CO_2 emission intensity. Another effective strategy to reduce CO_2 emissions is fuel switching. For example, replacing a conventional coal-fired steam power plant with a natural gas-fired combined cycle (NGCC) can reduce the CO_2 emission intensity by half.

5.5.3 CO_2 Capture and Sequestration Systems

The ultimate solution to combust fossil fuels to generate electricity without emitting GHG emissions is using **CO_2 capture and sequestration or storage (CCS)** systems. In these systems, the captured CO_2 can be stored by the injection of pressurized CO_2 into **geological formations**. Alternatively, the captured CO_2 can be used to improve recovery in oil and natural gas reservoirs. The storage facility can be at the power plant site or far from it. In the latter case, the pressurized CO_2 is transferred to the storage site by pipelines. However, there are serious questions on the viability of using geological formations for long-term CO_2 storage. For example, for how long CO_2 can be kept in the formations and what the environmental impacts are, particularly in case the stored CO_2 leaks. An important issue is who is responsible (and liable) if a CO_2 storage system fails. In small scale systems, the separated CO_2 can be used for industrial applications such as in the food industry.

CCS systems can be divided into three categories:

- pre-combustion CO_2 capture systems
- CO_2 capture after combustion with pure oxygen (oxygen firing)
- post-combustion CO_2 capture systems

The CO_2 capture process from flue gas heavily depends on the **concentration of CO_2 in the flow**. The lower the concentration, the more difficult and costlier the removal of CO_2 will be. In traditional combustion systems, using air to provide oxygen for the combustion process introduces an enormous amount of nitrogen to the combustion process (3.76 moles of nitrogen for every mole of oxygen, Section 4.3.1). This nitrogen gas does not participate in the combustion process (except a negligible amount that forms thermal NO_x). This nitrogen dilutes the flue gas flow and makes CO_2 capture more difficult. To eliminate this problem two approaches can be used, which will be elaborated in the next two sections.

5.5.4 Pre-combustion CO_2 Capture Systems

In this approach, fuel (primarily coal) is first gasified and converted to a gaseous fuel flow, called synthesis gas or syngas, composed of mainly carbon dioxide, carbon monoxide, and hydrogen. Due to the high concentration of CO_2 in the syngas flow, it can be easily captured and stored. Then, nearly pure hydrogen can be burned with little to no CO_2 emissions. Understandably this approach is referred to as the pre-combustion CO_2 capture. This process is typically used in CCPPs where hydrogen is burned in a gas turbine and then the combustion products are used to raise steam for a steam cycle. This cycle is known as an integrated gasification combined cycle (IGCC) and will be covered in Chapter 10.

5.5.5 Oxygen-Firing Systems

Another approach to address the problem of low concentration of CO_2 in flue gas is to use relatively pure oxygen in the combustion process instead of air. This technique is referred to as **oxygen firing**, **oxy firing**, or **oxy-fuel combustion**. These systems are not currently being used in commercial power plants. Several pilot plants are currently being tested as proof-of-concept demonstration plants. While it might seem counterintuitive, combustion of fuels with pure oxygen is not desirable. The presence of nitrogen from air in conventional combustion systems limits the maximum flame temperature. In the combustion process with relatively pure oxygen, with **95% to over 99% purity**, the flame temperature is much higher than that in the combustion with air. The high flame temperature can cause ash slagging and fouling in the downstream heat-transfer surfaces in steam generators, particularly in coal-fired steam generators (refer to Chapter 4). Also, this may promote NO_x generation (more on this below). To address this problem, a portion of the flue gas is recycled and is mixed with pure oxygen. The mixture of oxygen and recycled flue gas (RFG) is then used for combustion in the furnace by means of either the primary or secondary air flow (or more accurately the primary or secondary mixture of oxygen and RFG flow). Note that the flue gas is primarily composed of carbon dioxide and is free of nitrogen. The flue gas is typically extracted after the particulate matter removal system and after or before the scrubber depending on the concentration of sulfur oxides in the flow. The ratio of the RFG flow to the oxygen flow depends on the system design. The volumetric concentration of oxygen in the mixture is typically higher than that in the atmospheric air (21%) and can be up to 40%. In the design of oxygen-fired furnaces, special attention should be paid to the fact that the combustion properties of carbon dioxide are different from nitrogen. Therefore, the furnaces designed for air combustion may not be used for oxygen combustion and some modifications are typically required. The unrecycled portion of the flue gas is compressed and sent to the storage facility. If the flow needs to be transported through a piping system, the gas may need treatment to avoid corrosion in the pipes.

A major advantage of oxy-firing systems is that due to the lack of nitrogen from air, thermal NO_x production is almost entirely eliminated (except for a small amount of possible air infiltration to the system). While fuel NO_x production is still present, its magnitude is significantly reduced (perhaps due to the presence of NO_x from the recycled flue gas). As a result, NO_x removal systems (typically SCR) can be eliminated. These systems also have lower sulfur oxide emissions for the same fuel compared to conventional air-fired combustion systems.

An obvious problem with this technology is a need for pure oxygen which requires an **air separation unit**. The extraction of oxygen from air can be achieved by commercially available processes such as a cryogenic process. However, these processes are very energy-intensive and can consume a significant portion of the power produced by the power plant. Currently, a process called the chemical looping combustion process is being investigated to reduce this energy consumption.

5.5.6 Post-combustion CO_2 Capture

As the name implies, in post-combustion systems CO_2 is captured after the combustion process from the flue gas flow. For this purpose, CO_2 is chemically absorbed from combustion products by means of a scrubbing process. Currently, an **aqueous solution of amines**, typically monoethanolamine (MEA), is commonly used as a solvent. Amines are compounds based on ammonia (NH_3) where one or more hydrogen atoms are replaced by one or more organic groups. For example, monoethanolamine has a chemical formula of C_2H_7NO where a hydrogen atom in ammonia is replaced by a $HOCH_2CH_2$ group (Figure 5.9). The concentration of amines in aqueous solutions varies depending on the type

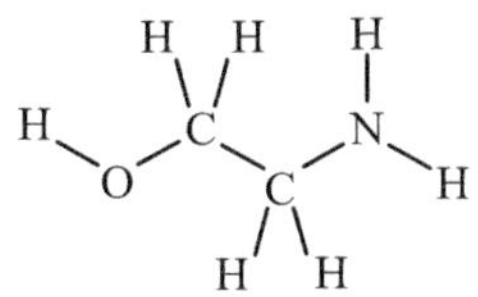

FIGURE 5.9
Molecular structure of monoethanolamine (MEA).

of amines and the application, between 20% and 50%. Other solvents such as ammonium carbonate are also currently being investigated.

The removal of CO_2 from flue gas in a post-combustion system is typically achieved by means of two-stage processes in two separate vessels: an **absorption column** and a **stripping column** (Figure 5.10). In the absorption column, the solvent closely interacts with the flue gas and absorbs most of CO_2 from the flow. The solvent can also remove some SO_2 and NO_2 from the

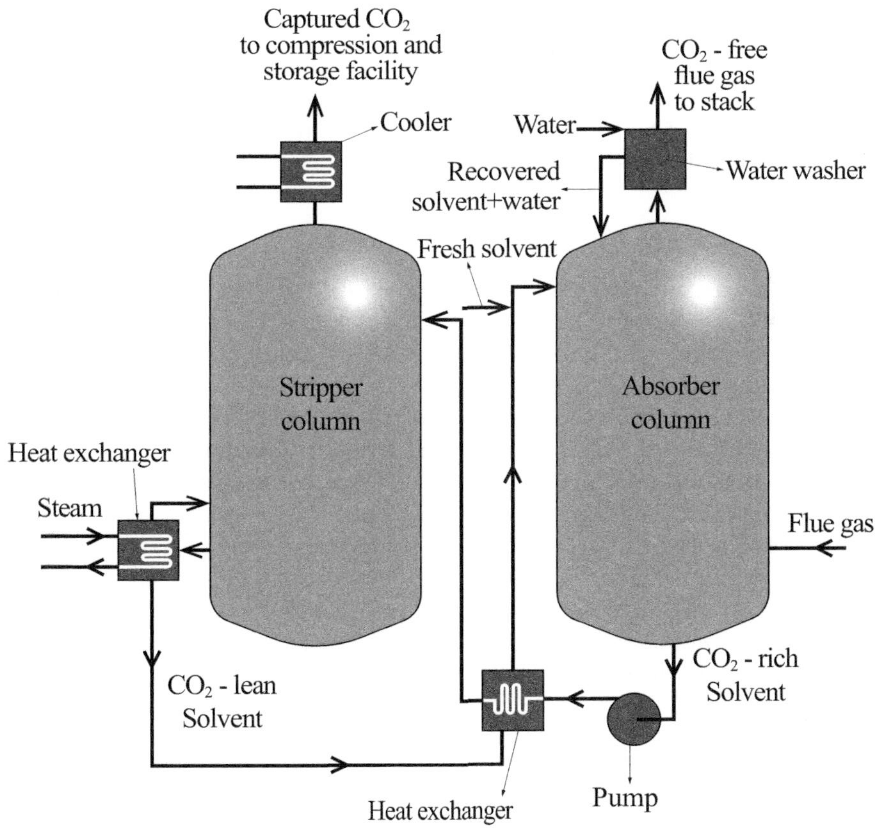

FIGURE 5.10
Schematic of a scrubbing post-combustion CO_2 capture.

flue gas. The solvent enters the top of the column and flows downward while the flue gas enters at the bottom and flows upward. To prevent the solvent carryover from the absorber by the flue gas flow, a water washer is used in the outlet flue gas flow. The mixture of the washing water and recovered solvent is returned back to the absorber. If the temperature of the flue gas flow is higher than a certain limit, the solvent can be degraded. To avoid the loss of the solvent, the inlet temperature of the flue gas and solvent must be controlled and maintained within an acceptable range.

The spent solvent is then sent to the stripping column (also referred to as a **reboiler**) where the absorbed CO_2 is released from the solvent by heating it. The CO_2-rich solvent enters at the top of the column and the regenerated solvent, which is CO_2-lean, is sent back to the absorption column from the bottom. The heater of the column is fed by the steam extracted from the steam cycle. The extraction of this steam from the power cycle reduces the overall efficiency of the cycle. To reduce the required heat, a heat exchanger is typically used to recover some of the thermal energy from the high-temperature lean solvent leaving the stripping column. This recovered heat is used to increase the temperature of the relatively low-temperature solvent entering the stripping column. Some solvents can be carried over by the outlet CO_2 flow from the stripping column. This solvent can be recovered by cooling the outlet CO_2 flow. The released CO_2 is sent to the compression and storage facility. The degraded solvent is separated and replaced by the fresh solvent (Figure 5.10).

5.5.7 CO_2 Capture by Microalgae

The above-discussed systems are costly because they use expensive chemical scrubbers to remove CO_2. In addition, they merely capture CO_2 from flue gas and the storage of the sequestered CO_2 is a major challenge. As noted before, various solutions have been presented, ranging from storage in rock formations to deep-ocean storage, both of which have their own potentially catastrophic ramifications. A major advantage of CO_2 capture by microalgae is that the captured CO_2 is in the form of a marketable product. The algae grown in these CCS systems can be used in a variety of ways, such as in food supplements and animal feeds, bio-plastics and pigments, fertilizers, or biofuels, thus negating the need for a special storage mechanism. Moreover, algae-based systems require less energy than other CCS systems. Microalgae are so effective at sequestering CO_2 that algal biomass can grow by up to threefold in a single day. Carbon capture systems using photosynthetic microalgae can be a simpler and more environmentally friendly alternative to traditional forms of CO_2 capture and sequestration.

In microalgae-based CCS systems, the flue gas flow is injected into a container full of solution where microalgae grow. Microalgae are exposed to either solar radiation or artificial light that promotes the photosynthesis process in which carbon dioxide, water, and minerals are converted into organic compounds (making more algae) and oxygen. The produced algae can be harvested and used as an animal food source or used to produce biofuel or other industrial products.

Microalgae growth containers can be open or closed systems. Open systems are typically concrete water channels with or without plastic liners. They are usually shallow to increase sunlight exposure area especially considering the fact that microalgae in the solution do not allow deep penetration of solar radiation. They are typically equipped with guiding baffles and a slow-moving paddle wheel to reduce the stress level on microalgae which otherwise can kill or damage them. The paddle wheel slowly circulates the water to avoid stagnation. While open systems are simple and relatively inexpensive and can be installed on a large scale, it is hard to measure the captured CO_2. Microalgae can also be contaminated by algae-killing bacteria.

Closed microalgae CO_2 capture systems typically consist of transparent containers, commonly referred to as photobioreactors (PBR) where light can easily penetrate. While they do not have the problems of open systems, they are more expensive to build on a large scale. However, they allow operational conditions, such as thermal, fluid, and light properties, to be effectively controlled to maximize the efficiency of CO_2 capture and the rate of microalgae growth. This results in smaller units and less land requirement compared to those of open systems. A variety of closed photobioreactors have been experimented with, including bagged, flat plate, horizontal tubular, helical tubular, stirred, and annular containers. These systems are currently being researched to determine their economic viability.

5.6 Other Emissions and Environmental Problems

While the above-discussed emissions are the most serious environmental problems associated with fossil fuel power generation plants, they are not the only ones. Some other issues are briefly discussed here.

5.6.1 Carbon Monoxide Emissions

Carbon monoxide (CO) is a colorless, odorless, and toxic gas that can be lethal if inhaled by humans or animals. Incomplete combustion of carbon can lead to the production of CO rather than CO_2 as follows:

$$2C + O_2 \rightarrow 2CO \tag{5.15}$$

To reduce CO emissions, the catalytic oxidation process is typically used. In this process, carbon monoxide molecules react with oxygen molecules in the presence of a catalyst.

$$2CO + O_2 \rightarrow 2CO_2 \tag{5.16}$$

The catalyst is typically platinum or other precious metals coated on stainless steel plates. To achieve optimum CO removal, the system is placed at a high-temperature section of steam generators.

5.6.2 Mercury Emissions

Coal can contain very small but sufficient amount of mercury to harm the environment and human health if released by combustion products. The mercury content of coal can be in the form of particulate matters or the vapor form as elemental mercury or oxidized mercury. Typically mercury emissions can be removed by a combination of flue gas desulfurization, SCR, and fabric filter systems. Sometimes to adequately remove all forms of mercury from combustion products, a sorbent such as activated carbon in powder form is also injected into the flue gas flow.

5.6.3 Sight and Noise Pollution

General **appearance and aesthetic aspects** of power plants have been increasingly scrutinized in the past few decades. Plant designers, owners, and operators have been using various strategies to reduce the intrusive physical appearance of power generation units. They use careful landscaping by growing lawns, trees, shrubs, and other plants for this purpose. Architects are included in the design teams to design pleasant buildings and structures. Also, some of the most intrusive facilities such as oil storage tanks and coal storage fields are hidden from public view by installing them underground or enclosing them by tall walls. Some components, such as fans, can be very noisy. They can be problematic for neighboring communities, particularly in urban areas. This problem can mostly be addressed by quieter equipment and acoustic barriers. A good public relation in general and proper communication and support of local communities in particular are major factors in hearing and addressing their concerns.

5.7 Questions and Problems

5.7.1 Introduction

5.1. Explain why the installation of emission control systems require careful study and tradeoff between all the involved factors.

5.2. Which acts protect air quality in the U.S.?

5.7.2 Nitrogen Oxide Emissions

5.3. Explain how the productions of NO_x and CO contradict each other.

5.4. Provide examples of environmental phenomena that NO_x emissions can contribute to.

5.5. Briefly explain how NO_x emissions contribute to smog, acid rains, and particulate matter emissions.

5.6. Explain how NO_x production can be reduced before, during, and after the combustion process.

5.7. What are two main post-combustion NO_x reduction technologies? Explain and compare these technologies.

5.8. Where is an SCR system installed for a high-ash fuel? What about a low-ash fuel? Why?

5.9. What are the differences and similarities between SCR and SNCR in terms of the regent, reactions, operation temperatures, installation location, and system configuration?

5.7.3 Sulfur Oxide Emissions

5.10. Describe how acid rain is formed by NO_x and SO_x.

5.11. Explain how pre-combustion NO_x and SO_x reduction can be achieved. Provide an example for each case.

5.12. Provide an example of combustion systems that have an inherent capability to remove the majority of sulfur oxide emissions during the combustion process.

5.13. Which substance is the most common reagent used in wet flue gas desulfurization systems?

  a. dolomite (calcium magnesium carbonate)
  b. lime (CaO)
  c. limestone ($CaCO_3$).
  d. caustic soda (sodium hydroxide)
  e. ammonia

5.14. Briefly explain the operation of wet flue gas desulfurization systems.

5.15. Why are desulfurization scrubbers typically installed right before the stack?

5.16. What is the difference between a wet and a semi-dry flue gas desulfurization system?

5.17. What are the typical reagents for wet scrubbers and dry scrubbers?

5.18. Explain why and how the sulfur trioxide (SO_3) emission in flue gas is harmful. How can it be removed from the flue gas flow?

5.7.4 Particulate Matter Emissions

5.19. Name several physical principles that have been employed for the separation of solid particles from combustion product gases.

5.20. Where are electrostatic precipitators and fabric filters typically installed?

5.21. Describe how electrostatic precipitators can separate particulate matters from the flue gas flow.

5.22. Describe several mechanisms that are being used to remove dust-cakes from bag filters.

5.23. What are two parameters to determine the appropriate time to clean filters?

5.7.5 Global Climate Change and Greenhouse Gas (GHG) Emissions

5.24. Briefly explain the greenhouse effect and how it causes climate change using the following figure.

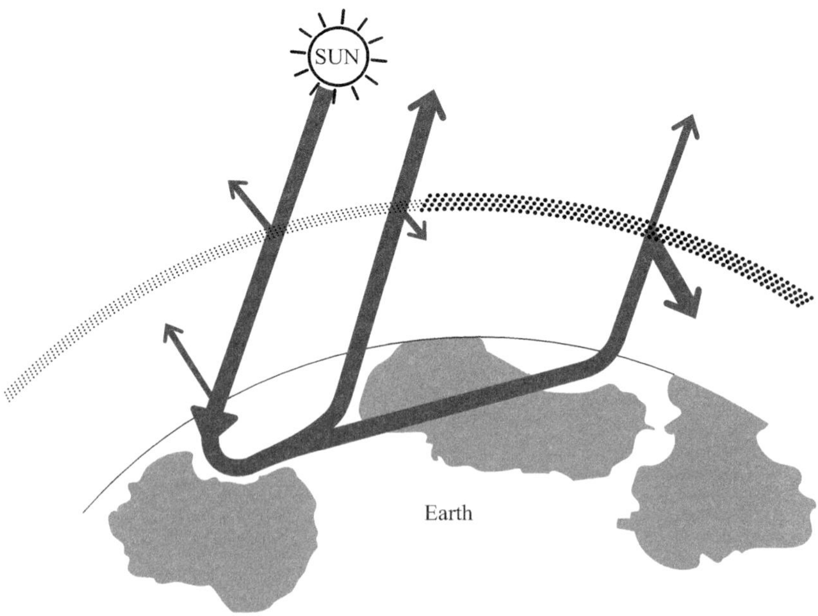

5.25. What are climate change feedbacks? Name a few of them.

5.26. Explain the impacts of water vapor and clouds on climate change. Why are they considered as both reinforcing and balancing feedbacks?

5.27. Provide several indications that we are not moving in the right direction regarding climate change.

5.28. List some greenhouse gases (at least five gases).

5.29. What sectors are the largest contributors to greenhouse gas emissions in the U.S.?

5.30. Compare the per capita CO_2 emissions in the U.S., China, India, and the European Union.

5.31. What are the two most common CO_2 emission control policies? Explain each.

5.32. What is the definition of the CO_2 emission intensity for power generation units?

5.33. What can be done with the captured CO_2 by CCS systems?

5.34. How does the concentration of CO_2 in the flue gas flow impact the CO_2 capture process?

5.35. Explain how oxygen-firing systems can capture CO_2?

5.36. Describe the post-combustion CO_2 capture systems by the scrubbing process. Use the following schematic diagram.

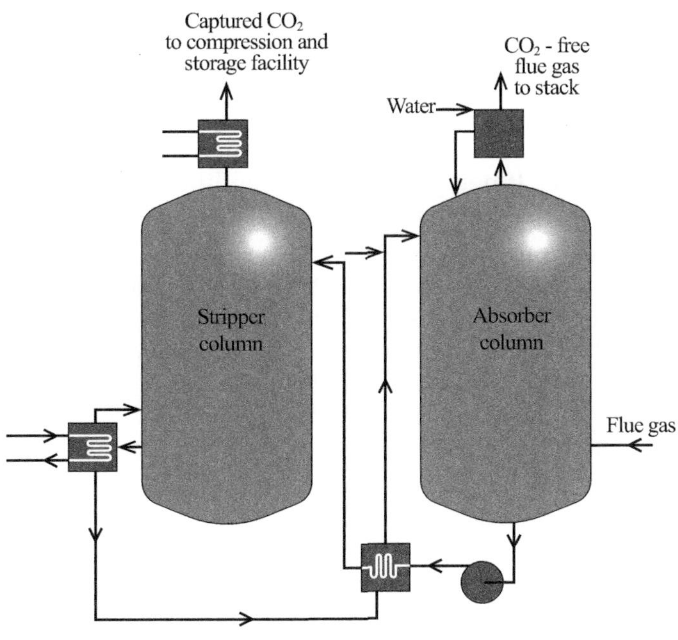

5.37. List the advantages of microalgae CO_2 capture systems over other CCS systems.

5.7.6 Other Emissions and Environmental Problems

5.38. Explain how carbon monoxide (CO) can be removed from combustion products.

5.39. How can mercury and its compounds be removed from flue gas?

5.40. List some ways to improve the appearance and aesthetic aspects of power plants.

5.7.7 Open-Ended Questions

5.41. Conduct research to identify several countries or regions where carbon tax and cap-and-trade policies have been implemented. What have been the outcomes so far?

5.42. Identify an underground CO_2 storage demonstrational project. Research the challenges the project has faced. Overall how successful was the project?

5.43. Conduct research to identify the pros and cons of the carbon tax and cap-and-trade policies.

5.44. Locate a fossil fuel power plant in the region you live (the closer to where you live, the better). What are some emissions released by the plant? What are their human health hazard, damage to the environment, and contribution to climate change? Have any of the emission reduction strategies discussed in this chapter been implemented in the plant? If yes, try to find some specifications of these systems. If not, propose some modifications to the plant to reduce its environmental impacts.

5.45. Locate the largest power plants in your country or state. What are some emissions released by the plant? What are their human health hazard, damage to the environment, and contribution to climate change? Have any of the emission reduction strategies discussed in this chapter been implemented in the plant? If yes, try to find some specifications of these systems. If not, propose some modifications to the plant to reduce its environmental impacts.

5.7.8 General

5.46. Provide an example of pre-combustion, during combustion, and post-combustion NO_x production control technologies.

5.47. What are the functions of an absorption column and a stripping column in a post-combustion CO_2 capture system?

5.48. How is pure oxygen provided for oxygen combustion?

5.49. Explain how melting glaciers and ice-caps impact climate change (two impacts).

5.50. Explain why Mars's surface is very cold while Venus's surface is very hot.

5.51. The lower half of an absorption tank in a wet scrubber is referred to as the … Why.

a. recirculation zone

b. reaction zone

c. oxidation zone

d. all of the above

5.52. What is the byproduct of wet scrubbers? What are its industrial applications?

5.53. List some ways that CO_2 is naturally released into the atmosphere and absorbed from it in significant quantities.

5.54. List three alkaline substances that can be used as a reagent in sulfur oxide reduction systems.

5.55. What are some of the questions on the viability of using geological formations for long-term CO_2 storage?

5.56. Which one has shorter wavelengths? Why?

a. Solar radiation

b. Radiation from Earth's surface

5.57. What are three general categories of CO_2 capture and sequestration or storage (CCS) systems?

5.58. What is the function of the catalyst in an SCR system?

5.59. Why is the study of climate change very complex?

5.60. Which of the following systems is the most effective approach for sulfur oxide reduction?

a. Wet systems

b. Semi-dry systems

c. Dry systems

5.61. What are the National Ambient Air Quality Standards (NAAQS)?

5.62. Explain pre-combustion CO_2 capture systems?

5.63. What is NO_x emission?

5.64. List countries with the largest contribution to the global GHG emissions.

5.65. What type of electricity is used in electrostatic precipitators, AC power or DC power? At what voltage?

5.66. Sulfur is one of the elements in fuels that can burn and slightly contribute to its heat content.

a. True

b. False

5.67. Identify where SCR and scrubbers are typically installed in steam power plants.

5.68. What are the most effective and affordable means to reduce GHG emissions in the energy sector?

5.69. What are the advantages of electrostatic precipitators?

5.70. What is a carbon trade market?

5.71. In an SCR system, the catalytic materials are not consumed but accelerate the reactions though they deteriorate with time.

a. True

b. False

5.72. What are the similarities and differences between WESPs and ESPs?

5.73. How is CO_2 naturally released into the atmosphere and absorbed from it? List at least four ways.

5.74. Which one of the following reactions is dominant in selective catalytic reduction systems?

a. $4NH_3 + 4NO + O_2 \rightarrow 4N_2 + 6H_2O$

b. $8NH_3 + 6NO_2 \rightarrow 7N_2 + 12H_2O$

c. $2NH_3 + NO + NO_2 \rightarrow 2N_2 + 3H_2O$

5.75. What are the conflicting feedbacks of tree and plant growth on climate change?

5.76. Which one is more effective in removing NO_x emissions? Which one is less costly?

a. Selective catalytic reduction

b. Selective non-catalytic reduction

5.77. Explain what happens when solar radiation is intercepted by Earth.

5.78. What is the problem with burning fuels with pure oxygen? How this problem can be addressed?

5.79. Compare open or closed microalgae CCS systems. List the pros and cons of each system.

6

Steam Turbines

Abstract

A turbine-generator set is where the ultimate product of a power plant, electricity, is generated. In that sense, they can be considered as the heart of the plant. In this chapter, you will learn how turbines and generators work. We will start this topic by looking at different types of blades, namely impulse and reaction blades. You will learn how to conduct basic relevant calculations for impulse blades. Then, you will use this knowledge to conceptually deduct similar understandings for reaction blades. Based on what you learned about turbine blades, we will discuss about various configurations of real-world steam turbines. Furthermore, we will review fundamental principles of other steam turbine components, such as rotors, casings, radial and axial bearings, gland steam seals, and many others. We will then briefly talk about electric generators and their components. We will conclude the chapter with a quick discussion on the control mechanisms of turbines and generators.

LEARNING OUTCOMES

After successful completion of this chapter, you will be able to:

- distinguish impulse and reaction turbine blades and describe characteristics of each type
- conduct basic relevant calculations for impulse blades
- conceptually deduct basic understandings for reaction blades
- explain various configurations of real-world steam turbines
- describe fundamental principles of other steam turbine components such as rotors, casings, radial and axial bearings, and gland steam seals
- explain operation of electric generators and their components
- describe control mechanisms of turbines and generators

6.1 Introduction

From a humble origin of just a few kW power generation capacity in the late 19th century, steam turbines have grown to large sophisticated units that can generate well over 1000 MW power. They can weigh thousands of tonnes with over 30 m (over 100 ft) in length. In Chapter 2, we discussed that in early steam power generation units, reciprocating engines were used to extract mechanical work from steam flows. In these engines, steam expanded in a cylinder and piston assembly and produced power by moving a piston. However, due to limitations in their power generation capacity, steam turbines soon replaced them. Due to the high speed of steam turbines, they can produce a large quantity of power in a relatively compact size. Their high speed also suits direct coupling with AC electricity generators. Nowadays electricity produced in power plants is predominantly generated in either steam or gas turbines.

It should be noted that steam turbines are typically designed for **maximum continuous rating** (**MCR**). As the name implies, this is the maximum power rating of the turbine so that it can operate continuously for long periods. However, the turbine might be able to operate for a short period above this maximum. The designer and manufacturer of a turbine determine these limitations.

Thermodynamically speaking, the objective of any turbine is to produce mechanical work by reducing the pressure and enthalpy of a fluid stream. Ideally, this process is an isentropic process (constant entropy) which can be seen in the expansion process in the Carnot cycle (process 3–4s in Figure 2.25). In this process, superheated steam is expanded to a saturated mixture of steam and liquid water, which is discharged to the condenser. The power production in the turbine depends on the length of the line connecting states 3 and 4s. This length can be increased by **increasing the temperature and pressure of the steam entering the turbine** from the steam generator (state 3), by **reducing the pressure of the flow leaving the turbine** toward the condenser (state 4s), or a combination of these two actions. The inlet steam conditions are limited by the **materials of turbine components**. The pressure of the turbine outlet flow and the condenser is determined by the temperature of the cooling water. The actual expansion process is always accompanied by some irreversibilities and losses. As a result of these losses, the expansion process is not isentropic, and the entropy increases in the process (process 3–4 in Figure 2.25). For more information on this topic, refer to Section 2.3.6.

Large utility steam turbines are commonly equipped with several turbines, which can typically be a combination of a high-pressure (HP) section, an intermediate-pressure (IP) section, and one or more low-pressure (LP) sections. They can be installed on a single shaft or double shaft. The high-temperature and pressure steam from the steam generator, is sometimes referred to as the live steam, first enters the high-pressure turbine and expands to an intermediate pressure. The outlet steam from this turbine, which is sometimes referred to as the cold reheat, is sent back to the steam generator to be reheated in the first heater. The reheated steam at a temperature equal or higher than that of the live steam is then fed to the intermediate-pressure turbine. If the plant is equipped with the second reheater, this process can be repeated for that reheater and the low-pressure turbine(s). The exit flow from this turbine is sent to the condenser for the condensation process.

In a steam turbine, the high-temperature and pressure superheated steam flow passes through a set of nozzles (or a series of fixed blades acting as nozzles) where the velocity of the flow increases. Then, the flow goes through a series of moving blades installed on a shaft. The force acted on the blades as a result of the steam expansion causes the shaft to rotate and produce mechanical work. In the next paragraph, the design and operation of steam turbines will be briefly reviewed. If you have difficulty understanding this paragraph, do not worry! We will elaborate on all of these terms and concepts in the rest of this chapter. To get an idea about the size and shape of steam turbines, Figure 6.1 shows a medium-size steam turbine.

The first step of the turbine design process is the thermodynamics analysis of the cycle. This analysis determines the states of the turbine inlet and outlet flows. This can determine the pressure and enthalpy drops of the working fluid through the turbine. This drop cannot be accomplished in a single step and should be divided into smaller pressure and enthalpy drops. Designers

FIGURE 6.1
A medium-size steam turbine. (Image by David Mark from Pixabay, https://pixabay.com/photos/speyer-germany-city-technik-museum-84394/.)

should decide on the number of smaller steps through which the expansion must occur. Each of these steps is called a **stage of the turbine** which is composed of a set of **stationary blades** (also known as **nozzles**) or a set of **rotating blades** (also known as **buckets**). These stages of expansion should be individually designed. These stages can be impulse or reaction stages. A flow of a fluid, either gas or liquid, can impose force on a solid surface that interacts with the flow in two manners. It can strike the surface and exert force on the surface because of the **change in its momentum**, such as when you try to remove dust and dirt using a water hose. This is called the **impact** or **impulse force**. Alternatively, the force can be exerted in the form of the **reaction force**, such as the force created when a balloon full of air is released to deflate and as a result, it spirals in air. We will start our discussion with impulse turbines.

6.2 Impulse Turbines

When two objects collide, they impose forces on each other called the impact force. Impulse blades operate based on the same concept. In these blades, the flow of the working fluid impinges the blades of the turbine. The blades, in turn, alter the direction of the flow. This results in the change in the momentum of the flow (Reminder Box 6.1). According to the second law of Newton ($Fdt = mdV$), this will create impulse force. In other words, in impulse turbines, the **kinetic energy of the stream is converted to the shaft work**.

REMINDER BOX 6.1

LINEAR MOMENTUM CONSERVATION

From fluid mechanics, the linear momentum conservation equation for a control volume can be written in the following form for a steady state flow with uniform properties across the inlets and outlets (Figure 6.2):

$$\sum \vec{F}_{CV} = \sum \vec{V}_{out}\dot{m}_{out} - \sum \vec{V}_{in}\dot{m}_{in} \tag{6.1}$$

where $\sum \vec{F}_{CV}$ is the resultant force of the body forces and surface forces (forces that act on the volume such as weight and surface such as pressure force, respectively) exerted on the control volume (N), $\vec{V}$ is the absolute velocity vector of the flows relative to an inertial reference frame (m/s), and $\dot{m}$ is the mass flow rates of inlet and outlet flows (kg/s).

Equation 6.1 indicates that the force exerted on the control volume is equal to the net rate of the linear momentum that accompanies the mass flows crossing the boundary of the control volume (the control surface). Note that in the blade analysis, we are interested in the force that is imposed by the flow on the blade, which is the reaction force of $\vec{F}_{CV}$. So from now on

$$\vec{F}_B = -\sum \vec{F}_{CV} = \sum \vec{V}_{\text{in}} \dot{m}_{\text{in}} - \sum \vec{V}_{\text{out}} \dot{m}_{\text{out}} \tag{6.2}$$

For a single inlet and single outlet control volume, Equation 6.2 can be simplified as follows:

$$\vec{F}_B = \dot{m}\left(\vec{V}_1 - \vec{V}_2\right) \quad \left(\text{vector form}\right) \tag{6.3}$$

$$F_{Bx} = \dot{m}\left(V_{1x} - V_{2x}\right) \quad \left(\text{for x-direction}\right) \tag{6.4}$$

Note that the difference between vector parameters, such as flow velocity, and scalar parameters, such as the components of the flow velocity on the main axes, are presented by a vector sign over the parameter.

6.2.1 Single Impulse Blades

Since the interaction of a fluid stream and blades is the building block of any turbomachine, such as turbines, pumps, compressors, and fans, in the next few examples this effect will be considered. First, let us assume a horizontal jet strikes a stationary blade and exits it at the angle θ (Figure 6.3). Then, the example will be repeated when the blade is moving with the velocity of V_B.

Example 6.1

Determine the force imposed by the mass flow of the fluid on the fixed blade in Figure 6.2.

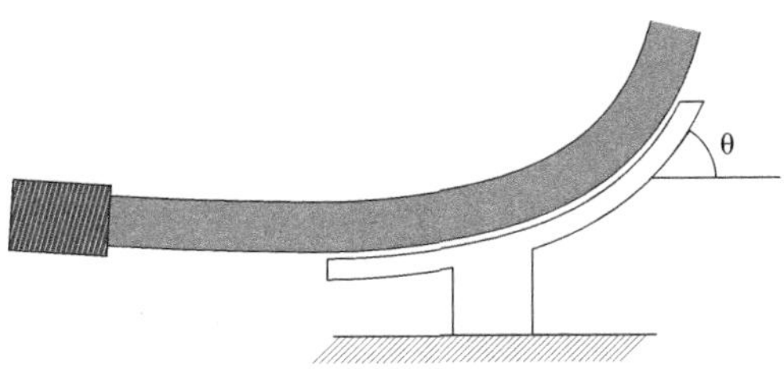

FIGURE 6.2
Interaction of a fluid stream and a fixed blade.

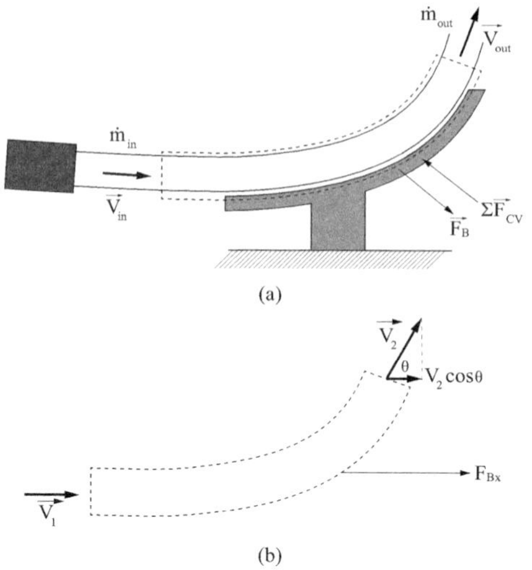

FIGURE 6.3
Control volume for a fluid stream and a fixed blade.

Analysis

First, an appropriate control volume for the problem should be selected (Figure 6.3).

Assuming a constant uniform external pressure on the control volume, a frictionless flow over the vane, no spreading of the stream, and a horizontal jet, from Bernoulli's equation,

$$\frac{V_1^2}{2g}+\frac{P_1}{\gamma}+Z_1=\frac{V_2^2}{2g}+\frac{P_2}{\gamma}+Z_2 \tag{6.5}$$

$$V_1 = V_2 \tag{6.6}$$

Equation 6.6 indicates that the magnitudes of the inlet and outlet stream velocities, V_1 and V_2 respectively, are equal but their directions are not. Neglecting the weight of the fluid in the control volume, the horizontal component of the force acting on the control volume can be calculated using Equation 6.4 (it will be shown later that the vertical component of the forces does not produce any work).

$$F_{Bx}=\dot{m}(V_{1x}-V_{2x})=\dot{m}(V_1-V_2\cos\theta)=\dot{m}V_1\ (1-\cos\theta) \tag{6.7}$$

This force is called the **impulse force**. Let us consider three specific cases for the angle θ:

If $\theta = 0°$, i.e., the blade is a plane plate:

$$F_{Bx}=0 \tag{6.8}$$

In this case, the fluid slides through the plate without exerting any force on the plate.

If $\theta=90°$, i.e., the blade is a vertical plate:

$$F_{Bx} = \dot{m}V_1 \tag{6.9}$$

If $\theta=180°$, i.e., the flow completely turns around:

$$F_{Bx} = 2\dot{m}V_1 \tag{6.10}$$

It is evident that the maximum force is when the angle is 180°. In this example, although the flow exerts a force on the blade, there is no work done because the blade is fixed and does not move.

Example 6.2

Determine the force imposed by the mass flow on the moving blade in Figure 6.4.

Analysis

Again in this case, the component of the force on the x-direction can be calculated using Equation 6.4. In this case, for the frictionless flow, the **magnitude of the stream velocity relative to the blade (the velocity of the stream seen by an observer riding on the blade) is constant** (but not its direction).

$$V_{r1} = V_{r2} \tag{6.11}$$

So in order to find the absolute velocity of the outlet flow, first the inlet flow velocity relative to the blade should be found.

$$\vec{V}_1 = \vec{V}_{r1} + \vec{V}_B \tag{6.12}$$

$$V_{r1} = V_1 - \vec{V}_B \text{ (in } x\text{-direction)} \tag{6.13}$$

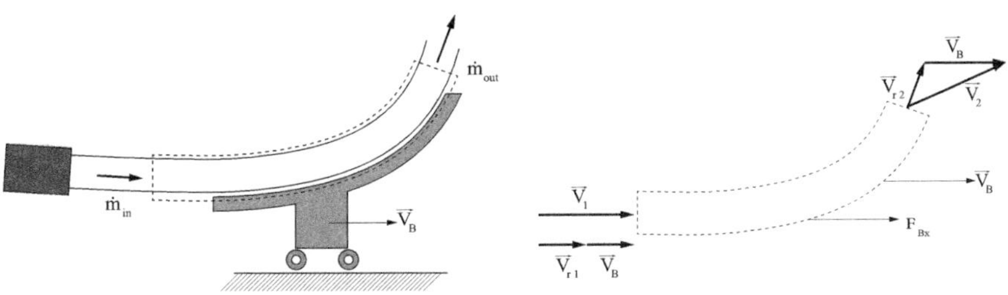

FIGURE 6.4
Interaction of a fluid stream and a moving blade.

For the outlet stream velocity vector,

$$\vec{V}_2 = \vec{V}_{r2} + \vec{V}_B \tag{6.14}$$

For x-direction,

$$V_{2x} = V_{r2}\cos\theta + V_B \tag{6.15}$$

Substituting Equation 6.11 and then Equation 6.13 into Equation 6.15,

$$V_{2x} = V_{r1}\cos\theta + V_B = (V_1 - V_B)\cos\theta + V_B \tag{6.16}$$

Substituting Equation 6.16 into Equation 6.4[1]

$$F_{Bx} = \dot{m}\left[V_1 - (V_1 - V_B)\cos\theta - V_B\right] = \dot{m}(V_1 - V_B)(1 - \cos\theta) \tag{6.17}$$

Again in this case if $\theta = 0°$,

$$F_{Bx} = 0 \tag{6.18}$$

If $\theta = 90°$,

$$F_{Bx} = \dot{m}(V_1 - V_B) \tag{6.19}$$

If $\theta = 180°$,

$$F_{Bx} = 2\dot{m}(V_1 - V_B) \tag{6.20}$$

Again in this case, the maximum force is exerted on the control volume when $\theta = 180°$.

Example 6.3

Determine the blade velocity (V_B) that maximizes the power output of the blade in Example 6.2.

---

1 In Equation 6.4, $\dot{m}$ is the mass flow rate of the stream that experiences the momentum change. In this case, since the blade is moving away from the stream, not all the mass flow of the stream enters the control volume and experiences the change in the momentum. The part of the mass flow that actually experiences the change in the momentum is

$$\dot{m}_r = \rho A(V_1 - V_B)$$

where $\dot{m}_r$ is the mass flow that actually experiences the momentum change (kg/s), ρ is the density of the fluid (kg/m^3), and A is the cross section of the stream (m^2). However, it will be shown later that for an actual turbine blade all the mass flow experiences the change in momentum. Therefore, for simplicity in this analysis, it is assumed that $\dot{m}_r = \dot{m}$. This assumption will not impact the analysis.

Analysis The power developed in the blade in Example 6.2 is the force exerted on the blade multiplied by the velocity of the blade. Note that there is no work at the *y*-direction because the component of the blade velocity in this direction is zero.

$$\dot{W} = F_{Bx}V_B = \dot{m}(V_1 - V_B)(1-\cos\theta)V_B \tag{6.21}$$

In Equation 6.21, the maximum power is developed when $\theta = 180°$. This was shown in Examples 6.1 and 6.2, and it can be easily shown for this example too.

$$\dot{W}_{(\theta=180°)} = 2\dot{m}(V_1 - V_B)V_B \tag{6.22}$$

Equation 6.17 shows that when the blade is stationary $(V_B = 0)$, the force is maximum. However, Equation 6.21 illustrates that the power developed for the stationary blade is zero. Equations 6.17 and 6.21 also show that when the velocities of the blade and the stream are equal $(V_B = V_1)$, the force and power are zero. This means that there is a blade velocity between 0 and V_1 in which the power is maximum. In order to find this optimum velocity, the power equation (Equation 6.21) should be differentiated with respect to V_B and equated to zero (this is left for practice), which results in $\dfrac{d\dot{W}}{dV_B} = 0$

$$V_{B,\text{opt}} = \frac{V_1}{2} \tag{6.23}$$

This indicates that the power is maximized when the speed of the blade is one-half of the speed of the inlet stream (for a frictionless flow). Substituting the optimum blade velocity (Equation 6.23) into Equations 6.17 and 6.21, the maximum force that can be exerted on the blade and the maximum power that can be developed in the blade are

$$F_{Bx}, max = \frac{1}{2}\dot{m}V_1(1-\cos\theta) \tag{6.24}$$

$$\dot{W}_{\max} = \frac{1}{4}\dot{m}V_1^2(1-\cos\theta) \tag{6.25}$$

The efficiency of the moving blade (later we will extend this to rotating blades) can be defined as the ratio of the developed power to the power of the inlet stream (due to its kinetic energy).

$$\eta_B = \frac{\dot{m}(V_1 - V_B)(1-\cos\theta)V_B}{\frac{1}{2}\dot{m}V_1^2} = 2(1-\cos\theta)\left[\left(\frac{V_B}{V_1}\right) - \left(\frac{V_B}{V_1}\right)^2\right] \tag{6.26}$$

For θ=90°, the maximum efficiency $\left(V_{B,\text{opt}} = \dfrac{V_1}{2}\right)$ is

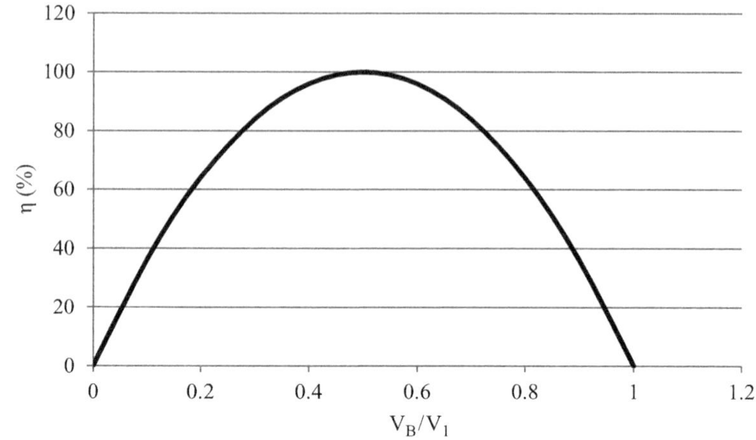

FIGURE 6.5
Efficiency of a moving blade as a function of the ratio of the blade velocity to the inlet stream velocity.

$$\eta_{B,\max\,(\theta=90^\circ)} = 50\% \tag{6.27}$$

For $\theta=180^\circ$

$$\eta_{B,\max\,(\theta=180^\circ)} = 100\% \tag{6.28}$$

As Equation 6.29 shows, in this particular case (Equation 6.28), the outlet stream's absolute velocity is zero. So the entire kinetic energy of the inlet stream is converted to shaft work (using Equation 6.16).

$$V_{2x} = (V_1 - V_B)\cos\theta + V_B = \left(V_1 - \frac{V_1}{2}\right)\cos 180 + \frac{V_1}{2} = 0 \tag{6.29}$$

The curve in Figure 6.5 illustrates the power developed in the blade as a function of the ratio of the blade velocity to the inlet stream velocity.

6.2.2 Actual Impulse Blades

In order for a turbine made of blades to be able to generate continuous power, there should be a series of blades that are mounted on the circumference of a disc. Furthermore, although the power is maximum at the stream exit angle of 180°, in practice this angle should be less than 180° so that the stream can actually leave the turbine (the angle θ_2 in Figure 6.6). Similarly, the inlet stream should have a shallow angle as well (the angle θ_1 in Figure 6.6).

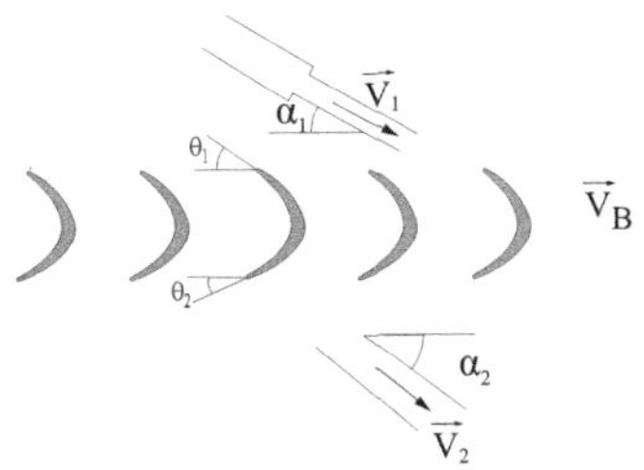

FIGURE 6.6
Series of impulse blades and the inlet and outlet flows.

In an actual turbine, the force exerted on a specific blade by a specific jet is zero until the blade is on the path of the jet and is struck by the jet. The force exerted on the blade gradually increases until it reaches a maximum when it is fully exposed to the jet flow. Then, while the blade moves away, the force decreases until it becomes zero again. The effect of these variable velocities (and hence the variable forces) can be averaged (Figure 6.7). Also, since in turbines the blades are not actually moving away from the inlet jet and the jet always strikes at least one blade, the momentum of all mass exiting the nozzle changes ($\dot{m} = \dot{m}_r$). Thus, the assumption made in footnote 1 is indeed valid.

To study the power generated in the blades, the absolute and relative velocities of the inlet and outlet streams should be studied. This can be done through a **velocity vector diagram** (or triangle). In a typical velocity vector diagram, the relative velocity of the inlet (and outlet) flow can be determined using the inlet (and outlet) flow's absolute velocity vector and the blade velocity. Ideally, to minimize the disturbance of the flow over the blade, the relative velocities of the inlet and outlet flows should be at the blade entrance and exit angles. For example, in Figure 6.7, V_{r1} and V_{r2} are at the directions θ_1 and θ_2, respectively. Therefore, the blade entrance and exit angles, which determine the shape of the blade, are dictated by the relative velocities of the inlet and outlet flows.

In the velocity vector diagram in Figure 6.7, the relationship between absolute and relative velocity of the flows is

$$\vec{V} = \vec{V}_r + \vec{V}_B \tag{6.30}$$

For the inlet stream, Equation 6.30 can be written on x and y directions as follows:

$$x\text{-direction}: V_{1x} = V_1 \cos\alpha_1 = V_B + V_{r1}\cos\theta_1 \tag{6.31}$$

$$y\text{-direction}: V_{1y} = V_1 \sin\alpha_1 = V_{r1}\sin\theta_1 \tag{6.32}$$

Similarly for the outlet stream

$$x\text{-direction: } V_{2x} = V_2 \cos\alpha_2 = V_B - V_{r2}\cos\theta_2 \tag{6.33}$$

$$y\text{-direction: } V_{2y} = V_2 \sin\alpha_2 = V_{r2}\sin\theta_2 \tag{6.34}$$

where V_1 is the absolute velocity of the inlet stream, V_{r1} is the relative velocity of the inlet stream with respect to the blade, V_B is the velocity of the blade, V_2 is the absolute velocity of the outlet stream, V_{r2} is the relative velocity of the outlet stream with respect to the blade, θ_1 is the blade entrance angle, θ_2 is the blade exit angle, α_1 is the fluid inlet angle and the nozzle angle, and α_2 is the fluid exit angle. In the design of blades, typically the inlet flow (nozzle) angle (α_1) and the inlet flow absolute velocity (nozzle velocity) are given, and other parameters need to be determined from the velocity vector diagram.

From Equation 6.31

$$V_{r1}\cos\theta_1 = V_1\cos\alpha_1 - V_B \tag{6.35}$$

$$V_{r1} = \frac{V_1\cos\alpha_1 - V_B}{\cos\theta_1} \tag{6.36}$$

From Equation 6.33, since $V_{r1} = V_{r2}$ (a frictionless flow)

$$V_{2x} = V_2\cos\alpha_2 = V_B - V_{r2}\cos\theta_2 = V_B - V_{r1}\cos\theta_2 = V_B - \left(\frac{V_1\cos\alpha_1 - V_B}{\cos\theta_1}\right)\cos\theta_2 \tag{6.37}$$

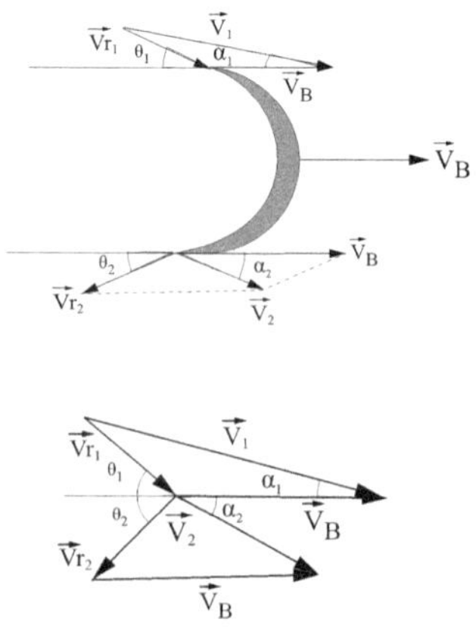

FIGURE 6.7
Velocity vector diagram for an impulse blade.

The force, power, and efficiency of the blade can be estimated by the following equations:

$$F_{Bx} = \dot{m}(V_{1x} - V_{2x}) = \dot{m}(V_1 \cos\alpha_1 - V_2 \cos\alpha_2) \tag{6.38}$$

$$\dot{W} = \dot{m}V_B(V_1 \cos\alpha_1 - V_2 \cos\alpha_2) \tag{6.39}$$

$$\eta_B = \frac{\dot{m}V_B(V_1 \cos\alpha_1 - V_2 \cos\alpha_2)}{\frac{1}{2}\dot{m}V_1^2} \tag{6.40}$$

Substituting Equation 6.37 into Equations 6.38 and 6.39

$$F_{Bx} = \dot{m}\left[V_1 \cos\alpha_1 - \left(V_B - \left(\frac{V_1 \cos\alpha_1 - V_B}{\cos\theta_1}\right)\cos\theta_2\right)\right] \tag{6.41}$$

$$\dot{W} = \dot{m}V_B\left[V_1 \cos\alpha_1 - \left(V_B - \left(\frac{V_1 \cos\alpha_1 - V_B}{\cos\theta_1}\right)\cos\theta_2\right)\right] \tag{6.42}$$

In impulse blades, the inlet and outlet angles are typically equal $(\theta_1 = \theta_2)$. Substituting this in Equations 6.37, 6.41, and 6.42

$$V_{2x} = V_2 \cos\alpha_2 = 2V_B - V_1 \cos\alpha_1 \tag{6.43}$$

$$F_{Bx} = 2\dot{m}(V_1 \cos\alpha_1 - V_B) \tag{6.44}$$

$$\dot{W} = 2\dot{m}V_B(V_1 \cos\alpha_1 - V_B) \tag{6.45}$$

In order to determine the blade velocity in which the power is maximum, the derivative of the power equation (Equation 6.45) with respect to V_B should be equated to zero.

$$\frac{d\dot{W}}{dV_B} = 0$$

$$V_{B,\text{opt}} = \frac{1}{2}V_1 \cos\alpha_1 \tag{6.46}$$

Note that this is for a frictionless flow turbine and for real turbines with friction the optimum blade velocity is slightly lower. Substituting Equation 6.46 into Equation 6.45

$$\dot{W}_{\max} = \frac{1}{2}\dot{m}(V_1 \cos\alpha_1)^2 \tag{6.47}$$

For the maximum efficiency

$$\eta_{B,\max} = \frac{\dot{W}_{\max}}{\frac{1}{2}\dot{m}V_1^2} = \frac{\frac{1}{2}\dot{m}(V_1 \cos\alpha_1)^2}{\frac{1}{2}\dot{m}V_1^2} = (\cos\alpha_1)^2 \tag{6.48}$$

Also, substituting Equation 6.46 into Equation 6.43

$$V_2 \cos\alpha_2 = 2\left(\frac{1}{2}V_1 \cos\alpha_1\right) - V_1 \cos\alpha_1 = 0 \text{ so} = 90° \tag{6.49}$$

This means when **the blade velocity is optimum, the exit absolute velocity is perpendicular to the blade velocity vector and is parallel to the axis of rotation of the turbine** (an axial flow).

Note that from the first law of thermodynamics (energy conservation), the produced power should be equal to the reduction in the kinetic energy (assuming no loss of energy and all other forms of energy are unchanged).

$$\dot{W} = \frac{1}{2}\dot{m}\left(V_1^2 - V_2^2\right) \tag{6.50}$$

Example 6.4

Superheated steam enters an impulse moving blade with the speed of 500 m/s at the angle of 15° (the angle between the blade velocity and the absolute velocity of the inlet flow). The blade is symmetrical and its velocity is optimized for power generation. Draw the velocity vector diagram for the inlet and outlet flows and determine the maximum output power of the blade per unit of mass flow rate and blade efficiency. The flow is frictionless.

AnalysisLet us determine the blade optimum velocity first.

$$V_{B,\text{opt}} = \frac{1}{2}V_1 \cos\alpha_1 = 0.5 \times 500 \times \cos 15 = 241.5 \ \frac{\text{m}}{\text{s}}$$

From Equations 6.31 and 6.32

$$x\text{-direction}: V_{1x} = V_1 \cos\alpha_1 = V_B + V_{r1}\cos\theta_1$$

$$V_{r1}\cos\theta_1 = V_1 \cos\alpha_1 - V_B = 500 \times \cos 15 - 241.5 = 241.5 \ \frac{\text{m}}{\text{s}}$$

$$y\text{-direction}: V_{1y} = V_1 \sin\alpha_1 = V_{r1}\sin\theta_1 = 500 \times \sin 15 = 129.4 \ \frac{\text{m}}{\text{s}}$$

Dividing the last two equations:

$$\tan\theta_1 = \frac{129.4}{241.5} = 0.536 \quad \text{so} \quad \theta_1 = 28.19°$$

$$V_{r1} = \frac{129.4}{\sin 28.18} = 274.0 \ \frac{\text{m}}{\text{s}}$$

Now all the information needed to draw the inlet velocity vector diagram is known. Let us determine similar information for the outlet velocity vector diagram.

Since the flow is frictionless,

$$V_{r2} = V_{r1} = 274.0 \frac{\text{m}}{\text{s}}$$

The blade is symmetrical.

$$\theta_1 = \theta_2 = 28.19°$$

From Equations 6.33 and 6.34

$$x\text{-direction}: V_{2x} = V_2 \cos\alpha_2 = V_B - V_{r2}\cos\theta_2 = 241.5 - 274.0\cos\ 28.19 = 0$$

$$y\text{-direction}: V_{2y} = V_2 \sin\alpha_2 = V_{r2}\sin\theta_2 = 274.0\ \sin\ 28.19 = 129.4 \frac{\text{m}}{\text{s}}$$

$$\tan\alpha_2 = \frac{129.4}{0} = \infty \, \text{so} \, \alpha_2 = 90$$

$$V_2 = \frac{V_{r2}\sin\theta_2}{\sin\alpha_2} = \frac{274.0\sin 28.18}{\sin 90} = 129.4 \ \frac{\text{m}}{\text{s}}$$

Using Equation 6.39, the power output can be determined as follows for the unit of mass flow rate $(\dot{m} = 1\ \text{kg/s})$:

$$\dot{W} = \dot{m}V_B(V_1\cos\alpha_1 - V_2\cos\alpha_2) = 1 \times 241.5\ (500\ \times \cos 15 - 129.39\ \times \cos 90)$$
$$= 116{,}627\ W = 116.6\,\text{kW}$$

Alternatively, from Equation 6.47

$$\dot{W}_{\max} = \frac{1}{2}\dot{m}(V_1\cos\alpha_1)^2 = 0.5 \times 1\ \times (500\ \times \cos 15)^2 = 116{,}626\ W = 116.6\ \text{kW}$$

Also from Equation 6.50

$$\dot{W} = \frac{1}{2}\dot{m}\left(V_1^2 - V_2^2\right) = 0.5 \times 1 \times \left(500^2 - 129.39^2\right) = 116{,}627\ \text{W} = 116.6\ \text{kW}$$

For the efficiency of the blade

$$\eta_B = \frac{\dot{W}}{\frac{1}{2}\dot{m}V_1^2} = \frac{116{,}627}{0.5 \times 1 \times 500^2} = \frac{116{,}627}{125{,}000} = 0.933 \text{ or } 93.3\%$$

This example clearly shows that in the ideal impulse turbine, the velocity of the flow relative to the blades in the inlet and outlet streams remains constant but their directions change. However, the absolute velocity of the flow reduces significantly. In actual turbines, the relative velocity of the flows also slightly reduces due to fluid friction. In order to take the friction into account, the **velocity coefficient**, k_v, is defined as the ratio of the outlet relative velocity to the inlet relative velocity.

$$k_v = \frac{V_{r2}}{V_{r1}} \quad V_{r2} < V_{r1} \tag{6.51}$$

Example 6.5

Repeat Example 6.4 for the velocity coefficient of 95%.

AnalysisThe inlet flow conditions are the same and the blade is symmetrical, so

$$\theta_1 = \theta_2 = 28.19^\circ$$

$$V_{r1} = 274.0\ \frac{\text{m}}{\text{s}}$$

$$V_{B,\text{opt}} = 241.5\,\text{m/s}$$

Since the flow has friction, the relative velocity of the flow in the outlet is smaller than the relative velocity in the inlet. The exit relative velocity can be found using the velocity coefficient.

$$V_{r2} = k_v V_{r1} = 0.95 \times 274.0 = 260.3\frac{\text{m}}{\text{s}}$$

$$x\text{-direction}: V_{2x} = V_2 \cos\alpha_2 = V_B - V_{r2}\cos\theta_2 = 241.5 - 260.3\ \cos\ 28.19 = 12.1\frac{\text{m}}{\text{s}}$$

$$y\text{-direction}: V_{2y} = V_2 \sin\alpha_2 = V_{r2}\sin\theta_2 = 260.3\ \sin\ 28.19 = 122.9\ \frac{\text{m}}{\text{s}}$$

$$\tan\alpha_2 = \frac{122.9}{12.1} = 10.2 \quad \text{so} \quad \alpha_2 = 84.4°$$

$$V_2 = \frac{V_{r2}\sin\theta_2}{\sin\alpha_2} = \frac{260.3\sin 28.19}{\sin 84.4} = 123.5\ \frac{\text{m}}{\text{s}}$$

$$\dot{W} = \dot{m}V_B\left(V_1\cos\alpha_1 - V_2\cos\alpha_2\right) = 1\times 241.5\ \left(500\ \times\ \cos 15 - 123.5\times\cos 84.4\right)$$

$$= 113{,}711\,W = 113.7\ \text{kW}$$

Note that neither Equation 6.47 nor Equation 6.50 can be used because of the loss.

For the efficiency of the blade

$$\eta_B = \frac{\dot{W}}{\frac{1}{2}\dot{m}V_1^2} = \frac{113{,}711}{0.5\times 1\ \times 500^2} = \frac{113{,}711}{125{,}000} = 0.9097 \text{ or } 90.97\%$$

As expected, the efficiency of the blade reduces due to the friction loss.

All these examples indicate that the force exerted on a blade and its output power are functions of the inlet flow velocity. So in order to maximize the power output of an impulse blade, the inlet velocity should be increased as much as possible. This means that the inlet steam pressure should be converted to the velocity (the kinetic energy) before it strikes the blades. For this purpose, there are **nozzles**, typically in the form of a set of stationary blades acting as nozzles, ahead of the rotating blades where the velocity of the flow increases while the pressure reduces (Figure 6.8). The maximum flow velocity is in the outlet of the nozzles. In the case of impulse blades, the entire steam expansion, which is manifested in the pressure drop of the stage, occurs in the stationary blades (nozzles). Note that the enthalpy of the steam also reduces in the nozzles while it remains unchanged through the rotating impulse blade. Figure 6.8 illustrates that the pressure of the steam remains unchanged through the rotating blades, which is why these types of turbines (or stages of turbines) are called **equal pressure turbines** (or stages). In order to keep the pressure unchanged in the rotating blades, the area between the blades should remain unchanged; otherwise the blades will act as a nozzle or diffuser, depending on whether the area reduces or increases. This is the reason for the symmetrical shape of impulse blades. This is also why the entrance and exit angles of these blades are equal.

The efficiency of fixed blades, rotating blades, and fixed and rotating blades combined (a stage) of an impulse turbine is defined as follows:

$$\eta_{\text{fixed}} = \frac{\Delta h_{\text{actual}}}{\Delta h_{\text{isentropic}}} \tag{6.52}$$

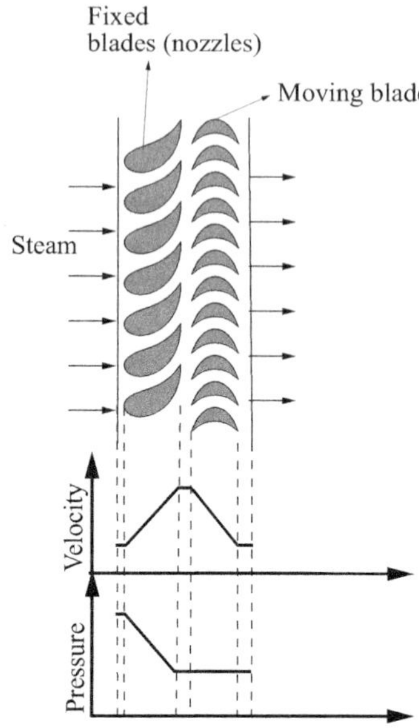

FIGURE 6.8
Stationary and rotating blades in an impulse blade, the pressure and absolute velocity variation across an impulse turbine (note that the real-world variation may not be linear as shown here for simplicity).

$$\eta_{\text{rotat}} = \frac{\dot{W}}{KE_{\text{outlet fixed blade}}} = \frac{\dot{W}}{KE_{\text{inlet rotating blade}}} \tag{6.53}$$

$$\eta_{\text{stage}} = \frac{\dot{W}}{\Delta \dot{H}_{\text{stage}}} = \frac{\dot{W}}{\Delta \dot{H}_{\text{fixed}}} = \frac{\dot{W}}{\dot{m}\Delta h_{\text{fixed}}} \tag{6.54}$$

6.2.3 Single-Stage Impulse Turbines

Karl Gustaf Patrik de Laval (1845–1913, a Swedish engineer) first introduced the idea of impulse turbines in 1882 and built a small single-stage demonstrational impulse turbine, known as the de Laval turbine, in 1887. In order to convert the steam pressure to velocity, he developed a supersonic nozzle for his turbine. One or several of these nozzles fed the high-velocity steam to the rotating blades (Figure 6.9). This meant that the nozzles did not cover all blades, and not all the blades were struck by the jets at a given time. The nozzles were also used to control the steam flow rate through the turbine by shutting some of the nozzles. Both the steam turbine and the nozzle bear his name. In order to withstand the engines' high rotational speed (up to 30,000 rpm), he employed oil-fed bearings. This speed was reduced to 3000 rpm by a gearing system. Then, he developed a centrifugal separator to separate steam from oil. In the early 1890s, he successfully built commercial steam turbines

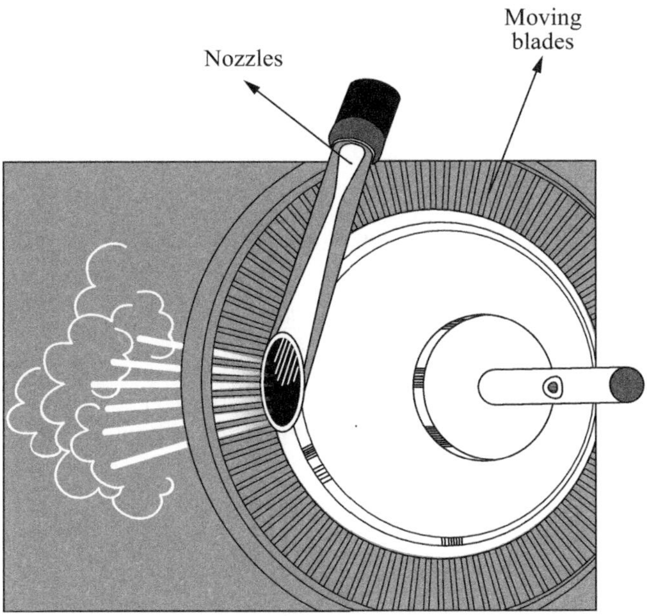

FIGURE 6.9
Single-stage impulse turbine or de Laval turbine.

with a capacity of first 3.7 kW (5 hp) and then 11.2 kW (15 hp) for marine applications. The problem with his engines was that due to their high rotational speed and thus high centrifugal forces, materials available then (and now) were not strong enough for large power production units. This problem limited the size and capacity of his turbines.

Example 6.6

1200 kg/s superheated steam from a steam generator enters a hypothetical single-stage impulse turbine at the pressure of 150 bar and the temperature of 500°C. The turbine discharges the exit steam to the condenser at the pressure of 0.2 bar. The isentropic efficiency of the turbine is 88% and the velocity coefficient of the blade is 98%. The outlet angle of the stationary blades (nozzles) is 18°. The rotating blades are symmetrical and are rotating at the optimum velocity. Draw the velocity vector diagram for the inlet and outlet flows and determine the blade entrance and exit angles, the maximum output power of the blade, the rotating blade efficiency, and the stage efficiency.

Stationary Let us determine the blade optimum velocity first. For the inlet steam at the pressure and temperature of 150 bar and 500°C, respectively, the enthalpy and entropy of the steam can be found from the steam property table (superheated steam table, note that depending on the table that you use, you may get slightly different values):

$$h_{\text{in}} = 3310.8 \frac{\text{kJ}}{\text{kg}}$$

$$s_{\text{in}} = 6.3479 \frac{\text{kJ}}{\text{kg K}}$$

The turbine discharges the steam to the condenser at the pressure of 0.2 bar. The isentropic efficiency of the turbine is defined as

$$\eta_{\text{turbine}} = \frac{\Delta h}{\Delta h_s} = \frac{h_{\text{in}} - h_{\text{out}}}{h_{\text{in}} - h_{\text{out}_s}} \tag{6.55}$$

where the subscript s represents the isentropic expansion process between the inlet flow and the outlet flow of the turbine. So,

$$p_{\text{out}} = 0.2\,\text{bar}$$

$$s_{\text{out}_s} = s_{\text{in}} = 6.3479 \frac{\text{kJ}}{\text{kg K}}$$

By interpolation using the saturated steam property table,

$$h_{\text{out}_s} = 2089.4 \frac{\text{kJ}}{\text{kg}}$$

Note that the quality of steam in this state will be about 78% which is well below the acceptable level. The steam reheating in real steam cycles can partially address this problem.

$$0.88 = \frac{\Delta h}{\Delta h_s} = \frac{3310.8 - h_{\text{out}}}{3310.8 - 2089.4}$$

$$h_{\text{out}} = 3310.8 - (3310.8 - 2089.4) \times 0.88 = 2236 \frac{\text{kJ}}{\text{kg}}$$

Since in an impulse turbine, the pressure and enthalpy of the steam in the rotating blade do not change, the pressure and enthalpy of the stationary blade exit steam are the same as the pressure and enthalpy of the turbine exit flow.

Writing the first law of thermodynamics for the fixed blades (nozzles):

$$h_{f1} + \frac{V_{f1}^2}{2} = h_{f2} + \frac{V_{f2}^2}{2} \tag{6.56}$$

$$\frac{V_{f2}^2}{2} = h_{f1} + \frac{V_{f1}^2}{2} - h_{f2} \tag{6.57}$$

Ignoring the velocity of the inlet flow to the turbine, which is typically an acceptable assumption,

$$\frac{V_{f2}^2}{2} = h_{f1} - h_{f2}$$

$$V_{f2} = \sqrt{2(h_{f1} - h_{f2})} = \sqrt{2(3310.8 - 2236)\left[\frac{\text{kJ}}{\text{kg}}\right] \times \frac{1000\ [\text{J}]}{1\ [\text{kJ}]}} = 1375.4\ \frac{\text{m}}{\text{s}}$$

The outlet flow leaving the nozzle is the inlet flow to the rotating blades, so

$$V_1 = V_{f2} = 1375.4\ \frac{\text{m}}{\text{s}}$$

Also, the angle between the blade velocity and the absolute velocity of the inlet flow is

$$\alpha_1 = 18°$$

The blade optimum velocity is

$$V_{B,\text{opt}} = \frac{1}{2} V_1 \cos\alpha_1 = 0.5 \times 1375.4 \times \cos 18 = 654.0\ \frac{\text{m}}{\text{s}}$$

$$x\text{-direction}: V_{1x} = V_1 \cos\alpha_1 = V_B + V_{r1}\cos\theta_1$$

$$V_{r1}\cos\theta_1 = V_1\cos\alpha_1 - V_B = 1375.4 \times \cos 18 - 654.0 = 654.0\ \frac{\text{m}}{\text{s}}$$

$$y\text{-direction}: V_{1y} = V_1 \sin\alpha_1 = V_{r1}\sin\theta_1 = 1375.4 \times \sin 18 = 425.0\ \frac{\text{m}}{\text{s}}$$

Dividing last two equations:

$$\tan\theta_1 = \frac{425.0}{654.0} = 0.650 \quad \text{so} \quad \theta_1 = 33.02°$$

$$V_{r1} = \frac{425.0}{\sin 33.02} = 780.0\ \frac{\text{m}}{\text{s}}$$

Now all the information needed to draw the inlet velocity vector diagram (on a real scale) is known. Let us determine similar information for the outlet velocity vector diagram. Since the flow has friction, the relative velocity of the flow in the outlet is smaller than the relative velocity in the inlet. The exit relative velocity can be found using the velocity coefficient.

$$V_{r2} = k_v V_{r1} = 0.98 \times 780.0\frac{\text{m}}{\text{s}} = 764.4\frac{\text{m}}{\text{s}}$$

$$x\text{-direction}: V_{2x} = V_2 \cos\alpha_2 = V_B - V_{r2}\cos\theta_2 = 654.0 - 764.4 \cos 33.02 = 13.1 \frac{\text{m}}{\text{s}}$$

$$y\text{-direction}: V_{2y} = V_2 \sin\alpha_2 = V_{r2}\sin\theta_2 = 764.4 \sin 33.02 = 416.5 \frac{\text{m}}{\text{s}}$$

$$\tan\alpha_2 = \frac{416.5}{13.1} = 31.8 \quad \text{so} \quad \alpha_2 = 88.20°$$

$$V_2 = \frac{V_{r2}\sin\theta_2}{\sin\alpha_2} = \frac{764.4 \sin 33.02}{\sin 88.20} = 416.7 \frac{\text{m}}{\text{s}}$$

$$\dot{W} = \dot{m} V_B \left(V_1 \cos\alpha_1 - V_2 \cos\alpha_2\right)$$

$$= 1200 \times 654.0 \left(1375.4 \times \cos 18 - 416.7 \times \cos 88.20\right)$$

$$= 1{,}016{,}371{,}203 \ W = 1{,}016.4 \text{ MW}$$

For the efficiency of the blade,

$$\eta_B = \frac{\dot{W}}{\frac{1}{2}\dot{m}V_1^2} = \frac{1{,}016{,}371{,}203}{0.5 \times 1200 \times 1375.4^2} = 0.8955 \text{ or } 89.55\%$$

For the efficiency of the stage,

$$\eta_{\text{stage}} = \frac{\dot{W}}{\dot{m}\Delta h_{\text{fixed}}} = \frac{1{,}016{,}371{,}203 \ W}{1200 \left[\frac{\text{kg}}{\text{s}}\right] (3310.8 - 2236) \left[\frac{\text{kJ}}{\text{kg}}\right] \frac{1000 \ [\text{J}]}{1 \ [\text{kJ}]}}$$

$$= 0.7880 \text{ or } 78.80\%$$

Question: Determine the rotational speed of the rotor if the above calculations were for the blade with diameter of 1 m (including the shaft diameter and the blade height).

If the diameter of the blade is assumed to be 1 m, the rotational speed of the shaft will be

$$\omega = \frac{V_B}{r} = \frac{654.0}{0.5} = 1308 \frac{\text{rad}}{\text{s}} \tag{6.58}$$

$$\text{Rotational speed} = \frac{\omega}{2\pi} = \frac{1308}{2\pi} = 208.2 \frac{\text{revolutions}}{\text{s}} = 12491.2 \text{ rpm} \tag{6.59}$$

The blade velocity and the shaft rotational speed that were calculated for the previous example are well beyond what a typically modern turbine can tolerate. Due to high stresses caused by centrifugal forces as well as other considerations including the generator rpm and the frequency of electricity, the rotational speed of steam turbines is typically limited to less than about 4000 rpm.[2] Furthermore, flow friction losses are proportional to the square of the flow velocity. So such a high flow velocity (Example 6.6) will result in unacceptable efficiency reduction due to high friction losses. As noted earlier, these problems severely limited the size and capacity of the de Laval turbines. To address these problems, two methods have been used: **velocity compounding** and **pressure compounding**.

6.2.4 Velocity-Compounded Impulse Turbines

In this type of turbines, there are **two or more stages of impulse rotating blades**. So unlike de Laval turbines, the kinetic energy (and velocity) reduces through more than one stage, which reduces the blade velocity of each stage. This can potentially eliminate the problem of unacceptably high blade velocity that de Laval turbines faced. Charles Gordon Curtis (1860–1953, an American engineer) developed the first velocity-compounded impulse stage, which is named after him, Curtis stage turbines.

In velocity-compounded impulse steam turbines, there are **a nozzle stage at the entrance of the turbine, several stages of rotating blades**, and **a stage of fixed blades between rotating stages** (Figure 6.10). In the nozzle stage, the steam pressure and enthalpy are reduced and its velocity is increased. In fact the **entire pressure reduction of the turbine occurs in this nozzle stage,** and the **steam pressure remains unchanged in the rest of the turbine**. The velocity and kinetic energy are maximum at the outlet of the nozzle stage. This high velocity of steam is a major problem for this type of turbine. As noted before, friction losses in a steam flow are proportional to the square of the flow velocity. Therefore, the high velocity in the outlet of the nozzle results in high friction losses, which reduces the overall turbine efficiency. After the nozzles, the steam is fed to the first stage of the rotating blades where some power is produced because of the momentum exchange between the blades and the steam flow (Figure 6.10). While the velocity reduces in the rotating blades, the pressure remains constant. The relative velocity of the steam also remains unchanged in the rotating blades for a frictionless flow. For an actual flow, there is a slight reduction in the relative velocity due to the friction that can be accounted for by the velocity coefficient for each rotating blade. For example, the velocity coefficient for the first stage of the rotating blades in Figure 6.10 is defined as $k_{v1} = \dfrac{V_{r2}}{V_{r1}}$.

The only function of the fixed blades between the rotating stages is to redirect the flow so it can enter the second stage of the rotating blades at a correct angle. So for a frictionless flow, the kinetic energy, absolute velocity,

[2] Note that there are turbines, such as micro gas turbines, that operate at much higher rpm, even over 100,000 rpm. But they are typically limited to small turbines.

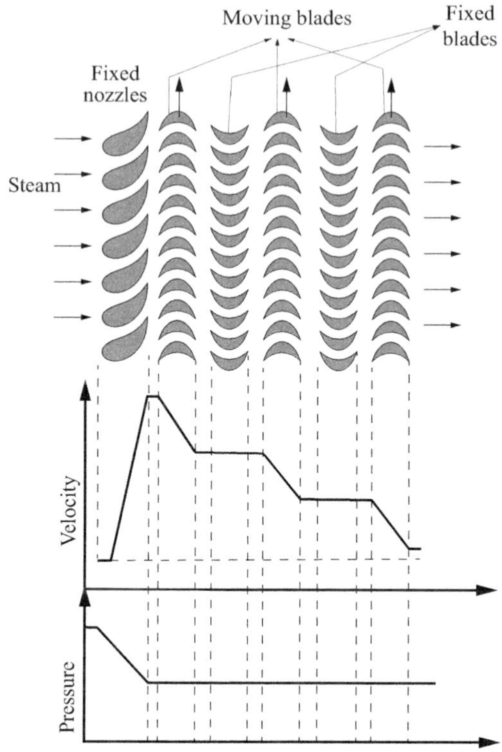

FIGURE 6.10
Schematic diagram of a three-stage velocity-compounded impulse turbine along with pressure and velocity diagrams.

and pressure remain constant in these fixed stages. For a flow with friction, the absolute velocity slightly reduces in these blades. This reduction can be taken into account by the velocity coefficient defined for the fixed blade stage. For example, for the fixed blade stage after the first rotating blades in Figure 6.10, the velocity coefficient can be written as $k_{v2} = \frac{V_{s3}}{V_{s2}}$. The remaining kinetic energy of the steam is extracted in the remaining stages of the rotating blades until the velocity reduces to the turbine exit velocity.

Theoretically, in these turbines, there can be any number of moving stages. The early rotating and fixed blades where the flow has high velocity are curvier and thicker compared to those in the later low velocity blades that are flatter and thinner.

For a velocity-compounded turbine with n rotating impulse stages since the kinetic energy of the steam is extracted in n stages, inspired by Equation 6.46 for single-stage impulse turbines, the optimum blade velocity for the frictionless flow turbine can be determined by the following equation:

$$V_{B,\text{opt}} = \frac{1}{2n} V_1 \cos \alpha_1 \tag{6.60}$$

Note that in this case, the absolute velocity of the final turbine exit flow (and not exit from each rotating blade) is perpendicular to the blade velocity vector and is parallel to the axis of rotation of the turbine (an axial flow).

In the velocity-compounded impulse turbines, the total power production of the turbine is divided among the rotating stages. However, the share of each stage is not equal. The earlier high-velocity stages extract much more power than those of the later low-velocity stages. It can be shown that the ratio of the power extracted in each stage are as follows (form high to low velocity blades):

Two-rotating-stage turbines: 3:1

Three-rotating-stage turbines: 5:3:1

Four-rotating-stage turbines: 7:5:3:1

This indicates that the lower velocity stages contribute little to the total power produced in the turbine. In fact, there is no economical justification to add stages beyond a certain number of stages, perhaps over two stages. This problem significantly limits the applications of the velocity-compounded turbines. Also, as noted earlier, friction losses are high, particularly in the early stages. The pressure-compounded impulse turbines can address some of these problems.

6.2.5 Pressure-Compounded Impulse Turbines

Similar to velocity-compounded impulse turbines, pressure-compounded impulse turbines are composed of several rotating impulse stages (Figure 6.11). However, unlike velocity-compounded turbines, in pressure-compounded turbines, all stationary stages (and not just the first one) are nozzles. This means the steam is expanded as well as redirected in these stages. This also means all velocity increase and pressure (enthalpy) drop do not occur in the nozzles at the entrance of the turbine. Instead, the entire pressure (enthalpy) drop of the turbine is divided among several stages of the fixed nozzles. In fact in this configuration, each stage is like an independent impulse stage. This configuration eliminates both problems of velocity-compounded turbines. Since the velocity increase occurs in several stages, the velocity of the steam after the nozzle at the turbine entrance does not increase to such a high level that it can create high friction losses. Furthermore, the power produced in each stage can be controlled by the designer. Auguste Rateau (1863–1930,

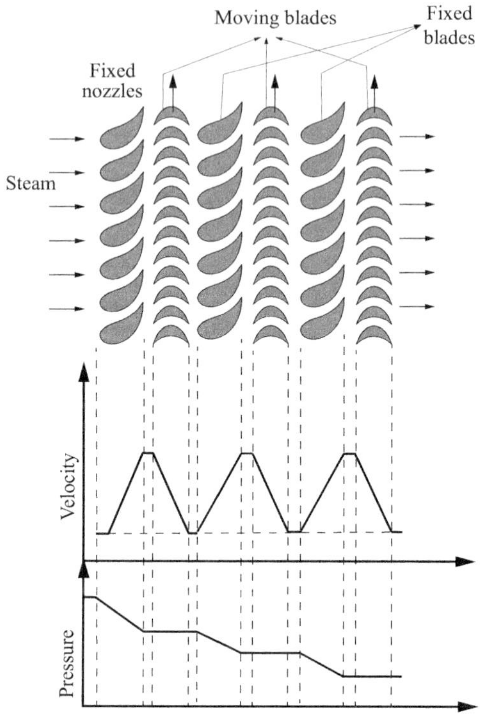

FIGURE 6.11
Schematic diagram of a three-stage pressure-compounded impulse turbine along with pressure and velocity diagrams.

a French engineer) invented this type of multistage impulse turbines, which was named after him **Rateau turbines** or Rateau-stages.

Typically, the enthalpy drop in pressure-compounded impulse turbines is divided equally among the stages. This means the inlet absolute velocity to each rotating stage is almost equal. However, the pressure drop among the stages is not equal. Using the first law of thermodynamics for the nozzles (neglecting the inlet absolute velocity to the first nozzle), the inlet velocity to the moving stages can be calculated by the following equation:

$$V_1 = V_2 = \cdots = V_n = \sqrt{\frac{2\Delta h_{\text{total}}}{n}} \tag{6.61}$$

where V_1, V_2, ..., V_n are the absolute velocities of the nozzle outlet flows and the absolute velocities of the inlet flows to the rotating stages, Δh_{total} is the entire specific enthalpy drop of the turbine, and n is the number of stages in the turbine. Note that in the actual design process, this equation cannot be used because the inlet absolute velocity is required for the flow to enter the nozzles and cannot be ignored. However, Equation 6.61 can provide a very good starting point for calculations.

Essentially, each stage (combined stationary and rotational blades) is similar to a single-stage impulse turbine but with a much smaller enthalpy drop. Therefore, all characteristics of single-stage impulse turbines can be applied for each stage of pressure-compounded turbines. For instance, ideally, the absolute velocity of the turbine exit flow from each rotating blade is perpendicular to the blade velocity vector and is parallel to the axis of rotation of the turbine. Another compounding possibility is to have a composite compounding where several velocity-compounded stages are combined through pressure compounding.

In summary, the advantages of pressure-compounded impulse turbines over velocity-compounded and single-stage impulse turbines can be listed as follows:

- The steam absolute velocity is lower throughout the turbine, particularly in the early stages, which results in less friction losses.
- The blade velocity, i.e., the rotational speed of the turbine, is lower.
- The share of each stage in the output power can be controlled by designers and is commonly roughly equal among stages.
- These turbines have higher efficiency.

6.3 Reaction Turbines

Unlike impulse forces that are based on the collision of objects, in reaction forces, there is no need for direct impact. We encounter devices and equipment operating based on reaction forces in our daily life from whirling water sprinklers to jet engine airplanes and space rockets. In these devices, the flow pressure drops and its velocity increases through the device, which create the reaction forces on the flow. The corresponding force exerted on the device in the opposite direction and equal in magnitude (according to the third law of Newton) is called the **reaction force**.

The first recorded turbine in history was a pure reaction turbine, known as an aeolipile, built by Hero of Alexandria (Greek mathematician and engineer who lived in Roman Egypt, c. 10–c. 70 AD) in the 1st century AD. It was similar to a lawn sprinkler but steam tangentially exited the nozzles instead of water (Figure 6.12). The steam was generated in the closed container at the bottom and was transferred to the nozzles via tubes that acted as the support for the turbine as well.

Much later in the 16th and 17th centuries, several authors described devices, mainly roasting jacks (a machine to rotate meat roasting on a spit), operating based on the reaction steam turbine principle. There were several attempts to build steam turbines similar to water turbines but all of them failed except the one developed by William Avery (an American mechanic). He patented a device similar to the Hero's reaction turbine. It was made of

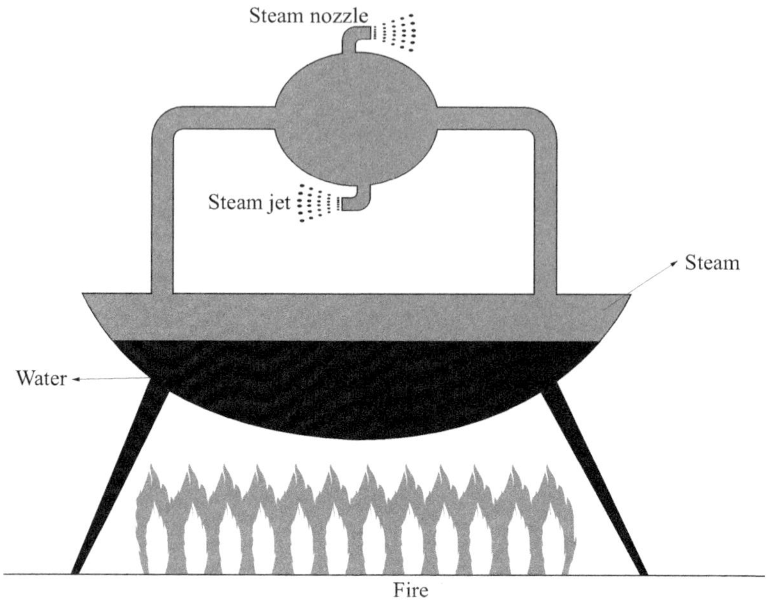

FIGURE 6.12
Hero of Alexandria's steam reaction turbine (note that the steam jets are perpendicular to the page, one outward and one inward).

a hollow cylinder that could rotate around its axis. Steam exited through two arms that discharged it at a right angle relative to the cylinder axis at opposite directions (Figure 6.13). For practical reasons, such as high noise and maintenance, his steam engine was never widely used.

Sir Charles Algernon Parsons (1854–1931, a British engineer) developed the first modern reaction turbine in 1884 with a capacity of 7.5 kW and an efficiency of less than 2%. In order to address the problem of high rotational

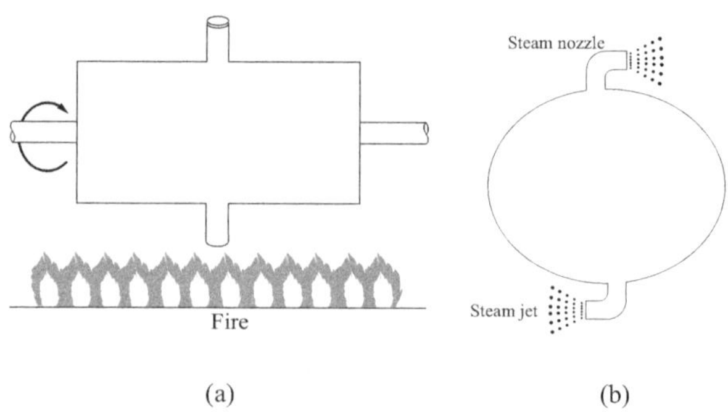

FIGURE 6.13
Schematic of Avery's steam reaction turbine. (a) Front view, (b) side view.

speed, which limited the capacity of de Laval's impulse turbines, Parsons' turbines had multiple stages, which divided the expansion process to small steps, so the steam speed never increases to an unacceptable level. The capacity and performance of Parsons' turbines were quickly improved. In 1890, the first power plant equipped with two generators each with 75 kW electricity generation capacity was installed. In 1899, the first-megawatt power plant was installed in Germany employing a Parsons' turbine. George Westinghouse (1846–1914, an American entrepreneur and engineer) installed the first turbine-based power plant in Pennsylvania. These turbines were also successfully installed in many ships and broke many records.

Reaction turbines, similar to impulse turbines, have both fixed and moving blades. It was discussed that in impulse turbines all enthalpy (and pressure) drop takes place in the fixed blades and the enthalpy and pressure are constant over the moving blades (except for the pressure drop due to friction). In reaction turbines however **pressure reduces both through the fixed and moving blades**. The pressure drop in fixed nozzles creates the impulse force when the flow strikes the moving blades. The pressure drop in the moving blades, which causes the flow acceleration (increase in the velocity), creates the reaction force. Note that if there is no momentum change in the blades, which means the direction of the flow does not change in the blades (as in Hero's turbine), the turbine will be a pure reaction turbine. In reality, there is **no commercial pure reaction turbine**. Although we call them reaction turbines, they are in fact a combination of reaction and impulse turbines. A 50% reaction turbine stage is one where the total enthalpy drop is equally divided between the fixed and moving blades. Sometimes turbine stages are named according to the variation of pressure through them, **equal pressure for impulse stages** and **unequal pressure for reaction stages.**

Figure 6.14 demonstrates the variation of pressure and absolute velocity throughout a series of reaction blades. While the pressure of the steam decreases, the absolute velocity increases as the steam passes through the fixed blades. In the moving blades, the pressure continues to decrease and the absolute velocity also drops (note the similarity of the velocity change in the reaction stages and pressure-compounded turbines in Figure 6.11). As shown in Figure 6.14, since both the fixed and moving blades are acting as nozzles, the passage areas between the blades should reduce along both stages. This results in **asymmetrical blades for both moving and fixed blades**. Note that they are curved in the **opposite direction** because the fixed blades should also redirect the steam so that it can enter the next stage of the moving blades at an appropriate entrance angle.

Note that although the relative velocity increases in the moving blades (as the flow accelerates), the absolute velocity reduces, which is in agreement with the first law of thermodynamics. Using a similar approach that was already used for impulse turbines, it can be shown that the blade velocity for the maximum power outlet and the maximum power generation in reaction

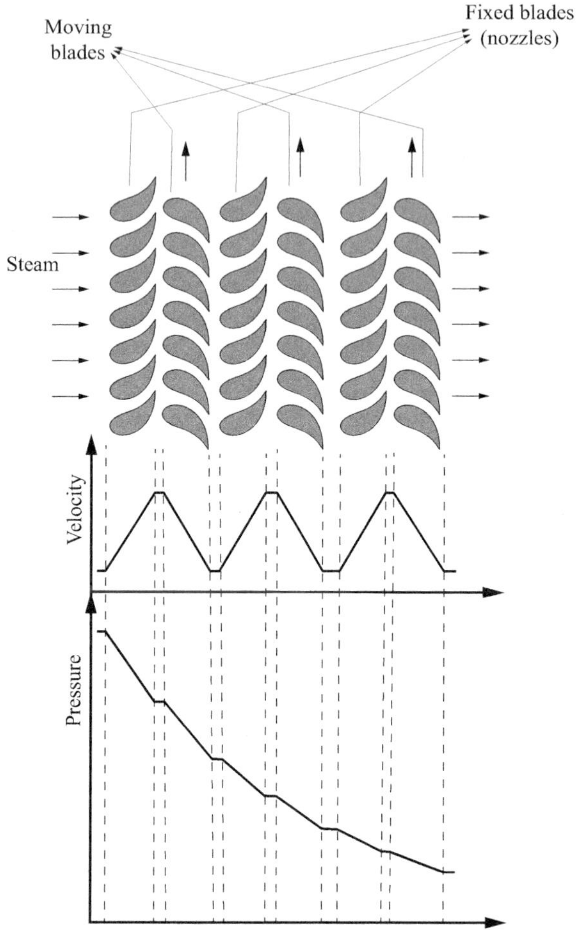

FIGURE 6.14
Variations of pressure and velocity of steam in a three-stage reaction turbine.

turbines can be determined by the following equations (the terms in these equations are presented in Figure 6.7):

$$V_{B,\text{opt}} = V_1 \cos\alpha_1 \tag{6.62}$$

$$\dot{W}_{\text{max}} = \dot{m} \tag{6.63}$$

Comparing Equations 6.62 and 6.63 for reaction turbines with Equations 6.46 and 6.47 for impulse turbines indicates that the reaction turbines' maximum rotational speed and maximum power generation are twice those in impulse turbines for the same given inlet flow conditions.

6.4 Modern Utility Steam Turbines

To understand the reasons behind the configuration of modern steam turbines, which is typically a combination of impulse and reaction stages, we need to first compare and contrast the characteristics of these two types of stages. There are several distinguishing factors that differentiate impulse and reaction stages. In this section, we will review these factors.

6.4.1 Impulse Stages versus Reaction Stages

The first distinction is the pressure variation throughout moving blades. We noted that while the steam pressure remains constant through the moving blades in impulse stages, it reduces in reaction stages. This means in an impulse stage the steam pressure before and after the moving stage is equal. However, in a reaction stage, the steam pressure after the moving stage is lower than that before the moving stage. This pressure gradient creates a potential for steam leakage. In actual turbines, there are always some gaps between the tip of moving blades and the turbine casing to avoid their contact, which can be catastrophic and can destroy the turbine. Due to the pressure difference between before and after the moving stage in reaction stages, some steam will always go through this gap bypassing the blades. This steam leakage negatively impacts the output power and efficiency of the stage since the leaked steam does not contribute to the stage power generation. The magnitude of the steam leakage depends on the **blade tip clearance** and the **pressure gradient across the stage**. While designers and manufacturers have tried to reduce this gap to as little as possible, there is a limit for this reduction, particularly considering temperature variations and thermal expansion between cold start-up and full load operation. To reduce the steam leakage, the pressure drop across a reaction stage should be limited. For this reason, for the same power output, reaction turbines need more stages than that required for impulse turbines.

Now let us consider the optimum rotational speed and maximum output power of impulse and reaction turbines. From dividing two sides of Equations 6.46 and 6.62 and Equations 6.47 and 6.63,

$$\frac{V_{(B,\text{opt})\,\text{Im}}}{V_{(B,\text{opt})\,\text{Re}}} = \frac{1}{2} \times \frac{V_{1,\text{Im}}}{V_{1,\text{Re}}} \times \frac{\cos\alpha_{1,\text{Im}}}{\cos\alpha_{1,\text{Re}}} \tag{6.64}$$

$$\frac{\dot{W}_{(\max)\,\text{Im}}}{\dot{W}_{(\max)\,\text{Re}}} = \frac{1}{2}\left(\frac{V_{1,\text{Im}}}{V_{1,\text{Re}}} \times \frac{\cos\alpha_{1,\text{Im}}}{\cos\alpha_{1,\text{Re}}}\right)^2 \tag{6.65}$$

These two equations indicate that for the given inlet steam conditions (the same V_1 and $\cos\alpha_1$), the optimum blade velocity (and the optimum turbine rotational speed) and the maximum power generated in the reaction stage double those in the impulse stage.

$$V_{1,\text{Im}} = V_{1,\text{Re}} \quad \text{and} \quad \cos\alpha_{1,\text{Im}} = \cos\alpha_{1,\text{Re}} \tag{6.66}$$

$$\frac{V_{(B,\text{opt})\ \text{Im}}}{V_{(B,\text{opt})\ \text{Re}}} = \frac{1}{2} \tag{6.67}$$

$$\frac{\dot{W}_{(\text{max})\ \text{Im}}}{\dot{W}_{(\text{max})\ \text{Re}}} = \frac{1}{2} \tag{6.68}$$

However, in practical applications, the limiting parameters are the blade speed and the turbine rotational speed due to the centrifugal forces that the blade and shaft materials can withstand. For a given blade speed, the steam velocity and the maximum power generated in the impulse stage double those in the reaction stage. From Equations 6.64 and 6.65,

$$V_{(B,\text{opt})\ \text{Im}} = V_{(B,\text{opt})\ \text{Re}} \tag{6.69}$$

$$\frac{V_{1,\text{Im}}}{V_{1,\text{Re}}} \times \frac{\cos\alpha_{1,\text{Im}}}{\cos\alpha_{1,\text{Re}}} = 2 \tag{6.70}$$

$$\frac{\dot{W}_{(\text{max})\ \text{Im}}}{\dot{W}_{(\text{max})\ \text{Re}}} = 2 \tag{6.71}$$

This is another reason why reaction turbines need more stages than impulse turbines to extract the same amount of power. This means that impulse turbines can produce the same amount of power in **fewer stages** and in a shorter turbine compared to those needed for reaction turbines. Note that in Equations 6.64–6.71 the fluid is assumed to be frictionless. For real fluids, these values might be different. However, the similar general trends can be observed in real steam turbines too.

For the stage efficiency, it can be shown that for a frictionless fluid, **reaction stages have much higher efficiency than that of impulse stages**. Since the velocity of the steam flow is lower in reaction stages (Equation 6.70), the friction losses are lower in these stages, which further improves their efficiency.

6.4.2 Configuration of Modern Steam Turbines

Now we can discuss the configuration of modern steam turbines. Typically, a combination of impulse and reaction turbines is used in modern steam turbines. The first few stages are typically impulse stages followed by many reaction stages. To explain the reasons for these configurations, first it should be noted that the **length of blades increases with the progression of steam** in the turbine. In early stages (high-pressure stages), the blades are short while the longest blades are in the last stage of the turbine. The reason is that when the steam flow moves along the stages of the turbine, it expands. As a result, its specific volume increases and its density decreases. This is reflected in the progressively increasing volumetric flow rate of the steam in the turbine. For the same mass flow rate, to keep the velocity of the steam flow at an acceptable range,[3] the area perpendicular to the flow should increase (according to the equation $\dot{m} = \rho AV$). Clearly, the larger area can only be achieved with longer blades. This is important for the selection of the type of a stage. We recalled from our earlier discussion that in reaction stages, there are always some steam leakages from the clearance between the tip of the blades and the casing. For short blades, the gap can be significant compared to the total length of the blade. This means for the same operational conditions the percentage of the total flow that bypasses short blades is much larger than that of longer blades. Furthermore, the fact that shorter blades are located at early stages of the turbine operating at high pressures, where the density of the steam is higher, intensifies this problem (again according to the above mass flow rate equation). This means that if reaction stages are used in the high-pressure early stages of the turbine, a considerable percentage of steam will bypass the blades. The leakage can severely impact the efficiency of these stages. This is why the first few stages of large steam turbines are typically impulse stages. These stages are followed by many reaction stages due to their higher efficiency. The degree of reaction in the reaction stages progressively increases. Note that the longer length of reaction blades increases their power generation, which can help to make up for the inherently lower power generation of reaction stages (Equations 6.47 and 6.63).

In summary, in utility steam turbines the high-pressure end is composed of few stages of short symmetrical impulse blades followed by increasing longer and more asymmetrical reaction blades at intermediate and low-pressure sections. It should be noted that the above discussion is rather a general one and each manufacturer can have a different design strategy on the degree of reaction for each stage of their turbine. Some manufacturers for example use low percentage reaction stages (perhaps 10% reaction and 90% impulse) at the high-pressure end of their turbines. Finally, it should be remembered that this configuration is only for large utility steam turbines. In small steam turbines, typically just a couple of impulse stages are used due to their compactness and simplicity.

6.4.3 Twisted Blades

In the study of velocity vector diagrams, we noticed that the blade velocity (V_B) has a significant impact on the shape of a blade and its performance.

[3] It should be noted that generally the objective is not necessarily to keep the steam velocity fixed but to distribute the total turbine power generation approximately uniform among all the turbine stages. However, these two effects are interrelated.

We know that the velocity of a point on a blade can be determined by the following equation:

$$V_B = r\omega = r(2\pi N) \tag{6.72}$$

where r is the radius of the rotation of the point or the vertical distance between the point and the axis of rotation (m), ω is the angular velocity of the turbine (rad/s), and N is the number of revolutions of the turbine per second (1/s). In this equation, the angular velocity of the turbine is the same for the entire turbine. However, the radius and as a result the blade velocity vary throughout the blade length. The farther the point of interest is from the axis of rotation, the greater the blade velocity at that point is. This indicates that the velocity vector diagram varies for different sections of a blade throughout its length. This, in turn, dictates that the **optimum profile of each section of the blade varies depending on how far it is from the base of the blade**. In general, **the base of a blade should be more impulse than its tip** and the maximum degree of reaction should be in the blade's tip. This can be typically achieved by a **twisted blade**. Longer blades experience a greater variation in their velocity along the blade's length, which, in turn, require more twisting. For short blades, i.e., the symmetrical impulse blades at the high-pressure end of turbines, the change in the blade velocity can be small enough to be negligible. That is why the blades of early turbine stages have typically a fixed profile for the entire length of the blade with no twist. But the blades of reaction stages are twisted, which progressively increases for longer blades. Due to limited manufacturing tools and nonexistence of computerized design tools, the blades of early turbines were usually uniform. However, most modern steam turbines (for that matter most large turbines[4]) have twisted blades, especially at lower pressure sections.

6.4.4 Blade Materials and Sealing

Steam turbine blades, particularly at the high-pressure sections of the turbine, are exposed to high-temperature and pressure superheated steam. Since the steam is highly purified, corrosion is not a major problem for blades. However, blades are susceptible to **creep**[5] and **fatigue**.[6] With the increasing inlet flow temperature and pressure of modern turbines, creep is particularly a major challenge. A low coefficient of thermal expansion is also a desirable attribute for turbine blade materials. Steam turbine blades are typically made of **nickel superalloys** formed by nickel (Ni) alloyed with titanium (Ti) and aluminum (Al) as well as sometimes ruthenium (Ru) and rhenium (Re). They may also

---

[4] Next time that you see a wind turbine, pay attention to how twisted its blades are.

[5] Creep is permanent deformation of material when exposed to long-term stresses, which are less than its yield strength, typically at high temperatures.

[6] Fatigue is material failure due to cyclic stresses.

be coated with a thermal protective layer of zirconium (Zr)-based ceramic to reduce their exposure to the high-temperature steam. The research and development activities are underway to develop blades materials that can operate with steam at temperatures around and above 750°C (about 1400°F).

In the previous section, we noted that the clearance between a moving stage and the turbine casting is a critical parameter in determining the steam leakage in reaction stages. Several sealing mechanisms have been used to minimize this gap. An important consideration in the design of these systems is that the minimum required clearance needed to avoid any rubbing between the blade tips and the casing is larger during start-up and shutdown than that during steady-state operation. In the past, these seals had a spring on the back to create a fixed clearance. In these seals, the clearance had to be large enough to accommodate all operation conditions. This resulted in larger than needed gaps for steady-state operation and unnecessary availability losses. In one of the latest designs, called **active clearance control (ACC) seals**, the clearance is controlled by steam pressure applied at the seal rings. Other common sealing techniques are leaf seals and abradable seals.

6.4.5 Turbine Rotors

A steam turbine rotor is a very large and heavy metallic component subject to high physical, thermal, and cyclic stresses. The rotor of a steam turbine from the functional point of view (and not necessarily from physical appearance) is composed of two parts: **shaft** and **disk assemblies**. The discs are where turbine blades are attached to the rotor.

There are several approaches to manufacture a steam turbine rotor. In the first method, which is sometimes referred to as **integral rotors**, the rotor is manufactured from a single forged block (Figure 6.15a). In this mono-block design, the discs are formed by machining. Due to its high material needs and machining time as well as the high possibility of manufacturing deficiencies, this approach compared to other methods is expensive. However, when manufactured, they are very robust during operation. Also, the overall length of the rotor is generally shorter in this design.

In the second technique, sometimes referred to as the **composite construction**, **built-up rotors**, or **shrunk-on disk design**, the shaft and discs are forged and machined separately (Figure 6.15b). After construction of the components, the discs are heated and then shrunk and keyed on the shaft. The forging, machining, and inspection processes in this technique are relatively simple and certainly less demanding than the previous method. However, there is a possibility of discs becoming loose during operation, particularly at the high-temperature sections of turbines. This problem can be intensified in long term due to the effects of creep.

The last method can be seen as a combination of the first two methods. In this approach, similar to built-up rotors, the manufacturing process is divided into several segments. Each disk and its corresponding shaft are forged and

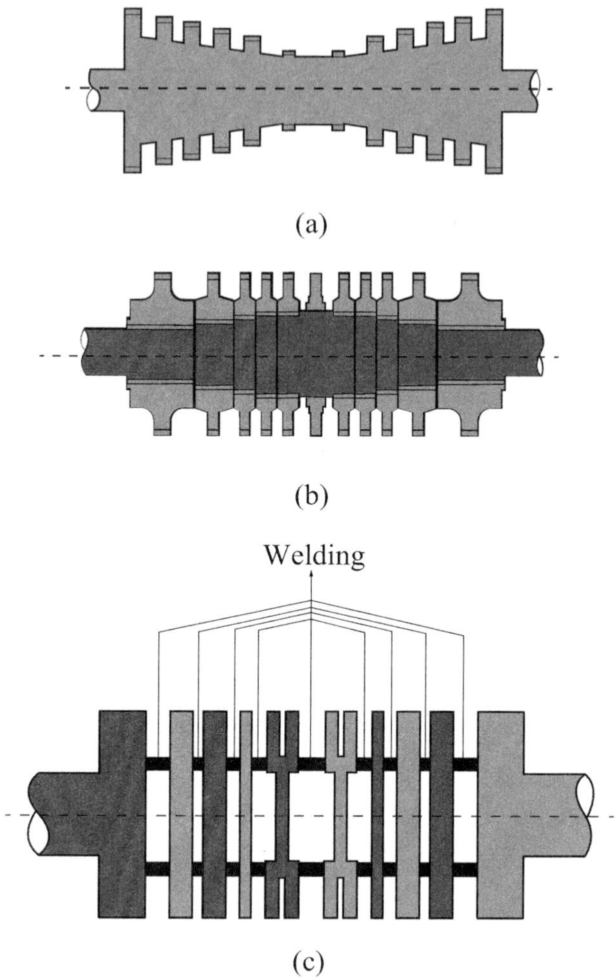

FIGURE 6.15
Different rotor manufacturing methods. (a) Integral or mono-block rotor, (b) built-up rotor, (c) welded disk rotor.

machined together but separated from the rest of the rotor (Figure 6.15c). These separate segments are then welded together to form a solid rotor, similar to integrated rotors, which in this case is called the welded disk rotor. This method requires advanced welding techniques that only became available in the 1920s. The first rotor manufactured using this method was built in Europe and installed in Japan in 1929. Since then this approach to construct rotors have been the most popular one, especially for large turbines. An advantage of this method that others cannot compete with is that different segments can be made up of different metals according to their operational requirements. While high-temperature sections can be fabricated from high-temperature resistant materials, low-temperature sections can be manufactured from low-carbon steels. In this method, the rotor should be stress relieved by a heat treatment process.

As noted above, turbine blades are attached to the shaft by means of disk assemblies. There are three common mechanisms to fasten blades to discs: **dovetail, fir-tree**, and **T-root** attachments (Figure 6.16a–c). Due to their high stress, particularly for long low-pressure blades, and the probability of high stress concentrations at sharp corners, the design of these joints is very important.

A steam turbine rotor has another function that can be critical in certain circumstances. If for whatever reason the flow of steam to the turbine is interrupted, for example due to an emergency shutdown, the quick cooling of turbine components especially blades can create such a large temperature gradient that may distort these components. In these situations, the huge amount of heat that is stored in the mass of the rotor gradually dissipates to allow other parts of the turbine a timely gradual cooling process. However, this heaviness of the rotor along with long bearing spans (below) can cause a problem. If during the cooling process the rotor stops rotating, the shaft can bend under the heavy load. The same can happen if the turbine does not operate for a long period. For this reason, the rotor in large units is typically equipped with a turning gear that rotates the rotor very slowly during the cooling process or during a prolong storage. The same system can be used at the beginning of a cold start-up. This system can be driven by an AC motor. However, special provisions should be considered for when the unit stops due to a blackout, which is called a **black trip**, and no grid electricity is available to rotate the rotor. This can be a battery-powered DC motor or an emergency power source.

6.4.6 Turbine Casings and Bearings

Turbine casings are enclosures that contain turbine's internal components, most notably moving blades. Turbines' fixed blades are also attached to casings. The shape of a casing is dictated by the size of blades. Since the blades

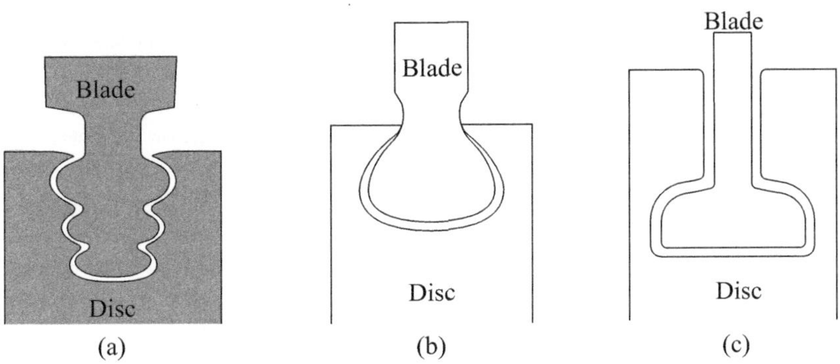

FIGURE 6.16
Turbine blade attachment to disk assemblies. (a) Dovetail, (b) fir-tree, (c) T-root.

get increasingly longer along the turbine from the high-pressure end to the low-pressure one, the casing should follow the same pattern. This means the **steam enters at the small diameter end and exits at the large diameter end.** Even when all turbine sections are on the same shaft, each high-pressure (HP), intermediate-pressure (IP), and one or more low-pressure (LP) sections have their own casings. The stress that a casing is exposed to is a direct function of the operating pressure of the turbine it encloses. Higher pressure turbines require walls with a higher thickness. Moreover, different parts of a casing are exposed to different temperatures since the temperature of the steam reduces along the turbine. This temperature gradient can cause different expansions at two ends of the casing, which can distort it. Special considerations should be given to counter these thermally induced stresses including using specially designed ribs. Turbine casings are typically made in two pieces: upper and lower sections. These two sections are connected by heavy bolts. The lower section is fixed to the foundation. During turbine maintenance, the upper section is removed to give access to the internal components of the turbine. In many designs, the HP and IP turbines have an inner and an outer casing.

In Section 2.3.4, we learned that in regenerative steam cycles we heat feedwater in feedwater heaters before it enters the steam generator. The steam for this purpose should be extracted from the turbine. The steam can also be extracted from the turbine for the turbine driving the feedwater pump (Chapter 7) and for other auxiliary steam needs. The turbine casing should be equipped with **various outlets at different pressures for steam extractions.** These extraction lines are all equipped with check valves (non-return valves) to prevent the reverse flow back to the turbine. These valves are closed when the turbine is shutdown. The lines also have isolating valves and condensate drain lines.

The bearings that support the turbine shaft are also enclosed within the casing. In this section, we will briefly discuss about the radial bearings but the axial (thrust) bearings will be discussed later in this chapter. The turbine shaft is supported by **journal bearings**, also known as **plain bearings** or sliding bearings. These bearings do not have any rolling element; instead the shaft (or the journal) slides over a film of lubrication oil formed between the journal and bearing sleeve (or shell). While the construction of these bearings may seem simple, their theory, design, and operation to ensure safety and reliability can be complicated. Obviously, it is critical to make sure that the oil film is reliably present during turbine operation. The loss of this layer can be disastrous and extremely costly. While in many applications the bearings are self-lubricating, in steam turbines an oil pump is used to pump the lube to the bearings. This pump is typically a positive displacement oil pump driven by the turbine shaft. This ensures that the pump operates and the oil is available whenever the shaft is rotating. However, a backup oil pump is needed for start-up and emergencies. Similar to the turning gear discussed in the previous section, special consideration should be given to the system's safe operation during a black trip. The lubrication system also includes an oil tank to provide makeup oil and an oil cooler to reduce the oil temperature heated by the bearings.

6.4.7 Gland Steam Seal Systems

An important aspect of turbine casings is their sealing. The turbine shaft almost always extends out of the casing on both ends. This can leave a gap between the shaft and the casing. At high and intermediate turbine sections the steam pressure is higher (often much higher) than the atmospheric pressure, which can lead to steam leakage out of the casing to the atmosphere. At low-pressure turbine sections, the steam pressure can drop below the atmospheric pressure (note that a condenser typically operates at vacuum pressures). In this case, atmospheric air can leak into the casing, which means the loss of vacuum and efficiency as well as the introduction of air into the steam that later should be removed in the condenser (Chapter 7). Clearly both of these phenomena are undesirable and negatively impact the performance of the turbine. Furthermore, the leakage of air into the turbine can cause thermal stress when the cold air gets in contact with hot blades.

The most universally used sealing system to avoid leakage is a **gland steam sealing system**. A gland seal is made of several **labyrinth seals**. Labyrinth seals create a **high-resistance path** against the flow of steam or air by forming a maze. To create this maze small groves and thin metallic strips are machined on the shaft and fixed support, respectively (Figure 6.17a). These strips are slightly pressed into the grooves by a spring system behind the fixed section. The result is the creation of a high-resistance flow path (Figure 6.17b).

A typical gland steam seal is made of three labyrinth seals (Figure 6.18). Let us first discuss how the gland steam seal system works when the turbine shaft crosses the casing where the pressure inside the turbine is higher than the atmospheric pressure. Technically, during the turbine full load operation, this is the case for all sealing points except the low-pressure end of the low-pressure turbine right before the condenser. In these high pressure sealing points, the first labyrinth seal significantly reduces steam leakage by restricting its

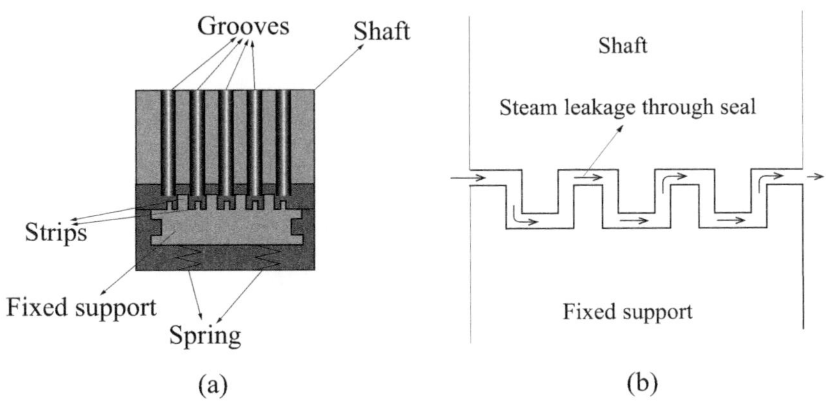

FIGURE 6.17
Schematic of a labyrinth seal for a gland steam seal system. (a) Grooves and stripes on the shaft and support, (b) high-resistance flow path.

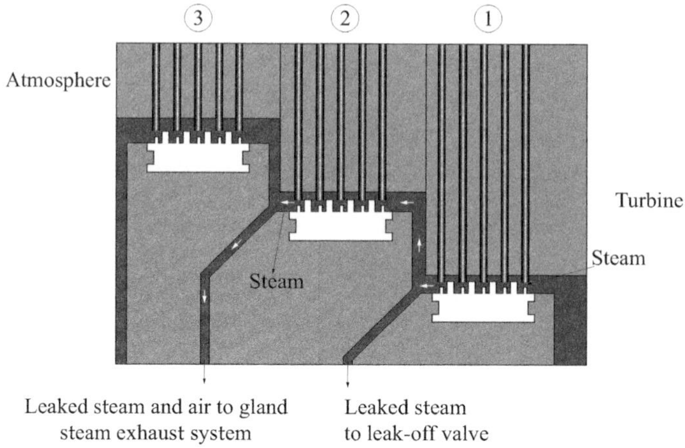

FIGURE 6.18
Gland steam seal operation when the turbine pressure is above the atmospheric pressure.

path (Figure 6.18). However, some steam still leaks through this seal. Note that the amount of the leaked steam in this case is not comparable with that without the seal. This steam is high-quality steam and is collected and sent to low-pressure sections of the cycle, typically the main condenser (the bottom right outlet in Figure 6.18). The pressure of this leaked steam is maintained slightly above the atmospheric pressure, perhaps 0.5 bar higher, by a valve called the leak-off valve (Figure 6.19). Since the pressure of the steam after the first seal is still higher than the atmospheric pressure, a small amount of steam leaks through the second labyrinth seal. This steam is collected (the bottom left outlet in Figure 6.18) by the **gland steam seal exhaust system** (Figure 6.19). In this system, a blower, called the **gland steam exhauster,**

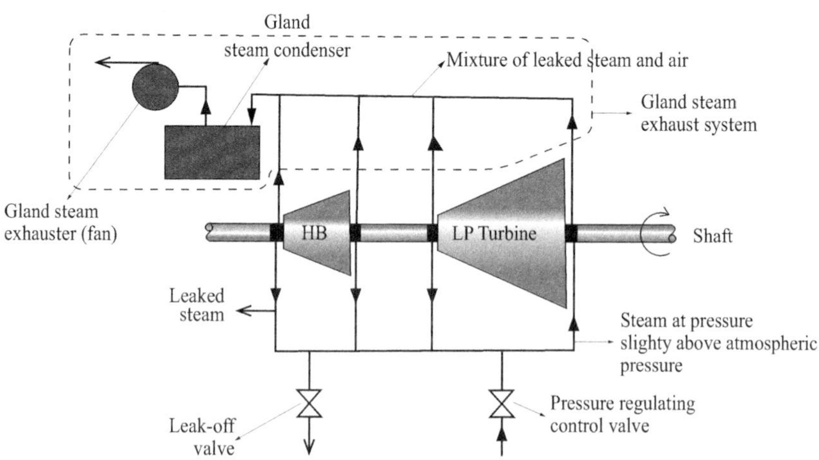

FIGURE 6.19
Schematic of a complete gland steam seal system including the exhaust system.

collects the leaked steam by creating slight vacuum. A small amount of air may also leak to this point through the third labyrinth seal. This leaked air is also collected by the gland steam exhaust system. This mixture of steam and air cannot be sent to the main condenser due to the presence of air in it (we will talk about why this is undesirable in Chapter 7). Therefore, a dedicated gland steam condenser is used to condense and remove steam from the mixture. To recover the thermal energy of the gland steam, the relatively cold condensate (condensed water) from the main condenser is used as the cooling water in the gland steam condenser. The condensed steam leaving the gland steam condenser is then sent back to the main condenser. The remaining air and other noncondensable gases are then discharged to the environment via the exhauster (Figure 6.19).

Now let us review how the gland steam seal works when the turbine shaft crosses the casing where the pressure inside the turbine is lower than the atmospheric pressure. As noted above, in the turbine full load operation is only the case for the low-pressure end of the low-pressure turbine; however, in part load operation and particularly during start-up and shutdown, this operating condition may be the case for more sealing points. In this case, since the steam within the turbine is in vacuum, an external source of steam at the pressure of slightly above the atmospheric pressure is needed. This steam can be provided from the steam collected at the high-pressure gland steam seals right after the first labyrinth seal (Figure 6.18). As you recall from the previous discussion, this steam is pure steam (and not mixed with the leaked air) and its pressure is maintained at slightly above the atmospheric pressure by the leak-off valve. If this steam is not sufficient, which can be the case in start-up and low load operation, some steam can be extracted from the high-pressure turbine for this purpose. The pressure of this fresh steam is reduced to the required pressure by a control valve (Figure 6.19). To prevent air leakage into the turbine, this low-pressure steam is fed to the gland seal between the first and second labyrinth seals (Figure 6.20). Since the pressure of this steam is higher than both atmospheric air and the steam inside the turbine, this flow is divided into two flows: one going through the first seal toward the turbine and the other going through the second seal toward the third seal. The former is safely mixed with the steam inside the turbine. The latter is the flow that prevents air from leaking into the turbine. This portion of the steam along with some air that leaked across the third seal is collected between the second and third seals by the gland steam exhaust system in the same manner previously explained (Figure 6.20). As noted, gland steam seal systems are the most common sealing mechanisms in utility turbines. However, smaller turbines may be equipped with other sealing systems such as carbon seals or water seals.

6.4.8 Axial Thrust in Steam Turbines

When the steam flow passes through the blades of a turbine, it imposes both tangential and axial forces on the blades. The tangential forces (F_x in Equation 6.4) cause the rotation of the turbine rotor. The axial forces

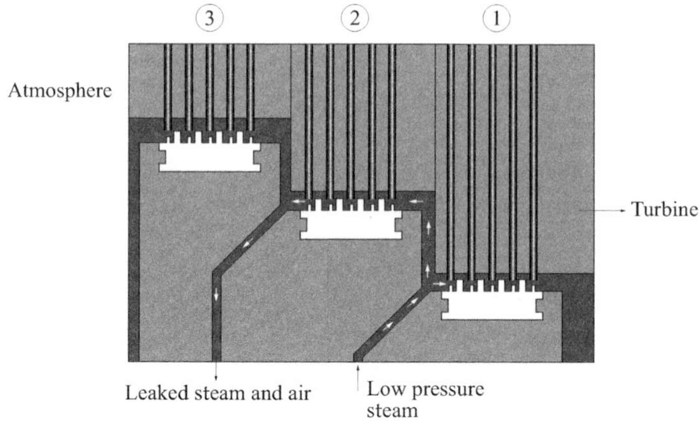

FIGURE 6.20
Gland steam seal operation when the turbine pressure is below the atmospheric pressure.

create the axial thrust that must be dealt with. If unopposed, it would be difficult, if not impossible, to maintain the axial position of the rotor and to balance its rotation. In the previous section, we discussed how journal bearings are used to radially keep the rotor in place. Unfortunately, journal bearings cannot absorb any axial thrust and other means are needed for this purpose. In this section, the sources of the axial thrust in impulse and reaction stages will be reviewed. Then, some strategies to counter them will be discussed.

The axial thrust in turbines can be created for two reasons: the change in the axial momentum and the pressure drop across a turbine stage or more specifically across the moving blades of a stage. Let us first consider the axial momentum change in an impulse turbine blade. In Equation 6.4, we showed that the x-component of the force imposed by the flow on a blade generates output power. We did not discuss the y-component of the force because it does not contribute to power generation; rather it creates the axial thrust. By writing the equation for the y-component of the force (similar to Equation 6.4) and applying it to the blade in Figure 6.7 (similar to Equation 6.38) the following equation can be derived for the axial force on an impulse blade:

$$F_{\text{Axial}} = F_y = \dot{m}\left(V_{1y} - V_{2y}\right) = \dot{m}\left(Vr_1 \sin\theta_1 - Vr_2 \sin\theta_2\right) \tag{6.73}$$

You recall from Section 6.2.2 that in impulse blades $\theta_1 = \theta_2$ and if the flow is assumed frictionless, $Vr_1 = Vr_2$. This indicates that the axial thrust due to momentum change in impulse stages is zero. Similarly for a reaction stage, it can be shown that the axial component of the momentum is unchanged and the axial thrust due to momentum change is zero. It should be noted that in both cases for real steam with friction, the velocity and thus the momentum slightly decrease, which means the actual axial thrust due to momentum change is not zero but is still relatively small.

Now let us take a look at how the pressure drop across a turbine stage contributes to the axial thrust. The moving blades on a turbine stage and its disk assembly can be treated as a plane perpendicular to the turbine's axis of rotation. The flow of steam creates a pressure field before and after this plane. The difference between these two pressure fields is manifested in the form of the axial thrust. The magnitude of this axial thrust depends on the overall pressure difference and the area of the plane. Note that all these discussions are about the moving blades. The fixed blades are attached to the turbine casing and cannot impose any thrust on the rotor.

In Section 6.2.2 (Figure 6.8), we learned that in an impulse stage the pressure only varies in the nozzles (fixed blades) and remains unchanged on the moving blades. Therefore, the axial thrust as a result of the pressure change through the moving blades in impulse stages is negligible. This means that since the contribution of both of the above discussed axial thrust sources is small, for an impulse stage the axial thrust is generally small and does not create a major problem. However, this is not the case for reaction stages. In Section 6.3 (Figure 6.14), we discussed that in reaction stages the steam pressure reduces both in the fixed blades and moving blades. This pressure gradient creates a significant axial thrust in reaction stages. For a given pressure drop across a stage, the true magnitude of the axial thrust depends on the length of the turbine blades. We know that the blades in high-pressure turbine stages are relatively short due to the low specific volume of the steam in these stages. As a result, the axial thrust in high-pressure stages is relatively small regardless of the pressure drop through the moving blades. However, in intermediate-pressure stages and particularly low-pressure stages the combination of long blades and the pressure drop across them creates a substantial axial thrust.

To counteract axial thrust in steam turbines two methods are commonly used: **dummy pistons** and **double-flow turbines**. A dummy piston is a disk (or a piston) that is integrated as a part of the turbine rotor, typically at the high-pressure end of the turbine (Figure 6.21). The disk is exposed to steam with two different pressures on two sides. The force excreted on the disk by this pressure gradient balances the axial thrust. In the design of a turbine, the diameter of the disk can be varied to create a sufficient force to cancel out the axial thrust. The two sides of the disk are separated by a labyrinth seal (as in Figure 6.17). This method is typically used for turbines with a relatively small axial thrust such as impulse turbines, high-pressure turbine sections, and small turbines. It may also be used to supplement the second approach discussed below in large turbines.

In the second method, the turbine is designed as two identical turbines that are facing each other at their high-pressure end (Figure 6.22a). The steam enters at the center of the turbine and is divided into two flows. Each flow expands through one half of the turbine. The two steam flows move in the opposite directions and exit the turbine at two opposing ends. These two turbines create two essentially identical axial thrusts but in the opposite directions, which can cancel each other out. Moreover, this design reduces the flow rate of each half, which, in turn, reduces the length of the blades. These shorter blades not only make the operating conditions of the blades less challenging

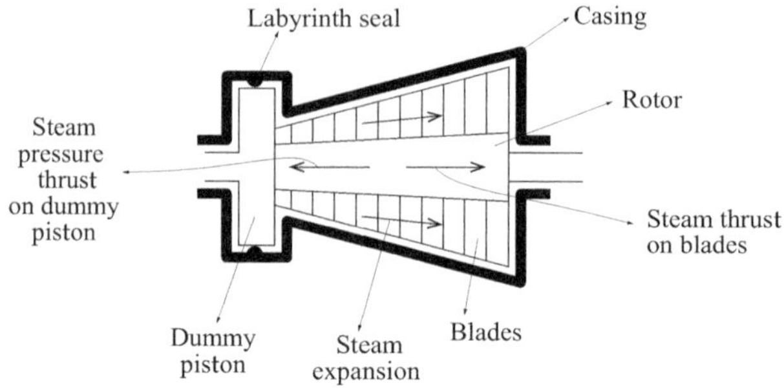

FIGURE 6.21
Schematic of a dummy piston (disc).

but also reduce the overall axial thrust of each turbine. Understandably this turbine design is known as a **double-flow, split-flow, or two-flow turbine**. Sometimes a combination of a single-flow and double-flow turbines are also used (Figure 6.22b). These turbines are typically used for large LP and IP steam turbines.

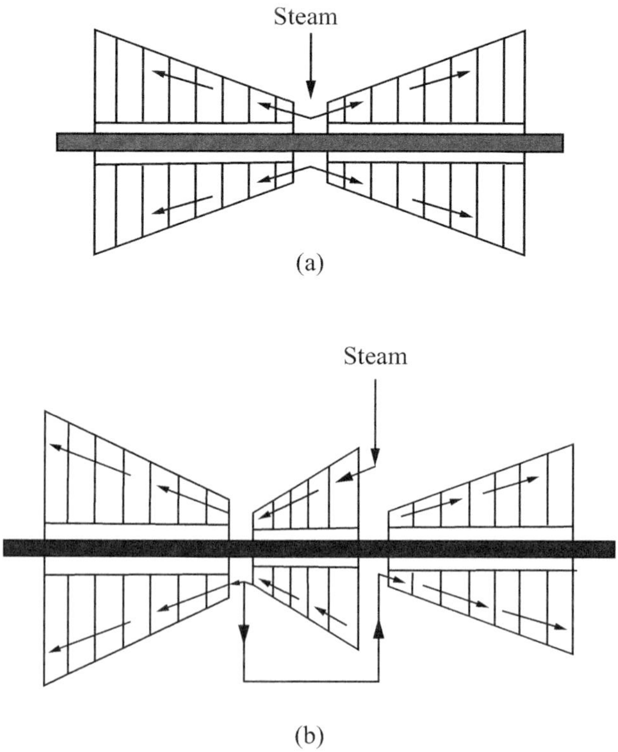

FIGURE 6.22
Schematic two configurations for double-flow steam turbines.

6.4.9 Turbine Arrangements

There is a wide variety of steam turbine arrangements. Various manufacturers design their turbines according to their design philosophy and the specific requirements of each project. In this section, we will review some common utility turbine configurations.

We learned that modern utility steam turbines are typically composed of a combination of impulse and reaction stages. The first few stages, typically the first two or three stages, are impulse. They are followed by many reaction stages. These stages are typically divided into either high-pressure (HP) and low-pressure (LP) sections or high-pressure (HP), intermediate-pressure (IP), and low-pressure (LP) sections, modules, or cylinders. These divisions are for practical reasons such as extractions for reheating or other auxiliary steam requirements, size limitations, and manufacturing constraints. These turbine sections typically have their own casings. However, sometimes HP and IP or IP and LP sections are housed within the same casing. These turbine sections can be single- or double-flow arrangements. Particularly low-pressure sections are almost universally double-flow arrangements to reduce their axial thrust.

We have already learned that the specific volume and as a result the volumetric flow rate of a turbine increase with pressure reduction along the turbine. This is especially the case for low-pressure turbine sections. This can result in a high flow velocity in low-pressure turbines, which can increase friction losses (recall that these losses are a function of velocity square). Moreover, since these turbine's exhaust flow is discharged to the condenser, a high flow velocity results in the loss of high kinetic energy of the flow to the condenser. In extreme cases, the steam velocity may reach the speed of sound and chocking can occur. Likewise, the very low steam flow speed is undesirable because of increase in the turbulence and possible steam recirculation both of which increase losses. To add to these complexities, turbine operational conditions can be heavily impacted by ambient conditions. This is due to the dependency of the turbine backpressure, which is provided by the condenser, to the temperature of the cooling water (much more on this topic in Chapter 8). Therefore, there is a delicate balance for an optimum design of low-pressure turbine sections. For all these reasons and also to manage the height of the blades in LP turbine sections, one, two, or even three parallel double-flow low-pressure turbine sections can be employed in utility turbines. Furthermore, a few last stages of LP sections typically operate at the pressure lower than the saturation pressure at their operating temperature. This means they have to deal with the two-phase flow of a saturated mixture of water and steam. The condensed water droplets and dissolved impurities can cause erosion and solid deposition on the blades. One way to reduce these problems is to add moisture removal orifices to the casing to extract the condensed steam, which is then bled to the condenser or one of the low-pressure feedwater heaters. In any case, the moisture content at the turbine outlet flow is typically limited to a maximum 12% (i.e., the mixture with minimum 88% quality).

In terms of the number of shaft, the most common design, referred to as the **tandem compound** design, is a single-shaft design where all turbine sections and the generator are on the same shaft. Alternatively, different turbine sections can be on two or more shafts coupled with several generators. This design is known as the **cross compound** and is less common than the tandem compound design.

6.5 Turbine Control

We learned that the steam pressure reduces in each stage of a steam turbine. In impulse stages, this pressure reduction takes place in fixed blades (nozzles) while in reaction stages both fixed and moving blades contribute to the reduction in the steam flow pressure. The magnitude of the pressure drop depends on the turbine load. The maximum pressure drop in a turbine happens at the MCR or the full load operation. When the turbine load reduces, the pressure drop across the turbine also reduces. Assuming a constant condenser pressure (more on this assumption in Chapter 7), this means that with a reduction in the load the pressure in the inlet of the turbine reduces. Depending on how the steam generator responds to this pressure change, there are two types of operating modes: **sliding-pressure** and **constant-pressure**.

In the sliding-pressure operating mode, there is no pressure control between the steam generator and the turbine. In this mode whenever there is a need to increase the turbine load, the fuel flow rate of the steam generator is increased. The higher firing rate increases the mass flow rate of the steam generated in the steam generator, which, in turn, increases the steam mass flow rate through the turbine. In this mode, the **steam generator pressure can vary with load variation**. This pressure is typically controlled and maintained by a variable-speed feedwater pump. In Chapter 7, we will see that variable-speed feedwater pumps are driven by either a variable-speed motor or a steam turbine, which allows them to change the output pressure of the pump by adjusting its rotational speed. In this operating mode, the feedwater pump output pressure at low loads can be reduced, which can significantly decrease the power consumption of the pump. Since in this mode load changes are achieved by manipulating the steam generator's operating conditions and the turbine follows the steam generator's lead, it is sometimes referred to as the **turbine-follow operating mode**. During the start-up and shutdown procedures, this operation mode allows a gradual increase in the turbine pressure while the steam generator heats up. This reduces the thermal stress on the turbine blades, which can increase their life expectancy and reduce required maintenance. Due to the time needed to adjust the fuel flow rate, especially for coal-fired steam generators, and more importantly the massive thermal inertia of steam generators, it takes a relatively long time to adjust the mass flow rate and the pressure of the generated steam. This means the response of the sliding-pressure operating mode to turbine load change is **slow**. Regardless of this problem, many supercritical once-through steam cycles and combined cycle power plants (Chapter 10), particularly those that are not used for peak-load electricity generation, are operated in this mode. The sliding-pressure operating mode is very stable, safe, and easy to implement.

In the constant-pressure operating mode, as the name implies, the pressure of the steam generator is kept constant at the level needed for the MCR. In order to control the pressure and mass flow rate at the turbine inlet, a control valve is used at the entrance to the turbine. In this mode, even at part load operation, the steam generator operates at its full load pressure. Depending on the load, the control valve throttles the steam pressure to the needed pressure at the turbine inlet. In this case, since the high-pressure steam is always available, in order to increase the load, the control valve at the turbine inlet can be opened and the turbine load can be almost immediately increased. In essence at part load operation, the steam at high pressure is stored in the steam generator and whenever needed, the turbine can quickly tap into the stored steam. While this characteristic improves the rapid load-following capabilities, it **negatively impacts the cycle efficiency at part load operation**. The reason is that a significant amount of energy is consumed in the feedwater pump to pressurize the water only to throttle the steam later to drop the pressure. Note that after the initial load increase, the steam generator should catch up by increasing the firing rate to increase the steam mass flow rate. For this reason, the constant-pressure operating is commonly called the **steam generator-follow (or boiler-follow) operating mode**. If the load increase is administered too quickly, due to the sluggish nature of the steam generator response, the steam pressure and mass flow rate can drop before the steam generator can catch up. This delay can create instability in the cycle. The stable operation under this mode during load changes requires careful implementation. During start-up, the generated steam in the steam generator is not allowed to enter the turbine until its temperature and pressure are approximately at their design values. This rapid exposure of the turbine blades to high-temperature steam causes a considerable thermal stress on them, which can reduce their life. The constant-pressure operating mode is typically used in drum-type steam generators since a significant amount of steam can be stored in the drum. Some cycles use a combination of these two approaches, which is referred to as the **coordinated or combined operating mode**.

We just learned that in the constant-pressure operating mode a control valve is used to regulate the steam mass flow rate to the turbine. Before we conclude this section, let us briefly discuss how these control strategies work. There are two general approaches for this purpose: **full arc admission** and **partial arc admission**. In full arc admission systems, all inlet steam to the turbine is distributed along the entire circumference of the first stage. To control the steam flow in this case, the entire flow should be throttled. This, as noted above, results in a significant energy loss. In partial arc admission systems, steam is admitted to different segments (arcs) of the first stage separately via several independent control valves. This system allows some valves to be closed in part loads while others are fully open. This control strategy results in only some segment of the first stage to be exposed to the entering steam. This control system is more efficient since it is not required to throttle the entire steam flow. Note that generally all sliding-pressure systems use full arc admission systems, and no control valve is needed (since the steam mass flow rate is controlled by the firing rate in the steam generator).

6.6 Generators

Electric generators are devices where the ultimate product of a power plant is generated. They receive mechanical energy from the turbine and convert that energy to electricity. They generate **three-phase AC power** at the **frequency of 50 or 60 Hz**, depending on the grid in which they are connected to. The generated electricity has to go through one more component before it is ready in its final form. Depending on their power generation capacity, the generated electricity in the generator can have a voltage of 10,000–27,000 V. The voltage of power transmission lines is typically much higher than the output voltage of the generator. Therefore, the electricity generated by the generator is transferred to a step-up transformer via the generator terminals. Finally, the electricity at an appropriate voltage is delivered to the grid. Power plants in general and turbines in particular are rated based on the output of their generator. In utility power plants, generators are directly connected to the same shaft as the turbine. That is why they are sometimes referred to as **turbogenerators (or turbine generators)**. In smaller units, they can be connected to the turbine via a speed-reducing gearing system. Modern generators are very efficient machines with an efficiency **between 98% and 99%**.

Like any other equipment that converts mechanical energy to electrical energy or vice versa, a generator needs two main components. One component generates **electric and magnetic forces** (**EMF**) to form an **electromagnetic field**, which induces electricity in the other component. Generally speaking, the electromagnetic field can be created by a permanent magnet or an electrical coil (known as a field coil). The process of electromagnetic field creation is called the **excitation process**. In electricity generators, this field is formed by a **rotor,** and electricity is generated in a **stator**. The field can typically have two or four poles.

The rotor is the rotating part of a generator. It includes the rotor winding (field winding) that is formed on the shaft. It creates a rotating electromagnetic field for the stator. For this purpose, the rotor needs an excitation power supply. This is DC power that passes through the rotor winding to generate the needed electromagnetic field. There are two common types of **exciters**: **static** and **rotating exciters**.

In static exciters, electricity to power the rotor winding is taken from the generator's terminals (before the electricity is sent to the main transformer). The voltage of the tapped electricity is first reduced in a small transformer. Then, AC power is converted to DC power in a **rectifier**. This electricity is transferred to the rotor winding via the graphite brushes and slip rings. In rotating exciters, a small additional generator is connected to the generator rotor. In this case, since this generator rotates with the rotor winding, it does not need brushes. That is why they are also known as **brushless exciters**. While static exciters do not need a dedicated generator, the graphite brushes should be periodically replaced. Both these excitation methods are being used for all generator sizes.

The stator or armature[7] is the stationary portion of generators. It is attached to the casing, which is bolted to the foundation. Similar to the rotor, the stator has a winding called the stator winding or armature winding where electricity is generated. However, this winding is actually composed of three independent windings, one for each phase of generated electricity. The rotating electromagnetic field created by the rotor winding induces alternative voltage in the stator windings. Each of these windings is connected to a generator terminal.

In order to connect a generator to the grid, the voltage of the generated electricity and the grid voltage should be exactly the same. Recalling that the voltage of AC power as a function of time is typically sinusoidal, i.e., $V = V_{\max}\sin(\omega t + \theta)$, the grid voltage and the generated electricity must have the same **shape**, **maximum voltage** $(V_{\max})$, **frequency** (ω), and **phase** (θ). In fact these conditions should be satisfied to connect any two AC electricity producing devices to a shared load. The process of connecting a generator to the grid is referred to as **synchronization** and can be done automatically by a device known as synchroscope (the details of this process are out of the scope of this text).

The passage of electricity from the rotor and stator windings creates significant thermal energy (which is a source of energy loss in generators). This thermal energy should be removed from the generator. The generator cooling can be achieved by circulating ambient air through both windings, which is known as an open ventilation system. However, dust and contaminations in air can damage the winding in long term. Alternatively, air can be used as a cooling medium in a closed cycle. In this system, air is cooled in a heat exchanger and recirculated back to the windings. The most common cooling system in large generators is **hydrogen cooling systems**. In these systems hydrogen, typically at the pressure of about 5 bar, is circulated through the windings. Hydrogen is used as the cooling medium due to its high cooling capacity and low friction losses. In both closed air and hydrogen cooling systems, the generator should be equipped with proper sealing to avoid the cooling medium loss. The required hydrogen for the system is provided by hydrogen bottles. To avoid any contact between hydrogen and air, which may create fire or explosion hazards, CO_2 gas is used to purge air before hydrogen injection and to remove hydrogen before air introduction to the generator.

[7] In general an armature is the component in motors and generators that carries AC electricity.

6.7 Control of Generators

The frequency of electricity produced in a generator is directly proportional to the rotational speed of the generator, according to the following equation:

$$f = \frac{P \times N}{120} \tag{6.74}$$

where f is the frequency (Hz), P is the number of generator poles (always an even number equal to or greater than 2), and N is the generator rotational speed (rpm). For example, the rotational speed of a generator with two poles should be 3600 or 3000 rpm, respectively, to produce electricity with the frequency of 60 or 50 Hz. Large generators typically have two poles. They are known as **full-speed generators.** Smaller generators can have four poles typically with a reduction gearbox. Their rotational speed is 1800 or 1500 rpm for the frequency of 60 or 50 Hz, respectively, which is why they are referred to as **half-speed generators.** On a side note, since the rotation speed of the turbine-generator set can only vary in a tight margin (just a few percent, perhaps ±5), **units operating with the frequency of 60 and 50 Hz are not interchangeable.**

Since the **frequency of electricity in the grid must be maintained in a very tight range,** e.g., ±0.1 Hz, the rotational speed of generators must be closely controlled. Speed control systems of generators can be categorized into two groups based on how they respond to load increase: **isochronous** and **droop.** In the former, the rotational speed of the generator is kept constant with load variations while in the latter the speed can reduce when load increases. In utility turbine-generator units that are synchronized with the grid, the droop speed control is used. Turbines that are used as mechanical drives or isolated units for electricity generation (also referred to as the island-operating mode when the generator is not connected to the grid) are controlled by the isochronous control mode. The difference between these two-speed control systems is in how they maintain the rotational speed of a generator when load changes.

Since load (power) is equal to torque times rotational speed, for a fixed input torque from the turbine if the load of a generator is increased, the rotational speed should reduce. Obviously, this is not acceptable for a generator. In the isochronous rotational speed control system, when load increases, the control system immediately detects this and tries to maintain the rotational speed (hence the frequency of the produced electricity). For this purpose, a governor is used. This can be achieved by increasing the input torque from the turbine. In a steam turbine, this is accomplished by a higher steam flow rate to the turbine.[8] In other words, the set point of the speed controller is fixed by the frequency of electricity. When the actual rotational speed differs from the set point, the

[8] In a gas turbine, this is done by a higher fuel flow rate to the combustor (Chapter 9).

turbine compensates the difference in the speed by changing the inlet flow of energy (in this case the steam flow rate). Usually this control system does not operate smoothly during a load change. The cruise speed control of your automobile is an isochronous speed control system. One can observe the unsmooth operation of the system when the vehicle's cruise speed control tries to maintain the speed in an uphill drive by applying too much pressure on the engine.

Generally, the isochronous speed control works fine for an isolated generator. However, if two or more parallel generators are all controlled by this method, load variations will be challenging. Since each generator tries to maintain its rotational speed, when load increases, each generator will try to respond first. They will "fight" for frequency control, and eventually one generator will take the entire load. This will result in the frequency oscillation and instability in the grid. In the grid where many generators are connected in parallel, only one generator, typically the largest baseload generator, operates in the isochronous speed control mode to fix the grid frequency and the rest of the generators follow the frequency set by the leading generator.

In a synchronized generator, the frequency of produced electricity and hence the actual rotational speed of the generator rotor are fixed by the frequency of the grid (i.e., the generator rotational speed is locked to the grid power frequency) and the operator has no control over it. Instead, the operator can control the set point for the speed controller. Therefore, to increase the load, the operator can increase the speed set point. The speed controller detects that the actual speed (which corresponds to the grid frequency) is less than the set point and tries to compensate it by allowing more input energy (steam) to the system. However, the actual rotation speed cannot be varied, so the load will increase. Note that the rate of change in the set point determines the rate of load change and the turbine power output change.

From the electrical point of view, when load increases, the current drawn from the stator increases. If no action is taken, the voltage in the generator terminals will drop. This is absolutely unacceptable and the output voltage should always remain constant. To prevent this from happening, all generators are equipped with an automated voltage regulator (AVR). The function of this regulator is to measure, monitor, and control the generator terminal voltage. The AVR measures the terminal voltage and compares it with the set point. When a discrepancy is detected, the AVR takes a corrective action by changing the rotor current. For example, when the load increases, the voltage in the terminals drops. So the AVR immediately increases the rotor winding current to maintain the voltage in the stator winding at the set value.

Before we conclude this chapter, let us briefly discuss some other measurement and control devices in a turbine-generator set. Considering the complexity of both the turbine and the generator and the importance of their connectivity to other plant components, it is not surprising that the safe and efficient operation of these systems requires many sensors and measurement systems. The parameters that these devices measure are used to monitor and control the entire system. Here is the list of just some of these devices/systems and/or parameters they measure: load control, speed control, vibration monitoring, steam pressure gauges, thermometers to measure steam temperature in various locations, thermometers to measure lube oil temperatures, condenser pressure gauge, synchronization system, governor, stress monitoring sensors, component remaining-life determination systems, and many more. The performance of steam turbines is governed by ASME PTC 6. For generators, a variety of standards developed by the Institute of Electrical and Electronics Engineers (IEEE) and the International Electrotechnical Commission (IEC) are used.

6.8 Questions and Problems

6.8.1 Introduction

6.1. How can the power production in a turbine be increased?

6.2. How and why are the inlet and outlet flow conditions to a turbine limited?

6.3. Briefly explain the steps of the steam turbine design process.

6.8.2 Impulse Turbines

6.4. Write the conservation of momentum equation for a control volume.

6.5. In an impulse blades, which parameter remains unchanged?

a. The magnitude of the stream absolute velocity
b. The magnitude of the stream relative velocity
c. The direction and magnitude of the stream absolute velocity
d. The direction and magnitude of the stream relative velocity

6.6. Derive the equation for the force imposed by the mass flow on the moving blade in the following figure. When is this force zero?

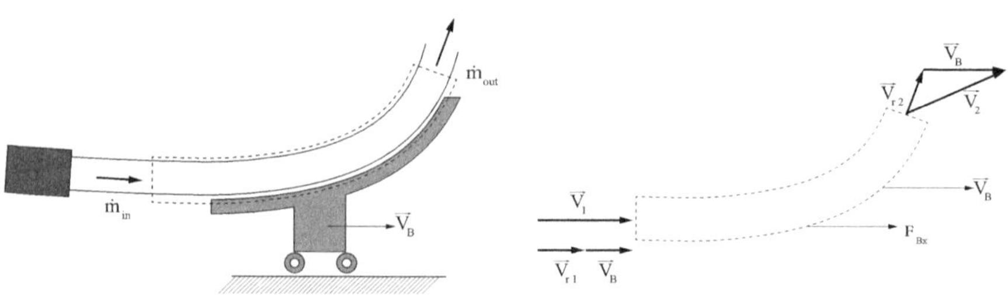

6.7. For a single moving blade in the above figure, for what velocity of the blade (V_B) as a function of the inlet stream velocity the efficiency is maximum? What is the value of this maximum efficiency?

6.8. Explain why in an actual turbine blade the stream exit angle of 180° is impractical and this angle should be less than 180°.

6.9. What is a velocity vector diagram? Why is this diagram important for the design of turbine blades?

6.10. In the following velocity vector diagram, the inlet flow angle (α_1) and the inlet flow absolute velocity are given. Derive the equations for the maximum power developed by the blade and its maximum efficiency. Assume that the blades' inlet and outlet angles are equal ($\theta_1 = \theta_2$).

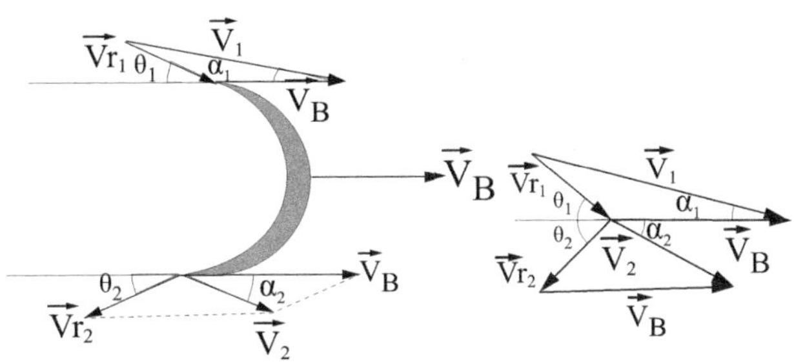

6.11. Superheated steam enters an impulse moving blade with the mass flow rate of 500 kg/s and the speed of 800 m/s at the angle of 13° (the angle between the blade velocity and the absolute velocity of the inlet flow). The blade is symmetrical. Draw the velocity vector diagram for the inlet and outlet flows and determine the maximum output power of the blade (i.e., the blade velocity is optimum) and the blade efficiency. The flow is frictionless.

6.12. Repeat Problem 6.11 for the velocity coefficient of 96%.

6.13. Why are impulse turbines also referred to as equal pressure turbines?

6.14. Why are impulse blades symmetrical in shape?

6.15. Draw the pressure and velocity diagrams for a single-stage impulse turbine.

6.16. What was the major problem with de Laval turbines that prevented their applications in large-size turbines? How could this problem be addressed?

6.17. What are the similarities and differences between de Laval turbines, Curtis turbines, and Rateau turbines?

6.18. Explain how velocity-compounded impulse turbines work. What are their drawbacks? How can pressure-compounded impulse turbines address these problems?

6.19. Similar to velocity-compounded turbines, in pressure-compounded turbines, all stationary stages are nozzles.

a. True
b. False

6.20. Which statement is not correct about pressure-compounded impulse turbines?
a. The enthalpy drop is divided equally among the stages.
b. The inlet absolute velocity to each rotating stage is almost equal.
c. The pressure drop among the stages is equal.

6.21. Draw the pressure and velocity diagrams for a two-stage pressure-compounded impulse turbine.

6.22. Which statement is not the advantage of pressure-compounded impulse turbines over velocity-compounded and single-stage impulse turbines?
a. In pressure-compounded impulse turbines, the steam absolute velocity is lower throughout the turbine, particularly in the early stages, which results in less friction losses.
b. In pressure-compounded impulse turbines, the blade velocity, i.e., the rotational speed of the turbine, is higher.
c. In pressure-compounded impulse turbines, the share of each stage in the output power can be controlled by designers and is commonly roughly equal among stages.
d. Pressure-compounded impulse turbines have higher efficiency.

6.8.3 Reaction Turbines

6.23. Explain how early pure reaction turbines worked.

6.24. How do reaction and impulse turbines differ in terms of pressure variation through each stage?

6.25. Which statement is correct (select all the correct ones)?
a. In reaction turbines, pressure reduces through both fixed and moving blades.
b. The pressure drop in fixed nozzles creates the impulse force when the flow strikes the moving blades.
c. The pressure drop in the moving blades creates the reaction force.
d. All of the above.

6.26. In a 50% reaction turbine stage, the total enthalpy drop is …
a. equally divided between the fixed and moving blades
b. is greater in the moving blades
c. is greater in the fixed blades
d. not equally divided between the fixed and moving blades

6.27. Draw the variation of pressure and absolute velocity diagrams throughout the following reaction blades.

6.28. What are the similarities of the velocity change in the reaction stages and pressure-compounded turbines?

6.29. Determine if the following blades are asymmetrical or symmetrical.

- Fixed blades in a reaction turbine
- Moving blades in a reaction turbine
- Fixed blades in a pressure-compounded impulse turbine
- Moving blades in a pressure-compounded impulse turbine
- Fixed blades in a velocity-compounded impulse turbine
- Moving blades in a velocity-compounded impulse turbine (except the first fixed stage)

6.30. What are the function(s) of fixed blades in a velocity-compounded impulse turbine?

6.31. The reaction turbines' maximum rotational speed and maximum power generation are ... those in impulse turbines for the same given inlet flow conditions.

a. same as

b. half

c. twice

d. just a fraction of

6.8.4 Modern Utility Steam Turbines

6.32. Why is there a potential for steam leakage in a reaction stage across the moving stage?

6.33. Why are there always some gaps between the tip of moving blades and the turbine casing?

6.34. The magnitude of the steam leakage across a moving stage in reaction stages depends on ....

6.35. For the same power output, reaction turbines need ... stages than that for impulse turbines.

a. more

b. less

c. same

d. more, less, or same

6.36. Why do reaction stages have less friction losses compared to those for impulse stages?

6.37. Reaction stages have much ... efficiency than that of impulse stages.

a. lower

b. higher
c. same
d. depends on the design and operating conditions

6.38. Why some blades are twisted?

6.39. Which statement is not correct?
a. The radius and as a result the blade velocity vary throughout the blade length.
b. The optimum profile of each section of the blade varies depending on how far it is from the base of the blade.
c. The base of a blade should be more impulse than its tip.
d. The maximum degree of reaction should be in the blade's base.

6.40. In two sentences briefly explain what creep and fatigue are. Why are they important for steam turbine blades?

6.41. What is the main characteristic of active clearance control (ACC) seals?

6.42. What are three techniques to construct a steam turbine rotor? Briefly explain each method. What are the pros and cons of each?

6.43. What is the function of a turbine rotor during an emergency shutdown?

6.44. In the following figure, identify three methods for a turbine blade attachment to disk assemblies, i.e., dovetail, fir-tree, and T-root.

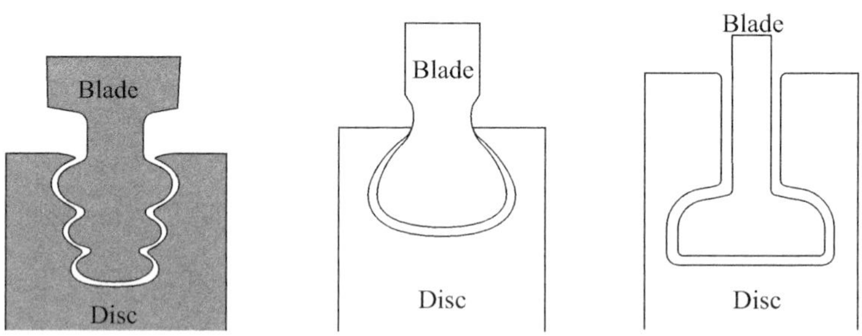

6.45. What is the function of a turning gear?

6.46. What is a black trip? Why is it important for a steam turbine?

6.47. Which statement is not correct about a steam turbine casing?
a. The shape of a casing is dictated by the size of blades.
b. The casings enclose turbine's internal components.
c. The fixed blades are attached to the casing.
d. The steam enters at the large diameter end and exits at the small diameter end of the casing.

6.48. Why do steam turbines have steam extraction outlets?

6.49. What type of bearing is used to support the main shaft of a steam turbine? Briefly explain how they work.

6.50. Explain how and why the sealing between the turbine shaft and the casing is important for high-pressure and low-pressure steam turbine sections. What are the problems that leakage can cause in each case?

6.51. Explain how gland steam seal systems work when the steam pressure in the turbine is above and below the atmospheric pressure.

6.52. Why does a gland steam seal system require a dedicated condenser?

6.53. What are two mechanisms that generate axial thrust?

6.54. Discuss the magnitude of the axial thrust in the following cases:

- Axial thrust due to the change in the axial momentum in impulse stages
- Axial thrust due to the pressure drop across a turbine stage in impulse stages
- Axial thrust due to the change in the axial momentum in reaction stages
- Axial thrust due to the pressure drop across a turbine stage in reaction stages

6.55. Explain two methods to deal with axial thrust in steam turbines.

6.56. Compare the magnitude of axial thrust between high-pressure and low-pressure stages of a reaction turbine.

6.57. What are two methods commonly used to counteract axial thrust in steam turbines? Briefly explain the operational principles of each method.

6.58. How can double-flow turbines substantially reduce axial thrust (two reasons)?

6.59. Explain why commonly several parallel low-pressure turbine sections are employed in large utility power plants.

6.60. Compare tandem compound and cross compound steam turbines.

6.61. Describe what happens to the inlet pressure of a turbine when its load reduces.

6.62. What are two methods to control the inlet pressure to a turbine? Briefly explain each method.

6.63. Which statement is not accurate for steam cycles with the sliding-pressure operating mode?

a. There is no pressure control between the steam generator and the turbine.

b. The steam generator pressure is typically controlled and maintained by a variable-speed feedwater pump.

c. During the start-up procedure, this operation mode allows a rapid increase in the turbine pressure while the steam generator heats up.

d. It takes a relatively long time to adjust the mass flow rate and the pressure of the generated steam.

6.64. Why is the response of the sliding-pressure operating mode to turbine load change slow?

6.65. Which statement is correct for the constant-pressure operating mode of steam turbines?

a. In order to control the pressure and mass flow rate at the turbine inlet, a control valve is used at the entrance to the turbine.

b. Even at part load operation, the steam generator operates at its full load pressure.

c. Since the high-pressure steam is always available, in order to increase the load, the control valve at the turbine inlet can be opened and the turbine load can be almost immediately increased.

d. All of the above.

6.66. Which statement is not correct for the constant-pressure operating mode of steam turbines?

a. The pressure of the steam generator is kept constant at the level needed for the MCR.

b. Depending on the load, the control valve throttles the steam pressure to the needed pressure at the turbine inlet.

c. The rapid load-following capability negatively impacts the cycle efficiency at part load operation.

d. At part load operation, the steam at high pressure is stored in the steam generator and whenever needed, the turbine can quickly tap into the stored steam.

6.67. Which one is more stable, safer, and easier to implement?

a. Constant-pressure operating mode

b. Sliding-pressure operating mode

6.68. Why is the life expectancy of blades shorter in the constant-pressure operating mode compared to that in the sliding-pressure operating mode?

6.69. Which is more efficient in part load operation?

a. Full arc admission

b. Partial arc admission

6.8.5 Generators

6.70. The generated electricity in a utility generator is ….
 a. three-phase AC power
 b. at the frequency of 50 or 60 Hz
 c. the voltage of 10–27 kV
 d. all of the above

6.71. Which part of a generator creates the electromagnetic field?
 a. Rotor
 b. Stator
 c. Exciter
 d. Permanent magnet

6.72. What are the two common types of exciters? Briefly explain how they work. What are their pros and cons?

6.73. What is the function of a rectifier in the excitation process?

6.74. Which one has three independent windings
 a. Rotor winding
 b. Stator winding

6.75. In the synchronization process, the grid voltage and the generated electricity must have the same ….
 a. shape
 b. maximum voltage
 c. frequency
 d. phase
 e. all of the above

6.76. What are the three generator cooling methods? Which one is the most common in utility generators?

6.8.6 Control of Generators and Turbines

6.77. A generator producing 60 Hz electricity has four poles. Determine the rotational speed of the generator (in rpm).

6.78. Explain why generators operating with the frequency of 60 and 50 Hz are not interchangeable.

6.79. The frequency of electricity in the grid must be maintained in a very tight range.
 a. True
 b. False

6.80. What does distinguish isochronous and droop speed control systems?

6.81. Describe how the rotational speed of a generator and hence the frequency of the produced electricity is maintained in droop and isochronous rotational speed control systems.

6.82. Explain why the isochronous rotational speed control mode is not appropriate for generators that are synchronized to the grid.

6.8.7 General

6.83. A generator producing 50 Hz electricity has two poles. What is the rotational speed of the generator (in rpm)?

6.84. Explain why in an actual turbine blade, the stream inlet angle of 180° is impractical and this angle should be less than 180°.

6.85. Why do static exciters require more maintenance?

6.86. The voltage of the generated electricity in a utility generator is … the voltage of the grid it is connected to.

a. greater than

b. smaller than

c. equal to

d. equal to or smaller than

6.87. Compare the magnitude of axial thrust between an impulse stage and a reaction stage.

6.88. Explain why typically impulse blades are not twisted.

6.89. Explain what the synchronization process is and how it can be achieved.

6.90. Even when all turbine sections are on the same shaft, each high-pressure (HP), intermediate-pressure (IP), and one or more low-pressure (LP) sections have their own casings.

a. True

b. False

6.91. What metal alloy is steam turbine blades typically made of?

6.92. Why can the sliding-pressure operating mode increase the life expectancy and reduce required maintenance of steam turbines?

6.93. Which one can more rapidly change the turbine load?

a. Constant-pressure operating mode

b. Sliding-pressure operating mode

6.94. Explain how the pressure drop across a turbine stage and the change in the axial momentum can create axial thrust.

6.95. In the constant-pressure operating mode during part-load operation, a significant amount of energy is consumed in the feedwater pump to pressurize feedwater only to throttle the steam later to drop the pressure.

a. True

b. False

6.96. Why are steam turbine casings vulnerable to thermal gradient-induced distortion?

6.97. Describe how labyrinth seals work.

6.98. Why is axial thrust very small in impulse turbine stages?

6.99. In a pressure-compounded impulse turbine, each stage is similar to a single-stage impulse turbine but with a much smaller enthalpy drop.

a. True

b. False

6.100. Draw the pressure and velocity diagrams for a two rotating-stage Curtis turbine.

6.101. Why is the design of a turbine blade attachment to disk assemblies challenging?

6.102. Which part of a generator produces electricity?

a. Rotor

b. Stator

c. Exciter

d. Permanent magnet

6.103. How can a dummy piston absorb axial thrust?

6.104. Why do generators require cooling?

6.105. What is the excitation process in generators?

6.106. What is the source of electricity in static and rotating exciters?

6.107. Which type of rotational speed control mode is appropriate for isolated generators (those that are not synchronized to the grid)? Why? What is the other application of this mode for grid-connected generators?

6.108. Turbines' fixed blades are attached to …

a. shafts

b. casings

c. disk assemblies

d. bearings

6.109. How is a turning gear powered during a normal shutdown and a black trip?

6.110. Explain double-flow turbines.

6.111. In the following figure, identify three manufacturing methods for steam turbine rotors, i.e. integral or mono-block rotor, built-up rotor, and welded disk rotor.

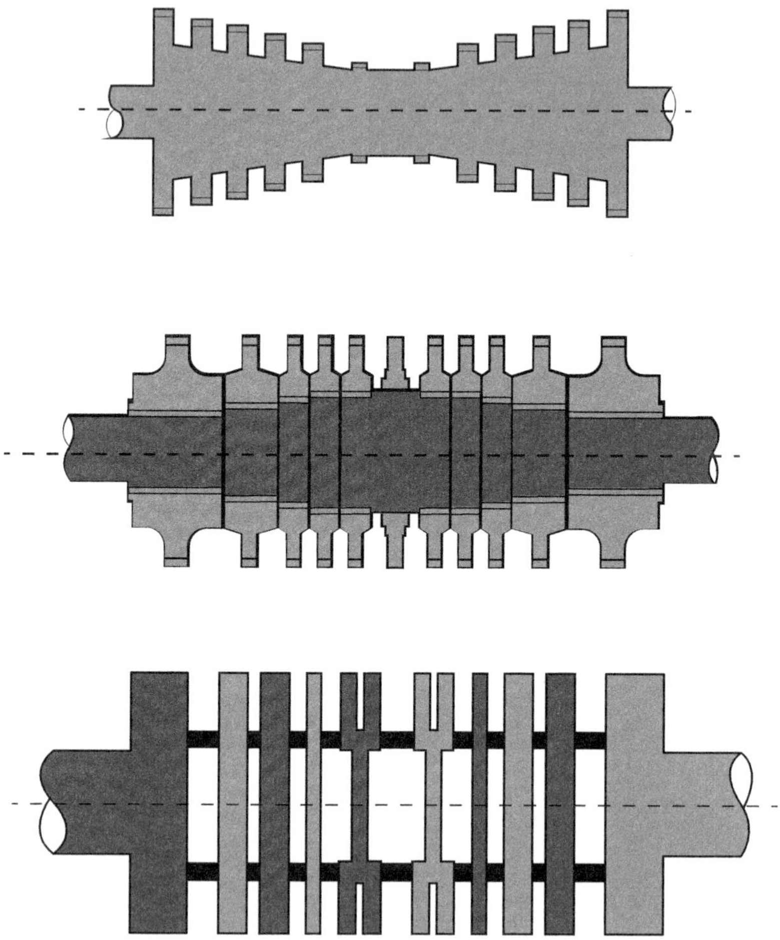

6.112. What are the functions of a steam turbine rotor?

6.113. Explain a typical utility turbine configurations in terms of impulse and reaction stages.

6.114. Describe why corrosion is not a major problem for steam turbine blades.

6.115. Which statement is not accurate for steam cycles with the sliding-pressure operating mode?

a. The steam generator pressure is fixed with load variation.
b. Whenever there is a need to increase the turbine load, the fuel flow rate of the steam generator is increased.
c. The feedwater pump output pressure at low loads can be reduced, which can significantly decrease the power consumption of the pump.
d. Load changes are achieved by manipulating the steam generator's operating conditions and the turbine follows the steam generator's lead.

6.116. What is the function of a thrust bearing?

6.117. Explain how the frequency of generated power is maintained in droop or isochronous rotational speed control systems.

6.118. How is the fire hazard of using hydrogen as the cooling medium in generators mitigated?

6.119. Why do gland steam seal systems need a source of steam when the turbine pressure is below the atmospheric pressure? What are the sources of the steam?

6.120. Which statement is not correct?

a. The farther the point on a blade is from the axis of rotation, the greater the blade velocity at that point is.
b. The velocity vector diagram varies for different sections of a blade throughout its length.
c. Longer blades experience a greater variation in their velocity along the blade's length.
d. Longer blades require no twisting.

6.121. Why can steam turbine blades be coated with a thermal protective layer?

6.122. Explain how journal bearings, also known as plain bearings or sliding bearings, work.

6.123. Why is the axial thrust small in high-pressure turbine stages?

6.124. In which type of steam turbines dummy pistons cannot be used to counteract axial thrust?

a. Utility turbines
b. Impulse turbines
c. High-pressure turbine sections
d. Small turbines

6.125. What is the difference between a full-speed generator and a half-speed generator?

6.126. The reaction turbines' maximum rotational speed and maximum power generation are equal to those in impulse turbines for the same given inlet flow conditions.

a. True

b. False

6.127. Explain why some steam always bypasses moving blades in a reaction stage. How does this impact the stage efficiency?

6.128. Name three components of an extraction line?

6.129. What is the maximum continuous rating (MCR) of a turbine?

6.130. The minimum required clearance needed to avoid any rubbing between the blade tips and the casing is larger during start-up and shutdown than that during steady-state operation.

a. True

b. False

6.131. Why is the clearance between a moving stage and the turbine casting a critical parameter?

6.132. What are the two techniques to regulate the steam mass flow rate to the turbine?

6.133. Typically the enthalpy drop in pressure-compounded impulse turbines is divided equally among the stages.

a. True

b. False

6.134. In a de Laval turbine what were the two functions of nozzles?

6.135. Why is hydrogen gas used as a cooling medium in generators?

6.136. Explain why in an impulse stage the steam pressure before and after the moving stage is equal.

6.137. While the steam pressure … through the moving blades in impulse stages, it … in reaction stages.

a. increases – remains constant

b. reduces – remains constant

c. remains constant – increases

d. remains constant – reduces

6.138. What are the function(s) of fixed blades in a pressure-compounded impulse turbine?

6.139. Why is the response of the constant-pressure operating mode to turbine load change faster compared to that in the sliding-pressure operating mode?

6.140. Compare velocity-compounded impulse turbines with pressure-compounded impulse turbines.

6.141. In reaction turbines pressure reduces only through moving blades.

a. True

b. False

6.142. Explain how the reduction in the flow velocity due to fluid friction is taken into account in turbine blade calculations (compared to a frictionless fluid).

6.143. What are the two methods to create an electromagnetic field in generators?

6.144. There is no commercial pure reaction turbine and although we call them reaction turbines, they are in fact a combination of reaction and impulse turbines.

a. True

b. False

6.145. Which statement is correct about Hero's turbine?

a. There is no momentum change in the turbine.

b. The direction of the flow does not change in the nozzles.

c. The turbine is a pure reaction turbine.

d. All of the above.

6.146. How many stages of nozzles are there in velocity-compounded impulse turbines? What about moving blade stages?

6.147. Why are impulse stages sometimes called equal pressure and reaction stages called unequal pressure?

6.148. For short symmetrical impulse blades at the high-pressure end of turbines, the change in the blade velocity through the blade length can be small enough to be negligible.

a. True

b. False

6.149. Why did turbines replace early reciprocating steam engines?

6.150. Describe why in a reaction stage the steam pressure after the moving stage is lower than that before the moving stage.

6.151. Why is the constant-pressure mode for the operation of steam turbines commonly called the steam generator-follow (or boiler-follow) operating mode?

6.152. Which statement is correct?

a. The blades of early turbine stages have typically a fixed profile for the entire length of the blade with no twist.

b. The blades of reaction stages are twisted.

c. The twist of blades progressively increases for longer blades.

d. Most modern steam turbines have twisted blades, especially at lower pressure sections.

e. All of the above.

6.153. Describe how Hero's and Avery's reaction turbines worked.

6.154. Both velocity-compounded impulse turbines and pressure-compounded impulse turbines have several rotating impulse stages.

a. True

b. False

6.155. Why is the cycle efficiency at part load operation low in the constant-pressure operating mode for steam turbines?

6.156. In the following velocity vector diagram, the inlet flow angle (α_1), the inlet flow absolute velocity, and the blade velocity are given. Derive the equation for the power developed by the blade. Assume that the blades' inlet and outlet angles are equal ($\theta_1 = \theta_2$).

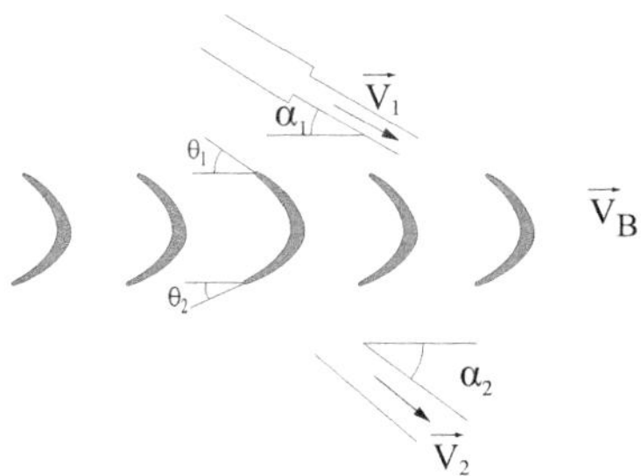

6.157. What is the objective of any turbine? How do they achieve this goal?

6.158. In an impulse stage with a frictionless flow, when the blade velocity is optimum, the exit absolute velocity is perpendicular to the blade velocity vector and is parallel to the axis of rotation of the turbine (axial flow).

a. True

b. False

6.159. Explain the configuration of a de Laval turbine.

6.160. A horizontal jet strikes a stationary blade and exits it at the angle θ (figure below Figure 6.2). Derive the equation for the force imposed on the blade. For what value of θ is the force maximum?

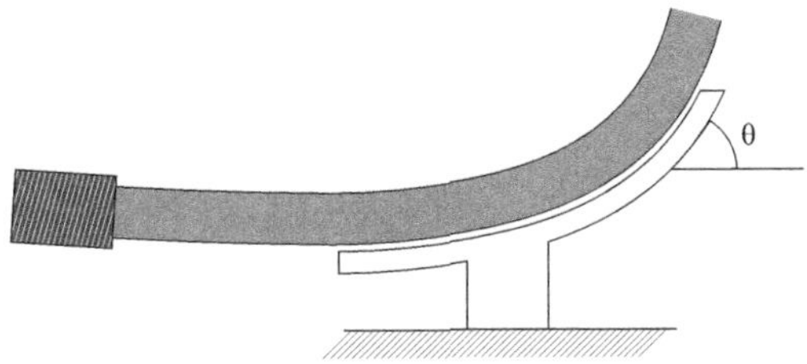

6.161. The following equation can be used to determine the power developed when a horizontal jet strikes a moving blade (figure above). Determine for which values of $V_B$the power is zero and maximum.

$$\dot{W} = \dot{m}(V_1 - V_B)(1 - \cos\theta)V_B$$

6.162. In the above question derive the equation for the efficiency of the blade. Determine the efficiencies when the blade angle is 90° and 180°.

6.163. Why are nozzles (nozzles stages) used in steam turbines?

6.164. In an impulse blade the magnitudes of the inlet and outlet stream absolute velocities remain unchanged.

a. True

b. False

6.165. Provide an example of impulse forces and reaction forces from your daily life.

6.166. Superheated steam enters an impulse moving blade with the mass flow rate of 700 kg/s and the speed of 600 m/s at the angle of 17° (the angle between the blade velocity and the absolute velocity of the inlet flow). The blade is symmetrical. Draw the velocity vector diagram for the inlet and outlet flows and determine the maximum output power of the blade (i.e., when the blade velocity is optimum) and the blade efficiency. The flow is frictionless.

6.167. Repeat the previous problem for the velocity coefficient of 94%.

7

Condensate-Feedwater Systems

Abstract

In this chapter, we will complete our journey following the working fluid in a steam power cycle. We will look at the cycle components from the outlet of the turbine to the inlet of the steam generator. The first piece of equipment that will be discussed is a condenser. You will learn that while in principle a steam power plant can operate without a condenser, why all modern plants are equipped with one. We will extensively cover different types of condensers and their components. Then, after a brief general discussion on pumps, we will talk about condensate pumps and feedwater pumps. Then, we will review how closed feedwater heaters and deaerators help to improve the performance of a steam cycle and to protect the steam generator. Finally, we will learn about water treatment systems for makeup water and feedwater.

LEARNING OUTCOMES

After successful completion of this chapter, you will be able to:

- list components of condensate-feedwater systems from the outlet of the turbine to the inlet of the steam generator
- explain the functions of a condenser, its types, and components
- describe the principle of pumps and the procedure for their sizing
- express the functions of condensate pumps and feedwater pumps and important considerations for their design
- describe the functions of open feedwater heaters and deaerators as well as their components
- explain various water treatment systems for makeup water and feedwater

7.1 Introduction

We have been following the flow of the working fluid in the steam cycle. We started with the steam generator (both water side and fire side) and then the turbine. In this chapter, we will continue following the working fluid from the outlet of the turbine to the inlet of the steam generator around the power plant cycle. Unlike previous chapters where we dealt with very specific components, namely steam generators and turbines, in this chapter, we will consider a wide variety of equipment and systems. Some of the components that we will evaluate are condensers, feedwater heaters, deaerators, feedwater and condensate pumps, and water treatment systems. The cooling side of condensers will be studied in the next chapter.

7.2 Condensers

We noted in Section 2.3 that the outlet flow of the turbine cannot be pumped back to the steam generator because we do not have a compressor capable of pressurizing a two-phase flow of saturated vapor and liquid. Even if we had, such a compressor would have consumed most of the work produced in the turbine (recall that work required in a compressor is determined by $w = \int v\,dp$).

We can reject the steam to the environment and use freshwater as the inlet to the steam generator. These cycles are referred to as **non-condensing** steam power plants. But high-quantity water to make up for the discharged steam may not be available. Even if the water is available, the cycle needs well-treated high-quality water, which needless to say requires expensive treatments. Some early steam power plants used this strategy. But none of modern steam power plants are non-condensing with only one exception. In some cogeneration or combined heat and power (CHP) plants (more on these plants in Chapter 10) steam (typically near-saturated steam) is sent for process needs and no condensation is needed. So generally we need to **recycle and condense high-quality valuable steam to liquid water**. These plants, which are essentially modern steam power plants of all types (with the exception discussed above), are referred to as condensing steam power plants.

So the condenser's main function is to **condense the saturated mixture of steam and water leaving the turbine and to recover highly treated high-value feedwater and recycle it back to the cycle**. While accomplishing this objective, a condenser can perform an even more important task, which can significantly improve the efficiency of the system. We know that the power produced in a turbine depends on the enthalpy difference between the turbine inlet and outlet flows. Lowering the enthalpy of the turbine outlet flow would increase the power output. This can be achieved by reducing the pressure at the outlet of the turbine. The pressure in the condenser is equal to the saturation pressure of the condensing steam, which, in turn, is a function of the condenser cooling water temperature. If the temperature of the cooling water is low enough, which is commonly the case, the low-saturation pressure provides a **backpressure that is lower than the atmospheric pressure** (vacuum) for the turbine to exhaust to. For example, at the temperature of 100°F the saturation pressure is about 1 psia. Now if we release the steam to the environment, the outlet pressure would be about 14.7 psia but if we use a condenser with the condensation temperature of 100°F, the outlet pressure would be about 1 psia. Which one produces more work? Clearly the lower pressure steam at the outlet can produce more work in the turbine. However, one may say a few tens of kPa pressure decrease in the turbine outlet flow is very small compared to many thousands of kPa pressure in the turbine inlet and may not make a big difference! But it is not over yet! At low pressures, the enthalpy change per unit of pressure change is much higher than that at high pressures. This means the power production in the turbine (and cycle efficiency) per unit of pressure change can be increased much more effectively by reducing the pressure of the condenser than by increasing the pressure of the steam generator. When the steam pressure is reduced from 0.35 to 0.06 bar at the temperature of 80°C, the enthalpy reduces from 2645.6 to 2650.1 kJ/kg. This indicates that by this pressure reduction the available enthalpy in the turbine will increase by 15.52 kJ/kg/bar. If the steam pressure is increased from 160 to 180 bar at the temperature of 480°C, the enthalpy increases from 3234.4 to 3203.2kJ/kg. In this case, the available enthalpy in the turbine will increase by 1.56 kJ/kg/bar. This is about an order of magnitude lower than the previous case. A similar effect can be seen in Example 2.8 (Section 2.3.2.4). Therefore, providing a **lower backpressure for the turbine** (i.e., vacuum in the condenser) can significantly improve the output work in the turbine and its efficiency.

Considering the definition of the efficiency of a power cycle, reducing the temperature and pressure of the condenser, and hence improving the efficiency of the cycle, reduces the amount of heat that needs to be rejected to the environment. This means a smaller and less capital intensive condenser and circulating water system. Therefore, reducing the pressure of the condenser not only improves the cycle efficiency but also reduces the capital cost of the system.

In summary, based on what is discussed here and in Section 2.3.2.4 for a Rankine cycle, to improve the power output and efficiency of a steam power plant, we should try to reduce the backpressure of the turbine by reducing the temperature of the condensation process in the condenser. This requires lowering the temperature of the condensation, which, in turn, means lowering the temperature of the cooling water to as low as possible temperature considering the availabilities and limitations of a specific site. In the next chapter, we will discuss how the temperature of the condensation process in the condenser can be minimized. In this section, we will focus on how the condensation process takes place in the condenser.

In addition to condensing steam to be reused in the cycle and providing a low backpressure for the turbine, condensers are where **noncondensable gases are removed from the condensate water.** The condensed steam in a condenser is called condensate. Also, condensate drains from various sources, such as feedwater heaters, turbine driving the feedwater pump, and ejectors (more on these later in this chapter), are collected in the condenser.

From a practical point of view, a condenser is simply a heat exchanger where thermal energy is transferred from spent steam from the last stage of the lowest pressure turbine to cooling water (there are some exceptions to this statement that will be discussed in Chapter 8). As a result of this heat transfer, the latent heat of steam is transferred to cooling water, and steam is condensed. In this section, we will only consider the steam part of the condenser. The entire Chapter 8 is dedicated to cooling water and its components. It should be noted that due to the large latent heat of water, the magnitude of the cooling water flow rate at the condenser is enormous.

Condensers can be divided into three general categories: **direct-contact condensers**, **surface condensers**, and **air-cooled condensers**. In direct-contact condensers, the steam and cooling water flows are directly mixed while in surface condensers heat transfer occurs through a surface/wall that separate two flows. In air-cooled condensers the latent heat of steam is directly rejected to the atmosphere in an air-cooled heat exchanger with no intermediate cooling water. Since this type of condenser has more commonalities with cooling in circulating water systems, it will be covered in the next chapter.

7.2.1 Direct-Contact Condensers

In **direct-contact condensers**, also known as **open condensers**, the condensation is achieved by **mixing the spent steam flow and the cooling water flow**. This type of condenser is not usually used in conventional modern utility steam power plants. However, there are some power generation units that use these condensers such as dry-cooling tower steam power plants (refer to Chapter 8), geothermal plants (Chapter 15), and ocean thermal energy Conversion (OTEC) (Chapter 14).

In direct-contact condensers, the mixing of cooling water and steam occurs either by spraying water into the flow of steam through nozzles or by allowing water to flow through a series of trays while steam rises through water curtains. The former is referred to as **spray-type condensers** (Figure 7.1). In this case, the water flows downward, and, depending on the direction of the steam flow, the condenser can be either counterflow or parallel flow. The latter is referred to as **tray-type condensers**. In both cases, the idea is to increase the surface area of interaction between cooling water and steam. As noted in the previous section, the saturation pressure of water at the condenser temperature is well below the atmospheric pressure. While this is very desirable for the performance of the turbine, the vacuum condenser pressure means that the mixture of the cooling water and condensed steam cannot flow out of the condenser unassisted.

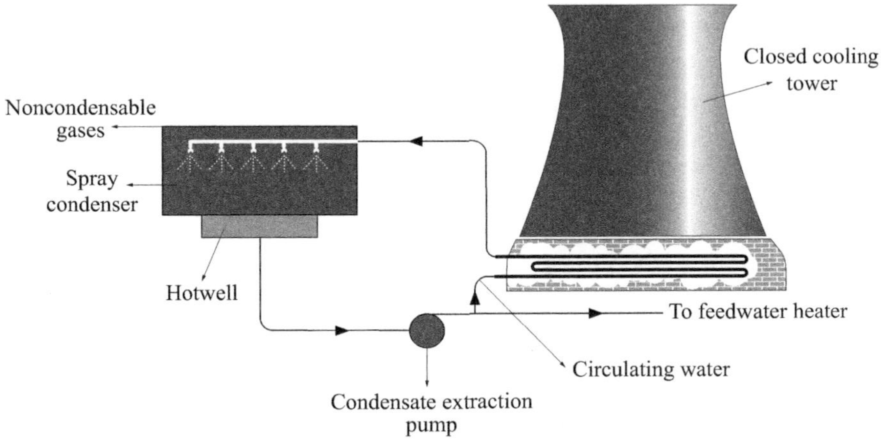

FIGURE 7.1
Schematic of a direct-contact spray condenser along with a closed cooling tower.

One way to compensate for this pressure difference is using gravity. In this method, the condenser should be installed at a high evaluation, hence the name **high-level barometric condensers**. The condenser is connected to a vertical tailpipe. The pipe is long enough that at the bottom of the pipe the pressure of the condenser plus the pressure created by the water column in the pipe minus the pressure drop of the flow in the pipe is slightly greater than the atmospheric pressure. The tailpipe drains saturated or nearly saturated water to a container called the **hotwell**. Alternatively, a pump called a **condensate extraction pump** can be used to pressurize the flow (Figure 7.1). In this case, the system is called a **low-level barometric condenser**. Barometric condensers are not used in current utility power plants while they are still common in some other industrial applications. Note that these condensers, similar to other types of condensers, must be equipped with a mechanism to remove noncondensable gases (discussed later in this chapter).

In all types of direct-contact condensers, the condensate flow after the condenser is divided into two flows (Figure 7.1). Some of the condensate is sent to the power cycle (to be more accurate to feedwater heaters). The mass flow rate of this flow is typically equal to the mass flow rate of the turbine outlet. The rest of the condenser outlet flow is sent to a cooling tower to be cooled before being returned to the condenser. This flow is commonly known as the **circulating water**. We already know that the ratio of the latent heat of water to its sensitive heat is very high. Because of this property of water, the ratio of the circulating water mass flow rate to the feedwater mass flow rate is very high, perhaps about 25.

Since the cooling water flow and the exhausted steam flow from the turbine are mixed in direct-contact condensers, the cooling water should be as pure as the feedwater. This is why a **closed cooling tower** (Figure 7.1) should be used with direct-contact condensers to maintain the quality of the cooling water (more on these cooling towers later in the next chapter).

7.2.2 Surface Condensers

Almost all conventional utility steam power plants use **surface condensers** where there is no direct mixing of the spent steam flow and the cooling water flow. In these condensers, thermal energy is transferred through metal walls (i.e., tube walls). Surface condensers used in steam power plants are typically **shell-and-tube heat exchangers**. The **cooling water flows inside tubes** and the **condensing steam flows outside tubes**. The cooling water can be from natural or human-made bodies of water, such as a river, ocean, lake, or pond, or cooling tower. We will deal with the cooling water system in the next chapter.

A surface condenser is composed of a **cylindrical or cubical carbon-steel vessel** (shell) and many tubes within that shell along with some other components all integrated to a single unit (Figure 7.2). Tubes are supported to **tube sheets** on both ends. These tube sheets also separate the flow of the steam and the cooling water flow. These two flows must be always kept separate; otherwise, the cooling water, with little to no treatment, would contaminate the steam flow, which can cause fouling and corrosion in boiler tubes and turbine blades. Similarly, any air leakage to the steam flow is undesirable and should be stopped (the reason will be later discussed in this section when we talk about noncondensable gases). To create a leak-tight connection, the tubes are rolled to the tube sheets at both ends by physically expanding them into the holes on the tube sheets (Figure 7.3). To ensure no leakage, some groves can be made in the tube sheet holes that the tubes can expand into. Also, the tubes can be welded to the tube sheets.

Since the tubes are exposed to untreated cooling water, their operating environment is more challenging than that of the shell. The tube material

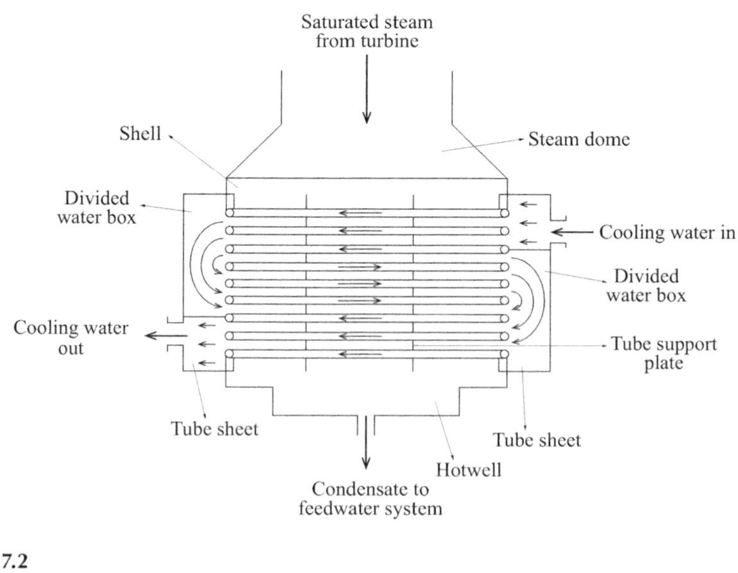

FIGURE 7.2
Schematic of a three-pass surface condenser.

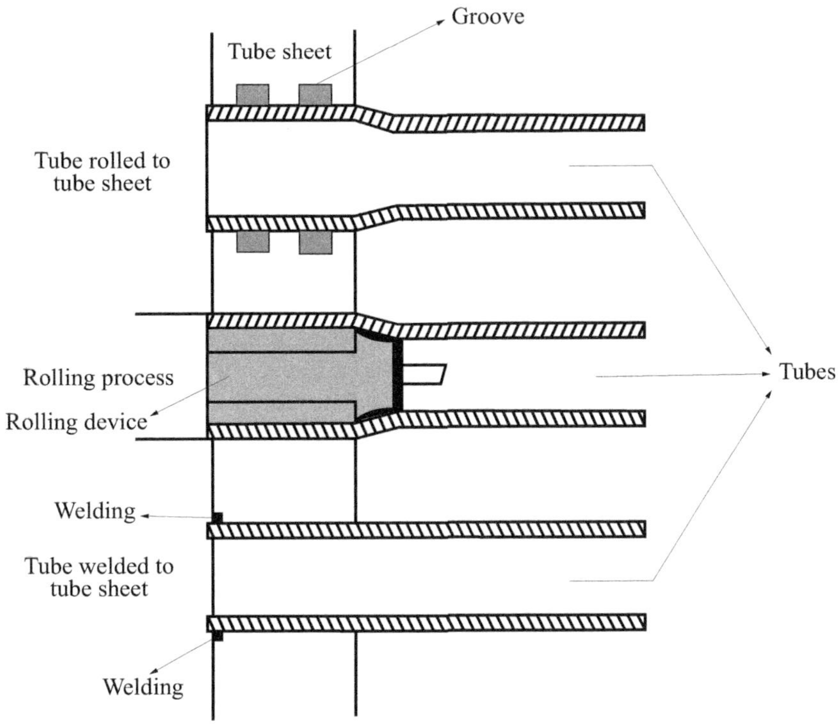

FIGURE 7.3
Tube attachments to a tube sheet.

depends on the type and quality of cooling water. The worst case is when seawater is used for a once-through system (discussed in the next chapter) in which titanium tubes are used. For other less-challenging environments, stainless steel tubes are popular. The leakage of cooling water to condensate in the condenser is not uncommon in steam power plants. It is hard to identify the site of the leakage and to fix it. Therefore, it is critical to choose the tube material very carefully considering the quality of the current and possible future cooling water.

Early condensers in the first half of the 20th century had <10,000 m^2 (about 100,000 ft^2) of heat-transfer surface areas. The tubes were about 4 m (about 13 ft) long and were uniformly and tightly distributed over tube sheets. Modern condensers for large steam power plants can have over 100,000 m^2 (about 1,000,000 ft^2) of heat-transfer surface areas with a tube length of 10 to over 15 m (about 30–50 ft). The tube arrangement is also different in modern condensers (more on these differences later in this section).

On two sides of the condenser, there are water boxes. The cooling water enters the water box on one side of the condenser. Figures 7.2 and 7.4 illustrate several configurations of condensers in terms of the number of passes for the cooling water flow. The number of water box divisions is a function of the number of passes. Depending on the number of passes and configuration, the cooling water may leave the condenser on the same water box that it entered or on the opposite water box. In one-pass condensers, cooling water enters at one water box and leaves on the opposite water box (Figure 7.4a). In this case, both water boxes are undivided. Two or more pass condensers have one or both water boxes divided. In a two-pass condenser, the cooling water enters at the divided water box and is distributed among one half of tubes (Figure 7.4b). The flow is then turned around at the undivided water box and is returned to the divided water box at the opposite direction through the other half of the condenser tubes.

A condenser can typically have up to three passes of cooling water. The number of passes impacts the effectiveness of the condenser and its dimensions. Increasing the number of passes reduces the required flow rate of cooling water, which, in turn, reduces the pumping power and the size of the

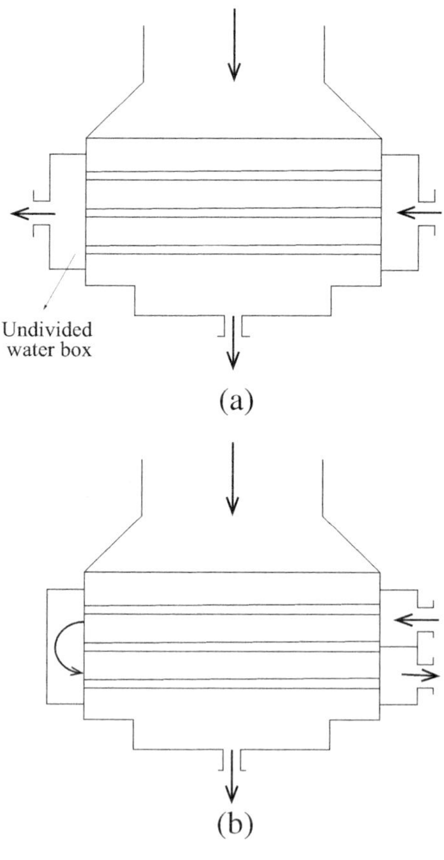

FIGURE 7.4
Schematic of a one-pass (a) and a two-pass (b) surface condenser (for name of components and flows refer to Figure 7.2).

water transportation system from the cold water source (a cooling tower or a natural body of water) to the condenser. However, it increases the temperature of the cooling water leaving the condenser and as a result the temperature of the condensing steam. This means a higher backpressure for the turbine and a lower overall efficiency of the cycle. It also increases the thermal pollution of the natural body of water in once-through cooling systems. In order to allow the condenser partial operation while cleaning or repairing, the water boxes are typically partitioned to two parallel lines. The tubes on each part can be independently operated while the tubes on the other one are out of service. It should be noted that the inner walls of tubes require periodic cleaning, particularly for poor quality cooling water.

Now let us consider the steam flow in a surface condenser. Surface condensers typically receive the exhausted steam at the top from the bottom of the lowest pressure turbine. The steam flow enters the space above the tube bundles, called the **steam dome,** via a hood (Figure 7.2). The steam entrance to the condenser is typically equipped with a perforated impaction plate to protect tubes from the impact of the high-speed two-phase saturated mixture of water and steam flow from the turbine.

The condenser can be supported on the ground and connected to the turbine by means of an expansion joint to absorb the thermal expansion of the system. Alternatively, the condenser can be bolted to the turbine outlet duct. In essence, in this case, the condenser is **hanging from the turbine**. In this arrangement, the turbine support system should be able to handle the stress due to the load of the condenser. The condenser is connected to the ground via a spring support mechanism.

As noted above, early condensers had a uniform and tight tube arrangement. The uniform arrangement however did not allow steam to penetrate deep into the tube bundle and only front tubes were fully exposed to steam flow and participated in the condensation process. This arrangement also caused a high pressure drop in the steam flow, which, in turn, increases the turbine backpressure. To address these problems, the tubes in modern condensers are divided into several separate bundles. Each bundle is formed with some passages through which steam can reach inner tubes in the bundles (Figure 7.5). Also, most condensers are designed to guide the steam flow around the shell so that steam can penetrate the bundle from all directions and converge toward the center. These condensers are known as **radial flow condensers**. When the steam moves deeper in the bundle, more steam condense on tube outer surfaces and drip to the hotwell. This reduces the steam volumetric flow rate but since the area of the heat transfer also reduces while the steam flow moves toward the center of the shell (due to a smaller area perpendicular to the flow), the steam velocity remains relatively unchanged. This is important for a balanced pressure drop, hence uniform steam flow throughout the condenser, and to avoid cross-flow. To assist with the distribution of steam along the tube bundle, condensers are generally equipped with some cross baffles.

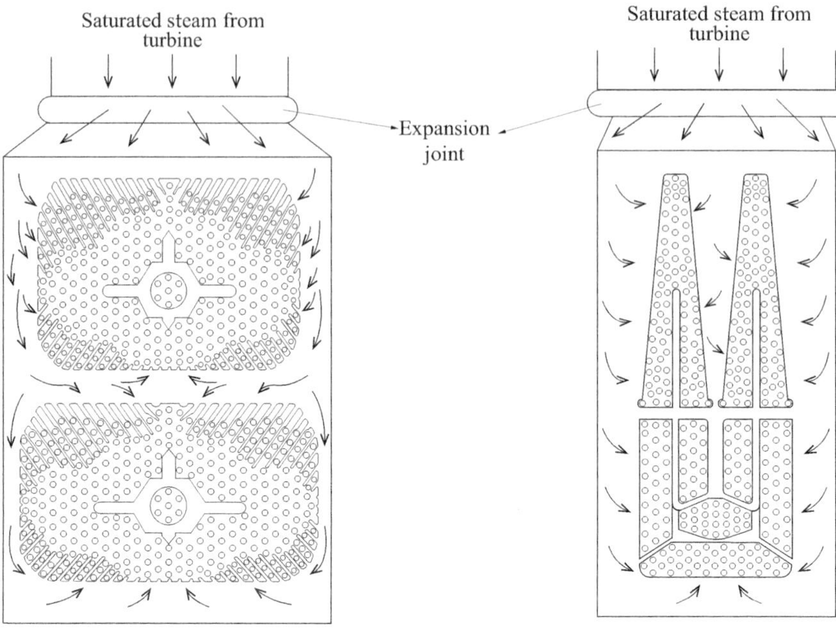

FIGURE 7.5
Schematic of tube arrangement in modern surface condensers.

All condenser tubes can be located in a single shell where the entire exhausted steam from the turbine is received through a hood. In this case, the entire condenser operates at a single pressure. While the cooling water flows through condenser tubes, its temperature increases. The pressure of a single-shell condenser is equal to the saturation pressure of steam at the highest temperature of the cooling water. This is a lost potential for a lower backpressure that could have been achieved by the lower temperature cooling water. Many large steam power plants are equipped with multiple identical shells (typically up to three), hence the name multiple-shell condensers. Each shell receives the steam through a separate hood from the exhaust of the low-pressure turbine. Depending on the cooling water arrangement, these shells can operate at same pressure or different pressures. In a parallel arrangement, the cooling water is divided among the shells. Since all shells receive the cooling water at the same temperature, they all have the same pressure. Usually in this design the steam domes and hotwells are also connected to balance the pressure. In a series arrangement, the entire cooling water flow goes through the first shell, then the second one, and so on. Since the temperature of cooling water increases while advancing from one shell to the next, the shell pressures are accordingly increased. To distinguish between these two arrangements, the former is referred to as a **multiple-shell condenser**, while the latter is known as a **multiple-pressure condenser**.

The steam side of the condenser is typically equipped with several steel tube support plates known as **baffles** to prevent tube vibration (induced by high velocity) and to provide several passages of steam to distribute the steam flow through tube surfaces. Finally, the condensate is accumulated in a water reservoir at the bottom of the condenser called a **hotwell**. The hotwell should be large enough to provide some storage capacity for the condensate. The stored condensate should be sufficient to allow the cycle to operate for a few minutes (around 2–5 minutes). This storage capacity is to accommodate the short-term variation of the condensate flow rate. The condensate from the hotwell is extracted by a condensate pump.

7.2.3 Noncondensable Gases and Deaeration Process in Condensers

A small percentage of **noncondensable gases** always accompanies the exhausted steam leaving the turbine. They can be formed by the **decomposition of water** or by water reaction with system components. Air can also leak to the flow in the condenser or low-pressure stages of turbine where there is vacuum. When the steam condenses in the condenser, these gases remain in the shell side. If they are not properly removed, they can interrupt the performance of the condenser in several ways. The heat-transfer surfaces, that is the outer surface of tubes, can be **covered by gas bubbles**. Since the thermal conductivity of the gases is several orders of magnitude lower than that of water, this layer can significantly reduce the overall coefficient of heat transfer of the condenser tubes. This can negatively impact the performance of the condenser by raising the temperature and pressure of the condensation process, which, in turn, increases the backpressure of the turbine. Furthermore, raising the concentration of noncondensable gases **increases their partial pressure**. For a given condensation temperature and pressure, increase in the partial pressure of noncondensable gases increases the total pressure of the condenser (and the backpressure of the turbine).

The process of the removal of noncondensable gases is called the **deaeration process**. In steam power plants, the deaeration process takes place in two equipment: the **condenser** and the **open feedwater heater** (the latter will be discussed later). Modern condensers typically have a maximum oxygen content of 10–20 ppm at the outlet condensate water from the condenser when operating at full load.

In radial flow condensers when the steam flow penetrates the tube bundle toward its center, there is less steam in the flow; therefore, the concentration of noncondensable gases increases. At the center of the condenser, the flow is primarily composed of noncondensable gases. To reduce the specific volume of these gases and the power needed to pump them out of the condenser, they need to be further cooled. Some of the tubes at the center of the bundle are dedicated to the cooling of noncondensable gases. These gases are separated from the steam by means of some baffles, which guide them to the venting system. Since the pressure of the gases in the condenser is lower

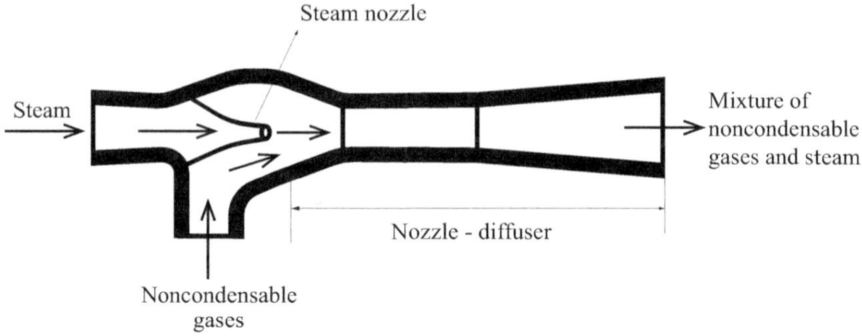

FIGURE 7.6
Schematic of a steam jet air ejector (SJAE).

than the atmospheric pressure, in order to discharge them to the atmosphere, they should be pressurized. For this purpose, either a **compressor** or an **ejector** can be used, the latter being more common in large steam power plants. **Steam jet air ejectors** (SJAE), sometimes also referred to as jet pumps, educators, or jet compressors, utilize high-pressure steam to pressurize a flow. Figure 7.6 illustrates the schematic diagram of an ejector. It is composed of a steam nozzle and a nozzle-diffuser assembly (similar to a Venturi). The steam nozzle injects the incoming high-pressure steam at high velocity into the nozzle-diffuser. In the nozzle, the velocity is increased and pressure is reduced (according to the Bernoulli equation). This low pressure draws the noncondensable gases from the condenser and the exchange of momentum between the high-velocity steam flow and noncondensable gases accelerates the flow of gases. Before the flow leaves the ejector the kinetic energy of the flow is converted to the pressure in the diffuser. Ejectors are self-priming, simple, and reliable devices. They have no moving parts and so require little maintenance and no lubrication. They can be made from a wide variety of materials from stainless steel to PVC and fiberglass. To provide an adequate pressure increase, up to three stages of ejectors can be used. Typically after each ejector there is a heat exchanger to cool the flow and condense the steam, which is returned to the condenser.

7.3 General Discussion on Pumps

In this chapter, several important pumps used in steam power plants will be discussed. Since pumps are one of the most commonly used components in steam power plants, here we will first have a brief discussion about pumps in general. Then, we will review their applications in steam power stations. Some of the most important pumps used in a typical steam power plant are the **condensate pump**, **condensate booster pump**, **feedwater pump**, and **circulating water pump** along with many other auxiliary and service pumps.

Pumps are devices that **add energy to a flow of liquid to increase its pressure**. While pumps can be categorized based on several characteristics, such as their applications, geometry, or pumped liquid, the most important categorization is based on how energy is imparted to the fluid. Generally, pumps can be either dynamic or displacement. In dynamic pumps, the energy addition to the flow is **continuous** and the pressure of the flow is increased **indirectly**. In these pumps, the added energy accelerates the flow and increases its velocity. This velocity is then converted to pressure. Unlike dynamic pumps, in displacement pumps, the energy addition to the flow is **intermittent** where the flow is **directly** pressurized in a closed space with moving boundaries. Displacement pumps can be reciprocating, such as piston, plunger, or diaphragm pumps, or rotary types such as vane, piston, screw, gear, lobe, or peristaltic pumps.

Dynamic pumps (Figure 7.7) depending on the direction of the flow through the pump relative to the direction of the shaft rotation, can be **radial**, **axial**, or **mixed flow**.[1] These pumps can have a **single stage** or **multiple stages**. Each stage has moving blades (vanes), typically referred to as **impellers**, and fixed blades or **diffusers**. An impeller imparts energy to the flow and accelerates it. Impellers can be with or without a shroud (also known as enclosed or open impellers, respectively). Impellers can

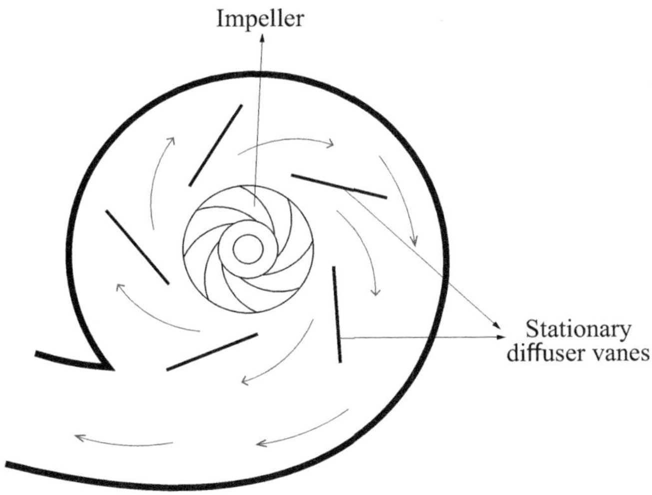

FIGURE 7.7
Schematic of a dynamic radial pump.

[1] While some special effect pumps, such as jet pumps, are also considered as dynamic pumps, we will not cover them in this section.

receive the flow from one side (single suction) or both sides (double suctions). A diffuser converts the resulting kinetic energy from the impeller to pressure. In radial flow pumps, the flow from the diffuser is typically discharged to a casing with an increasing area, called a volute, to maintain a constant flow velocity. Pumps can have a horizontal or vertical shaft. Pump casings are usually manufactured in two or more parts and depending on how these parts are connected, a pump can have an axially or radially split casing.

In the design and selection of pumps, special attention should be paid to avoid the cavitation problem. In a liquid flow, if the flow pressure drops below the saturation pressure of the fluid at the operating temperature (the vapor pressure), the liquid starts to boil. For pumps, due to the suction effect of the impellers, the lowest pressure is in the inlet of the pump. If the flow pressure drops below a certain limit, some liquid is evaporated and vapor bubbles are formed in the inlet flow. The formation of these small vapor-filled cavities is called **cavitation**. When the flow gets close to the impeller, its pressure increases and the vapor bubbles collapse, which can generate shock waves. The cavitation and the subsequent shock waves increase noise, vibration, and erosion of surfaces in pumps, particularly impellers. A parameter called the **net positive suction head (NPSH)** is used to avoid cavitation. The required net positive suction head ($NPSH_R$) is the pressure needed at the pump inlet to avoid cavitation. The available net positive suction head ($NPSH_A$) of an application should be always above the $NPSH_R$ with a safe margin.

The design principles of these pumps are similar to those of steam turbines (refer to Chapter 6) and involve using velocity vector diagrams. While essential for designing and manufacturing of pumps, for most users what they need is the pump selection (or sizing). For this propose, first a parameter called the **specific speed** is used to identify the type of the required pump. The specific speed is defined as follows:

$$N_s = \frac{N\sqrt{Q}}{(H)^{3/4}} \tag{7.1}$$

where N is the rotational speed of the pump in revolutions per minute (rpm), Q is the flow rate (m^3/s), and H is the total head (m). The specific speed is actually the simplified version of a dimensionless parameter, $\frac{N\sqrt{Q}}{(gH)^{3/4}}$, that can be found from the dimensional analysis of parameters impacting the performance of a pump. The value of the specific speed can help to identify the geometry (shape) of the pump impeller. A high value of the specific speed, which indicates a high flow rate and/or a low head, indicates a requirement for an **axial pump**, while a low value of the specific speed indicates a need for a **radial pump**.

When the geometry of the impeller is determined, pump performance curves that are provided by pump manufacturers can be used to identify the specific appropriate pump and its operating conditions. A pump performance curve in the simplest form illustrates the relationship between the head that a pump can create as a function of the volumetric flow rate (Figure 7.8). The curve shows that with increase in the volumetric flow rate, the head decreases (means less pressure rise in the flow). Typically the performance curve comes with the efficiency, required power input, and the required NPSH curves as a function of the flow rate (Figure 7.8). Pump manufacturers typically group the performance curves of a series of pumps with the same geometry but different sizes into a single diagram. These curves are typically presented according to the diameter of the impellers. Note that when two or more pumps operate in parallel, as is very common in steam power plants, the performance curve of the system can be created by adding up the flow rate of all pumps for each given head.

There are a variety of standards issued by several institutions that regulate pump manufacturing and applications, including the American Petroleum Institute (such as API 610/ISO 13709), American Society of Mechanical Engineers (such as ASME B73 series), and Hydraulic Institute (HI) standards and guidebooks. We will occasionally refer back to this introduction to pumps later whenever we talk about a specific pump used in a steam power cycle.

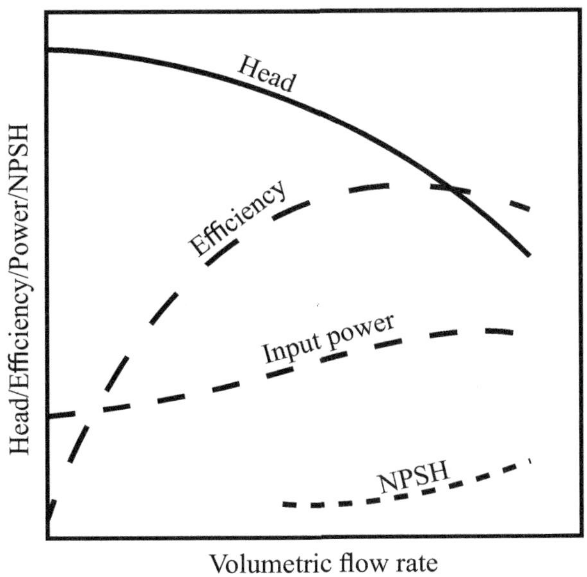

FIGURE 7.8
Performance curve of a typical pump.

7.4 Condensate Pumps

It was noted in Section 2.3.4.1 that steam power plants are equipped with one open feedwater heater. Since in an open feedwater heater the flows of steam and water are mixed, the two flows should be at the same pressure. This means that the pressure of the working fluid cannot be increased to the pressure of the steam generator in a single pump. The pressurization of the condensed steam (condensate) in a steam power plant takes place in two pumps. In this section, we will cover the first pump called the condensate pump and later we will discuss about the second pump called the feedwater pump.

A **condensate pump**, also known as a **condensate extraction or circulating pump** or simply **extraction pump**, receives saturated or slightly subcooled condensate water from the hotwell of the condenser at vacuum pressures and delivers the flow to the open feedwater heater at the pressure of steam bled from the turbine. The pressure increase in the condensate pump should be sufficient to compensate the pressure drop through the low-pressure feedwater heaters and other components in between.

Since the condensate in the condenser's hotwell is saturated water, the slight drop in the pressure of the flow will cause the condensate to re-evaporate. This means unless the required NPSH of the condensate pump is provided, the pump can suffer from cavitation in its inlet. While thermodynamically speaking the outlet flow of the condenser should be saturated liquid, to prevent cavitation, the outlet condensate is typically slightly subcooled. This will increase the condenser load and the required heat addition in feedwater heaters and will slightly reduce the overall efficiency of the cycle.

To meet the NPSH requirement, condensate pumps are typically located **under the condenser below the floor of the turbine hall**. The elevation difference between the condenser outlet and the pump inlet should be high enough so that the static pressure in the pump inlet due to the column of water can exceed the required NPSH with a safe margin at all operating conditions. To create the needed level difference, the condensate pump can be installed in a pit underneath the condenser. The pit can be dry or wet. In a dry-pit condensate pump, the condensate from the hotwell is transferred to the pump inlet by a piping system. To minimize the pressure drop in the condensate pump inlet flow, large diameter pipes are used to reduce the flow velocity. Also, a direct route with minimum fittings is used. Alternatively, the condensate pump can be submerged in a wet-pit filled with condensate water. Another configuration that is very common in modern steam power plants is a can-type condensate pump. In this design, a vertical pump is submerged in the condensate water collected in a suction can (or barrel), which is vertically installed at the bottom of the hotwell.[2] This design requires the least floor space. Furthermore, submerged condensate pumps do not need a

[2] Similar to pumps used to extract water from water wells.

priming system. In all designs, the hotwell is equipped with a level sensor. When the water level in the hotwell reaches a predetermined low level, to avoid cavitation the system is tripped.

Condensate pumps, depending on the needed head, can be single stage or multistage. They can be vertical or horizontal pumps with a single- or double-suction impellers. An impeller is made of a material with high erosion resistance, such as stainless steel, sometimes with a coating to resist possible cavitation. To maximize the available head, the inlet of the pump is at its bottom. To reduce the required NPSH, condensate pumps are typically low rotational speed (rpm) pumps. The efficiency of these pumps is at mid to high 80%. The flow rate of condensate pumps can be controlled by means of a **variable speed drive**, a **control valve in the outlet flow**, or the **recirculation of the flow to the condenser from the pump discharge**. To avoid air leakage to the inlet flow, where pressure is below the atmospheric pressure, the inlet is equipped with a specially designed sealing. While some plants may have only one condensate pump, typically there is a stand-by pump as a backup so that the cycle can operate at the full capacity while one of the pumps is out of service. In the latter case, the plant can be equipped with two, three, or four pumps each capable of handling the entire, one-half, or one-third of the designed condensate flow rate, respectively. The condensate flow is distributed among these pumps by a suction header and collected by a discharge header.

Many modern steam power plants, particularly supercritical ones, are equipped with a unit called a condensate polishing plant (CPP, refer to Section 7.8.5). The process requirements of this unit favor dividing the function of a single condensate pump between two pumps. In these plants, most of the needed pressure head is provided by a second pump called a **condensate booster pump**. This pump is usually located after the CPP. Since the condensate pump provides a sufficient NPSH at the inlet of the condensate booster pump, the condensate booster pump is typically a conventional centrifugal pump.

7.5 Feedwater Heaters

In Section 2.3.4 we discussed that heating subcooled water from the outlet of the feedwater pump to saturated water in an economizer involves significant irreversibility. This irreversibility that can negatively impact the efficiency of the cycle is due to the large temperature difference between the working fluid and combustion products. While ideally a regenerative steam cycle can be used to minimize this irreversibility, this cycle is not practical. Therefore, steam power plants are equipped with several feedwater heaters to reduce this loss. Since their introduction in the 1920s, feedwater heaters have gained steady popularity and now are the major components in majority of modern steam power plants.

The function of feedwater heaters is to step by step increase the temperature of feedwater before being fed to a steam generator. Although even with feedwater heaters the inlet feedwater to the steam generator is not saturated liquid and an economizer is still needed, the size of the economizer is much smaller than what it would have been if no feedwater heater is used. In all feedwater heaters, heat is transferred from steam extracted (bled) from the turbine to the working fluid. The steam is bled from the turbine and is piped to the appropriate feedwater heater. The steam bled from low-pressure stages of the turbine is sent to low-pressure feedwater heaters. This steam increases the temperature of the condensate water. To provide sufficient temperature difference between the bled steam and the feedwater in high-pressure feedwater heaters, the high-pressure steam at high temperature is sent to high-pressure feedwater heaters. To avoid any possibility of backflow of water or steam from entering the turbine through these extraction points, the piping systems are equipped with non-return check valves and automatic valves that are closed in abnormal operating conditions such as a turbine trip or a high water level in heaters. Depending on the mechanism for the heat transfer, feedwater heaters can be **closed** or **open**. Before covering the details of these feedwater heaters you are advised to review the relevant materials in Section 2.3.4.

7.5.1 Closed Feedwater Heaters

Closed feedwater heaters are **shell-and-tube heat exchangers**[3] where condensate/feedwater flows inside the tubes and steam from the turbine flows outside of them. The steam from the turbine enters the shell at one end and leaves at the other end. Depending on the pressure of the bled steam, it can be superheated steam at high pressure or saturated steam. Due to heat transfer to the water, most of the extracted steam condenses and the outlet flow can be a mixture of condensate, vapor, and noncondensable gases.

Some closed feedwater heaters are located right after the condensate pump and before the open feedwater heater. They operate at low pressures, hence the name **low-pressure (LP) feedwater heaters**. The extracted steam in these feedwater heaters can be saturated or slightly superheated steam. The rest of the feedwater heaters are located after the feedwater pump and before the steam generation. They operate at high pressures and accordingly are called **high-pressure (HP) feedwater heaters**. If the cycle is equipped with a condensate booster pump, feedwater heaters located between this pump and the feedwater pump operate at intermediate pressures. These feedwater

[3] For a useful introduction to shell and tube heat exchangers, please refer to http://thermopedia.com/content/1121/.

heaters are referred to as **intermediate pressure (IP) feedwater heaters**. The extracted steam for HP and IP feedwater heaters is typically superheated.

Most components of closed feedwater heaters are similar to the components of a condenser. The shell is typically cylindrical in shape capped by divided or undivided water boxes on one end or both ends (Figure 7.9). These heat exchangers can be installed horizontally or vertically. The orientation of the unit is usually determined by space availability. Tubes are usually made of stainless steel and can be U-shaped or straight, the former being more popular. The U-tube feedwater heaters are two-pass heat exchangers connected by a single tube sheet on one side of the shell (Figure 7.9). In the straight tube configuration, tubes typically have two passes.

To provide several passes in the steam flow and allow the steam interaction with the entire surface of tubes, closed feedwater heaters are equipped with several **baffles**. Baffles are plates that fit inside the shell and cover most of the shell intersection except one segment cut or two opposing segments cuts (Figure 7.10). These cuts are alternatively rotated to induce a zig-zag flow pattern in the steam flow. Since tubes go through baffles via tightly fit holes,

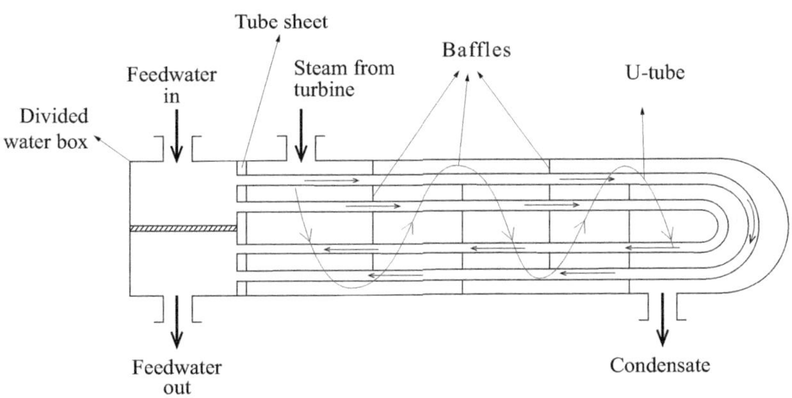

FIGURE 7.9
Schematic of a U-tube feedwater heater.

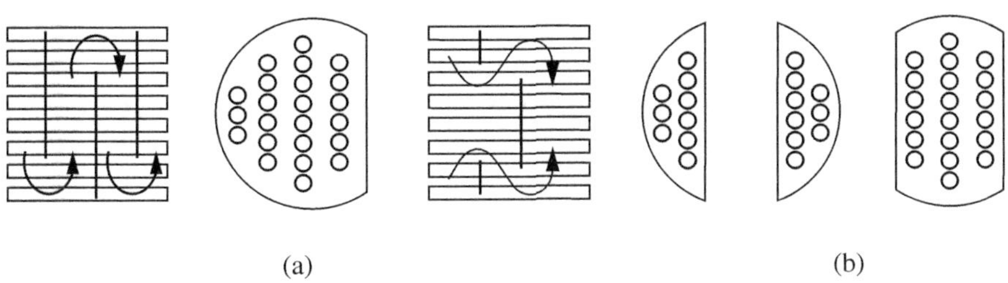

FIGURE 7.10
Schematic of baffle arrangements in closed feedwater heaters

they also support tubes and prevent the flow induced vibration in the tubes. Increasing the number of baffles increases the heat--transfer coefficient at the expense of increasing the flow pressure drop.

The heat-transfer mechanism in the shell side of a feedwater heater primarily depends on the state of the bled steam. In all feedwater heaters, steam is condensed. If the bled steam is superheated, the flow is also desuperheated. In many feedwater heaters, the condensed bled steam (often referred to as the heater drain) is cooled down to subcooled water by further heat transfer to the working fluid. In very low-pressure feedwater heaters, since the drain is at a very low temperature, there is not much opportunity for this heat transfer and the drain subcooling may be skipped. Depending on the dominant heat-transfer mechanism, a feedwater heater can be divided into three sections: **desuperheater**, **condenser**, and **drain cooler**. While all closed feedwater heaters have a condenser section, they may have either, both, or none of the other two sections. Typically the desuperheating and drain-cooling zones are separated from the condensing zone by a shroud or wrapper. Sometimes the drain cooling takes place in a separate heat exchanger.

Now let us consider the feedwater side of closed feedwater heaters. In a typical feedwater heater, the feedwater enters the lower water box and is distributed among the tubes fixed to the tube sheet. The feedwater is first heated in the drain cooler by the condensed steam accumulated at the bottom of the shell. Then, the feedwater receives heat from condensing steam and superheated steam in the condensing and desuperheating zones, respectively. Finally, the heated feedwater returns to the upper section of the water box and exits the feedwater heater. In the shell side, the bled steam enters the feedwater heater through one or more inlets. Upon entering, the steam is guided to the tube bundle by impingement plates. These plates also protect the tubes from steam flow erosion. The steam passes through the desuperheater and condenser zones and gives up its energy to the feedwater inside the tubes. The condensed steam falls to the bottom of the shell to the drain-cooling zone. To remove noncondensable gases, the condensing zone of a closed feedwater heater should be equipped with a vent. The lowest pressure feedwater heater is sometimes located in the inlet neck of the condenser in the transition duct between the turbine and the condenser.

As noted in Section 2.3.4.2, the heater drain in the shell side of a closed feedwater heater is at a pressure lower than that of the feedwater/condensate flow in the tubes. This means the outlet flows of the shell and the tubes cannot be directly mixed. To address this problem, two approaches can be used: drains pumped forward or drains cascaded backward. In the former, the condensed bled steam is pressurized in a small pump to the pressure of the feedwater flow and mixes with the tube-side outlet flow in a mixing chamber. In a drain cascaded backward system, the condensed bled steam is throttled through a steam trap (Section 2.3.4.2) to a lower pressure and drained to either the next lower pressure closed feedwater heater, the open feedwater heater, or the condenser. Drains pumped forward feedwater heaters slightly improve the cycle efficiency because they avoid the irreversibility of the throttling process. However, drains cascaded backward feedwater heaters are simpler and less expensive. For this reason, drains cascaded backward feedwater heaters are much more common. Some designs use the backward cascading system for all feedwater heaters, so each feedwater heater successively drains to the next lower pressure heater except for the lowest pressure feedwater heater in each LP and HP feedwater heater stages (Figure 7.11). This last heater is a drain pumped forward feedwater heater that pumps all the collected drains to the feedwater or condensate lines. In this case, the heat content of the drains is not lost to the condenser. The drain pump operating condition is severe since the inlet water is near-saturated liquid (similar to the condensate pump). To provide a sufficient NPSH, vertical can-type pumps are typically used for this application.

The decision about the number of closed feedwater heaters is primarily made based on economic considerations, particularly the balance between the capital cost and the plant efficiency. The size of the plant is a significant

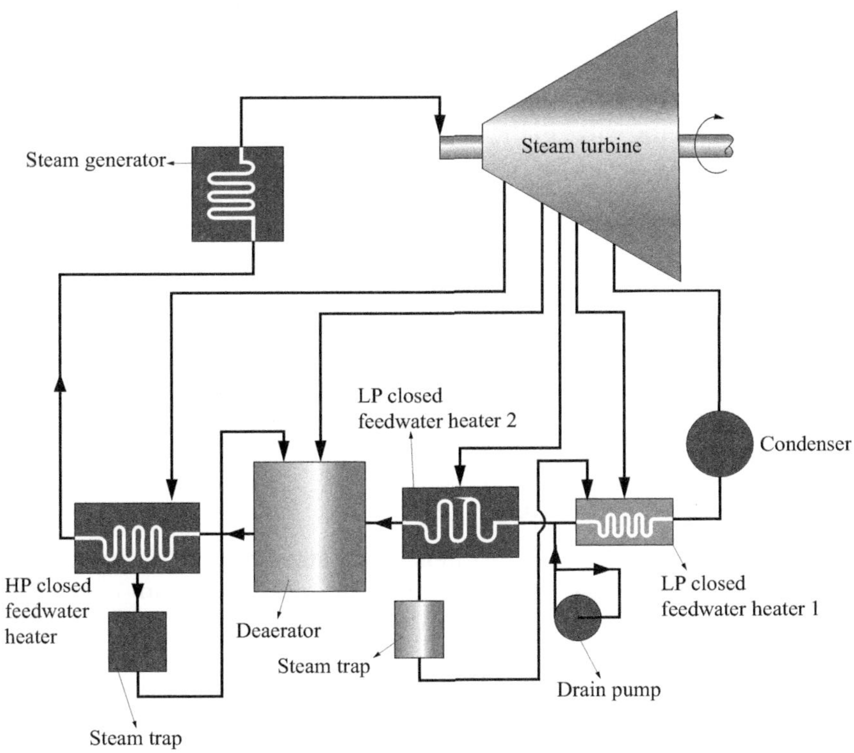

FIGURE 7.11
Schematic of a typical feedwater heater arrangement (condensate and feedwater pumps are not shown).

factor in this balance. Typically smaller units have fewer feedwater heaters. Large utility scale steam power plants can have up to eight closed feedwater heaters. To determine the pressure of the bled steam from the turbine for each feedwater heater an optimization process is required. However, a rule of thumb for a rough estimate is to divide the temperature difference between the condenser outlet and the steam generator inlet into equal increments. Then, the steam is extracted from the turbine at the pressure equal to the saturation pressure of each temperature increment.

Feedwater heaters are typically designed and manufactured according to the **Tubular Exchanger Manufacturers Association (TEMA)** standards.[4] The performance of closed feedwater heaters is evaluated using the ASME PTC 12.1 Feedwater Heater Standard.

7.5.2 Steam Trap

As noted in the previous section, a steam trap is used to throttle the outlet condensed steam from a drain cascaded backward closed feedwater heater to lower pressure. As the name implies, a **steam trap** not only reduces the flow pressure but also traps the steam accompanying the condensate. A steam trap allows the flow of condensate and noncondensable gases to pass but holds on to the steam until it condenses. A steam trap in essence is composed of a valve that opens when a certain condition(s) is met. Steam traps typically operate in cycles. That means the valve in the trap is closed until the sufficient condensate is collected and then the valve is opened to allow the condensate to discharge. To determine the condensate discharge timing, the following three principles or a combination of them can be used: mechanical traps, thermodynamic traps, or thermostatic traps. These traps use the condensate level, thermodynamic and fluid property differences between steam and condensate, and the temperature difference between steam and condensate, respectively, to determine the amount of accumulated condensate and to respond by opening the valve.[5]

7.6 Open Feedwater Heaters

As noted in Section 2.3.4.1, in an **open feedwater heater**, also known as a **direct-contact feedwater heater**, the feedwater and bled steam are directly mixed to increase the temperature of the feedwater. Because of this direct

[4] The Tubular Exchanger Manufacturers Association, Inc. (TEMA), established in 1939, is the trade association of shell and tube heat exchanger manufacturers. The latest (Tenth Edition) of the standards were published in 2019. http://www.tema.org/.

[5] For more information on steam traps, you can refer to the following freely available handbook: "The industrial steam trapping handbook" https://www.emerson.com/documents/automation/yarway-steam-traps-en-us-3587464.pdf or the following video https://www.youtube.com/watch?v=IiRyxcCBTa0.

mixing, the pressure of two flows, the extracted steam and the feedwater, must be equal. This means in an open feedwater heater the pressure of the feedwater cannot be higher than that of the extracted steam. Therefore, for each open feedwater heater, a pump is needed to pressurize the feedwater. Because of this, all steam power plants (except some nuclear ones) are equipped with only one open feedwater heater, which follows by the main feedwater pump. Even small industrial units, which may not have any closed feedwater heater, have an open feedwater heater. This is because open feedwater heaters also function as the site where gases are removed from feedwater. In fact, these heaters are typically known as **deaerating heaters** or simply **deaerators** (DA). In large utility steam power plants, the open feedwater heater is typically located in the middle of closed feedwater heaters, between LP and HP heaters. The exact location of a deaerator relative to other feedwater heaters and its operational conditions are determined to maximize the condensate deaeration (next paragraph).

Previously we discussed how noncondensable gases are removed from the condenser. While this process effectively removes almost all noncondensables in gaseous form, it cannot do anything with gases dissolved in the condensate. These gases are mainly oxygen and carbon dioxide with much less amount of nitrogen, argon, hydrogen, methane, and ammonia. They can enter the working fluid by makeup water, cooling water leakage in the condenser, or the air leakage to the system where the fluid pressure is lower than the atmospheric pressure at piping connections (e.g., in flanges) or sealing of rotating devices. **The solubility of these gases in water reduces with temperature increase**. For example, while the oxygen solubility is about 0.07 g of oxygen per kg of water at the temperature of 0°C, at the temperature of 100°C the solubility reduces to about zero. This means the condensate at the condenser has the maximum amount of dissolved gases. With gradual increase in the water temperature in low-pressure feedwater heaters, the solubility of the condensate reduces and some of the dissolved gases are released. The presence of these gases in the feedwater is undesirable. Particularly oxygen is problematic because it can react with metals like iron and copper and can corrode metal surfaces. The reactions can also generate metal ions that are soluble in water and can contaminate feedwater. Also, carbon dioxide increases water acidity, which again increases corrosion. Furthermore, some of these gases, such as hydrogen and methane, are combustible and their presence in the steam generator can be hazardous. An open feedwater heater can provide ideal conditions for the removal of gases, especially considering the mixing process that takes place within the heater. The operating temperature of the heater, at around 100°C, is selected to maximize the release and removal of noncondensable gases. Physically speaking, the deaerator is installed at high elevation in the boiler room to provide a sufficient head at the inlet of the following feedwater pump to avoid cavitation (refer to Section 7.7).

An open feedwater heater can have several other functions. The makeup water is added to the cycle with the open feedwater heater. Chemicals that are needed for water treatment are commonly added to the feedwater within the heater or somewhere nearby. Moreover, the heater, sometimes in connection with the feedwater tank, functions as the container in which the fluctuation in the needed feedwater flow rate is accommodated.

Condensate heating and deaeration can be achieved by three mechanisms. In the first mechanism, the water is sprayed into a container filled with steam. The interaction between water droplets and steam results in heat transfer from the steam to the water and the removal of dissolved noncondensable gases from the feedwater. The large surface area of water droplets immensely facilitates both processes. In the second mechanism, steam is injected into the feedwater collected at the bottom of the deaerator. The upward moving of steam bubbles agitates the water which results in water heating and deaeration. In the last mechanism, feedwater descends through a stack of horizontal trays while steam rises in the opposite direction. The interaction between the cascading water flow and the rising steam increases the water temperature and removes most noncondensable gases. Based on these three mechanisms, three types of deaerators are common in modern steam power plants: the spray deaerators, spray-scrubber deaerators, and spray-tray deaerators. In these deaerators, some combinations of the abovementioned mechanisms are used for water heating and deaeration.[6]

7.6.1 Spray Deaerators

A **spray deaerator** is a cylindrical horizontal pressure vessel (Figure 7.12). Inside the vessel there is a small shrouded area at the top of the heater. The bottom of the shrouded area is typically (but not always) made of a perforated plate. Steam enters through the shrouded area via this plate from the main body of the container. The condensate from the last low-pressure feedwater heater is sprayed into the steam within the shrouded area. The mixing of steam and water increases the temperature of the condensate and releases some noncondensable gases. Water droplets fall on the perforated plate and fall down to the preheating area of the heater through the holes on the plate. While dropping, water droplets pass through the steam filling the container and continue heating up. Water droplets first land in an area of the heater called the preheater, which is separated from the deaerating area of the heater by a baffle (Figure 7.12). The extracted steam from the turbine is distributed by pipes to the submerged steam sprays that inject steam into the entire body of water at the lower part of

[6] For an excellent presentation of how deaerators work, please watch the following visual discussion at "How Deaerators Work (Engineering)" https://www.youtube.com/watch?v=M_jOsTWVIH8.

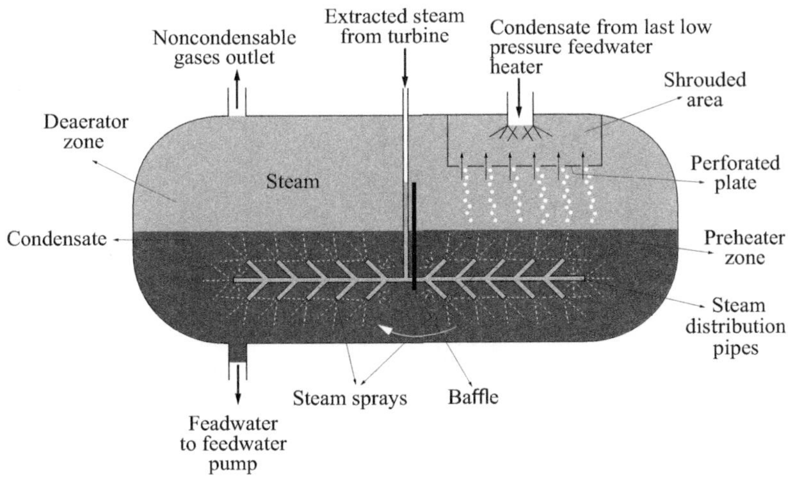

FIGURE 7.12
Schematic of a spray deaerator.

the heater. The steam bubbles up through the water by which the heating and deaerating processes are accomplished. After the preheating, the water flows under the baffle to the other side of the container, called the deaerator zone, for further heating and deaerating. As noted, in both the preheater and deaerator zones the steam is injected into the water. After sufficient heating and deaerating, the water, which is now called the feedwater, leaves the heater via an outlet at the bottom of the deaerator area. The released noncondensable gases along with some steam are vented to the atmosphere through an outlet (Figure 7.12).

7.6.2 Spray-Scrubber Deaerators

In **spray-scrubber deaerators**, the inlet water, usually just the fresh makeup water, is first sprayed into the steam accumulated at the upper portion of the heater (Figure 7.13). The sprayed water is then collected by a tray right below the sprays. The condensate is typically fed to the tray directly. Both condensate and makeup water are then directed by a pipe to the bottom of a can called a scrubber. The extracted steam enters the heater at the top and is transferred to sprays submerged at the bottom of the scrubber. The steam is sprayed into the water inside the scrubber and bubbles up through the water. The rising steam violently agitates the water inside the scrubber and scrubs the water, facilitating the heat-transfer process and the removal of noncondensable gases. The remaining steam leaves the scrubber and is collected at the upper section of the heater. The heated and deaerated feedwater is collected at the lower section of the heater and is discharged through an outlet at the bottom to the feedwater pump (Figure 7.13).

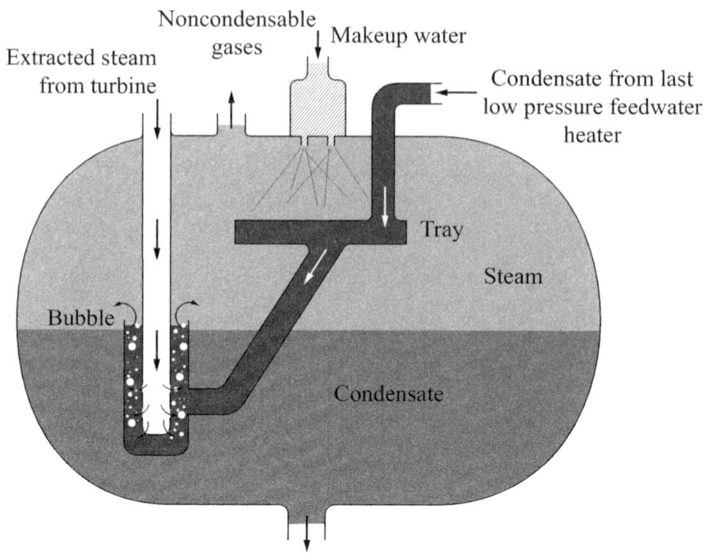

FIGURE 7.13
Schematic of a spray-scrubber deaerator.

7.6.3 Spry-Tray Deaerators

Spray-tray deaerators are composed of a series of horizontal stainless steel trays. The top trays are referred to as the heating trays or first-stage trays and the bottom ones are known as the air separating, deaeration, or secondary trays. The inlet water flow is sprayed on the top tray through the steam filling the container. The extracted steam flow is directed to the bottom of the tray stack. The steam ascends through trays while the water descends (counterflow). The high surface area between two flows facilitates the heat transfer and noncondensable gas removal processes. This type of deaerator is typically installed above (or sometimes inside) a much larger feedwater tank (the function of this tank is discussed below). The heated and deaerated feedwater falls to the feedwater tank below. The feedwater tank itself may have inlet steam and submerged distribution sprays for further heating and deaeration within the tank. Sometimes the deaerator system is used to refer to both the deaerator and the feedwater tank.

7.6.4 Other Functions of Deaerators

As discussed above, dissolved noncondensable gases are removed from condensate by a deaerator. These gases are vented to the atmosphere through a dedicated outlet. Unlike the condenser, the pressure at the deaerator is above the atmospheric pressure, so no pressure increase device, such as a steam ejector or compressor, is needed to vent the gases. When these gases are discharged to the atmosphere, steam cannot be separated effectively. So it is inevitable that some steam escapes through the deaerator along with noncondensable gases. The escaped steam forms very visible plume upon leaving the system. While the released steam should be minimized because it is a source of thermal loss, the effective removal of gases should also be ensured. Typically a deaerator is equipped with a safety relief valve installed at the highest elevation of the system.

We have already discussed some components and processes that result in losses of the working fluid. For example, saturated water is continuously blown down from the drum to maintain the quality of feedwater. Also, steam leaves the plant accompanying noncondensable gases released to the atmosphere from the condenser and in the open feedwater heater. Therefore, to maintain the required feedwater flow rate throughout the power cycle and to compensate for the lost feedwater, makeup water must be added to the feedwater. This makeup water is supplied by the water treatment facility (refer to Section 7.8.3). This water either directly enters the deaerator or is mixed with the condensate before being fed to the deaerator. The makeup water contains a significantly higher level of dissolved noncondensable gases than that of the condensate. So the deaeration process in the open feedwater heater is even more critical for this fresh working fluid. In some deaerator designs, such as some spray-scrubber deaerators, the makeup water goes through extra steps of heating and deaeration to properly treat the water.

The deaerator can remove almost all CO_2 and most oxygen from the feedwater. However, the deaerated water still contains about 7 part per billion (ppb) of dissolved oxygen. This remaining oxygen cannot be removed mechanically in the deaerator and must be removed chemically. For this purpose, volatile oxygen absorbers (also known as oxygen-scavenging chemicals or oxygen scavenger agents), such as sodium sulfite with the chemical formula of Na_2SO_3 (for low-pressure steam generators), hydrazine with the chemical formula of N_2H_4,[7] or erythorbic acid with the chemical formula of $C_6H_8O_6$, can be added to the deaerated feedwater. The makeup water may also require some chemical treatments depending upon the water treatment mechanism (refer to Section 7.8.3). The addition of the abovementioned chemicals can take place on the makeup water line, at the deaerator, after the CPP (refer to Section 7.8.5), at the feedwater tank, or after the deaerator before the steam generator.

The final function of the deaerator is to accommodate the variation of the feedwater flow due to plant load changes. Consider a sudden increase in the load. The increase in the load requires a higher steam flow rate through the turbine and a higher rate of steam generation in the steam generator. In the absence of an accumulator, the extra water flow rate should be drawn from the condenser hotwell. While the hotwell is sized to provide some

[7] Hydrazine is considered a carcinogen, which limits its applications.

margin for this purpose, if the load change is large enough, the hotwell may not be able to cope and get dry. This may starve the steam generator and may damage some components. This condition will continue until the excessively drawn condensate is returned to the condenser and condensed back to the hotwell and the cycle reaches a new equilibrium condition. The opposite can happen when the plant's load reduces. To avoid these conditions, a large enough tank should be located before the steam generator to temporarily accommodate the fluctuations in the feedwater flow rate. To perform this function the deaerator should be either sufficiently large or, as noted for the case of spray-tray deaerators, installed above a large feedwater tank. The feedwater tank is where the highly treated water from the deaerator is stored.

7.7 Feedwater Pumps

A **steam generator feedwater pump**, also known as a **boiler feed pump**, takes the heated and deaerated feedwater from the outlet of the deaerator (the open feedwater heater) and pressurizes it. At the outlet of the feedwater pump, the feedwater pressure is at its highest throughout the entire cycle. In cycles with a drum-type steam generator, this pressure should be high enough to overcome all pressure drops between the pump and the steam generator drum and deliver the feedwater to the drum at the required pressure. In cycles with a once-through steam generator, the pressure after the pump should be sufficient to deliver superheated steam to the turbine at an appropriate pressure.

Since the deaerator outlet is saturated or near-saturated water, the inlet pressure to the feedwater pump is approximately equivalent to the saturation pressure of water at the operating temperature of the deaerator. Therefore, similar to the condensate pump, the feedwater pump is susceptible to cavitation. To provide the required NPSH in the inlet of the pump, there should be a sufficient elevation difference between the deaerator and the feedwater pump inlet. For this purpose, the deaerator is typically installed at a high elevation in the turbine room structure and the feedwater pump is installed at the floor level. The available NPSH for the feedwater pump that is equal to the height difference between the water level in the deaerator heater or the feedwater tank and the pump inlet pipe centerline lessens the pressure drop in the piping system between two components. The head should be higher than the required NPSH of the pump with an acceptable safety margin for transient operating conditions. In some designs to reduce the severity of the main feedwater pump operation conditions, another pump is used upstream of the main feedwater pump. This pump is referred to as the **steam generator feedwater booster pump** (also the **boiler feed booster pump**). Its function is to provide sufficient pressure in the inlet of the main feedwater pump to meet its required NPSH. This is a low head pump with a low rotational speed (rpm), which reduces its required NPSH. The booster pump can be powered by the same driving system as the main pump through a speed reduction gear.

Feedwater pumps are typically horizontally installed centrifugal pumps. They can be single stage or multistage pumps. Similar to condensate pumps, several feedwater pumps can be used in parallel with one pump as a reserve. In some steam power plants, this pump is driven by a **variable speed electric motor**. But for some large units, the pump is so large that its power consumption can significantly increase the internal electricity consumption of the plant and reduce its net power generation. To increase net electricity production and to improve the overall efficiency of the cycle, in most medium- and large-size steam power plants, the feedwater pump is driven by a steam turbine. This steam turbine should not be mistaken with the main steam turbine that is coupled with the generator. This turbine is dedicated to driving the feedwater pump and is a relatively small turbine. This turbine is typically a variable speed steam turbine so that the feedwater flow rate can be controlled without throttling the flow; thus the irreversibilities associated with the throttling process are avoided. The turbine can be an impulse or impulse-reaction multistage turbine. For more effective speed control it may receive steam at different pressures. The flow rates of these steam flows are controlled by control valves to achieve the desired speed and pump flow rate. The discharge flow of the turbine can be exhausted to either the main condenser, a small dedicated condenser, or a low-pressure closed feedwater heater. Note that during the start-up when steam from the steam generator is not available yet, an alternative approach to starting the turbine should be considered such as an auxiliary steam generator. Many aspects and components of this turbine are similar to the main steam turbines covered in Chapter 6.

7.8 Treatment of Feedwater and Makeup Water

While steam power plants operate in a closed cycle and the working fluid circulates over and over, the chemical properties and the composition of the working fluid continuously vary throughout the cycle. This variation is due to the fact that water and steam can dissolve or chemically react with what they contact. Water can also decompose to positive and negative ions (H^+ and OH^-). Dissolved solids in water, expressed as water hardness, can be deposited in cycle components when the temperature and velocity of flows change. This can be particularly problematic during phase change processes. The deposition of scale forming impurities on heat-transfer surfaces can form an insulating scale layer, which negatively impacts the rate of heat transfer. Similarly, depositions can also happen in turbines, pumps, and piping systems. The presence of these layers reduces the efficiency of components. Their removal and surface cleaning increase maintenance costs and plant outages. Low-quality feedwater can also accelerate corrosion and erosion processes in system components, which again increase required maintenance and the frequency of component replacements. Therefore, it is critical to ensure that the working fluid quality is always within acceptable limits. The hardness of feedwater can be typically expressed by its **electric conductivity** with the unit of microsiemens per centimeter (μS/cm). The required conductivity of feedwater in a steam generator depends on many factors, including steam generator type and operating pressure, turbine steam requirements, and the rate of blowdown. The feedwater quality is more critical for once-through steam generators than that for drum-type steam generators, because they do not have a drum to remove dissolved and suspended solids. In these steam generators, all impurities in the feedwater would be deposited on tubes' inner surfaces. Both ASME and Electric Power Research Institute (EPRI) have guidelines for feedwater and steam chemistry control; however, the best source of information in this regard is equipment manufacturers.

Some of the feedwater quality factors are dissolved and suspended solids, dissolved oxygen, pH, and oil content. The control of water chemistry in a steam power plant can be achieved by many processes in many components and can be grouped as follows:

- Blowdown in the drum
- Mechanical and chemical deaeration in the deaerator heater
- Chemical control of pH
- Makeup water treatment
- Condensate polishing

The first two items have already been covered. In this section, the remaining items will be discussed.

7.8.1 Chemical Control of pH

The chemical control of pH can be achieved by adding chemicals, such as alkalizers including ammonium hydroxide, to the feedwater before the steam generator. This is commonly referred to as an **all-volatile treatment (AVT)**. The addition of chemicals is commonly done at the condensate pump outlet or at the downstream of the CPP (refer to Section 7.8.5). Alternatively, chemicals, such as certain phosphate compounds, can be added to the drum. Drum-type steam generators have an option to chooose between adding chemicals both to the upstream of the steam generator and in the drum or just to the upstream of the steam generator. However, since once-through steam generators do not have a drum, the chemicals can only be added to the upstream of the steam generator. The pH level can be determined by measuring pH or conductivity. The latter is more accurate and is preferred in steam power plant applications. The measurements for the feedwater pH control are typically taken from the condensate pump outlet and the economizer inlet flows.

7.8.2 Makeup Water Treatment

In our discussion on deaerator heaters, we learned that makeup water is added to feedwater in the deaerator. The makeup water should compensate for the working fluid losses throughout the cycle. These losses can include the steam generator blowdown, leakage from fittings and sealing, and steam accompanying noncondensable gases discharged from the condenser and the deaerator heater. In typical steam power plants, the majority of the feedwater is composed of recycled condensate. A very small portion of the feedwater, typically about 1% of the total feedwater flow rate depending on the system design and its age, is provided by the makeup water. However, for cogeneration units, when steam is directly used in industrial processes, the share of makeup water may increase all the way to fully replace the entire feedwater. To maintain the quality of the feedwater, it is imperative to properly treat the makeup water.

The most important factor in determining the characteristics of the makeup water treatment facility is the source of raw water. The availability of a water source is a critical factor in identifying a site to build a steam power plant in the feasibility study phase. Water sources can be the city potable water supply, recovered water from other uses, surface water (such as lake or river), underground water (from shallow or deep wells), or rainwater. The available water should be analyzed to determine the range of all its impurities, including dissolved and suspended solids and gases. Dissolved solids are primarily composed of various salts of sodium, magnesium, calcium, and silica oxide as well as some carbonates, bicarbonates, and hydrates. The required makeup water treatment facility and its specifications depend on the required makeup water quality and the composition and properties of raw water. In large utility steam power plants, the water treatment facility can be so extensive that a building, the water treatment building, typically adjacent to the boiler room is dedicated to containing all needed equipment.

7.8.3 Makeup Water Treatment Systems

The objective of any makeup water treatment facility is to produce water free of dissolved and suspended solids. This water is referred to as **demineralized water**. Note that the removal of gases takes place later in the deaerator heater. In this section, we will briefly review the makeup water treatment processes.

First raw water goes through the pre-treatment processes. This starts with the clarification process where the water is placed in **settling tanks**, also referred to as **clarifier vessels**. In these tanks, coarse suspended solids, such as sand, mud, dirt, and debris, are separated and deposited by gravity. To remove finer particles and to accelerate the sedimentation process, special chemicals, such as iron sulfide, may be added to coagulate particles, which are then deposited. The water is also chlorinated to kill microorganisms and avoid the biofouling of system components. The **filtration process** is another method to remove suspended solid particles, which is used to supplement the abovementioned method. Filters with different media have been used for the filtration process. **Sand filters** are the most common filters where solid particles in the water are trapped by the fine pores within a sand bed. These filters are vertical cylindrical vessels typically fabricated from carbon steel. In these filters, the water flow enters the filter at the top and leaves at the bottom. To stop the media (sand) from being carried away by the exit flow, a strainer or trap is installed at the exit. As with any other filter, after operating for a long enough time, the filter will be nearly saturated with captured solids and will not be able to perform effectively. With increase in the trapped solid particles, the pressure drop across the filter increases. To determine how saturated the filter is, the pressure difference between the inlet and outlet flows of the filter is measured. When this pressure drop reaches a predetermined value, it is time to clean the filter. For this purpose, the flow of water is reversed. In this process, known as the backwashing process, the bed is expanded by the reverse water flow and the collected solid particles are released, carried away, and discharged to a waste collection system. At this point, the clean filter is ready to be operated again. Another type of filter is activated charcoal filters. They are used to remove microorganisms, organic compounds, and residential chlorine from the previous stages of water treatment.

The next step is the water softening stage where water hardness is reduced by removing calcium and magnesium salts. Also, the water alkalinity level is reduced by removing carbonates and bicarbonates. Some other impurities, such as silica, are reduced too. Several processes can be used for **water softening**, including the **lime-soda softening**. In the lime-soda softening process, lime (calcium carbonate, $CaCO_3$) and soda ash (Sodium carbonate, Na_2CO_3) are added to the water. They react with abovementioned impurities and form products that are insoluble in water and are deposited. If the temperature of the water is near the ambient temperature, the process is called the cold lime-soda softening. At elevated temperatures, above 100°C, the process is more effective and is referred to as the hot lime-soda softening. The cold process is more common in steam power plants. The lime-soda processes are labor intense and handling of the produced mud (sludge) is a challenge. Other processes for water softening are sodium or hydrogen ion exchange. Ion exchange processes will be explained next for the demineralization process.

7.8.4 Water Demineralization

The next step in the makeup water purification process is the **demineralization process** where almost all dissolved solids in the water are removed and the demineralized water is produced. There are several approaches for this purpose: **ion exchange demineralizers, reverse osmosis units,** and **evaporators.**

7.8.4.1 Ion Exchange Demineralizers

Ion exchange units are one of the main components of any demineralization facility. An ion exchange unit is a vessel that contains **resins.** Ion exchange resins are chains of hydrocarbons with some sites where actual ion exchange agents are attached to the polymer matrix. Resins are white or yellowish porous beads with diameters less than 1 mm. Based on the charge of ion exchange agents, there are two types of resins: **cation resins** and **anion resins.** Cation resins have positive charges and anion resins have negative charges. The types of ions in resins depend on their applications. For example, the cation and anion resins used for water softening typically have sodium ions (Na^+) and chloride ions (Cl^-), respectively. This process is referred to the sodium ion exchange. However, the cation and anion resins used for the water demineralization process typically have hydrogen ion (H^+) and hydroxide ion (OH^-), respectively. The cation resins can remove positively charged calcium, magnesium, and sodium ions from the water by replacing them with hydrogen ions in the resins and releasing H^+. For example, sodium and calcium ions are removed from the water according to the following reactions:

$$(H)_{Resin} + Na^+ \rightarrow (Na)_{Resin} + H^+ \tag{7.2}$$

$$2(H)_{Resin} + Ca^{++} \rightarrow (Ca)_{Resin} + 2H^+ \tag{7.3}$$

Similarly, the anion resins can capture negatively charged ions of chlorides, bicarbonates, sulfates, and silicate by releasing hydroxide ions (OH^-). The cation resins can be either strong-acid cation (SAC) exchange resins or weak-acid cation (WAC) exchange resins. Similarly, the anion resins can be either strong-base anion (SBA) exchange resins or weak-base anion (WBA) exchange resins.

Each type of resins is placed in a vertical vessel to form a resin bed. A cation ion exchange bed is followed by an anion ion exchange bed. To remove any remaining dissolved solids from the water, the individual ion exchange vessels are typically followed by a mixed-bed ion exchange vessel where both cation and anion resins are simultaneously present (Figure 7.14). A combination of vessels containing WAC, SAC, WBA, and SBA resins, as well as mixed-bed vessels are used in a typical makeup water demineralization unit (Figure 7.14).

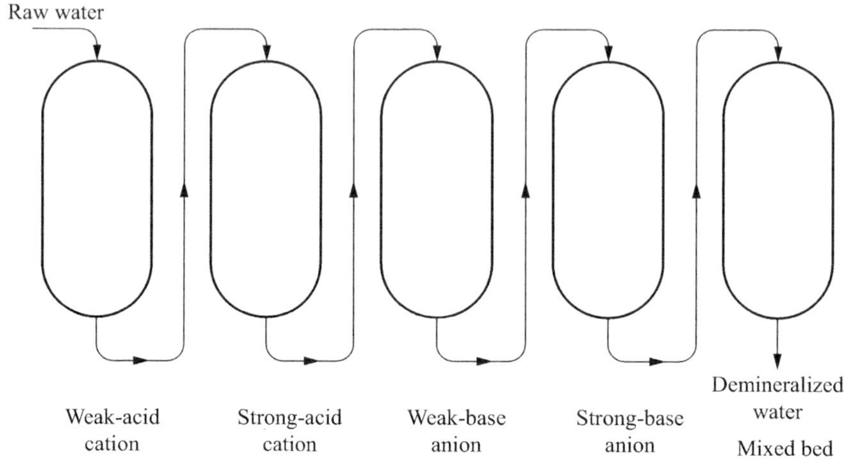

FIGURE 7.14
Schematic of a typical water demineralization plant.

Resins' capability to bond with ions in water is not unlimited and it is depleted after capturing a certain amount of ions. Fortunately, the reactions in both types of resins are reversible and resins can be regenerated to return to their initial form to be reused. To determine when the resins are exhausted and need regeneration, the conductivity of the outlet flow is measured. When this conductivity exceeds a certain limit, it is an indication that resins are not able to effectively remove dissolved solids and they should be regenerated. For the regeneration of cation resins, the flow of hydrochloric or sulfuric acid passes through the bed. For anion resins, caustic soda (sodium hydroxide) flow is used. Throughout the regeneration process, ions that are removed from the water by resins are exchanged with the original ions. In mixed-bed ion exchange units, before regeneration, resins must be separated and are transferred to different vessels. The separation process takes advantage of the different densities of cation and anion resins. A backwash water flow is used to separate resins hydraulically. The velocity of the water flow should be selected carefully to reduce the cross-contamination of resins. A resin regeneration facility should be equipped with the chemical storage tanks as well as metering pumps and appropriate piping systems to produce chemicals with proper concentrations. The facility may have the dedicated vessels to regenerate resins outside of the ion exchange vessels and the tanks containing fresh resins ready to replace depleted resins.

Similar to sand-bed filters discussed previously, ion exchange beds can trap suspended solid particles from the water flow. These particles along with fractured resin beads can reduce the porosity of the bed and increase its pressure drop. To detect this problem, the pressure difference across the bed is measured. If the water conductivity exceeds its limits before an excessive pressure drop is detected, the regeneration process will also physically clean the resins.

If the pressure drop exceeds its limits earlier, some beds are equipped with an ultrasonic cleaner to clean the resins by means of a mechanical process.

As noted earlier, resins are made of polymers and as a result, they are very sensitive to temperature. The high temperature of the water flow can damage or even melt the resins. Therefore, the water flow is allowed to pass through ion exchange beds up to a certain temperature limit. The operating temperature of resins depends on the resin type and is specified by the manufacturer. It can be as low of 50°C or as high as 130°C.

Typically several demineralizer units operate in parallel with one unit always in standby. As soon as a unit requires regeneration, the standby unit replaces that unit. A makeup water purification unit needs many supporting components, including a tank containing new resins to replace damaged resins, a resin transportation system, and resin traps to avoid resins from being carried over by the water flow. Also, the discharge flows from the regeneration processes should be neutralized with a balanced pH before being sent to a wastewater collection and processing unit. All these systems are automatically operated in harmony by a control system.

7.8.4.2 Reverse Osmosis (RO) Units

To understand the reverse osmosis (RO) process, let us first briefly review the direct osmosis process. This phenomenon happens when two water solutions with different concentrations of dissolved maters are separated by a **semipermeable membrane** (Figure 7.15). By its nature, the semipermeable membrane does not permit dissolved solids to cross and only water can pass through the membrane. The direction of this water flow is to equalize the concentration of the water at the two sides of the membrane. This means the water will flow from the solution with a low concentration to the one with a high concentration. The osmosis process takes place due to the partial pressure difference of the water between the two sides of the membrane. Note that the total pressures of the solutions are equal. This pressure difference is referred to as the **osmotic pressure** (Figure 7.15). In essence, the osmotic pressure is the pressure needed to stop the osmosis process. The fundamental base for reverse osmosis units is that if we apply a pressure higher than the osmotic pressure at the opposite direction, the water will flow from the high concentration solution to the low concentration solution through the semipermeable membrane. Using this approach, pure water can be accumulated on one side of the membrane while impurities are collected on the other side.

Note that in this water purification method the entire raw water cannot be purified and some water with high concentration of dissolved solids should be rejected. The inlet water is referred to as the feed. The portion of the feed that passes through the membrane (purified water) is called permeate. The water that does not pass through the membrane and is rejected from the system is known as the retentate or concentrate. The high rate of the retentate is

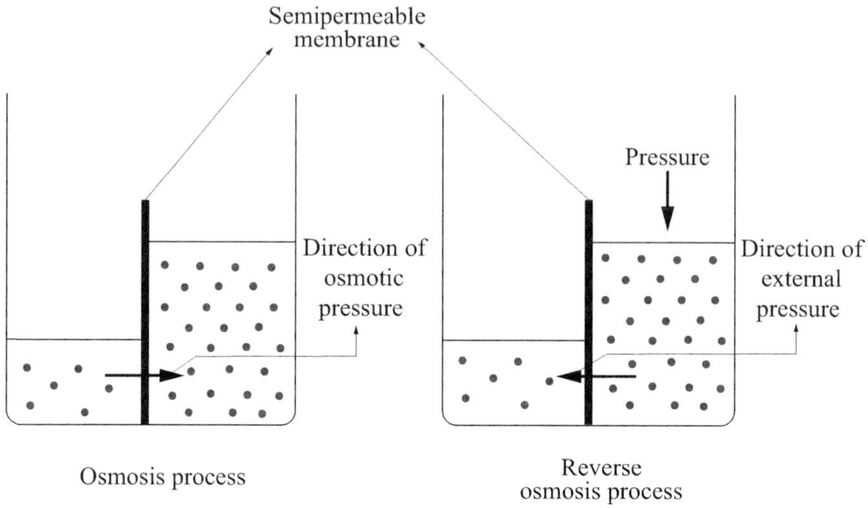

FIGURE 7.15
Schematic of principles of osmosis and reverse osmosis processes.

a major disadvantage of RO systems because a significant portion of the feed water is wasted as the retentate.

The concept of water purification using reverse osmosis principle has been known for a while, at least since the 1950s. However, it has only gained popularity in the past couple of decades, mostly due to the affordability and improved longevity of membranes produced by modern manufacturing methods. An ideal membrane should be able to handle a high flow rate of water while maintaining the solid crossover at a minimum level. It should also withstand the high-pressure gradient between its two sides. For this purpose, in many membranes a thin layer of membrane is placed on a thicker layer of a porous substrate. Most membranes are currently manufactured from synthetic polymers. Membranes can be also made of metal plates with very fine pores or microporous ceramic.

Membranes can be arranged in four configurations: plate and frame, tubular, spiral wound, and hollow. An important factor in selecting the membrane configuration is the ratio of the area of the membrane to the overall size of the system. The spiral wound configuration is commonly used in utility applications (Figure 7.16). In this configuration, the membrane is sandwiched between two supporting porous layers. The membrane sheet is sealed on three sides and rolled. The membrane is then placed in a cylindrical pressure vessel in a way that the unsealed edge of the membrane is next to one end of the vessel. The raw water at high pressure enters the shell side of the vessel. The permeate passes through the membrane and leaves the system at the unsealed edge of the membrane either at the side of the feed entrance or on the opposite side. The retentate is also collected and leaves the system at the opposite side of the feed entrance (Figure 7.16).

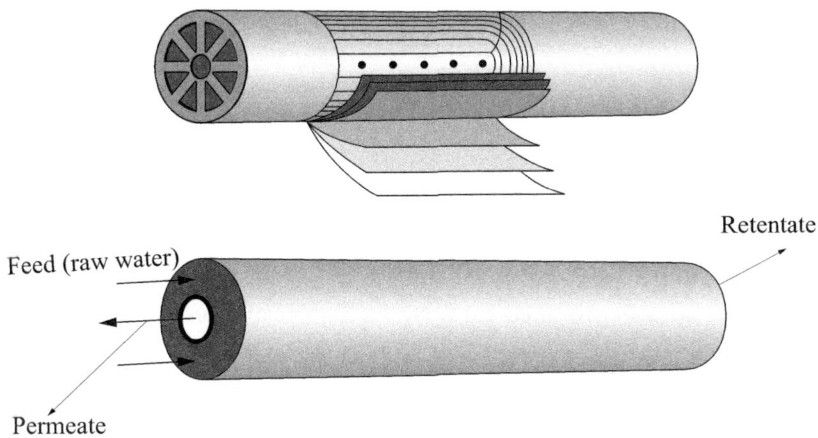

FIGURE 7.16
Schematic of a reverse osmosis system with spiral wound configuration.

When saturated by impurities, membranes can be cleaned by automatic flushing cycles. Membranes in RO units are susceptible to plugging if the feed water contains suspended solids or living organisms. To protect membranes, the feed water to RO units should be pretreated. The pre-treatment processes may include some of the following processes: sand filtration, cartridge filtration, activated carbon filtration, pH control, water softening, and killing biological organisms using an ultraviolet lamp. To ensure that all dissolved solids are removed, usually a mixed bed ion exchange demineralizer is used after RO units. Finally, the treated makeup water by RO units can be acidic and may require pH treatment before being added to the feedwater in the deaerator.

7.8.4.3 Evaporators

In some old utility power plants and nuclear-powered ships and submarines, evaporators are used to produce makeup and drinking water. These units may use steam from the main steam generator to evaporate water and then re-condense the generated steam. Alternatively, they may use flash-type steam generators (refer to Chapter 15).

7.8.4.4 Other Components of Makeup Water Systems

The makeup water system may have other components. The extent of these components and their specifications varies in different units and depends on the characteristics of the available raw water source and the power cycle requirements. If there is uncertainty involved with the availability of raw water, it can be necessary to have a raw water storage tank along with a raw water transfer pump. This tank is usually made of carbon steel. Typically there is also a makeup water storage tank (also referred to as a demineralized water storage tank). This typically stainless steel tank is commonly located next to the water treatment building. At downstream of the tank, the demineralized makeup water pumps are located to send makeup water to the deaerator as needed. This tank is equipped with level indicator sensors. When the water level is low, the makeup water treatment facility starts to produce demineralized water.

7.8.5 Condensate Polishing

We know that both feedwater and makeup water are highly purified. While feedwater can be contaminated by metallic ions from components, generally the feedwater quality is high enough for many steam power cycles, particularly those with drum-type steam generators. However, some cycles, such as those with once-through steam generators or some nuclear reactors, require higher feedwater quality. In once-through steam generators, feedwater quality is critical because the entire feedwater evaporates in tubes, with no drum to collect dissolved solid, and any impurity would end up on the inner surface of tubes. Moreover, sometimes cooling water leakage in the condenser can contaminate feedwater to an extent that its quality is not acceptable anymore, particularly if seawater is directly used as cooling water. In these cases, feedwater itself needs to be treated. This process is called the **condensate polishing process** and takes place in the **CPP**. Condensate polishing plants are typically equipped with mixed-bed ion exchange units to remove both dissolved and suspended solids. As noted previously, ion exchange resins are made of polymers and can be damaged or melted at high temperatures. Therefore, the placement of the CPP on the pass of the feedwater and the specifications of resins are very important factors to ensure that the feedwater temperature is appropriate for resins. Also, the vessels must be equipped with a trap or strainer to prevent resin escape along with the outlet feedwater which can deposit on steam generator tubes and cause damage. A CPP may be equipped with a cartridge filter to further remove suspended solids. Depending on the requirements of the power cycle, either the entire feedwater or a portion of it may be directed to the CPP.

7.9 Questions and Problems

7.9.1 Condensers

7.1. Why cannot the outlet flow of the turbine be directly pumped back to the steam generator (at least two reasons)?

7.2. You have two options: to increase the pressure of the steam generator by 3 psia or reduce the pressure of the condenser by 3 psia. In order to improve the efficiency of the power plant, which one will be your choice? Why?

7.3. Explain why the pressure in the condenser is partial vacuum and how this affects the power output of the turbine.

7.4. Name three types of condensers.

7.5. What are two mechanisms for mixing cooling water and steam in direct-contact condensers?

7.6. The saturation pressure of water at the condenser temperature is well above the atmospheric pressure.

a. True

b. False

7.7. Why should barometric condensers be installed at high elevations?

7.8. What is the difference between high-level and low-level barometric condensers?

7.9. Why is the ratio of the circulating water mass flow rate to the feedwater mass flow rate very high in direct-contact condensers?

7.10. Why are direct-contact condensers always used along with dry cooling towers?

7.11. In surface condensers the cooling water flows .... tubes and the condensing steam flows .... tubes.

a. inside – outside

b. outside – inside

7.12. In the following schematic, label the main components.

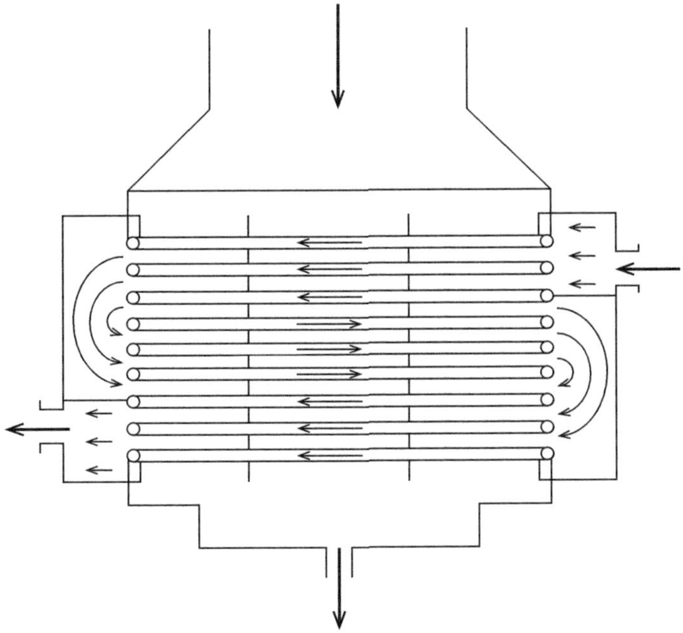

7.13. What are the functions of the following condenser components? Where are they located?

a. Tube sheets
b. Hotwell
c. Water boxes
d. Steam dome
e. Perforated impaction plate
f. Baffles

7.14. What were the problems with a uniform and tight tube arrangement in early surface condensers? How are these problems addressed in modern surface condensers?

7.15. How is the capacity of the hotwell in a condenser determined?

7.16. How can noncondensable gases interrupt the performance of a condenser if they are not properly removed?

7.17. What are the two equipment in steam power plants where the deaeration process takes place?

7.9.2 Application of Pumps in Steam Power Plants

7.18. Name types of pumps according to how energy is imparted to the fluid. Briefly explain each type.

7.19. In which type of pumps, the energy addition to the flow is intermittent where the flow is directly pressurized in a closed space with moving boundaries?

a. Dynamic pumps
b. Displacement pumps
c. Both dynamic and displacement pumps

7.20. What are the functions of the following pump components?

a. Impellers
b. Diffusers
c. Volute

7.21. Label the parts of a dynamic radial pump in the following schematic.

7.22. Explain why, when, and where cavitation takes place in a pump. Why is it undesirable? Which parameter is used to avoid cavitation?

7.23. What is the specific speed of a pump? Explain how the specific speed can be used to identify the geometry (shape) of the pump impeller.

7.24. A high value of specific speed indicates that a/an ...... pump is required while a low value indicates a need for a/an ...... pump.

a. axial – radial

b. radial – axial

c. radial – radial

d. axial – axial

7.25. Describe what parameters are presented in pump performance curves? How are they used to size a pump?

7.9.3 Condensate Pumps

7.26. What is a condensate pump?

7.27. Why are condensate pumps vulnerable to cavitation in their inlet? How can this problem be addressed?

7.28. How can the flow rate of condensate pumps be controlled?

7.29. What is the function of a condensate booster pump?

7.9.4 Feedwater Heaters

7.30. What is the function of feedwater heaters?

7.31. How is steam for feedwater heaters provided?

7.32. Explain how the possibility of backflow of water or steam from entering the turbine through extraction points is avoided.

7.33. What are the two types of feedwater heaters?

7.34. Where are the locations of high-pressure (HP), intermediate pressure (IP), and low-pressure (LP) feedwater heaters in a steam plant?

7.35. How can baffles improve heat transfer in closed feedwater heaters? What other effects, positive and negative, do they have?

7.36. What are the three sections of closed feedwater heaters? Explain heat-transfer processes in each section.

7.37. How can the drain from a closed feedwater heater be handled? Explain both approaches. Compare them in terms of the cycle efficiency and capital cost.

7.38. What is the function of a steam trap? Where is it installed?

7.39. What are the three mechanisms used in steam traps?

7.9.5 Open Feedwater Heaters

7.40. Explain why for each open feedwater heater a pump is needed to pressurize feedwater.

7.41. In large utility steam power plants, where is an open feedwater heater typically located? Why?

7.42. List some of the problems that the release of dissolved gases in feedwater can cause (three problems).

7.43. List five functions of an open feedwater heater.

7.44. What are the three mechanisms for feedwater deaeration? Briefly describe each mechanism.

7.45. Name three common deaerators and briefly explain how they work.

7.46. What is the function of a feedwater tank?

7.9.6 Feedwater Pumps

7.47. What is the function of a feedwater pump? Where is it located?

7.48. Why is a feedwater pump susceptible to cavitation?

7.49. Why are some power plants equipped with a steam generator feedwater booster pump? What is the main characteristic of this pump?

7.50. How can a feedwater pump be powered?

7.51. In a turbine-driven feedwater pump, how is the exhaust steam from the turbine handled?

7.9.7 Treatment of Feedwater and Makeup Water

7.52. Why does the quality of feedwater vary throughout a steam cycle?

7.53. Briefly explain the importance of feedwater quality control for long-term operation of a steam cycle.

7.54. List five feedwater quality control strategies.

7.55. Where can the chemicals to control pH be added to feedwater?

7.56. How and where can the pH level be measured?

7.57. What are some water sources that can be used as makeup water?

7.58. What are three water demineralization technologies?

7.59. What are ion exchange resins? Name two types of resins. How can they remove dissolved solids from water?

7.60. What is a mixed-bed ion exchange vessel? How does it work?

7.61. Explain the direct osmosis and reverse osmosis phenomenon.

7.62. In the following figure identify which one is an osmosis process. and which one is a reverse osmosis process. Describe each process.

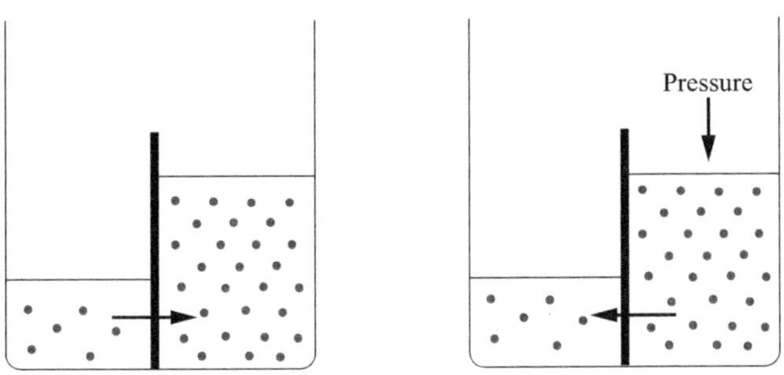

7.63. Name three desirable characteristics of a semipermeable membrane.

7.64. Using the following figure explain how a spiral wound reverse osmosis system works.

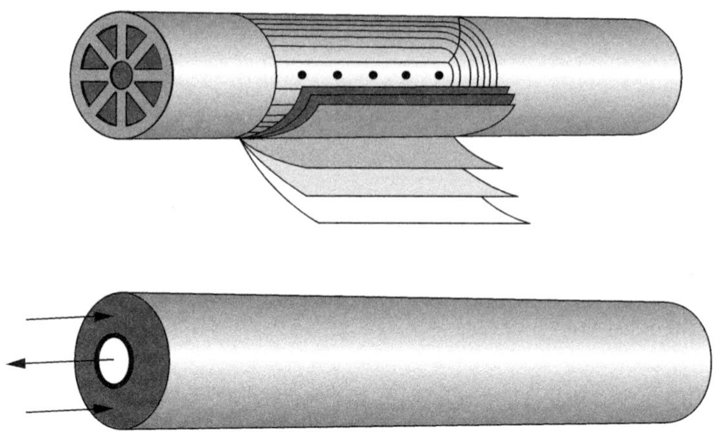

7.65. What is the objective of a condensate polishing plant? Explain why drum-type steam generators do not typically need it but once-through steam generators need one.

7.9.8 General

7.66. Name the equipment that immediately follows the open feedwater heater in a typical steam cycle.

7.67. Explain the objective of jet pumps or steam jet air ejectors (SJAEs) in a large steam power plant.

7.68. Explain the configuration of the tubes in the following condenser.

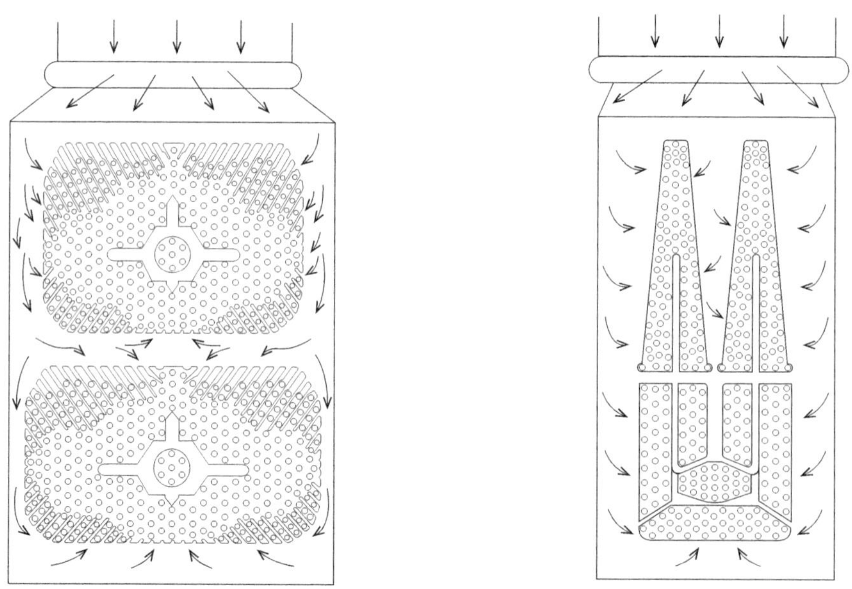

7.69. Why is an extraction pump right after the condenser typically located deep in a pit in the plant's floor?

7.70. Explain why the steam outlet from turbine ....

- is not discharged to the environment.
- is not pumped to the inlet of the steam generator.

7.71. Which plant controls and maintains the quality of the feedwater in the power plant with a once-through steam generator?

7.72. Which one is correct about closed feedwater heaters?

a. They are shell-and-tube heat exchangers.

b. They usually have U-tube bundles.

c. They can be horizontal or vertical.

d. All of the above.

7.73. List two components that the deaeration process occurs in a large steam power plant.

7.74. Explain why the rejection of the outlet flow of the turbine to the environment and using freshwater as the inlet to the steam generator is not a good idea.

7.75. What are two support mechanisms for surface condensers?

7.76. What are the differences between a makeup water treatment system and a condensate polishing plant?

7.77. Name three types of open feedwater heaters.

7.78. What is the function of a condensate polishing plant? When is it used? Where is it located?

7.79. Explain what happens when resins in an ion exchange unit are saturated. How can the system performance be restored?

7.80. Why is only one open feedwater heater used in steam power plants?

7.81. What is the difference between a deaerator and an open feedwater heater?

7.82. Which one is more efficient, an open feedwater heater or a close feedwater heater? Why?

7.83. Why are feedwater heaters used in steam power plants?

7.84. What is the component that sometimes accompanies an open feedwater heater (literally)? What is the function of this component?

7.85. Explain what all-volatile treatment (AVT) of water is.

7.86. Where does the deaeration process of feedwater take place?

a. Closed feedwater heater

b. Open feedwater heater

c. Feedwater pump

d. Steam trap

7.87. What are the main sources of steam/water losses in a steam power plant?

7.88. Due to the large latent heat of water, the magnitude of the cooling water flow rate at the condenser is enormous.

a. True

b. False

7.89. What are non-condensing steam power plants?

7.90. What are the characteristics of dynamic or displacement pumps?

7.91. In a reverse osmosis water treatment system, what are the following flows?

a. Permeate flow

b. Retentate or concentrate flow

c. Feed flow

7.92. Why do we have two types of resins, cation resins and anion resins?

7.93. Why are in a surface condenser the operating environment of the tubes more severe than those of the shell?

7.94. What are the main equipment in a water demineralization facility?

7.95. Ion exchange resins are made of polymers and as a result they are very sensitive to temperature.

a. True

b. False

7.96. Why are condensate pumps typically low rotational speed pumps?

7.97. Why is an open feedwater heater used as a deaerator too?

7.98. A high value of specific speed indicates a …... flow rate and/or a …… head.

a. high – low

b. high – high

c. low – high

d. low – low

7.99. Describe how a leak-tight connection between condenser tubes and tube sheets can be created.

7.100. Using the following figure explain the principles of reverse osmosis water treatment systems.

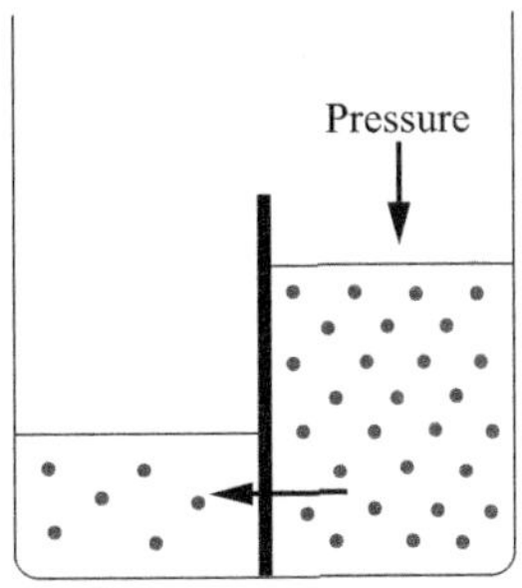

7.101. What are the required net positive suction head ($NPSH_R$) of a pump and the available net positive suction head ($NPSH_A$) of an application? How are they used in the designing/sizing of pumps to avoid cavitation?

7.102. The poor removal of noncondensable gases negatively impacts the performance of condensers by ……

a. raising the temperature and pressure of the condensation process.

b. covering the outer surface of tubes with gas bubbles.

c. raising the concentration of noncondensable gases, which increases their partial pressure.

d. all of the above.

7.103. What is the difference between multiple-shell condensers and multiple-pressure condenser?

7.104. What is the function of a condensate or extraction pump and where is it located?

7.105. What are the sources of dissolved gases in the condensate water?

7.106. Explain how a deaerator can temporarily accommodate fluctuations in the feedwater flow rate.

7.107. How does the jet pump or steam jet air ejector (SJAE) in a large steam power plant operate?

7.108. How can ion exchange resins be regenerated?

7.109. List at least three mechanisms of heat transfer in a condenser.

7.110. What is the difference between high-pressure (HP) and low-pressure (LP) feedwater heaters? Where are they located?

7.111. List at least three components of the condensate-feedwater system.

7.112. What is the source of the steam required for feedwater heaters (which equipment)?

7.113. What are the sources of noncondensable gases in steam in the condenser?

7.114. In a surface condenser, the cooling water is ……. tubes and condensing steam is …….. tubes (inside or outside).

7.115. How is the temperature of an open feedwater heater determined (thermodynamically)?

7.116. Why is the flow of steam radial (the steam flows from all sides of the tube-nest to the center) in modern condensers?

7.117. List two types of feedwater heaters. Which type works as a deaerator as well as a feedwater heater?

7.118. Name some of the most important pumps used in a typical steam power plant.

7.119. In a steam power plant equipped with feedwater heaters, the inlet feedwater to the steam generator is saturated liquid and an economizer is no longer needed.

a. True

b. False

7.120. Why do not some cogeneration or combined heat and power (CHP) plants need the condensation process?

7.121. Explain why an open feedwater heater is an ideal site for the deaeration process.

7.122. What is the main difference between open and closed feedwater heaters in terms of the number of pumps required in the cycle?

7.123. What is the function of a boiler feedwater pump and where is it located?

7.124. What are the main components of a water pre-treatment facility?

7.125. In which type of pumps, the energy addition to the flow is continuous and the pressure of the flow is increased indirectly?

a. Dynamic pumps

b. Displacement pumps

c. Both dynamic and displacement pumps

7.126. Explain how radial flow condensers work.

7.127. Why are condensate pumps typically located under the condenser below the floor of the turbine hall?

7.128. Which statement is not correct about surface condensers?

a. In surface condensers there is a direct mixing of the spent steam flow and the cooling water flow.

b. Surface condensers are typically shell-and-tube heat exchangers.

c. In surface condensers the cooling water flows inside tubes and the condensing steam flows outside tubes.

d. In surface condensers the cooling water can be either from natural or human-made bodies of water, such as a river, ocean, lake, or pond, or cooling tower.

7.129. Explain why low-level barometric condensers do not need to be installed at a high elevation?

7.130. Explain why typically there is only one open feedwater heater in a steam power plant.

7.131. Explain two important functions of the condenser.

7.132. What are the two main functions of an open feedwater heater?

7.133. What is the net positive suction head (NPSH) of a pump? How is it used in the designing/sizing of pumps to avoid cavitation?

7.134. In surface condensers, why are water boxes typically partitioned?

7.135. Which statement is not accurate about the function of condensers?

a. The condenser's main function is to condense the saturated mixture of steam and water leaving the turbine and to recover highly treated high-value feedwater and recycle it back to the cycle.

b. If the temperature of the cooling water is low enough, which is commonly the case, the low saturation pressure provides a backpressure that is lower than the atmospheric pressure for the turbine to exhaust to.

c. Providing a high backpressure for the turbine can significantly improve the output work in the turbine and its efficiency.

d. Condensers are where noncondensable gases are removed from the condensate water.

e. Condensate drains from various sources, such as feedwater heaters, the turbine driving the feedwater pump, and ejectors, are collected in the condenser.

7.136. In a steam power generation plant, what are the two advantages of using the condenser instead of discharging the low-pressure steam exhausted from the low-pressure steam turbine to the environment?

7.137. What is the function of a pump?

7.138. What is the objective of the hotwell in a condenser?

7.139. What type of condenser is more common in utility steam power plants?

a. Direct-contact condenser

b. Surface condenser

8

Cooling Systems (Circulating Water Systems)

Abstract

In this chapter, we will take a look at the last process in steam power plants. We will learn how low-temperature heat extracted from steam during the condensation process in a condenser is eventually transferred to the environment. We will start with once-through cooling systems and their components. We will investigate how these systems can harm the environment, and why despite being the most effective of all cooling systems, they are the least employed in new steam power plants. We will then discuss wet (evaporative) cooling towers. Due to their popularity, we will extensively cover various types of wet cooling towers and their related components with particular attention to natural draft and mechanical draft cooling towers. Finally, we will discuss about dry cooling towers and some hybrid cooling systems. We will conclude with a brief discussion on why dealing with the environmental impacts of these systems is complicated, and there is no simple straightforward solution.

LEARNING OUTCOMES

After successful completion of this chapter, you will be able to:

- compare and contrast various cooling systems to remove low-temperature thermal energy from a condenser and discharge the heat to the environment
- explain once-through cooling systems, their components, and environmental impacts
- describe wet (evaporative) cooling towers and their components
- express operational principles of various mechanical draft and natural draft cooling towers
- explain dry cooling towers and various hybrid cooling systems
- describe inherent complexities of dealing with environmental effects of cooling systems

8.1 Introduction

In Chapter 7, we learned that a condenser is where the low-temperature heat is removed from the working fluid. We discussed that this heat is typically transferred to the cooling water, which flows inside the condenser tubes. Also, it was discussed that how reducing the temperature and pressure of the condenser can improve the efficiency of a steam power plant. In this chapter, we will learn how this heat is eventually transferred to the environment. This system is typically referred to as the **circulating water system**. The naming might be somewhat confusing and should not be mistaken for the working fluid that is repeatedly circulated around the main cycle. Also, not all steam power plants have a separate cooling water loop. As we will discuss later, some power plants directly condense steam without a circulating water system. That is why the main title for this chapter is "cooling systems" rather than "circulating water systems".

The function of a steam cycle cooling system and a circulating water system is to provide cooling for the condenser at the lowest possible temperature and to discharge the extracted heat from the condenser to the environment while meeting all relevant environmental regulations. Generally speaking, the heat rejection to the environment can be achieved by means of the following different approaches:

- once-through circulating water systems
- closed circulating water systems – wet cooling towers
- combined once-through and wet cooling towers
- closed circulating water systems – dry cooling towers
- hybrid wet and dry cooling towers

In **once-through circulating water systems**, as the name implies, the cooling water is withdrawn from a natural body of water such as ocean, river, or lake. This water is circulated through the condenser and absorbs the latent heat of the exhaust steam flow from the turbine. The heated water is then returned to the original source of water at a higher temperature than that when it was withdrawn. While this approach can typically provide the lowest temperature possible in the condenser, it can cause substantial damages to the environment. In many countries, new power plants are moving away from this approach.

In **closed circulating water systems**, the cooling water exiting the condenser is cooled down in a cooling tower and is returned to the condenser. This type of circulating water system is the most popular system in utility steam power plants. The cooling tower in closed systems can be **wet or**

dry. A particular case for dry systems is air-cooled condensers. As the name implies, **air-cooled condensers** are systems where the condenser and the cooling tower are combined and the latent heat of the condensing steam is directly discharged to the environment with no intermediate cooling water. In a combined system, the once-through cooling system is assisted or at certain periods is replaced by a cooling tower.

8.2 Once-Through Circulating Water Systems

In this cooling system, the power plant is built near a sea, ocean, large river, large lake, or aquifer. The water source can be freshwater or saltwater. A large amount of water is taken from the water source and is pumped through a surface condenser. The temperature of the cooling water increases due to the turbine exhaust steam condensation. Then, almost the same amount of water is returned to the source at a few degrees Celsius higher temperature. Among all cooling systems, once-through circulating water systems, also known as **direct cooling systems**, are the simplest, least expensive, and most effective in minimizing the condenser temperature. These advantages made this type of cooling system very popular in older power plants. Until the 1970s, a majority of steam power plants used a once-through cooling system.[1] However, these systems can severely damage local ecosystems on various levels. Fish, shellfish, and other relatively large marine life can be captured and killed by being trapped in fish screens in their intake structures. Microorganisms, fish eggs, fish larvae, and smaller marine life that can pass through screens can be hurt or killed by chemicals added to clean cooling water or by thermal shock within the condenser due to a sudden temperature change. They can also be physically hurt going through equipment, particularly pumps. Furthermore, increasing the temperature of water in the source can degrade an entire local ecosystem by reducing the oxygen dissolved in water (for more on the relationship between the oxygen level in water and its temperature refer to Section 7.6). The higher temperature of water can also alter marine life behaviors, including their growth rates and reproduction patterns, particularly for cold-blooded living things since their body temperature depends on water temperature. Moreover, it has become increasingly more difficult to find suitable sites near natural bodies of water to build power plants. Finally, climatic changes due to global climate change make many power plants equipped with once-through cooling systems vulnerable to water shortage due to short-term and long-term draught.

[1] https://www.eia.gov/todayinenergy/detail.php?id=14971.

In the U.S., the **Clean Water Act**[2] is to regulate water pollutions and protect water resources. Section 316 of the act regulates water thermal pollution and intake structures of cooling water. The latest addition to the law is Section 316(b) that became law in 2014 and states that "the location, design, construction, and capacity of cooling water intake structures (CWIS) at regulated facilities …. [must] reflect the best technology available (BTA) for minimizing adverse environmental impact".[3]

Due to all the above-mentioned problems and the regulatory requirements, in the U.S. new power plants rarely use once-through cooling systems and some existing power plants have moved to replace their once-through systems with other alternatives.

8.2.1 Components of Once-Through Systems

The first part of a once-through cooling water system is the intake structure where water is extracted from a natural body of water. We know that the density of water is at its maximum at the temperature of 4°C and with temperature increase water density reduces. This means the higher temperature water stays on the surface and the deeper we go, the water temperature deceases. As a result, the water intake structure, which is typically made of reinforced concrete, is designed to withdraw inlet water from the bottom of the water source usually by means of a curtain wall (Figure 8.1). The configuration of the structure and its design heavily depend on the site. Shoreline structures should be able to withstand possible local storms. The intake is always equipped with a screening system to intercept debris from the inlet water flow and stop them from entering the system. This mechanism typically includes a bar rack and a rotating screen.

The vertical bar rack is to remove large debris. It is equipped with a traversing trash rake to clear its pass from debris (Figure 8.1). The vertical rotating (or traveling) screen is a continuously linked metal mesh frame that loops around two rollers at the top and bottom. At any given time, a certain number of panels are exposed to the cooling water flow. After they catch certain

[2] The Federal Water Pollution Control Act (FWPCA), typically referred to as the Clean Water Act (CWA), was first enacted in 1948 as the first major law to protect water resources. But what we know today as the foundations of the current CWA was passed in the U.S. Congress in 1972. At the time about 70% of the nation's bodies of water were unsafe for swimming or fishing. The act passed the House by a vote of 366 to 11. The vote in the Senate was unanimously and all 74 Senators voted to approve it. President Richard Nixon vetoed the act on the ground that it is too expensive (the act came with a $24 billion price tag!). In an excellent example of bipartisanship, in a matter of few hours both the Senate and the house overrode the president's veto making the act the law of the land (in the Senate the vote was 52 to 12 and in the House the vote was 247 to 23). The U.S. Environmental Protection Agency (EPA) is in charge of implementing, enforcing, and revising/updating the act (to read more on the history of this act, refer to https://waterkeeper.org/news/bipartisan-beginnings-of-clean-water-act/).

[3] https://www.federalregister.gov/documents/2014/08/15/2014-12164/national-pollutant-discharge-elimination-system-final-regulations-to-establish-requirements-for.

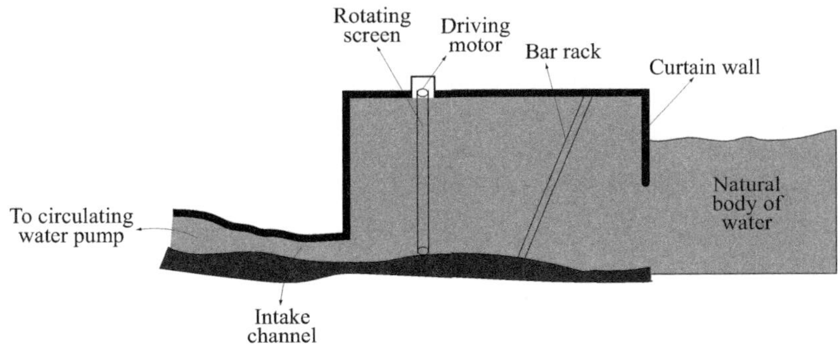

FIGURE 8.1
Schematic of an intake structure to a once-through circulating water system.

amount of debris (determined by the pressure drop across the screens), the motor-driven roller rotates to bring clean screens in front of the inlet water. Alternatively, they can run slowly and continuously. The screens already covered with debris are backwashed to clean them up. A dedicated screen-wash pump(s) provides water needed for this purpose. While these screens are typically equipped with a fish collection system, inevitably some fish are trapped and die in the system. Considering the size of these meshes, typically about 1 cm (about 0.4 in), small organisms can easily pass through screens. They are designed to only remove pieces that can clog condenser tubes.

After the screening process, the cooling water free from relatively large debris is fed to a series of parallel water pumps through intake channels. These motor-driven pumps have a relatively low head and high flow rate. They are typically single-stage vertical centrifugal pumps. The outlets of the pumps are combined in a single flow and are sent the condenser through a large diameter conduit (as large as 2 m diameter, about 6 ft). The cooling water enters the condenser via its inlet water box and is distributed among condenser tubes. Since a large increase in the cooling water temperature is undesirable (and usually restricted by environmental regulations), the condenser in power plants with a once-through cooling system is typically **single-pass**. The cooling water picks up the thermal energy released by the condensing steam in the condenser shell side and leaves the condenser at a higher temperature by about 5°C–15°C (about 10°F–30°F). The heated water is returned to the original body of water downstream of the extraction point via a discharge tunnel. In order to avoid local hot spots, the returned water should be properly distributed and mixed with the water within the reservoir. Depending on the specific conditions of the water source, mostly dictated by environmental requirements, the returned water may be distributed by means of long pipes equipped with nozzles on the water surface or submerged in the water. The thermal energy of the returned water is dissipated by mixing with water in the source or by surface evaporation.

If the inlet cooling water has a high concentration of microorganisms, some chemicals are added to avoid the growth of these organisms on the surfaces of components. Finally, it should be noted that the materials that components are made of are heavily dependent on the inlet cooling water quality. For example, due to the corrosiveness of seawater, all components dealing with this water should be made up of materials that can withstand the salinity of the water.

8.2.2 Cooling Ponds and Lakes

An alternative for once-through cooling systems that to some degree can avoid their environmental consequences is using artificially made bodies of water. In the simplest form, hot water can be just discharged to a manmade lake where its temperature is reduced by heat transfers to the environment by means of natural evaporation, convection (by wind), or radiation (to the sky). Cold water is then pumped back to the condenser. This cooling process requires no mechanical components (other than a makeup water pump). Due to the slow nature of heat-transfer processes, a large lake is needed to provide a large enough heat-transfer area and a long water residence time. In some designs, water is directed to a series of canals where water slowly moves and its temperature reduces. To assist heat transfer and to accelerate the cooling process, water can be sprayed into air by nozzles to form water droplets to increase the surface area. The efficiency of the cooling process can be increased by making finer droplets; however, this increases the carryover of water droplets by wind (drift). While these systems require more equipment, they can substantially increase the heat-transfer rate and thus reduce the area required for the lake. The lake in these systems can be used for a secondary utilization, for example as a recreational facility. An example of this type of power plants is 1600 MW Mount Storm Power Plant in Grant County, West Virginia with a 4.9 km^2 (1,200 acres) cooling lake (Figure 8.2).

8.3 Closed Circulating Water Systems – Wet Cooling Towers

In **closed (or closed-loop) circulating water systems**, the heated cooling water in the condenser is cooled down in a cooling device before being returned to the condenser. The cooling device is typically a **cooling tower**. Closed-loop cooling towers can be **dry-type**, **wet-type**, or some **combination of two types**. Furthermore, a once-through cooling system can be combined with a cooling device. In this section and next we will review all these systems.

As the name implies, in this type of cooling tower, the cooling process of circulating water is achieved by bringing the water into intimate contact with ambient air. These cooling towers are also known as **evaporating cooling**

FIGURE 8.2
Mount Storm Power Plant in Grant County, West Virginia. (By Raeky – Own work, Public Domain, https://commons.wikimedia.org/w/index.php?curid=3658660.)

towers. Similar to once-through cooling systems, evaporative closed-loop circulating water systems are always used with surface condensers. Generally speaking, regardless of its type in a typical wet cooling tower, hot water from a condenser enters the tower. The water is distributed uniformly over a medium, called the fill, which facilitates the mixing of air and water (Figure 8.3). The hot water falls by gravity through the fill. In the fill, the air flow and the hot water are intimately mixed and the water temperature reduces. The primary heat-transfer mechanism in these cooling towers is the **evaporation of a portion of the circulating cooling water**. The latent heat of the evaporated water is provided by lowering the temperature of the remaining circulating water. The share of this heat-transfer mechanism from the total heat transfer depends on the ambient air humidity. If the temperature of the ambient air is lower than that of the inlet circulating water, sensible heat transfer from the cooling water to the air also assists the cooling process. The share of this secondary heat-transfer process depends on the temperature difference between the circulating water and the ambient air and can be up to 40% of the total heat transfer. The cold water is collected at the bottom of the tower, which can be pumped back to the condenser. On the other hand, the ambient air enters the side of the tower through louvers and mixes with hot water throughout the fill. The air then leaves at the top of the tower. So in wet cooling towers the air at the ambient temperature and humidity enters the tower and the hotter and more humid air leaves the tower (Figure 8.3).

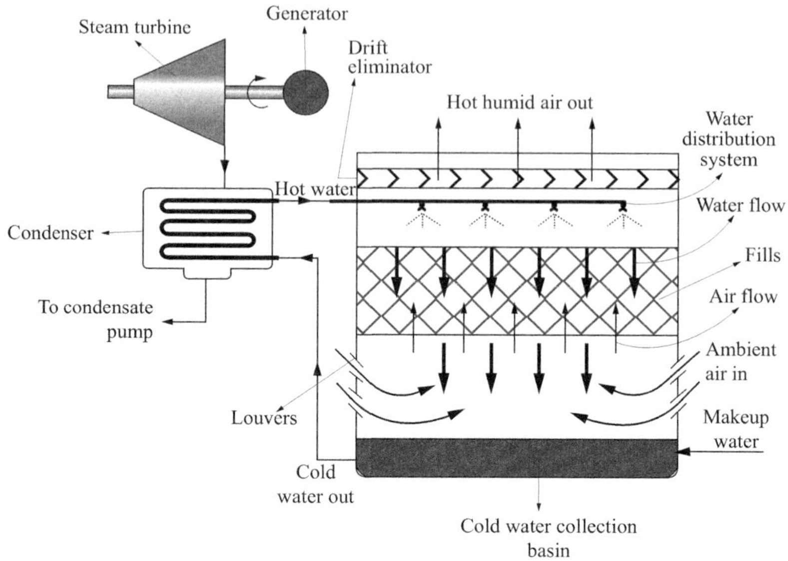

FIGURE 8.3
Schematic of a counterflow wet cooling tower.

While wet cooling towers certainly do not need nearly as much water extraction from a natural body of water as once-through systems need, still a natural source of water is needed to provide the makeup water to compensate for water evaporation. Furthermore, while in once-through systems almost the entire extracted water is returned to the source, albeit at a higher temperature, in evaporative systems some water is lost during the evaporation process and cannot be returned to the source.

Wet cooling towers are more costly, both initial and operation costs. However, due to the reasons explained in the previous section, most new steam power plants of all types use these cooling systems. In the U.S., they have been the dominant cooling system since the 1980s.

8.3.1 Water Flow-Related Components in Wet Cooling Towers

In this section, the components of wet cooling towers that are needed in the cooling water circuit will be discussed.

8.3.1.1 Water Transportation and Distribution System

First, the hot water exiting the condenser must be transferred to the cooling tower via a piping system. The specifications of the piping system depend on the type of the tower, the layout of the site, and the ground topography at the site. The piping system can be elevated or buried. Buried piping systems are protected from elements that substantially reduce piping loadings, thermal expansions, and freezing problems. On the contrary, elevated piping systems are easier and less expensive to install and repair.

At the tower, the hot water should be elevated by a riser to the roof of the tower or a high enough level that it can be distributed over the fill. This can be achieved by a single or double riser. A double riser includes two risers for the easier distribution of the incoming hot water flow. For rectangular towers, the risers are typically located at the two ends of the structure. The risers can be external pipes or internal concrete channels. The hot water can be delivered to either open or closed hot water distribution basins (Figure 8.4a) or a water distribution piping system (Figure 8.4b). In the former system, commonly used in crossflow towers (discussed below), the water is evenly distributed over the fill through metering orifices by gravity. In the latter, typically used in counterflow towers, a network of pipes delivers the hot water to distribution headers and eventually to spray nozzles. These nozzles shower the water over the fill (Figure 8.4c). While cast iron, steel, redwood stave, or concrete pipes can be used in the system, fiber-reinforced plastic pipes are more common.

8.3.1.2 Fill bank

A fill bank is critical components in the performance of wet cooling towers as it is the site where the hot water from the condenser and the air flow interact and heat and mass transfers take place. While there are some cooling towers with no fill where the water is directly spayed into the air flow for non-power plant applications, all cooling towers for utility power plants are typically equipped with the fill. A well-designed fill should be able to maximize the contact surface area and contact time between the water and air flows while minimizing the air pressure drop. Depending upon the relative direction of the water and air flows, a fill bank (and generally cooling towers) can be crossflow or counterflow. In both types, the water flow is vertical and downward. In the **counterflow** configuration, the air flow is vertical and upward at the opposite direction to the water flow (Figure 8.3). In the **crossflow** configuration however the air flow is horizontal (perpendicular to the water flow) (Figure 8.5).

There are two categories of the fill: the film type and the splash type. The **film-type fill** creates a large air and water contact surface by forming vertical sheets of water (Figure 8.6). This type of the fill is more effective in cooling the water. As a result, it can generate the same amount of cooling in a smaller volume compared to other types. The fill is made of vertical surfaces that water can adhere to. Due to its high resistance to corrosion and fouling, high strength, lightness, fire resistance, and formability, PVC (polyvinyl chloride) is the dominant material to make the film-type fill (Figure 8.6). This type of the fill is susceptible and sensitive to uneven distribution of air and water flows, hence may not be suitable when the cooling water is contaminated.

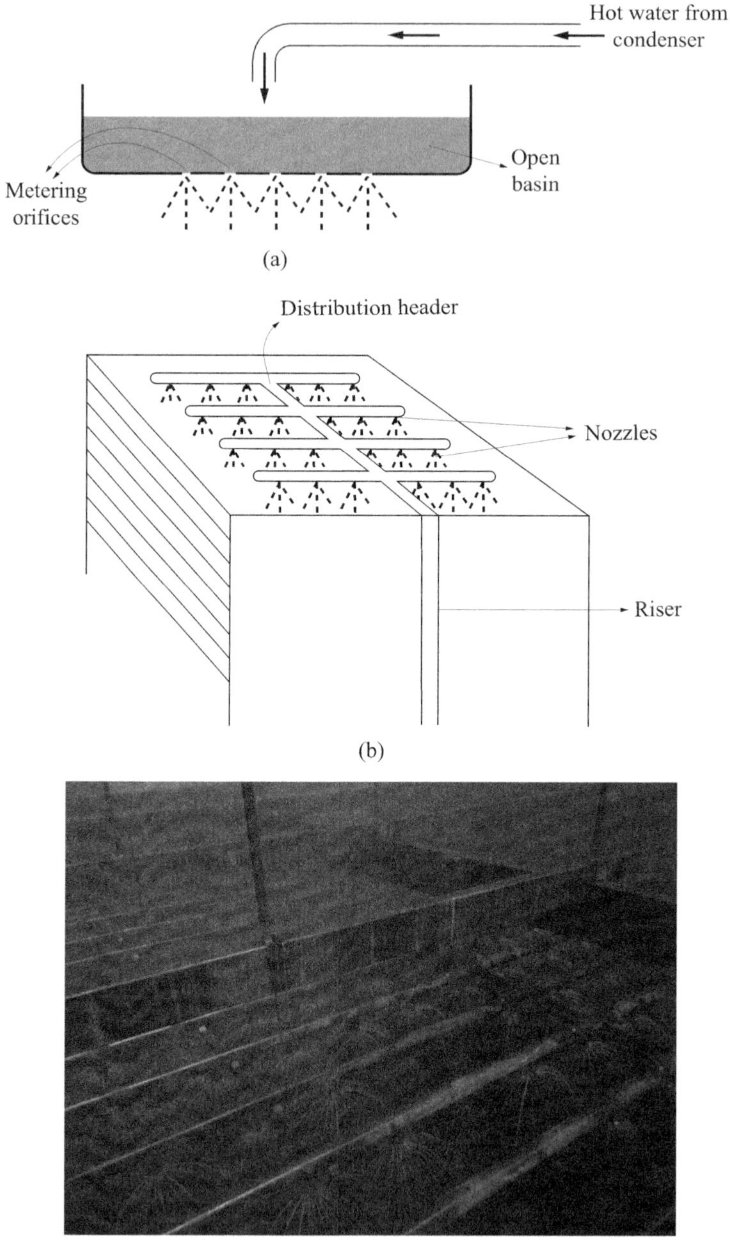

FIGURE 8.4
Hot water distribution systems in a wet cooling tower: (a) schematic of a gravity distribution system, (b) schematic of a spray distribution system, (c) an actual spray water distribution system. (By Jan Beránek – Own work, CC BY-SA 4.0, https://commons.wikimedia.org/w/index.php?curid=79463485.)

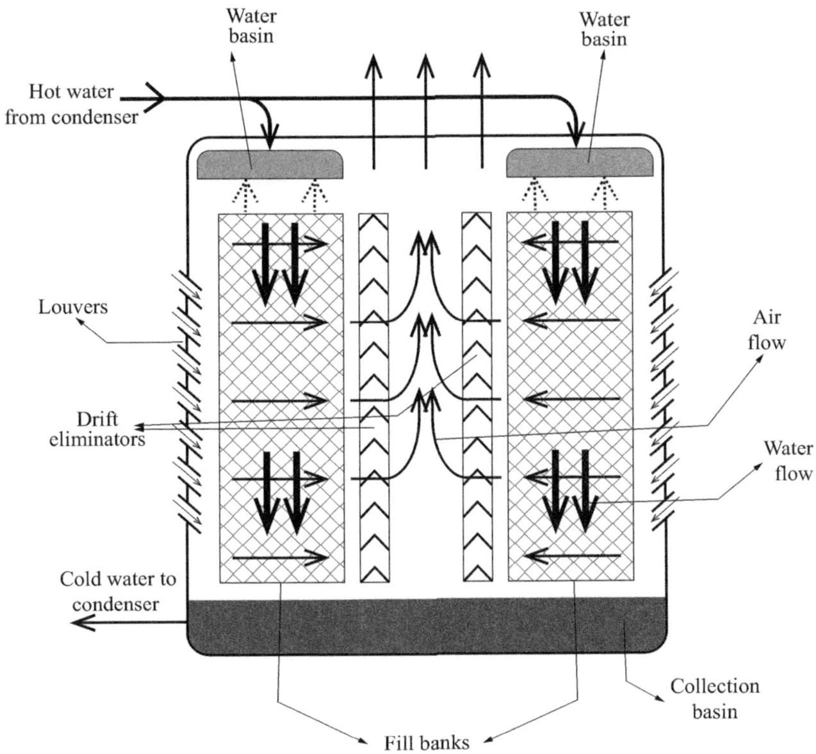

FIGURE 8.5
Schematic of a counterflow wet cooling tower.

In the **splash-type fill**, the large contact surface area is formed by breaking up the hot water flow into small droplets (Figure 8.7). To create this effect, the splash fill is composed of horizontal parallel bars, called splash bars, at different levels. The hot water vertically cascades through these bars under gravity and splashes into pieces within the air flow. The mass and heat transfers take place between the air flow and water droplets as well as between the air and bars' wet surfaces. The splash fill has a low air pressure drop, needs a low water head, and is not susceptible to blockage. However, it is very sensitive to bars' alignments. A slightest slope or sagging in the bars can interrupt the splashing process. The most common material for bars is reinforced plastic or plastic-coated metals. However, bars made of treated redwood, stainless steel, or aluminum are also being used.

Theoretically, both types of the fill can be used for counterflow and crossflow cooling towers. However, in practice, crossflow cooling towers are most commonly used with the film type while counterflow cooling towers can use both types. Generally speaking, counterflow cooling towers are more efficient. Both types are commonly used in utility steam power plants.

FIGURE 8.6
Film-type fill along with the water distribution system and water nozzles in a cooling tower for a steam power plant. (By Juhan Harm – originally posted to Flickr as Iru EJ, #4, CC BY 2.0, https://commons.wikimedia.org/w/index.php?curid=9485786.)

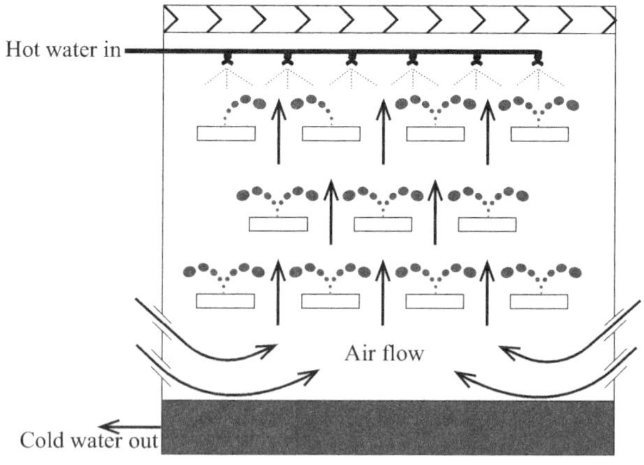

FIGURE 8.7
Schematic of a splash-type fill.

8.3.1.3 Cold Water Basin

After passing through the fill, the now cold circulating water is collected at the **cold water basin** at the bottom of the tower. This concrete basin has several other functions. One of these functions is to feed the cold water to the circulating water pump inlet so that the water can be pumped back to the condenser. A screen is placed at the outlet from the basin to stop debris from entering the pumps. The basin also forms the main foundation of the tower structure. Moreover, the basin functions as the site for a temporary circulating water storage. The basin size is determined in a way that it can store the sufficient water for a few hours of system operation if the flow of makeup water is interrupted. Also, the basin should be able to accommodate water flow fluctuations during transient operations. When the cooling tower is shutdown, the entire circulating water in the system as well as possibly the back flow of the circulation water pumps is accumulated in the basin, so it should be large enough to accommodate all these flows. Furthermore, typically water needed for the fire protection system of the power plant is taken from the basin and it acts as the water storage tank for the purpose.

When water and air interact in the fill, the circulating water washes out fine solid particulates in the air. These materials are settled at the bottom of the basin and should be continuously or periodically removed. For this purpose, some basins are equipped with a concrete sump pit with a dedicated silt removal pump. Some chemical agents can be added to accelerate this process. Other chemicals may also be added to control corrosion and the growth of algae and slime. The amount of chemical addition should be carefully monitored because excessive chemicals in the water can evaporate and negatively impact the surrounding environment, particularly local vegetation.

The basin is equipped with a blowdown (or bleeding) outlet. The reason why we need blowdown in an evaporative cooling tower is exactly the same as why we need it in a steam generator drum. When circulating water evaporates in a tower, it leaves most suspended and dissolved solids behind. These solids are accumulated in the basin and if not treated, they can reach an unacceptable level. The blowdown flow may require appropriate treatments before it can be released to the environment.

The basin is where circulating water makeup is added to compensate for all water losses. These losses are primarily from water evaporation (about 1%–3% of the total circulating water flow rate), blowdown (about 0.2%), and water drift (about 0.001%) (we will talk about the last one in the next section). The flow rate of the makeup water is controlled by a water level sensor, typically a float switch and an electric probe. The basin is also equipped with connections for overflow, drain, cleanout drain, and quick-fill. There is also a bypass pipe to send the hot circulating water returning from the condenser directly to the basin. This line is used in very cold conditions to reduce ice formation in the tower.

8.3.2 Air Flow-Related Components in Wet Cooling Towers

So far we have learned about the flow of circulating water in a cooling tower and the related components. In this section, we will consider the air flow through a cooling tower and its components. The first thing that we need to explain and can impact all other components is how the air flow is driven through a wet cooling tower. In this regard, cooling towers can be **mechanical**

draft, **natural draft**, or **hybrid draft**. In mechanical draft cooling towers, the pressure drop of the air flow through the cooling tower is compensated by fans. In natural draft cooling towers, the driving force for the air flow is provided by the density difference between hot moist air inside the cooling tower and outside ambient air. In hybrid draft cooling towers, a combination of these two approaches is used to provide the air flow through the tower.

8.3.2.1 Mechanical Draft Cooling Towers

Historically, mechanical draft cooling towers were the workhorse of steam power plants for many decades (until the commercial introduction of hyperbolic natural draft cooling towers which will be covered next). In these cooling towers, a single fan or more often several fans provide the driving force needed to move the required volume of air through the tower. Compared to natural draft cooling towers, mechanical draft cooling towers are typically more compact with a smaller footprint and hence less land requirements. They also have less initial and construction costs. Since the driving force is provided by a controllable external source independent of ambient conditions, these cooling towers are less sensitive to changes in ambient conditions. However, running the fans consumes a huge amount of power and they require frequent maintenance, which mean a very high operating cost. Fans are also very noisy.

Depending on the location of the fans, mechanical draft cooling towers can be forced draft or induced draft. In **forced draft (FD)** cooling towers, fans (known as **FD fans**) are located upstream of the cooling tower where they blow the ambient air flow through the tower (Figure 8.8a). These towers can suffer from several inherent problems. In these cooling towers, air velocity in the inlet is much higher than that of the outlet. This can create a low-pressure

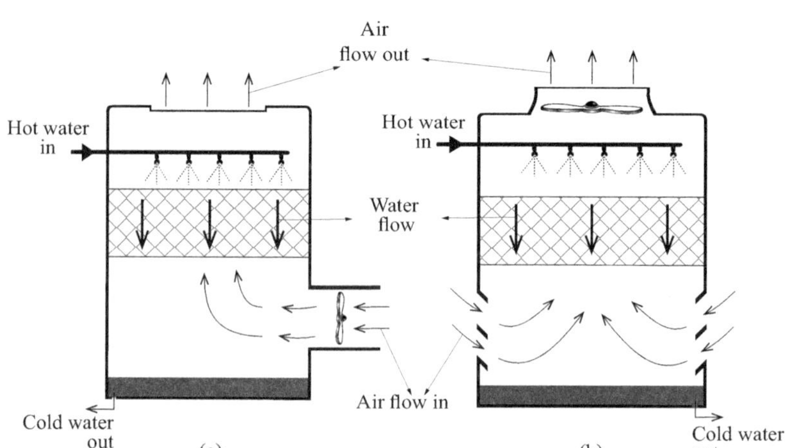

FIGURE 8.8
Schematic of forced draft (a) and induced draft (b) cooling towers.

zone in the inlet that can draw the moist outlet air back into the tower. This air recirculation can severely deteriorate the thermal performance of the tower. It is also challenging to achieve a uniform air distribution in these towers. Since these towers operate at pressures slightly higher than the atmospheric pressure, the air can leak from the tower before fully participating in the cooling process, which can increase the load of the fans. Finally, in cold ambient conditions, since the fans are located in the inlet cold flow, they are prone to icing when the inlet air is moist (either natural moisture or moisture in the recycled air). However, since the fans operate with dry and relatively cold air (compared to the outlet air flow from the tower), they consume less power (recall that the required power is directly proportional to the air specific volume and cold air has a lower specific volume) and require less maintenance.

In **induced draft** (**ID**) cooling towers, fans (known as **ID fans**) are installed on the top of the tower and draw the air flow through the tower (Figure 8.8b). The air flow enters at the lower sides of the tower. In these towers, since the outlet air velocity is several times higher than the inlet air velocity, the air recirculation is unlikely to happen. In these towers, the pressure of the air inside the tower is slightly lower than the atmospheric pressure, so air leakage from the tower is not a problem. Also, since the fans are installed on the exiting hot air stream, the fans are protected from icing and the air distribution is more uniform than that in FD towers. However, since the fans are working with the hot humid air flow, they consume more power and require more frequent maintenance (Figure 8.8b). Regardless of these drawbacks, a large majority of steam power plants that are equipped with mechanical draft cooling towers use induced draft fans (Figure 8.9).

FIGURE 8.9

An induced draft cooling tower. (By Cenk Endustri – Own work, CC BY-SA 3.0, https://commons.wikimedia.org/w/index.php?curid=16049857.)

8.3.2.2 Components of Induced Draft Cooling Towers (Air Flow)

The water flow-related components of induced draft cooling towers were discussed in the previous sections. In this section, we will focus on the air flow-related components of these cooling towers. The air flow enters the cooling tower horizontally at its sides. If the air flow enters the tower from both sides through a fill bank on each side, it is called the double-flow tower. Sometimes one side is too close to an obstacle, such as a building, to allow free air flow. In these cases, a single-flow cooling tower is used.

In many designs, the air flow enters the tower through **louvers**. Louvers are closely spaced horizontal slates sloped downward at the side of the tower (Figure 8.8b). Their function is to retain the water inside the tower and avoid water droplets splashing out of the tower. All splash-type cooling towers are equipped with louvers while film-type fill banks may or may not have louvers. The air flow then passes through the fill banks.

Right above the fill bank in counterflow towers and next to the side of the fill bank in crossflow towers, there are **drift eliminators**. The function of these drift eliminators is to remove tiny unevaporated water droplets carried by the air flow from the fill. If uncaptured, these droplets (known as drift) escape the tower without contributing to the cooling process within the tower and reduce the effectiveness of the tower. Moreover, they can not only create plume over the tower but also land on nearby structures and make them wet. In some cases, such as when drift is deposited on power lines, it can create operational hazard. In winter weather they can form excessive icing problems. To remove drift from the air exiting the fill bank, drift eliminators are designed to impose one or a few sudden changes in the air flow direction (called passes) (Figures 8.3 and 8.5). While the air and vapor flows can easily follow the directional changes created by the drift eliminator passes, heavier water droplets, due to their greater momentum, cannot follow the pattern and impinge on the surfaces of the drift eliminator. The separated water is then returned to the tower. Drift eliminators can be made of the similar material as that of the fill, commonly a variety of plastics. They can be installed separate from the fill or they can be integrated into the fill. The design of both louvers and drift eliminators should be in a way that they can provide the desirable functionalities while keeping air pressure drops at as low as possible. Modern drift eliminators can remove the great majority of the drift from the exiting air flow, allowing very little amount of water droplets, as low as 0.001% of the total circulating water flow rate, leaving the tower.

In induced draft cooling towers for utility steam power plants, typically several **fans** are installed on the roof of the tower. Since these fans should handle a high volume of air at a low-pressure head, in a majority of towers large multiblade propellers are used. These fans can have blades with the radius of 0.3–5 m (about 1–15 ft). In essence, they are the larger version of ceiling fans. They are exposed to the hot humid and corrosive air flow leaving the tower. To tolerate this challenging environment, the fans used in cooling towers are custom designed for this particular application. While cast aluminum has been the most common material for the blades (and are still being used), other materials, such as fiberglass, plastic, and stainless steel, are also being used.

These fans are driven by electric motors. Their rotational speed depends on their blade length (larger fans turn slower). To achieve the desired rotational speed, a speed reduction is needed on the power transmission system between the motor shaft and the fan shaft. Some units use a single speed motor to drive the fans. In this case, regardless of the ambient conditions and load, the same volumetric flow rate is drawn from the tower. This means at part-load operations or low ambient temperatures, the fans provide more volumetric air flow than what is really needed, which can unnecessarily increase power consumption in the motors. There are several strategies to regulate the air flow rate through the fans. Some cooling towers use **multiple-speed** (typically two-speed) motors (again similar to ceiling fans). For example, the fans can operate at high speed during the day but at night, when typically both the load and the ambient temperature are low, they can be switched to low-speed operation. The fan operation at low speed also reduces the noise level, which can be particularly problematic during night. Some units are equipped with **variable speed** motors to accurately control the air flow based on the need and to minimize power consumption. Furthermore, in some cases fans with **adjustable pitch blades** are used alone or in combination with two-speed motors. The ability to control the air flow rate is also important for icing control in cold weather. Since ice formation depends on the air flow rate, it can be reduced to address the icing problem.

Typically, the fans are installed inside a fan cylinder made of fiber-reinforced plastic to provide a smoother air flow pattern with less noise as well as for personnel protection. If the height of the cylinder is not high enough, a fan guard may be required. The proper selection of the cylinder height can help to reduce the possibility of air recirculation. To transfer fan loads to the tower structure, the fans are installed on a fan deck.

The typically concrete structure of cooling towers must be able to withstand a variety of heavy loads. The casing of cooling towers is designed to enclose the tower and keep water within the tower and reduce water drift. The casing is typically made of corrugated steel or reinforced plastic plates. These plates are overlapping and are sealed to avoid air and water leakage. The side air entry plenums are integrated into the casing. Note that not all cooling towers have a casing. The structure should be equipped with access ladders or stairways, safety handrails, and access doors to internal components.

Typically, induced draft cooling towers are designed and manufactured in single units or cells. For utility power plants, several cells are combined to form a single multicell tower. The cells can be arranged in a rectilinear, rectangular, round, or octagonal configuration. Typically, these cooling towers are field erected. The configuration, orientation, and site layout should be selected carefully to minimize the air recirculation phenomenon and/or the interference of other local thermal sources, such as other cooling towers or stacks, on the performance of the tower. The local wind speed and its direction as well as local topography must be taken into consideration in the process of the system site placement.

8.3.2.3 Natural Draft Cooling Towers

As the name implies, in **natural draft cooling towers** there is no mechanical device to move air through the tower. Instead, the driving force to overcome the pressure drop in the air flow pass is provided by the density difference between the hot air inside the tower (with lower density) and relatively the cold ambient air (with higher density). These units are also known as **atmospheric cooling towers**. Previously we showed that the driving force due to the density difference can be determined by the following equation:

$$\Delta P = (\rho_{\text{Out}} - \rho_{\text{in}})gh \tag{8.1}$$

where ΔP is the driving force created by the density differential (Pa), ρ_{Out} is the average ambient air density (kg/m^3), ρ_{in} is the average hot air density inside the tower (kg/m^3), g is the gravitational acceleration (m/s^2), and h is the height of the column of hot air inside the tower (the distance between the fill top and the highest point of the tower) (m). In this equation, the magnitude of the density difference $(\rho_{\text{Out}} - \rho_{\text{in}})$ is small. Therefore, to create an enough pressure difference to overcome the pressure drop on the air flow through the tower, the height of the tower should be sufficiently high. For this reason, it is not unusual to have a cooling tower taller than 150 m (about 500 ft) (Figure 8.10).

FIGURE 8.10
Sheer size of a natural draft cooling tower (use the high-voltage power lines in foreground as reference). (Image by Marc Pascual from Pixabay, https://pixabay.com/photos/nuclear-energy-central-electricity-4600565/.)

The idea of using natural draft cooling towers is not new. There is some evidence that even as early as the early 1900s it was known that tall towers can be used for evaporative cooling purposes. However, economically building a tall enough tower that could withstand severe loads was a major challenge. A breakthrough came with the introduction of **hyperbolic structures** to natural draft cooling towers. In the 1880s, Vladimir Shukhov (a Russian polymath, 1853–1939) discovered that hyperbolic structures can provide extraordinary structural strength. A hyperbolic structure is a surface formed by rotating a hyperbola around its central axis (Figure 8.10). In the late 1910s, a Dutch engineer named Frederik van Iterson (1877–1957) patented the idea of using hyperbolic structures in natural draft cooling towers. At the same time the first hyperbolic natural draft cooling tower was built in the Netherlands. With the advancements in construction technologies, concrete hyperbolic cooling towers gradually became popular in Europe because they could provide the needed structural integrity with minimal construction materials. However, they were not commonly used in the U.S. until the early 1970s. Note that the shape of these towers does not have much to do with the flow of air inside the tower though the shape is desirable when it comes to the effect of wind on the tower (wind load). Nowadays, most large utility size steam power plants (both fossil fuel and nuclear plants) are equipped with this type of cooling towers. Due to their sheer size and height, they are the most visible component of steam power plants (Figure 8.11). In fact, with the typical column of plume rising from them, they have become the public

FIGURE 8.11
Natural draft cooling towers as the most visible structures of steam power plants. (Image by Bruno/Germany from Pixabay, https://pixabay.com/photos/industry-power-plant-2663191/.)

FIGURE 8.12
Similarities of plume from a cooling tower and smoke from a stack. (Image by Monika P. from Pixabay, https://pixabay.com/photos/power-plant-industry-flame-2012377/.)

symbol (both positive and negative) of these power stations.[4] Speaking of plume, it should be noted that when the ambient relative humidity is high (close to saturation), typically in cold humid winter days, some of the exiting water vapor can be condensed to fine water droplets that appear as visible fog or plume. The fog will remain visible until the hot moist air flow is sufficiently mixed with the ambient air and becomes sub-saturated. In many sites, this is considered sight pollution and undesirable[5] (Figure 8.12). If released in low elevations, fog can create icing problems on structures located downwind of the tower. To reduce or eliminate this effect, the relative humidity of the exiting air should be reduced. For this purpose, the less humid air can be mixed with the exiting air and/or it can be heated before being discharged. In some units, where combustion products have gone through an extensive cleaning process and flue gas is free of contaminations, the stack structure is eliminated. Instead, the flue gas is discharged into the natural draft cooling tower (Figure 8.13). This strategy can not only save the cost of stack construction but also heat up the moist air in the tower, which can reduce the plume problem.

[4] While for some they are a symbolic sign of industrial development, for others they are purely a source of sight pollution. It might be difficult to find another empty industrial structure (!) that can provoke such emissions (good or evil) depending on who you are talking to!

[5] The author has even heard from some people unfamiliar with steam power plants that pointed to the plume rising above a natural draft cooling tower as an evidence of how polluting these power plants are (mistaking water vapor with smoke from combustion, Figure 8.11).

FIGURE 8.13
Clean flue gas exhaust into a natural draft cooling tower. (By Regionalplaner – Own work, CC BY-SA 3.0, https://commons.wikimedia.org/w/index.php?curid=26638533.)

While there are some natural draft cooling towers constructed by steel structures that are used in steam power plants, a majority of them are made of reinforced concrete (Figure 8.14). Using a steel structure in cooling towers can be due to the long construction time of concrete towers, high cost of concrete, or low strength of local soils. Inside these towers at the bottom, the water-cooling components are located. The internal components of natural draft cooling towers, including hot water distribution systems, water nozzles, fill banks, drift eliminators, and cold water basin, are very similar to those of the previously discussed mechanical draft cooling towers (Figure 8.15). The fill can be crossflow or counterflow. The fill in crossflow units is located outside the tower around the periphery of its base. In counterflow units, the fill is located inside the tower. The rest of the tower, which is a great majority of its volume, is just a round empty shell (Figure 8.16). These cooling towers are superbly reliable. Due to their great height, these towers are equipped with aircraft warning lights on their top.

Compared to mechanical draft cooling towers, natural draft cooling towers require much greater initial investments. However, they consume substantially less power to operate and need much less maintenance. This means they require a long amortization period, so a higher capital cost can be paid off by a lower operational cost. The nature of their operation dictates that the air flow rate cannot be varied with a change in load and ambient conditions.

FIGURE 8.14
A natural draft cooling tower with a steel structure. (Image by 2427999 from Pixabay, https://pixabay.com/photos/power-plant-power-plant-electricity-1367316/.)

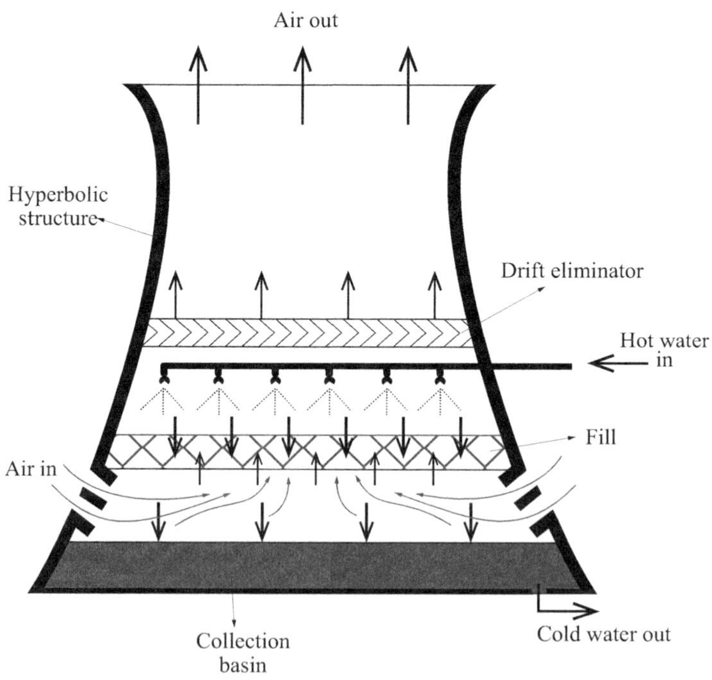

FIGURE 8.15
Schematic of the internal components of an evaporative natural draft cooling tower.

FIGURE 8.16
Vast empty volume inside a natural draft cooling tower. (Image by MichaelMep from Pixabay, https://pixabay.com/photos/architecture-cooling-tower-3321995/.)

That is why these cooling towers are more suitable for large units where usually large load fluctuations are not expected. They are also more effective in climates with relatively high relative humidity. At very cold climate to avoid ice formation in the air inlet duct to the tower and around it, the hot water flow rate to the pipes near the air inlet can be increased.

8.3.2.4 Hybrid Hyperbolic Cooling Towers

Hybrid hyperbolic cooling towers, also known as **fan-assisted hyperbolic cooling towers** or simply **hybrid draft cooling towers**, are a combination of hyperbolic natural draft cooling towers and mechanical draft cooling towers. From far they look like hyperbolic cooling towers, just shorter and with smaller base diameters. This lower height comes at the expense of required additional fans. These typically forced draft fans, which are installed around the tower base, assist the tower to achieve the needed driving force with a shorter height (Figure 8.17). The fans are usually operated when the plant produces its peak power and/or at high ambient temperatures. The design also eliminates the possibility of moist air recirculation and provides some

FIGURE 8.17
A fan-assisted hyperbolic cooling tower (hybrid hyperbolic cooling tower) (the photo shows an abandoned one!). (Image by Peter H. from Pixabay, https://pixabay.com/photos/fan-cooling-propeller-cooler-turn-3645379/.)

air flow rate control. Hybrid towers are typically used when the building of a tall hyperbolic tower is not acceptable or the release of the plume leaving the cooling tower at low elevations is not permissible.

8.4 Combined Once-Through and Wet Cooling Towers

Once-through cooling systems can be combined with wet cooling towers to reduce their environmental impacts and/or extend their operation period. In one such combined system, sometimes referred to as an **open wet cooling tower**, a cooling tower is placed at the pass of the hot water leaving the condenser. In this design, similar to a once-through cooling system, the cooling water is withdrawn from a natural body of water. However, the temperature of the hot water from the condenser is first reduced in the evaporative cooling tower before being returned to the water source. While this combined system cannot reduce marine life fatality within the cooling system, it can substantially reduce the thermal pollution in the natural body of water. Another strategy for combined cooling systems, known as a **helper cooling tower**, is to divide the cooling load between a once-through system and a wet cooling tower. The contribution of each unit generally can be from 0% to 100% of the load. When the once-through cooling system cannot be used, for example during a fish reproduction season, or its use is limited, for example when available water in the source is limited due to drought, the cooling tower can take a larger share of the cooling load.

8.5 Closed Circulating Water Systems – Dry Cooling Towers

Unlike wet cooling towers, where circulating water interacts with air and its evaporation is a major cooling mechanism, in dry cooling towers circulating water does not come in contact with ambient air at all. The only heat-transfer mechanism is sensible heat transfer by conduction and convection. In these cooling towers, the circulating water flows inside tubes while air passes outside. Since the heat-transfer coefficient of the water flow is much higher than that of the air flow, the total heat transfer is dictated and limited by the low heat transfer at the air side. To accelerate the heat-transfer process, typically finned tubes are used in dry cooling towers to extend heat-transfer surfaces on the air side.

The most important advantage of dry cooling towers is that they do not need a major water source to directly transfer heat to (like once-through cooling systems) or to make up for evaporated water (like evaporative cooling systems). This means the site for steam power stations equipped with a dry cooling system can be selected independent of the availability of water resources. These power plants can be located in regions where water is not abundantly available or environmental regulations restrict the usage of available water for cooling purposes. This opens up increasingly large regions for the installation of new steam power plants or for the rehabilitation or extension of existing power plants that would be unavailable otherwise, particularly sites close to consumption centers or fuel resources. Compared to wet cooling systems, dry cooling systems need less maintenance, no chemicals for water treatment, and no equipment cleaning. Furthermore, they face neither the icing problem nor produce plume. Their greatest disadvantage is that the condenser temperature is not nearly as low as that of once-through or evaporative cooling systems. The outlet water temperature of dry cooling towers can be up to about 20°C (about 35°F) higher than that of evaporative cooling towers. This lower condenser temperature leads to a lower cycle efficiency, which, in turn, means that more thermal energy needs to be discharged in the condenser and eventually in the cooling system. This results in larger cooling towers requiring larger equipment and more land for the installation. For this reason, the initial costs of dry cooling towers are substantially higher than those of wet cooling towers. Another major problem with dry cooling systems is their sensitivity to ambient conditions, particularly ambient temperature.

Until the 1990s there were just a handful of steam power plants with dry cooling systems. However, since then the number of power plants using dry cooling towers has considerably increased primarily due to increased environmental restrictions and water shortages in many regions. In fact, since 2000 substantially more new power plants have been built with dry cooling towers than with once-through cooling systems. Similar to wet cooling towers, dry cooling towers can be mechanical draft or natural draft types. There are two types of dry cooling towers: **direct cooling towers** and **indirect cooling towers**.

8.5.1 Direct Dry Cooling Towers or Air-Cooled Condensers

In this type of cooling systems, the condenser and the cooling tower are integrated into a single component. That is why these systems are commonly referred to as **air-cooled condensers**. In these systems, many air-cooled finned tubes condense the exhausted steam from the turbine (Figure 8.18). The steam is distributed among the tubes by a very large diameter header to reduce the pressure drop in the steam flow. Since the cooling system also acts as the condenser, it should provide some of its functions, including the separation and removal of noncondensable gases. The condensate is collected by the condensate header and is returned to the power cycle by the condensate pump. The air flow through the tubes can be provided by mechanical draft fans (typically forced draft) or natural draft cooling towers (Figure 8.18).

8.5.2 Indirect Dry Cooling Towers

Indirect dry cooling towers are equipped with a condenser as well as a cooling tower. These cooling towers can be used with either a direct-contact condenser or a surface condenser. In the former, as described previously, the cold water from the cooling tower is mixed with the saturated steam from the turbine. This is possible only because the cold water is not contaminated in the cooling tower by direct contact with the ambient air. The condensate from the condenser can be pumped by a single condensate pump, which is then divided between the flow to the cooling tower (the larger portion) and back to the feedwater system (the smaller portion) (Figure 8.19). Alternatively,

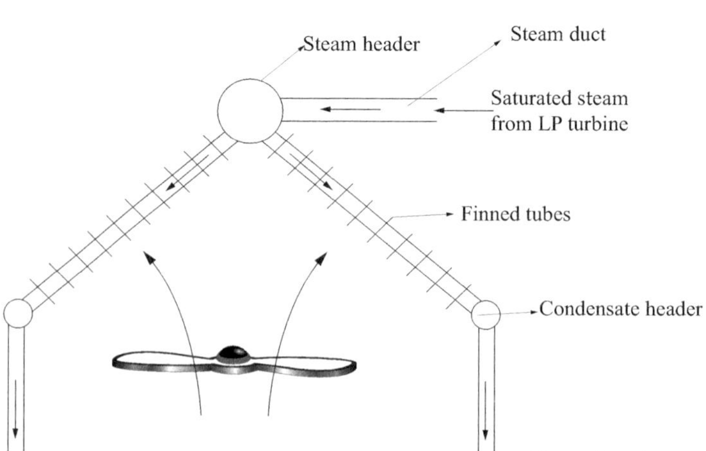

FIGURE 8.18
Schematic of a mechanical draft air-cooled condenser (or a direct dry cooling tower).

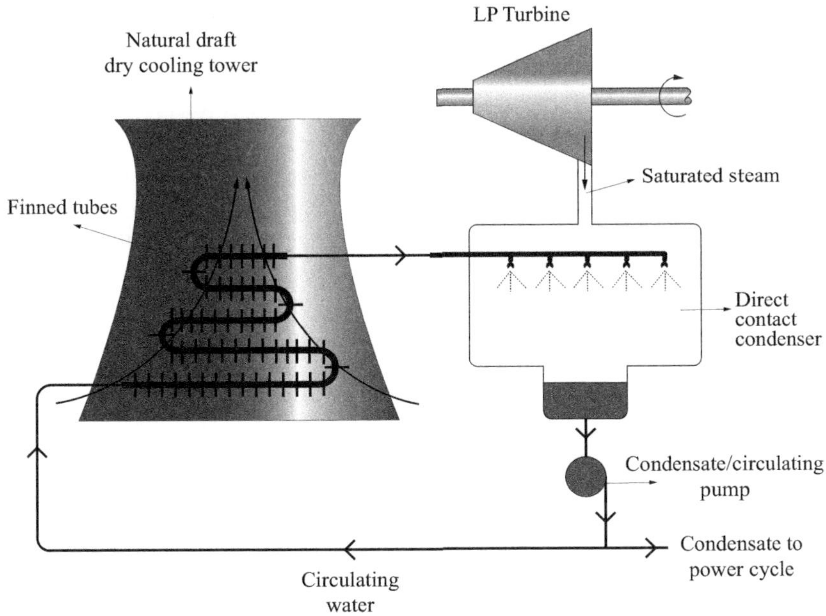

FIGURE 8.19
Schematic of an indirect dry cooling tower.

two pumps, the condensate pump and the circulation pump, can pressurize the two flows individually.

When an indirect dry cooling tower operates with a surface condenser, there is an intermediate cooling water loop but there is no mixing between the condensate and the cooling water. The irreversibilities due to two indirect heat-transfer processes, one in the condenser and one in the dry cooling tower, negatively impact the efficiency of the cycle. In order to reduce these irreversibilities and improve the efficiency of the cooling system and the overall cycle, a refrigerant, such as ammonia, can replace water as the secondary heat-transfer medium. The refrigerant evaporates in the condenser and condenses in the cooling tower. The phase change process can significantly improve the heat-transfer coefficients in both the condenser and the cooling tower and reduce the condenser temperature. In both the above-mentioned types of indirect dry cooling towers, the tower is composed of finned tubes with the air flow provided either by a natural draft tower or induced draft fans.

As noted, dry cooling towers are not efficient and are particularly vulnerable to high ambient temperatures. To assist these units on hot summer days, a supplementary evaporative strategy can be included in these systems such as spraying water over the outer surface of tubes in the cooling tower or adding a standby wet cooling tower.

8.6 Hybrid Wet and Dry Cooling Towers

A **hybrid wet and dry cooling tower** integrates these two cooling systems into a single unit with the objective of reducing water consumption and/or plume. Depending on the primary objective, these units may also be referred to as **water conservation cooling towers** or **plume abatement cooling towers**. In hybrid cooling towers, the dry cooling section is located at the top of the wet cooling section (Figure 8.20). Hot water first enters the dry section, which is made of a series of sloped parallel finned tubes. The temperature reduction in this section is achieved by means of sensible heat transfer and is limited by the ambient dry bulb temperature. If the ambient temperature is low enough that the water temperature can be reduced to the desired temperature, the water is directly bypassed to the cold water basin without going through the wet section of the tower. Otherwise, the water leaving the dry section is fed to the wet section for further cooling. In the former case, the water consumption is almost zero (other than little evaporation from the cold basin) and in the latter case water consumption can be substantially reduced because of precooling of the water in the dry section. The inlet air flow is admitted separately to the dry and wet sections. The air leaving the wet section is saturated with humidity. When this flow is mixed with hot dry air leaving the dry section (Figure 8.20), it is desaturated and the air leaving the cooling tower will have the relative humidity of <100% which is less likely to create visible fog (plume) than that in a wet cooling tower. The share of each

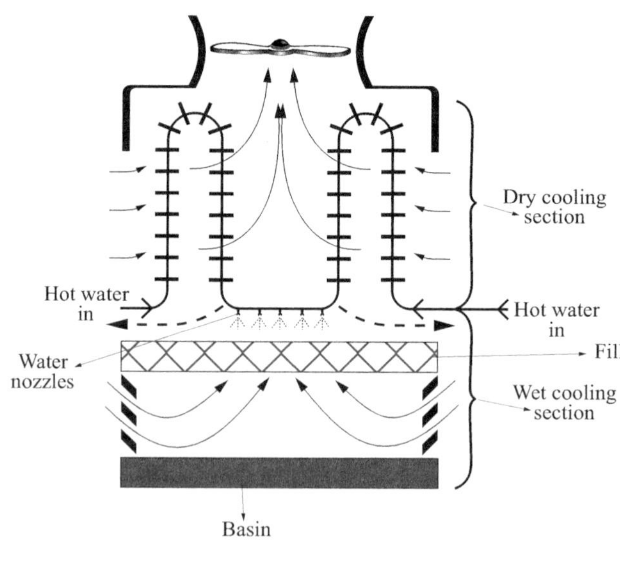

FIGURE 8.20
Schematic of a hybrid wet and dry cooling tower.

section of a hybrid tower from the total heat-transfer load depends on the ambient conditions and the primary concern of the system (water consumption reduction or visible fog reduction).

8.7 Challenge of Environmental Problems

Cooling towers are an excellent example to demonstrate how complicated it is to deal with environmental problems and how they need a holistic approach. We learned that once-through cooling systems can hurt the local environment and ecosystem by directly killing marine life and thermally and possibly chemically polluting local water sources. By using wet cooling towers, water extraction from natural sources and return of heated water are eliminated; however, this comes with the loss of a certain percentage of water to evaporation, which must be made up by a natural source of water. While this percentage is relatively small (about 1%–3% of the total circulating water flow), considering a large volume of circulating water, it is unbelievably significant. According to the U.S. Geological Survey, **thermal power plants (both fossil and nuclear plants) use about 41% of the total water withdrawals from natural water resources for all uses.** This accounted for 34% of total freshwater withdrawals, and 48% of fresh surface-water withdrawals. A majority of this extracted water is used for cooling in evaporative cooling towers. To eliminate any water consumption, dry cooling towers can be used. But as noted, a power plant with a dry cooling tower has low efficiency. This means for the same power generation, the power plant requires to combust more fuel in the steam generator. So although no water is used for cooling purposes, more flue gas and heat are rejected to the environment. Similar arguments can be made for most problems and there is not really a single solution. For every problem, a compromise between many conflicting factors should be reached to optimize the system in question.

8.8 Questions and Problems

8.8.1 Introduction

8.1. What is the objective of a circulating water system in a steam power cycle?

8.2. Name the approaches for the heat rejection to the environment in steam power plants.

8.3. What is the function of a cooling tower in a steam power cycle?

8.8.2 Once-Through Circulating Water Systems

8.4. Name some water sources that can be used in once-through circulating water systems.

8.5. How are debris removed from inlet cooling water to a once-through circulating water system? How are these devices cleaned?

8.6. Label the main components of a once-through circulating water system in the following figure.

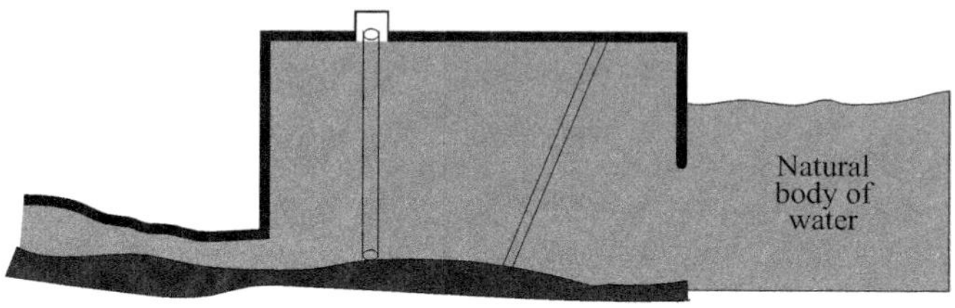

8.7. Why is the condenser in power plants with a once-through cooling system typically single-pass?

8.8. Explain why in a once-through cooling system the returned water should be properly distributed and mixed with the water within the reservoir.

8.9. Describe how heat is transferred to the environment in cooling ponds and lakes. How can the effectiveness of the cooling process be improved?

8.8.3 Closed Circulating Water Systems – Wet Cooling Towers

8.10. Which one has higher humidity?

a. Air entering a wet cooling tower

b. Air exiting a wet cooling tower

8.11. Which cooling system requires a higher flow of cooling water and which one consumes more water?

a. Once-through cooling systems – evaporative cooling systems

b. Once-through cooling systems – once-through cooling systems

c. Evaporative cooling systems – once-through cooling systems

d. Evaporative cooling systems – evaporative cooling systems

8.12. Which cooling system is the most common in new steam power plants of all types?

a. Evaporative cooling systems

b. Once-through cooling systems

c. Combination of two systems

8.13. What is the difference between evaporating cooling towers and dry cooling towers?

8.14. What are the primary and secondary heat-transfer mechanisms in wet cooling towers? Explain how the share of each heat-transfer mechanism from the total heat transfer is impacted by the ambient air properties.

8.15. Name two ways to distribute the hot water flow over the fill in a wet cooling tower. Briefly describe each method.

8.16. What are the pros and cons of elevated systems for hot water transportation systems from a condenser to a cooling tower? What about buried piping systems?

8.17. What are the functions of the cold water basin in evaporative cooling towers?

8.18. List three ways the air flow can be driven through a wet cooling tower. Briefly explain each method.

8.19. What are the advantages and disadvantages of mechanical draft cooling towers?

8.20. Compare forced draft and induced draft cooling towers. Which one is the dominant type in utility steam power plants?

8.21. What is the function of the following components of induced draft evaporative cooling towers?

a. Louvers

b. Drift eliminators

c. Fans

8.22. Why are some fans in mechanical draft cooling towers multiple-speed or variable speed?

8.23. Explain how the pressure drop in the air flow pass is overcome in natural draft cooling tower.

8.24. Why do natural draft cooling towers always require a tall structure?

8.25. Describe why natural draft cooling towers have hyperbolic structures.

8.26. The shape of natural draft cooling towers does not have much to do with the flow of air inside the tower.

a. True

b. False

8.27. When and why do wet cooling towers create visible fog or plume? When and why this plume disappears in the ambient air? What problems can plume create? How can the effects be reduced?

8.28. In what conditions can the flue gas flow be discharged into the natural draft cooling tower?

8.29. When are fan-assisted hyperbolic cooling towers used?

8.30. How is the physical appearance of natural draft cooling towers and fan-assisted hyperbolic cooling towers different?

8.8.4 Combined Once-Through and Wet Cooling Towers

8.31. Briefly explain open wet cooling towers and helper cooling towers.

8.32. Provide two examples of circumstances when helper cooling towers are used.

8.8.5 Closed Circulating water Systems – Dry Cooling Towers

8.33. In dry cooling towers, which flow side is equipped with finned tubes? Why?

8.34. What is the most important advantage of dry cooling towers? List other advantages.

8.35. Why has the number of power plants equipped with dry cooling towers substantially increased recently?

8.36. Explain how air-cooled condensers work.

8.37. Describe why the cold water from the cooling tower can be mixed with the saturated steam from the turbine in indirect dry cooling towers.

8.38. Which type of cooling systems can be used with direct-contact condensers? Why?

a. Once-through cooling systems

b. Evaporative cooling towers

c. Indirect dry cooling towers

8.39. Which types of dry cooling towers can be equipped with a condenser?

a. Indirect dry cooling towers

b. Direct dry cooling towers

8.8.6 Hybrid Wet and Dry Cooling Towers

8.40. Explain how a hybrid wet and dry cooling tower work. Describe the process at low and high ambient temperatures.

8.41. What are two possible objectives of hybrid wet and dry cooling towers?

8.42. Explain how a hybrid wet and dry cooling tower can reduce plume.

8.43. The share of dry and wet sections of a hybrid tower from the total heat-transfer load depends on the ambient conditions and the primary concern of the system (water consumption reduction or visible fog reduction).

a. True
b. False

8.8.7 General

8.44. A steam power station is to be built near a river with sufficient water for directly cooling of the plant's condenser. Which type of circulating water system do you recommend for this plant (if there is no limit on water removal/return from/to the river)? Draw a schematic diagram of this system showing main components and streams.

8.45. A steam power station is to be built near a river with sufficient water for directly cooling of the plant's condenser. However, during fall seasons due to the reproduction of a certain type of fish and their sensitivity to water temperature variations, the environmental regulation prohibits increasing the water temperature of the river. Which type of circulating water system do you recommend for this plant?

8.46. Draw a schematic diagram of a direct-contact spray-type condenser and indirect dry cooling tower showing all main components and steams. Explain which one is higher: the recalculating water mass flow rate or the water mass flow rate of the cycle (the cycle working fluid)? Why?

8.47. List advantages and disadvantages of the mechanical force draft (FD) system in a wet cooling tower. Which of the force draft (FD) or induced draft (ID) systems is used in modern mechanical draft cooling systems?

8.48. How do evaporative cooling towers work?

8.49. Identify which of the following schematics represent mechanical force draft (FD) system and induced draft (ID) system in a wet cooling tower.

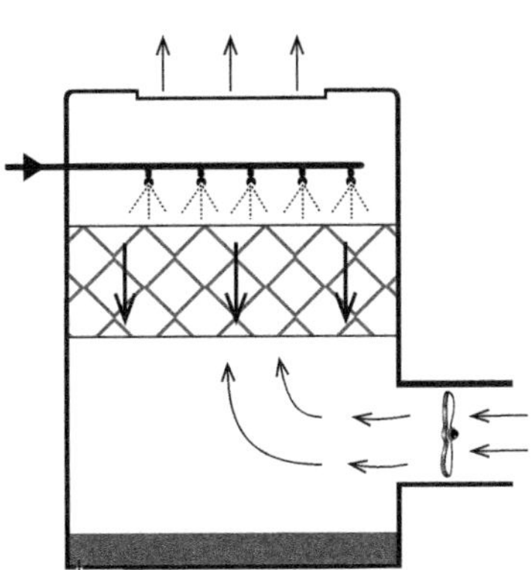

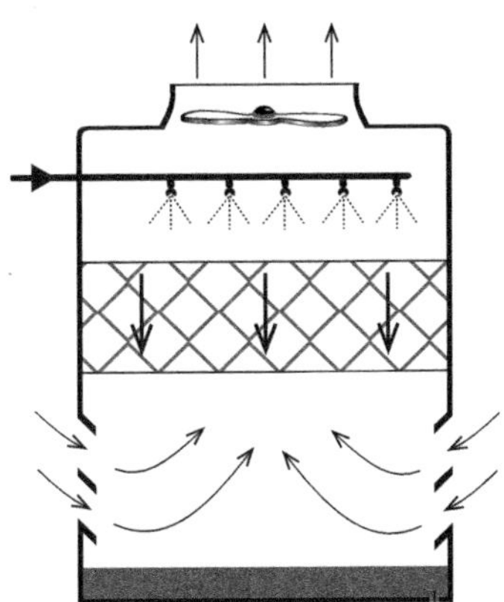

8.50. Why are splash-type fills sensitive to bars' alignments?

8.51. List three sources of water losses in a wet cooling tower.

8.52. All natural draft cooling towers are made of concrete structures.

a. True

b. False

8.53. Explain a hot water transportation system from a condenser to a cooling tower.

8.54. Compare once-through cooling systems, wet cooling towers, and dry cooling towers in terms of their environmental impacts.

8.55. Compare natural draft, mechanical draft, and fan-assisted hyperbolic cooling towers.

8.56. The share of the evaporative heat-transfer mechanism from the total heat transfer in wet cooling towers depends on the ambient air ...... (check all correct items.)

a. humidity

b. temperature

c. pressure

d. composition

8.57. Which circulating water system is the simplest, least expensive, and most effective in minimizing the condenser temperature?

8.58. In a wet cooling tower, explain:

- Why the drain (bleed off) is needed.
- Why the drift eliminator is needed.

8.59. Which federal act protects water resources in the U.S.?

8.60. Explain how a hybrid wet and dry cooling tower can reduce water consumption.

8.61. List problems that unevaporated water droplets can create if uncaptured by drift eliminators in induced draft cooling towers.

8.62. How can supplementary evaporative strategies be used to assist dry cooling towers?

8.63. Why are some natural draft cooling towers constructed with steel structures?

8.64. Explain why some fans in mechanical draft cooling towers have adjustable pitch blades?

8.65. Compare dry and wet cooling towers.

8.66. Explain how the natural circulation cooling towers operate. Why are they typically the tallest structure in a steam power plant? Why are they hyperbolic in shape?

8.67. Name two hot water distribution systems over the fill in a wet cooling tower.

8.68. What are two types of the fill used in evaporative cooling towers?

8.69. What are the pros and cons of forced draft fans and induced draft fans?

8.70. How is the large contact surface area formed in splash-type wet cooling towers?

8.71. Name two types of the fill in terms of the relative direction of the water and air flows. Briefly explain each type.

8.72. Explain why dry cooling towers are not efficient and particularly vulnerable to high ambient temperatures?

8.73. Determine which one of the following schematics is a counterflow configuration and a crossflow configuration.

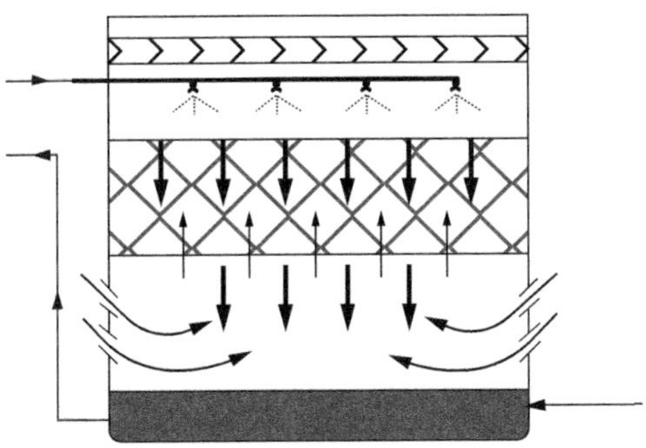

(a)

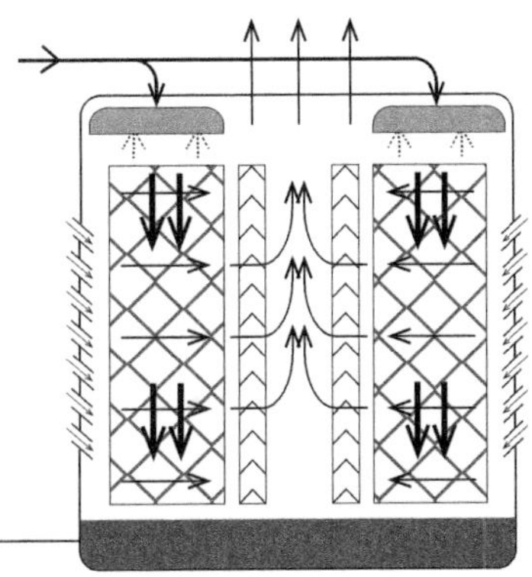

(b)

8.74. Explain how atmospheric cooling towers work.

8.75. Explain how direct dry cooling towers work.

8.76. Which one has lower temperature?

a. Air entering a wet cooling tower

b. Air exiting a wet cooling tower

8.77. What is the function of a cooling tower?

8.78. What are the advantages and disadvantages of natural draft cooling towers?

8.79. Explain why in some indirect dry cooling towers operating with a surface condenser as a refrigerant, such as ammonia, can replace water as the secondary heat-transfer medium. What are its benefits?

8.80. The share of the sensible heat-transfer mechanism from the total heat transfer in wet cooling towers depends on the ambient air ……

a. humidity
b. temperature
c. pressure
d. composition

9

Gas Turbines

Abstract

Gas turbine power generation units are an essential technology for utility electricity generation. Their contribution to electricity generation is second only to steam power cycles. They are also critical for many other industries, most notably the aviation industry. We will start this chapter with an extensive study of thermodynamic principles of the Brayton cycle as an ideal thermodynamic cycle representing gas turbine units. We will consider the impacts of various parameters on the performance of the cycle and how the cycle can be improved by adding auxiliary components. We will also study how actual gas turbine units deviate from the ideal cycle. We will conduct many calculations of simple and modified ideal and non-ideal Brayton cycles by means of air-standard and cold air-standard analyses. Then, we will briefly review the history of gas turbine developments and their applications in various industries. We will discuss components of gas turbine units for electricity generation, including compressors, combustion chambers, turbines, and compressor inlet air cooling systems. We will conclude the chapter with evaluating start-up and shutdown sequences, normal operation, and control of gas turbine units.

LEARNING OUTCOMES

After successful completion of this chapter, you will be able to:

- describe ideal Brayton cycle
- explain impacts of operating conditions on the performance of Brayton cycle
- describe various modifications in the simple Brayton cycle to improve its performance
- identify distinctions between ideal Brayton cycle and actual gas turbine power plants
- calculate the thermodynamic performance of various forms of Brayton cycles
- explain the history of gas turbine developments

- list gas turbine applications in various industries and identify how their characteristics match the requirements of each industry
- explain types of compressors for gas turbine units and their performance
- explain various combustor types and their emission control technologies
- describe gas turbine blade specifications and their cooling systems
- identify various control modes for gas turbines

9.1 Introduction

Gas turbines, sometimes also referred to as **combustion turbines** or **fire turbines** (the latter were common in older publications), are power generation units with a gaseous working fluid involving steady-state flow processes. In gas turbine cycles, a rotary machine (which also confusing can be called a gas turbine) extracts thermal energy of a high pressure and temperature flow of combustion products and converts it to mechanical rotational power, electricity, and/or thrust. They are commonly used in the **aviation industry** to provide propulsion for commercial and military airplanes. They are also widely used to generate electricity in **utility power plants**, **backup power units**, and **emergency power units**. Historically speaking, gas turbines are considered a new technology among conventional power generation technologies. Unlike steam power generation units and internal combustion engines that were introduced in the 18th century and developed in the 19th century, gas turbines were not introduced until the 20th century and early commercial units started to appear in the late 1930s and the early 1940s. Gas turbine technology advanced quickly in the second half of the 20th century. However, the introduction and widespread utilization of combined cycle power plants (CCPPs) in the late 20th century led to the significant contribution of gas turbines to the electricity generation industry. A more comprehensive study of the history of gas turbines and their applications will be presented later in this chapter.

Gas turbines have a high power-to-weight ratio, which makes them ideal power systems to propel aircrafts. For utility electricity generation, gas turbines' quick start-up and shutdown, and rapid load-following characteristics as well as their relatively low capital costs perfectly suit the needs of peak-load power generation units. However, they can also be used as baseload power plants due to the relatively high efficiency of modern gas turbines as well as their low maintenance, high availability, and high capacity factor. Their other favorable characteristics are no need for water (which is critically important in regions where water is not readily available), simple design and construction, short delivery and installation time, and remote control capability.

Despite these positive attributes, standalone gas turbines have the following several inherent disadvantages: high-temperature exhaust flow, strong dependency of both power output and efficiency on ambient conditions, and inability to consume solid fuels. The high temperature of the gas turbine exhaust flow means high-grade energy is wasted. This problem can be overcome by using modified gas turbines or more commonly by using the exhaust flow to raise steam (Chapter 10). The power output and efficiency of gas turbines and other cycles based on gas turbines strongly depend on ambient conditions, particularly ambient temperature. The power output and efficiency of a gas turbine reduce when the ambient temperature increases. This problem can be addressed by using compressor inlet air cooling systems (Section 9.5.4.1). Gas turbines are most commonly fueled by natural gas, but they can also be fueled by other fuels, such as heavy oil (fuel oil), diesel, propane, or manufactured fuels, such as hydrogen, biogas, and syngas (synthesis gas). The latter fuel is produced by gasification of coal or biomass to enable gas turbines to be fueled indirectly by solid fuels (Chapter 10).

9.2 Thermodynamic of Gas Power Cycles

As the name implies, unlike phase-changing working-fluid power cycles, in gaseous working-fluid power cycles, the phase of the working fluid does not change, and the working fluid always remains in the gaseous phase. These power cycles can be of two types: internal combustion engines and external combustion engines. In the former, the working fluid participates in the combustion process, and hence, the composition of the working fluid changes throughout the system. In the latter, the heat is transferred from an external source of thermal energy, possibly combustion or nuclear reactions, to the working fluid so the composition of the working fluid does not change. External combustion engines have little applications in large power generation systems with gaseous working fluids. An example of such a cycle is the application of gas turbines in nuclear power plants (refer to closed gas turbine cycles below). Although most gaseous working-fluid power cycles are internal combustion engines, the term usually refers to reciprocating internal combustion engines typically used in transportation applications.

While in this section the thermodynamic principles to study a gas turbine cycle will be covered, the initial discussion about the air-standard power cycles is general for all gaseous working-fluid power cycles, including piston and cylinder systems involving boundary movement work, such as spark- and compression-ignition internal combustion engines. The main difference between a gas turbine unit and a piston and cylinder unit is that in gas turbines the processes are continuous, while in piston and cylinder units, the processes are intermittent.

Gas turbine power generation cycles need **four processes**: a compression process in a compressor, a heat transfer from a high-temperature thermal energy source to the working fluid in a combustor or a heat exchanger, an expansion process in a turbine to produce work, and a heat transfer from the working fluid to the environment (a low-temperature thermal reservoir).

Gas turbines (and gaseous working-fluid power cycles in general) can be operated based on open or closed cycles. Open systems involve intake and exhaust of the working fluid (air). The great majority of gas turbines for all applications are open power systems. In these systems, the **compressor** receives fresh air from the atmosphere (stream 1 in Figure 9.1) and pressurizes it (stream 2). The temperature of the compressed air is then increased in the **combustor**, also known as the **combustion chamber**, by the injection of fuel (stream 3) and its combustion. The high-temperature and pressure combustion products (stream 4) then enter the **turbine** (gas turbine) where they expand and produce work. Finally, the outlet of the turbine is exhausted to the atmosphere (stream 5). If the gas turbine cycle is used as a power generation unit, a portion of the work produced in the turbine is used to run the compressor and the rest is available for electricity generation in the **generator** or for an external mechanical work utilization. If the gas turbine is for propulsion, e.g., in an aircraft, the turbine produces enough power to run the compressor and the rest of the combustion products go through a nozzle, which convert the high-temperature and pressure flow to a high-velocity flow to produce thrust. It is evident that none of these systems need a heat exchanger to reject heat to the environment. The heat rejection to the environment is achieved by discharging the turbine outlet to the atmosphere.

In closed-cycle gas turbines, the working fluid, which theoretically can be any gas, is recirculated and there is no inlet or exhaust of the working fluid (Figure 9.2). In these cycles, the turbine exhaust stream goes through a heat exchanger where heat is rejected to the surrounding from the working fluid before returning to the compressor. Note that closed-cycle gas turbines are not internal combustion engines because the working fluid does not participate in the combustion process.

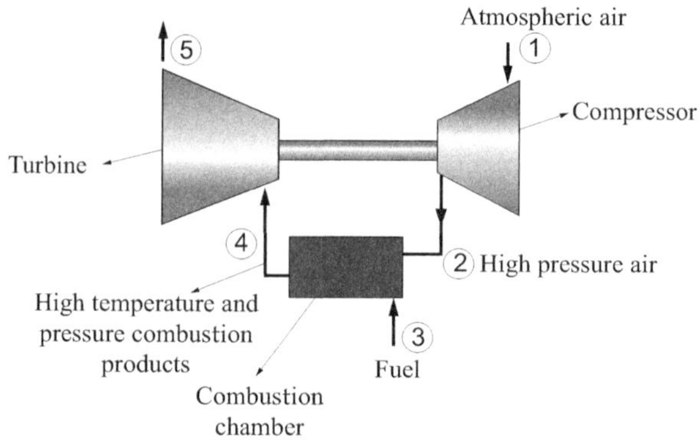

FIGURE 9.1
Schematic of an open direct gas turbine system.

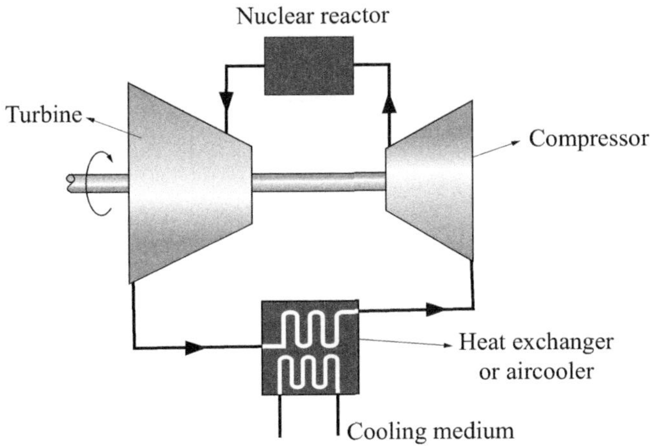

FIGURE 9.2
Schematic of a closed direct gas turbine cycle.

In Figures 9.1 and 9.2, the source of high-temperature thermal energy is a combustion process in the combustion chamber and a gas-cooled nuclear reactor, respectively. In both systems, the outlet of the combustion chamber or the reactor is directly fed to the turbine. These systems are called **direct** systems since the heat is directly transferred to the working fluid from the heat source.

A gas turbine cycle can also be an indirect cycle (Figures 9.3 and 9.4). In this cycle, the outlet flow of the combustor or the reactor, which is called the secondary fluid, is used to heat the working fluid of the cycle (the primary fluid) through a heat exchanger (an air heater). The secondary fluid can be

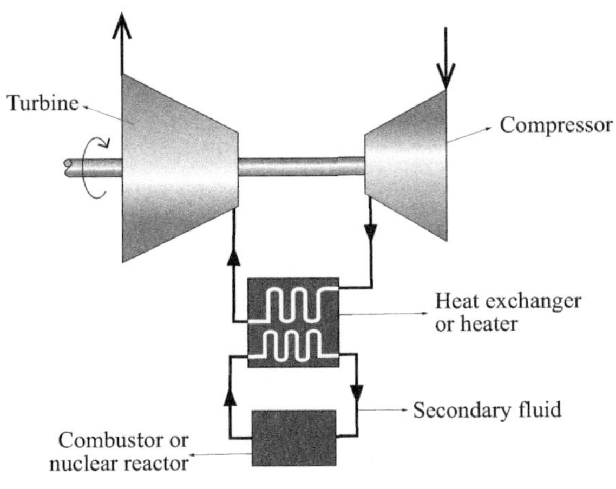

FIGURE 9.3
Schematic of an open indirect gas turbine system.

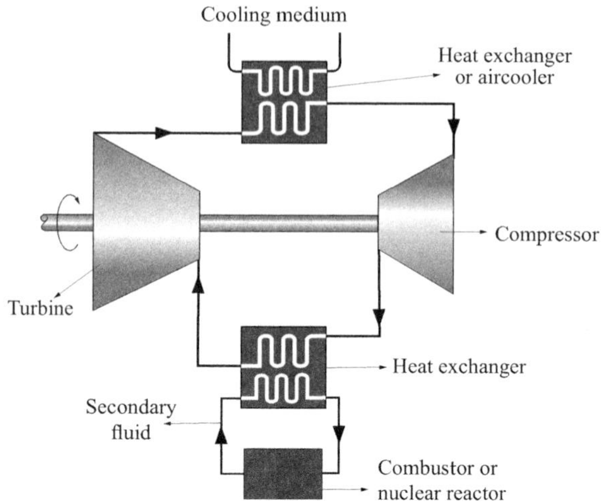

FIGURE 9.4
Schematic of a closed indirect gas turbine system.

water, liquid metal, or helium among others. Indirect cycles are used to prevent the release of contaminants, e.g., radioactive matters, to the environment if the source of thermal energy is a nuclear reactor. Alternatively, they can help the cycle to use fuels, such as coal or crude oil, which are not suitable for direct combustion in gas turbines. This concept was heavily investigated in the 1950s and 1960s with several large-scale proof-of-concept plants installed. However, they could not compete with conventional gas turbines and steam cycles, and commercial units were never materialized.

In summary, in indirect cycles, the working fluid is separated from the high-temperature heat source. In closed cycles, the working fluid is separated from the atmosphere (as the low-temperature heat sink).

In all configurations of gas turbine plants, several factors, such as the combustion process, heat transfer, temperature and pressure gradients, and friction, make the study of actual systems very complicated, which requires detailed computer numerical simulations. An elementary thermodynamic study of these power systems requires significant simplifications. An **air-standard analysis**, also called an air-standard power cycle analysis, is used for this purpose, which is based on a set of simplifications.

REMINDER BOX 9.1

AIR-STANDARD ANALYSIS

In order to qualitatively examine the effects of various parameters on the performance of gaseous working-fluid cycles, the air-standard analysis can be conducted with the following assumptions:

- The working fluid is air which always behaves according to the ideal gas model.
- All processes are internally reversible.
- They operate in a closed cycle.

The last assumption means that there are no inlet and exhaust flows, and a fixed amount of air goes through the cycle as the working fluid. There is also no combustion process. The combustion process is replaced by a reversible heat transfer process from the high-temperature heat source, and the exhaust process is replaced by a reversible heat transfer process to the low-temperature heat sink.

It should be noted that these assumptions, particularly avoiding combustion, introduce significant errors to analyses. The estimated properties and system performance indicators such as efficiency and specific work may differ substantially from those of actual systems operating at the same conditions. Therefore, these analyses should be only used for qualitative comparisons of cycles. Since all irreversibilities are absent in these idealized cycles, their performance should be treated as an upper limit for the performance of actual systems that cannot be achieved but can be approached.

In a particular kind of air-standard cycle called a **cold air-standard analysis**, the specific heats of air are assumed to remain constant throughout the cycle at their ambient temperature values (300 K). Obviously, this assumption introduces further errors to analyses and is commonly avoided for noneducational purposes.

REMINDER BOX 9.2

IDEAL GAS LAW

In Reminder Box 4.1, we learned that for ideal gases, we can write the following equations:

$$pV = n\bar{R}T = mRT \tag{9.1}$$

$$pv = RT \tag{9.2}$$

$$R = \frac{\bar{R}}{M} \tag{9.3}$$

$$n = \frac{m}{M} \tag{9.4}$$

$$\bar{R} = 8.314\,\text{kJ/kmol.}K = 1.986\ \text{Btu/lbmol.}^\circ R$$

$$= 1545\ \text{ft lbf/lbmol.}^\circ R$$

$$R_{\text{air}} = 0.2870\ \text{kJ/kg}\ \ K = 0.06855\ \ \text{Btu/lb}\,R\ (\textit{Table A}.5)$$

For these gases, specific internal energy (u) and specific enthalpy (h) depend only on temperature. The specific heat at constant pressure (C_p) and volume (C_v) can be defined as

$$C_p = \frac{dh}{dT} \tag{9.5}$$

$$C_v = \frac{du}{dT} \tag{9.6}$$

Both C_p and C_v are affected by the molecular structure of the gas. For monatomic gases, such as argon (Ar) and helium (He), the specific heats can be assumed constant with the temperature. But they increase with an increase in the gas temperature for diatomic and triatomic (faster for the latter).

It can be shown that the relationship between C_p and C_v is

$$C_p - C_v = R \tag{9.7}$$

Also, the specific heat ratio, k, is

$$k = \frac{C_p}{C_v} \tag{9.8}$$

The changes in specific enthalpy and internal energy from state 1 to state 2 are

$$h_2 - h_1 = \int_{T_1}^{T_2} C_p(T)\,dT \tag{9.9}$$

$$u_2 - u_1 = \int_{T_1}^{T_2} C_v(T)\,dT \tag{9.10}$$

In order to determine the integrals in Equations 9.9 and 9.10, the specific heats as a function of temperature are needed. There are three common methods to determine the changes in specific enthalpy and internal energy.

Method 1

The values of specific heats as a function of temperature are presented in the form of equations or sometimes tables or graphs (you can find them in many thermodynamic textbooks). For instance, for air, the following equation estimates the variation of the specific heat at constant pressure for the temperature range of 300 and 1000 K:

$$\frac{\bar{C}_p}{\bar{R}} = 3.653 - 1.337 \times 10^{-3} \times T + 3.294 \times 10^{-6} \times T^2 - 1.913 \times 10^{-9} \times T^3 + 0.2763 \times 10^{-12} \times T^4 \quad (9.11)$$

By substituting Equation 9.11 into Equation 9.9 and finding the integral, the change in the specific enthalpy can be determined.

Method 2

Many thermodynamic textbooks present the specific enthalpy and internal energy of various ideal gases as a function of temperature in tables, such as Tables A.7 (SI) and A.7 (typically using equations similar to Equation 9.11).

Method 3

By assuming constant specific heats (as in cold air-standard analyses), Equations 9.9 and 9.10 can be simplified as

$$h_2 - h_1 = C_p (T_2 - T_1) \quad (9.12)$$

$$u_2 - u_1 = C_v (T_2 - T_1) \quad (9.13)$$

This assumption is acceptable for monoatomic gases and cases that the variation of the gas temperature is small so the changes in specific heats are negligible. But for other cases, this assumption may introduce a significant error to calculations. However, it is very helpful in developing qualitative expressions for cycles under investigation.

Similarly, if the specific heats are taken as constants, the change in specific entropy is

$$s_2 - s_1 = C_v \ln \frac{T_2}{T_1} + R \ln \frac{v_2}{v_1} = C_p \ln \frac{T_2}{T_1} - R \frac{p_2}{p_1} \quad (9.14)$$

Since constant entropy (isentropic) processes play a major role in the analysis of gas turbine cycles, it is important to review the relationship among pressure, temperature, and specific volume in such processes when C_p and C_v are constant.

$$\frac{p_2}{p_1} = \left(\frac{T_2}{T_1} \right)^{k/(k-1)} \quad (9.15)$$

$$\frac{p_2}{p_1} = \left(\frac{v_1}{v_2} \right)^{k} \quad (9.16)$$

$$\frac{T_2}{T_1} = \left(\frac{v_1}{v_2}\right)^{k-1} \tag{9.17}$$

The process 1-2 is an isentropic process. Similarly, if air property tables are used for retrieving properties, the following equation can be used for the isentropic process 1-2:

$$\frac{p_2}{p_1} = \frac{pr_2}{pr_1} \tag{9.18}$$

where p_r is the relative pressure.

9.2.1 Brayton Cycle as an Ideal Cycle for Gas Turbine Power Systems

The idealized cycle for simple gas turbines is called a **Brayton cycle** (sometimes also referred to as a **Joule cycle**). This cycle is an air-standard power cycle that involves four reversible processes (Figures 9.5 and 9.6): two constant-pressure (isobaric) processes and two reversible adiabatic (and hence constant entropy or isentropic) processes. Figure 9.6 shows the T-s and P-v diagrams of a Brayton cycle. Process 1-2 is an isentropic compression in the compressor, and process 3-4 is an isentropic expansion in the turbine. Process 2-3 is a constant-pressure (isobaric) heat transfer from the high-temperature heat source to the working fluid, which typically replaces the combustion process. Process 4-1 is an isobaric heat transfer from the working fluid to the low-temperature heat sink, which typically replaces the exhaust of combustion products to the environment. The heat transfer processes are at constant pressure because all irreversibilities,

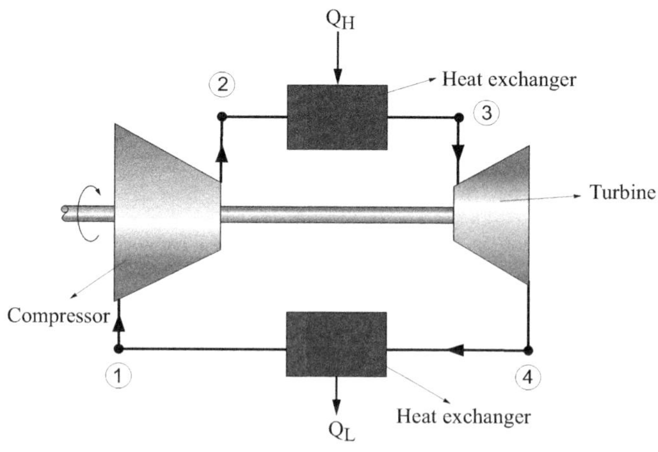

FIGURE 9.5
Schematic of a Brayton cycle.

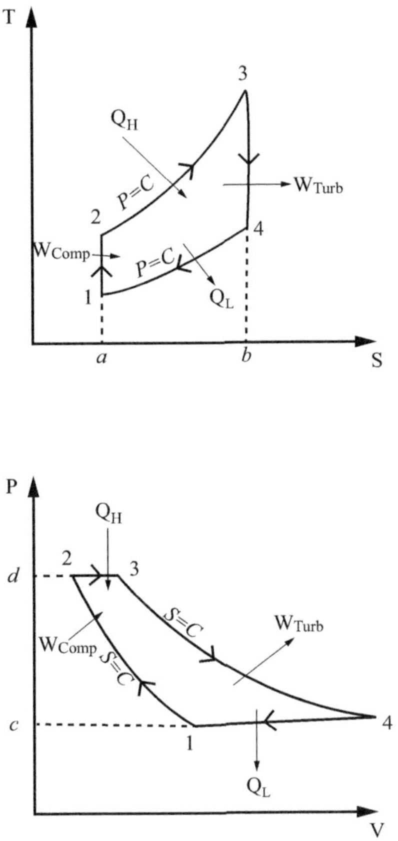

FIGURE 9.6
T-s and P-v diagrams of a Brayton cycle.

including pressure drops in air flows, are ignored. Since all processes in this cycle are reversible, the areas in the diagrams in Figure 9.6 represent work and heat transfer per unit of mass involved in the cycle as follows:

T-s diagram:
q_L = area 1-4-b-a-1
q_H = area 2-3-b-a-2
$q_{Net} = q_H - q_L = w_{Net}$ = area 1-2-3-4-1

P-v diagram:
w_{Comp} = area 1-2-d-c-1
w_{Turb} = area 4-3-d-c-4
$w_{Net} = w_{Turb} - w_{Turb} = q_{Net}$ = area 1-2-3-4-1

The work produced and consumed in the turbine and the compressor, respectively, can be estimated by reducing the energy rate balance equation (the first law of thermodynamics) for a control volume assuming adiabatic processes (no heat losses), and negligible change in potential and kinetic energies.

$$\dot{W}_{\text{Turb}} = \dot{H}_3 - \dot{H}_4 = \dot{m}(h_3 - h_4) \quad w_{\text{Turb}} = h_3 - h_4 \tag{9.19}$$

$$\dot{W}_{\text{Comp}} = \dot{H}_2 - \dot{H}_1 = \dot{m}(h_2 - h_1) \quad w_{\text{Comp}} = h_2 - h_1 \tag{9.20}$$

where $\dot{W}_{\text{Turb}}$ is the power production in the turbine, $\dot{W}_{\text{Comp}}$ is the power consumption in the compressor (both in W or Btu/h), $\dot{H}$ is the rate of enthalpy change in the gas flow (W or Btu/h), h is the specific enthalpy of the working fluid (J/kg or Btu/lb$_m$), and $\dot{m}$ is the mass flow rate of the working fluid (kg/s or lb$_m$/h). Similarly, the heat transfers to and from the cycle are

$$\dot{Q}_H = \dot{H}_3 - \dot{H}_2 = \dot{m}(h_3 - h_2) \quad q_H = h_3 - h_2 \tag{9.21}$$

$$\dot{Q}_L = \dot{H}_4 - \dot{H}_1 = \dot{m}(h_4 - h_1) \quad q_L = h_4 - h_1 \tag{9.22}$$

From the balance of energy rate equation

$$\dot{W}_{\text{Net}} = \dot{W}_{\text{Turb}} - \dot{W}_{\text{Comp}} = \dot{Q}_H - \dot{Q}_L = \dot{Q}_{\text{Net}} \tag{9.23}$$

The thermal efficiency of a power cycle can be found as follows:

$$\eta_{\text{Th}} = \frac{\dot{W}_{\text{Net}}}{\dot{Q}_H} = \frac{w_{\text{Net}}}{\dot{q}_H} = \frac{\dot{q}_H - \dot{q}_L}{\dot{q}_H} = 1 - \frac{\dot{q}_L}{\dot{q}_H} = 1 - \frac{h_4 - h_1}{h_3 - h_2} \tag{9.24}$$

The back work ratio (BWR) of gas turbine plants is defined as the ratio of the power consumption in the compressor to the power production in the turbine.

$$\text{BWR} = \frac{\dot{W}_{\text{Comp}}}{\dot{W}_{\text{Turb}}} = \frac{h_2 - h_1}{h_3 - h_4} \tag{9.25}$$

This parameter is an indication of how much of the work produced in the turbine is consumed internally in the cycle and how much is available externally. As we will see in the following examples, the back work ratio of gas turbine power systems is much higher than that of steam power plants.

In steady-state devices involving shaft work, such as turbines and compressors, the work per unit of mass can be calculated by

$$w = \int v \, dp \tag{9.26}$$

The magnitude of the difference between specific volume during the compression and expansion processes determines the output power and efficiency of the cycle. In gaseous working-fluid cycles, the difference between specific volume in the compression and expansion processes is a result of a limited temperature difference of these two processes (unlike steam cycles where the main reason for the different specific volumes is the phase change of the working fluid). Therefore, in gas turbines, the power consumption in the compression process is significant. In fact, in gas turbine power plants, one-third to three-quarters of the power produced in the turbine can be consumed in the compressor (compared to 1%–2% in steam cycles). Thus, the isentropic efficiencies of the turbine and the compressor in gas turbine cycles are very important because the lower efficiencies mean less power is produced in the turbine and more power is consumed in the compressor.

For isentropic processes 1-2 and 3-4, using air property tables

$$\frac{p_2}{p_1} = \frac{p_{r2}}{p_{r1}} \tag{9.27}$$

$$\frac{p_3}{p_4} = \frac{p_{r3}}{p_{r4}} \tag{9.28}$$

where P_r is the relative pressure for isentropic processes given in ideal gas property tables.

Using the assumption of constant specific heat,

$$\frac{T_2}{T_1} = \left(\frac{P_2}{P_1}\right)^{\frac{k-1}{k}} \tag{9.29}$$

$$\frac{T_3}{T_4} = \left(\frac{P_3}{P_4}\right)^{\frac{k-1}{k}} \tag{9.30}$$

where k is the specific heat ratio of air. Note that in a cold air-standard cycle, k is constant because C_p and C_v are constant. In above equations, the ratio of the compressor outlet pressure to the compressor inlet pressure $\left(\frac{P_2}{P_1}\right)$ is called the **compressor pressure ratio (PR)**, also known as the **compression ratio (CR)**. Since the processes 2-3 and 4-1 are isobaric processes (assuming no pressure losses in the heat transfer processes):

$$P_2 = P_3 \quad \text{and} \quad P_1 = P_4 \Rightarrow \frac{P_2}{P_1} = \frac{P_3}{P_4} \tag{9.31}$$

This means that in the absence of frictional pressure drops, the pressure ratios of the compressor and turbine are equal. In a cold-standard air Brayton cycle using Equations 9.29 and 9.30

$$\frac{T_2}{T_1} = \frac{T_3}{T_4} \rightarrow (T_2T_4) = (T_1T_3) \tag{9.32}$$

Example 9.1

Currently, one of the world's largest and most efficient gas turbines is model 9HA.02 introduced by General Electric (GE) in 2014 with the power production of 470 MW and the efficiency of 41.5%. The compressor pressure ratio is 21.8. The turbine inlet temperature (TIT) is 1700 K. Air enters the cycle with the mass flow rate of 978 kg/s at the temperature of 300 K and the pressure of 100 kPa. Assuming that the gas turbine operates as an air-standard Brayton cycle, determine the compressor and turbine outlet temperatures (in K), the energy transfer per second in each component (in kW), the net power generation (in kW), the back work ratio, and the efficiency of the cycle (a) using the air property table data and (b) assuming constant specific heats for air. Compare your results from parts (a) and (b) with the same data provided by the manufacturer.

Solution

Given: An air-standard Brayton cycle operates at the given compressor inlet air temperature, pressure, and mass flow rate; turbine inlet air temperature; and CR.

Find: Determine the compressor and turbine outlet temperatures, energy transfer per second in each component, the net power generation, the back work ratio, and the efficiency of the cycle.

Schematic diagrams: Figure 9.7.

Assumptions:

1. The cycle is an air-standard Brayton cycle; therefore, all processes are internally reversible. The heat transfer processes are at constant pressure, the expansion and compression processes are isentropic, and the working fluid is air and behaves as an ideal gas.
2. The steady-state control volume mass and energy balance analyses with no kinetic and potential energy changes are applied to each component.

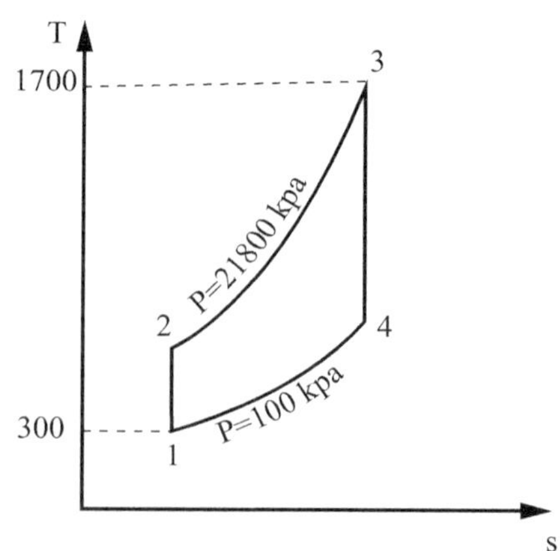

FIGURE 9.7
Schematic diagram of the Brayton cycle in Example 9.1.

3. For part (b), the cycle is a cold air-standard Brayton cycle (constant C_p and k).

Analysis:

(a) Using air property table data

Table A.7 (SI) can be used to find the properties of air at four states of the cycle.

State 1: For the air temperature 300 K, from Table A.7 (SI)

$$h_1 = 300.19 \text{ kJ/kg} \quad \text{and} \quad P_{r1} = 1.3860$$

State 2: For the isentropic compression in the compressor with the pressure ratio $\frac{p_2}{p_1} = 21.8$ from Equation 9.18

$$\frac{p_2}{p_1} = \frac{p_{r2}}{p_{r1}} \rightarrow p_{r2} = (21.8)(1.3860) = 30.21$$

For $p_{r2} = 30.21$ and interpolating between temperatures of 700 and 710 K in Table A.7 (SI)

$$h_2 = 722.91 \text{ kJ/kg} \quad \text{and} \quad T_2 = 708.95 \text{ K}$$

State 3: For the air temperature 1700 K, from A.7 (SI)

$$h_3 = 1880.1 \text{ kJ/kg} \quad \text{and} \quad P_{r3} = 1025$$

State 4: For the isentropic expansion in the turbine from Equation 9.18

$$\frac{p_3}{p_4} = \frac{p_{r3}}{p_{r4}} \rightarrow p_{r4} = \frac{1025}{21.8} = 47.02$$

For $p_{r4} = 47.02$ and interpolating between temperatures of 790 and 800 K in Table A.7 (SI)

$$h_4 = 818.30 \text{ kJ/kg} \quad \text{and} \quad T_4 = 796.67 \; K$$

The heat transfer rates are

$$\dot{Q}_L = \dot{m}(h_4 - h_1) = (978) \; [\text{kg/s}] \; (818.30 - 300.19) \; [\text{kJ/kg}]$$
$$= 506,712 \text{ kW}$$

$$\dot{Q}_H = \dot{m}(h_3 - h_2) = (978)[\text{kg/s}](1880.1 - 722.91)[\text{kJ/kg}]$$
$$= 1,131,732 \text{ kW}$$

The power consumption in the compressor and the power production in the turbine are

$$\dot{W}_{\text{Comp}} = \dot{m}(h_2 - h_1) = (978)[\text{kg/s}](722.91 - 300.19)[\text{kJ/kg}]$$

$$= 413,\ 420 \text{ kW}$$

$$\dot{W}_{\text{Turb}} = \dot{m}(h_3 - h_4) = (978)[\text{kg/s}](1880.1 - 818.30)[\text{kJ/kg}]$$

$$= 1,038,440 \text{ kW}$$

The correctness of the answers can be checked by the balance of energy (the first law of thermodynamics).

$$\text{inlet energy} = \text{outlet energy} \rightarrow \dot{Q}_H + \dot{W}_{\text{Comp}} = \dot{Q}_L + \dot{W}_{\text{Turb}}$$

$$1,131,732 + 413,420 = 506,712 + 1,038,440 \rightarrow 1,545,152$$

$$= 1,545,152$$

From the balance of energy rate

$$\dot{W}_{\text{Net}} = \dot{W}_{\text{Turb}} - \dot{W}_{\text{Comp}} = 1,038,440 - 413,420 = 625,\ 020 \text{ kW}$$

The back work ratio (BWR) is

$$\text{BWR} = \frac{\dot{W}_{\text{Comp}}}{\dot{W}_{\text{Turb}}} = \frac{413,\ 420}{1,038,440} = 0.398$$

Finally, the efficiency of the cycle is

$$\eta_{\text{Th}} = \frac{\dot{W}_{\text{Net}}}{\dot{Q}_H} = \frac{625,\ 020}{1,131,732} = 0.552$$

(b) Using constant specific heats for air

The specific heat ratio of air and various terms that are used in Equation 9.15 as well as the C_p of air is

$$C_p = 1.004 \text{ kJ/kg}\,K, k = 1.4, k - 1 = 0.4, \frac{k}{k-1} = 3.5, \frac{k-1}{k} = 0.2857$$

First, the temperature of air at states 2 and 4 should be determined using Equation 9.15

$$\left(\frac{T_2}{T_1}\right) = \left(\frac{P_2}{P_1}\right)^{\frac{k-1}{k}}$$

$$\left(\frac{T_2}{300}\right) = (21.8)^{0.2857} \rightarrow T_2 = 723.6\text{K}$$

$$\left(\frac{T_4}{T_3}\right)=\left(\frac{P_4}{P_3}\right)^{\frac{k-1}{k}}$$

$$\left(\frac{T_4}{1700}\right)=\left(\frac{1}{21.8}\right)^{0.2857} \rightarrow T_4=704.8\,\text{K}$$

The energy transfer rates in the components are as follows:

$$\dot{Q}_L=\dot{m}(h_4-h_1)=\dot{m}C_p(T_4-T_1)$$
$$=(978)[\text{kg/s}](1.004)[\text{kJ/kg K}](704.8-300)[\text{K}]=397,478\,\text{kW}$$

$$\dot{Q}_H=\dot{m}(h_3-h_2)=\dot{m}C_p(T_3-T_2)$$
$$=(978)[\text{kg/s}](1.004)[\text{kJ/kg K}](1700-723.6)[\text{K}]=958,739\,\text{kW}$$

$$\dot{W}_{\text{Comp}}=\dot{m}(h_2-h_1)=\dot{m}C_p(T_2-T_1)$$
$$=(978)[\text{kg/s}](1.004)[\text{kJ/kg K}](723.6-300)[\text{K}]=415,938\,\text{kW}$$

$$\dot{W}_{\text{Turb}}=\dot{m}(h_3-h_4)=\dot{m}C_p(T_3-T_4)$$
$$=(978)[\text{kg/s}](1.004)[\text{kJ/kg K}](1700-704.8)[\text{K}]=977,199\,\text{kW}$$

$$\dot{W}_{\text{Net}}=\dot{W}_{\text{Turb}}-\dot{W}_{\text{Comp}}=977,199-415,938=561,\,261\,\text{kW}$$

Finally, the back work ratio and efficiency are

$$\text{BWR}=\frac{\dot{W}_{\text{Comp}}}{\dot{W}_{\text{Turb}}}=\frac{415,938}{977,199}=0.426$$

$$\eta=\frac{\dot{W}_{\text{Net}}}{\dot{Q}_H}=\frac{561,261}{958,739}=0.585$$

Table 9.1 compares all the results of the air-standard Brayton cycle (part a) and the cold air-standard Brayton cycle (part b).

The comparison of the results indicates the significant effect of the constant specific heat assumption for the cold air-standard analysis on the calculated cycle parameters. The result for the back work ratio points out that about 40% of the power produced in the turbine is consumed in the compressor, which is much higher than that of steam power plants (about 1%–2%). The reason for this was explained earlier in this chapter.

TABLE 9.1

Comparison of the Results of Parts a and b in Example 9.1

Parameter	Air-Standard Brayton Cycle	Cold Air-Standard Brayton Cycle
T_2 (K)	709	724
T_4 (K)	797	705
$\dot{Q}_L$ (MW)	507	397
$\dot{Q}_H$ (MW)	1132	959
$\dot{W}_{Comp}$ (MW)	413	416
$\dot{W}_{Turb}$ (MW)	1038	977
$\dot{W}_{Net}$ (MW)	625	561
BWR (%)	39.8	42.6
η_{Th} (%)	55.2	58.5

TABLE 9.2

Comparison of Selected Parameters of the Air-Standard Brayton Cycle and the Cold Air-Standard Brayton Cycle with the Actual Gas Turbine Specifications in Example 9.1

Parameter	Net Power Output (MW)	Efficiency (%)	Exhaust Temperature (K)	Exhaust Energy (MW)
Actual gas turbine	470	41.5	892	712
Air-standard Brayton cycle	625 (33%)	55.2 (33%)	797 (–11%)	507 (–29%)
Cold air-standard Brayton cycle	561 (19%)	58.5 (40%)	705 (–21%)	397 (–44%)

Table 9.2 compares some selected parameters of the air-standard Brayton cycle and the cold air-standard Brayton cycle with the actual gas turbine specifications reported by the manufacturer. The percentages of the error of the theoretical results compared to the actual parameters are presented in parentheses.

Table 9.2 indicates that errors introduced by the air-standard analysis are substantial, and the results of the analysis should be always used qualitatively. These errors are partly due to the assumptions of the air-standard analysis, especially ignoring irreversibilities, and partly due to the simplifications in the gas turbine cycle and the absence of some components of the real cycle in the simplified cycle. Moreover, surprisingly, the percentage of the error in the net power output is greater for the air-standard analysis compared to that for the cold air-standard analysis. The reason is that the greater errors in the calculation of $\dot{W}_{Turb}$ and $\dot{W}_{Comp}$ for the cold air-standard cycle were canceled out when the net power was calculated so incidentally more accurate result is obtained in the case of the cold air-standard analysis.

9.2.2 Effects of Compressor Pressure Ratio on Efficiency and Specific Work Output of Brayton Cycle

As stated, the results of the cold air-standard analysis of an ideal Brayton cycle can be qualitatively applied to the practical applications of gas turbine power plants. In this section, the effects of the compressor pressure ratio on the efficiency and specific power output of the cycle will be evaluated. For a cold air-standard cycle, q_L and q_H are

$$q_L = (h_4 - h_1) = C_p (T_4 - T_1) \tag{9.33}$$

$$q_H = (h_3 - h_2) = C_p (T_3 - T_2) \tag{9.34}$$

Substituting Equations 9.33 and 9.34 into Equation 9.24 and rearranging it

$$\eta_{Th} = 1 - \frac{C_p (T_4 - T_1)}{C_p (T_3 - T_2)} = 1 - \tag{9.35}$$

Considering Equations 9.29 and 9.30

$$\left(\frac{T_2}{T_1}\right) = \left(\frac{P_2}{P_1}\right)^{\frac{k-1}{k}} \quad \text{and} \quad \left(\frac{T_4}{T_3}\right) = \left(\frac{P_4}{P_3}\right)^{\frac{k-1}{k}}$$

Since (Equation 9.31)

$$\frac{P_2}{P_1} = \frac{P_3}{P_4}$$

Thus,

$$\frac{T_2}{T_1} = \frac{T_3}{T_4} \tag{9.36}$$

Rearranging Equation 9.36

$$\frac{T_4}{T_1} = \frac{T_3}{T_2} \tag{9.37}$$

$$\frac{T_4}{T_1} - 1 = \frac{T_3}{T_2} - 1 \tag{9.38}$$

Replacing Equation 9.38 in Equation 9.35:

$$\eta_{Th} = 1 - \frac{T_1}{T_2} \tag{9.39}$$

Replacing Equation 9.29 in Equation 9.39:

$$\eta_{\text{Th}} = 1 - \frac{1}{\left(\frac{P_2}{P_1}\right)^{\left(\frac{(k-1)}{k}\right)}} \text{(based on the cold air – standard analysis)} \tag{9.40}$$

Equation 9.40 shows that the efficiency of the cold air-standard Brayton cycle is only a function of the CR of the compressor and the specific heat ratio of the working fluid. The higher CR results in the higher efficiency of the cycle.

Example 9.2

Using Equations 9.39 and 9.40, determine the efficiency of the Brayton cycle in Example 9.1.

Solution

Given, find, schematic diagrams, and assumptions are exactly similar to Example 9.1.

Analysis

Substituting the pressure ratio into Equation 9.40

$$\eta_{\text{Th}} = 1 - \frac{1}{\left(\frac{P_2}{P_1}\right)^{\left(\frac{(k-1)}{k}\right)}} = 1 - \frac{1}{21.8^{0.2857}} = 0.585$$

Similarly, from Equation 9.39

$$\eta_{\text{Th}} = 1 - \frac{T_1}{T_2} = 1 - \frac{300}{723.6} = 0.585$$

The result is the same as what was found in Example 9.1 based on the cold air-standard analysis but it is different from the air-standard analysis.

This can also be shown graphically. Figure 9.8 illustrates two Brayton cycles, 1-2-3-4-1 and 1–2′-3′-4-1. The CR of the latter is greater than that of the former. For the same q_L, the latter has higher q_H and thus higher efficiency (according to Equation 9.24). Note that the area under the T-s diagram for a reversible process is equal to the heat transfer during the process.

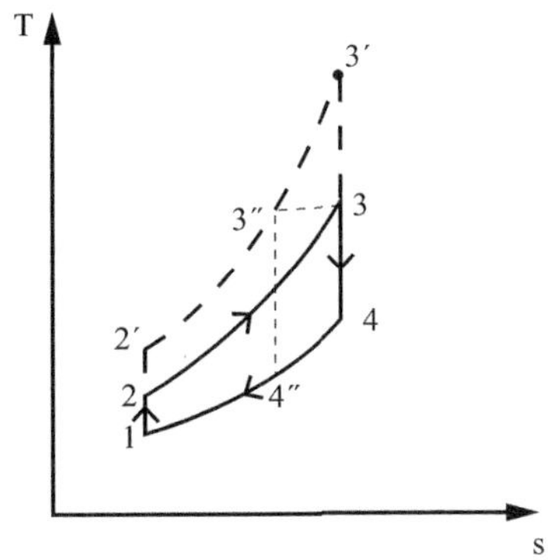

FIGURE 9.8
Effect of the compressor pressure ratio on the efficiency of the cycle.

The improvement of efficiency with an increase in the pressure ratio of the compressor can also be explained based on the fact that the average temperature of the heat transfer to the cycle from the high-temperature heat source is higher in the cycle with the higher CR. Note that the area enclosed by the cycle in the T-s diagram represents the net work per kilogram of the working fluid (specific work) produced by the cycle. So in this case, since the area enclosed by the cycle increases, the net specific work of the cycle also increases.

Obviously, the CR of a compressor is technologically limited; however, even before this limitation is reached, another parameter limits the increase in the CR (currently in modern compressors between 20 and 40). As Figure 9.8 illustrates, increasing the compressor outlet pressure from P_2 to P_2' increases the **TIT** from T_3 to T_3'. But in actual gas turbines, the maximum temperature of the cycle (TIT) is limited by metallurgical considerations of the blades of the initial stages of the turbine. Thus, for a fixed maximum temperature of the cycle, if the pressure ratio is increased, instead of the cycle 1-2-3-4-1, we will have the cycle 1-2′-3″-4″-1 (not 1-2′-3′-4-1). This new cycle has a higher average temperature of the heat transfer to the cycle, which in turn means a higher efficiency of the cycle. However, the area enclosed by the cycle 1-2′-3″-4″-1 is smaller than that of cycle 1-2-3-4-1 (Figure 9.8). This means the Brayton cycle's specific work reduces with increasing pressure ratio of the compressor for a constant TIT. This will result in larger equipment for the same net power output and a higher initial cost of the system.

Example 9.3

Determine the net specific power output of a Brayton ideal cycle as a function of the inlet temperature of the compressor and the maximum temperature of the cycle (TIT) using the cold air-standard assumptions. Note that in a Brayton cycle, both the inlet temperature of the compressor (T_1) and the maximum temperature of the cycle (T_3) are independent variables of a given Brayton cycle.

Solution

Given
The cycle operates based on a cold air-standard Brayton ideal cycle.

Find
Determine the net specific power output of the cycle.
Schematic diagrams: Figure 9.8

Assumptions

The cycle is a cold air-standard Brayton cycle operating at the steady-state condition.

Analysis

The power outputs of the turbine and compressor are

$$\dot{W}_{\text{Turb}} = \dot{H}_3 - \dot{H}_4 = \dot{m}(h_3 - h_4) = \dot{m}C_p(T_3 - T_4)$$

$$= \dot{m}C_pT_3\left(1 - \frac{T_4}{T_3}\right) = \dot{m}C_pT_3\left(1 - \frac{1}{\dfrac{T_3}{T_4}}\right) = \dot{m}C_pT_3\left(1 - \frac{1}{\left(\dfrac{P_3}{P_4}\right)^{(k-1)/k}}\right)$$

$$\dot{W}_{\text{Comp}} = \dot{H}_2 - \dot{H}_1 = \dot{m}(h_2 - h_1) = \dot{m}C_p(T_2 - T_1)$$

$$= \dot{m}C_pT_2\left(1 - \frac{T_1}{T_2}\right) = \dot{m}C_pT_2\left(1 - \frac{1}{\left(\dfrac{P_2}{P_1}\right)^{(k-1)/k}}\right)$$

The net power output of the cycle is

$$\dot{W}_{\text{Net}} = \dot{W}_{\text{Turb}} - \dot{W}_{\text{Comp}} = \dot{m}C_pT_3\left(1 - \frac{1}{\left(\dfrac{P_3}{P_4}\right)^{(k-1)/k}}\right) - \dot{m}C_pT_2\left(1 - \frac{1}{\left(\dfrac{P_2}{P_1}\right)^{(k-1)/k}}\right)$$

Since $\dfrac{P_2}{P_1} = \dfrac{P_3}{P_4}$

$$\dot{W}_{\text{Net}} = \dot{m}C_p(T_3 - T_2)\left(1 - \frac{1}{\left(\dfrac{P_2}{P_1}\right)^{(k-1)/k}}\right)$$

Let us rearrange the above equation as a function of the inlet temperature of the compressor (T_1) and the maximum temperature of the cycle (T_3).

$$T_2 = T_1\left(\frac{P_2}{P_1}\right)^{(k-1)/k}$$

$$w_{\text{Net}} = \frac{\dot{W}_{\text{Net}}}{\dot{m}} = C_p\,(T_3 - T_2)\left(1 - \frac{1}{\left(\frac{P_2}{P_1}\right)^{(k-1)/k}}\right)$$

$$= C_p\left(T_3 - T_1\left(\frac{P_2}{P_1}\right)^{(k-1)/k}\right)\left(1 - \frac{1}{\left(\frac{P_2}{P_1}\right)^{(k-1)/k}}\right)$$

$$w_{\text{Net}} = C_p\left[T_3\left(1 - \frac{1}{\left(\frac{P_2}{P_1}\right)^{(k-1)/k}}\right) + T_1\left(1 - \left(\frac{P_2}{P_1}\right)^{(k-1)/k}\right)\right] \tag{9.41}$$

Note: The efficiency of the cycle can be found as

$$\eta_{\text{Th}} = \frac{W_{\text{Net}}}{Q_H} = \frac{W_{\text{Turb}} - W_{\text{Comp}}}{Q_H} = \frac{\dot{m}C_p\,(T_3 - T_2)\left(1 - \frac{1}{\left(\frac{P_2}{P_1}\right)^{(k-1)/k}}\right)}{\dot{m}C_p\,(T_3 - T_2)}$$

$$= 1 - \frac{1}{\left(\frac{P_2}{P_1}\right)^{(k-1)/k}} \tag{9.42}$$

As expected, this equation for the cold air-standard Brayton cycle efficiency is the same as Equation 9.40.

Equation 9.41 illustrates that unlike the efficiency of the cycle, the net specific work output of the cycle depends on T_1, T_3, the pressure ratio of the compressor, and k and C_p of the working fluid. Also, a careful investigation of Equation 9.41 shows that the compressor pressure ratio appears in both $1-\frac{1}{\left(\frac{P_2}{P_1}\right)^{(k-1)/k}}$ and $1-\left(\frac{P_2}{P_1}\right)^{(k-1)/k}$ terms. The higher-pressure ratio increases the specific work in the former term and decreases it in the latter term. Everything else being equal, this indicates that there should be a pressure ratio that maximizes the net specific work output of the cycle. To find the pressure ratio that maximizes the net specific work output of the cycle for a fixed T_1, T_3, and the working fluid, the derivative of the net specific work output with respect to the pressure ratio in Equation 9.41 should be equated to zero, which results in

$$\left[\left(\frac{P_2}{P_1}\right)^{2(k-1)/k}\right]_{\text{opt}} = \frac{T_3}{T_1} \tag{9.43}$$

This means that other factors being equal, the **pressure ratios that the optimum efficiency and specific work output can be achieved are different**. This has a profound impact on the design of gas turbines. If a gas turbine is designed and operated for a baseload power generation, the optimization of the cycle efficiency is the most important factor to consider. In this case, the compressor pressure ratio should be as high as possible (limited by the TIT). However, for peak-load power generation plants, since the cycle operates for a relatively short period of time, the cycle specific work is more important than the cycle efficiency. In this case, the compressor pressure ratio should be determined using Equation 9.43. This requires a design compromise and trade-off to determine the desired pressure ratio of the cycle. Note that while the cycle efficiency is directly related to the operational cost (i.e., fuel cost) of the system, the net specific work output is proportional to the size and hence the capital cost of the system. This discussion can also be applied to gas turbines in transportation applications, especially aircraft propulsion, where the power-to-weight ratio is one of the most important design considerations. Furthermore, substituting Equation 9.29 into Equation 9.43

$$\left(\frac{T_2}{T_1}\right)^2 = \frac{T_3}{T_1} \rightarrow (T_2)^2 = T_1 T_3 \;\; (\text{at the maximum net specific work output}) \tag{9.44}$$

Substituting Equation 9.32 into Equation 9.44

$$T_2 = T_4 \, (\text{at the maximum net specific work output}) \tag{9.45}$$

This means that the specific work output is optimum when the output temperature of the compressor is equal to the output temperature of the turbine.

Example 9.4

Repeat Example 9.1 for the compressor pressure ratio of 5–40 with the increment of 5, and draw the diagram of each parameter as a function of the pressure ratio. Consider the cycle as a cold air-standard Brayton cycle.

Solution

Given: Same as Example 9.1
Find: Same as Example 9.1
Schematic diagrams: Same as Example 9.1
Assumptions: Same as Example 9.1.

Analysis

The best way to do this example is to create an Excel spreadsheet for Example 9.1. Table 9.3 and Figure 9.9 illustrate the efficiency, net power output, and back work ratio as a function of the compressor pressure ratio in Example 9.1.

As expected, the efficiency continuously increases with increasing compressor pressure ratio, while the net power output reaches a maximum at the pressure ratio of about 20 and then gradually reduces with further increase in the pressure ratio. This can be verified using Equation 9.43 to find the compressor pressure ratio with the maximum net power output.

$$\left[\left(\frac{P_2}{P_1}\right)^{2(k-1)/k}\right]_{\text{opt}} = \frac{T_3}{T_1}$$

$$\left[\frac{P_2}{P_1}\right]_{\text{opt}} = \left(\frac{T_3}{T_1}\right)^{k/2(k-1)}$$

TABLE 9.3

Efficiency, Net Power Output, and Back Work Ratio of the Brayton Cycle as a Function of the Compressor Pressure Ratio in Example 9.1

Pressure Ratio	T_2 (K)	T_4 (K)	$\dot{Q}_L$ (MW)	$\dot{Q}_H$ (MW)	$\dot{W}_{\text{Comp}}$ (MW)	$\dot{W}_{\text{Turb}}$ (MW)	$\dot{W}_{\text{Net}}$ (MW)	BWR (%)	η_{Th} (%)
5	475	1073	759	1203	172	615	443	27.9	36.9
10	579	881	570	1101	274	805	530	34.1	48.2
15	650	784	475	1031	344	899	555	38.3	53.9
20	706	722	415	976	399	960	561	41.5	57.5
25	753	678	371	930	444	1004	559	44.3	60.1
30	793	643	337	891	484	1038	554	46.6	62.2
35	828	616	310	856	519	1065	546	48.7	63.8
40	861	593	287	824	551	1087	537	50.6	65.1

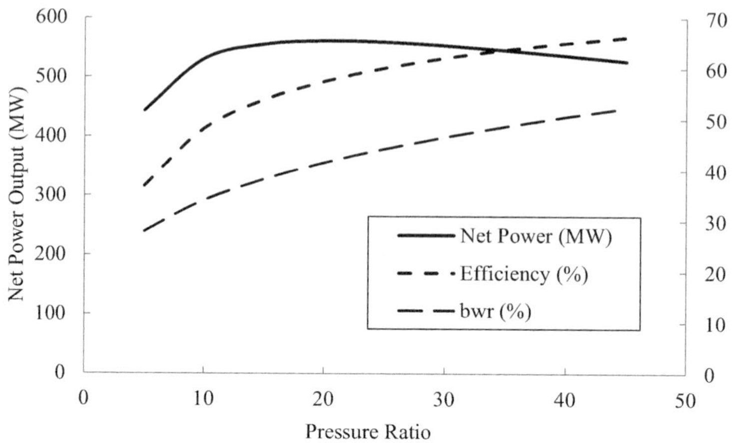

FIGURE 9.9
Variation of efficiency, net power output, and back work ratio as a function of compressor pressure ratio in Example 9.1.

$$\left[\frac{P_2}{P_1}\right]_{opt} = \left(\frac{1700}{300}\right)^{1.4/2(1.4-1)} = 20.81$$

Also, the diagram shows that the net power output varies a little when the pressure ratio changes about the maximum point and the diagram is almost flat around this point. This characteristic of the gas turbine plants gives designers a wide range of the compressor pressure ratio to choose from while the power output is almost at its optimum point.

Figure 9.10 illustrates the variation of other parameters as a function of the compressor pressure ratio. The explanation of these diagrams is left as an exercise.

Now let us answer this question: Considering the values of the compressor pressure ratio, is the gas turbine under investigation in this example designed for peak-load or baseload operation?

Comparing the compressor pressure ratio of the actual gas turbine plant and the compressor pressure ratio to achieve the maximum power output (21.8% versus 20.8, respectively) clearly indicates that the unit is designed for the maximum power output and not for optimum efficiency. However, these units are typically utilized for baseload power generation and not peak-load power generation. In order to understand this apparent contradiction, it should be kept in mind that these large gas turbines are usually installed as a part of CCPPs (Chapter 10). Therefore, the objective of designers is to optimize the overall efficiency of the whole plant rather than individual components.

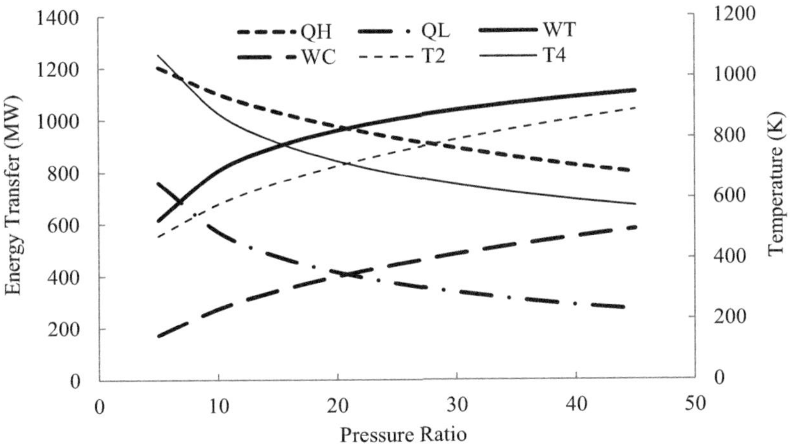

FIGURE 9.10
Variation of various operating parameters as a function of the compressor pressure ratio in Example 9.1.

9.2.3 Brayton Cycle with Irreversibilities and Losses

The main differences between an actual gas turbine plant and an ideal Brayton cycle are

- Irreversibilities due to the combustion process in the combustor
- Irreversibilities in the compressor and turbine
- Frictional pressure drops in flow passages and the combustion chamber

The effects of combustion irreversibilities, though the most important cause of departure from an ideal cycle, cannot be evaluated by the air-standard analysis and were covered in Chapter 4. The irreversibilities in the compressor and turbine, the second most important losses in gas turbines, are mainly due to frictional effects and are manifested in entropy increase. In real gas turbines (Figure 9.11), process 1-2 replaces the isentropic process 1-2_s and process 3-4 replaces the isentropic process 3-4_s. In order to take these irreversibilities into account, the isentropic (also known as adiabatic) efficiencies of the turbine and the compressor are defined by comparing the work or power involved in the nonideal cycle with that of the ideal cycle.

$$\eta_{\text{Comp}} = \frac{(w_{\text{Comp}})_s}{w_{\text{Comp}}} = \frac{(\dot{W}_{\text{Comp}})_s}{\dot{W}_{\text{Comp}}} = \frac{h_{2s} - h_1}{h_2 - h_1} \tag{9.46}$$

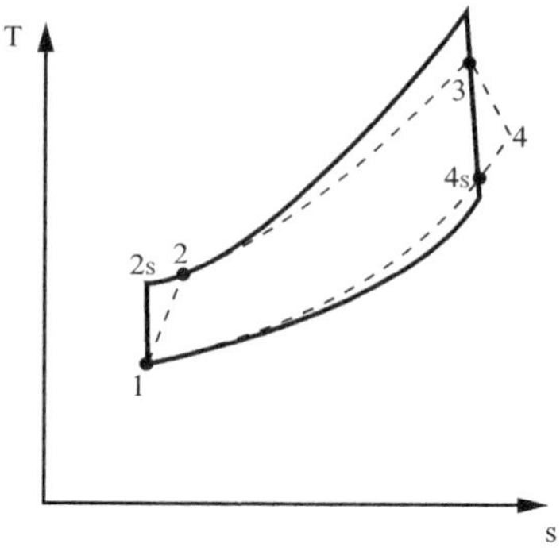

FIGURE 9.11
Ideal Brayton cycle versus nonideal gas turbine cycle (with irreversibilities).

$$\eta_{\text{Turb}} = \frac{w_{\text{Turb}}}{(w_{\text{Turb}})_s} = \frac{\dot{W}_{\text{Turb}}}{(\dot{W}_{\text{Turb}})_s} = \frac{h_3 - h_4}{h_3 - h_{4s}} \tag{9.47}$$

The isentropic efficiencies of the compressor and the turbine are very important factors in the gas turbine performance because low efficiencies mean less power production in the turbine and more power consumption in the compressor. The combined effect of these two can significantly reduce the net power output and efficiency of the system. As will be discussed later in the history of gas turbines, this was one of the reasons that the development of gas turbines did not progress until the 1930s. Today's modern compressors and turbines can have isentropic efficiencies of 80%–90%. The irreversibilities in the compressor and turbine are the only ones that are considered in the nonideal gas turbine analysis in this chapter.

Due to flow friction, the pressure drops across the two heat transfer processes, from P_2 to P_3 and from P_4 to P_1, as shown in Figure 9.11. This means the pressure ratio across the compressor (the ratio of the outlet flow pressure to the inlet flow pressure) is higher than the pressure ratio across the turbine (the ratio of the inlet flow pressure to the outlet flow pressure). These losses are less important than the two aforementioned irreversibilities and will not be taken into account in analyses presented in this chapter.

Other losses such as mechanical losses, bearing friction, heat losses from the combustion chamber and other components, and air used for turbine blade cooling (discussed later in this chapter) are typically less important and are not considered in the presented analyses.

It can be shown that the specific power output and efficiency of nonideal Brayton cycles are functions of the compressor pressure ratio, compressor inlet temperature, maximum cycle temperature (TIT), isentropic efficiencies of the turbine and the compressor, and properties of the working fluid. This is unlike the efficiency of the ideal Brayton cycle, which is only a function of the pressure ratio and properties of the working fluid.

Example 9.5

Reconsider Example 9.1, but this time instead of the isentropic compression and expansion processes, the compressor and turbine have the isentropic efficiency of 90%. Compare the results with the results from Example 9.1.

Solution

Given

An air-standard Brayton cycle operates at the given compressor inlet air temperature, pressure, and mass flow rate; turbine inlet air temperature; CR; and isentropic efficiencies of compressor and turbine.

Find

Determine the compressor and turbine outlet temperatures, energy transfer per second in each component, the net power generation, the back work ratio, and the efficiency of the cycle.

Schematic diagrams: Figure 9.11.

Assumptions

1. The cycle is an air-standard Brayton cycle; therefore, the heat transfer processes are at constant pressure, and the working fluid is air and behaves as an ideal gas.
2. The expansion and compression processes are not isentropic, and entropy increases through the processes.
3. The steady-state control volume mass and energy balance analyses with no kinetic and potential energy changes are applied to each component.
4. For part (b), the cycle is a cold air-standard Brayton cycle (constant C_p and k).

Analysis

(a) Using air property table data

From Example 9.1

$$h_1 = 300.19 \text{ kJ/kg}$$

$$h_{2s} = 722.91 \text{ kJ/kg}$$

$$h_3 = 1880.1 \text{ kJ/kg}$$

$$h_{4s} = 818.30 \text{ kJ/kg}$$

Note that states 2 and 4 are now 2s and 4s. Using Equations 9.46 and 9.47, the specific enthalpy of states 2 and 4 can be found.

$$\eta_{\text{Comp}} = \frac{h_{2s} - h_1}{h_2 - h_1} \rightarrow 0.90 = \frac{722.91 - 300.19}{h_2 - 300.19} \rightarrow h_2 = 769.88 \text{ kJ/kg}$$

$$\eta_{\text{Turb}} = \frac{h_3 - h_4}{h_3 - h_{4s}} \rightarrow 0.90 = \frac{1880.1 - h_4}{1880.1 - 818.3} \rightarrow h_4 = 924.48 \text{ kJ/kg}$$

The temperature of the compressor and turbine outlet flows can be found by interpolating in Table A.7 (SI).

$$T_2 = \mathbf{752.4\,K}$$

$$T_4 = \mathbf{892.4\,K}$$

As expected, due to irreversibilities, both temperatures have increased compared to the isentropic processes in Example 9.1 (Table 9.1).

The heat transfer rates are

$$\dot{Q}_L = \dot{m}(h_4 - h_1) = (978)(924.48 - 300.19) = \mathbf{610,556}\ \text{kW}$$

$$\dot{Q}_H = \dot{m}(h_3 - h_2) = (978)(1880.1 - 769.88) = \mathbf{1,085,795}\ \text{kW}$$

The power consumption in the compressor and the power production in the turbine are

$$\dot{W}_{\text{Comp}} = \dot{m}(h_2 - h_1) = (978)(769.88 - 300.19) = \mathbf{459,357}\ \text{kW}$$

$$\dot{W}_{\text{Turb}} = \dot{m}(h_3 - h_4) = (978)(1880.1 - 924.48) = \mathbf{934,596}\ \text{kW}$$

Alternatively, the power developed in the turbine and the power input to the compressor can be calculated as follows:

$$\eta_{\text{Comp}} = \frac{\left(\dot{W}_{\text{Comp}}\right)_s}{\dot{W}_{\text{Comp}}}$$

$$\eta_{\text{Turb}} = \frac{\dot{W}_{\text{Turb}}}{\left(\dot{W}_{\text{Turb}}\right)_s}$$

From Example 9.1

$$\left(\dot{W}_{\text{Comp}}\right)_s = 413,420\ \text{kW}$$

$$\left(\dot{W}_{\text{Turb}}\right)_s = 1,038,440\ \text{kW}$$

So,

$$\dot{W}_{\text{Comp}} = \frac{\left(\dot{W}_{\text{Comp}}\right)_s}{\eta_{\text{Comp}}} = \frac{413,420}{0.9} = \mathbf{459,356\ kW}$$

$$\dot{W}_{\text{Turb}} = \eta_{\text{Turb}}\left(\dot{W}_{\text{Turb}}\right)_s = 0.9 \times 1,038,440 = \mathbf{934,596\ kW}$$

$$\dot{W}_{\text{Net}} = \dot{W}_{\text{Turb}} - \dot{W}_{\text{Comp}} = 934,596 \ - 459,357 = \mathbf{475,239\ kW}$$

The back work ratio (BWR) is

$$\text{BWR} = \frac{\dot{W}_{\text{Comp}}}{\dot{W}_{\text{Turb}}} = \frac{459,357}{934,596} = \mathbf{0.492}$$

Finally, the efficiency of the cycle is

$$\eta_{\text{Th}} = \frac{\dot{W}_{\text{Net}}}{\dot{Q}_H} = \frac{475,239}{1,085,795} = \mathbf{0.438}$$

(b) Using constant specific heats for air

From Example 9.1

$$T_{2s} = 723.6\ K$$

$$T_{4s} = 704.8\ K$$

The temperatures of the compressor and turbine outlet flows are

$$\eta_{\text{Comp}} = \frac{T_{2s} - T_1}{T_2 - T_1} \rightarrow 0.90 = \frac{723.6 - 300}{T_2 - 300} \rightarrow \boldsymbol{T_2 = 770.7\ \mathrm{K}}$$

$$\eta_{\text{Turb}} = \frac{T_3 - T_4}{T_3 - T_{4s}} \rightarrow 0.90 = \frac{1700 - T_4}{1700 - 704.8} \rightarrow \boldsymbol{T_4 = 804.3\ \mathrm{K}}$$

The rest of the calculations are exactly similar to Example 9.1. Table 9.4 lists and compares the results based on the air-standard and cold air-standard analyses with the compressor and turbine isentropic efficiencies of 90%, and the results for the air-standard analysis when the compression and expansion processes are isentropic.

The comparison of the air-standard Brayton cycle with isentropic and non-isentropic compression and expansion processes indicates that the temperatures of the compressor outlet and turbine outlet are higher in the non-isentropic cycle. This is expected because in the non-isentropic processes, the entropy of the working fluid increases due to irreversibilities. These increases in entropy are manifested in an increase in the temperatures. The most interesting difference is the reduction of $\dot{Q}_H$ in the nonideal cycle, which means less heat input (i.e., less fuel consumption) is needed in the combustor. The reason is the higher temperature of the combustor inlet flow, which requires less heat to reach the same T_3. However, the overall performance of the cycle is markedly degraded as is evident in the significant reduction of the efficiency and net power output.

TABLE 9.4

Comparison of the Results for the Air-Standard and Cold Air-Standard Analysis with the Compressor and Turbine Efficiency of 90%, and the Air-Standard Analysis with the Isentropic Compression and Expansion

Parameter	Air-Standard Brayton Cycle with Isentropic Compression and Expansion Processes	Air-Standard Brayton Cycle with $\eta_{Comp} = \eta_{Turb} = 90\%$	Cold Air-Standard Brayton Cycle with $\eta_{Comp} = \eta_{Turb} = 90\%$
T_2 (K)	709	752	771
T_4 (K)	797	892	804
$\dot{Q}_L$ (MW)	507	611	495
$\dot{Q}_H$ (MW)	1132	1086	912
$\dot{W}_{Comp}$ (MW)	413	459	462
$\dot{W}_{Turb}$ (MW)	1038	935	879
$\dot{W}_{Net}$ (MW)	625	475	417
BWR (%)	39.8	49.2	52.6
η_{Th} (%)	55.2	43.8	45.7

Table 9.5 compares some selected parameters of the air-standard Brayton cycles with isentropic and non-isentropic expansion and compression processes with the actual gas turbine specifications reported by the manufacturer. The percentage of error of theoretical results compared to the actual parameters is presented in parentheses. The comparison indicates a remarkably lower error in the results when the non-isentropic expansion and compression processes are used.

Note: To investigate the effects of the isentropic efficiencies of the compressor and turbine, this example can be repeated for the isentropic efficiencies of 85% for both compressor and turbine. The comparison of the results in Table 9.6 indicates that by increasing the efficiencies of the

TABLE 9.5

Comparison of Results of Selected Parameters for the Actual Gas Turbine, Nonideal Brayton Cycle, and Ideal Brayton Cycle

Parameter	Net Power Output (MW)	Efficiency (%)	Exhaust Temperature (K)	Exhaust Energy (MW)
Actual gas turbine	470	41.5	892	712
Air-standard Brayton cycle – isentropic processes	625 (33%)	55.2 (33%)	797 (–11%)	507 (–9%)
Air-standard Brayton cycle with $\eta_{Comp} = \eta_{Turb} = 90\%$	475 (1%)	43.8 (6%)	892 (0%)	611 (–14%)

turbine and compressor from 85% to 90%, the cycle net power output and efficiency can be improved by 20% and 16%, respectively. Similarly, the back work ratio significantly reduces.

Figure 9.12 illustrates the cycle efficiency and net power output as a function of the isentropic efficiencies of the compressor and turbine at the range of 65%–100% for a cold air-standard nonideal Brayton cycle. The diagram shows that how quickly the efficiency and net power of the cycle drop with a reduction of the isentropic efficiencies of the

TABLE 9.6

Comparison of the Results for Nonideal Brayton Cycle with Isentropic Efficiencies of 85% and 90%

Parameter	Air-Standard Brayton Cycle with $\eta_{Comp} = \eta_{Turb} = 85\%$	Air-Standard Brayton Cycle with $\eta_{Comp} = \eta_{Turb} = 90\%$
T_2 (K)	778	752
T_4 (K)	940	892
$\dot{Q}_L$ (MW)	682	611
$\dot{Q}_H$ (MW)	1058	1086
$\dot{W}_{Comp}$ (MW)	486	459
$\dot{W}_{Turb}$ (MW)	883	935
$\dot{W}_{Net}$ (MW)	396	475
BWR (%)	55.1	49.2
η_{Th} (%)	37.5	43.8

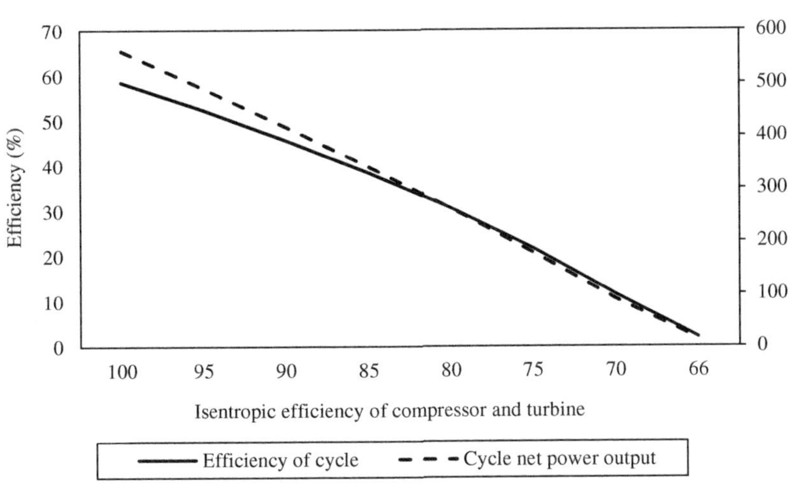

FIGURE 9.12
Efficiency and net power output of nonideal Brayton cycle as a function of the isentropic efficiencies of the compressor and turbine (cold air standard analyses).

compressor and turbine. When the isentropic efficiencies of the compressor and turbine reach about 65%, the cycle cannot produce any output power, and almost all power produced in the turbine is consumed in the compressor.

The performance of the Brayton cycle can be improved by reducing Q_H and W_{Comp} as well as increasing W_{Turb}. By incorporating a regenerator, intercoolers, and reheaters, respectively, these objectives can be achieved. In the following sections, the modified Brayton cycles by integrating these components will be explained.

9.2.4 Brayton Cycle with Regenerator

As previously mentioned, one of the problems of gas turbine power systems is that the high-temperature thermal energy is exhausted to the environment. Using a regenerator is one way of dealing with this problem. In gas turbine cycles, the temperature of the turbine exhaust (T_4 in Figure 9.6) is typically higher than the temperature of the compressor outlet (T_2). For instance in Example 9.1, T_4 and T_2 are 797 and 709 K, respectively. A heat exchanger, commonly counter-flow surface-type known as a **regenerator** or **recuperator**, can be used to transfer heat from the turbine outlet to the combustor inlet. Increasing the inlet temperature of the combustor means less fuel is required for the combustion process, which in turn improves the efficiency of the cycle.

In a gas turbine with a regenerator (also known as a **regenerative gas turbine cycle**), the turbine exhaust gas enters the hot section of the regenerator at T_4, and as a result of heat transfer, its temperature reduces to T_y. In the cold section of the regenerator, the high-pressure air from the compressor enters at T_2, and its temperature increases to T_x across the regenerator. Having a higher temperature difference between the outlet flow of the turbine and the outlet flow of the compressor provides a better heat transfer potential in the regenerator. Recalling Equation 9.45 for the Brayton cycle with a maximum net power output ($T_2 = T_4$), it can be concluded that in a Brayton cycle designed for the maximum net power output, there is no or little benefit for integrating a regenerator to the cycle.

In a regenerative Brayton cycle with an ideal counter-flow heat exchanger (Figure 9.13), the outlet temperature of the turbine (T_4) and the inlet temperature of the combustor (T_x) are equal. So the combustor only requires to increase the temperature from T_x to T_3 rather than T_2 to T_3 in a simple non-regenerative cycle. In the T-s diagram in Figure 9.13, by using a regenerator, the heat transfer from the high-temperature source reduces from area 2-3-d-a-2 to area x-3-d-c-x and the heat rejected to the low-temperature source

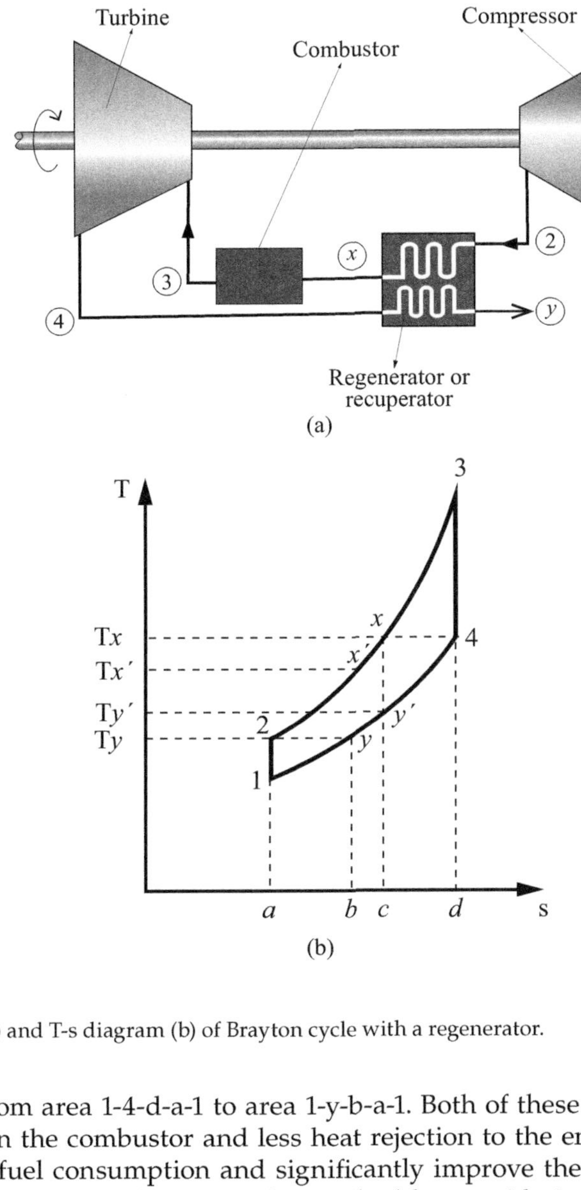

FIGURE 9.13
Schematic (a) and T-s diagram (b) of Brayton cycle with a regenerator.

reduces from area 1-4-d-a-1 to area 1-y-b-a-1. Both of these effects (less heat required in the combustor and less heat rejection to the environment) help to reduce fuel consumption and significantly improve the efficiency of the cycle. The same conclusion can be reached by considering the increase in the average temperature of the heat transfer from the high-temperature heat source and the decrease in the average temperature of the heat transfer to the low-temperature heat sink. It should be noted that theoretically the net power production of the cycle remains unchanged because the area enclosed by the cycle does not change. However, in reality, the pressure drop in the heat exchanger slightly reduces the power output.

It can be shown that the efficiency of the Brayton cycle with a regenerator can be calculated by Equation 9.48:

$$\eta_{th} = 1 - \frac{T_1}{T_3}\left(\frac{P_2}{P_1}\right)^{\left(\frac{(k-1)}{k}\right)} \tag{9.48}$$

In this modified cycle, the efficiency depends on both the compressor pressure ratio and the ratio of the minimum temperature of the cycle to the maximum temperature of the cycle (the compressor and turbine inlet flow temperatures, respectively) as well as the properties of the working fluid (k).

Equation 9.48 shows that increasing the compressor pressure ratio reduces the efficiency of the regenerative Brayton cycle (unlike the simple Brayton cycle). This can be seen in Figure 9.13, where increasing compressor pressure ratio results in a lower-temperature difference between outlet flows of the turbine and compressor. This means there is less opportunity for heat recovery in the regenerator.

Due to the limited size of real heat exchangers, it is impractical to have the exit temperature of the cold flow (T_x) to be equal to the temperature of the incoming hot flow (T_4). Similarly, it is impossible to have $T_2 = T_y$. The higher temperature difference between the flows causes more irreversibilities in the heat transfer process. In order to account for this irreversibility, the effectiveness of the regenerator can be defined as

$$\eta_{Reg} = \frac{h_{x'} - h_2}{h_4 - h_2} \tag{9.49}$$

Note that $h_{x'} - h_2$ is the change of enthalpy in the cold stream in the real heat exchanger and $h_4 - h_2$ is the change of enthalpy in the cold stream in the ideal heat exchanger. A similar relationship can be expressed for the effectiveness of the regenerator based on the change of enthalpy in the hot stream.

$$\eta_{Reg} = \frac{h_4 - h_{y'}}{h_4 - h_2} \tag{9.50}$$

For a cold-standard air cycle (constant specific heat)

$$\eta_{Reg} = \frac{T_{x'} - T_2}{T_4 - T_2} = \frac{T_4 - T_{y'}}{T_4 - T_2} \tag{9.51}$$

The effectiveness of the regenerator is an indication of the departure of the real regenerator from the ideal one. It should be noted that in a real heat exchanger, T_4 and T_x cannot be equal so the effectiveness of the regenerator cannot be 1 (100%) and typically is within the range of 60%–80%. A larger heat exchanger will result in a closer T_4 and T_x and higher regenerator effectiveness but a higher initial cost of the heat exchanger. Also, a larger heat exchanger causes a higher flow pressure drop across the regenerator, which adversely affects the overall performance of the system.

Example 9.6

In the cycle explained in Example 9.1, a regenerator with the effectiveness of 70% is incorporated into the cycle. Determine the energy transfer per second in each component (in kW), the net power generation (in kW), the back work ratio, and the efficiency of the cycle using data from the air property table.

Solution

Given

A regenerative air-standard Brayton cycle operates at the given compressor inlet air temperature, pressure, and mass flow rate; turbine inlet air temperature; CR; and regenerative effectiveness.

Find

Determine the energy transfer per second in each component, the net power generation, the back work ratio, and the efficiency of the cycle for the regenerator effectiveness of 70%.

Schematic diagrams: Figure 9.13.

Assumptions

1. The cycle is an air-standard Brayton cycle; therefore, all processes are internally reversible. The heat transfer processes are at constant pressure, the expansion and compression processes are isentropic, and the working fluid is air and behaves as an ideal gas.
2. The steady-state control volume mass and energy balance analyses with no kinetic and potential energy changes are applied to each component.

Analysis:

From Example 9.1

$$h_1 = 300.19 \text{ kJ/kg}$$

$$h_2 = 722.91 \text{ kJ/kg}$$

$$h_3 = 1880.1 \text{ kJ/kg}$$

$$h_4 = 818.30 \text{ kJ/kg}$$

The first step is to determine h_x' and h_y' using Equations 9.50 and 9.51.

$$\eta_{\text{Reg}} = \frac{h_{x'} - h_2}{h_4 - h_2} = \frac{h_4 - h_{y'}}{h_4 - h_2}$$

$$0.7 = \frac{h_{x'} - 722.91}{818.30 - 722.91} \rightarrow h_{x'} = 789.68\ \text{kJ/kg}$$

$$0.7 = \frac{818.30 - h_{y'}}{818.30 - 722.91} \rightarrow h_{y'} = 751.53\ \text{kJ/kg}$$

The heat transfer rates are

$$\dot{Q}_L = \dot{m}\left(h_y - h_1\right) = (978)\left[\text{kg/s}\right](751.53 - 300.19)\left[\text{kJ/kg}\right] = \mathbf{441,411\ kW}$$

$$\dot{Q}_H = \dot{m}\left(h_3 - h_x\right) = (978)\left[\text{kg/s}\right](1880.1 - 789.68)\left[\text{kJ/kg}\right] = \mathbf{1,066,431\ kW}$$

Comparing $\dot{Q}_L$ and $\dot{Q}_H$ of the regenerative Brayton cycle with those of the simple cycle (506,712 and 1,131,732 kW, respectively, for the simple cycle) shows about 15% and 6% reduction, respectively.

The power consumption in the compressor and the power production in the turbine, and hence the net power output and back work ratio, remain unchanged.

$$\dot{W}_{\text{Comp}} = \mathbf{413,420}\ \text{kW}$$

$$\dot{W}_{\text{Turb}} = \mathbf{1,038,440}\ \text{kW}$$

$$\dot{W}_{\text{net}} = \mathbf{625,020}\ \text{kW}$$

$$\text{BWR} = \mathbf{0.398}$$

The efficiency of the cycle is

$$\eta_{\text{th}} = \frac{\dot{W}_{\text{net}}}{\dot{Q}_H} = \frac{625,020}{1,066,431} = \mathbf{0.586}$$

which is higher than the efficiency of the simple cycle at 0.552.

9.2.5 Brayton Cycles with Compressor Intercoolers

Before we discuss about compressor intercooling and turbine reheating, it will be very informative and helpful to review the Ericsson cycle.

REMINDER BOX 9.3

ERICSSON CYCLE

In the Brayton cycle, the turbine and compressor are both reversible and adiabatic (isentropic) processes. If we replace these two components with two reversible and isothermal processes (Figure 9.14), the cycle is called the **Ericsson cycle** (after John Ericsson, 1803–1899, a Swedish engineer who first proposed the cycle), which consists of two constant-temperature processes and two constant-pressure processes (all reversible). The efficiency of this cycle is higher than that of the Brayton cycle and with a regenerator is equal to the Carnot cycle operating between the same high- and low-temperature thermal energy sources. However, this cycle is not practical because both compressor and turbine are high flow rate devices that involve work. Therefore, it is practically impossible to have a large quantity of heat transfer in these devices (to keep the temperature constant). They are both adiabatic processes by nature.

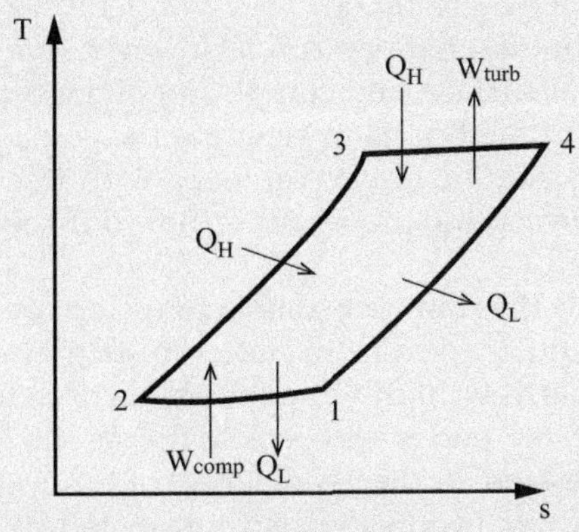

FIGURE 9.14
T-s diagram of Ericson cycle.

To modify the Brayton cycle to get closer to the Ericsson cycle, multiple steps of compression with intercooling and multiple steps of expansion with reheating are used. The efficiency of this cycle is closer to that of the Ericsson cycle (thus the Carnot cycle) than that of the simple Brayton cycle.

The objective of incorporating intercoolers into gas turbine plants is to reduce power input to the compressor that will increase both the net power output and efficiency while reducing the back work ratio of the cycle. The work of equipment involving a steady-state flow process, e.g., a compressor and turbine, can be calculated using $w = \int v dp$. For an air-standard power cycle with an ideal gas as the working fluid $(Pv = RT)$,

$$w = \int RT \frac{dP}{P} \tag{9.52}$$

This equation indicates that in a compressor, higher-temperature air requires more power input than lower-temperature air for the same CR (due to the higher specific volume of air flow). According to the ideal gas equation, the temperature of air flow increases as it goes through the various stages of the compressor. This means to create the same air pressure increase, more power input is needed at the later stages of the compressor (stages with higher air pressure and temperature) than that at the earlier stages of the compressor (stages with lower air pressure and temperature). Therefore, it is desirable to keep the air flow temperature as low as possible while it proceeds through the compressor and its pressure increases. Theoretically, it is ideal to lower the air flow temperature by continuous cooling of the flow in the compressor (as in the Ericson cycle where the air flow temperature is maintained constant at the compressor inlet temperature). However, due to the high air flow rate, small heat transfer surface, and complexity of the compressor, this is not practical.[1] The practical method is to separate the compression and cooling processes by compressing air in several steps with the air cooling process between the steps. The cooling process is achieved through heat exchangers known as **intercoolers**.

Figure 9.15 shows the Brayton cycle with a two-step compressor intercooler where the air flow is compressed from state 1 to state 2 in the first step of the compressor. Then, relatively high-temperature air is sent to the intercooler and cooled down to state 3 and is sent back to the second step of the compressor for further compression. In the T-s diagram of the cycle in Figure 9.15, the increase in the net power production is represented by the area 2-3-4-2′-2. Since the constant-pressure lines in the T-s diagram converge slightly with reducing entropy,

$$h_{2'} - h_1 > (h_2 - h_1) + (h_4 - h_3) \tag{9.53}$$

which again indicates that less power input is required for the compression process in the cycle with the intercooler.

The diagram also shows that along with an increase in the net power production of the cycle, the heat transfer to the cycle from the high-temperature reservoir increases (area a-4-2′-b-a) because the temperature of the combustor inlet flow is lower. This means more heat (and fuel) is needed to increase the air flow temperature to the desired TIT. However, the lower

[1] The same reasons that in steam turbines continuous regeneration is not practical and feedwater heaters are used.

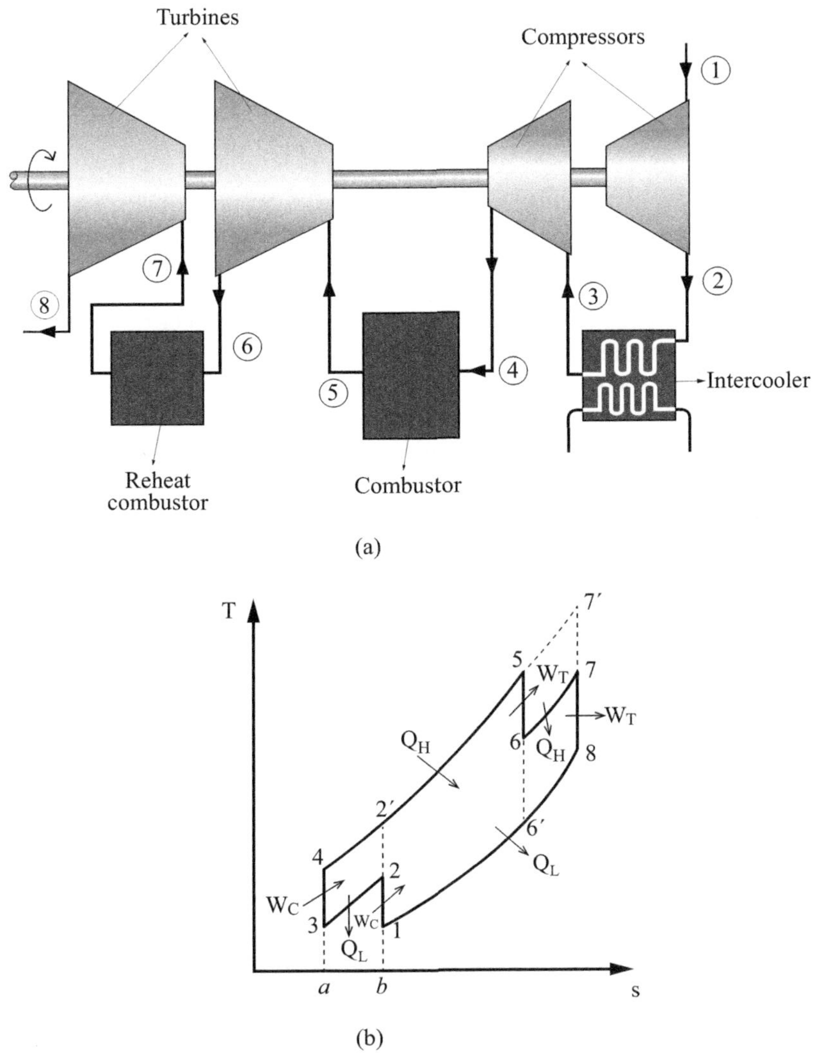

FIGURE 9.15
Schematic of modified Brayton cycle with two steps of compression with intercooling and two steps of expansion with reheat (a) and T-s diagram of the cycle (b).

air temperature at the outlet of the compressor improves the potential for the waste heat recovery from the turbine exhaust to the compressor outlet through a regenerator.

In practice, intercoolers can be air-cooled or water-cooled heat exchangers, the latter being more common. Large gas turbines can be equipped with several steps of compression with intercoolers in between. The decision about the number of intercoolers and their operational conditions is based on economic considerations.

9.2.6 Brayton Cycles with Turbine Reheaters

Theoretically, what a reheater does in the turbine is similar to what the intercooler does in the compressor but in this case to increase the air specific volume across the turbine. Note again that the power generation in the turbine is given by $w = \int v\,dp$. In this case, it is desirable to increase the average temperature of the expansion process in the turbine. The best scenario is to heat the air flow continuously while it is going through the turbine to maintain the air temperature at the TIT (as in the Ericsson cycle). But this is not practically possible so the expansion and heating processes should be separated. For this purpose, the expansion process occurs in steps with reheating between the steps. In Figure 9.15, the high-pressure air from the compressor enters the combustion chamber at state 4 and is heated to state 5 before being fed to the high-pressure turbine. Then, the air flow at moderate pressure (state 6) is extracted from the high-pressure turbine and sent to the reheat combustion chamber where it is reheated to state 7 before entering the low-pressure turbine. Because the constant-pressure lines in the T-s diagram diverge slightly with increasing enthalpy,

$$h_5 - h_{6'} < (h_5 - h_6) + (h_7 - h_8) \tag{9.54}$$

which again indicates that more power is produced in the turbine in the cycle with the reheater. Note that the performance of the cycle would have been even better if the air could be heated all the way to state 7′ in the combustor but this is not practical due to the limitations in the TIT. Therefore, turbine reheaters improve the specific power output of the cycle at the expense of additional heating. Note that the additional combustion in the reheat combustor requires excess air; therefore, the outlet flow of the first combustor should contain enough air for this purpose. This means that the overall air-to-fuel ratio in reheat gas turbines should be high enough to support combustion in both combustors. Since the temperature of the turbine exhaust is higher in the reheat gas turbine cycle compared to that of the simple cycle ($T_8 > T_{6'}$), there is more potential for waste heat recovery through a regenerator or the integration of a gas turbine into CCPP (Chapter 10). Furthermore, dividing the combustion processes (and fuel injection) between two combustors helps in the reduction of NO_x production. In Example 9.7, we will explore at what pressure of the reheater the specific net work output of the turbine will be maximized.

Example 9.7

In an ideal Brayton cycle, the two-step turbine is equipped with a reheater that increases the temperature of the low-pressure turbine inlet flow to the temperature of the high-pressure turbine inlet flow. For the fixed given states of the first turbine inlet and the second turbine outlet, determine the pressure of the reheater that maximizes the specific net work output of the turbine (both steps combined). The turbines operate according to the cold air-standard assumptions.

Solution

Given

An ideal two-step turbine with a reheater operating at the fixed first turbine inlet and second turbine outlet states.

Find

Determine the pressure of the reheater that maximizes the power output of the turbine.

Schematic diagrams: Figure 9.15.

Assumptions

1. The turbine analysis is based on the cold air-standard assumptions; thus, the reheater operates at constant pressure, the expansion processes in both steps are isentropic, and the working fluid is air and behaves as an ideal gas with constant C_p and k.
2. The steady-state control volume mass and energy balance analyses with no kinetic and potential energy changes are applied to both steps of the turbine.
3. The temperature of the low-pressure turbine inlet flow is equal to the temperature of the high-pressure turbine inlet flow ($T_5 = T_7$).

Analysis:

The turbine total specific work output is

$$w_{\text{Turb}} = (h_5 - h_6) + (h_7 - h_8)$$

$$w_{\text{Turb}} = C_p(T_5 - T_6) + C_p(T_7 - T_8) = C_pT_5\left(1 - \frac{T_6}{T_5}\right) + C_pT_7\left(1 - \frac{T_8}{T_7}\right)$$

Since $T_5 = T_7$

$$w_{\text{Turb}} = C_pT_5\left[\left(1 - \frac{T_6}{T_5}\right) + \left(1 - \frac{T_8}{T_7}\right)\right] \tag{9.55}$$

Since the expansion processes are isentropic and k is constant

$$\left(\frac{T_6}{T_5}\right) = \left(\frac{P_6}{P_5}\right)^{\frac{k-1}{k}} \quad \text{and} \quad \left(\frac{T_8}{T_7}\right) = \left(\frac{P_8}{P_7}\right)^{\frac{k-1}{k}}$$

Substituting in Equation 9.55

$$w_{\text{Turb}} = C_pT_5\left[\left[1 - \left(\frac{P_6}{P_5}\right)^{\frac{k-1}{k}}\right] + \left[1 - \left(\frac{P_8}{P_7}\right)^{\frac{k-1}{k}}\right]\right]$$

Since there is no pressure drop in the reheater so $P_6 = P_7$.

$$w_{\text{Turb}} = C_p T_5 \left[\left[1-\left(\frac{P_6}{P_5}\right)^{\frac{k-1}{k}}\right]+\left[1-\left(\frac{P_8}{P_6}\right)^{\frac{k-1}{k}}\right]\right] \tag{9.56}$$

For the given states of the first turbine inlet and the second turbine outlet, all the parameters in this equation are fixed except P_6. So in Equation 9.56, the specific work of the turbine is only a function of the reheater pressure. To find the reheater pressure that maximizes the turbine specific work, the derivative of Equation 9.56 with respect to p_6 should be equated to zero.

$$\frac{\partial w_{\text{Turb}}}{\partial P_6} = \frac{\partial}{\partial P_6} C_p T_5 \left[\left[1-\left(\frac{P_6}{P_5}\right)^{\frac{k-1}{k}}\right]+\left[1-\left(\frac{P_8}{P_6}\right)^{\frac{k-1}{k}}\right]\right]$$

$$= C_p T_5 \left(\frac{k-1}{k}\right)\left[-\left(\frac{P_6}{P_5}\right)^{\frac{-1}{k}}\left(\frac{1}{P_5}\right)+\left(\frac{P_8}{P_6}\right)^{\frac{-1}{k}}\left(\frac{P_8}{p_6^2}\right)\right]$$

$$= C_p T_5 \left(\frac{k-1}{k}\right)\frac{1}{P_6}\left[-\left(\frac{P_6}{P_5}\right)^{\frac{k-1}{k}}+\left(\frac{P_8}{P_6}\right)^{\frac{k-1}{k}}\right] = 0$$

$$-\left(\frac{P_6}{P_5}\right)^{\frac{k-1}{k}}+\left(\frac{P_8}{P_6}\right)^{\frac{k-1}{k}} = 0$$

Thus,

$$\frac{P_6}{P_5} = \frac{P_8}{P_6} \quad \text{or} \quad P_6 = \sqrt{P_5 P_8}$$

$$\text{So} \quad \frac{P_6}{P_5} = \frac{P_8}{P_7} = \sqrt{\frac{P_8}{P_5}} \tag{9.57}$$

$$(\text{Pressure Ratio})_{\text{stage}} = \sqrt{(\text{Pressure Ratio})_{\text{total}}} \tag{9.58}$$

It can be shown that a similar relationship exists for a higher number of turbine steps and reheaters. The pressure ratio of each step can be estimated by

$$(\text{Pressure Ratio})_{\text{stage}} = \sqrt[N]{(\text{Pressure Ratio})_{\text{total}}} \tag{9.59}$$

where $(\text{Pressure Ratio})_{\text{stage}}$ is the pressure ratio of each individual step, $(\text{Pressure Ratio})_{\text{total}}$ is the pressure ratio of the whole turbine (combined steps), and N is the number of steps in the turbine. Note that the number of reheaters is equal to N–1. So in order to have a pressure ratio of 16 in a two-step turbine, two turbines with a pressure ratio of 4 $\left(=\sqrt{16}\right)$ are required.

Also, from $T_5 = T_7$ and $\frac{P_6}{P_5} = \frac{P_8}{P_7}$, it can be concluded that $T_6 = T_8$. This means that in the optimum condition, the temperature decrease and enthalpy drop in the steps of the turbine are equal and the power produced in steps is equally distributed.

It can be shown that if the assumption of $T_5 = T_7$ is relaxed and the temperature of the low-pressure turbine inlet flow is allowed to be lower than or equal to the temperature of the high-pressure turbine inlet flow, Equation 9.57 will become

$$P_6 = \sqrt{P_5 P_8 \left(\frac{T_7}{T_5}\right)^{\frac{k}{k-1}}} \tag{9.60}$$

A similar cold air-standard analysis can be conducted for a two-step compressor with an intercooler. In this case, the temperature of the inlet flow to the high-pressure compressor is greater than or equal to the temperature of the inlet flow to the low-pressure compressor. For the fixed given states of the first compressor inlet and the second compressor outlet, it can be shown that the pressure of the intercooler that minimizes the power input of the compressor can be calculated by the following equation:

$$P_2 = \sqrt{P_1 P_4 \left(\frac{T_3}{T_1}\right)^{\frac{k}{k-1}}} \tag{9.61}$$

If $T_3 = T_1$

$$P_2 = \sqrt{P_1 P_4} \tag{9.62}$$

So,

$$\frac{P_2}{P_1} = \frac{P_4}{P_3} = \sqrt{\frac{P_4}{P_1}} \tag{9.63}$$

$$(\text{Pressure Ratio})_{\text{stage}} = \sqrt{(\text{Pressure Ratio})_{\text{total}}} \tag{9.64}$$

It can be shown that a similar relationship exists for a higher number of compressor steps and intercoolers. The pressure ratio of each step can be estimated by

$$(\text{Pressure Ratio})_{\text{stage}} = \sqrt[N]{(\text{Pressure Ratio})_{\text{total}}} \tag{9.65}$$

where $(\text{Pressure Ratio})_{\text{stage}}$ is the pressure ratio of each individual step, $(\text{Pressure Ratio})_{\text{total}}$ is the pressure ratio of the whole compressor (combined steps), and N is the number of steps in the compressor. So in order to have a pressure ratio of 16 in a two-step compressor, two compressors with a pressure ratio of $4\left(=\sqrt{16}\right)$ are required. Similar to the turbine with a reheater, in a compressor with an intercooler operating at the optimum condition, the temperature and enthalpy increases in the steps of the compressor are equal and the power input for each step is equal.

Example 9.8

In Example 9.1, a two-step compressor with an intercooler between two steps and a two-step turbine with a reheater between the steps are integrated into the Brayton cycle instead of the single-step compressor and turbine, respectively. All steps of the compressor and turbine are isentropic. Assuming $T_3 = T_1$ and $T_5 = T_7$ (Figure 9.16), conduct a cold air-standard analysis on this modified Brayton cycle to determine the net power output and efficiency of the cycle if the pressures of the intercooler and reheater are selected so the cycle operates at the optimum power output. Conduct the analysis once without a regenerator and once with an ideal regenerator (with an effectiveness of 100%).

Solution

Given

An ideal air-standard Brayton cycle is modified by incorporating a two-step turbine with a reheater and a two-step compressor with an intercooler. The cycle operates at the given conditions. The pressures of the intercooler and reheater are selected so the cycle operates at the optimum power output.

Find

Determine the net power output and efficiency of the cycle with and without an ideal regenerator.

Schematic diagrams: Figure 9.16

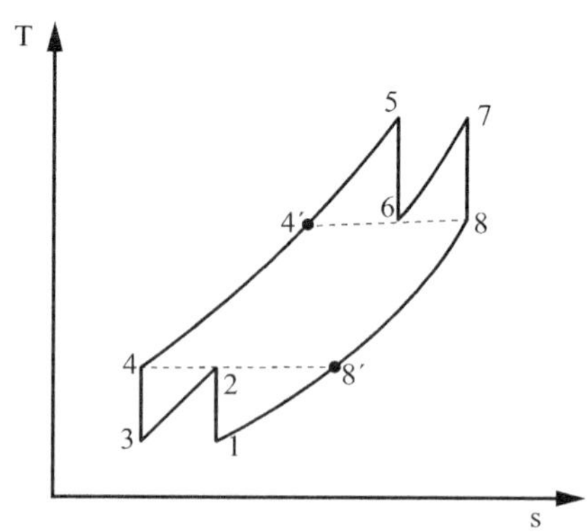

FIGURE 9.16
T-s diagram of a Brayton cycle with two steps of compression with intercooling, two steps of expansion with reheat, and a regenerator for Example 9.8.

Assumptions:

1. The cycle analysis is based on the cold air-standard assumptions; thus, the reheater and intercooler operate at constant pressures, and all steps of the compression and expansion processes are isentropic. Also, the working fluid is air and behaves as an ideal gas with constant C_p and k.
2. The steady-state control volume mass and energy balance analyses with no kinetic and potential energy changes are applied to both steps of the turbine and the compressor.
3. The temperature of the low-pressure turbine inlet flow is equal to the temperature of the high-pressure turbine inlet flow ($T_5 = T_7$).
4. The temperature of the high-pressure compressor inlet flow is equal to the temperature of the low-pressure compressor inlet flow ($T_1 = T_3$).
5. For the cycle with the regenerator, the effectiveness of the regenerator is 100%.

Analysis:

Cycle with No Regenerator
Let us first consider the compressor. Since $T_1 = T_3$ (Figure 9.16) and the pressure of the intercooler is selected for minimum power input to the compressor, the pressure ratios of the two steps of the compressor can be estimated by Equation 9.63

$$\frac{P_2}{P_1} = \frac{P_4}{P_3} = \sqrt{\frac{P_4}{P_1}} = \sqrt{21.8} = 4.7$$

$$\left(\frac{T_2}{T_1}\right) = \left(\frac{P_2}{P_1}\right)^{\frac{k-1}{k}}$$

$$\left(\frac{T_2}{300}\right) = (4.7)^{0.2857} \rightarrow T_2 = 466.8\,\text{K}$$

Also,

$$T_4 = T_2 = 466.8\,\text{K}$$

The outlet temperature of the compressor in this case is much less than that of Example 9.1 (723.6 K).

The power input required for the compression is

$$\begin{aligned}\dot{W}_{\text{Comp}} &= \dot{m}(h_2 - h_1) + \dot{m}(h_4 - h_3) = \dot{m}C_p\left[(T_2 - T_1) + (T_4 - T_3)\right] \\ &= (978)(1.004)\left[(466.8 - 300) + (466.8 - 300)\right] \\ &= 327,566\,\text{kW}\end{aligned}$$

Comparing this result with the result of Example 9.1 (415,938 kW) shows a significant reduction in the power input of the compressor.

Similarly for the turbine, since $T_5 = T_7$ and the pressure of the intercooler is selected for a maximum power output of the turbine, the pressure ratios of the two steps of the turbine can be estimated by Equation 9.57.

$$\frac{P_6}{P_5} = \frac{P_8}{P_7} = \sqrt{\frac{P_8}{P_5}} = \sqrt{\frac{1}{21.8}} = 0.214$$

$$\left(\frac{T_6}{T_5}\right) = \left(\frac{P_6}{P_5}\right)^{\frac{k-1}{k}}$$

$$\left(\frac{T_6}{1700}\right) = (0.214)^{0.2857} \quad \rightarrow T_6 = 1094.3\,\text{K}$$

Also,

$$T_8 = T_6 = 1094.3\,\text{K}$$

The total power produced in the turbine is a sum of the power produced in each step.

$$\dot{W}_{\text{Turb}} = \dot{m}(h_5 - h_6) + \dot{m}(h_7 - h_8) = \dot{m}C_p\left[(T_5 - T_6) + (T_7 - T_8)\right]$$

$$= (978)(1.004)\left[(1700 - 1094.3) + (1700 - 1094.3)\right]$$

$$= 1,189,488\,\text{kW}$$

Comparing this result with the result of Example 9.1 (977,199 kW) shows a significant increase in the power output of the turbine. The net power output of the cycle is

$$\dot{W}_{\text{Net}} = \dot{W}_{\text{Turb}} - \dot{W}_{\text{Comp}} = 1,189,488 - 327,566 = 861,922\,\text{kW}$$

As expected, the net power output of the modified cycle is much higher than that in the simple Brayton cycle in Example 9.1 (561,261 kW).

To determine the efficiency of the cycle, the heat added to the cycle is required.

$$\dot{Q}_H = \dot{m}(h_5 - h_4) + \dot{m}(h_7 - h_6) = \dot{m}C_p\left[(T_5 - T_4) + (T_7 - T_6)\right]$$

$$= (978)(1.004)\left[(1700 - 466.8) + (1700 - 1094.3)\right]$$

$$= 1,805,638\,\text{kW}$$

The heat added to the cycle in the modified cycle is much higher than the simple cycle in Example 9.1 (958,739 kW) because of the lower temperature of the inlet flow to the combustor and the heat required in the reheater. Finally, the back work ratio and efficiency are

$$\text{BWR} = \frac{\dot{W}_{\text{Comp}}}{\dot{W}_{\text{Turb}}} = \frac{327{,}566}{1{,}189{,}488} = \mathbf{0.275}$$

$$\eta = \frac{\dot{W}_{\text{Net}}}{\dot{Q}_H} = \frac{861{,}922}{1{,}805{,}638} = \mathbf{0.477}$$

It is evident that the back work ratio is significantly reduced in this cycle compared to the cycle in Example 9.1 (0.426). The results show that the compressor outlet temperature in the modified cycle (466.8 K) is much lower than that of the simple cycle (723.6 K). Also, the turbine outlet temperature in the modified cycle (1094.3 K) is much higher than that of the simple cycle (704.8 K). This indicates that there is a much higher potential for waste heat energy recovery through a regenerator in the modified cycle.

Cycle with Regenerator

In the regenerative cycle with a regenerator's effectiveness of 100%, the temperature of the combustor inlet is equal to the temperature of the turbine exhaust

$$T_{4'} = T_8 = 1094.3 \text{ K}.$$

The heat added to the cycle in this case is

$$\begin{aligned}\dot{Q}_H &= \dot{m}(h_5 - h_{4'}) + \dot{m}(h_7 - h_6) = \dot{m}C_p\left[(T_5 - T_{4'}) + (T_7 - T_6)\right] \\ &= (978)(1.004)\left[(1700 - 1094.3) + (1700 - 1094.3)\right] \\ &= 1{,}189{,}488\,\text{kW}\end{aligned}$$

Recall that in the regenerative gas turbine, the net power output remains unchanged, so the efficiency of the regenerative modified cycle is

$$\eta = \frac{\dot{W}_{\text{Net}}}{\dot{Q}_H} = \frac{861{,}922}{1{,}189{,}488} = \mathbf{0.725}$$

As these examples point out, utilizing a multistep turbine with reheaters and a multistep compressor with intercoolers increases the net power output of the cycle and improves the potential for heat recovery for regeneration. Thus, incorporating a regenerator can significantly improve the efficiency of these cycles.

Using large number of compression and expansion steps helps the Brayton cycle to get closer to the Ericsson cycle (Figure 9.17) but economic considerations (increase in the capital costs), compressors and turbines losses, and pressure drops limit the number of these steps to two or three steps.

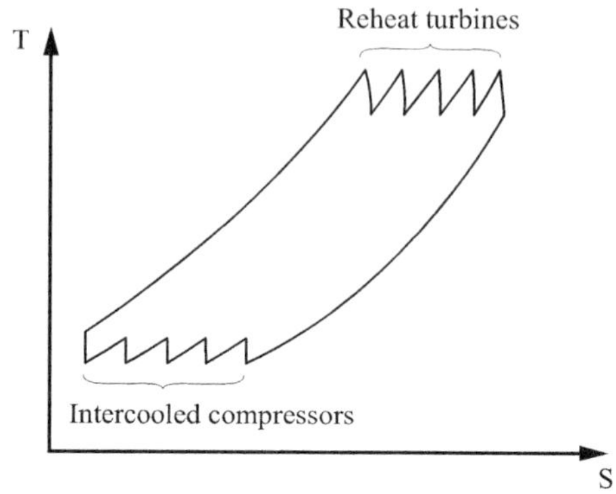

FIGURE 9.17
T-s diagram of a Brayton cycle with a large number of compression and expansion steps (approaching an Ericsson cycle).

9.3 Brief History of Gas Turbines

As noted in the introduction of this chapter, gas turbines are the youngest power generation technology among conventional power generation systems with less than a century of development history. While the concept of gas turbine cycles was known for a long time, the practical efforts to build a working gas turbine were proved to be futile until the early 20th century. Regardless of this short history, gas turbines have been able to achieve impressive improvements in their power generation capacity and efficiency, from 4 MW and 18% of the first-utility gas turbine to over 500 MW and over 40% efficiency of modern heavy-duty gas turbines.

As with many other devices and engineering ideas, the first known idea of a device that resembled a gas turbine appeared in Leonardo da Vinci's (1452–1519, Italian polymath, perhaps the greatest mind of whole history) drawing of a chimney jack (also known as a smoke or roasting jack) in his Codex Atlanticus.[2] His design used hot combustion products from fire to

[2] The Codex Atlanticus (Atlantic Codex) is the largest collection of Leonardo's famous handwritings and notes bonded in 12 volumes, currently preserved at the Biblioteca Ambrosiana in Milan. The collection included 1119 pages that Leonardo created in the span of over 30 years. Many of them dealt with the engineering design of tools and machines, one of his many outstanding talents. The collection was arranged on large pages in the 16th century to protect the original documents. That large size was at the time used for atlases and that is how the collection got its name. You can view all pages of the Codex at http://www.codex-atlanticus.it. Other surviving notes can be found at the Royal Library at Windsor Castle, the Louvre, the Biblioteca Nacional de España, the Victoria, Albert Museum, and several other locations. Probably the most well-known collection is the Codex Leicester currently owned by Bill Gates. He purchased the collection in 1994 for about $30 million. The collection is displayed annually in cities around the world.

run a turbine rotor mounted in the chimney of a fireplace. The single-stage reaction axial turbine had four overlapping blades that looked like modern propellers. The power produced by the turbine is transferred by a gear and pulley mechanism to turn a roasting spit. The design had an excellent inherent self-control characteristic. With a higher intensity of the fire, the turbine could produce more power and turn the roasting meat faster, which was what one needed to avoid burning of the meat. Reportedly, this device was commonly used in Europe after Leonardo.

The first patent for a true gas turbine was registered in 1791 in England by John Barber (1734–1793, an English inventor) to power a carriage.[3] However, the early gas turbine designs were not able to produce any net power. Franz Stolze (1836–1910, a German inventor and explorer) introduced the idea of using an axial turbine and compressor on a single shaft for gas turbines. Charles Algernon Parsons (1854–1931, a British engineer) continued the work done by his predecessors and patented his design in Britain. The first gas turbine patent in the U.S. was registered in 1899 by Charles Gordon Curtis (1860–1953, an American engineer).[4] Regardless of these efforts, none of these designs was able to generate a meaningful output power primarily due to the lack of proper understanding of compressor and turbine blade aerodynamics. While several experimental units were built and tested with varying degrees of success, none of these attempts resulted in building a gas turbine for continuous operation.

The first gas turbine that could produce net power was built by Jens William Ægidius Elling (1861–1949, a Norwegian inventor) in 1903 (Figure 9.18). Between this first experimental gas turbine and when the first commercial gas turbine was built, various "hybrid" gas turbines were tested and widely used in several industries. In one such design, a steam generator was integrated into a gas turbine. One of the main challenges in the early gas turbines was that while the combustion chamber could produce high-temperature gas flow, due to material limitations of turbine blades at the time, most of gas energy was wasted via the turbine cooling system. To address this problem, a steam generator was integrated into the combustion chamber. In this design, air was compressed in the compressor before being sent to the combustor. The combustion products were used to raise steam in an evaporator and superheater. The gas was then fed to the gas turbine at a relatively low temperature. The outlet of the turbine was finally used in an economizer before being exhausted. Tens of these units were deployed in the 1930s and 1940s. Other efforts included integrating gas turbines to industrial facilities in the form of turbochargers. In these units, air was compressed in a compressor and was sent to industrial processes where high-temperature

[3] For an excellent detailed history of gas turbines, refer to Eckart, Dietrich. Gas Turbine Powerhouse: The Development of the Power Generation Gas Turbine at BBC-ABB-Alstom. Walter de Gruyter, 2013.

[4] Both Parsons and Curtis are best known for their innovations on steam turbines (refer to Chapter 6).

FIGURE 9.18
The first gas turbine that could produce net power by Jens William Ægidius Elling. (By Photographer anonymous – Norsk Teknisk Museum, Public Domain, https://commons.wikimedia.org/w/index.php?curid=13255410).

air was needed, such as blast furnaces in steel manufacturing factories or some processes in refineries. Then, the exhaust high-temperature gas flow from these processes was expanded in the turbine which in turn powered the compressor.

The idea of using gas turbines as a sole source of propulsion for an aircraft was first patented in 1930 by Frank Whittle (1907–1996, an English engineer and an air officer of the Royal Air Force (RAF) of Britain). In August 1939, a few days before the beginning of World War II, the first gas turbine-powered airplane was tested. It was a single-engine He178 military fighter that took off at Rostock-Marienche, Germany.

The first gas turbine for utility power generation was installed in Switzerland in July 1939 for an emergency power station by Brown, Boveri & Compagnie (BBC), with an electricity generation capacity of 4 MW and an efficiency of 17.4%[5] (all the performance data are for the inlet air temperature of 20°C). While there were more mature and proven technologies available at that time, two factors that were taken into account before making the decision about the plant were space needed for the system and required maintenance. Both factors were pushed by the concerns about war right before World War II. The low space was needed so that the system could be installed in a bomb-proof cavern underground. The proposed gas turbine was able to meet both requirements which overweighed

[5] The world's first industrial gas turbine set – GT Neuchâtel, A historic mechanical engineering landmark, https://www.asme.org/wwwasmeorg/media/resourcefiles/aboutasme/who%20we%20are/engineering%20history/landmarks/135-neuchatel-gas-turbine.pdf.

its low efficiency and the risk involved with the new technology. The system was a single-shaft simple cycle fueled with light fuel oil. The system had a TIT of 550°C and a back work ratio of about 75%. To reduce the high temperature of the combustion to the acceptable TIT, 400% excess air (500% theoretical air) was used. The pressure ratio of the 23-stage air compressor was about 4.4 with an impressive isentropic efficiency of about 85%. The 7-stage turbine had an efficiency of about 88%. The unit remained in service for 62 years before retiring in 2002. This unit is currently at display in Birr, Switzerland, and is designated as a Historic Mechanical Engineering Landmark by ASME (American Society of Mechanical Engineers) entitled "World's first successful electricity-generating gas turbine to go into commercial operation."[5]

A decade later in 1949, the first-utility gas turbine was installed in the U.S. by General Electric Company (GE) with the capacity of 3.5 MW in Oklahoma. This unit is also designated as the ASME historic landmark and is currently on display at Greenville (SC) where GE, the largest gas turbine manufacturer in the world, is currently manufacturing its latest gas turbines.

After World War II, BBC[6] was the leading company in the small but profitable gas turbine market; however, gradually GE took over. The major northeast blackout of 1965 that left tens of millions of people without power for hours in several states in the U.S. and Ontario in Canada provided an unexpected boost for the gas turbine industry in North America. Furthermore, the prefabrication of gas turbines and the use of standard components substantially reduced the manufacturing and installation time of gas turbine units as well as their costs. Later, the introduction of remote operation of units helped to increase the popularity of gas turbines. In the meantime, advancements in high-temperature alloys and turbine blade cooling technologies resulted in an increase in the TIT and compressor pressure ratio, which in turn increased the power production capacity and efficiency of gas turbine units. By the early 1960s, the TIT was improved to about 800°C. In the following 20 years, this temperature increased by about 300°C and by another over 200°C a decade later. By the 1970s, hundreds of units with the capacity of several GW were sold every year. The electricity generation capacity of units increased by more than fivefolds between the early 1970s and the 1990s. Larger and more efficient gas turbine units were built using two-shaft units equipped with reheater, intercooler, and regenerator. In the 1980s and 1990s, the growing concerns on air quality forced gas turbine manufacturers to move toward low-emission designs, particularly low NO_x units. The greatest development for the modern gas turbine technology has been the introduction and widespread utilization of CCPPs. These cycles will be extensively

[6] BBC merged with the Swedish company ASEA in the 1980s to form ABB. About 2000, Alstom, a French company, acquired power generation section of ABB. Eventually in 2015, General Electric (GE) purchased Alstom power generation business.

discussed in Chapter 10. While the first combined cycle was installed in 1967, their popularity came in the 1980s and the 1990s. Since then, natural gas fuel CCPPs have been one of the fastest growing technologies for utility electricity generation. Modern large heavy-duty gas turbines are mostly designed for integration to CCPPs. In the 21st century, gas turbines with the capacity of over 500 MW, TIT of 1600°C, and efficiency of over 40% have been built. A historic breakthrough for the power generation industry in general and the gas turbine technology in particular was in 2012 when a CCPP unit built by Siemens in Germany was able to achieve an **efficiency of 60.75%, breaking the "magic" efficiency barrier of 60%**. Currently, the aim of the industry is to increase the TIT to 1700°C to achieve the efficiency of above 43% in simple gas turbines and over 64% in CCPPs.

9.4 Applications of Gas Turbines

As noted earlier, gas turbines are the driving force in several industries. Generally, the gas turbine market is divided to **aviation** and **non-aviation** sectors. Aviation gas turbines are further divided into commercial and military applications. Non-aviation gas turbines are also divided into **electricity generation units**, **mechanical drives**, and **marine** applications. Currently, the majority of manufactured gas turbines, in terms of both the number of produced gas turbines and their monetary values, are for the aviation industry. While small in size, the military aviation sector is considered as a driving force for research and development in the field. In the non-aviation sector, the majority of gas turbines (about 80% in the monetary value) are for electricity generation, followed by mechanical drive applications (about 17%) and marine applications. It is predicted that the market value of gas turbines sold in both aviation and non-aviation applications will rise at least for the next decade.[7] In the rest of this section, various applications of gas turbines will be briefly reviewed before focusing on the gas turbines for electricity generation.

9.4.1 Transportation Applications

As noted in the introduction to this chapter, gas turbines have an exceptionally **desirable power-to-weight ratio**. This ratio is a critical characteristic for propulsion systems in transportation applications. The high power-to-weight ratio of gas turbines means that for the same power output, they are lighter,

[7] http://www.fi-powerweb.com/.

smaller, and more compact than other systems, which leaves more space and weight for cargo and passengers. This makes them very popular not only for airplane propulsion systems but also to provide power for civilian and military helicopters, ships, and trains. While in this text we are primarily interested in gas turbines for electricity generation applications, it is beneficial to have some insights to other applications, particularly aviation gas turbines for two main reasons. First, many advancements in the field, such as a higher TIT and compressor pressure ratios, are initially developed for and introduced to aviation gas turbines. Second, some aviation gas turbines have been modified to be used for electricity generation, which are commonly referred to as aeroderivative gas turbines.

Unlike other gas turbine applications, the ultimate objective of aviation gas turbines is not to generate shaft work, rather to produce thrust, albeit sometimes via shaft work production. There are three common configurations for aviation gas turbines: turboprop, turbojet, and turbofan jet engines. **Turboprop** engines are in principle similar to electricity generation gas turbines where the objective is to generate shaft work. In these engines, combustion products from the combustion chamber are expanded in the turbine to the atmospheric pressure (Figure 9.19). A portion of the produced work is used to power the compressor, and the remaining is used to run a propeller via a reduction gear to generate thrust for the aircraft. These engines are suitable for airplanes flying at relatively low speed (about 700 km/h) and low altitude. At these conditions, they can achieve better fuel efficiency per passenger than other engines. While many turboprop engines are made every year, due to their usually small size, their share of the aviation gas turbine market, in terms of the monetary value, is very small.

In **turbojet** engines, the turbine extracts sufficient energy from combustion products to run the compressor. Therefore, the turbine exhaust flow is at substantially higher pressure than the atmospheric pressure, which is

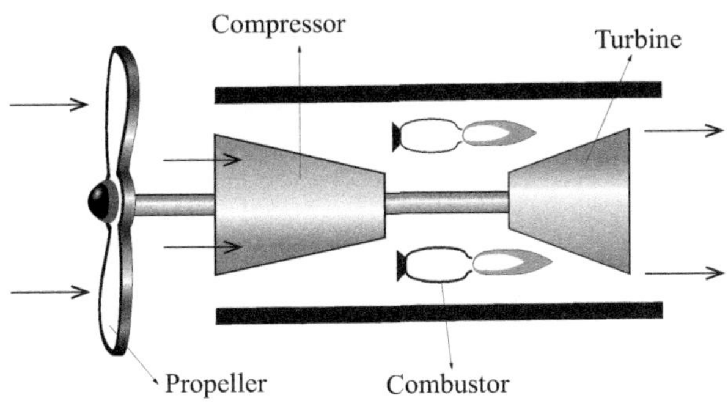

FIGURE 9.19
Schematic of a turboprop engine.

used to produce the entire thrust for the airplane. For this reason, besides a gas turbine (also called the gasifier or the gas generator section), these engines are equipped with a nozzle, common known as a propelling nozzle, downstream of the turbine to accelerate the flue gas leaving the turbine (Figure 9.20). The high-velocity flow upon leaving the nozzle can produce thrust. Furthermore, since the inlet flow to the compressor is at high velocity (due to the relative velocity of the aircraft with respect to surrounding air), a diffuser is added to the upstream of the compressor. The diffuser decelerates the inlet air flow and converts its kinetic energy to pressure, which in turn reduces the load of the compressor. To produce even greater thrust, some turbojet engines are equipped with an afterburner between the turbine and the nozzle to add more energy to combustion products before the expansion process in the nozzle. While turbojet engines powered many early jet airplanes, due to their low efficiency, they are not used in the majority of modern airplanes.

In essence, **turbofan** engines, also known as **fanjet** engines, are a combination of turbojet and turboprop engines. Similar to turbojet engines, they have a gas turbine engine at the core. However, they also have a large-diameter ducted fan at the front powered by the gas turbine (Figure 9.21). The fan accelerates air which bypasses the core gas turbine and directly contributes to the produced thrust. While typically the core gas turbine provides the needed thrust during cruising, the combination of both the core engine and the fan provide the thrust during takeoff. Turbofan engines are more efficient and quieter than turboprop engines and are used in the majority of commercial and military airplanes. They may also be equipped with an afterburner. Most advanced turbofan engines are equipped with two or three shafts, commonly referred to as 2-spool and 3-spool engines, respectively. In 2-spool turbofan engines, the fan and low-pressure stages of the compressor are on

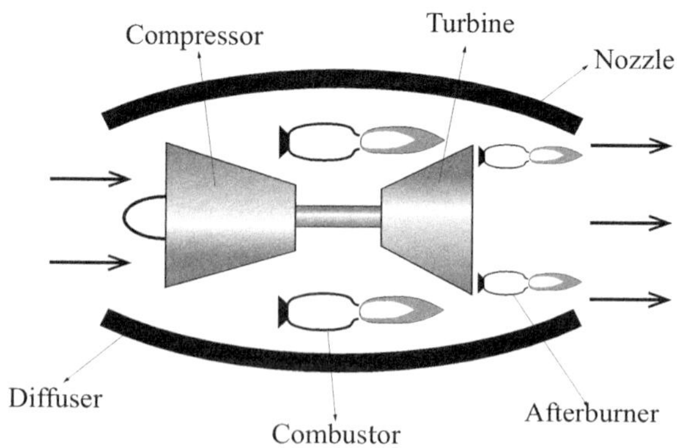

FIGURE 9.20
Schematic of a turbojet engine.

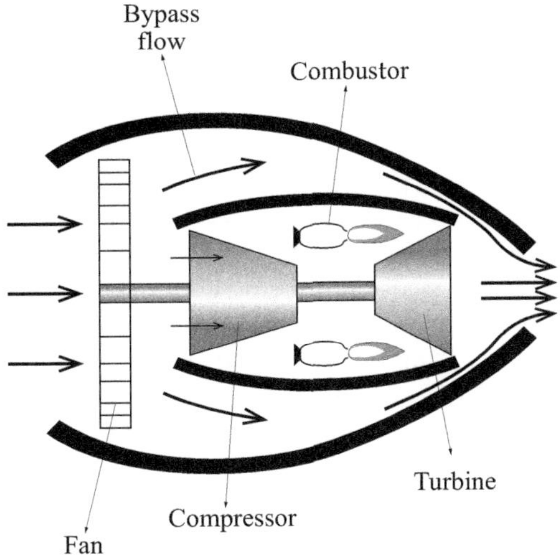

FIGURE 9.21
Schematic of a turbofan engine.

the same shaft as the low-pressure turbine, while the high-pressure stages of the compressor and turbine are on a different shaft. In this design, since the fan and low-pressure stages of the compressor have drastically different diameters, they have different optimal rotational speeds (rpm). This means either the fan or the low-pressure stages of the compressor operate at nonoptimal conditions. To address this problem, in 3-spool turbofan engines, the fan is separately driven by the low-pressure turbine. While 3-spool turbofan engines are more efficient, they are more complex, heavier, and more expensive. Currently, about one-half of the manufactured aviation gas turbines in terms of the number of produced engines and over 90% in terms of their value are of this type.[8]

Turboshaft engines are gas turbines that generate shaft work for transportation applications, such as helicopters, locomotive, auxiliary power units (APUs), tanks, hovercrafts, and marine applications including ships and boats. In marine applications, gas turbines are also used for shipboard use. Sometimes the usage of the term "turboshaft engines" is extended to include power generation applications.

Finally, gas turbines are also very common in military applications. Other than military jet airplane, they are used in marine applications, helicopters, and even in tanks (Figure 9.22).

[8] http://www.fi-powerweb.com/Aviation-Gas-Turbines.html.

FYI BOX 9.1

UTILIZATION OF GAS TURBINES IN TANKS

Leonardo da Vinci has been credited as the first person to conceive the concept of a mobile-armored military fighting vehicle in 15th century although there is no record that he actually made one. Tanks were first used in World War I by British and German armies. These tanks were powered by spark-ignition internal combustion engines fueled by gasoline, thus the name "gasoline or petrol engines". For example, in 1917–1918, German A7V tanks used two four-cylinder petrol engines with the power of 75 kW (about 100 hp) each. In World War II, tanks were main tools in Germany's Blitzkrieg tactics. During and after World War II, diesel engines were the primary power source of tanks (with the output power of up to about 1100 kW (about 1500 hp)). Even now, diesel engines are used to power many modern tanks. The idea of using gas turbines to power tanks (Panther tanks in particular) was first introduced in Germany in 1944, but the end of World War II stopped further development. First actual use of gas turbines in tanks was in a Swedish tank, Stridsvagn 103B. In the initial design (in the mid-1950s), the tank was powered by a diesel engine but the tank was proved to be underpowered. So a complementary gas turbine (made by Boeing) was added, though later it was replaced by a second diesel engine. By the 1980s, the dominant mentality in the U.S. Army was that diesel engines reached their technological limits. The gas turbine technology was the main candidate to replace diesel engines mainly due to its desirable power density (power-to-weight ratio), as it was proven in the aviation industry. As a result, M1 Abrams tanks began to be produced in the U.S. in 1980 (Figure 9.22). They were powered by AGT-1500 gas turbines capable of producing 1120 kW (about 1500 hp).

FIGURE 9.22
M1 Abrams tank with gas turbine propulsion system. (Image by Alexander Lesnitsky from Pixabay, https://pixabay.com/photos/tank-abrams-m1-us-tank-abrams-tank-1530098/.)

These gas turbines significantly improved the reliability and maneuverability of tanks. These tanks were able to reach the top speed of about 70 km/h (about 45 mph) in a few second. There were also much quieter than diesel-powered tanks (they were nicknamed "whispering death"). They also enjoyed other advantages such as their capability to be fueled by a range of fuels (unlike diesel engines which are very sensitive to fuel type and its quality), excellent cold-starting performance, light weight which allowed added armor and more protection, less moving parts which resulted in less maintenance and repair, and lower cost. However, their efficiency was lower than that of diesel engines so their fuel consumption was higher. To improve the efficiency of the cycle, they were equipped with a recuperator to recover some wasted thermal energy from the exhaust flow to heat compressed air and reduce fuel consumption. This tank has been improved in the proceeding models, including M1A1 Abrams and M1A2 Abrams, using the new generation of gas turbines (LV 100–50), which are lighter, more efficient, and more reliable. Gas turbines are still being produced and are in service in the U.S. Army.

9.4.2 Mechanical Drive Applications

Gas turbines are commonly used to provide shaft work for a variety of applications. These units are known as **mechanical drive applications** but sometimes are referred to as **industrial gas turbines**. They are widely used in petrochemical plants and other industrial plants to power large pumps and compressors. They are also commonly used in offshore drilling rigs fueled with flare gas. Today, many nations have a national pipeline network to distribute natural gas. In order to have a continuous flow of gas in the pipeline, the pressure of natural gas needs to be boosted in a compression station every 80 km (about 50 miles) or so. Natural gas-fueled gas turbines are used to drive large compressors in these stations. These are usually small- or medium-sized gas turbines with the power production capacity of 2–25 MW. Since their objective is to maximize the shaft output power, their design principles are very similar to electricity generation gas turbines. As a result, commonly aeroderivative or small heavy-duty gas turbines (these gas turbines will be discussed later) are modified for these applications. Since in many applications, output power is needed at a wide range of speeds or units need to be operated at part-load conditions, gas turbines in mechanical drive applications often have split-shaft designs. Some stages of the turbine are connected to the compressor, and the remaining stages provide output power. The former section is called the gasifier section, and the latter is referred to as the power section. This way the gasifier section operates at optimal conditions for the compressor and the power section operates according to needs of the external load. In some cases to improve the unit efficiency, they are equipped with a heat recovery heat exchanger (recuperator) to provide a required processing steam or as a regenerator.

9.4.3 Gas Turbines for Electricity Generation

From its first conception, gas turbines have been a major player in the electricity generation industry. The largest segment of non-aviation applications of gas turbines is the utility electricity generation industry. Their quick start-up (as low as several minutes) and fast load-changing capability make them particularly suitable as utility power plants for peak-load power generation. However, nowadays they are being used for intermediate and some baseload duties, as well as CCPPs. They are also popular as backup and emergency units as well as APUs. More recently, they are being utilized in distributed power generation. Gas turbines are also used for starting large power plants when the grid cannot provide initial electricity needed for start-up. Similarly in this situation, when electricity from the grid is not available, gas turbines can provide electricity needed for the safe plant shutdown, for example, for lubrication and cooling of sensitive components. Based on the basic design of gas turbines, they can be categorized into two groups: **aeroderivative** and **non-aeroderivative** gas turbines.

9.4.3.1 Aeroderivative Gas Turbines

Aeroderivative gas turbines, as the name implies, are gas turbine units for ground applications that use the modified design and components that are originally created for aviation applications. As noted earlier, most research and development activities on gas turbines are dedicated to aviation applications and most innovations are first introduced to these applications. Many of these developments are eventually used by other gas turbine applications; however, it makes sense to directly adapt these advancements to electricity generation gas turbines. The concept of using airplane gas turbines for ground applications is not new, and since the 1960s, aeroderivative gas turbines have been used for electricity generation. Recent tremendous improvements in turbofan engines have just accelerated this trend.

To convert a turbojet and turbofan engines to an aeroderivative ground gas turbine, several modifications should be implemented. The fan, diffuser, and bypass flow duct should be removed. A few extra stages of blades should be added to the low-pressure section of the compressor (in front of the compressor) to make up for the compression effects of the fan and the diffuser. A few stages of expansion should be added. This is typically achieved by adding new stages at the low-pressure section of the turbine. The new turbine section is connected to the generator by a shaft that is independent of the existing turbine sections from the original turbofan engine (referred to as the gas generator). For this reason, the added turbine is usually referred to as a free-power turbine. This configuration helps the compressor and turbine from the original design and the free-power turbine to operate at their optimum rotational speeds. The combustion chamber should be modified so it can burn the fuel available for the ground application, commonly natural gas or diesel fuel. However, these units cannot burn heavy fuels, and if these are the only available fuels, heavy-duty gas turbines should be used (below).

Aeroderivative gas turbines have typically high efficiency of up to about 46% and can be typically used for the power output range of between 20 and 100 MW. These units have high potential for the incorporation of regenerators or heat recovery systems. Since aeroderivative gas turbines can start in a very short time, in a couple of minutes from standstill to their full load, and can easily handle rapid load changes, they are ideal as peak-load units to stabilize the grid as well as for emergency and backup units. Their other favorable characteristics are their low capital cost, low maintenance, and simple remote operation.

9.4.3.2 Non-Aeroderivative Gas Turbines

Unlike aeroderivative gas turbines, these gas turbines are specifically designed for ground electricity generation applications. Therefore, these gas turbines are designed with requirements of electricity generation applications in mind. For example, for these units, size and weight are not considered design constraints. However, other parameters, such as long life (durability), high availability, high efficiency, low emissions, fuel flexibility, and less noise level, are considered critical for this type of gas turbines. Unlike aeroderivative gas turbines, in these gas turbines, the external diameter of the unit (its frontal area) is not limited so they can have a large frontal area. This reduces the velocity of the inlet air flow, which in turn lowers the noise level of these units compared to aeroderivative gas turbines. As noted in the historical review of gas turbines, these units were developed parallel to aviation and aeroderivative gas turbines. In their early development stages, they commonly used technologies that were developed for steam power plants but gradually they moved toward aeroderivative gas turbine technologies. Nowadays they share and exchange latest technologies with aeroderivative gas turbines.

The capital cost per kilowatt of these units is a function of their electricity generation capacity. Therefore, they are categorized according to their size to the following groups:

- Large-sized gas turbines
- Medium-sized gas turbines
- Small-sized gas turbines
- Micro gas turbines

Large-sized gas turbines are commonly referred to as **heavy-duty** or **frame-type gas turbines**. Generally, these gas turbines have an electricity generation capacity of above 100 MW. Currently, the largest units in this category have the rating of above 500 MW. These units are extensively used for peak-load and sometimes baseload electricity generation. They can achieve high pressure ratios of 40 with up to 25-stage axial compressors. They can also have a TIT up to low 1600°C using the latest blade cooling technologies. The components of these gas turbines will be extensively discussed in the coming sections.

Medium-sized gas turbines have a rating of roughly between 5 and 100 MW. For the higher range of power generation capacity, roughly between 20 and 100 MW, this type of gas turbines shares the market with aeroderivative gas turbines. Small-sized gas turbines commonly have the rating below 5 MW. Both medium- and small-sized gas turbines share many aspects of large gas turbines. They can be used for peak-load utility applications, distributed electricity generation, emergency and backup power units, and axillary power units. Generally speaking, smaller gas turbines are less complex but less efficient than larger gas turbines due to lower TITs and less-efficient components. To improve their efficiency, they may be used as cogeneration units.

9.4.3.3 Micro Gas Turbines

Micro gas turbines, or simply **microturbines**, have a power rating of less than 500kW. They can be fueled by a variety of fuels, including natural gas, diesel, biodiesel, petrol, and propane. In the late 1990s, there were great expectations that microturbines can revolutionize the distributed power generation sector. However, they have not been able to meet those expectations. For widespread applications for distributed electricity generation, microturbines must operate efficiently with low emissions at reasonable capital costs. However, they are still expensive relative to other technologies. Their efficiency is about 15%–18% (in simple configuration with no regeneration or heat recovery), which results in high emissions. The emission level is critical for urban applications since these units need to operate close to population centers. Other requirements for microturbines are low maintenance, rapid start-up and load change, quiet operation, and compactness. To improve their efficiency, they can be equipped with a regenerator or a heat recovery system. They commonly have a radial compressor and turbine on single- or double-shaft rotating at very high speed, in some cases above 100,000 rpm. Regardless of these disappointing results, there are hopes that in the future, microturbines can greatly contribute to electricity generation.

9.5 Gas Turbine Components

In this section, we will review the most important components of gas turbines for electricity generation.

9.5.1 Compressors

Compressors are one of the three main components of gas turbines where the inlet air flow is pressurized. As previously noted, a significant portion of the power generated in the turbine is used to drive the compressor so the slightest inefficiency in the compressor design and manufacturing can have substantial impacts on the overall performance of the system. Compressors used in the gas turbine units can be either axial or centrifugal. Almost all large- and medium-sized gas turbines, including aeroderivative units, use axial compressors, while centrifugal compressors are more common for small-sized gas turbines and microturbines.

9.5.1.1 Axial Compressors

In **axial compressors**, as the name implies, the direction of the inlet and outlet flows, and the flow through the compressor are all **parallel to the axis of rotation** (the axial direction). As noted in Section 9.3, from the beginning, gas turbines used axial compressors. The very first-utility gas turbine, installed in Switzerland in 1939, was equipped with a 23-stage axial compressor with the pressure ratio of 4.4. In modern gas turbines, the pressure ratio of compressors can be as high as 45. However, large-utility compressors typically have more moderate pressure ratios of 14–35 (refer Section 9.2.2 for the reason). The isentropic efficiency of these compressors is from high 80s to low 90s.

Similar to axial steam turbines (Chapter 6), axial compressors have alternative rows of moving blades (or vanes), known as the rotor, and fixed blades, known as the stator. An axial compressor for a gas turbine is made of several stages (for large gas turbines between 14 and 25), and each stage has one moving blade row attached to the main shaft and one fixed blade row attached to the casing. The air pressure is slightly increased in each stage. The moving blades impart energy to the passing air flow, which accelerates the flow and increases its velocity and kinetic energy. Then, the diverging fixed blades deaccelerate the flow by acting like a diffuser that converts the kinetic energy of the flow to pressure (unlike turbine fixed blades that are converging and act as a nozzle). The fixed blades also adjust the direction of the flow so it can enter the next stage of moving blades at an optimum angle. The pressure ratio of each stage of modern compressors can be between 1 and 1.5. Aeroderivative compressors are typically operated on the higher end of the range and utility compressors on the lower and middle end of the range. This incremental pressure increase helps these compressors to achieve very high efficiency. Besides all the fixed blade rows between the moving blade rows, there is a row of fixed vanes, commonly known as **inlet guide vanes (IGVs)**, before the first moving row to induce a swirl motion in the inlet air flow so the flow enters the first stage at an optimum angle. The pitch angle of these vanes is typically variable. Similar to the IGV, the initial few rows of fixed blades, up to five rows, are typically equipped with variable pitch blades.

This capability helps them to adjust the flow direction so the air flow enters moving rows at an optimum angle during part-load operation and when the flow density varies due to change in ambient conditions. Finally, after the last row of moving blade, there is another row of fixed vanes, known as **exit guide vanes (EGVs)**, to provide one final diffusion process and also to control the air flow so it can enter the combustion chamber at desired conditions, especially a proper velocity.

Fundamentally, the design of compressor blades is very similar to the design of steam turbine blades, as discussed in Chapter 6. The main difference is that compressor blades impart energy to the flow rather than extracting energy from it. Figure 9.23 shows the variation of velocity and pressure throughout various fixed and moving rows of an axial compressor. This is very similar to Figure 6.14 for a reaction steam turbine; only here the pressure increases while the flow advances axially through the compressor. Note that, similar to a reaction steam turbine, the velocity decreases in the fixed rows and pressure continuously increases in both moving and fixed rows. Also note that since no energy is added in fixed rows, the total pressure (which is the sum of pressure and kinetic energy) remains constant, while

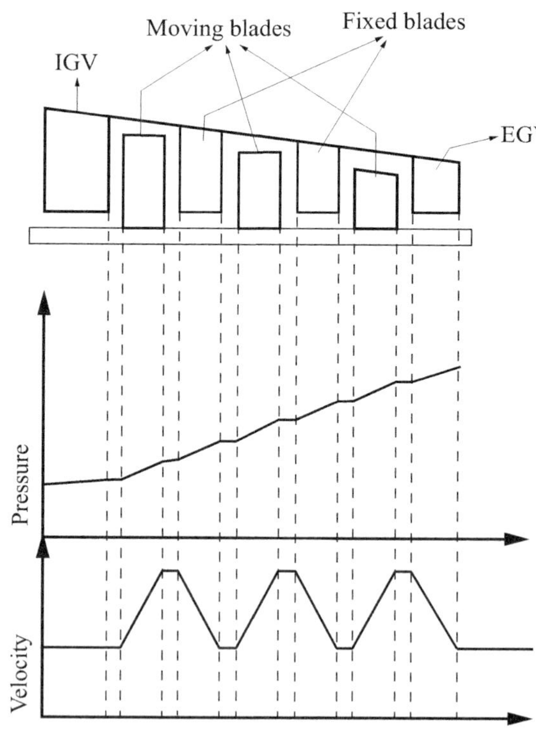

FIGURE 9.23
Variation of velocity and pressure throughout various fixed and moving rows of an axial compressor.

the pressure increases even in fixed blades because of conversion of kinetic energy to pressure (via the diffusion process). The rule of thumb on the distribution of the load on various stages of the compressor is that the total energy (enthalpy) that needed to be added to the flow should be (roughly) equally distributed among stages. This may not be reflected in equal pressure increase in stages.

To evaluate the distribution of pressure increase in fixed and moving rows within each stage, a parameter called the degree of reaction is used. The degree of reaction is defined as the ratio of increase in the flow pressure in the moving blades to the pressure increase in the entire stage (i.e., the sum of pressure increase in the moving and fixed blades). A 100% degree of reaction means that the entire pressure increase of the stage occurs in the rotor, i.e., an impulse stage. A common degree of reaction for axial compressors used in aeroderivative gas turbines is 50%. For heavy-duty gas turbines, higher degree of reaction is typically used. A higher degree of reaction means that more kinetic energy is converted to pressure in the rotor so the flow leaving the rotor and entering the stator is at lower velocity. Note that the fixed blades should change the direction of the flow so the flow can enter the moving blades at an optimum angle. Therefore, a lower velocity is more desirable in the stator because it reduces turbulence in fixed blades and improves the overall efficiency of the compressor. However, this requires more stages of compression for a given pressure increase, which in turn increases the size and weight of the compressor. For ground gas turbines, the size and weight are not critically important factors, so high degrees of reaction are common, whereas for aviation gas turbines, these two factors are critical, so lower degrees of reaction are more common.

As it is clear in Figure 9.23, the height of blades increasingly reduces along the axis of the compressor (the higher-pressure blades are shorter). This is to maintain the velocity of the air flow at various stages of the compressor. Since the pressurized air is denser, to keep velocity constant, the area of the flow and as a result the height of blades should reduce (according to this equation, $V = \dot{m}/\rho A$).

The early axial compressors had untwisted blades with a symmetrical profile that did not change throughout various stages. However, this has substantially changed in the past few decades. Particularly, the applications of computational fluid dynamics (CFD) have provided tools to design optimum blades for axial compressors. Modern compressor blades are on one side concave and on the other side convex (Figure 9.24). Since the compressor rotates toward the concave side, this side is the front of blades and is called the pressure side, while the convex side is behind blades and is called the suction side. These blades are typically made of special alloys of stainless steel and usually are coated by metal and/or ceramic protective coatings.

To design the profile of an airfoil (aerofoil) for an axial compressor, velocity vector diagrams can be used. The principles of velocity vector diagrams for axial compressor blades are similar to those described for steam

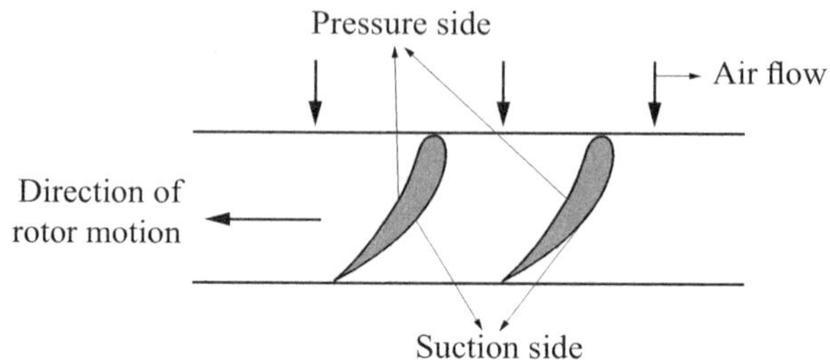

FIGURE 9.24
Pressure and suction sides of a moving blade row.

turbine blades (in Chapter 6). Note that nowadays these diagrams are usually used for preliminary design steps and detailed designs of modern compressor blades are based on the radial equilibrium theory.[9]

Traditionally, the shape of compressor blades is designed based on preexisting airfoils. The most well-known and common airfoil profile categorization is called NACA. In the first half of the 20th century, the National Advisory Committee for Aeronautics (NACA)[10] developed and tested many airfoils. The curvature of these airfoils is expressed in the form of several numerical values. Accordingly they can be recreated using these parameters. These profiles have been published, and their properties, particularly their drag and lift coefficients, have been used to design airfoils in many industries, including axial compressor blades. For each airfoil, these parameters, most notably drag and lift coefficients, vary with the angle of attack of the airfoil with respect to the direction of the unaltered flow, the Reynolds number, and the roughness of the airfoil. Nowadays, there are many other airfoil databases that are available for this purpose. Note that the drag and lift coefficients are defined by the following equations:

$$C_D = \frac{F_D}{\frac{1}{2} A \rho V^2} \tag{9.66}$$

$$C_L = \frac{F_L}{\frac{1}{2} A \rho V^2} \tag{9.67}$$

where C_D and C_L are the drag and lift coefficients, respectively; F_D and F_L are the drag and lift forces, respectively; ρ is the fluid density; A is the characteristic area of the airfoil; and V is the velocity of the fluid flow.

[9] This topic is out of scope of this text. For more information on the topic, refer to turbomachinery design textbooks.

[10] The National Advisory Committee for Aeronautics (NACA) was an agency in charge of the U.S. federal government's activities related to aeronautic in the first half of the 20th century. In the late 1950s, it was replaced by the newly formed National Aeronautics and Space Administration (NASA).

9.5.1.2 Centrifugal Compressors

In **centrifugal compressors**, the flow of air through the compressor is radial (hence the other common name **radial compressors**). This type of compressor is common for small gas turbines and microturbines. A combination of axial and centrifugal compressors, sometimes referred to as axi-centrifugal compressors, are also used in some small gas turbines (typically for units at the higher end of the capacity range for small gas turbines) where the air flow first goes through several axial stages followed by a centrifugal one. While centrifugal compressors are generally less efficient than axial compressors, they have a higher-pressure ratio per stage, which makes it possible to achieve high enough pressure for the combustion chamber with even one-stage compression. They can also be safely operated with a wide range of operating conditions, which is an important attribute for load-following units. Furthermore, they are relatively simpler to design and manufacture, more reliable, and shorter in length compared to axial compressors with the same pressure ratio. Finally, they are less sensitive to solid deposit buildup on blades.

In centrifugal compressors, the axial air flow enters the center (known as the eye) of the rotor (also known as the impeller) (Figure 9.25). These compressors may or may not be equipped with IGVs. The flow direction is rotated by 90°, from the axial flow to the radial flow, while going through the inducer section of the impeller. The flow is then accelerated due to the energy imparted to the flow on the exducer section of the rotating impeller. The impeller also acts as a diffuser since with an increase in the radius, the area normal to the air flow increases. The impeller can be either shrouded (also referred to as open)

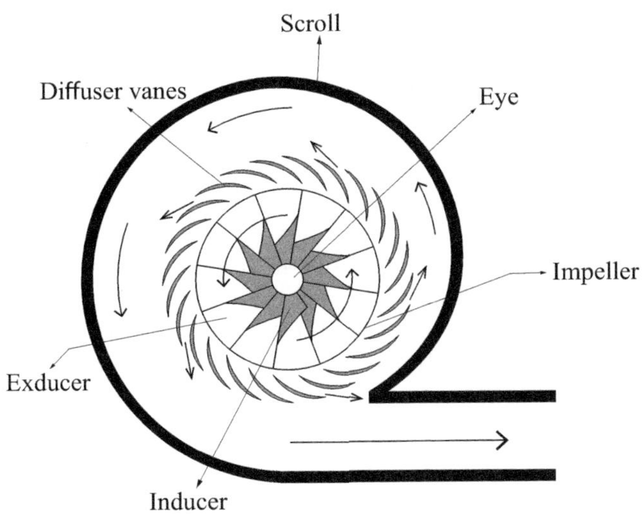

FIGURE 9.25
Schematic of a centrifugal compressor.

or unshrouded (closed). Depending on the orientation of vanes, impellers can be radial, backward-curved (or backward-swept), or forward-curved (or forward-swept) (Figure 4.38). Backward-curved impellers are more common than the other two types. The flow leaving the impeller is further diffused through the diffuser, which is made of diverging vanes. The conversion of kinetic energy to pressure is often equally divided between the impeller and the diffuser. The flow from the diffuser is exhausted to the scroll, volute, or collector where it is led to the outlet of the compressor and eventually to the combustion chamber.

9.5.1.3 Compressor Performance

Like every other component, compressors must be designed to operate with their maximum efficiency at design conditions and operate safely at off-design conditions. Most other systems can safely operate at off-design conditions, albeit at lower efficiency than their design conditions. However, if not properly designed, the operation of compressors at certain off-design conditions can lead to catastrophic consequences. Off-design conditions in a compressor can be encountered not only during part-load or overload operation of gas turbine units, but also during start-up and shutdown sequences (discussed later in this chapter), variation in ambient conditions (particularly ambient temperature), or any other unexpected operational conditions. Note that the discussions in this section are applicable for both axial and centrifugal compressors.

In order to understand the performance of a compressor at off-design conditions, we need to introduce the performance curve of compressors. The performance curve of a compressor shows the relationship between the pressure rise (head) across the compressor and the mass flow rate of the gas passing through the compressor (similar to that of pumps discussed in Section 7.3). The axes in performance curves are sometimes expressed as dimensionless numbers or percent of design values relevant to each parameter. It can be shown that the relationship between pressure head and mass flow rate at an ideal condition with no losses is linear (Figure 9.26). To get an actual performance curve, clearly flow losses should be taken into account. The most important losses are frictional, incident, and clearance losses. The frictional losses are due to fluid friction while passing through the compressor and are a function of the square of the air flow velocity. Note that even in the optimum operating condition, friction losses are present. The magnitude of these losses rises with an increase in velocity (and mass flow rate). The **incident losses** are due to misalignment of the gas flow with respect to the surfaces of the blades (both moving and fixed blades). They are sometimes referred to as **shock losses**. This naming might be misleading because there is no relationship between these losses and the shock phenomena that occur in supersonic compressible flows. Since in the optimum condition the blade contours are designed to be parallel to the flow direction, these losses are

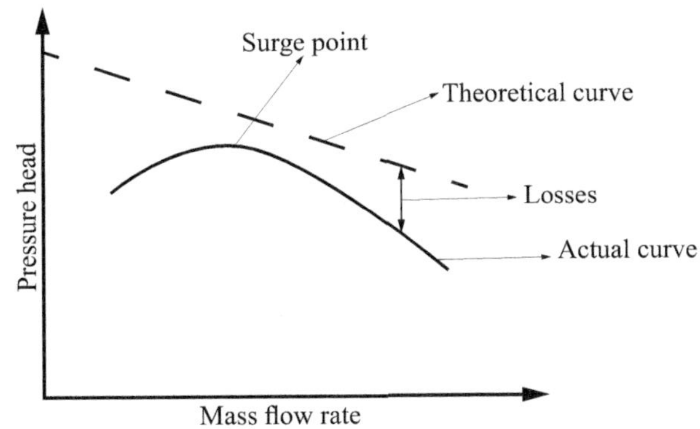

FIGURE 9.26
Performance curve for an ideal and actual compressor.

minimal at the design condition. The clearance or leakage losses are due to the gas flow through the gap between the blade tips and the casing (scroll). The actual typical performance curve, also known as the performance map, is also shown in Figure 9.26. Due to the complexity of the flow in real compressors, these curves are always created from experimental values from compressor performance tests. The performance curves are typically created for various shaft speeds and include the islands of constant efficiencies.

Before continuing our discussion on the compressor performance curve, let us consider incident losses in more detail and how they contribute to stall in compressors. Figure 9.27 shows an airfoil in the flow of air with varying attack angles (the angle between the unaltered flow and the blade) and the diagram of the lift and drag forces as a function of the attack angle. As the

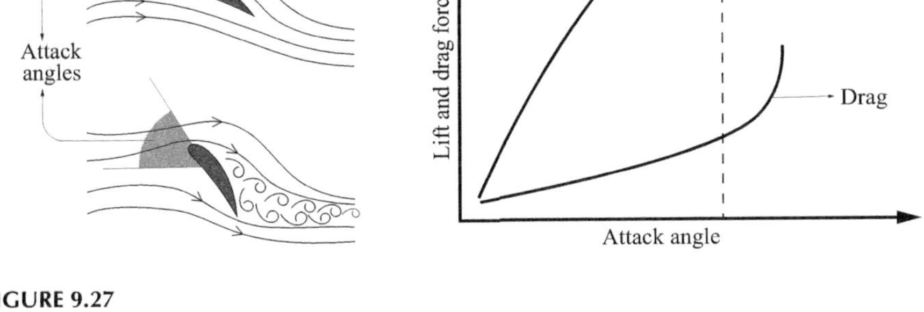

FIGURE 9.27
Relationship between attack angles and lift and drag forces.

diagram shows the increase in the attack angle, both lift and drag forces increase. At a certain attack angle (perhaps around 20°), the lift force reaches a maximum. If the attack angle is further increased, the flow separation occurs, which results in a sharp decline in the lift force and a significant rise in the drag force. **This separation limits the pressure increase in each stage of a compressor and is the reason for compressors needing many stages of compression.** This phenomenon is known as the **blade stall**. Stall can happen whenever an airfoil and gas are in relative movements. Stall has been the cause of some high-profile airplane accidents, including Air France Flight 447 in 2009 killing all 228 passengers and crew on board and Colgan Air Flight 3407 crashed in New York in 2009 killing 50 people.

In compressor blades, stall typically happens locally and is due to a mass flow rate either much higher or much lower than the design mass flow rate. In Figure 9.28, let us assume blade A is the first one to stall due to the excessive angle between the blade and the air flow. Note that stall can occur in the entire length of the blade or just a portion of it. Due to the flow separation in the suction side of the blade, this stalled blade cannot impart sufficient kinetic energy into the air flow. This means the air pressure cannot be increased as expected and air cannot flow through the passage between this blade and the adjacent one. The blocked air has to be diverted around the blades next to the stalled blade, which in turn will stall them. This mechanism moves stall around the stage at the rotation rate of about half of the rotation speed of the shaft. This stall mechanism can cover several blades of a stage at once and is called **rotating or propagating stall.**

It should be noted that a compressor can have one or more stages stalled, while other stages are unstalled. For example, consider a case when the air flow rate reduces compared to the design flow rate. In the first stage of compression, for almost the same power induced to the flow, there will be less

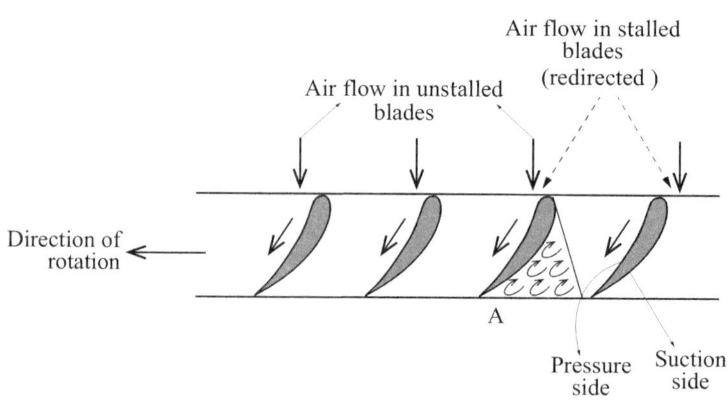

FIGURE 9.28
Rotating or propagating stall of a compressor stage.

air flow. This results in a higher-pressure rise in the stage than that in the design condition, which in turn increases the air density and reduces the air volumetric flow rate. This will result in even more pressure rise in the second stage, and so on. The accumulation of this effect causes the final stages of the compressor to stall, while the early stages are unstalled. Similar phenomena can occur when the air flow rate is higher than that in the design point (but in the opposite direction). While stall negatively impacts the efficiency of a compressor and may cause some noise, a compressor can operate with one or more stages stalled.

As noted, stall is due to local flow separation in blades. If the flow separation is expanded to an entire stage, it is known as the **compressor surge**. Similar to stall, surge hinders the ability of the stage to pressurize the flow. However, unlike stall that the air flow could be diverted to other blades, surge in a compressor stage results in most of the air flow not being able to pass through the stage. As a consequence of this temporary blockage, the pressure of the flow right after the surged stage abruptly drops. If this pressure is lower than the pressure of the already-compressed air downstream of the surged stage, the air flow will reverse. This means after the surged stage the air flow will be from the compressor outlet toward its inlet. But this effect is temporary because the discontinuation of the air flow and the air flow reversal gradually reduces the pressure downstream of the surged stage. This will eventually rebuild a sufficient pressure gradient to restore the air flow. However, if corrective actions are not taken, this will lead to another cycle of surge. As you can imagine, in this situation the compressor cannot deliver the required compressed air. But the consequences of compressor surge can be much more serious. As a result of the air flow reversal, all the forces imposed on the blades are reversed, which can create excessive vibration and noise. It can also damage compressor bearings. In a worst-case scenario, continuous surge can overwhelm bearings that results in the axial movement of the shaft and the collision of fixed and moving blades and the destruction of the compressor.

Beyond the compressor, the shortage of oxygen in the combustion chamber can cause flameout or incomplete combustion, which results in unburned fuel in the combustion products. The visible effect of the presence of unburned fuel in the exhaust is sudden explosive combustion in the exhaust flow where there is enough air available to complete the combustion process.[11] Compressor surge can have very serious consequences for aviation gas turbines. In some cases, the engine stall was initiated by the engine striking a bird. For example, the U.S. Airways Flight 1549 accident in 2009, which later became known as the "Miracle on the Hudson", was caused by gas turbine stall due to a Canadian geese striking the engine.

[11] This is more significant in aviation gas turbines. There are many videos in youtube.com showing this phenomenon in airplane engines.

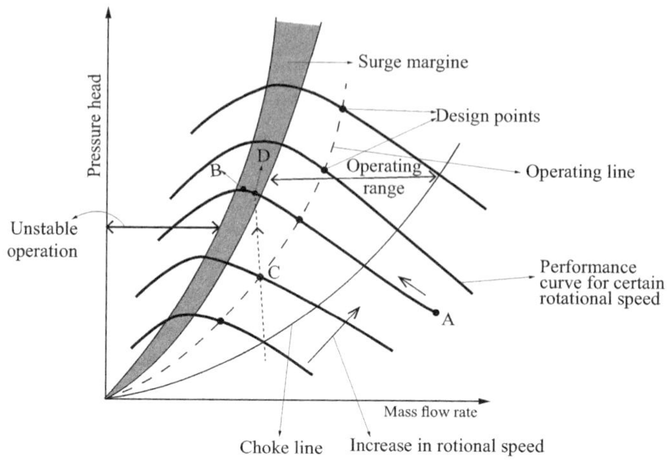

FIGURE 9.29
Compressor performance curve, surge margin and choke line.

The compressor surge can also be explained using the compressor performance curve (Figure 9.29). Starting from the bottom-right corner of the curve (point A), if we reduce the air mass flow rate, the outlet pressure of the compressor increases. As long as the pressure increases, the compressor can push the inlet air out and can continue a stable operation. However, at point B, the pressure is maximum, and if the air mass flow rate reduces beyond this point, the outlet pressure starts to reduce. Theoretically, for mass flow rates lower than the mass flow rate of point B, the compressor can surge. When the mass flow rate further reduces beyond point B, the mass flow rate and the outlet pressure reduce simultaneously. This means the compressor cannot push the air out because the pressure that it can generate is lower than the pressure of the air already pressurized. This can result in the air flow reversal in the compressor. In reality, considering many parameters that can impact the performance of the compressor, there is a surge margin. Manufacturers include a surge limit in the performance curve of compressors, which is at or near the point where the slope of the curve is zero. As noted, the performance map includes the performance curves for the compressor operating at various shaft rotational speeds. The surge line can be created by connecting surge points for various performance curves. Similarly, connecting design operating points at various shaft speeds will form the compressor operating line. The operating line of a compressor must be always to the right of the surge line with a certain safety margin.

The compressor performance curve in Figure 9.29 indicates that for a given rotational speed of the compressor, reducing the air mass flow rate can result in the compressor surge. Similarly for a given mass flow rate, increasing the rotational speed can cause surge (moving from point C to point D in Figure 9.29).

So far we learned that the performance of a compressor is limited at a low mass flow rate by compressor surge. On the other end, the performance of a compressor is limited at high air mass flow rates by **compressor choke**. The choke phenomenon occurs for a compressible fluid flowing through a nozzle or any other restrictions. When the pressure difference between the upstream and downstream of the restriction rises, the flow rate through the restriction, and hence the flow velocity, increases. This continues until the fluid velocity reaches the speed of sound (Mach number = 1). This is the maximum flow and velocity achievable for a given condition and reducing the pressure of downstream will not increase the flow rate anymore. The flow in this condition is called the choked flow.

In the compressor performance curve, the choke line shows when the flow through the compressor is choked (Figure 9.29). For a given compressor operating at a given rotational speed, the fluid mass flow rate is maximum at the choke point and it cannot be increased beyond this point. This point is also referred to as **stone wall**. While operating a compressor at choke conditions can significantly reduce its efficiency, unlike compressor surge, it does not lead to a physical damage to the compressor.

Referring back to Figure 9.29, we can observe that the surge margin defines the lower limit of the air mass flow rate through a compressor and the choke line defines its upper limit. The operating line of a compressor can be located between these two lines. For a given rotational speed, the compressor's stable operating range can be only between the stall and choke points. Note that designers try to choose a compressor in which the operating line is as close to the surge margin as possible while keeping a safe distance from the surge margin. Remember that when the operating line is away from the surge margin, it means the pressure increase of the flow is less than what the compressor can potentially provide, which in turn indicates oversizing the compressor (i.e., lower loading per stage).

Using multiple-shaft (twin- or triple-spool) gas turbine configurations, variable inlet guide vanes (VIGVs), and variable stator vanes (VSVs) can help to reduce the possibility of surge. However, even with these provisions and an appropriate compressor design, at least during the system start-up and shutdown, the compressor would cross the surge margin and will temporarily operate at the unsafe region of the performance map. To avoid surge during start-up and shutdown, compressors are equipped with bleed valves. These valves are located on the casing in the intermediate stages of the compressor. Their function is to extract (bleed) air half a way through the compressor. We already noted that how at low mass flow rates, overcompression of the air flow at the early stages of a compressor causes later stages to stall. Using bleed valves, we allow the air mass flow rate at the early stages of the compressor to be higher than that in later stages, which will delay surge. To remove dirt and other fouling materials from the surface of blades and to reduce the possibility of surge, the compressor blades are frequently washed with a specific solvent and demineralized water at a low rotational speed.

9.5.2 Combustion Chambers

A combustion chamber or a combustor is where the high-pressure air from the compressor is mixed with fuel and is combusted. The heat generated in the combustion process increases the temperature of combustion product gases, which are then sent to the turbine to produce power. Combustion chambers are designed to operate at the highest possible efficiency so that they can transfer the maximum possible thermal energy to the air flow for a given fuel rate while satisfying the emission limits imposed by appropriate regulations. Accordingly, the efficiency of a combustion chamber is defined as the ratio of the increase in the enthalpy of the flow through the combustor to the heat content of the fuel. Either the lower or higher heating value of the fuel can be used to determine its heat content depending on the predetermined standard or convention.

$$\eta_{\text{Comb}} = \frac{\Delta h_{\text{Gas}}}{\text{Heat content of inlet fuel}} = \frac{(\dot{m}_{\text{Air}} + \dot{m}_{\text{Fuel}})h_{\text{Outlet hot gases}} - \dot{m}_{\text{Air}}h_{\text{Inlet air}}}{\dot{m}_{\text{Fuel}}LHV} \quad (9.68)$$

The pressure drop in combustion chambers should be reduced to the lowest possible so the pressure ratio in the expansion process in the turbine can be as close to the pressure ratio of the compressor as possible. The pressure drop in a combustion chamber can be up to 10% of the compressor outlet pressure and is proportional to the square of air velocity through the chamber. Furthermore, the temperature uniformity in the outlet flow from the combustion chamber is critical because an uneven distribution of flow temperature in the inlet of the turbine can cause severe thermal stress on the turbine. Also, we know that the TIT is limited by the maximum temperature of the combustion chamber. This means in an uneven temperature distribution, the average TIT will be less than what it could have been if the temperature distribution was more uniform. Combustion chambers should also be capable of operating at a wide range of operating conditions while maintaining acceptable combustion stability and efficiency. Moreover, effective use of available space is important, particularly in aviation applications and by extension aeroderivative gas turbines.

As noted in the examples in Section 9.2.1, the temperature of the air flow increases throughout the compressor. This temperature which is the inlet temperature to the combustion chamber depends on the pressure ratio of the compressor. Furthermore, the combustion chamber inlet temperature will be higher if the unit is equipped with a regenerator. Due to these factors, the combustion chamber inlet temperature can vary significantly. However, the outlet temperature of the combustion chamber is limited by the TIT and can be somewhere between 900°C and 1700°C.

The combustion temperature within the combustion chamber is typically above 1800°C and can be as high as 2000°C. This high-temperature flow cannot be directly sent to the turbine since its temperature exceeds the TIT.

To reduce the temperature of the combustion products to the temperature acceptable for the turbine, the flow rate of the inlet air must be much higher than that required for the stoichiometric combustion process. This results in excess air of as high as 300% or even higher. This high air flow rate means the very high velocity of air (as high as 180 m/s). This extremely high air velocity can create two problems for combustion chambers. First, as noted above, the pressure drop within the combustor is proportional to the square of air velocity. The high air flow rate means the high air velocity, which results in a significant and unacceptable pressure loss in the combustor. To avoid this undesirable effect, a diffuser is needed in the inlet to the combustion chamber to reduce the air velocity. The second problem of high air velocity is that, even with the diffuser (which can reduce the air velocity to about 25 m/s), the air velocity is still high enough to blow away the flame (think of extinguishing the flame of a candle by blowing). In order to sustain stable combustion, the flame should be protected from the whole air flow and allow only a fraction of the air flow to come in contact with the combustion process. This can be achieved by creating a region where the air velocity is in an acceptable magnitude.

Based on their emission level, combustion chambers for gas turbines are generally categorized as diffusion-type combustors and low-emission/NO_x combustors. In the following section, we will cover the working principles of diffusion-type combustors and their different types. Then, we will discuss how low-emission/NO_x combustors have evolved from diffusion-type combustors.

9.5.2.1 Diffusion-Type Combustion Chambers

There are several types of diffusion combustion chambers but they share the same fundamental operation principles. So we first discuss how a generic diffusion-type combustion chamber works, and then we will evaluate the unique aspects of each specific type. In order to achieve an acceptable TIT while maintaining stable combustion, the incoming air to the combustor is divided to three main parts: primary air, secondary air, and dilution air. The primary air directly participates in the combustion process. The secondary air creates recirculation needed to sustain the flame. Finally, the dilution air is used for cooling the chamber and to reduce the temperature of combustion products. The combustion process takes place in a container called the **flame tube** (or **combustion liner**). Figure 9.30 shows an old combustor design that we will use to explain the operation of diffusion-type combustion chambers. Later, we will explain how modern combustion chambers have evolved from this design.

The primary air enters the flame tube at the left end of the tube, possibly after going through swirl vanes to impart circulation into the air flow. The primary air is typically sufficient for the near stoichiometric combustion process. Depending on the level of excess air, the primary air can be one-fifth to

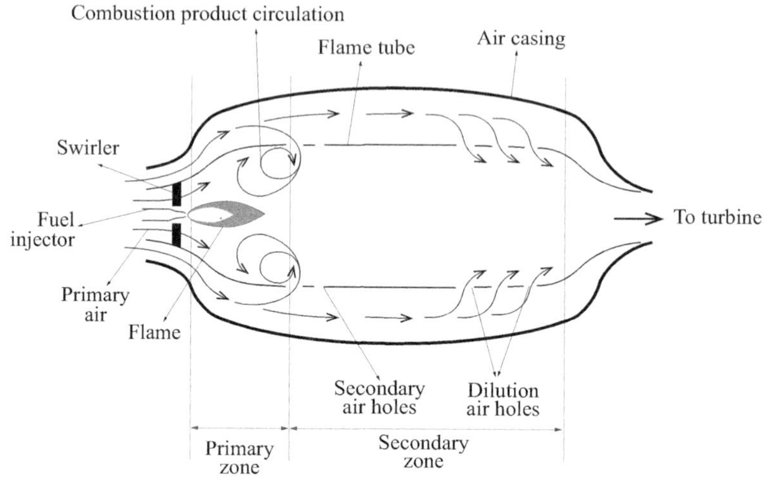

FIGURE 9.30
Schematic of a straight-through diffusion-type combustion chamber.

one-third of the total inlet air flow. The fuel is injected into the primary air to form the primary zone where the flame is anchored and most of the combustion process takes place. Note that in these combustors, air and fuel are mixed in the primary zone (we will later discuss that in modern combustors, the air and fuel flows are premixed before entering the combustion area). The rest of the inlet air is directed to the space between the flame tube and the air casing. In the case of Figure 9.30, the air casing is a concentric container enclosing the flame tube.

To have a stable and self-sufficient flame, a mechanism should be devised to protect the flame; otherwise, combustion products will be quickly carried away by the air flow, and a permanent presence of an ignitor is required. For this purpose in the downstream of the primary zone, some of the remaining air (about 20% of the total air flow rate) is admitted into the flame tube as the secondary air via specifically designed secondary air holes. The secondary air interacts with the primary air flow to create a pattern of vortexes to recirculate a portion of the high-temperature combustion products back to the primary zone to ignite the fresh air and fuel mixture and to sustain the flame.

As noted earlier, the temperature of the combustion products is much higher than the TIT. The next stage is the dilution zone (sometimes referred to as the secondary zone) where the remaining air is used to moderate the temperature of the combustion gases before being sent to the turbine. The admission of most of the inlet air flow in the dilution zone helps to provide a more uniform flow temperature at the inlet of the turbine. It also shortens the length of the flame, which in turn reduces the length of the flame tube and helps in having a more compact combustor. Over half of the entire air flow is introduced to the flame tube in this zone. About one-third of this air is directly admitted through one or more rows of holes or slots and mixes with the combustion products. The rest of the air flow first provides cooling for the walls of the flame tube before mixing with the combustion products. The cooling of the walls is critical since the intense combustion within the flame tube can increase the temperature of the walls, via radiation, to temperatures well above those the tube material can tolerate. For this purpose, air is passed through specially designed holes to form a layer of cooling air along the inner walls of the flame tube to separate them from the hot combustion products. This is called film cooling. To reduce the air flow required for cooling, sometimes the cooling air passes through passages built into the wall before being injected into the chamber for film cooling. Later, we will see that this is desirable because it leaves more air for emission control processes. Furthermore, in modern combustors, the walls may be coated with ceramic materials or covered with refractory bricks or metallic tiles to provide more protection against high-temperature combustion products. Also, some newer combustors use steam flow in the passages to provide cooling.

It is important to make sure that all fuel is completely burned before the secondary zone since the cooling effect of dilution air will mostly stop progress in the combustion reaction. All unburned fuel, primarily in the form of CO, can remain in the flow if they reach the secondary zone. This is undesirable since it reduces the combustor efficiency and contributes to emissions in the exhaust. To avoid these problems in some designs, primarily older gas turbines, when the combustion process cannot be completed in the primary zone, a small portion of the remaining air is injected into the intermediate zone (also known as the burning zone) to complete the combustion reaction.

Before explaining various types of combustion chambers, it should be noted that generally combustors can be straight-through flow or reverse flow. In the straight-through-type combustors, the air flows parallel to the flame in the same direction (Figure 9.30). In reverse flow-type combustors, the air flow is in the opposite direction of the flame. This means in Figure 9.31 that the

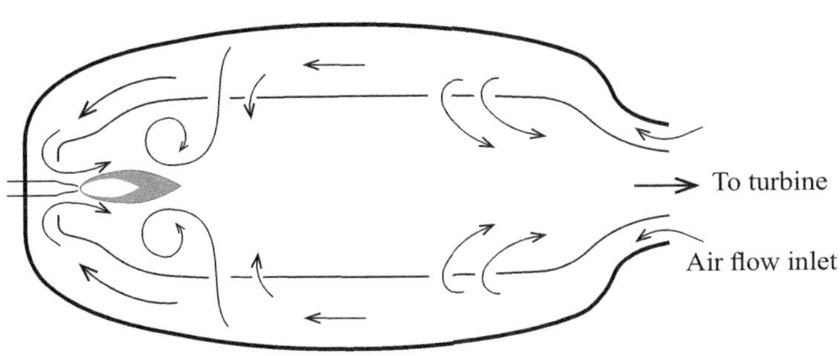

FIGURE 9.31
Schematic of a reverse flow diffusion-type combustion chamber.

air flow enters the combustor at its right end. While the former is popular in large aviation and aeroderivative gas turbines, both configurations are used in electricity generation units. There are four types of diffusion combustion chambers: can, annular, can-annular, and silo-type. We will cover these combustors in the following sections.

9.5.2.2 Can Combustors

Can combustors are sometimes referred to as **tubular combustors** or simply **multiple-chamber** (or multiple-can) combustors. In these combustors, several individual cans (between 6 and 16) are placed between the compressor and the turbine around the shaft connecting them (Figure 9.32). Each can is made of a flame tube surrounded by a concentric air casing. The air flow from the compressor is divided into several flows and directed to each can via a dedicated duct. Each can receives fuel from a common fuel line and has its own fuel injection system. The flame tubes are connected by connecting tubes (also referred to interconnectors, crossover, or cross-fire tubes) to equalize the pressure between cans. Also, typically only two of the cans are equipped with an igniter so the connecting tubes help to initiate the combustion among all cans during start-up. The basic working principles of these combustors were discussed in Section 9.5.2.1. Most early aviation gas turbines used this type of combustors. The main reason is that these combustors are relatively easy and inexpensive to test. Note that due to the complexity of combustors, their development was and to some degree still is primarily based on experimentations. For these combustors, tests can be conducted on a single can, and then, the results can be extended to the whole system. Furthermore, they provide easy maintenance access to cans and their fuel injection systems. However, they need a long flame tube, which results in

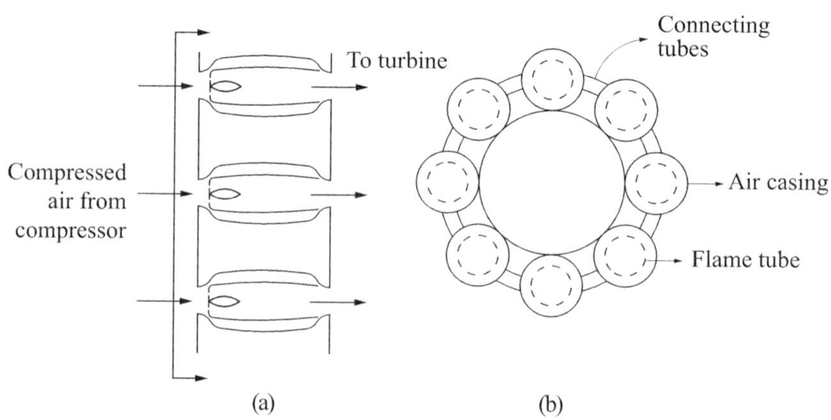

FIGURE 9.32
Schematic of a can combustor a) side view b) front view.

high weight and volume with a large frontal area. These combustors are not used in modern gas turbines and are currently only used for some small gas turbines with centrifugal compressors.

9.5.2.3 Annular Combustors

Similar to can combustors, annular combustors are located between the compressor and the turbine, and placed around their common shaft. However, unlike can combustors, they have a **single flame tube** created by an inner and outer concentric casing forming a ring-shaped combustor around the shaft (Figure 9.33). These combustors are open to the compressor on one end and to the turbine on the other end. Their combustion process and air distribution are similar to what was explained in Section 9.5.2.1 except that they have many fuel injectors creating flames in a single continuous space. They can achieve excellent combustion performance at a very short length, which makes them compact, light, and relatively inexpensive to manufacture. These attributes make them an ideal choice for aviation gas turbines. Due to their streamlined shape, they have a low-pressure drop. Moreover, they have a small surface area, compared to other combustors. This means they require less air for cooling, which in turn leaves more air for a complete combustion process and emission control system. Furthermore, since there is only a single flame tube, they do not require a propagation system, similar to the interconnecting tubes for can combustors explained in the previous section, and can avoid problems associated with it. Due to all these positive attributes, they are very common in airplane gas turbines and are also used in some electricity generation gas turbines. However, they have some drawbacks. They are structurally weak due to heavy buckling loads on the large inner and outer ring-shaped walls.

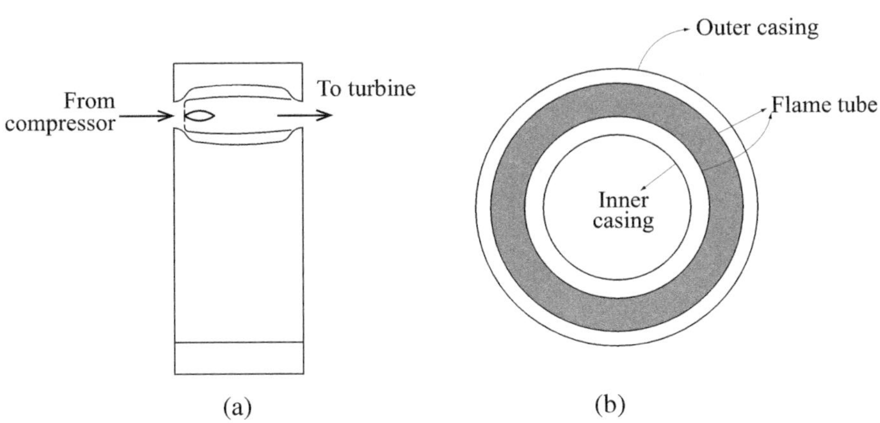

FIGURE 9.33
Schematic of an annular combustor: (a) side view and (b) front view.

It is also difficult to create an even fuel distribution around the flame tube, which makes the temperature distribution in the outlet flow to the turbine less uniform. Finally, these combustors are more expensive to develop since to test them, unlike can combustors, an entire combustor should be built and full-load air and fuel flows should be provided. These combustors are typically straight-through.

9.5.2.4 Can-Annular Combustors

Can-annular combustors, also known as tubo-annular combustors, are a crossover between can combustors and annual combustors. Similar to can combustors, they have several (usually 5–16) cans acting as the flame tubes placed uniformly around the shaft between the compressor and the turbine. Each can have its own fuel injection system and is connected to adjacent cans by interconnectors to equalize the pressure among the cans and to propagate the flame during start-up (note that typically only two cans are equipped with igniters). However, can-annular combustors, similar to annual combustors, have a common annular air casing. This hybrid nature of can-annular combustors helps them to enjoy mechanical strength and ease of maintenance of can combustors and compactness of annular combustors. Many large-utility gas turbines use this type of combustion chamber. The operation principles of these combustors are similar to what was explained in Section 9.5.2.1. These combustors can be straight-through or reverse flow. The former is used for some aircraft gas turbines, while the latter is more popular with electricity generation gas turbines.

9.5.2.5 Silo-Type Combustors

This type of combustor is the oldest technology used for gas turbines. The first-utility gas turbine, explained in Section 9.3, used this type of combustor. Unlike other types of combustors discussed so far where the combustor is physically located between the compressor and the turbine, these combustors are made of one or two large cylindrical chambers placed next to the rest of the gas turbine system. The high-pressure air from the compressor is transferred to the combustor via air ducts, and high-temperature combustion products are transferred from the combustor to the turbine through combustion product ducts. Early combustors had vertical chambers, which resembled silos, hence the name. However, later both vertical and horizontal combustors became popular so they are sometimes referred to as side combustors too. Their operational principles are similar to what was described in Section 9.5.2.1. They may have a louver cooling system where the dilution air flow enters the flame tube via overlapping cylindrical tube segments. Alternatively, they may have holes on the walls where the air flow is admitted. Historically, these combustors have been predominantly used in European gas turbine designs.

Since these combustors require large space, they are only used for large ground applications where the size of the system is not an important factor. They can be straight-through flow or reverse flow type. Due to the large diameter of combustors, they have a relatively low velocity of the flow, which results in a low-pressure drop in these combustors. They are also simple to design and manufacture, easy to maintain, and durable. They can also burn low-quality fuels. Regardless of these advantages, they were typically used in old European gas turbines, and **a great majority of modern gas turbines are equipped with either annual or can-annual combustors**. The main reason for abandoning silo combustors was the high air flow needed for the cooling of the walls which limited employment of the NO_x emission reduction technologies.

9.5.2.6 Low Emission and Low NO_x Combustors

As explained in Section 9.5.2.1, in diffusion combustors, the fuel combustion process in the primary zone is under near stoichiometric air. This means a very high flame temperature, which in turn requires a high quantity of mixing air flow in the delusion zone and for the cooling of the walls. As noted in Chapter 5, one mechanism for NO_x generation is thermal NO_x production, which is promoted at high-temperature environments. Therefore, the high-temperature combustion in diffusion combustors means high NO_x production. Until the late 1970s, there was no meaningful regulation to limit gas turbine emissions. In 1977, in the U.S. the Environmental Protection Agency (EPA) started to put a legal cap on emissions that gas turbines can release to the environment under the Clean Air Act (CAA). Initially, only NO_x emission was regulated but later with an increase in other emissions as a result of NO_x control technologies (discussed below), carbon monoxide and unburned hydrocarbons were also regulated. The early NO_x limit was 75 volumetric parts per million (vppm) at 15% oxygen (dry basis) for single gas turbines and the gas turbine portion of combined cycles. This was substantially lower than about 500 vppm NO_x level of the gas turbine exhaust flows for pre-1970s units. Since then, these regulations have become more and more stringent and currently are a part of the National Emission Standards for Hazardous Air Pollutants (NESHAP). By the mid-1980s, the NO_x level of about 30 vppm was common. Currently, large modern-utility size gas turbines have a single-digit NO_x emission, as low as 5 vppm. The goal is to reduce it to lower than 2 vppm. Before continuing this section, it might be a good idea to review Section 4.6.4.1 on the mechanisms of NO_x generation in the combustion process. Note that since the majority of gas turbines are fueled with natural gas, the primary concern is thermal NO_x production and fuel NO_x is not typically an important factor. Thus, most NO_x reduction approaches for gas turbines are focused on reducing the flame temperature. This is a dilemma for gas turbine designers because we know that a higher TIT improves the efficiency of a gas turbine. So the objective here is to reduce the maximum temperature of the combustion process without compromising the outlet temperature of the combustor.

Gas turbine NO_x reduction approaches can be divided into two groups: those that reduce NO_x production during the combustion process and those that remove NO_x from combustion products after its production during the combustion process. The most common approach to remove NO_x from combustion products is the selective catalytic reduction (SCR) system that was discussed in Chapter 5. The most common techniques to reduce NO_x production during combustion are as follows:

- injection of water or steam into the combustor
- dry low NO_x (DLN) combustion
- catalytic combustion

In the following sections, we will discuss these technologies. But before it should be noted that there are other NO_x reduction technologies for gas turbines but they are not common. One such method is to combust a rich air and fuel mixture in the primary zone followed by rapid cooling of the gases and finally the lean combustion of remaining fuel by the addition of more air in the dilution zone. This method is sometimes referred to as the rich-burn/quick-quench/lean-burn (RQL) technique. The shortage of oxygen in the primary zone does not allow complete combustion, which reduces the peak flame temperature. Another method is to burn a very lean air and fuel mixture in the primary zone, which again reduces the peak flame temperature. However, the method that may result in flame instability. In another approach, exhaust combustion products can be recirculated into the combustion process. This, however, needs either compression of high-temperature exhaust gases or cooling and then compression of gases, both of which can be costly. Finally, in the late 1970s, sequential or staged combustion was used to reduce NO_x. In this method, the combustion process is divided into two separate processes in two different combustors. These systems are very similar to reheat gas turbines where each combustor was followed by a turbine but in this case, no fresh fuel is injected into the second combustor. In the first high-pressure combustor, the fuel was partially burned for a moderate increase in the outlet flow temperature. This flow is then expanded in the high-pressure turbine to an intermediate pressure. The combustion process is completed in the second low-pressure combustor, and the outlet flow is then expanded in the low-pressure turbine. This two-stage combustion is able to substantially reduce the maximum flame temperature. This technique was first implemented in the world's first large-scale compressed air energy storage system (CAES) in Germany in 1977 (refer to Chapter 16 compressed air energy storage (CAES) systems). This system was able to reduce the overall NO_x production by about half.

9.5.2.6.1 *Injection of Water or Steam to Combustors*

The injection of water or steam into the primary zone to reduce the flame temperature, also known as **wet NO_x control**, was the first method employed to reduce NO_x production and was used even before regulatory restrictions. In early gas turbines, the steam generated during the water cooling of the combustion chamber was injected into the flame to increase the combustion product mass flow rate and thus power generation. It was the most popular method to reduce NO_x in the 1980s and 1990s. The injection of a massive amount of water or steam (the injected flow rate can get close to the mass flow rate of fuel) into the flame reduces the flame temperature, which in turn reduces NO_x production. This technique can reduce NO_x emissions by almost an order of magnitude. The injection of water or steam also increases the mass flow rate of the gas passing through the turbine, which can increase its power generation. However, this method introduces thermodynamic losses, which reduce the efficiency of gas turbines due to the reduction in the TIT, incomplete combustion, and thermodynamic irreversibilities. The incomplete combustion not only reduces the system efficiency but also increases other emissions, particularly CO. In fact, before the introduction of the wet NO_x reduction method, the level of CO emission in the exhaust of gas turbines was so low that CO emission was not initially regulated by EPA in the late 1970s. However, with a significant increase in CO emission as a result of water or steam injection, the CO emission was also regulated later. This method also needs a huge amount of demineralized water. This volume of water may not be available on many sites. Even if water is available in a site, a water purification plant requires a significant capital investment and increases the operational cost. This method also increases maintenance requirements and the system noise. All these problems led to the development of DLN combustors. Wet NO_x reduction is still used in some gas turbines, usually as a power augmentation technique as well as a NO_x control method, especially in some applications such as peak-load power plants where output power is more important than efficiency.

9.5.2.6.2 *Dry Low Emission Combustors*

As noted in the previous section, simply reducing the flame temperature on its own cannot satisfactorily address the emission problem in gas turbines. While the reduction of the combustion temperature may reduce NO_x production,it increases CO emission. So when evaluating the emission reduction strategies, NO_x, CO, and unburned hydrocarbon emissions must be taken into account, typically in the form of the CO/NO_x ratio. The solution came in the form of a combination of several techniques collectively initially referred to as DLN combustors but later, with the inclusion of the reduction of other emissions, it is now more commonly called **dry low-emission (DLE) combustors** or **DLN/DLE combustors**. These combustors achieve low emissions by employing lean combustion of a premixed air and fuel mixture in a staged combustion process.

We have already discussed the benefit of lean combustion to reduce the flame temperature. In these combustors, the excess air of about 100% is usually used. The premixing of air and fuel before the combustion process is to provide a homogeneous air and fuel mixture to avoid local high-temperature spots. Otherwise, these hot spots can promote NO_x production. The mixing can take place either before the air and fuel mixture is fed into the combustor or inside the combustor. While a combustion chamber can be designed and manufactured to provide an acceptable CO/NO_x ratio to meet regulatory restrictions at the design condition, it is quite a challenge to design a combustor that can meet the requirements at all possible operating conditions, including part- and transient-load operation, start-up and shutdown, and varying ambient temperatures. Meeting emission limitations while keeping combustion stable and accommodating all operating conditions require the design of special components with appropriate operational procedures.

It should be noted that generally speaking, gas turbines have a constant volume air flow rate since the rotational speed of the compressor is constant (at least in a single-shaft configuration). This means when the fuel flow rate changes to accommodate the required load, the air-to-fuel ratio changes. Therefore, the combustion chamber should be able to operate with a varying air-to-fuel ratio. For this purpose, staged combustion is used. Depending on the type of the combustor, different companies came up with different designs to address this problem. However, they all share some common design features. All DLN/DLE combustors are equipped with multiple fuel nozzles. At various operating conditions, a predetermined mixture of air and fuel is directed to different fuel nozzles to ensure a stable flame with an acceptable emission level. Note that for dual-fuel combustors, these requirements must be achieved for both gaseous fuel (typically natural gas) and liquid fuel (commonly diesel or heavy oil). Other factors that should be considered in designing operational procedures, other than emission levels, are ensuring combustion stability, minimizing noise level, and optimizing maintenance requirements. It is important to remember that unlike the water or steam injection system, in DLN/DLE combustors the NO_x control mechanism does not introduce significant thermodynamic losses.

It should be noted that having a very lean air and fuel mixture means more air should be directed toward the combustion process. As a consequence, there is less air available for the cooling of the flame tube walls. This is why the DLN/DLE method favors combustors with fewer surfaces that require cooling. This is one of the main reasons that silo-type combustors are no longer used in modern gas turbines, whereas annular and can-annular combustors are ideal in this regard.

9.5.2.6.3 Catalytic Combustion Chambers

As noted earlier, a very lean air and fuel mixture can cause an unstable combustion process. One solution to this problem is to have the combustion process in the presence of a catalyst. This type of combustion chambers is known as **catalytic combustors** (sometimes also somewhat misleadingly is referred to as flameless combustors). The catalyst facilitates the combustion process without being consumed during the process. This allows the stable combustion of a leaner air and fuel mixture than that of current commercially available technologies. The technology should not be mistaken with catalytic converters that can be used to remove emissions from the exhaust flow of gas turbines. This technology is currently under development. While it is being utilized in some small gas turbines and some experimental units, the technology has not entered the mainstream gas turbine market. However, it is a promising method to achieve extremely low NO_x combustion, as low as 2 vppm, in gas turbines. In current designs, only an initial portion of fuel combustion (less than half of the combustion) takes place on the catalyst stage, and most of the fuel burns (60%–80%) in the following combustion zone.

9.5.2.7 Other Components of Combustion Chambers

For a combustion chamber to operate properly, many components should be incorporated into the design. In this section, some of these components will be briefly discussed.

In the fuel system, first the fuel flow should be pressurized and delivered to the combustor. Then, it should be distributed among fuel injectors where the fuel is injected into the flame. Natural gas is typically delivered in high enough pressure that can be directly injected into the primary zone with no need for a compressor. If natural gas pressure is not high enough, a fuel compressor is used to pressurize the fuel flow. The fuel pressure should be higher than the air pressure in the combustor, by about 3–5 bar, so the natural gas flow can be easily injected into the air flow.

Liquid fuels are first pressurized, typically in positive displacement pumps. The fuel flow is then delivered to fuel injectors by a common fuel supply line. The fuel distribution system is typically equipped with control valves to ensure an even distribution of fuel among all fuel injectors. In order to control combustion temperature at various loads, the fuel supply line is commonly equipped with an accurate flowmeter.

Fuel injectors are used to supply fuel into the flame tube. The design of fuel injectors heavily depends on fuel phase, liquid or gaseous. Gaseous fuel injectors are made of orifices that can impart swirl in the fuel flow. For liquid fuels, the fuel may be vaporized in a heated vaporizer before being injected to the combustor. Alternatively, the fuel can be atomized through a fuel spray nozzle. In the latter case, the fuel is converted to fine droplets so they can be quickly evaporated and combusted in the primary zone. The fuel atomization can be achieved by pressure nozzles where the high-pressure fuel is injected through fine orifices or by air-spray nozzles where a high-velocity air flow atomizes the fuel (also known as aero liquid fuel spray nozzles).

Many combustors for ground gas turbines should be able to operate using both gaseous and liquid fuels, separately or simultaneously. They can switch between fuels while running (called duel fuel combustors) or after changing the nozzles (called double fuel combustors). The former is typically similar to air-spray nozzles.

During gas turbine start-up, an **igniter** is needed to start the combustion process and to support the flame until the combustor can sustain the flame. All igniters use electric sparks to initiate combustion. They may use the similar principle as typical automotive spark plugs though there are other designs too. They may create sparks with the frequency of 1–2 Hz at 2000–25,000 V. In annular combustors, there are typically two igniters at the opposite sides of the combustor. Similarly, can-annular combustors have two ignitors at two opposing cans and the fire propagates via cross-fire tubes to other cans. Flame detectors are used to ensure the presence of the flame during start-up, shutdown, and normal operation by detecting ultraviolet radiation from the flame. Typically, there are two detectors placed opposite each other.

Another component that is related to a combustor, though may not be considered as a part of it, is a duct that transfers combustion products from the combustor to the turbine. This is more challenging for can-annular combustors where the circular intersection of the cans must be converted to the rectangular one of the turbine. These connectors are subject to high thermal fatigue, buckling stress, and cracking and should be manufactured in a shape and from a material that can stand this harsh environment. They may be coated for thermal protection and have cooling air holes on their body.

9.5.3 Gas Turbines

Like steam turbines, the function of a turbine in a gas turbine cycle is to extract energy from the flow of the working fluid. It should be noted that the term "gas turbine" may confusingly refer to the entire cycle or just the turbine itself. In this section, we will typically use the term "turbine" or "gas turbine" when we are referring to the single component and use the term "gas turbine cycle" or "unit" when we are referring to the entire cycle. In gas turbines, the turbine receives the high-pressure and temperature combustion products from the combustion chamber, extracts energy from the flow, and discharges the relatively low pressure and temperature flow to the atmosphere in the case of single gas turbines or to the bottoming cycle in the case of CCPPs. In electricity-generating gas turbines, the turbine extracts as much power as possible from the gas flow. This power is used to run the compressor and the generator. In aviation gas turbines, depending on the type of the engine, the turbine may extract the maximum power from the gas flow (in turboprop engines) or just enough to run the compressor and fan (in turbojet and turbofan engines).

Historically, gas turbine designs were an extension of steam turbine designs. They followed the similar design principles. However, the operating conditions of modern gas turbines are much more challenging than those of steam turbines. In fact, gas turbine blades are arguably exposed to one of the harshest environments in any engineering field. Unlike steam turbines that are exposed to extremely pure steam, gas turbines have to directly deal with corrosive combustion products. Gas turbines also operate at much higher temperatures than steam turbines. The tip of gas turbine blades can rotate at the velocity of over 400 m/s (roughly about 1500 km/h or 900 mph). Turbines used in gas turbine units can be axial or radial turbines. Generally speaking, axial turbines are more efficient than radial ones. The majority of large- and medium-sized gas turbines use axial flow turbines, while radial turbines are used in some small gas turbines.

Radial-inflow turbines, also known as inward-flow radial turbines, are in principle similar to centrifugal compressors but expand the flow rather than compressing it. Many early gas turbines were equipped with this type of turbines. But currently they are only used in some small gas turbines, micro gas turbines, and turbochargers besides being shorter, these turbines can generate much more power in a single-stage than axial turbines.

9.5.3.1 Axial Gas Turbine Blades

In axial flow gas turbines, the flow throughout the turbine is parallel to its axis of rotation. Similar to axial flow steam turbines, these turbines are composed of several stages. Each stage has a row of fixed blades (also known as nozzles or vanes) and a row of moving blades (also referred to as rotor or buckets). These rows can be impulse, reaction, or a combination of the two. As explained in Chapter 6 for steam turbines, in impulse turbine stages, the entire conversion of pressure (enthalpy) to velocity (kinetic energy) occurs within the fixed blades so in the moving blades velocity drops (due to the extraction of energy from the flow), while the pressure remains constant. Therefore, the fixed blades are formed in nozzle shape, and the moving blades are symmetrical. In pure reaction stages, the entire pressure drop and velocity drop occur in the moving blades. So the fixed blades are symmetrical with the only function of redirecting the flow to the proper angle for the moving blades, while the moving blades are converging to form nozzles. In the combined impulse and reaction stages, the pressure reduction is divided between the moving and fixed blades. Since there is little pressure drop across the moving blades in impulse stages, there are little blade tip leakage losses. This makes impulse stages ideal for the high-pressure stages of steam turbines. However, in gas turbines, the flow pressure is much lower than that in steam turbines (typically up to 40 bar for gas turbines versus sometimes above 300 bar for supercritical steam turbines), so **typically impulse stages are not used in gas turbines**. Gas turbine blades are typically a combination of reaction and impulse. The blades, both moving and fixed, are twisted to accommodate the varying velocity of the blades along their length. This requires different blade profiles to ensure all sections along the blade length are extracting a similar magnitude of energy from the flow. The degree of reaction is used to identify a stage type. The degree of reaction varies from stage to stage in gas turbines. Also in a single blade, the degree of reaction increases from the hub (root) of the blade to its tip. However, generally speaking, the blades are designed to have the degree of reaction of about 50% at the mean diameter of the blade. In the inlet to the turbine, there are nozzle guide vanes where the inlet flow is directed to the moving blades at a correct angle. The moving blades are attached to the turbine shaft, while the fixed blades are installed on the casing. Both moving and fixed blades are arranged to form nozzles (converging passage) between adjacent blades.

Modern turbines used in gas turbine cycles are very efficient machines. They typically have a higher than 90% efficiency. The major losses in turbines are friction flow losses in the moving and fixed blades, gas leakage from blade tips, and energy loss through the exhaust flow.

As noted above, in reaction stages, the pressure drops along the moving blades, which results in a pressure gradient across the moving stage. This pressure difference can cause some gas flow to bypass the blades (or leak) via the gap between the blade tips and the turbine casing. This leakage, sometimes referred to as the tip clearance loss, not only results in lost power generation potential but also can disturb the main flow through the stage. Note that the tip clearance varies due to different thermal expansion coefficients of the blades and the casing. One way to reduce this leakage is to add a shroud to the blade tips. Generally, turbine stages can be shrouded or shroudless. The shrouds are interlocked pieces that connect the blade tips to form a ring around the blade tips. The shrouds also substantially reduce the vibration of blades that otherwise can lead to blade failure, particularly for low-pressure long blades.

Since turbine blades operate at a high-temperature environment, they are exposed to creep. At low temperatures, the strain (deformation) of a material is only a function of the stress level in the material and is independent of time (duration of stress). However, at high temperatures, the strain can be a time-dependent parameter. What this means is that blades can experience failure in the long-term even if the stress is below the allowable stress of their material. The turbine blades should be made of low creep alloys. These alloys should also have high ductility (ability to undergo plastic elongation before failure), low coefficient of thermal expansion, high thermal conductivity, as well as high corrosion, oxidization, erosion, and thermal fatigue resistance. The blade corrosion can be accelerated at high temperatures in the presence of sulfates or vanadium compounds. The blades can be eroded by tiny particles that passed through the filters or pieces separated from the combustor walls. One of the most important causes of blade failure is thermal fatigue as a result of cyclic temperature variations, such as start-up, shutdown, or major load changes. The maintenance and inspection intervals are heavily impacted by the number of these thermal cycles. On top of all these properties, the alloys must have manufacturability and machinability characteristics so the delicate complex blade surfaces needed for the optimal aerodynamic performance of blades can be achieved. Modern turbine blades, especially early stages that are exposed to the highest temperature and the harshest environment, are being made of nickel-based superalloys by casting techniques such as directionally solidified and single-crystal alloys. In directionally solidified casting, alloy crystals are formed along the blade length, which improves its desired properties, particularly its creep resistance. Currently, blades made of nonmetallic materials, such as reinforced ceramics, are being developed.

The blades are either directly or via the turbine disk connected to the turbine shaft. The shaft and the disks are cooled by the high-pressure air flow through air channels. The blades are attached to the turbine disks by a specially designed mechanism discussed in Section 6.4.5 (Figure 6.15). This attachment mechanism requires an accurate machining process to ensure an even load distribution among all surfaces. In some mechanisms, the blades are loose when the turbine is not rotating, but when the turbine is spinning, the centrifugal forces fix the blades in place.

9.5.3.2 Turbine Blade Cooling

So far we have established that increasing the TIT and the compressor pressure ratio improves the power output and efficiency of gas turbine units. The early stages of turbines are the most important limit for increasing the TIT. There are two main strategies to increase the TIT without damaging the turbine blades. Clearly, one strategy is to fabricate blades using high-temperature-resistant materials. While materials used to manufacture blades have been progressively able to tolerate higher inlet gas flow temperatures, there is always a limit on this temperature from the material point of view. Another strategy is to allow the gas flow temperature to increase while protecting the blades from being exposed to this high-temperature flow. Two common methods for this purpose are coating blades with protective surfaces and using cooling techniques to reduce the temperature of blades. These two methods keep the blade temperatures within the limits while the gas flow temperature can exceed those limits.

The protection coating, such as ceramic coating, is applied to the surface of blades to protect them from **chemical and corrosion attacks** as well as to act as the **thermal barrier**. This allows further increase in the TIT and the blade life expectancy. A wide variety of cooling techniques are employed to cool down the high-temperature parts of turbines. The cooling can be applied to moving and fixed blades, shrouds, and all components that are exposed to the high-temperature gas flow. The medium for the cooling system can be high-pressure air, steam, or water. In air-cooled blades, the high-pressure air flow is bled from the compressor (in some systems from several stages of the compressor) and directed toward the components that need cooling.

Note that the high-pressure air leaving the compressor is at a high temperature, so it may need cooling before being sent to the turbine. For cooling of moving blades, the high-pressure air flow passes through dedicated channels along the shaft and is distributed among the blades. For cooling of fixed blades, the air is distributed via the turbine casing. The cooling process impacts the flow through the blades and the flow temperature. The magnitude of impacts depends on the cooling medium, its temperature and flow rate, and the cooling mechanism. Overall, the cooling process can slightly reduce the efficiency of the turbine. However, the loss due to this lower turbine efficiency is more than compensated by the overall cycle efficiency improvement as a result of the higher TIT. The cooling process can be achieved through the following several cooling mechanisms: convection cooling, film cooling, transpiration cooling, multiple small-hole cooling, and water or steam cooling. Note that small gas turbine units commonly use uncooled turbine blades, which is one of the reasons for their low efficiency.

Convection cooling systems require channels inside the component that needs cooling, often serpentine ones, where high-pressure air flow cools down the component by convective heat transfer. This method is commonly used in current gas turbines. Another method is film cooling. This technique works similar to a protective coating layer where air is injected into the blade surface via specially designed nozzles to form an insulating film of air to separate the blade surface and the high-temperature gas flow. Often a combination of convection cooling and film cooling is utilized. In transpiration cooling systems, the blade is made of a porous material and cooling air passes through the holes to cover the entire blade surface and keep it cooled. While this is an effective cooling method, the small holes are susceptible to clogging. A similar concept is used in multiple small-hole cooling systems where the cooling air is injected into the blade surface via many holes. These holes are larger than pores in the transpiration cooling, which is why this method is less vulnerable to hole clogging. The most recent blade cooling technology is using water or steam as the cooling medium. In these techniques, tubes (commonly copper tubes) are embedded in the blades. The water or steam flow, usually from the bottoming steam cycle in CCPPs, passes through these tubes and provides very effective cooling. The water flow may need preheating to avoid thermal shock in the blades. The outlet cooling medium, always steam since water also evaporates while passing through the tubes, can be either returned to the steam cycle (known as closed-loop cooling) or injected into the gas flow (open-loop cooling).

9.5.3.3 Gas Turbine Arrangements

As noted, turbine blades extract energy from the gas flow and transfer the energy to the turbine shaft. The rotational power of the turbine should be used to drive the compressor and either the generator (in electricity generation applications) or other external power uses. There are several arrangements on how the power is transferred from the turbine to the compressor and the generator. The objectives of these different configurations are to allow various components to operate at their optimum rotational speeds and/or to improve the load-following capability of the power generation unit.

In the single-shaft configuration, the turbine, compressor, and generator are all on the same shaft spinning with the same rotational speed. This fixed rotational speed is dictated by the constant frequency of the electricity generated by the generator or the specification of the required external power. Note that in this discussion when we refer to the turbine and the compressor being on the same shaft, it means either the turbine and the compressor share a common shaft or their shafts are connected via a self-aligning coupling. While inherently simple, in this configuration the compressor and the turbine cannot operate at their optimum rotational speeds. On the contrary, in multiple-shaft configurations, there is much more flexibility on the rotational speed of various components. There are a variety of arrangements for multiple-shaft configurations. For example, the compressor and the high-pressure turbine can be on the same shaft, while the low-pressure turbine and the generator (or the external load that may require a variable rotational speed) can be on the separate shaft (Figure 9.34a). The combination of the former two components and the combustor is often referred to as the gas generator, while the latter turbine is referred to as the free-power turbine or simply the power turbine. In this configuration, the high-pressure turbine and compressor can be operated at their optimum rotational speed, while the rotational speed of the low-pressure turbine is fixed by the requirements of the external load (e.g., the generator). In another configuration, the high-pressure turbine and the high-pressure compressor can be on one shaft, and the low-pressure turbine, the low-pressure compressor, and the generator on another shaft. So each compound can run with the rotational speed that optimizes its performance. Another advantage of multiple-shaft configuration is that during system start-up, only the gas generator portion needs to be driven by the starter, such as an electric motor or a diesel engine, which reduces the inertia that should be overcome. Again note that in all these configurations, the generator can be replaced by an external load. High-pressure ratio gas turbines, particularly aeroderivative gas turbines, can have two- or three-shaft configurations, which are typically referred to as double- or triple-spool units. In these units, typically high-, intermediate-, and low-pressure sections of the turbine and the compressor are on the same shaft since their optimum rotational speeds match. Note that in all of the aforementioned configurations, to overcome radial and axial forces imposed on each portion of the system, shafts are mounted on a variety of bearings installed on casings. To provide lubrication for these bearings, several pumps, called the lubrication skid, are used.

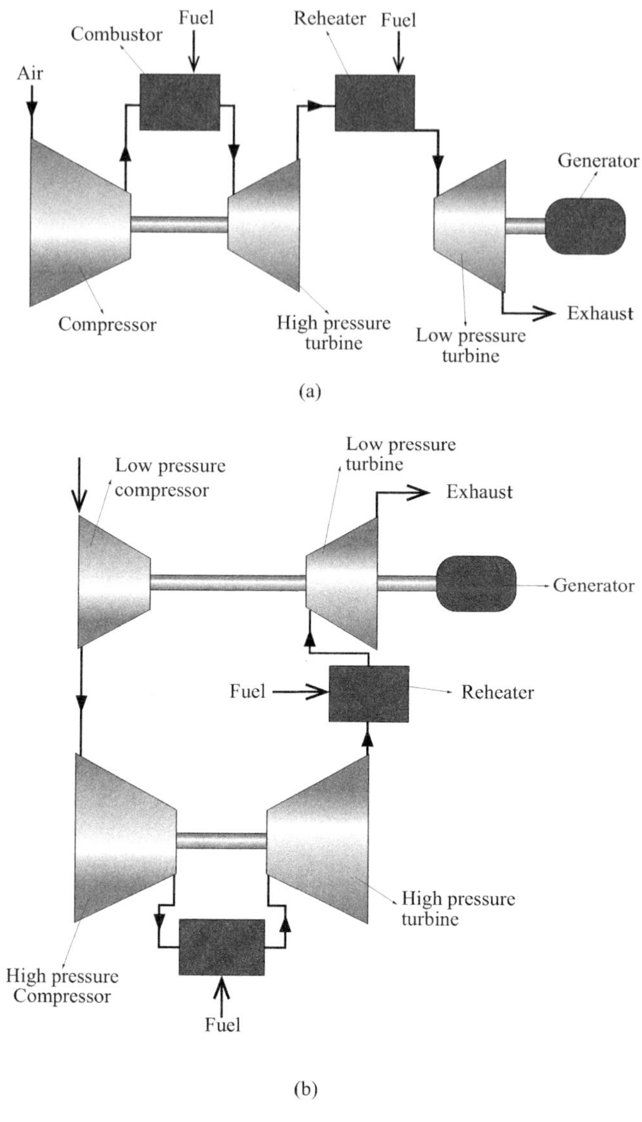

FIGURE 9.34
Schematic of various configurations of gas turbine units.

9.5.4 Auxiliary Systems and Components

In this section, some of the auxiliary components of gas turbine power-generating units will be discussed.

9.5.4.1 Compressor Inlet Air Cooling Systems

From our discussion earlier in this chapter, for the specific work consumption in a compressor, we can write (considering the ideal gas assumption)

$$w_{\text{Comp}} = h_2 - h_1 = C_p\left(T_2 - T_1\right) = C_pT_1\left(\frac{T_2}{T_1} - 1\right) = C_pT_1\left(\left(\frac{P_2}{P_1}\right)^{(k-1)/k} - 1\right) \tag{9.69}$$

This means that for a given compressor pressure ratio, the power required to pressurize the unit of mass of air increases if the inlet air temperature rises. Note that this is for an ideal compressor with 100% isentropic efficiency. For a nonideal compressor, the right side of the equation should be divided to the isentropic efficiency of the compressor. The same conclusion can be made using Equation 9.52.

The net total power generation in a gas turbine cycle and the mass flow rate through the system can be determined by the following equations:

$$\dot{W}_{\text{Net}} = \dot{W}_{\text{Turb}} - \dot{W}_{\text{Comp}} = \dot{m}\left(w_{\text{Turb}} - w_{\text{Comp}}\right) \tag{9.70}$$

$$\dot{m} = \rho\dot{V} \tag{9.71}$$

where $\dot{W}_{\text{Net}}$ is the net total power generation of the unit (kW), $\dot{W}_{\text{Turb}}$ is the power generation in the turbine (kW), w_{Turb} is the specific work generation in the turbine (kJ/kg), $\dot{W}_{\text{Comp}}$ is the power consumption in the compressor (kW), w_{Comp} is the specific work consumption in the compressor (kJ/kg), $\dot{m}$ is the air mass flow rate (kg/s), ρ is the air density (kg/m^3), and $\dot{V}$ is the air volumetric flow rate (m^3/s).

We know that in a compressor for the given operating conditions, the volumetric flow rate of the fluid through the compressor is constant. Therefore, the air mass flow rate only varies with changes in the air density as follows:

$$pv = \frac{p}{\rho} = RT \tag{9.72}$$

$$\rho = \frac{p}{RT} \tag{9.73}$$

This indicates that with an increase in the compressor inlet temperature, the mass flow rate of the air through the compressor reduces. So an increase in the compressor inlet temperature, i.e., the ambient temperature, will reduce the air mass flow rate through the compressor and will increase the power needed to pressurize a unit of mass of air in the compressor.[12] Both these effects negatively impact the power generation and efficiency of the gas turbine unit. In actual gas turbines with every 1°C increase in the temperature of the compressor inlet flow, the power output of gas turbines can reduce by 0.5%–1%. To worsen the problem, this loss of power happens during the

[12] Note that this is true for all air-breathing engines.

hottest time of the day, when in hot regions, the demand is maximum and the price of electricity is the highest.

Like every other situation when high temperature is a problem, the solution is the cooling process. The utilization of compressor inlet air cooling systems started in the early 1990s in some pioneering units. However, they are nowadays standard options that one can choose when purchasing a gas turbine unit. There are two ways for this purpose: **refrigeration systems** and **evaporative cooling systems**. In refrigeration systems, sometimes referred to as indirect cooling systems, cooling coils are installed in the compressor inlet duct, inside of which a coolant flows and cools down the inlet air flow. The cooling can be provided by an absorption or mechanical (vapor compression) refrigeration system. The refrigerant can be directly circulated in the coils or water as a secondary cooling fluid can receive cooling from the refrigerant in a heat exchanger and then flows through the cooling coils. Alternatively, for gas turbines that burn liquid natural gas (LNG) as fuel, the thermal energy needed to evaporate the fuel can be extracted from the compressor inlet air flow. These systems can reduce the inlet air flow temperature to below the ambient wet-bulb temperature. However, it should be noted that in this case, air humidity will be condensed, which requires significant energy. Also, the compressor inlet temperature cannot be lower than 8°C (about 45°F) because at lower temperatures there is a possibility of ice formation, which can damage compressor blades. These refrigeration systems have high capital costs, high costs of operation and maintenance (O & M), and high power consumption. Furthermore, the cooling coils in the compressor inlet duct can impose a small but significant pressure drop in the inlet air flow. This pressure drop negatively impacts the performance of the system, even when the cooling system is not operational, for example, during winter. A variation of refrigeration cooling systems is thermal energy storage systems. These systems have all parts of conventional refrigeration systems plus a container to store ice, cold water, or other heat transfer liquids. In these systems, the refrigeration system operates during off-peak periods (commonly at night). The generated ice or cold water (or another liquid) is stored in the storage system. Then, in mid-day peak periods, when the ambient temperature goes up, the energy storage media are used for the compressor inlet air cooling. This system usually can only provide a few hours of cooling during hot hours of the day.

A much more common compressor inlet air cooling technology is evaporate cooling system. As the name implies, in **evaporate** (or **direct cooling**) systems, water is evaporated by the compressor inlet air flow. The latent heat required for this evaporation is drawn from the air. This process produces air at lower temperatures with higher humidity at the outlet compared to those in the inlet. The minimum theoretical air temperature can be achieved when the outlet flow is saturated air. This temperature is equal to the web-bulb temperature of the ambient air. In practice, this temperature cannot be reached. The effectiveness of the system is defined to indicate how close the outlet temperature is to the lowest possible temperature. This is one of the most important limitations of evaporative cooling systems. The level of cooling that they can produce depends on the ambient humidity. Their capacity to cool down the compressor inlet flow diminishes with an increase in the ambient humidity. These systems are ideal for hot dry regions. However, generally speaking, in these regions, water is a very valuable commodity. Evaporative systems are significantly less expensive to install and operate than those for refrigeration cooling systems.

There are two main types of evaporative cooling systems, **media-type evaporative systems** and **fog cooling systems**. In media cooling systems, blocks of honeycomb-like medium made of fibrous corrugated materials (similar to the fill in wet cooling towers) are assembled in the inlet duct of the compressor, commonly after but sometimes before the air filter. Water is distributed among them at the top via the water distribution piping system. Water and air flows pass through the media in a cross-flow fashion, and water evaporates into the air flow. The internal channels of the media maximize the surface area of the interaction between water and air flows to maximize water evaporation. This process is similar to what we described for a wet cooling tower in Chapter 8. The difference is that in this case, the objective is to reduce the temperature of the air flow rather than the temperature of the water flow. The unevaporated water is discharged into a tank at the bottom of the media block. Then, the collected water is pumped back again to the top of the media. To compensate for the evaporated water, makeup water is continuously added to the tank. To maintain water quality, a certain percentage of the water in the tank is bled. Also, to avoid carryover of water droplets, a drift eliminator is typically installed right after the media. The media cause some pressure drop in the compressor inlet air flow. Some systems are designed in a way that during winter, the media can be removed to eliminate this pressure drop when the system is not operational. These systems have a modest cooling effectiveness of about 85%.

In the past couple of decades, **fog cooling systems** have gained popularity as a compressor inlet air cooling technology. In order to increase the surface area of water and to facilitate its rapid evaporation, in fog cooling systems water is atomized to extremely small droplets. To achieve fast enough evaporation, the size of droplets is commonly 5–20 μm.[13] This level of water atomization can be achieved using compressed air nozzles. Alternatively and more commonly, extremely high-pressure water can be sprayed through specially designed nozzles into the compressor inlet air flow. These nozzles can atomize water either by creating swirl in high-pressure water or by passing it through an orifice followed by impacting the water flow to a pin. The water pressure is increased, typically by piston pumps, to 70–200 bar (about 1000–3000 psi) before being sent to nozzles. The body of these nozzles are commonly made of stainless steel but the orifice is made of an erosion-resistant material, such as ruby. A network of these nozzles is created by assembling

[13] $1\ \mu m = 1\times10^{-6}$ m.

horizontal and vertical piping systems in the compressor inlet duct after filters. The exact location of the system depends on the air passage before the compressor and the time required for full evaporation of water droplets. The analysis of the two-phase flow and the evaporation process is quite complicated, and CFD methods are used for this purpose.

In evaporative systems, the control unit receives the data from the ambient temperature and humidity sensors and – based on the gas turbine operating conditions – determines the amount of water needed to be injected into the inlet air flow. There are two mechanisms to control the water flow rate: first, using several pumps in which the number of operating pumps determines the water flow rate; and second, a limited number of variable speed pumps. The latter system can provide smoother and more continuous control.

Fog cooling systems can achieve the lowest temperature possible with evaporative systems due to their high effectiveness of 95% or higher. These systems also enjoy low-pressure drop in the compressor inlet, minimal alternation required in the existing structure, and low installation time. Overall, fog cooling systems typically have the highest return for the investment and the shortest payback period among all cooling systems. They can even have a positive impact on the NO_x reduction of the gas turbine unit.

In fog cooling systems, the injected water directly goes through the compressor and the rest of the gas turbine system. Therefore, to avoid compressor blade fouling, the injected water needs to be demineralized. This is also needed to avoid the plugging of orifices in nozzles. If the power plant does not have a sufficient excess capacity to produce demineralized water (in the case of CCPPs), a dedicated water purification system, commonly a reverse osmosis system (refer to Chapter 7), is needed. In early fog cooling systems, there were some concerns about the effects of unevaporated water droplets on the compressor blades and possible erosion in the blades. However, long-time experience with the operation of these systems proved their safe operation. These systems may even have some positive impacts due to the compressor blade washing effect.

Recently, some cooling systems have even taken a step forward by excess fogging to provide compressor intercooling. For this purpose, water is oversprayed into the air flow. This means some water droplets enter the compressor unevaporated. When the air flow is compressed in the early stages of the compressor, the air temperature goes up. The high-temperature air has more capacity to absorb water vapor so the remaining water droplets will evaporate, which can provide further cooling within the compressor. Another approach with a similar effect is to inject water between compressor stages.

Many criteria should be considered when choosing the technologies and design parameters for the compressor inlet cooling system. Besides installation and operating costs, other factors including local climatic conditions, desired power gain, and NO_x reduction requirements should all be taken into account.

9.5.4.2 Regenerators

As discussed in Section 9.2.4, a regenerator can be used in a gas turbine cycle to recover some of the thermal energy in the turbine exhaust flow by heating the compressed air from the compressor before entering the combustor. This can reduce the fuel consumption in the combustor for the same power output, which leads to a higher overall efficiency. A variety of heat exchangers can be used as a regenerator, including shell-and-tube, plate-fin, and ceramic heat wheel. The latter, in principle, is similar to a Ljungström air preheater (Chapter 4). While regenerators have been used since the beginning of gas turbine development, modern large-utility gas turbines are not typically equipped with a regenerator. There are several reasons for this. First, with an increase in the compressor pressure ratio, the temperature difference between the turbine outlet flow and the combustor inlet flow reduces. Therefore, in gas turbines with a high compressor pressure ratio, there is less potential for heat recovery by a regenerator. Also, most current large gas turbines are installed as combined cycle power plants where the thermal energy leaving the gas turbine unit is fully recovered by the bottoming steam cycle. So there is no need for a regenerator. Finally, regenerators are bulky and expensive, and require frequent maintenance.

9.5.4.3 Other Auxiliary Components

In this section, a few other auxiliary components that are typically used in gas turbine units will be covered. Let us start with the air inlet to the compressor. Typically, the inlet air flow is not directly in front of the compressor. There might be several reasons for this, including size consideration particularly the length of the system and the fact that the inlet should be sufficiently away from the stack to prevent the recirculation of exhausted combustion products to the compressor. The inlet air first goes through a filter house. The filteration process is to remove foreign particles, both solid and liquid particles, in the inlet air flow. If these particles are allowed to enter the compressor, they can cause erosion, corrosion, and fouling in compressor blades and other downstream components. Several filteration mechanisms are utilized in gas turbine units. This mechanism along with the type of the filter, its fiber size, and the packing density of the media that the filter is made of determines the characteristics of the filter. The design of a filter is always a compromise between the effectiveness of particle removal and the air pressure loss across the filter. Depending on how the filtering media are attached to the fixed structure, filters are divided into three types: rectangular, bag, and cartridge filters. In rectangular-type filters, the continuous filtering media are attached to a flat rectangular supporting structure (in principle similar to a typical residential air conditioner filter but much larger with metal-supporting structure). In bag filters, as the name implies, the filtering media are formed in the shape of a garbage bag, and then many of these bags are connected to a structure to form the filter. In cartridge or cylindrical filters, individual cylindrical filters (very similar to automobile air filters but larger) are connected to a vertical or horizontal supporting plate.

The filtration process in these filters can take place on the surface of the media or deep within the fabric of the media (surface- and depth-loaded filters, respectively). Rectangular and bag filters are typically depth-loaded, while cartridge filters can be depth- or surface-loaded. This distinction is very important to determine how to revive a filter when it is saturated by collected particles after a long period of operation. In a new filter, air can easily pass through clear passages within the filter media with little pressure drop. With the operation of the filter, the collected particles gradually plug these passages, and as a result, the pressure drop increases. This pressure drop across the filter is measured by a pressure gauge. When the measured pressure drop reaches a sufficiently high predetermined value, the filter is saturated and can no longer function as intended. For depth-loaded filters, the only option is to replace the filters. Surface-loaded filters can be cleaned and reused. For this purpose, a flow of relatively high-pressure air is sent through the filter by a series of dedicated air nozzles in the opposite direction of the normal air flow. The cleaning air flow continues for a fraction of a second, which is why these self-cleaning filters are also referred to as pulse filters. To allow a continuous flow of combustion air during the cleaning process, typically a small portion of the filters is cleaned at any given time. Since typically a single filtering stage cannot be effective in capturing particles of all sizes, gas turbine filters can be composed of two or three stages of different types of filters. The decision about the number of stages, their type, and specifications of each stage depends on the level of particles in the local air that needed to be removed, including sand, dust, and organic materials.

Typically, weather louvers (also referred to as weather hoods) are installed in the inlet to the filter house to stop very large pieces and water droplets during rain from entering the filters. After the filter house, there is typically a trash screen to protect the compressor from filters and its parts in an unlikely event that a filter gets loose and is dislodged. The filter house is connected to the compressor via an air duct, sometimes a relatively long one with several changes in the direction of the air flow. To reduce the noise and vibration induced by the air flow and possibly to reduce the pressure drop in the duct by streamlining the flow, the duct may be equipped with intake silencers.

The outlet of the turbine must be sent to the bottoming cycle in combined cycles or discharged to the environment. The exhaust duct may also be equipped with some types of a silencing device. Even gas turbines in combined cycles may have a dedicated stack independent of the bottoming cycle in case they are operated without the bottoming cycle in a single cycle mode. In this case, there is a damper that directs the combustion products to either the stack or the bottoming cycle. The design principle of gas turbine stacks is similar to that of steam cycles but about an order of magnitude shorter.

9.6 Operation of Gas Turbine Units

The combustion process and its control is the most important task in every fossil fuel power plant. This is even more important for internal combustion power plants, such as gas turbines, because the combustion process is not only the source of thermal energy for the system but also formulates the working fluid of the cycle. In general, the control of combustion can be achieved by manipulating the flow rates of fuel and air. Since gas turbines are essentially constant volumetric flow rate machines, the control of the combustion is primarily achieved by changing the fuel flow rate. Therefore, regardless of the type of the gas turbine and its application, the fuel flow rate is the most important factor to control the load. The fuel flow rate is controlled under three modes or loops: frequency or speed control mode, temperature control mode, and start-up mode. The fuel flow rate of a gas turbine unit is always controlled under one of these modes. In the start-up (and shutdown) mode, the start-up sequence controls the fuel flow rate. In the speed mode, the fuel flow rate is set to achieve the required load at the predetermined rotational speed of the shaft (which is tied to the frequency of the generated electricity), while in the temperature mode, the TIT controls the fuel flow rate. At any given conditions, the fuel flow rate is estimated by one of these control modes and then the setpoint of the fuel control valve is set to the minimum of these estimated values.

9.6.1 Start-up Sequence

We know that in order for a gas turbine unit to operate, the compressor should pressurize the inlet air, which is followed by the combustion process in the combustor.[14] Then, the high-temperature combustion products expand in the turbine to generate shaft power. A part of this power is used to run the compressor. What this means for the start-up process is that to run the compressor, turbine power is needed but the turbine cannot generate power unless the compressor runs. Therefore, gas turbine units are not self-starting machines. Like your car's engine, gas turbines need a starter (an external source of power) to start. The starter for gas turbine units is typically an electric motor or a diesel engine. The former can be either a dedicated electric motor or a generator, which can function as a motor during start-up. In either case, obviously, electricity is required to run the starter. In a normal start-up, this electricity is provided by the grid. But if the electricity is not available from the grid, i.e., during a blackout, units with an electric motor as the starter cannot be started. Starting a gas turbine while no electricity from the grid is available is called a **black start**. Therefore, only gas turbines with

[14] Before reading this section on the gas turbine control mechanisms, please make sure you review Section 6.6 on the operation of a generator and its synchronization to the grid.

a diesel engine starter can have a black start. So while most gas turbines have an electric motor starter, in every power grid there are several gas turbines with a diesel engine as the starter. In a blackout, these units are first started to provide the electricity needed for the other units to start. This is one of the main reasons that recovering from a blackout can take many hours if not days.

The first step in a start-up sequence is to activate all the systems required for starting the unit, such as the lubrication oil system, fuel pump (for liquid fuels), fuel compressor (if needed for gaseous fuels), igniters, valves, sensors and instruments, computers, control systems, and other electronics. In a black start, these systems must be powered by a backup battery pack or a small power generation unit. After completion of this step, the starter is engaged to the main shaft to crank the unit. In multiple-shaft units, the gas generator (the portion of the unit with the compressor, combustor, and high-pressure turbine) is cranked. At a certain rotational speed, the fuel is injected into the combustion chamber and igniters start the combustion process. As noted earlier, typically only two igniters initiate the fire on the opposite corners of the combustor, which is then propagated to the entire combustor. The ignition process is completed when the flame detectors detect fire in the entire combustor. Otherwise, if the flame is not detected in a predetermined time period, about 1 minute, the system is purged by the air flow to ensure all unburned fuel is removed before trying again. After the flame establishment, the combustion continues for about 1 minute for the system to warm up. Then, the rotational speed is increased by increasing the fuel flow rate. To prevent compressor surge, the bleed valves are opened. This continues until a certain point when the turbine can run the system without the starter. This is usually referred to as a self-sustaining condition. At this point, the starter is disengaged. The acceleration continues until the rotational speed of the system reaches its design value. The next step is to connect the generator to the grid. To achieve this connection, which is typically referred to as the **generator synchronization** (refer to Section 6.6), the electricity generated by the generator and the grid electricity should have the same characteristics. These characteristics are manifested in frequency, voltage, and phase. Since the frequency of the electricity generated by the generator is a function of the rotational speed of the generator, the synchronization takes place at the generator's design rotational speed by closing the main breaker. After the synchronization of the generator, the start-up procedure is completed and the unit is under the normal operation control system. As will be discussed in the next section, in the next step the generator can be loaded to the desired power generation level.

These steps are commonly known as the **start-up sequence** and can be implemented manually or automatically through a local or remote control system. The sequence can be normally completed in less than 20 minutes, as low as a couple of minutes in some systems. In an emergency and when quick power generation is needed, gas turbines can be started much faster.

However, the fast start-up should be only reserved for exceptional circumstances since it can increase the required system maintenance due to much higher thermal stress involved in the processes. After the completion of the start-up sequence, the unit enters "normal operation" mode. But before discussing the normal operation of gas turbines, let us briefly cover the shutdown sequence.

The normal shutdown procedure is similar to the start-up sequence but in a reverse order: the generator load reduction, the generator desynchronization, closing of fuel valves, opening of bleed valves to avoid compressor surge, reduction of system rotational speed, and the starter engagement. The starter runs the system at a very low rotational speed to allow uniform cooling of the shaft and avoid it bowing (refer to Section 6.4.5 for the same process for steam turbines). The shaft cooling process is very slow and can take many hours. The system can be restarted during the cooling process. Note that in the case of an emergency shutdown, the removal of load, generator desynchronization, and fuel valve closure take place at the same time at once.

9.6.2 Normal Operation

As explained in the previous section, when the generator is synchronized to the grid, the start-up sequence is considered complete and the unit is operated under the normal operation control mode. The most important part of the normal operation control mode is the mechanism to increase or decrease the load or electricity generation in the generator. In Section 6.5, we learned that in steam power cycles, the change in the load is directly achieved by the change in the steam flow rate (via the steam control valve), which in turn is indirectly controlled by the fuel flow rate. However, in gas turbine units, the load change is directly achieved by varying the fuel flow rate. The fuel flow rate can be varied by changing the degree of openness of the fuel valves. The increase in the power generation by opening the fuel valves is commonly termed "raise" in power by operators. The increase in the fuel flow rate can continue until the desired electricity generation is reached and the "raise" order is stopped automatically by the control system or manually by the operator. This is called the frequency or speed control mode.

We know that adding more fuel to the combustion process in the combustor increases the TIT. As discussed earlier, the TIT is limited by the maximum allowable temperature of the blade material of the early stages of the turbine. If the desired load is high enough, the speed control mode can increase the fuel flow rate to the extent that the TIT reaches its limit. Further raise in power generation can result in the TIT exceeding its limit. Clearly, this cannot be allowed. As soon as the TIT exceeds a predetermined setpoint, the temperature control mode takes over the control of the unit from the speed control mode. The temperature control mode maintains the TIT at its setpoint. It is evident that the gas turbine in normal operation is always controlled by the speed mode unless the TIT reaches the maximum allowable temperature (the setpoint). A gas turbine can generate its maximum power when it is controlled under the temperature control mode.

In some units, especially units with heat recovery systems, the air flow rate through the compressor is used to supplement the control of the system. The air flow rate can be controlled either by the variation of the compressor rotational speed (in multirotor units) or by changing the VIGVs of the compressor. The variation of the air flow rate is mainly employed to maintain the gas turbine exhaust temperature at part-load operations of the system to optimize the overall efficiency of the combined gas turbine and heat recovery system. Moreover, VIGVs are used to reduce the air flow rate during start-up to prevent compressor surge.

9.6.3 Gas Turbine Control Systems

We have learned so far that a gas turbine is a sophisticated machine with many interlinked components. The control system acts like the brain of the machine. It receives and processes all the information and data from a variety of sources, including instruments measuring the performance parameters of components, and uses complex algorithms to make decisions on how to operate the gas turbine safely, economically, and within environmental regulations while optimizing its performance. The control system performs start-up, shutdown, and normal operation procedures (both full-load and part-load operations). While the control system operates the unit automatically, it also interacts with operators and provides them with information to monitor the performance of components and the system, particularly any abnormal conditions, potential problems, and alarms. When and if needed, operators can override the automatic operation to manually control the unit. A gas turbine control system is typically a single integrated system that can perform many other tasks besides the control of the unit operation. It can detect problems or potential problems, perform diagnostic procedures to identify the root of problems, protect the system and its component, and create maintenance schedules. Finally, a modern control system can determine the operating conditions for short-term, long-term, and life cycle overall optimization of the unit, usually in real time. The optimization process takes into account many factors, including fuel costs, time-dependent price of generated electricity, labor costs, financial factors (such as interest rates), maintenance, and many other factors with the objective of the maximum system profitability.

9.6.4 Instruments and Sensors

It would be beneficial if you review our discussion on common instrumentations in the power generation industry in Section 3.6. While that section focused on steam power plants, almost all those instruments are equally applicable and commonly used in gas turbine units. In this section, we will focus on how those instruments are used in gas turbines. The readings from these instruments are used to operate, protect, and optimize the unit and its components. In many systems, the protective systems operate independently and have a set of dedicated instruments, while other systems may share some sensors. One attribute of these protective systems is the high degree of redundancy incorporated to them to improve their reliability. It is not uncommon to see double or even triple redundancy in these systems. Too low or too high readings in these sensors can result in alarms to operators, unit unloading, or even emergency shutdown (trip) of the unit. Some of the most important protective systems are for protection against shaft overspeed and excessive TIT. The former protects the shaft rotation speed from exceeding typically 110% of the design shaft rotational speed. The importance of the TIT protection system cannot be overstated, and because of that, a set of dedicated thermocouples (other than those that are used for the fuel flow rate control in the temperature control mode) are used for this purpose. Some other protective systems are to protect the unit against flame loss, excessive vibration in bearings, extreme lubrication oil temperature and pressure, compressor surge, and fuel loss.

Some other important temperatures that need to be measured and monitored are the compressor inlet and outlet flows, lubrication oil, and bearings. Pressure transducers are used to measure the pressure of air or combustion products at the inlet and outlet of the filter, compressor, and turbine. The pressure of fuel and lubrication oil is also monitored in several locations. All major bearings are equipped with vibration measurement sensors to detect any unusual and excessive vibration and provide timely warning to operators to take necessary corrective actions. Gas turbines can also be equipped with many other sensors, including flame detectors, flowmeters, humidity meters, torque meters, and emission meters.

9.6.5 Maintenance of Gas Turbines

Like every complicated machine, power plants in general and gas turbines in particular need continuous planned and unplanned maintenance. The maintenance procedure of every power plant is highly customized to its requirements, and only manufacturers can provide such a guideline. Here, we will very briefly review how the maintenance philosophy has evolved in the past few decades. While this discussion is presented for gas turbines, it is equally applicable to other power generation technologies too.

The most fundamental form of maintenance is to operate a system until something goes wrong or something breaks down and needs repair or replacement. This maintenance strategy is referred to the **corrective maintenance, breakdown maintenance**, or the **fix-as-fail strategy**. The main idea in this maintenance strategy is to minimize the maintenance cost by allowing components to operate as much as possible until they must be repaired or replaced. While power plant operators should be always prepared for corrective maintenance, it is not used as a maintenance strategy in modern plants because an untimely failure of a part can severely reduce the availability of the unit, which can have significant financial ramifications. Also, the failure of a component can destroy or damage other components that can increase costs beyond replacing or repairing just one component.

Experience manufacturers and operators gained by corrective maintenance led to the development of the **preventive maintenance** strategy.[15] In this strategy, specific guidelines are developed for routine inspections of components and systems so a potential problem can be detected as early as possible before an actual failure happens with consequences mentioned in the previous paragraph. This strategy, while much more effective than corrective maintenance, is still a reactionary strategy. This means even in the preventive maintenance strategy, we wait for a sign of trouble before we take any action to mitigate it, which can be often too late. Also, both these strategies aim at reducing maintenance costs and increasing the unit availability and not the optimization of the system performance.

The problems with preventive maintenance strategy led to the **predictive maintenance** strategy, which is currently being used in almost all modern power plants. This strategy is based on the fact that there is a correlation between the performance of the unit and the possibility of the failure of system components. The long maintenance history of units helps to create an algorithm that establishes these correlations. To implement this strategy, the operation of the unit is continuously monitored and used for maintenance predictions. An example of this strategy was explained earlier for the correlation between turbine blade life expectancy and the number of unit start-up, shutdown, and load changes. Modern preventive maintenance systems are now integrated into the overall unit control system. These systems take into account a wide variety of factors to optimize the performance of the system and maximize its profitability.

9.6.6 Standards for Gas Turbine Performance Tests

There is no standard for the design and manufacturing of gas turbines, and each manufacturer has developed its own approach based on its unique experience. However, there are national standards for performance tests of gas turbines. These standards are to ensure the safe and reliable operation of systems and the comparability of reported specifications by manufacturers. These standards define a set of detailed rules and conditions under which the performance tests for gas turbine units must be conducted and reported. In the U.S., these standards are developed and maintained by the ASME and the American Petroleum Institute (API). The ASME B133 standard series includes many volumes on a wide variety of gas turbine-related topics, such as fuels, control and protection systems, exhaust and sound emissions, auxiliary equipment, and electrical equipment. The ASME PTC 22 standard is focused on the evaluation of the units' thermal performance tests.

[15] The regular inspections for your car recommended by the manufacturer is an example of the preventive maintenance strategy.

Similarly, ISO 2314 can be used for the acceptance tests of gas turbines. The ASME PTC 51 is for the testing of gas turbine inlet air-conditioning equipment. Generally, ASME standards are applicable for both utility and industrial gas turbines. The API standard 616 can be only applied to gas turbines that are used in the petroleum, chemical, and gas industry.

9.7 Questions and Problems

9.7.1 Introduction

9.1. List some of the applications of gas turbine units.

9.2. List the three disadvantages of gas turbine engines and the means to address these problems.

9.3. Why are gas turbines ideal power systems to propel aircrafts?

9.4. Which characteristics of gas turbines make them to perfectly suit the needs of peak-load power generation units?

9.5. Gas turbine units are only suitable for peak-load power generation.

a. True

b. False

9.6. Which item is not a characteristic of gas turbine cycles?

a. quick start-up and shutdown

b. rapid load-following

c. low power-to-weight ratio

d. relatively low capital costs

9.7. Which one is an inherent disadvantage of standalone gas turbines?

a. high-temperature exhaust flow

b. inability to consume solid and liquid fuels

c. relatively low efficiency

d. high fuel cost

9.7.2 Thermodynamic of Gas Power Systems

9.8. Compare open versus closed and direct versus indirect gas turbine cycles.

9.9. List assumptions for an air-standard analysis and a cold air-standard analysis.

9.10. Describe the four processes of a Brayton cycle. Draw T-s and P-v diagrams for a simple Brayton cycle.

9.11. In the following T-s diagram for a Brayton cycle, name the four processes and show them in the cycle.

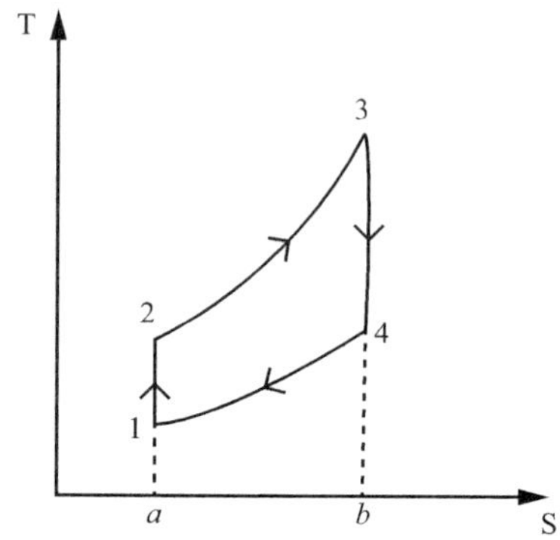

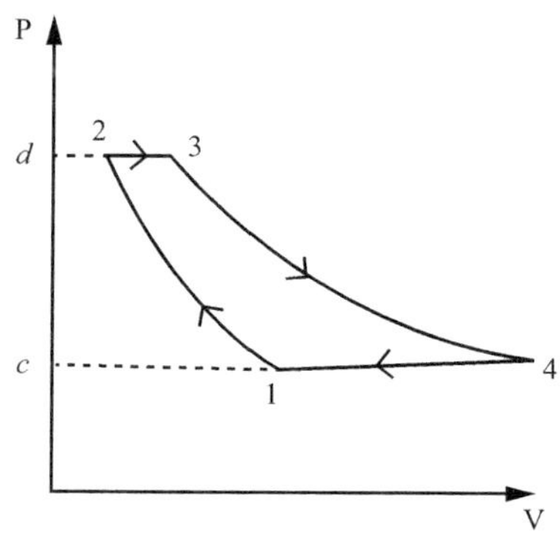

9.12. In the above P-v diagram for a Brayton cycle, show the areas that represent the work per unit of mass involved in the compressor and turbine.

9.13. In the above T-s diagram for a Brayton cycle, explain how the efficiency of the cycle can be determined.

9.14. Why is the difference between specific volume in the compression and expansion processes important in gas turbines?

9.15. Why is the back work ratio of gas turbines much higher than that of steam cycles?

9.16. The results of the cold air-standard analysis of an ideal Brayton cycle can be only qualitatively applied to the practical applications of gas turbine power plants.

a. True

b. False

9.17. The efficiency of the cold air-standard Brayton cycle is only a function of ….

a. the compression ratio of the compressor

b. the specific heat ratio of the working fluid

c. both a and b

d. neither a or b

9.18. What does TIT stand for? Why is it important?

9.19. Increasing the compressor outlet pressure …. the turbine inlet temperature (TIT).

a. increases

b. decreases

c. does not change

d. can increase or decrease

9.20. Determine the efficiency of a simple Brayton cycle with the compressor pressure ratio of 10 and 20. The working fluid of the cycle is air.

9.21. Air enters a compressor of a gas turbine at the temperature of 25°C and leaves the compressor at 400°C. If the cycle operates as an ideal Brayton cycle, determine the efficiency of the cycle.

9.22. The area under the T-s diagram for a reversible process …. equal to the heat transfer during the process.

a. is

b. is not

9.23. Explain what happens to the efficiency and specific work of a Brayton cycle if we increase the compressor pressure ratio with and without a limit on the TIT.

9.24. Increasing the compressor pressure ratio of a Brayton cycle with a limited TIT, …. the cycle efficiency and …. the cycle specific work.

a. increase – decrease

b. decrease – increase

c. increase – increase

d. decrease – decrease

9.25. Using the following two equations justify how the efficiency and net specific work of a simple Brayton cycle change as a function of the compression ratio of the compressor.

$$w_{\text{Net}} = C_p\left[T_3\left(1-\frac{1}{\left(\frac{P_2}{P_1}\right)^{(k-1)/k}}\right)+T_1\left(1-\left(\frac{P_2}{P_1}\right)^{(k-1)/k}\right)\right]$$

$$\eta_{\text{Th}} = 1 - \frac{1}{\left(\frac{P_2}{P_1}\right)^{(k-1)/k}}$$

9.26. The above equations indicate that everything else being equal, there should be a pressure ratio, which maximizes the net specific work output of the cycle.

a. True

b. False

9.27. If a gas turbine is designed and operated for baseload power generation, the optimization of the cycle …. is the most important factor to consider.

a. efficiency

b. mass flow rate

c. specific work

d. TIT

9.28. Explain why for peak-load power generation plants the cycle specific work is more important than the cycle efficiency.

9.29. The specific work output is optimum when the output temperature of the compressor is equal to the output temperature of the turbine.

a. True

b. False

9.30. A gas turbine with a turbine inlet temperature of 1200°C operates at a site with an ambient temperature of 25°C. Determine the compressor pressure ratio that maximizes the net power output of the cycle.

9.31. List the three most important irreversibilities in gas turbines. Rank them according to their magnitude.

9.32. In a Brayton cycle designed for the maximum net power output, there is a limit for integrating a regenerator to the cycle.

a. True

b. False

9.33. What are the similarities and differences between the Ericsson cycle and the Brayton cycle?

9.34. An Ericsson cycle is composed of … and ….

a. two reversible and adiabatic processes – two reversible and constant-pressure processes

b. two isentropic processes – two reversible and isothermal processes

c. two reversible and constant-pressure processes – two reversible and isentropic processes

d. two reversible and isothermal processes – two reversible and constant-pressure processes

9.35. The efficiency of the Ericsson cycle is ….. that of the Brayton cycle and with a regenerator is …. to the Carnot cycle operating between the same high- and low-temperature thermal energy sources.

a. higher than – equal
b. equal – lower than
c. higher than – lower than
d. lower than – equal

9.36. Explain how and why compressor intercooling and turbine reheating can improve the performance of a gas turbine cycle.

9.37. In the following T-s diagram for a gas turbine with an intercooler, explain why

$$h_{2'} - h_1 > (h_2 - h_1) + (h_4 - h_3).$$

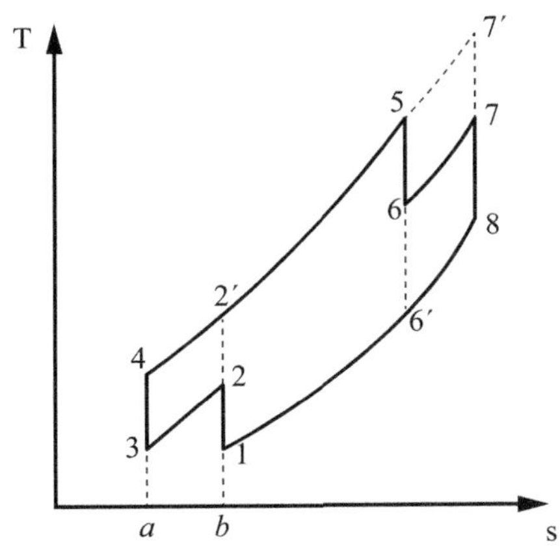

9.38. In the above diagram, describe why heating all the way to state 7′ in the combustor is beneficial but impractical.

9.39. Explain why reheat gas turbines have more potential for waste heat recovery through a regenerator or the integration of the gas turbine into a combined cycle power plant.

9.40. Why can using reheaters in gas turbines help in the reduction of NO_x production?

9.41. In an intercooled gas turbine with one intercooler, determine the pressure ratio of each stage of the compressor that minimizes the power input of the compressor if the total combined pressure ratio of the compressor is 25. The temperature of the inlet flow to the high-pressure compressor is equal to the temperature of the inlet flow to the low-pressure compressor.

9.42. In a reheat gas turbine with three steps of expansion, determine the pressure ratio of each stage of the turbine that maximizes the specific net work output of the turbine if the total combined pressure ratio of the turbine is 27. The turbine inlet flow temperatures are equal.

9.7.3 Applications of Gas Turbines

9.43. What are the three types of non-aviation gas turbines?

9.44. What is the critical characteristic for propulsion systems for transportation applications? Why?

9.45. What are the three common configurations for aviation gas turbines? Explain the principle operation of each type. What are their applications?

9.46. Identify the type of each of the following aviation gas turbines. Which one is used in the majority of commercial and military airplanes?

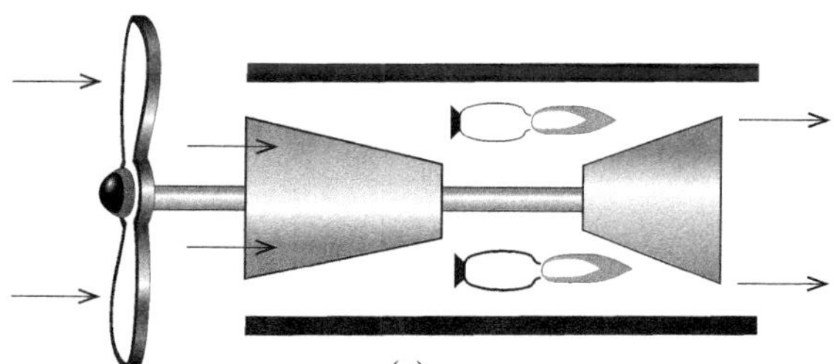

(a)

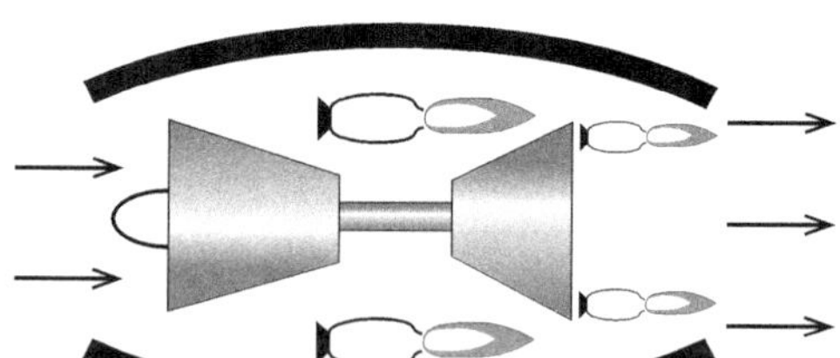

(b)

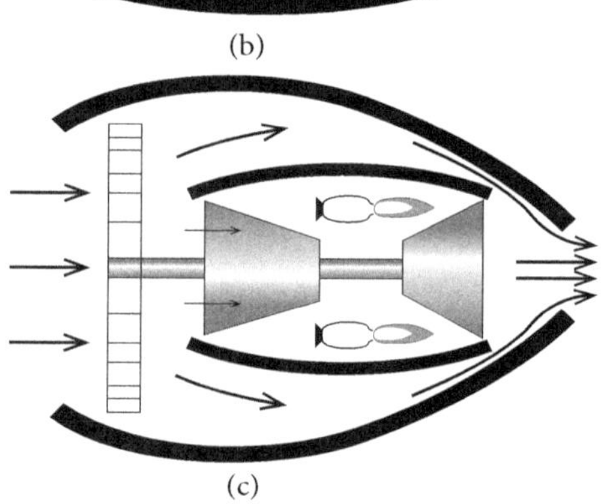

(c)

9.47. What are the advantages of turbofan engines over turbojet engines?

9.48. Why are advanced turbofan engines equipped with two or three shafts?

9.49. Explain how mechanical drive gas turbines are used in pipeline networks to distribute natural gas.

9.50. What is the main difference between aviation gas turbines and stationary gas turbines? What are their similarities and differences in terms of their components?

9.51. Describe why gas turbines in mechanical drive applications often have split-shaft designs.

9.52. What are the differences between aeroderivative and non-aeroderivative gas turbines? Explain the advantages and disadvantages of each type.

9.53. Explain the required processes to convert an aviation gas turbine to an aeroderivative gas turbines.

9.54. What are the design objectives specific for non-aeroderivative gas turbines?

9.55. List the criteria that microturbines should meet for their successful widespread utilization.

9.7.4 Gas Turbine Components

9.56. What is the difference between axial and centrifugal compressors? What type of gas turbine units each type is commonly used?

9.57. What are the functions of the rotor and the stator in an axial compressor?

9.58. Why is an axial compressor for a gas turbine made of several stages? What is the pressure ratio of each stage in modern compressors?

9.59. What are the functions of inlet guide vanes (IGVs) and exit guide vanes (EGVs) in compressors in modern gas turbine units?

9.60. Draw the velocity and pressure diagrams throughout various fixed and moving rows of an axial compressor.

9.61. Define the degree of reaction for an axial compressor. Explain why heavy-duty gas turbines have a higher degree of reaction compared to aeroderivative gas turbines.

9.62. In the following diagram, show the pressure and suction sides.

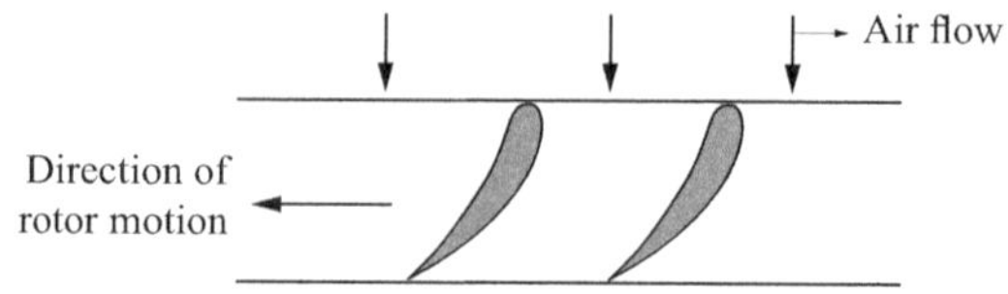

9.63. What are the advantages and disadvantages of centrifugal compressors compared to axial compressors?

9.64. Name various components of a centrifugal compressor in the following schematic.

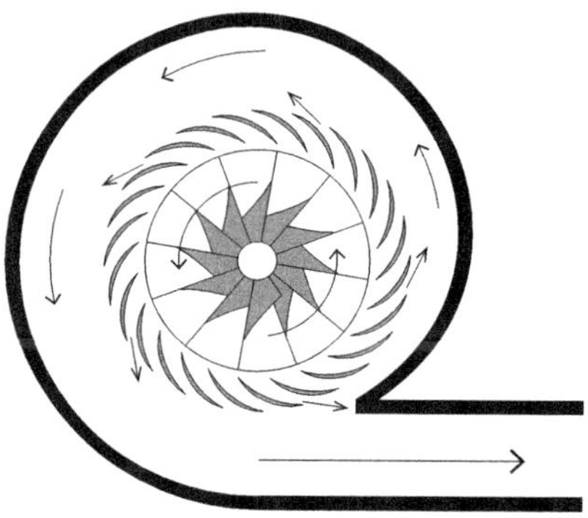

9.65. List some conditions that can lead to the off-design operation of a compressor.

9.66. Draw a simple performance curve of an ideal and real compressor. What factors differ these two curves?

9.67. What are three types of losses in an actual compressor?

9.68. Explain what the blade stall phenomenon is in compressors. How does it limit the pressure increase in each stage of a compressor?

9.69. A compressor can have one or more stages stalled, while other stages are unstalled.

   a. True

   b. False

9.70. Explain how and why compressor surge happens. Use the following compressor performance curve in your explanation.

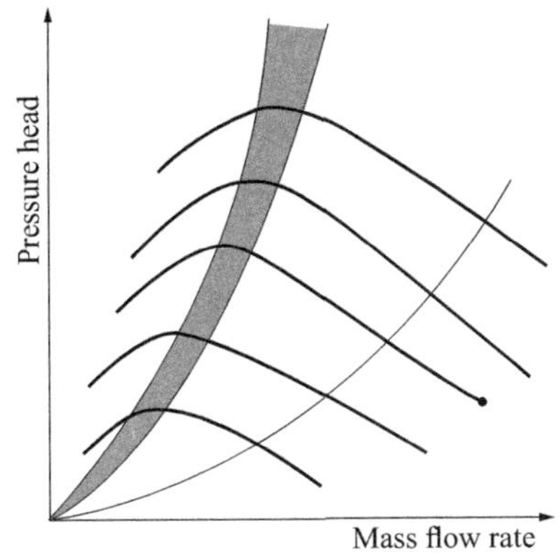

9.71. In an air compressor … has the most damaging effect. Why?

a. stall

b. surge

c. choke

9.72. How can compressor surge be avoided during gas turbine start-up?

9.73. What is compressor choke? When does it happen?

9.74. Why do gas turbines generally need high excess air? Explain how this high excess air impacts the operation of combustors.

9.75. Explain how diffusion-type combustion chambers work.

9.76. What is the difference between straight-through-type and reverse flow-type diffusion combustion chambers?

9.77. What are the functions of primary, secondary, and dilution air flows in diffusion-type combustion chambers?

9.78. Name various parts of a diffusion-type combustion chamber in the following schematic.

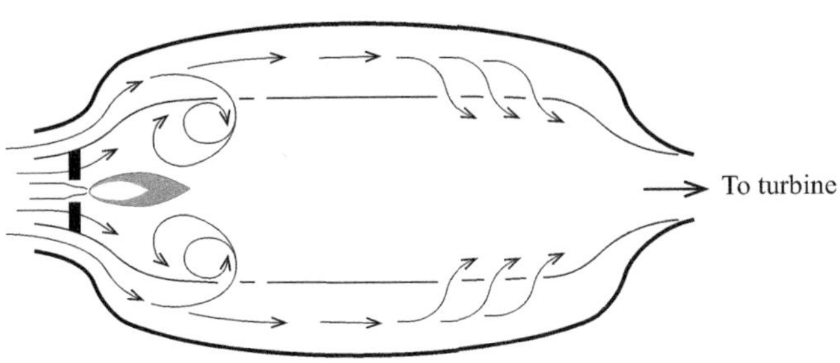

9.79. What are two functions of connecting tubes in a can combustor?

9.80. Explain why can combustors were very popular with early gas turbines and now rarely used in modern large gas turbines.

9.81. How are can combustors and annular combustors similar? What are their differences?

9.82. Explain how silo combustors are different from other combustion chambers. What are their pros and cons? Why are they uncommon in modern gas turbines?

9.83. Name the three most common NO_x control technologies during the combustion process for gas turbines?

9.84. What are the advantages and disadvantages of wet NO_x reduction technologies?

9.85. Briefly explain the operation of DLN/DLE combustors.

9.86. In DLN/DLE combustors why are air and fuel premixed before the combustion process?

9.87. What is the difference between duel fuel combustors and double fuel combustors?

9.88. Generally speaking, gas turbines have a constant volume air flow rate. Why?

a. True

b. False

9.89. How many igniters do can-annular combustors have? How does the flame propagate to other cans?

9.90. Explain why typically impulse stages are not used in gas turbines.

9.91. Explain the tip clearance losses in gas turbines.

9.92. Why are some turbine stages shrouded?

9.93. What are the two main strategies to increase the TIT without damaging the turbine blades?

9.94. What are the two functions of turbine blade coating?

9.95. What are the cooling medium for turbine blade cooling systems?

9.96. Explain how each of the following turbine blade cooling systems works.

a. convection cooling

b. film cooling

c. transpiration cooling

d. multiple small-hole cooling

e. water or steam cooling

9.97. Explain the effect of ambient temperature on the performance of a gas turbine cycle and its reason(s) and how the problems can be overcome.

9.98. An increase in the compressor inlet temperature, i.e., the ambient temperature, will reduce the air mass flow rate through the compressor and will increase the power generation in the turbine.

a. True

b. False

9.99. Compare evaporative cooling systems and refrigeration systems for compressor inlet air cooling systems.

9.100. What is the most important limitation of evaporative cooling systems when used for compressor inlet air cooling of gas turbines?

9.101. Describe the following two evaporative compressor inlet air cooling systems. List their pros and cons.

a. media-type evaporative cooling systems

b. fog cooling systems

9.102. Explain why in fog cooling systems, the injected water needs to be demineralized water.

9.103. Explain why compressor inlet air cooling systems are used in gas turbine units. Name the two types of the compressor inlet air cooling. Explain their operation principles. List pros and cons of each system.

9.104. Explain why modern-utility gas turbine units are not typically equipped with a regenerator.

9.105. Briefly explain how each of the following filtering mechanisms works.

a. Rectangular filters

b. Bag filters

c. Cartridge filters

9.7.5 Operation of Gas Turbine Units

9.106. Explain why in gas turbine units the load change is primarily achieved by the fuel flow rate control.

9.107. List the three gas turbine control modes. Describe the load control mechanism for each mode.

9.108. Explain why some gas turbines are equipped with an electric motor as starter, while others have a diesel engine for this purpose.

9.109. What is a black start of a power generation unit?

9.110. Explain why recovering from a blackout can take many hours if not days.

9.111. Briefly explain a general gas turbine unit start-up sequence.

9.112. Why is a starter needed during the gas turbine unit shutdown sequence?

9.113. Explain how the control system of a typical gas turbine unit changes the unit load.

9.114. Compare the speed control mode and TIT control mode of a gas turbine.

9.115. Which one is not a function of a gas turbine unit control system?

a. the control of the unit operation
b. detect problems or potential problems
c. perform diagnostic procedures to identify the root of problems
d. protect the system and its component
e. create maintenance schedule
f. short-term, long-term, and life cycle optimization
g. all of the above are the functions of a control system

9.116. Name some protective systems in a gas turbine unit.

9.117. List the three maintenance strategies that can be used in a power generation unit. Explain the evolution of these strategies. What is the objective of each strategy? What are their advantages and disadvantages? Which one is currently used in modern power plants?

9.7.6 General

9.118. List inherent disadvantages of standalone gas turbines. How can they be addressed?

9.119. Describe how catalytic combustors operate. Why are they being investigated for modern combustors?

9.120. What are the typical upper limits for the pressure ratio and TIT of large gas turbines?

9.121. Using the following schematic diagram, explain the combustion process in the combustion chamber of a gas turbine.

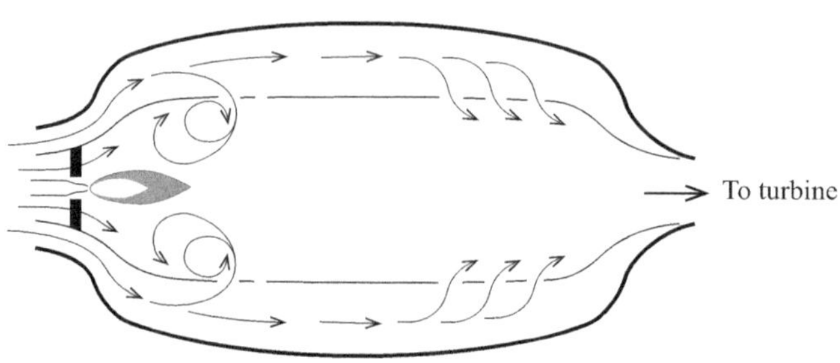

9.122. Using the following equation, explain how the ambient temperature impacts the specific work consumption in a compressor.

$$w_{Comp} = C_p T_1 \left(\left(\frac{P_2}{P_1} \right)^{(k-1)/k} - 1 \right)$$

9.123. Explain how combustors can work.

9.124. Which type of gas turbines has the largest share of gas turbine market?

a. aviation

b. non-aviation

c. military

9.125. What are the functions of the gasifier section and the power section in an industrial gas turbine?

9.126. Why is corrective maintenance the least effective maintenance strategy? How can the preventive maintenance address some of these problems?

9.127. Explain how single-shaft and split-shaft gas turbines are different (be specific about equipment connected to shafts).

9.128. Choose items that describe the characteristics of a gas turbine plant correctly (there are more than one correct choice).

a. quick start-up, shut down, and load change

b. high capital cost per unit of power output

c. short delivery and installation time

d. low power output-to-weight ratio

e. simple design and construction

f. relatively low efficiency

g. high exhaust temperature

h. power output and efficiency independent of ambient conditions

i. no need for water

9.129. Explain how and why rotating or propagating stall occurs in air compressors.

9.130. Is the following statement correct? Why?

"All inlet air is not directly used in the combustion process in a gas turbine".

9.131. Explain why military applications are so important for gas turbine development while it is a very small portion of the global market.

9.132. Which type of compressor inlet air cooling systems needs water?

a. Evaporative cooling

b. Mechanical refrigeration system cooling

9.133. Using the following figure, explain the relationship between the angle of attack and the drag and lift forces.

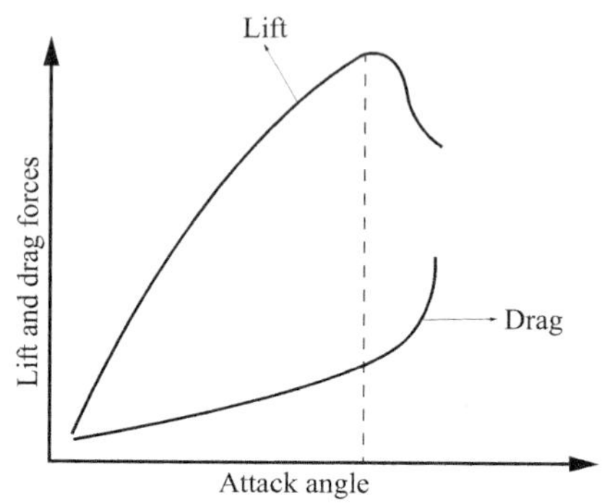

9.134. Explain how can-annular combustors are a crossover between can combustors and annular combustors.

9.135. A Brayton cycle is an air-standard power cycle that involves four reversible processes: two constant-pressure (isobaric) processes and two reversible adiabatic and hence isentropic processes

a. True

b. False

9.136. Define the drag and lift coefficients for an airfoil.

9.137. What is the definition of compression ratio (CR)?

9.138. Name the two main applications of stationary gas turbines.

9.139. Evaporative systems are significantly less expensive to install and operate than refrigeration cooling systems.

a. True

b. False

9.140. Explain fuel supply systems of gas turbines for liquid fuels.

9.141. Describe aeroderivative gas turbines for power generation.

9.142. Is corrective maintenance the least expensive maintenance strategy in long term? Why?

9.143. What type of gas turbines has very high shaft rotational speed, in some cases above 100,000 rpm?

a. Large-sized gas turbines

b. Medium-sized gas turbines

c. Small-sized gas turbines

d. Micro gas turbines

9.144. In the following T-s diagram, explain how using a large number of compression and expansion steps helps the Brayton cycle to get closer to the Ericsson cycle.

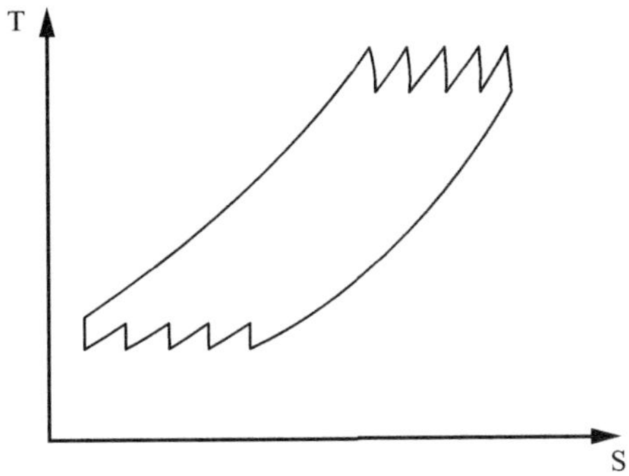

9.145. If a gas turbine is designed and operated for baseload power generation, the compressor pressure ratio should be as high as possible (limited by the TIT).

a. True

b. False

9.146. Typically in what type of gas turbines are centrifugal compressors used?

9.147. What are the pros and cons of the preventive maintenance strategy? How can predictive maintenance address these problems?

9.148. Explain incident or shock losses in a compressor. How are these losses intensified at off-design conditions?

9.149. List some ways and components that can be used to avoid compressor surge during gas turbine operation.

9.150. Briefly explain a general gas turbine unit shutdown sequence.

9.151. In the following T-s diagram, which cycle has the lowest compressor pressure ratio?

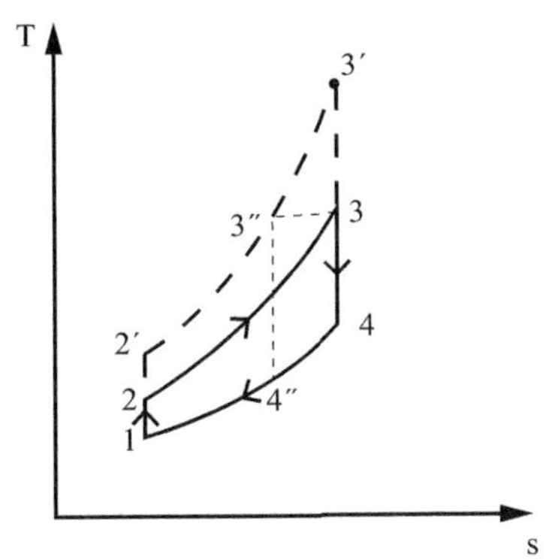

a. 1-2-3-4-1

b. 1-2′-3″-4″-1

c. 1-2′-3′-4-1

9.152. In the above diagram, which cycle has the lowest efficiency?

a. 1-2-3-4-1

b. 1-2′-3″-4″-1

c. 1-2′-3′-4-1

9.153. In the above diagram, which cycle has the lower specific work?

a. 1-2′-3″-4″-1

b. 1-2′-3′-4-1

9.154. In the following T-s diagram for a gas turbine with an intercooler, explain why

$$h_5 - h_{6'} < (h_5 - h_6) + (h_7 - h_8).$$

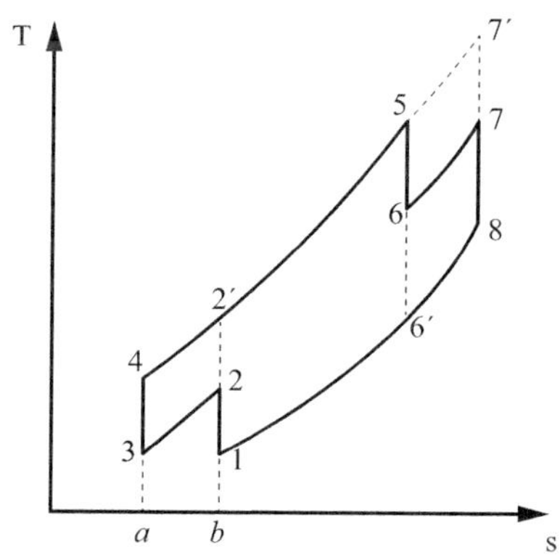

9.155. One of the problems of gas turbine power systems is that the low-temperature thermal energy is exhausted to the environment.

a. True

b. False

9.156. What is a regenerator or recuperator in regenerative gas turbines and what is its function?

9.157. Define the efficiency of a combustor.

9.158. Explain why wet NO_x reduction technologies are not commonly used in modern gas turbines.

9.159. If improperly designed, the operation of compressors at certain off-design conditions can lead to catastrophic consequences.

a. True

b. False

9.160. What is the ultimate objective of aviation gas turbines?

9.161. Describe clearance losses in compressors.

9.162. Other factors being equal, the pressure ratios that the optimum efficiency and specific work output can be achieved are different.

a. True

b. False

9.163. What is the function of a compressor in a gas turbine plant?

9.164. Explain refrigeration systems (indirect cooling systems) for the compressor inlet flow cooling.

9.165. Draw the T-s diagram of a gas turbine with the compressor intercooling system, and compare it with a simple gas turbine cycle.

9.166. Increasing the compressor outlet pressure decreases the turbine inlet temperature (TIT).

a. True

b. False

9.167. How are isentropic (also known as adiabatic) efficiencies of the turbine and compressor defined?

9.168. Name the two major types of gas turbines in terms of their applications.

9.169. Operating at similar conditions, theoretically, which power cycle generally has a greater efficiency, a steam power plant or a gas turbine plant?

9.170. If in the Brayton cycle the isentropic turbine and compressor are both replaced with reversible and isothermal processes, the cycle is called the …. cycle.

a. Ericsson

b. Rankine

c. Carnot

d. regenerative Brayton

9.171. In a reheat gas turbine with one reheater, determine the pressure ratio of each stage of the turbine that maximizes the specific net work output of the turbine if the total combined pressure ratio of the turbine is 16. The temperature of the low-pressure turbine inlet flow is equal to the temperature of the high-pressure turbine inlet flow.

9.172. What is the function of an afterburner?

9.173. Compare evaporative cooling systems and refrigeration systems for compressor inlet air cooling systems in terms of their costs.

9.174. The magnitude of the difference between specific volume during the compression and expansion processes determines the output power and efficiency of a cycle.

a. True

b. False

9.175. Compare the efficiency of two simple Brayton cycles: one with the compressor pressure ratio of 5 and the other with the compressor pressure ratio of 25. The working fluid of the cycles is air.

9.176. Name some military applications of gas turbines.

9.177. The comparison of the air-standard Brayton cycle with isentropic and non-isentropic compression and expansion processes indicates that the temperatures of the compressor outlet and turbine outlet are higher in the non-isentropic cycle.

a. True

b. False

9.178. In the following P-v diagram for a Brayton cycle, name the four processes and show them in the diagram.

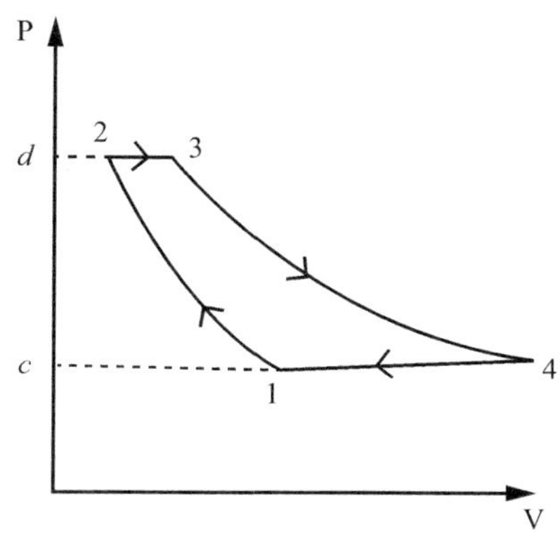

9.179. In the above P-v diagram for a Brayton cycle, show the areas that represent the net work per unit of mass of the cycle and the compressor and turbine work per unit of mass.

9.180. The higher compression ratio of a gas turbine cycle results in the lower efficiency of the cycle.

a. True

b. False

9.181. Why is an Ericson cycle practically impossible?

9.182. Draw the T-s diagram for a regenerative gas turbine.

9.183. Why is it beneficial for the study of gas turbines for electricity generation applications to have some insights to other applications particularly aviation gas turbines?

9.184. The area under the T-s diagram for a reversible process is …. to the heat transfer during the process.

a. equal
b. greater
c. smaller
d. not equal

9.185. Explain why gas turbines with intercoolers have more potential for waste heat recovery through a regenerator.

9.186. What are turboshaft engines?

9.187. What is creep? Why is it important for gas turbine blades?

9.188. Gas turbine units are being used for both baseload and peak-load power generation.

a. True
b. False

9.189. Explain why the continuous cooling of the air flow in the compressor is not practical.

9.190. What are the two methods to keep turbine blade temperatures within their limits while the gas flow temperature can exceed those limits?

9.191. Draw the schematic diagram and the T-s diagram of a modified Brayton cycle with multiple stages of compression with intercooling and multiple stages of expansion with reheaters and a regenerator.

9.192. What is the main difference between aviation gas turbines and stationary gas turbines, in terms of their components?

9.193. Which one is an inherent disadvantage of standalone gas turbines?

a. High-temperature exhaust flow
b. Strong dependency of both power output and efficiency on ambient conditions
c. Inability to consume solid fuels
d. All of the above

9.194. What is the definition of the back work ratio? Why is it important?

9.195. Explain how 2-spool and 3-spool gas turbine engines work.

9.196. Describe the impact of fatigue on gas turbine blades.

9.197. Select all items that indicate the characteristics of gas turbine cycles.

a. No need for water
b. Simple design but high construction cost
c. Short delivery and installation time

d. Low availability
e. Remote control capability
f. Relatively high efficiency
g. High fuel cost
h. Quick start-up and shutdown
i. Rapid load-following
j. Low power-to-weight ratio
k. Relatively low capital costs
l. Low maintenance
m. High capacity factor

9.198. What is the difference between an air-standard analysis and a cold air-standard analysis?

9.199. In an intercooled gas turbine with three steps of compression, determine the pressure ratio of each stage of the compressor that minimizes the power input of the compressor if the total combined pressure ratio of the compressor is 27. The temperature of the inlet flow to the high-pressure compressor is equal to the temperature of the inlet flow to the low-pressure compressor.

9.200. Why is the power-to-weight ratio the most important characteristic for propulsion systems in transportation applications?

9.201. What alloys are modern turbine blades made of?

9.202. An Ericsson cycle consists of two constant-temperature processes and two constant-pressure processes (all reversible).

a. True
b. False

9.203. What is the function of a compressor intercooler?

9.204. What are groups of non-aeroderivative gas turbines according to their power ratings? What is their approximate size range?

9.205. Increasing the compressor pressure ratio of a Brayton cycle with no limit in the TIT, … the cycle efficiency and ….. the cycle specific work.

a. increase – decrease
b. decrease – increase
c. increase – increase
d. decrease – decrease

9.206. Air enters a compressor of a gas turbine at the temperature of 35°C and leaves the compressor at 425°C. If the cycle operates as an ideal Brayton cycle, determine the efficiency of the cycle.

9.207. For a given inlet temperature to the compressor, the inlet temperature to the turbine, and the working fluid, derive the equation for the compressor pressure ratio that maximizes the net specific work output of a cold air-standard Brayton cycle.

9.208. In a cold air-standard Brayton cycle, the compressor inlet air temperature and pressure are 280 K and 100 kPa, respectively. The pressure ratio of the compressor is 10. If the turbine inlet temperature (TIT) is 1400 K, determine (1) the temperature of the compressor outlet and the turbine exhaust, (2) net specific work produced in the cycle, (3) the thermal efficiency of the cycle, and (4) the back work ratio of the cycle, once using the air property table data and then assuming the constant specific heats for air.

9.209. Repeat the previous problem for the compressor pressure ratio between 5 and 40 with an increment of 5. Draw the diagram of the efficiency, net specific work output, and back work ratio as a function of the compressor pressure ratio. Consider the cold air-standard Brayton cycle.

9.210. Repeat the cold air-standard analysis for Problem 9.208 with the isentropic efficiencies of the turbine and compressor equal to 80%.

9.211. Repeat Problem 9.208 for the regenerative cold air-standard Brayton cycle with 100% efficient regenerator.

9.212. Repeat Example 9.1 if the compressor and turbine have the isentropic efficiency of 85%. Compare the results with the results from Example 9.1.

9.213. In the cycle explained in Example 9.1, a regenerator with the effectiveness of 100% is incorporated to the cycle. Determine the energy transfer per second in each component, the net power generation, the back work ratio, and the efficiency of the cycle using data from air property table and the cold air-standard analysis.

9.214. In the nonideal air-standard Brayton cycle, explained in Example 9.5, a regenerator is incorporated to the cycle with the effectiveness of 100%. Determine the heat transfer per second in heat exchangers and the efficiency of the cycle using constant specific heats for air. Repeat your calculation for the regenerator effectiveness of 70%, and compare your results for two cases.

9.215. In an ideal Brayton cycle, a two-step turbine is equipped with a reheater. The temperature of the low-pressure turbine inlet flow is lower than or equal to the temperature of the high-pressure turbine inlet flow. For the fixed given states of the first turbine inlet and the second turbine outlet, derive the equation for the pressure of the reheater that maximizes the power output of the turbine (both steps combined). The turbines operate according to the cold air-standard assumptions.

9.216. In an ideal Brayton cycle, a two-step compressor is equipped with an intercooler. The temperature of the inlet flow to the high-pressure compressor is greater than or equal to the temperature of the inlet flow to the low-pressure compressor. For the fixed given states of the first compressor inlet and the second compressor outlet, derive the equation for the pressure of the intercooler that minimizes the power input of the compressor (both steps combined). The compressors operate according to the cold air-standard assumptions.

9.217. A 200-MW gas turbine operates based on the Brayton cycle. The ambient pressure is 1.1 bar, and the combustion chamber pressure is 22 bar. Determine the thermal efficiency of the cycle (%) and the rate of heat rejection to the environment (MW). Assume the working gas is an ideal gas and the ratio of the specific heats of the working gas is 1.4. Compare the results with those of the Carnot cycle with the same power generation capacity operating between the heat source and sink at temperatures 1200°C and 25°C, respectively.

9.218. In a steam power plant, the difference between the specific volume of the working fluid in the compression and expansion processes is greater than that in a gas turbine (remember that the specific volume of steam is in the order of 10^3 times greater than that of liquid water). Therefore, generally, everything else being equal, the efficiency of steam power plants is higher than that of gas turbines:

a. True

b. False

10

Combined Cycle Power Plants and Cogeneration Plants

Abstract

In the previous chapters, we learned about steam power plants and gas turbine units. In this chapter, we will investigate how we can combine them to achieve an efficiency that is unimaginable for individual plants. First, we will discuss about how a steam cycle and a gas turbine unit can be combined to form a combined cycle power plant (CCPP). We will learn how to calculate the efficiency of a CCPP using efficiencies of its steam cycle and gas turbine unit. Then, we will briefly review various components and configurations of combined cycles. We will place a particular emphasis on the functions and various types of heat recovery steam generators (HRSG) as the interface between the steam cycle and the gas turbine. Finally, we will conclude the chapter by briefly discussing cogeneration plants, also known as combined heat and power (CHP) plants, and integrated gasification combined cycles (IGCC).

LEARNING OUTCOMES

After successful completion of this chapter, you will be able to

- describe combined cycle power plants (CCPPs) and their working principles
- calculate efficiency of a CCPP using efficiencies of its steam cycle and gas turbine unit
- explain various components and configurations of combined cycles
- explain functions and various types of heat recovery steam generators (HRSGs)
- describe cogeneration plants (combined heat and power, CHP)
- explain integrated gasification combined cycles (IGCC)

10.1 Introduction

Technically speaking, combined cycle power plants (CCPPs) can refer to any combination of two or more power generation cycles. However, in the power generation industry, the term specifically refer to when a gas turbine power unit (Brayton cycle) and a steam power plant (Rankine cycle) work together (Figures 10.1). The term "combined cycle" might be somewhat misleading. In CCPPs, the two cycles are not combined, as being integrated, rather they work alongside each other. The exhaust flow of one cycle is the inlet flow of the other cycle. Since in a great majority of cases they are fueled by natural gas, they are also commonly referred to as **natural gas combined cycles** (**NGCC**). Other less common terms that are used for these plants are combined cycle gas turbine (CCGT) or the binary of the Brayton and Rankine cycles.

In Chapter 9, we learned that the temperature of the gas turbine exhaust flow is high, which results in the loss of high-grade energy. Depending on the size and design of a gas turbine (Figure 10.1a), its exhaust temperature can be typically between 450°C and 650°C (about 840°F–1200°F) with the mass flow rate of up 1000 kg/s. On the other hand, the temperature of the superheated steam generated in a typical fossil fuel burning steam generator (Figure 10.1b) can be typically between 500°C and 650°C (about 930°F–1200°F). The principl idea of combined cycles is to take advantage of this temperature overlap of the working fluids of the two cycles to avoid the high-grade energy loss in the exhaust of gas turbines (Figure 10.1c). From the Carnot cycle point of view, a combined cycle takes advantage of a high-temperature heat addition (T_H) of a Brayton cycle and a low-temperature heat rejection (T_L) of a Rankine cycle. In a combined cycle, one or more gas turbine units function as a **topping cycle**, and a steam cycle works as a **bottoming cycle** (Figure 10.2). To recover thermal energy from the gas turbine exhaust flow, the steam generator of the Rankine cycle is replaced by a **heat recovery steam generator** (**HRSG**), also known as a heat recovery boiler. Combined cycles are capable of achieving extraordinarily high efficiency. The combination of gas turbines, which are typically peak-load power generation units, and steam cycles, which are generally baseload power generation units, provides flexibility for these cycles to perform in a variety of demand conditions, both as peak-load and baseload power plants. They have also relatively low capital costs, small footprint, rapid installation period, and low emissions.

Due to their superior efficiency and since they typically burn natural gas as fuel, combined cycles have extraordinarily low-carbon emissions. Their low emissions and small footprint provide them with the flexibility to be installed close to electricity consumption centers even within cities.

The concept of a combined cycle or its variation can be used for replacing, retrofitting, or repowering existing or retiring power plants. These can take place in several levels and to different extents. For example, gas turbine

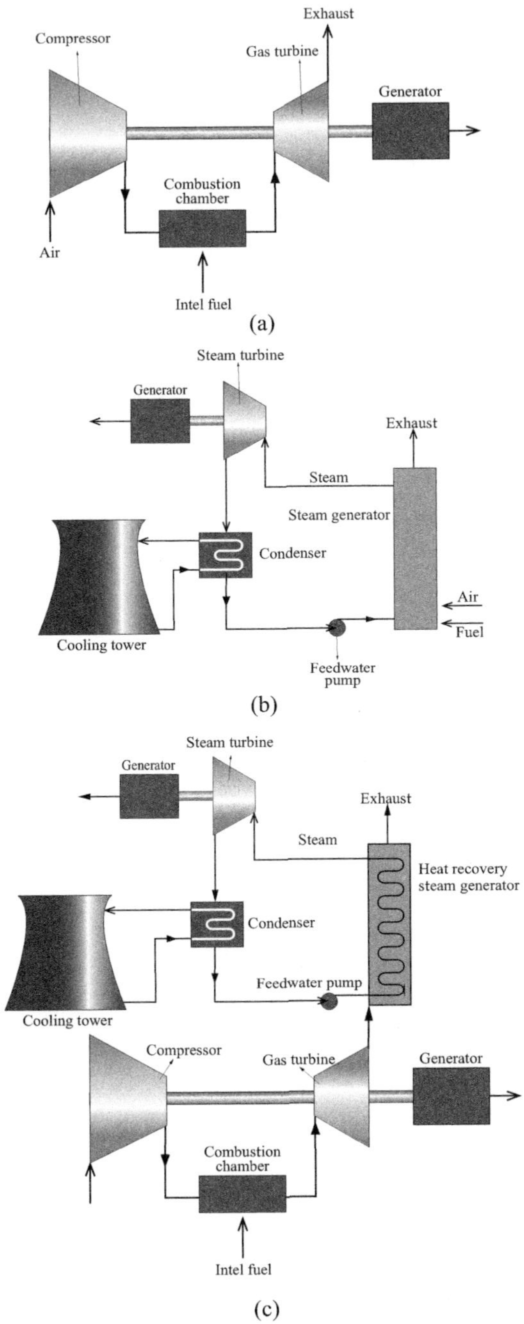

FIGURE 10.1
Schematic of (a) single gas turbine unit, (b) single steam power plant, and (c) combined cycle power plant (CCPP).

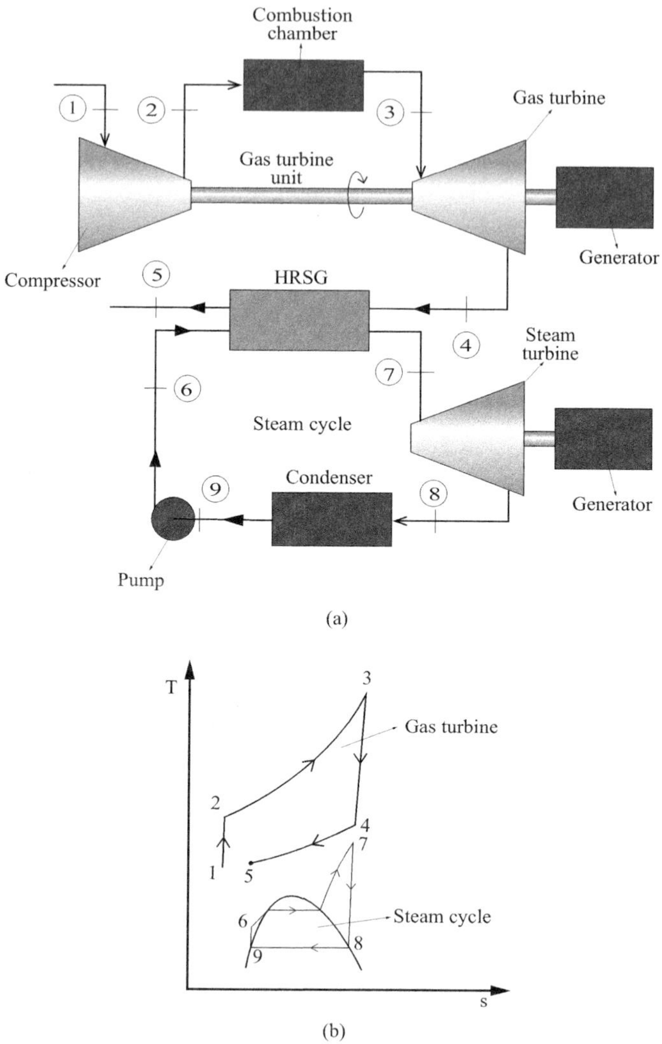

FIGURE 10.2
Schematic (a) and *T-s* diagram (b) of a combined cycle power plant.

units can be added to an existing steam power plant to use their exhaust flow to fully or partly replace the ambient air for the combustion process in the steam generator. Alternatively, the entire steam generator can be replaced by a gas turbine-HRSG assembly with the rest of the steam cycle intact. Even the entire power cycle can be replaced by a combined cycle but using an existing cooling system, power transmission lines, etc.

Generally speaking, in a CCPP, the atmospheric air is first pressurized in the compressor (Figure 10.1c). The high-pressure air flow is then heated in the combustion chamber by burning fuel, typically natural gas. The high temperature and high-pressure combustion products are expanded in the gas turbine to generate electricity in the generator. The exhaust flow of the gas turbine enters the HRSG. In the steam generator, the thermal energy is transferred from the gas turbine exhaust flow to the working fluid of the steam cycle, i.e., feedwater. The superheated steam from the steam generator then expands through the steam turbine and generates further electricity. The saturated mixture of steam and liquid water is condensed in the condenser and is pumped back to the HRSG.

Sadi Carnot (1796–1832, a French engineer and one of the founders of thermodynamics as a branch of science) was the first person who laid the foundation of combined cycles when he wrote about how the hot outlet air flow of an engine can be used to create low-temperature steam instead of being thrown out. A more specific concept was introduced in the early 20th century. Early experimental CCPP units were installed in the 1950s. Early commercial units were commissioned in the early 1960s, which was followed by the installation of tens of units in the 1970s. This was primarily due to the fact that improvements in the gas turbine blade technology allowed a significant increase in the turbine inlet temperature (TIT). This in turn resulted in a high enough temperature of the turbine exhaust, which made an effective integration of the HRSG economically feasible. However, a breakthrough in the widespread utilization of CCPPs came in the 1990s when very large and efficient gas turbines with power generation capacities of up to about 250 MW (extraordinarily large units with those days' standards) were introduced. This pattern still continues with gas turbine units with capacities up to about 500 MW. Currently, CCPPs are quickly replacing other forms of fossil fuel power generation systems particularly when natural gas is available as fuel. Even with the high contribution of electricity generated from renewable energy sources, combined cycles are playing and will continue to play a vital role in providing power to smooth their intermittent nature (more on this topic in Chapter 16). They are even being integrated into solar thermal energy power plants to supplement these units.

10.2 Efficiency of Combined Cycle Power Plants

The efficiency of a CCPP as a function of the efficiency of its gas turbine unit and steam cycle can be determined as follows (Figure 10.3):

$$\eta_{CCPP} = \frac{\dot{W}_{CCPP}}{\dot{Q}_{GT,\,in}} = \frac{\dot{W}_{GT} + \dot{W}_{St}}{\dot{Q}_{GT,in}} \tag{10.1}$$

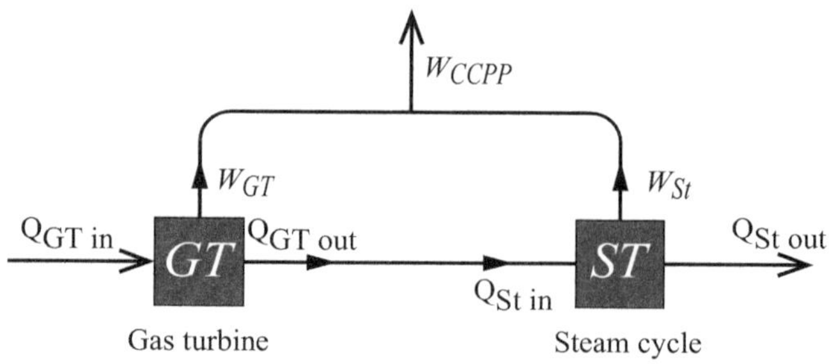

FIGURE 10.3
Flows of heat and work in a combined cycle power plant.

where η_{CCPP} is the efficiency of the combined cycle, $\dot{W}_{GT}$ is the output power of the gas turbine unit (W), $\dot{W}_{St}$ is the output power of the steam cycle (W), and $\dot{Q}_{GT,\,in}$ is the time rate of the input thermal energy of fuel in the combustion chamber of the gas turbine (W).

The efficiency of the gas turbine and steam cycle are

$$\eta_{GT} = \frac{\dot{W}_{GT}}{\dot{Q}_{GT,\,in}} \tag{10.2}$$

$$\eta_{St} = \frac{\dot{W}_{St}}{\dot{Q}_{St,\,in}} \tag{10.3}$$

where η_{GT} is the efficiency of the gas turbine unit, η_{St} is the efficiency of the steam cycle, and $\dot{Q}_{St,\,in}$ is the time rate of the input thermal energy to the steam cycle from the outlet of the gas turbine (W). Assuming that no energy is lost between the exhaust of the gas turbine and the inlet of the HRSG (which is a very reasonable assumption considering their physical proximity),

$$\dot{Q}_{St,\,in} = \dot{Q}_{GT,\,out} \tag{10.4}$$

Substituting Equation 10.4 in 10.3,

$$\eta_{St} = \frac{\dot{W}_{St}}{\dot{Q}_{GT,\,out}} \tag{10.5}$$

Rearranging Equations 10.2 and 10.5,

$$\dot{W}_{GT} = \eta_{GT}\dot{Q}_{GT,\,in} \tag{10.6}$$

$$\dot{W}_{St} = \eta_{St}\dot{Q}_{GT,out} \tag{10.7}$$

From the energy conservation law (the first law of thermodynamics) for the gas turbine cycle,

$$\dot{W}_{GT} = \dot{Q}_{GT,in} - \dot{Q}_{GT,out} \tag{10.8}$$

Substituting Equation 10.8 in 10.2,

$$\eta_{GT} = \frac{\dot{Q}_{GT,in} - \dot{Q}_{GT,out}}{\dot{Q}_{GT,in}} \tag{10.9}$$

Rearranging Equation 10.9,

$$\eta_{GT}\dot{Q}_{GT,in} = \dot{Q}_{GT,in} - \dot{Q}_{GT,out} \tag{10.10}$$

$$\dot{Q}_{GT,out} = \dot{Q}_{GT,in} - \eta_{GT}\dot{Q}_{GT,in} = \dot{Q}_{GT,in}\left(1 - \eta_{GT}\right) \tag{10.11}$$

Substituting Equation 10.11 in 10.7,

$$\dot{W}_{St} = \eta_{St}\dot{Q}_{GT,in}\left(1 - \eta_{GT}\right) \tag{10.12}$$

Substituting Equations 10.6 and 10.12 in Equation 10.1,

$$\eta_{CCPP} = \frac{\eta_{GT}\dot{Q}_{GT,in} + \eta_{St}\dot{Q}_{GT,in}\left(1 - \eta_{GT}\right)}{\dot{Q}_{GT,in}} \tag{10.13}$$

$$\eta_{CCPP} = \eta_{GT} + \eta_{St} - \eta_{GT}\eta_{St} \tag{10.14}$$

As noted, CCPPs can achieve an extraordinarily high efficiency. Until recently, the efficiency of above 60% seemed unachievable for any type of thermal power generation technology. In May 2012, a 570 MW CCPP designed and built by Siemens in southern Germany, called Irsching Power Station, was eventually able to break the 60% efficiency barrier by achieving the efficiency of 60.75%.

Example 10.1

The largest and most efficient gas turbine that is currently available in the market is the GE 9HA.01 gas turbine with an efficiency of 41% manufactured by GE Power. When combined with a steam cycle, the efficiency of the cycle increases to 61%. Determine the efficiency of the steam cycle that is used in this CCPP.

$$\eta_{CCPP} = \eta_{GT} + \eta_{St} - \eta_{GT}\eta_{St}$$

$$0.61 = 0.41 + \eta_{St} - 0.41 \times \eta_{St}$$

$$\eta_{St} = 0.3390 = 33.90\%$$

Example 10.2

A 500 MW CCPP has an efficiency of 55%. If the efficiency of the gas turbine cycle is 36%, determine the power output of the gas turbine and the steam cycle (in MW), the mass flow rate of natural gas burned in the combustion chamber of the gas turbine (in kg/s), and the thermal energy exhausted by the HRSG (in MW). The heating value of natural gas is 45 MJ/kg.

$$\dot{Q}_{GT,\,in} = \frac{\dot{W}_{CCPP}}{\eta_{CCPP}} = \frac{500}{0.55} = 909.1\,\text{MW}$$

$$\dot{W}_{GT} = \eta_{GT} \times \dot{Q}_{GT,\,in} = 0.36 \times 909.1 = 327.3\,\text{MW}$$

$$\dot{W}_{St} = \dot{W}_{CCPP} - \dot{W}_{GT} = 500 - 327.3 = 172.7\,\text{MW}$$

$$\dot{Q}_{GT,\,in} = \dot{m}_{Fuel} \times HV$$

$$\dot{m}_{Fuel} = \frac{\dot{Q}_{GT,\,in}}{HV} = \frac{909.1}{45} = 20.2\ \text{kg/s}$$

$$\dot{Q}_{St,\,out} = \dot{Q}_{GT,\,in} - \dot{W}_{GT} - \dot{W}_{St} = 909.1 - 327.3 - 172.7 = 409.1\,\text{MW}$$

10.3 Configuration of Combined Cycle Power Plants

As noted, any CCPP is composed of a gas turbine unit and a steam cycle. Since the HRSG is the interface between these two cycles, we will treat it separate from the cycles. In this section, we will briefly discuss the two cycles since they are very similar to standalone steam cycles and gas turbines that were covered extensively in the previous chapters. Then, because of the unique characteristics of HRSGs, in the next section, we will discuss them in more detail. In terms of how these two cycles are assembled, there are several common configurations. For example, each gas turbine can have a dedicated HRSG and steam cycle, or a pair of gas turbines each can have an HRSG but share a single steam cycle.

The gas turbine section of combined cycles is exactly the same as standalone gas turbines. In fact, most gas turbine manufacturers offer their systems as standalone units or combined cycles. During the system operation, the backpressure imposed by the pressure drop in the flue gas flow in the HRSG slightly reduces the efficiency of the gas turbine compared to that of the standalone unit. Some cycles have a stack right after the gas turbine and before the HRSG (this stack is separate from the main stack after the HRSG). This stack allows the gas turbine to operate without the steam cycle being in service. There is a diverter damper that controls whether the flue gas flow is sent to the stack or the HRSG.

Between the gas turbine and the HRSG there is an interconnection that connects the exhaust of the gas turbine unit to the inlet duct of the HRSG. The function of the inlet duct is to uniformly distribute the flue gas at the full length of heat-transfer surfaces. The flue gas flow passes through the HRSG and raises steam before being discharged to the atmosphere. The steam from the HRSG is sent to the turbine-generator assembly to generate electricity. The gas turbine and steam turbine can be connected to the same generator or have their dedicated ones. The saturated steam from the turbine is condensed in the condenser by the cold water from the cooling system, typically a cooling tower. The condensate from the condenser is pumped by the condensate (extraction) pump to the deaerator (open feedwater heater). The deaerated feedwater is pumped back to the HRSG by the feedwater pump. Combined cycles are typically equipped with one or more closed feedwater heaters before or after the deaerator. It should be noted that the structures and foundations of the gas turbine, HRSG, and stack are separated by expansion joints.

10.4 Heat Recovery Steam Generators (HRSGs)

Since the ultimate objective of both HRSGs and conventional fossil fuel-steam generators is to absorb the maximum amount of thermal energy from a flue gas flow, they share many common components. However, they have some significant differences too. The first and most consequential one is the lack of the furnace in HRSGs. Although they may have a supplementary firing system, they do not have a dedicated space for fuel burning. From the heat-transfer point of view, this means that the main heat-transfer mechanism for heat transfer from the flue gas flow to the water or steam flow inside tubes is convection and the contribution of radiation is small. One result is that HRSGs do not have waterwall tubes as the boiler (evaporator). Instead, a certain section of heat-transfer surfaces (tube bundles) is dedicated for this purpose. While some combined cycles are fueled with fuel oil or synthetic gases (more on this topic later in this chapter), a majority of CCPPs are natural gas-burning units. This simplifies some of the required components.

HRSGs can be drum-type or once-through. While the former is currently more common, the latter is becoming popular due to their rapid start-up and better load-following characteristics. These steam generators can be natural or force circulation often used in horizontal and vertical configurations, respectively.

HRSGs can be equipped with a supplementary firing system, called a **duct burner**, to increase the temperature and the mass flow rate of the flue gas before or while going through the HRSG. Accordingly, a duct burner can be installed at the entrance of the HRSG or within the HRSG. Commonly, the flue gas flow from the gas turbine(s) is rich in oxygen and contains sufficient oxygen to support the secondary combustion process in the duct burner. Otherwise, some fresh air may be added to the flue gas for this purpose. Using a duct burner provides more control over the operation of the steam generator. This allows to compensate for the variations of temperature and mass flow rate of the gas turbine outlet flow due to changes in the ambient conditions or during part-load operation. This is particularly critical if the temperature of the generated steam has a narrow window of acceptable variations. However, adding a duct burner not only increases the capital cost of the plant, but also the direct burning of fuel reduces the overall efficiency of the cycle.

In terms of the direction of the flue gas through the steam generator, HRSGs can be horizontal or vertical. In the horizontal configuration, the vertical heat-transfer surfaces are located on the horizontal flue gas flow (Figure 10.4). The tubes can be supported from the top or bottom. This configuration provides a simpler and more accessible (for maintenance) layout. In the vertical configuration, the heat-transfer surfaces are arranged horizontally with the flue gas flow passing through them at the vertical direction. The tubes are supported vertically by a support structure hanging from top. While slightly more compact, the vertical configuration requires more steel structure and more complex layout of heat exchangers. For these reasons, the horizontal configuration is more common in current HRSGs.

According to the steam pressure in the HRSG, combined cycles can be divided to two different types: single-pressure and multiple-pressure (double pressure or triple pressure) HRSGs. In the following sections, we will discuss these HRSGs.

10.4.1 Single-Pressure HRSGs

Single-pressure drum-type HRSGs are in principle very similar to conventional fossil fuel-steam generators discussed in Chapter 3. Their major components are the economizer, boiler (evaporator), drum, and superheater. They may or may not be equipped with reheaters. The superheater and reheater are close to the flue gas entrance. This means the lowest level heat-transfer surfaces in the vertical configuration or the ones closest to the gas turbine in the horizontal configuration. The economizer is typically right before the stack.

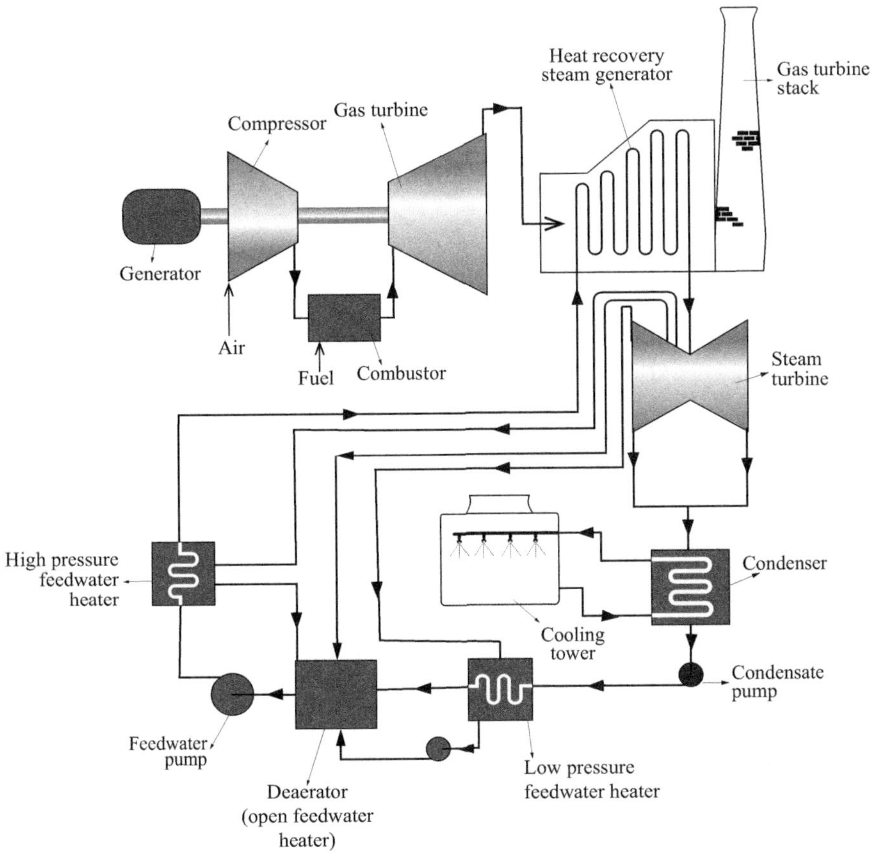

FIGURE 10.4
Schematic of a combined cycle power plant.

The temperature as a function entropy diagram of the water flow inside the tubes and the flue gas flow outside of the tubes are shown in Figure 10.5. As can be observed in the diagram, in the economizer compressed liquid from the feedwater heaters is heated to near-saturated liquid before being fed to the drum. The temperature of the water leaving the economizer and entering the drum is slightly lower than the saturation temperature of the feedwater (the temperature of the boiling water in the drum). This temperature difference, called the **approach temperature**, is to ensure that even at part-load operation the boiling process does not occur in the economizer, which can damage the heat-transfer surfaces of the economizer. The approach temperature can vary between 5°C and 20°C (about 10°F and 35°F), with about 10°C (about 18°F) being a common value. A larger approach temperature results in less heat-transfer surface area in the economizer, which must be compensated by more heat-transfer area in the evaporator.

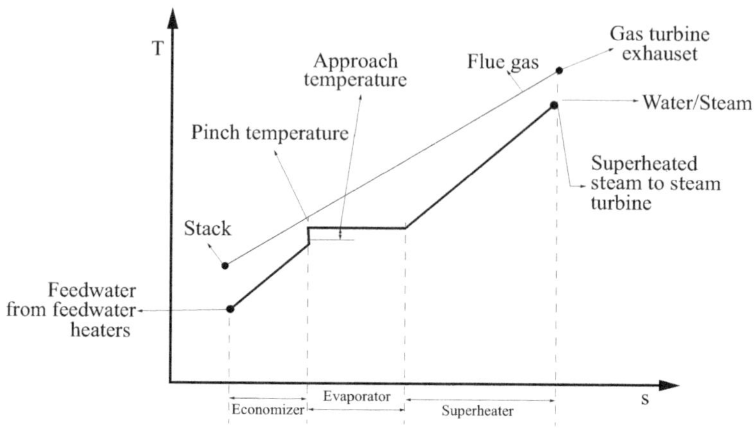

FIGURE 10.5
T-s diagram of a single-pressure HRSG.

The boiling process takes place in the evaporator-drum assembly followed by the steam superheating process in the superheater. The superheated steam is then sent to the turbine at the pressure up to about 180 bar and the temperature of up to about 600°C. The superheated steam temperature is controlled by an attemperator (refer to Section 3.6.2), which in this case is typically referred to as a **desuperheater**.

The temperature of the flue gas flow outside the tubes continuously decreases. As expected, the temperature of the flue gas is always higher than that of the water. These two temperatures are closest at the saturation temperature of the water at a point, called the **pinch point** (refer to Section 2.3.5.2). The difference between the flue gas temperature leaving the evaporator and the water boiling temperature is called the **pinch temperature**. Reducing the pinch temperature decreases the temperature difference between the two flows, which in turn reduces the irreversibilities in the heat-transfer processes and increases the efficiency. However, a steam generator with a lower pinch temperature requires a larger heat-transfer area (and a more expensive one) to achieve same heat transfer compared to that of a steam generator with a higher pinch temperature. Therefore, the selection of an appropriate pinch temperature, typically 5°C–17°C, involves a tradeoff between the system size versus the efficiency or more to the point the initial cost versus the operating cost.

10.4.2 Multiple-Pressure HRSGs

A close observation of Figure 10.5 indicates that there is a significant temperature difference between two flows in most of single-pressure HRSGs. This temperature difference causes a high level of irreversibility in the steam generator and a high temperature of the flue gas leaving the steam generator both of which negatively impact the efficiency. To reduce these negative impacts, multiple-pressure HRSGs can be utilized in CCPPs. Figure 10.6 shows that in a multiple-pressure HRSG, the temperature difference between the two flows is much smaller than that in a single-pressure HRSG (Figure 10.5) throughout the steam generator. These HRSGs have significantly higher efficiencies compared to those of single-pressure HRSGs. Dual- and triple-pressure HRSGs are currently being employed, with the latter being the most common among the three types of HRSGs. While having more pressure levels can slightly increase the cycle efficiency, the gain cannot justify the increase in the capital cost and increased in the complexity of the system. The number of pinch points for these steam generators is equal to the number of pressure levels.

Triple-pressure HRSGs have three loops with high pressure (HP), intermediate or medium pressure (IP or MP), and low pressure (LP). Increasing the high pressure and/or decreasing the low pressure improve the efficiency of the HRSG. However, the latter increases the required heat-transfer surfaces and piping size, which in turn increases the capital cost of the unit. This means there are economic limits on how low this pressure can be. In the most complex configuration, each pressure level has a dedicated economizer, boiler (including evaporator and drum), superheater, feedwater pump, and possibly reheater. These cycles also typically have three steam turbines each corresponding to a pressure level. The outlet flows from the HP and/or IP turbines are typically mixed with the inlet flow of the next lower pressure steam turbine. Furthermore, in some plants a single high-pressure feedwater pump is used, where the feedwater for the low-pressure circuit is provided by reducing the water pressure in a valve.

There are a wide variety of configurations for HRSGs. In some designs, the LP economizer is eliminated, or the LP and IP levels share the same

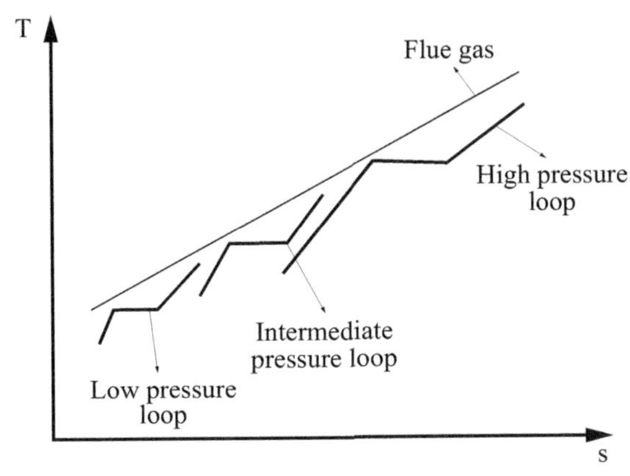

FIGURE 10.6
T-s diagram of a triple-pressure HRSG.

economizer while in others the HP and/or LP economizer, superheater, or reheater are divided into two or even three sections. The overall objective is to have a closed match between the flue gas temperature, and those of the water/steam flows across the steam generator in a most economically sound way. The layout of these heat-transfer surfaces is according to their outlet temperature. The surfaces with higher temperatures are closer to the flue gas inlet flow from the gas turbine. For example, the superheaters of intermediate and low-pressure circuits may be located downstream of the high-pressure boiler (evaporator). It is not uncommon to have over 20 heat-transfer surfaces to be distributed along an HRSG.

10.5 Cogeneration Plants

As the name implies, **cogeneration plants**, also known as **combined heat and power (CHP)**[1] plants, are able to generate both electricity and heat, in the form of steam or hot water, using a single source of fuel (Figure 10.7). The thermal efficiency (but not the electrical efficiency) of these plants is much higher than that of individual systems. Their efficiency can be determined by the following equation:

$$\eta_{\text{Cogen}} = \frac{\dot{W}_{\text{Elec}} + \dot{Q}_{\text{Cogen}}}{\dot{Q}_{\text{In}}} \tag{10.15}$$

where η_{Cogen} is the overall efficiency (the thermal and electrical efficiencies combined) of the cogeneration plant, $\dot{W}_{\text{Elec}}$ is the generated electrical power in the plant (W), $\dot{Q}_{\text{Cogen}}$ is the time rate of the generated thermal energy in the plant (W), and $\dot{Q}_{\text{In}}$ is the time rate of the input thermal energy (from fuel combustion) of the plant (W). The thermal energy generated in these plants can be used to provide either steam or hot water for industrial applications, such as food processing and paper manufacturing, or thermal energy for building heating systems. Small- and medium-size cogeneration plants have gained remarkable popularity in the past few decades to provide superheated or saturated steam for indusial processes or to provide heat (and of course electricity) for space heating in large complexes, such as airports or university campuses.

Technically speaking, cogeneration plants can be based on gas turbine power plants, steam power plants, internal combustion engines, or fuel cells. The main design objective of these plants can be different from that of typical

[1] Recently, the idea of trigeneration or combined cooling, heat, and power (CCHP) are also being investigated.

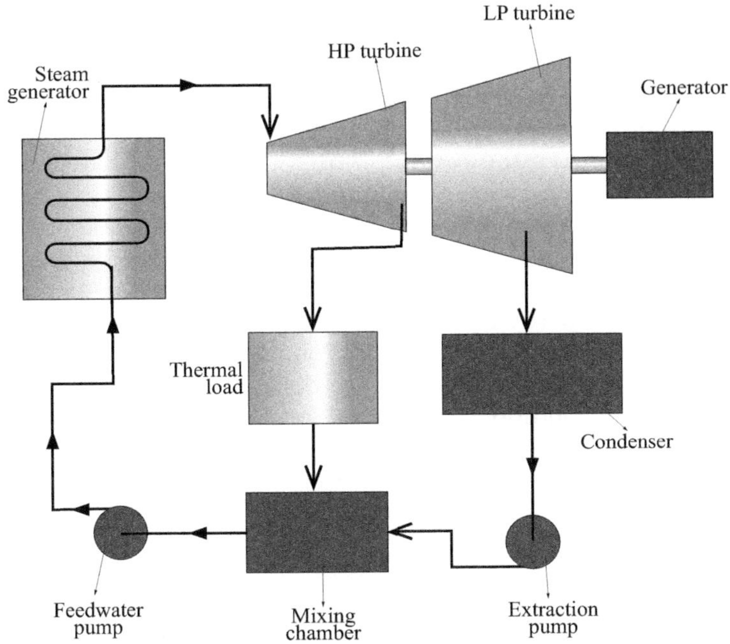

FIGURE 10.7
Schematic of a typical extraction cogeneration plant.

power plants. In some applications, the main objective is to provide thermal energy (typically steam) at desired conditions at all times rather than to maximize electricity generation and its efficiency. The design of these units can be challenging since the electrical and thermal loads may not coincide. For this reason, there is a wide variety of configurations depending on the requirements of a specific application.

These plants can be topping cycles or bottoming cycles. In topping cycles, electricity is generated first and then the remaining thermal energy is recovered from the plant exhaust flow for the thermal load. In these plants the thermal needs can be provided by the outlet flow of a steam turbine (known as back-pressure CHP plants, below), can be extracted from the stages of a steam turbine (referred to as extraction CHP plants, below), or can be recovered from the flue gas from the combustion process of a power generation unit. The former two types are based on steam power plants while the latter is used in gas turbines or internal combustion engines.

In bottoming cycles, the thermal energy for electricity generation is recovered from the exhausted energy of the thermal load. In these plants, the thermal load is directly provided by an HRSG (or boiler) or a furnace. This is typically used when thermal energy at a high temperature is needed for process requirements.

10.5.1 Back-Pressure CHP Plants

In these cycles, either the outlet flow of the turbine is used as a source of thermal energy or thermal energy rejected in the condenser is used for this purpose. In the former configuration, there is no condenser to create low pressure at the outlet of the turbine, which significantly reduces the efficiency of the power generation portion of the unit. Similarly, if a condenser is used as a source of thermal energy, the temperature of the condenser and as a result its pressure increases, which again increases the backpressure of the steam turbine. Increasing the temperature and pressure of the needed thermal energy will increase the backpressure of the turbine, which in turn will reduce its efficiency. In these plants, typically some provisions should be considered to ensure that the heat load is always met regardless of the turbine operational conditions. For example, a bypass line can be included across the turbine to directly send the superheated steam to the thermal load. Commonly, a desuperheater is employed to control the temperature of the steam outlet flow before being sent to the external thermal load.

10.5.2 Extraction CHP Plants

This type of CHP plant is very similar to conventional steam power plants, but in this case, the steam required to provide thermal energy is extracted from the steam turbine (Figure 2.7) (similar to the steam extracted for feedwater heaters). Obviously, extracting steam from the turbine reduces the power output of the turbine. This configuration is very flexible for varying thermal energy requirements.

10.6 Integrated Gasification Combined Cycles

As noted in Chapter 9, one of the limitations of gas turbine units is that they cannot be fueled by solid fuels, such as coal, biomass, solid municipal waste, or even a mixture of solid fuels. **Integrated gasification combined cycles (IGCC)** are a way to overcome this limitation by integrating a gasifier and a combined cycle. In these cycles, solid fuel is first converted to gaseous fuel, called **syngas** (synthesis gas), by the gasification process (Figure 10.8). In the simplest term, the gasification process involves partial combustion of the solid fuel with oxygen in the presence of steam. The required oxygen is produced in an air separation unit (ASU), which is a part of the plant. The exact composition of the produced syngas depends on the type of feedstock (fuel), the used gasification process, and its conditions. The gases that contribute to the heating value of the syngas are hydrogen, carbon monoxide, and to lesser degree methane. The generated syngas is then fed to the combustion

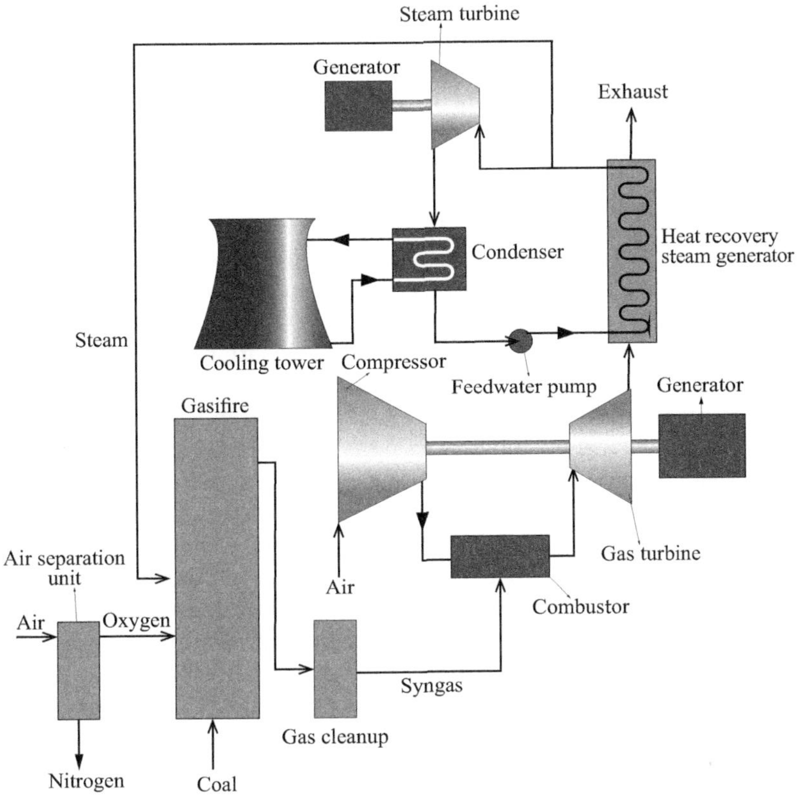

FIGURE 10.8
Schematic of an integrated gasification combined cycle (IGCC).

chamber of the gas turbine in a CCPP. These units have a great potential to use precombustion carbon dioxide capture systems with much lower costs compared to that of conventional post-combustion carbon dioxide capture systems (for more information refer to Chapter 5). A typical plant can include several stages of syngas cleaning such as the particulate, mercury, and sulfur removal processes. The processes in a gasifier require certain temperature ranges. To achieve these temperatures several heat recovery heat exchangers are typically employed in the gasifier.

IGCC plants can achieve efficiency at the low 40%, which is significantly higher than that of other coal-based power generation units, particularly if carbon dioxide capture systems are included. However, their high capital cost is a substantial obstacle for their widespread use. Several proof-of-concept and demonstrational plants have been constructed since the 1990s with varying degrees of success. The performance tests of these plants can be conducted according to ASME PTC 47 standard (IGCC Power Generation Plants).

10.6.1 Gasification Process

Syngas is a gaseous fuel mainly made of carbon monoxide (CO), hydrogen (H_2), carbon dioxide (CO_2), nitrogen (N_2), and some hydrocarbons (such as CH_4, C_2H_4, and C_2H_6), and trace amounts of H_2S, NH_3, and tar. The use of syngas as a fuel is not a new technology. For example, during World War II, the shortage of transportation fuels led to the conversion of millions of vehicles to syngas-fueled engines in Europe where wood or coal was used as the feedstock. Syngas can be generated from biomass, coal, or natural gas. These feedstocks are primarily composed of carbon, hydrogen, and oxygen with lesser amounts of nitrogen and sulfur. The exact chemical composition of syngas depends on the feedstock and the gasification process. Syngas produced from solid feedstocks typically has a low concentration of hydrogen, while syngas from gaseous feedstocks is typically rich in hydrogen.

The main approaches to generate syngas are the gasification and reforming processes. Gasification is a series of thermochemical processes that convert solid feedstocks to syngas whereas reforming is to produce syngas from gaseous feedstock, such as natural gas, or liquid feedstock. Confusingly, the term gasification is commonly used to refer to both an entire conversion and one of its processes. In this text, we use the term as it represents the entire process. Generally, gasification includes drying, pyrolysis, combustion, cracking, and possibly water-gas shift processes. Each of these processes takes place at a specific temperature range, from 100°C to 1500°C (about 210°F–2700°F). The overall processes in gasification require heat (endothermic). The heat can be provided by the partial combustion of the feedstock (known as autothermal or partial oxidation) or by an external heat source (allothermal).

In the drying or dehydration process, the moisture is removed from the feedstock. In the pyrolysis or devolatilization process, the feedstock is decomposed to CO, CO_2, CH_4, C_2H_4, steam, char, and tar. This process takes place at the temperature of 200°C–300°C (about 400°F–570°F) in an oxygen-free environment. The combustion process is, as noted, to provide the heat required for the processes. To avoid complete combustion, less than stoichiometric oxygen is provided to burn some volatiles and char. In the next step, the tar produced in the pyrolysis process decomposes to CO, CO_2, CH_4, C_2H_4, and carbon. Tar is a dark brown liquid primarily composed of hydrocarbons and water (with a broad chemical formula of CH_xO_y). Finally, the produced gases may go through the water-gas shift reaction to increase the concentration of hydrogen by carbon monoxide conversion. The unit where these processes occur at is referred to as the **gasifier**. Some of the typical reactions that take place during the various stages of gasification are

$$C + O_2 \rightarrow CO_2 \text{ (char combustion)}$$

$$C + \frac{1}{2}O_2 \rightarrow CO \text{ (char partial combustion)}$$

$$CO + \frac{1}{2}O_2 \rightarrow CO_2 \text{ (carbon monoxide combustion)}$$

$$H_2 + \frac{1}{2}O_2 \rightarrow H_2O \text{ (hydrogen combustion)}$$

$$CH_4 + \frac{1}{2}O_2 \rightarrow CO + 2H_2 \text{ (methane partial combustion)}$$

$$C + CO_2 \rightarrow 2CO \text{ (Boudouard reaction)}$$

$$C + H_2O \rightarrow CO + H_2 \text{ (water-gas reaction or steam reforming reaction)}$$

$$CO + H_2O \rightarrow CO_2 + H_2 \text{ (water-gas shift reaction)}$$

$$C + 2H_2 \rightarrow CH_4 \text{ (methane production reaction)}$$

$$CH_4 + H_2O \rightarrow CO + 3H_2 \text{ (steam methane reforming reaction)}$$

$$CH_4 + CO_2 \rightarrow 2CO + 2H_2 \text{ (dry reforming reaction)}$$

Some of these reactions take place at the same section of the gasifier, while others are physically separated. While there are several types of gasifiers, the most common ones are fluidized-bed and fixed-bed gasifiers. The former types can be bubbling, circulating, or dual fluidized beds (refer to Section 4.6.4.11 for more information on this topic). In fixed-bed gasifiers, the feedstock enters at the top of the container. Depending on the location of the air flow entrance and its direction relative to the feedstock flow, fixed-bed gasifiers can be updraft or downdraft.

10.6.2 Pressurized Fluidized-Bed Combustion Combined Cycle Power Plants

Another proposed approach to use solid fuels in a gas turbine and a steam cycle is pressurized fluidized-bed combustion (PFBC). In this approach, a bubbling fluidized bed (Section 4.6.4.11.3) is fed by a pressurized air flow from the compressor of a gas turbine unit. The fluidized bed produces steam as in conventional fluidized beds, which is sent to a steam cycle. The high-pressure and temperature combustion products are sent to a gas turbine after cleanup processes to generate further electricity. While both a steam cycle and a gas turbine are present in this design, it is not a combined cycle in a sense discussed so far. The steam cycle and the gas turbine unit operate independently using a shared steam generator-combustion chamber. This technology has not been commercialized yet.

10.6.3 Binary Vapor Cycles

For the sake of completeness, let us very briefly review one last combined cycle. A **binary vapor power plant** is composed of two Rankine cycles, one working at high-temperature end of the cycle (topping cycle) and the other at low-temperature end of the cycle (bottoming cycle). These two cycles operate using different working fluids, possibly water and organic fluids, each suitable for the operating temperature of one of the Rankine cycles. The condenser of the high-temperature Rankine cycle is the vapor generator of the low-temperature Rankine cycle. Binary vapor cycles are able to take advantage of exchanging heat with a high-temperature thermal energy source at the topping cycle and exchanging heat with a low-temperature energy source at the bottoming cycle. A variation of this concept is very popular in geothermal power plants (Chapter 15).

10.7 Questions and Problems

10.7.1 Introduction

10.1. Explain the principl idea of combined cycle power plants.

10.2. In a combined cycle power plant, which cycle functions as a topping cycle and which one as a bottoming cycle?

- a. Brayton cycle – Rankine cycle
- b. Brayton cycle – Brayton cycle
- c. Rankine cycle – Brayton cycle
- d. Rankine cycle – Rankine cycle

10.3. To recover thermal energy from the gas turbine exhaust flow, the steam generator of the Rankine cycle is replaced by a ......

10.4. Combined cycle power plants are suitable for ....

- a. peak-load power generation
- b. baseload power generation
- c. both peak-load and baseload power generation
- d. neither peak-load nor baseload power generation

10.5. Explain why even with the high contribution of electricity generated from renewable energy sources, combined cycles are playing and will continue to play a vital role in the power generation industry.

10.6. Draw a schematic diagram of a simple combined cycle power plant showing all main components and streams.

10.7.2 Efficiency of Combined Cycle Power Plants

10.7. The first thermal power cycle that was able to exceed the efficiency of 60% was a combined cycle power plant designed and built by Siemens in southern Germany in May 2012 (Irsching Power Station). The efficiency of this combined cycle was 60.75%. If the efficiency of the gas turbine cycle was 40%, determine the efficiency of the steam cycle.

10.8. In the previous problem, the output power of Irsching Power Station is 570 MW. Determine the power output of the gas turbine and the steam cycle (in MW), the mass flow rate of natural gas burned in the combustion chamber of the gas turbine (kg/s), and the thermal energy exhausted from the HRSG (MW). The heating value of natural gas is 43 MJ/kg.

10.9. In a combined cycle with an efficiency of 57%, the efficiency of the steam cycle is 29%. Determine the efficiency of the gas turbine cycle.

10.10. Which one of these combinations of a gas turbine and a steam cycle can form a combined cycle with an efficiency of 56%?

a. A gas turbine with an efficiency of 38% and a steam cycle with an efficiency of 28%
b. A gas turbine with an efficiency of 38% and a steam cycle with an efficiency of 29%
c. A gas turbine with an efficiency of 39% and a steam cycle with an efficiency of 28%
d. A gas turbine with an efficiency of 33% and a steam cycle with an efficiency of 23%

10.11. A 160 MW gas turbine has the efficacy of 34%. This gas turbine operates once standalone and once combined with a steam cycle with an efficiency of 25%. Determine the exhausted thermal energy in each case.

10.7.3 Configuration of Combined Cycle Power Plants

10.12. In CCPPs, the backpressure imposed by the pressure drop in the HRSG slightly reduces the efficiency of the gas turbine compared to that of the standalone unit. True or false? Why?

10.13. In the following schematic of a combined cycle power plant, name the major components of the cycle.

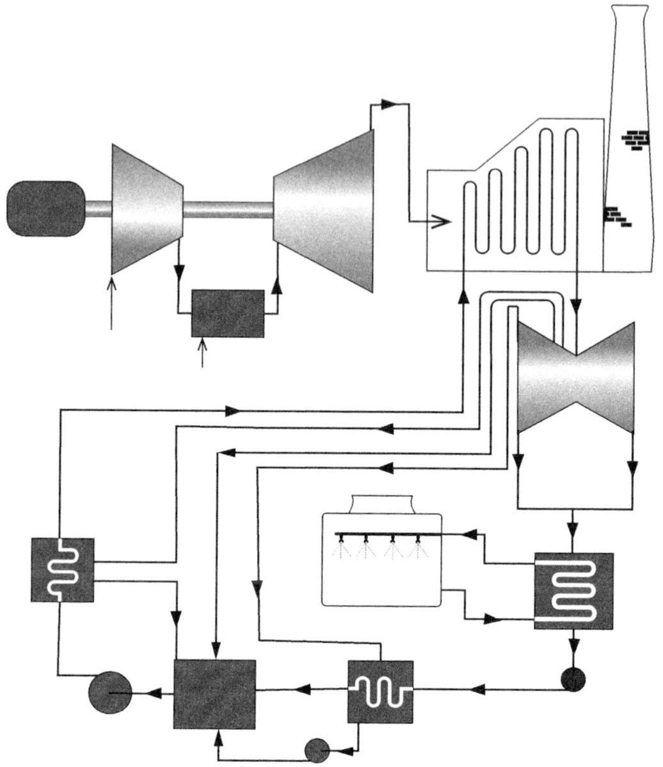

10.7.4 Heat Recovery Steam Generators (HRSGs)

10.14. What is the most important difference between heat recovery steam generators and conventional fossil fuel-steam generators?

10.15. What is the dominant heat-transfer mechanism in HRSGs? Why?

10.16. Which component does not exist in HRSGs?

a. Superheater
b. Economizer
c. Evaporator tube bundles
d. Waterwall tubes

10.17. Which type of HRSGs is currently more popular (drum-type or once-through)? Which type is gaining momentum? Why?

10.18. What is the function of a duct burner in an HRSG?

10.19. What are the advantages and disadvantages of adding a duct burner to an HRSG?

10.20. What is the approach temperature in an HRSG? Why is it important to consider this temperature difference in the design?

10.21. What is the pinch temperature? From an efficiency point of view, which one is better, a higher pinch temperature or a lower one? What about the size of the HRSG?

10.22. Which statement is not correct about the pinch temperature?

a. Reducing the pinch temperature, decreases the temperature difference between the flue gas and steam flows.

b. Reducing the pinch temperature, increases the irreversibilities in the heat-transfer processes and reduces the efficiency.

c. A steam generator with a lower pinch temperature requires a larger heat-transfer area.

d. The selection of an appropriate pinch temperature involves a tradeoff between the system size versus the efficiency.

10.23. Explain why many HRSGs have multiple pressures?

10.7.5 Cogeneration Plants

10.24. What does CHP stand for?

10.25. How can the design objective of a cogeneration plant be different from a conventional power plant?

10.26. Define the thermal efficiency and the electrical efficiency of a cogeneration plant.

10.27. List some applications of thermal energy in a cogeneration plant.

10.28. What is a topping CHP plant? How is it different from a bottoming CHP plant?

10.29. What are the differences between back-pressure CHP plants and extraction CHP plants?

10.7.6 Integrated Gasification Combined Cycles

10.30. Explain how an integrated gasification combined cycle works.

10.31. What is the major advantage of an IGCC plant over a conventional power plant in terms of the integration of carbon dioxide capture systems?

10.32. Describe the configuration of the following integrated gasification combined cycle (IGCC) power plant.

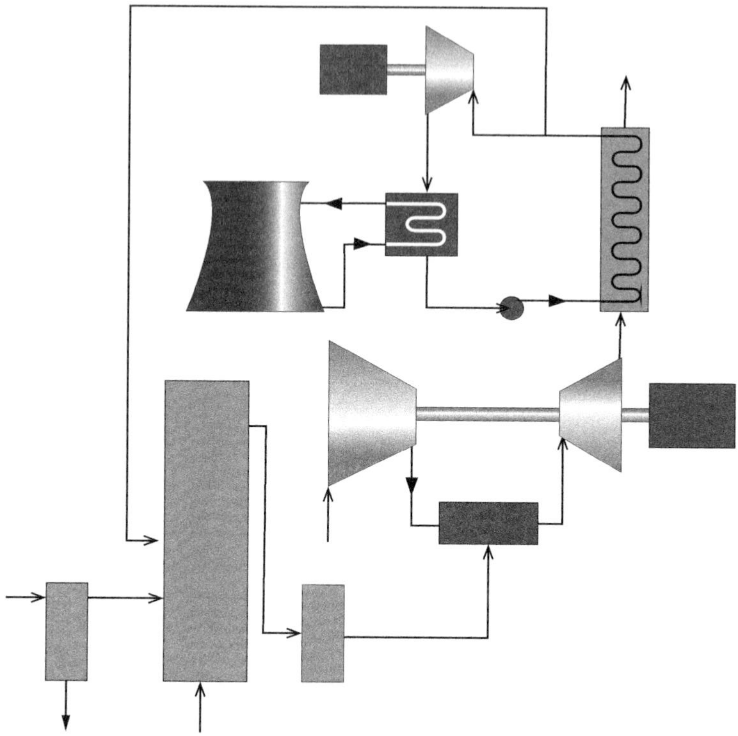

10.33. Describe the gasification of a solid feedstock. What are the processes involved?

10.34. Why does an IGCC plant need an air separation unit (ASU)?

10.35. Explain how a pressurized fluidized-bed combustion unit can be integrated into a gas turbine unit and a steam cycle.

10.36. What are the different types of gasifiers?

10.37. Explain how a binary vapor power plant works.

10.7.7 General

10.38. Why is using the term "combined cycle" somewhat misleading?

10.39. What is syngas composed of?

10.40. A typical gas turbine and steam cycle have an efficiency of 40% and 30%, respectively. Determine the efficiency of a combined cycle made of these two cycles.

10.41. In combined cycle power plants, the two cycles are truly integrated.

a. True

b. False

10.42. List some of the advantages of combined cycle power plants.

10.43. What is the advantage of multiple-pressure HRSGs over single-pressure ones?

10.44. Explain combined cooling, heat, and power (CCHP) plants.

10.45. What is the function of a desuperheater in an HRSG?

10.46. Provide some examples of how combined cycles can be used for replacing, retrofitting, or repowering existing or retiring power plants.

10.47. Describe the configuration of the following combined cycle power plant (CCPP).

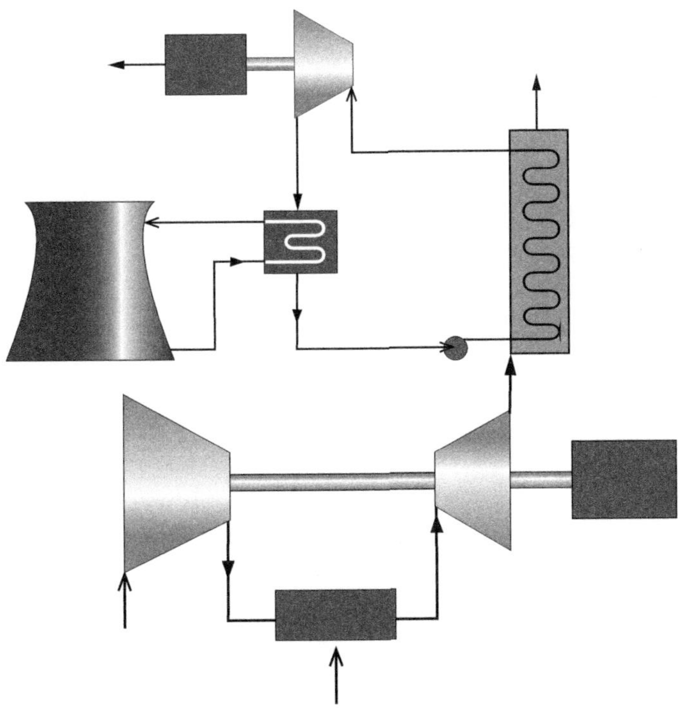

10.48. A large approach temperature results in less heat-transfer surface area in the economizer, which must be compensated by more heat-transfer area in the evaporator.

a. True

b. False

10.49. Can solid fuels be used to fuel gas turbines? Based on your answer, explain integrated gasification combined cycles (IGCC).

10.50. Name three types of power plants in the following figure.

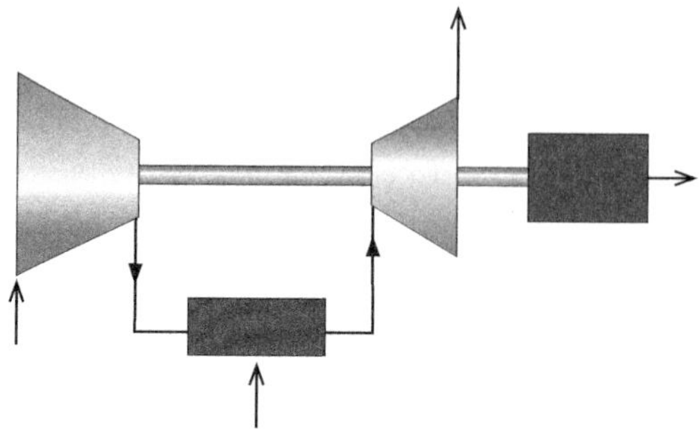

(a)

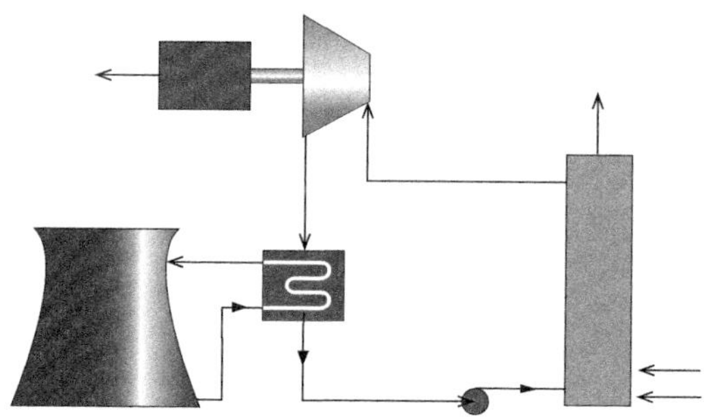

(b)

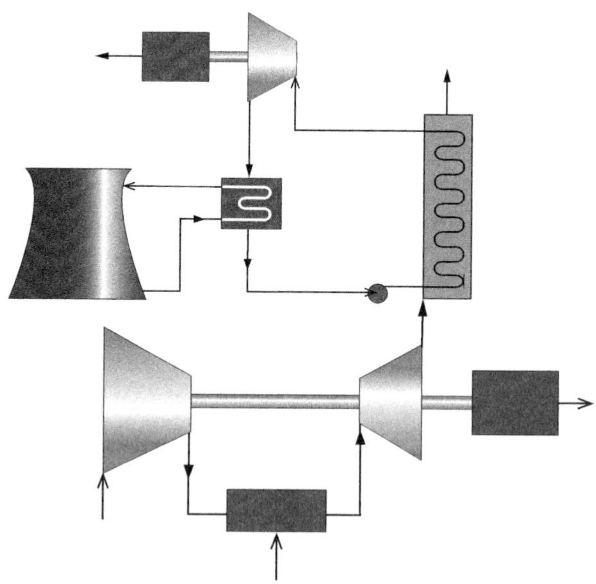

(c)

10.51. The efficiency of a combined cycle composed of a gas turbine with the efficiency of 35% and a steam cycle with the efficiency of 27% is equal to ……

a. 9.45%

b. 52.55%

c. 62%

d. 71.45%

10.52. Why do some CCPPs have a stack right after the gas turbine and before the HRSG?

10.53. Combined cycle power plants are capable of achieving extraordinarily high efficiency.

a. True

b. False

10.54. Draw the *T-s* diagram of a combined cycle power plant.

10.55. Why do combined cycles have extraordinarily low-carbon emission?

10.56. What are the functions of the inlet duct in an HRSG?

10.57. Explain combined heat and power (CHP) plants.

10.58. A gas turbine with the efficiency of 41% is combined with a steam cycle with the efficiency of 31% to form a combined cycle power plant. Determine the efficiency of the combined cycle.

10.59. A heat recovery steam generator is the interface between the gas turbine unit and the steam cycle in a CCPP.

a. True

b. False

10.60. Very briefly explain the principle of power generation by integrated gasification combined cycle (IGCC) power plants.

10.61. The combination of gas turbines, which are typically peak-load power generation units, and steam cycles, which are generally baseload power generation units, provides flexibility for these cycles to perform in a variety of demand conditions, both as peak-load and baseload power plants.

a. True

b. False

11

Solar Energy and Photovoltaic Solar Units

Abstract

Solar energy is the origin of most of the energy resources that our life literally depends on. In this chapter, we will learn a great deal about the direct use of solar energy. Generally speaking, the direct use of solar energy can be divided into three main categories, solar photovoltaic units, high temperature solar thermal energy systems, and low temperature solar thermal energy systems. In this chapter, we will discuss about the first application in the list, solar photovoltaic systems. This chapter will start with why and how solar energy varies both spatially and temporally on Earth. Then, the conventional and emerging solar photovoltaic systems will be extensively covered. In the first half of the chapter, we will follow solar energy from its creation on the Sun, crossing the void between the Sun and Earth's atmosphere, and traveling through the atmosphere until its interception by the surface of Earth. We will particularly focus on how the geometric relationship between the Sun and Earth and the atmospheric conditions impact solar energy on Earth.

In the second half of the chapter, we will study solar photovoltaic electricity generation. We will start with a historical review of these systems and their applications in various industries. Then, we will discuss about semiconductors as the building blocks of solar cells and their unique properties as well as the internal operation of solar cells from the microscopic point of view (at the subatomic level). We will learn how solar radiation and electricity generation in solar cells are interrelated. We will then focus on the macroscopic point of view of solar cells, and how their microscope characteristics can explain their macroscopic behaviors. We will learn how to determine solar cell characteristics using I-V diagrams and calculate their performance. Next, financial models to support solar PV units will be explained. We will discuss how the size of grid-connected solar PV arrays can be calculated. Finally, we will learn different types of solar photovoltaic cells of first, second, third generations and their characteristics. We will conclude the chapter by studying the manufacturing processes for currently dominant crystalline silicon-based PV units.

LEARNING OUTCOMES

After successful completion of this chapter you will be able to:

- explain factors that impact solar irradiance variation on Earth
- describe parameters that impact irradiance at the certain location including latitude, time of day and year, and atmospheric conditions
- determine solar irradiance outside the atmosphere and in a given location on Earth
- explain semiconductors and their behaviors at the subatomic level
- describe solar cell internal working principles at the subatomic level
- determine the relationship between light and power generation in solar cells
- calculate the performance of solar cells and their characteristics using I-V diagram
- list balance of plant for PV systems and their specifications, including batteries, inverters, mounting systems, etc.
- explain financial models to support solar PV units
- calculate the size of grid-connected solar PV arrays
- name different types of solar photovoltaic cells of first, second, and third generations and their characteristics
- describe manufacturing processes for crystalline silicon-based PV cells

11.1 Nature of Solar Energy

The Sun is, directly or indirectly, the origin of a great majority of energy resources that we use everyday, from biomass, wind energy, and ocean wave energy to fossil fuels and of course solar energy.

In order to identify suitable sites for the installation of solar energy systems, we need to evaluate how much solar energy a specific region receives. For this purpose, **solar irradiance**, also called **solar energy flux or intensity**, is defined as the incident solar energy or power, also referred to as insolation, received by a unit of area. The amount of solar irradiance varies with both time and location. In this section, we will study the factors that affect solar irradiance. Solar radiation is generated on the Sun, travels through space between the Sun and Earth, passes through the atmosphere, and shines on a

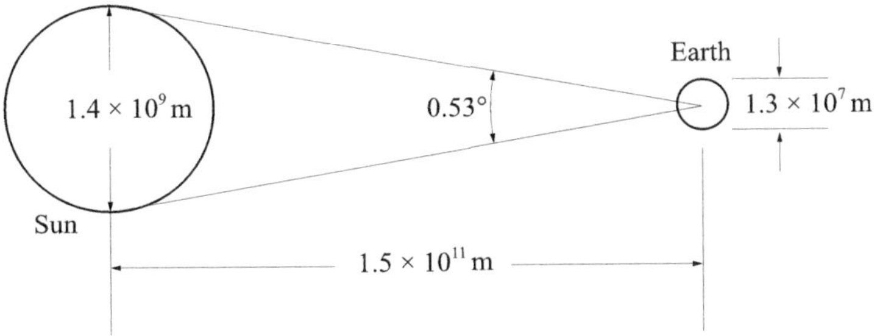

FIGURE 11.1
Dimensional relationship between Earth and the Sun.

specific area. Therefore, to understand how solar irradiance varies on Earth, we will evaluate the following factors:

- Solar energy production on the Sun
- Solar radiation traveling from the Sun to Earth (extraterrestrial radiation as received above the atmosphere)
- Geometric relationship between the Sun and Earth
- Solar radiation traveling through the atmosphere

11.1.1 Solar Energy Production on the Sun

The Sun is located at the center of the solar system where the majority of the energy needed to sustain life on Earth is originated. This energy is generated by the thermonuclear reactions at its core where hydrogen atoms are fused into helium. The estimated rate of this reaction is about 600 billion kg per second, of which 4 billion kg per second is converted into energy (Example 11.1). The diameter of the Sun is about 110 times greater than that of Earth and it is 330,000 heavier than Earth[1] (Figure 11.1). While most known elements exist in the Sun, the great majority of the Sun is made of hydrogen and helium (with the ratio of about three to one). The energy produced at the center of the Sun is estimated at about 3.8×10^{26} W. The Sun is so massive that it takes tens of thousands of years for the thermal energy created at its core to transfer to its surface. While it is estimated that the temperature of the core of the Sun is over 15 million degree Celsius, this temperature gradually reduces through many layers of gases. On the surface of the Sun, the temperature is about 5500°C. This high-temperature surface is where the solar radiation is originated from.

[1] About 99% of the mass of the Solar System is concentrated on the Sun.

Example 11.1

If the energy generated in the fusion reaction on the Sun is 3.8×10^{26} W, determine the rate of the conversion of mass to energy on the Sun in kg/s.

Solution

According to the mass-energy equivalence equation (Albert Einstein's famous formula),

$$\dot{E} = \dot{m}c^2 \tag{11.1}$$

where c is the speed of light, $c = 3 \times 10^8$ m/s

$$\dot{m} = \frac{P_{\text{Sun}}}{c^2} = \frac{3.8 \times 10^{26}}{(3 \times 10^8)^2} = 4.22 \times 10^9 \text{ kg/s}$$

The Sun was formed about 4.5 billion years ago and is expected to continue to shine for another 5 billion years. While some theories have related the historic climate changes on Earth, such as ice ages, to the variation of solar radiation,[2] in the past few thousand years the Sun has been calm. The Sun has a cycle of solar magnetic activity (referred to as solar sunspot activity) roughly every 11 years. However, the variation of solar radiation as a result of this cycle is <0.1%.[3] Overall, while the variation of energy released by the Sun can have significant impacts on the long-term climate change on Earth, for the practical purposes of solar energy utilization on Earth **it can be safely assumed constant.**

11.1.2 Extraterrestrial Radiation

Solar radiation generated by the surface of the Sun travels the void between the Sun and Earth unaltered. The average distance between the Sun and Earth is about 1.5×10^{11} m (93 million mi), which is also referred to as one astronomical unit (1 AU). The value of solar irradiance outside Earth's atmosphere when the distance between the Sun and Earth is approximately one AU is called the **solar constant** (which includes all solar spectrum not just visible light). Note that the solar constant (sometimes referred to as G_{SC}) is not a fixed physical constant (such as the gravitational constant or the speed of light) and a range of values has been reported for it, from 1360 to 1370 W/m^2. Because of this variation, some references stop using the term solar constant

---

[2] Ehrlich, R. "Solar resonant diffusion waves as a driver of terrestrial climate change." *Journal of Atmospheric and Solar-Terrestrial Physics* 69.7 (2007): 759–766.

[3] https://www.nasa.gov/centers/goddard/news/topstory/2008/solar_variability.html.

and now use the term **total solar irradiance** (**TSI**) instead. The following example shows how we can determine the value of the solar constant. The current estimated value for the solar constant by NASA is 1360.8±0.5 W/m².[4] This value is relatively constant with a small variation due to the solar sunspot activity.

Example 11.2

We know that the Sun releases 3.8×10^{26} W of radiation. If the average distance between the Sun and Earth is 1.5×10^{11} m, determine the solar power received on Earth per unit of area (solar irradiance in W/m²).

Solution

When solar radiation arrives at Earth's atmosphere, it is uniformly distributed on the surface of a sphere with the radius of the average distance between the Sun and Earth.

$$I = \frac{\text{Total solar radiation}}{\text{Surface of sphere}} = \frac{P_{\text{Sun}}}{4\pi r^2} = \frac{3.8 \times 10^{26}}{4 \times 3.14\left(1.5 \times 10^{11}\right)^2} = 1345\ \text{W/m}^2 \quad (11.2)$$

If we repeat the calculations with more accurate values of solar radiation and the Sun and Earth distance, a much more accurate value of $1361.8\ \text{W/m}^2$ can be achieved.

Example 11.3

Jupiter is about 5.2 times farther away from the Sun than Earth. Determine the solar constant on Jupiter.

Solution

$$I = \frac{\text{Total solar radiation}}{\text{Surface of sphere}} = \frac{P_{\text{Sun}}}{4\pi r^2} = \frac{3.828 \times 10^{26}}{4 \times 3.14\left(5.2 \times 1.496 \times 10^{11}\right)^2} = 50.4\ \text{W/m}^2$$

This means the solar constant on Jupiter is 3.7% of that on Earth. This was a major challenge for the designers of the spacecraft Juno, which was launched in 2011 to orbit Jupiter. To address this problem Juno is equipped with the largest solar array ever installed on a spacecraft.

In the above examples, we assumed Earth rotates around the Sun in a circle with the radius of the average distance of the Sun and Earth. However, Earth's orbit around the Sun is slightly elliptical (with the eccentricity of 0.0167), which means Earth is sometimes farther away from the Sun than the

[4] https://atmospheres.gsfc.nasa.gov/climate/index.php?section=136.

average distance and sometimes closer. The seasonal changes in weather patterns may be mistakenly attributed to these differences in the distance that solar radiation should travel. Clearly, this is not correct for several reasons. First, if this was the case, both northern and southern hemispheres should have summer or winter at the same time. However, we know that when it is summer in the northern hemisphere, in the southern hemisphere it is winter and vice versa. Also, it turns out that Earth is closest to the Sun in January (perihelion) when it is winter in the northern hemisphere and is farthest in July (aphelion). The following example determines the impact of the elliptical orbit of Earth around the Sun on solar irradiance outside the atmosphere.

Example 11.4

The ratio of the longest distance between the Sun and Earth to the shortest distance is 1.034. Determine the ratio of solar irradiance outside the atmosphere for these two cases.

Solution

$$\frac{I_1}{I_2} = \left(\frac{r_2}{r_1}\right)^2 = \left(\frac{1}{1.034}\right)^2 = 0.935$$

This means the variation of solar irradiance on Earth due to the changes in the distance from the Sun to Earth is only about 6.5%, which clearly cannot cause drastic seasonal changes.

To determine the solar constant, solar irradiance is measured outside Earth's atmosphere by satellites and then adjusted for one AU. Now we can determine how much of the Sun's total radiation is intercepted by Earth.

Example 11.5

Determine what percentage of the Sun's total radiation is intercepted by Earth. Use the solar constant of 1361.8. Earth's mean radius is 6371 km (3958.8 mi).

Solution

The area that intercepts solar radiation on Earth is equal to the area of a disk with the radius of Earth.

$$\text{Area of a disk with radius of Earth} = \pi r_{\text{Earth}}^2 = 3.14\left(6.371\times10^6\right)^2 = 1.27\times10^{14}\ \text{m}^2$$

$$\text{Radiation intercepted by Earth} = 1361.8\times1.27\times10^{14} = 1.74\times10^{17}\ \text{W}$$

Percentage of the Sun's total radiation intercepted by Earth

$$= \frac{\text{Radiation intercepted by Earth}}{\text{Total solar radiation}} = \frac{1.74\times10^{17}}{3.828\times10^{26}}$$

$$= 4.53\times10^{-10} = 4.53\times10^{-8}\,\%$$

Example 11.6

The area of the solar array installed on Juno orbiting Jupiter is 72 m². Determine the amount of solar radiation intercepted by this satellite when it orbits Jupiter. What if it orbits Earth, Mars, or Saturn? The Sun releases 3.828×10^{26} W of radiation. The average distance between the Sun and Jupiter, Earth, Mars, and Saturn is 5.2, 1, 1.52, and 9.5 AU, respectively.

Solution

If the satellite orbits Jupiter:

$$I = \frac{\text{Total solar radiation}}{\text{Surface of sphere}} = \frac{P_{\text{Sun}}}{4\pi r^2} = \frac{3.828\times10^{26}}{4\times3.14\left(5.2\times1.496\times10^{11}\right)^2} = 50.4 \text{ W/m}^2$$

$$\text{Solar radiation intercepted by solar array} = I\times A = 50.4\times72 = 3626.1 \text{ W}$$

If the satellite orbits Earth:

$$I = \frac{\text{Total solar radiation}}{\text{Surface of sphere}} = \frac{P_{\text{Sun}}}{4\pi r^2} = \frac{3.828\times10^{26}}{4\times3.14\left(1.496\times10^{11}\right)^2} = 1361.8 \text{ W/m}^2$$

$$\text{Solar radiation intercepted by solar array} = I\times A = 1361.8\times72 = 98050.9 \text{ W}$$

If the satellite orbits Mars:

$$I = \frac{\text{Total solar radiation}}{\text{Surface of sphere}} = \frac{P_{\text{Sun}}}{4\pi r^2} = \frac{3.828\times10^{26}}{4\times3.14\left(1.52\times1.496\times10^{11}\right)^2} = 589.4 \text{ W/m}^2$$

$$\text{Solar radiation intercepted by solar array} = I\times A = 589.4\times72 = 42438.9 \text{ W}$$

If the satellite orbits Saturn:

$$I = \frac{\text{Total solar radiation}}{\text{Surface of sphere}} = \frac{P_{\text{Sun}}}{4\pi r^2} = \frac{3.828\times10^{26}}{4\times3.14\left(9.5\times1.496\times10^{11}\right)^2} = 15.1 \text{ W/m}^2$$

$$\text{Solar radiation intercepted by solar array} = I\times A = 15.1\times72 = 1086.4 \text{ W}$$

This means the solar radiation received on Mars, Jupiter, and Saturn is 43.3%, 3.4%, and 1.1% of that on Earth. Now you can explain why it is very challenging to use solar photovoltaic (PV) arrays to power space missions in the solar system beyond Jupiter.

11.1.3 Solar Radiation and Geometric Relationship of the Sun and Earth

So far we learned about the variation of the energy released by the Sun and how the distance between Earth and the Sun impacts the irradiance outside Earth's atmosphere. The actual irradiance that a certain location on Earth receives depends on the following four factors:

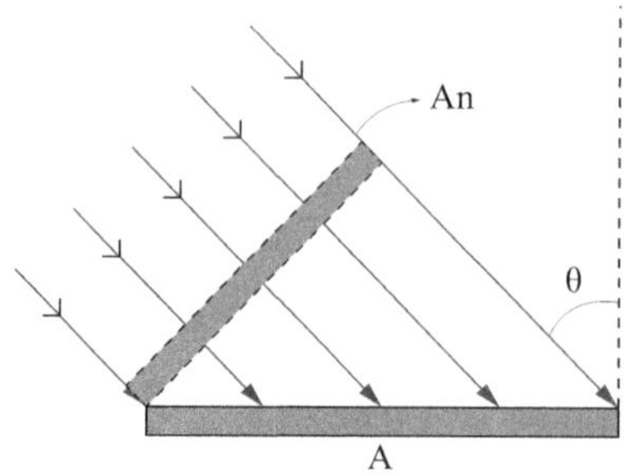

FIGURE 11.2
Illustration for the cosine emission law.

- Latitude of the location
- Time of the day
- Time of the year
- Atmospheric conditions

In this section, we will learn about the first three factors while the last one will be discussed in the next section. The topic of this section is very important because it will help to identify suitable regions to install solar energy units as well as to determine the best orientation and installation conditions.

Let us start with the **cosine emission law**, also known as Lambert's cosine law, named after its discoverer Johann Heinrich Lambert (1728–1777, a Swiss polymath[5]). It states that the irradiance a surface actually receives is directly proportional to the cosine of the angle between the direction of the light and the direction perpendicular to the surface (Figure 11.2).

$$I_s = I\cos\theta \tag{11.3}$$

It simply means when a surface is not normal to the direction of the light, the irradiance (radiation per unit of area) received by the surface reduces because the same amount of radiation will be distributed on a larger surface (surface A rather than A_n in Figure 11.2 where $A_n = A\cos\theta$). This law is critical for us because for the rest of this section we will try to determine the spatial and temporal factors that impact the angle θ.

We know that Earth orbits around the Sun, which takes 365.256 days for every complete rotation. Also, Earth rotates around its own axis every 86,400 (=24 × 60 × 60) seconds, which results in the day and night cycle. If the axis of rotations for these two rotations were parallel, there would be no season

[5] A polymath is a person with a wide variety of knowledge in many fields such as art, science, engineering, and philosophy. The best example of a polymath is probably Leonardo Da Vinci, the all-time genius.

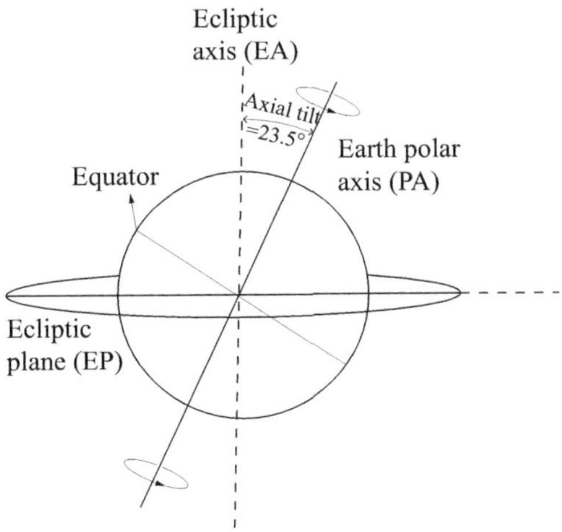

FIGURE 11.3
Rotation of Earth around itself and the Sun.

on Earth and the duration of day and night would always be fixed and equal (12 hours each). Fortunately for us humans, this is not the case[6] and the axis of rotation of Earth around itself, sometimes referred to as **Earth's polar axis (PA)**, is tilted. In order to explain this, let us represent the axis of rotation of Earth around the Sun, called the **ecliptic axis (EA)**, by a line passing through the center of Earth and perpendicular to the orbit of Earth around the Sun, called the **ecliptic plane (EP)**, as shown in Figure 11.3. This line, the ecliptic axis, is parallel to the axis of the rotation of Earth around the Sun. The angle between the polar axis and the ecliptic axis is a fixed angle currently at about **23.5°**.[7] This angle is referred to as **Earth inclination, axial tilt**, or **obliquity**. When Earth rotates around the Sun, the direction of the ecliptic axis does not change but the axis of rotation of Earth around itself rotates around the ecliptic axis. As a result, **Earth is sometimes tilted toward the Sun and sometimes titled away from it**.

Typically on June 20–22 each year, Earth is most tilted toward the Sun, which is called the **June solstice**. In this day, the northern hemisphere is most exposed to solar radiation, which has two major impacts. First, the duration of the exposure of all regions in the northern hemisphere is the longest. That means the longest day and the shortest night of the year. Second, the angle between solar rays and the normal line to the surface of Earth (horizon) is the smallest possible. The actual value of this angle depends on the location (the latitude) of each site, but everywhere it is the closest solar radiation gets to the normal line. This indicates that according to the cosine emission law,

[6] Not having seasons would have been catastrophic for human civilization, just search the internet.

[7] This angle varies between 22.1° and 24.5° on a 41,000-year cycle.

the **highest solar irradiance** can be observed. Both these impacts increase the solar energy gain in the northern hemisphere and that is why this day is the beginning of summer in the northern hemisphere and is referred to as the **summer solstice** in the northern hemisphere (the winter solstice in the southern hemisphere).

After the summer solstice, with the continued rotation of Earth around the Sun, Earth is less tilted toward the Sun. On a day between September 21 and 24, which is called the **autumnal equinox** (also fall, September, or Southward equinox), Earth is tilted neither toward nor away from the Sun. In this day, the length of day and night is almost the same. From this day on Earth starts to tilt away from the Sun. In the northern hemisphere, this means the end of summer and the beginning of fall.

Earth is the most tilted away from the Sun usually on December 21 or 22, which is called the **December solstice** (also the winter solstice in the northern hemisphere or the summer solstice in the southern hemisphere). In the northern hemisphere, the longest night and the shortest day of a year are experienced in this day. Also, solar rays are the farthest from the normal line to the surface of Earth. For these two reasons, in the **northern hemisphere, solar energy gain is the lowest on this day**, which is why the winter solstice marks the beginning of winter. Finally, usually on March 20 or 21, called the **vernal equinox** (also spring or March equinox), Earth orientation is similar to the autumnal equinox but in the opposite direction, which marks the beginning of spring in the northern hemisphere.

As it can be clearly observed in Figure 11.4, the exposure of the southern hemisphere to the Sun is opposite to that of the northern hemisphere so, as

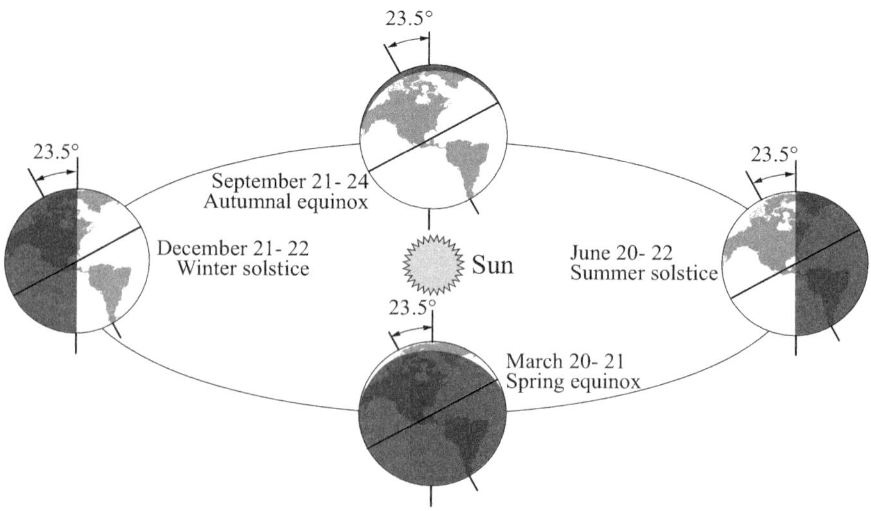

FIGURE 11.4
Schematic of the Sun and Earth relative positions during solstices and equinoxes.

noted earlier, when it is winter in the northern hemisphere, it is summer in the southern hemisphere and vice versa. Also notice that the orientation of the equator in June and December solstices is similar (but opposite direction), so there is no meaningful seasonal variation on the equator and the length of day and night does not change much. However, the farther we go toward the North or South Poles, there are more dramatic changes in seasons and the duration of day and night. Now you should be able to explain why the North and South Poles are the coldest places on Earth and why they do not experience 24-hour cycle of day and night.

11.1.4 Apparent Movement of the Sun across the Sky

All the concepts introduced so far are being used in theoretically determining the relative position of the Sun and Earth (which are beyond the scope of this text). For qualitative discussions of this text, it is more useful to consider the apparent movement of the Sun across the sky from the perspective of an observer on Earth (this view is referred to as the geocentric model).

Figure 11.5 illustrates the apparent movement of the Sun across the sky from the view point of an observer in the northern hemisphere. As shown in the figure, the Sun rises in the east or south-east. With the passage of time, it moves toward south and upward on the horizon. At solar noon, it reaches the highest point on the horizon due south. Then, in afternoon, the Sun again moves downward and to west. The Sun's path on the horizon depends on the season. It appears higher in summer and lower in winter (Figure 11.5). The Sun's path appears the highest in the summer solstice and the lowest in the winter solstice. In the two equinoxes, the Sun's path appears in the middle of the two extremes in the two solstices. The actual height in these cases

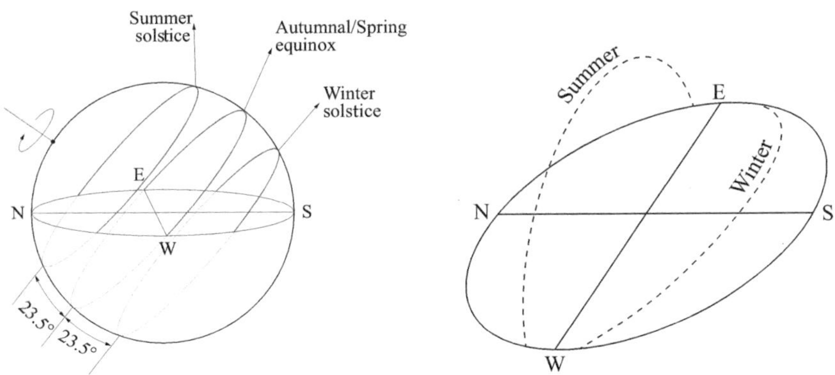

FIGURE 11.5
Apparent movement of the Sun across the sky from the view point of an observer in the northern hemisphere.

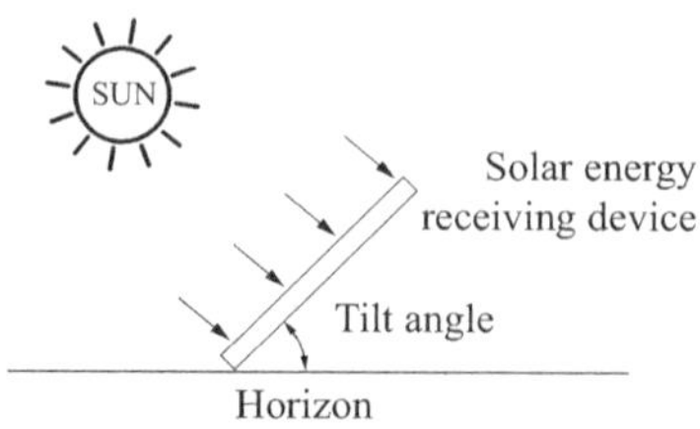

(a)

(b)

FIGURE 11.6
Schematic of the tilt angle of a solar energy receiving device (a) and tilted solar panels in a solar farm (b). (Image by Andreas Troll from Pixabay, https://pixabay.com/photos/solar-field-solar-photovoltaic-191683/.)

depends on the latitude of the location.[8] Clearly, **the maximum solar incident occurs when the Sun is more directly overhead during the summer solstice**. In contrary, the minimum solar incident occurs during the winter solstice. During the two equinoxes, the solar incident is about average. Again note that all these discussions are for the northern hemisphere and the opposite happens for the southern hemisphere.

All these discussions lead us to the main question of this section: what is the orientation and tilt angle that maximize the solar energy intercepted by a device. The **tilt angle** for a solar energy receiving device is the angle between the surface of the device and the horizon in a given location (Figure 11.6).

[8] The latitude is the angle between the line that connects a location to the center of Earth and the equatorial plane. It is 0° for the equator, 90° N or +90° for the North Pole, and 90° S or −90° for the South Pole. It is expressed in degrees, minutes (1′ = 1/60°), and seconds (1″ = 1/60′). It is the first number in the latitude-longitude coordinate. For example, the coordinate of Sacramento, California, is 38.5816° N, 121.4944° W.

First of all, it should be noted that **the solar incident will be maximum when the device is always directly facing the Sun and its surface is always perpendicular to solar rays** (according to the cosine emission law). Considering the motion of the Sun, this can only happen if the device can follow the Sun across the sky. This requires a Sun tracking device, which will be discussed later. For fixed solar energy receiving devices, such as most solar photovoltaic (PV) panels and solar water heaters, **the device should be always orientated toward true south in the northern hemisphere** (true north in the southern hemisphere). This is because the Sun's entire path across the sky, from sunrise to sunset, takes place in the south direction albeit at different heights depending on the time of the year and the latitude of the location. In order to understand the concept of true north, we need to know the difference between the Geographic Poles and the Magnetic Poles. The points where the axis of rotation of Earth around itself intersects the surface of Earth in the northern and southern hemispheres are called **North Pole and South Pole**, respectively. These poles are slightly off the **Magnetic Poles** and that is why they are sometimes referred to as the **Geographic Poles** or **Terrestrial Poles**. Therefore, if you use a compass to identify the northern direction, what you have is the magnetic north, which is different than the true north. This difference can be significant. It depends on the location and varies with time. To take this difference into account, the **magnetic declination** is defined as the angle between the direction of the true north and the magnetic north. For example, the magnetic declination in the state of California is between +11° to +15° and in the state of New York is between –12° and –13°. Here a positive magnetic declination means to find the magnetic north, you need to turn counter clockwise from the true north (Figure 11.7). You can find the magnetic declination of your location at http://www.magnetic-declination.com/ or a simple internet search.

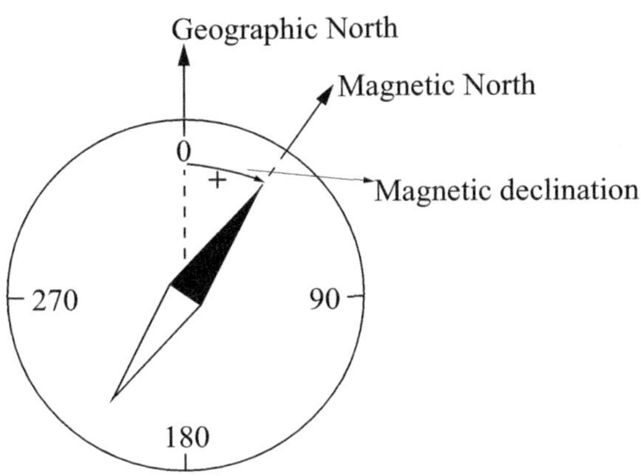

FIGURE 11.7
Definition of the magnetic declination and its positive direction.

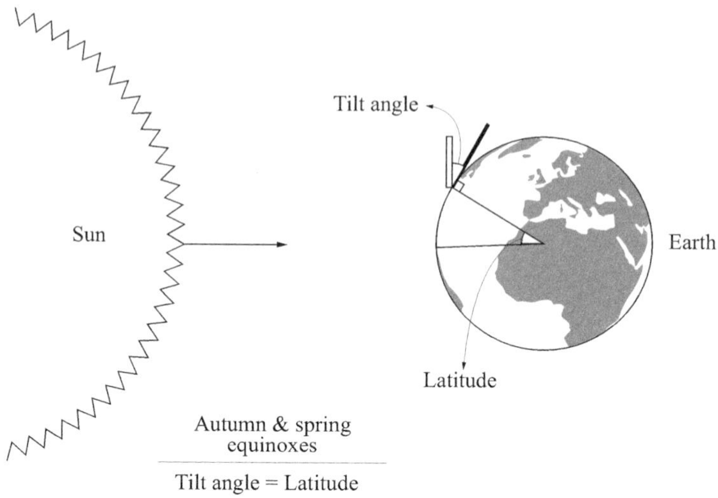

FIGURE 11.8
When the tilt angle is equal to the latitude of the site, the surface of the device will be perpendicular to Sun's rays on the two equinoxes.

The tilt angle of a device is a little more complicated. **If the objective is to maximize the annual intercepted solar energy, the best tilt angle is equal to the latitude of the site**. In this case, the surface of the device will be perpendicular to Sun's rays on the two equinoxes (Figure 11.8). But **if it is desired to have more energy in summer, for example, for photovoltaic panels installed in a hot region where the peak load happens in summer, the tilt angle of the device should be smaller than the latitude**. On the contrary, **if more energy is required in winter, for example, for a solar water heater in a cold region where more thermal energy is required in winter, the tilt angle should be greater than the latitude**. As a rule of thumb, for summer the tilt angle of the latitude minus 15° and for winter the tilt angle of the latitude plus 15° is typically used. Also, a seasonal change of the tilt angle is recommended to improve the intercepted energy by up to about 5%.[9] Note that while these suggested values can roughly maximize the performance of solar energy systems, all sorts of deviations from them are inevitable in many installations. Although in these cases the outputs will not be optimum, some deviations will not significantly reduce the system outputs. This is important since in some applications these parameters are out of control of designers. For example, the orientation of a building and the slope of its roof cannot be altered to install a solar energy unit. There are some other factors that can influence the orientation and tilt angle of solar energy receiving devices and we will get back to this topic once again at the end of the next section.

[9] You can find a lot of useful information about the tilt angle of solar devices at https://www.solarpaneltilt.com/.

11.1.5 Traveling of Solar Radiation through the Atmosphere

So far we learned how we can determine the solar irradiance outside the atmosphere and how the geometric relationship of the Sun and Earth impacts the irradiance in a given location. Now solar radiation should enter the atmosphere and travel through it before striking the surface of Earth. Throughout this passage, solar energy is attenuated by means of three mechanisms, namely **reflection**, **absorption**, and **scattering**. About one-third of extraterrestrial solar energy is reflected by the atmosphere back to the sky. Some of the radiation that actually enters the atmosphere is absorbed by the gases that make up the atmosphere. When solar radiation travels through the atmosphere, it interacts with various gases and each gas selectively absorbs certain wavelengths in the solar spectrum depending on the specific absorption bands of the gas (we will discuss about the nature of solar radiation and its spectrum later). Most absorption is due to the contribution of water vapor and carbon dioxide, each absorbing light at several wavelength ranges. Ozone (O_3) in the atmosphere is mostly responsible for the absorption of ultraviolet solar radiation, while nitrogen and oxygen absorb very short wavelength portion of solar radiation. These absorptions create gaps in the solar spectrum received on the surface of Earth (Figure 11.9).

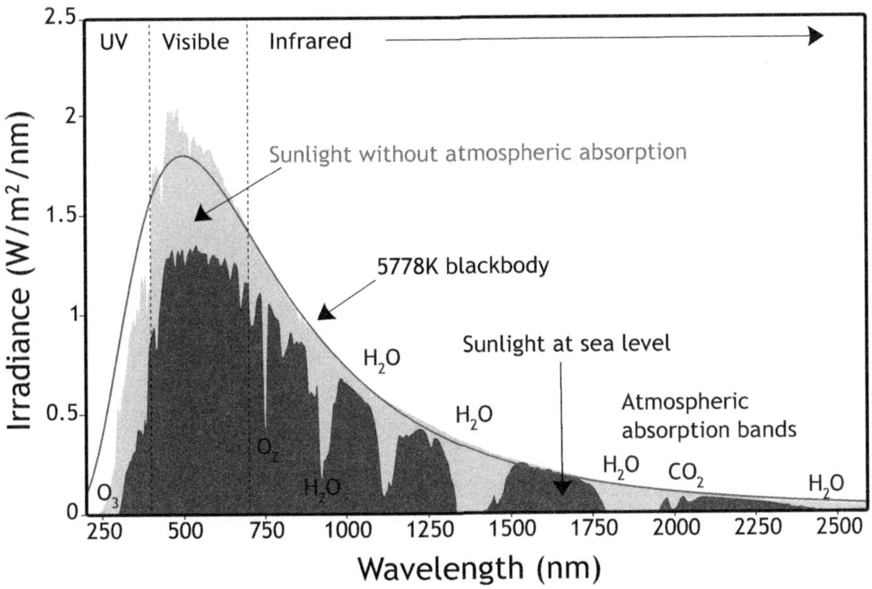

FIGURE 11.9
Solar spectrum received on the surface of Earth and gaps due to atmospheric absorptions (the light area represents the solar spectrum above the atmosphere and the dark area represents the solar spectrum at sea level). (By Nick84- File:Solar_spectrum_ita.svg, CC BY-SA 3.0, https://commons.wikimedia.org/w/index.php?curid=24648395.)

The magnitude of the absorption of solar radiation by the atmosphere is directly proportional to the **distance that the radiation travels through the atmosphere**. To take this factor into account we define a parameter called the air mass coefficient or simply **air mass (AM)**. In the definition of air mass, we consider the vertical thickness of the atmosphere as a benchmark. The air mass is the ratio of the actual distance radiation travels through the atmosphere to the thickness of the atmosphere. This ratio determines the solar spectral power distribution and partly the solar irradiance received on the surface of Earth. A **larger air mass means more attenuation of solar radiation and a lower solar irradiance**. Air mass 0 or AM0 refers to the solar spectrum above the atmosphere with irradiance equal to the solar constant (from Example 11.2 about 1361.8 W/m^2). AM1 is the solar spectrum when radiation passes through the atmosphere normal to Earth's surface. This is the case when the Sun is directly overhead in a location. When solar radiation travels through the atmosphere in any pass other than normal to Earth, the air mass will be larger than 1 and can be determined using the following equation (Figure 11.10):

$$\mathrm{AM} = \frac{H}{h} = \frac{H}{H\cos\theta} = \frac{1}{\cos\theta} \tag{11.4}$$

Sometimes the word "Global" (or G) is added, such as AM1.5 Global or AM1.5G, to indicate that diffuse radiation is also included (albeit typically its share is small). Later we will see that AM1.5 is used as the standard test condition to measure and express the performance of solar photovoltaic units.

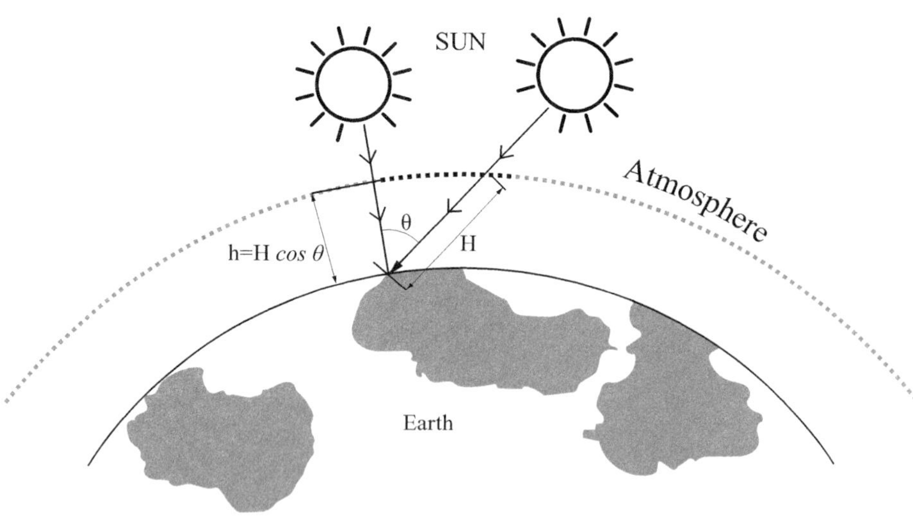

FIGURE 11.10
Schematic showing concept of air mass (AM).

Example 11.7

Determine the angle of solar radiation and the line perpendicular to the surface of the Earth for solar radiation at the standard test conditions (STC).

Solution

For the STC, we have AM1.5.

$$\mathrm{AM} = \frac{1}{\cos\theta}$$

$$\cos\theta = \frac{1}{\mathrm{AM}} = \frac{1}{1.5}$$

$$\theta = \cos^{-1}\left(\frac{1}{1.5}\right) = \cos^{-1}(0.67) = 48.2°$$

If we repeat the example for AM2.5, the angle will be 66.4°. This is expected because a larger angle of solar radiation and the line perpendicular to the surface of Earth result in the light traveling a longer distance through the atmosphere.

The last attenuation mechanism is the scattering of solar radiation by gases and aerosols such as vapor (cloud), dust, and particulate matters in the atmosphere. When solar rays reach Earth's atmosphere, they are nearly parallel (almost parallel but not quite, this topic will be covered later). While passing through the atmosphere, some of these rays are laterally scattered. The details for the scattering process depend on the size of particles and the wavelength of light. If the wavelength is larger than the particle size, the process is called the Rayleigh scattering and if the wavelength is smaller than the particles, the process is called Mie scattering. The scattering is stronger for the shorter wavelength portion of the solar spectrum. The scattering of the shorter wavelength light is the reason for the blue color appearance of the sky in clear days and reddish color when the Sun is low in the sky at dawn and dusk. The level of scattering depends on the number of particles in the atmosphere and their sizes. This in turn depends on the cloudiness of the region and the dust level, which is higher, for example, in large cities and industrial regions.

Solar radiation we observe on Earth can be divided into two types: direct radiation and diffuse radiation. **Direct radiation**, also referred to as **beam radiation**, is near parallel radiation that passes through the atmosphere unscattered. **Diffuse, indirect, or sky radiation** (also simply **skylight**) is the portion of solar radiation that is scattered by the atmosphere. Unlike direct rays that come straight from the actual Sun position, diffuse rays come from various directions from all over the sky. The sum of direct and diffuse radiation is called **global radiation**. The contribution of diffuse radiation to global radiation can vary widely from a small portion to more than a half depending on the location and time. Even in a clear day at noon about 10% of total radiation is diffuse radiation. Near sunrise and sunset most of solar incident is diffuse radiation. In a cloudy day, depending on how overcast the sky is, majority or almost entire radiation can be diffuse radiation. Note that in a cloudy day all three attenuation mechanisms are stronger so more radiation is reflected or absorbed and most of the remaining solar radiation is diffuse radiation. In winter, beside the reasons explained in the previous section, typically more cloud coverage also contributes to reduced solar radiation. In an annual basis, the share of diffuse radiation can typically vary from about 20% to a little over 50% of the total solar energy. Typically for regions with higher annual solar irradiance, the share of direct radiation is higher.

To take into account the impacts of the atmosphere on solar radiation, the **clearness index** is defined as the ratio of the average global irradiance on Earth's surface in a certain site during a given period to the corresponding irradiance outside the atmosphere. The clearness index can have any value from 0 to 0.70 depending on the aforementioned factors.

Now we can revisit our discussion on the orientation and tilt angle of solar energy receiving devices. It should be noted that most solar energy units, such as most solar photovoltaic units and water heaters, **can operate with both direct and diffuse radiation** (but not all, particularly concentrated solar energy systems can only operate with direct radiation). However, the optimum tilt angle for direct radiation for a given location is not the same as the optimum tilt angle for diffuse radiation. As noted above, diffuse radiation comes from all directions in the sky so if the receiving surface is tilted to one side, it will not be able to receive diffuse radiation from the part of the sky behind the surface. Therefore, **for pure diffuse light, the optimum case is when the surface is parallel to the ground (horizon) facing the sky**, because in this case it can "see" the entire sky and can receive diffuse radiation from all directions. However, since in regions with significant available solar energy, the majority of solar radiation is direct radiation, the rule of thumb previously discussed (the south-facing orientation and the tilt angle of latitude ±15°) is still valid.

11.1.6 Measurement of Solar Irradiance

Like the measurement of many other parameters, in order to measure solar irradiance, a detector is needed. This detector converts solar irradiance to a measurable parameter, which can be calibrated to determine the solar irradiance in the desired location. Typically two classes of detectors are used: thermopile and photovoltaic. In **thermopile detectors**, a series of thermocouples (called a thermopile) are exposed to solar radiation under a glass cover and the voltage between the two ends of the thermopile is measured. The measured voltage depends on the temperature difference between the junctions of the thermocouples, which in turn depends on the magnitude of incident solar radiation (for more on thermocouples refer to Chapter 3). **Photovoltaic detectors** are much simpler and can be based on a photovoltaic solar cell or

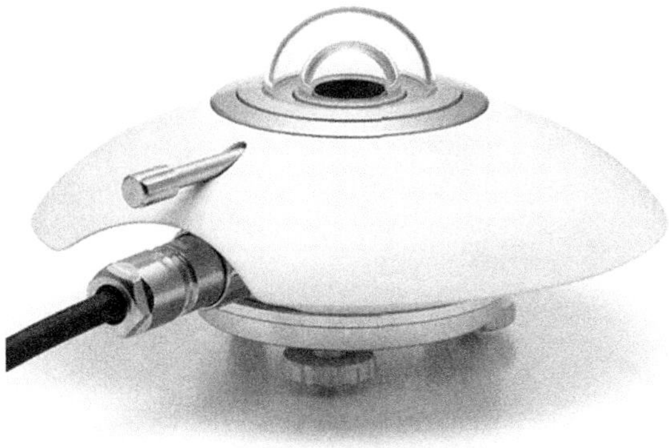

FIGURE 11.11
A pyranometer to measure GHI. (By Hukseflux Thermal Sensors – Hukseflux Thermal Sensors, CC BY 4.0, https://commons.wikimedia.org/w/index.php?curid=59851751.)

a photodiode. They are also referred to as **solarimeter** or **solar sensor** where the short-circuit current produced by a crystalline silicon solar cell under a glass cover is measured as an indication of solar irradiance (we will discuss much more on these cell and their characteristics later).

These detectors are used in two types of sensors: pyranometers and pyrheliometers. A **pyranometer** is typically installed horizontally facing the sky (Figure 11.11), so the detector measures global irradiance, which in this case is called **global horizontal irradiance** or **GHI**. A pyranometer can also be used to measure diffuse radiation (in this case the sensor is sometimes referred to as a diffusometer, diffusionmeter, or scatterometer) if it is equipped with a shadow ring or a shadow ball that casts a narrow strip or a disk of shadow on the sensor all day long using a Sun tracking system. The measured diffuse solar radiation is called **diffuse horizontal irradiance** or **DHI**.

Pyrheliometers are used to measure direct solar radiation. In a pyrheliometer the detector is installed at the end of a tube that limits the view of the detector to a small angle of about 5° (Figure 11.12). Pyrheliometers are also equipped with a Sun tracking system so the open-end of the tube is always directed at the Sun. Therefore, in a pyrheliometer the detector measures direct solar radiation along with diffuse light within its view angle. The measured solar radiation is called **direct normal irradiance** or **DNI**.

11.1.7 Insolation Maps

Thus far we learned that many factors influence incident solar irradiance in any location and learned, at least qualitatively, their impacts. But in order to start any practical design, we need to have the actual value of solar irradiance for the site of the project of interest. Of course, we can always use a pyranometer to measure local irradiance, but we need to have the annual

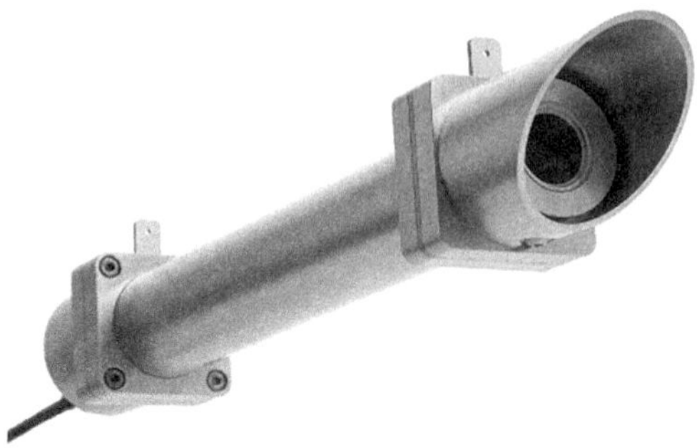

FIGURE 11.12
A pyrheliometer to measure DNI. (By Hukseflux Thermal Sensors – Hukseflux Thermal Sensors, CC BY 4.0, https://commons.wikimedia.org/w/index.php?curid=59851747.)

solar irradiance. Considering the uncertainties involved in the measurements and, to some degrees, unpredictable and random nature of local climate, we need a few years of measurements. Clearly this is beyond the scope of most real-world projects.

Fortunately, the solar irradiance at many locations have been measured and recorded for a long period of time. These data are reported as the **typical meteorological year** (**TMY**).[10] However, these data sets are not typically organized in a user friendly manner and typically used by software developers to generate more accessible information via specialized software tools, such as System Advisor Model (SAM, refer to Section 11.2.8). Besides these tools, solar irradiance for some cities can be found in many resources.[11] One of the best tools for this purpose, at least for initial designs, is **solar energy maps**, also known as **insolation maps**. One such map is presented in Figure 11.13, where average daily solar irradiance is presented in the global scale. As expected, generally regions close to the equator have higher solar incidents and regions further north or south have lower incidents. However, it can be observed that even for areas with similar latitudes, there are significant variations in solar irradiance.

A wide variety of similar maps are available at the National Renewable Energy Laboratory (NREL) website (https://www.nrel.gov/gis/solar.html) for the U.S. in annual and monthly bases as well as for each individual state. The insolation map in Figure 11.14 might be a good starting point (you can access

[10] In the U.S., the National Renewable Energy Laboratory (NREL) has published long-term solar data along with weather data as the typical meteorological year, version 3 (TMY3) at https://nsrdb.nrel.gov/data-sets/archives.html. The user manual for the TMY3 data sets can be accessed at https://www.nrel.gov/docs/fy08osti/43156.pdf

[11] NREL has published average irradiance information for many cities in the U.S. in a report entitled "Solar Radiation Data Manual for Flat-Plate and Concentrating Collectors" accessible at https://www.nrel.gov/docs/legosti/old/5607.pdf

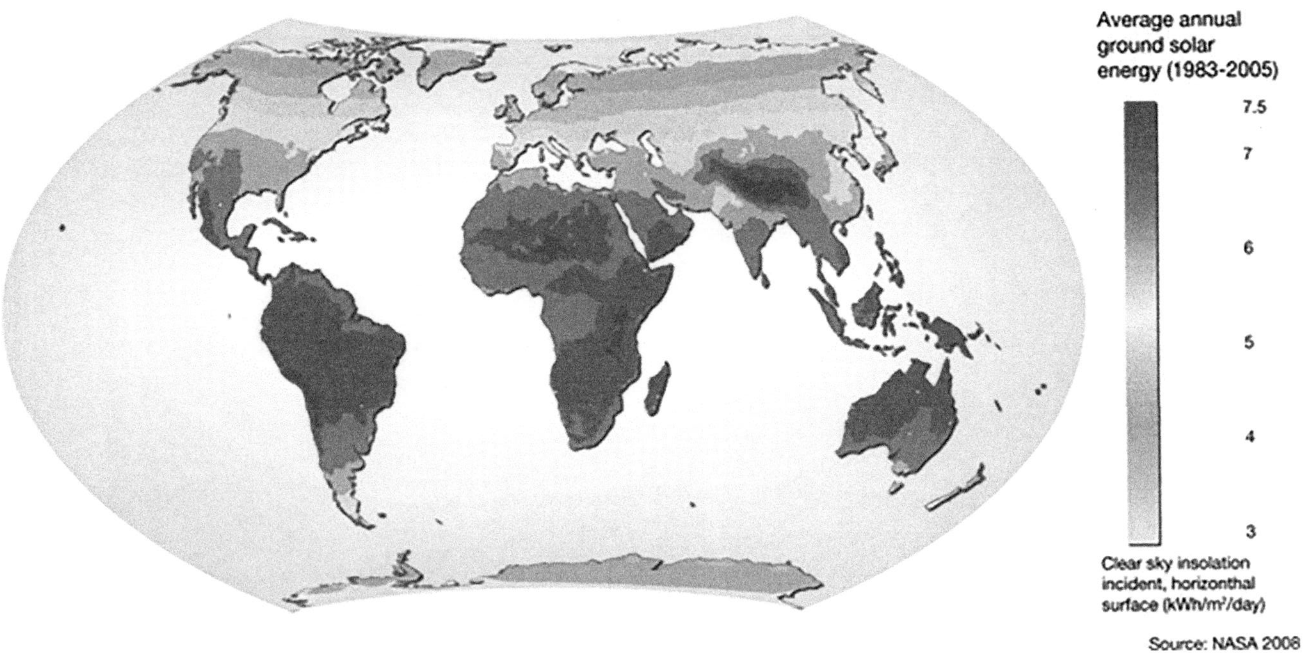

FIGURE 11.13

Global solar insolation map. (Source: The U.S. Energy Information Administration, https://www.eia.gov/energyexplained/solar/where-solar-is-found.php.)

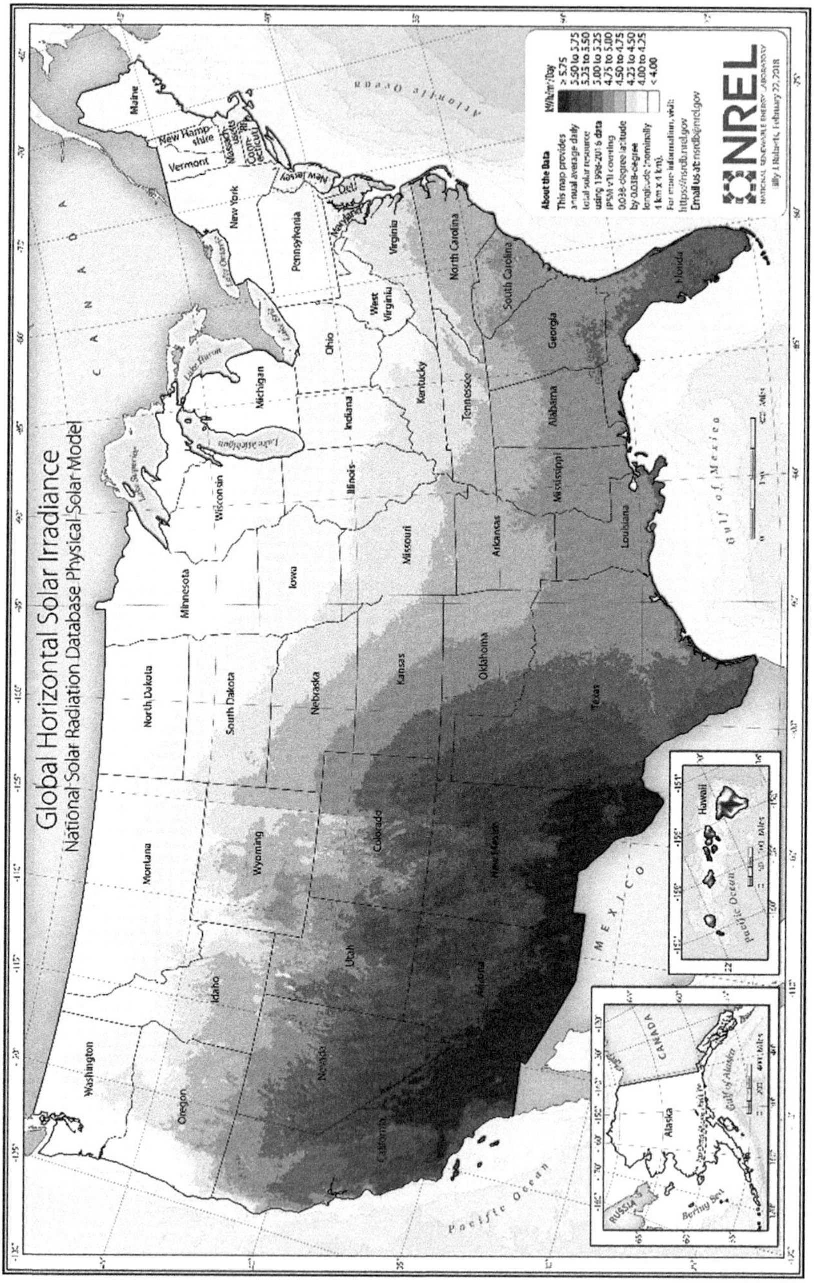

FIGURE 11.14

The U.S. average daily insolation map available at https://www.nrel.gov/gis/assets/images/solar-annual-ghi-2018-usa-scale-01.jpg. (Source: the National Renewable Energy Laboratory (NREL).)

the online map, which is much more readable, via the provided link). The map is color-coded from dark color for very high solar energy to light color for very low energy and many colors in between. You can also find an interactive insolation map at the Department of Energy (DOE) website.[12] Note that these maps are presenting daily average of solar irradiance in kWh/m^2/day and to find the annual irradiance the values should be multiplied by 365. However, some maps may directly provide annual irradiance so a particular attention should be paid to what a map presents, e.g., global horizontal irradiance (GHI) or direct normal irradiance (DNI), and the unit of solar irradiance in the map.

11.2 Solar Photovoltaic Power Generation

If we could envision an ideal electricity production technology, it could have been very much like solar photovoltaic systems. They directly convert solar radiation into electricity. The primary source of energy, i.e., solar radiation, is renewable and virtually limitless (considering the scale of human needs). Their operation does not emit any pollutants or produces any radioactive materials. Even if life-cycle emissions, including raw materials, manufacturing processes, operation, and end-of-life disposal, are all taken into account, they are one of the most environmentally friendly power generation technologies. The energy payback for manufacturing of solar panels is typically between 1 and 2 years. Solar panels are solid-state devices, and there are no working fluid or moving parts involved. This makes the technology very safe and low maintenance with no mechanical wear out. Therefore, they can be utilized in location where continuous and regular maintenance may not be available or convenient such as remote areas or residential buildings. This provides a great opportunity for electrification of rural remote areas where grid connection is not available, particularly in developing countries. Moreover, they are ideal for decentralized distributed power generation. They can also be built in a variety of capacities ranging from a few microwatt for small handheld devices to many hundred MW units for utility power production. They can be built and installed in modular units to meet the power demand from a small residential building to a large community. They are uniquely suitable for integration to buildings. In terms of financial viability, the cost of solar panels has reduced substantially in the past few years. Currently, there are dozens of countries where the cost of electricity from solar panels is less than other available sources even without any incentive. The current dominant solar panel technology in the market is silicon crystalline panels. Silicon is the second most abundant element on Earth. Simply put, the solar photovoltaic power generation can be described as simple, elegant, convenient, dependable, and in many cases affordable.

[12] https://www.energy.gov/maps/solar-energy-potential.

Regardless of all these extraordinary attributes, the contribution of solar photovoltaic energy to overall global electricity production is small. To significantly alter this picture, the cost of photovoltaic panels should be further reduced. Also, the variable nature of solar energy and its impacts on the electrical grid should be addressed. This includes predictable variations, i.e. the day and night cycle, which creates the problem of a duck curve (refer to Chapter 16), and local long- and short-term unpredictable events, from a few rainy days to a passing cloud and even a solar eclipse.

11.2.1 History of Solar Photovoltaic Power Generation

Forty years after Alessandro Volta (1745–1827, an Italian physicist and chemist) demonstrated that electricity can be generated by non-living things (refer to FYI Box 1.1 on electric fishes and human encounter with them) through electrochemical reactions using his voltaic piles, it was discovered that light energy can also produce electricity. This phenomenon was coined the **photovoltaic effect** by combining the word light in Greek (phôs) and Volta's name. The effect was first discovered by Alexandre-Edmond Becquerel (1820–1891, a French physicist[13]) in 1830. He was actually conducting experiments on a battery. In his experiments he noticed that when the electrode plates were illuminated by light, the current produced by the battery was increased. Therefore, while he did not really make an actual photovoltaic cell, he accidentally discovered that at right conditions light can have electrical effects. Later in 1873 Willoughby Smith (1828–1891, an English electrical engineer) observed that conductivity of selenium increased when exposed to light, showing selenium is a semiconductor. A few years later, in 1877, William Grylls Adams (1836–1915, an English scientist) and his student Richard Evans Day were able to produce electricity on a cell made of selenium and platinum for the first time. In 1883, Charles Fritts (1850–1903, an American engineer) built the first working solar cell that was able to generate a meaningful amount of electricity. His cell, known as the selenium cell, was made of selenium coated with a very thin layer of semi-transparent gold on one side and an iron plate on the other side. He connected several of his cells and installed them on a rooftop in New York, making it the first rooftop-installed solar photovoltaic device. His invention was never commercialized for electricity production due to its high cost and low efficiency, about 1%. However, it was used in light sensors, for example, in cameras. Just to understand the timing of his invention, when he announced his discovery the early coal steam power plants were just entering the market. From then to mid-20th century there were many

[13] Being a prominent physicist was a tradition in the Becquerel's family. His father and son were also very well-known physicists with many accomplishments. His son Henri received the 1903 Nobel Prize in Physics along with Marie and Pierre Curie for the discovery of radioactivity. Alexandre-Edmond Becquerel was 19 years old when he discovered photovoltaic effect. He conducted some pioneering research on photography, particularly colored photos, and solar spectrum.

patents filed related to solar photovoltaic cells.[14] In the meantime, many physicists studied the photovoltaic (photoelectric) effect and the theoretical understanding of the phenomenon increased significantly. Among them was Albert Einstein who received the 1921 Nobel Prize in Physics for his work on "the law of the photoelectric effect" (and not for relativity).[15] While these early progresses created great expectations, commercial solar photovoltaics were never materialized until the second half of the 20th century.

The major breakthrough on photovoltaic power generation came after the invention of transistors in the early 1950s because both transistors and solar cells work based on the properties of semiconductors especially silicon (we will extensively discuss about these material later). Transistors were first developed at the Bell Laboratories in New Jersey, then a subsidiary of AT&T which was founded by Alexander Graham Bell the famous inventor of telephone (1847–1922, a Scottish-born American scientist and engineer).[16,17] In the early 1950s, a team of Bell Labs scientists[18] accidentally reinvented the concept of photovoltaic solar cells while studying the impact of light on silicon-based semiconductors to improve the performance of transistors. They then tried to use their solar cells to power remote rural telephone amplifiers located in humid environments. The efficiency of the cell was initially about 2%, but after some experiments and optimizations, it was improved to about 6%. The systems was able to generate sufficient electricity to power typical electrical equipment of those days and was considered, albeit briefly, for electricity generation in remote rural areas. While the public announcement of the invention created a lot of excitements, the prohibitively high cost of the system eliminated any meaningful application (somewhere between $250 and $1000 per watt).[19] Even if we consider the lower range of the cost, a small typical 6 kW residential unit would have cost about $1,500,000. They were also very fragile. Later, one of the major players in the improvement of solar cells was Hoffman Electronics Corporation founded by Leslie Hoffman (1906–1971, an American entrepreneur). Between 1957 and 1960, they were able to improve

[14] Just search for solar cell in Google Patent (https://patents.google.com) by limiting the period to 1850–1940 to review some.

[15] At the time some committee members believed the relativity is still controversial and unproven. So they awarded him the prize for the work he published in 1905. Some have speculated that the committee wanted to award him the second Nobel Prize when the relativity was "properly" proven. The second prize was never awarded perhaps because he never accepted the well-established quantum theory. By the way, he did not attend the award ceremony partly because his name was in the target list of the Europe's anti-Semite groups. For more information refer to "Why Einstein never received a Nobel prize for relativity" https://www.theguardian.com/science/across-the-universe/2012/oct/08/einstein-nobel-prize-relativity.

[16] The Bell Laboratories are currently a subsidiary of the Finnish company Nokia.

[17] The three American physicists who invented transistors, John Bardeen, Walter Brattain, and William Shockley, received the 1956 Nobel Prize in Physics for this invention and its subsequent improvements.

[18] The team consisted of three American scientists, Daryl Chapin (1906–1995), Calvin Fuller (1902–1994), and Gerald Pearson (1905–1987).

[19] Compared it to current cost of less $1/W.

FIGURE 11.15
Hoffman KP706 Trans Solar Radio (1959) running on batteries or with a Hoffman solar cell array on the top (https://en.wikipedia.org/wiki/Les_Hoffman#/media/File:Hoffman_model_KP706_Transolar_Radio_1959.jpg). (By Engradio at English Wikipedia, CC BY-SA 3.0, https://commons.wikimedia.org/w/index.php?curid=31304864.)

the efficiency of solar cells from 8% to about 14% and markedly reduced the cost. The company manufactured and sold a solar cell powered radio called the Hoffman Trans Solar Radio in 1959 (Figure 11.15).[20]

By the late 1950s a niche new application appeared in which the hefty cost of early solar cells was not an issue. In 1957, the Soviet Union launched the world's first satellite, Sputnik 1. The transmitters on board of the satellite were powered with batteries, and they stopped operating when the batteries ran out 21 days after its launch. This was followed by Sputnik 2 and the first American spacecraft, Explorer 1. They both used batteries as the power source and stopped transmitting signal after 1 week and 4 months, respectively. In all these cases, the satellites stayed in the orbit for a duration much longer than they could transmit the signals.[21] To address this problem, a continuous supply of power was required. Considering the operating conditions of a satellite and its direct exposure to solar radiation, solar photovoltaic cells were an ideal solution. Solar cells were used to power a satellite for the first time in Vanguard 1 launched in March 1958.[22] This small 1.5 kg (3 lb) and 0.165 m (6.4 inches) diameter sphere made of aluminum was equipped with six solar cells manufactured by Hoffman Electronics Co. each 25 cm^2 (5 cm×5 cm) installed around the sphere. With the efficiency of about 10% (at the operating temperature of 28°C), the cells were made of single crystalline silicon and produced total power of about 1 W.[23] While the produced power was very small, it was sufficient to power the transmitter on board. The solar cells were able to power the transmitter for more than 6 years (Figure 11.16).[24]

---

[20] Reportedly the radio was sold $79.95 (those days dollar).

[21] Sputnik 1 and 2 stayed in the orbit for 3 months and 162 days, respectively. Explorer 1 was in the orbit for about 12 years.

[22] Vanguard 1 was the fourth satellite ever launched and the second American.

[23] https://nssdc.gsfc.nasa.gov/nmc/spacecraft/display.action?id=1958-002B.

[24] The satellite also had a battery powered radio transmitter though it stopped sending signal in <4 months of launching due to the battery depletion.

FIGURE 11.16
Vanguard 1 satellite equipped with six solar cells. (Source: https://nssdc.gsfc.nasa.gov/image/spacecraft/vanguard1.jpg.)

FIGURE 11.17
Explorer 6. (Source: https://nssdc.gsfc.nasa.gov/planetary/image/explorer_6.jpg.)

Since then many spacecraft was powered by solar cells,[25] for example:

- Explorer 6 (launched in 1959) had closed to 10,000 solar cells manufactured by Hoffman Electronics (Figure 11.17).
- Telstar (launched in 1962) had solar cells that produced about 14 W of power (Figure 11.18).
- The Hubble Space Telescope (launched in 1990) was equipped with enough solar cells to produce 5.5 kW peak and 2.1 kW average electrical power (in two wings each about 8 m or about 25 ft) (Figure 11.19).
- Finally and most notably, the International Space Station has 262,400 solar cells covering 2500 m^2 (27,000 ft^2) arranged in eight solar arrays with the wingspan of 73 m (240 ft) (Figure 11.20). They could generate up to 120 kW of electrical power (enough to provide power to more than 40 homes).

[25] But not all satellites were powered by solar cells. Probably the most notably, the Gemini and Apollo series did not have solar cells.

FIGURE 11.18
Telstar satellite. (Source: https://www.nasa.gov/sites/default/files/thumbnails/image/665926main_telstar_full.jpg.)

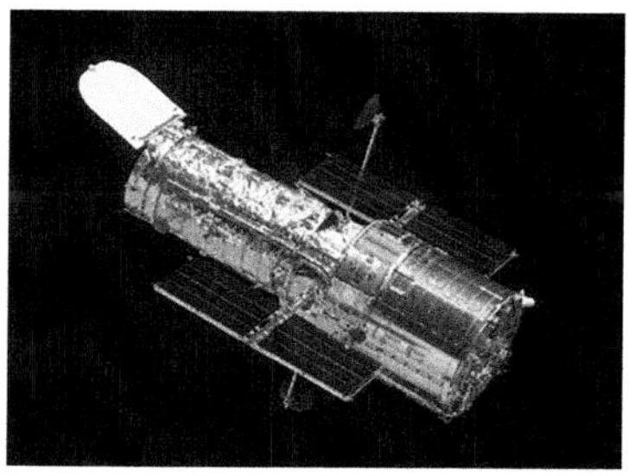

FIGURE 11.19
Hubble space telescope. (Source: https://www.nasa.gov/sites/default/files/thumbnails/image/hst-sm4.jpg.)

FIGURE 11.20
International space station. (Source: https://www.nasa.gov/sites/default/files/thumbnails/image/iss056e201248.jpg.)

Solar photovoltaic cells proved to be an ideal power source for extraterrestrial applications due to their high power-to-weight ratio, which is a critical criterion for this application. Furthermore, the high cost of early solar cells was not a major factor for this application. Since then solar cells have been the power source for many space exploration missions. This application provided a reliable driving force to push solar cell technologies forward.

FIGURE 11.21
A solar cell-powered calculator. (Image by DG-RA from Pixabay, https://pixabay.com/photos/algebra-arithmetic-calculator-casio-2154417/.)

In many cases (some will be discussed later), new (and expensive) technologies were first developed for this application and then after a substantial cost reduction was adapted for other applications.

The advancement in the solar cell manufacturing processes gradually reduced the cost of solar cells to the order of \$100/W. While this price was still too high for widespread use of solar cells, it helped their gradual deployments in remote applications where no other source of conventional electricity production was economically available, including road and railroad signs as well as offshore and telecommunication devices. The rapid advancement of semiconductors for electronic devices in the 1970s provided a significant boost for the solar cell industry. The oil crisis of the late 1970s helped to provide strong supports for research and development in the field, at least for a while. In the early 1980s, the global installed capacity of photovoltaic power generation units reached about 20 MW. In the 1980s another popular niche market opened up for the utilization of solar cells in low-powered devices particularly handheld calculators (Figure 11.21).

During the 1990s and 2000s, the efficiency of solar cells continuously improved and their cost reduced. In 1985, 1994, and 2006, the records for the efficiency of solar cells were broken at 20%, 30%, and 40%, respectively.[26]

[26] When dealing with these records, one should be very careful about the conditions they were recorded. For example, these reported records were for concentrated solar radiation. The highest recorded efficiency for unconcentrated solar radiation was 34.5% in 2016.

In 1992, Sacramento Municipal Utility District (SMUD) in California initiated the first residential distributed solar power program designed for general public. Many states and countries started to provide similar systematic incentives. In the meantime, the development in axillary systems, such as inverters, helped to lay the foundation for the applications of solar photovoltaic cells in areas with access to the grid, which reduced or eliminated the reliance on battery storage systems. Until 1999 when the capacity of the installed solar cell units reached 1.2 GW, almost the entire capacity was off-grid units. However, large-scale utility solar power generation units, which followed residential applications, had a much higher rate of increase. By 2004, the capacity of grid-connected units overtook that of off-grid units. Nowadays the great majority of the units are grid-connected. This created the cycle of increased installed units and increased unit production, which brought down the manufacturing costs due to the economy of scale. This in turn encouraged more installation. The reduction in the required raw materials (per unit of power production) and progress in manufacturing technologies in the past decade have pushed down the cost to the level that, even without any incentive, now the electricity generated by photovoltaic cells is the least expensive source of electricity in increasing number of countries and regions.

The diagram in Figure 11.22 illustrates the global installed solar photovoltaic unit capacity between 1996 and 2018, both existing and new capacities for each year. In 1996, the total global capacity of photovoltaic power generation was merely 0.7 GW. By 2006 this increased by 10 folds (it took 11 years). In order to increase the capacity by another 10 folds, only 5 years was needed.

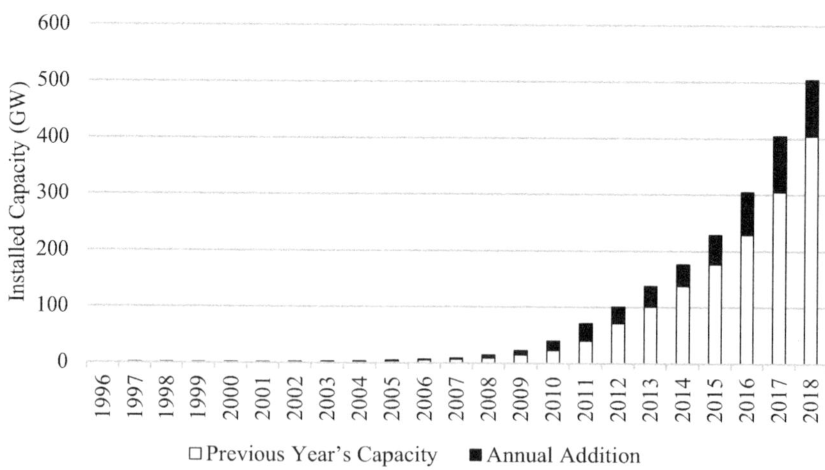

FIGURE 11.22
Global installed solar photovoltaic capacity between 1996 and 2018, both existing and new capacities for each year. (Based on the data provided by IEA, https://iea-pvps.org/wp-content/uploads/2020/04/IEA_PVPS_Snapshot_2020.pdf)

By 2011 the installed capacity increased to 71 GW, which was 78% greater than that of the prior year. By 2018, the global installed capacity increased to 505 GW in which 100 GW was new units installed in that year. This indicates an exponential growth of the installed solar photovoltaic power generation units with over 30% annual growth rate between 1996 and 2018. This means the installed capacity was doubled roughly every 2 or 3 years.

Considering the abundance of solar energy, California has always been a natural leader in the field. But in the international level, in the early 2010s, the leadership came from an unexpected country, Germany (considering the fact that Germany is located in northern Europe where solar energy is not among the highest in Europe) and Japan. Germany's progress was extraordinary in the early 2010s, but the increase in the installed capacity has been stagnant in the past few years since it seems that the country has reached to the saturation point in terms of economic viability. Since 2015 China has had the largest capacity of installed solar units, with very wide margins in the past couple of years. Currently China has over a third of the global installed capacity. In 2018, the six highest contributors to the new install capacity of solar cells were China (45%), India (11%), the U.S. (11%), Japan (7%), Australia (4%), and Germany (3%). However, Germany still has the highest installed capacity per capita (W/person). In 2018, there were eight countries that the share of photovoltaic solar units in their annual electricity production mix exceeded 5%. Regardless of all these rapid progresses, the global share of electricity from photovoltaic solar units is still <3%.[27]

It is expected that the current growth rate in the global installed solar photovoltaic units will continue in the future. The ongoing reduction in the cost of these units is the most important factor. The continued incentives provided for these power production units will further encourage their utilization. However, in many countries the price of the electricity from solar cells is already competitive with the electricity from traditional power plants (called the **grid parity**) even without the financial incentives. The significant investments by some of the largest emerging economies, particularly China and India, are a major driving force for this trend. The fact that solar panels are providing an alternative affordable path toward electrification of rural areas in some parts of the world is another encouraging factor. The global, national, and regional plans to combat global CO_2 emission and local pollutions are also major factors for promotion of this technology.

In the U.S., according to the Solar Energy Industries Association, the capacity of installed solar units was about 69 GW in 2019.[28] From the beginning of the industry, California has been the leading force with the installed capacity of about 26 GW. This followed by North Carolina and Arizona with about 6 and 4 GW capacity, respectively. Nevada, Florida, Texas, New Jersey, and Massachusetts followed all with about 3 GW capacity.

---

[27] https://www.iea.org/reports/solar-pv.

[28] https://www.seia.org.

11.2.2 Applications of Photovoltaic Solar Units

In the previous section, some applications of solar photovoltaic electricity generation were discussed from the historical point of view. In this section, various applications and some categorizations in modern units will be presented.

Solar electricity generation units can be distributed generation units or utility-scale plants. **Distributed generation** (also known as **decentralized**) units are located at or near the load so the produced electricity is directly transferred to the consumer without going through the grid. In contrary, **utility-scale** PVs (sometimes also referred to as **solar farms**) are much like central power plants where a huge amount of electricity is produced and then distributed among many consumers via the grid. For most of the history of the utilization of photovoltaic power, distributed units were dominant. However, in the U.S., since the early 2010s, utility-scale units have generated more power than distributed units.

Photovoltaic units can be **off-grid** or **on-grid** (or grid connected). As the name implies, off-grid units do not have any interconnection with the grid. In these units the generated electricity needs to be used or stored as soon as it is produced. Normally they are equipped with a battery storage system. When the power production exceeds the consumption, the excess electricity is stored in the battery bank. When the solar energy is not available (e.g., at night) or when the produced power is not sufficient to meet the demand, the electricity is pulled from the batteries. Note that the power produced by the solar cells and the power from the batteries are direct current (DC) but the power needed by consumers should be alternative current (AC). Therefore, an inverter is needed to convert DC power to AC power (refer to Section 11.2.5.2). While off-grid units were very popular in early solar systems, their current applications are limited to locations where the grid is not available, such as rural regions and remote applications, or situations when electricity from the grid is not available, for example in emergency and backup power units.

On-grid photovoltaic systems are interconnected to the grid. While they can be equipped with a battery storage unit, they are not required to have one. In these units, the grid acts as a gigantic virtually unlimited storage system. When the generated electricity is more than the consumption, the excess power is sent to the grid. When the generated power is not sufficient, the required balance is pulled from the grid. There are various financial mechanisms to determine the cost of power exchange with the grid (refer to Section 11.2.6). Also, an inverter is required to convert the produced DC power to AC power. As previously noted, currently the majority of solar photovoltaic power generation capacity is connected to the grid.

As noted in the introduction, one major advantage of photovoltaic cells is their modularity and scalability, from less than a watt to hundreds of megawatts. In the scale that is of the interest for this text, these units can be divided to small-scale residential applications, mid-size industrial or commercial applications, and many-megawatt utility units.

Solar photovoltaic units can be installed on a building structure or can be ground-mounted. While **building-mounted** units are typically in the form of distributed generation, **ground-mounted** units can be large-scale utility units. There are many potential lands to install ground-mounted solar panels, including deserts, semi-desert areas, brownfields, lands prone to shallow flooding, shades for parking lots and other areas, or even floating units. Ground installations have the advantage of easy maintenance.

Building-mounted units can be added to existing buildings or can be integrated to new building designs (also known as **building integrated photovoltaics** or **BIPV**). When rooftop-mounted solar units are added to existing buildings, their sole function is to generate electricity but in integrated buildings they may have several energetic and architectural functions. Building integrated units may also address the visual impacts and aesthetic aspects of rooftop modules that some may find objectionable. When PV units are installed on commercial or industrial buildings, they have an added benefit that the period of typical occupation of buildings, i.e., during daytime, matches the power production period from solar units.

Solar PV modules can be fixed or equipped with sun tracking systems. Building-mounted units are almost universally fixed, but ground-mounted units can be fixed or have tracking units. Other than utility-scale and building-mounted units, photovoltaic panels can be assembled on small-scale units with the power generation capacity of a few kW. They are typically used to replace diesel power generators in emergency units, for example, in disaster areas or in construction sites. Smaller micro-scale units can be used to power outdoor security cameras, road signs, etc., as well as personal military applications or outdoor recreational applications. These units are not typically equipped with inverters, and DC power is directly used by consuming devices.

11.2.3 Principles of Solar Photovoltaic Cell Operation

Now that you have good understanding of the historical background, the current state of solar photovoltaic systems, and the variety of their applications, we can discuss about physical principles behind their operation. These fundamentals are not only helpful in the understanding of internal operation of cells but also essential in the understanding of the impacts of various parameters on their performance. As noted in Section 11.2.1, the development of PV cells and transistors has been intertwined because both used similar building blocks, semiconductors. Thus, we will first discuss the semiconductors' structure and their photovoltaic properties. Then, we will explain how these properties help solar cells to generate electricity.

11.2.3.1 Semiconductors

Depending on how well a substance can conduct electrons (and heat), it can be categorized as a conductor, insulator, or semiconductor. At one end, there are metals with very high conductivity (and very low resistivity) and at the other end there are insulators with very low conductivity (and very high resistivity). The conductivity (and resistivity) of these two ends can differ by the factor of 10^{30}. Semiconductors have conductivity between conductors and insulators at the room temperature. At very low temperatures and high temperatures they behave as insulators and conductors, respectively.

In order to explain such a wide range of conductivity of materials, we need to study their subatomic structure. While atoms are made of differing number of electrons, protons, and neutrons, most properties of materials, particularly their conductivity, are determined by the number of electrons in their atom. We know that in every atom, protons and neutrons form the nucleus and electrons orbit this nucleus. From quantum mechanics, it is proven that for a single atom, electrons can only occupy defined orbits with certain energy levels. These energy levels are expressed according to their energy in electron-volt (eV). Electrons can only "jump" from one energy level to a higher or lower energy level by absorbing or releasing a photon. The energy difference between the two energy levels must be equal to the energy of the absorbed or released photon. When there are many atoms bonded together to form a bulk of a substance, due to the interaction of the atoms, we will have a range of energy levels that electrons can have. These ranges of acceptable energy levels for electrons are called **energy bands** (also known as **allowed bands** or simply **bands**) (Figure 11.23). The gaps between these bands are called **forbidden bands**, forbidden gaps, or band gaps. Note that electrons naturally fill these bands from the lowest energy level upward and are organized in a manner that results in the minimum energy. While an atom can have many bands depending on the number of electrons it has, the following two bands mostly determine the properties of the atom. The band that has the maximum energy level and is occupied by some electrons (i.e., fully or partially filled with electrons) is called the **valence band**. The next band,

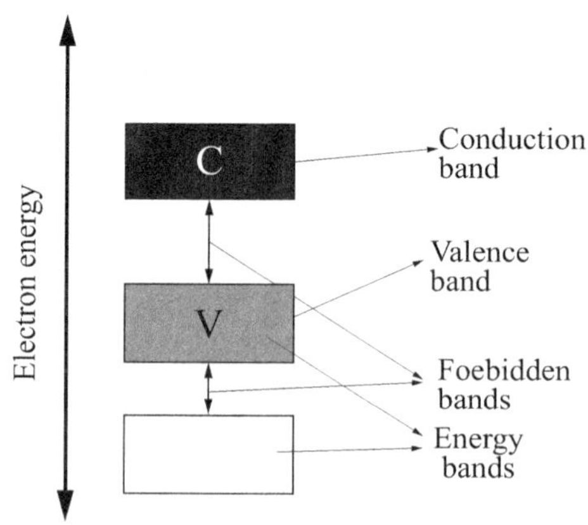

FIGURE 11.23
Electron energy levels in an atom.

which is empty (otherwise this band would have been the valence band), is called the **conduction band** (Figure 11.23). When two atoms bond, the electrons in the valence band of one atom form bonds with the electrons in the valence band of the other atom.

The electrons in the valence band are bonded to the nucleus and cannot move freely. Electrons can conduct electricity (and heat) only when they are elevated from the valence band to the conduction band. Technically speaking the electrons occupying the conduction band are not bonded to the nucleus of the atom so they belong to the whole bulk of the substance rather than to a particular atom. That means they can freely move across the body and transfer electricity and conduct heat. Therefore, the conductivity of a substance is a function of the relative range of energy levels of the valence and conduction bands. If these two bands overlap (i.e., when there is no forbidden gap) or the gap between them is very small (i.e., a very narrow forbidden gap), the substance is a good conductor (Figure 11.24). In this case, either all electrons of the valence band can freely move to the conduction band or they can be easily elevated. In insulators, the gap is so large (larger than 3 eV) that hardly any electron can be elevated so the conduction band is almost empty and there is no free electron in the material. In semiconductors, the gap is moderate (<3 eV), so some electrons can elevate to the conduction band. Therefore, semiconductors have medium conductivity. The band gap for some semiconductors that are used in the manufacturing of solar cells at the room temperature is as follows: silicon 1.1 eV, germanium 0.7 eV, gallium arsenide 1.4 eV, and cadmium telluride 1.4 eV.

At right conditions, such as high temperatures, sufficient electrons can be moved to the conduction band that the semiconductor behaves like a conductor. At very low temperatures, hardly any electrons can be elevated so it

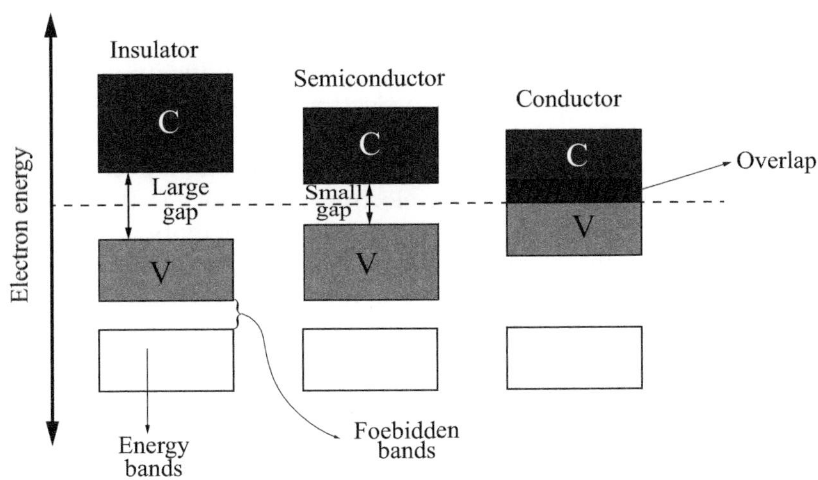

FIGURE 11.24
Band gap for conductors, semiconductors, and insulators.

becomes like an insulator. Semiconductors can be **elemental** or **compound**. While elemental semiconductors are made of a single element, compound semiconductors have two or more species. The examples of elemental semiconductors are the elements in Group IV of the periodic table: carbon (C), silicon (Si), germanium (Ge), and tin (Sn); and in Group VI: sulfur (S), selenium (Se), and tellurium (Te). The compound semiconductors can be composed of two elements (binary), e.g., gallium arsenide (GaAs), three elements (ternary), e.g., copper gallium diselenide (CuGaSe), or four elements (quaternary), e.g., copper indium gallium selenide (CuInGaSe). The binary semiconductors can be grouped according to the groups the elements belong in the periodic table; for example, gallium arsenide (GaAs) is a III-V semiconductor. Other examples are II-VI, I-VII, IV-VI, V-VI, and II-V semiconductors.

Most semiconductors have a crystalline structure. Let us consider silicon crystalline, which is the main building block of the most common type of solar cells. Silicon's atomic number (proton number) is 14, which means in an uncharged form a silicon atom has 14 electrons. These electrons occupy first, second, and a part of the third energy bands with 2, 8, and 4 electrons in each, respectively. That means the valence band contains four electrons. In its pure crystalline form each of the electrons in the valence band forms a bond with a similar electron from another atom. This creates a lattice crystalline structure where each silicon atom is surrounded with four other atoms and shares two electrons with each neighbor. Figure 11.25 shows this arrangement for the pure crystalline silicon in three-dimensional and two-dimensional drawings (Figure 11.25a and b, respectively). As the drawings illustrate, in this structure there is no free electron (i.e., electrons that occupy the conduction band). In reality this structure only exists at very low temperatures (close to absolute zero). However, due to a relatively small energy gap (about 1.1 eV), at room temperatures some electrons can be transferred to the conduction band (Figure 11.25c). The higher the temperature, the more electrons can be promoted, which in turn increases the conductivity of silicon.

As Figure 11.25c illustrates, when an electron is promoted to the conduction band, the bond between the two silicon atoms will have a shortage of an electron. This absence of an electron, which translates to a positive charge, is called a **hole**. Holes, similar to free electrons, can move in the crystalline structure and carry charges, which increase the conductivity of silicon. This process is known as the **electron-hole pair generation process**. When an electron and a hole rejoin, the process is known as the electron-hole pair recombination process. In order to understand how holes impact conductivity, imagine two consecutive shelves in your bookshelf with some books. In this analogy, the lower and upper shelves represent the valence and conduction bands, respectively, and the books represent the electrons. When the lower level shelf is full of books and the upper level shelf is empty, none of the books in the lower shelf can move. But if we move one book to the upper shelf, not only that book (representing an electron promoted to the conduction band) can move in the upper shelf, but also the books in the lower shelf

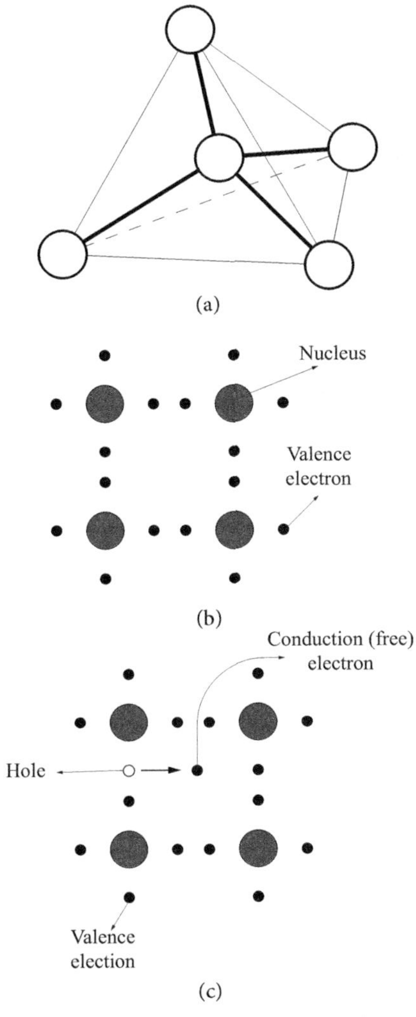

FIGURE 11.25
Atomic structure of silicon crystalline (a) three-dimensional structure with no electron promoted to conduction band (b) simplified two-dimensional representation with no promoted electron (c) simplified two-dimensional representation with one electron promoted to conduction band.

can move because of the space left by the removed book (representing the hole). Note that even in the latter case what is really moving is the electrons in the valence band (or books in the lower shelf). But it is simpler to assume the holes (the space left by the removed book) are moving.

Elements or compounds that exhibit an appropriate semiconductor ability in their pure form are called **direct or intrinsic semiconductors**; for example, gallium arsenide (GaAs) and cadmium telluride (CdTe) with the band gap of

1.4 and 1.5 eV, respectively, are considered direct semiconductors. The meaning of "appropriate semiconductor ability" will be discussed later. For now, just note that in order for a semiconductor to be suitable for solar cells, it should have a certain band gap (about 1.4 eV being the optimum value). Some other semiconductors do not have this favorable ability. For example, pure silicon crystalline has a band gap of 1.1 eV, which is too small (again we will later talk about why a too small band gap is not favorable). To remedy this problem, a very small and controlled amount of certain impurities can be added to a pure semiconductor. The process of adding the impurity is called the **doping process,** and this type of semiconductors is known as **indirect**, **extrinsic**, or **doped semiconductors**. Adding the impurity adds an energy level (and not an energy band), similar to the ones in individual atoms, between the valence and conduction bands. Depending on the type of the impurity, also known as the **doping agent** or the **dopant**, the added energy level can be close to the bottom of the conduction band or the top of the valence band (Figure 11.26). If the energy level is near the bottom of the conduction band (just below it), it is called the **donor level** and it helps to send

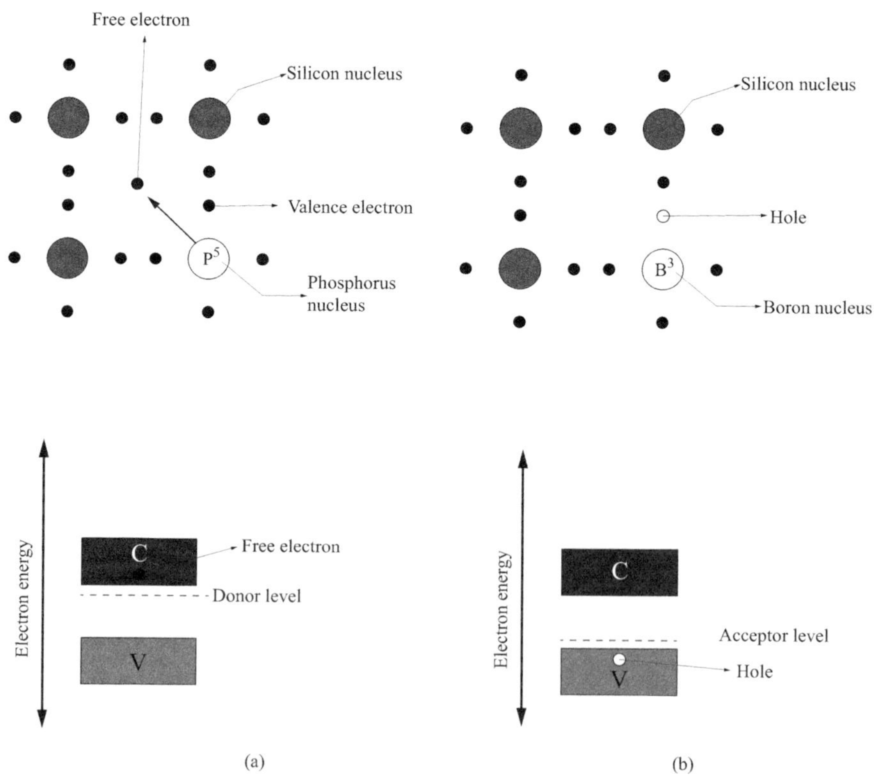

FIGURE 11.26
Doping silicon crystalline with (a) phosphorus to create an n-type semiconductor (b) boron to create a p-type semiconductor.

many more electrons to the conduction band than that of a pure semiconductor. If the energy level is near the top of the valence band, it is called the **acceptor level**. In this case, many electrons are promoted from the valence band to the acceptor level, which leaves positively charged holes (electron deficiencies) behind. These holes in turn can conduct electricity or heat. In indirect semiconductors, it is easier to control the properties of the semiconductor. Both direct and indirect semiconductors can be used to manufacture solar cells. However, indirect semiconductors need a much thicker layer to absorb the incoming light than that of the direct semiconductors (will be discussed later in this chapter).

In the case of silicon crystalline, we can dope the crystalline lattice with phosphorus, which has five electron in the valence band.[29] Four of these electrons bind with the electrons in the neighboring silicon atoms and one electron remains free and available for conduction (Figure 11.26a). Since the dominant charge carrier in this case is negatively charged electrons, it is called **n-type semiconductors**. In the same fashion, silicon can be doped with boron with three electrons in the valence band.[30] In this case, each of the three electrons binds with one electron from the neighboring silicon atoms, but the fourth bond will have one electron deficiency, i.e., a hole (Figure 11.26b). The holes are the dominant charge carriers in this case. Since they have positive charge, this type is called **p-type semiconductors**. In these two semiconductors, phosphorus and boron atoms are known as the **donor atoms** and **acceptor atoms**, respectively. The conductivity of doped semiconductors depends on the type and amount of the doping agent. Note that while only a trace amount of the dopant is added, the improvement in conductivity is dramatic. In terms of the number of atoms, the dopant-to-silicon ratio of $1{:}10^5$ is typical.

While in both n-type and p-type semiconductors light can create electron-hole pairs, the creation of the pairs will not be translated to electricity generation until we can separate electrons and holes and we can transfer them to different locations. This separate accumulation of holes and electrons creates a difference in charge (voltage) and the flow of electrons in an external circuit, i.e., electricity. In order to be able to separate electrons and holes in solar cells, a p-n junction is needed. The **p-n junction** is created when a layer of a p-type semiconductor is in physical contact with a layer of n-type semiconductor (Figure 11.27). When the connection is established, the negatively charged free electrons in the n-type semiconductor migrate to the p-type semiconductor where they join the positively charged holes. Note that since the distance the electrons and holes can diffuse is limited, the impacted area is limited to the area near the junction. The diffusion of the electrons across the junction will leave a positive net charge in the n-type semiconductor

[29] Phosphorus is in Group V (group 15 in the modern notation) of the periodic table and has an atomic number of 15 and the electrons arrangement in the energy bands is 2, 8, and 5.

[30] Boron is in Group III (group 13 in the modern notation) of the periodic table and has an atomic number of 5 and the electrons arrangement in the energy bands is 2 and 3.

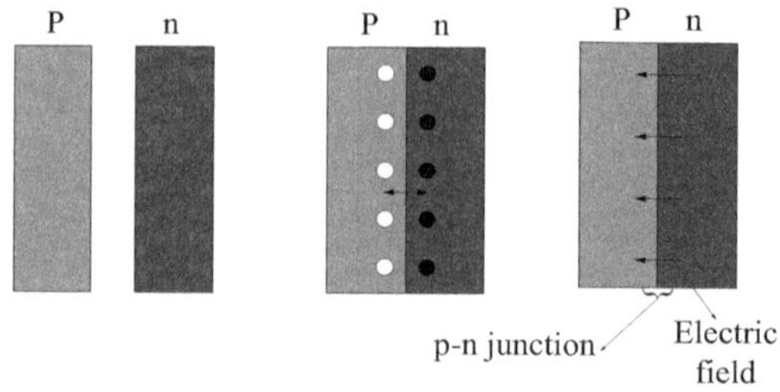

FIGURE 11.27
Schematic of a p-n junction to separate electrons and holes in electron-hole pairs.

and will create a negative net charge in the p-type semiconductor. These two fixed electrical charges in two opposite sides of the junction will create an electric field around the junction. If an electron-hole pair is created within this electric field, the electron will be immediately attracted to the n-type semiconductor (where there is a positive net charge) and the hole to the p-type semiconductor. Now we have all we need to explain how a solar cell convert solar radiation to electricity.

11.2.3.2 Operation of Solar Cells

In its simplest form a typical solar cell is a very thin layer of crystalline silicon with **two electrodes** (also known as the **contacts**) on the front and back surfaces (Figure 11.28). A p-n junction is developed within the single layer of silicon. So actual solar cells do not have two pieces of p-type and n-type semiconductors that are physically connected. Instead there is a single piece

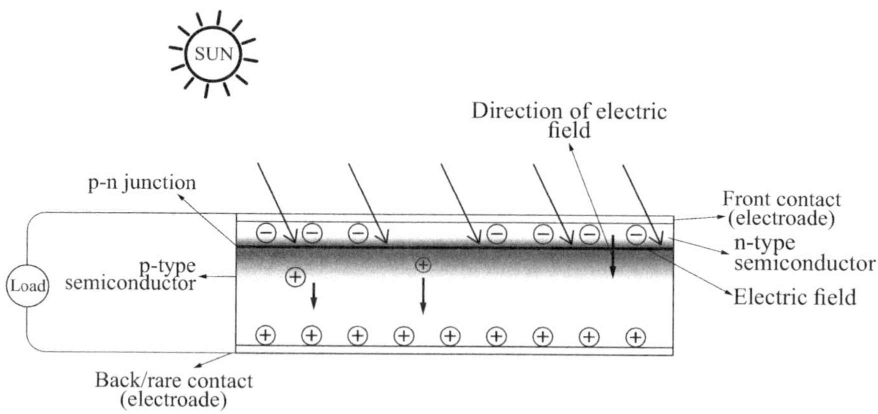

FIGURE 11.28
Schematic of what happens when sunlight strikes the surface of a solar cell.

of silicon where the doping agent changes from p to n to create the p-n junction. The n-side is much thinner than the p-side (typically about 1 μm vs. a few hundred μm). The n-side is on the front of the cell and is exposed to solar radiation. The electrodes' function is to collect the electric charges at two opposite surfaces of the cell. The electrode on the surface of the n-side collects the electrons to send them to the load, and the electrode on the p-side returns the electrodes to complete the circuit.

When a solar cell is exposed to sunlight, the light first strikes the n-side. Due to its thinness, the light can easily penetrate the n-type and strikes the p-n junction and beyond. The energy of photons of the light can be absorbed by some electrons, which in turn can excite and promote them to create free electrons in the n-type semiconductor and free holes in the p-type semiconductor (Figure 11.28). Another way to look at this phenomenon is that the photons create electron-hole pairs. If the photon absorption takes place in the electric field around p-n junction or near it, the electric field in the p-n junction separates the electrons from the holes. The electrons are forced toward the n-type semiconductor and the holes toward the p-type. As noted above, in the electric field within the p-n junction, the n-side has positive charges. The electrons diffuse through the n-type to the electrode on the front surface of the cell. Similarly the holes diffuse through p-type to the electrode on the back surface of the cell. This accumulation of positive and negative charges on the surfaces of the cell creates a potential difference (voltage). If connected to a load this voltage can create a flow of electrons, i.e., electrical current, in which the electrons flow from the electrode of the n-type to the electrode in the p-type where they recombine with the holes. If a photon is absorbed away from the electric field, it may or may not be able to contribute to power production depending on where the absorption takes place.

Now let us take a look at the mechanism of how light can promote electrons and create electron-hole pairs. We know that sunlight is made up of the flow of very small energy particles called photons. The energy of each photon is only a function of the light wavelength and independent of other light parameters including its intensity. For visible light, the wavelength depends on its color. For the same color light, the energy of the photons is the same but more intense light has simply more photons. The photon energy, typically between 1 and 3.5 eV, is inversely proportional to its wavelength (or directly proportional to its frequency). The energy of a photon as a function of its frequency and wavelength can be determined by the following equation:

$$E = hv = hc/\lambda \tag{11.5}$$

where E is energy of the photon (J), h is the Planck constant ($h = 6.626 \times 10^{-34}$ Js), v and λ are the frequency (Hz) and wavelength of the photon (m), respectively, and c is the light speed $c = 2.998 \times 10^{8}$ m/s.

Example 11.8

We know that the wavelength of solar radiation spectrum on Earth ranges from about 300 nm to about 2500 nm. For the light with the wavelength of 300, 1130, and 2500 nm,

a. determine the photon energy of the part of solar radiation with the above wavelength in J and eV.
b. determine the number of photons per second in each case if the light energy that is 0.1 W.

Solution

$$E_{300} = \frac{hc}{\lambda} = \frac{6.626\times10^{-34}\times2.998\times10^{8}}{300\times10^{-9}} = 6.62\times10^{-19}\ \frac{\text{J}}{\text{photon}}$$

$$= 6.62\times10^{-19}\times6.242\times10^{18}\left[\frac{\text{eV}}{\text{J}}\right] = 4.13\frac{\text{eV}}{\text{photon}}$$

$$E_{1130} = \frac{hc}{\lambda} = \frac{6.626\times10^{-34}\times2.998\times10^{8}}{1130\times10^{-9}} = 1.76\times10^{-19}\ \frac{\text{J}}{\text{photon}}$$

$$= 1.76\times10^{-19}\times6.242\times10^{18}\left[\frac{\text{eV}}{\text{J}}\right] = 1.10\ \text{eV/photon}$$

$$E_{2500} = \frac{hc}{\lambda} = \frac{6.626\times10^{-34}\times2.998\times10^{8}}{2500\times10^{-9}} = 7.95\times10^{-20}\ \frac{\text{J}}{\text{photon}} = 0.50\frac{\text{eV}}{\text{photon}}$$

$$\text{Number of photons for wavelernth of 300 nm} = \frac{0.1}{6.62\times10^{-19}} = 1.51\times10^{17}\ \frac{\text{photons}}{\text{s}}$$

$$\text{Number of photons for wavelernth of 1130 nm} = \frac{0.1}{1.76\times10^{-19}} = 5.69\times10^{17}\ \frac{\text{photons}}{\text{s}}$$

$$\text{Number of photons for wavelernth of 2500 nm} = \frac{0.1}{7.95\times10^{-20}} = 1.26\times10^{18}\ \frac{\text{photons}}{\text{s}}$$

We know from quantum mechanics that if an electron descends from a higher energy band to a lower energy band, it releases a photon with energy equal to the band gap between the two bands. Inversely, in order for an electron to be promoted from the valence band to the conduction band, it needs to receive energy equal to the band gap energy.

When a photon from sunlight strikes the p-n junction of a solar cell, depending on the energy of the photon and the band gap of the semiconductor, three scenarios may take place. If the photon energy is equal to the band gap, it dislodges and promotes an electron from the valence band to the conduction band creating an electron-hole pair. If the photon energy is greater than the band gap, an electron will be excited and the band gap energy will be absorbed from the photon's energy. However, the difference between the photon energy and the band gap will be wasted and converted to heat (known as the thermalization loss). If the photon energy is less than the band gap, the photon will not be able to excite any electron and its energy will be converted to heat without any contribution to power production (known as the transmission loss).

This indicates that to minimize the thermalization and transmission losses and to achieve the maximum possible efficiency in a solar cell, the best case scenario is to have incoming light that its photons have energy equal to the band gap of the semiconductor the solar cell is made of. The following example clearly shows this fact.

Example 11.9

Determine the maximum electrical power (in W) that can be generated if the light with the wavelength of 300, 1130, and 2500 nm strikes a solar cell made of crystalline silicon (with the band gap of 1.1 eV). Also, determine the maximum theoretical efficiency that can be achieved in the cell for each case.

Solution

We know that the photons of the light can excite electrons and create electron-hole pairs only if their energy is larger than the band gap of the semiconductor. So we first need to determine the photon energy of each light (Example 11.8).

$$E_{300} = 6.62\times10^{-19}\ \frac{\text{J}}{\text{photon}} = 4.13\ \frac{\text{eV}}{\text{photon}}$$

$$E_{1130} = 1.76\times10^{-19}\ \frac{\text{J}}{\text{photon}} = 1.10\ \frac{\text{eV}}{\text{photon}}$$

$$E_{2500} = 7.95\times10^{-20}\ \frac{\text{J}}{\text{photon}} = 0.50\ \frac{\text{eV}}{\text{photon}}$$

Therefore, the photons of the light with the wavelength of 300 and 1130 nm will be able to create electron-hole pairs, while the photons of the light with the wavelength of 2500 nm do not have sufficient energy to dislodge the electrons.

The maximum electrical energy for the light with the wavelength of 300 and 1130 nm will be generated if all photons are absorbed and each proton can excite one electron (and creates an electron-hole pair). The energy of each electron will be equal to the band gap of the crystalline silicon. So the maximum electrical energy will be equal to the number of photons in each case multiplied by the band gap of the semiconductor.

$$\text{Maximum possible generated electrical power}$$
$$= \text{Number of photons} \times \text{Band gap of the semiconductor}$$

We already determined the number of photons for each wavelength in the previous example.

$$P_{\text{max},300} = 1.51\times10^{17}\left[\frac{\text{photons}}{\text{s}}\right]\times1.1\left[\frac{\text{eV}}{\text{photon}}\right]\times1.6\times10^{-19}\left[\frac{\text{J}}{\text{eV}}\right] = 2.7\times10^{-2}\ \text{W}$$

$$P_{\text{max},1130} = 5.69\times10^{17}\left[\frac{\text{photons}}{\text{s}}\right]\times1.1\left[\frac{\text{eV}}{\text{photon}}\right]\times1.6\times10^{-19}\left[\frac{\text{J}}{\text{eV}}\right] = 0.1\ \text{W}$$

$$P_{\text{max},2500} = 0\ \text{W}$$

For the maximum efficiency in each case

$$\eta_{\text{max},300} = \frac{\text{Maximum power}}{\text{Light energy}} = \frac{0.027}{0.1} = 0.27 = 27\%$$

$$\eta_{\text{max},1130} = \frac{\text{Maximum power}}{\text{Light energy}} = \frac{0.1}{0.1} = 1 = 100\%$$

$$\eta_{\text{max},2500} = 0\%$$

The efficiency of the cell can also be calculated using the following equation:

$$\text{Light power} = \text{Number of photons}\times\text{Energy of each photon}$$

$$\text{Maximum possible generated electrical power}$$
$$= \text{Number of photons}\times\text{Band gap of the semiconductor}$$

$$\eta_{\text{max}} = \frac{\text{Maximum power}}{\text{Light energy}}$$
$$= \frac{\text{Number of photons}\times\text{Band gap of the semiconductor}}{\text{Number of photons}\times\text{Energy of each photon}}$$
$$= \frac{\text{Band gap of the semiconductor}}{\text{Energy of each photon}}$$

$$\eta_{\text{max},300} = \frac{\text{Band gap of the semiconductor}}{\text{Energy of each photon}} = \frac{1.1}{4.13} = 0.27 = 27\%$$

$$\eta_{\text{max},1130} = \frac{\text{Band gap of the semiconductor}}{\text{Energy of each photon}} = \frac{1.1}{1.1} = 1 = 100\%$$

In summary, for silicon solar cells, if the energy of light's photon is <1.1 (the wavelength of >1130 nm), the generated power and the efficiency of the cell will be zero. If the energy of photons is equal to the band gap (the wavelength equal to 1130 nm), the maximum possible efficiency of the cell will be 100%. If the energy of photons is higher than the band gap (the wavelength of <1130 nm), the maximum possible efficiency of the cell will be <100%. The more the difference between the photons' energy and the band gap, i.e., the shorter the wavelength, the lower the efficiency of the cell will be. Note that this is the maximum theoretical efficiency and in real solar cells cannot be achieved due to many forms of losses including various mechanisms for recombination of electron-hole pairs.

Since the spectrum of the solar radiation contains a range of wavelengths and is relatively constant, we cannot choose the light that its photons' energy matches the band gap. The only way we can control the effect of the difference between the photons' energy and the band gap is to choose a semiconductor with a more favorable band gap. If the band gap is too small, most photons will have sufficient energy to excite electrons and they will contribute to electricity generation in the cell. However, since the band gap is small, for most photons (particularly low-wavelength high-energy photons) only a small portion of their energy contributes to power generation, and most of their energy is wasted in the form of thermal energy. If the band gap of the semiconductor is too high, the opposite will take place. While less electrons will contribute to electricity generation (mostly high-energy photons will be able to excite electrons), for those that contribute a higher portion of their energy will be converted to electricity. These two opposing phenomena, i.e. the number of electrons that contribute to power production and the percentage of their energy that is converted, indicate that there must be an optimum point where their impacts are balanced to maximize energy conversion and efficiency. This **optimum band gap for the solar radiation spectrum is about 1.4 eV**.

Another important factor that significantly impacts the manufacturing and operation of solar cells is whether the semiconductor used in the fabrication of the cell is a direct or indirect semiconductor. Imagine we have two semiconductors one direct and the other indirect with the same thickness. If the same light strikes them both, the absorption of the light in the indirect semiconductor will be much lower than that of the direct semiconductor. In practice this means solar cells manufactured out of indirect semiconductors (such as crystalline silicon) must be thick enough to absorb the incoming light; otherwise, much of the light can simply pass through the cell without contributing to power production. This is particularly important for the absorption of long wavelength light because semiconductors are weaker absorber of these wavelengths. As a result, crystalline silicon solar cells are typically a few 100 μm thick, while direct semiconductor cells are typically just a few micron thick. In later sections we will come back to this point when we discuss recent high-efficiency solar cells.

11.2.3.3 Losses in Solar Cells

In this section, we will try to understand which factors can contribute to the loss of efficiency in a solar cell and how we can reduce the impacts of those parameters. These losses can be divided into two general groups. The first group is related to light absorption. If the light energy is not absorbed by the semiconductor atoms, the unabsorbed energy is considered as the lost energy. The second group of losses occurs after light is absorbed by the cell, but some other factors prevent its conversion to electricity. Understanding these losses can help us in minimizing them and also identifying the limits in their reductions.

Some of the incoming light to the surface of the solar cell will be reflected and some will be blocked by the electrodes on the surface. The reflection can be particularly problematic for silicon-based solar cells because the polished surface of silicon is very reflective and can mirror a substantial portion of the incoming light. The reflection of the surface of solar cells can be reduced by covering the surface of the cell with an antireflective coating or by texturing the surface. As will be discussed later, shadowing a solar unit, even a very small portion of it, can have profound and disproportional impacts on the efficiency of the unit. Furthermore, the electrodes on the surface of cells can block the incoming light from penetrating the semiconductor and contributing to power generation.

We previously discussed that in solar cells the incoming light creates electron-hole pairs. These pairs, in turn, when separated by the electric field in the p-n junction, generate electricity. So the efficiency of a solar cell highly depends on the creation and persistence of these pairs. However, these pairs can be recombined eliminating their possible contributions to power generation. Therefore, it is critically important to evaluate the mechanisms for the recombination of electron-hole pairs and minimize the possibility of the occurrence of this recombination. This phenomenon can be characterized by two parameters, carrier lifetime or diffusion length. These two parameters refer to the average time of existence and the average traveled distance of the pairs from when they are created until they are recombined, respectively. They can be in the order of 10^{-3} to 10^{-6} seconds and 50 to 500 μm, respectively, for silicon crystalline. They strongly depend on the impurity in both the semiconductor material and the dopant as well as imperfections in the crystalline structure of the semiconductor. They are also **very strong functions of the temperature of the cell**.

The diffusion length is a very important design and performance parameter for solar cells. Note that when a photon is absorbed and creates an electron-hole pair, the pair must travel to the electric field in the p-n junction. Therefore, the diffusion length determines the area that the produced pairs can effectively contribute to power generation. Typically the thickness of a solar cell should be in the order of the diffusion length of the semiconductor. A thicker cell will not improve power generation but rather will adversely impact the efficiency because some photons will be absorbed far away from the p-n junction where the produced pairs will have very little to no chance of survival before they reach the electric field in the p-n junction. On the other hand, it was discussed earlier that each semiconductor material requires a certain thickness to fully absorb the incoming light. The combined impact of this thickness and the diffusion length determines the suitability of a semiconductor for solar cell manufacturing and the limitations in the maximum efficiency they can achieve.

Other sources of losses include

- internal resistance of the cell
- imperfection in the semiconductor structure created during manufacturing processes
- slight electrical mismatch among cells and modules that causes losses when cells are connected
- electric losses in connecting cables
- losses in auxiliary components such as the inverter

These losses will be covered later. Furthermore, note that the efficiency of a PV module is typically lower than the efficiency of the cells that make up the module. This is because when the cells are organized to form a module, there are areas between the cells and between the cells and the frame that are left uncovered by cells. While these unused areas are taken into account in determining the area of the module, they do not contribute to the power generation of the unit. To evaluate the magnitude of this effect, a parameter called the packing factor is defined as the ratio of the total area of the cells on the panel to the total area of the panel (sum of both covered and uncovered areas).

11.2.4 Electricity Generated in Photovoltaic Solar Cells

So far we have learned about semiconductors and their unique properties as well as the internal operation of solar cells primarily from the microscopic point of view (at the atomic level). From now on we will mainly focus on the macroscopic point of view of solar cells and how their microscope characteristics can explain their macroscopic behaviors. In this section, we will discuss the characteristics of the generated electricity in solar cells. We will look at how the generated power and the efficiency of cells can be determined.

11.2.4.1 Current-Voltage Curve of Solar Cells

Previously we learned how photons can create electron-hole pairs that are separated in the electric field on the p-n junction. We also discussed how these electric charges are collected by the electrodes on the two surfaces of the cell to create electrical potential (voltage) that can power a load. In this section, we will discuss how load resistance can influence the power generation of the cell.

The performance of electricity generating devices is conventionally characterized by their **current-voltage curve** (also known as characteristic curve). The current-voltage diagram (often referred to as the I-V curve) illustrates the current drawn from a solar photovoltaic unit[31] as a function of the voltage measured between the two electrodes. This curve not only provides a detailed

[31] It can be for a solar cell or module, which will be discussed later.

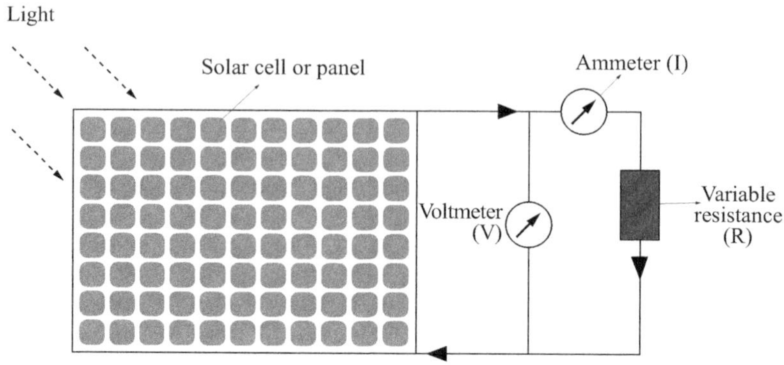

FIGURE 11.29
Schematic of a setup to create a characteristic curve (I-V curve) of a solar cell by varying the load's resistance and measuring current and voltage.

understanding of the system behavior, that is the relationship between current and voltage, when connected to a load but can also be used to determine the electricity generation by the system and its efficiency as well as the optimum operational current and voltage. In this diagram, the current is a function of the area of the cell. In order to eliminate this area dependency, sometimes the current is normalized by dividing the current to the area of the cell. In this case, the vertical axis is the current density with the unit of A/m^2.

A current-voltage curve of a solar cell or panel can be easily created by two simple multimeters, one as a voltmeter and the other as an ammeter (also known as an ampere meter), and a variable electrical resistance as the load assembled as shown in Figure 11.29. The I-V diagram can be drawn by changing the resistance and recording the current and voltage for the given solar photovoltaic system at the given solar irradiance and temperature of the experiment.

While using the resistive load is the most basic and least expensive method to develop the I-V curve, using a source measure unit (SMU) (also known as source-meter) can provide more data relevant to the performance of the solar cell under investigation with higher accuracy (but at a higher cost). Most SMUs can be programmed to create the I-V curve automatically.

Figure 11.30 shows an I-V curve for a typical solar cell. In this curve, two points of interest are when the curve intersects the vertical axis (current) and horizontal axis (voltage). The intersection of the curve and the voltage axis is the maximum voltage that a cell can produce. This voltage can be measured when the current drawn from the unit is zero. This is the current when the circuit is open (equivalent to a load with infinite resistance), thus the name the **open-circuit voltage** (V_{OC}). Practically the V_{OC} is measured by connecting a voltmeter directly to the electrodes of the cell (with no load).[32] The intersection

[32] Because the resistance of a voltmeter is very large, connecting a voltmeter directly to an electricity generating device will measure the V_{OC}.

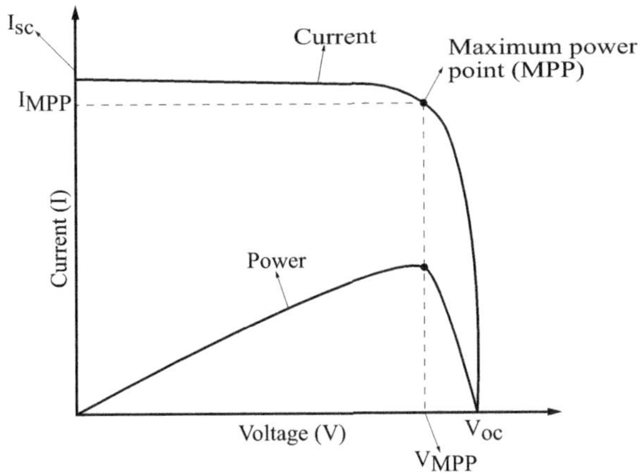

FIGURE 11.30
Current-voltage (I-V) and power-voltage curves for a solar photovoltaic power generating device showing V_{OC}, I_{SC}, MPP, P_{MPP}, V_{MPP}, and I_{MPP}.

of the I-V curve and the current axis represents the current when the resistance of the load is reduced to zero so it is called the **short-circuit current (I_{SC})**. This is the maximum current that the system can generate and can be measured by directly connecting an ammeter to the two electrodes of the system.

A close observation of the I-V curve in Figure 11.30 indicates that while the voltage varies with the load, the current is nearly constant for most of the voltage range. The fact that solar cells can maintain an almost constant flow of current regardless of the load is a unique aspect of solar cells operation.[33] However, at a certain voltage the current starts to rapidly decline.

Like any other power generating device, here we are mostly interested in the magnitude of electrical energy generated in the solar units. Recalling the following equation to determine the power as a function of the current and voltage, the generated power curve can be easily determined from the I-V curve (Figure 11.30).

$$P = \text{Current} \times \text{voltage} = IV \tag{11.6}$$

where P represents power (W), I is current (in A or Amp), and V is voltage (V). As expected while at the open-circuit voltage and short-circuit current the voltage and current are maximum, respectively, the generated power at these two points is zero. In the I-V curve, this starting with zero power generation and returning to zero again indicates that there must be a point where the power generation is maximum. In this curve starting from the short-circuit point, the voltage increases while the current is nearly constant (Figure 11.30).

[33] In later chapters, we will see that for example batteries are constant voltage devices and for fuel cells the current continuously varies with voltage.

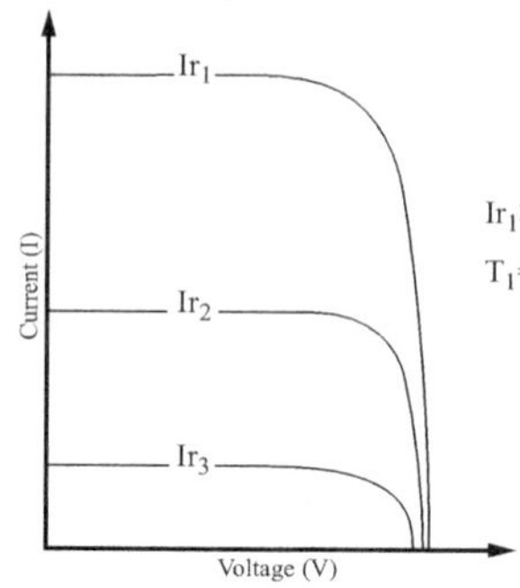

FIGURE 11.31
Impacts of solar irradiance on I-V curve.

This means the power almost linearly increases as a function of the voltage. This trend continues until the point where the current starts to reduce with increase in voltage. With the sharp decrease in the current, the increase in the power quickly stops and it reduces rapidly. The point with the maximum power generation is called the **maximum power point (MPP)**, and its current, voltage, and power are shown as I_{MPP}, V_{MPP}, and P_{MPP}, respectively (sometimes the subscript MP is used instead of MPP) (Figure 11.30). Note that in Figure 11.30 the power at the maximum power point is equal to the area of the rectangle shown by dashed line. Note that the **current generated by a solar cell depends on the size** of the cell but the **voltage is independent of the size** and is only a function of the material and operating conditions of the cell.

At this point let us evaluate the impacts of the incoming light intensity and the cell temperature on the I-V curve of a solar cell. The **short-circuit current is directly proportional to irradiance.** While the reduction in irradiance reduces the open-circuit voltage, the impact is not significant[34] (Figure 11.31). For example, if we double the light irradiance, the general shape of the I-V curve will remain unchanged. However, the height of the horizontal section of the curve will nearly double and the V_{OC} will slightly increase.

The **open-circuit voltage is inversely proportional to the cell temperature**. If the cell temperature decreases, the V_{OC} increases linearly (Figure 11.32). However, the variation of the short-circuit current with the temperature of the cell is typically negligible. The deterioration of the performance of a solar cell at higher operation temperatures can be explained based on the rate of the recombination of electron-hole pairs. The higher the cell temperature, the faster the rate of the recombination is. This more rapid rate of recombination will decrease the carrier lifetime and diffusion length, which in turn reduces the voltage of the cell.

In summary, both the solar irradiance and operating temperature of a solar cell have significant impacts on the performance of the system. These

[34] The change in the V_{OC} is logarithmically proportional to irradiance.

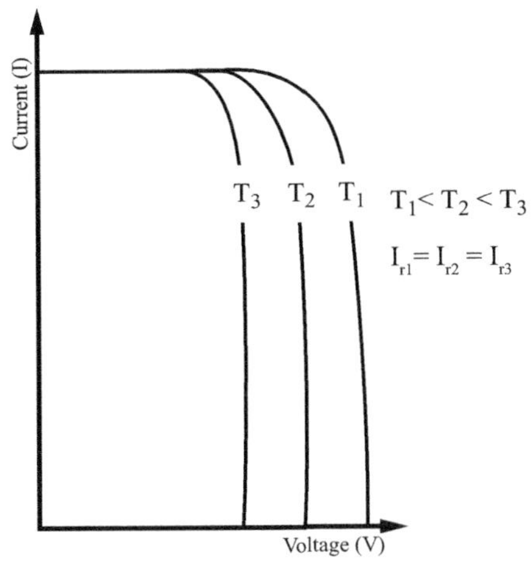

FIGURE 11.32
Impacts of cell temperature on I-V curve.

dependencies influence the operation of solar photovoltaic systems and how they are tested.

We know that the solar irradiance and the operating temperature of solar cells vary considerably both daily and seasonally. The I-V curves presented in Figures 11.31 and 11.32 indicate that the maximum power point of a solar unit changes with these two parameters. To ensure that the system is always operated around its maximum power point regardless of the load, the solar irradiance, and the operating temperature, many units are equipped with a **maximum power point tracking circuit**. This circuit automatically adjusts the output current and voltage so the system operates close to the MPP all the time.

Another implication of these variations is that in order to be able to compare the performance of various solar units, they need to operate at the comparable irradiance and operating temperature. This is more important with today's international trades. We need to have uniform standard conditions to test and present the performance of solar units. These conditions are called the **standard test conditions (STC)** and they include the following conditions:

- **solar irradiance**: 1000 W/m^2
- **operating temperature**: 25°C (77°F)
- **air mass**: AM 1.5 sunlight

The solar radiation with these specifications is referred to as **1 (one) Sun** (or 1 full sun). Note that the solar irradiance determines the intensity of the incoming light and the air mass determines the spectrum of the light. Since solar radiation is variable by nature, the standard test conditions must be recreated in the laboratory environment where the intensity of light, its spectrum, and the cell temperature can be readily controlled. To recreate an indoor light source that can mimic the natural solar light, typically the xenon, halogen, or tungsten lamps (the former being much more popular) and more recently light-emitting diodes (LEDs) with or without filters are used. These artificial light generating units are called **solar simulators**. The most important factors that impact the appropriate performance of solar simulators are their spectral match to the target light, and their ability to uniformly maintain this match through the entire surface of the cell under investigation during the entire time period of the experiment. Sometimes to improve the accuracy of the artificial light, the correction procedure is applied to reduce spectral mismatch. These factors are standardized via ASTM E927 or IEC 60904-9.[35]

The **maximum power generation (the power rating) at the standard test conditions is referred to as the peak power and is expressed with the unit of watt-peak (W_p)**. Notice that the power generation in the standard test conditions does not necessarily correspond to the actual power generation by the unit. The instantaneous power generation can be more and less than the peak power rating of a unit.

Sometimes for some solar systems, a specification sheet is provided instead of or along with an I-V curve by manufacturers. This specification sheet provides I_{MPP}, V_{MPP}, I_{SC}, V_{OC}, and P_{MPP} as well as the physical characteristics of the module including its weight and dimension. Note that all these parameters are referring to the DC power directly from the PV module and do not take into account the conversion of DC power to AC power. Furthermore, it can include some parameters to indicate how the cell performance parameters are affected by the operating temperature of the unit. These parameters include the temperature coefficient of the open-circuit voltage, temperature coefficient of the short-circuit current, and the temperature coefficient of the maximum power. They can be expressed with units of V/°C, A/°C, and W/°C, respectively. Alternatively, they may be presented as percentage of change per unit of temperature change (%/K) (this topic will be discussed later). A simple internet search for "solar panel spec sheet" can provide you with many samples of specification sheets for a variety of solar panels.

Note that these values are not reflecting the actual performance of the solar panel when it operates under real-world conditions. Other than the variation in the incoming solar irradiance and the cell temperature, the losses in the auxiliary systems such as wiring losses and the inverter loss (Section 11.2.5.2) as well as other miscellaneous losses such as dust covering the surface of the

[35] The ASTM E927 standard is titled as "Standard Classification for Solar Simulators for Electrical Performance Testing of Photovoltaic Devices" which was last revised and adapted in 2019. The International Electrotechnical Commission counterpart is IEC 60904-9 (Photovoltaic devices – Part 9: Solar simulator performance requirements).

panel further reduce the panel efficiency. These losses may add up to a significant loss, between 10% and 20% of the panel rating, and should be taken into account in sizing of solar photovoltaic systems.

Example 11.10

A solar PV module specification sheet lists the open-circuit voltage, short-circuit current, and maximum power as 22 V, 8 A, and 130 W, respectively. The temperature coefficient of these parameters is –0.3%/°C, 0.04%/°C, and –0.4%/°C, respectively. Determine the open-circuit voltage, short-circuit current, and maximum power if the module operating temperature is 55°C.

Solution

Note that the values that are reported in the specification sheet are all for STC. So the temperature difference is measured in reference to 25°C.

$$\frac{\Delta V_{OC}}{\Delta T} = \frac{V_{OC,2} - V_{OC,Ref}}{T_2 - T_{Ref}} = -0.003 V_{OC,Ref}$$

$$\frac{V_{OC,55°C} - V_{OC,25°C}}{55 - 25} = -0.003 V_{OC,25°C}$$

$$V_{OC,55°C} = -0.003 \times 22(55 - 25) + 22 = 20.02 \text{ V}$$

$$I_{OC,55°C} = 0.0004 \times 8(55 - 25) + 8 = 8.096 \text{ A}$$

$$P_{OC,55°C} = -0.004 \times 130(55 - 25) + 130 = 114.4 \text{ W}$$

11.2.4.2 Connecting Solar Cells

So far we learned that the current and voltage generated by a solar cell depend on the type of the semiconductor used to manufacture the cell (and its quality), the intensity of the incoming light and its spectrum, the size of the cell, and the load connected to the cell. Considering these factors, a typical single solar cell can produce a very small amount of power. For a typical single silicon solar cell, the generated current and voltage are in the order of 8 A and 0.5 V. This is equivalent to a third of the voltage of an AA battery and the power of about three AA batteries. Clearly, this is not sufficient for a majority of applications, and solar cells must be interconnected to increase their power generation capacity to a reasonably high level. Like any other electrical device, solar cells can be connected in parallel or series connections. In this section, we will see how these connections will change the current, voltage, and power generation of connected cells.

When solar cells are connected in series, it is called a **series string**. Figure 11.33 illustrates this configuration along with the I-V curves of a

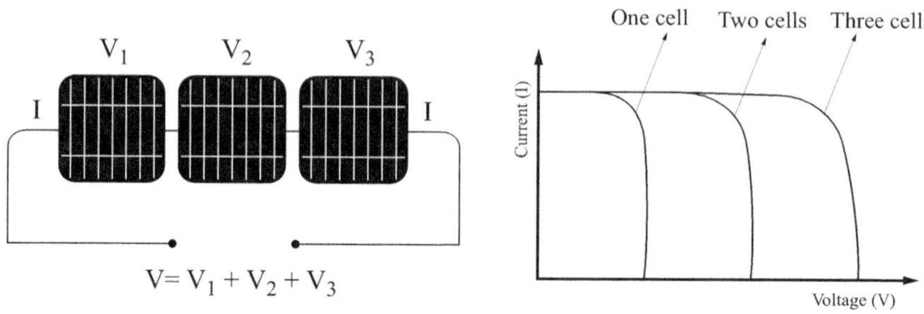

FIGURE 11.33
Solar cells connected in series and I-V curves of a single cell and when two or three similar cells connected in series.

single cell and when two or three similar cells are connected in series. In this connection, the current through all the cells remains the same but the total voltage is equal to the sum of the voltage of all cells, as the following equations indicate. Thus, this configuration is used when an increase in voltage is desirable. Note that if the cells are not similar, while the total voltage will be still the sum of the all voltages, the current will be primarily determined by the lowest current produced by the cells.

$$I = I_1 = I_2 = \cdots = I_n \tag{11.7}$$

$$V = V_1 + V_2 + \ldots + V_n = \sum_{m=1}^{m=n} V_m \tag{11.8}$$

In a parallel configuration, while voltage is the same for all cells, the total current is equal to the sum of the currents generated by all cells (equations below). Therefore, this configuration is used when an increase in current is desirable. Figure 11.34 illustrates this configuration along with the I-V curves of a single cell and when two or three similar cells are connected in parallel. Note that if the cells are not similar, while the current will be still the sum of all the currents, the voltage will be primarily determined by the lowest voltage.

$$V = V_1 = V_2 = \cdots = V_n \tag{11.9}$$

$$I = I_1 + I_2 + \cdots + I_n = \sum_{m=1}^{m=n} I_m \tag{11.10}$$

As noted earlier, a single solar cell produces a small voltage, so it is ideal to be able to increase the voltage by connecting several solar cells in series. A series string of typically up to 72 cells (usually 60 cells) is placed in an aluminum

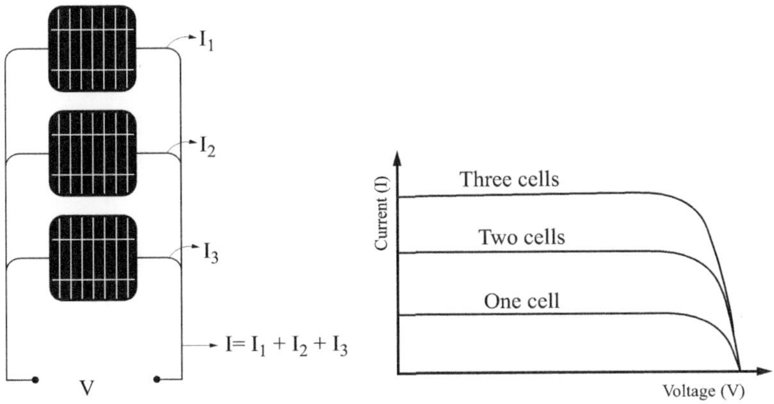

FIGURE 11.34
Solar cells connected in parallel and I-V curves of a single cell and when two or three similar cells connected in parallel.

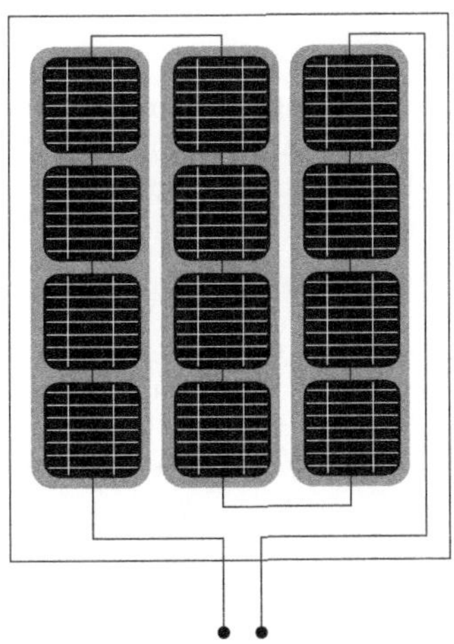

FIGURE 11.35
A solar module (panel).

frame and covered by a protective cover on the front, and a backing plate to form an individual unit called a **solar module** or simply a **solar panel** (Figure 11.35). The size of solar panels is usually about 1.5 m^2 (or sometimes larger) with the peak power output of between 120 and 300 W.

The solar modules can be connected in series, parallel, or a combination to form a system of modules with a desirable output, which is called a **solar array** (Figure 11.36). This collection of solar panels (or a solar array) is what you can see on the rooftop of residential applications.

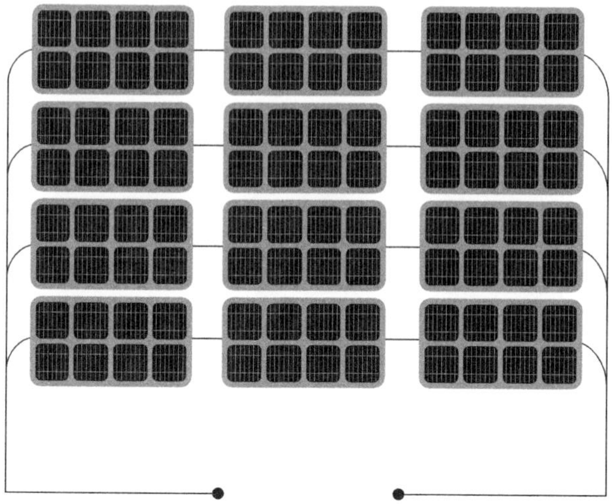

FIGURE 11.36
A solar array.

Example 11.11

Five series strings each containing four similar solar cells are connected in parallel. If each cell has the current and voltage of 8.5 A and 0.55 V, respectively, determine the current, voltage, and power generation of the entire system.

Solution

Each series string: $I_{string} = 8.5$ A and $V_{string} = 0.55 \times 4 = 2.2$ V
Entire system: $I_{system} = 5 \times 8.5 = 42.5$ A and $V_{system} = 2.2\ V$

$$P_{system} = 2.2 \text{ V} \times 42.5 \text{ A} = 93.5 \text{ W}$$

Or

$$P_{cell} = 8.5 \times 0.55 = 4.675 \text{ W}$$

$$P_{system} = (5 \times 4) \times 4.675 = 93.5 \text{ W}$$

Example 11.12

Two solar panels are connected once in series and once in parallel. The V_{OC} and I_{SC} of the cells are given as follows. Determine the V_{OC} and I_{SC} when they are connected in series and parallel configurations.

$$V_{OC,1} = 0.55 \text{ V}$$

$$I_{SC,1} = 8 \text{ A}$$

$$V_{OC,2} = 0.5 \text{ V}$$

$$I_{SC,2} = 7 \text{ A}$$

Solution

Series: (Note that in a series configuration, the short-circuit current is dictated by the minimum of the two short-circuit currents.)

$$V_{OC} = V_{OC,1} + V_{OC,2} = 1.05 \text{ V}$$

$$I_{SC} = \text{Minimum}\left(I_{SC,1} \text{ and } I_{SC,2}\right) = 7 \text{ A}$$

Parallel: (Note that in a parallel configuration, the open-circuit voltage is dictated by the minimum of the two open-circuit voltages.)

$$I_{OC} = I_{OC,1} + I_{OC,2} = 15 \text{ A}$$

$$V_{SC} = \text{Minimum}\left(V_{SC,1} \text{ and } V_{SC,2}\right) = 0.5 \text{ A}$$

11.2.4.3 Effects of Shadow on Performance of Solar Systems

Before we start to discuss the efficiency of solar systems, it might be appropriate to evaluate the effects of shading on the performance of solar panels and arrays. Let us assume we have a set of parallel solar cells. Shading a half of the cells in the set will approximately reduce the generated power by half compared to that of the unshaded set. That is because while the total current roughly reduces by half, the voltage remains mostly unchanged (remember that the open-circuit voltage of a solar cell is not significantly affected by the light intensity and the cell size). In this case, the percentage of power loss is nearly directly proportional to the number of cells in shadow. Basically in this case, the shaded cells will not contribute to the total current while the illuminated ones will be almost unaffected and continue to generate current.

Now let us assume a set of two solar cells that are connected in series. What will happen if one of the solar cells is shaded? Intuitively your answer might be: the system power generation will reduce by half compared to what it could generate when both cells were illuminated. It turns out that when even one cell is shaded in a set of many cells connected in series, the power generation drastically reduces. Remember that in solar cells in series, the current passing through all cells is the same. Basically the current of the entire system will be dictated by the cell with the lowest current. In this case, the shaded cell with the small current generation will severely limit the current of the entire block and substantially reduce the power generation of the system. This effect can also increase the temperature of the shaded cell significantly. This means when designing and installing solar photovoltaic systems, a careful attention should be paid to avoid partial shading of the system. To lesser extents, the partial covering of solar modules by dust, dirt, or snow can create similar problems.

To eliminate the adverse effect of shading in solar cells connected in series, a bypass diode can be used. If a diode is placed over the affected cell, it can provide a bypass for the current to avoid the severe limiting impact of the low current of the shaded cell. So in this case, only the power of the shaded cell will be lost and the rest of the system will remain intact. Due to the cost as well as manufacturing and operation problems, the bypass diodes are not mounted on each cell; rather, they are typically installed in the module level.

Example 11.13

Five series strings each containing four similar solar cells are connected in parallel. Each cell has a current and voltage of 8.5 A and 0.55 V. Determine the current, voltage, and power generation of the entire system if one of the cells is totally shaded.

Solution

One of the five legs of the parallel cells will not generate any power.

Each series string: $I_{string} = 8.5$ A and $V_{string} = 0.55 \times 4 = 2.2$ V

Entire system: $I_{system} = 4 \times 8.5 = 34$ A and $V_{system} = 2.2$ V

$$P_{system} = 2.2 \text{ V} \times 34 \text{ A} = 74.8 \text{ W}$$

Or

$$P_{cell} = 8.5 \times 0.55 = 4.675 \text{ W}$$

$$P_{system} = (4 \times 4)4.675 = 74.8 \text{ W}$$

11.2.4.4 Efficiency of Solar Photovoltaic Units

We know that the efficiency of a power generating unit is directly proportional to how effectively the input energy is converted to the desirable form of energy. The efficiency determines the amount and the cost of the consumed fuel in conventional power plants. However, in solar photovoltaic units the input energy, i.e., solar radiation, is free. So you may be wondering why we care about their efficiency. The efficiency of a power generation unit is also proportional to the size of the unit, in this case, the area of the solar modules, required to generate the given amount of electricity. The size of the unit not only is proportional to the material needed to manufacture the unit but also the cost of the balance of the plant (refer to Section 11.2.5), such as support structure, as well as the required land. We will discuss about these parameters more in depth later when we cover the cost of the solar units. For now, it suffices to realize that the efficiency of solar units is important. The efficiency of a solar photovoltaic unit is defined as follows:

$$\eta_{\text{Solar unit}} = \frac{\text{Generated electricity}}{\text{Incoming solar energy}} = \frac{I_{MPP} \times V_{MPP}}{P_{in}} \quad (11.11)$$

In this equation, the generated electricity and incoming solar energy are typically expressed per unit of area. In previous sections, we learned that the generated power in a solar unit is a function of many factors including the solar irradiance, light spectrum, cell temperature, and load. Therefore, to make the reported efficiencies for various units universally uniform and comparable, they are always determined at the STC and for the maximum power point. The following example will demonstrate how the efficiency of a solar panel can be determined.

Example 11.14

A real-world solar panel manufacturer claims that the efficiency of its solar panels is about 16%. Evaluate this claim if the voltage and current at the maximum power point for the panels are 17.6 V and 7.39 A, respectively. The size of the panel is 0.608 m × 1.367 m. All the given values are at the STC.

Solution

$$P_{MPP} = 17.6 \times 7.36 = 129.54 \text{ W}$$

This is the peak power generation per panel. The next step is to determine the power per unit of area of the panel.

$$A = 0.608 \times 1.367 = 0.8311 \text{ m}^2$$

$$\frac{P_{MPP}}{A} = \frac{129.54}{0.8311} = 155.8 \text{ W/m}^2$$

This is the power that one square meter of solar cell produces if the intensity of the incoming light is 1000 W/m^2 (at the STC), so the efficiency of the solar panel is

$$\eta_{\text{Solar unit}} = \frac{155.8}{1000} = 0.1558 \approx 16\%$$

This means as long as the given characteristics of the solar panels are accurate, the claim of the manufacturer is valid.

Note that again this is not the actual efficiency of the solar panel when it operates under real-world conditions since many real-world losses are not taken into account in its determination. To take these losses into account, an external efficiency of typically between 80% and 90% is used in the sizing of solar panels.

Example 11.15

The best efficiency for the commercially available solar panels is about 24%. Determine the required area of solar panels to generate 1 kW peak power using these panels.

Solution

$$\eta_{\text{Solar unit}} = \frac{\text{Generated electricity per area}}{1000} = 0.24$$

$$\text{Generated electricity per area} = 240\ \text{W/m}^2$$

$$\text{Required area} = \frac{\text{Required electricity}}{\text{Generated electricity per area}} = \frac{1000}{240} = 4.2\ \text{m}^2$$

Example 11.16

The typical efficiency of the commercially available solar panels is about 16%. Determine the required area of solar panels to generate 1 kW peak power using these panels. Compare the result with the answer for the previous example.

Solution

$$\eta_{\text{Solar unit}} = \frac{\text{Generated electricity per area}}{1000} = 0.16$$

$$\text{Generated electricity per area} = 160\ \text{W/m}^2$$

$$\text{Required area} = \frac{\text{Required electricity}}{\text{Generated electricity per area}} = \frac{1000}{160} = 6.25\ \text{m}^2$$

This type of solar panels requires much more area for the same power generation compared to the solar panels in the previous example. However, since the per-area cost of this type is much cheaper than that of the previous type, overall this type might be more economical.

Example 11.17

The estimated installed global solar photovoltaic panels in 2018 were 505 GW. Assuming the average efficiency of 17%, determine the total area of these panels.

Solution

$$\text{Area} = \frac{\text{Required electricity}}{1000 \times \eta_{\text{Solar unit}}} = \frac{505 \times 10^9}{1000 \times 0.17} = 3 \times 10^9\ \text{m}^2 = 3000\ \text{km}^2$$

This is less than the area of Rhode Island (at about 4000 km^2). This is about 0.03% of the total area of the U.S.

Example 11.18

As discussed in Section 11.2.1, the efficiency of early solar cells was very low, in the order of 1%. Determine the required cell area to power a house that consumes 10 kW using these early solar cells.

Solution

$$\text{Area} = \frac{\text{Required electricity}}{1000 \times \eta_{\text{Solar unit}}} = \frac{10,000}{1000 \times 0.01} = 1000\ \text{m}^2$$

As expected the required area is unreasonably high.

11.2.4.5 Fill Factor of Solar Photovoltaic Cells

Previously we learned that while the maximum current and voltage are at V_{OC} and I_{SC}, respectively, the current and voltage that provide the maximum power are lower than V_{OC} and I_{SC}. The parameter that related the short-circuit current and open-circuit voltage to the maximum power is called the **fill factor** (**FF**). The fill factor is defined as the ratio of the maximum power to the product of the open-circuit voltage and short-circuit current.

$$\text{FF} = \frac{P_{\text{MPP}}}{I_{\text{SC}} \times V_{\text{OC}}} = \frac{I_{\text{MPP}} \times V_{\text{MPP}}}{I_{\text{SC}} \times V_{\text{OC}}}$$

Using this parameter, the efficiency of a solar unit can be defined as

$$\eta_{\text{Solar unit}} = \frac{P_{\text{MPP}}}{P_{\text{in}}} = \frac{\text{FF} \times I_{\text{SC}} \times V_{\text{OC}}}{P_{\text{in}}}$$

The fill factor can be expressed as the ratio of the areas of two rectangles in the I-V curve (Figure 11.37). The fill factor is equal to the ratio of the area of the rectangle with the sides of I_{MPP} and V_{MPP} (the area of OADEO) to the area of the rectangle with the sides of I_{SC} and V_{OC} (the area of OBCFO). The fill factor is an indication of the quality of the material and the manufacturing process. While for high-quality solar cells the value of the fill factor can exceed 0.8, it can be as low as 0.4 for lower quality solar cells.

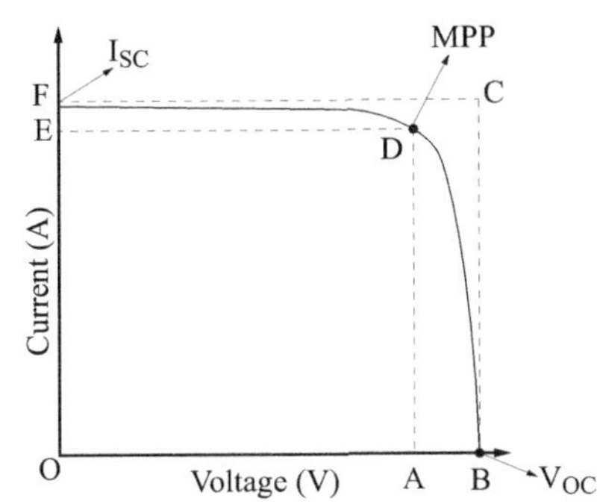

FIGURE 11.37
Fill factor definition on I-V curve.

Example 11.19

Determine the fill factor for a solar panel with the following I-V curve (Figure 11.38).

Solution

$$I_{\text{MPP}} = 7.4 \text{ A}$$

$$V_{\text{MPP}} = 17.5 \text{ V}$$

$$I_{\text{SC}} = 8 \text{ A}$$

$$V_{\text{OC}} = 22 \text{ V}$$

$$\text{FF} = \frac{I_{\text{MPP}} \times V_{\text{MPP}}}{I_{\text{SC}} \times V_{\text{OC}}} = \frac{17.5 \times 7.4}{8 \times 22} = 0.736 \text{ or } 73.6\%$$

Example 11.20

A solar module has the short-circuit current of 8 A and the open-circuit voltage of 30 V. Determine the maximum generated power for two fill factors of 0.4 and 0.75.

$$P_{\text{MPP}} = \text{FF} \times I_{\text{SC}} \times V_{\text{OC}}$$

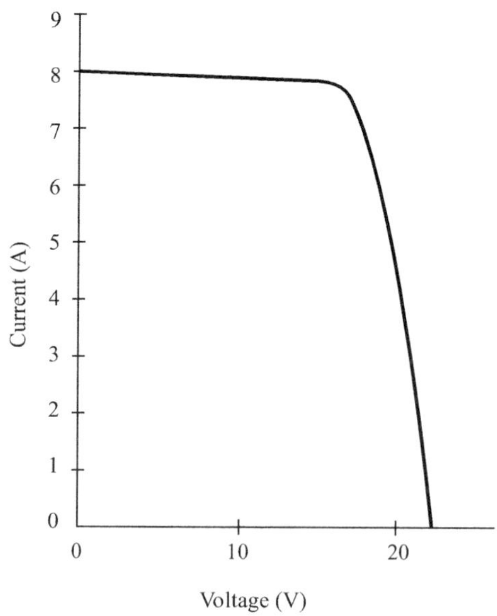

FIGURE 11.38
I-V curve for Example 11.19.

$$P_{MPP,1} = FF \times I_{SC} \times V_{OC} = 0.4 \times 8 \times 30 = 96 \text{ W}$$

$$P_{MPP,1} = FF \times I_{SC} \times V_{OC} = 0.75 \times 8 \times 30 = 180 \text{ W}$$

This shows how impactful is the fill factor in the power generation of the module.

11.2.4.6 Modeling of Solar Photovoltaic Cells

Being able to express the behaviors of solar cells in the form of mathematical equations is a very powerful tool for both understanding of their internal operation and for their numerical modeling. Solar cells are typically modeled in the form of an equivalent circuit, which includes a current source (with the generated current proportional to light intensity), a diode, and a parallel resistant that are connected with another resistant in series (Figure 11.39).

The two parallel and series resistances, which are sometimes referred to as the **parasitic resistances**, are called the **parallel or shunt resistance** (R_{SH}) and **series resistance** (R_S), respectively. The series resistance primarily represents the losses in the wiring system and electrical contacts. A poor solar cell design and manufacturing can contribute to high values for this resistance. The shunt resistance is mainly due to manufacturing defects from the low-quality manufacturing processes of semiconductor crystal, which causes the generated current in the cell to bypass the external circuit. These losses are sometimes referred to as the current leakage across the p-n junction. This equivalent circuit clearly indicates that the low values of R_S and the high values of R_{SH} are desirable. It can be shown that the value of R_S is equal to the slope of the I-V diagram at the open-circuit voltage and the value of R_{SH} is equal to the slope of the I-V diagram at the short-circuit current. The creation of the I-V curve using the measurements for a given solar cell or module and the determination of R_{SH} and R_S can be used as a diagnostic tool to identify the losses in the unit and ways to improve its performance.

The development of a mathematical representation of this model requires a clear understanding of a diode and its mathematical formula (the Shockley diode equation), which is beyond the scope of this text. It suffices to say that using this model it can be shown that the current generated by a cell

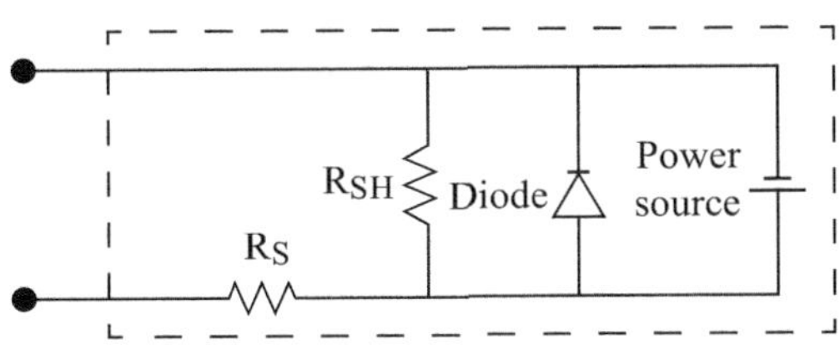

FIGURE 11.39
Schematic of a typical solar cell equivalent circuit including a current source (from photovoltaic phenomenon), a diode, and two resistances.

is linearly proportional to the incoming light intensity. On the contrary, the open-circuit voltage is only logarithmically proportional to the incoming light intensity while it is linearly dependent on the cell temperature.

11.2.5 Balance of Plant for PV Systems

Before we start our discussion on sizing PV units, it is useful to get familiar with some components of PV installations besides PV modules and arrays. We discussed that solar photovoltaic units can directly convert light to electricity. However, in order for PV units to function, they need some auxiliary equipment to support their performance. These components are sometimes referred to as the **balance of plant** (**BOP**) or the balance of system (BOS). The supporting components are substantially small in number, size, and sometimes cost compared to other power generation technologies. Nevertheless, they are critical for the proper deployment of these units for the majority of applications. The major auxiliary components are as follows:

- batteries to store electricity
- charge controller to control the battery charging process
- inverter to convert DC power to AC power
- wiring systems to transfer electricity from PV to other components and to consumers
- electrical monitoring and protection systems to ensure the safe operation of the array
- mounting systems to physically support modules' installation
- cooling and sun tracking systems

Note that not all these subsystems are required for a given PV system for a specific application.

11.2.5.1 Batteries and Charge Controllers

Batteries are used to store electricity when electricity generation exceeds consumption so they can supply electricity when there is no generation or generated electricity is less than demand. Batteries are devices that store electricity via electrochemical reactions. The different types of batteries and their internal operation are discussed in Chapter 16. Here, we will briefly review those aspects of batteries that are directly relevant to PV units.

As explained before, the grid-connected PV arrays practically use the grid as a giant storage system. So they are not typically equipped with batteries. However, if the solar system is intended to provide backup power, they need to have a battery storage system. Also with the introduction and increasing use of the time-of-day or time-of-use rate by utilities, it may financially make sense to add batteries to PV units to avoid the high cost of electricity during peak-load periods. In contrast, for off-grid PV units and those used to provide backup power, a battery storage system must be a part of the system.

While currently lead acid batteries are the dominant type for PV applications, lithium ion batteries are gradually being introduced to this market. Both sealed-type and flooded-type lead acid batteries are used for this application (different types of batteries will be discussed in Chapter 16). The stored energy-to-mass ratio is an important factor for determining the location of the battery bank. Since the bank is typically a heavy component, it is not usually installed on the roof or attic of buildings to avoid adding an excessive weight load to building structures. Instead battery banks are commonly located in basements or ground floors.

Another component that always accompanies a battery bank is a **charge controller**, also known as a **battery regulator** or **charge regulator**. As the name implies, a charge controller controls the level of charging or discharging of a battery bank. Both excessive charging and discharging a battery bank can reduce battery life. Overcharging a battery can also cause significant safety problems. Charge controllers are relatively inexpensive electronic components.

Previously we learned that power generated by a PV system depends on the load connected to it and is maximum at the maximum power point (MPP). Many PV solar arrays are equipped with an electronic component called the **maximum power point tracking** (**MPPT**) device that makes sure the PV unit always operates at or near MPP. Sometimes the maximum power point tracking system and the charge controller are integrated into a single system.

Earlier we discussed about a bypass diode to reduce the negative impact of partial shading of an array. Another diode that is typically used in PV units is called the blocking diode. This diode is used to stop the current from flowing to PV modules when they are not producing any power, for example, during night time. It should be noted that batteries have relatively low efficiency. So it is preferable to consume the generated power directly from a PV array (if possible) rather than storing it in a battery bank and then using it.

11.2.5.2 Inverters

While some appliances can be designed to operate with DC power, such as some pumps and fans, most electric devices need AC power to operate. This means DC power that is either generated in PV modules or drawn from a battery bank should be first converted to AC power before being supplied to consuming devices. An inverter is certainly required when generated power is sent to the utility grid.

Inverters are available in a wide range of sizes from large inverters that can support a big solar farm to medium size for solar arrays on residential and commercial buildings to very small units that can support individual panels. The latter type, which is sometimes referred to as **micro inverters**, is now the most popular one in building-mounted PV units. The inverter cost is a significant portion of the total cost of a PV array. Depending on the size of the unit, the inverter cost can be from one-fifth to half of the total cost. The larger the unit is, the lower the share will be. Sometimes a maximum power point tracking system and an inverter are integrated into a single system. Similarly inverters are typically equipped with a monitoring system to ensure the output power is at an appropriate voltage and frequency. This monitoring system is critical for grid-connected PV arrays to ensure that the voltage, frequency, phase, and waveform of generated power match those of the grid. This is called synchronization process. The monitoring system also disconnects the PV system from the grid whenever it is required, such as when there is no power in the grid. Inverters are very efficient devices with a typical efficiency of above 95%.

11.2.5.3 Wiring Systems

The wiring system provides electrical connections among modules within an array and between the array and the inverter, load, and grid. In building-mounted PV units the generated power in the PV array is connected to the existing wiring of the building via the main electrical panel and does not require a dedicated distribution system. In grid-connected units, a **bidirectional electric meter** is also needed to measure both the imported power from the grid and the power exported to the grid. This measurements are required to determine the annual financial dealing with the utility (refer to Section 11.2.6). In the U.S., all wiring systems must comply with the National Electric Code (NEC), article 690.

11.2.5.4 Mounting Systems

A solar PV mounting system is a series of mechanical components that keep PV modules in place and connect them to a main support structure, either a building structure or a concert foundation (in ground-mounted systems). For building-mounted units, regardless of their specific mechanism, they should be minimally intrusive and should not compromise the integrity of the building structure. Furthermore, when a solar array is added to an existing building, the impacts of the added weight of the system on the structure should be evaluated to ensure the building can withstand the added load. Whenever outer surfaces of the building are penetrated, a proper waterproofing technique should be used to protect the building from moisture damage.

Ground-mounted units are typically less challenging because they are typically designed for the unit and are independent of any existing structure and the weight management is less complicated. In this case, modules can be installed on a rack or on a single post. While the former systems are much more common, the latter systems can be used with sun tracking systems. Note that ground-mounted units have the added advantages of relatively easier control of the tilt angle as well as the relative ease of system maintenance.

11.2.5.5 Other Components

Solar modules need to be exposed to maximum solar radiation possible to generate maximum possible power; however, this exposure has an unintended consequence, i.e., heating of modules. With increase in the temperature of modules, their efficiency reduces and the rate of cell degradation increases. To address this problem, various active or passive cooling techniques can be used, including forced air circulation, forced water circulation, water spray, and water immersion cooling.[36] Sometimes the removed thermal energy is used for other purposes in combined heat and power (CHP) systems.[37] These technologies are typically used for large units particularly when light concentration systems are used (refer to Section 11.2.12.2).

In order to maximize the exposure of solar units to solar radiation, some units are equipped with **sun tracking systems.**[38] These units can be categorized according to the number of axes they have, namely **one-axis or two-axis** systems. A higher number of axes increases the received solar radiation at the expense of higher costs of the system.

11.2.6 Cost of Solar PV Units and Financial Models

In this section, we will take a look at some economic aspects of solar PV systems. Our focus will be primarily on the cost of PV units and financial models to exchange generated solar electricity from PV units with utilities in residential and commercial sectors (but not utility level power generation).

11.2.6.1 Cost of Solar PV Units

The cost of any power generation unit can be categorized into initial or capital costs and operating costs. Operating costs include fuel cost, maintenance costs, and fixed costs. For PV systems, the fuel cost is zero and the maintenance cost is typically very low. Nowadays manufacturers of typical commercial PV modules guarantee that their modules can generate at least 95% and 85% of their rated power when the module is new and after 25 years, respectively. However, there are some other operating costs such as cleaning modules to remove dust and dirt from their surfaces. The fixed costs (also known as soft costs) are associated with tax, insurance, permits, etc. They are becoming a major portion of the total cost of systems especially with steady reduction in the cost of other components.

---

[36] Siecker, J., K. Kusakana, and B.P. Numbi. "A review of solar photovoltaic systems cooling technologies." *Renewable and Sustainable Energy Reviews* 79 (2017): 192–203.

[37] Jia, Y., G. Alva, and G. Fang. "Development and applications of photovoltaic–thermal systems: A review." *Renewable and Sustainable Energy Reviews* 102 (2019): 249–265.

[38] Nadia, A.L.R., N.A.M. Isa, and M.K.M. Desa. "Advances in solar photovoltaic tracking systems: A review." *Renewable and Sustainable Energy Reviews* 82 (2018): 2548–2569.

The capital cost of a solar PV unit can be divided to the cost of PV modules and the costs of other components. The latter can be a significant portion of the total cost. Particularly with recent reduction in the cost of PV modules, the cost of the rest of the system can account to over a half of the total cost. The initial cost of PV systems can be divided to the area-dependent costs and area-independent costs. Some of the **area-dependent** costs are

- PV modules
- land (for ground-mounted units)
- foundation
- module support structure

Some of **area-independent** costs are

- inverter
- battery bank and charge controller
- wiring
- connection to the grid
- electrical protection devices
- metering and monitoring devices

When we are talking about the capital costs of PV units, it should be always remembered that due to their low utilization factors, these units are operational only for a few hours a day. Therefore, while the initial cost of PV units is expressed per peak watt, their actual cost is much higher. So when comparing various power generation technologies, it is more meaningful to compare the cost of generated power ($/kWh) rather than the capital cost ($/kW).

Historically the cost of PV modules has continuously reduced. The capital cost of commercial solar cells was over $200 per peak watt in the late 1950s. By the mid-1970s, this reduced by about ten folds. By the early 1980s, the solar module cost reached $10 per peak watt. In the early 2010s, the cost of PV modules was reduced by a half in just a couple of years. Currently this cost is as low as $\$0.50/W_p$. The cost of some other components such as inverters has also reduced, while others have remained relatively stable such as mounting and support systems.

So far much of PV cell cost reductions have been achieved by lowering the manufacturing costs. But this reduction has slowed down and new approaches are being pursued in using new materials (much more on this topic is discussed later in this chapter). The current dominant strategy is to install relatively low-efficiency inexpensive PV modules rather than more expensive but more efficient modules. But this may change in future.

Overall, all these cost reductions have helped to increase the utilization of PV units in a variety of applications. In many countries, they can compete with fossil fuel-based electricity generation units with the help of government incentives. But in tens of countries, they have reached grid parity. That means the price of electricity from PV modules is equal or less than that of electricity from conventional power plants even without incentives.

11.2.6.2 Financial Models for Exchanging Generated Solar Electricity with Utilities

It was noted that government financial incentives are playing a crucial role in promoting solar PV units. For grid-connected units powering commercial and residential buildings, these incentives are typically offered in the form of financial models for exchange of generated electricity with the utility. Some of more common models will be discussed in this section. Note that all these discussions are for grid-connected PV units. For off-grid units, the owner has no relation with the utility.

There are currently several established financial models for the relationship between the owner of a residential or commercial building that is equipped with solar PV units and the utility that traditionally provides power for the building. The following list represents the most common models currently being used:

- net metering
- remote metering
- feed-in tariff
- power purchase agreement

In all these models, a PV unit generates electricity that is used in a building. When the produced energy is more than the needs of the building, the excess energy is sent to the grid managed by a utility. When the generation is not sufficient to cover the building needs, power is received from the grid. These models (except the power purchase agreement model) just differ on how this exchange of power between the building and the utility is accounted for and financial gains and losses are determined.

11.2.6.2.1 Net Metering

In this model the building or the site owner makes the investment on the design, purchase, installation, and operation of the PV system and generates electricity. After the building consumes its needs, the excess electricity is sold to the utility at the retail price. Also, the needed electricity for the building is purchased at the retail price. The balance evaluation is typically done annually, which determines the building's annual utility bill based on the credit received from selling electricity and the charge for the electricity pulled from the grid. Note that in a traditional market, the utility purchases electricity at a wholesale price, which is much lower than the retail price. So the provided incentive in this model is that the building owner can sell the power at the price much higher than what was possible otherwise. Also, the annual balance helps to even out high electricity production in summer with low power generation in winter. So the building does not need to meet the daily load for everyday of a year but just the daily annual average. Basically a credit from 1 month can roll over to cover the expenses in another month. The implementation of this model is relatively simple and just need a bidirectional meter to keep track of the exported electricity to the grid and the imported electricity from the grid. If annual credits and charges are equal, they cancel out each other.[39] However, note that the owner also pays for a connection fee to the utility to provide access to the grid and the grid maintenance. You may ask, "is it possible to make money if you install a large enough PV array?". The answer is typically no. The amount of power one can produce in this model is limited to the annual consumption of the building. This depends on the laws of a specific state or country. Sometimes one can generate a little more than what consumes, say 10% more. If a building produces more than what is allowed, the excess power can be sold to the grid but at the wholesale price and not the retail price. There are many varieties of this model that are being implemented in different states and countries. Due to the high cost of maintaining the two-way interconnection between the house and the grid as well as the loss of revenue because of local power generation, utilities are claiming that this model is not financially sustainable.

A derivative of the net metering model is called the **remote net metering model**. The only difference between the two models is that in the remote net metering model more than one building can be in the same account. That means the credit for generating solar electricity in one building that exceeds the consumption of the building can be used to offset the electricity bill of a different building owned by the same person or company.

11.2.6.2.2 Feed-in Tariff Model

In the feed-in tariff (FIT) model while the building owner purchases electricity from the utility at the retail price, the excess electricity is sold to the utility at a higher price compared to the retail price. This higher price is guaranteed for several years (typically up to 25 years) and then gradually the provided incentive reduces and the selling price goes done. The timing and rating of the reduction in the incentive is typically determined by the number or percentage of adapters. So the earlier the solar system is installed, the more incentive the owner can receive. The idea is that the early incentive can help owners to pay for the high cost of system installation. Also, the more popular the program becomes, the lower the capital cost of the system will be (hopefully!), due to the economy of scale. Therefore, more people can buy into the program even with less incentive. The main idea is that eventually solar PV units can compete

[39] Note that the owner may be required to pay a monthly bill for the monthly energy usage and then the total balance is settled annually.

with conventional power generation units (called the grid parity or socket parity) even without any incentive. Unlike the net metering model where the energy generation of the system is limited, in this model the limitations are for the generation capacity of units. Another major side benefit of this model is that it encourages energy conservation efforts so the building owner has more electricity to sell. While this model was first developed in the U.S. in the late 1970s, it is currently more popular outside the U.S., particularly in Europe.

11.2.6.2.3 Power Purchase Agreement

Unlike the previous two models that the building owner invests on the system (and owns the PV system) and the contract is between the owner and the utility, in the power purchase agreement (PPA) the leasing contract is between the building owner and a third-party company (and not necessarily the utility). The company is responsible for all aspects of the development and operation of the PV unit with no or little investment from the building owner. The building purchases its consumption from the PV unit managed by the developer at a cost lower than the grid retail price. The price is either fixed for a number of years (typically for 10–25 years) or increases at a rate lower than the predicted rate of increase for power from the grid. Typically in this model there is no limit on the size of the unit and the developer sells the excess power to the utility according to a separate contract. If the building consumption is more than what the PV unit generates, the excess energy is provided from the grid at the retail price (similar to if there was not PV unit). Note that all incentives for the system are collected by the developer. At the end of the contract, the building owner can extend it, ask for the system removal, or purchase it from the developer. In this model, building owners can reduce their electricity cost and increase the value of their property with no or little investment or risk. However, one certainly cannot make any money in this model and the saving level is much lower compared to that of the previous two models. Also, it should be noted that unlike the net metering and feed-in tariff models that are regulated by the state or national policies, power purchase agreements are contracts between two parties. The power purchase agreement model is not available in all regions. As you can see, unlike the previous two models that were bookkeeping methods only for financial aspects of the system, the PPA has much broader impacts.

The choice of a financial model in each region is typically limited to the option provided by the local and national policies. The choice between owning a PV units and leasing it (via a PPA) is all about a high gain at a high capital cost and risk or a low gain with no or little initial cost and risk.

11.2.7 Sizing of Solar PV Systems

The first step in designing any electricity generation system, including PV systems, is to identify required energy production by the system (in kWh) and possibly the system power (in kW). In utility level power generation units the required energy is determined by a complicated study of the current and predicted future demands in the target region. In residential and commercial units required energy is determined by an energy audit of the building of interest. The building energy audit field has grown to a large industry on its own. While the building energy audit topic is beyond the scope of this text, a very brief introduction is presented in the next section. After the energy consumption of the building is determined, the available solar incident must be calculated. The next stage is to size the solar PV array. Finally, the supporting auxiliary components are needed to be selected. These steps will be covered in this section.

11.2.7.1 Load Determination for Residential and Commercial Applications

There are three main methods to determine electricity consumption for residential and commercial buildings. In the first method, direct measurements are used. The electricity consumption of each appliance can be measured for a certain period of time using a wattmeter. The overall building energy consumption is determined by adding the consumption of all appliances. Alternatively, the consumption of each appliance can be determined by reading its electrical characteristics from its label along with the duration of its operation. For some appliances, the consumption can be identified from available online tools. Finally, the monthly utility bills can be used to determine the annual energy consumption of a building. Clearly this last method can only be used for grid-tied buildings because it does not provide any information regarding daily variation of the demand. One of these methods or a combination of them can be utilized to determine the daily, monthly, or annual electrical energy consumption and the maximum electrical power requirement of the building.

The first two methods not only help us to determine the energy consumption of the building, they can also be used for the building energy audit. While not directly part of a solar PV system design, typically the design process starts with an energy audit to understand the building energy performance, the energy consumption for the building air conditioning system and appliances, and the occupants' behaviors and habits relevant to energy consumption. The audit can help to identify areas where improvements are possible. Most often reducing the energy consumption is more economical than to invest in a PV system to produce the extra electricity. It is usually less expensive for example to replace an old refrigerator with a new more efficient one than to install enough PV modules to produce electricity to power the old one.

The energy consumption and power of an appliance can be directly measured using a device called **wattmeter**. The measurements can be directly read from the device (Figure 11.40) or, for more advanced units, the device can be connected to internet and the data can be recorded and accessed

FIGURE 11.40
A simple local wattmeter. (By Gareth.randall – Own work, Public Domain, https://commons.wikimedia.org/w/index.php?curid=7745495.)

remotely. Note that even when not operating, electrical equipment typically consume power in the stand-by mode. Just by plugging them in, they usually consume power. While the consumed energy in this case is small, due to the long period of consumption and the large number of devices, they can quickly add up to substantial energy consumption and cost. That is why they are sometimes referred to as **power vampires**.

Alternatively, the power and energy consumption of appliances can be calculated from their electrical specifications. A majority of electrical equipment come with a label showing their rated power consumption (Figure 11.41). Sometime instead of power, the rated voltage and current of the device are

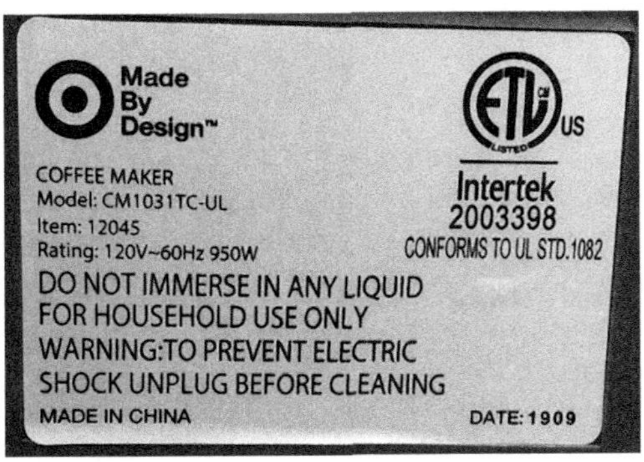

FIGURE 11.41
Electrical specification label for a typical coffee maker.

listed. In this case, the rated power can be easily determined by multiplying the voltage and current. For a device with a known rated power, we only need to know the duration of the use for the device to determine its energy consumption. The following example can help to understand how this method can be used to determine the annual energy consumption of a building.

Example 11.21

A building energy audit resulted in the following table. Determine the annual electricity consumption of the building.

Device	Number of Device	Rated Power (W)	Usage Hours per Day (h)
Fluorescent lights	6	20	4
Air condition unit	1	250	8
Computer	2	150	5
Microwave	1	450	0.5
Coffee maker	1	900	0.5

Analysis

The annual electricity consumption of each group of appliances can be determined by multiplying their rated power, usage hours, and number, as reflected in the table below.

Device	Number of Device	Rated Power (W)	Usage Hours per Day (h)	Annual Electricity Consumption (kWh/year)
Fluorescent lights	6	20	4	480
Air condition unit	1	250	8	2000
Computer	2	150	5	1500
Microwave	1	450	0.5	225
Coffee maker	1	900	0.5	450
				Total: 4655

For some home appliances, such as refrigerators and freezers, it is more difficult to determine their electricity consumption because it is not easy to determine how many hours the system actually operates. For these cases, the energy consumption can be determined from information provided by the manufacturer or other independent information providers, such as Energy Star website (energystar.gov).

To get the annual energy consumption of a building, the utility bills provide the easiest and most direct way. By adding the monthly consumptions, the annual electricity consumption can be determined. While for most grid-tied units the annual consumption rate is used, for designing units that are operational seasonally or off-grid units, monthly or even daily consumption might be more relevant.

Example 11.22

A house is equipped with ten light bulbs each with a rated power of 40 W.

a. If on average each light is on for four hours a day, determine the annual electrical energy consumption and the cost of the consumed electricity. The average cost of electricity for the building is \$0.15/kWh.
b. If the light bulbs are replaced with LED lights with the rated power of 9 W, determine the annual saving.

Solution

a.

$$\text{Annual electricty consumption of each light bulb} = 40[\text{W}]\times 4\left[\frac{h}{\text{day}}\right]\times 365\left[\frac{\text{day}}{\text{year}}\right]$$

$$= 58,400[\text{Wh}] = 58.4[\text{kWh}]$$

$$\text{Annual electricty consumption of all light bulbs} = 10\times 58.4[\text{kWh}] = 584\ [\text{kWh}]$$

$$\text{Annual electricty cost of all light bulbs} = 584\ [\text{kWh}]\times 0.15\left[\frac{\$}{\text{kWh}}\right] = \$87.6$$

b)

$$\text{Annual electricty consumption of each LED light} = 9[\text{W}]\times 4\left[\frac{\text{h}}{\text{day}}\right]\times 365\left[\frac{\text{day}}{\text{year}}\right]$$

$$= 13,140[\text{Wh}] = 13.14[\text{kWh}]$$

$$\text{Annual electricty consumption of all LED lights} = 10\times 13.14[\text{kWh}] = 131.4[\text{kWh}]$$

$$\text{Annual electricty cost of all lights} = 131.4[\text{kWh}]\times 0.15\left[\frac{\$}{\text{kWh}}\right] = \$19.71$$

$$\text{Annual saving} = \$87.6 - \$19.71 = \$67.89$$

Example 11.23

The following table shows the monthly electricity consumption of a residential building. Determine the annual and average monthly electricity consumption of the building.

Month	Monthly Electricity Consumption (kWh)
January	893.2
February	846.3
March	762.3
April	692.3
May	741.3
June	1053.5
July	1486.1
August	1393
September	1019.2
October	1003.8
November	940.1
December	764.4

Solution

Adding the right column and averaging it, you can find that the annual and average monthly electricity consumption of the building is 11,592 and 966 kWh, respectively. According to the U.S. Energy Information Administration (EIA), the average annual residential electrical energy usage in the U.S. was slightly <11,000 kWh per household in 2018.[40,41]

Example 11.24

A solar farm with the rated capacity of 3 MW can generate power at this rate for on average 5.5 hours a day. Determine the annual electricity generation of the system in MWh.

Solution

$$\text{Annual electricty generation} = 3[\text{MW}] \times 5.5\left[\frac{\text{h}}{\text{day}}\right] \times 365\left[\frac{\text{day}}{\text{year}}\right]$$

$$= 6022.5 \text{ MWh}$$

11.2.7.2 Sizing Grid-Connected Solar PV Arrays

Before we discussed about sizing of PV arrays using the calculated electricity demand, it should be noted that for grid-connected PV units, the design is typically based on the annual electricity consumption of the building. The daily and seasonal variations of the demand do not affect the design. This is because the grid can even out these variations. Also, typically the financial calculations with the utility are on the annual basis. Also, note that the

[40] https://www.eia.gov/tools/faqs/faq.php?id=97&t=3.

[41] The highest average annual consumption was reported in Tennessee at 15,394 kWh per household and the lowest was in Hawaii at 6213 kWh per household.

annual electricity generation of a building should not exceed its annual consumption. The reason for all these was discussed in Section 11.2.6, where we covered various financial models between a PV unit owner and the utility. Furthermore, depending on the specific conditions and requirements of each project, the solar array may be designed for the replacement of 100% demand of the building, which is an ideal case, or a portion of the load. Some factors that may limit the design load for a PV array are the limitation in the capital cost that the owner can or want to invest in the system and the availability of space for the installation of the modules.

From our discussion earlier, we know that solar panels are tested and their performance is reported for the STC. In this discussion, we assume that the solar intensity is constant and is equal to 1000 W/m^2. Let us assume that the annual electrical energy required to power a house is 10,000 kWh. While from the consumer point of view, the electricity generation is important, for sizing the system we need to determine the power (the time rate of energy consumption). Notice that the generated electricity depends on both the size of the system and the duration of its operation. If we could provide solar radiation every hour of every day, we could determine the required power generation from the PV array by dividing the annual energy by the number of hours in a year.

$$\text{Required power of PV array if solar radiation was always availble}$$

$$= \frac{10{,}000\left[\dfrac{\text{kWh}}{\text{year}}\right]}{24\left[\dfrac{\text{h}}{\text{day}}\right]\times 365\left[\dfrac{\text{day}}{\text{year}}\right]} = 1.14\,\text{kW}$$

However, clearly this is not the case and the solar radiation is not available all the time at 1000 w/m^2. This leads us to the concept of **sunlight hour**, also known as **full sun hour** or **sun hour**. This is the number of hours that if we had the solar radiation with the intensity of 1000 W/m^2, it would have the same amount of energy as the actual solar radiation in an average 24 hours. So if a region has an average daily solar intensity of 5500 Wh/(m^2×day), the sunlight hour of the region is

$$\text{Sunlight hour} = \frac{5500\left[\dfrac{\text{Wh}}{\text{m}^2\times\text{day}}\right]}{1000\left[\dfrac{\text{W}}{\text{m}^2}\right]} = 5.5\,\frac{\text{h}}{\text{day}}$$

As it is evident, the sun hour for a given location is numerically the same as the annual solar energy when it is expressed in $\dfrac{\text{kWh}}{\text{m}^2\times\text{day}}$ as in the isolation maps discussed in Section 11.1.7 and in Figures 11.13 and 11.14. Now we can determine the required power (capacity) of the unit as follows:

$$\text{Required power}(\text{capacity})\text{of the unit} = \frac{\text{Annual required energy generation}}{\text{Sunligth hour} \times 365}$$

$$= \frac{10{,}000\left[\frac{\text{kWh}}{\text{year}}\right]}{5.5\left[\frac{\text{h}}{\text{day}}\right] \times 365\left[\frac{\text{day}}{\text{year}}\right]} = 4.98\ \text{kW}$$

As expected, the actual required power of 4.98 kW is much higher than 1.14 kW we calculated above if the solar energy was always available. This is because you need a larger system so it can generate the same amount of energy at a much shorter operation period.

The next step is to determine the number of solar panels we need so we can form an array with the desired power production capacity. In order to do this, we first need to determine the solar module manufacturer. Then, from the specification sheet of a specific panel of interest that the manufacturer provides, the power generation capacity of each panel can be determined. Let us assume each solar panel generates 150 W of power. So theoretically about 33 solar panels are required (=4.98/0.15). However, this calculation neglects that the actual operating conditions may not match the testing laboratory environment. For instant, dust and stain on the surface of solar panels can prevent some of solar radiation from penetrating the cell. Also, not all the power generated by the array is available for the user. For example, some of the generated electricity is lost in the inverter and connecting cables. To take all these factors into account, an external efficiency is used in the calculations. This external efficiency can typically be between 75% and 90%, in this case, assuming an external efficiency of 80%

$$\text{Number of required solar panels} = \frac{\text{Required power}(\text{capacity})\text{of the unit}}{\text{Power of each panel} \times \text{external efficiency}}$$

$$= \frac{4.98}{0.15 \times 0.8} = 41.5$$

This number can be round down or up depending on the specific requirements of each project.

In designing a real-world PV array, some of the following factors also need to be taken into account, which makes calculation much more complicated:

- The availability of rooftop space, the orientation of each surface, and the roof angle

- The nearby trees, building, or any other obstacles that may cast a shadow on some part of the modules
- The long-term impact of modules' aging
- The local air pollution and dust in the atmosphere that may reduce the direct solar light and increase indirect (diffused) light
- The accumulation of dust and dirt on the surface of modules and the frequency of their cleaning
- The possible accumulation of snow in the winter and the possible snow removal schedule

Example 11.25

A PV array is expected to power a house with annual electricity consumption of 8000 kWh. If the house is located in San Francisco and the external efficiency of the array is 85%, determine the required nameplate power of the array in kW.

Solution

Based on the isolation map in Figure 11.14, the solar intensity in San Francisco is in the range of 5.00–5.25 kWh/m²/day. Let us consider the lower value of the range. The sunlight hour for San Francisco is

$$\text{Sunlight hour} = \frac{5000}{1000} = 5.0\frac{\text{h}}{\text{day}}$$

$$\text{Required power}\left(\text{capacity}\right)\text{of the unit}\left(\text{with no external losses}\right)$$

$$= \frac{\text{Annual required energy generation}}{\text{Sunligth hour} \times 365}$$

$$= \frac{8000}{5.0 \times 365} = 4.38\text{ kW}$$

Taking the external efficiency of the array into account,

$$\text{Required nameplate power}\left(\text{capacity}\right)\text{of the array} = \frac{4.38}{0.85} = 5.16\text{ kW}$$

Example 11.26

a. Determine the maximum rated power of a building-mounted PV array installed on a building with an annual electricity consumption of 12,500 kWh. The net metering contract with the regional utility limits the annual electricity generation by residential applications to 10% more than the annual electricity consumption of the building. The local sunlight hour is 4.5 hours and the external efficiency of the system is 86%. (b) How many solar PV modules are required if the panel specification indicated the power rating of each panel is 250 W? (c) Determine the power generation if the homeowner can only afford half of the panels. Round down the calculated number of the modules.

Solution

a. Maximum annual energy generation $= 1.1 \times 12,500 = 13,750$ kWh

$$\text{Maximum power}\left(\text{capacity}\right)\text{of the unit}$$

$$= \frac{\text{Annual required energy generation}}{\text{Sunligth hour} \times \text{external efficiency} \times 365}$$

$$= \frac{13,750}{4.5 \times 0.86 \times 365} = 9.73\text{ kW}$$

b. Number of required solar panels

$$= \frac{\text{Required power }\left(\text{capacity}\right)\text{of the unit}}{\text{Power of each panel}}$$

$$= \frac{9.73}{0.25} = 38.9 \approx 39\text{ modules}$$

Note that we have already taken the external efficiencies into account in the previous section.

c. You can simply half the number of modules $\left(\frac{38.9}{2} = 19.4\right)$. So 19 modules are needed in this array. We can also half the electricity requirement and then repeat the calculation.

11.2.7.3 Consideration in Sizing Off-Grid Solar PV Arrays

It was discussed that grid-tied arrays are designed based on the annual electricity consumption of the target consumer regardless of seasonal and daily variations. For off-grid PV units, however, the daily demand variation is an important design parameter. In this case, the design philosophy will dictate the sizing of the PV array and the battery bank to ensure that the maximum demand is always met. This typically means meeting both daily electrical energy consumption and maximum power demand. For example, assume an off-grid house with two appliances, one with rated power of 0.5 kW and the other 1 kW. Each appliance is operated on average 2 hours a day. If only one appliance is operated at a time, the PV array energy load is 3 kWh/day $(2 \times 0.5 + 2 \times 1)$ and the maximum power demand is 1 kW. So in this case, not only the average electrical energy consumption of each appliance should be determined but their peak electric power also needs to be taken into account. Even whether the consumption occurs during the day time or night should be considered. This is important because a PV array equipped with a battery bank can be operated on several modes: PV-to-load, PV-to-battery, or battery-to-load. The overall efficiency of the system is different in each of these modes, particularly because the efficiency of storing and releasing electricity in the battery bank needs to be accounted for (refer to Chapter 16 on round trip efficiency of batteries). To further complicate the design, some appliances may use AC power while others may operate using DC power.

Example 11.27

In an off-grid PV array designed for a remote house, both DC- and AC-powered appliances are utilized. If the efficiency of various parts of the system is as follows, determine the overall efficiency of the system in PV-to-DC load, PV-to-AC load, PV-to-battery-to-DC load, and PV-to-battery-to-AC load modes.

- the efficiency of the distribution system 95%
- the efficiency of the battery pack 78%
- the efficiency of the inverter 90%

Solution

The efficiency of PV-to-DC load mode = 0.95%

The efficiency of PV-to-AC load mode = 0.95 × 0.90 = 85.5%

The efficiency of PV-to-battery-to-DC load mode = 0.95 × 0.78 = 74.1%

The efficiency of PV-to-battery-to-AC load mode = 0.95 × 0.78 × 0.90 = 66.69%

11.2.8 Commercial Software Tools to Simulate Performance of Solar PV Units

Considering the complexity involved in designing and evaluating the performance of solar arrays, computational tools have long been used for these purposes. There are currently many commercial packages that are being used in the industry and academia. Some of these software tools are exclusively developed for PV units and some have many built-in components, which include PV units. They have a wide variety of approaches, complexity, and details. Some of the common software packages are as follows:

- System Advisor Model (SAM)[42]
- PVsyst
- HOMER
- RETScreen
- Helioscope

[42] This software was initially designed as Solar Advisor Model (SAM) by National Renewable Energy Laboratory (NREL), and later other renewable power generation systems were gradually added. SAM is available free of charge to the public at https://sam.nrel.gov/ and is commonly used for educational purposes.

- TRANSYS
- ESP-r
- PV F-Chart
- SolarDesignTool
- PV*SOL
- SOLARGIS
- INSEL
- SolarPro
- PV DesignPro-G
- SOLSIM
- PVLib
- iHOGA
- iGRHYSO

11.2.9 Different Types of Solar Photovoltaic Cells

In the previous sections, we learned about how solar PV cells work and their performance calculations. In the following sections, various types of solar cells will be discussed along with their manufacturing methods. There are several ways to categorize solar cells, one of which is the so-called generation of cells. There are three generations of solar cells. The **first generation** is **crystalline silicon-based cells** that are currently dominating the market. The **second-generation** cells, also referred to as **thin film solar cells**, are made of direct semiconductor. They are commercially available and have a small share of the market. The **third-generation** PV cells are a reference to many emerging technologies that are in various stages of development.

11.2.10 First Generation: Crystalline Silicon-Based PV Cells

As noted in Section 11.2.1, crystalline silicon solar cells were the first practical cells that were developed and they are still the **workhorse of the solar PV power generation industry**. Currently, over 90% of the manufactured solar modules are the crystalline silicon type. A majority of modules you may see installed in rooftops are of this type. Over many decades their efficiency has significantly improved and the price has remarkably reduced. In the early 1980s the highest efficiency of crystalline silicon cells in the laboratory environment was about 16% and the highest efficiency of commercial modules was about 10%. These efficiencies for modern solar units are in the range of 21%–26% and 17%–20%, respectively. These efficiencies are close to the maximum theoretical efficiency that can be achieved using silicon at the STC at about 28%. The life expectancy of more than 25 years and the cost of \$0.50\$1/W_p are currently the industry standard.

(a) (b)

FIGURE 11.42
Flaky appearance of a polycrystalline cell (a) compared to uniformly black surface of a monocrystalline cell (b). ((a) Image by Hans Braxmeier from Pixabay, https://pixabay.com/photos/solar-cells-technology-current-62559/, (b) Image by Valentin J-W from Pixabay, https://pixabay.com/photos/solar-cells-energy-solar-energy-4518991/.)

The reasons for their current dominance can be explained based on several determining factors. **Silicon is the second most abundant element in the Earth's crust** after oxygen and it is the most abundant semiconductor. This abundance provides virtually unlimited supply of the raw material to produce solar cells. Furthermore, silicon is a relatively safe element and poses little safety concerns during the entire life cycle of solar units, from manufacturing to operation to end-of-life disposal. More importantly, the band gap of silicon (about 1.1 eV) is suitable for photovoltaic electricity generation. Another reason for undisputable dominance of crystalline silicon cells is that their technology has a lot in common with the electronic components, such as transistors and integrated circuits (also referred to as ICs or microchips). Solar cells have immensely benefited from rapid advancements in the electronic industry. The crystalline silicon-based solar modules can provide power with good performance for a very long period of time without significant decline in their performance by environmental elements and other external factors. A long period of laboratory and field experiences have proved their high stability and reliability. However, they suffer from inflexibility and very energy-intensive manufacturing processes.

There are two types of crystalline silicon photovoltaic solar PV cells: monocrystalline silicon cells and polycrystalline silicon cells. **Monocrystalline** solar cells, also referred to as **single crystal silicon** cells, are made of a single crystalline of silicon. They are black in color and uniform in texture (Figure 11.42b). They are one of the most efficient types of PV cells with the best cell efficiency of about 26% in the laboratory environment and about 20% for typical commercial modules, which can go up to about 22% for high-end products. They have currently about 25% of the market share for the crystalline silicon PV market.[43]

Polycrystalline silicon cells, also known as **multi-crystalline** cells, are the dominant type with over three-quarters of the market share for the

[43]Shubbak, M.H. "Advances in solar photovoltaics: Technology review and patent trends." *Renewable and Sustainable Energy Reviews* 115 (2019): 109383.

crystalline silicon PV market. They are made of many small silicon crystals and this gives them a flaky look (Figure 11.42a). Since they are made of less pure silicon and their manufacturing process is less energy-intensive, they are less expensive than monocrystalline cells. For the same reason, they are less efficient than monocrystalline cells. For polycrystalline PV units, the best cell efficiency in laboratory conditions has been about 22% and the commercial modules have up to 17% efficiency.

11.2.10.1 Manufacturing of Crystalline Silicon PV Cells

The manufacturing processes of solar PV modules involve a wide variety of processes. These processes will be briefly reviewed in this section. Figure 11.43 illustrates these manufacturing steps.

The manufacturing of crystalline silicon requires extremely pure silicon, in the order of 99.9999% purity.[44] This level of purity is achieved in several stages. The first stage is to convert naturally bonded silicon to metallurgical-grade silicon. As noted earlier, silicon is the second most abundant element in the nature. Silicon is naturally found as **silicon dioxide** (SiO_2), also known as quartz crystal, silica, or more commonly **sand**. In order to remove oxygen from silicon dioxide (referred as the reduction process), sand and coal (coke) are heated in an electric arc furnace. At the temperature about 1800°C (about 3270°F), the following reduction reaction takes place:

$$2SiO_2 + 4C \rightarrow 2Si + 4CO \tag{11.12}$$

The product of this process is silicon with a purity of about 98%, called metallurgical-grade silicon (MGS or MG-Si). As noted above, this level of purity is not sufficient for the manufacturing of PV cells. To increase the purity of silicon, metallurgical-grade silicon is ground into fine particles and then dissolved in hydrochloric acid (HCl) in a fluidized-bed reactor to produce trichlorosilane ($SiHCl_3$) through the following silane process at the temperature of about 300°C (about 570°F) at the presence of a catalyst.

$$Si + 3HCl \rightarrow SiHCl_3 + H_2 \tag{11.13}$$

The purity of trichlorosilane is increased through the distillation process. Then, the highly pure trichlorosilane goes through the modified Siemens process, also known as the chemical vapor deposition (CVD) process, in a Siemens reactor. The reactor operates at partial vacuum and typically contains U-shaped slim rods of pure silicon (several meters long) that are heated to a temperature of around 1000°C (about 1830°F). Within the reactor, a series

[44] The fineness of metal is commonly represented by "number of 9s". For example, the 99.9999% pure silicon is shown as 6 nines or 6N. The purity of silicon required for the electronic industry is typically 9 nines or 9N.

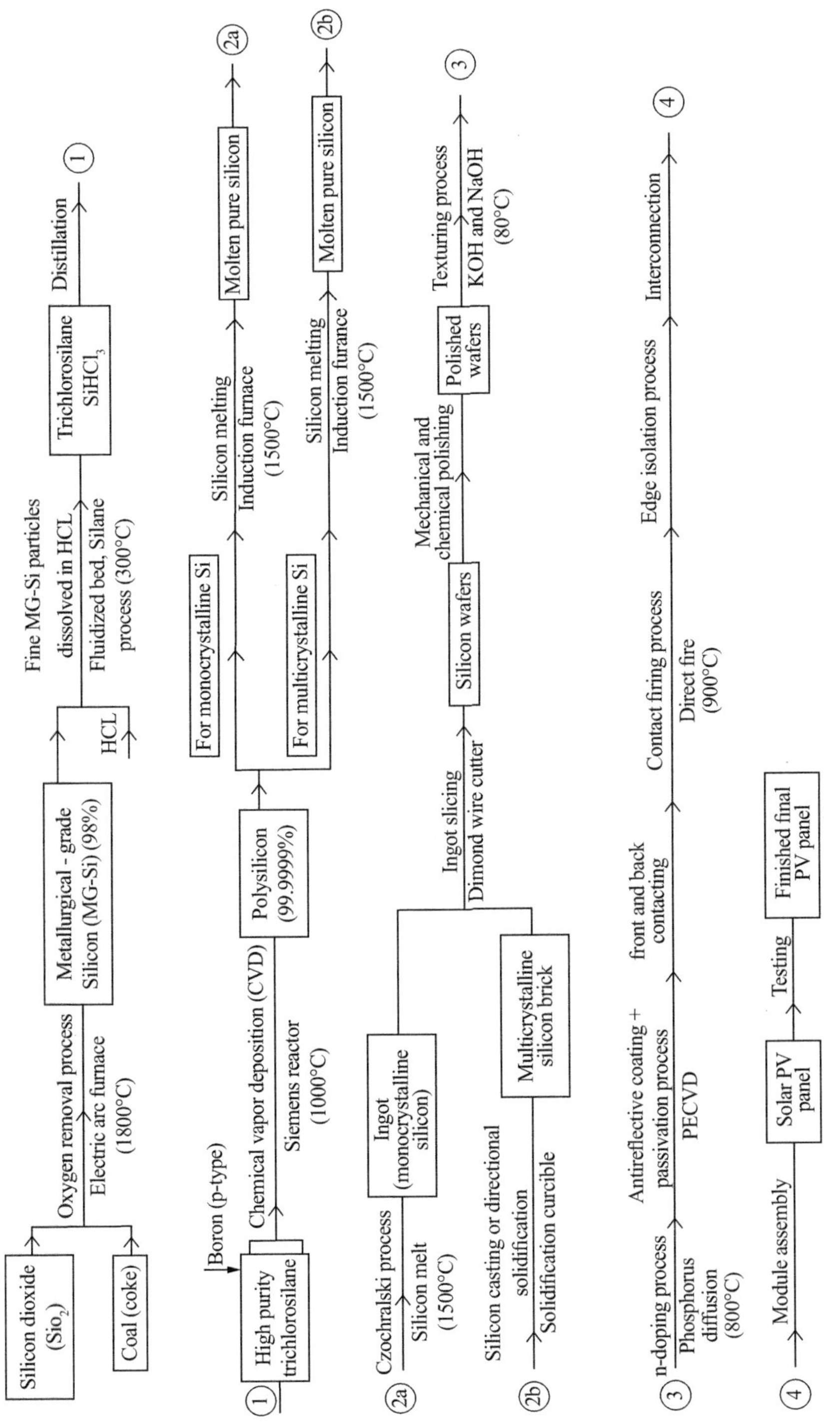

FIGURE 11.43

A general overview of manufacturing processes for crystalline silicon cells.

of complicated reactions take place in the presence of hydrogen, which eventually results in the deposition (recrystallization) of highly pure silicon on the rods. These reactions can be simplified in the form of the following reversed silane process.

$$SiHCl_3 + H_2 \rightarrow Si + 3HCl \tag{11.14}$$

In order to produce doped silicon, a very small but carefully controlled amount of gases containing the dopants are also injected into the reactor. In general, different dopants can be added depending on the type of the desired semiconductor. For n-type semiconductors, phosphorus (P) or arsenic (As) containing gases are added, while for p-type semiconductors boron (B) containing gases are added. For PV cell production typically **p-type is required**. The produced pure silicon, called **polysilicon** (not to be confused with polycrystalline silicon cells), is sufficiently pure for PV cell manufacturing. However, it does not have a suitable crystal structure because the structure is made of too many crystal grains; therefore, further processes are required to achieve a desirable structure. The processes to achieve this structure differentiate the monocrystalline silicon cells from multi-crystalline silicon cells. It should be noted that the described process to produce polysilicon is a very energy-intensive process and involves very toxic, explosive, and corrosive gases, handling of which adds to the cost of the process. There are other processes that require less energy but the purity of the produced silicon is generally lower than this process.

To produce monocrystalline silicon, first polysilicon is melted in an induction furnace using a very strong electromagnetic field at the temperature of 1400°C–1500°C (about 2250°F–2730°F) under an inert gas environment such as argon. To create a single crystal of silicon, a small seed crystal attached to a rod is inserted into the molten silicon. The crystal silicon is gradually formed around the seed crystal. To grow a uniform silicon rod, the seed crystal slowly rotates. The orientation of the seed crystal determines the orientation of the entire produced silicon rod, which is called an **ingot**. When the ingot is grown to the desired size, it is slowly polled out of the bath. This solidification process takes place under a controlled temperature distribution. The speed of the rotation and withdrawal is controlled to achieve an ingot with a desired size. Each ingot can grow up to 30 cm diameter and a couple of meters length and may weigh several hundred kilograms (Figure 11.44). The process may take several days and since the silicon must be kept at the molten condition for such a long time, the process is very energy-intensive. This process is called the Czochralski process after its inventor Jan Czochralski (1885–1953, a Polish chemist) who invented the process in 1918.[45] Another process to pro-

[45] He discovered the process when he mistakenly dipped his pen into molten tin rather than ink and quickly withdrew it. He observed a long thread of solid metal attached to the pen which he later confirmed as a single crystal.

duce highly pure, properly oriented single crystalline silicon is called the float-zone process where a vertical silicon rod is melted locally (in zones).

In the production of polycrystalline silicon, the slow and energy-intensive Czochralski process is avoided. Instead, the pure polysilicon produced from the Siemens process is melted and then solidified to cubical ingots or bricks (the size can be up to 70 × 70 × 25 cm, about 2.3 × 2.3 × 0.8 ft). This process can be achieved through silicon casting or directional solidification. In the casting process, silicon is melted in a graphite crucible and is then cast in a growth crucible. In the directional solidification process, both melting and solidification processes take place in the same crucible. The orientation of crystals is controlled by allowing silicon to slowly solidify from the bottom of the crucible. The final product is a cubical multi-crystalline ingot that is formed from many small crystal grains.

Both the monocrystalline and multi-crystalline silicon manufacturing processes result in forming silicon ingots. From this point on again the manufacturing processes for both types are similar. The next step is to mechanically

FIGURE 11.44
Monocrystalline silicon ingots. (CC BY-SA 3.0, https://commons.wikimedia.org/w/index.php?curid=314282.)

cut ingots into very thin slices of silicon called silicon wafers. For this purpose, diamond wire cutters are used. Every sawing process inevitably involves converting some material into dust. The magnitude of this loss is a function of the diameter of the wire, in this case between 100 and 140 μm. This may not look significant but if we compare it to the thickness of the produced wafers, typically between 150 and 200 μm, it is clear that a significant portion of silicon ingots will be converted to dust. This means about 50% of the silicon produced via very energy-intensive processes can be lost, which can constitute to more than 30% of the total cost. This mechanical sawing process is one of the bottlenecks of the manufacturing processes of crystalline silicon cells because any attempt to reduce the thickness of wafers to reduce the silicon consumption increases the possibility of breakage and increases the percentage of the sawing material loss. Later we will learn how new manufacturing processes try to eliminate this process. Note that the workers are not exposed to the produced dust and the highly pure silicon dust is recycled.

In the next step individual wafers with the side of about 15 cm (about 0.5 ft) are chemically and mechanically polished to remove contaminants and surface damages caused by the sawing process. The surface of wafers at this point is shiny and mirror-like that can reflect a significant portion of the incoming light. To reduce this effect, two processes are used: texturing the surface and adding an antireflection coating to the surface.

In the texturing process, the wafers are placed in a potassium hydroxide (KOH or caustic potash) or sodium hydroxide (NaOH or caustic soda) solution at a temperature of about 80°C (about 180°F). Through a chemically violent reaction, some of the silicon atoms are removed from the surface and the previously smooth surface is converted to a surface covered with microscopic pyramids. These pyramids significantly reduce the reflectiveness of the cell surface by so-called light trapping.

We learned that in the Silane process we typically add boron-containing gas so the wafer that we start the cell manufacturing process with is already a p-doped semiconductor. The next step in this process is to create the p-n junction. For this purpose, we need an n-doping process. This process takes place in a furnace where the batches of silicon wafers are exposed to a phosphorus-containing gas at the temperatures between 800°C and 900°C (about 1470°F–1650°F). In this environment, phosphorous can diffuse sufficiently into the front of the wafers to create the n-type silicon. Note that the thickness of the n-type semiconductor layer on the surface of cells is less than 1 μm so the phosphorus diffusion is effective for this region. Due to its high temperature, this is another relatively energy-intensive process.

The next step is to add an antireflective coating to the front surface of wafers to further reduce the reflectiveness of the cell surface and increase light absorption. The coating material is typically silicon nitride that is deposited by the plasma-enhanced chemical vapor deposition (PECVD) process. This coating gives PV cells their characteristic blue appearance (in polycrystalline cells). Since this coating process can also be integrated to the surface passivation process, let us briefly talk about this process.

The surface of crystalline silicon wafers contains some reactive silicon atoms. These silicon atoms, also referred to as unsaturated atoms, have only three bonds with their neighboring atoms and one electron remains unpaired (we learned earlier that silicon atoms have four electrons in their valence band). This unpaired electron (referred to as the dangling bond) can bond with other atoms, which changes the properties of silicon semiconductor and creates a defect that can cause recombination of electron-hole pairs and can negatively impact the performance of the cell. To avoid this problem, hydrogen atoms can form a bond with this unpaired electron. This process is called the surface passivation process. While there are several sources of hydrogen to achieve this effect, the most common one is to use hydrogenated silicon nitride for the antireflective coating process. Using this coating causes some hydrogen atoms to be trapped during the deposition process. In order for these hydrogen atoms to diffuse into the surface of silicon and react with unsaturated silicon atoms, a high-temperature environment is needed. This high-temperature process can be completed during the firing process to create the cell back electrode (will be discussed below). The advantage of this method for the passivation process is that no additional process is needed and all the required processes are needed for the manufacturing of cells anyway.

The next step is to apply the front and back electrodes (contacts). For the front contact, narrow strips of silver paste is screen printed and then dried. For the rear contact, first the surface is almost entirely covered with a thick layer of aluminum paste. Some small areas are left uncovered. These areas are covered with silver paste in the next step. They are later used as soldering pads to interconnect cells to form modules. Finally, cells go through the firing process where they are exposed to direct fire at the temperature of about 900°C (about 1650°F). This process serves several purposes. The paste for the contacts hardens and diffuses to the silicon wafer by fire to create good contact. The hydrogen passivation process is also completed. Another function of the firing process is to create the **back surface field**. To explain this field, we need to know that aluminum can act as the dopant to form a p-type silicon semiconductor. When aluminum (from aluminum paste) diffuses into the rear surface of the wafer, a thin layer of p-type is formed, which is called **p^+ (p-plus) layer**. This creates a p^+-p junction, which performs like a p-n junction. This new junction enhances the performance of the cell by repelling the small number of electrons that exist in the p-type silicon (known as minority charges)[46] from reaching the rear surface of the cell and recombining with the holes there. Finally, the firing process can help to improve the performance of cells by what is called the gettering process. While the silicon used to manufacture wafers is very pure, still there are very small about of impu-

[46]The electrons in the p-type and holes in the n-type are generally referred to as the minority charges.

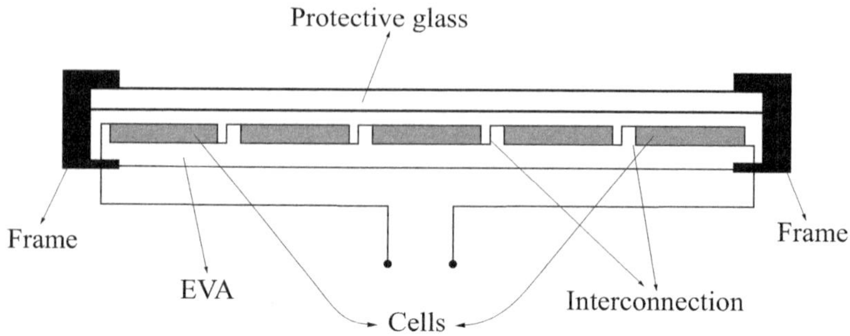

FIGURE 11.45
Schematic of a finished solar PV panel.

rity that can negatively impact the performance of cells. At the high temperature of the firing process these metal impurities can move freely in the silicon crystal structure. Since these metals have higher solubility in aluminum than silicon, some of them will diffuse to aluminum in the rear layer paste where they will not be harmful anymore. As you can see, while the firing is an energy-intensive process, it can help in several improvements of the cells.

The last stage of cell manufacturing is called the edge isolation process to eliminate any electrical contact (shortcut) between the front and rear surfaces so all electrons are forced to pass through the external load. To ensure the quality of manufactured cells, they pass through a quality control process before being sent to the module assembly line.

As noted earlier, in order to produce a meaningful about of power, cells are electrically connected to create a panel or module. This is also important for the protection of cells because individual cells are very fragile. The electrical connection of a cell string is accomplished by means of flat wires, called tabs, connecting the front contact of one cell to the rear contact of the next cell (Figure 11.45). The strings are then laminated between two sheets of ethyl vinyl acetate (ETA) on two sides. They are then further protected by a glass sheet on the front and a foil on the back. Finally, the entire unit is placed in an aluminum frame ready to be shipped to customers.

It should be noted that the explained manufacturing processes are a general description of what typically happens in the industrial manufacturing of solar PV panels; however, each manufacturer may have a different variation of some aspects of these processes.

11.2.10.2 Alternative Manufacturing Methods of Silicon Crystalline Cells

We learned in the previous section that a huge percentage of silicon ingots is lost during the ingot sawing process. There are alternative methods to produce silicon wafers that can avoid this process. The **ribbon silicon technology** is a technique that directly produces multi-crystalline silicon wafers at the desired thickness (so no need for sawing) by pulling a sheet (ribbon) of silicon from a silicon melt. There are several mechanisms to form ribbon silicon. In the edge-defined film-fed growth (EFG) technology molten silicon is pulled through an octagonal graphite mold or shaper. The produced silicon is in the shape of a long thin-walled octagonal shell with no free edge. This shell can be cut by laser to a desirable size. In the **string ribbon technology**, the ribbon is formed between two hot wires while being pulled out of the molten silicon crucible. In the **ribbon growth on substrate** (RGS) process, a sheet of silicon is formed on a substrate. Due to high rate of cooling, the solidification process is fast so a high rate of production can be achieved. The separation of silicon ribbon and the substrate takes place because of the different rate of contraction of the two surfaces.

In all these techniques, since almost the entire highly pure silicon is converted to silicon wafers, the cost can be significantly reduced. Some of these methods are already cost competitive with traditional methods; however, the quality of the produced wafers in some methods cannot compete with that of the conventional manufacturing methods and the efficiency of final PV modules is lower. Also, their widespread adaptation to the capital-intensive PV cell manufacturing industry may take a while.

11.2.10.3 Concluding Remarks on Crystalline Silicon PV Units and Their Future

Now that we learned about the difference in manufacturing processes for monocrystalline and polycrystalline silicon cells, we can briefly discuss why monocrystalline cells are more efficient. We learned that while monocrystalline cells have one single silicon crystal in their structure, polycrystalline cells are made of many crystals. This means that in polycrystalline cells there are many grain boundaries in the structure between crystals at various orientations. These defects can obstruct the flow of electrons and holes and provide sites for pair recombination. This is the main reason for poorer performance and lower efficiency of polycrystalline silicon PV cells compared to those for monocrystalline PV cells. Also, this is why larger crystalline sizes are more desirable in polycrystalline silicon wafers.

An important take away from our discussion on the manufacturing of silicon solar cells is that it involves several very energy-intensive processes, which means a lot of energy is consumed in the manufacturing of these cells. Therefore, it is a legitimate question to ask if produced cells can produce enough electricity in their lifetime to make up for the energy consumed during their manufacturing and if so, how many years this may take. The answer to this question leads us to the concept of the **energy payback period**, which is the number of years that an electricity generation system must operate to generate energy that was used to fabricate the system. For crystalline silicon solar units, of course, this time is a function of solar light intensity in the location of the system and actual manufacturing processes. The energy payback time for current crystalline silicon solar units is about 13 years. So considering the long life of solar modules (between 25 and 30 years) and the fact that their performance slightly degrades with time, electricity they can generate in their lifetime is many times more than the energy consumed in their manufacturing.

While the best efficiency of silicon-based cells in laboratory environments and commercial modules were about 17% and 10% in the early 1980s, respectively, these efficiencies for current units are 20% and 26%, respectively. As impressive as this improvement is, the current efficiencies have been stagnant for a long time and there is not much hope for further improvements since the best efficiency is very close to the maximum theoretical efficiency possible. So while there is room for some improvements in some aspects of manufacturing processes, no major breakthrough is expected. This is why new technologies are being researched and developed.

Overall it seems that considering the very well-established manufacturing infrastructure of the solar crystalline silicon technology, these units will continue their current dominance at least in near future. Currently, the lower cost of polycrystalline silicon units can make up for their inferior performance. So in most projects for a given capital investment, it economically makes sense to install more units of less efficient but inexpensive polycrystalline silicon units. However, this may change in the future if for example, the share of the cost of land in the total cost of installed systems increases dramatically so it is more profitable to have more efficient units for a given land area. Similarly, a change in the price of energy used in the manufacturing of units may alter the current balance.

11.2.11 Second Generation: Thin Film Solar PV Cells

Previously we learned that solar cells made of indirect (extrinsic or doped) semiconductors, such as crystalline silicon cells, need the thickness of a couple of 100 μm to be able to absorb most of incoming solar energy. So from the manufacturing point of view, at least in terms of the cost of raw materials, it is desirable to have thinner wafers. However, practically there is a limit on how thin these cells can be. This limitation leads us to investigate solar cells made of direct (intrinsic) semiconductors with the band gap of 1.4–1.8 eV. Direct semiconductor cells can achieve similar absorption as that in crystalline silicon cells with a much thinner absorber (typically <5 μm compared to about 200 μm in crystalline silicon cells). These cells are generally referred to as **thin film solar cells** or **second-generation** solar cells (2G).

In these cells, thin layers (or films) of semiconductor materials are deposited over a substrate. This substrate is commonly made of glass to achieve the best performance but can also be made of plastic or metal. Due to their thinness, depending on the substrate, these PV cells can be very light and flexible. Moreover, their performance is not heavily impacted by increase in the operating temperature. They also demonstrate better performance in low light intensity and indirect (diffused) light, such as light on cloudy days, than that in conventional solar cells. Their manufacturing processes consume much less material compared to that of crystalline silicon and are much less energy-intensive. That means their energy payback period is very short (less than a year). However, generally speaking, their efficiency is lower than that of conventional solar cells. While some thin film solar cells have achieved an efficiency comparable to or higher than that of polycrystalline silicon cells, their high efficiency has not been reflected in commercial cells yet.

While they can be used in any application that crystalline silicon modules are being used, they are particularly suitable for building integrated PV units (BIPV), semi-transparent cells, and applications where flexibility is important. They have been even proposed for wearable solar systems. While about a quarter of the global market was for thin film solar units in the mid-1990s, currently that share is <10%. This reduction is primarily due to the significant reduction in the cost of crystalline silicon modules in the past two decades.

The most common types of thin film solar cells are

- cadmium telluride (CdTe)
- copper indium gallium selenide (CIGS)
- amorphous silicon (a-Si)
- Gallium arsenide (GaAs)

In the following sections, we will review these cells.

11.2.11.1 Cadmium Telluride (CdTe) Solar PV Cells

Cadmium telluride is a direct II–VI compound semiconductor with a band gap of about 1.5, which is very close to the optimum band gap to absorb maximum energy from the solar spectrum. It is very stable and can operate for years without any meaningful degradation. Their efficiency is not significantly impacted by high operating temperatures, low light intensity, and diffused light. The manufacturing cost of CdTe cells is currently less than that of conventional PV cells due to lower material consumption, less energy-intensive processes, and manufacturing simplicity. Some very large solar farms used this type of module, such as the Desert Sunlight Solar Farm and the Topaz Solar Farm both in California, with 550 MW power generation capacity each.

Early cadmium telluride cells were developed in the 1970s and since then their performance and efficiency have continuously improved. Currently, their cell laboratory condition efficiency is about 22% exceeding that of polycrystalline silicon cells. However, the efficiency of commercial modules is much lower at about 11%.[47,48] While this type of solar cell is the most popular non-silicon PV cell type, its market share at about 6% is still very small compared to that of conventional silicon cells.[49]

[47] He, S., et al. "Study of CdTe/ZnTe composite absorbing layer deposited by pulsed laser deposition for CdS/CdTe solar cell." *Materials Science in Semiconductor Processing* 67 (2017): 41–45.

[48] Shubbak, M.H. "Advances in solar photovoltaics: Technology review and patent trends." *Renewable and Sustainable Energy Reviews* 115 (2019): 109–383.

[49] Sinha, T., D. Lilhare, and A. Khare. "A review on the improvement in performance of CdTe/CdS thin-film solar cells through optimization of structural parameters." *Journal of Materials Science* 54 (2019): 1–17.

One of the problems with CdTe units is that tellurium is a very rare element with very limited annual production, which can impose severe limits on widespread applications of this technology. The discovery of new sources of tellurium and recycling of old CdTe panels might to some extent address this problem.

Another problem typically associated with this type of PV cells is the toxicity of the materials used in their fabrication. Both cadmium and tellurium are highly toxic, particularly cadmium, which is a heavy metal. However, crystalline cadmium telluride used to manufacture solar units is much less toxic and is stable. Thus, the operation of these units poses little risk though safety considerations must be carefully integrated to manufacturing processes. End-of-life recycling of CdTe modules is currently a standard practice both for the recovery and reuse of valuable materials and for safety reasons. One may even argue that using cadmium in the manufacturing of solar units might be beneficial for the environment because cadmium is a byproduct of zinc production and if it is not used, it can be landfilled and can pollute the environment.

Unlike conventional solar cells that both p- and n-type semiconductors are made of the same crystalline silicon doped with different materials, CdTe cells are made of layers of dissimilar p- and n-type semiconductors. The p-n junction in these cases is referred to as **homojunction** and **heterojunction**, respectively.

The first layer (the surface that light strikes first) is a transparent superstrate commonly made of glass (Figure 11.46). The next layer is the front contact (electrode), which collects electrons. Since the incoming light needs to pass through this layer, it is referred to as a transparent conducting oxide (TCO). This layer is typically made of tin oxide (SnO_2), which replaces metal contacts in conventional cells. Then, there is a very thin layer (in the order of 0.1 μm) of doped cadmium sulfide (CdS), which forms the n-type semiconductor layer. To form the p-n junction, a layer of doped CdTe with a thickness of about 5 μm is added as the p-type semiconductor. The rear contact layer is then overlaid to collect positive charges. Finally, to protect the module from environmental elements, it is laminated and encapsulated.

The manufacturing processes used in the fabrication of CdTe are relatively inexpensive. For this reason along with low material and energy consumptions, CdTe cells are one of the least expensive PV modules in the market. The manufacturing processes generally involve various methods that deposit

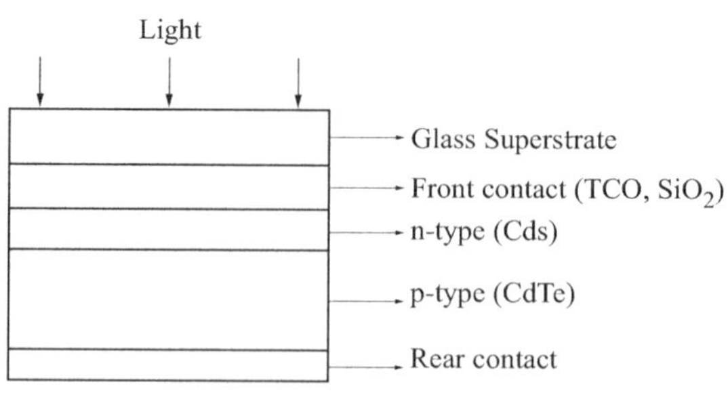

FIGURE 11.46
Schematic of a CdTe cell with various layers.

a thin layer of a substance on a surface, such as close-spaced sublimation, physical- or chemical vapor deposition, or sputter deposition.

11.2.11.2 Copper Indium Gallium Selenide (CIGS) Solar PV Cells

This type of thin film solar cells is a solid solution[50] of copper, indium, gallium, and selenium (diselenide), often referred to as **CIGS** solar cells. It is a I–III–VI semiconductor. Both indium and gallium are from the group III elements in the periodic table. The relative amount of indium and gallium in the solution determines the band gap of the semiconductor and can be used to control its precise value. The composition can vary from pure copper indium selenide ($CuInSe_2$) to pure copper gallium selenide ($CuGaSe_2$) with the band gap of about 1 and 1.7 eV, respectively, and any mixture in between. This is why these cells are sometimes referred to as CI(G)S or CIS. Sometimes even indium and gallium are replaced by aluminum (another group III element). Also, a compound of sulfur (disulfide), which is a group VI element (similar to selenium), can be added to form what is called CIGSS. Similar to CdTe, CIGS is a direct semiconductor with very high light absorption capability. With a thin layer of just a couple of microns, it can nearly absorb the entire solar light spectrum. This thinness of cells reduces the consumption of materials and energy required for the manufacturing processes. It also allows the manufacturing of flexible cells if the used substrate is flexible. CIGS cells have superior performance compared to crystalline silicon cells when exposed to indirect light. They are ideal for integration into building structures.

The development of CIGS cells started in the 1970s and their performance has improved over the following years. Their cell efficiency in the laboratory conditions is about 23%, but the efficiency of commercial modules is lower at about 15%. However, similar to CaTe cells, there are some concerns about the toxicity of some of the elements used in their manufacturing and the scarcity of indium. Recently, the share of CIGS units has been about 2% of the global PV market.

Similar to CdTe cells, CIGS cells have heterojunction and are made of several layers deposited separately. However, unlike CdTe cells, CIGS cells have a substrate made of glass, plastic, or metal. For cells that flexibility is not desired, typically soda-lime glass with a thickness of a couple of millimeters is used as the supporting substrate. A critical characteristic of the suitable substrate is the compatibility of its thermal expansion coefficient with other cell layers. A thin layer (about 1 μm) of molybdenum is added on the substrate as the rear contact (Figure 11.47). This layer is not only conductive enough to collect positive charges but also is a reflective layer that reflects unabsorbed light back into the cell. On top of that, the CIGS layer with the thickness of 2 to 4 μm is deposited as the polycrystalline p-type semiconductor. All elements can be deposited either at once in a single step (called co-evaporation) or in a two-step or even three-step process. To form the p-n junction, a very thin

[50] In a solid solution several solid elements, typically metals, are mixed, but they do not chemically react and atoms are present in their original forms.

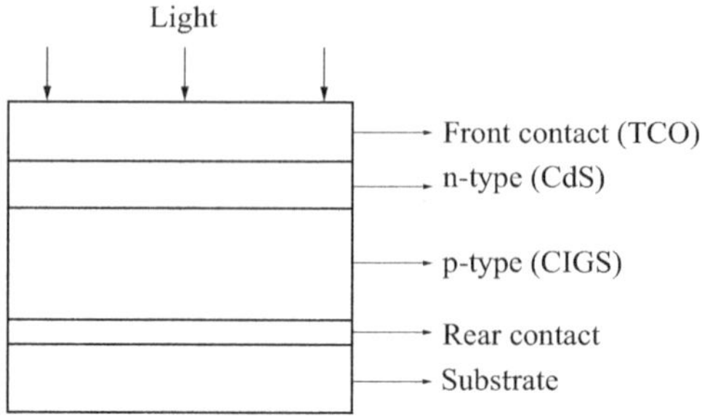

FIGURE 11.47
Schematic of a CIGS cell with various layers.

layer of cadmium sulfide (CdS) is deposited as the n-type semiconductor with a thickness of about 0.05 μm. The front contact made of aluminum doped zinc oxide is then overlaid, which later is covered with an antireflection layer. Both the CdS layer and the front contact should be transparent to allow light passage to the p-n junction. Finally, the module is laminated and encapsulated ready to be shipped to the installation site. Note that unlike crystalline silicon modules that are manufactured as individual cells and then connected in series to form a panel, thin layer units can be made in a single piece. For this reason, there are several laser-scribing processes between the above steps to internally form individual cells connected in series. A variety of manufacturing processes have been used to deposit different layers by manufacturers including co-evaporation, precursor reaction processes, various sputtering methods, screen-printing, spray coating, and electrodeposition. It should be noticed that some manufacturing processes may require some extra steps.

11.2.11.3 Amorphous Silicon (a-Si)

The third type of thin film solar cells is amorphous silicon cells. As the name implies, like crystalline silicon cells (sometimes referred to as **c-Si**), they are made of silicon with a non-crystalline structure (i.e., in small grains) with band gap of about 1.7 eV. Therefore, unlike other common thin film cells, they do not have the problem of using scarce and sometimes toxic materials. Since amorphous silicon is a direct semiconductor, it can be deposited in very thin layers of a few microns. Thus, the consumption of silicon is just a couple of percent of what is needed for conventional silicon cells. The manufacturing processes also consume much less energy. All these attributes result in very inexpensive and environmentally friendly cells. However, with the cell efficiency of about 14% in a laboratory environment and commercial unit efficiency of about 7%, their efficiency is much lower compared to other technologies. They also suffer from low stability during their initial operation. During the initial few months of their operation (roughly about the first 1000 hours of operation), their efficiency reduces by about 20% and then stabilizes. The reported performance data and module ratings are typically

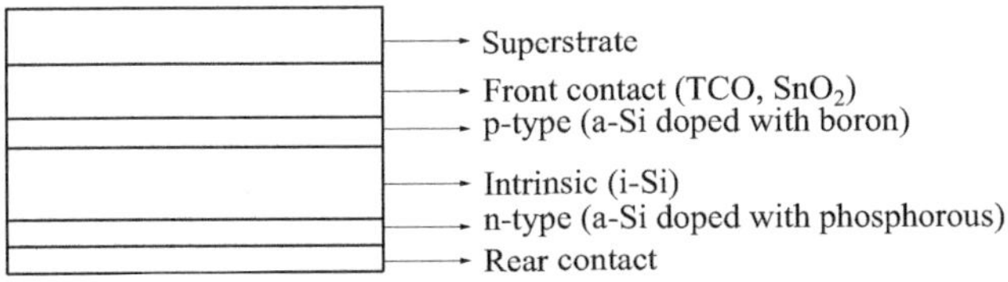

FIGURE 11.48
Schematic of an amorphous silicon cell with various layers.

for after this initial period and reflect the stable operation. One reason for this rapid degradation is that in amorphous silicon atoms some the valence electrons are not paired creating what is called dangling bonds. These bonds degrade the performance of cells. Similar to the passivation process explained earlier, hydrogen atoms can be used to reduce this problem.

This type of cell has been used since the 1970s for powering small electronic devices such as handheld calculators. In recent years, a-Si cells have had about 2% of the global solar PV market. Unlike CdTe and CIGS cells, a-Si cells are homojunction similar to c-Si cells. However, instead of p-n junction as in conventional silicon PV cells, a-Si cells have a p-i-n structure. In this case, an intrinsic (undoped) a-Si layer (i-Si) is sandwiched between very thin layers of p- and n-type doped a-Si and the electric field is created in the i-Si layer (Figure 11.48). To manufacture this type of cell, a transparent conducting oxide (TCO) layer, typically tin oxide (SnO_2), is deposited on the superstrate as the front contact. While a glass superstrate can provide the best performance, due to typically low-temperature manufacturing processes, plastic superstrates can be also used to make flexible cells. Then, several layers of silicon are deposited. Since there is no opportunity for the growth of crystalline grains, the deposited layers are amorphous silicon. The first layer, immediately after the TCO, is doped with boron to form the p-type layer. This layer is followed by an undoped silicon layer (intrinsic Si) and then the phosphorus doped n-type silicon layer. Finally, the rear contact is deposited, which also functions as the reflective surface. Various chemical vapor deposition (CVD) or sputtering methods have been used to deposit different layers.

11.2.11.4 Gallium Arsenide (GaAs)

Another type of thin film solar cell is made of gallium arsenide, which is a III–V direct compound semiconductor with a crystalline structure similar to c-Si. The efficiency of this type of PV technology far exceeds even monocrystalline silicon modules. With an efficiency of about 29%, they are the most efficient single-junction solar cells ever made. However, their manufacturing costs are very high and they only suit applications where efficiency and low weight are much more important factors than cost. For this reason and their resistance against space radiations, since the 1990s they have been used for space applications. They are also used in multi-junction solar cells.

11.2.12 Third Generation: Emerging Solar Cell Technologies

The third-generation solar cells are a combination of a wide variety of technologies that are currently under investigation. Some of them, such as multi-junction solar cells and concentrated solar cells, are extensions of the second-generation technologies, while some are based on entirely new concepts, such as organic solar cells, perovskite solar cells, dye-sensitized solar cells, and quantum dot solar cells. In this section, these emerging technologies will be briefly reviewed.

11.2.12.1 Multi-Junction Solar Cells

Previously we learned how the band gap of semiconductors has an opposing impact on which wavelength of photons can be absorbed and what portion of the absorbed photons can be converted to electricity. We also noticed that for the solar spectrum, there is an optimum band gap that can provide maximum efficiency for a single p-n junction cell. **Multi-junction solar cells** (also referred to as **tandem** or **cascade** cells), as the name implies, can help us to overcome this theoretical limitation by stacking several layers of p-n junctions. The technologies developed for thin film solar cells allow us to place several very thin layers of transparent semiconductors in a way that light can pass through them. Each p-n junction in the stack is made of different semiconductor materials and is designed to optimally absorb a certain range of the solar spectrum. While several technologies can be used to create the stack of p-n junctions, including organic solar cells (will be discussed later in this chapter), multi-junction solar cells typically refer to those cells made by thin film cell technologies. Currently, multi-junction solar cells are very expensive and are only used commercially in space applications. Some of the most high-profile space missions are using this type of cells including Mars rovers, Spirit and Opportunity, and Juno spacecraft exploring Jupiter.

In a multi-junction solar cell, the first layer is the one that absorbs the more energetic photons (with lower frequencies) in the blue portion of the spectrum by the semiconductor with the largest band gap. The photons with an energy level lower than the band gap of this semiconductor pass through mostly intact. This process continues with the next layers of semiconductors with the band gap lower than the previous layer but higher than the next layer to absorb the yellow and red/infra-red portions of the spectrum. There are various semiconductors proposed and used for different layers of multi-junction solar cells. One of these combinations is gallium indium phosphide ($GaInP_2$), gallium arsenide (GaAs), and germanium (Ge) (ordered from top to bottom) with the band gaps of about 1.9, 1.4, and 0.7 eV, respectively. This three-junction combination has achieved an efficiency of about 38% with non-concentrated solar radiation. The highest efficiency that has ever been achieved for a solar cell operating under non-concentrated solar radiation has been 39% for a four-junction cell.

The manufacturing of multi-junction solar cells is very challenging and expensive. Particularly, removing the generated current from each layer is a major challenge. There are two design concepts for stacking various layers. In one design the layers are deposited on top of each other so the middle layer functions as an n-type on one surface and p-type on the other surface. In this design, referred to as the **monolithic design**, the cell behaves similar to two cells in series. The challenge, in this case, is to match the generated current in various layers (recall that in cells connected in series, the overall current is determined by the lowest current in the string). Another design concept is to make the cell individually and then mechanically connect them. This design, sometimes referred to as **mechanical stacking**, does not face the problem of current mismatch but requires transparent contacts.

As noted above, multi-junction solar cells are very expensive to manufacture (per unit of area). One way to reduce the impact of this problem is to concentrate light before it strikes the cell surface. This way a cell with a small surface can generate much more than what it can produce with non-concentrated light. This leads us to our next topic.

11.2.12.2 Concentrating PV (CPV) Units

Exposing a solar cell to concentrated solar radiation can help to improve the performance of solar PV systems by increasing the power output per unit of area of the cell and increasing its efficiency. At the same temperature, the cell efficiency is much higher with concentrated light than that with non-concentrated light. Concentrating devices can be integrated into high-efficiency solar cells such as multi-junction cells. The highest efficiency ever achieved for a solar cell is 46%, which is recorded in a four-junction cell illuminated with concentrated light. Light concentration can be achieved by various designs of mirrors, such as dish mirrors, parabolic mirrors, or heliostat arrays (refer to Chapter 12), or lenses, such as array of small lenses or trough Fresnel lenses.[51] The concentration ratio of 500 (500 Sun) is not unusual. Based on their concentration ratio, these units can be divided into low concentration PVs (LCPV), medium concentration PVs, and high concentration PVs (HCPV). Note that by using concentrated light while the required area of the cell substantially reduces, the area of the concentrating device that is exposed to the incoming light remains unchanged compared to non-concentrating cells.

These systems have some major drawbacks other than their expensive price tag. When exposed to concentrated light, the temperature of PV cells can increase, which negatively impacts their performance. The impact on the cell performance depends on the magnitude of temperature increase, which is a function of the concentration ratio and the type of the cell. While low concentration ratio systems may be operated without cooling, for higher concentration ratios an active or passive cooling system is required that adds to the overall cost of the system. Another limitation is that in order to operate properly, these units must be always directed toward the Sun, which typically

[51] Fresnel lenses concentrate light gradually by thin glasses rather than one-step thick lens.

requires a one-axis solar tracking system for LCPVs or a two-axis solar tracking system for HCPVs. Furthermore, unlike non-concentrated units that can operate with both direct and diffused light, CPV units can only operate with direct sun radiation. Therefore, they are only effective for regions with mostly clear skies with high solar irradiance. For regions with low solar intensity, many cloudy days, or high air pollution, non-concentrated PV units may perform better than concentrated ones. Also, the high impact of optical losses means that they require more frequent cleaning to remove dust from the module surface. As a result of these problems, concentrated solar units have a very small share of the market at about 0.1%–0.2%.

Another mechanism to enhance the performance of solar cells is by using bifacial cells. While it is not a light concentration method, in this method the amount of light that a cell is exposed to increases. In these cells, the back layer of cells is transparent, similar to the front layers so they can produce power regardless of which side is receiving the incoming light. Mirrors can be used to reflect solar radiation to the back of these cells.

11.2.12.3 Organic Solar Cells

Organic solar cells, as the name implies, are made of organic materials (i.e., carbon-based compounds). In this case, materials are primarily long-chain polymers (plastics) that function as semiconductors. This technology has many advantages over other cell types, including very low energy and material consumption for manufacturing processes (i.e., very short energy payback period), very low negative environmental impacts, no scarcity or toxicity of materials, very high light absorption, lightweight, easy end of end-of-life disposal, and flexible structure. However, they suffer from low efficiency and stability. Their efficiency in the laboratory conditions is about 12%, and the efficiency of early commercial units is merely about 6%. These cells also age rapidly in a span of just a few years compared to a few decades for conventional technologies. While their manufacturing processes have not matured yet, there are great potentials for fast and inexpensive manufacturing processes for these solar cells. The major advantage of this technology is that it can be made from a solution, which makes it possible to manufacture them with inexpensive printing or coating methods on plastic substrates.

This technology has not been commercialized yet and still is in the basic research stage. However, there have been great progresses on all fronts, which have improved their efficiency, lifetime stability, and manufacturing methods. This technology is one of the most promising technologies for the future of the solar PV industry.

11.2.12.4 Perovskite Solar Cells

Perovskite solar cells (PSC) are made of materials with a perovskite structure. Perovskite materials are a group of crystallines with the general formula of ABX_3 (A, B, and X are elemental atoms or compound molecules where A has a much larger size than that of B). This is a new technology with impressive progress in the past decade. The efficiency of these cells in laboratory conditions is currently about 22%, which indicates an about 6 fold increase in less than a decade. Several perovskite crystallines have been used as the absorber in these cells, the most common of which is methylammonium lead trihalide, $CH_3NH_3PbX_3$ where X is a halogen element and can be iodine (I), bromine (Br), or chlorine (Cl).[52] This perovskite crystalline can be inexpensively manufactured. One of the major advantages of these cells is that their band gap can be controlled by the halogen atoms in the crystalline structure. The presence of lead (Pb) atom in the structure, which is a hazardous heavy metal, is one of the concerns for this specific crystalline. One of the main promises of this technology is that it can be manufactured by relatively simple and inexpensive printing or coating processes from a solution, similar to organic solar cells. However, perovskite solar cells suffer from a short lifetime and low stability under outdoor operating conditions (again similar to organic solar cells). While this technology has not been commercialized yet, it has a great potential to make a significant impact on the industry and currently many research groups are working to improve its performance.

11.2.12.5 Dye-Sensitized Solar Cells

Dye-sensitized solar cells (DSSC), sometimes referred to as Grätzel cells after its co-inventor Michael Grätzel (1944, a Swiss chemist) who invented the technology along with Brian C. O'Regan (an American chemist), can be made as translucent and flexible cells. These attributes open many new applications, particularly in building integrated photovoltaics (BIPV). The best efficiency they have achieved is about 15%. They can be manufactured using printing processes. While most of the materials needed in their manufacturing are inexpensive, they still need a small amount of platinum and ruthenium. They currently suffer from a short lifetime and involvement of some hazardous materials in their manufacturing processes. While they are far from commercialization, the progress in the field are expected to continue. This is another promising future technology for solar cells.

11.2.12.6 Quantum Dot Solar Cells

The quantum dot technology has been developed based on recent advancements in nanotechnology (based on the principles of quantum mechanics). Quantum dots are nanoparticle semiconductors. Unlike bulk semiconductors whose band gap is only determined by the material type, in quantum dots the band gap depends on both the material and the size of particles. This allows adjusting the band gap by changing the size of the particles, which provides flexibility to develop appropriate materials to fabricate solar cells. Some of the

---

[52] Al-Amri, A.M., B. Cheng, and J.-H. He. "Perovskite methylammonium lead trihalide heterostructures: Progress and challenges." *IEEE Transactions on Nanotechnology* 18 (2018): 1–12.

previously discussed semiconductors, such as cadmium telluride (CdTe) and cadmium sulfide (CdS), have been successfully utilized to make quantum dots. They have been used to manufacture multi-junction solar cells. While currently the highest efficiency they have achieved is about 13%, theoretically they can achieve extremely high efficiency with inexpensive fast printing manufacturing methods. While far away from commercialization, they can provide a pathway for future highly efficient affordable solar PV units.

11.2.12.7 Hybrid PV Systems

Solar PV units can be combined with a variety of other power generation systems to improve their performance. They can be integrated to other electricity generation systems to provide smoother power production. In off-grid applications, this can result in no electricity storage unit (i.e., battery bank) or smaller required storage capacity. They can be combined with diesel generators, fuel cells, or hydro turbines so these units can provide the required power when PV units are not operational. A hybrid solar PV array and wind turbine system can be used so they can complement each other.

Alternatively, solar PV units can be combined with other thermal units to use the solar thermal energy absorbed by solar arrays. These systems are referred to as **hybrid solar photovoltaic/thermal (PV/T)** units. While providing thermal energy, they also help to cool PV modules to improve their efficiency and reduce their rate of degradation. The produced thermal energy can be utilized for space heating or cooling, heat pumps, water desalination units, thermoelectric units, or solar greenhouses.

11.2.12.8 Comparison of Various Solar Cell Technologies and Their Future

Figure 11.49 clearly illustrates the efficiency of various solar cell technologies between 1975 and 2018. Please try to follow the trends in the improvement of the efficiency of each type that we have learned about. The three generations of solar cells along with multi-junction cells are distinguished with different colors. Also, notice the difference between the efficiencies when cells are exposed to concentrated or non-concentrated light and the fact that for the same technology, the former always has a higher efficiency. The diagrams indicate that generally speaking, the conventional cell technologies experienced a significant and continuous increase in their efficiency in the last quarter of the 20th century and then their efficiencies have been primarily flat. Many new technologies in the past decade have experienced similarly steep increase in their efficiency. This growth is expected to continue until these technologies are mature. However, this diagram does not show whether these improvements are accompanied by increase in the cost of systems. This is a very important factor that can determine the economic viability of each technology. Regardless of impressive improvements in new technologies, it is expected that in the short-term future, crystalline silicon units will continue to dominate the market. However, new technologies will gradually gain some

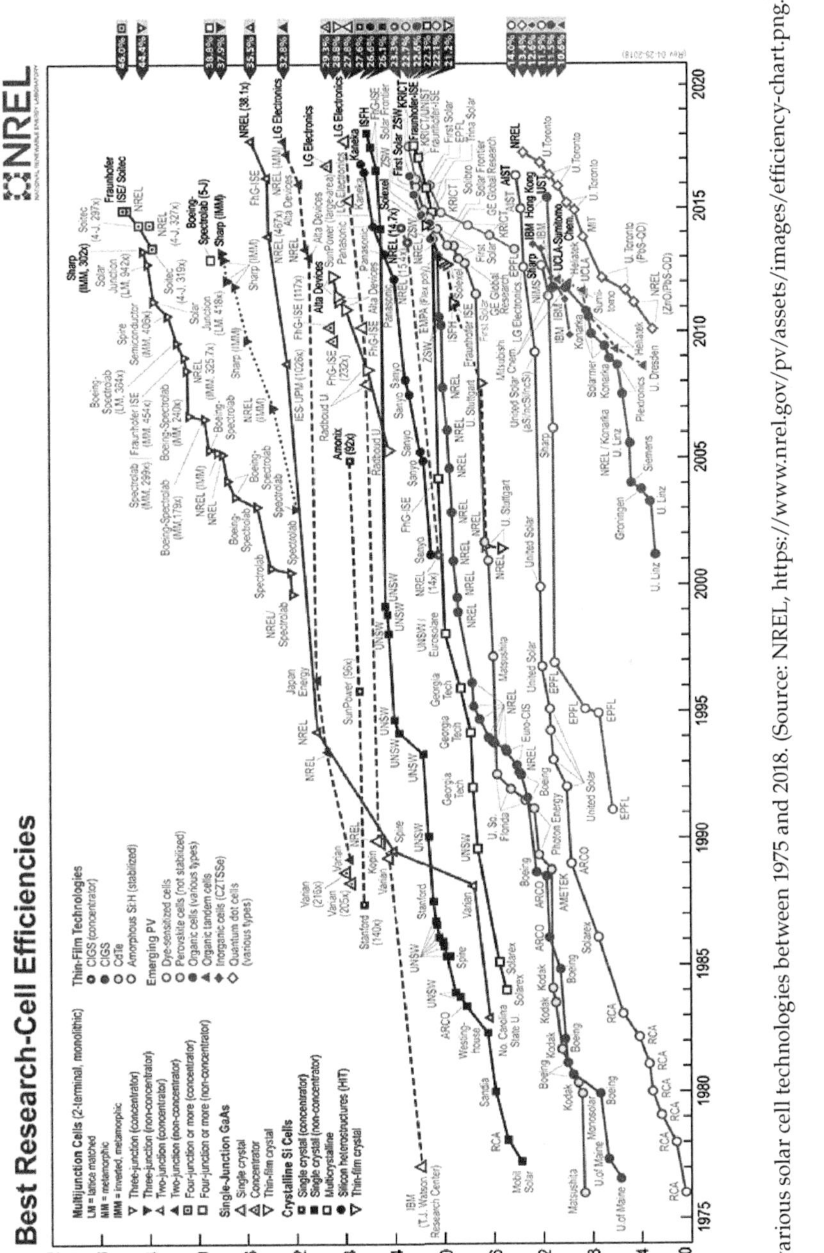

FIGURE 11.49
Efficiency of various solar cell technologies between 1975 and 2018. (Source: NREL, https://www.nrel.gov/pv/assets/images/efficiency-chart.png.)

small share of the market until one or several technologies can prove their technical viability and economic parity with traditional technologies.

11.3 Extra Learning Resources

11.3.1 Nature of Solar Energy

- Seeing Equinoxes and Solstices from Space (a video from NASA) https://www.youtube.com/watch?v=FmCJqykN2J0
- You can find the magnetic declination of your location at http://www.magnetic-declination.com/
- You can find a lot of useful information about the tilt angle of solar devices at https://www.solarpaneltilt.com/.
- You can find a lot of useful information about various aspects of solar energy at https://www.builditsolar.com/.
- In the U.S., National Renewable Energy Laboratory (NREL) has published the long-term solar data along with weather data as the typical meteorological year version 3 (TMY3) at https://rredc.nrel.gov/solar/old_data/nsrdb/1991-2005/tmy3/. The user manual for TMY3 data sets can be accessed at https://www.nrel.gov/docs/fy08osti/43156.pdf
- NREL has published irradiance data for many cities in the U.S. in a report entitled "Solar Radiation Data Manual for Flat-Plate and Concentrating Collectors accessible at https://www.nrel.gov/docs/legosti/old/5607.pdf
- A wide variety of insolation maps are available at NREL website, https://www.nrel.gov/gis/solar.html
- You can also find interactive insolation maps at the Department of Energy (DOE) website at https://www.energy.gov/maps/solar-energy-potential

11.3.2 Solar Photovoltaic Power Generation

- Courses on various aspects of solar PV systems:
  - Solar Energy Basics (Coursera.org)[53]
  - Introduction to solar cells (Coursera.org)

[53] This course provides useful information on the structure of the industry and certification process.

 - Photovoltaic solar energy (Coursera.org)
 - Physics of silicon solar cells (Coursera.org)
 - Organic Solar Cells – Theory and Practice (Coursera.org)
 - Series of courses on Solar Energy (edx.org offered by TUDelft)
- This website contains an amazing collection of videos and photos of industrial production processes: https://**www.pveducation.org/**
- This website contains incredibly informative virtual instruments to illustrate impacts of various parameters on the performance on PV cells: https://**infinitypv.com/learn/virtual-tools**
- If you are interested in the certification processes for the PV industry, here are some resources to start with:
 - National Roofing Contractors Association (NRCA) http://www.nrca.net/
 - North American Board of Certified Energy Practitioners (NABCEP) Associate Handbook: http://www.nabcep.org/wp-content/uploads/2018/01/ASSOCIATE-HANDBOOK-1.2018.1.pdf
 - North American Board of Certified Energy Practitioners (NABCEP) Photovoltaic Associate™ (PVA™) Program: https://www.nabcep.org/certifications/associate-program/, http://www.nabcep.org/wp-content/uploads/2018/01/NABCEP-PV-Associate-JTA-9-13-17.pdf
- Some useful short YouTube videos (there are plenty more and counting. Try your luck on each topic):
 - How Solar Panels Work: https://www.youtube.com/watch?v=JxOTulMExWU&
 - How Does a Transistor Work? https://www.youtube.com/watch?v=IcrBqCFLHIY&feature=youtu.be&t=1m0s
 - How It's Made – Solar Panels: https://www.youtube.com/watch?v=qZgWC-Cxd44
 - Wafer manufacturing process: https://www.youtube.com/watch?v=3TOpg1niATg
 - Solar cell manufacturing and solar panel production by Suntech: https://www.youtube.com/watch?v=fZ1SC-vUe_I
 - Juno: Making Solar Power History: https://www.youtube.com/watch?v=ka6OERznXh4
 - How It's Made – Foldable Solar Panels: https://www.youtube.com/watch?v=Lk--SiDQU40
 - Roll and roll to roll coater – making a solar cell: https://www.youtube.com/watch?v=qe1l3kRMeeU
 - Constructing a Dye-Sensitized Solar Cell: https://www.youtube.com/watch?v=17SsOKEN5dE

11.4 Questions and Problems

11.4.1 Nature of Solar Energy

11.1. Define solar irradiance.

11.2. What is the origin of solar energy?

11.3. The energy released by the Sun for the practical purposes of solar energy utilization on Earth can be safely assumed constant.

a. True

b. False

11.4. The solar constant is a fixed physical constant.

a. True

b. False

11.5. We know that the Sun releases 3.828×10^{26} W of radiation. If the average distance between the Sun and Earth is 1.496×10^{11} m, determine the solar power received on Earth per unit of area (solar irradiance in W/m^2).

11.6. The solar irradiance outside Earth's atmosphere is measured by a satellite to be 1363.4 W/m^2. If at the moment of the measurement, the satellite is 149.5 million kilometers from the Sun, determine the total radiation released by the Sun.

11.7. Explain why the seasonal changes in weather patterns cannot be attributed to the distance that solar radiation should travel between the Sun and Earth.

11.8. Explain why it is very challenging to use solar PV arrays to power space missions in the solar system beyond Jupiter.

11.9. Earth is closest to the Sun in January and is farthest in July.

a. True

b. False

11.10. The axes of rotation of Earth around itself and around the Sun are parallel.

a. True

b. False

11.11. What is the ecliptic axis (EA)?

11.12. Show Earth's polar axis (PA), ecliptic axis (EA), ecliptic plane (EP), and inclination in the following figure.

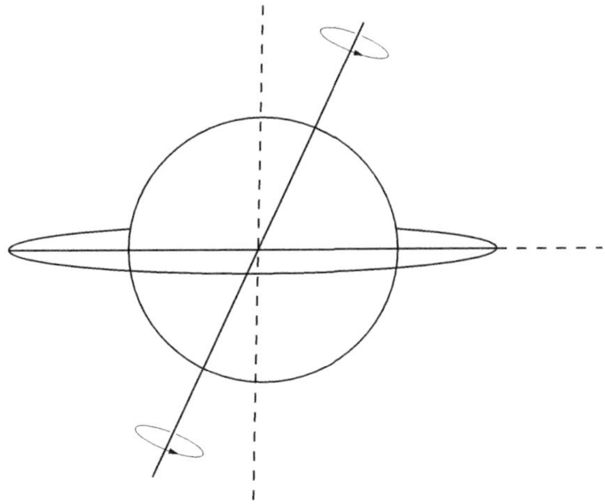

11.13. What is Earth's axial tilt? What is its numerical value?

11.14. Explain why Earth is sometimes tilted toward the Sun and sometimes titled away from it.

11.15. When is the northern hemisphere most exposed to solar radiation?

11.16. Which one is not correct about the June solstice in the northern hemisphere?

a. On this day, the northern hemisphere is most exposed to solar radiation.
b. Earth is most tilted away from the Sun.
c. It is the longest day and the shortest night of the year.
d. The angle between solar rays and the normal line to the surface of Earth (horizon) is the smallest possible.
e. It is when the highest solar irradiance can be observed.

11.17. When is Earth tilted neither toward nor away from the Sun?

11.18. If the axial tilt of Earth was 0°, ……

a. days would be always longer than nights
b. days would be always shorter than nights
c. the duration of days and nights would be varied
d. days and nights would be always 12 hours

11.19. Explain why the duration of day and night changes throughout a year.

11.20. When it is winter in the northern hemisphere, it is ….. in the southern hemisphere.

a. spring

b. fall

c. winter

d. summer

11.21. On which days are the length of day and night almost the same?

11.22. Explain why the North and South Poles are the coldest places on Earth.

11.23. The maximum solar incident occurs when the Sun is directly overhead during the …….

a. summer solstice

b. winter solstice

c. vernal equinox

d. autumnal equinox

11.24. What is the tilt angle for a solar energy receiving device? Show it in a schematic diagram.

11.25. For fixed solar energy receiving devices, the device should be always orientated toward true ….. in the northern hemisphere

a. west

b. east

c. north

d. south

11.26. What is the difference between the magnetic north and the true north? How can each one be determined?

11.27. Solar collectors are either non-concentrating or concentrating.

a. True

b. False

11.28. Define the magnetic declination.

11.29. If the objective of a solar device is to maximize the annual intercepted solar energy, the best tilt angle is equal to the latitude of the site.

a. True

b. False

11.30. For photovoltaic panels installed in a hot region where the peak load happens in summer, the tilt angle of the device should be ……. the latitude.

a. smaller than

b. larger than

c. equal to

11.31. Define the air mass coefficient or air mass (AM).

11.32. Air mass 0 or AM0 refers to the solar spectrum on the surface of Earth with irradiance equal to the solar constant.

a. True

b. False

11.33. The solar spectrum when radiation passes through the atmosphere normal to Earth's surface is …..

a. AM0

b. AM1

c. AM1.5

d. AM2

11.34. In the following figure which sun shows AM1?

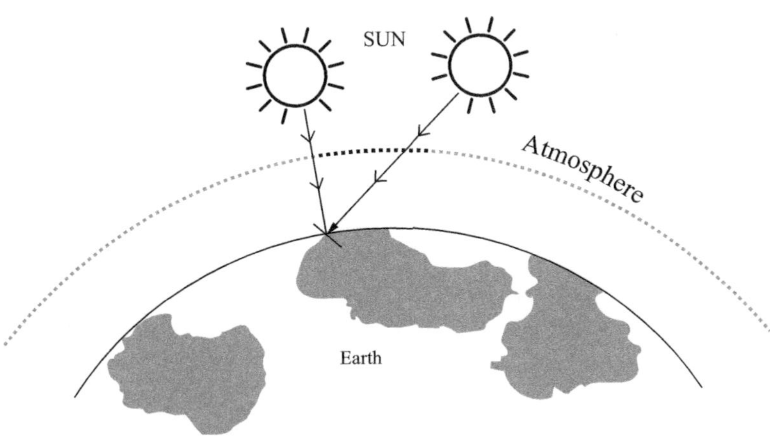

11.35. What is the air mass coefficient for the STC?

a. AM2

b. AM1.5

c. AM1

d. AM0

11.36. What is the angle of solar radiation and the line perpendicular to the surface of Earth for AM2.5?

11.37. What is direct radiation or beam radiation? How is it different from diffuse or indirect radiation?

11.38. What is global radiation?

11.39. Define the clearness index.

11.40. Explain why for pure diffuse light, the optimum tilt angle of solar energy receiving devices is when the surface is parallel to the ground (horizon) facing the sky.

11.41. Name two classes of detectors used to measure solar irradiance. Explain the working principles of each type.

11.42. What is global horizontal irradiance (GHI)? How is it measured?

11.43. What is diffuse horizontal irradiance (DHI)? How is it measured?

11.44. How can direct normal irradiance (DNI) be measured?

11.45. Using the insolation maps in this chapter, identify the regions in the U.S. that are receiving the highest and lowest irradiance.

11.46. Describe what solar energy maps or insolation maps are and how they can be used.

11.4.2 History of Solar Photovoltaic Power Generation

11.47. What makes solar photovoltaic power generation an ideal electricity production technology?

11.48. Why are solar photovoltaic cells an ideal power source for extraterrestrial applications?

11.49. The first spacecraft powered by solar cells was ....

a. Sputnik 1 – USSR
b. Explorer 1 – U.S.A.
c. Vanguard 1 – U.S.A.
d. Explorer 1 – U.S.A.

11.50. Which one is not a factor in the growth of solar photovoltaic power generation in the past two decades?

a. The reduction of manufacturing costs.
b. The increase of the cost of electricity generated by conventional fossil fuel power plants.
c. The incentives provided in federal (national) and state (regional) levels.
d. Growing environmental concerns about fossil fuel-based electricity generation.

11.51. Which country has the largest added installed capacity of photovoltaic power generation units in 2018?

11.52. Which country has the largest total installed capacity of photovoltaic power generation units in 2018?

11.53. Which country has the largest total installed capacity of photovoltaic power generation units per capita in 2018?

11.4.3 Applications of Photovoltaic Solar Cells

11.54. Explain the concept of grid parity for an electricity generation unit?

11.55. What is distributed or decentralized electricity generation?

11.56. Which one has the largest contribution to the electricity produced from photovoltaic cells in the U.S.?

a. Grid-connected units
b. Off-grid units

11.57. Which type of PV units currently has the greatest share of the market?

a. Distributed generation
b. Utility-scale generation

11.58. Which one is not a characteristic of ground-mounted photovoltaic units?

a. They can be utility-scale power generators.
b. They are easier to maintain.
c. They can be installed in otherwise unused lands.
d. They are typically at or near the load.
e. They can be equipped with sun tracking systems.

11.59. Which ones are characteristics of building-mounted photovoltaic units?

a. They can be utility-scale power generators.
b. They are easier to maintain.
c. They can be installed on otherwise unused rooftop of buildings.
d. They are typically at or near the load.
e. They can be equipped with sun tracking systems.

11.60. Which one must have a battery storage system?

a. An on-grid unit
b. An off-grid unit
c. Both a and b
d. Neither a nor b

11.61. Which one can have a battery storage system?

a. An on-grid unit
b. An off-grid unit
c. Both a and b
d. Neither a nor b

11.62. Which one needs an inverter in a residential application?

a. An on-grid unit
b. An off-grid unit
c. Both a and b
d. Neither a nor b

11.4.4 Principles of Solar Photovoltaic Cell Operation

11.63. Define the following terms.

- Energy bands
- Forbidden bands
- Valence band
- Conduction band

11.64. Which band has higher energy?

a. Conduction band
b. Valence band

11.65. Label the following diagram using energy bands, forbidden bands, conduction band, and valence band.

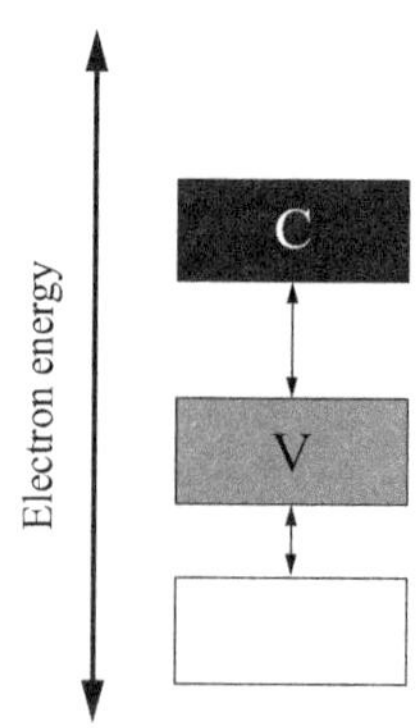

11.66. Which of the following electron energy diagrams shows conductor, semiconductor, and insulator?

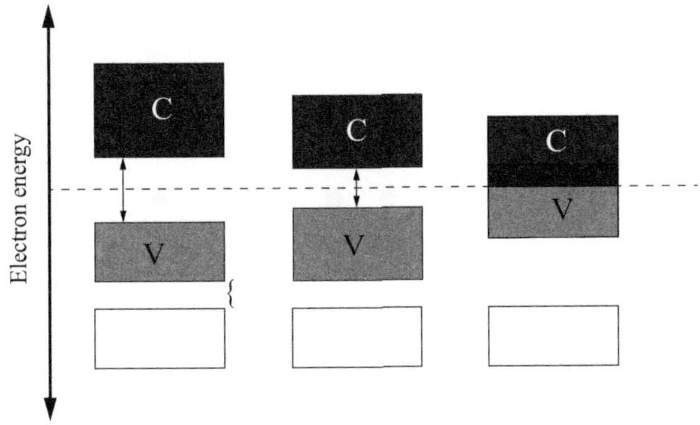

11.67. Which type of material has a large band gap between the conduction and valance bands?

a. Conductor
b. Semiconductor
c. Insulator

11.68. In which type of material the conduction and valance bands can overlap?

a. Conductor
b. Semiconductor
c. Insulator

11.69. Describe the atomic structure and conductivity of silicon using the following figure.

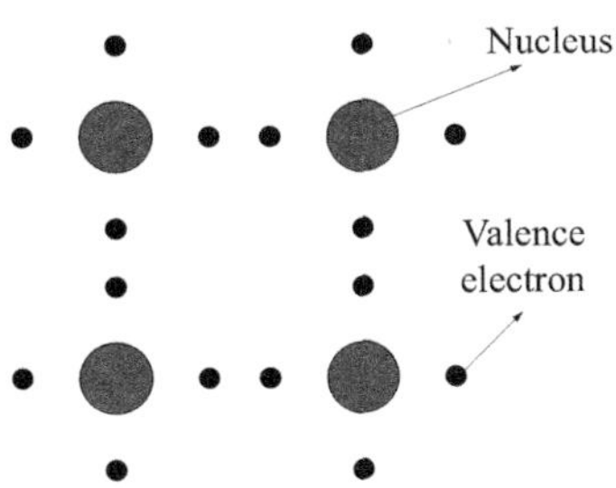

11.70. Explain the electron-hole pair generation process.

11.71. Describe the difference between direct and indirect semiconductors.

11.72. Explain how n-type and p-type semiconductors are different.

11.73. What type of semiconductors is shown in the following schematic?

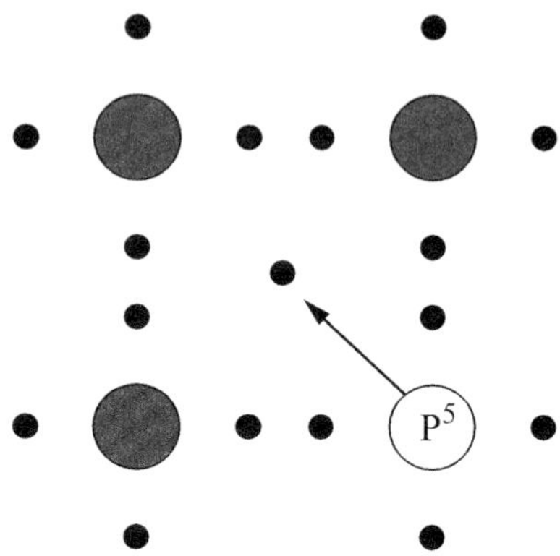

a. n-type

b. p-type

11.74. What is the typical ratio of the number of atoms of silicon to that of dopants in a doped silicon semiconductor?

a. 10

b. 10^3

c. 10^5

d. 10^6

11.75. An electron-hole pair is created on the p-n junction of a solar cell. The electron moves toward ……

a. n-side

b. p-side

c. does not move

d. can move on either side

11.76. A substance is a semiconductor if its band gap is ……

a. zero

b. very small

c. very large

d. within a certain range

11.77. Why do we need a p-n junction in a solar cell?

11.78. Explain how an electric field is formed in a p-n junction in a solar cell. What is the function of this electric field?

11.79. In the following schematic identify which side is the n-type semiconductor. Why?

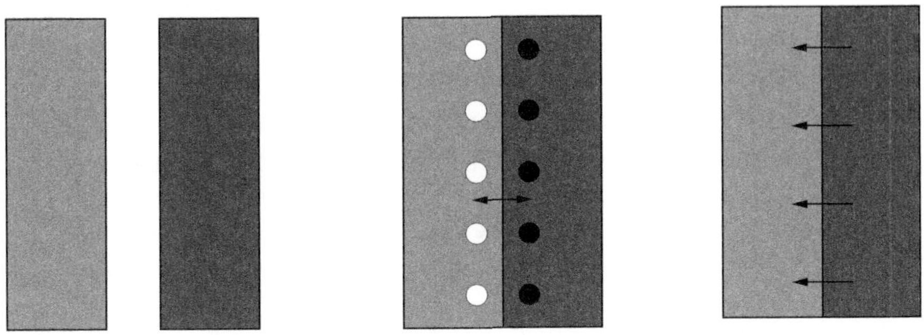

11.80. Describe the internal structure and components of a solar PV cell.

11.81. What are the functions of the two electrodes (contacts) in a solar PV cell?

11.82. Which one is thinner in a silicon PV cell, the n-type semiconductor or the p-type semiconductor? Why?

11.83. Explain what happens when sunlight strikes a solar PV cell.

11.84. Which side of a solar PV cell does light strike first?

11.85. The energy of each photon is only a function of the light ......

a. wavelength
b. intensity
c. size
d. origin

11.86. A photon from sunlight strikes the p-n junction of a solar cell. Depending on the energy of the photon and the band gap of the semiconductor, three scenarios may take place. Describe these scenarios. Which one is the best case scenario (minimal energy losses)?

11.87. Photon A has higher energy than photon B. Which one has a larger wavelength?

a. A
b. B
c. Either a or b

11.88. Explain how the band gap of a semiconductor that a solar cell is made of and the energy of the light photons impact the power production and the efficiency of the cell.

11.89. A photon strikes a solar cell, when can the cell produce maximum output?

a. When the photon energy is larger than the band gap.

b. When the photon energy is smaller than the band gap.

c. When the photon energy is equal to the band gap.

d. When the photon energy is not equal to the band gap.

11.90. A helium-neon laser beam has a wavelength of 633 nm. Determine the photon energy of solar radiation in J and eV. Determine the number of photons per second in the light if the light energy is 1 W.

11.91. In the previous problem, determine the maximum possible electric power that can be generated if a solar cell made of crystalline silicon is illuminated with this light (with the band gap of 1.1 eV). Also, determine the maximum theoretical efficiency possible in the cell.

11.92. Derive an equation to directly determine the energy of photons in eV as a function wavelength in μm.

11.93. Using the equation from the previous problem, determine the maximum electrical power (in W) that can be generated if the light with the wavelength of 0.3 and 2.5 μm strikes a solar cell made of crystalline silicon (with the band gap of 1.1 eV). Also, determine the maximum theoretical efficiency that can be achieved in the cell for each case.

11.94. Determine the wavelength of a photon that has energy equal to the band gap of crystalline silicon in μm. The band gap of crystalline silicon is 1.1 eV.

11.95. Determine the wavelength of a photon in μm that has energy equal to the optimum band gap of the semiconductor that operates under the solar spectrum (about 1.4 eV).

11.96. Determine the number of photons per second in the light with the wavelength of 2.5 μm if the light energy is 10 W.

11.97. In the previous problem, determine the maximum possible electric power that can be generated if a solar cell made of crystalline silicon (with the band gap of 1.1 eV) is illuminated with this light. Also, determine the maximum theoretical efficiency possible in the cell.

11.98. Determine the number of photons per second in the light with the wavelength of 880 nm if the light energy is 1 kW. Also, determine the maximum theoretical efficiency that can be achieved in the cell for two cases; when the cell is made of germanium (with the band gap of 0.7 eV) and gallium arsenide (with the band gap of 1.4 eV).

11.99. What are two general groups of losses in solar cells?

11.100. Why using a very low band gap semiconductor may not improve the efficiency of a solar cell while it can help to absorb photons with a wide range of wavelengths?

11.101. Why using a very large band gap semiconductor may not improve the efficiency of a solar cell while it can help to convert more energy of the photons that are absorbed?

11.102. How do the surface reflection and blockage contribute to losses in solar cells?

11.103. What are two factors that contribute to the recombination of electron and hole pairs in a solar cell?

11.104. The carrier lifetime or diffusion length of a solar cell is very strong functions of the temperature of the cell.

a. True

b. False

11.105. Why should typically the thickness of a solar cell be in the order of the diffusion length of the semiconductor?

11.106. Define the packing factor of a solar panel. Why is this factor important?

11.4.5 Electricity Generated in Photovoltaic Solar Cell

11.107. Explain how the load's resistance can influence the power generation of a PV cell.

11.108. What is current density? What is its unit? Why is it often used in creating solar cell characteristic curves?

11.109. Why is the current-voltage curve of a solar cell important? What can we learn from them (be specific on the parameters)?

11.110. What is a source measure unit (SMU)?

11.111. How can the open-circuit voltage (V_{OC}) be measured?

11.112. Explain what the short-circuit current is and how it can be measured.

11.113. Which statement is correct about the short-circuit current?

a. The current when the load is maximum.

b. The current when the voltage is maximum.

c. The current when power generation is maximum.

d. It is the maximum current of a solar cell.

11.114. In the following I-V curve for a solar cell, label points A to E.

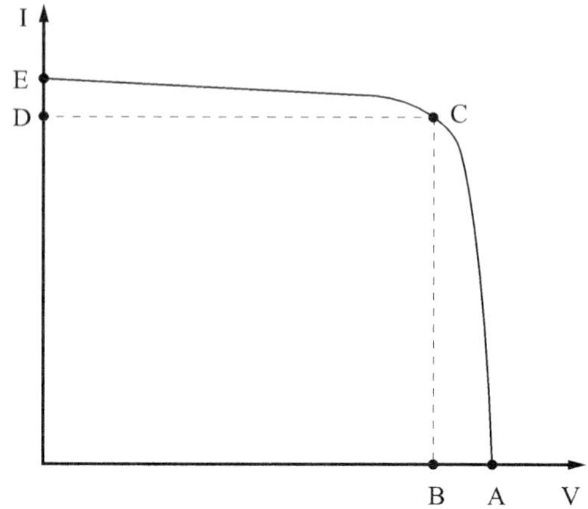

Answer the following questions, 11–115 to 11–117, using the above curve.

11.115. Which parameter is associated with point E?

a. open-circuit voltage (V_{OC})

b. short-circuit current (I_{SC})

c. current of the maximum power point (I_{mpp})

d. voltage of the maximum power point (V_{mpp})

e. maximum power point (MPP)

11.116. Which points correspond to the current and voltage of the maximum power point, respectively?

a. A and C

b. B and D

c. C and E

d. A and D

e. D and A

11.117. Which point corresponds to the short-circuit current?

a. A

b. B

c. C

d. D

e. E

11.118. What is the function of a maximum power point tracking circuit?

11.119. What are solar simulators? Why and when are they used?

11.120. The maximum power point is a state of a solar cell operation when .....

a. both the current and voltage are maximum.
b. either the current or voltage is maximum.
c. the term current × voltage (IV.is maximum.
d. the term current/voltage (I/V) is maximum.

11.121. Explain why in the power-voltage curve the generated power varies nearly linearly between $V = 0$ and $V = V_{MPP}$.

11.122. Which statements are not correct?

a. The power generation at I_{SC} and V_{OC} is zero.
b. I_{SC} is the maximum possible current that a solar cell can achieve.
c. The maximum power that a solar cell can produce is equal to $I_{SC} \times V_{OC}$.
d. The maximum power that a solar cell can generate is either at I_{SC} or V_{OC}.

11.123. The following diagram shows the I-V curve of a solar photovoltaic system at an operating temperature of 45°C. Determine the following parameters: the short-circuit current (A), the open-circuit voltage (V), the maximum power generation (W), the current of the maximum power point (A), and the voltage of the maximum power point (V). Also, draw the generated power curve in the diagram.

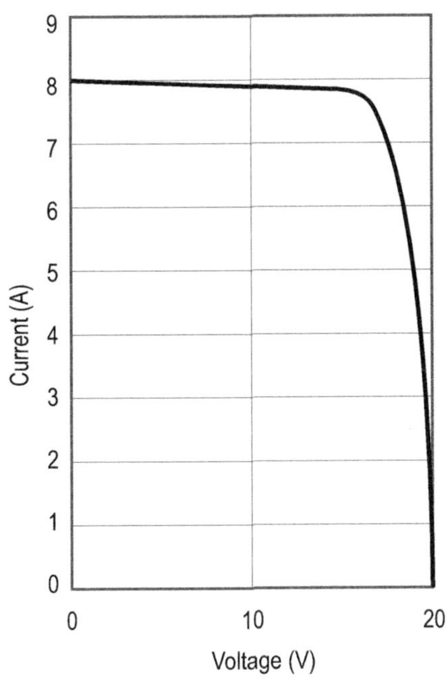

11.124. If the light intensity reduces, the current generated by a solar cell will ….

a. increase

b. decrease

c. mostly remain unchanged

11.125. If the light intensity reduces, the voltage of the cell will ….

a. increase

b. decrease

c. mostly remain unchanged

11.126. A solar PV module specification sheet lists the open-circuit voltage and short-circuit current as 22 V and 8 A, respectively. The temperature coefficient of these parameters are −0.08 V/°C and 0.003 A/°C, respectively. Determine the open-circuit voltage and short-circuit current if the module operating temperature is 55°C.

11.127. The following values are provided from a specification sheet provided by a manufacturer of a solar PV module at the STC. Determine the open-circuit voltage, short-circuit current, and maximum power if the module operating temperature is 42°C.

$$V_{MPP} = 30 \text{ V}$$

$$I_{MPP} = 8.5 \text{ A}$$

$$V_{OC} = 37 \text{ V}$$

$$I_{SC} = 9 \text{ A}$$

Temperature coefficient of $V_{OC} = -0.3\%$
Temperature coefficient of $I_{SC} = 0.05\%$
Temperature coefficient MPP = −0.4%

11.128. Which of the following conditions is set by the standard test conditions (STC) for solar cells? (check all applicable conditions)

a. 0.85 kW/m^2

b. 1 kW/m^2

c. AM 2

d. AM 1.5

e. 25°F

f. 25°C

11.129. Select all the important factors in the accuracy of a solar simulator.

a. Intensity match
b. Temperature match
c. Spectral match
d. Electricity generation match
e. Temporal uniformity
f. Short-circuit uniformity
g. Spatial uniformity

11.130. For a solar simulator, a light intensity match is a sufficient factor to achieve an accurate result from the efficiency test.

a. True
b. False

11.131. What are the generated current and voltage for a typical single silicon solar cell?

11.132. Five similar solar cells each with the current and voltage of 8 A and 0.5 V are connected in series. What is the current and voltage of the entire system, respectively?

a. 8 A and 0.5 V
b. 8 A and 2.5 V
c. 40 A and 0.5 V
d. 40 A and 2.5 V

11.133. Five similar solar cells each with the current and voltage of 8 A and 0.5 V are connected in series. What is the power generation of the entire system?

a. 4 W
b. 10 W
c. 20 W
d. 40 W

11.134. Eight series strings each containing five similar solar cells are connected in parallel. If each cell has a current and voltage of 8 A and 0.5 V, determine the current and voltage of the entire system.

11.135. Eight series strings each containing five similar solar cells are connected in parallel. If each cell has the current and voltage of 8 A and 0.5 V, what is the power generation of the entire system?

a. 4 W
b. 20 W
c. 32 W
d. 160 W

11.136. A solar panel with 36 similar solar cells produces 18 V. How many cells are required in the panel if the desired voltage is 120 V (note that in a solar panel, the cells are typically connected in series)?

a. 18
b. 36
c. 60
d. 72

11.137. Two solar panels are made of similar cells, one with 60 cells and the other with 72 cells. The two solar panels are connected once in series and once in parallel. The V_{OC} and I_{SC} of each individual cell are given as follows. Determine the V_{OC} and I_{SC} when they are connected in series and in parallel configurations.

$$V_{OC} = 0.55 \text{ V}$$

$$I_{SC} = 8 \text{ A}$$

11.138. How can the adverse effects of partial shadowing be eliminated?

a. By connecting the cells in series
b. By placing the entire system in the shadow
c. By mounting a bypass diode
d. By mounting a bypass resistor

11.139. Explain the function of a bypass diode in a solar panel.

11.140. A bypass diode can be mounted in a solar cell or panel to ......

a. increase current
b. increase voltage
c. increase power generation
d. reduce the effect of high temperature
e. reduce the effect of shadow

11.141. Eight series strings each containing five similar solar cells are connected in parallel. Each cell has a current and voltage of 8 A and 0.5 V. Determine the current and voltage of the entire system if one of the cells is totally shaded.

11.142. Eight series strings each containing five similar solar cells are connected in parallel. Each cell has a current and voltage of 8 A and 0.5 V. What is the power generation of the entire system if one of the cells is totally shaded?

a. 128 W
b. 140 W
c. 156 W
d. 160 W

11.143. Ten similar solar cells each with the current and voltage of 7 A and 0.6 V are once connected in parallel and once in series. What are the rough power generation of the entire system if two of the cells are shaded in the parallel and series configuration, respectively?

a. 42 and 0 W

b. 0 and 42 W

c. 0 and 33.6 W

d. 33.6 and 0 W

11.144. Determine the efficiency of a solar panel if the voltage and current at the maximum power point for the panel are 30 V and 8 A, respectively, at the STC. The area of the panel is 1.5 m^2.

11.145. The solar cells on Vanguard satellite could generate about 1 W electricity with an area of about 10 cm^2. Determine the efficiency of the cells if the extraterrestrial solar irradiance is assumed 1360 W/m^2.

11.146. A solar panel with an area of 1.8 m^2 has been rated 250 W at the STC. What is the efficiency of the panel?

a. 45%

b. 25%

c. 18%

d. 13.9%

11.147. Determine the efficiency of a 100 cm^2 solar cell that can generate 0.5 W electricity if the intensity of the incoming light is 0.7 kW/m^2.

11.148. What is the definition of the efficiency of a solar cell or panel?

11.149. Determine the efficiency of a solar cell if the short-circuit current, open-circuit voltage, and fill factor at the STC are 15 A, 40 V, and 0.8, respectively. Assume the area of the panel to be 2.5 m^2.

11.150. In a solar cell, the current is ...... dependent on incoming light density.

a. linearly

b. logarithmically

c. exponentially

d. They are independent.

11.4.6 Balance of Plant for PV Systems

11.151. List components of the balance of plant (BOP) for a solar PV system?

11.152. Explain the function of a charge controller in an off-grid solar PV installation.

11.153. What is the function of an inverter?

11.154. Which component is not required in a grid-tied PV unit?

a. Inverter
b. Battery
c. Mounting system
d. Monitoring system

11.155. The cost of inverters can be between ..... of the total system cost.

a. 5%–20%
b. 15%–25%
c. 20%–50%
d. 40%–65%

11.156. List different types of invertors.

11.157. What are the functions of a bidirectional electric meter?

11.158. Why are some PV systems equipped with a cooling system?

11.159. What is the function of sun tracking systems? List two types of sun tracking systems.

11.4.7 Financial Models for Exchanging Generated Solar Electricity with Utilities

11.160. In the net metering model ......

a. we always need two electric meters
b. we do not need any electric meter
c. we always need a bidirectional meter
d. we always need one meter that cannot be a bidirectional meter

11.161. In the net metering model, the annual electricity generation of a PV unit in a building ......

a. is unlimited
b. is limited by the annual electricity consumption of the building
c. is limited by an agreed upon amount fixed in the contract
d. is limited by the size of the unit

11.162. The price of electricity sold to the utility in a net metering model is ... (select all correct items).

a. fixed and guaranteed for a number of years
b. fixed at a value less than the retail electricity price
c. increased at a rate less than the predicted rate of increase for the electricity provided by the utility
d. equal to the retail price of electricity

11.163. A manufacturing firm has an office in a high-rise building with no possibility of installing a PV array and a manufacturing facility in suburb with a large space to install a PV array. Which financial model provides the best economic return for the company? Explain how the model works in this situation.

11.164. Who owns the PV unit in a power purchase agreement?

a. The building owner

b. The third-party developer

c. The state or federal government

11.165. Which factor is guaranteed in all three common financial models for PV systems?

a. the price of the electricity sold to the utility

b. the price of the electricity purchased from the utility

c. the access to the grid to exchange electricity

d. the ownership of the PV system

11.166. Which of these financial models requires a battery storage system?

a. Feed-in tariff model

b. Net metering model

c. Power purchase agreement

d. None of the models

11.4.8 Sizing of Solar PV Systems

11.167. A house is equipped with 10 LED lights with the rated power of 10 W. If on average each light is on for 4 hours a day, determine the annual electrical energy consumption and the cost of the consumed electricity. The average cost of electricity for the building is \$0.15/kWh.

11.168. Determine the annual electrical energy consumption and the cost of the consumed electricity for a coffee maker with the rated voltage and current of 120 V and 9 A, respectively. The coffee maker is operated on average 45 minutes a day and the average cost of electricity for the building is \$0.15/kWh. Compare the cost with that of the previous problem and comment on the results.

11.169. The following table shows the specifications of appliances in a house resulted from a building energy audit. Determine the annual electricity consumption of the building.

Device	Number of Device	Rated Power (W)	Usage Hours per Day (h)
LED lights	10	8	3.5
Air condition unit	1	400	6
Computer	2	200	4
Microwave	1	500	0.5
Coffee maker	1	950	0.5

11.170. The following table shows the monthly electricity consumption of a residential building. Determine the annual and average monthly electricity consumption of the building.

Month	Monthly Electricity Consumption (kWh)
January	1276
February	1209
March	1089
April	989
May	1059
June	1505
July	2123
August	1990
September	1456
October	1434
November	1343
December	1092

11.171. A solar PV module manufacturer has listed an efficiency of 20% in the published specification. But the installed PV system had an actual efficiency of 17%. Assuming that the reported efficiency by the manufacturer is honest, explain the discrepancy and list some of the important factors.

11.172. A solar panel has an efficiency of 19% listed in its specification sheet. What would be the likely efficiency of the unit in actual operating conditions?

a. 25%

b. 22%

c. 15%

d. 8%

11.173. An annual electricity consumption of a house is 11,000 kWh. If the owner wants to produce 70% of that energy from a solar array with an external efficiency of 77%, determine the annual electrical energy that the solar array needs to generate.

11.174. How much energy a 5 kW array can generate weekly if it is located in an area with an average solar intensity of 6.5 kWh/m^2/day? Assume the external losses are negligible.

11.175. A residential unit in Kentucky is equipped with a 6 kW solar array. What is the annual electricity generation if the external efficiency of the unit is 82% (in kWh/year)? Use the sunlight hour of 4 h/day.

a. 244.9
b. 1197.2
c. 1795.8
d. 7183.2
e. 8760

11.176. In an off-grid PV array designed for a remote house, both DC- and AC-powered appliances are utilized. If the losses of generated electricity in various parts of the system are as follows, determine the overall efficiency of the system in PV-to-DC load, PV-to-AC load, PV-to-battery-to-DC load, and PV-to-battery-to-AC load modes.

- 3% loss in the distribution system
- 25% loss in the storage and recovering of power in the battery bank
- 7% loss in the inverter

11.4.9 First Generation: Crystalline Silicon-Based PV Cells

11.177. What type of solar cells is referred to as the first generation?

11.178. Which ones are the characteristics of crystalline silicon PV modules? (check all the correct ones)

a. High reliability
b. Abundance of raw materials
c. High flexibility
d. High efficiency
e. High stability
f. Energy-efficient manufacturing process

11.179. Name two types of crystalline silicon photovoltaic solar cells.

11.180. Which type of crystalline silicon PV cells has higher efficiency?

a. Polycrystalline
b. Monocrystalline

11.181. What type of solar modules is shown in this figure?

a. Monocrystalline silicon modules
b. Polycrystalline silicon modules
c. Amorphous silicon modules
d. Cadmium telluride modules

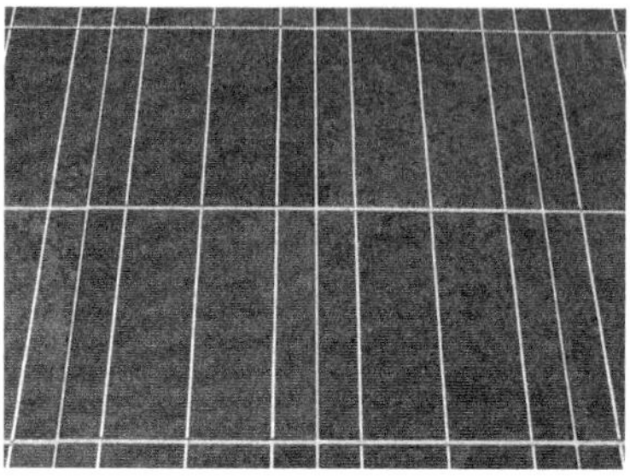

(Image by Hans Braxmeier from Pixabay, https://pixabay.com/photos/solar-cells-technology-current-62559/)

11.182. Briefly explain the manufacturing processes for polycrystalline silicon and monocrystalline silicon solar cell.

11.183. Explain why metallurgical-grade silicon is not suitable to directly manufacture silicon wafers.

11.184. Explain why polysilicon is not suitable to directly manufacture silicon wafers and what the further required processes are.

11.185. Briefly explain the similarities and differences of the manufacturing processes for polycrystalline silicon and monocrystalline silicon solar PV modules.

11.186. Explain why silicon wafer thickness cannot be reduced below certain limits from both performance and manufacturing points of view.

11.187. Explain how the ribbon silicon technology reduces losses in the manufacturing of multi-crystalline silicon wafers.

11.188. Define the energy payback period. Why is it important? What is its value for current crystalline silicon solar units?

11.189. Which technology has higher semiconductor consumption during the manufacturing process?

a. First generation
b. Second generation

11.4.10 Second Generation: Thin Film Solar PV Cells

11.190. Which technology has lower energy consumption during the manufacturing process?

a. First generation
b. Second generation

11.191. The energy payback period for thin film solar cells or second-generation solar cells is very short.

a. True
b. False

11.192. Select the characteristics of thin film solar cell technologies.

a. Low thickness
b. Flexibility
c. Short energy payback period
d. Low cost
e. High efficiency

11.193. Which one of the following solar PV types is considered a second-generation technology? (check all correct answers)

a. Monocrystalline silicon cells
b. Polycrystalline silicon cells
c. CIGS
d. CdTe
e. a-Si

11.194. What is the difference between a heterojunction and a homojunction (p-n junction)?

11.195. CdTe cells are considered a .... PV technology.

a. first-generation
b. second-generation
c. third-generation

11.196. Identify the p-n junction of the following CIGS solar cell.

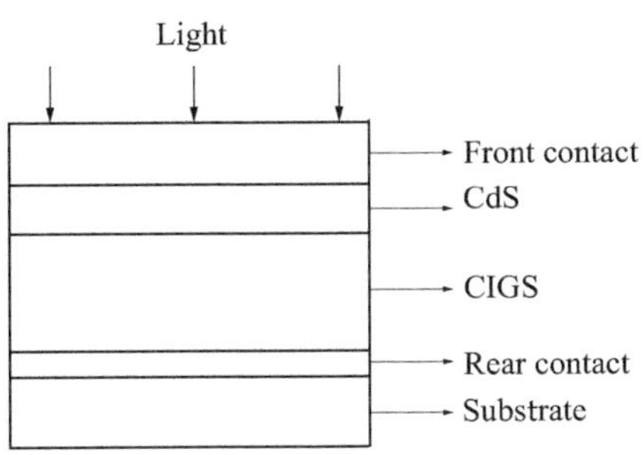

11.197. The manufacturing of which type of solar cells is more energy intensive?

a. Monocrystalline silicon modules
b. Polycrystalline silicon modules
c. Cadmium telluride modules

11.198. Which type of solar cells is more efficient?

a. Monocrystalline silicon modules
b. Polycrystalline silicon modules
c. Amorphous silicon modules

11.199. Rank the following solar PV cells based on their efficiency (highest to lowest).

a. Monocrystalline silicon modules
b. Polycrystalline silicon modules
c. Amorphous silicon modules

11.200. Which one of the following materials is used in the second-generation PV units? (check all correct answers)

a. Monocrystalline silicon cells
b. Polycrystalline silicon cells
c. CIGS
d. CdTe
e. a-Si

11.201. Which one of the following solar PV types has a heterojunction (p-n junction)? (check all correct answers)

a. Monocrystalline silicon cells
b. Polycrystalline silicon cells
c. CIGS
d. CdTe
e. a-Si

11.202. Rank the following thin film solar cell technologies according to the efficiency of commercial modules.

a. CIGS
b. CdTe
c. a-Si

11.203. Rank the following type of solar cells according to their share of market from the highest to lowest.

a. Monocrystalline silicon
b. Multi-crystalline silicon
c. CIGS
d. CdTe
e. Amorphous silicon

11.4.11 Third Generation: Emerging Solar Cell Technologies

11.204. A multi-junction solar cell is made of the following semiconductors

Semiconductor A with a band gap of 0.7 eV

Semiconductor B with a band gap of 1.4 eV

Semiconductor C with a band gap of 1.9 eV

What is the required order of the semiconductors (top to bottom)?

a. A-B-C
b. C-B-A
c. A-C-B
d. C-A-B

11.205. What are the major benefits of organic solar cells?

a. Rapid manufacturing
b. Inexpensive manufacturing
c. High efficiency
d. Flexibility
e. Short energy payback period
f. Stability

11.206. Which solar PV cell type currently has the highest efficiency in the cell level in laboratory conditions?

a. Multi-crystalline silicon
b. Monocrystalline silicon
c. Amorphous silicon
d. CdTe
e. CIGS
f. Multi-junction
g. Organic
h. Perovskite
i. Quantum dot

11.4.12 Open-Ended Questions

11.207. Conduct a search for solar cell in Google Patent (https://patents.google.com) by limiting the period to 1850–1940. Report at least three of them. How many patents did you find? What is the oldest patent? Pick one patent, read through it and then prepare a short report on the subject of the patent. Compare your choice with a recent relevant patent.

11.208. Based on what you learned in this chapter, how do you think the future of commercial solar PV technologies will look like? In the next 10 years? Next 20 years? Next 50 years? By the end of the 21st century?

11.209. You are asked to design a PV array for a recreational vehicle (RV). What type of solar cells will you choose? Justify your answer.

11.210. You are asked to design a PV system for on foot disaster relief first responders. What type of solar cells will you choose? Justify your answer.

11.4.13 General

11.211. The following curves show the current of a solar cell versus its voltage. Determine the short-circuit current, open-circuit voltage, maximum power, and filling efficiency of the cell. How the output power can be improved?

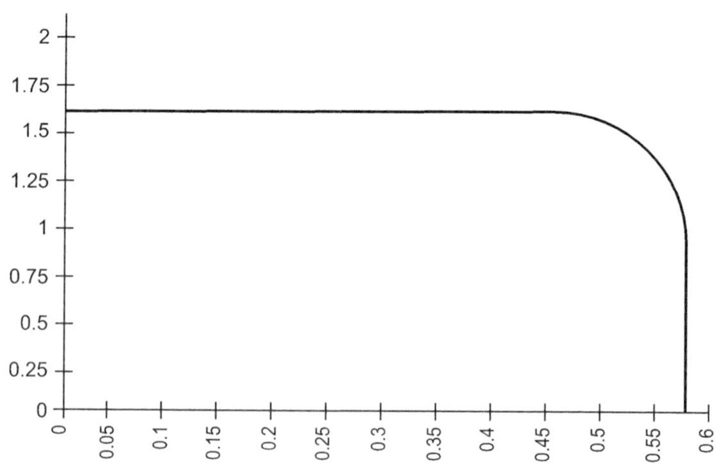

11.212. Explain why about half of silicon is lost during the manufacturing of monocrystalline and polycrystalline silicon solar cells. How does the ribbon silicon manufacturing process address this problem?

11.213. The efficiency of solar cells ….. with increase in the operating temperature.

a. increases

b. decreases

c. remains unchanged

d. independent

11.214. Identify the p-n junction(s) of the following a-Si solar cell.

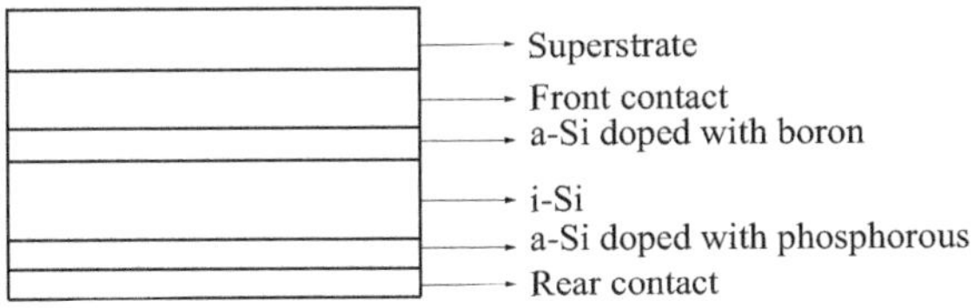

11.215. What type of solar modules is shown in this figure?

a. Monocrystalline silicon module

b. Polycrystalline silicon module

c. Amorphous silicon module

d. Cadmium telluride module

(Image by Valentin J-W from Pixabay, https://pixabay.com/photos/solar-cells-energy-solar-energy-4518991/)

11.216. Which solar cell type currently has the highest efficiency for commercial modules for residential applications?

a. Multi-crystalline silicon

b. Monocrystalline silicon

c. Amorphous silicon

d. CdTe

e. CIGS

f. Multi-junction

g. Organic

h. Perovskite

i. Quantum dot

11.217. What are the typical efficiencies for commercial multi-crystalline silicon and monocrystalline silicon modules? Explain why they are different.

11.218. In the below I-V curve for a solar cell identify the points that correspond to the following parameters:

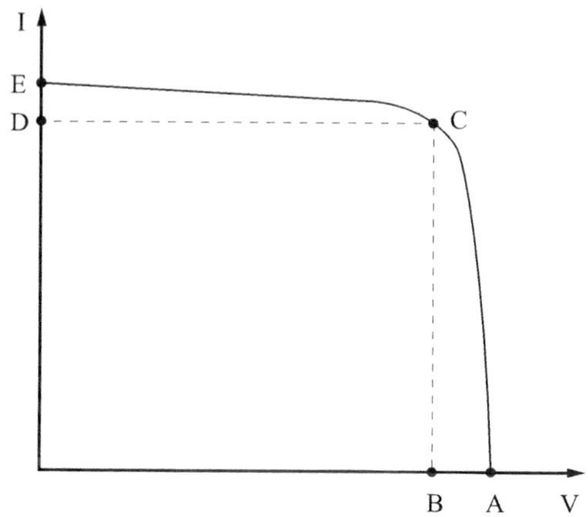

a. V_{oc}:

b. I_{SC}:

c. MPP:

d. V_{MPP}:

e. I_{MPP}:

11.219. In the above I-V curve which point corresponds to the open-circuit voltage?

a. A

b. B

c. C

d. D

e. E

11.220. In the above I-V curve which parameter is associated with point B?

a. open-circuit voltage (V_{OC})

b. short-circuit current (I_{SC})

c. current of the maximum power point (I_{MPP})

d. voltage of the maximum power point (V_{MPP})

e. maximum power point (MPP)

11.221. In the above I-V curve which point corresponds to the maximum current?

a. A

b. B

c. C

d. D

e. E

11.222. In the above I-V curve which points correspond to the maximum current and voltage, respectively?

a. A and C

b. B and D

c. C and E

d. A and E

e. D and A

11.223. Five series strings each containing four similar solar cells are connected in parallel. If each cell has the current and voltage of 8.5 A and 0.55 V, what is the power generation of the entire system?

a. 4.675 W

b. 18.7 W

c. 23.375 W

d. 93.5 W

11.224. For a solar water heater in a cold region where more thermal energy is required in winter, the tilt angle of the device should be .... the latitude.

a. smaller than

b. larger than

c. equal to

11.225. Explain the difference between elemental and compound semiconductors.

11.226. What type of solar units would you recommend for a residential application where the rooftop space is limited but financing the project is relatively easy?

a. Multi-crystalline silicon modules

b. Monocrystalline silicon modules

c. Amorphous silicon

d. CdTe modules

e. CIGS modules

f. Multi-junction modules

11.227. Which model is designed to encourage early adaptors?

a. Feed-in tariff model
b. Net metering model
c. Power purchase agreement
d. None of the models

11.228. How much usable energy an 80 kW array can monthly generate, if it is located in an area with an average solar intensity of 4.5 $kWh/m^2/day$ and it has an external efficiency of 88%?

11.229. List the benefits of using silicon to manufacture solar modules.

11.230. Which one of the following solar PV types has a homojunction (p-n junction)? (check all correct answers)

a. Monocrystalline silicon cells
b. Polycrystalline silicon cells
c. CIGS
d. CdTe
e. a-Si

11.231. Monocrystalline silicon cells are considered a …. PV technology.

a. first-generation
b. second-generation
c. third-generation

11.232. With an increase in the operating temperature of a solar module, its efficiency ….

a. increases
b. decreases
c. remains unchanged
d. becomes zero

11.233. A solar array has a rated power generation capacity of 125 kW. If the area of the panels is 800 m^2, calculate the generated electricity per area for the array (in W/m^2).

11.234. The power production of a solar cell at $I = I_{SC}$ is …….

a. maximum
b. zero
c. nonzero
d. negative

11.235. Define the carrier lifetime and diffusion length of electron-hole pairs in a solar cell.

11.236. What type of silicon solar modules has bluish flakey surface?

a. Multi-crystalline silicon
b. Monocrystalline silicon
c. Amorphous silicon

11.237. Which financial model might be most suitable for a homeowner with limited or no financial means to invest in a PV system?

a. Feed-in tariff model
b. Net metering model
c. Power purchase agreement
d. None of the models

11.238. Which application provided an early market for solar cell technology and still is a driving force for the most advanced modern solar cells?

a. Residential applications
b. Distributed power production
c. Low-powered devices such as calculators
d. Space exploration

11.239. What is the June solstice?

11.240. Which type of crystalline silicon PV cells is less expensive?

a. Polycrystalline
b. Monocrystalline

11.241. What is the intensity of light at one full sun?

11.242. Which one of the following solar PV types is considered a first-generation technology? (check all correct answers)

a. Monocrystalline silicon cells
b. Polycrystalline silicon cells
c. CIGS
d. CdTe
e. a-Si

11.243. A solar panel with 36 similar solar cells produces 135 W at 18 V and 7.5 A. Determine the current, voltage, and generated power if the number of cells in the panel increases to 60 cells (note that in a solar panel the cells are typically connected in series).

11.244. Explain various types of losses in solar cells.

11.245. What type of silicon solar modules has uniform black surface?

a. Multi-crystalline silicon
b. Monocrystalline silicon
c. Amorphous silicon

11.246. Five series strings each containing four similar solar cells are connected in parallel. Each cell has a current and voltage of 8.5 A and 0.55 V. What is the power generation of the entire system if one of the cells is totally shaded?

a. 70.125 W

b. 74.8 W

c. 88.825 W

d. 93.5 W

11.247. Which statement is not correct about the open-circuit voltage?

a. The voltage when the load is zero.

b. The voltage when the current is zero.

c. The voltage when power generation is maximum.

d. It is the maximum voltage of a solar cell.

11.248. Explain the difference between the net metering model and the feed-in tariff model in terms of the limits for electricity generation and/or the capacity of the PV unit.

11.249. What type of solar units is more suitable for a residential application where there is ample rooftop space but the capital investment is limited?

a. Multi-crystalline silicon modules

b. Monocrystalline silicon modules

c. Amorphous silicon modules

d. CdTe modules

e. CIGS modules

f. Multi-junction modules

11.250. List the drawbacks of organic solar cells.

11.251. The optimum tilt angle of solar energy receiving devices for direct radiation for a given location is the same as the optimum tilt angle for diffuse radiation.

a. True

b. False

11.252. How can a pyranometer be converted to a diffusionmeter?

11.253. Which one of the following materials are used in the first-generation PV units? (check all correct answers)

a. Monocrystalline silicon

b. Polycrystalline silicon

c. CIGS

d. CdTe

e. a-Si

11.254. Why is it important for a solar simulator to have the spectrum that matches the spectrum of natural light?

11.255. The power production of a solar cell at $V = V_{OC}$ is ....

a. maximum

b. zero

c. nonzero

d. negative

11.256. Explain what holes are in the atomic structure of semiconductors.

11.257. Which one can directly generate AC power?

a. Utility-scale PV units

b. Ground-mounted PV units

c. Building rooftop PV units

d. Micro-size PV units

e. All of the above

f. None of the above

11.258. What are conductor, insulator, and semiconductor?

11.259. Which of the following thin film solar cell technologies has the highest efficiency for commercial modules?

a. CIGS

b. CdTe

c. a-Si

11.260. An electron-hole pair is created on the p-n junction of a solar cell. To which side the hole will be attracted?

11.261. In the southern hemisphere, the solar devices orientated toward true ......

a. west

b. east

c. north

d. south

11.262. Explain why there are seasons.

11.263. If the cost of installing a multi-crystalline solar array is $2 per peak watt, determine the total cost of installing a 5 kW array.

11.264. Which of the following thin film solar cell technologies has the lowest efficiency for commercial modules.

a. CIGS

b. CdTe

c. a-Si

11.265. Explain why the generated electricity in a solar cell is zero at the open-circuit and short-circuit conditions.

11.266. The photovoltaic effect was first observed by ...... in ....

a. Alessandro Volta 1790

b. Alexandre-Edmond Becquerel – 1830

c. Willoughby Smith – 1873

d. Albert Einstein 1921

e. Bell Labs – 1948

f. Hoffman Electronics Corporation – 1957

11.267. What are the desirable characteristics of silicon that make it an ideal choice for the manufacturing of solar cells?

11.268. What is a doping agent or a dopant in an indirect semiconductor?

11.269. When it is summer in the northern hemisphere, in the southern hemisphere it is winter and vice versa.

a. True

b. False

11.270. What is the air mass coefficient for a solar PV panel installed on a moon landing spacecraft (when it is landed on the moon)?

a. AM2

b. AM1.5

c. AM1

d. AM0

11.271. Is the solar constant a physical constant? In other word, can the solar constant vary?

11.272. CIGS cells are considered a .... PV technology.

a. first-generation

b. second-generation

c. third-generation

11.273. Explain if there is a limit to annual electricity generation for a PV unit in the net metering model.

11.274. List the major advantages of organic solar cells over conventional cells.

11.275. Which band has lower energy?

a. Conduction band

b. Valence band

11.276. Explain why the sky is blue.

11.277. What is extraterrestrial solar radiation?

11.278. The first-generation solar cells use direct semiconductors.

a. True

b. False

11.279. Explain why in the I-V curve, the generated power varies nearly linearly over the section of the curve where the current is constant.

11.280. Utility-scale PV units ….. (check all the correct statements).

a. are always connected to the grid

b. generate DC power

c. currently generate the majority of solar electricity

d. always require a battery

11.281. Provide two examples of elemental semiconductors and compound semiconductors.

11.282. Eight similar solar cells each with the current and voltage of 7 A and 0.6 V are connected in parallel. What are the current and voltage of the entire system, respectively?

a. 7 A and 0.6 V

b. 7 A and 4.8 V

c. 56 A and 0.6 V

d. 56 A and 4.8 V

11.283. Eight similar solar cells each with the current and voltage of 7 A and 0.6 V are connected in parallel. What is the power generation of the entire system?

a. 4.2 W

b. 8.4 W

c. 33.6 W

d. 268.8 W

11.284. AM0 refers to the solar spectrum above the atmosphere.

a. True

b. False

11.285. AM1 is when the Sun is directly overhead in a location.

a. True

b. False

11.286. Explain what the solar constant is.

11.287. The efficiency of initial solar cells made in Bell Lab was at best 6%. Calculate the electricity that 1 m^2 of these cells could generate at the STC.

11.288. What are the rated voltage and current of a typical commercial solar module with 72 cells? (Select the closest choice to what you expect.)

a. 35 V and 8 A

b. 72 V and 4 A

c. 30 V and 16 A

d. 8 V and 30 A

11.289. Which type of crystalline silicon solar modules consumes more energy to manufacture? Explain why.

11.290. What are the typical current and voltage of a single cell made of crystalline silicon?

a. 120 V and 8 A

b. 40 V and 16 A

c. 5 V and 4 A

d. 0.5 V and 8 A

11.291. In the following figure identify the Sun's summer and winter passes.

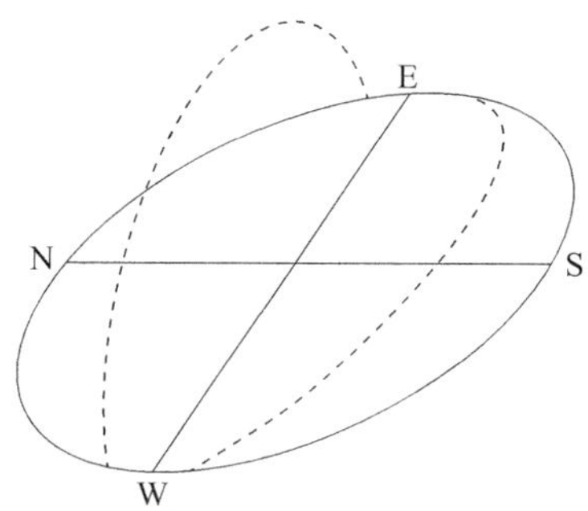

11.292. There is no meaningful seasonal variation on the equator and the length of day and night does not change much.

a. True

b. False

11.293. Explain why the North and South Poles do not experience the 24-hour cycle of day and night.

11.294. Albert Einstein received the 1921 Nobel Prize in Physics for his work on the ......

a. general relativity

b. special relativity

c. photoelectric effect

d. development of the first solar cell

11.295. When it is autumn in the southern hemisphere, it is ….. in the northern hemisphere.

a. spring
b. fall
c. winter
d. summer

11.296. What percentage of the global electricity production is currently generated by solar photovoltaic units?

11.297. Select the characteristics of organic solar cells.

a. Rapid and inexpensive fabrication
b. High efficacy
c. Long lifetime
d. Low stability
e. Flexibility
f. Short energy payback period

11.298. The winter solstice in the northern hemisphere is the summer solstice in the southern hemisphere.

a. True
b. False

11.299. The distance between the Sun and Mars is about 1.52 times greater than that of the Sun and Earth.[54] Determine the solar constant on Mars.

11.300. When are the longest day and the shortest night of the year in the northern hemisphere?

11.301. In the northern hemisphere, solar devices are orientated toward ……

a. west
b. east
c. north
d. south

11.302. What is the cosine emission law?

11.303. On-grid PV units ….. (check all the correct statements).

a. are always connected to the grid
b. generate DC power
c. currently generate the majority of solar electricity
d. always require a battery

[54] https://www.nasa.gov/audience/foreducators/k-4/features/F_Measuring_the_Distance_Student_Pages.html.

11.304. When was the first spacecraft powered by solar cells launched?

a. 1957

b. 1958

c. 1959

d. 1962

11.305. What is the difference between beam radiation and diffuse radiation (sky radiation)? How they impact solar power generation systems?

11.306. AM2 refers to the solar spectrum for solar radiation that has traveled through the atmosphere for a distance twice the thickness of the atmosphere.

a. True

b. False

11.307. The minimum solar incident occurs during the ……...

a. summer solstice

b. winter solstice

c. vernal equinox

d. autumnal equinox

11.308. Explain why there is no meaningful winter or summer on the equator?

11.309. Which one has the largest contribution to the electricity produced from photovoltaic cells?

a. Distributed solar electricity production

b. Utility-scale solar electricity production

11.310. What is Earth's polar axis (PA)?

11.311. Earth's orbit is slightly elliptical, which causes various seasons.

a. True

b. False

11.312. If the axial tilt of Earth was 0°, ...

a. winters would be much longer

b. there would be no season

c. there would be only two seasons

d. the duration of seasons would be unstable

11.313. When is the angle between solar rays and the normal line to the surface of Earth (horizon) the smallest possible in the northern hemisphere?

11.314. The solar irradiance in most of the northeast U.S. is in the range of 4–5 kWh/m^2/day. What is the range of the annual full sun hours for the region?

a. 1640–1825

b. 1460–1825

c. 1046–1285

d. 1604-1528

11.315. A solar array with a installed capacity of 12 kW is located in western Arizona. Assuming the system efficiency of 84%, how many kWh electrical energy the array can annually generate?

11.316. The second-generation solar cells use direct semiconductors.

a. True

b. False

11.317. Earth's orbit is slightly elliptical with the ratio of the major axes to the minor axes of 0.0167.

a. True

b. False

11.318. Air mass 0 or AM0 refers to the solar spectrum above the atmosphere with irradiance equal to the solar constant.

a. True

b. False

11.319. Which type of PV units currently has the greatest share of the market?

a. On-grid

b. Off-grid

11.320. The solar array installed on Juno has the rated power of 435 Watt when operated on the orbit of Jupiter. Determine its rated power at the STC if all operational characteristics remain constant.

11.321. Determine the size of a solar array for a satellite with a power requirement of 125 W if the satellite orbits Earth, Mars, Jupiter, and Saturn. The efficiency of the cells used to build the array is 28%. Ignore all losses.

11.322. The area of the solar array installed on Juno orbiting Jupiter is 72 m^2 and has the rated power of 435 W when operated on the orbit of Jupiter. Determine the efficiency of the array.

11.323. In the following diagram, what are A and B representing, respectively?

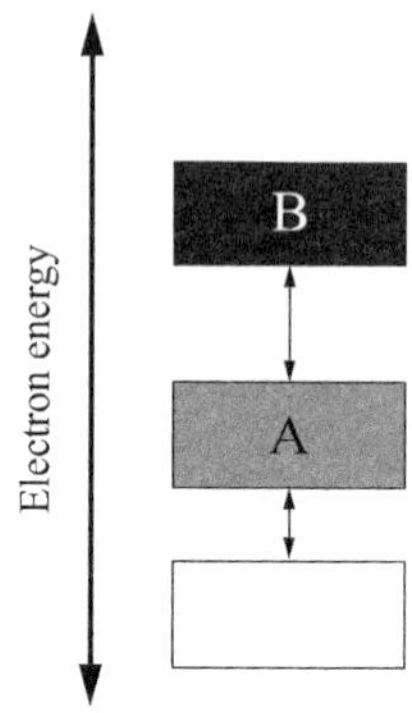

a. Valence band – Band gap
b. Conduction band – Band gap
c. Band gap – Conduction band
d. Conduction band – Valence band

12

Solar Thermal Energy

Abstract

In this chapter, we will study how solar thermal energy can be utilized. We will start with a brief discussion on low-temperature solar thermal energy systems for water heating. We will introduce unglazed, flat plate, and evacuated tube water heaters. Then, after a brief review of the history of solar thermal energy utilization, we will learn about different types of high-temperature solar thermal energy systems. We will consider two types of solar thermal power plants and their components. Next, we will focus on solar thermal power plants with a central receiver and their components, including heliostats, the arrangement of heliostat fields, and central receivers. Then, we will learn about solar thermal power plants with distributed or dispersed collectors and their types and components, including line-focus and point-focus collectors. Finally, we will consider heat transfer fluids and power generation units for solar thermal power plants. We will conclude the chapter with a brief discussion on the negative environmental effects of these power plants.

LEARNING OUTCOMES

After successful completion of this chapter, you will be able to

- explain low-temperature solar thermal energy systems: unglazed, flat plate, and evacuated tube water heaters
- describe high-temperature solar thermal energy systems, their types and components
- identify different types of high-temperature solar thermal energy systems and their components
- explain solar thermal power plants with a central receiver and their components, including heliostats, arrangement of heliostat fields, and central receivers

- explain solar thermal power plants with distributed or dispersed collectors and their components, including line-focus and point-focus collectors
- describe negative environmental effects of solar thermal power plants

12.1 Introduction

We know that the direct use of solar energy can be divided into three main categories, solar photovoltaic units, high-temperature solar thermal energy systems, and low-temperature solar thermal energy systems. In the previous chapter, we learned how solar energy can be directly converted to electrical energy using photovoltaic cells. We also learned that these units have many advantages, which have facilitated their tremendous growth in recent years. However, they have some significant drawbacks the most important of which is their inherently low efficiency. Probably, many of you have a pleasant memory of the Sun warmth on your skin on a chilly day. In this section, we will explore how solar thermal energy can be exploited to generate useful energy particularly electricity. Aside from the benefits of all forms of renewable resources, solar thermal energy systems enjoy two distinguishing advantages. First, their integration into thermal energy systems is relatively simple and inexpensive. Also, they share many common components with well-developed fossil fuel conventional power plants, which allows significant technology transfer between the two industries. In this chapter, we will focus on high-temperature solar thermal energy systems. Low-temperature solar thermal energy systems for buildings are not for electricity generation and technically beyond the scope of this text. However, for the sake of completeness of the coverage of various solar thermal energy applications and since using these units reduces the consumption of electricity, we will start this section with a very brief review of this application. Then, we will focus on the applications of solar thermal energy in electricity generation.

12.2 Low-Temperature Solar Thermal Energy Applications for Buildings

In the U.S., over a half of energy usage in residential applications is for space heating and air conditioning and a quarter is used for water heating, lighting, and refrigeration.[1] Solar energy can help to substantially reduce these consumptions. Building energy systems can be passive or active. In **passive systems**, building structures, envelopes, and windows are used to absorb solar thermal energy and to allow light to enter the interior of buildings to reduce energy usage for heating and artificial lighting. Blinds or overhangs also can be used to reduce heating load in summer. Passive solar energy methods are beyond the scope of this text and are studied in building and air conditioning system design fields.

Active solar thermal energy systems use a collector to absorb solar energy. The collected heat is then transferred to the desired space by a working fluid. They may have a pump or fan to circulate the working fluid and a storage tank to store thermal energy for when solar energy is not available.

The working fluid in active systems can be air or liquid. Air-based systems can be advantageous because air as a nontoxic working fluid does not evaporate, freeze, or degrade. However, due to the low density, heat-transfer coefficient, and specific heat of air, it requires a larger system with more fan power for the same thermal performance than those of liquid-based systems. For this reason, air solar collectors have a small share of the market.

Solar water heaters (SWHs) use water or a mixture of water as the working fluid. This technology has been around for over a century, so it is a well-developed and mature technology. These units can be **domestic hot water (DHW)** systems for standalone buildings or central heating systems (or district heating systems) for several (or many) buildings. These units are very popular in some countries where a significant portion of buildings are equipped with these units including Greece, Turkey, Israel, Australia, Japan, and China.

The categorization of SWHs is typically based on their solar collector. These collectors can be **nonconcentrating** or **concentrating**. In general, solar collectors can be divided into the following four groups (along with the temperature increase in them and whether they are nonconcentrating or concentrating)

- Unglazed water heaters, up to 10°C (18°F) temperature increase (nonconcentrating)
- Flat-plate water heaters, up to 50°C (90°F) temperature increase (nonconcentrating)
- Evacuated tube water heaters, up to 155°C (280°F) temperature increase (nonconcentrating or concentrating)
- Parabolic trough water heaters, up to 280°C (500°F) temperature increase (concentrating)

[1] https://www.eia.gov/energyexplained/use-of-energy/homes.php#:~:text=Electricity%20and%20natural%20gas%20are%20the%20most%2Dused%20energy%20sources%20in%20homes&text=Natural%20gas%2C%20which%20was%20used,the%20residential%20sector%20in%202019.

The most important factors in selecting an appropriate collector are the desired temperature of the produced hot water and the range of the ambient temperature. With an increase in the temperature of produced hot water and/or decrease of the ambient temperature, the temperature difference between the working fluid and the surrounding increases, which in turn increases the heat loss via various heat-transfer mechanisms. There is an operating temperature that balances the heat gain and loss. The operating temperature of a solar heater cannot increase beyond that temperature. In the following sections, we will briefly review these SWHs.

12.2.1 Unglazed Solar Water Heaters

Unglazed collectors are typically made of flexible (foldable) plastic collectors with no protective cover. They can be usually rolled out to provide relatively hot water for applications where low-temperature water is required. The lack of protective cover while reduces optical losses, makes unglazed collectors vulnerable to heat losses when the temperature difference between water and surrounding air increases. This limits the usage of these very inexpensive units to very low-temperature rise applications, such as water heaters for swimming pools.

12.2.2 Flat-Plate Solar Water Heaters

Flat-plate collectors are the most common type of collectors used in SWHs for residential and commercial buildings mainly due to their simple, reliable, safe, and low-maintenance design. They are used to provide domestic hot water and thermal energy for space heating. While air can be used as the heat-transfer medium for space heating applications, a majority of current systems use liquid for this purpose.

As the name implies, these SWHs have a copper, aluminum, or steel flat-plate absorber (collector) that receives and absorbs direct and diffuse solar radiation, which increases the temperature of the absorber. The parallel rows of copper or steel tubes are embedded in the plate. They are in good contact and are thermally bonded with the absorber. The heat-transfer fluid flows in the tubes to remove thermal energy from the plate. The fluid is distributed among the tubes at the top by a header and collected from them at the bottom by another header though there might be more than one pass. To maximize absorption and minimize reflection and re-radiation of solar radiation, the absorber is coated with a dark-colored selective coating. The entire unit is placed in a typically metallic or plastic housing to provide mechanical strength and to protect it from environmental elements. The front of the housing is covered by a transparent surface made of special glass or plastic to reduce heat loss by convection while allowing solar radiation to pass through albeit with some reflection loss. The back and sides of the housing are insulated to reduce conduction heat loss. If freezing is not a major problem in winter, typically water is used as the heat-transfer fluid due to its availability, low cost, nontoxicity, good thermal characteristics (conductivity and specific heat), low corrosion, and low viscosity. Otherwise, water is mixed with antifreeze liquid (typically propylene or ethylene glycol). In the latter case, a heat exchanger is needed to separate the heat-transfer fluid from domestic water. The produced hot fluid can be either used directly or stored in a storage tank to be used when solar radiation is not available. The circulation of the heat-transfer fluid can be achieved by natural circulation or forced circulation. In **natural circulation**, also known as **thermosiphon or passive** SWHs, water is circulated because of the density difference between unheated cold water and hot water. In this type, a storage tank (sometimes more than one tank) is added to the top of the unit (Figure 12.1). In forced-circulation systems, a pump is used for this purpose.

Typically, solar heaters are designed for a maximum daily average solar irradiance. This means they can only provide the required hot water in the day with the maximum daily average solar irradiance. In this way, they are fully functional throughout the year but require a backup or auxiliary heat source, such as an electric heater or a natural gas burning heater for when solar radiation is not sufficient to meet the demand. When designing these systems, special attention should be paid to the protection of the system when there is high solar irradiance but no hot water consumption, which can result in very high temperatures, well above the boiling temperature of the heat-transfer fluid (this circumstance is commonly referred to as the stagnation).

FIGURE 12.1
A natural-circulation solar water heater (SWH). (By Gilabrand at English Wikipedia, CC BY 2.5, https://commons.wikimedia.org/w/index.php?curid=7954929.)

12.2.3 Evacuated Tube Solar Water Heaters

In these systems, to practically eliminate conductive and convective heat losses, the absorber and tubes are enclosed in a glass vacuum tube (Figure 12.2). They are able to raise the temperature of water to much higher than that of flat-plate systems. Usually, solar thermal energy absorbed in the absorber can be transferred out of the tube by means of either a direct flow system or a heat pipe. In direct flow systems, the heat-transfer liquid flows inside copper tubes attached to the collector plate and carries the thermal energy to the storage tank. In heat pipe systems, the thermal energy is transferred by a phase changing volatile fluid. This fluid receives heat from the absorber and evaporates. Then, it moves upward to the condenser section of the pipe, which is located in the header that the evacuated tube is connected to. The heat is transferred from the vapor to the circulating water and the condensed fluid flows down the evacuated tube by gravity for further heat transfer. We will study parabolic trough water heaters later in this chapter since they are also used for power generation systems.

12.2.4 Efficiency and Operation of Solar Water Heaters

The efficiency of SWHs is defined as the ratio of the annual thermal energy supplied by the system to the annual solar radiation intercepted by the collector. Another parameter that is used to express the performance of SWHs with regards to the building they serve is called the solar fraction. It is defined as the ratio of the thermal energy supplied by the solar thermal energy system to the total corresponding thermal energy needs of the building.

FIGURE 12.2
An evacuated tube SWH. (Image by Vijaya Narasimha from Pixabay, https://pixabay.com/photos/solar-water-heater-shimoga-india-331314/.)

Solar thermal energy systems are almost universally stationary (with no sun tracking system). Therefore, similar to other fixed solar energy systems, they can operate most efficiently facing true south in the northern hemisphere. While the tilt angle for a year-round operating system (for optimum annual output) should be equal to the latitude of the site, these units often need the maximum output in winter so the tilt angle equal to the site's latitude plus 15° is common (refer to Chapter 11).

Example 12.1

In a building, a solar air heater provides 8000 kWh of thermal energy needed for space heating and a natural gas burning furnace provides the remaining 2500 kWh. Determine the solar fraction of the system.

Solution

$$\text{Solar fraction} = \frac{\text{Thermal energy supplied by solar water heater}}{\text{Total annual thermal energy consumption}}$$

$$= \frac{8000}{8000 + 2500} = 0.762$$

12.3 Introduction and Brief History of High-Temperature Solar Thermal Energy Utilization

Using solar thermal energy is not a new idea. Probably the first recorded use of solar thermal energy for an industrial application (or more precisely a military one) goes back over 22 centuries ago to ancient Greece. Legend has it that during the Siege of Syracuse[2] by the Roman navy (about 213 BC) Archimedes of Syracuse (287 – 212 BC, a Greek polymath) designed and built a device to reflect and focus sunlight (similar to concentrated solar power plants) perhaps using polished metal shields on Roman ships surrounding the port of Syracuse. Apparently, this resulted in some ships catching fire and being destroyed. There have been many debates on the feasibility of the device with those days' technology. Some experiments have been conducted to repeat the performance of the weapon with mixed results. Regardless of the accuracy of the legend, the city defenders were defeated, and the city remained an important Roman city for many centuries. As for Archimedes, he was killed by a Roman soldier when the city fell.

In more recent history, Horace de Saussure (1740–1799, a Swiss scientist and explorer) built the first modern solar oven in 1767. Augustin Mouchot (1825–1912, a French inventor) in the 1860s and 70s designed a parabolic collector, steam generator, and steam engine that could power a variety of

[2] The city of Syracuse is currently part of Sicily, Italy.

machines. He also wrote one of the early books on solar thermal energy entitled "Solar Heat and its Industrial Applications". However, the low efficiency of his designs and increasingly more available inexpensive coal eventually domed his inventions. Similarly, in 1912 Frank Shuman (1862 –1918, an American engineer) used semi-circle troughs to raise steam to power a water pump in Africa. The first large-scale solar collector was the Odeillo solar furnace built in southern France on the border with Spain in 1969 (Figure 12.3). The collector included over 60 steerable heliostats that reflected solar rays on a parabolic mirror installed on a relatively tall building (Figure 12.4). The mirror with a diameter of about 50 m, in turn, reflected and concentrated the

FIGURE 12.3
Odeillo solar furnace in France. (Image by Marc Pascual from Pixabay, https://pixabay.com/photos/solar-oven-odello-odeillo-france-921116/.)

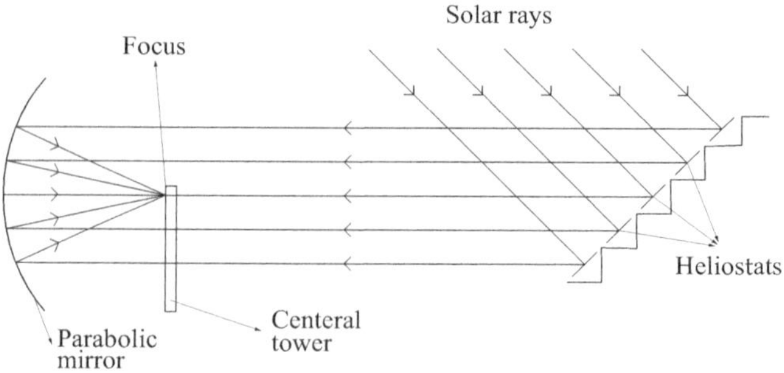

FIGURE 12.4
Schematic of Odeillo solar furnace.

rays on a 0.4 m diameter area on a central tower. The facility was not designed to raise steam or to produce power but rather has been used as a furnace to research materials at very high temperatures (up to 3500°C, about 6300°F).

The first experimental solar thermal power plant, Solar One, was built in Barstow, California, in 1981 followed by its redesign and modification as Solar Two in 1995. A major breakthrough came from Europe where the first commercial solar thermal power plants, Planta Solar 10 (PS10) and Planta Solar 20 (PS20), were constructed in Seville in southern Spain, in 2007 and 2009, respectively. Since then, many large-scale commercial plants have been completed all over the world. The largest plant in the world is a 510 MW plant in Morocco followed by the Ivanpah Solar Power Facility in California with a power generation capacity of 392 MW. The cost of solar thermal power plants per kWh has substantially reduced in the past few decades. This continuous cost reduction along with the possibility of integrating relatively inexpensive thermal energy storage systems make solar thermal energy power plants a major candidate to play a meaningful and important role in the future renewable energy infrastructure.

Generally speaking, solar thermal power plants consist of the following main components: **solar collectors**, **receivers**, **heat-transport systems**, **energy storage systems**, and **energy conversion systems**. Collectors directly receive solar rays and redirect them toward a receiver (Figure 12.5). Since they have to always face the receiver regardless of the position of the Sun, collectors are universally equipped with a tracking mechanism. The receiver, as the name implies, receives concentrated solar rays from the

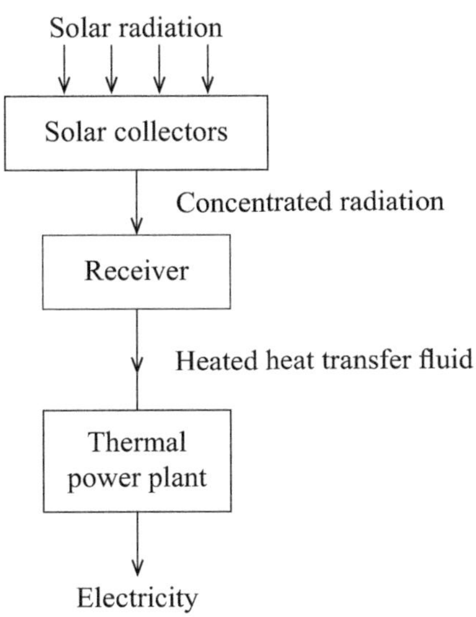

FIGURE 12.5
Parts of a solar thermal energy power plant.

collectors. This thermal energy is then transferred to the heat-transfer fluid. The heat-transfer fluid provides interconnection between the solar thermal energy collection section of the plant and its electricity generation section. The power generation unit in solar thermal power plants can be similar to conventional power plants. A solar thermal power plant may or may not have a thermal energy storage system. If the plant is equipped with a thermal energy storage system, the excess solar thermal energy is stored in the daytime and then is used to produce electricity when direct solar energy is not available.

We know that to have an efficient power generation unit, we require a high-temperature thermal reservoir. Solar radiation in its natural form cannot produce nearly high enough temperature that can be meaningfully exploited. For that reason, all solar thermal energy power plants are equipped with devices to collect and concentrate solar radiation. This concentrated solar radiation can provide a high-temperature heat source for power generation. That is why these plants are commonly referred to as **concentrated solar power (CSP)**, concentrating solar power, or concentrated solar thermal power plants.

For the concentration of solar radiation, either mirrors or lenses can be used. Due to their lower costs and easy assembly in large numbers, mirrors are much more commonly used in thermal solar power plants. The temperature of thermal energy when solar radiation is concentrated depends on the **concentration ratio** of the solar collector. The concentration ratio of a device is the ratio of the surface area of the device perpendicular to solar rays to the area of the Sun's image created by the device (i.e., the area that receives the concentrated radiation).[3] If rays in solar radiation were perfectly parallel to each other, they could be concentrated on a single point, as shown in Figure 12.6a. In this case, based on the above definition of the concentration

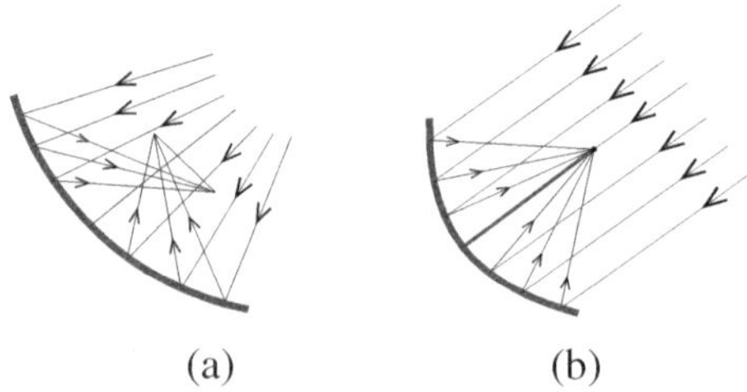

FIGURE 12.6
Schematic of concentration of parallel rays (a) and diffusing real solar radiation (b).

[3] This definition is called the geometric concentration ratio. There is another definition called optical concentration ratio, which is based on solar radiant flux. Here we will use the geometric concentration ratio.

ratio, the maximum concentration ratio could approach infinity (note that the area of a point is zero). But in reality, Sun's rays are not perfectly parallel, so concentrated Sun's rays are distributed in an area (not a point) as shown in Figure 12.6b. This means there is a theoretical limit to the concentration ratio that a solar collector can achieve. However, practical concentration ratios for typical solar collectors are much lower than this limit. Some reasons for the lower than the maximum concentration ratio are

- diffused portion of incoming solar radiation
- lower than perfect reflection coefficient of reflecting surfaces
- imperfection on the surface of reflectors
- errors in the tracking device

There are two design philosophies for the design of solar energy collection and concentration systems both requiring a large number of solar collectors. What distinguishes these two philosophies is how they transfer the collected high-temperature thermal energy to the heat-transfer fluid so the energy can be sent to the energy conversion system. In one design approach, many solar collectors reflect and concentrate solar radiation to a **central receiver**, typically on a **central tower**, where the energy is transferred to the heat-transfer fluid. These plants are referred to as **central receiver** or **tower power plants**. In the other approach, each solar collector unit collects and concentrates solar radiation on a receiver dedicated to that unit and transfer the energy to the heat-transfer fluid. The high-temperature heat-transfer fluid is collected from all receivers and sent to the power conversion system. These systems are referred to as **distributed** or **dispersed solar thermal energy collection systems**. Note that in both cases, the amount of solar energy in the concentrated form is not greater than its initial form (as expected from the first law of thermodynamics). Just the same amount of energy is concentrated in a smaller area. For the next several sections, we will focus primarily on the major components of solar thermal power plants. We will start with power plants with a central receiver followed by distributed plants.

12.4 Solar Thermal Power Plants with a Central Receiver

These plants are composed of a large number of individual mirrors, called **heliostats,** that reflect solar radiation to a central receiver on the top of a tower, hence the name **central receiver** or **tower solar thermal power plants**.

12.4.1 Heliostats

Heliostats[4] are steerable reflecting mirrors that receive solar radiation directly from the Sun. Considering the sheer number of heliostats and their size, their cost can be a significant portion of the total cost a solar thermal power plant (even as high as one-half of the total cost). Therefore, their appropriate design is not only critical for the performance of the plant but also for its cost. A heliostat is composed of one or several reflective surfaces, a support structure, and a Sun tracking system.

Traditionally, the reflective surfaces of heliostats are made of glass with a thin layer of silver on the back as the reflective layer. This layer is covered with a metal coating, typically copper, for the protection of the reflective layer. These mirrors are in principle similar to household mirrors and are referred to as **second-surface mirrors** since light has to pass through the glass before reaching the reflective layer (which is the second layer, hence the name). In second-surface mirrors, the passage of light and its reflection through the glass layer distort the reflected light (Figure 12.7). Also, a slight reflection of light by the front of the glass creates a faint secondary image (sometimes referred to as a "ghost image" or "ghosting effect"). While these problems are not usually important for mirrors used in heliostats, for some applications, they are not acceptable. In these cases, front-surface or first-surface mirrors are used where the reflective layer is applied to the front surface of the glass.

To reduce light absorption by the glass layer of heliostat mirrors, low-iron glass (also called extra-clear glass) is used. The reflective surface of a heliostat can be a few square meters to over $100\,m^2$. For the ease of manufacturing, transportation, and replacement (in the case of glass breakage), large heliostats are made of a series of smaller mirrors. Ideally mirrors used in heliostats should be slightly curved (concave). However, since the difference is not significant, to reduce cost, typically flat mirrors are used. These

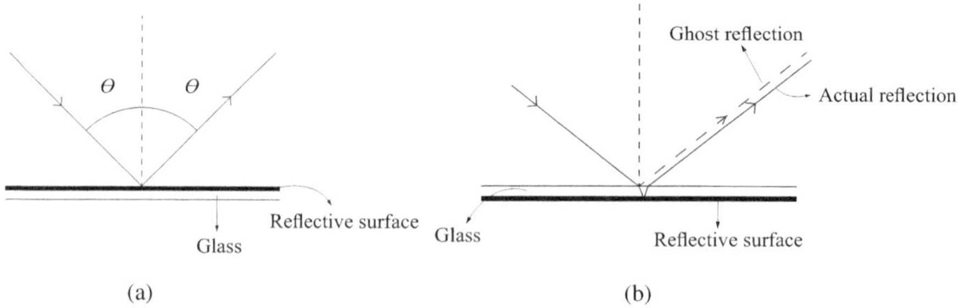

FIGURE 12.7
Schematics of first-surface (a) and second-surface (b) mirrors.

[4] The word Heliostat is derived from "helios", the Sun in Greek, and stat, a short form for the word stationary.

glass mirrors should be robust and have excellent reflectance and durability. However, they are generally expensive and vulnerable to breakage. They are also heavy, which not only increases their installation cost but also requires a stronger and more expensive heliostat support structure and tracking system. To address these problems and reduce the cost and weight of mirrors, heliostat reflective surfaces can be formed by silvered polymer mirrors or aluminum mirrors. These new designs however still suffer from lower reflectivity, lack of robustness, deterioration after prolong exposure to sunlight, and/or high manufacturing costs. There is ongoing research on improving these alternative reflective surfaces. Another factor that impacts the effectiveness of these mirrors is their cleanliness. The excessive amount of dirt on these surfaces can significantly increase their reflective losses. This problem can be intensified by dusty environments and occasional sandstorms, which are typical in hot arid regions suitable for these plants.

Aside from reflective surfaces, each heliostat has a support structure that maintains the integrity of mirrors and transfers the load to the pedestal. The pedestal in turn transfers the load to the foundation. These components should be designed to not only stand heavy mirrors but wind, snow, and seismic loads.

12.4.1.1 Heliostat Control Systems

As noted, heliostats typically reflect solar radiation to a central receiver although this is not always the case, for example, the Odeillo solar furnace discussed in Section 12.3. Unlike solar PV panels, the function of heliostats is not to **collect** the maximum amount of solar energy on the surface of the mirror but to **reflect** the maximum amount of solar energy to the central receiver. For this reason, unlike solar PV panels, the reflecting surfaces of heliostats are not always perpendicular to the incoming solar radiation (though this may occasionally happen), as shown in Figure 12.8. According to the cosine emission law (Chapter 11), this significantly reduces the effective

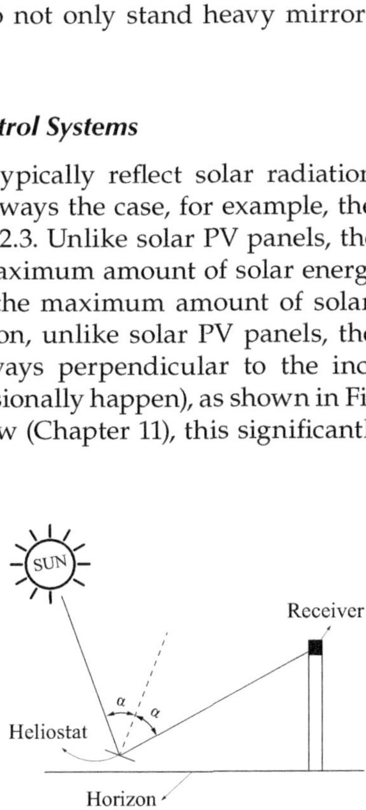

FIGURE 12.8
Schematic of reflection of solar radiation on a heliostat.

area of heliostats and decreases the intercepted solar radiation. It should be noted that solar thermal energy systems **can only use direct solar radiation**. Due to the random direction of diffused solar rays, they cannot be accurately reflected to the receiver. The optimal orientation of each heliostat depends on the position of the Sun relative to Earth (which in turn depends on the date and time), the location of the site, and the position of the heliostat relative to the tower. Geometrically speaking, the process of heliostat orientation control is simple. The mirror should be always perpendicular to the bisector of the angle between the line connecting the mirror to the central receiver and the Sun's rays (Figure 12.8).

This simple geometrical principle is the backbone of all computerized heliostat control systems. In older control systems, light sensors were used to detect the direction of the Sun. However, most modern systems use computational methods to determine the direction of the Sun rays based on the day, time, and location. Then, the location of the central receiver is used to determine the optimum orientation of each mirror. Based on the difference between the current and optimum orientations of the heliostat, the computer sends corrective signals to motors, typically step motors, to turn the heliostat in an appropriate direction. This procedure is repeated at a predetermined interval to ensure that heliostats are always optimally orientated. Typically a closed-loop control system is used to detect all possible errors and correct them (for more on closed-loop control systems refer to Chapter 3).

The control mechanism of a solar collector can be generally **single-** or **two-axis** (dual-axis) with one or two degrees of freedom, respectively. Typically, a single-axis system only follows the Sun vertically, while a two-axis system follows the Sun both vertically and horizontally. Since heliostats must always reflect solar radiation to the central receiver accurately, they are universally equipped with two-axis control systems.

The accuracy of the heliostat control system is critical. In a case of even slight error in the angle of reflection, considering the relatively large distance between heliostats and the central receiver, the target can be missed with a large margin. This not only results in the loss of efficiency but also can cause damage to equipment and structure or hurt personnel (considering the very high temperature of the point where the reflection hits a surface). Another challenge is to maintain the accuracy of the control system during windy conditions. In a case of emergency, the control system should be able to rapidly redirect solar beams from the receiver to a safe direction.

12.4.1.2 Heliostat Field

In the previous section, we learned that a large number of heliostats should be used to reflect sunlight toward the central receiver (Figure 12.9).

The number of needed heliostats depends on the power output of the plant and the desired concertation ratio. The arrangement of these heliostats in a **heliostat field** is a surprisingly challenging task. In the first thought, it might

FIGURE 12.9
A heliostat field and central tower. (Image by Samuel Faber from Pixabay, https://pixabay.com/photos/sun-solar-energy-solar-power-plant-4020112/.)

be logical to place them in a circular pattern around the central receiver each with a small angle to the next one. However, we recall from our discussion in Chapter 11 that north-facing heliostats, even with a sun tracking system, receive much less energy than south-facing ones. Clearly, this is only valid in the northern hemisphere, and in the southern hemisphere it is the opposite. For the rest of this section, all discussions are for sites in the northern hemisphere. Therefore, it makes sense to try to place all heliostats in the northern part of the field facing south (Figure 12.10). In fact, in most small plants this

FIGURE 12.10
Arrangement of heliostats in the heliostat field for the PS20 power plants with 1255 heliostats. (By Koza1983 – Own work by the original uploader, CC BY 3.0, https://commons.wikimedia.org/w/index.php?curid=7008971.)

is the case. However, in large plants with a larger number of heliostats, placing all heliostats in the northern section of the field results in more of them being farther away from the central receiver. The sunlight reflected from these far-away heliostats must travel more distance through air to reach to the receiver, which causes more attenuation of their energy. We can reach the point that placing heliostats in the eastern, western, and eventually southern parts of the field can yield more energy than just placing them farther on the northern side (Figure 12.11). The optimum solution can be achieved by balancing the energy losses due to the cosine emission law and the energy losses due to the placement of the heliostats far from the receiver. It should be noted that the latter losses depend on the local air quality (especially particulates and smoke in the atmosphere) and the possibility of sandstorm, fog, etc., in the region.

Other important considerations in determining the arrangement of heliostats are their shadowing and blocking effects on each other. At a certain time of a day, some heliostats can cast shadow on other heliostats by not allowing solar radiation to reach them. Also, some heliostats can block the reflection of other heliostats from reaching the receiver. These two losses can be reduced by increasing the spacing between rows and columns of heliostats. However, this strategy increases the distance of heliostats and the receiver with its own negative impacts discussed above as well as increase in the area of the required land. Similar to many other engineering problems, there are conflicting parameters that require an optimum compromise. One strategy to reduce these problems is to place heliostat in a staggered pattern (Figure 12.12).

Another important factor in the size and shape of the field is the height of the receiver's tower. A taller tower can increase the visibility of the receiver by the heliostats but can increase the distance the reflected lights have to travel

FIGURE 12.11
Arrangement of heliostats in a heliostat field. (Image by Christo Ras from Pixabay, https://pixabay.com/photos/solar-farm-solar-photovoltaic-4683339/.)

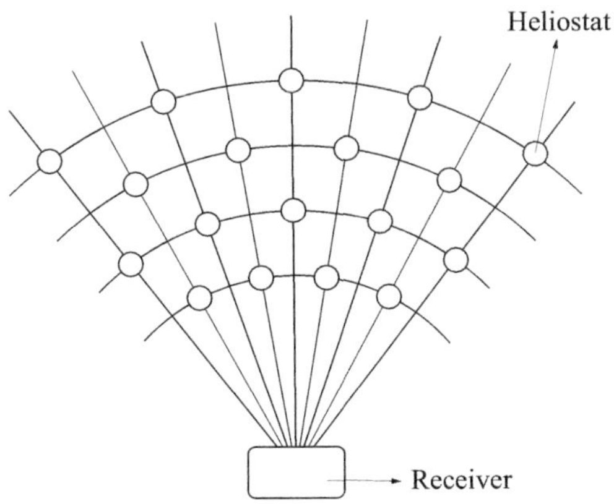

FIGURE 12.12
Schematic of heliostats in a staggered pattern.

and increase the negative impact of the tower shadow on the performance of the heliostats. Furthermore, it should be noted that the above discussion is for when there is a flat land surrounding the central tower, which is not uncommon considering suitable sites are typically located in hot deserts. However, sometimes, the local topography can significantly impact the possible arrangement.

It can be seen that there are many factors, including costs, to take into account in the layout design of a heliostat field. An optimum design can be often reached by a computerized optimization model considering all these sometimes-conflicting parameters.

12.4.2 Central Receivers

The functions of a central receiver are to receive concentrated solar radiation from heliostats, absorb the thermal energy, and transfer that heat to the heat-transfer fluid. As noted above, to perform the first function, to intercept reflected solar radiation, a receiver should be located on a tall tower. Central receivers can be divided into three categories: external, cavity, and volumetric receivers. In the following sections, we will review these types of receivers.

12.4.2.1 External Receivers

In external receivers, the heat-transfer surfaces are openly placed on the top of the tower. The heat exchanger surfaces, typically tubes (that is why they are also referred to as **external tubular receivers**), are completely exposed to the elements. In essence, they are similar to furnaces of fossil fuel steam generators where the furnace is enclosed by waterwall tubes. However, in this case, external surfaces (the outer perimeter of the enclosure) are exposed to the heat source. The absorbed heat from concentrated solar radiation is transferred to the heat-transfer fluid flowing inside the tubes. They are typically composed of flat or curved panels with parallel vertical tubes interconnected

FIGURE 12.13
Schematic of a cylindrical external central receiver.

by the fluid distribution and collection headers at two ends (Figures 12.13 and 12.14). These panels are typically assembled in a cylindrical or cubical shape though other configurations have also been proposed.[5] Due to their simplicity, they are most commonly used receivers. Their shape helps these receivers to intercept concentrated light from all directions, which makes them suitable for heliostat fields that surround the central tower. However, they suffer from various forms of heat loss, mainly radiation, reflection, and convection. The former two losses can be reduced by applying special coating or painting to the heat-transfer surfaces. These coverings can absorb a wide spectrum of solar energy but radiate little heat. The radiation losses are

[5] Wang, W.-Q., et al. "Optical efficiency improvement of solar power tower by employing and optimizing novel fin-like receivers." *Energy Conversion and Management* 184 (2019): 219–234.

FIGURE 12.14
External central receivers. (Image by Michael4Wien from Pixabay, https://pixabay.com/photos/sun-electricity-production-renewable-4535891/.)

a function of the operating temperature of the receiver. The latter loss is particularly intensified on windy days although it occurs by natural convection too. They have also more limitations on how much thermal energy they can receive and the temperature they can tolerate compared to other receivers. Ivanpah plant (discussed in Section 12.3) used this type of receivers.

12.4.2.2 Cavity Receivers

Cavity receivers, as the name implies, are in the form of a cavity, the inside of which is lined with tubes (Figures 12.15 and 12.16). The concentrated solar radiation from heliostats enters the cavity via an opening in the front of the receiver. The thermal energy is absorbed by the heat-transfer surfaces and is transferred to the heat-transfer fluid flowing inside the tubes. Due to the enclosed nature of the receiver, the energy reflected or radiated by the heat-transfer surfaces is more likely to be intercepted and absorbed by other surfaces rather than being lost. Also, since the heat-transfer surfaces are protected from environmental elements (particularly wind), these receivers have less convectional losses. For a given projected area, they can provide more heat-transfer surface area than external receivers. A limitation of cavity receivers is that a single cavity can only intercept solar radiation from one side (the direction of the opening) so they best suit fields with heliostats located in the north part of the field. For other types of fields, more than one cavity receivers are needed to cover the entire field. The opening of the receiver should be large enough to accommodate the entire reflected radiation from the farthest heliostats.

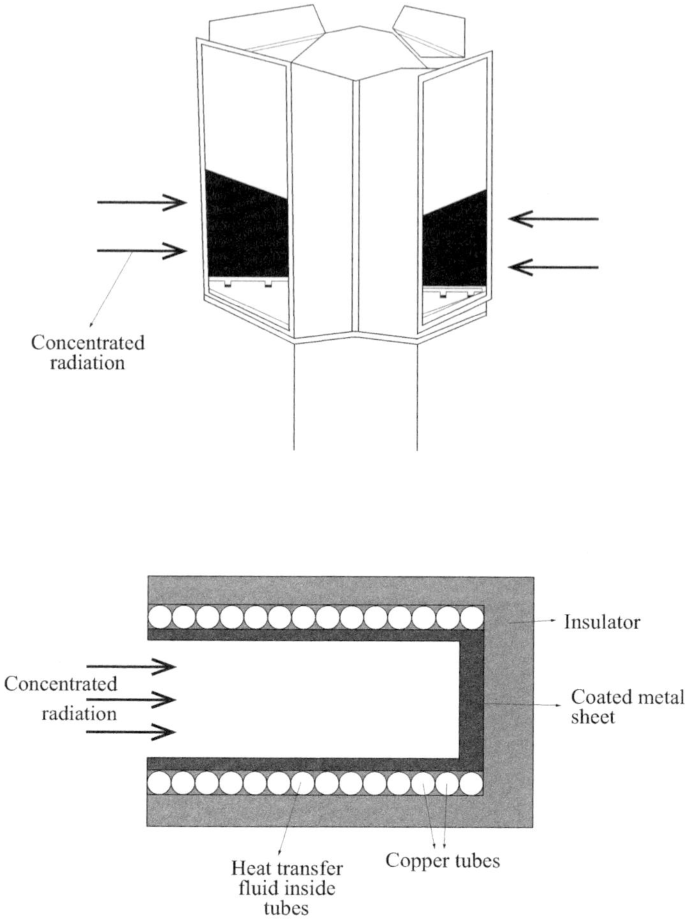

FIGURE 12.15
Schematic of a cavity receiver.

12.4.2.3 Volumetric Receivers

Unlike the previous two receiver types where the solar radiation is absorbed by heat-transfer surfaces and is then transferred to the heat-transfer fluid on the other side of the surfaces, volumetric receivers absorb concentrated radiation by the entire body of the receiver medium (Figure 12.17). These receivers are made of a porous medium that is exposed to solar radiation and absorbs heat. The heat-transfer fluid (typically air) flows through the medium at the same time to transfer heat away from the receiver. Several media have been used as the heat absorber in these receivers, including honeycomb, foam, and wire meshwork made of metals or ceramics. These receivers can be open or

FIGURE 12.16
A cavity receiver. (Image by sgrunden from Pixabay, https://pixabay.com/photos/solar-energy-solar-tower-sun-5366189/.)

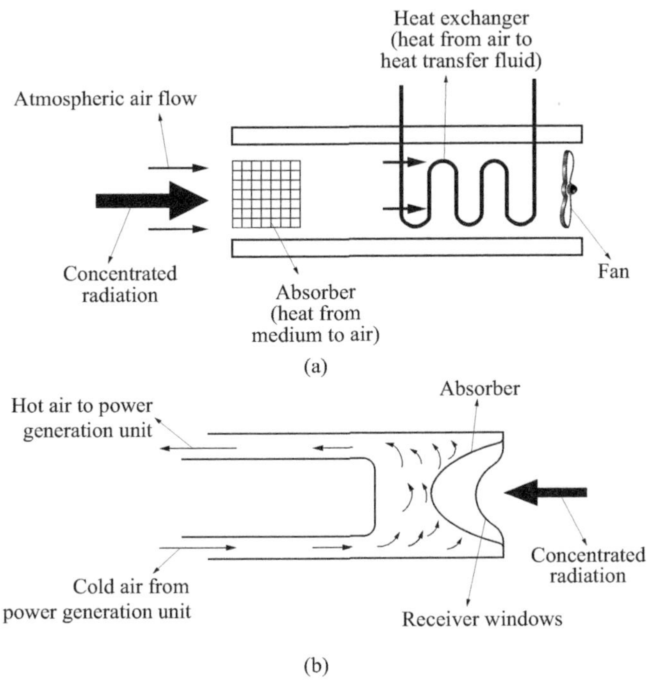

FIGURE 12.17
Schematics of volumetric receivers. (a) open receiver, (b) closed receiver.

closed systems. In open systems, the ambient air directly flows through the medium, with no pressurization, by a fan (Figure 12.17a). In closed systems, the pressurized air flow is used as a coolant in the receiver (Figure 12.17b). In this case, the medium is placed behind a transparent window to allow solar light in while avoiding air leakage. Due to manufacturing limitations in the size of these windows, commonly, several windows are used in a single receiver. These receivers are the youngest of the three types of receivers, and currently, a wide variety of concepts and designs are being suggested, analyzed, and tested.

12.4.2.4 Thermal Losses in Receivers

The losses in receivers can be divided into three groups:

- failure to intercept concentrated radiation from heliostats
- failure to absorb the intercepted light (by reflection losses)
- losing the absorbed thermal energy (by radiation or convection losses)

We have already discussed the latter two losses for each type of receiver. The losses because of failure to intercept concentrated radiation from heliostats, also referred to as **spillage losses**, are technically a part of collector losses. These losses occur because the concentrated radiation reflected from heliostats misses their target, which is heat-transfer surfaces of the receiver. These losses are primarily due to errors in the heliostat control system but can be intensified by wind impacts. Increasing the size of the part of the receiver that directly incepts the reflected light can reduce spillage losses but can increase heat-transfer losses.

12.5 Solar Thermal Power Plants with Distributed or Dispersed Collectors

Similar to power plants with a central receiver, solar thermal power plants with distributed or dispersed collectors have a large number of collectors. However, distributed collectors do not reflect and concentrate solar radiation to a central receiver, rather they have local dedicated receivers for each collector or a series of collectors. So solar radiation collection, concentration, and absorption all take place in the same unit. The heat-transfer fluid collects thermal energy from all receivers and transfers the energy to a power generation unit. Alternatively, each collector can have its own power generation unit.

There are two types of collectors-concentrators that can be used in these plants:

- line-focus collectors-concentrators
- point-focus collectors-concentrators

12.5.1 Line-Focus Collectors

A line-focus concentrator is a long trough or conduit with a parabolic or semicircle cross section (Figure 12.18). They are also known as trough, parabolic, or parabolic trough concentrators. The inside of the trough is lined with reflective surfaces. The reflective surfaces can be made of a single-piece glass mirror, strips of parallel mirrors, polymers with a silver layer inside, and most commonly polished metal sheets. A parabolic trough system is made of a series of modules, each a few meters to over 10 m, connected to form rows each hundreds of meters long. These mirrors concentrate solar rays at their optical focal line. The concentration ratio of these reflective surfaces is substantially lower than that for central receiver systems, which results in a lower temperature on the receiver. To receive this concentrated light, a straight pipe with a special coating to maximize absorption runs along the focal line. The heat-transfer fluid functions as a coolant for this pipe and transfers the thermal energy to a power generation unit (or an energy storage system).

Since in parabolic trough concentrators the receiver is fixed relative to the reflective surface, they can even operate without a sun tracking system (Figure 12.19). In this case, a row of collectors is installed on a west-east axis facing south. The direction of reflective surfaces can be seasonally adjusted according to the height of the Sun in the sky. While simple and inexpensive, fixed parabolic trough concentrators are inefficient in intercepting incoming solar rays and are not typically used. More commonly these concentrators

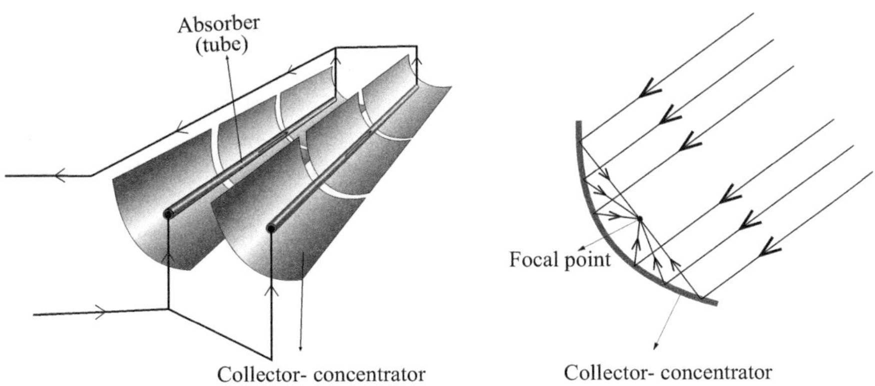

FIGURE 12.18
Schematic of a parabolic trough concentrator.

FIGURE 12.19
A parabolic trough concentrator. (By Z22 - Own work, CC BY-SA 3.0, https://commons.wikimedia.org/w/index.php?curid=27881587.)

are equipped with a single-axis sun tracking system (with one degree of freedom). In this case, the trough is installed on a north-south axis so the collectors can follow the Sun as it moves across the sky from east to west.

These concentrators have been around for over a century. The plant proposed by Frank Shuman discussed in Section 12.3 was an early example. They have also been extensively used for nonelectricity generation solar thermal energy applications where relatively high-temperature energy is needed. Parabolic trough solar thermal power plants have been commercially in service for many years. Currently, a majority of major solar thermal power plants are of this type, including some of the largest plants in the world, including most of the Ouarzazate Solar Power Station, the largest solar thermal power plant in the world with a total capacity of 510 MW in Morocco, the 361 MW Solar Energy Generating Systems (SEGS), and the Genesis Solar Energy Project and the Mojave Solar Project each with the capacity of 280 MW, all in California as well as the 280 MW Solana Generating Station in Arizona. Similarly, the Extresol, Andasol, and Solnova solar power stations each with about 150 MW power generation capacity and the 100 MW Valle Solar Power Station, all in Spain, used this technology.

Another line-focus concentrator, called **compact linear Fresnel reflectors** (CLFR), uses long narrow strips of mirror to reflect and concentrate solar radiation on two points (Figure 12.20). Each strip is individually controlled to reflect light to the receivers. The main advantage of these systems is that receivers can be fixed and do not need to move with reflective surfaces, as in

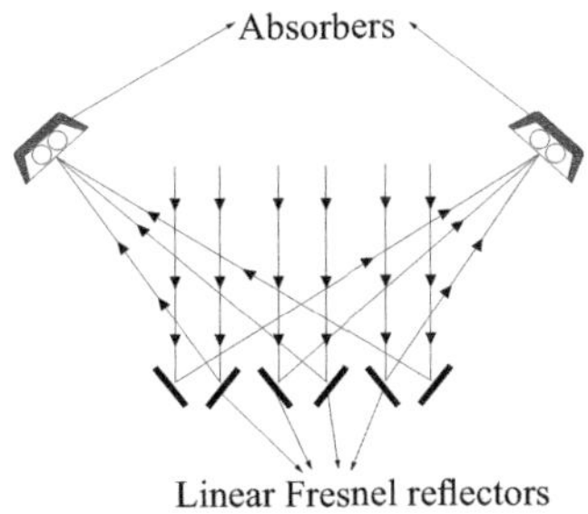

FIGURE 12.20
Schematic of a line-focus compact linear Fresnel reflector-collector-concentrator.

conventional trough collectors. Also, they have an inherently low resistance to wind, which substantially reduces the cost of their support system. Several designs of these systems have been tested but no large-scale unit has been built yet.

12.5.2 Point-Focus Collectors

Point-focus concentrators are typically in the shape of a parabolic dish in which the inside of the dish is covered by reflective surfaces (Figure 12.21). These surfaces focus the incoming solar rays on its focal point, hence the name. They are also referred to as **dish concentrators**. Fresnel lenses have also been used with a similar objective, i.e., to concentrate solar rays in a point. These collectors can achieve a high concentration ratio, which in turn means they can provide a high temperature at the receiver. These systems are typically equipped with a two-axis solar tracking system, which also promotes a high solar gain.

It should be noted that in both line- and point-focus concentrators, reflective surfaces should be directly pointed at the Sun (similar to PV panels but unlike heliostats). The receiver in these units is placed on the focal point of

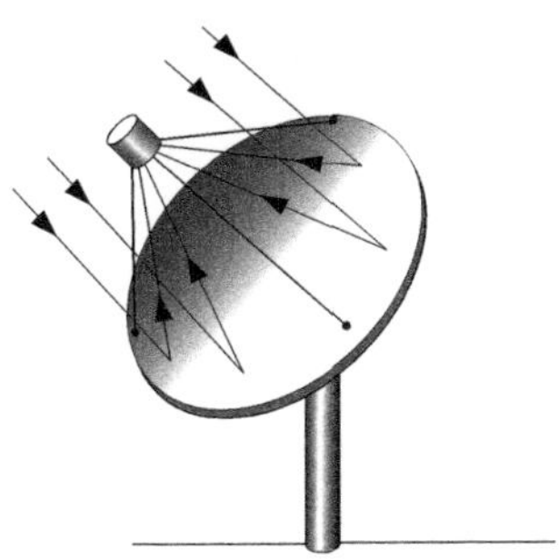

FIGURE 12.21
Schematic of a dish concentrator.

the dish and is fixed to the dish. This practically eliminates relative movements of the receiver with respect to the dish, which reduces spillage losses to almost zero. The receiver in dish concentrators can be in the forms discussed earlier for central receivers. Cavity receivers particularly suit these collectors. However, in these units the power generation unit, a Stirling engine (more on this engine below), is commonly installed as a receiver.

Each disk and receiver along with their associate systems is mounted on an individual pedestal and foundation. This modular nature of point-focus solar thermal systems, especially if the system is equipped with an engine, can give significant flexibility in the construction and operation of these units. Similar to heliostats, the size of these units is limited by forces the support structure can tolerate. While several companies have developed and tested individual modules for dish solar thermal power units, a large-scale plant has not been built yet.

12.6 Heat-Transfer Fluid

After reflected radiation is intercepted and absorbed by the receiver, it should be transferred to the heat-transfer fluid. The heat-transfer fluid functions as the coolant for the receiver and the heat source for the power generation system. The heat-transfer fluid can be the working fluid of the power generation unit or an intermediate fluid that transfer thermal energy to the working fluid of the cycle. Several fluids have been used for this purpose, including pressurized water, molten salt, liquid metal, or air. The properties of the fluid, such as its heat-transfer properties, specific heat, viscosity, and density, can impact the design and size of the receiver. Aside from the heat-transfer surfaces and the fluid itself, the system includes a network of pipes or ducts as well as pumps, fans, or compressors.

Using **pressurized water** as the heat-transfer fluid provides some unique benefits. In this case, the heat-transfer fluid and the working fluid of the power cycle are the same. This reduces the number of required equipment. In this approach, the pressurized liquid is heated in the receiver and is converted to superheated or saturated steam, which is then sent to the steam turbine. In principle, the receiver functions as a steam generator. A major advantage of this system is that the already existing technologies developed in more than a century for conventional steam power plants can be adapted for the new application. The receiver can raise steam using either a drum-type steam generator or a once-through steam generator (refer to Chapter 3 for more information on these steam generators). A major disadvantage of these systems is that steam cannot be typically used for energy storage with a high efficiency. Also, the operation of the steam turbine is challenging since the variable solar gain in the receiver can result in ever-changing conditions of the generated steam. Regardless of these disadvantages some of the most recent concentrated solar power systems are using this technology, including Ivanpah power facility and Mojave Solar Project in California, and PS10 and PS20 in Spain.

Another option for the heat-transfer fluid is using a **molten salt mixture**. For this purpose, a mixture of different nitrate salts (referred to as a eutectic mixture), such as sodium nitrate, potassium nitrate, or calcium nitrate, can be used. The melting temperature of the mixture can be controlled by varying the ratio of various salts to match the requirements of the optimum operation of the receiver and the steam generator. A mixture of 60% sodium nitrate and 40% potassium nitrate has been commonly used. The relatively high specific heat of these molten salts helps them to carry and store a large amount of thermal energy for a given mass and volume. Unlike receivers that use pressurized water/steam as a coolant/heat-transfer fluid, molten salt systems can operate at near atmospheric pressures. In these units, the "cold" molten salt flow at the temperature of about 300°C (about 570°F) enters the heat exchanger in the receiver and is heated up to the temperature of about 600°C (about 1110°F). A major advantage of these systems is that they can be readily integrated into an energy storage system. The high-temperature molten salt flow can be directly sent to the steam generator or can be stored for later use. A major drawback for these systems is the high melting temperature of the salt mixture, which means the fluid will be solidified below this temperature, for example when the plant is not operational. To avoid this problem, all components and pipes that are in direct contact with the molten salt mixture are equipped with electric heaters to maintain their temperature above the freezing temperature of the mixture at all times. The Solar Two project in California and more recently the Solana Generating Station, a solar thermal power plant with the nameplate capacity of 250 MW in Arizona, and Crescent Dunes plant, a 110 MW unit in Nevada, have used this system. Andasol solar power station and Extresol plant in Spain, each with about 150 MW power generation capacity, also used this technology.

Liquid metals, particularly liquid sodium, has been demonstrated to be effective heat-transfer fluid for solar thermal power plants. These systems have the advantages and disadvantages very similar to molten salt systems. In the case of liquid sodium, the fact that sodium can violently react with air and water and create fire hazards should be carefully considered in the system design.

Heat-transfer oils, such as synthetic oils, have been also used as the heat-transfer fluid in some plants. While they enjoy most benefits of molten salts, they have an added benefit of lower corrosiveness. However, they have a tendency to decompose at high temperatures, which limits the maximum temperature in which they can operate. They are also flammable and have low viscosity at low temperatures. To address the latter problem, electric heating elements should be installed on piping systems.

Finally, gases, mainly air, can be used to transfer thermal energy from the receiver to the power generation unit. While the heated air can raise steam in a steam generator to power a steam power plant, this system is more aligned with gas turbines (Brayton cycles) where the flow of hot air directly runs a gas turbine. Aside from being freely available, stable, and nontoxic, a major advantage of air is that there are no physical limits on the operation temperature of the air flow. The air temperature in the receiver can exceed 700°C (about 1300°F). While technically all types of receivers can use air as the coolant, this system is typically associated with volumetric receivers. The air flow to the receiver can be at near atmospheric pressure or can be pressurized. A drawback of this system is that due to the low specific heat and density of air, a large volume flow rate of air is required.

12.7 Power Generation Units

We learned that generally speaking, solar collectors-concentrators gather solar radiation and concentrate it on the receiver. The receiver intercepts and absorbs the concentrated radiation and transfers it to the heat-transfer fluid. The fluid transfers the heat away from the receiver. In this section, we will discuss how this energy can be converted to electricity in a central power generation unit. This unit can operate based on a Rankine cycle, Brayton cycle, or a combined cycle. The basic thermodynamic principles of all these cycles are as discussed for fossil fuel energy sources in previous chapters. The main difference is that the source of thermal energy is not the combustion of fossil fuels but solar thermal energy. Clearly, this difference will influence the design of some components, particularly the steam generator or combustion chamber; however, most of the cycle will remain unchanged. In this section, we will review the major differences for each cycle.

For a Rankine cycle used in a solar thermal energy plant, the important difference between this cycle and the cycle used for a fossil fuel power plant is in the steam generator. The configuration and design of a steam generator in a solar thermal plant depend on the heat-transfer fluid. If pressurized water is used as the heat-transfer fluid, the same water is also the working fluid of the Rankine cycle. As a result, the receiver also functions as the steam generator. For central receiver plants, commonly the entire power generation unit, except the cooling tower, is located inside the central tower. If the heat-transfer fluid other than water is used, a separate steam generator is needed where the thermal energy of the hot heat-transfer fluid is used to raise superheated or saturated steam. As noted earlier, either a drum-type or once-through steam generator can be used. For example, in a parabolic trough system, most collectors can act as the boiler (evaporator) that provides boiling water to the drum and the rest of the collectors work as the superheater. The design and operation of the condenser, turbine, and feedwater system are similar to their counterparts in fossil fuel power plants although they have to be costume designed for the operating conditions of each case. As for the cooling tower, both wet and dry cooling towers have been used. However, considering environments that these units are typically installed, dry cooling towers are more common. Note that using a dry cooling tower will inevitably reduce the efficiency of the cycle (as discussed in Chapter 8), which requires larger components than those for a cycle with a wet cooling tower for the same capacity.

As previously noted, steam does not suit energy storage systems. Therefore, when water/steam is used as the heat-transfer fluid, typically the plant does not include an energy storage facility. In plants with steam cycles to overcome problems caused by the intermittent nature of solar energy, the plant is typically equipped with a supplementary natural gas steam generator. Whenever power is needed to be generated while solar energy is not available, natural gas is combusted to produce steam for the power plant. While burning fuel reduces the environmental benefits of using renewable solar energy, it substantially increases the reliability of the plant in meeting the peak demands of the grid (this topic will be extensively discussed in Chapter 16). Currently, the Rankine cycle is the only cycle that has been used in large commercial solar thermal power plants.

Due to the simplicity of the Brayton cycle, there is currently a great deal of research on adapting this technology for solar thermal energy systems. In these systems, the air flow is first pressurized in a compressor. The high-pressure air flow is then sent to the receiver, typically a volumetric one, where it absorbs solar thermal energy from the collectors. Finally, the high-temperature and pressure air expands through the gas turbine and generates power. A major advantage of these plants is that they do not need any water, a typically scarce commodity in deserts where most solar thermal power plants are located, to operate. Note that even with a dry cooling tower, steam power plants need a significant amount of water as makeup feedwater.

There have been proposals to use a combined cycle as the power generation unit in a solar thermal power plant, but no actual system has been built so far based on this concept. It has also been suggested to use the solar thermal energy technology to repower retiring or even currently operating fossil fuel steam power plants to reduce their fossil fuel consumption.

12.7.1 Stirling Engines

In all the above power generation cycles, a central unit is used to convert thermal energy gathered by the entire solar collection system to electricity. There is one important exception for these central power units where every individual solar collector has its own power generation unit (engine). This engine is known as a Stirling engine, which is typically installed on the focal point of a dish collector.

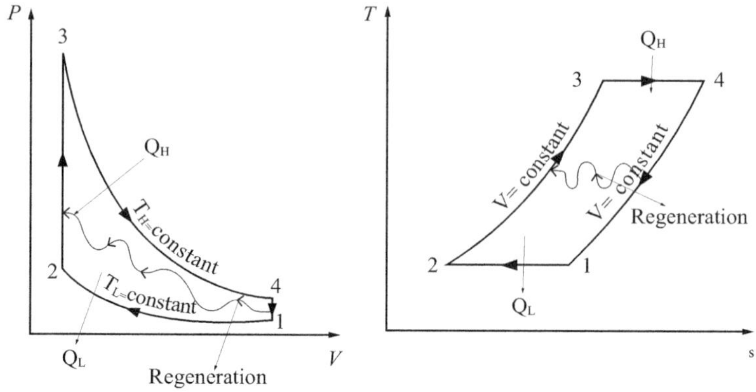

FIGURE 12.22
T-s and P-v diagrams of a Stirling thermodynamic cycle.

Stirling engines are external "combustion"[6] engines that operate on a closed regenerative cycle. They operate based on a Stirling thermodynamic cycle (Figure 12.22). Thermodynamically, a Stirling cycle is composed of two isothermal processes (constant temperature) and two isochoric processes (constant volume). The engine was invented by Robert Stirling (1790–1878, a Scottish inventor and church minister). In an ideal case, the efficiency of a regenerative Stirling cycle is equal to that of a Carnot cycle operating between the same thermal reservoirs. They are efficient and relatively simple with low-maintenance requirements. These characteristics make them an ideal choice for dish solar thermal power units, where the receiver and the power generation unit (including the generator) are integrated into a single unit. They may also have an integrated natural gas burner to generate electricity when solar radiation is not available.

12.7.2 Thermal Energy Storage

The topic of energy storage systems is critical in the context of renewable energy sources, particularly solar systems. For this reason, an entire chapter (Chapter 16) is dedicated to this topic (especially the concept of the duck curve). However, since **thermal energy storage** (**TES**) systems are currently associated with solar thermal energy systems, this particular energy storage system will be briefly discussed in this section. Here it suffices to note that in order for solar systems to make a significant impact on our electricity generation mix, we need to produce electricity from these systems even when solar energy is not available, for example at nighttime or cloudy days. This also includes early morning and evening peak periods when solar energy is not

[6] In solar applications, "combustion" is irrelevant and here is only used in reference to the difference between Stirling engine and internal combustion engines. In this context, what really matters is that the same working fluid goes through the cycle over and over (closed cycle).

fully available. For this purpose solar thermal energy power plants should be either equipped with thermal energy storage systems or have a backup fossil fuel burning (typically natural gas) combustion system. We have already briefly discussed the latter.

As the name implies, thermal energy storage systems store thermal energy when there is excess thermal energy available. This excess thermal energy can be at least partially collected at off-peak periods during the day. However, typically the solar collection system, the receiver, and the heat-transfer system should be designed with more capacity than that of the power generation unit. In this way, the extra thermal energy that is collected but not used by the power generation unit can be stored.

12.8 Environmental Impacts of Solar Thermal Energy Plants

While solar thermal energy is considered a renewable and sustainable energy source, it can have some adverse environmental impacts. Heliostat fields can occupy tens of square kilometers of land, which can cause visual/sight pollution particularly if the site is close to national parks or protected regions. These plants are typically installed in hot deserts that may look like a great place to install a power plant because of the lack of wildlife but in fact, many of these environments have a fragile ecosystem that can be easily disturbed. For example, when the Ivanpah Solar power project was being constructed in the Mojave Desert in California, there was a serious concern that it can impact desert tortoises, an endangered species. These problems can be significantly worse for central receiver power plants. The central receiver in these plants can get extremely hot, and it can emit very bright light (Figures 12.14 and 12.16). This light can attract insects, which can not only kill them but can also attract birds to the area. Highly concentrated rays between heliostats and the receiver can easily burn and kill birds that collide with them. In the case of the Rankine cycle, the usage of water for makeup water and wet cooling towers in dry regions is another concern. However, these problems can be addressed, at least to some degrees, by careful planning and in-depth environmental studies before or during the construction phase.

12.9 Questions and Problems

12.9.1 Low-Temperature Solar Thermal Energy Applications for Buildings

12.1. Explain how nonconcentrating collectors for solar water heating systems work.

12.2. Solar systems for heating water or air usually have … (nonconcentrating/concentrating) collectors.

12.3. Describe the following solar water heaters.

a. Unglazed collector

b. Flat plate

c. Evacuated tube

d. Parabolic trough

12.4. Provide an application of unglazed and flat-plate collectors.

12.5. Explain how the fluid is circulated in a natural-circulation SWH.

12.6. What are the two types of evacuated tube collectors? Explain how each one works.

12.7. A total annual thermal energy needed for water heating in a building is 2500 kWh. If a SWH annually provides 1800 kWh of that energy, determine the solar fraction of the system.

12.9.2 Solar Electricity Generation

12.8. What are the main components of a solar thermal power plant?

12.9. Name the main components of a solar thermal power plant in the following schematic diagram.

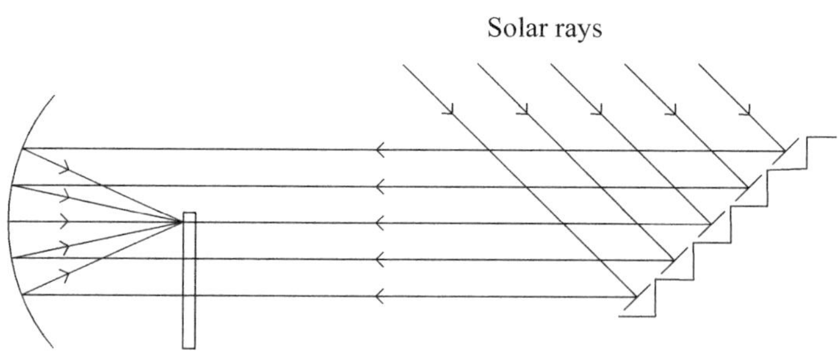

12.10. What is the definition of the concentration ratio of a device?

12.11. Explain why in reality concentrated Sun's rays are distributed in an area (not a point).

12.12. List factors that contribute to the reduced concentration ratio of real solar collectors.

12.13. Which of the following schematics illustrates a real solar collector?

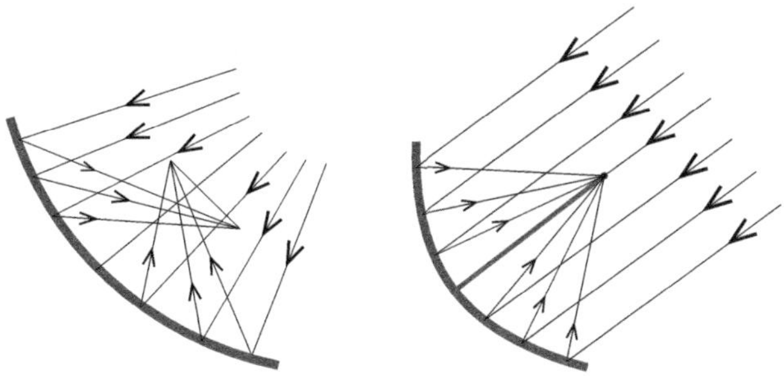

12.14. What are two design philosophies for the design of solar energy collection and concentration systems? Explain them.

12.15. What is the function of a heliostat?

12.16. What are the functions of the following components of a heliostat?

- Mirror
- Support system
- Pedestal
- Control system

12.17. What is the difference between first-surface and second-surface mirrors? What are the pros and cons of each?

12.18. What are mirrors in heliostats made of?

12.19. Explain how PV solar panels and heliostats are different in terms of their orientation relative to solar radiation.

12.20. Describe how the control system of a heliostat determines its orientation.

12.21. What are single- and two-axis (dual-axis) control mechanisms of solar energy collectors? How do they work? Which one is used for heliostats?

12.22. Why is the accuracy of the heliostat control system critical?

12.23. What is the ideal location and direction of heliostats in a heliostat field? Explain why in large plants we may need to move away from this ideal.

12.24. Explain various parameters that impact the arrangement of heliostats in a heliostat field.

12.25. Compare the heliostat fields in Figures 12.10 and 12.11.

12.26. What are various losses in heliostat fields?

12.27. Why are heliostats typically placed in a staggered pattern?

12.28. How can the height of the central tower impact the performance of the solar radiation collection system?

12.29. What are the different types of central receivers? Explain each type and pros and cons of each?

12.30. Name each of the following receiver.

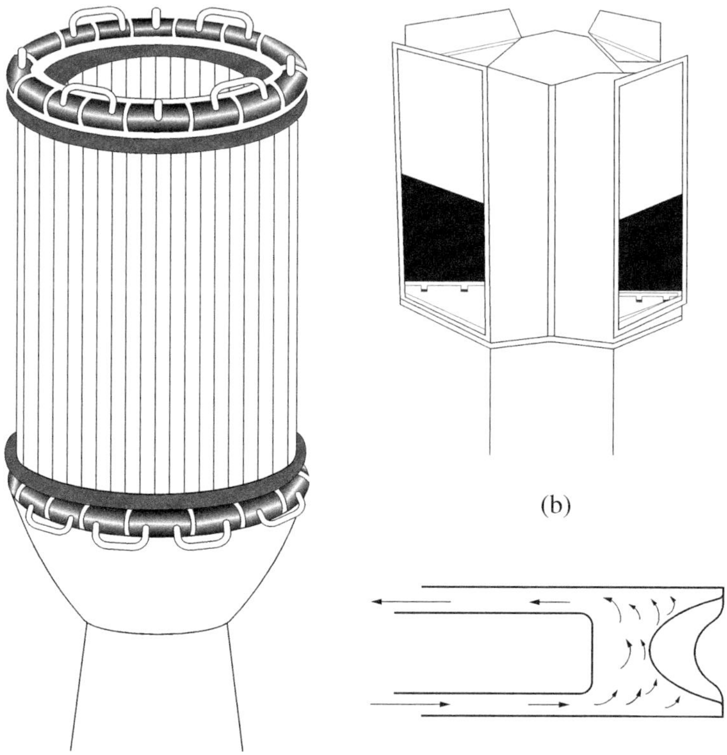

(a) (b) (c)

12.31. Can thermal power plants operate with diffuse solar radiation? Why?

12.32. Which one of the following solar collectors can operate with a single-axis (with one degree of freedom) sun tracking system?

a. Point-focus collectors-concentrators

b. Line-focus collectors-concentrators

c. Heliostats

12.33. What are thermal losses in a receiver? Explain how the reduction of one loss can increase another.

12.34. What are two types of collectors-concentrators that can be used in distributed collector plants?

12.35. Which one of the following solar collectors has the lowest temperature? Why?

a. point-focus collectors-concentrators
b. line-focus collectors-concentrators
c. heliostats

12.36. Compare line-focus and point-focus concentrators. What are their similarities and differences?

12.37. Which one of the following solar collectors is the most common one?

a. Point-focus collectors-concentrators
b. Line-focus collectors-concentrators
c. Heliostats

12.38. Explain how compact linear Fresnel reflectors (CLFR) work.

12.39. Which one is not correct for dish concentrators?

a. They can achieve a high concentration ratio.
b. They can provide a high temperature at the receiver.
c. They are typically equipped with a two-axis solar tracking system.
d. They are line-focus concentrators.

12.40. List some fluids that have been used as the heat-transfer fluid in solar thermal power plants. Compare the characteristics of each fluid.

12.41. What are the advantages and disadvantages of using high-pressure water as the heat-transfer fluid in a solar thermal power plant?

12.42. What are the advantages and disadvantages of using a molten salt mixture as the heat-transfer fluid in a solar thermal power plant?

12.43. Explain how air can be used as the heat-transfer fluid in a solar thermal power plant. What power cycle is typically used in this case?

12.44. Explain how a Rankine cycle can be used as a power generation unit in a solar thermal power plant. Describe how steam can be generated for the power cycle. Can energy storage systems be readily integrated to the system? Why?

12.45. What are the advantages and disadvantages of using a Brayton cycle as a power generation unit in a solar thermal power plant?

12.46. Which type of solar thermal power plants best suits for a Stirling engine?

a. Point-focus collectors-concentrators

b. Line-focus collectors-concentrators

c. Heliostats

12.47. How can the intermittent nature of solar energy be addressed in solar thermal power plants? (two approaches)

12.48. List some negative environmental impacts of solar thermal power plants. How can they be addressed?

12.9.3 General

12.49. Which one of the following solar collectors has the lowest concentration ratio?

a. Point-focus collectors-concentrators

b. Line-focus collectors-concentrators

c. Heliostats

12.50. Explain why the radially staggered layout is the most cost-effective layout for heliostats to minimize losses.

12.51. How can the local air quality (especially particulates and smoke in the atmosphere) and the possibility of sandstorm, fog, etc., in a region impact arrangement of a heliostat field?

12.52. Flat-plate collectors are used mainly for producing domestic hot water and, in some cases, building space heating.

a. True

b. False

12.53. Explain the shadowing and blocking as two sources of energy losses between heliostats and the receiver.

12.54. What are the typical maximum temperatures that each of the following systems is best suited to provide?

- Unglazed collector
- Flat plate
- Evacuated tube
- Parabolic trough

12.55. What are the advantages of using a Stirling engine in a solar thermal power plant?

12.56. What are the major sources of energy losses between heliostats and the receiver?

12.57. Space heating and hot water heating collectively can typically amount to about … of the average homeowner's energy bill.

a. 10%

b. 55%

c. 90%

d. 95%

12.58. Solar thermal energy can serve a wide variety of applications. Provide some examples.

12.59. In nonconcentrating collectors for solar water heating systems, the collector area (the area that intercepts the solar radiation) is the same as the absorber area (the area absorbing the radiation).

a. True

b. False

12.60. Name the following distributed collectors.

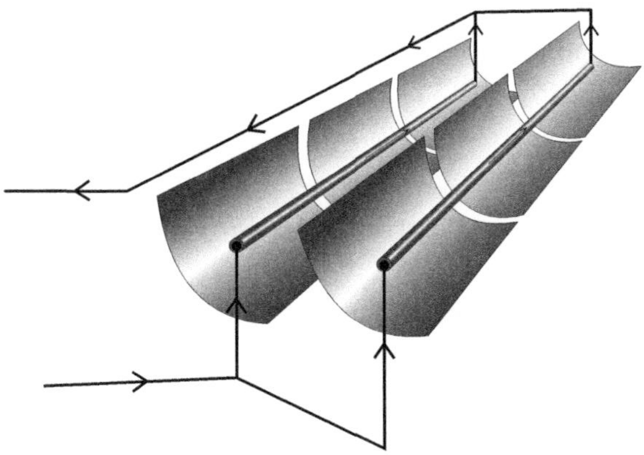

(a)

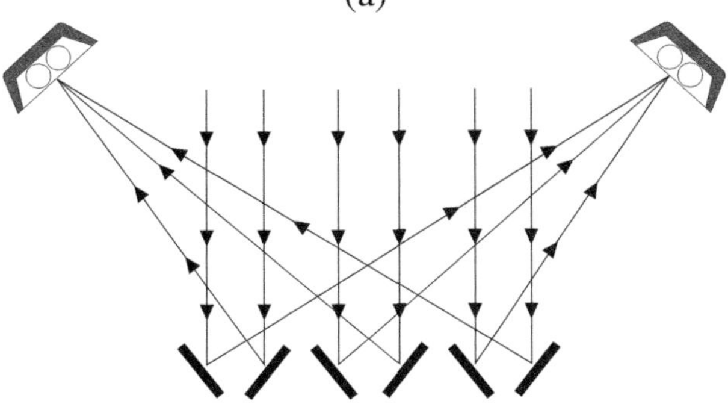

(b)

12.61. How and why is the orientation of solar PV panels and heliostats different (or similar)? Why?

12.62. Why typically slightly parabolic heliostats are not used for solar thermal power plants while they have a better performance?

12.63. Describe passive solar heating systems.

12.64. A solar thermal power plant always has a thermal energy storage system.

a. True

b. False

12.65. The number of needed heliostats for a solar thermal power plant depends on ….

12.66. What is the spillage loss?

12.67. Illustrate how heliostats are positioned around a central receiver in a small solar thermal power plant. What about a large power plant? What is the determining factor in the shape and size of the heliostat field?

12.68. For a real solar collector, the maximum concentration ratio could approach infinity. Why?

a. True

b. False

12.69. How are central and distributed solar thermal energy collection systems different?

12.70. Explain the thermodynamic cycle for a Stirling engine.

12.71. Which one of the following solar collectors has most spillage losses?

a. Point-focus collectors-concentrators

b. Line-focus collectors-concentrators

c. Heliostats

12.72. List various components of a heliostat.

12.73. In concentrating collectors the area intercepting the solar radiation is greater, sometimes hundreds of times greater, than the absorber area.

a. True

b. False

12.74. What is the spillage energy loss in central receivers? How can this loss be minimized?

12.75. Solar thermal power plants use … solar collector systems because they can produce high-temperature heat.

a. Concentrating

b. Nonconcentrating

c. Flat

12.76. Why unglazed collectors are more suitable for low-temperature applications?

12.77. Compare natural-circulation systems and forced-circulation systems for SWHs.

12.78. Solar water heaters (SWHs) represent a particularly cost-effective means of reducing one's energy costs.

a. True

b. False

12.79. What are the advantages and disadvantages of using liquid metals and oils as the heat-transfer fluid in a solar thermal power plant?

12.80. Explain the opposing impacts of the size of the receiver on the spillage loss and thermal losses of a central receiver of a solar thermal power plant.

12.81. Passive solar space heating happens when the Sun shines through the windows of a building and warms the interior.

a. True

b. False

12.82. Draw T-s and P-v diagrams of a Stirling cycle.

12.83. Why are thermal energy storage (TES) systems critical for solar thermal power plants?

12.84. Explain concentrating collectors for solar thermal systems.

12.85. Why do we need a storage system in a solar thermal power plant?

12.86. Active solar heating systems use a collector and a fluid that absorbs solar radiation. Fans or pumps circulate air or heat-absorbing liquids through collectors, which in turn transfer the heated fluid directly to a room or to a heat storage system.

a. True

b. False

12.87. List major components of a solar thermal power plant. Describe the function of each component.

12.88. Which one can produce higher temperature, point-focus systems or line-focus systems? Why? Which one is better suited for integration with a Stirling engine?

12.89. In concentrating collectors, the area intercepting the solar radiation is …. than the absorber area.

a. greater

b. smaller

c. equal

12.90. List several types of receivers. What are the advantages and disadvantages of each?

12.91. In a solar collector, the amount of solar energy in the concentrated form is greater than its initial form.

a. True

b. False

12.92. What is the functions of a central receiver?

12.93. Describe the operation and advantages of linear Fresnel collectors.

12.94. Solar collectors are either nonconcentrating or concentrating.

a. True

b. False

12.95. What are the various power production systems in a solar thermal power plant?

12.96. Why is the absorber surface in flat-plate systems often black painted?

13

Wind Energy

Abstract

Wind energy has been one of the fastest-growing energy sources for electricity generation. In this chapter, we will learn about the evaluation of wind energy, the operation of wind turbines, and their components. After a brief review of the history of utilizations of wind energy, we will learn how to determine wind energy using wind properties, particularly wind speed. Then, we will derive Betz's law to determine the upper theoretical limit for the efficiency of wind turbines. Next, we will discuss about some instruments that can be used to measure wind properties and will get familiar with statistical analysis of wind data collected through field measurements. Then, we will consider different categorizations of wind turbines and study each of them. We will learn what distinguishes lift or drag force wind turbines. We will discuss the characteristics of vertical axis and horizontal axis wind turbines and their applications. We will also review land-based and offshore wind turbines, grid-tied and stand-alone wind turbines, and small and hybrid wind turbines. Next, we will learn how we can use wind turbine power curves and the information they can provide. We will discuss how wind turbine power output can be controlled by stall-regulated, pitch-regulated, and variable speed control systems. Finally, we will discuss the major components of wind turbines. We will conclude the chapter with a brief discussion on environmental impacts of wind turbines.

LEARNING OUTCOMES

After successful completion of this chapter, you will be able to

- describe a brief history of utilization of wind energy
- determine wind energy using wind properties and wind maps
- derive and use Betz's law to determine the maximum theoretical efficiency of wind turbines
- explain wind property measurement instruments and their characteristics
- conduct simple statistical analyses of real wind data using the Weibull distribution

- determine wind speed as a function of height
- identify characteristics of lift and drag force wind turbines and conduct relevant simple calculations
- explain characteristics of vertical-axis and horizontal-axis wind turbines, land-based and offshore wind turbines, grid-tied and stand-alone wind turbines, and small and hybrid wind turbines
- describe power curves for wind turbines and information they can provide
- explain different methods to control wind turbine power output
- describe major components of wind turbines and their functions

13.1 Introduction

Probably, two decades ago, even the most optimistic observers could not predict that in such a short period, wind energy can become a nonnegotiable part of the global electricity generation industry. The study of incredible and successful evolution of wind energy can provide a road map for other renewable energy sources. To a certain degree, wind energy has moved from a novelty in the power generation industry to a form of a conventional power generation technology. In many parts of the world, one cannot have a day without seeing them integrated to landscapes. According to the IRENA (the International Renewable Energy Agency), the global installed capacity of wind turbines has increased from about 180 GW in 2010 to over 600 GW in 2019, over 330% growth in a decade.[1] The installed capacity of wind turbines in the U.S. in 2019 was 2.6 and 42 times greater than what it was in 2010 and 2000, respectively.[2] China, the U.S., Germany, and India have the largest wind turbine installations in the world.

Wind energy is a domestic renewable source of energy that is available in most of the world. Wind turbines can be installed relatively quickly in farms and lands that are already being used for other purposes. They do not need water for power generation. Most important of all, they are cost-effective and, in many cases, can compete with electricity generated from other sources. On the negative side, like most other renewable energy sources, wind energy has a variable nature and can be unpredictable. Regions with high wind energy are commonly far away from consumption centers, which require expansion of the grid to transmit the generated electricity (particularly challenging for offshore units). They also have real and perceived negative environmental impacts as well as aesthetic and noise pollution. This has led to so-called

[1] https://www.irena.org/wind.
[2] https://www.eia.gov/electricity/data.php#gencapacity.

"NIMBYism" (not in my back yard) in some communities where members of the community like the general concept, just not in their region.

13.2 History of Wind Energy

Wind energy is one of the oldest sources of energy that human being has been able to harness for practical applications to grind grain, pump water for irrigation or flood prevention, or run sawmills. Wind energy has been used to propel boats and ships for thousands of years. The first documented practical design for nontransportation applications of wind energy was called the Persian windmill (Figure 13.1a). This windmill was made of vertical wooden sails connected to a vertical shaft and was driven by the drag force of wind (Figure 13.1b). The system was enclosed in a building, which covered one half of the wheel from the incoming wind flow to prevent its effect on slowing the motion of the plates moving toward the wind. These windmills and their Chinese counterparts were used for grain grinding and water pumping.

The developments of windmills in Europe started in the late 13th century. Unlike their eastern earlier counterparts, early European windmills had a horizontal shaft.[3] In the next 500 years, these windmills evolved to relatively

FIGURE 13.1
A Persian windmill. (a) An actual windmill from southeastern Iran, (b) schematic top view of the wheel. (By MorvaridiMeraj – Own work, CC BY-SA 4.0, https://commons.wikimedia.org/w/index.php?curid=73206407.)

[3] Perhaps, the most romanticized appearance of this type of windmills in the literature was in Don Quixote (by Miguel de Cervantes Saavedra, Spanish novelist and poet, 1547–1616), when Don Quixote attacked windmills that he believed to be wild giants.

FIGURE 13.2
Windmills in Kinderdijk in the southern Netherlands (built in the 1730s for water drainage). (Image by Ellen26 from Pixabay (https://pixabay.com/photos/mills-kinderdijk-holland-3888276/).)

efficient power production units that included many features of modern wind turbines. Along with watermills, they powered many industrial plants in pre-industrial Europe. One advantage of these windmills compared to watermills was that their operation was not stopped due to water freezing. Many of these units are still operational, primarily as tourist attractions (Figure 13.2).

Wind energy was one of the main sources of power until the Industrial Revolution. The utilization of these windmills gradually declined after the introduction of steam engines. The next step in the development of windmills was in the U.S. in the 19th century. The application of wind energy to pump water was always one of the most common ways to use wind energy. However, small-size windmills were perfected in the U.S. in the second half of the 19th century (Figure 13.3). They were first wind-powered machines that were manufactured industrially and took advantage of the economy of scale. These multi-blade units were usually 1–3 m (about 3 to 9 ft) in diameters with typically less than 0.75 kW (about 1 hp) output power, although very large units were also built with the diameter up to 18 m (about 60 ft). They were initially made of wood but later using steel blades significantly improved their performance. Probably one of the most important reasons for the success of these units was their robust control system that used vanes to orientate the rotor with no human attention.

FIGURE 13.3
A windmill to pump water. (Image by Lisa Johnson from Pixabay, https://pixabay.com/photos/windmill-turbine-wind-water-pump-1110079/.)

Using wind energy to generate electricity was first tried almost simultaneously in Scotland and the U.S. in the late 1880s. Charles Brush (1849–1929, an American engineer and inventor) in Cleveland, Ohio, and James Blyth (1839–1906, a Scottish electrical engineer) in Scotland both successfully produced electricity using wind energy. The devices they developed were a horizontal-axis wind turbine and a vertical-axis wind turbine, respectively. The wind turbine built by Brush had a rotor with a diameter of 17 m (56 ft) and 144 blades (Figure 13.4). The unit generated the output power of 12 kW (DC), which was used to charge a battery bank or to provide electricity for his laboratory. While the unit output power was impressive for its time, it was small compared to the output power of over 100 kW for the same size in modern wind turbines. This unit featured a large tail to follow the wind direction or turn it out of the wind if needed and a gearbox to increase the shaft rotational speed.

Another important development was in 1891 when Dane Paul la Cour (a Danish scientist, 1846–1908) used electricity produced by wind energy to produce hydrogen through electrolysis. He was the first one to recognize the favorable effects of improving the aerodynamic efficiency of blades and increasing their circumferential velocity.

FYI BOX 13.1

A WIND ENERGY PIONEER

Charles F. Brush (1849 –1929) was born and raised in a farm in Cleveland, Ohio. At the age of 12, he was experimenting with simple electric equipment. He built his first arc lamp, which produced light by generating an electric arc between two electrodes when he was in high school. He designed and patented his electric generator (dynamo) in the mid-1870s to power his arc lamp. By the early 1880s, his arc lamps provided public light in many North American cities, including New York, Philadelphia, and Montreal before Edison's incandescent lamps. However, eventually, his lamps could not compete with more advanced lamps and in 1889 he sold his company, which eventually became part of General Electric. He built a mansion in Cleveland where he lived and worked. He tested and showcased his inventions in his home, including his wind turbine. His wind turbine provided electricity for his residence for 20 years.

Regardless of the relative success of these attempts, none of them became commercial mainly due to inexpensive electricity from fossil fuels. In the 20th century, research and development on wind turbines continued, although the pace of progress was directly proportional to the cost of fossil fuels. In this period, several experimental units were tested, including a 100 kW unit in 1931 in the former Soviet Union, a 1.25 MW unit in 1941 in the U. S. (world's first megawatt-size unit), an 800 kW unit in 1956 in France, and a 200 kW unit in the 1960s in Denmark. These units were mostly for remote off-grid applications. The oil crisis of 1973 significantly boosted the development of wind turbines through government supports. This resulted in various designs with a capacity of 1 kW to 3.2 MW, both horizontal and vertical axes. These designs took advantage of developments in the design of military aircrafts. The highlight of these designs was NASA's 100 kW MOD-0 built in 1975 to 3.2 MW MOD-5B built in 1997. While this golden period did not last long mainly due to the rapid decline of oil price, it set stage for many of technologies that are used in modern wind turbines.

California in the U.S. and Denmark in Europe were the pioneers in the development of wind farms in the early 1980s with turbine sizes between 20 and 100 kW. Then, Germany followed by Spain were major players in the early 2000s but later the U.S. took back its leadership position. Nowadays

FIGURE 13.4
Charles Brush's wind turbine in Cleveland, Ohio (Note the large tail and 144 blades. Compare the size of the unit with the size of the person standing on the ground on the right side of the unit.). (By Unknown author – Robert W. Righter (1996) Wind Energy in America: A History, University of Oklahoma Press, p. page44 Retrieved on 27 December 2008. ISBN: 0806128127, Public Domain, https://commons.wikimedia.org/w/index.php?curid=5599364.)

China and the U.S. are leading the world with the largest wind turbine farm installations in the world.

13.3 Wind Formation

In this section, we will discuss about the origin of wind and how we can evaluate its energy. Wind is the flow of air from a region with a higher atmospheric pressure to a region with a lower atmospheric pressure. This pressure gradient in the atmosphere is as a result of the rising and sinking of air in the atmosphere. The rising of air causes a low pressure near the surface of Earth and the sinking of air causes a high pressure at the ground level. The effect of air movement on pressure at high elevations is opposite, i.e., when air is rising, the pressure at high elevations increases and vice versa. The question is why air rises or sinks. To answer this question, imagine two regions A and B are next to each other and the temperature of the surface of Earth in region A is higher than that in region B. The temperature of air over region A increases as a result of heat transfer from Earth's surface to air. The density of air decreases when its temperature increases so the hot air rises in region A and the pressure on the surface decreases. The opposite process takes place in region B, so air sinks and the pressure increases on the ground level. Thus, as a consequence of this pressure difference, cold air flows (or wind blows) from region B (high pressure) to region A (low pressure). It should be noted that this process at higher elevations causes the flow of hot air in the opposite direction (from region A to region B). As long as the temperature difference between the surface of Earth in regions A and B exists, wind will continue to blow. In short, **wind is generated by uneven heating of Earth's crust by solar radiation** and is considered as **an indirect form of solar energy**. The temperature of the Earth's crust in a region depends on the amount of solar energy received by the region. This effect can be global or local.

Earth's surface at the equator receives a greater amount of solar radiation compared to North or South poles. This uneven heating of the Earth's crust on the global scale causes a flow of cold air from the poles to the equator on the surface of Earth (also known as atmospheric circulation or **planetary wind**) and the flow of hot air at the upper atmosphere at the opposite direction. It should be noted that the direction of planetary wind is affected by the rotation of Earth and the Coriolis effect.

The non-uniform temperature distribution between land and water as well as hills and mountainsides causes **local wind**. Due to the higher specific heat of water compared to that of ground, during the day, the temperature increase of land surface caused by solar radiation is greater than that of water surface. This process causes **sea breeze** or onshore breeze where wind blows from water toward shore. At the night, the faster cooling of land compared to water causes wind in the reverse direction (called **land breeze** or offshore breeze). A similar process (i.e., the faster heating of mountains than that of valleies) causes wind toward mountains during the day and cold wind toward lowlands at night (called valley breeze and mountain breeze, respectively).

13.4 Power of Wind

The kinetic energy of an object with a mass of m moving with the velocity of V is

$$KE = \frac{1}{2} m\ V^2 \tag{13.1}$$

where KE is kinetic energy (kg.m^2/s^2 or Joule), m is mass (kg), and V is velocity (m/s). In the case of wind energy, since we deal with the mass flow rate of air (mass per unit of time) instead of a fixed mass, Equation 13.1 can be rewritten in the following form for the time rate of kinetic energy (energy per unit of time) or power:

$$\dot{KE} = P = \frac{1}{2}\dot{m}V^2 \tag{13.2}$$

where $\dot{KE}$ is the time rate of the kinetic energy (J/s or W), P is wind power (W), and $\dot{m}$ is mass flow rate (kg/s). For wind with uniform velocity over area A perpendicular to the wind direction, assuming a constant and uniform air density (incompressible flow), the mass of air crossing area A can be calculated from the continuity equation.

$$\dot{m} = \rho AV \tag{13.3}$$

where ρ is the uniform density of air (kg/m^3) and A is the area perpendicular to the wind direction (m^2). Substituting Equation 13.3 in Equation 13.2,

$$P = \frac{1}{2}\rho AVV^2 = \frac{1}{2}\rho AV^3 \tag{13.4}$$

A more common parameter to express wind power is **wind power density**, which is defined as wind power per unit of area.

$$PD = \frac{P}{A} = \frac{1}{2}\rho V^3 \tag{13.5}$$

It is easier to compare the wind power of various locations using this parameter. Note that although the units of power density for wind energy and irradiance for solar energy are the same, the reference areas are different. In solar energy, the area is on Earth's surface, while in wind energy, the area is perpendicular to the wind direction.

Example 13.1

A horizontal-axis wind turbine with a radius of 10 m operates in a location with a uniform wind speed of 10 m/s. Determine the power of wind for an area equal to the swept area of the turbine and per unit of area.

Analysis

For wind at the standard temperature and pressure (STP), 1 bar (14.5 psi) and 15°C (59°F),

$$P = \frac{1}{2}\rho AV^3 = \frac{1}{2}\times 1.22\times \pi\times (10)^2\times 10^3 = 191540\,\text{W}$$

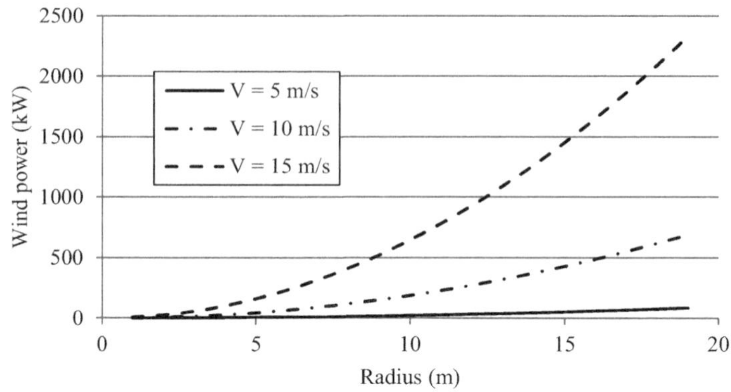

FIGURE 13.5
Diagram of wind power as a function of the radius of area perpendicular to the wind direction.

$$\frac{P}{A} = \frac{1}{2}\rho V^3 = \frac{1}{2} \times 1.22 \times 10^3 = 610 \text{ W/m}^2$$

Figure 13.5 illustrates wind power as a function of the radius of the area perpendicular to the wind direction for three wind speeds.

13.5 Maximum Theoretical Efficiency of Wind Turbines

In the previous section, we learned how we can use Equation 13.2 to determine the kinetic energy of wind. In this section, we will evaluate what portion of this energy can be theoretically extracted by a wind turbine. If a wind turbine could extract all wind energy, the kinetic energy and as a result the velocity of wind after the wind turbine become zero (according to the first law of thermodynamics). This means the mass flow rate of air through the wind turbine becomes zero (according to the conservation of mass law), which in turn means the output power of the wind turbine is zero. Therefore, the outlet velocity of wind from the turbine has two opposite effects. While reducing the exit velocity will increase the output power per unit of mass of the air going through the turbine, it will reduce the air mass flow rate. This means there should be an optimum air outlet velocity that maximizes the output power of the wind turbine. This optimum velocity, which places a limit to the maximum efficiency of wind turbines, can be determined by **Betz's law**. To derive this law, the concepts of the conservation of mass, momentum, and energy are used to find the upper limit for the efficiency of an ideal rotor.

FYI BOX 13.2

HISTORY OF BETZ'S LAW

Three scientists discovered the maximum efficiency of a wind turbine independently in a short period of time. Frederick W. Lanchester (1868–1946, an English scientist and engineer) in 1915, Nikolay Yegorovich Zhukovsky (1847–1921, a Russian scientist), and Albert Betz (1885–1968, a German scientist) both in 1920. All three of them had significant contributions to the science of aerodynamics. In fact, Zhukovsky sometimes is called the founding father of modern aerodynamics, while Betz was a pioneer of wind turbines.

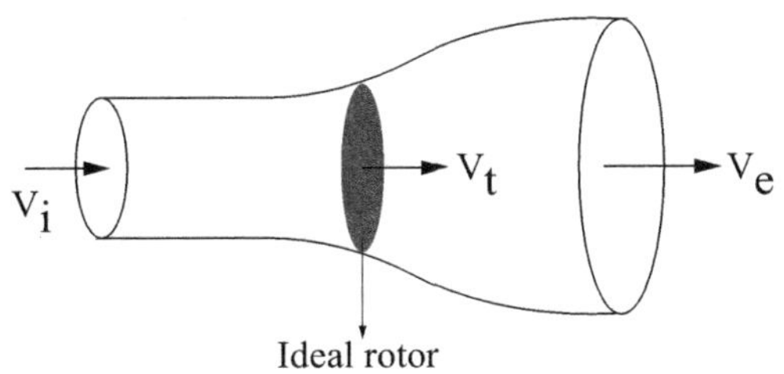

FIGURE 13.6
Schematic for the derivation of Betz's law.

In the following analysis to derive Betz's law, the wind turbine is a horizontal-axis turbine; however, this assumption will be relaxed later. Assume that an ideal turbine is placed in a flow of air to extract energy from the air flow (Figure 13.6). An ideal rotor, also called an actuator disk, has no hub with an infinite number of blades with no drag force. In this analysis, the aerodynamics of the blades are not taken into account. A control volume can be placed around the disk where air enters the control volume at the velocity of V_i and exits at the velocity of V_e. The flow of air in the inlet, exit, and throughout the control volume is axial and no flow crosses the boundary of the control volume except in the inlet and exit. The inlet and exit to the control volume are far upstream and downstream of the turbine, respectively, where the air pressure is unaffected by the presence of the turbine and they are equal to the ambient pressure $(P_i = P_e = P_{\text{amb}})$.

The velocity of the outlet flow is less than the velocity of the inlet flow and the difference between the kinetic energy of the inlet and outlet flows is equal to the energy captured by the wind turbine (according to the first law of thermodynamics). Assuming that the changes in potential energy and heat transfer are negligible and no work is added or extracted expect in the rotor, using Equation 13.2,

$$P = \frac{1}{2}\dot{m}\, V_i^2 - \frac{1}{2}\dot{m}\, V_e^2 = \frac{1}{2}\dot{m}\left(V_i^2 - V_e^2\right) \tag{13.6}$$

Since the air flow is assumed frictionless, and there is no nonconservative force, the second law of Newton can be used to find the force exerted on the rotor.

$$F = ma = m\frac{\Delta V}{\Delta t} = \frac{m}{\Delta t}\Delta V = \dot{m}(V_i - V_e) \tag{13.7}$$

The power extracted by the rotor because of this force is

$$P = \mathrm{FV} = \dot{m}(V_i - V_e)V_t \tag{13.8}$$

where V_t is the air velocity when crossing the rotor. Equating Equations 13.6 and 13.8 gives

$$V_t = \frac{1}{2}(V_i + V_e) \tag{13.9}$$

Thus, the velocity of air at the turbine can be calculated by averaging the inlet and exit air flow velocities. Substituting the mass flow rate of the air through the rotor from Equation 13.3 in Equation 13.6 gives:

$$P = \frac{1}{2}\rho A V_t\left(V_i^2 - V_e^2\right) \tag{13.10}$$

Combining Equations 13.9 and 13.10 gives

$$P = \frac{1}{2}\rho A\left(\frac{1}{2}(V_i + V_e)\left(V_i^2 - V_e^2\right)\right) \tag{13.11}$$

Equation 13.11 reconfirms the discussion at the beginning of this section. The velocity of the air flow at the outlet appears as a positive value in one term and a negative value in the other, which means there should be an optimum for this parameter that maximizes the output power of the turbine. In order to find the optimum value of the exit velocity, the derivative of P with respect to V_e should be equal to zero $\left(\frac{dp}{dV_e} = 0\right)$, which gives

$$3V_e^2 + 2V_eV_i - V_i^2 = 0 \tag{13.12}$$

Factoring out V_i^2 gives

$$3\left(\frac{V_e}{V_i}\right)^2 + 2\left(\frac{V_e}{V_i}\right) - 1 = 0 \qquad (13.13)$$

Solving the quadratic Equation 13.13 gives the exit to inlet air velocity ratio that maximizes the output power of the wind turbine.

$$\left(\frac{V_e}{V_i}\right)_{\text{opt}} = \frac{1}{3} \quad \text{or} \quad V_{e,\text{opt}} = \frac{1}{3} V_i \qquad (13.14)$$

Substituting Equation 13.14 in Equation 13.11 gives the maximum turbine output power.

$$P_{\max} = \frac{8}{27} \rho A V_i^3 \qquad (13.15)$$

The efficiency, also known as the **power coefficient** (C_p) of the wind turbine can be calculated by dividing this output power of the turbine to the power of the wind in Equation 13.4, which gives

$$\eta_{\text{wind turbien}} = C_p = \frac{P_{\text{wind turbine}}}{P_{\text{wind}}} \quad \text{or} \quad P_{\text{wind turbine}} = C_p P_{\text{wind}} = \frac{1}{2} C_p \rho A_i^3 \qquad (13.16)$$

The maximum power coefficient of the ideal wind turbine can be calculated as follows:

$$\eta_{\max} = C_{p,\max} = \frac{P_{\max}}{P_{\text{wind}}} = \frac{\frac{8}{27} \rho A V_i^3}{\frac{1}{2} \rho A V_i^3} = \frac{16}{27} \sim 0.593 \qquad (13.17)$$

Figure 13.7 illustrates the power coefficient of the ideal wind turbine as a function of the ratio of upstream wind velocity to downstream wind velocity with the maximum efficiency identified.

This means that no ideal rotor disc wind turbine can convert more than 59.3% of wind kinetic energy to mechanical energy. In reality, there are other losses in real wind turbines, e.g., viscous drag forces on blades, creation of vertices in the flow of air, mechanical losses in the internal components of the turbine, and losses during the conversion of mechanical energy to electrical energy, that further reduce the efficiency. Understanding this maximum theoretical power coefficient can help to appreciate the power coefficient of modern wind turbines that is in the range of 0.4–0.5, which is 65%–85% of

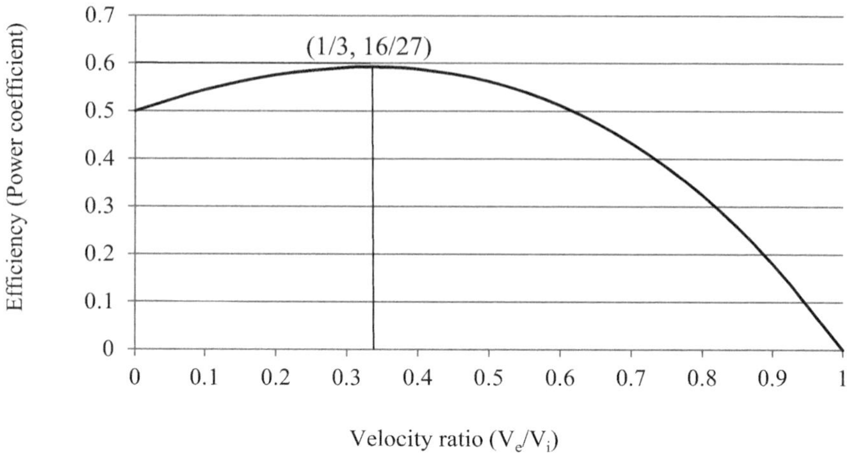

FIGURE 13.7
Efficiency (power coefficient) of an ideal wind turbine as a function of the ratio of upstream wind velocity to downstream wind velocity.

the Betz limit.[4] Although in deriving the Betz limit the configuration of a horizontal-axis wind turbine is used, this law is universal and can be used for any type and size wind turbine.

Example 13.2

Determine the blade diameter of a turbine with the rated (design) output power of 7.5 MW when the rated (design) wind speed is 12 m/s. The power coefficient of the turbine is 0.55. Repeat your calculation for the power coefficient equal to the Betz limit and 0.45.

Analysis

From Equation 13.16

$$P_{\text{rated}} = C_p P_{\text{wind}} = \frac{1}{2} C_p \rho A V^3 = \frac{1}{2} C_p \rho \left(\frac{\pi D^2}{4} \right) V^3$$

$$D = \sqrt{\frac{8 P_{\text{rated}}}{C_P \rho \pi V^3}}$$

For the density of air equal to 1.23 kg/m^3

[4] It should be noted that in this analysis the area is the cross-section area of the rotor perpendicular to the wind direction. If the wind turbine is equipped with a cowl that acts as a funnel to concentrate the air flow before guiding into the rotor, the inlet area of the cowl should be used.

$$D=\sqrt{\frac{8\times 7{,}500{,}000}{0.55\times 1.23\times 3.14\times 12^3}}=127.85\ \text{m}$$

An actual 7.5 MW wind turbine has a typical rotor diameter of about 125 m. The second section of the example is left for practice.

13.6 Wind Energy Evaluation

Every wind energy project starts with the evaluation of wind energy in the region of interest and the identification of suitable site(s). In this section, we will first introduce the wind classification to identify when wind energy is good enough for power generation. Then, we will study how wind energy can be evaluated.

13.6.1 Wind Classification

Wind can be classified based on the wind speed or power density. Table 13.1 shows four wind categories based on wind speed at 10 and 50 m above the ground.

13.6.2 Wind Property Measurements

Equation 13.4 shows that wind energy and as a result the output power of wind turbines are proportional to a cube of wind speed. This emphasizes the importance of accurate measurements of wind speed and direction for the analysis of wind energy potentials. In wind speed measurements, the horizontal component of wind velocity is the most important parameter in determining wind energy. However, for some wind turbines the vertical component of the wind velocity is also important because it may negatively affect the wind turbine performance. Note that during the measurement of wind properties, their spatial and temporal variation should be taken into account. For example, when a cup anemometer is used to measure wind speed, it does not measure the wind speed in a single point but measures

TABLE 13.1

Classification of Wind Speed at the height of 10 m and 50 m

Wind Power Category	Wind Speed at Height of 10 m (33 ft) in m/s (mph)	Wind Speed at Height of 50 m (164 ft) in m/s (mph)
poor	0–4.5 (0–10)	0–5.5 (0–12.3)
Moderate	4.5–5.5 (10–12.3)	5.5–7.0 (12.3–15.7)
Good	5.5–6 (12.3–13.4)	7–7.5 (15.7–16.8)
Excellent	>6 (>13.4)	>7.5 (>16.8)

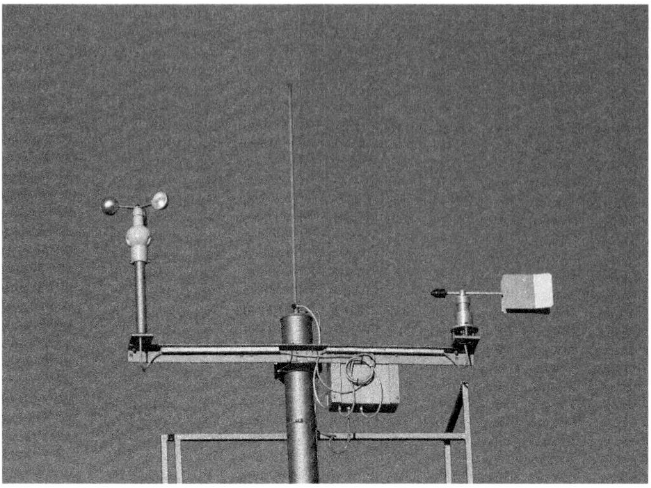

FIGURE 13.8
A meteorological tower. (Image by Hans Braxmeier from Pixabay, https://pixabay.com/photos/weather-station-anemometer-5580/.)

the average wind speed of the area that is covered by the cups. Instruments to measure wind properties can be in situ measurement instruments, such as anemometers, or remote sensing instruments, such as sonic detection and ranging (or SODAR, based on sound waves) and light detection and ranging (or LIDAR, based on light waves). Instruments used for in situ measurements are typically installed on a meteorological tower. The common instruments on a meteorological tower are instruments to measure wind speed and direction as well as those to measure ambient conditions including temperature, pressure, and relative humidity (Figure 13.8). Typical meteorological towers are 60 m (about 200 ft) in height.

Anemometers are the most popular instruments to measure wind speed. There are three types of anemometers: cup, propeller, and sonic anemometers. The principle operation of cup (Figure 13.9) and propeller anemometers (Figure 13.10) is similar to vertical and horizontal wind turbines, respectively. Unlike the former that only measures wind speed and needs a wind vane to measure wind direction (Figure 13.9), the latter can measure both speed and direction of wind. In both instruments, the frequency of the generated AC signal in the device is proportional to the wind speed. Sonic anemometers deploy ultrasound waves to create a three-dimensional velocity vector of wind and directly measure its direction.

The classification of anemometers is based on their accuracy and technology according to the standard published by the International Electrotechnical Commission (IEC 61400-12-1 standard). Note that a wind speed and direction measurement instruments are used not only for estimating wind energy in a certain area at the preliminary stage of any wind turbine project but for providing data for controling the later operation of wind turbines.

FIGURE 13.9
A cup anemometer. (Image by Dragan Stanojevic from Pixabay, https://pixabay.com/photos/anemometer-apparatus-wind-5532291/.)

FIGURE 13.10
A handhold propeller anemometer. (Image by Bernabe Colohua from Pixabay, https://pixabay.com/photos/anemometer-kelner-temperature-wind-5160777/.)

13.6.3 Statistical Analysis of Wind Data

Thus far we learned that in real-world wind velocity varies both temporally (from sudden gusts to seasonal variations) and spatially and is a stochastic property. To take into account these random variations, the first answer that may come to mind is to use the average wind speed to represent a specific wind. But this approach has a major problem. The cube of the average velocity is not equal to the average of the velocity cube.

$$\left(V_{\text{avg}}\right)^3 \neq \left(V^3\right)_{\text{avg}} \tag{13.18}$$

If we use the average wind velocity to estimate wind energy, it can be very misleading. If we have n readings for the velocity of wind, the average wind velocity is

$$\bar{V} = \frac{\sum_1^n V_i}{n} \tag{13.19}$$

But the mean velocity that correctly represents wind energy (known as the **mean energy velocity,** $\bar{V}_E$) can be estimated from the following equations:

$$\bar{V}_E = \left(\frac{\sum_1^n V_i^3}{n}\right)^{1/3} \tag{13.20}$$

Example 13.3

In a region for points *A* to *E*, wind velocity is measured ten times. The wind velocity data in the region are listed in Table 13.2. For each set of data determine the average wind velocity and the mean energy velocity.

As the table illustrates in all cases using average wind velocity will underestimate the wind energy sometimes with significant errors.

TABLE 13.2

Wind Velocity Data for Example 13.3

Wind Velocity	V_1	V_2	V_3	V_4	V_5	V_6	V_7	V_8	V_9	V_{10}	$\bar{V}$	$\bar{V}_E$
A	0	2	4	6	8	10	12	14	16	18	9.0	11.7
B	0	0	0	0	0	18	18	18	18	18	9.0	14.3
C	0	0	0	0	0	0	0	0	0	18	1.8	8.4
D	0	18	18	18	18	18	18	18	18	18	16.2	17.4
E	6	4	10	8	2	4	18	12	12	10	8.6	10.7

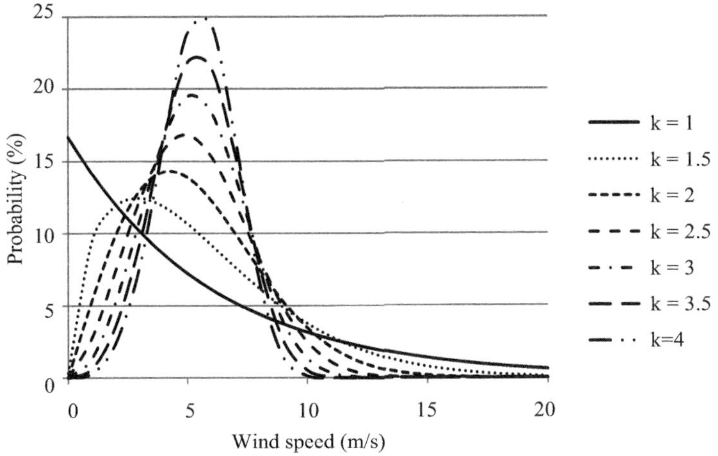

FIGURE 13.11
Weibull probability density function for $\lambda = 6\,\text{m/s}$.

In typical actual wind data, the wind average velocity is about 25% less than the mean energy velocity. This translates to the actual wind energy being twice the energy calculated by this average velocity (1.25^3). Therefore, wind energy depends on the nature of the time variation of the wind speed and the average wind speed is not the right parameter to evaluate wind energy.

Although wind speed is seemingly random and unpredictable, in reality to large extent it follows a predictable pattern that can be expressed by a probability distribution. The most common continuous probability distribution used to model wind speed is the **Weibull distribution** (after Waloddi Weibull (1887–1979) Swedish engineer, scientist, and mathematician and winner of the American Society of Mechanical Engineers gold medal in 1972). The probability density function (PDF) of the Weibull distribution is

$$f(V;k,\lambda)=\left(\frac{k}{\lambda}\right)\left(\frac{V}{\lambda}\right)^{k-1} e^{-\left(\frac{V}{\lambda}\right)^k} \quad \text{for } V>0 \tag{13.21}$$

where V is wind speed (m/s), k is the shape factor (dimensionless), and λ is the scale factor (m/s). In this equation, k and λ depend on the type of wind and its characteristics. For a given k and λ, the probability density function (f) can predict the likelihood of occurrence of a certain wind speed.

Example 13.4

For wind in a region, the shape factor and scale factor of the Weibull distribution that fits the distribution of the wind are 2 and 6 m/s, respectively. Determine the probability of having the wind speed of 10 m/s and how many hours a year the wind with the speed of 10 m/s is likely to blow.

Analysis

In this case $V = 10\,\text{m/s}, k = 2, \lambda = 6\,\text{m/s}$, substituting these values in the probability distribution function in Equation 13.21

$$f(10;2,6) = \left(\frac{2}{6}\right)\left(\frac{10}{6}\right)^{2-1} e^{-\left(\frac{10}{6}\right)^2} = 0.0345$$

Hours a year with the wind speed of $10\,\text{m/s} = 0.0345 \times 24 \times 365 = 302.22\,\text{h}$

That means for this wind there is only 3.45% chance of having wind with the speed of 10 m/s or about 302 hours a year (about 12.5 days a year). You can use WEIBULL(V, k, λ, FALSE) function in Excel to calculate the probability distribution function (in this case WEIBULL(10, 2, 6, FALSE)).

Note: A specific case of the Weibull distribution when $k = 2$ is called the **Rayleigh distribution** (after John W. S. Rayleigh (1842–1919) English physicist and winner of Nobel Prize for Physics in 1904) and is commonly used as a simpler approximation of the wind speed distribution.

Figure 13.11 illustrates the Weibull probability density function for $\lambda = 6\,\text{m/s}$ and k between 1 and 4 with an increment of 0.5. As the name implies, the shape factor determines the shape of the curve. In the same way, Figure 13.12 illustrates the effect of the scale factor on the Weibull probability density function for $k = 2$. Again, as the name suggests, in this case, the general shape of the curve is fixed and only the size (scale) of the curve changes.

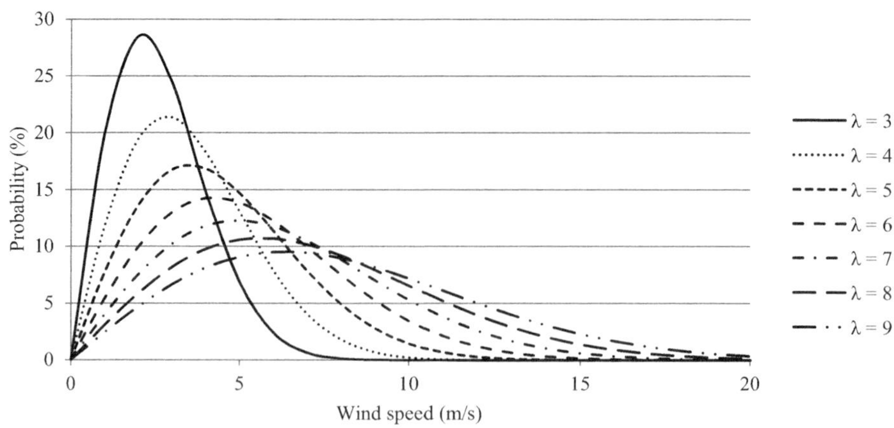

FIGURE 13.12
Weibull probability density function for $k = 2$.

For the study of wind energy, it is of special interest to be able to predict the probability of having wind at a certain range of the wind speed is. This is important because wind turbines can produce their full power (the rated power) above a certain wind speed (the rated wind speed). So it is vital to be able to determine what fraction of the wind distribution occurs above the rated wind speed. For this purpose, the probability density curve (Figures 13.11 and 13.12) can be used. The area under the curve between the upper and lower limits in this curve is equal to the probability of having wind at the specified range. A simpler method is to use the cumulative distribution function (CDF) of the Weibull distribution:

$$F(V;k,\lambda) = 1 - e^{-\left(\frac{V}{\lambda}\right)^k} \quad (13.22)$$

where F is the probability of wind with a speed of smaller than or equal to V.

Example 13.5

In Example 13.4, determine the probability of having the wind speed between 3 and 10 m/s and how many hours a year wind with the speed of 3 and 10 m/s is likely to blow.

Analysis

$$F(3;2,6) = 1 - e^{-\left(\frac{3}{6}\right)^2} = 0.221$$

$$F(10;2,\ 6) = 1 - e^{-\left(\frac{10}{6}\right)^2} = 0.938$$

$$\text{Probability of having wind speed between 3 and 10 m/s}$$

$$= F(10;2,\ 6) - F(3;2,\ 6) = 0.938 - 0.221 = 0.717$$

$$\text{Hours a year with the wind speed of between 3 and 10 m/s}$$

$$= 0.717 \times 24 \times 365 = 6280.92\,\text{h}$$

That means it is likely that about 262 days a year, the wind speed would be between 3 and 10 m/s. You can use WEIBULL(V, k, λ, TRUE) function in Excel to calculate the CDF (for example in this case WEIBULL(10, 2, 6, TRUE) − WEIBULL(3, 2, 6, TRUE)).

To determine wind energy in a specific location, wind speed is typically recorded every 1–3 seconds for a given elevation (the height of the measuring point from the ground level). Then, the data are typically divided into 10-minute (or sometimes an hour) intervals. In each interval, the average speed, standard deviation, maximum, and minimum values are determined and recorded, and the raw information is discarded. These averages are used to find the factors of the Weibull distribution. In this process, it is assumed that in each interval the wind speed is steady, and its value is equal to the average of the interval. The data are recorded for two to three years for an acceptable accuracy. For various analyses, different time periods might be needed, for example, hourly, daily, monthly, seasonal, or even annual. For each period of interest, wind speed histograms are generated and used to determine the wind speed distribution. The details of this calculation is beyond the scope of this text.

13.6.4 Variation of Wind Speed as a Function of Height

As explained in the previous section, the wind speed measurement is usually done on a single elevation (at least in preliminary stages of wind power assessment). On many occasions, this elevation is not at the level of the hub of the intended turbine. In this section, we will explain how wind speed information in one elevation can be used to predict the same properties in other elevations. The variation of wind speed as a function of height is known as **wind shear**. Generally speaking, the wind speed increases with an increase in height, but the magnitude of this variation depends on the ground roughness and obstacles, such as buildings and trees. The most widely used method for this purpose is the wind power law.

$$\frac{V_2}{V_1} = \left(\frac{h_2}{h_1}\right)^{\gamma} \tag{13.23}$$

where V_2 and V_1 are the wind speeds at the heights of h_2 and h_1, respectively, and γ is the wind shear exponent (also known as the Hellmann exponent). This exponent depends on the roughness of the terrain and its obstacles. The most commonly used wind shear exponent when wind speed is available at a single height in open flat terrains is 0.14 $\left(\text{or } \frac{1}{7}\right)$. This exponent can be as high as 0.6 for rough terrains with many obstacles. Also, the wind shear exponent changes at different hours of the day. During night hours, the exponent is higher. This means during a night when power demand is lower, wind turbines produce more power. This can be a problem because the price of electricity is lower at off-peak hours. Figure 13.13 illustrates the relationship between wind speed and height for various Hellmann exponents.

Example 13.6

A wind turbine can be installed at the hub height of 15 m or 60 m. If the location of the installation is open flat terrain, determine the power ratio of the turbine at two elevations.

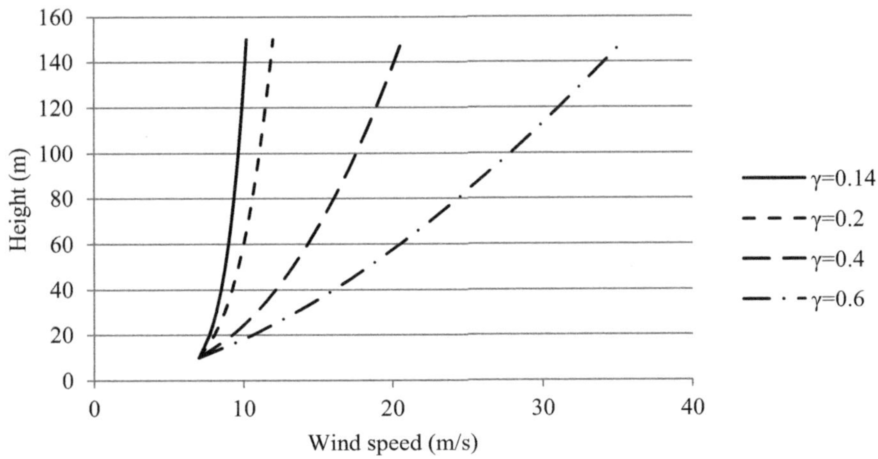

FIGURE 13.13
Relationship between wind speed and height for various Hellmann exponent.

Solution

Using Equation 13.23, the ratio of the average wind speed at the hub height of 60 m to that of the height of 15 m can be estimated as follows (for an open flat terrain $\gamma = 0.14$):

$$\frac{V_{avg,2}}{V_{avg,1}} = \left(\frac{60}{15}\right)^{0.14} = 1.21$$

Using Equations 13.4 and 13.16

$$P = \frac{1}{2} C_p \rho A V^3$$

$$\frac{P_2}{P_1} = \frac{C_{P,2}}{C_{P,1}} \times \frac{\rho_2}{\rho_1} \times \frac{A_2}{A_1} \times \left(\frac{V_2}{V_1}\right)^3$$

Since the same wind turbine is compared (constant C_p and A) and the density of the air can be considered constant in two elevations,

$$\frac{P_2}{P_1} = \left(\frac{V_2}{V_1}\right)^3 = \left(\frac{h_2}{h_1}\right)^{3\gamma} \tag{13.24}$$

As previously stated generally $(V_{avg})^3 \neq (V^3)_{avg}$; however, in many cases the only known information about wind speed is the average wind speed, so this average wind speed is used to estimate the wind energy. Obviously, using this approximation can introduce significant errors to the estimation. The magnitude of the error depends on the parameters of the Weibull distribution. Typically, this approximation is used to compare wind energy in the same area, such as this example, so:

$$\frac{P_2}{P_1} = \left(\frac{V_2}{V_1}\right)^3 = \left(\frac{V_{avg,\,2}}{V_{avg,\,1}}\right)^3 = 1.21^3 = 1.79$$

13.6.5 Effect of Variations of Density and Relative Humidity of Air

Equation 13.4 shows that the density of air directly affects wind power density. The density of air varies with the elevation and temperature. This effect should be particularly taken into account when the power law (Equation 13.23) is used. The density of air decreases approximately by 1% when the elevation increases from 10 m to 100 m. As a rule of thumb, the power of wind drops by 10% for every 1000 m increase in the elevation. Also, the density of air is affected by relative humidity of air. When the ambient relative humidity increases from 10% to 90%, the density of air decreases by about 0.5%.

13.6.6 Wind Energy Maps

The results of wind property measurements have been reflected in the global, national, and local wind power density maps. These maps are prepared using weather data, terrain topography, and computer modeling. The National Renewable Energy Laboratory (NREL) and the U.S. Department of Energy provide detailed wind energy maps for the U.S. for utility-scale (land-based and offshore), community-scale, and residential-scale wind turbines at various heights at national and state levels (Figure 13.14). Many nations and regions have similar wind resource maps available to the public. These maps should be used with caution as they are usually software-driven and just provide a rough estimation of wind energy in general. The wind energy of areas that are identified as suitable in these maps should be evaluated with local wind measurements. Thus, these maps should be only used as a tool to identify potential regions with good wind resources. The next step before local wind property measurements can be to use wind data from a neighboring airport or weather station to determine the rough values of wind energy.

13.7 Wind Turbine Categorization

Wind turbines can be categorized in several ways according to

- interaction of blades with wind (lift force or drag force)

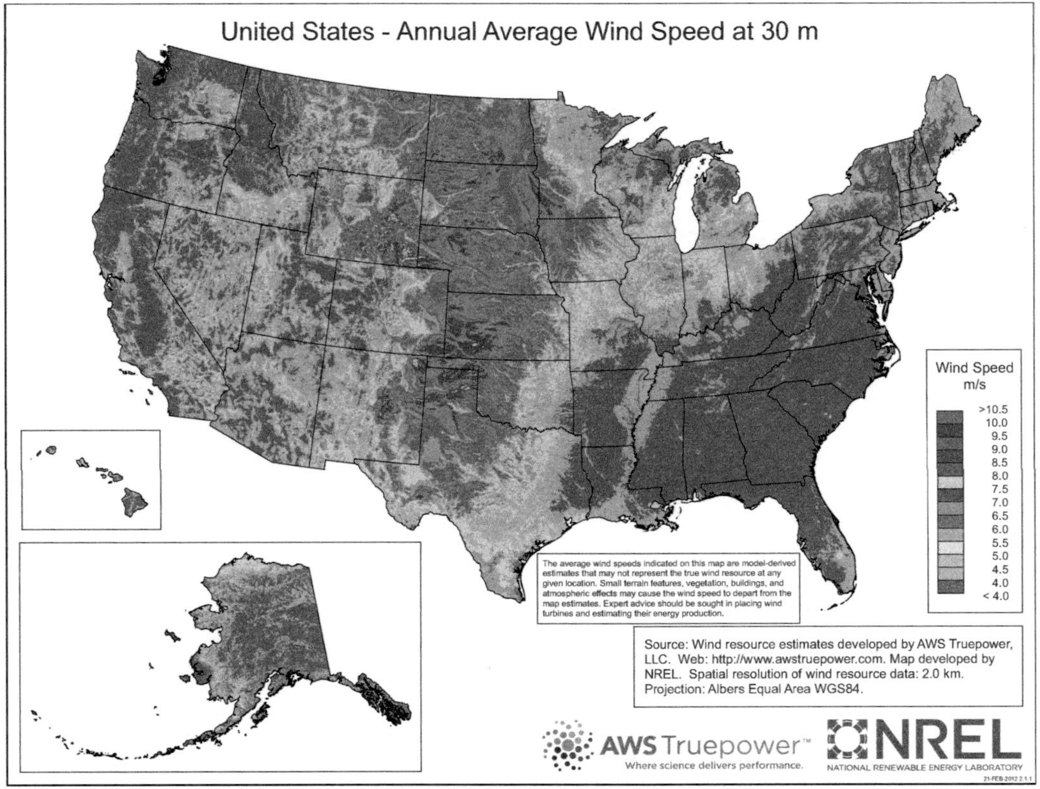

FIGURE 13.14

Wind power density map for the U.S. (The National Renewable Energy Laboratory (NREL) and the U.S. Department of Energy, https://windexchange.energy.gov/maps-data/325.)

- axis of rotation (horizontal- or vertical-axis wind turbines)
- connection to the grid (grid-connected or stand-alone)
- nature of rotational speed (constant or variable)
- power output capacity (small or large)
- number of blades
- location (land-based or offshore)

In the following sections, some of these categories will be discussed.

13.7.1 Lift-Force Dominant Wind Turbines and Drag-Force Dominant Wind Turbines

When a fluid flows over an object or an object moves through a fluid (or both), the fluid exerts a net force on the object. The net force acting on the object can be divided into two components, parallel and perpendicular to the velocity of the fluid relative to the object (Figure 13.15). As a convention, the former is called the drag force, and the latter is called the lift force, though this does not mean that the lift force is always upward. Note that both forces can have normal components to the axis of rotation of blades, which creates torque to rotate them. Typically, both of these components exist in a turbine and based on which one is dominant, turbines are called lift or drag force turbines.

In order to simplify the study of these two forces, the following two dimensionless parameters are defined:

$$C_D = \frac{F_D}{\frac{1}{2}\rho A V_{\text{rel}}^2} \tag{13.25}$$

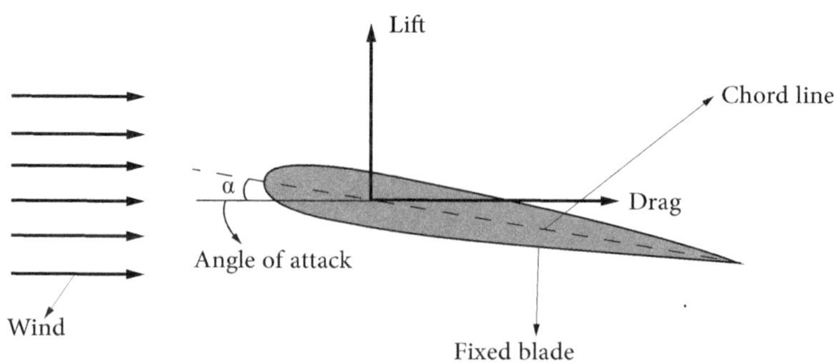

FIGURE 13.15
Schematic of lift and drag forces.

$$C_L = \frac{F_L}{\frac{1}{2}\rho A V_{rel}^2} \tag{13.26}$$

where C_D and C_L are **coefficients of drag and lift**, respectively, F_D and F_L are the drag and lift forces, respectively, and V_{rel} is the velocity of the fluid relative to the object. For a stationary object, V_{rel} will be equal to the velocity of the flow (wind).

In drag-force dominant devices, the component that interacts with wind, in this case blades, stand against the wind and the wind force produces torque. Since these devices move with the flow, their **speed relative to wind cannot exceed the wind speed**. They have low efficiency (power coefficient, see Example 13.7) and material required to build blades is substantial. As a result nowadays there is no commercial large drag force wind turbine for electricity production. But Persian turbines discussed in Section 13.2 (Figure 13.1) were purely drag force turbines.

Lift-force dominant devices do not stand against the wind but interact with it like an airplane wing. Their speed relative to wind can exceed wind speed and thus they are more efficient. They need much less material for fabrication than that for drag force wind turbines. Almost all modern wind turbines are lift force dominant.

Example 13.7

A flat plate is vertically placed on a cart (Figure 13.16). The cart is placed in a wind in a way that the plate is perpendicular to the wind direction. If as a result of the wind force the cart moves with a speed equal to one-third of the wind speed, determine the maximum power coefficient of this device. Assume the drag coefficient of the flat plate is 1.2.

Analysis

From Equation 13.25,

$$F_D = \frac{1}{2} C_D \rho A V_{rel}^2$$

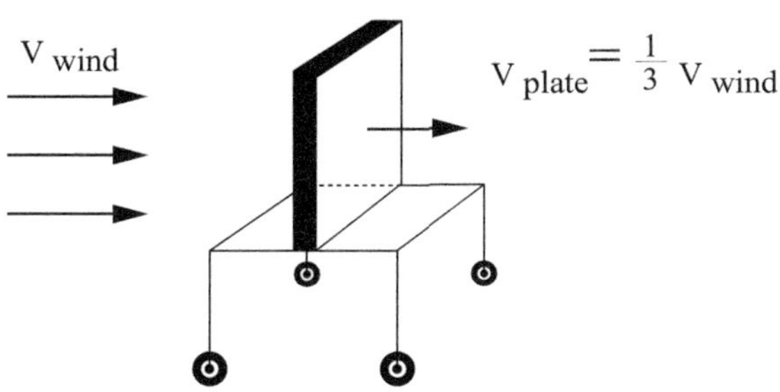

FIGURE 13.16
Schematic of a flat plate in wind for Example 13.7.

Since the direction of the wind and the plate movement are the same,

$$V_{rel} = V_{wind} - V_{plate}$$

If $V_{wind} = V_{plate}$, the force will be zero, and if $V_{Plate} = 0$, the force is maximum.

$$P = F_D V_{plate} = \frac{1}{2} C_D \rho A \left(V_{wind} - V_{plate}\right)^2 V_{plate}$$

Since $V_{plate} = \dfrac{V_{wind}}{3}$,

$$P = \frac{1}{2} C_D \rho A \left(V_{wind} - \frac{V_{wind}}{3}\right)^2 \times \frac{V_{wind}}{3} = \frac{2}{27} C_D \rho A V_{wind}^3$$

From Equation 13.16 for the power coefficient (C_p),

$$C_P = \frac{\frac{2}{27} C_D \rho A V_{wind}^3}{\frac{1}{2} \rho A V_{wind}^3} = \frac{4}{27} C_D$$

Assuming the drag coefficient of the flat plate is 1.2,

$$C_P = \frac{4}{27} C_D = 0.178$$

This power coefficient (efficiency) is significantly lower than those of modern wind turbines in the range of 0.4–0.5.

Example 13.8

Persian turbines, discussed in Section 13.2, are purely drag force turbines. For a Persian windmill shown in Figure 13.17, determine the drag force and the torque produced by a wind with the speed of 10 m/s. The

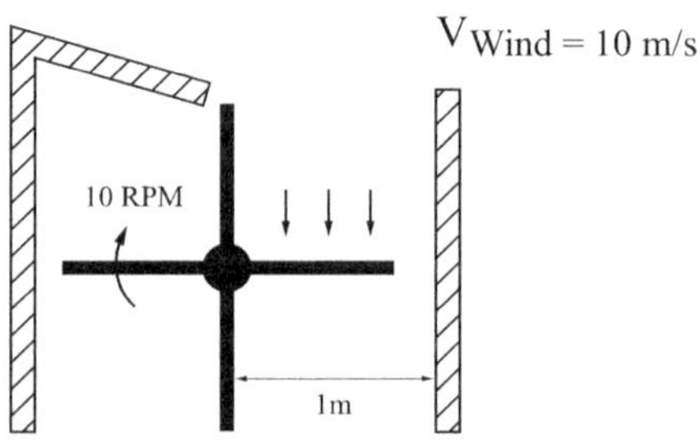

FIGURE 13.17
Schematic of the Persian windmill in Example 13.8.

height and radius of the wheel are 2 and 1 m, respectively, the rotational speed of the wheel is 10 rpm, and C_D is 1.2.

Analysis

For the sake of simplicity, let us find the drag force only in the state shown in Figure 13.17. In this case, there is no lift force and the net force exerted by the wind on the windmill is a pure drag force. From Equation 13.25,

$$F_D = \frac{1}{2} C_D \rho A V_{\text{rel}}^2$$

The drag coefficient for a rectangular plate is about 1.2. In this case, the wind velocity and the velocity of the wheel are at the same direction, so

$$V_{\text{rel}} = V_{\text{wind}} - V_{\text{wheel}} = V_{\text{wind}} - r\omega$$

where r is the radius of the wheel at a certain point on the plate and can vary from $r=0$ to $r=1$ m. For simplicity, let us assume r is equal to one half of the wheel radius. This means we assume the entire wind drag force is imposed on the middle of the plate.

$$V_{\text{rel}} = V_{\text{wind}} - \frac{R\omega}{2}$$

$$F_D = \frac{1}{2} C_D \rho A \left(V_{\text{wind}} - \frac{R\omega}{2} \right)^2$$

$$\omega = \frac{2\pi \times \text{rpm}}{60} = \frac{2\pi \times 10}{60} = 1.05 \text{ rad/s}$$

$$F_D = \frac{1}{2} \times 1.2 \times 1.22 \times (2 \times 1) \times \left(10 - \frac{1 \times 2\pi \times 10}{2 \times 60} \right)^2 = 131.5 N$$

Assuming the uniform effect of wind on the plate, the torque can be calculated as follows:

$$T = F_D \times \frac{R}{2} = 131.5 \times 0.5 = 65.7 \ N.m$$

Note that this is the maximum torque on the plate. When the plate moves forward, the amount of torque exerted on the plate reduces; however, at the same time the next plate moves toward the wind and the drag force of the wind starts to push that plate.

Now you can see why one half of the wheel is covered (Figure 13.17). If the whole wheel was placed on the wind, the torque exerted on the opposite plates would cancel each other, and the wheel would not move. So the main idea in this design is to have unbalanced torque on two sides of the wheel. In a cup anemometer, this effect is produced by different shapes of the two sides of the cup that is exposed to the wind, when it moves toward the wind and away from it (next example).

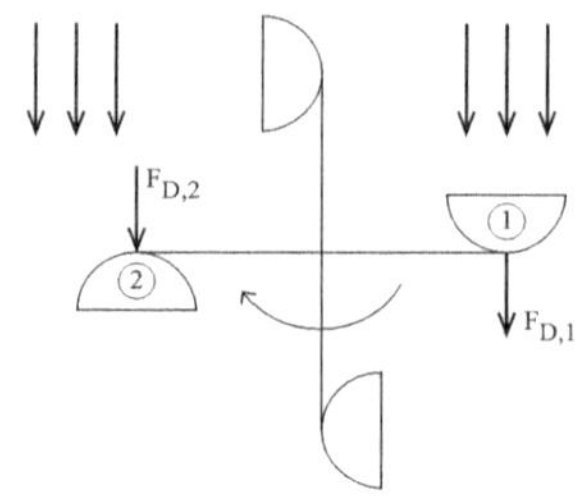

FIGURE 13.18
Schematic of an anemometer for Example 13.9.

Example 13.9

In an anemometer shown in Figure 13.18, determine the drag forces on cups 1 and 2 and discuss how this anemometer can measure wind speed. Assume when the cup moves toward the wind (the back of the cup is toward the wind), C_D is 0.38 and when it moves away from the wind (the front of the cup is toward the wind), C_D is 1.4.

Analysis

Again, let us study the problem in the state shown in Figure 13.18. In this case, while the wind exerts forces on both cups at the same direction, the produced torques by the cups on the opposite sides of the arm are against each other. The forces on the cups can be calculated using Equation 13.25. The only difference is the different values of the drag coefficients for the two sides of the cups and the velocity of wind relative to the cups. Since cup 2 moves toward the wind (the back of the cup is toward the wind), its C_D is 0.38 and cup 1 moves away from the wind (the front of the cup is toward the wind), its C_D is 1.4. The forces acting on two cups can be estimated as follows:

$$F_{D,1} = \frac{1}{2} C_{D,1} \rho A (V_{\text{wind}} - R\omega)^2 = \frac{1}{2} \times 1.4 \times \rho A (V_{\text{wind}} - R\omega)^2$$

$$F_{D,2} = \frac{1}{2} C_{D,2} \rho A (V_{\text{wind}} - R\omega)^2 = \frac{1}{2} \times 0.38 \times \rho A (V_{\text{wind}} + R\omega)^2$$

Note that for cup 2, the directions of the wind and the cup rotation are opposite. It can be shown that there is a direct relationship between ω and V, which makes anemometers useful in measuring wind speed. The manufacturers of anemometers provide an accurate empirical relationship between ω and V, which can be used to calibrate them.

Before concluding this section, let us introduce an important parameter called the **blade tip-speed ratio**. For an observer on the tip of a blade looking at wind, the wind relative velocity (V_{rel}) has two components, the wind velocity parallel to the axis of rotation (V_{wind}) and the velocity due to the motion of the observer itself (because of the blade rotation), which is perpendicular to the wind velocity ($V_{tip} = \omega R$). The tip-speed ratio (λ) of a blade is the ratio of the speed of the tip of the blade to the speed of the free wind (the wind unaffected by the blade).

$$\lambda = \omega R / V_{wind}$$

where R is the radius of the blades (m) and ω is the angular velocity of the blades (rad/s). It has been shown that the power coefficient of a wind turbine strongly depends on its tip-speed ratio. If the tip-speed ratio is too small, the turbine cannot effectively extract wind energy. If it is too large, the blades induce turbulence to the flow of air, which disturbs the wind flow and reduces the output power. Lift blades typically have a high tip-speed ratio (about 7).

Example 13.10

For a wind turbine with a blade radius of 10 m that rotates with the rotational speed of 15 rpm at wind with a speed of 12 m/s, draw the velocity diagram at the tip of the blade and determine the velocity of wind relative to the blade.

Analysis

$$\text{Horizontal component} = V_{wind} = 12\frac{\text{m}}{\text{s}}$$

$$\text{Vertical component} = V_{tip} = 10 \times 15 \times 2\pi \times \frac{1}{60} = 5\pi = 15.7\frac{\text{m}}{\text{s}}$$

$$V_{rel} = \sqrt{\left(V_{wind}^2 + V_{tip}^2\right)} = \sqrt{\left(12^2 + 15.7^2\right)} = 19.8\frac{\text{m}}{\text{s}}$$

$$\gamma = \tan^{-1}\left(\frac{V_{tip}}{V_{wind}}\right) = \tan^{-1}\left(\frac{15.7}{12}\right) = 52.6°$$

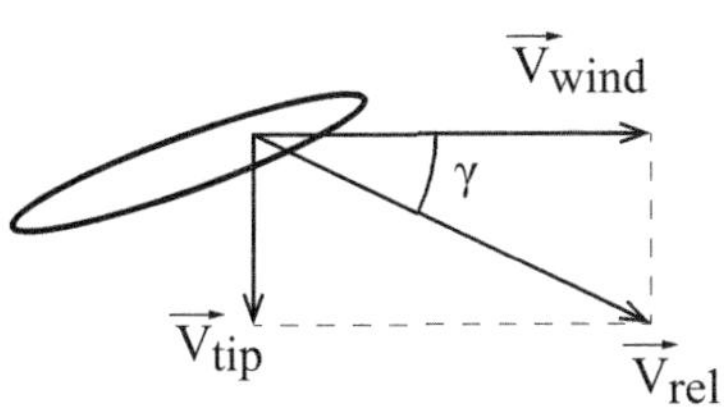

FIGURE 13.19
Velocity diagram for Example 13.10.

13.7.2 Vertical-Axis Wind Turbines

In **vertical-axis wind turbines** (**VAWT**), as the name implies, the axis of rotation of turbines is perpendicular to the direction of the wind and the local horizon. They were very popular in the early days of wind turbines both for the large-scale and small-scale applications, such as Persian windmills and James Blyth's turbine (Section 13.2). But nowadays, mainly due to their low efficiency, there are not many commercial large-scale VAWTs while a few small-scale VAWTs are commercially available. The most important advantage of VAWTs is their insensitivity to the direction of wind. These wind turbines do not need to be aligned at the wind direction and can produce power regardless of the wind direction. In contrast, horizontal-axis wind turbines need to be oriented to face wind, which is accomplished by yaw adjustments (Section 13.10). Normally, the yaw mechanism can follow the wind direction efficiently; however, the performance of horizontal-axis wind turbines can deteriorate when the wind direction changes rapidly. VAWTs are not sensitive to a rapidly changing wind direction, which makes them a suitable candidate for certain applications, particularly when turbulent and gusty winds are involved. Also in VAWTs the generator and other components are on the ground level, which makes them easy to install and maintain. Furthermore, the elimination or at least reduction of cyclic stresses due to change in the direction of the gravity relative to the direction of the blade motion, which can be damaging in long-term for horizontal-axis wind turbines, is another important feature of VAWTs. Moreover, typically VAWTs do not require a pitch adjustment mechanism, which simplify their control. Some VAWT designs are easier to manufacture than their horizontal-axis counterparts. They can be also packed in smaller areas, even beneath horizontal-axis wind turbines.

Traditionally, VAWTs are considered to have a lower efficiency compared to horizontal-axis wind turbines; however, some new VAWT designs have efficiency close to horizontal-axis wind turbines. Other than their lower theoretical and practical efficiencies, VAWTs suffer from the following disadvantages. Since they are commonly installed at low elevations, they are exposed to lower speed winds, which reduces their output power. The continuous change in the angle of attack causes the torque on blades to pulsate, which can cause high stress in components as well as undesirable variations in produced electricity. They require an extensive support system, which increases their footprint and limits their utilization in agricultural lands. They are also vulnerable to fatigue failure and stall. While in the past few decades, large capacity VAWTs have not been popular, recently there has been a resurgence of VAWTs particularly for small wind turbines. Also R&D activities on developing large size VAWTs have significantly increased. There are many types of VAWTs, but the three most popular ones are Savonius (Figure 13.20), Darrieus (Figure 13.21), and H rotor (or Giromill) VAWTs.

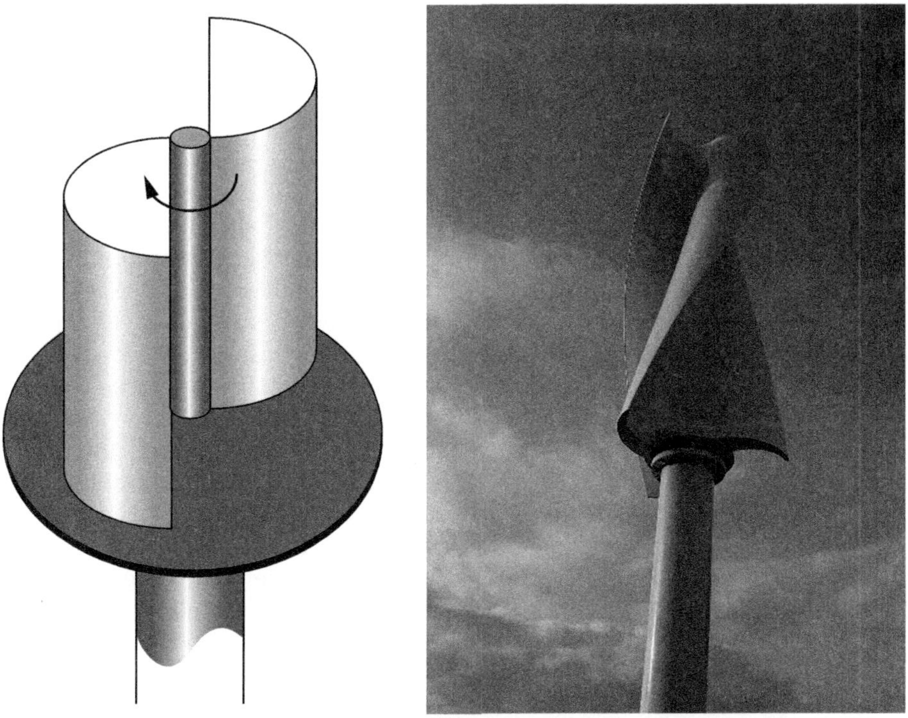

FIGURE 13.20
Schematic of a Savonius VAWT and an actual unit. (Image by Bernd Wolf from Pixab7ay, https://pixabay.com/photos/savonius-rotor-vertical-wind-turbine-3084781/.)

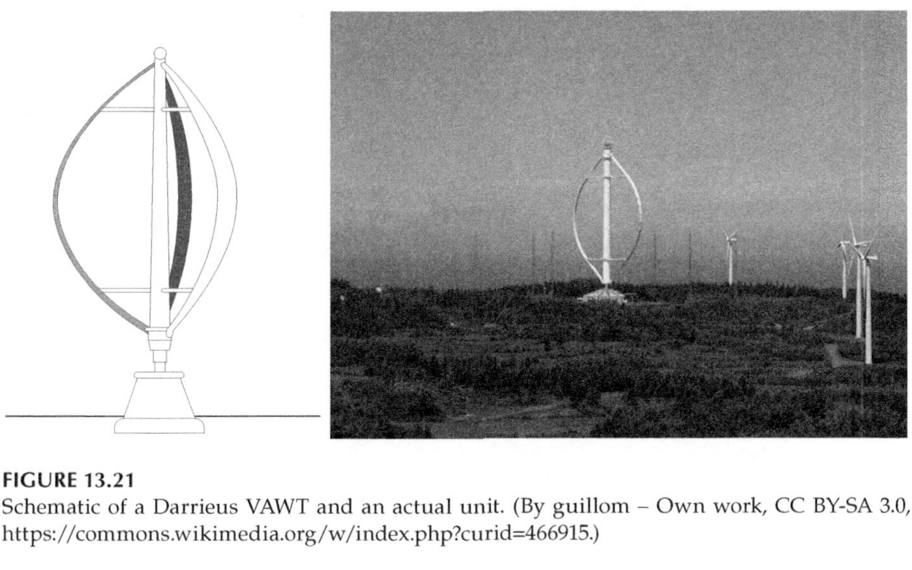

FIGURE 13.21
Schematic of a Darrieus VAWT and an actual unit. (By guillom – Own work, CC BY-SA 3.0, https://commons.wikimedia.org/w/index.php?curid=466915.)

13.7.3 Horizontal-Axis Wind Turbines

In **horizontal-axis wind turbines** (**HAWTs**), the axis of rotation of turbines is parallel to the wind direction and the local horizon. Almost all modern wind turbines, large or small, are horizontal axis. Their two main advantages are high efficiency and tall towers. The efficiency (or power coefficient) of HAWTs is extraordinarily high. This is primarily because they are able to extract wind energy through the entire blade length of all blades at all times (unlike VAWTs that blades in their return path oppose the rotation of the rotor). Also with their tall towers, they can be installed at high elevations where more powerful winds are available. While they are relatively expensive and are highly visible, their advantages overwhelmingly compensate for their disadvantages. For the rest of this chapter, we will mainly focus on this type of wind turbines.

These turbines almost universally have three blades (although there are some exceptions such as turbines in Figure 13.22). Theoretically, there is no limit for the number of blades in a wind turbine and wind turbines with one, two, three, four, or even more blades have been built (Figures 13.3, 13.4, 13.22 and 13.23); however, the turbines with three blades are the most popular ones due to their higher efficiency and lower vibration as well as from aesthetic standpoint. For HAWTs, the number of blades is typically an

FIGURE 13.22
Horizontal-axis wind turbines with two blades. (Image by skeeze from Pixabay, https://pixabay.com/photos/windmills-farm-technology-energy-705282/.)

FIGURE 13.23
A horizontal-axis wind turbine with six blades. (Image by Shutterbug75 from Pixabay, https://pixabay.com/photos/air-battery-blade-breeze-charge-1238449/.)

odd number. The main reason is an uneven force of wind on blades if the number of blades is even. When the two opposite blades are in the vertical position, the lower blade is affected by the tower, which reduces the wind force on the blade, while the upper blade is fully exposed to the wind. This effect can cause a cyclic stress on the structure of the blades that can damage them in long run.

While a great majority of existing horizontal-axis wind turbines are **land-based units, offshore wind turbines** are quickly gaining popularity. Until 2019 almost 30 GW offshore wind turbines were installed globally. While this installed capacity was small compared to the installed capacity of land-based wind turbines at about 600 GW, the rate of growth for offshore wind turbines has to be exponential in the past few years. They are one of the most promising power generation technologies in near future. They do not need any land for installation while enjoying faster and steadier winds that blow over oceans. Also, they are not subjected to scrutiny for sight and noise pollutions since they are typically installed far from shores. However, since they operate at harsh environments, it is more difficult to design, build, install, and maintain them. Also, it is more challenging to transmit generated electricity to shores and consumption centers. All of these factors make them more expensive than land-based wind turbines.

13.7.4 Grid-Connected and Stand-Alone Wind Turbines

Electricity produced by wind turbines can be easily supplied to the grid as long as the quality of produced electricity in terms of frequency and voltage meets the country-specific requirements. However, the highly intermittent nature of power produced from wind energy can be problematic when the share of renewable energy sources, particularly wind and solar energy, increases. The high degree of penetration of wind energy may require substantial energy storage systems (refer to Chapter 16) and a significant upgrade of the grid. Stand-alone wind turbines have fewer electrical constraints. They can produce AC electricity just like grid-connected wind turbines. Alternatively, they may produce variable frequency power, which can be used to charge a battery bank. The DC output of the batteries can be directly used in certain applications or it can be converted to AC power through an inverter.

13.7.5 Small and Hybrid Wind Turbines

Installing a small wind turbine for distributed, community, and residential applications can be very tempting and is becoming popular, especially in regions with frequent power outages. However, the economic feasibility of small wind turbines should be carefully assessed. In a region with apparently significant wind, the effects of natural and manmade obstacles can significantly reduce wind speed and thus its energy at low elevations. As a result, the wind turbine can produce its rated power only for a short period. Figure 13.24 shows a residential wind turbine. These turbines can have rated power between several hundred watts and 100 kW. They may be also used for telecommunication and water pumping. Also, local height limits for the installation of wind turbines should be taken into account. An attractive option is integrating a wind turbine and some solar PV panels in a **hybrid unit**. Figure 13.23 shows a small wind turbine and a solar PV panel in a hybrid unit. Alternatively, wind turbines can be used with mini- and micro-hydro units, bioenergy systems, diesel engines, or gas turbines.

13.8 Operation of Wind Turbines and Power Curves

The most common way of presenting the performance of wind turbines is through a **power performance curve** (or simply **power curve**). In this curve,

FIGURE 13.24
A residential wind turbine. (By Tobi Kellner – Own work, CC BY-SA 3.0, https://commons.wikimedia.org/w/index.php?curid=25275689.)

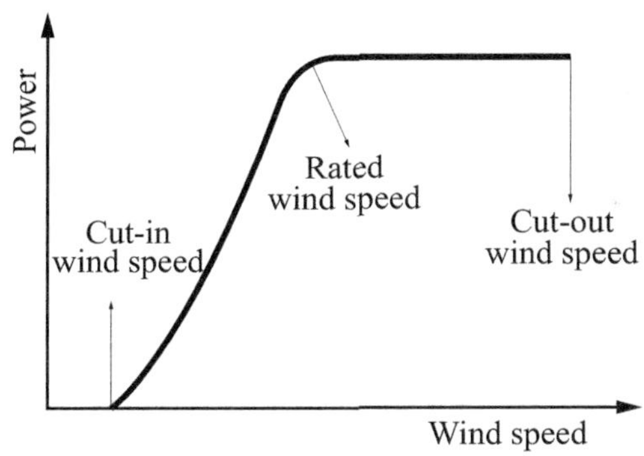

FIGURE 13.25
A typical power curve of a wind turbine.

the power output of a wind turbine is presented as a function of wind speed (Figure 13.25). The following three wind speeds are of main interest and importance in power performance curves.

Cut-in (start-up) wind speed: This is the wind speed that the wind turbine starts to produce electricity and is usually between 3 and 5 m/s (about 7–11 ft/s). At wind speeds lower than the cut-in speed, the force imposed by the wind is not sufficient to overcome the internal inertia of the wind turbine and the wind turbine will not turn.

Rated wind speed: This is the minimum wind speed that a wind turbine can produce its rated power and is commonly between 10 and 15 m/s (about 22–34 ft/s). A lower rated speed is more desirable because it means the turbine can operate at its rated power for a longer period of time and results in a larger capacity factor and more power generation.

Cut-out wind speed: At this speed, the turbine stops operation to protect its components from overloading and is typically about 25 m/s (about 55 ft/s). This can be done either by increasing the pitch of blades in pitch-regulated wind turbines (Section 13.9.2) or by turning the turbine 90° along its vertical axis (the tower that the turbine is installed on), so the turbine's rotational axis is perpendicular to the wind direction. The latter method was used in early wind turbines and is still being used in small turbines. Alternatively, blades can be turned to the feather position by a spring-loaded mechanism to protect the turbine. In this case, the turbine blades are parallel to the wind direction and the turbine is not operational. Another important wind speed is the **survival wind speed**. This is the wind speed above which the entire structure of the wind turbine, particularly the tower, may be at the risk of failure. This speed is typically between 50 and 60 m/s (about 110 and 135 ft/s).

The maximum or near-maximum wind speed at a given site should not be selected as the rated wind speed of the wind turbine because in this case the wind turbine will produce its rated power for a short period and most of the time its full capacity will remain unused. From an economic standpoint, the rated wind speed should be selected at a lower than maximum wind speed, which results in a smaller turbine and generator.

13.9 Control of Wind Turbine Power Output

In order to understand the control mechanisms of wind turbines, we need to recognize the objectives of these mechanisms. For this purpose, let us revisit the power curve of a typical wind turbine and discuss how the function of the control system changes with the variation of wind speed. The power curve of a typical wind turbine in Figure 13.25 indicates that between the rated and cut-out wind speeds the wind turbine should roughly produce its rated power. However, according to Equation 13.16, the power output of the turbine will increase with the cube of wind speed. To keep the output power of the turbine at a constant level and avoid overheating of the generator at this range, the power coefficient of the turbine should be reduced so that the turbine can produce its rated power (since A and ρ are fixed in Equation 13.16). A similar case can be made when the wind speed is between the cut-in and rated wind speeds but this time the objective is to increase C_p so that more power can be produced. In summary, the turbine operates at variable power between the cut-in wind speed and the rated wind speed and at constant power above the rated wind speed up to the cut-out wind speed. The wind turbine control system should be able to maximize the output power in the former range and maintain the power in the latter range.

Generally speaking, for a given wind turbine, the power coefficient can be controlled by adjusting tip-speed ratio, blade pitch, or rotor rotational speed. Using these three parameters have led to the introduction of the following three control strategies for wind turbines: stall-regulated, pitch-regulated, and variable speed (for tip-speed ratio, blade pitch, and rotor rotational speed, respectively). In the following sections, we will briefly review these control strategies.

13.9.1 Stall-Regulated and Pitch-Regulated Wind Turbines

Before discussing these control systems, we need to define the pitch of blades. The pitch or blade angle of a turbine blade is the angle between the chord line and the plane of the rotor (the plane of the rotation). Note that the chord of a blade is a line connecting the leading edge and the trailing edge of the blade. Figure 13.26 illustrates the pitch and the angle of attack of the blade. The attack angle is the angle between the chord line and the wind velocity vector with respect to the blade. The pitch is only a function of the design and geometry of the blade but the angle of attack depends on the design of the blade, its rotational speed, and the wind speed.

In terms of the pitch of blades, wind turbines can be either fixed pitch or variable pitch. Fixed-pitch wind turbines are simple and inexpensive because their blades are locked to the turbine rotor and they do not need a mechanism to change the pitch of the blades. These turbines are **stall-regulated**, also known as the Danish Concept, after where it was first pioneered.

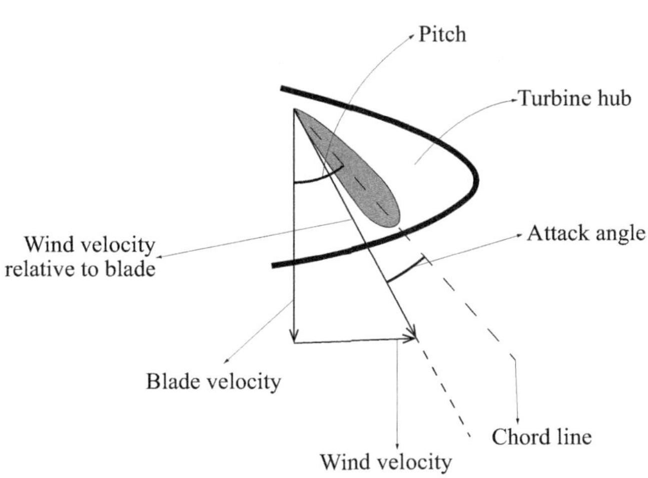

FIGURE 13.26
Schematic of a wind turbine blade along with pitch and attack angle.

They are the oldest and simplest design for wind turbine control. The rotational speed in these turbines is held nearly constant. At higher wind speeds, the attack angle of blades changes due to the change in the vector diagram shown in Figure 13.26. This results in the separation of the air flow from the blades. This separation increases the drag force and decreases the power output, which protects the generator from overloading. At very high wind speeds, the turbine eventually stalls so the name **stall-regulated** wind turbines (for more on stall refer to Chapter 9). Basically in these turbines, the output power of the turbine at wind speeds higher than the rated speed is controlled (reduced) passively. This concept was first developed in 1957 and most wind turbines until the 1980s were based on this design. Later this design was further improved by controlling stall by means of varying the blade pitch angle by a few degrees, which is called the **active-stall control**. Although stall-regulated wind turbines were robust and inexpensive, the output power of the turbine reduces at wind speeds higher than the rated speed. The active-stall and pitch-control concepts were introduced to overcome this problem.

In pitch-regulated wind turbines, the pitch of the blades varies greatly with a change in the wind speed to achieve the rated power output at above the rated wind speed by reducing C_P. This is accomplished by increasing the pitch, which reduces the angle of attack. In large wind turbines, the pitch is controlled by a mechanism that rotates the blades based on the signals received from the control system. The control system estimates the required pitch angle based on the wind speed measured by the wind speed measurement instrument (an anemometer). In smaller wind turbines, the pitch is controlled using spring-loaded blades that react to centrifugal forces. The pitch-regulated wind turbines were able to maintain the output power at the rated power when the wind speed was higher than the rated wind speed and they had also good performance at wind speed lower than the rated speed; however, they had a problem in gusty winds. In a sudden gust, the pitch-control system could not react fast enough to compensate for the effect, and this caused a short strong variation in produced power, which reduced the quality of electricity and imposed a shock to the structure. To overcome the problem, variable speed wind turbines were introduced.

13.9.2 Constant Speed and Variable Speed Wind Turbines

In terms of rotational speed, wind turbines can have constant or variable rotational speeds. Most modern large wind turbines are variable speed. The nature of the rotational speed of a wind turbine is determined by the type of electric generator connected to it. If the electric generator is directly connected to the grid, in order to maintain the constant frequency of produced electricity, the rotational speed of the wind turbine should be kept constant. The rotational speed of the generator depends on the desired frequency of generated electricity (60 Hz in the U.S. and Japan and 50 Hz in Europe and Asia) and the number of poles in the generator (for more on this refer to Chapter 6).

$$\text{Rotational speed of generator (rpm)} = \frac{f \times 60}{\frac{1}{2} \times \text{number of poles in generator}} \quad (13.27)$$

The number of poles in a generator is the number of magnetic north and south poles in the generator and is always an even number, commonly two or four. Therefore, the rotational speed of the generator can be either 3000 or 1500 rpm for electricity with the frequency of 50 Hz and 3600 or 1800 rpm for electricity with the frequency of 60 Hz. Obviously, a wind turbine cannot rotate with this rotational speed, and a gearbox is needed to convert the slow and constant rotation of the rotor to the higher speed of the generator.

Since in this configuration, generated electricity is directly fed to the grid and no power electronics is required, it is simple and inexpensive. However, even with pitch angle regulation, the turbine cannot smooth out the sudden variation of wind speed above the rated wind speed. The concept of variable rotational speed wind turbines can fix this problem.

There are two dominant concepts for this purpose. In the first design, the generator is still coupled to the grid and the rotational speed of the turbine can be adjusted by about 10% to smooth out the sudden variation of wind speed along with pitch control. This design however has a complete variability of rotational speed at wind speeds below the rated wind speed. The rotational speed can be doubled between the cut-in speed and rated speed, which results in smooth operation and reduced noise. More than half of newly manufactured wind turbines follow this design.

In the second design, the generator is completely decoupled from the grid. In some designs, the gearbox has been eliminated altogether and the rotor shaft is directly connected to the generator shaft. In this design, the rotational speed can triple between the cut-in speed and the rated speed. The electricity generated in the generator has a variable frequency and it should be converted to DC power and then back to AC power at the grid frequency through the frequency transverter, which requires a considerable amount of power electronics. This design improves the quality of produced power and reduces the sensitivity of the turbine to a disturbance in the grid. A great majority of the new large wind turbines are designed based on one of these two concepts.

13.10 Wind Turbine Components

In this section, some major components of wind turbines (mainly in HAWTs) will be covered. Typically, a large wind energy conversion system (WECS) is composed of the blades, hub, gearbox, generator, control system, nacelle, and tower. Figure 13.27 illustrates the main components of a wind turbine. We have already discussed some of these systems.

The rotor system of a wind turbine mainly consists of the blades, hub, and pitch motors. Blades of very large wind turbines can be over 100 m in length (Figure 13.28). These enormous components should endure extremely harsh atmospheric conditions. Also, they should be able to tolerate the cyclic stress caused by the bending force of the blades' own weight for hundreds of million cycles in the turbine's lifetime. To design these engineering marvels, the most advanced materials such as carbon-fiber composites are used.

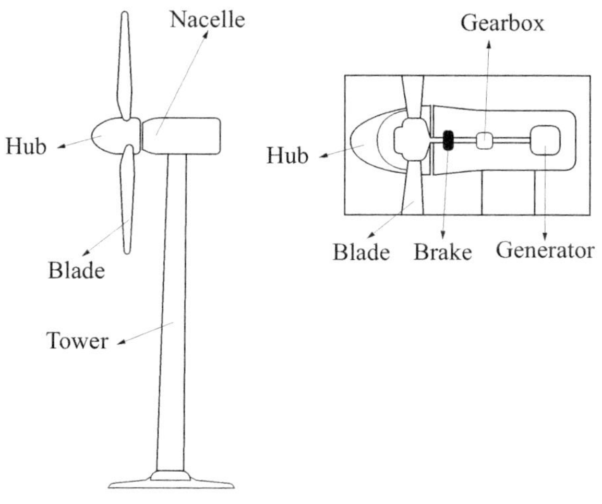

FIGURE 13.27
Schematic of main components of a large wind turbine.

FIGURE 13.28
A wind turbine blade at construction stage. (Image by Bishnu Sarangi from Pixabay, https://pixabay.com/photos/wind-mill-turbine-rolling-stock-3674248/.)

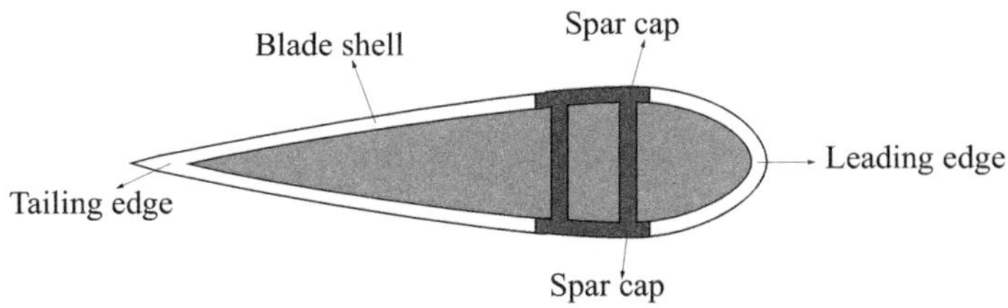

FIGURE 13.29
Cross-section of a wind turbine blade.

Wind turbine blades are composed of a central beam called the **spar** that supports the blade along its length. The spar can have an enclosed profile. Alternatively, it can have one or two shear webs with two spar caps at two ends (Figure 13.29). Similar to steam turbine blades (discussed in Chapter 6), the blade velocity ($r\omega$) varies along the blade length and is directly proportional to the radius. This variation results in different velocities of the wind relative to the blade and different angles of attack. As a result, the wind turbine blades are twisted.

This central piece is covered with a shell made of fiberglass and epoxy resin on both upwind and downwind toward the leading edge and tailing edge, respectively. The blade is connected to the **rotor hub** by a metallic cylinder either directly (in fixed pitch) or through the pitch-control mechanism (Figure 13.30). In the latter, the pitch motors are located inside the hub. The blades are also equipped with a lighting system (to warn aircrafts) and various sensors to measure stress, strain, noise, etc.

(a) (b)

FIGURE 13.30
Connection of blades to the hub before (a) and after installation (b). (a: Image by Hans Linde from Pixabay, https://pixabay.com/photos/windr%C3%A4der-pinwheel-site-2759645/. b: Image by Steppinstars from Pixabay, https://pixabay.com/photos/windmill-wind-wind-turbine-electric-62257/.)

The hub is connected to the generator via a shaft. Depending on if the wind turbine is fixed or variable speed, a gearbox may be placed between the hub and the generator to regulate the rotational speed of the generator. The gearbox converts the low rotational speed of the rotor to the high rotational speed required by the generator. Also, the shaft is typically equipped with a brake to stop the rotation of the shaft (and blades) for maintenance or when the wind speed is above the cut-out speed.

In HAWTs, it is very important to align the turbine so that its axis of rotation is always parallel to the wind direction. The angle between the axis of rotation of the turbine and the wind direction is called the **yaw angle**. If this angle is not zero, the effective wind velocity will be equal to the normal component of the wind speed to the rotor plane. For small turbines, the yaw angle is usually controlled passively by a vane. In large wind turbines, the yaw angle is controlled by a motor. The yaw control mechanism is in charge of aligning the wind turbine according to the wind direction measured by wind sensors. The yaw motor and its related system are located at the top of the tower, which turns the whole turbine, the rotor, and the nacelle, according to the wind direction about the tower axis.

The **nacelle** is a streamlined enclosure at the top of the tower that contains most mechanical and electrical components of the wind turbine, including the generator and gearbox (Figure 13.30). The turbine tower supports the entire system at the desired height. Typically for large HAWTs, the diameter of the turbine blades is roughly equal to the height of the hub. The bottom of the wind turbine blades should be at least several times higher than any obstacles near the turbine.

FYI BOX 3.3

AN INNOVATIVE DESIGN FOR WIND TURBINES BASED ON MAGNUS EFFECT

The Magnus effect, which was discovered by a German scientist Heinrich Gustav Magnus (1802–1870), states that if a cylinder or sphere rotates in the wind blowing at the direction perpendicular to the axis of rotation of the cylinder, a lift force will be generated that is perpendicular to both the axis of rotation of the cylinder and the wind direction. Have you ever wondered why the ball in soccer, golf, baseball, volleyball, tennis, and other games played with a ball curves away from its principal projectile path? Now you know. The first design based on this concept was introduced in 1912 but still, there is no commercial product based on this concept. However, several prototypes have been fabricated. These machines look like HAWTs but instead of airfoil blades, they have cylindrical blades. The cylinders rotate about their axis of symmetry by an electrical motor in the hub. This electric motor uses the power generated in the generator. For the start-up, there is a battery that can run the cylinders for several seconds.

13.11 Environmental Impacts of Wind Turbines

While wind energy is considered a renewable and sustainable source of electricity, like any other technology, it has some negative environmental impacts. The most important impact of wind turbines on wildlife is their effects on flying animals, such as birds and bats. They can kill or hurt these animals due to direct collision with blades or the variation in air pressure near turbines. Using advanced technologies and adjustments in operational procedures have proved successful in reducing these impacts. It should be noted that the life-cycle carbon emission of wind turbines is one to two orders of magnitudes less than that for fossil fuel power plants. These pollutions could possibly have killed much more birds and bats compared to those by wind turbines. Also, this number of casualties is small compared to those killed by other human activities, such as collision with buildings or even cats. Noise and sight pollutions have also caused some concerns in some people and communities. However, they have not been proven to have real harmful impacts on people's health. Complaints about noise seem to be a direct function of how friendly the relationship between residents and turbine owners is. Finally, the land use for wind turbine installations needs to be taken into account in the design of large projects. The installation of wind turbines in agricultural lands, where the space between turbines can be effectively used, and in brown lands where other activities are not possible have been proposed and successfully implemented.

13.12 Questions and Problems

13.12.1 Introduction

13.1. List pros and cons of wind energy.

13.2. Which countries have the largest wind turbine farm installations in the world?

13.12.2 Wind Formation

13.3. Explain how wind is created.

13.4. Why is wind considered as an indirect form of solar energy?

13.5. What are the similarities and differences between planetary wind and local wind?

13.6. Explain how sea breeze and land breeze are created? Why are they in the opposite direction?

13.12.3 Power of Wind and Wind Energy Evaluation

13.7. What is wind power density? How is it different from wind power?

13.8. A horizontal-axis wind turbine with a radius of 15 m operates in a location with a uniform wind speed of 5 m/s. Determine the power of wind for an area equal to the swept area of the turbine and per unit of area. Compare your answer with the value you read from the diagram in Figure 13.5.

13.9. Determine the blade diameter of the turbine with a rated (design) output power of 1.5 MW when a rated (design) wind speed is 10 m/s. The power coefficient of the turbine is 0.50. Repeat your calculation for the power coefficient equal to the Betz limit and 0.45. Compare your results with the size of an actual wind turbine with the same capacity.

13.10. Why is an accurate measurement of wind speed critically important for wind energy estimation?

13.11. In wind speed measurements, which component of wind velocity is the most important parameter in determining wind energy? Why?

   a. Horizontal component
   b. Vertical component

13.12. Name two types of instruments to measure wind properties.

13.13. In the following meteorological tower, what types of wind speed measurement instruments are used?

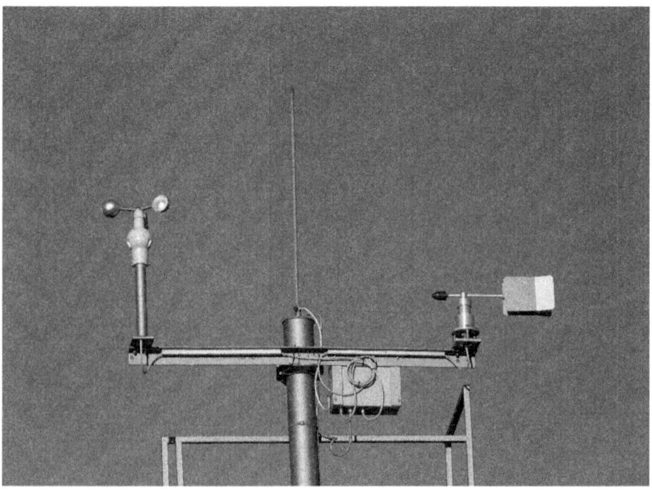

13.14. What are three types of anemometers? Explain how each one works.

13.15. Define the mean energy velocity.

13.16. In a region for points A to C wind velocity is measured five times. The wind velocity data in the region are listed in the following table. For each set of data determine the average wind velocity and the mean energy velocity.

Wind Velocity	V_1	V_2	V_3	V_4	V_5	$\bar{V}$	$\bar{V}_E$
A	1	2	5	6	8		
B	2	10	9	6	3		
C	12	7	3	9	4		

13.17. If the wind average velocity is 20% less than the mean energy velocity, the actual wind energy will be ... times ..... (greater/smaller) than the energy calculated by the average velocity.

13.18. Which continuous probability distribution is commonly used to model wind speed?

13.19. Explain what the shape factor and scale factor of the Weibull distribution represent.

13.20. For wind in a region, the shape factor and scale factor of the Weibull distribution that fits the distribution of the wind velocity are 3 and 8 m/s, respectively. Determine the probability of having the wind speed of 12 m/s and how many hours a year the wind with the speed of 12 m/s is likely to blow.

13.21. Describe the process to determine the shape factor and scale factor of the Weibull distribution for the measured wind speeds for a given region (you need to search for this question).

13.22. What is wind shear?

13.23. What is the wind power law? Why do we need it?

13.24. A wind turbine can be installed at the hub height of 100 m or 150 m. If the location of the installation is open flat terrain, determine the power ratio of the turbine at the two elevations.

13.25. Two different wind turbines with diameters of D_1=7.5 m and D_2=30 m are installed at the hub height of 15 m or 60 m, respectively. If the location of the installation is open flat terrain, determine the power ratio of the turbines at two elevations. The power coefficients of the two turbines are the same.

13.26. Show that wind power density doubles from the height of 10 m–50 m for the wind shear exponent of 0.14.

13.27. Explain the effects of variations of density and relative humidity of air in wind power.

13.28. What type of information can wind maps provide?

13.12.4 Wind Turbine Categorization

13.29. Explain lift force and drag force in a blade. Explain how lift force and drag force are different.

13.30. Show lift force and drag force in the following diagram.

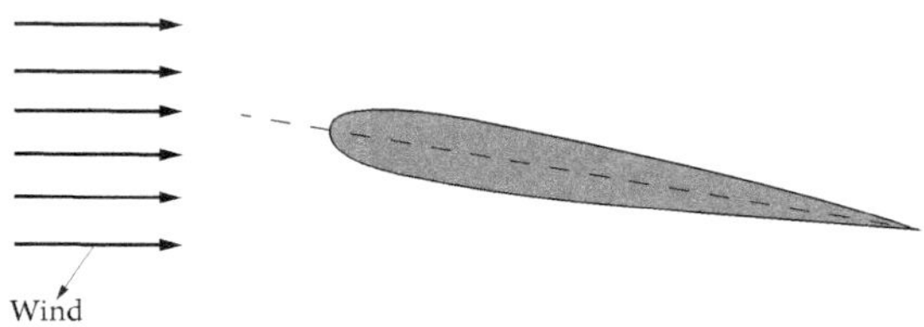

13.31. The speed of drag-force dominant devices relative to wind cannot exceed the wind speed. Why?

a. True

b. False

13.32. Explain why drag-force dominant wind turbines have low efficiency (power coefficient).

13.33. Persian turbines were purely …. force turbines.

a. lift

b. drag

13.34. Which one is more efficient?

a. Drag-force dominant wind turbines

b. Lift-force dominant wind turbines

13.35. For a Persian windmill, determine the drag force and the torque produced by wind with a speed of 8 m/s. The height and radius of the wheel are 3 and 1.5 m, respectively, the rotational speed of the wheel is 7 rpm, and C_D is 1.1 (make appropriate assumptions).

13.36. Define the tip-speed ratio of a blade. Explain how this ratio impacts the performance of the blade.

13.37. For a wind turbine with a blade radius of 25 m that rotates with the rotational speed of 12 rpm at wind with a speed of 9 m/s, draw the velocity diagram at the tip of the blade and determine the velocity of wind relative to the blade.

13.38. What are positive and negative attributes of vertical-axis wind turbines?

13.39. What are the three most popular types of VAWTs?

13.40. What are the two main advantages of HAWTs?

13.41. Explain why the efficiency (or power coefficient) of HAWTs is high.

13.42. Why do HAWTs almost universally have three blades?

13.43. A great majority of existing horizontal-axis wind turbines are … wind turbines.

c. land-based

d. offshore

13.44. Describe the pros and cons of land-based wind turbines compared to offshore wind turbines.

13.45. Compare grid-tied and stand-alone wind turbines.

13.46. What are the advantages of hybrid wind turbines and solar PV panels?

13.12.5 Operation and Control of Wind Turbines

13.47. Define the following wind speeds in power curves for wind turbines.

a. Cut-in (start-up) wind speed

b. Rated wind speed

c. Cut-out wind speed

13.48. In the following power curve, for a wind turbine, show cut-in, rated, and cut-out speeds.

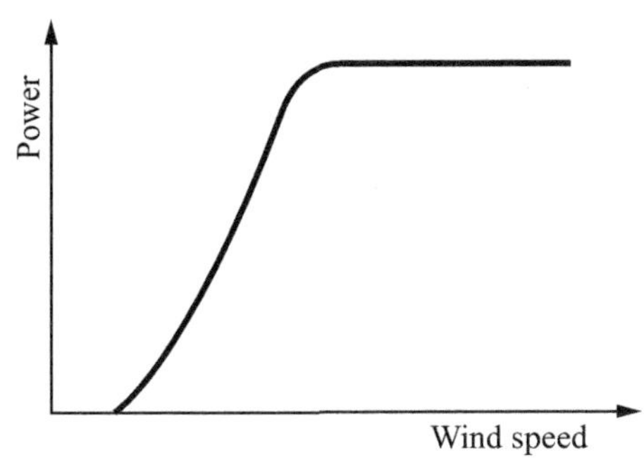

13.49. Explain how a wind turbine can be turned off at wind speeds above the cut-out wind speed.

13.50. What is the survival wind speed?

13.51. Describe why the maximum or near-maximum wind speed at a given site should not be selected as the rated wind speed of the wind turbine.

13.52. Explain how the function of a wind turbine control system varies between the cut-in wind speed and the cut-out wind speed.

13.53. How can the power coefficient be controlled (three methods)?

13.54. Define the pitch and the angle of attack of a blade.

13.55. In the following schematic, show the pitch and the angle of attack.

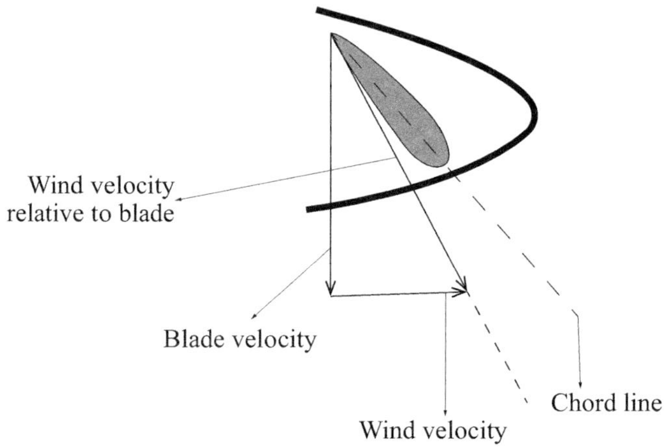

13.56. Name three wind turbine power control strategies.

13.57. Explain each of the following wind turbine power control strategies and their characteristics: stall-regulated, pitch-regulated, and variable speed.

13.12.6 Wind Turbine Components

13.58. What is the function of the spar in a wind turbine blade?

13.59. Why are wind turbine blades twisted?

13.60. How are blades connected to the hub?

13.61. What is the function of the brake? When is it used? Where is it?

13.62. What is the yaw angle of a wind turbine? Why is it important? How is it controlled in small and large wind turbines?

13.63. List some negative environmental impacts of wind turbines. How can they be addressed?

13.64. Name the components of a HAWT in the following schematic.

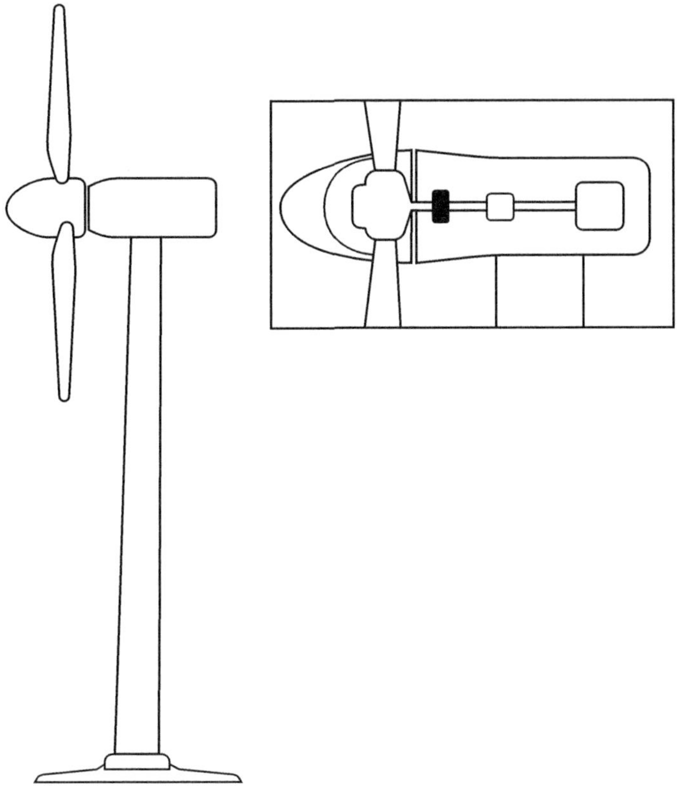

13.12.6 General

13.65. Compare planetary wind and local wind. How are they created?

13.66. Describe how stall-regulated wind turbines control the power output of the turbine. What are their pros and cons?

13.67. At wind speeds, lower than the rated wind speed, the force imposed by the wind is not sufficient to overcome the internal inertia of the wind turbine and the wind turbine will not turn.

a. True

b. False

13.68. Name two types for each of the following categories of wind turbines.

- Interaction of blades with wind
- Axis of rotation
- Connection to the grid

- Nature of rotation speed
- Power output capacity
- Location

13.69. Define coefficients of drag and lift.

13.70. What is the main reason that VAWTs are not used in modern wind turbines?

13.71. A wind turbine can be installed at the hub height of 75 m or 100 m. If the location of the installation is open flat terrain, determine the power ratio of the turbine at two elevations.

13.72. Where can wind maps for the U.S. be found?

13.73. Explain how the variable frequency power of a stand-alone wind turbine can be used if the system is equipped with a battery bank.

13.74. Describe how pitch-regulated wind turbines control the power output of the turbine. What are their pros and cons?

13.75. Which type is more expensive?

a. land-based wind turbines

b. offshore wind turbines

13.76. Theoretically, there is no limit for the number of blades in a wind turbine and wind turbines with one, two, three, four or even more blades have been built.

a. True

b. False

13.77. The cube of the average velocity is not equal to the average of the velocity cube.

a. True

b. False

13.78. Explain how valley breeze and mountain breeze are created? Why are they in the opposite direction?

13.79. Why wind energy varies at different locations and elevations?

13.80. How are vertical-axis and horizontal-axis wind turbines different? Compare their positive and negative characteristics.

13.81. Determine the blade diameter of a turbine with the rated (design) output power of 10 MW when the rated (design) wind speed is 14 m/s. The power coefficient of the turbine is 0.55. Repeat your calculation for the power coefficient equal to the Betz limit and 0.45. Compare your results with the size of an actual wind turbine with the same capacity.

13.82. The maximum or near-maximum wind speed at a given site should be selected as the design wind speed of the wind turbine.

a. True

b. False

13.83. Describe how variable speed wind turbines control the power output of the turbine. What are their pros and cons?

13.84. Wind is generated by uneven heating of Earth's crust by solar radiation.

a. True

b. False

13.85. Two different wind turbines with diameters of $D_1 = 70\,m$ and $D_2 = 140\,m$ are installed at the hub height of 100 m and 150 m, respectively. If the location of the installation is open flat terrain, determine the power ratio of the turbines at two locations. The power coefficients of the two turbines are the same.

13.86. What are wind classifications? What is the wind speed range for each class?

13.87. What is the Betz limit? Determine the value of this limit.

14

Ocean Renewable Energy

Abstract

Oceans are gigantic energy storage systems with a variety of renewable energy sources, namely wave energy, tidal energy, ocean thermal energy, ocean current energy, and salinity gradient energy. We will start our discussion with wave energy. We will learn about the origin of wave energy and conduct simple calculations to determine wave energy. Then, we will review various types of wave energy converters, including point absorbers, oscillating water columns, attenuators, overtopping devices, oscillating wave surge converters, and submerged pressure differential devices. Then, we will cover tidal energy systems, including gravitational potential energy and current energy, and their simple energy calculations. We will conclude the chapter with a brief discussion on ocean thermal, current, and salinity gradient energy.

LEARNING OUTCOMES

After successful completion of this chapter, you will be able to

- explain origin of wave energy and conduct simple calculation to determine wave energy and power
- describe various types of wave energy converters, including point absorbers, oscillating water columns, attenuators, overtopping devices, oscillating wave surge converters, and submerged pressure differential devices
- explain different tidal power generation plants and their characteristics
- determine tidal energy potential by simple calculations
- explain working principles of ocean thermal, current, and salinity gradient energy

14.1 Introduction[1]

Oceans cover over 70% of Earth's surface that act as a giant energy storage system. Among various forms of renewable energy sources, the energy potentials of oceans and seas are special because they have stored a great amount of energy in different forms, such as wave, tidal, thermal, and current energy. Practical experiences in the commercial applications of sea energy, such as tidal electricity generation, has proved that, in long term, these technologies can compete with conventional power plants.[2] In this chapter, ocean energy sources are classified and reviewed in five groups, namely wave energy, tidal energy, ocean thermal energy, ocean current energy, and salinity gradient energy. These sources of ocean energy will be discussed in the following sections.

14.2 Wave Energy Sources

Waves are the most well-known symbol of ocean energy. Waves are produced by wind effects on the water surface. That is why wave power is indirectly considered as a type of solar energy. The idea of harnessing wave energy has been around for a long time. Early patents on wave energy conversion devices started in the late 18th century. However, **wave energy converters** (**WECs**) were first seriously considered in the 1970s during the oil crisis. Those efforts were gradually ceased due to decrease in the oil price and disappointing results of experimental units. During the two past decades, the interest in WECs has been renewed. One of the most important advantages of wave energy compared to other forms of energy derived from solar energy is the high density of the wave energy. In fact, wave energy can be considered as a concentrated form of solar and wind energy. For instance, in latitude 15° N, the average density of direct solar, wind, and wave energies are 0.17, 0.58, 8.42 kW/m^2, respectively.[3] Also, the availability of wave energy is much higher than other forms of renewable energy. In addition, unlike solar and wind energy, wave energy does not require large lands and it is more available where it is required, such as remote islands that are not connected to the grid and where the cost of electricity is usually high.

[1] This chapter is adapted from the following paper published by the author with the permission from the copyright holder (Elsevier).
Zabihian, F. and A.S. Fung. "Review of marine renewable energies: Case study of Iran." *Renewable and Sustainable Energy Reviews* 15.5 (2011): 2461–2474, with permission from Elsevier.

[2] Andre, H. "Ten years of experience at the "La Rance" tidal power plant". *Ocean Management* 4(2–4) (1978): 165–178.

[3] McCormick, M. *Ocean Wave Energy Conversion*. First ed., New York: Wiley (1981).

The predictability of waves is another important factor to consider. While wind can be predicted only hours in advance, waves are predictable days in advance.

14.2.1 Energy and Power of Wave

The amount of wave energy depends on the wavelength (or period) and height. Wave energy is a combination of potential energy due to its height and kinetic energy arising from the movements of water particles. Wave energy converters can produce electricity from both forms. The characteristics of waves highly depend on wind conditions that created them, such as its speed, period, and fetch, as well as on seabed status.

The wave height is the most important factor for wave energy determination. While typically the height of wave may refer to its peak-to-peak (trough-to-crest) height (H in Figure 14.1), for wave energy calculations typically the significant wave height is used. The **significant wave height** (SWH) is defined as the average wave height (peak-to-peak) of the highest one-third of the waves. It may be shown as H_s, $H_{1/3}$, or H_{m0} (here we will use H_s). Using the significant wave height provides more realistic energy calculations. For a sinusoidal wave with the peak-to-peak height of H, H_s is equal to $\frac{H}{\sqrt{2}}$.

The average energy of a wave per unit of area in deep water, where the water depth is greater than one-third of the wavelength, can be estimated by the following equation[4]:

$$E = \frac{\rho g H_s^2}{16} \tag{14.1}$$

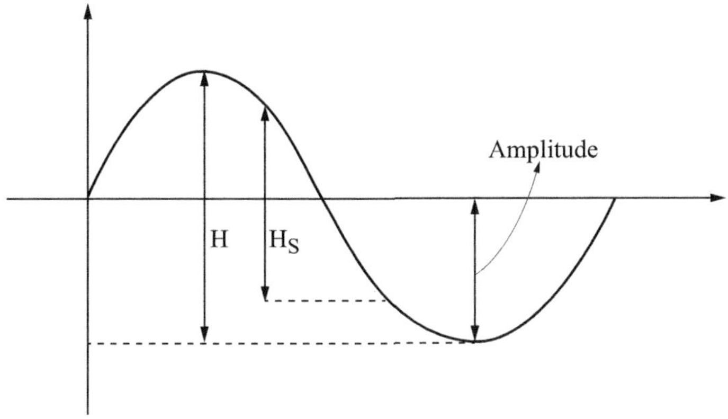

FIGURE 14.1
A sinusoidal wave.

[4] Falnes, J. "A review of wave-energy extraction". *Marine Structures* 20 (4) (2007): 185–201.

where E is the average energy of the wave per unit of area (J/m^2), ρ is the density of seawater (kg/m^3), g is the acceleration of gravity (m/s^2), and H_s is the significant wave height (m). This energy is equally divided between the wave's kinetic and potential energy. For the power of a wave per unit of wavefront width, the equation is[3]

$$P = \frac{\rho g^2 T H_s^2}{64\pi} \tag{14.2}$$

where P is the power of the wave per unit of wavefront width (W/m) and T is the wave period (s). When evaluating wave power in a certain area, it is important to find the ranges of significant wave height and period at which most wave power occurs, because these ranges should be considered for determining the design point of a WEC.

Example 14.1

For a sinusoidal wave with an amplitude of 2 m and a period of 9 s, determine wave energy and power. Assume the density of water is 1000 kg/m^3.

Solution

$$\frac{H}{2} = 2\text{ m} \rightarrow H = 4\text{ m}$$

$$H_s = \frac{H}{\sqrt{2}} = \frac{4}{\sqrt{2}} = 2.83\text{ m}$$

The wave energy and power can be calculated by Equations 14.1 and 14.2 as follows:

$$E = \frac{\rho g H_s^2}{16} = \frac{1000 \times 9.8 \times 2.83^2}{16} = 4905.5\text{ J/m}^2$$

$$P = \frac{\rho g^2 T H_s^2}{64\pi} = \frac{1000 \times 9.8^2 \times 9 \times 2.83^2}{64\pi} = 34447\text{ W/m}$$

14.2.2 Different Wave Energy Converters

The wave energy application has a long history. The first patent about WECs was recorded in 1799 in France.[5] The first actual unit was constructed in 1910 in France using an oscillating water column system. Since then, a great number of systems and mechanisms has been proposed. Many of these concepts are in the R&D stage, which means they require considerable time before they can be commercialized. However, there are a few systems that

[5] Clement, A., P.F.A. McCullen, A. Fiorentino, F. Gardner, K. Hammarlund, G. Lemonis, et al. "Wave Energy in Europe: Current status and perspectives". *Renewable and Sustainable Energy Reviews* 6(5) (2002): 405–431.

are mature enough to be in the precommercialization stage or even the early market entry level. The two leading countries in this field are Portugal and the U.K. In addition, Australia, Denmark, and Ireland are heavily investing in this sector. As noted, in the 1970s, there was great interest in renewable energy sources in general and wave energy in particular. However, this interest did not result in any system with acceptable technical and economic characteristics. One of the main reasons was ambitious targets that were set for programs. For instance, in the U.K. the objective was to develop 2 GW wave power generation capacity by 1983, which resulted in oversize designs with unacceptably high costs. Based on this experience, recently developed systems have been designed as small individual units, usually <500 kW, which can be installed in large numbers to generate a large amount of electricity. These devices can be divided into three main categories: **shoreline**, **near shore**, and **offshore** systems. Shoreline systems require neither sophisticated mooring mechanisms nor underwater electrical cables or high-pressure pipelines to transfer generated power to the shore. Their installation and maintenance are also easier. However, suitable sites to implement such plants are limited and the available wave energy in those sites is lower than that in offshore sites. In addition, due to the unique conditions of each shoreline, most likely the mass production and economy of scale cannot be applied to shoreline systems, which may cause higher installation costs.

Nearshore systems can be helpful for coastal protection since after wave energy is extracted, the waves are calmer and less harmful for coastal regions. For a long time, shoreline plants attracted attention, because they were less vulnerable to the wild nature of ocean storms. However, the real potential energy of waves can only be achieved in offshore locations. That is why the trend of progress in this industry is toward offshore locations, where not only there is more power, but also much more suitable sites. Nevertheless, the reliability and accessibility of offshore plants should be significantly improved before their commercialization can be realized.

WECs can be also identified by how they convert wave energy to the desired product. The extracted energy from waves has low velocity and high force. One method to apply this energy is to transform it to high-speed rotating motion before it can be used in a generator to produce electricity. This requires a complicated mechanical system that increases capital and maintenance costs of the plant and reduces system reliability due to many moving parts, which decreases the capability of the system to withstand strong storms. Alternatively, wave energy can be converted to high-pressure fluid. This high-pressure fluid can be then used to either generate electric power in a turbine-generator or produce desalinated water in conventional reverse osmosis (RO) membranes.

Another method is to use a linear generator directly driven by a WEC. In such a system, the reciprocating motion of a buoy is transferred to a permanent magnet, which is placed inside a stator. The reciprocating motion of the magnet induces electric current in the stator and produces electric power. Since in this system, no mechanical or hydraulic conversion system is required, it is mechanically less complicated and requires less maintenance. However, due to variation in both frequency and voltage of the generated electricity, it requires a more complex electrical system before it can be connected to the grid.

Based on how wave energy is extracted, WECs can be classified into six main categories, namely a point absorber, oscillating water column, attenuator, overtopping device, oscillating wave surge converter, and submerged pressure differential device. Although this categorization is not exhaustive, it includes a great majority of WECs. In this section, some approaches that are deemed to be more promising are briefly presented. Then, some actual systems developed based on each approach are explained.

14.2.2.1 Point Absorbers

One of the most commonly used approaches to exploit wave energy at or near water surface is using the **height variation** in waves by **floats**. The floating part interacts with passing waves in all directions and converts wave energy to other forms of energy by heaving and/or pitching motions (Figure 14.2). Depending on the used mechanism, the produced energy can be in the forms of shaft power, air pressure, or liquid pressure, which can be directly converted to electricity or converted to electrical energy in a shoreline or offshore facility. Also, the extracted energy can be used to pressurize seawater (Figure 14.2). Then, the generated high-pressure seawater can be used to run reverse osmosis water desalination systems.

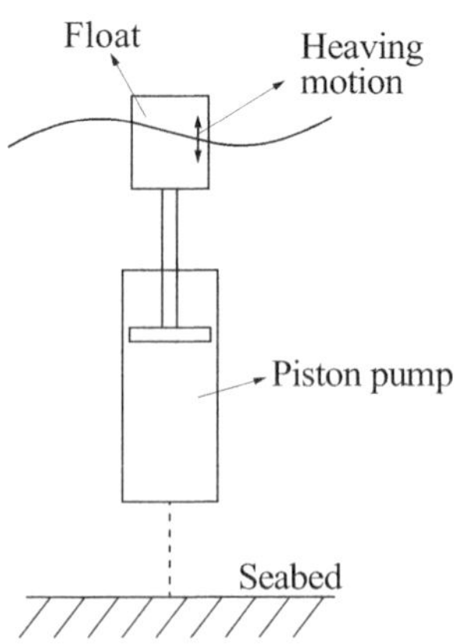

FIGURE 14.2
Schematic of a point absorber wave energy converter.

14.2.2.2 Oscillating Water Columns/Chambers

Another very common concept is an **oscillating water column or chamber (OWC)**. This system has had one of the longest development periods among all WECs. It consists of a partially submerged cell, which is open to seawater at the bottom. When waves pass through the system, the water column rises and falls inside the cell, acting as a piston, causing the air above the water surface to oscillate. This oscillation in turn compresses and depressurizes the enclosed air above the water column (Figure 14.3). Some designs use concentrating walls to concentrate wave energy into the air chamber. The high-pressure air can be then used to run a turbine to generate electricity. The cross section of the air passage is narrowed before it enters the turbine to accelerate the air flow and increase system efficiency. In this system, the air alternatively flows in two directions. Originally the air flow was rectified by a system of check valves. This system was replaced by a Wells turbine when it was invented by Alan Wells (1924–2005, a British structural engineer) in the late 1970s. The Wells turbine spins in the same direction regardless of the air stream direction due to its symmetrical blades. The disadvantages of this turbine are its relatively low efficiency and poor starting characteristics. Alternatively, some designs use a high-efficiency bi-directional airflow turbine with variable pitch blades. OWCs can be also integrated to breakwaters and coastal defenses.

14.2.2.3 Attenuators

Wave energy can be exploited by the relative motion of two or more floating structures. In this system, relative yaw and/or pitch motions between sections can be used to move double-action pumps. In one design, the system

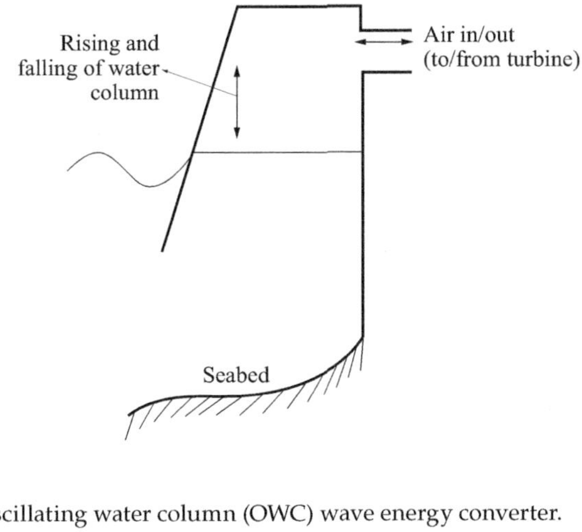

FIGURE 14.3
Schematic of an oscillating water column (OWC) wave energy converter.

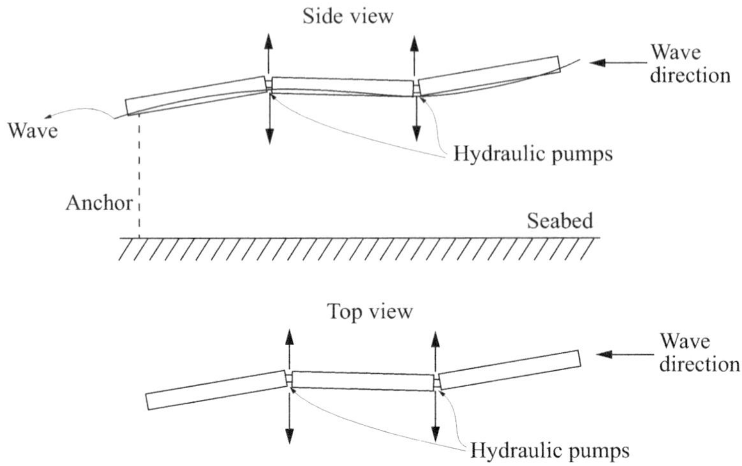

FIGURE 14.4
Schematic of an attenuator wave energy converter.

consists of several cylindrical segments. The relative motion of the segments, induced by waves, is converted to high-pressure fluid by hinged joints. These joints can capture both horizontal and vertical motions of the segments (Figure 14.4). This high-pressure fluid can be then smoothed in accumulators to be used to drive electrical generators via hydraulic motors. The generated electricity can be transmitted to the shore through a seabed cable. These machines operate parallel to the wave direction and should be able to effectively ride the waves. Since they have a lower area against the wave, they are subjected to lower force induced by the waves. This technology is one of the most promising WEC technologies. These systems are expected to be one of the market leaders in the future. The first full-scale grid-connected system and the first multi-MW wave energy Converter farm were tested using this type in 2004 and 2008, respectively (Figure 14.5).

Figure 14.6 illustrates a similar system that makes use of the relative motion of its parts. However, in this case, the hydraulic pumps are activated due to the motion of two pontoons relative to a fixed central part.

The schematic of another type of attenuator wave energy converters is shown in Figure 14.7. This system is based on a very old concept, and the first patent for this type of WECs was registered in the late 19th century. This nearshore multi-point absorption system consists of a series of hemisphere-shaped floats that are connected by arms to a central fixed section. The heaving motions of the half-submerged floats are transferred to pistons by arms to compress oil into the machine's common transmission system. The high-pressure oil drives a hydraulic motor and a generator to generate electricity. Since the system length covers several wavelengths, the wave energy can be harnessed continuously.

FIGURE 14.5
An attenuator wave energy converter. (By P123 – Own work, Public Domain, https://commons.wikimedia.org/w/index.php?curid=4859717.)

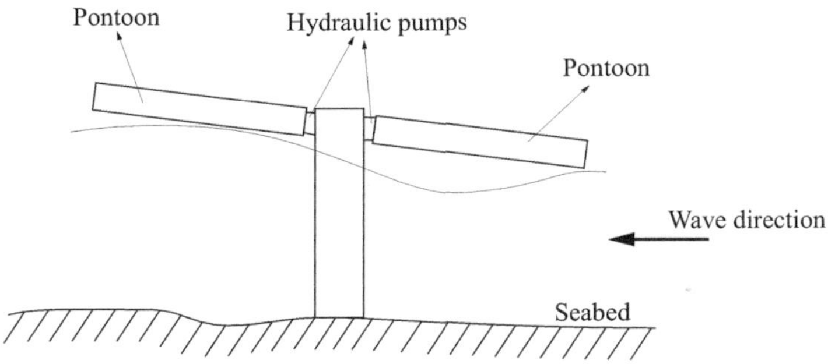

FIGURE 14.6
Schematic of an attenuator wave energy converter with two pontoons.

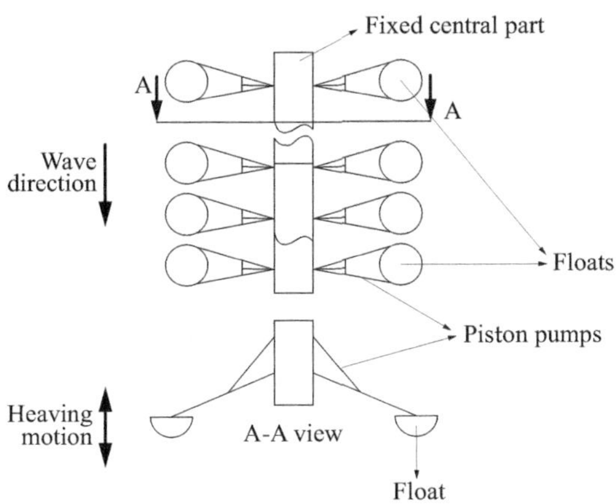

FIGURE 14.7
Schematic of a multi-point absorption system wave energy converter.

14.2.2.4 Overtopping Devices

The problem with all aforementioned systems is that, because of having many movable components, they are vulnerable to storms. Therefore, systems with minimal movable parts are desirable. One of the most successful WECs in this regard is overtopping devices. They can capture water from waves and hold it in a reservoir. The electricity can be then generated when the water from the reservoir returns to the sea via a low-head turbine. The only moving component is the turbine. This system makes use of the well-developed technology employed in conventional hydropower stations. Overtopping devices can be installed onshore or offshore and can be used in all weather conditions.

In onshore systems, waves are guided into a gradually narrowing channel, which amplifies waves, until they spill over to an above-the-sea level reservoir (Figure 14.8). The water stored in the reservoir can then produce electricity when it returns to the sea via a low-head turbine.

Figure 14.9 illustrates a schematic of an offshore overtopping device where a water head is generated when seawater overtops a reservoir. In some designs, there is no reservoir, and the overtopped water is guided directly to the turbine. Some designs may also integrate wave reflecting wings to increase the quantity of overtopped water (Figure 14.10). In some systems, an artificial beach in front of the device is used to catch water at different elevations through various inlets, which guides the water to the turbine through divided levels.

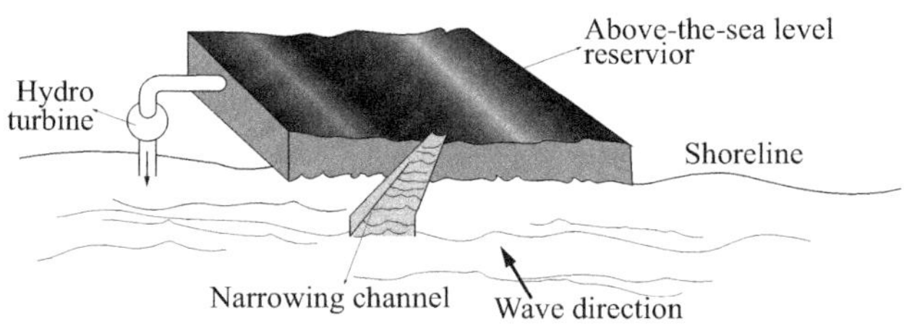

FIGURE 14.8
Schematic of an onshore overtopping device.

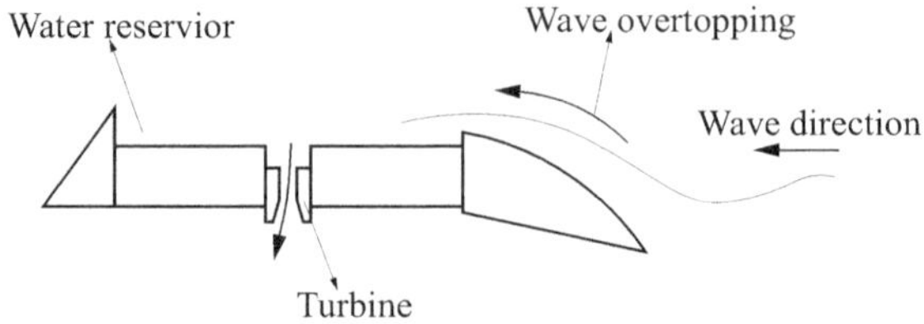

FIGURE 14.9
Schematic of an offshore overtopping device.

FIGURE 14.10
A small-scale prototype of an offshore overtopping device with reflecting wings. (By Erik Friis-Madsen at English Wikipedia, CC BY 3.0, https://commons.wikimedia.org/w/index.php?curid=11336506.)

14.2.2.5 Oscillating Wave Surge Converters

These machines rely on extracting wave energy from wave surges and the movement of water particles. Figure 14.11 shows a fully submerged and anchored array of buoys that are moved by passing waves. They are connected to pumps that convert wave energy to pressurized seawater. The high-pressure seawater is then transferred to shore to produce either electricity or desalinated water. In a similar design, inspired by the swaying motion of sea plants, an ocean wave energy system captures wave energy by the movement of buoyant blades as waves pass.

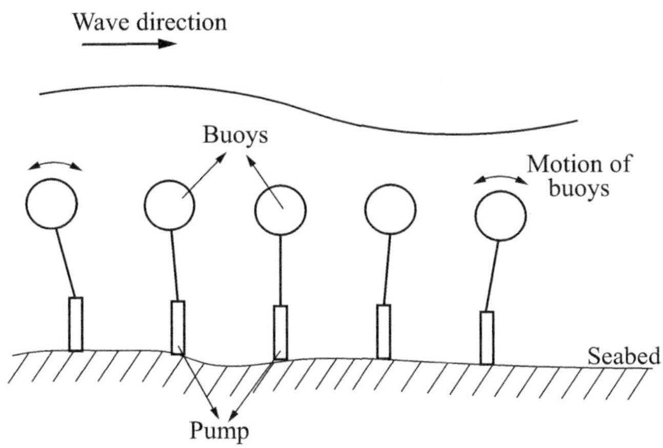

FIGURE 14.11
Schematic of an oscillating wave surge converter using buoys.

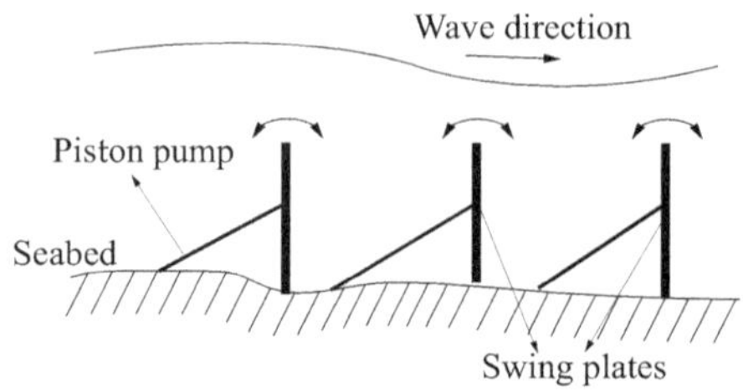

FIGURE 14.12
Schematic of an oscillating wave surge converter using plates.

In another design with the same concept, a series of plates are anchored on the seabed or shoreline (Figure 14.12). The wave energy is captured by the oscillation of the plates around their base, which then drives a piston pump to produce high-pressure oil. Being submerged provides some advantages for these devices over other systems, including protection from storms, self-tuning to the wave conditions and directions, and no visual impact. Another proposal uses a cam (called Salter cam), which vibrates due to the wave passage and can theoretically capture up to 90% of the wave energy.

14.2.2.6 Submerged Pressure Differential Converters

The rise and fall of the water level due to waves can create a differential pressure at the seabed. The devices based on the submerged pressure differential concept make use of this gradient to pump fluid to generate electricity. They are typically installed nearshore on the seabed. An example of a system based on this concept consists of a large air tank (with 10–20 m diameter) with a flexible upper section. The wave motions create a variable pressure, which in turn causes the upper section of the tank to move up and down and create high-pressure air flow.

It should be noted that some of the existing and proposal devices cannot be fitted to the aforementioned categorizations. Also, newer ideas to extract wave energy are being proposed from time to time, for instance specially designed turbines that are turned directly by waves and shape/volume changing flexible structures. Furthermore, many of the abovementioned systems have been designed to combine a WEC and an offshore wind turbine to simultaneously harness wave and wind energy.

14.2.3 Concluding Remarks on Wave Energy Converters

The main issue with the development of wave energy systems is their high cost. In order to reduce the cost of WECs, these systems need more demonstrations to gain the needed real-world practical experience in installation, operation, and maintenance, particularly in harsh ocean environments. In terms of environmental impacts, these systems have minimal negative effects on the ecosystem of installation sites. One study showed that neither species abundance nor biodiversity was reduced in the region after installation of WECs.[6] There were also positive signs that the area could recover after a disturbance during construction, even with ongoing research activities. Moreover, there was not any concern about the extinction of local marine organisms. However, long-term environmental analysis is required to evaluate the impacts of these systems. Some methods are being tested to improve the environmental performance of WECs. For instance, the differences between the concrete foundation with and without holes were investigated. The idea was that holes could provide shelter for many marine mobile species and could have a positive impact on colonizing species.[5]

14.3 Tidal Energy

Tidal energy is one of the most available forms of energy from oceans. Unlike most other renewable energy sources, which directly or indirectly are derived from solar energy, this energy is created by the gravitational forces of the Moon and the Sun on the waters of Earth and the rotation of Earth. One of the advantages offered by tidal energy over solar and wind energy is its predictability. While tidal energy in the world bodies of water is significant, only a fraction of this energy is exploitable because it occurs in gulfs and estuaries. The leading countries in harnessing tidal energy are Canada, France, the U.K., South Korea, and the U.S. Basically, tidal energy can be in the form of either the gravitational potential energy of the water-level difference during ebb and flood or the kinetic energy of tidal currents, which are briefly explained in the following sections.

14.3.1 Tidal Gravitational Potential Energy

Traditionally, tidal potential energy can be extracted by constructing a barrage against seawater and using the height difference of seawater and the water behind the barrage, caused by tides, to operate turbines installed in the barrage wall. The power that a tidal power generation plant can produce depends on both the tidal range and the capacity of the basin. The tidal energy is directly proportional to the square of the tidal range. This difference is almost 1 m (about 3.3 ft) in open oceans. But in coastal areas, it depends on shore conditions and can be as high as 16 m (about 52 ft). Also, with a higher capacity of basins, available energy will increase. Furthermore,

[6] Leijon, M., C. Boström, O. Danielsson, S. Gustafsson, K. Haikonen, O. Langhamer, et al. "Wave energy from the North Sea: Experiences from the Lysekil research site." *Surveys in Geophysics* 29(3) (2008): 221–240.

if basins are large enough, the problem of interrupted electricity generation can be partially solved by their proper arrangement. There are other factors to consider when selecting a suitable site, like the distance of the site from the electricity grid, the current application of the basin, and distance from sea waves and storms.

Tidal power has been around since at least the tenth century and was used to move mills in England, France, and several other countries. Although there have been many modern designs and/or experimental pilot projects to harvest tidal energy, only about a dozen or so commercial tidal electricity generation plants have been constructed in the world. The Rance Tidal Power Station in France is the world's first and one of the largest tidal power plants with a capacity of 240 MW and an annual output of 600 GWh and it has been operating since 1966.[2] North America's only tidal power plant, the Annapolis Royal Generating Station with a capacity of 20 MW and an annual output of 30 GWh, was opened in Nova Scotia, Canada in 1984.

The concept of the utilization of tidal potential energy is similar to hydro dams, where electricity is generated by the static pressure of water. However, unlike hydro dams, due to different possibilities for the flow direction in turbines, several different schemes have been employed or proposed for exploiting energy from the tides. These schemes are categorized as follows:

- single-pool ebb system
- single-pool flood system
- single-pool two-way operation system
- two-pool ebb and flood tide system
- two-pool one-way system (high and low pools)

In the first design, the basin is filled during the flood through gates. When the basin is full, the gates are closed until the desired water head is created due to the decline in the sea level. Then, the basin is emptied by the water flowing through turbines, which in turn generates electricity (Figure 14.13).

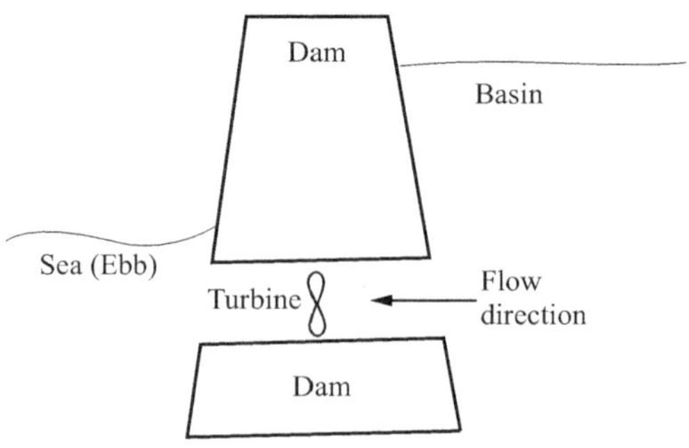

FIGURE 14.13
Schematic of a single-pool ebb system.

This method has been used in tidal mills for a long time and some of them are still operational. The same scheme has been employed in the Annapolis Royal Tidal Power Generating Station. In this plant, the minimum required water head is 1.6 m. When this head is available, ~400 m^3/s of water flow passes through the turbines. The plant generates electricity for 5 hours and then it is off for 7 hours.

The system and mechanism for the second scheme is similar to the first one. The only difference is that the water flows through the turbine when the basin is filling (Figure 14.14). The efficiency of this scheme is much lower than that of the first one.

In the third scheme, by taking advantage of a bi-directional turbine, electricity can be generated both when the water flows into the pool at flood and when it flows out of the pool at ebb. This method has been successfully employed in the Rance Tidal Power Station (Figure 14.15). At this power plant, the amplitude between low and high tides can reach up to 13.5 m. A 750 m structure was constructed to block a 22 km^2 area with the capacity of 180 million m^3 of seawater. Twenty-four specially designed axial flow turbines (also known as bulb sets) were used, each unit with a capacity of 10 MW. Although the capital cost of this plant was high, the overall average cost of electricity has been lower than the Electricite de France's (EDF) average generation costs.[2] One of the features of this power station is that when the high pool is nearly full, additional water can be pumped into it using excess electricity in the grid. While this water is pumped into the basin at a few meters of head, later falls through a much larger head generating energy and returns several times the energy required for pumping. This also increases system flexibility to meet the requirements of the grid. Essentially, in this case, the plant acts as both an energy storage system and a power generation unit.

In the fourth scheme, a combination of the first and second schemes are used, where one of the two pools operates based on the single-pool ebb system and the other one based on the single-pool flood system. The fifth scheme requires two pools, high and low, which are physically adjacent. The

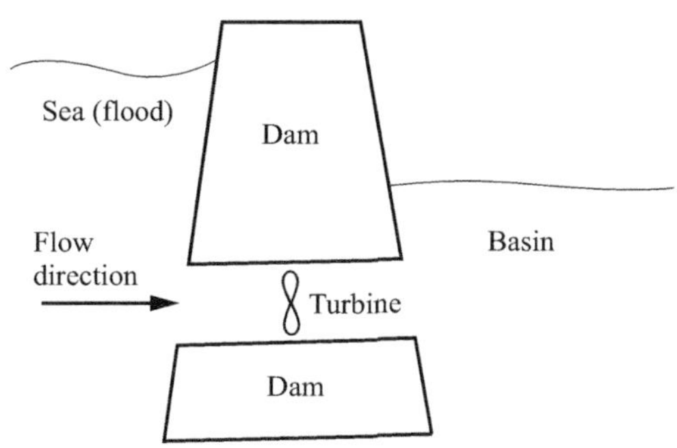

FIGURE 14.14
Schematic of a single-pool flood system.

FIGURE 14.15
The Rance Tidal Power Station in France. (Image by Gwenaële Moignic from Pixabay, https://pixabay.com/photos/rancid-ille-et-vilaine-st-malo-352956/.)

high pool is filled each high tide and the low pool is emptied each low tide. Electricity is generated when water flows from the high pool to the low pool through the turbine (Figure 14.16). This type of tidal power station can generate electricity almost continuously and has high flexibility in the time of electricity generation.

It should be noted that using the aforementioned systems have significant barriers. Firstly, these methods require a river estuary or a bay with an extremely high tidal range. Sites with such a high tidal range are very limited. Also, the ecosystem in river estuaries is very sensitive and constructing a barrage to exploit tidal energy may cause considerable negative impacts on the environment. Another problem with traditional tidal power generation that should be considered is that the neap tidal average range is typically about one-third of the tidal range of average spring tides. At the Rance River tidal power plant, electricity generation during neap tides is only

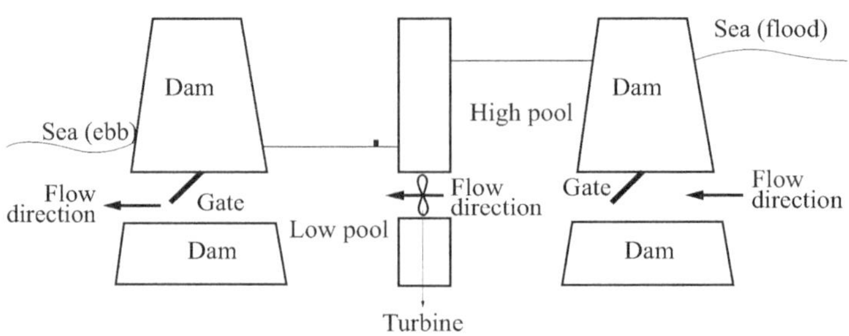

FIGURE 14.16
Schematic of a two-pool one-way system.

80,000 MWh/day, compared to 1,450,000 MWh/day during an equinoctial spring tide, a ratio of 18:1, which causes a disturbance in the grid control.[1]

Tidal power plants can be operated in either grid-tied or off-grid modes. In the former, the selection of the plant configuration and its operational conditions depends on grid requirements, either peak load or baseload. If these power plants are operated as peak-load electricity producers, the best scheme is a two-pool one-way system (Scheme 5) with a flexible power generation time. Also, during off-peak hours water can be pumped to store energy and produce more power later when needed most.

The required conditions for economic power generation from tidal potential energy include a high tidal range and a basin with a narrow enough mouth so that it can be dammed and deep enough so that turbines can be set below the level of the low tide. In order to estimate tidal power generation potentials, many parameters should be considered. However, the potential energy of a tidal energy system in each tidal cycle (one ebb and one flood) can be simplified and estimated by the following equation:

$$E = \eta \rho g h q \tag{14.3}$$

where E is the potential energy of the tidal system in each tidal cycle (J/cycle), η is the efficiency of the system (hydro turbine), ρ is the density of seawater (kg/m^3), g is the acceleration of gravity (m/s^2), h is the mean level difference between water in the basin and sea (m), and q is the volume of seawater that flows through the turbine at each cycle (m^3). It should be noted that Equation 14.3 should be applied each time the basin is filled or emptied through the turbine and power is generated. For instance, in the first scheme, the water goes through the turbine only when the basin is emptied, but in the third scheme power is generated in both directions. The volume of seawater that flows through the turbine at each cycle of ebb and flood in a simple case when the basin area is uniform can be calculated using Equation 14.4.

$$q = AH \tag{14.4}$$

where A is the area of the basin (m^2) and H is the maximum tidal range (m). Finally, the average tidal range is equal to one half of the maximum tidal range $h = \frac{H}{2}$. This is due to the fact that as the flow passes through the turbine, the difference between water in the basin and the sea level reduces. Therefore, there is less hydraulic head available for the turbine.

Example 14.2

In a site with a basin area of 170 km^2 and a tide range of 3.9 m, determine the tidal power generation potential of the site for the first scheme (in MW). Assume the efficiency of the turbine is 100%.

Solution

The basin capacity and the average tidal range can be estimated as follows:

$$q = AH = 170 \times 10^6 \times 3.9 = 6.63 \times 10^8 \text{ m}^3$$

$$h = \frac{H}{2} = \frac{3.9}{2} = 1.95 \text{ m}$$

The tidal power generation potential can be estimated by using Equation 14.3 as follows:

$$E = \eta \rho g h q = 1 \times 1000 \times 9.8 \times 1.95 \times 6.63 \times 10^8$$

$$= 1.27 \times 10^{13} \text{ J/Cycle} = 1.27 \times 10^7 \text{ MJ/Cycle}$$

Since there are two complete cycles of high and low tides every day, the total potential energy per day is equal to 2.54×10^7 MJ/day. Then, the mean tidal power generation potential can be estimated as follows:

$$\text{Power generation potential} = 2.54 \times 10^7 \text{ MJ/day} = \frac{2.54 \times 10^7}{86,400} = 293 \text{ MW}$$

As noted earlier, this is the simplest estimation with several assumptions. This is a good method to obtain some rough estimations of the tidal power generation potential, especially for sites with relatively large basins. But real-world evaluations require more accurate calculations by considering the actual conditions for each specific site. Due to the aforementioned problems, currently the main focus in tidal power generation is on the kinetic energy of tidal currents.

14.3.2 Tidal Kinetic Energy

During floods and ebbs, water flows from oceans to coastal regions and back to oceans, respectively. These currents have substantial kinetic energy. Tidal current generators in principle are similar to wind turbines. But because of the much higher density of water in comparison to air, they can operate with much lower current velocity. In order to use tidal kinetic energy, the current speed of between 2 and 3 m/s is required.[7] A current speed of lower than 2 m/s does not have significant energy and it is not economical to extract this energy. Handling the current speed of higher than 3 m/s, on the other hand, can be problematic for turbines. An energy flux of a tidal current with the speed of 3 m/s is ~14 kW/m^2.[8] Most suitable sites to harvest tidal cur-

[7] Rourke, F.O., F. Boyle, and A. Reynolds. "Tidal energy update 2009." *Applied Energy* 87(2) (2010): 393–409.

[8] Bryden, I.G. and S.J. Couch. "ME1-marine energy extraction: Tidal resource analysis". *Renewable Energy* 31(2) (2006): 133–139.

rent power are typically located in water depths of 20–30 m, which are about 1 km from shores.[9] In these sites, the tidal current speed is magnified by topographical features, such as the shape of seabed and a narrow channel.

Since the kinetic energy of tidal currents is directly proportional to the square of the current speed, it is essential to measure and analyze this speed to evaluate the suitability of candidate sites. In the tidal current energy evaluation, the variation of current velocities during a tidal cycle, which is strongest in the middle of the cycle, should be considered. In addition, the neap and spring tidal current velocities can vary considerably.

Tidal current energy converters can be divided into three main categories: **horizontal axis turbines**, **vertical axis turbines**, and **oscillating hydrofoils**. These systems can be either fixed to a seabed-mounted structure or floating. In the former systems, the turbine can be supported by a gravity structure or can be mounted on a pile penetrating into the seabed (similar to a off-shore wind turbine). Floating systems can be either fixed or flexible. Flexible mechanisms enjoy an advantage of being self-aligned to the tidal current direction. Almost all technologies used for tidal current kinetic energy extraction are similarly used for run-of-river hydroelectricity (ROR) units where the flow of rivers is used to produce power.

14.3.2.1 Horizontal Axis Turbines

In these devices, the rotational axis of the turbine is parallel to the current stream (axial flow turbine) and it consists of several blades mounted on a rotor. The rotational power extracted from tidal current is transferred to a generator via a gearbox to generate electricity. This type of tidal energy converters can be housed in a duct to concentrate the stream and increase the current speed. Some designs of these systems include several features, such as a large open-center turbine designed to provide a safe passage for marine life, lubricant-free (oil, grease, and other lubricating fluid) designs, and very low levels of mechanical noise to ensure minimal impacts on marine life. Many designs typically have three blades (very similar to HAWTs in Chapter 13). While some systems are fixed to the seabed, others are floating devices. The mooring system of the latter allows an optimal orientation of the system relative to the tidal current direction. Some designs use bi-directional turbines. Horizontal axis turbines are the most popular approach to harness tidal current energy.

14.3.2.2 Vertical Axis Turbines

The rotational axis of these devices is perpendicular to the tidal current stream. These systems can be similar to VAWTs (Chapter 13). In some designs, the oscillating motion of a hydrofoil connected to an arm is converted to hydraulic pressure, which in turn can generate electricity.

---

[9] Westwood, A. "Ocean power: Wave and tidal energy review". *Refocus* 5(5) (2004): 50–55.

14.4 Ocean Thermal Energy

About 70% of solar energy reaching Earth is absorbed by oceans. This causes the ocean surface water to be warmer than that at the bottom. This means the process of collecting and storing solar energy is free. On the other hand, because ocean water temperature at night, day, and in different seasons only slightly varies, this energy is more stable than other forms of solar energy and can be used for baseload power generation systems. The most important parameter in the application of this energy is the temperature difference between surface water and deep water. At the depth of about 1000 m, the water temperature can reduce by as much as 20°C. The essence of using this energy is to run a thermal engine between this heat sink and source to generate power. Due to the small temperature difference between warm and cold water, the efficiency of such an engine is very low. For instance, if the temperature difference is considered to be 20°C, the ideal cycle (Carnot cycle) efficiency can vary from about 6% to 7%. The actual efficiency of a real system could be around 3%. Thus, a high rate of flow is required to produce a reasonable amount of energy. Therefore, for a temperature difference of <20°C, this energy cannot be considered, since the required energy for pumping the water is more than the produced electrical energy. In addition, the depth at which cold water is available should be considered, because this water must be pumped to the surface to be used. Usually <1000 m is a suitable depth.[10] There are two methods to bring cold deep seawater to the surface. Deep water can be pumped to the surface or it can be desalinated at the sea bottom. Since the density of desalinated water is lower, it will float up to the surface automatically. The other important factor is the distance of this deep water to the coast. Since produced electricity should be transferred to the coast, this distance should be as short as possible. Alternatively, the plant can be located on the coast. In this case, cold and warm water should be transferred to the plant. In another design, the plant's output can be used to produce hydrogen or other fuels, which can be transferred to the shore. In all these cases, the distance of deep water from the shore is an important parameter. Furthermore, there are other factors like tide and storm impacts as well as ship movement patterns in the area that should be considered.

Ocean thermal energy conversion (**OTEC**) systems have some other positive attributes, like potable water production, utilization of output cold water for air conditioning systems, utilization of nutrient-rich deep water in aquaculture to raise seafood, and mineral extraction. Furthermore, commercial fishing is expected to benefit from these systems. Cold water from the depths of oceans is rich in nutrients and living organisms (planktons) and

[10] Pelc, R. and R.M. Fujita. "Renewable energy from the ocean". *Marine Policy* 26(6) (2002): 471–479.

when brought to the surface, would attract a wide range of sea life. However, there are some concerns about the ecological effects of OTECs.

First attempts to exploit ocean thermal energy were started in the 1800s. The first plant was proposed in 1881 and a 22 kW plant was constructed in 1930. Since then, many plants have been proposed and some of them have been built as pilot plants. Currently, Japan, the U.S., India, Cuba, and Indonesia are actively involved in the development of this technology.

OTEC plants based on their locations are categorized as floating, land-based, or shelf-based plants. They can be also categorized based on their operating system, closed or open cycles. The main difference between an open cycle and a closed cycle is their operating fluid. In a closed cycle, a low-boiling temperature fluid, like ammonia, is used in a Rankine cycle. On the other hand, in an open cycle, ocean warm surface water is evaporated in a low-pressure vessel (vacuum chamber) and used as working fluid. The main advantage of the latter system is the production of drinking water at the end of the cycle. In addition, closed and open cycles can be integrated to form hybrid cycles. These open, closed, and hybrid cycles that are used in OTEC systems are very similar to those used in geothermal plants and will be discussed in more details in Chapter 15.

14.5 Ocean Current Energy

Ocean currents are considered to be a form of indirect solar energy because they are generated by wind and temperature differences in oceans due to solar heating. The Gulf Stream and Florida Straits Current are a few examples of ocean currents. In this approach, ocean currents are used to run turbines. These turbines are usually installed in the flow path, so they do not need any dam. Similar to tidal stream generators, these systems, in principle, are similar to wind turbines. However, ocean current power has two major advantages over tidal current power, namely relatively constant and one directional flow. Laboratory tests have proven that this technology is feasible.

14.6 Salinity Gradient Energy

In this scheme, the osmotic pressure difference between fresh and salty water, equivalent to up to 240 m of hydraulic head, is used to produce energy. This concept has the highest energy density among ocean-based power resources. Also in another approach, the different rate of evaporation of fresh and salty water can be used to move a low-pressure turbine. Although the high cost is an obstacle, the feasibility of these systems has been confirmed in laboratory conditions. In addition, there have been considerable developments in their commercial application due to the introduction of new less expensive and cost-effective membranes to the market (for RO water treatment systems, Chapter 7). In order to employ salinity gradient energy, water with very high salinity density (up to 280 g/L) as well as freshwater is required.

14.7 Questions and Problems

14.7.1 Introduction

14.1. Why are the energy potentials of seas and oceans special among various forms of renewable energy sources?

14.2. Name five forms of the energy potentials of seas and oceans.

14.7.2 Wave Energy Sources

14.3. List advantages of ocean wave energy sources.

14.4. Why is wave energy considered an indirect form of solar energy?

14.5. Why can wave energy be considered as a concentrated form of solar and wind energy?

14.6. Rank the flowing energy sources according to their average density (descending order).

a. Solar energy
b. Wind energy
c. Wave energy

14.7. What are wave properties that impact its energy? What are two forms of wave energy? Which one can be exploited by wave energy converters?

14.8. The average energy of a wave per unit of area in deep water depends on ……...

a. wave speed
b. wave height
c. wave period
d. fetch
e. seabed status
f. all of the above

14.9. What is the difference between wave height and significant wave height? Which one is used to determine wave energy?

14.10. Determine the significant wave height for a wave with the (trough-to-crest) height of 3.5 m.

14.11. For a sinusoidal wave with an amplitude of 1.5 m and a period of 6 s, determine wave energy and power. Assume the density of water is 1000 kg/m^3.

14.12. Explain why the interest in wave energy in the 1970s did not result in any system with acceptable technical and economic characteristics.

14.13. What are three types of WECs in terms of their distance from the shoreline? What are the pros and cons of each type?

14.14. How can the extracted energy from waves be converted to useful forms of energy?

14.15. What are different forms of WECs according to how wave energy is extracted?

14.16. Explain how point absorber WECs work. How can the extracted wave energy be used?

14.17. How does an oscillating water column (OWC) wave energy converter work?

14.18. Explain how attenuator WECs work.

14.19. Using the following schematics explain how these attenuator wave energy converters work.

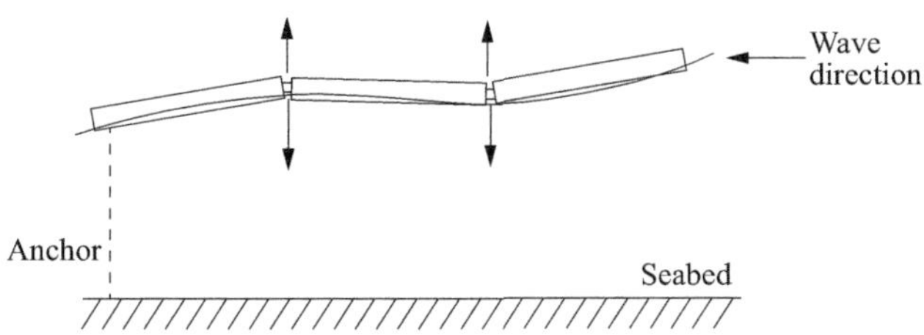

14.20. What is the unique characteristic of overtopping devices? Explain how these systems work.

14.21. Explain how oscillating wave surge converters operate with two examples.

14.22. Describe the concept behind submerged pressure differential converters.

14.7.3 Tidal Energy

14.23. Explain the origin of tidal energy.

14.24. What are the advantages of tidal energy over other sources of renewable energy?

14.25. Explain why unlike most other renewable energy sources, tidal energy is not directly or indirectly derived from solar energy.

14.26. What are two forms of tidal energy?

14.27. Explain how tidal potential energy can be traditionally extracted.

14.28. What are the most important factors in determining a suitable site for tidal energy?

14.29. How do the size of the basin and tidal range impact the amount of available tidal energy?

14.30. How are hydro dams and traditional tidal power plants similar and different?

14.31. What are five types of tidal power plants according to the number of pools and their arrangements? Explain the operational principle, advantages, and disadvantages of each type.

14.32. Explain how a tidal power plant can also function as an energy storage system.

14.33. What are the barriers for traditional tidal power plants?

14.34. How does the seasonal variation of tide impact traditional tidal power plants?

14.35. List different types of tidal current energy systems.

14.36. In a site with a basin area of 50 km^2 and a tide range of 7 m, determine the tidal power generation potential of the site. Assume the efficiency of the turbine is 95%.

14.7.4 Ocean Thermal, current, and salinity gradient Energy

14.37. Explain the essence of ocean thermal energy systems.

14.38. Why can ocean thermal energy systems be used for baseload power generation?

14.39. Why is the efficiency of ocean thermal energy systems low?

14.40. List important parameters for the suitability of a site for ocean thermal energy.

14.41. What are two methods to bring cold water from deep water to the surface?

14.42. Which water is provided from deep water for an OTEC system?

a. Cold water

b. Hot water

14.43. How can OTEC systems benefit ecosystems?

14.44. How can OTEC systems be categorized based on their operating system? Explain each system.

14.45. Explain how ocean currents are created. How can their energy be used?

14.46. What are the advantages of ocean currents over tidal currents?

14.47. Explain how salinity gradient energy systems work.

14.7.5 General

14.48. The amount of wave energy depends on the wavelength and its height.

a. True

b. False

14.49. For a sinusoidal wave with an amplitude of 7 m and a period of 10 s, determine wave energy and power. Assume the density of water is 1000 kg/m^3.

14.50. Tidal energy is a solar energy-base source of energy.

a. True

b. False

14.51. How are tidal current energy systems similar to wind turbines?

14.52. In a site with a basin area of 350 km^2 and a tide range of 5 m, determine the tidal power generation potential of the site. Assume the efficiency of the turbine is 94%.

14.53. Wave energy is a combination of potential energy due to its height and kinetic energy arising from the water particle's movements.

a. True

b. False

14.54. Determine the significant wave height for a wave with (trough-to-crest) height of 175 cm.

14.55. Unlike most of the other renewable energy sources, which directly or indirectly are derived from solar energy, tidal energy is created by the gravitational forces of the Moon and the Sun on the waters of Earth and the rotation of Earth.

a. True

b. False

15

Geothermal Energy

Abstract

We will first learn about the origin of Earth, its different inners layers and their characteristics, and the sources of its energy. We will then discuss how geothermal energy can be assessed. We will take a look at geothermal energy applications, particularly for heating purposes either directly or via heat pumps. Next, we will briefly review the history of geothermal energy utilizations and their current status. We will discuss different types of geothermal resources and their characteristics. Then, we will learn about various methods employed to find geothermal resources followed by a discussion on the drilling process to access geothermal energy resources and equipment used for this purpose. Next, we will study how geothermal energy is converted to electricity in power generation units. We will discuss about vapor-dominated, liquid-dominated, and binary cycles. Finally, we will learn about enhanced or engineered geothermal systems (EGS). We will conclude the chapter with a brief review of the environmental impacts of geothermal energy systems.

LEARNING OUTCOMES

After successful completion of this chapter, you will be able to

- explain the origin of Earth, its inner structure, and its energy
- describe the application of geothermal energy for heating purposes either directly or via heat pumps
- characterize geothermal resources
- identify various methods used in the exploration of geothermal resources
- describe drilling methods and equipment to access geothermal resources
- explain different types of power generation units for the conversion of geothermal energy to electricity (vapor-dominated, liquid-dominated, and binary cycles)

- explain enhanced or engineered geothermal systems (EGSs)
- identify the environmental effects of geothermal power generation systems

15.1 Introduction

Geothermal energy is a source of thermal energy provided by the inner core of Earth. Unlike most other energy sources, particularly renewable energy sources, this energy is not originated from the Sun. This characteristic of geothermal energy provides a unique advantage for geothermal-based energy systems that other renewable energy systems may find hard to compete. Geothermal energy sources are not intermittent, and unlike many other renewable energy sources, they do not suffer from daily, seasonal, and unpredictable variations. They are available around the clock all year. As a result, they can be used as baseload power generation units with a capacity factor of typically between 70% and 90%.

While the thermal energy flow from the Earth's interior to the surface, at about 4.5×10^{13} W (or about 1.4×10^{21} J/year), is huge,[1] the energy density on the Earth's surface is not significant. On average, the geothermal energy density on the Earth's surface is about 0.09 W/m^2, which is very small compared to the average solar energy density on the Earth's surface at about 1,000 W/m^2 (Chapter 11). However, in many locations, this energy is concentrated to the level that can provide a usable energy source. Overall, considering the magnitude and distribution of geothermal energy, it is not capable of single handedly addressing the world energy problem, but it can make a significant contribution in many locales. While, broadly speaking, geothermal energy is inexhaustible in the human scale, in a particular site, this may not be the case. The rate of thermal energy extraction from a geothermal energy reservoir is typically higher than the rate of energy replenishment from the Earth's interior. The experience has shown that after a few decades of operation, a geothermal field can be exhausted although the energy can be resorted if the field is left unused for some time after the exhaustion or used intermittently. A major advantage of this energy source is that it does not need any substantial new technology development. Geothermal power plants share most components with conventional steam power plants. While geothermal power plants have some negative environmental impacts, compared to fossil fuel power stations, these impacts, particularly CO_2 emission, are very small even if their construction is taken into account. Geothermal energy can be used not only to generate electricity but also can provide heating for residential, commercial, and industrial applications. In the past, this energy

[1] This is about twice the total annual global energy consumption at about 6×10^{20} J.

was only economically available in few sites, but with development of new technologies, currently and certainly in the near future, it will be available more widely (more on this later).

Example 15.1

The thermal energy flow from the Earth's interior to its surface is about 4.5×10^{13} W (or about 1.4×10^{21} J/year). Determine the density of thermal energy transfer to the surface of Earth. Earth's diameter is about 12,800 km.

Solution

$$\text{Earth's surface area} = 4\pi r^2 = 4\times3.14\left(12,800\times10^3/2\right)^2 = 5.1\times10^{14}\ \text{m}^2$$

$$\text{Density of thermal energy on the surface of Earth} = \frac{4.5\times10^{13}\ \text{W}}{5.1\times10^{14}\ \text{m}^2}$$

$$= 8.8\times10^{-2}\ \text{W/m}^2$$

15.2 Origin of Thermal Energy of Earth and Its Potentials

Human beings have always fantasized about what is underneath us almost as much as what is above us.[2] However, almost everything that we know about the Earth's interior comes from indirect measurements. The current deepest artificial hole in the world is the 12,262 m deep Kola Superdeep Borehole in Russia drilled in 1989 with the diameter of 23 cm. This is <0.2% of the Earth's radius and about one-third of the Earth's outer crust.

The current scientific consensus is that Earth was formed about 4.5 billion years ago (the age of the universe is estimated to be about 13.8 billion years). Like most stars, planets, or even galaxies, Earth was formed by clouds of dust and gas (called nebula) that originated from the Sun. These clouds gradually attracted more matters (particles) by their gravity. This increased their mass and gravity and the process continued. The gravitational energy of these particles was converted to heat and was stored in the molten body of matter. The originally molten surface of Earth was gradually cooled by thermal energy radiation to space and eventually formed the hard solid outer crust of Earth. Most of this energy from the early days of Earth was stored in the Earth's interior, which is still being dissipated to the surface. However, this energy source (and several others) has a relatively small contribution to the overall geothermal energy. The greatest source of Earth's thermal energy is the nuclear reactions of radioactive components of Earth, such as uranium,

[2] Think of the fact that Jules Verne (1828–1905, French novelist) wrote both "Journey to the Center of the Earth" (1864) and "From the Earth to the Moon" (1865).

potassium, and thorium. There is significant uncertainty on the total energy of Earth and the contribution of each of these mechanisms. But two things are certain: the temperature of the Earth's core is very high and there are sufficient radioactive materials on the Earth's interior that geothermal energy will be available for a very long time, probably for billions of years.

Earth has three main layers that each can be divided into sublayers (Figure 15.1). In the center, there is the **core**, which is divided into two distinct layers, the inner core and the outer core. The **inner core** is the central part of Earth with the radius equal to about 20% of the Earth's radius. The inner core is made of mostly iron and some nickel in a solid form at the temperature of about 5500°C (almost as high as the temperature of the surface of the Sun) and pressure of several million atmospheres. The **outer core** is made of similar materials but in a liquid form at a temperature of about 6000°C. The thickness of this layer is about 35% of the Earth's radius. The movement of molten metal in this layer provides violent convection[3] so effective that there is a small temperature gradient in this layer. The next layer is the **mantle**, which is the thickest layer of Earth with a thickness equal to about 45% of Earth's radius (roughly 85% of the Earth's volume). The current scientific understanding is that the mantle is made of circulating viscous fluid. The temperature and fluid viscosity in the mantle decrease with increase in the radius. Close to the hard Earth's surface, the mantel is in the form of molten rock (lava). Finally, the surface layer of Earth, called the crust, is the thinnest layer of all with the thickness of typically between 5 and 50 km (<1% of the Earth's radius and <1% of the Earth's volume). The crust is solid (but not in a single piece) and made of rocks and sediments. The crust is divided to several tectonic plates (or lithospheric plates). These plates gradually move relative to each other[4]

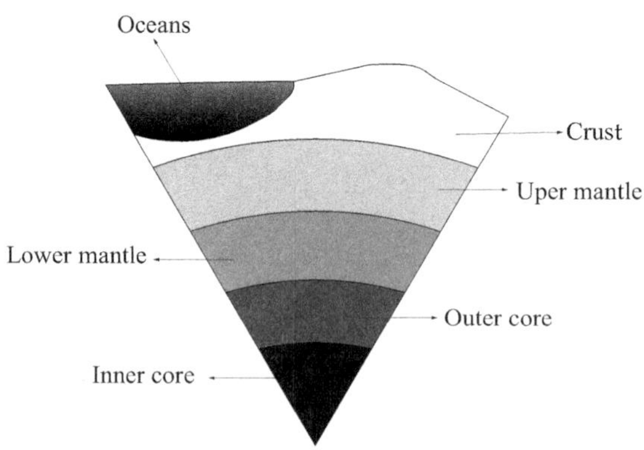

FIGURE 15.1
Schematic of different layers of Earth (not to scale).

[3] The movement of the outer core also causes Earth's magnetic field.
[4] You can visualize the tectonic plates as several icebergs floating on the viscous fluid of the upper mantel layer.

with a velocity of a few centimeters per year. While in the lower layers of the mantel, similar to the outer core, the dominant heat-transfer mechanism is convection, in the upper mantel and in the crust heat is transferred primarily by conduction. It should be emphasized here that since we have never been able to reach to the lower layers of the crust, let alone the mantel or the core, most of what we know about Earth comes from indirect measurements and their interpretations by scientists. Therefore, there is no single universally accepted theory on how the Earth's interior behaves and why.

One may ask that if we cannot directly observe the inner working of Earth, how we know what we know. The answer is the science of seismology, which will be discussed later in this chapter.

Considering the temperature of the Earth's core (about 6000°C) and Earth's radius (6400 km), the numerical average of the temperature gradient in the Earth's interior is about 1°C/km. If this numerical average was also practically true, considering our limited ability to dig into the Earth's crust, no geothermal energy source would be within our reach. However, unlike the lower mantle and upper core layers where the heat transfer is primarily convection, in the upper mantel and crust layers, the heat transfer is dominantly conduction. This means close to the Earth's surface, the temperature gradient is generally much higher than the rest of the Earth's interior. However, even this is not sufficient to bring geothermal resources to a depth available with current drilling technologies. As noted, below the Earth's crust, there are molten rocks (magma) and the thickness of the crust varies. Wherever this magma can get close to the surface, there is a potential for a higher temperature gradient and the possibility of reaching high-temperature geothermal reservoirs close to the Earth's surface. In The Geysers (we will extensively talk about it later in this chapter), the magma is believed to be at least 7 km below the surface.[5] These locations are typically on the boundaries between tectonic plates. These are typically sites where boiling mud pool, hot springs, fumaroles, and/or geysers exist. These boundaries are also the sites of a majority of earthquakes and volcanic activities. It should be noted that there are some local hot spots that do not fit this general rule.

15.3 Geothermal Energy Assessments

Broadly speaking, geothermal resources can be categorized to two major groups, deep and shallow geothermal energy resources. Deep resources are typically at high temperatures and located below 1 km underground, while shallow resources are typically at low temperatures and closer to the surface.

[5] https://geysers.com/The-Geysers/Steam-to-Electricity.

While the temperature of a geothermal reservoir is critically important in evaluating the available energy (considering thermal energy availability according to the second law of thermodynamics), it is not the only factor to be considered. The precise evaluation of available geothermal energy in a specific site depends on several factors, including temperature gradient as a function of the depth, the volume of the reservoir, the characteristic of the reservoir rock, and the depth that can be reached using available drilling technologies considering the temperature increase with the increasing depth. In its simplest form, the equation to determine the available geothermal energy in a specific site can be expressed as follows:

$$Q_{\text{Res}} = mC(T_{\text{Res}} - T_O) = \rho VC(T_{\text{Res}} - T_O) \tag{15.1}$$

where Q_{Res} is the available geothermal energy in a specific reservoir (kJ), m is the mass of the rocks within the reservoir (kg), V is the volume of the rocks from the depth that the minimum usable temperature is reached to the maximum depth that can be reached economically using available drilling technologies (m^3), ρ is the density of the rocks (kg/m^3), C is the specific heat of the rocks (kJ/K-kg), T_{Res} is the temperature of the reservoir (°C or K), and T_O is the temperature specified by the end-user (°C or K). For example, in electricity generation applications, this is the temperature of the condenser, while in space heating applications, this is the target space temperature. As noted, this is a simplified equation that involves several fundamental assumptions including no temperature gradient in the reservoir (the temperature does not vary with depth) and a uniform rock formation in the entire reservoir (so a constant density and specific heat). This equation indicates that the temperature of the reservoir is not the only factor in specifying a suitable geothermal reservoir. Particularly the volume (or more accurately the mass) of the rocks making up the reservoir have a substantial impact on energy availability. It should be noted that estimating geothermal energy involves a significant uncertainty depending on the accuracy of measurements and changes as more advanced technologies become economically available. Also, the entire thermal energy of a reservoir may not be available for extraction. For these reasons, similar to fossil fuel resources discussed in Chapter 4, one may encounter terms such as geothermal energy resources, reserves, etc.

15.4 Applications of Geothermal Energy

Geothermal energy resources are being used for several distinct applications. These applications require different source temperatures, which in turn determines the type (and to a large extent the depth of required resources). Clearly, geothermal energy can be used as the source of thermal energy for electricity-generating plants. This application needs the highest temperature reservoirs among all other applications. Geothermal energy can be also used where thermal energy is needed without conversion to other energy forms. For this purpose, geothermal energy can be directly used, or can be used via a heat pump. Geothermal electricity generation applications will be discussed for the rest of this chapter. While using geothermal energy to provide heat is out of the scope of this text, a brief discussion will be presented here.

While high-temperature heat sources are essential for some applications, most importantly for electricity generation, a significant portion of the energy needs of a society is for low-temperature thermal energy. Actually in some countries, most energy consumption is in this sector. In the U.S., about 45% of the residential energy consumption is being used for space heating while the share of water heating is about 20%.[6] Geothermal energy can play a significant role in providing this low-grade thermal energy. Geothermal energy stored in relatively shallow underground water sources (typically up to 100 m below surface) can be hot enough to be used for direct space heating. This water can be directly pumped to a heat exchanger in the target space. Alternatively, if the circulation of underground water in the living space is not desirable, for example, due a high level of contaminations in water, an intermediate fluid can be used. This fluid receives thermal energy from underground water in a heat exchanger and then transfers that heat to the space. In both cases, using a district heating approach is common. Similarly, geothermal energy can be used to produce hot water for residential and commercial applications. A similar approach can be used to provide hot water for industrial processes, heating large greenhouses, or feeding geothermal baths (as tourist attractions).

The direct use of geothermal energy is not available in many locations. Even if it is available, it may need a substantial initial investment both to get access to the reservoir and to pump underground water to surface applications. An alternative approach is to use **heat pumps**. The operation and components of heat pumps are very similar to air conditioning systems. Similar to air conditioners, they have four main components: compressor, condenser, expansion valve, and evaporator.

In a refrigerator, the vapor refrigerant is pressurized in the compressor, and its temperature increases. Then, the vapor is cooled and condensed in the condenser located in an open area out of the space needed for cooling by transferring heat to the atmosphere. The pressure of the condensed refrigerant is then reduced in the expansion valve through the throttling process. Finally, the low-pressure liquid refrigerant is evaporated in the evaporator located inside the space and takes the latent heat of evaporation from the space and provides the desired cooling. Heat pumps have the same components and operate exactly in the same fashion. The only difference is that

[6] https://www.eia.gov/energyexplained/use-of-energy/homes.php.

in heat pumps the evaporator is located outside and the condenser is inside the space. In winters, heat pumps remove heat from the cold outside air and transfer the heat to the air inside the space. In majority of cases, a heat pump and an air conditioner are integrated into a single unit where the function of indoor and outdoor heat exchangers as the condenser and evaporator can be switched based on the season. These conventional heat pumps are also referred to as **air source heat pumps**. The performance of heat pumps is expressed by the coefficient of performance (COP).

$$\text{COP} = \frac{Q_H}{W} = \frac{\dot{Q}_H}{\dot{W}} \tag{15.2}$$

where COP is the coefficient of performance, Q_H (or $\dot{Q}_H$) is the heat transferred (or its time rate) to the indoor space (kJ or kW), and W (or $\dot{W}$) is the work (or power) consumed in the compressor (kJ or kW). The COP of a heat pump is a function of the temperature difference between the inside space and the ambient temperature. This means when the ambient temperature drops and more heating is needed, heat pumps have less capability to do so. This is where geothermal heat pumps can help. We know that the temperature of the ground below about 6 m (about 20 ft) is almost constant regardless of the ambient temperature at a given time. The exact value of this temperature varies from location to location and can be between 10°C and 16°C, depending on the average annual ambient temperature of the site. In geothermal heat pumps, also known as **ground-source heat pumps**, the evaporator is in the form of a heat exchanger buried underground where it is protected from ambient temperature variations. The pipes of the heat exchanger can be buried in a shallow horizontal pattern or in a vertical deep borehole. Sometimes underground aquifers may be used as a heat source. An attractive option is using water-flooded abandoned mines as the source of geothermal energy. Note that these systems not only help to provide heating in winter with less input electricity but also can reduce electricity consumption for cooling when the system operates as an air conditioner. Since in winter the system extracts heat from the reservoir and in summer adds heat to the reservoir, the reservoir can sustain its function for a long period.

15.5 Past and Present of Geothermal Energy

15.5.1 History of Geothermal Energy Utilizations

Geothermal energy has been used for thousands of years in the form of thermal baths for recreation or their perceived therapeutically properties. There are evidence that Romans used thermal spas in the city of Hierapolis (now is known as Pamukkale, literally meaning "cotton castle", in Denizli in southwestern Turkey) since the 7th century BC.[7] The existence of hot springs in Larderello (south of Florence) in Tuscany in central Italy were reported by Romans over 2000 years ago. Similarly, the city of Bath, Somerset, England was founded and named by Romans for its thermal baths. Similar sites existed in many other cultures. Probably the early practical applications of geothermal energy were for space heating and cooking over 700 years ago. Much later, hot springs were used to extract chemicals.

The first attempts to produce mechanical work from geothermal energy were in Larderello, Italy, in the late 19th century. The site is extraordinary because the temperature of about 200°C can be reached in just a couple hundred meters underground. In the early 20th century, for the first time, electricity was generated from a geothermal source in the same site.[8] Its capacity was gradually increased to 127 MW by the mid-1940s. Until the late 1950s, the plant remained the only electricity-generating geothermal plant. With the capacity of about 800 MW, it is still one of the largest geothermal power plants in the world and the largest in Europe.

In the U.S., The Geysers geothermal complex, located in northern California (116 km north of San Francisco), houses the largest geothermal power generation facility in the world with a capacity of about 1600 MW. The site was first discovered by modern-day Americans in the mid-19th century[9] and was named The Geysers.[10] For many decades, the site was used as a resort. In the 1920s, some attempts to use the field for electricity generation failed until 1960 when the first geothermal power plant with the capacity of 12 MW was built. By 1988, over 30 power generating units were installed in the field with the total capacity of almost 2000 MW. Since then nine units have been nonoperational (seven units has been dismantled all together). Currently 22 units with the capacity of about 1600 MW are operational.

15.5.2 Current Status of Geothermal Energy

Figure 15.2 shows the global and the U.S. installed geothermal electricity generation capacity (MW) between 2010 and 2019. The global installed

[7] The carbonate mineral formations in the shape of cascading pools created by the flow of mineral-rich water is still a major tourist attraction. Nowadays Turkey's largest geothermal power plant, Kızıldere Geothermal Power Plant with the electricity generation capacity of 95 MW, is located nearby.

[8] The generated electricity was reportedly used to light four or five light bulbs (depending on the source). The pioneering project was led by a local prince named Piero Ginori Conti (1865–1939, an Italian businessman and politician).

[9] Native Americans lived in the region for thousands of years and used the hot water in thermal baths and for cooking before the arrival of European settlers.

[10] The name "The Geysers" might be inappropriate and misleading. Unlike real geysers, such as those in Yellowstone National Park, where the mixture of boiling water and steam is rejected to the air, there are no such eruptions in The Geysers. The more appropriate term might be "The Fumarole" because in this field steam vents though openings.

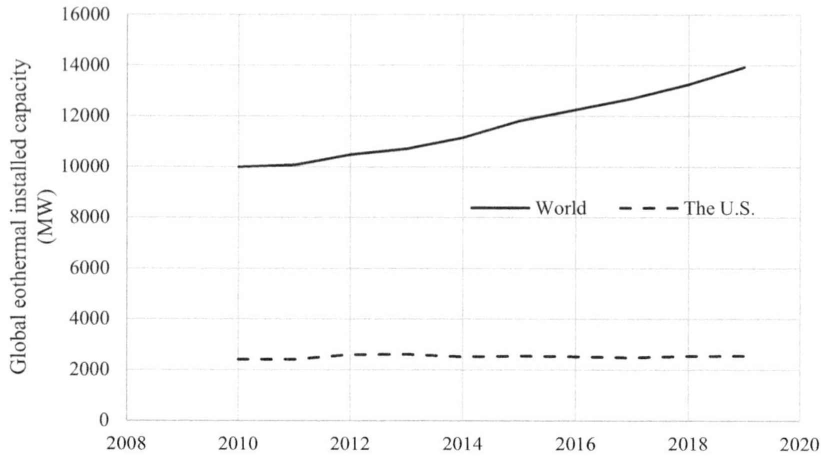

FIGURE 15.2
Global and the U.S. installed geothermal electricity generation capacity (MW) between 2010 and 2019. (Based on the data provided by International Renewable Energy Agency (IRENA) and the U.S. EIA.)

geothermal electricity generation capacity is currently about 14 GW, which is much higher than what it was just a decade ago at about 10 GW (an average 4% annual increase).[11] While this capacity compared to those of the wind and solar electricity production capacity (each at roughly about 600 GW) is small, in certain countries and regions it can make a significant contribution to overall electricity generation. In 2017, globally about 80 billion kWh of electricity was generated by geothermal power plants. Geothermal power plants are distributed among 27 countries in all five continents. The top ten countries with the largest installations are (in 2019), the U.S. (2555 MW), Indonesia (2131 MW), Philippines (1928 MW), Turkey (1515 MW), New Zealand (965 MW), Mexico (936 MW), Kenya (823 MW), Italy (800 MW), Iceland (753 MW), and Japan (525 MW). Among these countries, Kenya and Iceland are very important on how geothermal energy can make a significant contribution to electricity generation in certain favorable circumstances. Kenya is the first and only African country with any meaningful geothermal power production facilities. Geothermal electricity generation makes up over 30% of the country's installed electricity generation capacity and produces about half of the country's electricity. If you refer back to Table 1.1, it is clear that Kenya has one of the lowest per capita electricity consumptions in the world with the annual consumption of 197 kWh/capita. The fact that one half of this electricity comes from geothermal energy indicates how impactful this

[11] Renewable Capacity Statistics 2020. "International Renewable Energy Agency (IRENA)." (2020). https://www.irena.org/-/media/Files/IRENA/Agency/Publication/2020/Mar/IRENA_RE_Capacity_Statistics_2020.pdf.

energy is in Kenya. A similar case, in term of the widespread utilization of geothermal energy, is Iceland where about a quarter of its installed capacity is geothermal power plants. Interestingly, regarding electricity consumption per capita Iceland is at the other end of the spectrum (compared to Kenya). With the annual electricity consumption of 53,301 kWh/capita, Iceland has the highest per capita electricity consumption in the world.

In the U.S., the capacity of geothermal power generation was 2555 MW in 2019, which indicates about 6% increase compared to a decade ago. While the U.S. has the largest installed geothermal power generation capacity in the world, the geothermal electricity generation contribution to the total electricity generation in the country is <0.5%.[12] California and Nevada are responsible for over two-thirds and a quarter of the total geothermal electricity generation in the U.S., respectively. The rest of the country produces about 5% of the electricity generated from geothermal resources.[13]

15.6 Characteristics of Geothermal Resources

We learned that in order to have an exploitable geothermal energy source, magma (melted rocks) should be able to penetrate the Earth's crust to heat rocks close to the Earth's surface. Depending on the presence of water or lack of it, geothermal reservoirs can be divided into two categories: **hydrothermal** and **dry rocks**.

15.6.1 Hydrothermal Geothermal Reservoirs

In these reservoirs, the ground is made of porous rocks, such as sandstone, volcanic ash, and limestone, where water can be stored to form an aquifer (Figure 15.3). The water is heated by deep hot rocks. The hot water should be able to circulate to heat the entire reservoir. For this reason, rock porosities should be interconnected to allow water permeation through the rocks. So in the aquifer, the water is heated indirectly by magma. Then, due to the density difference between hot and cold water, the heated water is driven upward. This process is referred to as the hydrothermal convective circulation.

So to have a hydrothermal geothermal reservoir, we need hot rocks heated by magma and an aquifer with permeable rocks sufficiently close to the Earth's surface that is accessible by drilling. However, these are not enough and one more element is needed. The aquifer should be sandwiched between two layers of impermeable rocks above and below the reservoir so water or steam cannot escape downward or upward (Figure 15.3). The top layer,

[12] https://www.eia.gov/energyexplained/geothermal/use-of-geothermal-energy.php.
[13] https://www.eia.gov/energyexplained/geothermal/use-of-geothermal-energy.php.

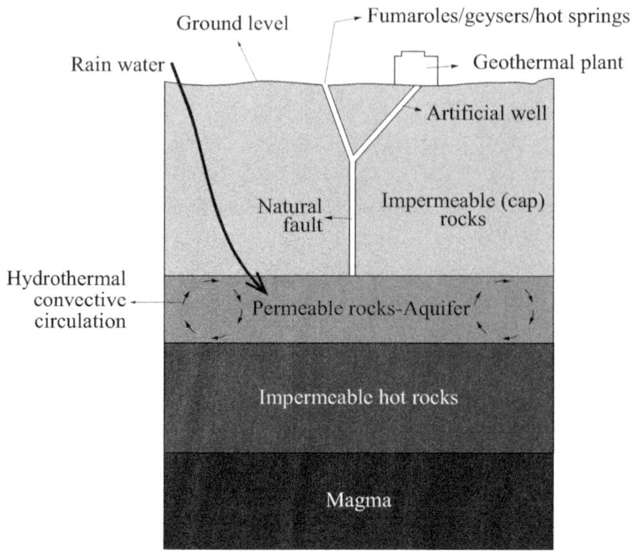

FIGURE 15.3
Schematic of a hydrothermal geothermal reservoir.

known as the **cap rocks** or seal, also acts as an insulating layer, due to the rocks' low conductivity, to keep thermal energy within the reservoir. Natural faults in the cap rocks can release some of the thermal energy to the atmosphere. Depending on the location of the fault, the released energy can be in the form of hot water (in hot springs), steam (in fumaroles), or a high-pressure mixture of vapor and liquid (in geysers).

Hydrothermal geothermal reservoirs can be at the temperature of 100°C400°C accessible within kilometers from the surface. At The Geysers, the deepest well is about 4 km deep, while the average well depth is about 2.6 km.[14,15] Hydrothermal geothermal reservoirs are the only ones that are currently commercially being exploited. While the temperature of a reservoir is critical, the pressure of the water is also very consequential. We know in a column of water, deeper waters are at higher hydrostatic pressures.[16] The same principle applies to hydrothermal reservoirs. The deeper reservoirs have higher pressures. There is a certain depth below which hydrostatic pressure exceeds the saturation pressure of water at the given temperature. For that temperature, if the depth of the reservoir is more than that specific depth, the water will be at the compressed liquid state (rather than vapor phase). Otherwise, for shallower reservoirs, the water will be at a superheated (or saturated) steam state.

[14] https://geysers.com/The-Geysers/Geysers-By-The-Numbers.
[15] Note that in this text when we refer to The Geysers, we actually refer to the portion of the field that is operated by Calpine, the company which operates majority of the plants.
[16] For every 10 m column of water, the hydrostatic pressure increases by about 1 bar or 100 kPa.

Example 15.2

For water at the temperatures of 200°C and 300°C, determine the depth below which the water will be compressed liquid.

Solution

The pressure and specific volume of saturated liquid water at 200°C and 300°C can be found from the water property tables:

$$P_{\text{Sat}@T=200^\circ\text{C}} = 1554.9\ \text{kPa}$$

$$v_{\text{Sat Liq}@T=200^\circ\text{C}} = 0.001157\ \text{m}^3/\text{kg}$$

$$P_{\text{Sat}@T=300^\circ\text{C}} = 8587.9\ \text{kPa}$$

$$v_{\text{Sat}@T=300^\circ\text{C}} = 0.001404\ \text{m}^3/\text{kg}$$

The hydrostatic pressure of a column of water can be found from the following equation:

$$P = \rho g h = \frac{gh}{v} \rightarrow h = \frac{Pv}{g}$$

$$h_{T=200^\circ\text{C}} = \frac{1554.9 \times 10^3\,[\text{Pa}] \times 0.001157 \left[\dfrac{\text{m}^3}{\text{kg}}\right]}{9.8 \left[\dfrac{\text{m}}{\text{s}^2}\right]} = 183.5\ \text{m}$$

$$h_{T=300^\circ\text{C}} = \frac{8587.9 \times 10^3\,[\text{Pa}] \times 0.001404 \left[\dfrac{\text{m}^3}{\text{kg}}\right]}{9.8 \left[\dfrac{\text{m}}{\text{s}^2}\right]} = 1230.3\ \text{m}$$

So for pure water at the temperature of 300°C, if the height of the column of water is higher than about 1.2 km, the water will be at the compressed liquid state. Clearly, this calculation is over simplified. For example, the geothermal water is never pure water. However, it provides a conceptual understanding for the following topic.

Depending on the pressure of a reservoir and the phase of water in it, hydrothermal geothermal fields can be categorized as vapor-dominated or liquid-dominated. Compared to vapor-dominated reservoirs, liquid-dominated reservoirs are generally located deeper underground.

15.6.2 Hot Rock Geothermal Reservoirs

In the previous section we learned that in order to have a hydrothermal geothermal reservoir, the minimum requirements are to have a hot underground formation and an aquifer. These requirements severely limit suitable sites to exploit geothermal energy. In contrast, in most parts of the Earth's surface hot rocks at high temperatures can be found where other requirements for hydrothermal reservoirs are not met, particularly the presence of water and permeable rocks. In early days of the conception of the idea of using these resources they were called **hot dry rocks** (**HDR**) or petrothermal geothermal reservoirs. But later explorations indicated that in a majority of cases they are not really dry but just impermeable. They are currently referred to **enhanced or engineered geothermal systems** (**EGSs**). The methods to use these resources will be covered later in this chapter.

15.7 Exploration for Geothermal Resources

The first and probably the most important step to utilize geothermal resources for electricity generation is to locate a suitable site. This includes determining the magnitude of the thermal energy, ensuring the presence of water in the sufficient quantity, and identifying if the ground has adequate permeability. The ultimate objective is to determine if the potential geothermal source can be economically exploited with reasonable environmental effects (determined by considering all regulations) with an acceptable certainty. The process of identifying geothermal resources is not a straightforward and linear process. Considering potentially unlimited areas to explore and the high cost of a complete exploration process, exploring for geothermal resources is conducted in several steps. In each step, more promising sites are picked for further comprehensive and more expensive exploration techniques. For this purpose, geochemical and geophysical measurement techniques are extensively used in coordination with statistical and numerical tools. The maps of geothermal resource can also provide some insight to potential available resources (Figure 15.4).

The very first step in this process is to identify a potential site for a detailed investigation by field surveys and geochemical methods. This identification can be as simple as looking for typical surface signs of geothermal fields, such as geysers, hot springs, boiling mud pools, fumaroles, etc. While this was how some of the most prominent geothermal sites, like The Geysers and Larderello, were discovered, there are many more sites that do not demonstrate such visible signs. A good place to start is regions with recent or distant past volcanic activities. However, it should be remembered that not all geothermal sites have a history of volcanic activities. The records of past geological mapping and drill records can be also used for this purpose. One can also look for the evidence of the presence of other manifestations of geothermal energy. For example, the composition of local geological formations can

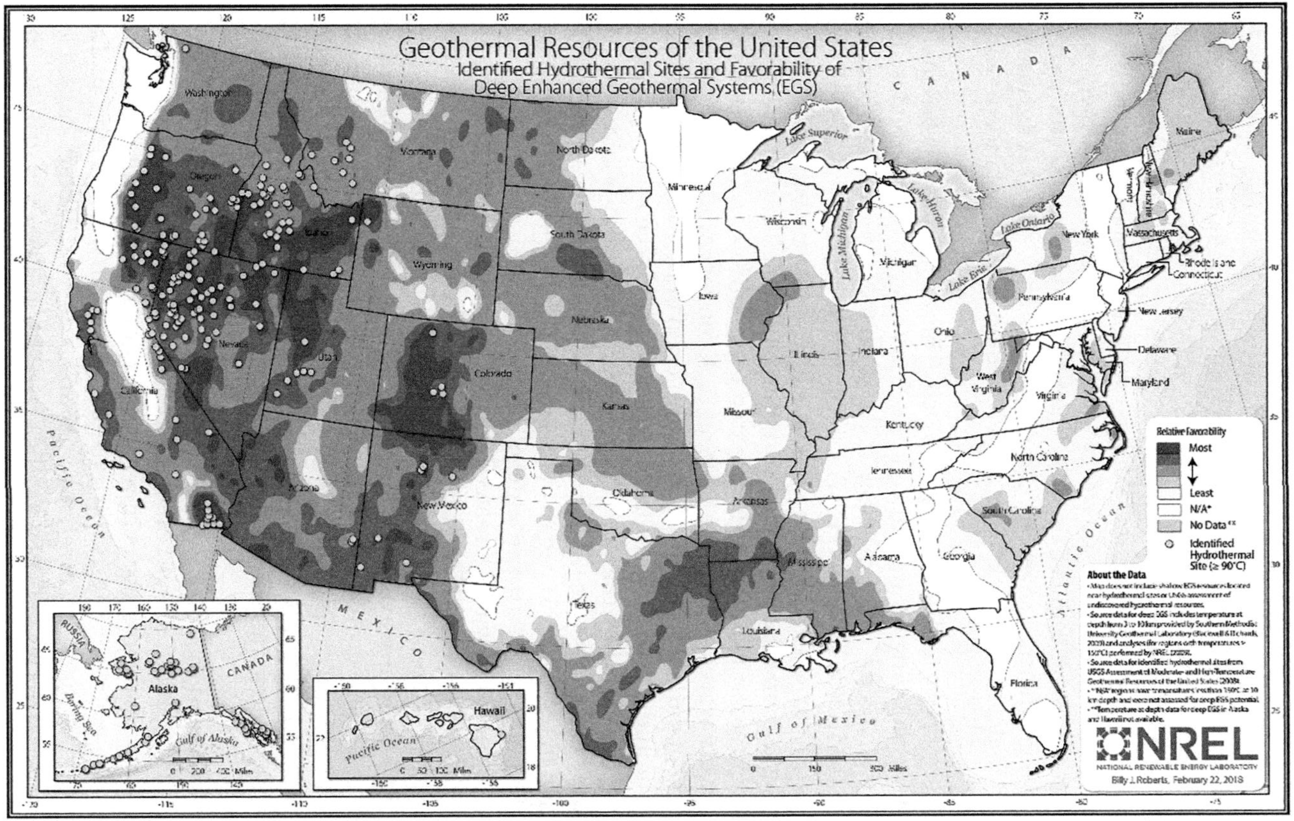

FIGURE 15.4

A geothermal energy map of the U.S. (Source: The National Renewable Energy Laboratory, NREL, https://www.nrel.gov/gis/assets/images/geothermal-identified-hydrothermal-and-egs.jpg.)

indicate the existence of hot springs in the past.[17] The composition of surface water streams and the presence of some substances, such as boron and certain isotopes, can also indicate their interaction with hot rocks. All abovementioned methods are inexpensive tools to create a preliminary picture of underground conditions. If the overall picture is sufficiently promising, it can trigger more extensive and expensive geophysical studies. One of these tools is magnetic surveys of magnetic anomalies of local rocks by magnetometers mounted on airplanes, helicopters, or drones (hence the name aeromagnetic studies). When rocks interact with geothermal water sources, their magnetic activities are reduced, and this characteristic can be an indication of a potential geothermal reservoir. Similarly, resistivity and gravity studies can be used to determine rocks' electrical resistance and gravitational fields, respectively.

Another important tool is a seismic analysis. This technique was originally developed for exploring oil and natural gas sources. Since this method is also used to understand the inner structure of Earth (as the science of seismology), we will briefly discuss how it works. For this purpose seismic waves with low frequencies are created at a certain point by either explosive seismic sources, such as gelatin dynamite explosion, or nonexplosive seismic sources, such as seismic vibrators. They can generate two types of waves, P-waves (pressure or primary waves) and S-waves (shear or secondary waves). In P-waves, the vibration of particles is parallel to the direction of the wave propagation. They are longitudinal waves, similar to sound waves, and move by compression and expansion. They can pass through both solids and liquids. In contrast, S-waves are transverse waves where particles vibrate perpendicular to the direction of the wave propagation (similar to waves created in a string of a guitar). They can only pass through solids.

The low-frequency waves created by the wave generator can easily propagate throughout Earth interior and are reflected and refracted while going through different environments and discontinuities between them. The velocity of these waves depends on the nature of the medium they are traveling in. These waves after passing through the Earth's interior are recorded by some recording devices, such as seismometers and accelerometers in land and hydrophones in water. The timing and characteristics of the recordings help to determine the structure and properties of the Earth's interior.

It should be noted that the results of none of these techniques is conclusive and a holistic approach should be employed for their interpretation. When and if there is convincing evidence with high enough probability that a site has a potential geothermal reservoir, one can proceed to the next step. In this step, many small diameter boreholes (<15 cm or about 0.5 ft) are drilled to the depth of about 2 km. Considering the extremely high cost of drilling, the

[17] The evidence that in the past certain bacteria, which thrive in hot springs, lived in a region can be used for this purpose.

number of holes and their layout should be strategically chosen. The subsurface temperature gradient is measured by lowering a temperature sensor into the hole and the specific heat of rocks is determined in the laboratory using samples taken during drilling. These two parameters help to determine the local heat flow rate.[18] When it is confirmed that a geothermal reservoir meets all the requirements with a very high certainty, the next step is to drill a well to gain access to geothermal energy.

15.8 Drilling for Geothermal Resources

While most power generation components of a geothermal power plant are similar to other steam power generation units, drilling boreholes is the most distinguishing feature of geothermal power units. With typically about one half of the total cost of geothermal power stations, drilling boreholes are the most expensive part of geothermal projects. Drilling is not only needed to extract steam or hot water but is required for reinjection of water to the filed. Furthermore, as noted earlier, drilling test boreholes is an important part of identifying and evaluating potential geothermal sites. To a large extent the technology and equipment used for geothermal systems are similar to those for oil and gas industry, although the environment is more challenging for geothermal well drilling. While this similarity allows a considerable technology sharing between the two industries, it also creates competition between them.

Every conventional drilling system requires several major processes. The heart of any drilling system is the part that interacts with ground and cut into rocks and soil. This part that is always located at the end of the drilling mechanism and is placed at the bottom of the well is called the **drill bit**, **rock bit**, or just **bit**. The breaking and penetrating mechanism can be hammering, jackhammering, or rotary cutting. The latter can use helical screws or rolling cutters (Figures 15.5 and 15.6). In any case the bit should be made of a hardened alloy or other extremely hard materials, such as tungsten carbide or diamond. During the drilling process, a particular attention should be paid to avoiding blow out when a high pressure and temperature pocket of gases is hit and uncontrollably discharged.

The power needed to break rocks should be generated and then is transmitted to the bit by a drill string. The drill string is an interlinked shaft that receives rotational energy on its surface end and delivers the energy to the bit

[18] Please note that the methods presented in this section are by no means complete. For more information on this topic, one may refer to Chapters 6 and 7 of the following book: Glassley, W.E. *Geothermal Energy: Renewable Energy and the Environment*. Boca Raton, FL: CRC Press (2014).

FIGURE 15.5
A drill bit. (By Rama – Own work, CC BY-SA 2.0 fr, https://commons.wikimedia.org/w/index.php?curid=346128.)

FIGURE 15.6
A drill rig with auger for shallow wells. (Image by Hans Braxmeier from Pixabay, https://pixabay.com/photos/drill-construction-equipment-site-57557/.)

so that it can break rocks and penetrate ground. This means the entire shaft, which can be potentially a few kilometer long, should rotate. This requires a huge amount of power, which can be in the order of a few MWs. The magnitude of this power depends on the type of rocks and how deep the well is. It should be noted that it is more difficult and takes more energy to break rocks as the depth of the well increases. This is true even if the type of rocks does not change. This is due to increase in the pressure imposed on rocks (called confining pressure) at deeper parts of the well. The time needed to drill a unit of length also increases the deeper the bit penetrates the ground. It should be also noted that the diameter of the bit and consequently the size of the borehole reduces in several steps the deeper the bit penetrates the ground, possibly from 90 to 25 cm (about 3–0.8 ft). In some drilling rigs, instead of using a drill string, the high-pressure **drilling fluid** (discussed next) is used to run a screw-like mechanism that powers the bit.

The pieces of broken rocks, sand, and dirt then should be transported out of the well. This can be achieved by the rotation of cutting mechanism, high-pressure air, or drilling fluid. The fluid, also known as **drilling mud**, is typically pumped to the bottom of the well within the drill pipe (string) and is returned back along with drill cuttings via the space between the string and well walls. The drilling fluid also functions as the lubricant and coolant for the drill string and the bit. Furthermore, the mud fluid stabilizes the wells by forming a thin protective layer on them. The fluid is typically a mixture of water, clay, and some additives. The returned mud fluid is discharged into a storage pit where after deposition and settling of drill cuttings, it is recycled to the well. A pump is needed to circulate the drilling fluid with a special design to handle the mud fluid. For deep wells where the returned fluid can be at high temperatures, a fluid cooling system may be also needed.

After the well drilling process is completed, pipes or casings (liners), typically made of steel, are placed in the well. Then, the space between them and the walls are filled with concrete or similar materials. This process is known as grouting. The grouting process is to stabilize the walls to ensure the structural integrity of the borehole and to physically and thermally seal the well from it surrounding. This is to prevent the contamination of local aquifers in the case of leakage. It also helps to avoid cross contamination between aquifers at different levels if they are not naturally interconnected, particularly when one is used as a source of drinking water. The diameter of the casing reduces with increase in the well depth, probably from 75 to 20 cm (about 2.5–0.7 ft). For the geothermal water production section of the well, depending on the type of rocks, special provision should be made to allow hot water or steam to enter the casing. If the rocks are sufficiently stable, the section is left open with no linings, otherwise perforated pipes are used to permit the water to enter the pipe while also supporting the walls. On the ground level, the pipe is connected to the wellhead, typically by welding.

Drilling equipment used for wells with the depth of a few 100 m is the same as those used for water well drilling (Figure 15.6). For this purpose, typically augers with a helical screw and a hardened bit mounted on a truck is used. The rotation of the screw moves the removed pieces upward to the surface.

Drilling deeper wells is obviously more challenging and requires more advanced drilling systems. The most visible part of a drilling rig is its tower that is used to lower, withdraw, or add to the drilling string and its pipe (Figure 15.7). Other components of drilling rigs are discussed above. Early geothermal wells were all vertical but with the advancement of the directional drilling technology, wells can be initially vertical and then can be expanded horizontally at desired directions.

Almost every aspect of the drilling process becomes more demanding with increased well depth: it is more difficult to shatter rocks, a longer drill string should be rotated, and the mud fluid needs to be pumped for a longer distance. All these effects as well as temperature increase, not only increases the time and energy needed per unit of length of drilling, but also increases the maintenance and part replacement frequency, which in turn increases the drilling time and cost. All these indicate that **the cost of drilling a unit of length of a well is an exponential function of the well depth**. This dependency severely limits how deep a geothermal well can be for economic exploitation of its energy. There are some ideas and patents that are being investigated to eliminate or reduce this exponential

FIGURE 15.7
A drill rig and its tower. (Image by Bruce Tunget from Pixabay, https://pixabay.com/photos/onshore-drilling-rig-geothermal-1928358/.)

dependency and make it more linear. In one such idea, the bit is replaced by an intense flame jet where thermal stress rather than mechanical stress breaks rocks. Another idea is to rehabilitate abandoned oil and gas wells for geothermal power production.

15.9 Geothermal Power Generation Units

In real world, a single geothermal well cannot produce enough hot water or steam to feed a power plant with more than a few MW capacity. Typically several wells are drilled to form a **geothermal field** to provide sufficient fluid for a sizable power station. Geothermal power plants typically have power generation capacity between 20 and 120 MW. This range of capacity is small compared to fossil fuel or nuclear steam power plants, with typical capacity of many hundreds of MWs. However, building larger units require a large number of wells scattered in a large field. Collecting and transferring hot water or steam from such a field is typically more costly than building several smaller units (rather than one large unit).

In the previous section, we discussed that the drilling process is completed by placing a wellhead over the well on the ground level. This wellhead is connected to an above-ground piping network that collects hot water or steam from several wells and transfers them to the power plant. The flow rate of fluid from each well is controlled by a valve right after the wellhead. Due to the high pressure of water or steam in the reservoir, the flow can exit the well with no pump required. The pressure of the reservoir when the valve after the wellhead is closed is called the shut-off pressure or the closed-in pressure. When the valve is opened steam flows through the piping network. Due to low density and high specific volume of the extracted steam, relatively large diameter pipes are needed to transfer the flow in the network.

The type and configuration of a geothermal power plant depend on the temperature, pressure, and composition of the geothermal water or steam. Generally there are three types of power generation units that are currently being used to generate electricity from geothermal energy: vapor-dominated, liquid-dominated, and binary cycle units (Figure 15.8). In both vapor-dominated and liquid-dominated systems the working fluid of the power cycle (the flow passing through the turbine) is the geothermal steam extracted from the well. Their difference is the state that the fluid is harvested from the well. In vapor-dominated systems, also known as dry steam power plants, the geothermal fluid is in the vapor state. In liquid-dominated or wet steam power plants, the geothermal fluid is in the liquid state. In binary cycles, the power cycle is a closed one where the working fluid receives heat from the hot water flow from the geothermal well.

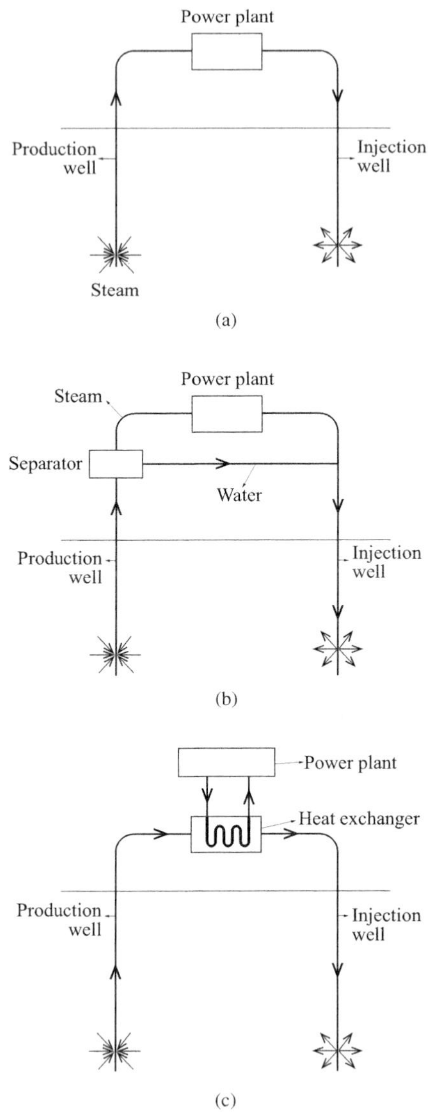

FIGURE 15.8
Schematic of (a) vapor-dominated (b) liquid-dominated (c) binary cycle units.

15.9.1 Vapor-Dominated Geothermal Units

As noted in the previous section, in vapor-dominated or dry geothermal power plants, the geothermal reservoir generates dry steam (with no water droplets in it). Depending on the temperature and pressure of the reservoir, this could be superheated steam or supercritical steam (if the pressure is above about 221 bar). This type of geothermal fluid is the most desirable type both in terms of the processes (and the equipment) needed in the power generation unit and the efficiency of the unit. However, there are very few locations on Earth that have vapor-dominated geothermal reservoirs and most of these sites have already been exploited. Most high-profile geothermal fields are vapor-dominated, such as The Geysers (the U.S.), Larderello (Italy), and Matsukawa (the first geothermal power plant in Japan built in 1966), though there are others in these countries and in Mexico and Indonesia. The steam temperature entering the power generation system can be between 180°C and 250°C and the pressure of typically <8 bar. For example, the average steam temperature and pressure in The Geysers are about 190°C and 6 bar, respectively.[19] In one of the geothermal units at The Geyser, to generate 55.1 MW of electrical power, about 380,000 kg/h of steam is extracted from the wells at the temperature of 170°C and the pressure of 7.8 bar. This is about 105 kg/s of the steam flow or about 6.9 kg of steam for each kWh of generated electricity. The average steam production per well at The Geysers is about 16,300 kg/h,[20] which indicates that on average each well can support about 2.4 MW units. Compared to conventional fossil fuel and nuclear steam power plants, the efficiency of geothermal power plants is relatively low, rarely exceeding 20%, due to low temperatures of the inlet steam to the turbine.

Dry steam geothermal power plants are similar to simple steam power plants operating based on a Rankine cycle except there is neither a steam generator nor a feedwater pump. In the reservoir, steam coexists with water, so the steam leaves the reservoir and enters the pipe within the well as saturated steam, typically at the temperature between 180°C and 250°C and pressure between 35 and 80 bar (flow 1 in Figures 15.9 and 15.10). While steam ascends the well and passes through the valve after the wellhead and the piping network in the field, its pressure drops (a throttling process with a constant enthalpy). At the entrance to the power generation unit (flow 2), the steam pressure reduces to below 8 bar and it is **superheated** steam. If the flow is contaminated with solid particles, a separator, typically a centrifugal one, is needed to separate the particles from the steam. The steam then enters a turbine (flow 3), typically an axial multistage one, where it expands and generates mechanical work, which is converted to electricity in a generator. The saturated mixture of steam and liquid water (flow 4) leaves the low-pressure stage of the turbine and enters the condenser, which is typically located right at the bottom of the turbine. The condenser can be a direct contact type or a surface one (Figures 15.9 and 15.10, respectively).

In a direct contact condenser, the exhausted steam from the turbine (flow 4 in Figure 15.9) is mixed with the cold water from the cooling tower (flow 9) (for more information refer to Chapter 8). The saturated water is collected at the bottom of the condenser in the hotwell (flow 5). The condensate is then pumped to the cooling tower by the condensate pump (flow 6). The cooling

[19] https://geysers.com/The-Geysers/Geysers-By-The-Numbers.
[20] https://geysers.com/The-Geysers/Geysers-By-The-Numbers.

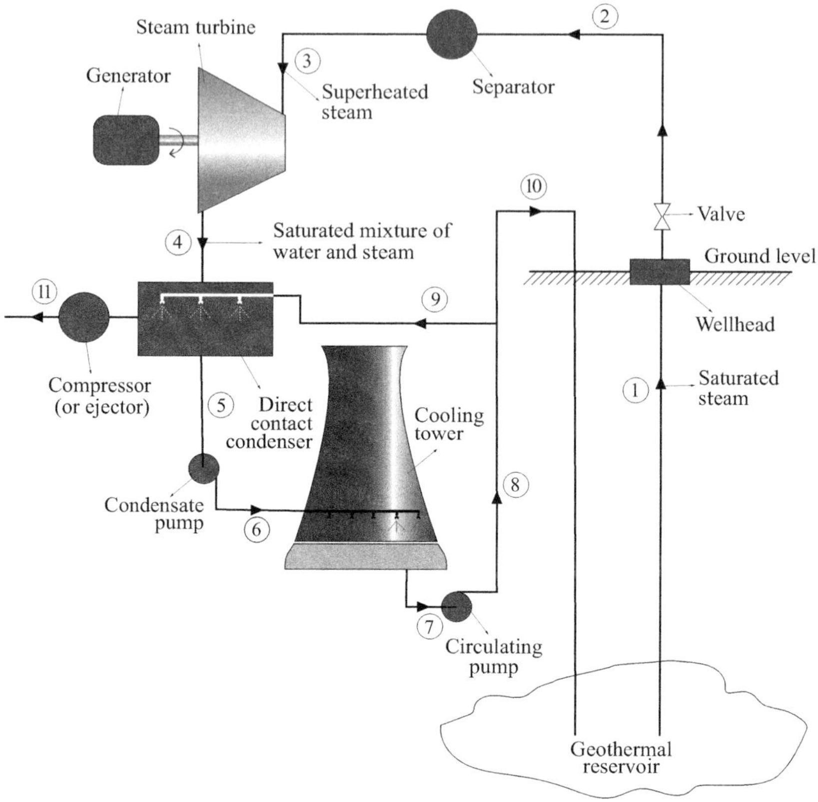

FIGURE 15.9
Schematic of a vapor-dominated geothermal power plant with a direct contact condenser.

tower is typically a wet (evaporative) tower, though for area where makeup water is not readily available, a dry cooling tower can also be used. Both natural draft and mechanical draft cooling towers are widely used in geothermal power plants (refer to Chapter 8). Note that in units with a wet cooling tower, most of the condensate is evaporated in the cooling process to provide the required cooling for the condensation of the turbine exhaust flow. Therefore, most of the condensate (flow 8), as high as 80% of the total flow, is pumped by the circulating pump(s) back to the condenser (flow 9). The leftover of the condensate (flow 10), as low as just 20% of the total condensate (and the extracted steam from the wells), is reinjected to the well after (as in Figure 15.9) or before the cooling tower. While direct condensers are less expensive and more effective in reducing the condenser pressure compared to that for surface condensers, they release all noncondensable gases to the atmosphere (flow 11). These gases include hydrogen sulfide (H_2S), which is a highly toxic and flammable gas, and its release to the atmosphere can harm

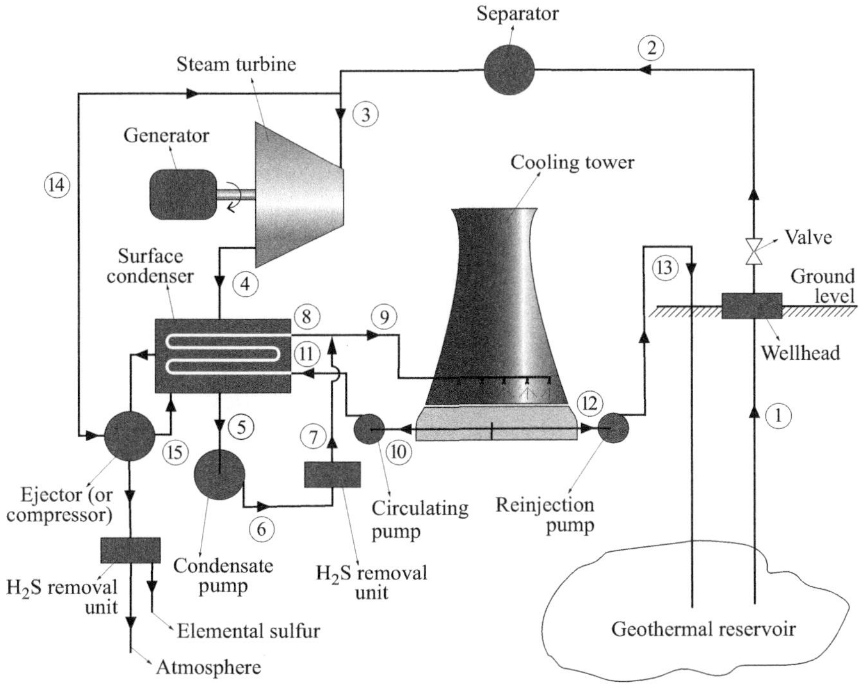

FIGURE 15.10
Schematic of a vapor-dominated geothermal power plant with a surface condenser.

the environment and is prohibited in most regions. So most modern geothermal units are equipped with a surface condenser.

If a surface condenser is used, similar to this type of condensers in fossil fuel steam power plants (refer to Chapter 7), the cooling water flows inside the condenser tubes (flow 11, in Figure 15.10) and the exhausted steam from the turbine (flow 4) flows outside the tubes. The cooling water from the basin of the cooling tower (flow 10) is pressurized in the circulating water pump(s) and is sent to the condenser (flow 11). The outlet flow of the turbine (flow 4) is condensed by losing its latent heat to the cooling water and is collected at the bottom of the condenser in the hotwell (flow 5). This condensate is then pressurized by the condensate pump to the pressure of the cooling water (flow 6) and after removal of H_2S (will be discussed below) (flow 7) is mixed with the cooling water leaving the condenser (flow 8). The mixture of the cooling water and the condensate (flow 9) is then recycled back to the hot water distribution system of the cooling tower. In this case, similar to the cycle with the direct contact condenser, a small portion of the flow is reinjected back to the reservoir (flow 13). For this purpose, the cooling tower basin is typically divided into two sections. The reinjection pumps are installed on one section of the cooling tower basin (flow 12). In Chapter 7, we discussed how the accumulation of noncondensable gases in the condenser can increase the back pressure of the turbine and significantly reduce the cycle efficiency. This problem is much more significant because the geothermal steam from the wells contains a substantial amount of noncondensable gases, perhaps 0.5% of the total mass of the extracted steam. To remove these gases, the condenser is always equipped with either a steam-jet ejector or a compressor. In the case of steam-jet ejectors, often a two-stage ejector is used. The steam needed for the ejectors is taken from the inlet of the turbine (flow 14). To partially condense and recover this steam, an intercondenser and an aftercondenser are typically placed between the two stages of the ejector and after the second ejector, respectively (not shown in Figure 15.10). A small portion of the cooling water is diverted to these condensers and their condensate is returned to the main condenser (flow 15). As noted above, noncondensable gases contain H_2S, which cannot be released to the atmosphere and should be removed. There are several processes for H_2S removal. Commonly, the Stretford process[21] is used for this purpose. In the Stretford process, hydrogen atoms are removed from H_2S by the oxidation process in which elemental sulfur is left behind. The process can remove over 90% of hydrogen sulfide from noncondensable gases and the marketable commercial grade sulfur is produced as a byproduct.

Generally speaking, the design principles and manufacturing methods for components of geothermal power plants are identical or very similar to those of conventional steam power plants. In fact, since the temperature and pressure that the components are exposed to are lower in geothermal power plants, they experience less physical stresses. However, since the concentrations of particulates, dissolved solids, and noncondensable gases are much higher in geothermal fluids than those in feedwater of conventional steam plants, material sections for the construction of components of geothermal systems are more challenging. Stainless steel and aluminum are common construction materials for various components of geothermal plants.

15.9.2 Liquid-Dominated Geothermal Units

As noted in the previous section, dry steam power plants are the simplest and most efficient geothermal units. However, the number of geothermal reservoirs that can directly provide dry steam are very limited throughout the world. Previously we discussed that if a geothermal reservoir is deep enough, it will contain compressed or saturated liquid rather than saturated steam. This type of geothermal reservoirs is much more common compared to dry steam reservoirs. In liquid-dominated geothermal reservoirs, liquid water is provided by a geothermal field typically at the temperature

[21] The process was first discovered in the late 1950s in a laboratory located at Stretford, a region in the Greater Manchester, England, hence the name.

of 150°C–300°C. This water is also referred to as brine due to its high dissolved solids. Geothermal water can flow through the well to the surface either naturally due to the pressure of the reservoir or by pumps. Before the geothermal fluid can be sent to the turbine, it should be converted to steam (at least a portion of it). The conversion takes place via a **flashing process**. The flashing process is technically a simple throttling process where the flow pressure reduces at a constant enthalpy process. This process can take place within the well when the fluid ascends through the well pipe. However, generally this is undesirable because when the fluid is flashed and is partly evaporated, the already high concentration of dissolved solids in the hot water further increases in the remaining liquid. This may result in excessive scaling of the pipe in the well, which can reduce the geothermal flow through the pipe or even plug the pipe altogether. To avoid this problem, usually the well operates under pressure (so no flashing process in the well) and most of the flashing process occurs on the surface inside a specially designed vessel. For this reason, the pressure drop of the flow in the well should be reduced. Since the pressure drop depends on the velocity of the flow in the pipe, reducing the pressure drop requires a lower velocity. However, this means a lower hot water flow rate from the well, which reduces the productivity of each well. Thus, a compromise is needed in the design of the system.

If the hot water is flashed in the well, the mixture of liquid and vapor (wet steam) flow enters the power plant. We know that the presence of water droplets in the turbine is very undesirable and can reduce the efficiency of the turbine and damage its blades. For this reason, the wet steam flow first passes through a separator, such as a cyclone separator. In the separator, the high-density liquid droplets are collected on the walls and accumulated at the bottom of the separator to be eventually reinjected to the well. The low-density dry steam stays in the middle of the separator and leaves through an outlet on the top toward the turbine.

If the flashing is needed to take place outside the well and within the power plant, both the flash process and the separation process are accomplished in a single container called the **flash tank** or separator. The function of a flash tank is similar to a boiler – to evaporate a portion of liquid water to steam. But how they achieve this objective is different between a boiler and a flash tank. In a boiler, the temperature of water is increased by heat addition until it reaches the saturation temperature at the pressure of the boiler and the water starts to boil. In contrary, in a flash tank, the pressure of water is reduced to below the saturation pressure of the water at its temperature, so it starts to boil. Lowering the pressure will result in more steam in the flow but at a lower temperature. The tank also functions as a separator of liquid and vapor (similar to a drum). The dry steam leaves from the top of the tank and is sent to the turbine while the saturated brine is sent to a collection system to be reinjected to the reservoir. The ratio of dry steam to saturated brine is typically low, perhaps 1:4. In some plants, this spent brine is used for low-grade energy uses, such as district heating systems. The rest of the power plant is very similar to dry steam power plants just with a lower temperature and pressure steam inlet to the turbine (commonly 150°C–300°C and 5–8 bar), which in turn means a lower efficiency of the plant. Similar to dry steam plants, these units are equipped with brine treatment facility to remove dissolved solids and solid particulates. Also, to remove entrained water droplets, the steam flow may pass through a steam scrubber.

Example 15.3

Saturated geothermal water (assume the water is pure) at the temperature of 220°C is flashed in a flash tank at the pressure of 7 bar. Determine the steam to liquid water ratio in the tank and the temperature of the tank. Also, determine how much hot water should be extracted to generate 1 kg of steam. Determine the ratio of the enthalpy of steam to the enthalpy of liquid water in the tank. What percentage of the total energy of the hot inlet water is converted to steam energy which is sent to the turbine?

Analysis

For the inlet brine from thermodynamic property tables:

$$T_1 = 220°\text{C}$$

$$h_1 = h_{f@T=220°\text{C}} = 943.6 \text{ kJ/kg}$$

Since the water and steam are at the saturated condition in the tank,

$$T_{\text{Tank}} = T_{\text{Sat}@P=7 \text{ bar}} = 164.9°\text{C}$$

Since the flash process in the tank is thermodynamically a throttling process,

$$h_2 = h_1$$

$$h_{f2@\,P=7 \text{ bar}} = 697.0 \text{ kJ/kg}$$

$$h_{fg2@\,P=7 \text{ bar}} = 2065.8 \text{ kJ/kg}$$

$$h_{g2@\,P=7 \text{ bar}} = 2762.8 \text{ kJ/kg}$$

$$x_2 = \frac{(h_1 - h_{f2})}{h_{fg2}} = \frac{(943.6 - 697.0)}{2065.8} = 0.119$$

As per the definition of the steam quality, for every kg of the inlet flow, 0.119 kg of steam is generated, and the rest is spent brine that needs to be returned to the reservoir. So

$$\text{Steam to liquid water mass ratio} = \frac{0.119}{1-0.119} = 0.136$$

$$\text{Spent brine to steam mass ratio} = \frac{1-0.119}{0.119} = 7.38$$

The amount of hot water that should be extracted to

$$\text{generate 1 kg of steam} = \frac{1}{0.119} = 8.38\ \text{kg}$$

$$\text{Ratio of enthalpy of steam to enthalpy of water} = \frac{0.119 \times 2762.8}{(1-0.119) \times 697.0} = 0.535$$

Ratio of the total enthalpy of steam to the total enthalpy of inlet water

$$= \frac{0.119 \times 2762.8}{0.119 \times 2762.8 + (1-0.119) \times 697.0} = 0.349 = 34.9\%$$

This indicates that about 34.9% of the energy of the hot water extracted from the well is sent to the turbine and the rest is reinjected to the reservoir.

Example 15.3 indicated that for the given conditions to generate 1 kg of steam, 8.38 kg of hot water should be extracted from the reservoir in which 7.38 kg should be reinjected back to the well. Similarly it showed that about 65% of energy that the extracted hot water contains is reinjected to the reservoir. While these values are specific for this case, generally similar values can be observed for most liquid-dominated geothermal power systems where most of extracted water (both in terms of mass and energy) is reinjected.

As noted above, the efficiency of this type of geothermal power plants is lower than that of dry steam power plants. Also, most of the extracted brine from the reservoir must be reinjected to the well as spent brine. Both these factors mean that flash steam geothermal power plants require much larger equipment and piping systems than those for a dry steam power plant, which significantly increase the initial cost of the system. One way to increase the rate of energy that can be extracted from the hot water is to reduce the pressure of the flash process.

Example 15.4

Repeat Example 15.3 for the flash tank pressure of 7, 5, and 3 bar. For each case, determine the temperature of the tank, the quality of steam and water mixture in the tank, the steam to liquid water ratio in the tank, the mass of the extracted hot water to generate 1 kg of steam, the specific enthalpy of the generated steam for 1 kg of the inlet hot water, the percentage of the total energy of the hot inlet water that is converted to steam energy, which is sent to the turbine.

Analysis

TABLE 15.1
Results for Example 15.4

Pressure of Tank	Temperature of Tank (°C)	Quality of Mixture	Steam-Specific Enthalpy (kJ/kg)	Steam to Liquid Ratio	Hot Water Mass to Generate 1 kg Steam (kg)	Enthalpy of Steam (kJ)	Extracted Energy (%)
7	164.9	0.119	2763	0.136	8.38	330	35.0
5	151.8	0.144	2748	0.168	6.95	396	41.9
3	133.5	0.177	2725	0.215	5.66	481	51.0

Example 15.4 shows that reducing the pressure of the flash tank increases the mass of produced steam. However, the temperature of the generated steam is reduced. Overall, this pressure reduction improves the energy of the extracted steam but the reduction in the steam temperature can reduce the positive impact of this increase in the extracted energy.

One way to address this problem is to use double-flash systems. Double-flash liquid-dominated geothermal power stations are equipped with two flash tanks, one high-pressure tank and one low-pressure tank. The geothermal hot water is first flashed in the high-pressure tank (similar to single flash systems). The high-pressure dry steam is sent to the high-pressure turbine. The unflashed water (in liquid phase) then enters the low-pressure flash tank for further flashing. The low-pressure dry steam is sent to the low-pressure turbine and low-pressure unflashed water is reinjected to the reservoir. These cycles are commonly equipped with two turbines designed to operate with high- and low-pressure steam flows. Double-flash units can be typically used for geothermal reservoirs with relatively high-quality water.

15.9.3 Binary Cycles

While globally the number of liquid-dominated geothermal reservoirs is much greater than vapor-dominated reservoirs, they still require relatively high-temperature resources. There are even a larger number of liquid-dominated geothermal reservoirs that have temperature between 80°C and 170°C, which is too low for power plants operating based on the water flashing process. At these low-hot water temperatures, the portion of the extracted hot water that can be flashed into dry steam is so low that producing a meaningful dry steam flow requires a prohibitively large flow of the geothermal fluid.

The challenge of using vast low-grade thermal energy reservoirs of all sorts, such as low-temperature geothermal resources, ocean thermal energy (Chapter 14), solar energy (Chapter 12), or waste thermal energy, can be addressed by closed cycles operating based on a Rankine cycle. While these cycles are known as **binary cycles** in the geothermal industry, they are more generally known as **Organic Rankine Cycles** (**ORC**). Needless to say that these cycles are only used in association with low-temperature liquid-dominated geothermal resources. Their operational principle is exactly similar to that of conventional steam power cycles, but they **do not use water as the working fluid**. The most important characteristic of the working fluid for these cycles is their low boiling temperatures. The low evaporation temperature of the working fluid allows the boiling and superheating processes in the steam generator to be achieved using low-temperature thermal energy sources. As the name implies, the working fluid of ORCs is an organic fluid, i.e., a carbon-based fluid. Various hydrocarbons, such as isopentane (C_5H_{12}), isobutane (C_4H_{10}), and propane (C_3H_8) with the boiling temperatures of about 28°C, –12°C, and –42°C, respectively, have been used. A major problem with these fluids is their inflammability, which can create hazardous operating conditions. Alternatively, hydrofluorocarbon refrigerants, such as R-134a (Freon 134a, CF_3CH_2F) or R245fa ($C_3H_3F_5$), with the boiling temperatures of about –26°C and 15°C, respectively, can be used.[22] A particularly interesting concept is to use a working fluid that is a mixture of two compounds with different boiling temperatures, such as an ammonia and water solution. Since two fluids boil at different temperatures, they can receive heat at much wider range of temperatures. For example, ammonia (NH_3) with a boiling temperature of –33°C can be evaporated at the lower temperature end of the heat exchanger while the water can boil at the high-temperature end of the heat exchanger. This cycle is called the **Kalina Cycle** (named after its inventor Alexander Kalina, a Russian-American scientist and engineer) and has a potential to achieve an efficiency substantially higher than typical ORCs. Several small-scale units based on the concept have been built and operated with varying level of successes. Kalina Cycles are not currently commercial and further developments are needed before their widespread use.

In ORCs, the low-grade thermal energy source, in this case, low-temperature water from a geothermal reservoir, passes through series of heat exchangers, typically shell-and-tube ones, and transfers heat to the working fluid typically flowing inside the tubes. In one particular design, the inlet and outlet temperatures of geothermal water were 135°C and 60°C, respectively. To minimize the possibility of scaling, typically these heat exchangers are designed as single-pass and counterflow heat exchangers with no baffle. Considering the very limited temperature difference between the working

[22] In older experimental units Chlorofluorocarbons (CFCs), such as R-11 (Freon 11, CCl_3F) and R-12 (Freon 12, CCl_2F_2) with the boiling temperatures of 24°C and –30°C, respectively, and hydrochlorofluorocarbons (HCFCs), such as R-21 ($CHCl_2F$) and R-22 ($CHClF_2$) with the boiling temperatures of 9°C and –41°C, respectively, were used in ORCs. However, their use has been banned by the Montreal Protocol and the Copenhagen Protocol, respectively.

fluid and the hot water, the key to the success of these cycles is high-efficiency heat exchangers. It is important that the well and the water surface transportation system are designed and operated to eliminate possibility of water flashing before the heat exchanger. The superheated working fluid is then fed to a multistage axial or radial turbine-generator assembly followed by a surface condenser. The cooling load of the condenser can be provided by any method discussed in Chapter 8. The condensed working fluid is pumped back to the heat exchanger by a feed-fluid pump. The design and operation of all these components are similar to those of conventional steam power plants only at much lower temperatures and pressures. Their construction materials should be compatible with the working fluid.

As noted, these power units operate on a completely closed cycle, which can provide them with several major advantages. Since the hot geothermal water is not the working fluid for the power cycle, it can simply pass through the heat exchanger and then directly reinjected to the reservoir by a pump without coming to contact with the environment or releasing any noncondensable gases. This minimizes any possible negative environmental impact of the water extraction process and eliminates a need for a water treatment facility. This also protects all components of the cycle, except the heat exchanger, from the corrosive nature of the geothermal water. This is why binary cycles are most suitable for geothermal reservoirs with a very high concentration of impurities. The closed cycle operation of the unit also provides an opportunity to pressurize the working fluid, either at subcritical or supercritical pressures, for the optimal thermodynamic operation of the system. Another advantage of this cycle is that units can be built in a modular form. Moreover, there are currently attempts to combine dry steam and flash-type geothermal power generation units with binary cycles to further improve the overall efficiency and power output of the system by using the output of the turbine and the liquid product of the flash tank as the sources of thermal energy for the binary cycle.

15.10 Enhanced Geothermal Systems

We discussed that if we dig deep enough (typically 3–10 km), we can reach a ground temperature of above 130°C in almost every location and above 250°C in many locations. However, even if the technological challenges and cost of drilling to such a depth is addressed, these resources may not be readily available in most sites for two reasons. First, the required water to absorb geothermal energy and transfer it to the surface may not be naturally available. Second, the natural interconnected rock porosity to create needed permeability may not be in place. A way to exploit these resources is to use

enhanced or engineered geothermal systems (EGSs). Among various types of geothermal reservoirs, these resources are largest and most dominant type with availability almost all over the world. The most readily available sources are those that have temperature of 150°C–290°C reachable at moderate depth.

Since the ground in these resources are typically impermeable, first rocks should be hydraulically stimulated to become permeable before water can pass through them. Depending on the conditions of rocks, either the **hydraulic shearing** (hydro-shearing) or **hydraulic fracturing** (hydro-fracturing or simply **fracking**) processes can be used to enhance their porosity and permeability. The hydraulic shearing is used when the rock formation has sufficient existing permeability, and no new fracture is needed. This process only activates preexisting fractures. When the rocks are impermeable, the hydraulic fracturing process is used to induce new fractures. Both methods are based on a similar physical phenomenon to create pathways within rock formations so water can flow and absorb their thermal energy.

The first step in applying these techniques is to drill a well in the site. Then, the fracturing fluid, typically water, is pumped to the rock formation where the potential geothermal reservoir exists. The fluid pressure should be high enough either to create fractures in the rocks (in hydraulic fracturing) or to force the rocks to slip along the existing discontinuities in their structure (in hydraulic shearing). However, one cannot simply inject the high-pressure water flow to the entire wellbore. This not only increases the cost of the process, but also can induce factures on portions of the well where permeability is undesirable. For example, if fractures are formed in the sections of the well where no geothermal energy exists, it can lead to loss of water or dilution of hot water with cold water flows. So, the stimulation process should be targeted on a specific and predetermined region. For this to occur, the targets should be first determined. There are several physical and optical methods for this purpose.[23] When the segment of the borehole is identified, several techniques can be used to target the fracturing process to that specific region. In one technique, some devices, known as packers, are placed right below and above the region. By inflating or pressurizing the packers, the target area can be isolated. Another method is to cover the borehole with a solid casing except the target area that is covered with perforated casing, which only allows that region to be exposed to the high-pressure fluid.

An important consideration during the process is to ensure that the pathways created by the induced fractures stay open after the pressure is removed after the completion of the fracking process. In some cases, loose

[23] These methods are out of the scope of this text. But one can find more information about these methods in Chapter 13 of the following book: Glassley, W.E. *Geothermal Energy: Renewable Energy and the Environment*. Boca Raton, FL: CRC Press (2014).

rock fragments formed during the process keep the new fractures open but commonly sand, ceramic spherules, or other materials, known as a proppant (propping agents), with a specific size (in the order of 1 mm) and hardness are mixed with the fracturing fluid for the purpose.

The required volume of the stimulated region depends on the size of the desired power output and the rock temperature. Considering the typically low conduction coefficient of rocks, a stimulated zone should be large enough, typically in the order of several km^3, to provide a sufficiently large heat-transfer surface area. Also, the volume should be large enough that it can provide sufficient geothermal energy for the intended life of the system, typically 20–30 years. Such a large area cannot be stimulated by a single well and several wells should be drilled in appropriate distances. The layout of the wells should be selected very carefully. The wells should be spaced in a way that their stimulated zones are interconnected while they can cover maximum volume of the geothermal field. When the entire volume is stimulated and the geothermal field is ready, one or more of the wells are used to inject water (or less commonly brine or seawater) into the field (hence the name the **injection wells**). The water flow passes through pathways within the rocks and is heated by heat transfer from them. The hot water is then collected at one or more wells, known as the **production wells**, and pumped to the surface. The power production units can be a flash steam power plant or an ORC exactly similar to those of liquid-dominated geothermal reservoirs. In order to avoid over-exhaustion of the field and to sustain its thermal energy for the entire life of the system, the injected water flow rate should be controlled to minimize the rock temperature reduction while providing enough thermal energy at the highest possible temperature for the electricity generation unit.

The concept of enhanced geothermal systems (then was called hot dry rocks, HDR) was first introduced in the 1970s. While several pioneering EGS projects were attempted, they were not seriously considered until the 1990s when the fracking technologies were perfected for the oil and gas industry. The geothermal industry has been taking advantage of many advancements in the technologies developed for the oil and gas industry. The two industries share most equipment and techniques. However, the conditions for EGSs are much more challenging than those for oil and gas projects. For example, wells are deeper, the rock temperature is higher, and rocks are harder in geothermal projects. Due to their high costs, there are currently just a handful of commercial/pilot EGSs operating worldwide. Much more research and development are needed before their commercial widespread use can become a reality. However, if and when this happens, considering their availability almost everywhere on the planet it can be a game changer. EGSs are very important for the continuation of the growth in the geothermal industry, especially since many of sites with readily available geothermal energy have already been developed or are being developed.

15.11 Other Ideas and Concepts for Geothermal Energy

So far we covered the methods that are either currently being commercially used or seriously being investigated. There have been many other ideas and concepts that have been introduced to use geothermal energy. Some of these ideas have been completely abandoned and some are still being considered at least on the paper. One of the more feasible ideas is to integrate geothermal energy into a conventional fossil fuel steam power plant. Two approaches can be used for this purpose. One approach, referred to as geothermal-preheat hybrid plants, is to use geothermal fluid (water or steam) to heat feedwater in feedwater heaters of a conventional steam power plant to reduce the steam bleeding from the turbine for the preheating of feedwater. Depending on the temperature of the geothermal fluid, it can be used for high-pressure feedwater heaters after the deaerator and/or low-pressure feedwater heaters before the deaerator. Another approach, referred to as geothermal-fossil superheated hybrid plants, is to take steam coming either directly from the reservoir (in vapor-dominated reservoirs) or from the flash tank (in liquid-dominated reservoirs) and superheat it in a fossil fuel-fired superheater before sending it to the turbine.

There are also several proposals to integrate geothermal energy plants with oil and gas production facilities. They include a variety of systems to coproduce oil and gas along with geothermal energy as well as using already existing exhausted oil and gas wells for geothermal applications. The latter could be particularly intriguing considering millions of abandoned wells all over the world including some deep ones. There have been also some suggestions to integrate geothermal units with CO_2 sequestration facilities.

Geopressured (geopressurized) resources are brines with very high dissolved solids trapped very deep underground (3–9 km) below Earth's surface at the temperature of 150°C–200°C. Due to chemical processes during decomposition of organic matters at a confined space, their pressure is extremely high, up to 1000 bar, and they are saturated with methane (the main component of natural gas). There are vast geopressured resources in the Gulf of Mexico and its northwestern coasts. While these resources are not currently commercially being used and are in the research and development stage, they have a great potential because of their temperature, pressure, and stored natural gas.

There are some ideas that are not being pursued now due to their technical and/or environmental challenges. One of these ideas was called total-flow geothermal systems where geothermal hot water was directly fed to a specially designed turbine. Such a turbine that can tolerate extremely corrosive two-phase flow has never been developed and there is serious doubt that it can ever be developed.

In the early 1970s, a bizarre idea was introduced to use a nuclear explosion to create fractures in rock formations. The idea was a part of an overall project called the Project Plowshare to use nuclear explosions for nonmilitary applications. There was even a more strange idea of creating an artificial geothermal reservoir as a thermal energy storage system by an underground nuclear explosion in deep salt formations to create molten salt. Fortunately, none of these ideas was materialized due to environmental concerns, though many nuclear explosions were experimented for other applications such as large excavation projects.

15.12 Environmental Impacts

One of the major concerns both in terms of the geothermal plants' long-term operations and their environmental impacts is the management of reinjected water. It was noted in earlier that over 80% of the extracted dry geothermal steam can be consumed primarily in the cooling tower (due to evaporation and drift) and also in the steam-jet ejector. Only about 20% of the total extracted flow may be reinjected back to the reservoir (a similar pattern can be observed for most other technologies). No reservoir can sustain such a large rate of net steam or hot water extraction in long-term. Eventually the water storage within the reservoir will be depleted and the reservoir's pressure will drop. Note that a useful life of a geothermal reservoir can span for many decades so it may take a few decades to observe this pressure drop. The reservoir pressure reduction can also affect local hot springs and geysers, which can negatively affect the local tourism industry and create public resistance against geothermal installations. Furthermore, the reduced reservoir pressure can cause geothermally induced land surface subsidence, particularly in liquid-dominated reservoirs. The Geyser geothermal complex is a perfect example in this regard. Throughout the first two decades of its operation, The Geyser geothermal field was relentlessly exploited with no regards to the magnitude of the net steam production (the difference between steam extraction and water reinjection) as it was customary in many plants at the time. The situations in The Geysers were particularly dire since several independent companies operated the power plants without much coordination. By the late 1980s, this led to a severe steam pressure drop of the reservoir, which threatened the operation of the complex. To address the problem, through a well-coordinated project known as the Southeast Geysers Effluent Pipeline (SEGEP), treated wastewater from nearby communities were transferred to the field via a 65 km pipeline and was continuously injected to the field. After the implementation of the project between 1995 and 1997, the steam pressure was recovered relatively fast. Nowadays the field has 322 production wells and 54 injection wells[24] where about 35 million

[24] https://geysers.com/The-Geysers/Geysers-By-The-Numbers.

liters of wastewater[25] is annually injected to the reservoir instead of being released to the environment. This solution has another positive impact that will be discussed next. This is an example of how a technical and environmental problem is addressed by a creative solution.

As noted earlier, geothermal water and steam are originated from typically ancient water entrapped underground and interacted with rock formations for a very long time. As a result, geothermal fluids contain extremely high concentrations of noncondensable gases and dissolved solids. A great majority of noncondensable gases is carbon dioxide with lower concentrations of methane (CH_4), hydrogen, nitrogen, ammonia (NH_3), sulfur dioxide (SO_2), radon, and hydrogen sulfide (H_2S). The latter has been specially publicized due to its rotten-egg-like smell and to lesser extent ammonia due to it toxic and noxious effects. While dissolved solids are typically reinjected to the reservoir, noncondensable gases are released in the condenser and must be dealt with. We previously discussed how hydrogen sulfide is removed and a commercial byproduct is produced from it. Other gases are typically released to the environment. While a substantial amount of CO_2 is released to the environment, in most cases it is a fraction of CO_2 emission by conventional fossil fuel power plants. Radon emission is also troublesome since it is known that it causes lung cancer. One of the unintended positive effects of using wastewater to recharge The Geysers field has been substantial improvement in the quality of extracted steam, which has reduced both the environment effects of the operation and equipment maintenance requirements. It has also improved the effectiveness of the condenser (due to lower amounts of noncondensable gases) and has increased the output of the turbine while reducing the steam requirement of the ejector.

While geothermal installations generally have much smaller footprint compared to most conventional steam power plants and especially other renewable resources, the well drilling process can cause significant surface disturbances, including the handling of the drilling fluid and drilling noises. The operation of the system can also sometimes be noisy. Particular attention must be paid to avoid contamination of underground aquifers, especially potable water resources, both during drilling and system operation.

A high-profile geological potential effect of geothermal systems, which has received intense public scrutiny, is geothermal-induced earthquakes. In some sites where geothermal projects were implemented a meaningful and sometimes significant spike in small earthquakes have been observed. At least one project was cancelled in Basel, Switzerland, in the fear of geothermal-induced earthquakes. The pressure reduction in a reservoir can intensify the possibility of geothermal-induced earthquakes. This can be particularly worrisome in EGSs based on the similar experiences in the oil and gas fracking industry. While never a major earthquake has been attributed to geothermal

[25] https://geysers.com/water.

activities, it is essential that seismic activities are monitored in the vicinity of any geothermal installation to observe any unusual increase in the number and/or magnitude of these activities.

In summary, geothermal systems are extremely complex and involve many uncertainties. Therefore, in the feasibility study, design, drilling, construction, and operation of these facilities a holistic approach must be exercised to minimize their potential negative impacts.

15.13 Questions and Problems

15.13.1 Introduction

15.1. What is the unique advantage for geothermal-based energy systems that other renewable energy systems may find hard to compete?

15.2. Geothermal energy in general is practically exhaustible in human time scale.

a. True

b. False

15.3. Why can geothermal power plants be used as baseload power generation units?

15.4. Geothermal power plants can be used as ...... units.

a. baseload power generation

b. peak-load power generation

c. both baseload and peak-load power generation

15.5. The geothermal energy density on the Earth's surface is .... compared to the average solar energy density.

a. small

b. large

c. similar

15.6. Considering the magnitude and distribution of geothermal energy, it is capable of addressing the world energy problem.

a. True

b. False

15.7. Geothermal energy in a particular site is inexhaustible in the human time scale.

a. True

b. False

15.8. The density of thermal energy transfer to the surface of Earth is about 9×10^{-2} W/m^2. Determine the total thermal energy flow from the Earth's interior to its surface. Earth's diameter is about 12,800 km.

15.9. Explain how Earth was formed.

15.10. What are the origins of geothermal energy?

15.11. Name the inner layers of Earth. What are their relative thicknesses? What are the dominant heat-transfer mechanisms in each layer?

15.12. Why does the temperature of the Earth's interior get very high very close to the Earth's surface in some locations?

15.13.2 Geothermal Energy Assessments

15.13. What are two major groups of geothermal sources in terms of their depth? What are their characteristics?

15.14. What are the factors for evaluating the available energy of a geothermal reservoir?

a. The temperature of the reservoir

b. The volume of the rocks

c. The mass of the rocks

d. The properties of rock formations

e. All of the above

15.15. Write the equation to determine the available geothermal energy in a specific site. What are the assumptions for the equation?

15.16. How can the end-user type impact the available geothermal energy in a specific site?

15.13.3 Applications of Geothermal Energy

15.17. What are the applications of geothermal energy?

15.18. How can the end user of geothermal energy impact the suitability reservoir for an application?

15.19. What are low-temperature thermal energy consumptions in residential applications?

15.20. What percent of energy consumption in a typical American house is used for low-temperature applications?

15.21. Explain two approaches for direct geothermal space heating.

15.22. Why are sometimes heat pumps used for geothermal heating applications?

15.23. Explain how heat pumps work. Compare heat pumps and refrigerators.

15.24. What is the definition of the coefficient of performance (COP) in a heat pump? How can it be determined? What factor impacts its value?

15.25. The COP of a heat pump is a function of the temperature difference between the inside space and the ambient temperature.

a. True

b. False

15.26. What is the main advantage of ground-source heat pumps over air source heat pumps?

15.27. How can an underground evaporator benefit a heat pump?

15.28. Ground-source heat pumps can reduce electricity consumption both in winter and in summer.

a. True

b. False

15.13.4 Characteristics of Geothermal Resources

15.29. What are two types of geothermal reservoirs in terms of the pressure of water?

15.30. Explain how hydrothermal geothermal reservoirs are formed.

15.31. What are the requirements for an underground formation to form a hydrothermal geothermal reservoir?

15.32. Describe how hot springs, fumaroles, and geysers are formed.

15.33. Why are cap rocks needed to form a hydrothermal geothermal reservoir?

15.34. Label various parts of a hydrothermal geothermal reservoir in the following schematic.

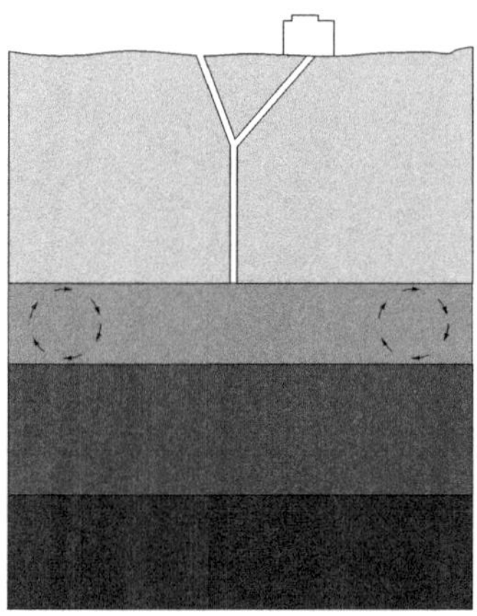

15.35. Explain why some hydrothermal geothermal reservoirs are in liquid state while others are in vapor state.

15.36. What are two types of hydrothermal geothermal reservoirs?

15.37. Explain enhanced or EGSs and how they are different from hydrothermal geothermal reservoirs. Why are they no longer called hot dry rocks (HDR)?

15.13.5 Exploration for Geothermal Resources

15.38. What are the objectives of exploration for geothermal resources?

15.39. The process of identifying geothermal resources is a straightforward and linear process.

a. True

b. False

15.40. What are typical surface signs of geothermal fields?

15.41. Where are good places to start exploration for geothermal resources?

15.42. List some methods for the exploration of geothermal sources.

15.43. What are some physical and geological evidence that may indicate the presence of a potential geothermal reservoir in a region?

15.44. Explain how seismic analyses can be used to understand the inner structure of Earth.

15.45. What are two types of seismic waves? What are their characteristics? How are they used in seismic studies?

15.46. Identify which types of seismic waves can pass through solid and liquid media.

a. S-waves

b. P-wave

15.47. Why does exploration for geothermal resources need a holistic approach?

15.13.6 Drilling for Geothermal Resources

15.48. To a large extent the technology and equipment used for the development of geothermal fields are similar to those used in the oil and gas industry.

a. True

b. False

15.49. Explain each of the following parts of a drilling rig.
- drill bit
- drilling fluid
- drill string
- casings (liners)
- tower

15.50. The power needed to break rocks depends on ....

15.51. What is the confining pressure on a rock formation? How does it change? How does it impact the power needed for drilling geothermal wells?

15.52. What are the functions of drilling mud? What is it made of? How is it recycled?

15.53. Explain why the cost of drilling a unit of length of a well is an exponential function of the well depth.

15.13.7 Geothermal Power Generation Units

15.54. Why are geothermal power plants relatively small in capacity (typically)? What is their capacity range?

15.55. What is shut-off pressure or the closed-in pressure of a geothermal well?

15.56. What are three types of power generation units that are currently being used to generate electricity from geothermal energy?

15.57. What is the working fluid of the power cycle operating based on vapor-dominated and liquid-dominated geothermal systems?

15.58. What is the main difference between vapor-dominated and liquid-dominated systems?

15.59. What is the typical range of efficiency for vapor-dominated geothermal power plants? Why is this efficiency relatively low?

15.60. What are the similarities and differences between dry steam geothermal power plants and simple steam power plants?

15.61. While steam in a geothermal reservoir is saturated, the inlet flow to the turbine is superheated steam. Why?

15.62. Which one is more common?

a. Liquid-dominated geothermal reservoirs

b. Vapor-dominated geothermal reservoirs

15.63. Explain how a vapor-dominated geothermal power plant works.

15.64. Describe the operation of liquid-dominated geothermal power plants.

15.65. In geothermal power plants with a wet cooling tower, explain why only a small fraction of extracted water or steam can be reinjected back to the geothermal reservoir.

15.66. In a liquid-dominated geothermal well, how can liquid water be converted to steam?

15.67. Why is the flashing process within the well undesirable?

15.68. What is the function of a separator before the turbine in a liquid-dominated geothermal system?

15.69. How are a flash tank and a boiler similar and different?

15.70. Hot water from a geothermal reservoir enters a flash tank at the temperature and pressure of 240°C and 35 bar, respectively. The pressure of the tank is 7 bar. Determine the ratio of mass of the spent brine to the mass of steam in the tank. Assume the water is pure.

15.71. In the previous problem, determine the ratio of the enthalpy of steam to the enthalpy of liquid water in the tank and the ratio of the enthalpy of steam to the enthalpy of inlet hot water from the well.

15.72. Hundred kilograms of saturated liquid geothermal water (assume the water is pure) at the temperature of 180°C is flashed in a flash tank at the pressure of 8 bar. Determine the steam to liquid water mass ratio in the tank. Also, determine the ratio of the total enthalpy of steam to the total enthalpy of liquid water in the tank. What percentage of the total energy of the hot inlet water is converted to steam energy, which is sent to the turbine?

15.73. Saturated geothermal liquid water (assume the water is pure) at the temperature of 180°C is flashed in a flash tank at the pressure of 8, 6, and 4 bar. For each case, determine the temperature of the tank, the quality of steam and water mixture in the tank, the steam to liquid water mass ratio in the tank, the mass of the extracted hot water to generate 1 kg of steam, the enthalpy of steam sent to the turbine for 1 kg of the inlet hot water, and the percentage of the total energy of the hot inlet water that is converted to steam energy, which is sent to the turbine.

15.74. How can the steam extraction in a flash tank be increased?

15.75. How can double-flash systems improve the performance of liquid-dominated geothermal power plants?

15.76. Explain why double-flash units can be typically used for geothermal reservoirs with relatively high-quality water.

15.77. Explain how binary cycles or Organic Rankine Cycles work. What are the advantages of these cycles?

15.78. What makes the Kalina Cycle a unique and attractive option for geothermal plants?

15.13.8 Enhanced Geothermal Systems

15.79. Explain why dry rocks cannot be used as a source of geothermal energy in their natural form. What needs to be done to use them for this purpose?

15.80. What is the essence of enhanced or EGSs?

15.81. What are two methods to hydraulically stimulate hot rocks for geothermal applications? Where is each method used?

15.82. Explain how the hydraulic shearing and hydraulic fracturing processes work.

15.83. Why is exposing the entire borehole to the high-pressure water flow during the hydraulic shearing and hydraulic fracturing processes undesirable? How can the high-pressure water flow be targeted only to the region with geothermal potentials? Explain two different methods.

15.84. What is the function of proppant (propping agents)?

15.85. What are the parameters to determine the size of the stimulated area in an enhanced or engineered geothermal field?

15.86. Explain why a single injection well and a production well are not typically sufficient to extract energy from an enhanced or engineered geothermal field.

15.87. What are the functions of injection and production wells in an enhanced or engineered geothermal field?

15.88. Why are EGSs very important for the continuation of the growth in the geothermal industry?

15.13.9 Other Ideas and Concepts for Geothermal Energy

15.89. Explain two approaches to form a hybrid geothermal and fossil fuel steam power plant.

15.90. What are geopressured resources? How are they formed? List their energy forms. What are challenges to exploit them?

15.91. What was the Project Plowshare? What were the geothermal energy related parts of the project? Explain why they were never implemented.

15.92. Explain some of the environmental impacts of geothermal energy resources. How can some of them be addressed?

15.93. How was the problem of steam pressure drop addressed in The Geyser geothermal field? What were the positive side impacts of the solution?

15.94. Explain the problem of geothermal-induced earthquakes.

15.13.10 General

15.95. Can geothermal power plants operate for peak-load power generation? Why?

15.96. How are seismic waves created?

15.97. Geothermal power plants have a small carbon footprint even when their construction is taken into account.

a. True

b. False

15.98. Describe the formation and geological structure of a typical geothermal field.

15.99. Where is the oldest geothermal power plant in the world located?

15.100. What are the similarities and differences between the following two types of hydrothermal geothermal systems?

- Vapor-dominated systems
- Liquid-dominated systems

15.101. Geothermal fields are installations with many boreholes.

a. True

b. False

15.102. Most high-profile geothermal fields are ………

a. vapor-dominated

b. liquid-dominated

c. binary systems

15.103. Explain how a flash tank works.

15.104. The nature of the Earth's center is relatively unknown. Why?

a. True

b. False

15.105. What is the depth of the deepest well on Earth?

15.106. Where is the oldest geothermal power plant in the U.S. located?

15.107. The complexity of geothermal systems requires a holistic approach for their exploration before a geothermal power plant can be built?

a. True

b. False

15.108. Which one has a higher efficiency: geothermal power plants or conventional fossil fuel power plants? Why?

15.109. What type of geothermal energy systems are largest source of geothermal energy of any kind?

15.110. How have scientists been able to study the nature of the Earth's inner space?

15.111. What is the purpose the fracturing process in EGSs?

a. Widens the existing fractures

b. Connect them

c. Creates new ones

d. All of the above

15.112. Scientists have been able to study the nature of Earth's inner space using the science of seismology.

a. True

b. False

15.113. Why geothermal power plants have steam pipes with larger diameters compared to those in conventional fossil fuel power plants?

15.114. What are hydrothermal geothermal systems?

15.115. What is the most important mechanism for the origin of geothermal energy?

15.116. The United States continues to lead the world in geothermal electricity production.

a. True

b. False

15.117. What are enhanced (engineered) geothermal systems (EGS)?

15.118. Geothermal energy is renewable because the Earth's interior will continue to cool and provide the source of energy.

a. True

b. False

15.119. What are the two types of seismic waves?

15.120. Which statement is not correct?

a. A major advantage of geothermal energy systems is that they do not need any substantial new technology development.

b. power plants share most components with conventional steam power plants.

c. Geothermal power plants have negative environmental impacts comparable to those of fossil fuel power stations.

d. Geothermal energy systems can be used not only to generate electricity but also can provide heating for residential, commercial, and industrial applications.

15.121. Explain how geothermal-fossil fuel superheated hybrid plants and geothermal-preheat hybrid plants work.

15.122. List two mechanisms for the origin of geothermal energy. Explain them.

15.123. The current well depth record is the Kola research borehole at 12.262 km in Russia.

a. True

b. False

15.124. List pros and cons for geothermal power plants.

15.125. Organic Rankine Cycles (ORC) have corrosion problem only in the heat exchanger and well casing. Why?

a. True

b. False

15.126. Geothermal power plants have …. capacity factors. Why?

a. relatively high

b. relatively Low

c. zero

d. 100%

15.127. Energy released by the decay of natural radioactive isotopes (uranium, thorium, and potassium) is the most important source of geothermal energy.

a. True

b. False

15.128. Briefly explain the principles of enhanced (engineered) geothermal systems.

15.129. Geothermal power plants use relatively similar technology to conventional steam power plants.

a. True

b. False

15.130. What is the difference between Organic Rankine Cycles (ORC) and binary power plants?

15.131. The cost of drilling a unit of length of a well is an exponential function of the well depth.

a. True

b. False

15.132. Why geothermal power plants have larger cooling towers compared to those of conventional fossil fuel power plants?

15.133. In liquid-dominated geothermal systems, water can flow …… out of wells.

a. naturally

b. by pump

c. both a and b

d. neither a nor b

15.134. In the following schematic, name various layers of Earth.

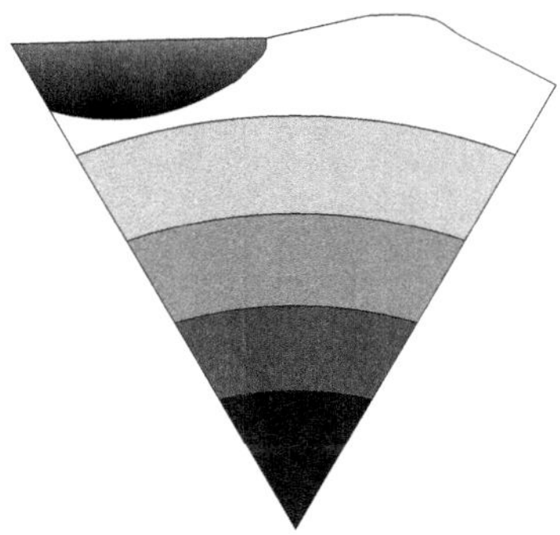

15.135. Which types of seismic waves can pass through solid media?

a. S-waves

b. P-wave

15.136. Hot water from a geothermal reservoir enters a flash tank at the temperature and pressure of 200°C and 30 bar, respectively. The pressure of the tank is 5 bar. Determine the ratio of mass of the spent brine to the mass of steam in the tank.

15.137. In the previous problem, determine the ratio of the enthalpy of steam to the enthalpy of liquid water in the tank and the ratio of the enthalpy of generated steam to the enthalpy of the inlet hot water from the well.

15.138. Considering the magnitude and distribution of geothermal energy, it is not capable of single handedly addressing world energy problems, but it can make a significant contribution in many locales.

a. True

b. False

15.139. Why geothermal power plants have larger turbine, condenser, etc., compared to those of conventional fossil fuel power plants?

15.140. While broadly speaking geothermal energy is inexhaustible in the human time scale in a particular site this may not be the case.

a. True

b. False

15.141. Which types of seismic waves can pass through liquid media?

a. S-waves

b. P-wave

15.142. In principle, geothermal energy is available everywhere and not just in active volcanic areas.

a. True

b. False

15.143. What type of hydrothermal geothermal plants are The Geysers and Larderello in Italy?

16

Energy Storage Systems

Abstract

Energy storage systems are a desirable addition for a conventional fossil fuel-based electricity generation infrastructure and an essential one for future renewable energy-based infrastructure. There are a wide variety of technologies to store energy depending on the form of energy that is being stored. In this chapter, the most important and promising technologies will be discussed. We will start the chapter with a discussion on why energy storage matters in the context of the current fossil fuel dominant electricity generation industry and the future renewable energy dominant industry. We will then introduce several parameters to characterize various energy storage technologies. The rest of the chapter is dedicated to explanations of different energy storage technologies. In each case, the operational principles of the technology and its advantages and disadvantages will be explained and in most cases, simple calculations will be presented. First, energy storage via raising mass and in pumped hydroelectric plants will be covered. Then, we will learn about compressed air energy storage systems. Next, flywheel energy storage systems will be explained. Battery storage systems, including lead-acid and lithium-ion batteries, as one of the most promising technologies, will be discussed next. Finally, supercapacitors and superconducting coils as direct electricity storage systems will be covered followed by a brief discussion on thermal energy storage systems.

LEARNING OUTCOMES

After successful completion of this chapter, you will be able to:

- explain the importance of energy storage in current fossil fuel-dominant electricity generation infrastructure and future renewable energy-dominant infrastructure
- characterize energy storage technologies
- explain gravitational energy storage systems in the form of raising mass and in pumped hydroelectric plants, and conduct simple relevant calculations

- explain flywheel energy storage systems, and conduct simple relevant calculations
- describe battery storage systems, including lead-acid and lithium-ion batteries, and conduct simple relevant calculations
- explain the principles of supercapacitors and superconducting coils as direct electricity storage systems
- describe thermal energy storage systems

16.1 Introduction

Like every other commodity, electricity should be available whenever and wherever it is needed by consumers. In Chapter 1, we discussed how the electrical grids, including high-voltage transmission lines and local distribution lines, are used to fill the spatial gap between electricity generation units and consumers. They have allowed us to locate power plants far away from consumption centers. However, we learned that electricity is probably the only product that cannot be readily and economically stored. Without energy storage systems, electricity must be produced at the moment of demand. So the ultimate objective of energy storage systems is the temporal movement of electrical energy. We also learned that electrical demand varies both regularly, such as predictable hourly, daily, and sessional variations, and irregularly (unpredictable variations). So the electricity generation industry must be able to collectively generate enough electricity to meet this varying demand all the time; otherwise, there will be a blackout or brownout.

Since the early days of widespread electrification, electricity storage systems have always been a desirable feature for the electricity generation infrastructure but not a necessity (after all humanity has enjoyed electricity for many decades without much (if any) meaningful electricity storage). But as will be elaborated in the next section, it is hard to envision a future renewable-based electrical infrastructure without large-scale energy storage systems.

16.1.1 Energy Storage: Fossil Fuel-Based versus Renewable Energy-Based Electrical Infrastructure

Before discussing the function of energy storage systems in a fossil fuel-based electrical infrastructure, we should keep in mind that fossil fuels are created from solar energy that millions of years ago was stored in ancient plants and other living organisms. Over the course of a long time, perhaps hundreds of millions of years, the remains of these organisms have evolved to chemical energy that we now use as fossil fuels. So in order for a fossil fuel-based electrical infrastructure to operate, we do not necessarily need energy storage systems since fossil fuels are energy storage media. Therefore, what we need are energy conversion systems that can convert chemical energy in fossil fuels to electricity in a sufficient quantity at the appropriate time. A fossil fuel-based electrical infrastructure only needs to have two attributes to function. First, its electricity generation capacity should be sufficiently large to meet the maximum demand (plus some extra capacity to cover planned and unplanned power plant outages). Second, it should have a proper mixture of baseload and peak-load power plants, where the former generate the minimum electricity requirements and the latter can rapidly follow the load to cover the difference between the baseload and the demand. So with an appropriate design and operating strategy, the entire infrastructure can operate perfectly with no storage facility. In fact, if peak-load power generation units were as efficient and inexpensive (in terms of the cost of a unit of the generated electricity, and not just capital cost) as baseload power plants, energy storage units would have been undesirable. This is because, as we will see later, the round trip efficiency of energy storage systems is always less than unity (for some systems significantly less). This means by storing energy and using it at a later time, we lose some portion of the energy. For everything else equal, it makes more sense to generate electricity in real-time rather than storing it. Of course, we know that the cost of a unit of electricity is much lower in baseload power plants than that of peak-load power stations.

In a real fossil fuel-based electrical infrastructure, while energy storage is not necessary, it can be desirable with two objectives. The excess inexpensive electricity that is generated during off-peak periods (typically at night) can be stored in an energy storage system and then can be recovered at peak periods. This plan only makes sense when the cost of the energy storage system and losses due to the round trip efficiency of energy storage are more than offset by the difference between peak and off-peak electricity costs. Another objective of energy storage systems is to reduce the overall required power generation capacity. In this strategy, instead of building a new power plant or using old inefficient plants, an energy storage facility is used to meet the peak demand. So the basic idea is to operate existing power plants for a longer period of time (at a higher capacity factor) rather than installing new power plants. Of course, this is only meaningful if the cost of building and operating an energy storage system is less than that of a new power generation unit. This is called **peak shaving**. Generally, peak shaving is any action or strategy that reduces the peak load (maximum demand). Peakshaving can be in the form of a reduction in the consumption during peak period, for example, by turning off some lights or adjusting the temperature setting for air-conditioning systems. Alternatively, it can be achieved by moving some electricity uses to off-peak periods without reducing the overall energy consumption, for example, by doing laundry during off-peak periods.

In a renewable energy-based electrical infrastructure, while demand variation (and its associated uncertainties) are still present, there is a new layer of variations and even greater uncertainties with electricity generation. Unlike fossil fuel power plants, in which the source of energy (i.e., fossil fuels) is available on demand, in renewable energy systems, with a few notable exceptions such as geothermal and biomass-fueled units, the source of energy is intermittent. This includes predictable and unpredictable variations. Electricity can only be generated if and when renewable energy sources are available. Generally, the availability of renewable energy sources and their intensity do not necessarily match the variations in the demand. The problem with the intermittent nature of renewable energy sources may not be tangible when their contributions to the total electricity generation are small. But with an increase in their share, the problem becomes evident. California is a perfect example in this regard. According to the Solar Energy Industries Association (SEIA), California has the largest solar photovoltaic installations in the U.S., with about 28,000 MW installed capacity, partly thanks to 40% cost reduction of PV units in the past five years.[1] In 2019, with over 1,100,000 installations, over 20% of electricity in the state was generated by these PV units. As discussed in Chapter 11, these units start to generate electricity in morning. Their power production increases until midday and then declines until they cease to produce electricity after sunset. While they significantly contribute to the summer midday peak period, they cannot generate electricity during the evening and early night peak periods. This problem is sometimes referred to as the so-called **duck curve** problem (the term is typically used in the industry[2]). This curve refers to a graph very similar to the load profile but instead of showing the total electricity demand, it shows the difference between the total demand and the electricity generation by renewable energy sources. In Figure 16.1, the dashed curve is the total demand, the dotted curve is the electricity generation by renewable sources (in this case primarily from solar PV units), and the solid curve is the duck curve (total demand minus power from renewable resources), which supposedly looks like the back of a duck! This curve actually shows the amount of electricity that needs to be generated by nonrenewable resources. This curve illustrates that PV units have been able to successfully eliminate the midday peak in California and reduce dependency on fossil fuel power generation during the day. However, the early morning peak and especially the late evening peak are intact, except for a small reduction due to wind, geothermal, and biomass-fueled units. While solar energy has been able to significantly reduce the overall fossil fuel consumption for electricity generation, it has almost no impact on the peak-load power generation. This means conventional fossil fuel power plants cannot be retired, and if they are retired, some replacement should be installed to cover peak demands. Power plants just operate for a shorter period of time (with lower capacity factors).

---

[1] https://www.seia.org/state-solar-policy/california-solar.

[2] The term is reportedly first used by the California Independent System Operator (California ISO or CAISO).

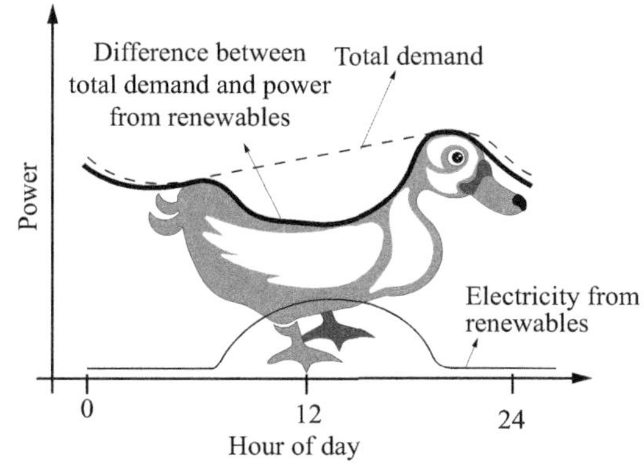

FIGURE 16.1
A duck curve.

This new dynamic can create major disturbance and financial problems for conventional power plants, particularly baseload power plants, including geothermal ones. This phenomenon began in California in 2013 when solar energy systems started to make a meaningful impact in the total electricity generation in the state, which has worsened every year since then. This problem is certainly going to become more intense with increased popularity of PV units as it is expected that about 15,000 MW more PV units is to be installed over the next 5 years.

A different problem but with a very similar nature is due to the unpredictability of renewable energy resources, for example, on cloudy days. A dramatic manifestation of this problem was materialized in the solar eclipse of August 21, 2017. California lost up to 6 GW of power at its peak and took several months of planning and collaboration of 17 western states to "survive solar eclipse unscathed", as described by a local news agency.[3]

While diversifying renewable energy sources can lessen this problem, the future society fully (or at least mostly) powered by renewable energy sources cannot be reached unless energy can be somehow stored. This is why the development of energy storage technologies is so important in shaping the future of the electricity generation industry.

Another reason that certain energy storage systems are critical for the future of the energy industry is their utilization in transportation applications. In Chapter 1, we learned that the transportation systems are one of the major energy consumers. These systems need to carry the energy required for their operation. The sustainable future of these systems depends on the development of affordable energy storage systems and/or renewable environmentally friendly fuels. While these systems are not the focus of this text,

[3] https://www.sfgate.com/business/article/California-power-grid-survives-solar-eclipse-11948034.php.

some energy storage systems that have the potential to be used in transportation applications will be covered in the coming sections.

16.2 Characterization of Energy Storage Technologies

As we will see in the coming sections, energy storage systems operate based on a wide variety of technologies and physical principles. So it is not surprising that there are many ways to characterize them. In this section, some of these categorizations will be discussed, which will help us to evaluate and compare various technologies.

The principle idea of energy storage systems is to store energy when energy production (e.g., electricity generation) or available energy (such as solar thermal energy) either is or can be more than the demand. For example, when electricity generated from PV units or wind turbines is more than the demand or when some capacity of a combined cycle power plant is unused due to the lack of demand, there is a potential for these systems to generate electricity in excess of the demand. The stored energy can be then used when the demand is more than the generation. Therefore, when characterizing energy storage systems, the efficiency, economy, and reliability of the energy storage process, the energy recovery process, and the entire process should be considered.

One of the most important parameters to evaluate an energy storage system is its efficiency or more accurately its **round trip efficiency**, also known as **turnaround efficiency** or **overall efficiency**. As the name implies, this is the **ratio of the recovered energy (energy output) to the energy input** in an energy storage system. While the definition seems simple and straightforward, there are some complexities that need to be clarified when determining this efficiency. One such factor is how the **system boundary** is defined and what components are included in the system when the efficiency is determined. For example, some electrical energy storage systems can only store DC power, such as batteries. The inclusion of the AC-to-DC and DC-to-AC inverters can have a significant impact on the value of the round trip efficiency. Another important factor in determining this efficiency is the **duration of the energy storage**. For example, while the short-term thermal energy storage can be quite efficient, long-term thermal energy storage can reduce the round trip efficiency due to thermal losses (even in well-insulated systems). While these issues are present for all energy storage technologies, their intensity strongly depends on each individual system. This leads us to another attribute of energy storage systems, the **timescale** they can effectively store energy.

Speaking of time, there are other factors while time-related; they have a totally different nature than the timescale as discussed above. As noted above, an energy storage system must be charged when excess energy is available and discharged when there is a need for energy. Since both the charge time and discharge time are typically in the same order, only the discharge time is used as their representative. This time is characterized by a parameter known as **self-discharge time** (**SDT**). This is the time takes for a fully charged device to release its energy up to a certain minimum energy level. We will see in the coming sections that most of the energy storage systems, for a variety of economic and/or technical reasons, either cannot be fully discharged or are undesirable to do so. The percentage of the total stored energy that can be typically discharged is referred to as the **depth of discharge** (**DOD**). If DOD of an energy storage device is 75%, it means that from its fully charged state, the device can only release 75% of the stored energy and the remaining 25% cannot be recovered. Note that this does not mean that the remaining energy is wasted because when recharging the device, one needs to provide sufficient energy to charge it from 25% to 100%. So a low DOD does not negatively impact the round trip efficiency. However, a low DOD reduces the energy density and specific energy of the device (will be covered next). It should also be noted that the SDT and DOD are not necessarily linearly related.

In terms of the time required to store and deliver energy, energy storage devices can be divided into three groups: **very short**, **short**, and **long storage periods**, which may refer to a fraction of a second, seconds to minutes, and hours to days, respectively. The discussion about each of these groups leads up to the various applications of energy storage systems. The first application of energy storage systems is to **maintain the quality of electrical power**. We know that ideally the voltage (and current) of electrical power as a function of time should be sinusoidal with a predetermined magnitude and frequency. The quality of electricity refers to how close the actual maximum voltage, frequency, and waveform of the power are to those of the ideal case. Some systems are sensitive to the quality of electricity, and low-quality power can negatively impact their performance or even physically damage them. For this reason, some consumers have specifically designed systems to condition the input power from the grid. To compensate for any instantaneous anomaly, electricity storage systems with very quick response time are needed. These systems are equipped with power-monitoring systems to detect any problem and react to rectify it in a split second by storing or discharging some electricity.

The second application of energy storage devices with a short discharge time is to **provide power when switching from one electricity source to another**. For instance, imagine a hospital that is equipped with a diesel generator as a backup power generator. If for whatever reason electricity from the grid is discontinued, for example, during a blackout, it will take a few seconds for the generator to start and speed up to the nominal rotational speed. During these few seconds, an energy storage system is needed to avoid discontinuity in the flow of electricity; otherwise, critical equipment in the hospital will shut down. The last group of energy storage systems is those that can take from a few hours to few days to discharge their stored energy. Many energy storage systems of interest in this text are of this type.

Other important parameters to characterize energy storage systems are their specific energy and energy density. The **specific energy** or **energy-to-mass ratio** of an energy storage system is the amount of energy that can be stored per unit of mass of the system (with the unit of kWh/kg or MJ/kg). The **energy density** or **energy-to-volume ratio** of an energy storage system is defined as the amount of energy that can be stored per unit of volume of the system (with the unit of kWh/m^3 or MJ/m^3). Similarly, the **specific cost** is the cost of an energy storage system to store the unit of energy (with the unit of $/kWh or $/MJ). The combination of these factors for a specific application determines the energy storage capacity (with the unit of kWh or MJ) achievable for the given conditions. The required capacity depends on a specific application, from a couple of tens of kWh for automotive applications to a few kWh to many tens of kWh for distributed storage applications, to several millions of kWh for utility-size storage applications. In the same way, the **specific power** (or power-to-mass ratio) with the unit of W/kg and the **power density** (or power-to-volume ratio) with the unit of W/m^3 can be defined. These two parameters depend not only on the energy stored in the system and its mass or volume but also on how quickly this energy can be discharged.

A critically important factor is the **system longevity**, which is typically expressed by the number of consecutive charge and discharge cycles before the system performance is significantly deteriorated. This parameter for some technologies is a strong function of the DOD. This life cycle for various energy storage technologies can be from a few hundreds to a few hundreds of thousands.

Finally, in terms of physical principles used to store energy, energy storage systems can be generally divided into three groups: **electrical**, **mechanical**, and **thermal** energy storage systems. They can directly store electricity in batteries, supercapacitors, or superconducting coils. Alternatively, they store energy in the form of kinetic energy (such as flywheels), gravitational potential energy (such as hydro power or elevated mass), high-pressure air, or spring potential energy. They can also store energy in the form of thermal energy. Sometimes nonthermal energy storage systems are collectively referred to as electrical–mechanical energy storage systems. In the remainder of this chapter, various energy storage technologies will be discussed.

16.3 Gravitational Energy Storage Systems

16.3.1 Energy Storage by Raising Mass

Probably the simplest way to store energy that one can come up with is to use the available energy to raise an object to a higher elevation. Then, whenever the energy is needed, the object can be brought to the original elevation and the stored energy can be recovered. An example of these systems is old antique gravity-powered clocks, also known as weight-driven clocks. The stored energy can be determined by the following equation:

$$\mathrm{PE} = mgh \tag{16.1}$$

where PE is the stored potential energy (J), m is the mass of the object (kg), g is the gravitational constant $\left(\frac{\mathrm{m}}{\mathrm{s}^2}\right)$, and h is the height that the object is raised (m). While simple, this approach may not be able to have a significant contribution to the large-scale energy storage due to their low energy density. Energy storage in springs also faces a similar problem. Because of this problem, these systems have been only used in very small applications, such as clocks and watches, sensors and instruments, and toys. The following example may provide some insight into the problem. However, recently there have been some innovative proposals for these systems.

Example 16.1

An object with a mass of 1500 kg is raised to the height of 100 m. Determine how much energy is stored in the object in the form of gravitational potential energy. Determine for how long that energy can power an average American household.

Analysis

The gravitational potential energy can be simply determined by Equation 16.1.

$$\mathrm{PE} = mgh = 1500 \times 9.81 \times 100 = 1,471,500\ J = 0.409\,\mathrm{kWh}$$

In 2018, an average American household consumed 10,972 kWh/year[4] of electricity.

$$\text{Average U.S. home electricity consumption} = 10,972 \frac{\mathrm{kWh}}{\mathrm{year}}$$

$$= \frac{10,972}{(365 \times 24 \times 60)\frac{\mathrm{min}}{\mathrm{year}}} \frac{\mathrm{kWh}}{\mathrm{year}} = 0.0209 \frac{\mathrm{kWh}}{\mathrm{min}}$$

Time the potential energy can power an average American household

$$= \frac{0.409\,\mathrm{kWh}}{0.0209 \frac{\mathrm{kWh}}{\mathrm{min}}} = 19.6\,\mathrm{min}$$

[4] https://www.eia.gov/tools/faqs/faq.php?id=97&t=3#:~:text=In%202018%2C%20the%20average%20annual,about%20914%20kWh%20per%20month.

Just to help you to visualize what we are talking about in this example, 1500 kg is approximately the mass of a midsize car and 100 m is a little taller than the Statue of Liberty. So if you raise your midsize car to the height of the Statue of Liberty, you can store sufficient energy to power your house for about 20 minutes or run your 1000-W iron for 24 minutes. To store sufficient energy to fully power an average house in the U.S., one needs to raise about 2.5 Boeing 737-MAX (with a mass of about 45,000 kg) to the height of 100 m. Alternatively, for this purpose, one can raise a midsize car to the height 16.6 times the highest point of the Empire State Building in New York City. Of course, this is with the assumption that all the processes are 100% efficient. It can be shown that to store electrical energy produced by a power plant with the capacity of 1000 MW (a midsize power plant) in an hour, almost 2.5 million midsize cars (with the mass of about 1500 kg each) are needed to be raised to the height of 100 m. A similar concept can be used for pumping water for energy storage.

16.3.2 Pumped Hydroelectric Energy Storage Systems

Pumped hydroelectric energy storage (**PHES**) systems, also referred to as **pumped storage hydroelectricity** (**PSH**), can be considered as an extension of the concept discussed in the previous section. Only in this case, the object that is being raised to store gravitational potential energy is water. This energy storage method requires two reservoirs: one lower-elevation water reservoir and one higher-elevation water reservoir. When excess electricity is available, usually inexpensive electricity during off-peak periods, or when excess renewable energy sources are available, water is pumped from the lower reservoir to the higher reservoir. When electricity is needed during peak periods, the water flows from the higher reservoir to the lower reservoir through a turbine, which converts the potential energy of the water flow to rotational kinetic energy and eventually electricity. Among various energy storage technologies, **PHES units are the only ones currently making a meaningful contribution to the electricity generation industry.** Currently, the energy storage capacity of pumped hydro plants in the world exceeds 125 GW in several hundred large plants (operating or under construction).[5] China, Japan, and the U.S. are leading countries in terms of the pumped hydro storage capacity, while China also has many new plants currently under construction. In the U.S., there are over 40 plants with a similar number being in the licensing process or under construction. The installed capacity of the pumped hydro plants in the U.S. is over 21 GW, accounting to about 2% of the country's electricity generation capacity. This is almost the entire large-scale utility energy storage capacity of the nation (less than 5% for other storage technologies). The early utilization of pumped hydro storage systems was reported in Europe. In the U.S., the first unit was installed in the 1930s

[5] https://www.hydro.org/wp-content/uploads/2018/04/2018-NHA-Pumped-Storage-Report.pdf.

in Connecticut. A majority of the plants were built in the 1970s. The largest currently operational pumped hydro storage facility in the world is the Bath County Pumped Storage Station on the border of Virginia and West Virginia with a power generation capacity of about 3000 MW. The plant is actually the second largest hydropower plant in the U.S. The second largest pumped hydro storage facility in the U.S. (and the fourth largest hydropower plant) is the Ludington Pumped Storage Plant in Michigan with a capacity of about 1900 MW. These two plants were both established in the 1970s, and both are exclusively energy storage facilities. However, pumped hydro systems can also be integrated to conventional dam hydropower plants. An example of such a systems is in the Grand Coulee Dam, in Washington on the Columbia River, the largest hydropower plant in the U.S., with an electricity generation capacity of about 6800 MW. The power generation capacity includes 314 MW of pumped hydro storage turbines.

Considering all the positive attributes of these plants, their popularity should not come as a surprise. These plants primarily share the same technologies that conventional hydropower plants are using for over a century. They can store a very large amount of energy using currently mature and widely available technologies. They have high round trip efficiency, easily exceeding 80% in some cases. They have a very long service life period sometimes over a century with a virtually unlimited number of "charging and discharging" cycles (borrowing the term from batteries; in fact, pumped hydro energy storage units are sometimes referred to as "water batteries"). Furthermore, they provide operational flexibility due to the flexible speed in which they store and release energy. Similar to hydro power plants, they can rapidly (in a matter of seconds) respond to load variations. This characteristic makes them ideal for peak-load power generation. They can also be a helpful tool in water management. Generally, these plants require high initial costs. However, overall, they are the only cost-effective and cost-competitive energy storage system for the large-scale energy storage. There are several major sources of energy loss in typical pumped hydro energy storage units. The most important source of energy loss is inefficiencies in the pump-turbine and motor-generator sets. Also, water evaporation and seepage to ground in the upper reservoir contribute to energy losses. These two losses also exist in the lower reservoir, but they do not directly contribute to energy losses, although the lost water should be compensated.

While they are more cost-effective in larger installations, they have been installed in smaller capacities, a few MWs, too. These systems have potentials to be used as decentralized energy storage systems in urban regions or even individual buildings. The concept can also be used to store energy in the ocean in underwater containers. When excess electricity is available, water is pumped out of the containers, and when energy is needed, ocean water is allowed to fill the containers via a turbine. These units are ideal to store electricity from offshore wind turbines.

16.3.2.1 Energy and Power in Pumped Hydroelectric Energy Storage Plants

The gravitational potential energy stored in the elevated water in the upper reservoir can be determined by the following equation (similar to Equation 16.1 but here m is the mass of the stored water):

$$\text{PE} = mgh = \rho V g h \tag{16.2}$$

where PE is the potential energy stored in the elevated water (J), m is the mass of the stored water (kg), V is the volume of the stored water (m^3), ρ is the density of water $\left(\frac{\text{kg}}{\text{m}^3}\right)$, g is the gravitational constant $\left(\frac{\text{m}}{\text{s}^2}\right)$, and h is the height difference between the upper and lower reservoirs (m). Clearly, this energy is not entirely available, and some portion of the energy will be lost to various losses (discussed below). The power that is needed to pump the water from the lower reservoir to the upper reservoir can be calculated by the following equation:

$$P_P = \frac{\rho \dot{Q}_P g h}{\eta_P} \tag{16.3}$$

where P_P is the power needed to pump the water (W), $\dot{Q}_P$ is the flow rate of water in the pumping mode $\left(\frac{\text{m}^3}{\text{s}}\right)$, and η_P is the efficiency of the system in the pumping mode. Similarly, the power generated by the flow of the water flowing from the upper reservoir to the lower reservoir can be determined by the following equation:

$$P_T = \eta_T \rho \dot{Q}_T g h \tag{16.4}$$

where P_T is the power generated by the flowing water (W), $\dot{Q}_T$ is the flow rate of water in the turbine mode $\left(\frac{\text{m}^3}{\text{s}}\right)$, and η_T is the efficiency of the system in the turbine mode. Alternatively, the power needed for pumping the water in the pumping mode and the power generated by the water in the turbine mode can be determined by the following equations:

$$P_P = \rho \dot{Q} g h_P \tag{16.5}$$

$$P_T = \rho \dot{Q} g h_T \tag{16.6}$$

$$h_P = h + h_{l,\,P} \tag{16.7}$$

$$h_T = h - h_{l,T} \tag{16.8}$$

where h is the height difference between the upper and lower reservoirs (m), h_P is the head of the pump (m), h_T is the head of the turbine (m), $h_{l,P}$ is the head loss during the pump mode (m), and $h_{l,T}$ is the head loss during the turbine mode (m). It should be noted that the system efficiencies during the pump mode and the turbine mode are not necessarily equal (the same applies to the head losses). Typically, the power generation capacity (the output power in the turbine mode) is greater than the input power during the pump mode. To make up for this imbalance, the time period of the energy storage (i.e., the time that the system operates in the pump mode) is typically longer than the power generation period (i.e., the time that the system operates in the turbine mode). For the sake of consistency and accuracy, these units are rated for **their power generation capacity.**

Example 16.2

The Bath County pumped hydro storage plant, the largest of its kind in the world, has a storage capacity of 24,000 MWh. Assuming a total efficiency of 100%, determine the mass (in kg) and volume (in m^3) of water if the elevation difference between the upper and lower reservoirs is 380 m.

Analysis

$$\text{PE} = 24{,}000\ \text{MWh} = 8.64\times10^{13}\ \text{J}$$

$$\text{PE} = mgh = \rho Vgh$$

$$m = \frac{\text{PE}}{gh} = \frac{8.64\times10^{13}}{9{,}8\times380} = 2.32\times10^{10}\ \text{kg}$$

$$V = \frac{\text{PE}}{\rho gh} = \frac{8.64\times10^{13}}{1000\times9{,}8\times380} = 2.32\times10^{7}\ \text{m}^3$$

In reality, the upper reservoir has a storage capacity of about 44 million m^3 and the lower reservoir has a storage capacity of about 35 million m^3. In this calculation, it is assumed that no energy loss is occurred in the system. Also, clearly, the reservoirs cannot be fully emptied and there is always some residue water in both reservoirs. On top of these assumptions, the actual calculations are more complicated since the turbine and pump heads vary when the reservoirs are being filled (or emptied). Let us consider this problem from a different angle in the next example.

Example 16.3

In the Bath County pumped hydro storage plant, the upper reservoir has a surface area of 1.07 km^2. Determine the change in the elevation of the surface of the reservoir when it is fully filled compared to that when it is empty. The volume of the pumped water to fill the reservoir is 2.32×10^7 m^3.

Analysis

$$V_{\mathrm{Up}} = A_{\mathrm{Up}} d_{\mathrm{Up}} \tag{16.9}$$

where V_{Up} is the volume change of the upper reservoir (m^3), A_{Up} is the surface area of the reservoir (m^2), and d_{Up} is the variation in the level of the reservoir (m).

$$A_{\mathrm{Up}} = 1.07 \text{ km}^2 = 1.07\times10^6 \text{ m}^2$$

$$d_{\mathrm{Up}} = \frac{V_{\mathrm{Up}}}{A_{\mathrm{Up}}} = \frac{2.32\times10^7}{1.07\times10^6} = 21.7 \text{ m}$$

Depending on how the level of water in the lower reservoir changes, the head available in the turbine mode varies. For example, in this case, the lower reservoir has a surface area of 2.25 km^2.

$$V_{\mathrm{Low}} = A_{\mathrm{Low}} d_{\mathrm{Low}}$$

$$d_{\mathrm{Low}} = \frac{V_{\mathrm{Low}}}{A_{\mathrm{Low}}} = \frac{2.32\times10^7}{2.25\times10^6} = 10.3\,\text{m}$$

This shows that the available heads in the upper and lower reservoirs are varying with different rates. The maximum turbine heat is available when the upper reservoir is full and the lower reservoir is at the lowest level. With water gradually flowing from the upper reservoir to the lower reservoir through the turbine, the level of water in the upper reservoir reduces and the level of water in the lower reservoir increases. The overall impact is the reduction in the available head. The rate of this reduction increases if the surface areas of the reservoirs become smaller. It should be noted that in this discussion, we assumed that the surface areas of reservoirs are constant. This assumption is generally not accurate. In many cases, the reservoir is like an inverted pyramid. So the variations in the water level are not linear and actual calculations are rather complex requiring the study of local topography.

Example 16.4

The Bath County pumped hydro storage plant is equipped with six turbine-pump units. Each unit can generate 500.5 MW of power in the turbine mode. The volumetric flow rate of the water in the turbine mode is 850 m^3/s. Determine the efficiency of the system in the turbine mode (h = 380 m).

Analysis

$$P_T = \eta_T \rho \dot{Q}_T gh$$

$$P_T = 500.5 \times 6 = 3003\,\text{MW} = 3.003 \times 10^9\ \text{W}$$

$$\eta_T = \frac{P_T}{\rho \dot{Q}_T gh} = \frac{3.003 \times 10^9}{1000 \times 850 \times 9.8 \times 380} = 0.949$$

16.3.2.2 Reservoirs in Pumped Hydroelectric Energy Storage Plants

As noted earlier, in order to have a pumped hydro storage plant, we need to have a high-elevation reservoir and a low-elevation reservoir. How these two reservoirs are created primarily determines the system operation and costs. The easiest and most straightforward approach is to integrate a pumped storage system to a conventional hydropower plant, typically an existing one. Since the storage system takes advantage of the already-existing components of the plant, a limited number of new components are required, which significantly reduces the cost associated with establishing the plant. These units are referred to as **pump-back plants**. While these hydropower plants operate like a conventional hydroplant by storing the naturally flowing river water, they can store excess electricity by pumping some water from downstream back to behind the dam. This increases the stored water behind the dam and allows the plant to operate for a longer period of time. Of course, this energy storage method cannot be used if the dam is already full by river water.

As noted in the previous section, the Grand Coulee Dam is equipped with a pumped storage facility. While the dam and its two powerhouses were completed between 1933 and 1942 (with the third powerhouse installed in 1973), the pumped storage units were added between 1973 and 1984. There are six pump-turbine units, each with about 50 MW power generation capacity. This capacity is about 5% of the total capacity of the facility (at about 6800 MW). At least in one tidal energy system, the Rance Tidal Power Station in France, this type of energy storage system was also incorporated into the unit (refer to Chapter 14).

Most major pumped hydro storage plants are exclusively designed and operated for energy storage purpose (and not integrated into conventional hydropower plants). In these plants, lower and upper reservoirs should be formed. These reservoirs should have a sufficient capacity to accommodate sufficient water corresponding to the desired energy storage. It should be noted that since these reservoirs are repeatedly filled and emptied, they are much smaller than those needed for conventional hydropower plants, where the entire water for a season must be stored. The most important characteristic of the reservoirs is their vertical and horizontal locations relative to each other. The vertical distance between two reservoirs should be increased as much as possible to maximize the energy storage for the unit of mass of water. On the contrary, the horizontal distance between two reservoirs should be minimized to reduce the pressure drop in the piping system.

They also need a reliable source of water to initially fill the reservoirs and provide the makeup water to compensate for water losses to ground and evaporation. The most important challenge that pumped hydro storage systems are facing is finding a site with desirable characteristics, as outlined above, which can be economically exploited. These suitable sites are generally limited and often located in areas with natural beauty and/or sensitive ecosystems, which makes receiving a license to construct the system challenging.

Either or both reservoirs can be natural or artificially made. Natural reservoirs can be in the form of rivers, lakes, seas, or oceans. These systems, where at least one of the reservoirs is a natural body of water, are referred to as **open-loop** systems. When none of the reservoirs is connected to a natural body of water and the water is only circulated between the two artificial reservoirs, the system is referred to as a **closed-loop** system.

Reservoirs can be aboveground or underground. Aboveground reservoirs can be created by excavating earth or by enclosing an area by building dams. Underground reservoirs can take an advantage of naturally existing underground cavities, abandoned mines, or underground military facilities as a lower reservoir. A wide variety of mining sites have been considered for this purpose, including metal mines and particularly coal mines considering the existence of a huge number of them.

In the most common configuration, the upper- and lower-level reservoirs are connected via a pipeline. In some designs, the water flow from the upper reservoir is first directed to a horizontal conduit or an open channel called a pressure tunnel. At the entrance to the intake from the upper reservoir, typically a trash bar or screen is installed to stop debris from entering the pipeline. The powerhouse, where the pump-turbine and motor-generator units are located, is placed at the lower end of the penstock, as close to the level of the lower reservoir as possible (Figure16.2). Somewhere along the penstock,

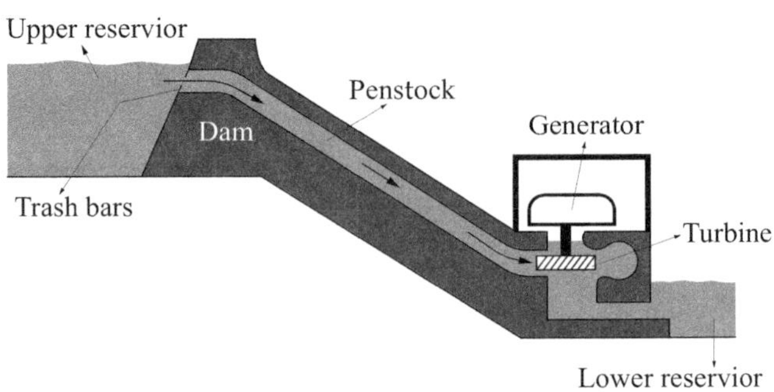

FIGURE 16.2
Schematic of a pumped hydro energy storage plant.

before the powerhouse, a valve house is located to control the water flow rate to the turbine. Also, a surge tank is installed to protect the line from the sudden change in the pressure.

16.3.2.3 Turbine-Pump Assembly

Before we talk about the turbine-pump assembly for pumped hydro plants, let us have a brief discussion on the different types of hydro turbines in general. Water wheels have provided power for human activities for centuries. During the Industrial Revolution, they were replaced by hydro turbines. Similar to steam turbines, hydro turbines can be either impulse or reaction. Water wheels are the oldest impulse turbines. The most common impulse hydro turbine is a **Pelton turbine**, named after its inventor Lester Allan Pelton (1829–1908, an American inventor). This turbine type is composed of many buckets attached to a runner (Figure 16.3). The water is directed toward buckets through one or more free nozzles (jets). The nozzles increase the velocity of the water flow so the flow can impart its kinetic energy to the buckets when it hits them. Note that these turbines are not submerged and spent water is collected at the bottom of the turbine casing and is then discharged. The flow rate of water is controlled by a spear installed inside the nozzles. By moving the spear forward or backward, the water flow rate can be decreased or increased, respectively. Pelton turbines can best operate with high heads and low flow rates with a capacity typically up to 400 MW. An alternative design that can operate with a medium head is called

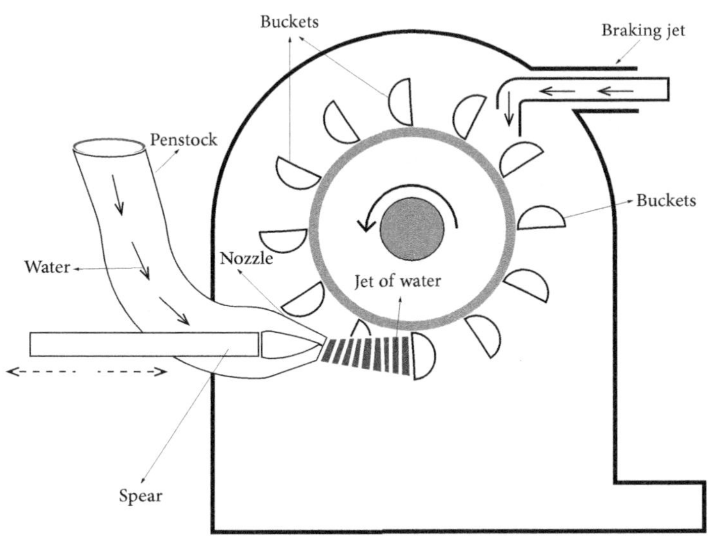

FIGURE 16.3
Schematic of a Pelton turbine.

Turgo turbine. Another design that operates based on the impulse principle is called cross-flow turbine, which is similar to "squirrel cage" blowers.

Reaction hydro turbines have two main types depending on the relative direction of the inlet and outlet flows and their designs. Both types are fully submerged. Perhaps the simplest design is **Kaplan turbines**, named after Viktor Kaplan (1876–1934, an Austrian engineer). These turbines are similar to large ship or boat propellers with three to six fixed or variable pitch blades. These axial flow turbines are suitable for low head and high flow rate applications.

Francis turbines, named after its inventor James Francis (1815–1892, a British-American engineer and inventor), are the most commonly used hydro turbines today. They can have a capacity ranging from several kWs to 800 MW with a diameter of up to 10 m. These turbines are typically installed in a vertical orientation where the water flow enters the turbine radially and leaves axially. These turbines are typically suitable for applications with intermediate heads, though the head range for their operation has a substantial overlap with those for Kaplan and Pelton turbines.

Early pumped hydro storage plants were equipped with a pump and a turbine to pump water to higher elevation and then to convert the stored energy back to electricity. Since the mid-20th century, these two components have typically been integrated into a single turbine-pump assembly. In the pump mode, these units operate as a motor-driven pump, and in the turbine mode, they function as a turbine generator. Typically, these units are single-speed, single-stage Francis pump turbines. After decades of operation history, these units have evolved to reliable systems with high efficiency at the full-load operation. However, they suffer from low efficiencies at the part-load operation. Even more importantly, in the pump mode, they cannot operate at the part load. Imagine that a pumped storage unit is equipped with two 50-MW single-speed pump-turbine units. If the excess electricity available for the storage is 70 MW, only 50 MW of the energy can be stored and the remaining 20 MW should be discarded. This is particularly problematic if the storage system is coupled with renewable energy sources with typically unpredictable intermittent nature. To address this problem, most new plants and almost all proposed units are equipped with variable-speed (or adjustable-speed) pump-turbine systems. These systems can store a wide range of power in the pump mode by adjusting the rotational speed of the motor pump. Also, they can use this capability to adjust the speed for maximum efficiency in the part-load operation in the turbine mode.

16.4 Compressed Air Energy Storage Systems

Compressed air energy storage (**CAES**) systems use high-pressure air as the medium to store energy. Using compressed air to store and deliver energy is not a new idea. In the late 1890s in some European cities, energy was distributed in urban regions via pressurized air pipelines. Similarly, for over a century, compressed air-powered transportation systems, such as cars and locomotives, have been proposed, tested, and even commercially employed. The first utility-scale CAES system was established in Huntorf, Germany, in 1978 with the electricity generation capacity of 290 MW and the energy storage capacity of 580 MWh. The second plant was commissioned in McIntosh, Alabama, in 1991 with an electricity generation capacity of 110 MW and the energy storage capacity of 2860 MWh. The third and latest plant was commissioned in Gaines, Texas, in 2012 with a power generation capacity of 2 MW.

In its simplest form, a CAES system requires a compressor to pressurize the air flow, a container (a very large one) to store the high-pressure air, and a turbine to expand the high-pressure air. In this system, when there is available excess electricity, the motor-driven compressor pressurizes air and stores the high-pressure air in the container. Whenever electricity is needed, the high-pressure air is expanded by flowing through the turbine to generate electricity in the generator. The functions of the motor and the generator are typically achieved by a single device, which is connected to the compressor and turbine via two clutches. These clutches allow the device to function as a motor for the pump or a generator for the turbine.

While the concept of increasing pressure of air to store energy (in the same way that pumped hydro plants increase water elevation to store energy) may look simple and straightforward, thermodynamically there are complications that need to be addressed.[6] In Chapter 9, we learned that the power generated during the expansion of a gas and the power needed to compress a gas depend on the specific volume of the gas (for the same pressure difference). We also learned that compressing a gas results in increase in its temperature. In essence, the electricity used to pressurize the air flow is converted to internal energy of the air flow, which is manifested in the form of increase in the air temperature. If the high-pressure air is stored for a long time and the air is cooled down by heat transfer to the environment, the cold air cannot generate electricity that is remotely close to the energy used to compress the same air. Thermodynamically, this problem can be addressed in several ways. The most efficient approach is to store high-pressure air in a very-well-insulated container so it can retain its internal energy at a constant temperature. However, as will be discussed later, underground conventional containers are challenging to insulate. An alternative approach but the similar concept is to remove the thermal energy from the hot compressed air and store the energy in a thermal storage medium. The thermal energy is then transferred back to the air before it is sent to the turbine for expansion. The thermal storage medium can be a packed bed of sand or other materials, hot

[6] Since thermodynamically and conceptually compressed air energy storage systems are to some degree similar to gas turbines, it might be useful to refresh your mind on gas turbines covered in Chapter 9.

oil, or molten salts (Chapter 12). The efficiency of these storage systems can be in the low to mid 70%. These storage systems also prevent freezing of the turbine during the expansion process. Note that in the expansion process, the air temperature reduces for the same reason that the air temperature increases during the compression process. This concept is in the research and development stage. The above two methods are referred to as the **adiabatic storage systems** since they avoid heat loss to the environment.

A more practical approach is to store the compressed air with no thermal energy storage. In this approach, the hot compressed air gradually losses its energy to the environment (typically rock formations surrounding the underground storage space). Alternatively, the partially pressurized air in the compressor can be cooled in intercoolers. When the electricity is needed, the compressed air is heated, typically by a natural gas furnace, before being sent to the turbine (Figure 16.4). In some plants, two compressor intercoolers (three stages of compression) and two heat addition processes (three stages of expansion) have been used. This system is thermodynamically very similar to a gas turbine cycle; just the compression and expansion processes are decoupled (timewise); and they do not take place simultaneously. You recall from Chapter 9 that in a gas turbine, a significant portion of the power generated in the turbine is consumed in the compressor. In CAES systems, the power needed to compress air is provided when extra electricity is available; therefore, the entire power generated by the turbine in the peak period is available for consumption. From a technical and implementation point of view, this approach is the most practical method. However, burning natural gas reduces the efficiency of the storage system and increases its operational

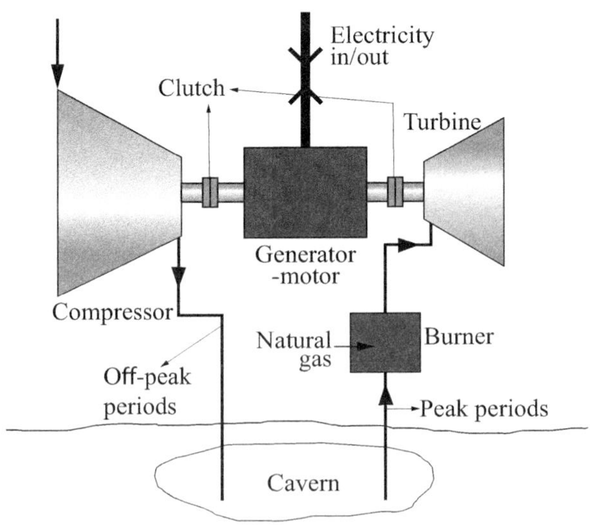

FIGURE 16.4
Schematic of a diabatic or hybrid compressed air energy storage system.

cost. Also, it partly eliminates the environmental benefits of storing electricity from renewable energy sources. Despite these problems, this technology is the only CAES system that has commercially been implemented. These plants are typically referred to as diabatic or hybrid CAES systems.

16.4.1 Compressed Air Containers

Unlike pumped hydro energy storage systems, CAES systems do not need two reservoirs with elevation difference and can be implemented with a single container in a flat region. However, they require enclosed air-tight containers. As the next example indicates, in order to have an even moderate energy storage capacity, a very large container is needed that can withstand high internal stress due to high air pressure. For this purpose, typically underground enclosed spaces, such as salt and hard-rock caverns, aquifers, or abandoned mines, have been proposed and used. In order to grasp the sheer size of containers needed to store compressed air in utility-size compressed air energy storage systems, let us have a real-world example.

Example 16.5

The world's first utility-size CAES system is the Huntorf plant built in 1978 in Germany (then West Germany). The plant has a power generation capacity of 290 MW and the energy storage capacity of 580 MWh. It takes 8 hours to fully charge the container. Assuming the ambient temperature and pressure of 25°C and 1 bar, respectively, the stored air pressure of 50 bar, and the turbine and compressor isentropic efficiency of 65%, determine how many hours the plant can operate at the full load, the temperature of the compressor and turbine outlet flows, the mass and volume of stored air, the mass flow rates of the compressor and the turbine, and overall efficiency of the system. Assume that the temperature of the compressed air is returned to its initial temperature before expansion by burning natural gas so the temperature of the compressor exit flow is equal to the temperature of the turbine inlet flow. Also, there is no pressure drop in the container and the compressed air temperature in the container is equal to the ambient temperature. Use the ideal gas assumption with the constant specific heat.

Givens

$$T_1 = 25°\text{C}$$

$$P_1 = 1\,\text{bar}$$

$$P_2 = 50\,\text{bar}$$

$$\text{Power generation capacity} = 290\,\text{MW}$$

$$\text{Energy storage capacity} = 580\,\text{MWh}$$

$$\eta_{\text{Turb}} = \eta_{\text{Comp}} = 0.65$$

Analysis

The equations and parameters used in this example are all explained in Chapter 9.

$$\text{Hours the plant can operate at full load} = \frac{\text{Energy storage capacity}}{\text{Power generation capacity}}$$

$$= \frac{580}{290} = 2\,\text{h}$$

The specific heat ratio and C_p of air are

$$C_p = 1004\,\text{J/kg K},\ k = 1.4,\ \frac{k-1}{k} = 0.2857$$

First, the temperature of air at states 2 and 4 should be determined

$$\left(\frac{T_2}{T_1}\right) = \left(\frac{P_2}{P_1}\right)^{\frac{k-1}{k}}$$

$$\left(\frac{T_{2s}}{25+273}\right) = \left(\frac{50}{1}\right)^{0.2857} \quad \rightarrow T_{2s} = 911.2\,\text{K}$$

$$\eta_{\text{Comp}} = \frac{h_{2s}-h_1}{h_2-h_1} = \frac{T_{2s}-T_1}{T_2-T_1} \rightarrow 0.65 = \frac{911.2-(25+273)}{T_2-(25+273)} \rightarrow T_2 = 1241.5\,\text{K}$$

Since it is assumed that the temperature of the compressor exit flow is equal to the temperature of the turbine inlet flow and since there is no pressure drop in the container,

$$T_2 = T_3 = 1241.5\,\text{K}$$

$$\left(\frac{T_{4s}}{T_3}\right) = \left(\frac{P_4}{P_3}\right)^{\frac{k-1}{k}}$$

$$\left(\frac{T_{4s}}{1241.5}\right) = \left(\frac{1}{50}\right)^{0.2857} \quad \rightarrow T_{4s} = 406.0\,\text{K}$$

$$\eta_{\text{Turb}} = \frac{h_3 - h_4}{h_3 - h_{4s}} = \frac{T_3 - T_4}{T_3 - T_{4s}} \rightarrow 0.65 = \frac{1241.5 - T_4}{1241.5 - 406.0} \rightarrow T_4 = 698.4\,\text{K}$$

$$W_{\text{Turb}} = m_{\text{Stored}} C_p \left(T_3 - T_4\right)$$

$$\text{Mass of stored air} = m_{\text{Stored}} = \frac{W_{\text{Turb}}}{C_p\left(T_3 - T_4\right)}$$

$$= \frac{580 \times 3600 \times 10^6}{1004(1241.5 - 698.4)} = 3.8 \times 10^6\,\text{kg}$$

$$\text{Volume of stored air} = V_{\text{Stored}} = \frac{mRT}{P}$$

$$= \frac{3.8 \times 10^6 \times 287 \times (25 + 273)}{50 \times 10^5} = 6.55 \times 10^4\,\text{m}^3$$

$$\text{Compressor air mass flow rate} = \dot{m}_{\text{Comp}} = \frac{m_{\text{Stored}}}{\text{Charging time}}$$

$$= \frac{3.8 \times 10^6}{8 \times 3600} = 133.0\,\text{kg/s}$$

$$\text{Turbine air mass flow rate} = \dot{m}_{\text{Turb}} = \frac{m_{\text{Stored}}}{\text{Charging time}}$$

$$= \frac{3.8 \times 10^6}{2 \times 3600} = 531.9\,\text{kg/s}$$

or

$$\dot{W}_{\text{Turb}} = \dot{m}_{\text{Turb}} C_p \left(T_3 - T_4\right) \rightarrow \dot{m}_{\text{Turb}}$$

$$= \frac{\dot{W}_{\text{Turb}}}{C_p\left(T_3 - T_4\right)} = \frac{290 \times 10^6}{1004(1241.5 - 698.4)} = 531.9\,\text{kg/s}$$

$$W_{\text{Turb}} = 580 \times 3600 \times 10^6 = 2.09 \times 10^{12}\,\text{J}$$

$$W_{\text{Comp}} = m_{\text{Stored}} C_p \left(T_2 - T_1\right) = 3.8 \times 10^6 \times 1004 \times (1241.5 - 298) = 3.63 \times 10^{12}\,\text{J}$$

$$\eta_{\text{Overall}} = \frac{W_{\text{Turb}}}{W_{\text{Comp}}} = \frac{2.09 \times 10^{12}}{3.63 \times 10^{12}} = 0.576 = 57.6\%$$

The actual Huntorf plant uses a salt cavern with a volume of approximately 300,000 m^3. It is equipped with a three-stage compressor with two intercoolers between stages and a two-stage turbine with a reheater between them. The compressed air is heated before the turbine by a natural gas burner. The reheater also uses a lower-pressure natural gas burner. The latest state-of-the-art components at the time from the gas turbine industry were utilized in this plant.

Note that in the above calculation, we did not include the energy of the combusted natural gas in our efficiency calculations. The actual round trip efficiency of the plant is about 42%. It should be noted that the volume of the needed container is a strong function of the compressed air pressure. For example, if we increase the air pressure in the container to 100 and 150 bar, the container volume reduces to 24,150 and 13,632 m^3, respectively. This means in this case, doubling and tripling the pressure reduce the volume by 2.4 and 4.8 times, respectively.

In the previous section, we discussed air temperature variations and how it can be handled. Now we can consider the air pressure and volume in air storage containers. A compressed air container can be constant volume or constant pressure. In a constant-volume container, as the name implies, the container's volume is constant while air pressure can vary. This means the boundaries of the container are fixed. These containers are typically simply underground voids with fixed walls. While these containers are physically simple, they can pose some challenges in the system operation. During charging of the container with high-pressure air, with an increase in the mass of the air stored in the container, the pressure of the air in the container increases. This means the compressor has to overcome a higher pressure in its outlet, which results in the compressor operation in off-design conditions with a lower-than-optimum efficiency. A similar case occurs when the system is discharging where the turbine has to operate with reducing pressure, the longer it runs. In these containers, the varying pressure also imposes a cyclic stress on the walls of the container effects of which should be carefully studied in the feasibility study for any such project.

Constant-pressure containers are typically connected to a water pond or lake on the surface. The air pressure in the space is kept constant by water displacement at the bottom of the space. During air storage, water is pushed out of the space, and during discharge, water enters the space.

When studying a potential underground space, parameters such as its size, geometry, and the possibility of air leakage, and the impacts of cyclic stress and fatigue on walls should be investigated. While existing underground spaces can provide convenient containers for compressed air storage systems, finding suitable potential sites, their characterization, and their modification can be challenging, including drilling to gain access to the space. Currently, the use of pipelines and underwater air energy storage systems are also under investigation. The latter are particularly attractive because they are just some flexible large air bags made of plastic-anchored underwater. A major advantage of these systems is that they operate under the constant pressure, the value of which depends on the depth of the storage system. These systems can be installed next to offshore wind farms to store their excess electricity generation.

16.5 Flywheel Energy Storage Systems

Flywheel energy storage (FES) systems are essentially made of a rotating disc (called a flywheel), and they store energy in the form of rotational kinetic energy. The same principle has been employed for centuries to smooth out the rotation of objects or generated power from potter's wheels and toys to early steam engines to internal combustion engines. In the latter application, a flywheel stores energy during the power stroke and releases some of that energy in the parts of rotation with no power production with the ultimate objective of creating a uniform power output. Having an engine with more cylinders (and thus more uniform power output) reduces the size of the required flywheel. There have been proposals (and actual vehicles) to entirely power a vehicle by a flywheel. In essence, they are like electric cars except that they store energy in a flywheel rather than a battery pack. In one such design that was implemented in Europe in the 1950s, a bus, called gyrobuses, was fully powered by a motor-driven flywheel. In this case, the flywheel was "charged" at terminals or stops by connecting the motor to an electric source. While this particular design was only used for a few years, since then several bus and locomotive prototypes have been designed and built in Europe and the U.S. Flywheels have been also used as regenerative braking systems in nonelectric hybrid cars as well as subway cars and elevators. NASA has developed FES units for space missions. In power generation industry, flywheels can be employed both by end-users and at the utility level. End-users can use flywheels as uninterrupted power supplies typically to supplement battery packs or to provide high load for a short period in some tests and research laboratories. In fact, some of the largest flywheels have been installed in physics laboratories around the world (Example 16.6). For utility-level applications, a few tens to several hundred flywheels have been employed to provide up to several tens of MW power. These units are typically used to maintain the quality of the grid power or as transitional power sources when the grid power is interrupted. There are only a handful of such plants typically as demonstrational and pilot plants.

FES systems are typically composed of four main components: flywheel, bearings, motor-generator set, and housing. Early systems typically used massive steel disks with mass ranging from many hundred kilogram to a few ten thousand kilogram rotating with the rotation speed of a few hundred to several thousand rpm. Newer more advanced systems are typically equipped with cylindrical rotors made of carbon-fiber composite connected to a shaft by a hub made of a metal, such as titanium. Carbon-fiber flywheels are lighter and stronger than metal disks and can store much higher energy.

The rotor or disk should be supported by bearings. Early flywheels used mechanical bearings, typically ball bearings. These bearings needed periodical inspection and replacement every few years. They also limited the maximum rotational speed that the rotor could spin. In modern systems, the flywheel is levitated by magnetic bearings. These bearings along with light carbon-fiber rotors and vacuumed housings (below) reduce friction to almost zero and substantially reduce system maintenance.

The FES system uses excess electricity to speed up the rotation of the rotor and extracts kinetic energy of the rotor when electricity is needed. For this purpose, all flywheels are equipped with a combined motor-generator set that can function as either a motor or a generator during charging and discharging periods, respectively. The flywheel and the motor-generator assembly can be connected via a central shaft. The entire system, typically (but not always) including the motor-generator assembly, is enclosed in a housing. The housing is typically vacuumed to eliminate the effect of drag force on the flywheel due to air friction. For this purpose, the housing should be properly sealed to avoid air leakage and is equipped with a vacuum pump. The housing should be strong enough to protect the adjacent environment (both personnel and equipment) from high-speed flying pieces of the system in case of system failure. With all technological advancements for the abovementioned components, modern flywheels can rotate at an extremely high speed of up to 100,000 rpm.

Generally, FES systems can have a round trip efficiency of about 90%. They are also very reliable. They can operate for several decades with a minimal required maintenance. More importantly, they can have several hundred thousands of charge–discharge cycles with no performance deterioration. They can also be quickly charged and discharged. Furthermore, their performance is not affected by ambient conditions, and they have minimal negative environmental impacts.

The kinetic energy of a rotating flywheel can be determined by the following equation:

$$\mathrm{KE} = \frac{1}{2} I\omega^2, \tag{16.10}$$

where *KE* is the kinetic energy stored in the flywheel (J), *I* is the moment of inertia of the object (kg.m^2), and ω is the angular velocity of the object (radian/s or rad/s). The moment of inertia for a solid disk (cylinder) and that for a hoop about their symmetry axes are

$$I_{\text{Disk}} = \frac{1}{2} mR^2 \tag{16.11}$$

$$I_{\text{Hoop}} = mR^2 \tag{16.12}$$

where m is the mass (kg) and R is the radius (m).

Example 16.6

The Joint European Torus (JET) in Oxfordshire, UK, is the world's largest operational magnetically confined plasma physics experiment to research on energy from nuclear fusion. To provide power needed for experiments (which can exceed 700 MW for a very short period), two flywheels are installed in the facility. Each flywheel weighs about 775,000 kg and has a diameter of 10 m. The flywheel spins with a rotational speed of up to 225 rpm.[7] Determine the maximum energy that can be stored in the flywheel (in GJ). During the experiment, the speed of the flywheel reduces to half of the maximum rpm in 20 seconds. Determine the energy released by the flywheel during the experiment (in GJ) and the average power (in MW).

Analysis

The maximum energy that can be stored in the flywheel is when it spins at the maximum speed of 225 rpm.

$$I_{\text{Disk}} = \frac{1}{2}mR^2 = \frac{1}{2}\times 775,000\times\left(\frac{10}{2}\right)^2 = 9,687,500 \text{ kg}\cdot\text{m}^2$$

$$\omega = \frac{\text{rpm}\times 2\pi}{60} = \frac{225\times 2\pi}{60} = 23.6\ \frac{\text{rad}}{\text{s}}$$

$$\text{KE} = \frac{1}{2}I\omega^2 = \frac{1}{2}\times 9,687,500\times 23.6^2 = 2.69\times 10^9\ J = 2.69\,GJ$$

During the experiment, the speed of the flywheel reduces by half, so

$$\omega_1 = 23.6\,\frac{\text{rad}}{\text{s}}$$

$$\omega_2 = 11.8\ \frac{\text{rad}}{\text{s}}$$

$$\Delta\text{KE} = \frac{1}{2}I\left(\omega_1^2 - \omega_2^2\right) = \frac{1}{2}\times 9,687,500\times\left(23.6^2 - 11.8^2\right) = 2.02\times 10^9\ J = 2.02\,\text{GJ}$$

$$\text{Average power} = \frac{\Delta\text{KE}}{t} = \frac{2.02\times 10^9}{20} = 0.1008\ \text{GW} = 100.8\,\text{MW}$$

The actual system can provide a maximum power of 400 MW.

[7] http://aei.pitt.edu/94542/1/jet_joint..__50695.pdf.

Example 16.7

A power plant with a capacity of 1500 MW operates for 30 minutes at the full load. Determine the number of flywheels from Example 16.6 that are needed to store this much energy.

Analysis

$$\text{Energy produced by the power plant} = 1500 \times 0.5 = 750\,\text{MWh} = 2.7 \times 10^{12}\ J$$

$$n = \frac{\text{Total energy storage}}{\text{Energy stored in each flywheel}} = \frac{2.7 \times 10^{12}}{2.69 \times 10^{9}} = 1004\,\text{Flywheels}$$

According to Equation 16.10 in order to increase the energy stored in a flywheel, one can increase the rotational speed of the flywheel and/or its moment of inertia. The moment of inertia can be increased by increasing the mass of the flywheel and more importantly its radius. In the first generation of FES systems, the strategy to increase the stored energy was to build massive rings using materials with high density. While this approach increased the moment of inertia of the flywheel, it severely limited the rotational speed of the flywheel. Using large disks made of high-density materials at high speed not only increases the load on bearings but also imposes an enormous centrifugal force on the disk, which can result in its disintegration. It can be shown that the maximum rotational speed and the maximum stored energy per unit of mass (maximum specific energy) for a flywheel are a function of the ratio of the allowable stress to the density of its material $\left(\frac{\sigma}{\rho}\right)$. This means a material with high strength and low density is desirable to maximize the ratio of σ/ρ. It should be noted that other parameters such as resistance to cyclic loads are also important.

Example 16.8

Compare high-strength steel alloy ASTM A514 with carbon nanotube (CNT) in terms of the maximum specific energy they can achieve when used to make FES systems. The tensile strengths of the steel alloy and CNT are 760 and 60,000 MPa, respectively, and their densities are 7800 and 1300 kg/m^3, respectively.

Analysis

The maximum specific energy these materials can achieve when used to make FES systems is a function of the ratio of the tensile stress to density.

$$\left(\frac{\sigma}{\rho}\right)_{\text{Steel}} = \frac{760,000}{7800} = 97.4$$

$$\left(\frac{\sigma}{\rho}\right)_{CNT} = \frac{60,000,000}{1300} = 46,153.9$$

This indicates why currently flywheels for FES systems are being made of CNT rather than steel.

16.6 Battery Storage Systems

Batteries are energy storage devices that **directly** store electricity by means of **electrochemical reactions.** In Chapter 1, we discussed about the Parthian battery, which has been speculated that was used to generate a very small amount of electricity. We also learned that in the 1800s, Alessandro Volta pioneered perhaps the first electricity generation device.

Generally, batteries can be **non-rechargeable** (also known as **primary**, single-use, or disposable batteries) or **rechargeable** (**secondary** batteries). Non-rechargeable batteries are those "charged" during manufacturing. When the stored energy is depleted, since the electrochemical reactions are irreversible, these batteries have no use and should be disposed of. Alkaline and zinc-carbon batteries are the most common type of primary batteries and are widely used to power many household and industrial small electronic devices, from sensors to toys. Since these batteries cannot be used to store and recover energy repeatedly, they are out of the scope of this text and will not be covered any further. There is a wide variety of rechargeable batteries in terms of size, shape, and technology with many applications in our daily life. Some common applications that we are all familiar with are cordless electric devices and tools; automotive applications such as starting units for conventional internal combustion engines and power source for electric and hybrid cars; portable electronic devices such as laptops, tablets, smartphones, and cameras; backup and emergency power applications; energy storage for renewable energy systems, particularly PV units in remote applications; and particularly of interests for this text large-scale utility-level energy storage.

The amount of energy that can be stored in a battery is referred to as the capacity of the battery. It is measured as the delivered power at the rated voltage of the battery with the unit of **amp-hour** (**A·h**). In any battery energy storage system, the fact that they store and generate DC power should be kept in mind because it means in the majority of applications, certainly those in connection with the grid need to be equipped with an inverter (Chapter 11).

The most common and the oldest type of these batteries are **lead-acid batteries.**[8] For this reason, we will start with this type of battery and explain

[8] The first lead-acid battery was invented by Gaston Planté (1834 – 1889, a French physicist).

the basic principles of their operation. Later, we will introduce other battery types and explain how they are similar to and different from lead-acid batteries.

16.6.1 Lead-Acid Batteries

Most batteries are composed of several typically identical units, called **cells**. Since each cell can generate a limited voltage and current, the cells can be connected in series or parallel to achieve the desired output voltage and current. Every battery cell needs three main components to operate: an **anode electrode** (the negative terminal), a **cathode electrode** (the positive terminal), and an **electrolyte**. The surfaces of the electrodes are the site of electrochemical reactions where negative and positive charges are created. The electrolyte is typically a liquid mixture of a salt in a solvent. The main characteristic of an electrolyte is that while it allows the passage of certain ions, it does not permit the flow of electrons. The choice of the electrolyte is directly a function of the desired electrochemical reactions. To increase the surface area of the sites of electrochemical reactions, both the anode and cathode electrodes are made of highly porous materials. In lead-acid batteries, the electrolyte solution consists of typically 63% water and 37% sulfuric acid (mass base) with a specific gravity of 1.28.[9] Note that when sulfuric acid and water are mixed, the solution is primarily made of the bisulfate negative ions $\left(HSO_4^-\right)$ and hydrogen positive ions (H^+ or protons). Batteries also need current collectors to guide electric charges to the terminals.

While dividing a cell into these three components can help in explaining the physical structure of a cell, to understand electrochemical reactions within a cell, it is more appropriate to divide it into two half-cells in series. Each half-cell is composed of one electrode and the electrolyte adjacent to it.

To explain the electrochemical reactions that take place in a lead-acid battery, let us consider a case when the battery is fully charged and is disconnected from any load. In this case, the anode is a lead (Pb) plate, the cathode is a lead oxide (PbO_2) plate, and the electrolyte is a solution of water and sulfuric acid (H_2SO_4). When a load is connected to the battery and the battery is discharging, lead in the anode reacts with the negative ions in the electrolyte and releases electrons via the following half-reaction:

$$Pb(s) + HSO_4^-(aq) \rightarrow PbSO_4(s) + H^+(aq) + 2e^- \text{ (anode half reaction)} \quad (16.13)$$

Note that in the above reaction equation, the parentheses indicate the phase of reactants and products ("s" for solid and "aq" for aqueous). The electrons generated in this electrochemical reaction give the lead electrode negative charge, which is why it is called the anode or the negative terminal. They flow through the external load to the cathode forming an electrical current.

[9] For a liquid, specific gravity, also known as relative density, is the ratio of the density of the liquid to the highest density of water, which is at the temperature of 4°C.

The generated positive ions pass through the electrolyte to the cathode where they react with the lead oxide via the following half-reaction:

$$PbO_2(s)+HSO_4^-(aq)+3H^+(aq)+2e^-$$
$$\rightarrow PbSO_4(s)+2H_2O \quad \text{(cathode half reaction)} \tag{16.14}$$

Note that if no load is connected to the battery, the negative charges will accumulate in the anode and will repel negative ions $\left(HSO_4^-\right)$ from the anode. This will stop the reaction from taking place until the load is connected and the electrons are removed from the anode. The voltage created by these accumulated electrons is referred to as the **open-circuit voltage (OCV)**. Combining the two half-reactions above, the overall cell reaction can be written as follows:

$$Pb\ (s)+PbO_2\ (s)+2H_2SO_4\ (aq)$$
$$\rightarrow 2PbSO_4\ (s)+2H_2O \quad \text{(overall cell reaction)} \tag{16.15}$$

The chemical energy of the reactants in this reaction is higher than that of the products. The difference between the energy level of the reactants and products is the electrical energy generated by the battery (according to the first law of thermodynamics). The amount of generated electricity depends on the electrochemical reaction that takes place within the battery and the number of molecules reacting. The voltage created per each released electron during the reaction is called the **electromotive force (EMF)**. This voltage is equal to the OCV. The EMF is only a function of the reaction taking place in the battery, which in turn is dictated by the material used for the electrodes and the electrolyte. However, it is independent of the size and shape of the battery. For lead-acid batteries, the EMF is about 2.1 V. We will see later that when a current is drawn from a battery, its voltage reduces.

Equations 16.13 and 16.14 indicate that with more energy extraction from the battery (battery being discharged), more Pb and PbO_2 are converted to $PbSO_4$ in the anode and cathode, respectively, and more sulfuric acid is converted to water in the electrolyte. The latter means while the battery is discharged, the electrolyte becomes more diluted (i.e., the mixture of more water and less acid). For each mole of the converted Pb (and PbO_2), two moles of sulfuric acid is converted to water. The lowest concentration will be when the battery is fully discharged. For this reason, the concentration of acid in the electrolyte can be used as an indication of the level of charge of the battery. For this purpose, the density (or more commonly specific gravity) of the solution is measured using a hydrometer.[10] Since the density of water is lower than that of the acid, a high density of the solution means there are more acid and less water in the solution. The composition of the electrolyte can play an important role

[10] For example, in battery-powered submarines to determine how much longer they can possibly stay under water.

in the operation of batteries at very low temperatures since a higher concentration of acid means the electrolyte will freeze at lower temperatures. So in these environments, it is advisable to keep a battery fully charged.

When a battery is partly or fully discharged, it can be recharged by connecting its terminals to an external source of electricity (called a **charger**). It should be noted that a DC power source is needed to charge a battery. During the charging process, the reactions expressed in Equations 16.13 and 16.14 are reversed. This results in the conversion of $PbSO_4$ to Pb in the anode and PbO_2 in the cathode. In the electrolyte, water is converted to sulfuric acid. Batteries should be protected from overcharging, which can result in water electrolysis and the generation of hydrogen and oxygen. If these gases have a way to escape (e.g., in unsealed batteries), the water level should be checked regularly and losses should be replaced.

The electrical power generated by a battery depends on the voltage difference between its terminals and the current drawn from it by the load. The voltage depends on the electrochemical reaction taking place in the battery. The current, however, depends on the quantity of the reaction, which in turn is a function of the surface area of the electrodes. To increase the surface area of the electrodes, as noted earlier, they are made of porous substances. In lead-acid batteries, the electrodes are formed by grids or matrixes made of hardened lead or a lead alloy filled with paste-like metallic Pb and PbO_2 (in a fully charged state). Several electrode plates are assembled in a parallel formation (physically) connected in series (electrically). They are housed in a compartment to form a cell. A collection of cells makes a battery. To avoid the electrodes touching each other and creating electrical short circuit, a separator is placed between them. In fact, batteries are typically manufactured using electrodes in their discharged state so that they do not degrade during their shelf time before they are actually installed for a certain application. So batteries you see in stores are typically discharged and must be recharged after or right before they are installed.

In old lead-acid batteries, the electrolyte was typically a liquid solution so they were referred to as **wet or flooded cells** or batteries. However, the electrolytes for more recent lead-acid batteries are gel-like paste and hence, these batteries are named **gel batteries** or **dry cells**. Alternatively, the electrolyte can be absorbed in a fiberglass substrate. In this case, the battery is known as **absorbed glass mat** (AGM). These batteries are typically sealed to avoid the escape of hydrogen and oxygen, for example, generated in the case of overcharging. The generated gases are recombined to form water, which means these batteries do not need the addition of water or other chemicals. That is why they are sometimes referred to as **maintenance-free batteries**. However, the excessive accumulation of gases can result in overpressurization of the battery casing and possible explosion. To release the gases in these situations, sealed batteries are typically equipped with a one-way valve, which is why they are also referred to as **valve-regulated lead-acid** (**VRLA**) batteries. They have a longer shelf life with less possibility of leakage. They are particularly suitable for applications where batteries operate under severe stress, vibration, and temperature variations.

As noted, when a battery is not connected to any load, the voltage difference between its terminals is equal to the OCV (and the EMF). However, this voltage reduces when the battery is used to power a device. This is due to the internal resistance of the battery. Let us consider a battery with the OCV of V_{OC} and the internal resistance of r that is connected to an external resistance (load) of R (Figure 16.5).

For the circuit in Figure 16.5,

$$V = V_{OC} - \mathrm{Ir} = \mathrm{IR} \tag{16.16}$$

$$I = \frac{V_{OC}}{(R+r)} \tag{16.17}$$

where V is the voltage difference between two terminals of the battery (V), V_{OC} is the open-circuit voltage (and the EMF) (V), I is the current (A), r is the internal resistance of the battery (Ω), and R is the external resistance of the load (Ω). Equation 16.16 indicates that the voltage difference between the two terminals of the battery is always less than the OCV (and the EMF) whenever a load is connected to the battery. The difference between these two voltages can be reduced by reducing r or increasing R. The current is maximum when $R=0$, which is the case in a short circuit. The capacity of a battery is the amount of electrical energy that it can provide. We know that electrical power (P in W or kW) and energy (E in Wh or kWh) can be calculated from the following equations:

$$P = VI \tag{16.18}$$

$$E = VIt \tag{16.19}$$

where t is the time that the battery can operate (hour). Technically speaking, E in the above equation is the energy storage capacity of a battery. However,

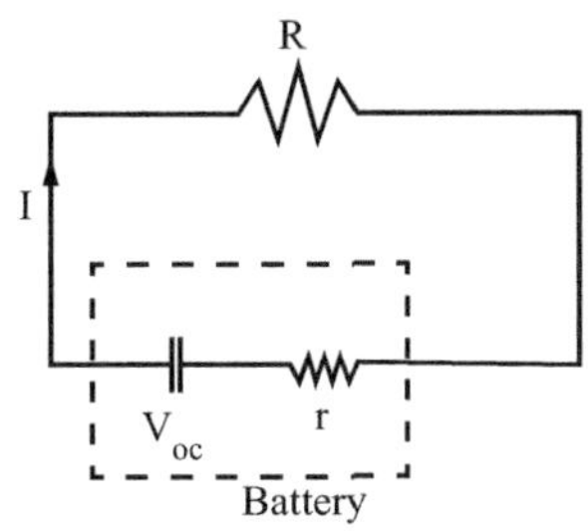

FIGURE 16.5
Schematic of a battery connected to a load.

as discussed before, the voltage of a battery is pretty much fixed by its electrochemical reactions and is known for each battery type. For this reason, typically the capacity of batteries is expressed by the current multiplied by time with the unit of A.h. It should be noted that the value of this parameter varies with the rate of battery discharge. A higher rate of discharge decreases the capacity of a battery. Also, the capacity of a battery reduces at lower temperatures. To have a uniform definition, the **capacity of batteries is typically measured at 20°C in a 20-hour interval**. This parameter, typically referred to as "**capacity 20hr**" (capacity at 20hr), can be determined by multiplying 20 hours to the maximum consistent current that a new battery can generate for a 20-hour period.

Example 16.9

For a battery with the OCV of V_{OC} and the internal resistance of r, determine the value of the external resistance R that can maximize the power drawn from the battery.

Analysis

We know that the power drawn from the battery is equal to the voltage difference between the battery terminals multiplied by the current ($P = VI$). From Equations 16.16 and 16.17, it is clear that increasing R decreases the current but increases the voltage. These two conflicting impacts indicate that there should be a value for R that maximizes the power. To determine this value, we can differentiate power as a function of R and equate that to zero

$$P = VI = I^2 R = \frac{R\, V_{OC}^2}{(R+r)^2}$$

$$\frac{dP}{dR} = \frac{V_{OC}^2 (R+r)^2 - 2(R+r) R\, V_{OC}^2}{(R+r)^2} = 0$$

$$V_{OC}^2 (R+r)^2 - 2(R+r) R\, V_{OC}^2 = 0$$

$$(R+r) - 2R = 0$$

$$R = r$$

Example 16.10

The voltage between two terminals of a battery is 24 V without a load and 23 V when a current of 20 A is drawn from the battery. Determine the internal resistance and OCV of the battery as well as the resistance and power consumption of the load. If the capacity of the battery is 100 A.h, for how long the battery can provide the given current?

Analysis

$$V_{OC} = 24\ V$$

$$V = IR\ R = \frac{V}{I} = \frac{23}{20} = 1.15\ \Omega$$

$$I = \frac{V_{OC}}{(R+r)}\ r = \frac{V_{OC}}{I} - R = \frac{24}{20} - 1.15 = 0.05\ \Omega$$

$$P = VI = 23 \times 20 = 460\ W$$

$$\text{Capacity} = \text{It} = 20t \quad t = \frac{100}{20} = 5\,\text{h}$$

Example 16.11

In the previous example, determine the resistance of the load that can draw the maximum power from the battery. What is the value of the maximum power? For how long the battery can provide the given current?

Analysis

$$R = r = 0.05\ \Omega$$

$$I = \frac{V_{OC}}{(R+r)} = \frac{24}{(0.05+0.05)} = 240\ A$$

$$V = V_{OC} - Ir = 24 - 240 \times 0.05 = 12\ V$$

$$P = VI = 12 \times 240 = 2880\ W$$

$$\text{Capacity} = \text{It} = 20t\,t = \frac{100}{240} = 0.42\,\text{h}$$

Example 16.12

In the previous example, determine the maximum current that the battery can provide.

Analysis

The maximum current can be achieved for a short circuit when $R=0$:

$$I = \frac{V_{OC}}{(R+r)} = \frac{24}{(0+0.05)} = 480 \text{ A}$$

Example 16.13

A multimeter is directly connected to two terminals of a battery. As a voltmeter, it reads 12 V, and as an ammeter, it reads 200 A. Determine the internal resistance and V_{OC} of the battery.

Analysis

$$V_{OC} = 12 \ V$$

$$I_{max} = \frac{V_{OC}}{r} \quad r = \frac{V_{OC}}{I_{max}} = \frac{12}{200} = 0.06\,\Omega$$

Example 16.14

A power plant with a capacity of 1500 MW operates for 30 minutes at the full load. This energy is stored in a battery pack made of batteries, each with a capacity of 90 A.h and a voltage of 24 V. Determine the number of batteries needed for this purpose.

Analysis

$$\text{Energy produced by the power plant} = 1500 \times 0.5 = 750 \text{ MWh} = 2.7 \times 10^{12} \text{ J}$$

$$\text{Energy stored in each battery pack} = V(\text{It}) = 24 \times 90 = 2160 \text{ Wh}$$

$$= 2160 \times 3600 = 7.78 \times 10^{6} \text{ J}$$

$$\text{Energy stored in the battery pack} = nV(\text{It})$$

$$n = \frac{\text{Energy stored}}{V(It)} = \frac{2.7 \times 10^{12}}{7.78 \times 10^{6}} = 3.47 \times 10^{5} \text{ Batteries}$$

This is the challenge with battery storage systems. In order to store energy generated in a midsize power plant in just half an hour, about 350,000 batteries are needed.

Example 16.15

In Example 16.6, how many batteries with a capacity of 65 A.h and the voltage of 12 V are needed to store the same amount of energy (for each flywheel when its speed is reduced by half).

Analysis

$$\text{Energy stored in each flywheel} = 2.02 \times 10^9 \text{ J}$$

$$\text{Energy stored in each battery pack} = V\,(\text{It}) = 12 \times 65 = 780 \text{ Wh}$$

$$= 780 \times 3600 = 2.8 \times 10^6 \text{ J}$$

$$\text{Energy stored in the battery pack} = nV\,(\text{It})$$

$$n = \frac{\text{Energy stored}}{V\,(It)} = \frac{2.02 \times 10^9}{2.8 \times 10^6} = 718 \text{ Batteries}$$

While these many batteries may seem manageable and certainly less expensive, they cannot replace the flywheel. The flywheel can release this much energy in a matter of seconds (10 seconds in this case), whereas it may take hours for the batteries to release their energy.

Lead-acid batteries can have a round trip efficiency of 50%–95%. They can also have 10–40 Wh/kg specific energy and 50–90 Wh/L energy density. These values are relatively low but the low cost of lead-acid batteries along with high current they can generate makes them popular in many applications. These batteries can have a very long service life too if operated properly.

In automotive applications, lead-acid batteries are used for starting, lighting, and ignition (SLI). These applications require a high surge current, particularly during an engine start-up. As noted above, a high-current density can be achieved when the internal resistance of the battery is low. For this reason, these batteries are made of thin electrode plates.

The service life of a battery strongly depends on the number of complete discharge (called **deep discharge**) and charge cycles the battery can go through. The thicker electrode plates generally can have more cycles since they contain more reacting substances. Thin electrolyte lead-acid batteries can only have a couple of hundreds of deep discharges. However, in automotive applications, batteries are typically recharged after slight discharge and they are rarely fully discharged. Therefore, thin electrolyte batteries perfectly fit these applications.

There are other applications where batteries have to go through many deep discharge cycles. These applications include batteries for electric cars and, more importantly for this text, energy storage systems. These batteries are generally referred to as **deep-cycle lead-acid batteries**. They have much thicker electrolyte plates compared to automotive batteries, which means they have more active electrode materials but lower peak currents (due to their high internal resistance). Some deep-cycle batteries have tubular or cylindrical positive electrodes to increase their surface area and power density.

Lead-acid batteries have been very well developed for many decades and are mature technologies for large-scale energy storage systems. However, other battery types are also being used or investigated with the potentials to reduce system initial and operation costs as well as the size with better efficiency and reliability. In the following sections, some of these batteries will be reviewed.

16.6.2 Lithium-Ion Batteries

Lithium-ion batteries, also known as **Li-ion batteries** or **LIBs**, were commercially developed in the early 1990s after decades of research and development. These batteries have a high round trip efficiency, very high energy density, long service life, and slow self-discharge. Their drawbacks are their high cost and some safety concerns.[11] Regardless of their disadvantages, due to their favorable characteristics, they soon became standard batteries in portable electronic devices, such as cell phones, laptops, and tablets, and later for electric and hybrid vehicles. These batteries have demonstrated promising potentials for utility-level energy storage systems.

In lithium-ion batteries, the cathode and anode electrolytes (the positive and negative electrolytes, respectively) are made of some lithium compounds, such as lithium cobalt oxide, $LiCoO_2$, and graphite, respectively. The electrolyte is a lithium salt solution that allows the passage of Li^+ ions while blocking electrons. The choice of the cathode material depends on the desirable characteristics of a battery, for example, energy density versus safety and service life. The operating cell voltage of these batteries is 3–4 V with the OCV of 4.1 V. Their energy density and specific energy are about 400 Wh/L and 150 Wh/kg, respectively, up to about an order of magnitude higher than those of lead-acid batteries. Their round trip efficiency is about 80%–90%. Typically, lithium-ion batteries are made in the form of standard cells and then are assembled in packs to achieve desirable current, voltage, and storage capacity.

While currently, lithium-ion batteries are too expensive for widespread applications for large-scale energy storage systems, there are great hopes and potentials for them. In the past few years, there have been some pilot plants of large-scale energy storage systems using these batteries. For example, in 2017 the world's largest lithium-ion battery storage system was installed in South Australia with a storage capacity of 129 MWh (100 MW power) sufficient to power 30,000 homes for an hour.[12] The facility was

[11] When these batteries are overheated, they can ignite or even explode. Probably the most high-profile manifestation of this problem was recalling of about 2.5 million Samsung products in Galaxy series.

[12] https://www.bbc.com/news/world-australia-42190358#:~:text=The%20world's%20largest%20lithium%20ion,caused%20by%20local%20hot%20weather.

designed and constructed by Tesla using the same technology that is used in electric cars' batteries. There are some unorthodox proposals on using this type of battery. In one such proposal, lithium-ion batteries used in electric and hybrid automobiles are repurposed for large-scale energy storage systems when vehicles are retired. Nowadays electric and hybrid cars are gaining popularity. The fact that most of today's electric and hybrid cars will be retired in a decade or so provides a unique source of batteries for this proposal. It should be noted that the large-scale energy storage industry does not require the level of reliability that is needed for electric-powered transportation systems. Another similar proposal is to use batteries currently in service electric and hybrid cars to provide distributed energy storage capability at residential units. In this concept, residents of a home can draw all or part of their electricity consumption during peak periods from energy stored in their car's batteries and then recharge the batteries during off-peak periods.

16.6.3 Molten-Salt Batteries

Molten-salt batteries are a group of batteries with molten salt as the electrolyte. This concept can be applied to both non-rechargeable and rechargeable batteries. In the non-rechargeable form, a molten-salt battery can be stored at room temperature in its solid inactive state for decades. To operate the battery, it should be heated to a certain temperature. For this reason, this type of battery is also known as **thermal batteries**.

Rechargeable molten-salt batteries are considered as a promising technology for large-scale energy storage systems. They generally have high efficiency, high energy and power densities, long service and cycle life, and potential for low cost. The two most developed batteries are sodium-sulfur and sodium-nickel chloride batteries. Here, the former will be discussed as a case study.

In sodium-sulfur (NaS) batteries, the anode, cathode, and electrolyte are made of sodium (liquid), sulfur (liquid), and alumina (solid), respectively. Sodium-sulfur cells are typically built in a large vertical cylinder shape. The molten sodium electrode is in the core of the cell surrounded by the solid electrolyte. The electrolyte is alumina ceramic (known as β-alumina or β-Al_2O_3) containing a variety of mobile ions. The outer layer is made of the molten sulfur electrode enclosed by the external metallic walls of the battery. To keep the electrodes molten, the cell operates at an elevated temperature of about 350°C although the most advanced low-temperature batteries can operate at a temperature of as low as 100°C. The operating cell voltage of these batteries is 1.8–2 V with the OCV of 2.1 V. Their energy density and specific energy can be as high as 350 Wh/L and 200 Wh/kg, respectively, which is comparable with those of lithium-ion batteries. Some of the largest battery energy storage installations in the world are of this type of battery. The technology is even being considered (and tested) for space applications.

16.6.4 Nickel-Based Batteries

In all nickel-based batteries, the cathode electrode is nickel oxide hydroxide (NiO(OH)) and the electrolyte is potassium hydroxide solution. However, using different anodes results in different types of batteries. In nickel-metal hydride (NiMH) batteries, the anode electrolyte can be selected from a wide variety of hydrogen-absorbing alloys, such as certain alloys of cerium, nickel, aluminum, vanadium, zirconium, and cobalt. Nickel-cadmium (NiCd) and nickel-zinc (NiZn) batteries have metallic cadmium and zinc, respectively, as the anode electrode. Some nickel-based batteries had a significant portion of the portable electronic device market until the 1990s. But they lost most of this market in competition with lithium-ion batteries although NiMH batteries are still being used in some devices and some hybrid vehicles. However, some of these batteries are being considered and used for large-scale energy storage systems, particularly NiCd batteries.

16.6.5 Other Battery Types

Considering a variety of materials that can be used as the electrodes and electrolyte in batteries, it should not come as a surprise that there are many concepts and battery types at different stages of research and development or commercialization. Here, we just list some of them in case one would like to look them up:

- Solid-state batteries
- Liquid-metal batteries
- Flow batteries, such as vanadium redox batteries (VRBs)
- Silver zinc batteries (AgZn)
- Lithium iron phosphate batteries (LFP)
- Lithium sulfur batteries (LiS)
- Aluminum ion batteries
- Magnesium ion batteries
- Carbon batteries
- Metal-air electrochemical cells

16.7 Direct Electricity Storage Systems

In all energy storage systems we discussed so far, electrical energy was converted to other forms of energy before being stored, for example, gravitational potential energy (in PHES systems), air pressure (in CAES systems), kinetic energy (in FES systems), and chemical energy (in batteries). Unlike these systems, in direct electricity storage systems, electrical energy is directly stored in electromagnetic fields with no need for the conversion of electrical energy to other forms of energy for storage. Two devices that can be used for this purpose are supercapacitors and superconducting coils.

16.7.1 Supercapacitor Energy Storage Systems

The operational principles of supercapacitors or ultracapacitors are similar to those of conventional capacitors that have been widely used for over a century in electronic devices. Every capacitor is composed of an electrical insulator, called a **dielectric**, sandwiched between two **positive or negative conductors**. The positive and negative charges accumulated on the conductors form an electric field in the space between them, which stores electrical energy. This electric field forces the molecules (or atoms) of the dielectric to align themselves at the opposite direction of the field, which weakens the field. This allows more charges to be stored in the conductors. A dielectric can be a layer of plastic, ceramic, or just air.

The ability of a capacitor to store electrical charges is characterized by its capacitance (C), measured in Farads (*F*),[13] and is expressed by the following equation (Figure 16.6):

$$C = \frac{Q}{V} \tag{16.20}$$

where *C* is the capacitance (*F*), *Q* is the charge stored in each conductor (positive or negative) (coulomb, C), and *V* is the voltage across two conductors (*V*). The electrical energy stored in a capacitor can be determined by the following equation:

$$E = \frac{1}{2}CV^2 \tag{16.21}$$

where *E* is the stored electrical energy (*J*). To increase the energy stored in a capacitor, we can increase the voltage in which the conductors are connected to. But any electrical insulator (used as a dielectric) will become a conductor if it is exposed to a strong enough electromagnetic field. So the voltage across two conductors is limited and the maximum voltage depends on the type of the dielectric and the distance between the conductors (or the thickness of the dielectric). So the only way that the energy stored in a capacitor can be realistically increased is by increasing its capacitance. Now let us see how we can increase the capacitance. For a parallel-plate capacitor (Figure 16.6), the capacitance is expressed by the following equation:

[13] After English scientist Michael Faraday (1791–1867) who was pioneer in the fields of electromagnetism and electrochemistry. He is considered one of greatest scientists of all time.

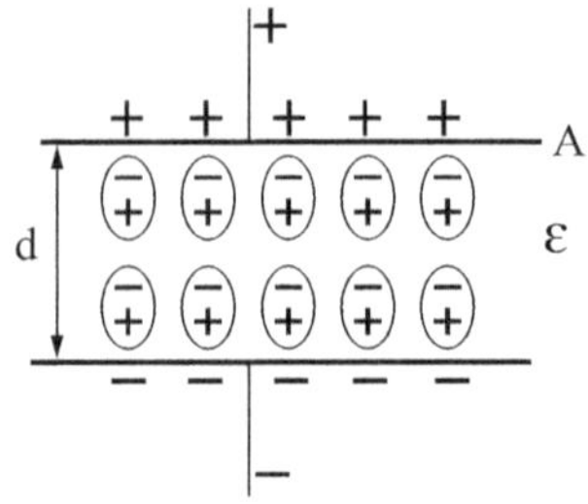

FIGURE 16.6
Schematic of a capacitor.

$$C = \frac{\varepsilon A}{d} \tag{16.22}$$

where ε is the permittivity of the dielectric (F/m), A is the area of the conductors (m^2), and d is the distance between the conductors (m). This equation indicates that to increase the capacitance of a capacitor, we can either increase the area of the conductors, increase the permittivity of the dielectric, or decrease the distance between the conductors.

In supercapacitors, conductors are composed of metallic thin plates coated with a porous material, typically a carbon-based material such as activated carbon, graphene, or CNTs. The pores significantly increase the effective area of the conductors. The most important phenomenon that causes a substantial increase in a supercapacitor's capacity to store electricity is how the gap between the conductors is filled. Unlike conventional capacitors that have a relatively thick dielectric, the space between conductors in supercapacitors is filled with an electrolyte. In the middle, there is a very thin separator made of a porous insulator, such as paper; polymer film; or carbon, glass, or ceramic fibers, to allow the passage of electrolyte's ions. The electrolyte is a liquid solution of salts in water or organic solvents that contains positive and negative ions. This electrolyte fills the pores of the conductors. When a supercapacitor is connected to a source of electricity and is charged, the electric field created between the conductors separates the positive and negative ions on two sides of the separator. This is called an electric double layer. Consequently, supercapacitors are commonly referred to as **double-layer capacitors** or **electric double-layer capacitors** (EDLC). Therefore, the energy storage process in supercapacitors involves **both electrostatic and electrochemical processes**. It should be noted that in supercapacitors, the gap between conductors is dramatically less than that in conventional capacitors. The thickness of an electric double layer can be just a few molecules. This not only increases the capacitance on its own, but also allows a huge area of conductors to be compressed in a small space, which further increases the capacitance. A side impact of the reduced gap is a lower permissible maximum voltage for supercapacitors compared to that for conventional capacitors. However, the overall impact is substantial increase in the electrical energy that can be stored in supercapacitors.

The overall impact of the abovementioned factors is that supercapacitors have much higher specific energy and energy density than conventional capacitors (we are talking about up to billions of times more!). However, even the best supercapacitors have specific energy substantially smaller than some other energy storage systems. The characteristic that distinguishes supercapacitors from some other energy storage systems is their ability to charge and discharge almost instantaneously. In other words, while their specific energy is low, their specific power is very high. Moreover, they can go through charging and discharging cycles hundreds of thousand times in the span of 10 years or more with an extraordinary round trip efficiency of up to 98%. These attributes make them an ideal choice for applications where **frequent and rapid charging and discharging are needed**, such as power quality management and regenerative braking. They are also useful as a voltage stabilizer for wind turbines and solar photovoltaic units when there is a brief fluctuation in power generation. It should be noted that supercapacitors require DC power to charge and they discharge DC power so AC/DC and DC/AC inverters are needed when they operate with AC power from the grid.

Example 16.16

Large commercial supercapacitors can have a capacitance of several thousand Farad. Consider a 0.15 kg supercapacitor with the capacitance of 4000 F that is charged with a 5 V electricity. Determine the energy stored in the supercapacitor (in J) and its specific energy (in Wh/kg). Compare the stored energy in the supercapacitor with that in a lead-acid battery with a capacity of 65 A.h and a voltage of 12 V.

Solution

$$E_{\text{Supercapacitor}} = \frac{1}{2}CV^2 = \frac{1}{2} \times 4000 \times 5^2 = 5 \times 10^4 \text{ J}$$

$$\text{Specific energy of supercapacitor} = \frac{E_{\text{Supercapacitor}}}{m_{\text{Supercapacitor}}} = \frac{5 \times 10^4}{0.15}$$

$$= 3.33 \times 10^5 \frac{\text{J}}{\text{kg}} = 92.6 \frac{\text{Wh}}{\text{kg}}$$

$$E_{\text{Battery}} = V(\text{It}) = 12 \times 65 = 780 \text{ Wh} = 780 \times 3600 = 2.8 \times 10^6 \text{ J}$$

This example shows that even the largest single supercapacitors can only store a small fraction of what a lead-acid battery can store. While the specific energy of this supercapacitor is greater than that of lead-acid batteries, it is still much smaller than that for lithium-ion batteries.

16.7.2 Superconducting Coils Energy Storage Systems

We know that electric coils store electrical energy when an electric current passes through them by means of creating an electromagnetic field. The magnitude of this energy depends on the current and the inductance of the coil, and can be found by the following equation:

$$E = \frac{1}{2} L I^2 \tag{16.23}$$

where E is the stored electrical energy (J), L is the coil inductance (Henry, H), and I is the current passing through the coil (A). A problem with conventional coils is that they lose a significant amount of energy due to their electrical resistance against the current. The efforts to reduce electrical resistance and thus the loss of energy in electric coils lead us to the concept of superconductivity. In the early 20th century, Kamerlingh Onnes (1853–1926, a Dutch physicist and Nobel laureate) discovered that when the temperature of some materials is brought to lower than a certain temperature, the resistance abruptly drops to almost zero. This temperature, called the transition or critical temperature, is typically in the cryogenic range, i.e., lower than –150°C.

In **superconducting magnetic energy storage** (**SMES**) systems, electrical energy can be stored in a coil made of a superconductor operating at a very low temperature. This device can store energy with very little to no loss. These systems enjoy many of the characteristics of supercapacitors, such as almost instantaneous charge and discharge, very high efficiency, long life span, and almost unlimited charge and discharge cycles with no performance degradation. For these reasons, they share some applications with supercapacitors. However, since they require a substantial cooling facility to reduce the temperature of the coil, they are not suitable for small-scale applications. Similar to supercapacitors, superconducting coils operate with DC power so inverters are needed when they are used to operate with power from the grid.

16.8 Thermal Energy Storage Systems

Thermal energy storage systems were briefly discussed in Chapter 12. In this section, we will cover these systems more extensively beyond just solar thermal power generation units. Thermal energy storage systems store the excess thermal energy whenever it is available and provide the stored energy when it is needed. The duration of the storage can be in the order of a few hours, days, or even months. An example of long-term thermal energy storage is when solar thermal energy in summer is stored for heating purposes in winter. However, in the power generation industry, thermal energy storage systems are used for the short-term applications, typically for hours or perhaps for a couple of days. Thermal energy storage is generally less expensive than storing energy after its conversion to electricity. Clearly, the proper insulation of piping systems and equipment is critical in an effective thermal energy storage system. Thermal energy storage systems also include the storage of "coldness" in the form of ice and cold water.

It should be noted that while these systems are currently used primarily in solar thermal energy systems, they are being seriously considered for other applications as well. One important example is baseload power plant where the load variation is very difficult if not impossible, such as geothermal and nuclear power plants. These units find it difficult to compete with inexpensive electricity generated by solar PV systems during the daytime. So the idea is to store the generated thermal energy when it is not financially viable to generate electricity and use that energy during peak periods at the late evening and early night. The technical and financial feasibilities of these applications are currently under investigation.

Thermal energy can be stored in a single-phase medium (in the form of sensible heat), a phase-changing medium (in the form of the latent heat of evaporation or melting), or a chemically active mixture (in the form of the heat of chemical reactions). In **single-phase thermal energy storage** systems, also referred to as **sensible heat energy storage** systems, heat is stored and recovered by changing the temperature of the storage medium. The specific energy and energy density of these systems can be increased by increasing the temperature difference between the fully charged and the fully discharged system. A drawback of this type of thermal energy storage system is that the temperature of the energy they release during the discharge varies. Also, the impact of the thermal expansion of the medium should be considered in the design. The heat storage medium of molten salts, organic thermal oils, liquid metals, large blocks of porous solid materials, and bed of solid materials, such as sand and rock, has been utilized for these systems.

In **phase-changing thermal energy storage** systems, also referred to as **latent heat energy storage** systems, heat is stored and recovered by changing the phase of the storage medium. This phase change can be from solid to liquid (melting), and vice versa (solidification), or from liquid to vapor (evaporation), and vice versa (condensation). Since typically for most materials, the latent heat (per unit of mass) is large, these systems can achieve high specific energy and energy density. A major advantage of latent heat energy storage systems is that since the phase change takes place at an isothermal process, they can provide thermal energy at a constant temperature during the discharge process.

In **steam accumulators** or the so-called steam storage steams, a combination of sensible and latent heat energy storage is used. In these systems, superheated steam enters a tank, called an accumulator, partly filled (commonly by about a half) with cold water. Some steam is accumulated on top of the water and increases the pressure of the accumulator. The rest of the steam condenses and increases the temperature of the water to the saturation temperature of water at the tank pressure. The condensed steam also increases the level of the water in the tank. By adding more steam, the amount of stored steam, the pressure of the tank, the temperature of the water, and the water level in the tank are all increased. The charging process continues until the pressure of the steam reaches a design pressure. When thermal energy in the form of steam is needed, saturated steam is discharged from the upper section of the accumulator (where steam is present). The removal of steam reduces the pressure of the tank to lower than the saturation pressure of the saturated water, which results in water boiling and production of more steam. The process continues until the accumulator is fully discharged and is returned to its initial conditions. The 11 MW PS10 Solar Power Plant, the world's first commercial solar thermal power plant, commissioned in 2007 in Spain used this type of thermal energy storage systems.

As noted, thermal energy storage can also be in the form of cold water (sensible energy) storage or ice/ice-slush (latent energy) storage. To produce cold water or ice, refrigeration systems operate during off-peak periods, commonly at night. The stored cold water or ice is used for cooling purposes during peak periods without adding to the peak load. In power generation, this approach has been used for compressor inlet air cooling systems for gas turbines.

In chemical reaction thermal energy storage systems, a reversible chemical reaction with high heat absorption and release at different directions of the reaction is used to store thermal energy. The thermal energy storage takes place during the endothermic reaction, and the thermal energy is released during the exothermic process. This method has a potential for high specific energy storage but has not been commercialized yet.

16.9 Questions and Problems

16.9.1 Introduction

16.1. How are spatial and temporal gaps between electricity generation and consumption filled?

16.2. Which statement is accurate?

a. Electricity is probably the only product that cannot be readily and economically stored.

b. Electricity should be available whenever and wherever it is needed by consumers.

c. The electricity generation industry must be able to collectively generate enough electricity to meet varying demand all the time; otherwise, there will be a blackout or brownout.

d. All of the above.

16.3. Why can a fossil fuel-based electrical infrastructure operate without electricity storage?

16.4. Explain why if peak-load power generation units were as efficient and inexpensive as baseload power plants, energy storage units would have been undesirable for a fossil fuel-based electrical infrastructure.

16.5. What are the two objectives of energy storage systems in a fossil fuel-based electrical infrastructure?

16.6. What is peak shaving? What are the two strategies for peak shaving? Provide an example for each strategy.

16.7. What makes the management of renewable energy-based electrical infrastructure more challenging than that of fossil fuel-based one?

16.8. What is the duck curve problem? What problem is manifested by this curve?

16.9. In California, while solar energy has been able to significantly reduce the … for electricity generation, it has almost no impact on the … power generation.

a. peak load – overall fossil fuel consumption

b. overall fossil fuel consumption – peak-load

c. baseload – peak-load

d. peak load – baseload

16.10. Explain why the development in energy storage technologies is so important in shaping the future of the electricity generation industry.

16.9.2 Characterization of Energy Storage Technologies

16.11. What is the principle idea of energy storage systems?

16.12. Define the round trip efficiency. Why is it important?

16.13. Explain why and how the system boundary definition and what components are included in the system impact the efficiency of an energy storage system. Provide an example.

16.14. How does the duration of the energy storage impact the round trip efficiency of the system? Provide an example.

16.15. What is the self-discharge time (SDT) of an energy storage system? How is it related to the depth of discharge (DOD)?

16.16. A low depth of discharge ... the round trip efficiency of some energy storage systems.

a. reduces

b. increases

c. does not impact

16.17. A low DOD ... the energy density and specific energy of an energy storage device.

a. reduces

b. increases

c. does not impact

16.18. In terms of the time required to store and deliver energy, energy storage devices can be divided into three groups. Name these groups. What is the time scale of each group? Provide an example for each group.

16.19. Explain how energy storage systems can be used to maintain the quality of electrical power.

16.20. What is the quality of electricity? What are the important factors for the quality of electricity?

16.21. What is the most important characteristic of energy storage systems used for maintaining power quality?

16.22. How can electricity storage systems provide power when switching from one electricity source to another?

16.23. What is the specific energy or energy-to-mass ratio of an energy storage system? What is its unit? What about the energy density or energy-to-volume ratio?

16.24. How is the system longevity defined for an energy storage system?

16.25. In terms of physical principles used to store energy, energy storage systems can be generally divided into ….

16.26. What form of energy is stored in the following energy storage technologies?

- Batteries
- Flywheels
- Supercapacitors
- Hydropower plants
- Springs
- Superconducting coils
- Elevated masses
- Thermal energy storage systems

16.9.3 Gravitational Energy Storage

16.27. Assume you live in the twentieth floor of an apartment complex. In order to reduce your electricity bill, you would like to store your peak-load period electricity consumption during the night time when the cost of electricity is substantially lower by raising $0.5 \times 0.4 \times 0.1$ m concrete blocks from ground to your apartment. Determine the number of blocks you need if your average peak-load electricity consumption is 10 kWh/day. Assume each floor is 3 m high and there is no energy loss. The density of concrete is 2400 kg/m^3.

16.28. A power plant with a capacity of 1000 MW (a midsize power plant) operates for an hour at the full load. How many midsize cars (with a mass of about 1500 kg each) are needed to be raised to the height of 100 m to store this much energy?

16.29. Among various energy storage technologies, ... energy storage units are the only ones currently making a meaningful contribution to the electricity generation industry.

16.30. Which energy storage technology shares similar technologies with conventional hydropower plants?

16.31. What are the advantages of pumped hydro storage plants?

16.32. What are the major sources of energy losses in pumped hydro storage plants?

16.33. In a pumped hydro plant, the system efficiencies during the pump mode and the turbine mode are equal.

a. True

b. False

16.34. Why is the energy storage time in a pumped hydro plant typically longer than the power generation period?

16.35. Pumped hydro plants are rated for their power ....

a. storage capacity

b. generation capacity

16.36. The Grand Coulee Dam is equipped with a pumped hydro storage system. The surface area of the lake behind the dam is 324 km^2. The water surface level is 116 m above the turbine. During the pump mode, enough water is pumped to raise the water level behind the dam by 5 cm. Determine the gravitational potential energy stored by this water in J and MWh.

16.37. In the previous problem, the Grand Coulee Dam is equipped with six pump-turbine assemblies with a total capacity of 314 MW. If the system operates for five hours in the pump mode, determine the rise in the level of water behind the dam (in cm). The efficiency of the system in the pump mode is 88%.

16.38. A fossil fuel power plant with a capacity of 2000 MW (a relatively large fossil fuel power plant) operates for two hours at the full load. This energy is stored in the Grand Coulee Dam. Determine the rise in the level of the upper reservoir.

16.39. A dam is equipped with a pumped hydro storage system. The surface area of the lake behind the dam is 250 km^2. The water surface level is 60 m above the turbine. During the pump mode, enough water is pumped to raise the water level behind the dam by 30 cm. Determine the gravitational potential energy stored by this water in J and MWh.

16.40. In the previous problem, the pump-turbine assembly has a total capacity of 220 MW. If the system operates for four hours in the pump mode, determine the rise in the level of water behind the dam (in cm). The efficiency of the system in the pump mode is 86%.

16.41. Explain how pumped hydro plants can be integrated to conventional hydropower plants. How do they operate?

16.42. What are pump-back plants?

16.43. What is the difference between open-loop and closed-loop pumped hydro plants?

16.44. Explain the components of the system connecting the upper and lower reservoirs in a pumped hydro plant.

16.45. What is the most common impulse hydro turbine? Explain its operational principle.

16.46. What are the components of a Pelton turbine? What is the function of each component?

16.47. At what heads and flow rates Pelton turbines operate their best?

16.48. Conduct research to understand and explain Turgo turbines and a cross-flow turbine.

16.49. Explain how Kaplan turbines work. At what heads and flow rates Kaplan turbines operate their best?

16.50. Which type of turbines is the most commonly used hydro turbine today?

16.51. What is the problem with single-speed pump-turbine units in pumped hydro units? How is this problem solved in newer units?

16.52. In the following schematic of a pumped hydro plant, name the four components.

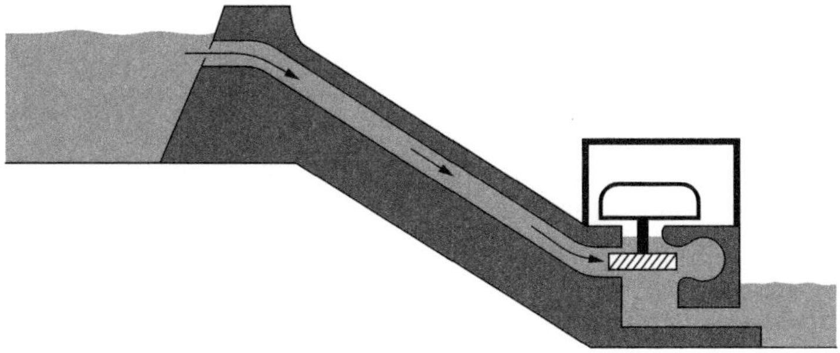

16.9.4 Compressed Air Energy Storage Systems

16.53. Explain a compressed air energy storage (CAES) system in its simplest form.

16.54. Explain why if the high-pressure air is stored for a long time and the air is cooled down by heat transfer to the environment, the cold air cannot generate electricity that is remotely close to the energy used to compress the same air. How can this problem be addressed?

16.55. What is the concept behind the adiabatic compressed air energy storage systems? Name the two approaches for this purpose. How do they work?

16.56. Explain how diabatic or hybrid compressed air energy storage systems work.

16.57. The world's second utility-size compressed air energy storage system was built in 1991 in McIntosh, Alabama. The plant has a power generation capacity of 110 MW and the energy storage capacity of 2860 MWh. Assuming the ambient temperature and pressure of 20°C and 1 bar, the stored air pressure of 45 bar, the turbine and compressor isentropic efficiency of 82%, determine how many hours the plant can operate at the full load after it is fully charged, the temperature of the compressor and turbine outlet flows, the mass and volume of the stored air, and the mass flow rate of the turbine. Assume that the temperature of the compressed air is returned at its initial temperature before expansion by burning natural gas so the temperature of the compressor exit flow is equal to the temperature of the turbine inlet flow. Also, there is no pressure drop in the container, and the compressed air temperature in the container is equal to the ambient temperature. Use the ideal gas assumption with a constant specific heat.

16.58. A compressed air energy storage system has a power generation capacity of 2000 MW and the energy storage capacity of 30,000 MWh. Assuming the ambient temperature and pressure of 20°C and 1 bar, the stored air pressure of 100 bar, and the turbine and compressor isentropic efficiency of 85%, determine how many hours the plant can operate at the full load after it is fully charged, the temperature of the compressor and turbine outlet flows, the mass and volume of the stored air, and the mass flow rate of the turbine. Assume that the temperature of the compressed air is returned to its initial temperature before expansion by burning natural gas so the temperature of the compressor exit flow is equal to the temperature of the turbine inlet flow. Also, there is no pressure drop in the container and the compressed air temperature in the container is equal to the ambient temperature. Use the ideal gas assumption with a constant specific heat.

16.59. Repeat the previous problem by varying the pressure from 20 to 200 bar with an increment of 20 bar, and determine the volume of the needed container for each case.

16.60. Explain how constant-volume or constant-pressure compressed air containers work. What are the advantages and disadvantages of each container?

16.61. Describe how underwater air energy storage systems work.

16.9.5 Flywheel Energy Storage Systems

16.62. Explain how flywheel energy storage systems work. What form of energy is stored in them?

16.63. List some of the applications of flywheel energy storage systems.

16.64. What are the main components of flywheel energy storage systems?

16.65. How is friction in bearings and friction due to air drag force eliminated in flywheels?

16.66. Why is a flywheel housing typically vacuumed?

16.67. In a concept car (prototype), the car manufacturer Volvo installed a 6-kg flywheel with a diameter of 20 cm and a spinning speed of 60,000 rpm to store energy through regenerative braking.[14] Determine the energy that the flywheel can store (in kJ).

16.68. A power plant with a capacity of 1500 MW operates for 30 minutes at the full load. Determine the number of flywheels from the previous problem that are needed to store this much energy.

16.69. As a designer, you have an option to choose between an aluminum alloy and concrete to manufacture a flywheel for an energy storage system. Which material will you choose? Justify your answer. The tensile strengths of aluminum alloy and concrete are 483 and

[14] https://newatlas.com/volvo-flywheel-kers-testing/27273/.

4 MPa, respectively, and their densities are 2800 and 2600 kg/m^3, respectively.

16.70. Explain why making flywheel disk out of high-density materials is not a good idea. What parameter should be considered for the selection of material? What type of materials are currently being used for this purpose?

16.9.6 Batteries

16.71. What are primary and secondary batteries?

16.72. What are some applications of rechargeable batteries?

16.73. How is the capacity of a battery expressed? What is its unit?

16.74. What are the three main components of every battery cell?

16.75. Why are the anode and cathode electrodes in a lead-acid battery made of highly porous materials?

16.76. What is the electrolyte in a lead-acid battery?

16.77. What are the compositions of the anode and cathode of a lead-acid battery when the battery is fully charged?

16.78. Using the following overall cell reaction, explain how a lead-acid battery works.

$$Pb\ (s) + PbO_2\ (s) + 2H_2SO_4\ (aq) \rightarrow 2PbSO_4\ (s) + 2H_2O$$

16.79. For each mole of the reacted Pb (and PbO_2), … moles of sulfuric acid is converted to water.

a. 0.5

b. 1

c. 2

d. 4

16.80. What is the open-circuit voltage (OCV) of a battery?

16.81. How can the density (or more commonly specific gravity) of the electrolyte be used to indicate the level of charge in a battery? How can the specific gravity of the electrolyte be measured?

16.82. In cold weather, why is it advisable to keep a lead-acid battery fully charged?

16.83. What is the electromotive force (EMF) of a cell?

16.84. The charging process of a lead-acid battery results in the conversion of … to … in the anode and to … in the cathode.

a. $PbSO_4 - PbO_2 - Pb$

b. $PbO_2 - Pb - PbSO_4$

c. $PbSO_4 - Pb - PbO_2$

d. $Pb - PbO_2 - PbSO_4$

16.85. In the charging process of a lead-acid battery, in the electrolyte, sulfuric acid is converted to water.

a. True

b. False

16.86. Why should batteries be protected from overcharging?

16.87. Explain why in some lead-acid batteries, the water level should be checked regularly.

16.88. Explain how electrodes are manufactured for lead-acid batteries.

16.89. Describe the internal structure of lead-acid batteries.

16.90. What is the form of the electrolyte in the following battery types?

- wet or flooded batteries
- gel or dry batteries
- absorbed glass mat batteries

16.91. Why are some batteries referred to as maintenance-free batteries?

16.92. Why are some batteries referred to as valve-regulated lead-acid (VRLA) batteries?

16.93. When a battery is used to power a device, its voltage compared to the open-circuit voltage ….

a. increases

b. decreases

c. remains unchanged

16.94. Explain how the capacity of batteries is typically expressed. What is its unit? Which factors impact this capacity?

16.95. What does the capacity at 20hr mean for a lead-acid battery?

16.96. A battery has an open-circuit voltage of 20 V, an internal resistance of 0.22 Ω, and a capacity of 150 A.h. Determine the potential difference between the terminals and the duration of the operation of the battery if the current is (a) 1 A, (b) 5 A, (c) 10 A, and (d) 50 A.

16.97. The maximum voltage a battery can provide is 30 V, and the maximum current it can provide is 100 A. Determine the internal resistance and V_{OC} of the battery.

16.98. A power plant with a capacity of 2000 MW operates for 2 hours at the full load. This energy is stored in a battery pack made of batteries each with a capacity of 70 A.h and a voltage of 12 V. Determine the number of batteries needed for this purpose.

16.99. What is the round trip efficiency of lead-acid batteries?

16.100. What types of lead-acid batteries are used in automotive applications? Why do they suit this application?

16.101. What is the definition of a deep discharge?

16.102. Why are thin electrolyte lead-acid batteries suitable for automotive applications? Explain why their relatively low number of complete cycles is not a problem for these applications.

16.103. How is the thickness of electrode plates in a battery related to its service life?

16.104. Provide the two examples for batteries that have to go through many deep discharge cycles.

16.105. What are deep-cycle lead-acid batteries? Which applications are they suitable for?

16.106. Deep-cycle lead-acid batteries have more active electrode materials but lower peak current. Why?

a. True

b. False

16.107. What are the advantages and disadvantages of lithium-ion batteries?

16.108. Describe how batteries from retired electric cars can be used for the large-scale energy storage.

16.109. How can currently in service electric and hybrid cars provide a distributed energy storage capability at residential units?

16.110. Why are molten-salt batteries sometimes referred to as thermal batteries?

16.111. What are the advantages of molten-salt batteries?

16.9.7 Direct Electricity Storage Systems

16.112. In direct electricity storage systems, how is electricity stored?

16.113. What are the components of a capacitor?

16.114. Explain how and why a dielectric can increase electrical energy stored in a capacitor.

16.115. Can we increase the energy stored in a capacitor by changing the voltage in which the conductors are connected to? What is the limit for increasing this voltage?

16.116. How can the capacitance of a capacitor be increased?

16.117. Explain why and how the capacitance of supercapacitors is substantially higher than that of conventional capacitors.

16.118. Which statement is not correct?

a. Even the best supercapacitors have specific energy substantially smaller than some other energy storage systems.
b. Supercapacitors can charge and discharge almost instantaneously.
c. While supercapacitors' specific energy is high, their specific power is very low.
d. Supercapacitors have a round trip efficiency of up to 98%.

16.119. For what type of applications are supercapacitors ideal for?

16.120. Why are inverters needed when supercapacitors operate with the grid?

16.121. Consider a 0.2 kg supercapacitor with the capacitance of 10000 F that is charged with a 4 V electricity. Determine the energy stored in the supercapacitor (in J) and its specific energy (in Wh/kg). Compare the stored energy in the supercapacitor with that in a lead-acid battery with a capacity of 200 A.h and a voltage of 24 V.

16.122. Explain how electrical energy can be stored in electric coils. What is the main source of energy loss in them? How can superconducting coils address this loss?

16.123. What are the pros and cons of superconducting magnetic energy storage (SMES) systems?

16.9.8 Thermal Energy Storage Systems

16.124. Thermal energy storage is generally less expensive than storing energy after its conversion to electricity.
a. True
b. False

16.125. Explain how and why thermal energy storage systems are being considered for some baseload power plants.

16.126. What are different types of thermal energy storage systems according to the medium used for storage?

16.127. How is thermal energy stored in single-phase thermal energy storage systems? What are some drawbacks of these systems?

16.128. List some media that have been used in single-phase thermal energy storage systems.

16.129. How is thermal energy stored in phase-changing thermal energy storage systems? What are some advantages of these systems?

16.130. Explain how steam accumulators store thermal energy in the form of both sensible and latent heat energy.

16.131. Explain how chemical reaction thermal energy storage systems work.

16.9.9 General

16.132. Why do we need energy storage?

- In a fossil fuel-based infrastructure
- In a renewable energy-based infrastructure

16.133. Why is the reversible pump-turbine and motor-generator set typically used in pumped hydro plants?

16.134. In most energy storage systems, electrical energy is converted to other forms of energy before being stored. In the following energy storage systems, determine the form of energy that is stored.

- Pumped hydroelectric energy storage systems
- Compressed air energy storage systems
- Flywheel energy storage systems
- Batteries
- Supercapacitors
- Superconducting coils

16.135. The power density and energy density of energy storage systems are the same.

a. True

b. False

16.136. Which one is a better choice to achieve the maximum specific energy in a flywheel energy storage system: high-strength steel alloy ASTM A514 or titanium. The tensile strengths of the steel alloy and titanium are 760 and 370 MPa, respectively, and their densities are 7800 and 4500 kg/m^3, respectively.

16.137. Determine how many midsize cars (with a mass of about 1500 kg each) should be raised to the height of 100 m to store sufficient energy to fully power an average house in the U.S. for a day. What about Boeing 737-MAX (with a mass of about 45,000 kg each)? Assume no energy losses.

16.138. A flywheel energy storage system uses a 2-kg disk with a diameter of 120 cm that spins with a rotational speed of 60,000 rpm. Determine the energy stored in the flywheel in kJ.

16.139. In a pump-back plant, the surface area of the lake behind the dam is 65 km^2. During the pump mode, enough water is pumped to raise the water level behind the dam by 20 cm. The water surface level is 95 m above the turbine. Determine the gravitational potential energy stored by this water in J and MWh.

16.140. In the previous problem, the pump-turbine assembly has a total capacity of 5 MW. If the system operates for three hours in the pump mode, determine the rise in the level of water behind the dam (in cm). The efficiency of the system in the pump mode is 89%.

16.141. Determine the energy stored in the supercapacitor with the capacitance of 1000 F that is charged with a 9 V electricity (in J).

16.142. A power plant with a capacity of 1500 MW operates for 30 minutes at the full load. This energy is stored in the dam in Problem 139. Determine the rise in the level of the upper reservoir.

16.143. A flywheel is made of carbon fiber with a density of 2000 kg/m^3. The flywheel is a cylinder with a diameter of 50 cm and a height of 40 cm. Determine the energy that can be stored in the flywheel if it spins with the rotational speed of 90,000 rpm (in MJ).

16.144. Any action or strategy that reduces the peak load is called ….

16.145. Explain the functions of energy storage systems in a renewable-based energy infrastructure.

16.146. Which one(s) is electrical–mechanical storage system?

a. Pumped hydro
b. Compressed air
c. Spring, torsion bars, mass elevation
d. Flywheels
e. Batteries
f. Thermal energy storage systems
g. Superconducting coils

16.147. Which one(s) is the sources of losses in pumped hydroplants?

a. Motor and pump losses
b. Seepage into ground
c. Leakage from pipes and equipment
d. Evaporation in upper reservoir
e. Generator and turbine losses
f. All of them

16.148. What is the oldest type of battery?

16.149. What is the difference between the power density and the energy density of energy storage systems?

16.150. In the utility level, two types of energy storage are electrical–mechanical storage and thermal storage.

a. True
b. False

16.151. Explain why the temperature of the outlet flow from a compressor is much higher than the temperature of the inlet flow. Why is it a good idea to store this thermal energy in CAES systems? How can the thermal energy be stored and used?

16.152. What are the three main components of every battery?

16.153. Provide an application of energy storage systems that requires short charging/discharging times (seconds to minutes).

16.154. The discharging process of a lead-acid battery results in the conversion of … to … in the anode and … to … in the cathode.

a. $PbSO_4 - PbO_2 - PbSO_4 - Pb$

b. $PbO_2 - PbSO_4 - Pb - PbSO_4$

c. $PbSO_4 - Pb - PbSO_4 - PbO_2$

d. $Pb - PbSO_4 - PbO_2 - PbSO_4$

16.155. What are the advantages of the adiabatic storage of compressed air?

16.156. In the discharging process of a lead-acid battery, in the electrolyte, sulfuric acid is converted to water.

a. True

b. False

16.157. The voltage of a battery depends on the … taking place in the battery. The current however depends on the …, which in turn is a function of the … of the electrodes.

16.158. What type of batteries is the most developed and used one?

16.159. Energy storage systems operate based on a wide variety of technologies and physical principles.

a. True

b. False

16.160. Why do we need a storage system in a solar thermal power plant?

16.161. In the following diagram, explain each component.

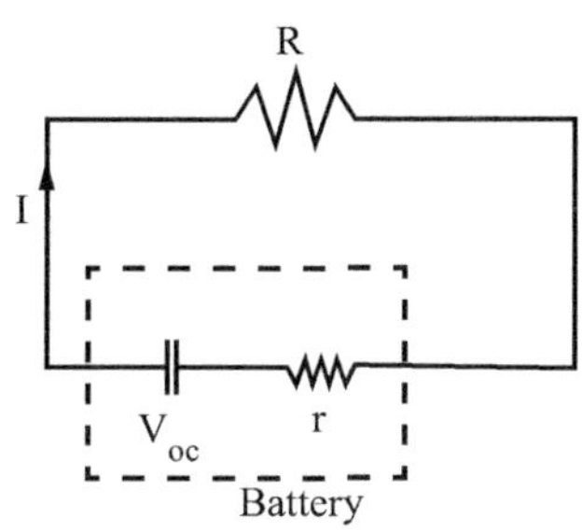

16.162. In the above circuit, what is the equation to find I?

16.163. What are deep-discharge lead-acid batteries?

16.164. The voltage produced by a battery when there is no load (open-circuit voltage) is 14 V. In the above circuit, if $r = R = 0.01\ \Omega$, determine the current I and the voltage across battery terminals.

16.165. There are several ways to characterize the numerous energy storage technologies. One of these characterizations is the time interval of storing energy or delivering it to a load (charging or discharging the device).

a. True

b. False

16.166. What is the adiabatic storage of compressed air?

16.167. Explain why batteries used in nonelectric vehicles are not suitable for energy storage systems.

16.168. Which one(s) is a direct electrical energy storage system?

a. batteries

b. superconducting coils

c. compressed air systems

d. flywheels

e. ultracapacitors

16.169. Pumped hydro plants are typically rated according to their power output (MW) not input power.

a. True

b. False

16.170. What does CAES stand for?

16.171. Explain the operation of lead-acid batteries.

16.172. Which one is not a characteristic of high-speed flywheels?

a. high energy density

b. high maintenance

c. long life

d. high efficiency

16.173. What are various types of energy storage systems in solar thermal power plants?

16.174. Why are battery cells connected in series or parallel inside a battery pack?

16.175. What form of energy is stored in the following energy storage systems?

- Pumped hydro

- Compressed air
- Spring and mass elevation
- Flywheels

16.176. How can the difference between the open-circuit voltage of a battery and its operating voltage be reduced?

16.177. The following reaction is given for a battery. What type of battery is this? Using the given electrochemical reactions, describe how this battery works.

$$Pb(s) + PbO_2\ (s) + 2H_2SO_4\ (aq) \rightarrow 2PbSO_4\ (s) + 2H_2O$$

16.178. Where is compressed air typically stored in a large-scale energy storage system?

16.179. Pumped hydro is a gravitational potential energy storage system.

a. True

b. False

16.180. Why do not we raise solid objects to store gravitational potential energy in them as an energy storage strategy?

16.181. What is the main application of deep-discharge lead-acid batteries?

16.182. In pumped hydro plants, why is a minimum horizontal distance between lower and upper reservoirs important?

16.183. Compressed air systems are suitable for large-scale energy storage systems.

a. True

b. False

16.184. Why do we need a vacuum pump in flywheel energy storage systems?

16.185. Explain how ultracapacitors can directly store electricity.

16.186. The voltage of a battery is fixed by its electrochemical reactions and is known for each battery type.

a. True

b. False

16.187. Describe the two main methods for thermal energy storage in solar thermal power plants.

16.188. Explain what a duck curve is. Why is it important? How is it related to energy storage systems?

16.189. Describe why in the past ultracapacitors were not suitable for energy storage. What has changed in this regard recently?

16.190. What is the main difference between the applications of batteries in nonelectric vehicles and energy storage systems? Why are deep-discharge lead-acid batteries used for the latter applications?

16.191. Explain the difference between energy and power density in energy storage systems.

16.192. In a fossil fuel-based energy infrastructure, why do we need energy storage?

16.193. Explain how superconducting coils can be used as direct electricity storage systems.

16.194. What is an example of electrochemical energy storage systems?

16.195. What are the pros and cons of pumped hydro plants?

16.196. The energy stored in a flywheel rotating at a speed of ω and moment of inertia I is ....

16.197. Energy management or load leveling allows energy generated when the demand is low to be released hours later at times of peak demand. This application requires an energy storage system with long charging/discharging times.

a. True

b. False

16.198. Why often pumped hydro plants are adjunct to standard hydropower plants?

16.199. What is the short circuit of a battery? What is its unit?

16.200. What are the three types of energy storage systems based on how long is required to charge and discharge them?

16.201. What is the unit of energy storage capacity of a battery?

16.202. In a pump hydro storage system, both or one of the lower and upper reservoirs can be natural (river or lake) or artificially excavated.

a. True

b. False

16.203. What is the most developed and used energy storage system for large plants?

16.204. A power plant with a capacity of 1000 MW (a midsize power plant) operates for an hour at the full load. How many Being 747 (with a mass of about 180,000 kg each) are needed to be raised to the height of the tip of Eifel tower (at 324 m) to store this much energy?

16.205. Provide an application of energy storage systems that requires long charging/discharging times (minutes to many hours).

16.206. Why do pumped hydro plants need two reservoirs: lower and upper ones?

Appendix 1: Supplemental Tables

TABLE A.1

Saturation Temperature (Water)

		Specific Volume (ft.^{3}/lb$_m$)		Internal Energy (Btu/lb$_m$)			Enthalpy (Btu/lb$_m$)			Entropy (Btu/lb$_m$·°R)		
Temp. °F	Press. psi	Sat. Liquid v_f	Sat. Vapor v_g	Sat. Liquid u_f	Evap. u_{fg}	Sat. Vapor u_g	Sat. Liquid h_f	Evap. h_{fg}	Sat. Vapor h_g	Sat. Liquid s_f	Evap. s_{fg}	Sat. Vapor s_g
32	0.08859	0.016022	3305	0.01	1021.2	1021.2	0.01	1075.4	1075.4	0.00003	2.1870	2.1870
32.018	0.08866	0.016022	3302	0.00	1021.2	1021.2	0.01	1075.4	1075.4	0.00000	2.1869	2.1869
35	0.09992	0.016021	2948	0.2.99	1019.2	1022.2	3.00	1073.7	1076.7	0.00607	2.1704	2.1764
40	0.12166	0.016020	2445	0.8.02	1015.8	1023.9	8.02	1070.9	1078.9	0.01617	2.1430	2.1592
45	0.14748	0.016021	2037	13.04	1012.5	1025.5	13.04	1068.1	1081.1	0.02618	2.1162	2.1422
50	0.17803	0.016024	1704.2	18.06	1009.1	1027.2	18.06	1065.2	1083.3	0.03607	2.0899	2.1259
60	0.2563	0.016035	1206.9	28.08	1002.4	1030.4	28.08	1059.6	1087.7	0.05555	2.0388	2.0943
70	0.3632	0.016051	867.7	38.09	995.6	1033.7	38.09	1054.0	1092.0	0.07463	1.9896	2.0642
80	0.5073	0.016073	632.8	48.08	988.9	1037.0	48.09	1048.3	1096.4	0.09332	1.9423	2.0356
90	0.6988	0.016099	467.7	58.07	982.2	1040.2	58.07	1042.7	1100.7	0.11165	1.8966	2.0083
100	0.9503	0.016130	350.0	68.04	975.4	1043.5	68.05	1037.0	1105.0	0.12963	1.8526	1.9822
110	1.2763	0.016166	265.1	78.02	968.7	1046.7	78.02	1031.3	1109.3	0.14730	1.8101	1.9574
120	1.6945	0.016205	203.0	87.99	961.9	1049.9	88.00	1025.5	1113.5	0.16465	1.7690	1.9336
130	2.225	0.016247	157.17	97.97	955.1	1053.0	97.98	1019.8	1117.8	0.18172	1.7292	1.9109
140	2.892	0.016293	122.88	107.95	948.2	1056.2	107.96	1014.0	1121.9	0.19851	1.6907	1.8892
150	3.722	0.016343	96.99	117.95	941.3	1059.3	117.96	1008.1	1126.1	0.21503	1.6533	1.8684
160	4.745	0.016395	77.23	127.94	934.4	1062.3	127.96	1002.2	1130.1	0.23130	1.6171	1.8484
170	5.996	0.016450	62.02	137.95	927.4	1065.4	137.97	996.2	1134.2	0.24732	1.5819	1.8293
180	7.515	0.016509	50.20	147.97	920.4	1068.3	147.99	990.2	1138.2	0.26311	1.5478	1.8109
190	9.343	0.016570	40.95	158.0	913.3	1071.3	158.03	984.1	1142.1	0.27866	1.5146	1.7932

(Continued)

TABLE A.1 (*Continued*)

Saturation Temperature (Water)

		Specific Volume ($ft.^3/lb_m$)		Internal Energy (Btu/lb_m)			Enthalpy (Btu/lb_m)			Entropy (Btu/lb_m·°R)		
Temp. °F	Press. psi	Sat. Liquid v_f	Sat. Vapor v_g	Sat. Liquid u_f	Evap. u_{fg}	Sat. Vapor u_g	Sat. Liquid h_f	Evap. h_{fg}	Sat. Vapor h_g	Sat. Liquid s_f	Evap. s_{fg}	Sat. Vapor s_g
200	11.529	0.016634	33.63	168.04	906.2	1074.2	168.07	977.9	1145.9	0.29400	1.4822	1.7762
210	14.125	0.016702	27.82	178.10	898.9	1077.0	178.14	971.6	1149.7	0.30913	1.4508	1.7599
220	17.188	0.016772	23.15	188.17	891.7	1079.8	188.22	965.3	1153.5	0.32406	1.4201	1.7441
230	20.78	0.016845	19.386	198.26	884.3	1082.6	198.32	958.8	1157.1	0.33880	1.3901	1.7289
240	24.97	0.016922	16.327	208.36	876.9	1085.3	208.44	952.3	1160.7	0.35335	1.3609	1.7143
250	29.82	0.017001	13.826	218.49	869.4	1087.9	218.59	945.6	1164.2	0.36772	1.3324	1.7001
260	35.42	0.017084	11.768	228.64	861.8	1090.5	228.76	938.8	1167.6	0.38193	1.3044	1.6864
270	41.85	0.017170	10.066	238.82	854.1	1093.0	238.95	932.0	1170.9	0.39597	1.2771	1.6731
280	49.18	0.017259	8.650	249.02	846.3	1095.4	249.18	924.9	1174.1	0.40986	1.2504	1.6602
290	57.53	0.017352	7.467	259.25	838.5	1097.7	259.44	917.8	1177.2	0.42360	1.2241	1.6477
300	66.98	0.017448	6.472	269.52	830.5	1100.0	269.73	910.4	1180.2	0.47320	1.1984	1.6356
310	77.64	0.017548	5.632	279.81	822.3	1102.1	280.06	903.0	1183.0	0.45067	1.1731	1.6238
320	89.60	0.017652	4.919	290.14	814.1	1104.2	290.43	895.3	1185.8	0.46400	1.1483	1.6123
330	103.00	0.017760	4.312	300.51	805.7	1106.2	300.84	887.5	1188.4	0.47722	1.1238	1.6010
340	117.93	0.017872	3.792	310.91	797.1	1108.0	311.30	879.5	1190.8	0.49031	1.0997	1.5901
350	134.53	0.017988	3.346	321.35	788.4	1109.8	321.80	871.3	1193.1	0.50329	1.0760	1.5793
360	152.92	0.018108	2.961	331.84	779.6	1111.4	332.35	862.9	1195.2	0.51617	1.0526	1.5688
370	173.23	0.018233	2.628	342.37	770.6	1112.9	342.96	854.2	1197.2	0.52894	1.0295	1.5585
380	195.60	0.018363	2.339	352.95	761.4	1114.3	353.62	845.4	1199.0	0.54163	1.0067	1.5483
390	220.2	0.018498	2.087	363.58	752.0	1115.6	364.34	836.2	1200.6	0.55422	0.9841	1.5383
400	247.1	0.018638	1.8661	374.27	742.4	1116.6	375.12	826.8	1202.0	0.56672	0.9617	1.5284

(*Continued*)

TABLE A.1 (*Continued*)

Saturation Temperature (Water)

		Specific Volume (ft.^{3}/lb$_m$)		Internal Energy (Btu/lb$_m$)			Enthalpy (Btu/lb$_m$)			Entropy (Btu/lb$_m$·°R)		
Temp. °F	Press. psi	Sat. Liquid v_f	Sat. Vapor v_g	Sat. Liquid u_f	Evap. u_{fg}	Sat. Vapor u_g	Sat. Liquid h_f	Evap. h_{fg}	Sat. Vapor h_g	Sat. Liquid s_f	Evap. s_{fg}	Sat. Vapor s_g
425	325.6	0.019014	1.4249	401.24	717.4	1118.6	402.38	802.1	1204.5	0.59767	0.9066	1.5043
450	422.1	0.019433	1.1011	428.6	690.9	1119.5	430.2	775.4	1205.6	0.6282	0.8523	1.4806
475	539.3	0.019901	0.8594	456.6	662.6	1119.2	458.5	746.4	1204.9	0.6586	0.7985	1.4571
500	680.0	0.02043	0.6761	485.1	632.3	1117.4	487.7	714.8	1202.5	0.6888	0.7448	1.4335
525	847.1	0.02104	0.5350	514.5	599.5	1113.9	517.8	680.0	1197.8	0.7191	0.6906	1.4007
550	1044.0	0.02175	0.4249	544.9	563.7	1108.6	549.1	641.6	1190.6	0.7497	0.6354	1.3851
575	1274.0	0.02259	0.3378	576.5	524.3	1100.8	581.9	598.6	1180.4	0.7808	0.5785	1.3593
600	1541.0	0.02363	0.2677	609.9	480.1	1090.0	616.7	549.7	1166.4	0.8130	0.5187	1.3317
625	1849.7	0.02494	0.2103	645.7	429.4	1075.1	654.2	492.9	1147.0	0.8467	0.4544	1.3010
650	2205	0.02673	0.16206	685.0	368.7	1053.7	695.9	423.9	1119.8	0.8831	0.3820	1.2651
675	2616	0.02951	0.11952	731.0	289.3	1020.3	745.3	332.9	1078.2	0.9252	0.2934	1.2186
700	3090	0.03666	0.07438	801.7	145.9	947.7	822.7	167.5	990.2	0.9902	0.1444	1.1346
705.44	3204	0.05053	0.05053	872.6	0	872.6	902.5	0	902.5	1.0580	0	1.0580

TABLE A.2

Saturation Pressure (Water)

		Specific Volume (ft.^{3}/lb$_m$)		Internal Energy (Btu/lb$_m$)			Enthalpy (Btu/lb$_m$)			Entropy (Btu/lb$_m$·°R)		
Press. psi	Temp. °F	Sat. Liquid v_f	Sat. Vapor v_g	Sat. Liquid u_f	Evap. u_{fg}	Sat. Vapor. u_g	Sat. Liquid h_f	Evap. h_{fg}	Sat. Vapor h_g	Sat. Liquid s_f	Evap. s_{fg}	Sat. Vapor s_g
0.50	79.56	0.016071	641.5	47.64	989.2	1036.9	47.65	1048.6	1096.2	0.09250	1.9443	2.0368
1.0	101.70	0.016136	333.6	69.74	974.3	1044.0	69.74	1036.0	1105.8	0.13266	1.8453	1.9779
1.5	115.65	0.016187	227.7	83.65	964.8	1048.5	83.65	1028.0	1111.7	0.15714	1.7867	1.9438
2.0	126.04	0.016230	173.75	94.02	957.8	1051.8	94.02	1022.1	1116.1	0.17499	1.7448	1.9198
3.0	141.43	0.016300	118.72	109.38	947.2	1056.6	109.39	1013.1	1122.5	0.20089	1.6852	1.8861
4.0	152.93	0.016358	90.64	120.88	939.3	1060.2	120.89	1006.4	1127.3	0.21983	1.6426	1.8624
5.0	162.21	0.016407	73.53	130.15	932.9	1063.0	130.17	1000.9	1131.0	0.23486	1.6093	1.8441
7.5	179.91	0.016508	50.30	147.88	920.4	1068.3	147.90	990.2	1138.1	0.26297	1.5481	1.8110
10	193.19	0.016590	38.42	161.20	911.0	1072.2	161.23	982.1	1143.3	0.28358	1.5041	1.7877
14.696	211.99	0.016715	26.80	180.10	897.5	1077.6	180.15	970.4	1150.5	0.31212	1.4446	1.7567
15	213.03	0.016723	26.29	181.14	896.8	1077.9	181.19	969.7	1150.9	0.31367	1.4414	1.7551
20	227.96	0.016830	20.09	196.19	885.8	1082.0	196.26	960.1	1156.4	0.33580	1.3962	1.7320
25	240.08	0.016922	16.306	208.44	867.9	1085.3	208.52	952.2	1160.7	0.35345	1.3607	1.7142
30	250.34	0.017004	13.748	218.84	869.2	1088.0	218.93	945.4	1164.3	0.36821	1.3314	1.6996
35	259.30	0.017078	11.900	227.93	862.4	1090.3	228.04	939.3	1167.4	0.38093	1.3064	1.6873
40	267.26	0.017146	10.501	236.03	856.2	1092.3	236.16	933.8	1170.0	0.39214	1.2845	1.6767
45	274.46	0.017209	9.403	243.37	850.7	1094.0	243.51	928.8	1172.3	0.40218	1.2651	1.6673
50	281.03	0.017269	8.518	250.08	845.5	1095.6	250.24	924.2	1174.4	0.41129	1.2476	1.6589
55	287.10	0.017325	7.789	256.28	840.8	1097.0	256.46	919.9	1176.3	0.41963	1.2317	1.6513
60	292.73	0.017378	7.177	262.06	836.3	1098.3	262.25	915.8	1178.0	0.42733	1.2170	1.6444
65	298.00	0.017429	6.657	267.46	832.1	1099.5	267.67	911.9	1179.6	0.43450	1.2035	1.6380

(Continued)

TABLE A.2 (*Continued*)

Saturation Pressure (Water)

		Specific Volume (ft.³/lb$_m$)		Internal Energy (Btu/lb$_m$)			Enthalpy (Btu/lb$_m$)			Entropy (Btu/lb$_m$·°R)		
Press. psi	Temp. °F	Sat. Liquid v_f	Sat. Vapor v_g	Sat. Liquid u_f	Evap. u_{fg}	Sat. Vapor. u_g	Sat. Liquid h_f	Evap. h_{fg}	Sat. Vapor h_g	Sat. Liquid s_f	Evap. s_{fg}	Sat. Vapor s_g
70	302.96	0.017478	6.209	272.56	828.1	1100.6	272.79	908.3	1181.0	0.44120	1.1909	1.6321
75	307.63	0.017524	5.818	277.37	824.3	1101.6	277.61	904.8	1182.4	0.44749	1.1790	1.6265
80	312.07	0.017570	5.474	281.95	820.6	1102.6	282.21	901.4	1183.6	0.45344	1.1679	1.6214
85	316.29	0.017613	5.170	286.30	817.1	1103.5	286.58	898.2	1184.8	0.45907	1.1574	1.6165
90	320.31	0.017655	4.898	290.46	813.8	1104.3	290.76	895.1	1185.9	0.46442	1.1475	1.6119
95	324.16	0.017696	4.654	294.45	810.6	1105.0	294.76	892.1	1186.9	0.46952	1.1380	1.6076
100	327.86	0.017736	4.434	298.28	807.5	1105.8	298.61	889.2	1187.8	0.47439	1.1290	1.6034
105	331.41	0.017775	4.234	301.97	804.5	1106.5	302.31	886.4	1188.7	0.47906	1.1204	1.5995
110	334.82	0.017813	4.051	305.52	801.6	1107.1	305.88	883.7	1189.6	0.48355	1.1122	1.5957
115	338.12	0.017850	3.884	308.95	798.8	1107.7	309.33	881.0	1190.4	0.48786	1.1042	1.5921
120	341.30	0.017886	3.730	312.27	796.0	1108.3	312.67	878.5	1191.1	0.49201	1.0966	1.5886
125	344.39	0.017922	3.588	315.49	793.3	1108.8	315.90	875.9	1191.8	0.49602	1.0893	1.5853
130	347.37	0.017957	3.457	318.61	790.7	1109.4	319.04	873.5	1192.5	0.49989	1.0822	1.5821
135	350.27	0.017991	3.335	321.64	788.2	1109.8	322.08	871.1	1193.2	0.50364	1.0754	1.5790
140	353.08	0.018024	3.221	324.58	785.7	1110.3	325.05	868.7	1193.8	0.50727	1.0688	1.5761
145	355.82	0.018057	3.115	327.45	783.3	1110.8	327.93	866.4	1194.4	0.51079	1.0624	1.5732
150	358.48	0.018089	3.016	330.24	781.0	1111.2	330.75	864.2	1194.9	0.51422	1.0562	1.5704
160	363.60	0.018152	2.836	335.63	776.4	1112.0	336.16	859.8	1196.0	0.52078	1.0443	1.5651
170	368.47	0.018214	2.676	340.76	772.0	1112.7	341.33	855.6	1196.9	0.52700	1.0330	1.5600

(*Continued*)

TABLE A.2 (*Continued*)

Saturation Pressure (Water)

		Specific Volume (ft.³/lb$_m$)		Internal Energy (Btu/lb$_m$)			Enthalpy (Btu/lb$_m$)			Entropy (Btu/lb$_m$·°R)		
Press. psi	Temp. °F	Sat. Liquid v_f	Sat. Vapor v_g	Sat. Liquid u_f	Evap. u_{fg}	Sat. Vapor. u_g	Sat. Liquid h_f	Evap. h_{fg}	Sat. Vapor h_g	Sat. Liquid s_f	Evap. s_{fg}	Sat. Vapor s_g
180	373.13	0.018273	2.533	345.68	767.7	1113.4	346.29	851.5	1197.8	0.53292	1.0223	1.5553
190	377.59	0.018331	2.405	350.39	763.6	1114.0	351.04	847.5	1198.6	0.53857	1.0122	1.5507
200	381.86	0.018387	2.289	354.9	759.6	1114.6	355.6	843.7	1199.3	0.5440	1.0025	1.5464
225	391.87	0.018523	2.043	365.6	750.2	1115.8	366.3	834.5	1200.8	0.5566	0.9799	1.5365
250	401.04	0.018653	1.8448	375.4	741.4	1116.7	376.2	825.8	1202.1	0.5680	0.9594	1.5274
275	409.52	0.018777	1.6813	384.5	733.0	1117.5	385.4	817.6	1203.1	0.5786	0.9406	1.5192
300	417.43	0.018896	1.5442	393.0	725.1	1118.2	394.1	809.8	1203.9	0.5883	0.9232	1.5115
350	431.82	0.019124	1.3267	408.7	710.3	1119.0	409.9	795.0	1204.9	0.6060	0.8917	1.4978
400	444.70	0.019340	1.1620	422.8	696.7	1119.5	424.2	781.2	1205.5	0.6218	0.8638	1.4856
450	456.39	0.019547	1.0326	435.7	683.9	1119.6	437.4	768.2	1205.6	0.6360	0.8385	1.4746
500	467.13	0.019748	0.9283	447.7	671.7	1119.4	449.5	755.8	1205.3	0.6490	0.8154	1.4645
550	477.07	0.019943	0.8423	458.9	660.2	1119.1	460.9	743.9	1204.8	0.6611	0.7941	1.4551
600	486.33	0.02013	0.7702	469.4	649.1	1118.6	471.7	732.4	1204.1	0.6723	0.7742	1.4464
700	503.23	0.02051	0.6558	488.9	628.2	1117.0	491.5	710.5	1202.0	0.6927	0.7378	1.4305
800	518.36	0.02087	0.5691	506.6	608.4	1115.0	509.7	689.6	1199.3	0.7110	0.7050	1.4160
900	532.12	0.02123	0.5009	523.0	589.6	1112.6	526.6	669.5	1196.0	0.7277	0.6750	1.4027
1000	544.75	0.02159	0.4459	538.4	571.5	1109.9	542.4	650.0	1192.4	0.7432	0.6471	1.3903
1250	572.56	0.02250	0.3454	573.4	528.3	1101.7	578.6	603.0	1181.6	0.7778	0.5841	1.3619
1500	596.39	0.02346	0.2769	605.0	486.9	1091.8	611.5	557.2	1168.7	0.8082	0.5276	1.3359

(*Continued*)

TABLE A.2 ***(Continued)***

Saturation Pressure (Water)

		Specific Volume (ft.^{3}/lb$_m$)		Internal Energy (Btu/lb$_m$)			Enthalpy (Btu/lb$_m$)			Entropy (Btu/lb$_m$·°R)		
Press. psi	Temp. °F	Sat. Liquid v_f	Sat. Vapor v_g	Sat. Liquid u_f	Evap. u_{fg}	Sat. Vapor. u_g	Sat. Liquid h_f	Evap. h_{fg}	Sat. Vapor h_g	Sat. Liquid s_f	Evap. s_{fg}	Sat. Vapor s_g
1750	617.31	0.02450	0.2268	634.4	445.9	1080.2	642.3	511.4	1153.7	0.8361	0.4748	1.3109
2000	649.20	0.02565	0.8813	662.4	404.2	1066.6	671.9	464.4	1136.3	0.8623	0.4238	1.2861
2250	652.90	0.02698	0.15692	689.9	360.7	1050.6	701.1	414.8	1115.9	0.8876	0.3728	1.2604
2500	668.31	0.02860	0.13059	717.7	313.4	1031.0	730.9	360.5	1091.4	0.9131	0.3196	1.2327
2750	682.46	0.03077	0.10717	747.3	258.6	1005.9	763.0	297.4	1060.4	0.9401	0.2604	1.2005
3000	695.52	0.03431	0.08404	783.4	185.4	968.8	802.5	213.0	1015.5	0.9732	0.1843	1.1575
3203.6	705.44	0.05053	0.05053	872.6	0	872.6	902.5	0	902.5	1.0580	0	1.0580

TABLE A.3

Properties of Superheated Steam

	Vapor (psia)											
P psi (*T* Sat. °F)	14.696 (211.99)				20 (227.96)				30 (250.34)			
T °F	*v*	*u*	*h*	*s*	*v*	*u*	*h*	*s*	*v*	*u*	*h*	*s*
Sat.	26.80	1077.6	1150.5	1.7567	20.09	1082.0	1156.4	1.7320	13.748	1088.0	1164.3	1.6996
150	*24.10*	*1054.5*	*1120.0*	*1.7090*	*17.532*	*1052.0*	*1116.9*	*1.6710*	*11.460*	*1047.3*	*1111.0*	*1.6185*
160	*24.54*	*1058.2*	*1125.0*	*1.7171*	*17.870*	*1056.0*	*1122.1*	*1.6795*	*11.701*	*1051.6*	*1116.5*	*1.6275*
170	*24.98*	*1062.0*	*1129.9*	*1.7251*	*18.204*	*1059.9*	*1127.2*	*1.6877*	*11.938*	*1055.8*	*1122.0*	*1.6364*
180	*25.42*	*1065.7*	*1134.9*	*1.7328*	*18.535*	*1063.8*	*1132.4*	*1.6957*	*12.172*	*1059.9*	*1127.5*	*1.6449*
190	*25.85*	*1069.5*	*1139.8*	*1.7405*	*18.864*	*1067.6*	*1137.4*	*1.7036*	*12.403*	*1064.0*	*1132.9*	*1.6533*
200	*26.29*	*1073.2*	*1144.7*	*1.7479*	*19.191*	*1071.4*	*1142.5*	*1.7113*	*12.631*	*1068.1*	*1138.2*	*1.6615*
210	*26.72*	*1076.9*	*1149.5*	*1.7553*	*19.515*	*1075.2*	*1147.5*	*1.7188*	*12.857*	*1072.1*	*1143.5*	*1.6694*
220	27.15	1080.6	1154.4	1.7624	*19.837*	*1079.0*	*1152.4*	*1.7762*	*13.081*	*1076.1*	*1148.7*	*1.6771*
230	27.57	1084.2	1159.2	1.7695	20.157	1082.8	1157.4	1.7335	*13.303*	*1080.0*	*1153.9*	*1.6847*
240	28.00	1087.9	1164.0	1.7764	20.475	1086.5	1162.3	1.7405	*13.523*	*1084.0*	*1159.0*	*1.6921*
250	28.42	1091.5	1168.8	1.7832	20.79	1090.3	1167.2	1.7475	*13.741*	*1087.9*	*1164.1*	*1.6994*
260	28.85	1095.2	1173.6	1.7899	21.11	1094.0	1172.1	1.7543	13.958	1091.7	1169.2	1.7064
270	29.27	1098.8	1178.4	1.7965	21.42	1097.7	1177.0	1.7610	14.173	1095.6	1174.2	1.7134
280	29.69	1102.4	1183.1	1.8030	21.73	1101.4	1181.8	1.7676	14.387	1099.4	1179.2	1.7202
290	30.11	1106.0	1187.9	1.8094	22.05	1105.0	1186.6	1.7741	14.600	1103.2	1184.2	1.7269
300	30.52	1109.6	1192.6	1.8157	22.36	1108.7	1191.5	1.7805	14.812	1106.9	1189.2	1.7334
310	30.94	1113.2	1197.4	1.8219	22.67	1112.4	1196.3	1.7868	15.023	1110.7	1194.1	1.7399
320	31.36	1116.8	1202.1	1.8280	22.98	1116.0	1201.0	1.7930	15.233	1114.4	1199.0	1.7462
330	31.77	1120.4	1206.8	1.8340	23.28	1119.7	1205.8	1.7991	15.442	1118.2	1203.9	1.7525
340	32.19	1124.0	1211.6	1.8400	23.59	1123.3	1210.6	1.8051	15.651	1121.9	1208.8	1.7586

(Continued)

TABLE A.3 (*Continued*)

Properties of Superheated Steam

	Vapor (psia)											
P psi (*T* Sat. °F)	14.696 (211.99)				20 (227.96)				30 (250.34)			
T °F	*v*	*u*	*h*	*s*	*v*	*u*	*h*	*s*	*v*	*u*	*h*	*s*
Sat.	26.80	1077.6	1150.5	1.7567	20.09	1082.0	1156.4	1.7320	13.748	1088.0	1164.3	1.6996
350	32.60	1127.6	1216.3	1.8458	23.90	1126.9	1215.4	1.8110	15.859	1125.6	1213.6	1.7646
360	33.02	1131.2	1221.0	1.8516	24.21	1130.6	1220.1	1.8168	16.067	1129.3	1218.5	1.7706
370	33.43	1134.8	1225.7	1.8574	24.51	1134.2	1224.9	1.8226	16.273	1133.0	1223.3	1.7765
380	33.84	1138.4	1230.5	1.8630	24.82	1137.8	1229.7	1.8283	16.480	1136.7	1228.1	1.7822
390	34.26	1142.0	1235.2	1.8686	25.12	1141.4	1234.4	1.8340	16.686	1140.3	1233.0	1.7880
400	34.67	1145.6	1239.9	1.8741	25.43	1145.1	1239.2	1.8395	16.891	1144.0	1237.8	1.7936
420	35.49	1152.8	1249.3	1.8850	26.03	1152.3	1248.7	1.8504	17.301	1151.4	1247.4	1.8047
440	36.31	1160.1	1258.8	1.8956	26.64	1159.6	1258.2	1.8611	17.709	1158.7	1257.0	1.8155
460	37.13	1167.3	1268.3	1.9060	27.25	1166.9	1267.7	1.8716	18.116	1166.1	1266.6	1.8260
480	37.95	1174.6	1277.8	1.9162	27.85	1174.2	1277.2	1.8819	18.523	1173.4	1276.2	1.8364
500	38.77	1181.8	1287.3	1.9263	28.46	1181.5	1286.8	1.8919	18.928	1180.8	1285.9	1.8465
520	39.59	1189.1	1296.8	1.9361	29.06	1188.8	1296.3	1.9018	19.333	1188.2	1295.5	1.8564
540	40.41	1196.5	1306.4	1.9457	29.66	1196.1	1305.9	1.9114	19.737	1195.5	1305.1	1.8661
560	41.22	1203.8	1315.9	1.9552	30.26	1203.5	1315.5	1.9210	20.140	1203.0	1314.8	1.8757
580	42.04	1211.2	1325.5	1.9645	30.87	1210.9	1325.2	1.9303	20.543	1210.4	1324.4	1.8851
600	42.86	1218.6	1335.2	1.9737	31.47	1218.4	1334.8	1.9395	20.95	1217.8	1334.1	1.8943
620	43.67	1226.1	1344.8	1.9827	32.07	1225.8	1344.5	1.9485	21.35	1225.3	1343.8	1.9034
640	44.49	1233.5	1354.5	1.9916	32.67	1233.3	1354.2	1.9575	21.75	1232.8	1353.6	1.9123
660	45.30	1241.0	1364.2	2.0004	33.27	1240.8	1363.9	1.9662	22.15	1240.4	1363.3	1.9211
680	46.12	1248.6	1374.0	2.0090	33.87	1248.3	1373.7	1.9749	22.55	1247.9	1373.1	1.9298

(Continued)

TABLE A.3 (*Continued*)

Properties of Superheated Steam

	Vapor (psia)											
***P* psi (*T* Sat. °F)**	**14.696 (211.99)**				**20 (227.96)**				**30 (250.34)**			
***T* °F**	*v*	*u*	*h*	*s*	*v*	*u*	*h*	*s*	*v*	*u*	*h*	*s*
700	46.93	1256.1	1383.8	2.0175	34.47	1255.9	1383.5	1.9834	22.95	1255.5	1383.0	1.9384
720	47.75	1263.7	1393.6	2.0259	35.07	1263.5	1393.3	1.9918	23.35	1263.2	1392.8	1.9468
740	48.56	1271.4	1403.4	2.0342	35.66	1271.2	1403.2	2.0001	23.75	1270.8	1402.7	1.9551
760	49.37	1279.0	1413.3	2.0424	36.26	1278.8	1413.0	2.0082	24.15	1278.5	1412.6	1.9633
780	50.19	1286.7	1423.2	2.0504	36.86	1286.5	1423.0	2.0163	24.55	1286.2	1422.5	1.9714
800	51.00	1294.4	1433.1	2.0584	37.46	1294.3	1432.9	2.0243	24.95	1294.0	1432.5	1.9793
850	53.03	1313.9	1458.1	2.0778	38.96	1313.8	1457.9	2.0438	25.95	1313.5	1457.6	1.9988
900	55.07	1333.6	1483.4	2.0967	40.45	1333.5	1483.2	2.0627	26.95	1333.2	1482.8	1.0178
950	57.10	1353.5	1508.8	2.1151	41.94	1353.4	1508.6	2.0810	27.95	1353.2	1508.3	2.0362
1000	59.13	1373.7	1534.5	2.1330	43.44	1373.5	1534.3	2.0989	28.95	1373.3	1534.0	2.0541
1100	63.19	1414.6	1586.4	2.1674	46.42	1414.5	1586.3	2.1334	30.94	1414.3	1586.1	2.0886
1200	67.25	1456.5	1639.3	2.2003	49.41	1456.4	1639.2	2.1663	32.93	1456.2	1639.1	2.1215
1300	71.30	1499.3	1693.2	2.2318	52.39	1499.2	1693.1	2.1978	34.92	1499.1	1692.9	2.1530
1400	75.36	1543.0	1747.9	2.2621	55.37	1542.9	1747.9	2.2281	36.91	1542.8	1747.7	2.1833
1500	79.42	1587.6	1803.6	2.2912	58.35	1587.6	1803.5	2.2572	38.90	1587.5	1803.4	2.2125
1600	83.47	1633.2	1860.2	2.3194	61.33	1633.2	1860.1	2.2854	40.88	1633.1	1860.0	2.2407
1800	91.58	1727.0	1976.1	2.3731	67.29	1727.0	1976.1	2.3391	44.86	1726.9	1976.0	2.2944
2000	99.69	1824.4	2095.5	2.4237	73.25	1824.3	2095.4	2.3897	48.83	1824.2	2095.3	2.3450
2200	107.80	1924.8	2218.0	2.4716	79.21	1924.8	2218.0	2.4376	52.81	1924.7	2217.9	2.3929
2400	115.91	2028.1	2343.4	2.5170	85.17	2028.1	2343.3	2.4830	56.78	2028.0	2343.3	2.4383

(Continued)

TABLE A.3 (*Continued*)

Properties of Superheated Steam

	Vapor (psia)											
P (*T* Sat.)	40 (267.26)				50(281.03)				60(292.73)			
T	*v*	*u*	*h*	*s*	*v*	*u*	*h*	*s*	*v*	*u*	*h*	*s*
Sat.	10.501	1092.3	1170.0	1.6767	8.518	1095.6	1174.4	1.6589	7.177	1098.3	1178.0	1.6444
200	9.346	1064.6	1133.8	1.6243	7.370	1060.9	1129.1	1.5940	6.047	1057.1	1124.2	1.5680
210	9.523	1068.8	1139.3	1.6327	7.519	1065.4	1135.0	1.6029	6.178	1061.9	1130.5	1.5774
220	9.699	1073.0	1144.8	1.6409	7.665	1069.9	1140.8	1.6115	6.307	1066.6	1136.6	1.5864
230	9.872	1077.2	1150.3	1.6488	7.810	1074.2	1146.5	1.6198	6.432	1071.2	1142.6	1.5952
240	10.043	1081.3	1155.6	1.6565	7.952	1078.6	1152.1	1.6279	6.556	1075.7	1148.5	1.6036
250	10.212	1085.4	1161.0	1.6641	8.092	1082.8	1157.7	1.6358	6.677	1080.1	1154.3	1.6118
260	10.380	1089.4	1166.2	1.6714	8.231	1087.0	1163.1	1.6434	6.797	1084.5	1160.0	1.6198
270	10.546	1093.4	1171.4	1.6786	8.368	1091.1	1168.5	1.6509	6.915	1088.8	1165.6	1.6275
280	10.711	1097.3	1176.6	1.6857	8.504	1095.2	1173.9	1.6582	7.031	1093.0	1171.1	1.6351
290	10.875	1101.2	1181.7	1.6926	8.639	1099.2	1179.2	1.6653	7.146	1097.2	1176.5	1.6424
300	11.038	1105.1	1186.8	1.6993	8.772	1103.2	1184.4	1.6722	7.260	1101.3	1181.9	1.6496
310	11.200	1109.0	1191.9	1.7059	8.904	1107.2	1189.6	1.6790	7.373	1105.4	1187.3	1.6565
320	11.360	1112.8	1196.9	1.7124	9.036	1111.2	1194.8	1.6857	7.485	1109.5	1192.6	1.6634
330	11.520	1116.6	1201.9	1.7188	9.166	1115.1	1199.9	1.6922	7.596	1113.5	1197.8	1.6700
340	11.680	1120.4	1206.9	1.7251	9.296	1119.0	1205.0	1.6986	7.706	1117.4	1203.0	1.6766
350	11.838	1124.2	1211.8	1.7312	9.425	1122.8	1210.0	1.7049	7.815	1121.4	1208.2	1.6830
360	11.996	1128.0	1216.8	1.7373	9.553	1126.7	1215.1	1.7110	7.924	1125.3	1213.3	1.6893
370	12.153	1131.7	1221.7	1.7432	9.681	1130.5	1220.1	1.7171	8.032	1129.2	1218.4	1.6955
380	12.310	1135.5	1226.6	1.7491	9.808	1134.3	1225.0	1.7231	8.139	1133.1	1223.5	1.7015

(*Continued*)

TABLE A.3 (*Continued*)

Properties of Superheated Steam

	Vapor (psia)											
P (T Sat.)	**40 (267.26)**				**50(281.03)**				**60(292.73)**			
T	*v*	*u*	*h*	*s*	*v*	*u*	*h*	*s*	*v*	*u*	*h*	*s*
390	12.467	1139.2	1231.5	1.7549	9.935	1138.1	1230.0	1.7290	8.246	1136.9	1228.5	1.7075
400	12.623	1143.0	1236.4	1.7606	10.061	1141.9	1235.0	1.7348	8.353	1140.8	1233.5	1.7134
420	12.933	1150.4	1246.1	1.7718	10.312	1149.4	1244.8	1.7461	8.565	1148.4	1243.5	1.7249
440	13.243	1157.8	1255.8	1.7828	10.562	1156.9	1254.6	1.7572	8.775	1156.0	1253.4	1.7360
460	13.551	1165.2	1265.5	1.7934	10.811	1164.4	1264.4	1.7679	8.984	1163.6	1263.3	1.7469
480	13.858	1172.7	1275.2	1.8038	11.059	1171.9	1274.2	1.7784	9.192	1171.1	1273.2	1.7575
500	14.164	1180.1	1284.9	1.8140	11.305	1179.4	1284.0	1.7887	9.399	1178.6	1283.0	1.7678
520	14.469	1187.5	1294.6	1.8240	11.551	1186.8	1293.7	1.7988	9.606	1186.2	1292.8	1.7780
540	14.774	1194.9	1304.3	1.8338	11.796	1194.3	1303.5	1.8086	9.811	1193.7	1302.6	1.7879
560	15.078	1202.4	1314.0	1.8434	12.041	1201.8	1313.2	1.8183	10.016	1201.2	1312.4	1.7976
580	15.382	1209.9	1323.7	1.8529	12.285	1209.3	1323.0	1.8277	10.221	1208.8	1322.2	1.8071
600	15.685	1217.3	1333.4	1.8621	12.529	1216.8	1332.8	1.8371	10.425	1216.3	1332.1	1.8165
620	15.988	1224.8	1343.2	1.8713	12.772	1224.4	1342.5	1.8462	10.628	1223.9	1341.9	1.8257
640	16.291	1232.4	1353.0	1.8802	13.015	1231.9	1352.4	1.8552	10.832	1231.5	1351.7	1.8347
660	16.593	1239.9	1362.8	1.8891	13.258	1239.5	1362.2	1.8641	11.035	1239.1	1361.6	1.8436
680	16.895	1247.5	1372.6	1.8977	13.500	1247.1	1372.0	1.8728	11.237	1246.7	1371.5	1.8523
700	17.196	1255.1	1382.4	1.9063	13.742	1254.8	1381.9	1.8814	11.440	1254.4	1381.4	1.8609
720	17.498	1262.8	1392.3	1.9147	13.984	1262.4	1391.8	1.8898	11.642	1262.0	1391.3	1.8694
740	17.799	1270.5	1402.2	1.9231	14.226	1270.1	1401.7	1.8982	11.844	1269.7	1401.2	1.8778
760	18.100	1278.2	1412.1	1.9313	14.467	1277.8	1411.7	1.9064	12.045	1277.5	1411.2	1.8860
780	18.401	1285.9	1422.1	1.9394	14.708	1285.6	1421.7	1.9145	12.247	1285.2	1421.2	1.8942

(*Continued*)

TABLE A.3 (*Continued*)

Properties of Superheated Steam

	Vapor (psia)											
P (T Sat.)	40 (267.26)				50(281.03)				60(292.73)			
T	v	u	h	s	v	u	h	s	v	u	h	s
800	18.701	1293.7	1432.1	1.9474	14.949	1293.3	1431.7	1.9225	12.448	1293.0	1431.2	1.9022
850	19.452	1313.2	1457.2	1.9669	15.551	1312.9	1456.8	1.9421	12.951	1312.7	1456.4	1.9218
900	20.202	1333.0	1482.5	1.9859	16.152	1332.7	1482.2	1.9611	13.452	1332.5	1481.8	1.9408
950	20.951	1352.9	1508.0	2.0043	16.753	1352.7	1507.7	1.9796	13.954	1352.5	1507.4	1.9593
1000	21.700	1373.1	1533.8	2.0223	17.352	1372.9	1533.5	1.9975	14.454	1372.7	1533.2	1.9773
1100	23.20	1414.2	1585.9	2.0568	18.551	1414.0	1585.6	2.0321	15.454	1413.8	1585.4	2.0119
1200	24.69	1456.1	1638.9	2.0897	19.747	1456.0	1638.7	2.0650	16.452	1455.8	1638.5	2.0448
1300	26.18	1498.9	1692.8	2.1212	20.943	1498.8	1692.6	2.0966	17.449	1498.7	1692.4	2.0764
1400	27.68	1542.7	1747.6	2.1515	22.138	1542.6	1747.4	2.1269	18.445	1542.5	1747.3	2.1067
1500	29.17	1587.4	1803.3	2.1807	23.332	1587.3	1803.2	2.1561	19.441	1587.2	1803.0	2.1359
1600	30.66	1633.0	1859.9	2.2089	24.53	1632.9	1859.8	2.1843	20.44	1632.8	1859.7	2.1641
1800	33.64	1726.9	1975.9	2.2626	26.91	1726.8	1975.8	2.2380	22.43	1726.7	1975.7	2.2179
2000	36.62	1824.2	2095.3	2.3132	29.30	1824.1	2095.2	2.2886	24.41	1824.0	2095.1	2.2685
2200	39.61	1924.7	2217.8	2.3611	31.68	1924.6	2217.8	2.3365	26.40	1924.5	2217.7	2.3164
2400	42.59	2028.0	2343.2	2.4066	34.07	2027.9	2343.1	2.3820	28.39	2027.8	2343.1	2.3618

(*Continued*)

TABLE A.3 (*Continued*)

Properties of Superheated Steam

	Vapor (psia)															
P (T Sat.)	**80 (312.07)**				**100 (327.86)**				**140 (353.08)**				**180 (373.13)**			
T	*v*	*u*	*h*	*s*	*v*	*u*	*h*	*s*	*v*	*u*	*h*	*s*	*v*	*u*	*h*	*s*
Sat.	5.474	1102.6	1183.6	1.6214	4.434	1105.8	1187.8	1.6034	3.221	1110.3	1193.8	1.5761	2.533	1113.4	1197.8	1.5553
250	4.902	*1074.5*	*1147.1*	*1.5720*												
260	4.998	*1079.3*	*1153.3*	*1.5806*												
270	5.093	*1083.9*	*1159.3*	*1.5890*												
280	5.186	*1088.5*	*1165.2*	*1.5970*												
290	5.277	*1092.9*	*1171.1*	*1.6049*												
300	5.367	*1097.4*	*1176.8*	*1.6125*	*4.228*	*1093.1*	*1171.4*	*1.5822*								
310	5.456	*1101.7*	*1182.5*	*1.6199*	*4.303*	*1097.7*	*1177.4*	*1.5900*	*2.978*	*1089.2*	*1166.4*	*1.5414*				
320	5.544	1106.0	1188.0	1.6271	*4.377*	*1102.3*	*1183.3*	*1.5976*	*3.037*	*1094.3*	*1173.0*	*1.5499*				
330	5.631	1110.2	1193.5	1.6341	4.449	1106.7	1189.1	1.6050	*3.094*	*1099.3*	*1179.4*	*1.5581*				
340	5.717	1114.3	1199.0	1.6409	4.521	1111.1	1194.8	1.6121	*3.150*	*1104.1*	*1185.7*	*1.5661*				
350	5.802	1118.5	1204.3	1.6476	4.592	1115.4	1200.4	1.6191	*3.205*	*1108.9*	*1191.9*	*1.5737*	*2.429*	*1101.7*	*1182.6*	*1.5368*
360	5.886	1122.5	1209.7	1.6541	4.662	1119.7	1205.9	1.6259	3.259	1113.5	1198.0	1.5812	*2.475*	*1106.9*	*1189.3*	*1.5450*
370	5.970	1126.6	1215.0	1.6605	4.731	1123.9	1211.4	1.6326	3.312	1118.1	1203.9	1.5884	*2.520*	*1111.9*	*1195.8*	*1.5529*
380	6.053	1130.6	1220.2	1.6668	4.799	1128.0	1216.8	1.6391	3.364	1122.6	1209.7	1.5954	2.563	1116.7	1202.1	1.5605
390	6.135	1134.6	1225.4	1.6730	4.867	1132.1	1222.2	1.6455	3.415	1127.0	1215.5	1.6022	2.606	1121.5	1208.3	1.5678
400	6.217	1138.5	1230.6	1.6790	4.934	1136.2	1227.5	1.6517	3.466	1131.4	1221.2	1.6088	2.648	1126.2	1214.4	1.5749
410	6.299	1142.5	1235.7	1.6850	5.001	1140.3	1232.8	1.6578	3.517	1135.7	1226.8	1.6153	2.690	1130.8	1220.4	1.5818
420	6.380	1146.4	1240.8	1.6908	5.068	1144.3	1238.1	1.6638	3.567	1139.9	1232.3	1.6217	2.731	1135.3	1226.3	1.5886

(Continued)

TABLE A.3 *(Continued)*

Properties of Superheated Steam

	Vapor (psia)															
P (T Sat.)	80 (312.07)				100 (327.86)				140 (353.08)				180 (373.13)			
T	v	u	h	s	v	u	h	s	v	u	h	s	v	u	h	s
430	6.460	1150.3	1245.9	1.6966	5.134	1148.3	1243.3	1.6697	3.616	1144.1	1237.8	1.6279	2.771	1139.8	1232.1	1.5951
440	6.541	1154.2	1251.0	1.7022	5.199	1152.3	1248.5	1.6755	3.665	1148.3	1243.3	1.6339	2.811	1144.2	1237.8	1.6015
450	6.621	1158.0	1256.0	1.7078	5.265	1156.2	1253.6	1.6812	3.713	1152.4	1248.6	1.6399	2.850	1148.5	1243.4	1.6078
460	6.700	1161.9	1261.1	1.7133	5.330	1160.1	1258.8	1.6868	3.762	1156.5	1254.0	1.6458	2.889	1152.8	1249.0	1.6139
470	6.780	1165.7	1266.1	1.7188	5.394	1164.1	1263.9	1.6923	3.810	1160.6	1259.3	1.6515	2.928	1157.0	1254.6	1.6199
480	6.859	1169.6	1271.1	1.7241	5.459	1168.0	1269.0	1.6978	3.857	1164.7	1264.6	1.6572	2.966	1161.3	1260.1	1.6257
490	6.938	1173.4	1276.1	1.7294	5.523	1171.8	1274.0	1.7032	3.904	1168.7	1269.9	1.6627	3.005	1165.4	1265.5	1.6315
500	7.017	1177.2	1281.1	1.7346	5.587	1175.7	1279.1	1.7085	3.952	1172.7	1275.1	1.6682	3.042	1169.6	1270.9	1.6372
600	7.794	1215.3	1330.7	1.7838	6.216	1214.2	1329.3	1.7582	4.412	1212.1	1326.4	1.7191	3.409	1210.0	1323.5	1.6893
620	7.948	1222.9	1340.6	1.7930	6.340	1221.9	1339.3	1.7675	4.502	1219.9	1336.6	1.7286	3.481	1217.9	1333.9	1.6990
640	8.102	1230.5	1350.5	1.8021	6.464	1229.6	1349.2	1.7767	4.592	1227.7	1346.7	1.7379	3.552	1225.8	1344.1	1.7084
660	8.255	1238.2	1360.4	1.8111	6.588	1237.3	1359.2	1.7857	4.682	1235.6	1356.8	1.7470	3.623	1233.8	1354.4	1.7177
680	8.408	1245.9	1370.4	1.8199	6.711	1245.0	1369.2	1.7946	4.771	1243.4	1367.0	1.7560	3.693	1241.7	1364.7	1.7268
700	8.561	1253.6	1380.3	1.8285	6.834	1252.8	1379.2	1.8033	4.860	1251.2	1377.1	1.7648	3.763	1249.6	1374.9	1.7357
720	8.714	1261.3	1390.3	1.8371	6.957	1260.5	1389.3	1.8118	4.949	1259.0	1387.2	1.7735	3.833	1257.5	1385.2	1.7445
740	8.866	1269.0	1400.3	1.8455	7.079	1268.3	1399.3	1.8203	5.037	1266.9	1397.4	1.7820	3.903	1265.4	1395.4	1.7531
760	9.018	1276.8	1410.3	1.8538	7.201	1276.1	1409.4	1.8286	5.125	1274.7	1407.5	1.7904	3.972	1273.4	1405.7	1.7615
780	9.170	1284.6	1420.3	1.8619	7.324	1283.9	1419.5	1.8368	5.214	1282.6	1417.7	1.7986	4.041	1281.3	1415.9	1.7699
800	9.321	1292.4	1430.4	1.8700	7.445	1291.8	1429.6	1.8449	5.301	1290.5	1427.9	1.8068	4.110	1289.3	1426.2	1.7781
1000	10.831	1372.3	1532.6	1.9453	8.657	1371.9	1532.1	1.9204	6.173	1371.0	1531.0	1.8827	4.793	1370.2	1529.8	1.8545

(Continued)

TABLE A.3 (*Continued*)

Properties of Superheated Steam

	Vapor (psia)															
P (T Sat.)	80 (312.07)				100 (327.86)				140 (353.08)				180 (373.13)			
T	v	u	h	s	v	u	h	s	v	u	h	s	v	u	h	s
1100	11.583	1413.5	1584.9	1.9799	9.260	1413.1	1584.5	1.9551	6.605	1412.4	1583.6	1.9176	5.131	1411.7	1582.6	1.8894
1200	12.333	1455.5	1638.1	2.0130	9.861	1455.2	1637.7	1.9882	7.036	1454.6	1636.9	1.9507	5.467	1454.0	1636.1	1.9227
1300	13.081	1498.4	1692.1	2.0446	10.461	1498.2	1691.8	2.0198	7.466	1497.7	1691.1	1.9824	5.802	1497.2	1690.4	1.9544
1400	13.830	1542.3	1747.0	2.0749	11.060	1542.0	1746.7	2.0502	7.895	1541.6	1746.1	2.0129	6.137	1541.2	1745.6	1.9849
1500	14.577	1587.0	1802.8	2.1041	11.659	1586.8	1802.5	2.0794	8.324	1586.4	1802.0	2.0421	6.471	1586.0	1801.5	2.0142
1600	15.324	1632.6	1859.5	2.1323	12.257	1632.4	1859.3	2.1076	8.752	1632.1	1858.8	2.0704	6.804	1631.7	1858.4	2.0425
1800	16.818	1726.5	1975.5	2.1861	13.452	1726.4	1975.3	2.1614	9.607	1726.1	1975.0	2.1242	7.470	1725.8	1974.6	2.0964
2000	18.310	1823.9	2094.9	2.2367	14.647	1823.7	2094.8	2.2121	10.461	1823.5	2094.5	2.1749	8.135	1823.2	2094.2	2.1470
2200	19.802	1924.4	2217.5	2.2846	15.842	1924.3	2217.4	2.2600	11.315	1924.0	2217.1	2.2228	8.800	1923.7	2216.8	2.1950
2400	21.294	2027.7	2342.9	2.3301	17.036	2027.5	2342.8	2.3054	12.169	2027.3	2342.5	2.2682	9.465	2027.0	2342.2	2.2404

(*Continued*)

TABLE A.3 (*Continued*)
Properties of Superheated Steam

	Vapor (psia)															
P (*T* Sat.)	200 (381.86)				300 (417.43)				350 (431.82)				400 (444.70)			
T	*v*	*u*	*h*	*s*	*v*	*u*	*h*	*s*	*v*	*u*	*h*	*s*	*v*	*u*	*h*	*s*
Sat.	2.289	1114.6	1199.3	1.5464	1.5442	1118.2	1203.9	1.511	1.3267	1119.0	1204.9	1.4978	1.1620	1119.5	1205.5	1.4856
400	2.361	1123.5	1210.8	1.5600	*1.4915*	*1108.2*	*1191.0*	*1.4967*								
410	2.399	1128.2	1217.0	1.5672	*1.5221*	*1114.0*	*1198.5*	*1.5054*								
420	2.437	1132.9	1223.1	1.5741	1.5517	1119.6	1205.7	1.5136	*1.2950*	*1112.0*	*1195.9*	*1.4875*				
430	2.475	1137.5	1229.1	1.5809	1.5805	1125.0	1212.7	1.5216	*1.3219*	*1117.9*	*1203.6*	*1.4962*	*1.1257*	*1110.3*	*1193.6*	1.4723
440	2.511	1142.0	1234.9	1.5874	1.6086	1130.3	1219.6	1.5292	1.3480	1123.7	1211.0	1.5045	*1.1506*	*1116.6*	*1201.7*	*1.4814*
450	2.548	1146.4	1240.7	1.5938	1.6361	1135.4	1226.2	1.5365	1.3733	1129.2	1218.2	1.5125	1.1745	1122.6	1209.6	1.4901
460	2.584	1150.8	1246.5	1.6001	1.6630	1140.4	1232.7	1.5436	1.3979	1134.6	1225.2	1.5201	1.1977	1128.5	1217.1	1.4984
470	2.619	1155.2	1252.1	1.6062	1.6894	1145.3	1239.1	1.5505	1.4220	1139.9	1232.0	1.5275	1.2202	1134.1	1224.4	1.5063
480	2.654	1159.5	1257.7	1.6122	1.7154	1150.1	1245.3	1.5572	1.4455	1145.0	1238.6	1.5346	1.2421	1139.6	1231.5	1.5139
490	2.689	1163.7	1263.3	1.6181	1.7410	1154.8	1251.5	1.5637	1.4686	1150.0	1245.1	1.5415	1.2634	1144.9	1238.4	1.5212
500	2.724	1168.0	1268.8	1.6239	1.7662	1159.5	1257.5	1.5701	1.4913	1154.9	1251.5	1.5482	1.2843	1150.1	1245.2	1.5282
510	2.758	1172.2	1274.2	1.6295	1.7910	1164.1	1263.5	1.5763	1.5136	1159.7	1257.8	1.5546	1.3048	1155.2	1251.8	1.5351
520	2.792	1176.3	1279.7	1.6351	1.8156	1168.6	1269.4	1.5823	1.5356	1164.5	1263.9	1.5610	1.3249	1160.2	1258.2	1.5417
530	2.826	1180.5	1285.0	1.6405	1.8399	1173.1	1275.2	1.5882	1.5572	1169.1	1270.0	1.5671	1.3447	1165.0	1264.6	1.5481
540	2.860	1184.6	1290.4	1.6459	1.8640	1177.5	1281.0	1.5940	1.5786	1173.7	1276.0	1.5731	1.3642	1169.8	1270.8	1.5544
550	2.893	1188.7	1295.7	1.6512	1.8878	1181.9	1286.7	1.5997	1.5998	1178.3	1281.9	1.5790	1.3833	1174.6	1277.0	1.5605
560	2.926	1192.7	1301.0	1.6565	1.9114	1186.2	1292.3	1.6052	1.6207	1182.8	1287.7	1.5848	1.4023	1179.2	1283.0	1.5665
570	2.960	1196.8	1306.3	1.6616	1.9348	1190.5	1297.9	1.6107	1.6414	1187.2	1293.5	1.5904	1.4210	1183.8	1289.0	1.5723
580	2.993	1200.8	1311.6	1.6667	1.9580	1194.8	1303.5	1.6161	1.6619	1191.6	1299.3	1.5960	1.4395	1188.4	1294.9	1.5781
590	3.025	1204.9	1316.8	1.6717	1.9811	1199.0	1309.0	1.6214	1.6823	1196.0	1304.9	1.6014	1.4579	1192.9	1300.8	1.5837

(Continued)

TABLE A.3 (*Continued*)

Properties of Superheated Steam

	Vapor (psia)															
P (*T* Sat.)	200 (381.86)				300 (417.43)				350 (431.82)				400 (444.70)			
T	*v*	*u*	*h*	*s*	*v*	*u*	*h*	*s*	*v*	*u*	*h*	*s*	*v*	*u*	*h*	*s*
600	3.058	1208.9	1322.1	1.6767	2.004	1203.2	1314.5	1.6266	1.7025	1200.3	1310.6	1.6068	1.4760	1197.3	1306.6	1.5892
620	3.123	1216.9	1332.5	1.6864	2.049	1211.6	1325.4	1.6368	1.7424	1208.9	1321.8	1.6172	1.5118	1206.1	1318.0	1.5999
640	3.188	1224.9	1342.9	1.6959	2.094	1220.0	1336.2	1.6467	1.7818	1217.4	1332.8	1.6274	1.5471	1214.8	1329.3	1.6103
660	3.252	1232.8	1353.2	1.7053	2.139	1228.2	1347.0	1.6564	1.8207	1225.8	1343.8	1.6372	1.5819	1223.4	1340.5	1.6203
680	3.316	1240.8	1363.5	1.7144	2.183	1236.4	1357.6	1.6658	1.8593	1234.2	1354.6	1.6469	1.6163	1231.9	1351.6	1.6301
700	3.379	1248.8	1373.8	1.7234	2.227	1244.6	1368.3	1.6751	1.8975	1242.5	1365.4	1.6562	1.6503	1240.4	1362.5	1.6397
720	3.442	1256.7	1384.1	1.7322	2.270	1252.8	1378.9	1.6841	1.9354	1250.8	1376.2	1.6654	1.6840	1248.8	1373.4	1.6490
740	3.505	1264.7	1394.4	1.7408	2.314	1261.0	1389.4	1.6930	1.9731	1259.1	1386.9	1.6744	1.7175	1257.2	1384.3	1.6581
760	3.568	1272.7	1404.7	1.7493	2.357	1269.1	1400.0	1.7017	2.0104	1267.3	1397.5	1.6832	1.7506	1265.5	1395.1	1.6670
780	3.631	1280.6	1415.0	1.7577	2.400	1277.3	1410.5	1.7103	2.0476	1275.6	1408.2	1.6919	1.7836	1273.8	1405.9	1.6758
800	3.693	1288.6	1425.3	1.7660	2.442	1285.4	1421.0	1.7187	2.085	1283.8	1418.8	1.7004	1.8163	1282.1	1416.6	1.6844
820	3.755	1296.6	1435.6	1.7741	2.485	1293.6	1431.5	1.7270	2.121	1292.0	1429.4	1.7088	1.8489	1290.5	1427.3	1.6928
840	3.818	1304.7	1446.0	1.7821	2.527	1301.7	1442.0	1.7351	2.158	1300.3	1440.0	1.7170	1.8813	1298.8	1438.0	1.7011
860	3.879	1312.7	1456.3	1.7900	2.569	1309.9	1452.5	1.7432	2.194	1308.5	1450.6	1.7251	1.9135	1307.1	1448.7	1.7093
880	3.941	1320.8	1466.7	1.7978	2.611	1318.1	1463.1	1.7511	2.231	1316.7	1461.2	1.7331	1.9456	1315.4	1459.4	1.7173
900	4.003	1328.9	1477.1	1.8055	2.653	1326.3	1473.6	1.7589	2.267	1325.0	1471.8	1.7409	1.9776	1323.7	1470.1	1.7252
920	4.064	1337.0	1487.5	1.8131	2.695	1334.5	1484.1	1.7666	2.303	1333.3	1482.5	1.7487	2.0094	1332.0	1480.8	1.7330

(*Continued*)

TABLE A.3 (*Continued*)

Properties of Superheated Steam

	Vapor (psia)															
P (*T* Sat.)	200 (381.86)				300 (417.43)				350 (431.82)				400 (444.70)			
T	*v*	*u*	*h*	*s*	*v*	*u*	*h*	*s*	*v*	*u*	*h*	*s*	*v*	*u*	*h*	*s*
940	4.126	1345.2	1497.9	1.8206	2.736	1342.8	1494.7	1.7742	2.339	1341.6	1493.1	1.7563	2.0411	1340.4	1491.5	1.7407
960	4.187	1353.3	1508.3	1.8280	2.778	1351.1	1505.3	1.7817	2.375	1349.9	1503.7	1.7639	2.0727	1348.7	1502.2	1.7483
980	4.249	1361.6	1518.8	1.8353	2.819	1359.3	1515.8	1.7891	2.411	1358.2	1514.4	1.7713	2.1043	1357.1	1512.9	1.7558
1000	4.310	1369.8	1529.3	1.8425	2.860	1367.7	1526.5	1.7964	2.446	1366.6	1525.0	1.7787	2.136	1365.5	1523.6	1.7632
1020	4.371	1378.0	1539.8	1.8497	2.902	1376.0	1537.1	1.8036	2.482	1375.0	1535.7	1.7859	2.167	1373.9	1534.3	1.7705
1040	4.432	1386.3	1550.3	1.8568	2.943	1384.4	1547.7	1.8108	2.517	1383.4	1546.4	1.7931	2.198	1382.4	1545.1	1.7777
1060	4.493	1394.6	1560.9	1.8638	2.984	1392.7	1558.4	1.8178	2.553	1391.8	1557.1	1.8002	2.229	1390.8	1555.9	1.7849
1080	4.554	1403.0	1571.5	1.8707	3.025	1401.2	1569.1	1.8248	2.588	1400.2	1567.9	1.8072	2.261	1399.3	1566.6	1.7919
1100	4.615	1411.4	1582.2	1.8776	3.066	1409.6	1579.8	1.8317	2.624	1408.7	1578.6	1.8142	2.292	1407.8	1577.4	1.7989
1200	4.918	1453.7	1635.7	1.9109	3.270	1452.2	1633.8	1.8653	2.799	1451.5	1632.8	1.8478	2.446	1450.7	1631.8	1.8327
1300	5.220	1496.9	1690.1	1.9427	3.473	1495.6	1688.4	1.8973	2.974	1495.0	1687.6	1.8799	2.599	1494.3	1686.8	1.8648
1400	5.521	1540.9	1745.3	1.9732	3.675	1539.8	1743.8	1.9279	3.148	1539.3	1743.1	1.9106	2.752	1538.7	1742.4	1.8956
1500	5.822	1585.8	1801.3	2.0025	3.877	1584.8	1800.0	1.9573	3.321	1584.3	1799.4	1.9401	2.904	1583.8	1798.8	1.9251
1600	6.123	1631.6	1858.2	2.0308	4.078	1630.7	1857.0	1.9857	3.494	1630.2	1856.5	1.9685	3.055	1629.8	1855.9	1.9535
1800	6.722	1725.6	1974.4	2.0847	4.479	1724.9	1973.5	2.0396	3.838	1724.5	1973.1	2.0225	3.357	1724.1	1972.6	2.0076
2000	7.321	1823.0	2094.0	2.1354	4.879	1822.3	2093.2	2.0904	4.182	1822.0	2092.8	2.0733	3.658	1821.6	2092.4	2.0584
2200	7.920	1923.6	2216.7	2.1833	5.280	1922.9	2216.0	2.1384	4.525	1922.5	2215.6	2.1212	3.959	1922.2	2215.2	2.1064
2400	8.518	2026.8	2342.1	2.2288	5.679	2026.1	2341.4	2.1838	4.868	2025.8	2341.1	2.1667	4.260	2025.4	2340.8	2.1519

(*Continued*)

TABLE A.3 (*Continued*)

Properties of Superheated Steam

	Vapor (psia)															
P (*T* Sat.)	450 (456.39)				500 (467.13)				550 (477.07)				600 (486.33)			
T	*v*	*u*	*h*	*s*	*v*	*u*	*h*	*s*	*v*	*u*	*h*	*s*	*v*	*u*	*h*	*s*
Sat	1.0326	1119.6	1205.6	1.4746	0.9283	1119.4	1205.3	1.4645	0.8423	1119.1	1204.8	1.4551	0.7702	1118.6	1204.1	1.4464
500	1.1226	1145.1	1238.5	1.5097	0.9924	1139.7	1231.5	1.4923	0.8850	1134.0	1224.1	1.4755	0.7947	1128.0	1216.2	1.4592
510	1.1417	1150.4	1245.5	1.5170	1.0106	1145.4	1238.9	1.4999	0.9056	1140.1	1232.0	1.4836	0.8118	1134.5	1224.6	1.4679
520	1.1605	1155.7	1252.3	1.5239	1.0283	1150.9	1246.1	1.5073	0.9196	1146.0	1239.6	1.4914	0.8283	1140.7	1232.7	1.4762
530	1.1788	1160.8	1258.0	1.5307	1.0456	1156.3	1253.1	1.5144	0.9361	1151.7	1246.9	1.4989	0.8443	1146.8	1240.5	1.4841
540	1.1969	1165.8	1265.5	1.5372	1.0625	1161.6	1259.9	1.5212	0.9522	1157.2	1254.1	1.5061	0.8598	1152.6	1248.0	1.4917
550	1.2146	1170.7	1271.9	1.5436	1.0792	1166.7	1266.6	1.5279	0.9679	1162.6	1261.1	1.5131	0.8749	1158.2	1255.4	1.4990
560	1.2320	1175.6	1278.2	1.5498	1.0955	1171.8	1273.1	1.5343	0.9834	1167.8	1267.9	1.5198	0.8896	1163.7	1262.5	1.5060
570	1.2492	1180.3	1284.4	1.5559	1.1115	1176.7	1279.5	1.5406	0.9985	1173.0	1274.6	1.5263	0.9040	1169.1	1269.5	1.5128
580	1.2662	1185.0	1290.5	1.5618	1.1273	1181.6	1285.9	1.5467	1.0133	1178.0	1281.1	1.5327	0.9181	1174.3	1276.3	1.5194
590	1.2830	1189.7	1296.5	1.5675	1.1429	1186.4	1292.1	1.5527	1.0280	1183.0	1287.6	1.5388	0.9320	1179.5	1282.9	1.5253
600	1.2996	1194.3	1302.5	1.5732	1.1583	1191.1	1298.3	1.5585	1.0424	1187.9	1293.9	1.5448	0.9456	1184.5	1289.5	1.532
610	1.3160	1198.8	1308.4	1.5788	1.1735	1195.8	1304.3	1.5642	1.0566	1192.7	1300.2	1.5507	0.9590	1189.5	1295.9	1.5381
620	1.3323	1203.3	1314.2	1.5842	1.1885	1200.4	1310.4	1.5698	1.0706	1197.4	1306.4	1.5565	0.9722	1194.4	1302.3	1.5440
630	1.3484	1207.8	1320.0	1.5896	1.2033	1205.0	1316.3	1.5753	1.0845	1202.1	1312.5	1.5621	0.9853	1199.2	1308.6	1.5497
640	1.3644	1212.2	1325.8	1.5948	1.2181	1209.5	1322.2	1.5807	1.0982	1206.7	1318.5	1.5676	0.9982	1203.9	1314.8	1.5554
650	1.3803	1216.6	1331.5	1.6000	1.2327	1214	1328.0	1.5860	1.1118	1211.3	1324.5	1.5730	1.0109	1208.6	1320.9	1.5609
660	1.396	1221	1337.2	1.6051	1.2472	1218.4	1333.8	1.5912	1.1252	1215.9	1330.4	1.5783	1.0235	1213.3	1326.9	1.5664
670	1.4116	1225.3	1342.9	1.6101	1.2615	1222.9	1339.6	1.5963	1.1386	1220.4	1336.3	1.5836	1.0360	1217.9	1332.9	1.5717

(*Continued*)

TABLE A.3 (*Continued*)

Properties of Superheated Steam

P (T Sat.)	Vapor (psia)															
	450 (456.39)				300 (417.43)				350 (431.82)				400 (444.70)			
T	*v*	*u*	*h*	*s*	*v*	*u*	*h*	*s*	*v*	*u*	*h*	*s*	*v*	*u*	*h*	*s*
680	1.4272	1229.6	1348.5	1.6151	1.2758	1227.3	1345.3	1.6014	1.1518	1224.9	1342.1	1.5887	1.0483	1222.5	1338.9	1.5769
690	1.4426	1233.9	1354.1	1.6199	1.2899	1231.7	1351.0	1.6063	1.1649	1229.4	1347.9	1.5938	1.0606	1227.0	1344.8	1.5821
700	1.4580	1238.2	1359.6	1.6248	1.3040	1236.0	1356.7	1.6112	1.1779	1233.8	1353.7	1.5987	1.0727	1231.5	1350.6	1.5872
720	1.4884	1246.7	1370.7	1.6342	1.3319	1244.7	1367.9	1.6208	1.2037	1242.6	1365.1	1.6085	1.0968	1240.4	1362.2	1.5971
740	1.5186	1255.2	1381.7	1.6435	1.3594	1253.3	1379.1	1.6302	1.2291	1251.3	1376.4	1.6180	1.1205	1249.3	1373.7	1.6067
760	1.5485	1263.7	1392.6	1.6525	1.3867	1261.8	1390.1	1.6394	1.2543	1260.0	1387.6	1.6273	1.1439	1258.1	1385.1	1.6161
780	1.5782	1272.1	1403.5	1.6614	1.4138	1270.3	1401.1	1.6483	1.2793	1268.6	1398.8	1.6364	1.1671	1266.8	1396.3	1.6253
800	1.6077	1280.5	1414.4	1.6701	1.4407	1278.8	1412.1	1.6571	1.3040	1277.1	1409.8	1.6452	1.1900	1275.4	1407.6	1.6343
820	1.6369	1288.9	1425.2	1.6786	1.4673	1287.3	1423.0	1.6657	1.3285	1285.7	1420.9	1.6539	1.2128	1284.1	1418.7	1.6430
840	1.666	1297.3	1436.0	1.6870	1.4938	1295.7	1433.9	1.6742	1.3528	1294.2	1431.9	1.6625	1.2353	1292.7	1429.8	1.6517
860	1.695	1305.6	1446.8	1.6952	1.5201	1304.2	1444.8	1.6825	1.377	1302.7	1442.9	1.6708	1.2577	1301.2	1440.9	1.6601
880	1.7238	1314.0	1457.5	1.7033	1.5463	1312.6	1455.7	1.6906	1.4010	1311.2	1453.8	1.6791	1.2800	1309.8	1451.9	1.6684

(*Continued*)

TABLE A.3 (Continued)

Properties of Superheated Steam

	Vapor (psia)															
	450 (456.39)				500 (467.13)				550 (477.07)				600 (486.33)			
P (T Sat.) T	v	u	h	s	V	u	h	s	v	u	h	s	v	u	h	s
900	1.7524	1322.4	1468.3	1.7113	1.5723	1321.0	1466.5	1.6987	1.4249	1319.7	1464.7	1.6872	1.3021	1318.4	1462.9	1.6766
920	1.7810	1330.8	1479.1	1.7191	1.5982	1329.5	1477.4	1.7066	1.4487	1328.2	1475.6	1.6951	1.3240	1326.9	1473.9	1.6846
940	1.8094	1339.2	1489.8	1.7269	1.6240	1337.9	1488.2	1.7144	1.4723	1336.7	1486.6	1.7030	1.3459	1335.5	1484.9	1.6925
960	1.8377	1347.6	1500.6	1.7345	1.6497	1346.4	1499.0	1.7221	1.4958	1345.2	1497.5	1.7107	1.3676	1344.0	1495.9	1.7003
980	1.8660	1356.0	1511.4	1.7420	1.6753	1354.9	1509.9	1.7296	1.5193	1353.7	1508.4	1.7183	1.3893	1352.6	1506.8	1.7079
1000	1.8941	1364.4	1522.2	1.7495	1.7008	1363.3	1520.7	1.7371	1.5426	1362.3	1519.3	1.7259	1.4108	1361.2	1517.8	1.7155
1020	1.9221	1372.9	1533.0	1.7568	1.7262	1371.8	1531.6	1.7445	1.5659	1370.8	1530.2	1.7333	1.4322	1369.7	1528.8	1.7230
1040	1.9501	1381.4	1543.8	1.7641	1.7515	1380.4	1542.4	1.7518	1.5890	1379.3	1541.1	1.7406	1.4536	1378.3	1539.7	1.7303
1060	1.9780	1389.9	1554.6	1.7712	1.7768	1388.9	1553.3	1.7590	1.6121	1387.9	1552.0	1.7478	1.4749	1386.9	1550.7	1.7376
1080	2.0058	1398.4	1565.4	1.7783	1.8020	1397.4	1564.2	1.7661	1.6352	1396.5	1562.9	1.7550	1.4961	1395.6	1561.7	1.7448
1100	2.034	1406.9	1576.3	1.7853	1.8271	1406.0	1575.1	1.7731	1.6581	1405.1	1573.9	1.7620	1.5173	1404.2	1572.7	1.7519
1150	2.103	1428.4	1603.5	1.8025	1.8896	1427.5	1602.4	1.7904	1.7152	1426.7	1601.3	1.7793	1.5699	1425.9	1600.2	1.7692
1200	2.172	1450.0	1630.8	1.8192	1.9518	1449.2	1629.8	1.8072	1.7720	1448.5	1628.8	1.7962	1.6222	1447.7	1627.8	1.7861
1250	2.240	1471.8	1658.3	1.8355	2.0137	1471.1	1657.4	1.8235	1.8285	1470.3	1656.5	1.8126	1.6742	1469.6	1655.5	1.8026
1300	2.308	1493.7	1685.9	1.8515	2.075	1493.1	1685.1	1.8395	1.8848	1492.4	1684.2	1.8286	1.7260	1491.7	1683.4	1.8186
1350	2.376	1515.8	1713.7	1.8670	2.137	1515.2	1712.9	1.8551	1.9409	1514.6	1712.2	1.8443	1.7775	1514.0	1711.4	1.8343
1400	2.444	1538.1	1741.7	1.8823	2.198	1537.6	1741.0	1.8704	1.9967	1537.0	1740.2	1.8596	1.8289	1536.5	1739.5	1.8497
1450	2.512	1560.6	1769.8	1.8972	2.259	1560.1	1769.2	1.8853	2.0525	1559.6	1768.5	1.8746	1.8801	1559.1	1767.8	1.8647
1500	2.580	1583.3	1798.2	1.9119	2.320	1582.8	1797.5	1.9000	2.1080	1582.3	1796.9	1.8892	1.9312	1581.9	1796.3	1.8794
1600	2.715	1629.3	1855.4	1.9403	2.442	1628.9	1854.8	1.9285	2.219	1628.4	1854.3	1.9178	2.033	1628.0	1853.7	1.9080
1800	2.983	1723.7	1972.1	1.9944	2.684	1723.3	1971.7	1.9827	2.440	1722.9	1971.2	1.9720	2.236	1722.6	1970.8	1.9622
2000	3.251	1821.3	2092.0	2.0453	2.926	1820.9	2091.6	2.0335	2.660	1820.6	2091.2	2.0229	2.438	1820.2	2090.8	2.0131
2200	3.519	1921.8	2214.9	2.0933	3.167	1921.5	2214.5	2.0815	2.879	1921.1	2214.2	2.0709	2.639	1920.8	2213.8	2.0612
2400	3.787	2025.1	2340.4	2.1388	3.408	2024.7	2340.1	2.1270	3.098	2024.4	2339.7	2.1164	2.840	2024.0	2339.4	2.1067

(Continued)

TABLE A.3 (*Continued*)

Properties of Superheated Steam

	Vapor (psia)							
P (*T* Sat.)	800 (518.36)				1000 (544.75)			
T	*v*	*u*	*h*	*s*	*v*	*u*	*h*	*s*
Sat.	.5691	1115.0	1199.3	1.4160	0.4459	1109.9	1192.4	1.3903
550	0.6154	1138.8	1229.9	1.4469	0.4534	1114.8	1198.7	1.3966
560	0.6287	1145.6	1238.6	1.4555	0.4669	1123.6	1210.0	1.4077
570	0.6415	1152.0	1247.0	1.4637	0.4795	1131.7	1220.5	1.4179
580	0.6539	1158.3	1255.1	1.4714	0.4915	1139.4	1230.4	1.4275
590	0.6659	1164.3	1262.9	1.4789	0.5030	1146.7	1239.8	1.4365
600	0.6776	1170.1	1270.4	1.4861	0.5140	1153.7	1248.8	1.4450
610	0.6890	1175.8	1277.8	1.4930	0.5245	1160.3	1257.4	1.4531
620	0.7002	1181.3	1285.0	1.4997	0.5348	1166.7	1265.7	1.4609
630	0.7111	1186.7	1292.0	1.5062	0.5447	1172.9	1273.7	1.4682
640	0.7218	1192.0	1298.9	1.5125	0.5543	1178.9	1281.5	1.4754
650	0.7324	1197.2	1305.6	1.5186	0.5637	1184.7	1289.1	1.4822
660	0.7428	1202.3	1312.3	1.5245	0.5730	1190.4	1296.5	1.4889
670	0.7530	1207.4	1318.8	1.5304	0.5820	1196.0	1303.7	1.4953
680	0.7631	1212.3	1325.3	1.5361	0.5908	1201.4	1310.8	1.5015

(*Continued*)

TABLE A.3 (*Continued*)

Properties of Superheated Steam

	Vapor (psia)							
P (*T* Sat.)	800 (518.36)				1000 (544.75)			
T	*v*	*u*	*h*	*s*	*v*	*u*	*h*	*s*
690	0.7730	1217.2	1331.7	1.5416	0.5995	1206.8	1317.7	1.5076
700	0.7829	1222.1	1338.0	1.5471	0.6080	1212.0	1324.6	1.5135
720	0.8023	1231.6	1350.4	1.5577	0.6247	1222.3	1337.9	1.5249
740	0.8212	1241.0	1362.6	1.5680	0.6410	1232.3	1350.9	1.5359
760	0.8399	1250.3	1374.6	1.5779	0.6569	1242.1	1363.7	1.5464
780	0.8583	1259.4	1386.5	1.5875	0.6725	1251.7	1376.2	1.5566
800	0.8764	1268.5	1398.2	1.5969	0.6878	1261.2	1388.5	1.5664
820	0.8943	1277.4	1409.8	1.6061	0.7028	1270.6	1400.6	1.5760
840	0.9120	1286.4	1421.4	1.6150	0.7176	1279.9	1412.7	1.5853
860	0.9295	1295.2	1432.8	1.6238	0.7323	1289.1	1424.6	1.5944
880	0.9468	1304.1	1444.2	1.6324	0.7467	1298.2	1436.4	1.6033

(*Continued*)

TABLE A.3 (*Continued*)

Properties of Superheated Steam

	Vapor (psia)							
P (*T* Sat.)	800 (518.36)				1000 (544.75)			
T	*v*	*u*	*h*	*s*	*v*	*u*	*h*	*s*
900	0.9640	1312.9	1455.6	1.6408	0.7610	1307.3	1448.1	1.6120
920	0.9811	1321.7	1466.9	1.6490	0.7752	1316.3	1459.7	1.6205
940	0.9980	1330.4	1478.2	1.6572	0.7892	1325.3	1471.3	1.6288
960	1.0149	1339.2	1489.4	1.6651	0.8031	1334.3	1482.9	1.6370
980	1.0316	1348.0	1500.7	1.6730	0.8169	1343.2	1494.4	1.6451
1000	1.0482	1356.7	1511.9	1.6807	0.8305	1352.2	1505.9	1.6530
1020	1.0647	1365.5	1523.1	1.6883	0.8441	1361.1	1517.3	1.6608
1040	1.0812	1374.2	1534.3	1.6959	0.8576	1370.0	1528.7	1.6684
1060	1.0975	1383.0	1545.5	1.7033	0.8710	1378.9	1540.1	1.6760
1080	1.1138	1391.7	1556.6	1.7106	0.8844	1387.9	1551.5	1.6834
1100	1.1300	1400.5	1567.8	1.7178	0.8976	1396.8	1562.9	1.6908
1120	1.1462	1409.3	1579.0	1.7249	0.9108	1405.7	1574.3	1.6980
1140	1.1623	1418.1	1590.2	1.7319	0.9240	1414.6	1585.6	1.7052
1160	1.1783	1426.9	1601.4	1.7389	0.9370	1423.6	1597.0	1.7122
1180	1.1943	1435.8	1612.6	1.7458	0.9500	1432.5	1608.3	1.7192
1200	1.2102	1444.6	1623.8	1.7526	0.9630	1441.5	1619.7	1.7261
1220	1.2261	1453.5	1635.9	1.7593	0.9759	1450.4	1631.0	1.7329
1240	1.2420	1462.3	1646.2	1.7659	0.9888	1459.4	1642.4	1.7396
1260	1.2578	1471.3	1657.4	1.7725	1.0016	1468.4	1653.8	1.7462

(*Continued*)

TABLE A.3 (*Continued*)

Properties of Superheated Steam

	Vapor (psia)							
P (*T* Sat.)	800 (518.36)				1000 (544.75)			
T	*v*	*u*	*h*	*s*	*v*	*u*	*h*	*s*
1280	1.2735	1480.2	1668.7	1.7790	1.0144	1477.4	1665.1	1.7528
1300	1.2892	1489.1	1680.0	1.7854	1.0272	1486.5	1676.5	1.7593
1350	1.3284	1511.6	1708.2	1.8013	1.0589	1509.1	1705.1	1.7753
1400	1.3674	1534.2	1736.6	1.8167	1.0905	1531.9	1733.7	1.7909
1450	1.4062	1556.9	1765.1	1.8319	1.1218	1554.8	1762.4	1.8061
1500	1.4448	1579.9	1793.7	1.8467	1.1531	1577.8	1791.2	1.8210
1600	1.5218	1626.2	1851.5	1.8754	1.2152	1624.4	1849.3	1.8499
1800	1.6749	1721.0	1969.0	1.9298	1.3384	1719.5	1967.2	1.9046
2000	1.8271	1818.8	2089.3	1.9808	1.4608	1817.4	2087.7	1.9557
2200	1.9789	1919.4	2212.4	2.0290	1.5828	1918.1	2211.0	2.0038
2400	2.1305	2022.7	2338.1	2.0745	1.7046	2021.3	2336.7	2.0494

TABLE A.4

Properties of Compressed Liquid (Water)

	Liquid (psia)											
P (T Sat.)	0				500 (467.13)				1000 (544.75)			
T	v	u	h	s	v	u	h	s	v	u	h	s
Sat.					0.019748	447.70	449.53	0.64904	0.021591	538.39	542.38	0.74320
32	*0.016022*	*0.01*	*0.01*	*0.00003*	0.015994	0.00	1.49	0.00000	0.015967	0.03	2.99	0.00005
50	*0.016024*	*18.06*	*18.06*	*0.03607*	0.015998	18.02	19.50	0.03599	0.015972	17.99	20.94	0.03592
100	*0.016130*	*68.05*	*68.05*	*0.12963*	0.016106	67.87	69.36	0.12932	0.016082	67.70	70.68	0.12901
150	*0.016344*	*117.95*	*117.95*	*0.21504*	0.016318	117.66	119.17	0.21457	0.016293	117.38	120.40	0.21410
200	*0.016635*	*168.05*	*168.05*	*0.29402*	0.016608	167.65	169.19	0.29341	0.016580	167.26	170.32	0.29281
250	*0.017003*	*218.52*	*218.52*	*0.36777*	0.016972	217.99	219.56	0.36702	0.016941	217.47	220.61	0.36628
300	*0.017453*	*269.61*	*269.61*	*0.43732*	0.017416	268.92	270.53	0.43641	0.017379	268.24	271.46	0.43352
350	*0.018000*	*321.59*	*321.59*	*0.50359*	0.017954	320.71	322.37	0.50249	0.017909	319.83	323.15	0.50140
400	*0.018668*	*374.85*	*374.85*	*0.56740*	0.018608	373.68	375.40	0.56604	0.018550	372.55	375.98	0.56472
450	*0.019503*	*429.96*	*429.96*	*0.62970*	0.019420	428.40	430.19	0.63798	0.019340	426.89	430.47	0.62632
500	*0.02060*	*488.1*	*488.1*	*0.6919*	*0.02048*	*485.9*	*487.8*	*0.6896*	0.02036	483.8	487.5	0.6874
510	*0.02087*	*500.3*	*500.3*	*0.7046*	*0.02073*	*497.9*	*499.8*	*0.7021*	0.02060	495.6	499.4	0.6997
520	*0.02116*	*512.7*	*512.7*	*0.7173*	*0.02100*	*510.1*	*512.0*	*0.7146*	0.02086	507.6	511.5	0.7121
530	*0.02148*	*525.5*	*525.5*	*0.7303*	*0.02130*	*522.6*	*524.5*	*0.7273*	0.02114	519.9	523.8	0.7245
540	*0.02182*	*538.6*	*538.6*	*0.7434*	*0.02162*	*535.3*	*537.3*	*0.7402*	0.02144	532.4	536.3	0.7372
550	*0.02221*	*552.1*	*552.1*	*0.7569*	*0.02198*	*548.4*	*550.5*	*0.7532*	*0.02177*	*545.1*	549.2	*0.7499*
560	*0.02265*	*566.1*	*566.1*	*0.7707*	*0.02237*	*562.0*	*564.0*	*0.7666*	*0.02213*	*558.3*	562.4	*0.7630*
570	*0.02315*	*580.8*	*580.8*	*0.7851*	*0.02281*	*576.0*	*578.1*	*0.7804*	*0.02253*	*571.8*	*576.0*	*0.7763*
580					*0.02332*	*590.8*	*592.9*	*0.7946*	*0.02298*	*585.9*	*590.1*	*0.7899*
590					*0.02392*	*606.4*	*608.6*	*0.8096*	*0.02349*	*600.6*	*604.9*	*0.8041*
600									*0.02409*	*616.2*	*620.6*	*0.8189*
610									*0.02482*	*632.9*	*637.5*	*0.8348*

(Continued)

TABLE A.4 (*Continued*)

Properties of Compressed Liquid (Water)

	Liquid (psia)											
P (*T* sat)	4000				18,000				20,000			
T	*v*	*u*	*h*	*s*	*v*	*u*	*h*	*s*	*v*	*u*	*h*	*s*
32	0.015807	0.10	11.80	0.00005	0.015188	0.54	50.05	0.00372	0.015116	0.65	55.30	0.00446
50	0.015821	17.76	29.47	0.03534	0.015227	16.32	67.04	0.03021	0.015154	16.14	72.23	0.02936
100	0.015942	66.72	78.52	0.12714	0.015372	62.83	114.03	0.11817	0.015298	62.37	118.99	0.11688
150	0.016150	115.77	127.73	0.21136	0.015572	109.64	161.51	0.19941	0.015497	108.91	166.26	0.19778
200	0.016425	165.02	177.18	0.28931	0.015813	156.62	209.29	0.27474	0.015736	155.62	213.86	0.27281
250	0.016765	214.52	226.93	0.36200	0.016096	203.57	257.19	0.34472	0.016015	202.28	261.55	0.34250
300	0.017174	264.43	277.15	0.43038	0.016422	250.58	305.28	0.41021	0.016333	248.96	309.41	0.40766
350	0.017659	314.98	328.05	0.49526	0.016791	297.80	353.73	.47197	.016693	295.83	357.61	0.46911
400	0.018235	366.35	379.85	0.55734	0.017207	345.32	402.63	0.53057	0.017096	342.97	406.24	0.52738
450	0.018924	418.83	432.84	0.61725	0.017669	393.15	452.00	0.58639	0.017541	390.39	455.31	0.58286
500	0.019766	472.9	487.5	0.6758	0.018183	441.3	501.9	0.6397	0.018031	438.1	504.8	0.6358
520	0.020161	495.2	510.1	0.6990	0.018404	460.6	521.9	0.6605	0.018242	457.2	524.7	0.6564
540	0.020600	517.9	533.1	0.7223	0.018636	480.0	542.1	0.6808	0.018461	476.4	544.7	0.6766
560	0.021091	541.2	556.8	0.7457	0.018879	499.5	562.4	0.7009	0.018689	495.6	564.8	0.6965
580	0.021648	565.2	581.2	0.7694	0.019134	519.0	582.8	0.7207	0.018928	514.9	585.0	0.7160
600	0.02229	590.0	606.5	0.7936	0.01940	538.7	603.3	0.7403	0.01918	534.3	605.2	0.7354
620	0.02304	616.0	633.0	0.8183	0.01968	558.4	623.9	0.7596	0.01944	553.7	625.6	0.7544
640	0.02394	643.3	661.1	0.8441	0.01998	578.2	644.8	0.7787	0.01972	573.2	646.1	0.7732
660	0.02506	672.7	691.2	0.8712	0.02030	598.1	665.7	0.7976	0.02001	592.7	666.8	0.7918
680	0.02653	704.9	724.5	0.9007	0.02064	618.2	686.9	0.8163	0.02031	612.4	687.6	0.8102
700	0.02867	742.1	763.4	0.9345	0.02099	638.4	708.3	0.8349	0.02063	632.1	708.5	0.8285
710	0.03026	764.3	786.7	0.9545	0.02118	648.5	719.1	0.8442	0.02080	642.1	719.1	0.8375

TABLE A.1 (SI)

Saturation Temperature (Water)

		Specific Volume (m^3/kg)		Internal Energy (kJ/kg)			Enthalpy (kJ/kg)			Entropy (kJ/kg.K)		
Temp. °C	Sat. Press. kPa	Sat. Liquid v_f	Sat. Vapor v_g	Sat. Liquid u_f	Evap. u_{fg}	Sat. Vapor u_g	Sat. Liquid h_f	Evap. h_{fg}	Sat. Vapor h_g	Sat. Liquid s_f	Evap. s_{fg}	Sat. Vapor s_g
0.01	0.6113	0.001000	206.14	0.00	2375.3	2375.3	0.01	2501.3	2501.4	0.0000	9.1562	9.1562
5	0.8721	0.001000	147.12	20.97	2361.3	2382.3	20.98	2489.6	2510.6	0.0761	8.9496	9.0257
10	1.2276	0.001000	106.38	42.00	2347.2	2389.2	42.01	2477.7	2519.8	0.1510	8.7498	8.9008
15	1.7051	0.001001	77.93	62.99	2333.1	2396.1	62.99	2465.9	2528.9	0.2245	8.5569	8.7814
20	2.339	0.001002	57.79	83.95	2319.0	2402.9	83.96	2454.1	2538.1	0.2966	8.3706	8.6672
25	3.169	0.001003	43.36	104.88	2304.9	2409.8	104.89	2442.3	2547.2	0.3674	8.1905	8.5580
30	4.246	0.001004	32.89	125.78	2290.8	2416.6	125.79	2430.5	2556.3	0.4369	8.0164	8.4533
35	5.628	0.001006	25.22	146.67	2276.7	2423.4	146.68	2418.6	2565.3	0.5053	7.8478	8.3531
40	7.384	0.001008	19.52	167.56	2262.6	2430.1	167.57	2406.7	2574.3	0.5725	7.6845	8.2570
45	9.593	0.001010	15.26	188.44	2248.4	2436.8	188.45	2394.8	2583.2	0.6387	7.5261	8.1648
50	12.349	0.001012	12.03	209.32	2234.2	2443.5	209.33	2382.7	2592.1	0.7038	7.3725	8.0763
55	15.758	0.001015	9.568	230.21	2219.9	2450.1	230.23	2370.7	2600.9	0.7679	7.2234	7.9913
60	19.940	0.001017	7.671	251.11	2205.5	2456.6	251.13	2358.5	2609.6	0.8312	7.0784	7.9096
65	25.03	0.001020	6.197	272.02	2191.1	2463.1	272.06	2346.2	2618.3	0.8935	6.937.5	7.8310
70	31.19	0.001023	5.042	292.95	2176.6	2469.6	292.98	2333.8	2626.8	0.9549	6.8004	7.7553
75	38.58	0.001026	4.131	313.90	2162.0	2475.9	313.93	2321.4	2635.3	1.0155	6.6669	7.6824
80	47.39	0.001029	3.407	334.86	2147.4	2482.2	334.91	2308.8	2643.7	1.0753	6.5369	7.6122
85	57.83	0.001033	2.828	355.84	2132.6	2488.4	355.90	2296.0	2651.9	1.1343	6.4102	7.5445
90	70.14	0.001036	2.361	376.85	2117.7	2494.5	376.92	2283.2	2660.1	1.1925	6.2866	7.4791
95	84.55	0.001040	1.982	397.88	2102.7	2500.6	397.96	2270.2	2668.1	1.2500	6.1659	7.4159
100	101.35	0.001044	1.6729	418.94	2087.6	2506.5	419.04	2257.0	2676.1	1.3069	6.0480	7.3549

(*Continued*)

TABLE A.1 (SI) (*Continued*)

Saturation Temperature (Water)

		Specific Volume (m³/kg)		Internal Energy (kJ/kg)			Enthalpy (kJ/kg)			Entropy (kJ/kg.K)		
Temp. °C	Sat. Press. MPa	Sat. Liquid v_f	Sat. Vapor v_g	Sat. Liquid u_f	Evap. u_{fg}	Sat. Vapor u_g	Sat. Liquid h_f	Evap. h_{fg}	Sat. Vapor h_g	Sat. Liquid s_f	Evap. s_{fg}	Sat. Vapor s_g
105	0.12082	0.001048	1.4194	440.02	2072.3	2512.4	440.15	2243.7	2683.8	1.3630	5.9328	7.2958
110	0.14327	0.001052	1.2102	461.14	2057.0	2518.1	461.30	2230.2	2691.5	1.4185	5.8202	7.2387
115	0.16906	0.001056	1.0366	482.3	2041.4	2523.7	482.48	2216.5	2699.0	1.4734	5.71	7.1833
120	0.19853	0.001060	0.8919	503.50	2025.8	2529.3	503.71	2202.6	2706.3	1.5276	5.6020	7.1296
125	0.2321	0.001065	0.7706	524.74	2009.9	2534.6	524.99	2188.5	2713.5	1.5813	5.4962	7.0775
130	0.2701	0.001070	0.6685	546.02	1993.9	2539.9	546.31	2174.2	2720.5	1.6344	5.3925	7.0269
135	0.313	0.001075	0.5822	567.35	1977.7	2545	567.69	2159.6	2727.3	1.687	5.2907	6.9777
140	0.3613	0.00108	0.5089	588.74	1961.3	25500	589.13	2144.7	2733.9	1.7391	5.1908	6.9299
145	0.4154	0.001085	0.4463	610.18	1944.7	2554.9	610.63	2129.6	2740.3	1.7907	5.0926	6.8833
150	0.4758	0.001091	0.3928	631.68	1927.9	2559.5	632.20	2114.3	2746.5	1.8418	4.996	6.8379
155	0.5431	0.001096	0.3468	653.24	1910.8	2564.1	653.84	2098.6	2752.4	1.8925	4.901	6.7935
160	0.6178	0.001102	0.3071	674.87	1893.5	2568.4	675.55	2082.6	2758.1	1.9427	4.8075	6.7502
165	0.7005	0.001108	0.2727	696.56	1876	2572.5	697.34	2066.2	2763.5	1.9925	4.7153	6.7078
170	0.7917	0.001114	0.2428	718.33	1858.1	2576.5	719.21	2049.5	2768.7	2.0419	4.6244	6.6663
175	0.892	0.001121	0.2168	740.17	1840.0	2580.2	741.17	2032.4	2773.6	2.0909	4.5347	6.6256
180	1.0021	0.001127	0.19405	762.09	1821.6	2583.7	763.22	2015.0	2778.2	2.1396	4.4461	6.5857
185	1.1227	0.001134	0.17409	784.1	1802.9	2587.0	785.37	1997.1	2782.4	2.1879	4.3586	6.5465
190	1.2544	0.001141	0.15654	806.19	1783.8	2590.0	807.62	1978.8	2786.4	2.2359	4.272	6.5079
195	1.3978	0.001149	0.14105	828.37	1764.4	2592.8	829.98	1960.0	2790.0	2.2835	4.1863	6.4698
200	1.5538	0.001157	0.12736	850.65	1744.7	2595.3	852.45	1940.7	2793.2	2.3309	4.1014	6.4323
205	1.723	0.001164	0.11521	873.04	1724.5	2597.5	875.04	1921.0	2796.0	2.3780	4.0172	6.3952
210	1.9062	0.001173	0.10441	895.53	1703.9	2599.5	897.76	1900.7	2798.5	2.4248	3.9337	6.3585

(*Continued*)

TABLE A.1 (SI) (*Continued*)

Saturation Temperature (Water)

		Specific Volume (m^3/kg)		Internal Energy (kJ/kg)			Enthalpy (kJ/kg)			Entropy (kJ/kg.K)		
Temp. °C	Sat. Press. MPa	Sat. Liquid v_f	Sat. Vapor v_g	Sat. Liquid u_f	Evap. u_{fg}	Sat. Vapor u_g	Sat. Liquid h_f	Evap. h_{fg}	Sat. Vapor h_g	Sat. Liquid s_f	Evap. s_{fg}	Sat. Vapor s_g
215	2.104	0.001181	0.09479	918.14	1682.9	2601.1	920.62	1879.9	2800.5	2.4714	3.8507	6.3221
220	2.318	0.001190	0.08619	940.87	1661.5	2602.4	943.62	1858.5	2802.1	2.5178	3.7683	6.2861
225	2.548	0.001199	0.07849	963.73	1639.6	2603.3	966.78	1836.5	2803.3	2.5639	3.6863	6.2503
230	2.795	0.001209	0.07158	986.74	1617.2	2603.9	990.12	1813.8	2804.0	2.6099	3.6047	6.2146
235	3.060	0.001219	0.06537	1009.89	1594.2	2604.1	1013.62	1790.5	2804.2	2.6558	3.5233	6.1791
240	3.344	0.001229	0.05976	1033.21	1570.8	2604.0	1037.32	1766.5	2803.8	2.7015	3.4422	6.1437
245	3.648	0.001240	.05471	1056.71	1546.7	2603.4	1061.23	1741.7	2803.0	2.7472	3.3612	6.1083
250	3.973	0.001251	0.05013	1080.39	1522	2602.4	1085.36	1716.2	2801.5	2.7927	3.2802	6.0730
255	4.319	0.001263	0.04598	1104.28	1496.7	2600.9	1109.73	1689.8	2799.5	2.8383	3.1992	6.0375
260	4.688	0.001276	0.04221	1128.39	1470.6	2599	1134.37	1662.5	2796.9	2.8838	3.1181	6.0019
265	5.081	0.001289	0.03877	1152.74	1443.9	2596.6	1159.28	1634.4	2793.6	2.9294	3.0368	5.9662
270	5.499	0.001302	0.03564	1177.36	1416.3	2593.7	1184.51	1605.2	2789.7	2.9751	2.9551	5.9301
275	5.942	0.001317	0.03279	1202.25	1387.9	2590.2	1210.07	1574.9	2785	3.0208	2.8730	5.8938
280	6.412	0.001332	0.03017	1227.46	1358.7	2586.1	1235.99	1543.6	2779.6	3.0668	2.7903	5.8571
285	6.909	0.001348	0.02777	1253.00	1328.4	2581.4	1262.31	1511.0	2773.3	3.1130	2.7070	5.8199
290	7.436	0.001366	0.02557	1278.92	1297.1	2576.0	1289.07	1477.1	2766.2	3.1594	2.6227	5.7821
295	7.993	0.001384	0.02354	1305.2	1264.7	2569.9	1316.3	1441.8	2758.1	3.2062	2.5375	5.7437
300	8.581	0.001404	0.02167	1332.0	1231.0	2563.0	1344.0	1404.9	2749.0	3.2534	2.4511	5.7045
305	9.202	0.001425	0.019948	1359.3	1195.9	2555.2	1372.4	1366.4	2738.7	3.3010	2.3633	5.6643
310	9.856	0.001447	0.018350	1387.1	1159.4	2546.4	1401.3	1326.0	2727.3	3.3493	2.2737	5.6230

(*Continued*)

TABLE A.1 (SI) (*Continued*)

Saturation Temperature (Water)

		Specific Volume (m³/kg)		Internal Energy (kJ/kg)			Enthalpy (kJ/kg)			Entropy (kJ/kg.K)		
Temp. °C	Sat. Press. MPa	Sat. Liquid v_f	Sat. Vapor v_g	Sat. Liquid u_f	Evap. u_{fg}	Sat. Vapor u_g	Sat. Liquid h_f	Evap. h_{fg}	Sat. Vapor h_g	Sat. Liquid s_f	Evap. s_{fg}	Sat. Vapor s_g
315	10.547	0.001472	0.016867	1415.5	1121.1	2536.6	1431.0	1283.5	2714.5	3.3982	2.1821	5.5804
320	11.274	0.001499	0.015488	1444.6	1080.9	2525.5	1461.5	1238.6	2700.1	3.4480	2.0882	5.5362
330	12.845	0.001561	0.012996	1505.3	993.7	2498.9	1525.3	1140.6	2665.9	3.5507	1.8909	5.4417
340	14.586	0.001638	0.010797	1570.3	894.3	2464.6	1594.2	1027.9	2622.0	3.6594	1.6763	5.3357
350	16.513	0.001740	0.008813	1641.9	776.6	2418.4	1670.6	893.4	2563.9	3.7777	1.4335	5.2112
360	18.651	0.001893	0.006945	1725.2	626.3	2351.5	1760.5	720.5	2481.0	3.9147	1.1379	5.0526
370	21.03	0.002213	0.004925	1844.0	384.5	2228.5	1890.5	441.6	2332.1	4.1106	0.6865	4.7971
374.14	22.09	0.003155	0.003155	2029.6	0	2029.6	2099.3	0	2099.3	4.4298	0	4.4298

TABLE A.2 (SI)

Saturation Pressure (Water)

		Specific Volume (m^3/kg)		Internal Energy (kJ/kg)			Enthalpy (kJ/kg)			Entropy (kJ/kg.K)		
Press. kPa	Sat. Temp. °C	Sat. Liquid v_f	Sat. Vapor v_g	Sat. Liquid u_f	Evap. u_{fg}	Sat. Vapor u_g	Sat. Liquid h_f	Evap. h_{fg}	Sat. Vapor h_g	Sat. Liquid s_f	Evap. s_{fg}	Sat. Vapor s_g
0.6113	0.01	0.00100	206.14	00.00	2375.3	2375.3	0.01	2501.3	2501.4	0.000	9.1562	9.1562
1.0	6.98	0.00100	129.21	29.30	2355.7	2385.0	29.30	2484.9	2514.2	0.1059	8.8697	8.9756
1.5	13.03	0.001001	87.98	54.71	2338.6	2393.3	54.71	2470.6	2525.3	0.1957	8.6322	8.8279
2.0	17.50	0.001001	67.00	73.48	2326.0	2399.5	73.48	2460.0	2533.5	0.2607	8.4629	8.7237
2.5	21.08	0.001002	54.25	88.48	2315.9	2404.4	88.49	2451.6	2540.0	0.3120	8.3311	8.6432
3.0	24.08	0.001003	45.67	101.04	2307.5	2408.5	101.05	2444.5	2545.5	0.3545	8.2231	8.5776
4.0	28.96	0.001004	34.80	121.45	2293.7	2415.2	121.46	2432.9	2554.4	0.4226	8.0520	8.4746
5.0	32.88	0.001005	28.19	137.81	2282.7	2420.5	137.82	2423.7	2561.5	0.4764	7.9187	8.3951
7.5	40.29	0.001008	19.24	168.78	2261.7	2430.5	168.79	2406.0	2574.8	0.5764	7.6750	8.2515
10	45.81	0.001010	14.67	191.82	2246.1	2437.9	191.83	2392.8	2584.7	0.6493	7.5009	8.1502
15	53.97	0.001014	10.02	225.92	2222.8	2448.7	225.94	2373.1	2599.1	0.7549	7.2536	8.0085
20	60.06	0.001017	7.649	251.38	2205.4	2456.7	251.40	2358.3	2609.7	0.8320	7.0766	7.9085
25	64.97	0.001020	6.204	271.90	2191.2	2463.1	271.93	2346.3	2618.2	0.8931	6.9383	7.8314
30	69.10	0.001022	5.229	289.20	2179.2	2468.4	289.23	2336.1	2625.3	0.9439	6.8247	7.7686
40	75.87	0.001027	3.993	317.53	2159.5	2477.0	317.58	2319.2	2636.8	1.0259	6.6441	7.6700
50	81.33	0.001030	3.240	340.44	2143.4	2483.9	340.49	2305.4	2645.9	1.0910	6.5029	7.5939
75	91.78	0.001 037	2.217	384.31	2112.4	2496.7	384.39	2278.6	2663.0	1.2130	6.2434	7.4564
MPa												
0.100	99.63	0.001 043	1.694	417.36	2088.7	2506.1	417.46	2258.0	2675.5	1.3026	6.0568	7.3594
0.125	105.99	0.001 048	1.3749	444.19	2069.3	2513.5	444.32	2241.0	2685.4	1.3740	5.9104	7.2844
0.150	111.37	0.001 053	1.1593	466.94	2052.7	2519.7	467.11	2226.5	2693.6	1.4336	5.7897	7.2233
0.175	116.06	0.001 057	1.0036	486.80	2038.1	2524.9	486.99	2213.6	2700.6	1.4849	5.6868	7.1717

(Continued)

TABLE A.2 (SI) (*Continued*)

Saturation Pressure (Water)

		Specific Volume (m³/kg)		Internal Energy (kJ/kg)			Enthalpy (kJ/kg)			Entropy (kJ/kg.K)		
Press. MPa	Sat. Temp. °C	Sat. Liquid v_f	Sat. Vapor v_g	Sat. Liquid u_f	Evap. u_{fg}	Sat. Vapor u_g	Sat. Liquid h_f	Evap. h_{fg}	Sat. Vapor h_g	Sat. Liquid s_f	Evap. s_{fg}	Sat. Vapor s_g
0.200	120.23	0.001 061	0.8857	504.49	2025.0	2529.5	504.70	2201.9	2706.7	1.5301	5.5970	7.1271
0.225	124.00	0.001 064	0.7933	520.47	2013.1	2533.6	520.72	2191.3	2712.1	1.5706	5.5173	7.0878
0.250	127.44	0.001 067	0.7187	535.10	2002.1	2537.2	535.37	2181.5	2716.9	1.6072	5.4455	7.0527
0.275	130.60	0.001 070	0.6573	548.59	1991.9	2540.5	548.89	2172.4	2721.3	1.6408	5.3801	7.0209
0.300	133.55	0.001 073	0.6058	561.15	1982.4	2543.6	561.47	2163.8	2725.3	1.6718	5.3201	6.9919
0.325	136.30	0.001 076	0.5620	572.90	1973.5	2546.4	573.25	2155.8	2729.0	1.7006	5.2646	6.9652
0.350	138.88	0.001 079	0.5243	583.95	1965.0	2548.9	584.33	2148.1	2732.4	1.7275	5.2130	6.9405
0.375	141.32	0.001 081	0.4914	594.40	1956.9	2551.3	594.81	2140.8	2735.6	1.7528	5.1647	6.9175
0.40	143.63	0.001 084	0.4625	604.31	1949.3	2553.6	604.74	2133.8	2738.6	1.7766	5.1193	6.8959
0.45	147.93	0.001 088	0.4140	622.77	1934.9	2557.6	623.25	2120.7	2743.9	1.8207	5.0359	6.8565
0.50	151.86	0.001 093	0.3749	639.68	1921.6	2561.2	640.23	2108.5	2748.7	1.8607	4.9606	6.8213
0.55	155.48	0.001 097	0.3427	655.32	1909.2	2564.5	655.93	2097.0	2753.0	1.8973	4.8920	6.7893
0.60	158.85	0.001 101	0.3157	669.90	1897.5	2567.4	670.56	2086.3	2756.8	1.9312	4.8288	6.7600
0.65	162.01	0.001 104	0.2927	683.56	1886.5	2570.1	684.28	2076.0	2760.3	1.9627	4.7703	6.7331
0.70	164.97	0.001 108	0.2729	696.44	1876.1	2572.5	697.22	2066.3	2763.5	1.9922	4.7158	6.7080
0.75	167.78	0.001 112	0.2556	708.64	1866.1	2574.7	709.47	2057.0	2766.4	2.0200	4.6647	6.6847
0.80	170.43	0.001 115	0.2404	720.22	1856.6	2576.8	721.11	2048.0	2769.1	2.0462	4.6166	6.6628
0.85	172.96	0.001 118	0.2270	731.27	1847.4	2578.7	732.22	2039.4	2771.6	2.0710	4.5711	6.6421
0.90	175.38	0.001 121	0.2150	741.83	1838.6	2580.5	742.83	2031.1	2773.9	2.0946	4.5280	6.6226
0.95	177.69	0.001 124	0.2042	751.95	1830.2	2582.1	753.02	2023.1	2776.1	2.1172	4.4869	6.6041
1.00	179.91	0.001 127	0.194 44	761.68	1822.0	2583.6	762.81	2015.3	2778.1	2.1387	4.4478	6.5865

(*Continued*)

TABLE A.2 (SI) (*Continued*)

Saturation Pressure (Water)

		Specific Volume (m³/kg)		Internal Energy (kJ/kg)			Enthalpy (kJ/kg)			Entropy (kJ/kg.K)		
Press. MPa	Sat. Temp. °C	Sat. Liquid v_f	Sat. Vapor v_g	Sat. Liquid u_f	Evap. u_{fg}	Sat. Vapor u_g	Sat. Liquid h_f	Evap. h_{fg}	Sat. Vapor h_g	Sat. Liquid s_f	Evap. s_{fg}	Sat. Vapor s_g
1.10	184.09	0.001 133	0.177 53	780.09	1806.3	2586.4	781.34	2000.4	2781.7	2.1792	4.3744	6.5536
1.20	187.99	0.001 139	0.163 33	797.29	1791.5	2588.8	798.65	1986.2	2784.8	2.2166	4.3067	6.5233
1.30	191.64	0.001 144	0.151 25	813.44	1777.5	2591.0	814.93	1972.7	2787.6	2.2515	4.2438	6.4953
1.40	195.07	0.001 149	0.140 84	828.70	1764.1	2592.8	830.30	1959.7	2790.0	2.2842	4.1850	6.4693
1.5	198.32	0.001 154	0.131 77	843.16	1751.3	2594.5	844.89	1947.3	2792.2	2.3150	4.1298	6.4448
1.75	205.76	0.001 166	0.113 49	876.46	1721.4	2597.8	878.50	1917.9	2796.4	2.3851	4.0044	6.3896
2.00	212.42	0.001 177	0.099 63	906.44	1693.8	2600.3	908.79	1890.7	2799.5	2.4474	3.8935	6.3409
2.25	218.45	0.001 187	0.088 75	933.83	1668.2	2602.0	936.49	1865.2	2801.7	2.5035	3.7937	6.2972
2.5	223.99	0.001 197	0.079 98	959.11	1644.0	2603.1	962.11	1841.0	2803.1	2.5547	3.7028	6.2575
3.0	233.90	0.001 217	0.066 68	1004.78	1599.3	2604.1	1008.42	1795.7	2804.2	2.6457	3.5412	6.1869
3.5	242.60	0.001 235	0.057 07	1045.43	1558.3	2603.7	1049.75	1753.7	2803.4	2.7253	3.4000	6.1253
4.0	250.40	0.001 252	0.049 78	1082.31	1520.0	2602.3	1087.31	1714.1	2801.4	2.7964	3.2737	6.0701

(*Continued*)

TABLE A.2 (SI) (*Continued*)

Saturation Pressure (Water)

		Specific Vo\lume (m³/kg)		Internal Energy (kJ/kg)			Enthalpy (kJ/kg)			Entropy (kJ/kg.K)		
Press. MPa	Temp. °C	Sat. Liquid v_f	Sat. Vapor v_g	Sat. Liquid u_f	Evap. u_{fg}	Sat. Vapor u_g	Sat. Liquid h_f	Evap. h_{fg}	Sat. Vapor h_g	Sat. Liquid s_f	Evap. s_{fg}	Sat. Vapor s_g
5	263.99	0.001 286	0.039 44	1147.81	1449.3	2597.1	1154.23	1640.1	2794.3	2.9202	3.0532	5.9734
6	275.64	0.001 319	0.032 44	1205.44	1384.3	2589.7	1213.35	1571.0	2784.3	3.0267	2.8625	5.8892
7	285.88	0.001 351	0.02737	1257.55	1323.0	2580.5	1267.00	1505.1	2772.1	3.1211	2.6922	5.8133
8	295.06	0.001 384	0.023 52	1305.57	1264.2	2569.8	1316.64	1441.3	2758.0	3.2068	2.5364	5.7432
9	303.40	0.001 418	0.020 48	1350.51	1207.3	2557.8	1363.26	1378.9	2742.1	3.2858	2.3915	5.6772
10	311.06	0.001 452	0.018 026	1393.04	1151.4	2544.4	1407.56	1317.1	2724.7	3.3596	2.2544	5.6141
11	318.15	0.001 489	0.015 987	1433.7	1096.0	2529.8	1450.1	1255.5	2705.6	3.4295	2.1233	5.5527
12	324.75	0.001 527	0.014 263	1473.0	1040.7	2513.7	1491.3	1193.6	2684.9	3.4962	1.9962	5.4924
13	330.93	0.001 567	0.012 780	1511.1	985.0	2496.1	1531.5	1130.7	2662.2	3.5606	1.8718	5.4323
14	336.75	0.001 611	0.011 485	1548.6	928.2	2476.8	1571.1	1066.5	2637.6	3.6232	1.7485	5.3717
15	342.24	0.001 658	0.010 337	1585.6	869.8	2455.5	1610.5	1000.0	2610.5	3.6848	1.6249	5.3098
16	347.44	0.001 711	0.009 306	1622.7	809.0	2431.7	1650.1	930.6	2580.6	3.7461	1.4994	5.2455
17	352.37	0.001 770	0.008 364	1660.2	744.8	2405.0	1690.3	856.9	2547.2	3.8079	1.3698	5.1777
18	357.06	0.001 840	0.007489	1698.9	675.4	2374.3	1732.0	777.1	2509.1	3.8715	1.2329	5.1044
19	361.54	0.001 924	0.006 657	1739.9	598.1	2338.1	1776.5	688.0	2464.5	3.9388	1.0839	5.0228
20	365.81	0.002 036	0.005 834	1785.6	507.5	2293.0	1826.3	583.4	2409.7	4.0139	0.9130	4.9269
21	369.89	0.002 207	0.004 952	1842.1	388.5	2230.6	1888.4	446.2	2334.6	4.1075	0.6938	4.8013
22	373.80	0.002 742	0.003 568	1961.9	125.2	2087.1	2022.2	143.4	2165.6	4.3110	0.2216	4.5327
22.09	374.14	0.003 155	0.003 155	2029.6	0	2029.6	2099.3	0	2099.3	4.4298	0	4.4298

TABLE A.3 (SI)

Properties of Superheated Steam

	$P = 0.010$ MPa (45.81)				$P = 0.050$ MPa (81.33)				$P = 0.10$ MPa (99.63)			
T	v	u	h	s	v	u	h	s	v	u	h	s
Sat.	14.674	2437.9	2584.7	8.1502	3.240	2483.9	2645.9	7.5939	1.6940	2506.1	2675.5	7.3594
50	14.869	2443.9	2592.6	8.1749								
100	17.196	2515.5	2687.5	8.4479	3.418	2511.6	2682.5	7.6947	1.6958	2506.7	2676.2	7.3614
150	19.512	2587.9	2783.0	8.6882	3.889	2585.6	2780.1	7.9401	1.9364	2582.8	2776.4	7.6134
200	21.825	2661.3	2879.5	8.9038	4.356	2659.9	2877.7	8.1580	2.172	2658.1	2875.3	7.8343
250	24.136	2736.0	2977.3	9.1002	4.820	2735.0	2976.0	8.3556	2.406	2733.7	2974.3	8.0333
300	26.445	2812.1	3076.5	9.2813	5.284	2811.3	3075.5	8.5373	2.639	2810.4	3074.3	8.2158
400	31.063	2968.9	3279.6	9.6077	6.209	2968.5	3278.9	8.8642	3.103	2967.9	3278.2	8.5435
500	35.679	3132.3	3489.1	9.8978	7.134	3132.0	3488.7	9.1546	3.565	3131.6	3488.1	8.8342
600	40.295	3302.5	3705.4	10.1608	8.057	3302.2	3705.1	9.4178	4.028	3301.9	3704.7	9.0976
700	44.911	3479.6	3928.7	10.4028	8.981	3479.4	3928.5	9.6599	4.490	3479.2	3928.2	9.3398
800	49.526	3663.8	4159.0	10.6281	9.904	3663.6	4158.9	9.8852	4.952	3663.5	4158.6	9.5652
900	54.141	3855.0	4396.4	10.8396	10.828	3854.9	4396.3	10.0967	5.414	3854.8	4396.1	9.7767
1000	58.757	4053.0	4640.6	11.0393	11.751	4052.9	4640.5	10.2964	5.875	4052.8	4640.3	9.9764
1100	63.372	4257.5	4891.2	11.2287	12.674	4257.4	4891.1	10.4859	6.337	4257.3	4891.0	10.1659
1200	67.987	4467.9	5147.8	11.4091	13.597	4467.8	5147.7	10.6662	6.799	4467.7	5147.6	10.3463
1300	72.602	4683.7	5409.7	11.5811	14.521	4683.6	5409.6	10.8382	7.260	4683.5	5409.5	10.5183

(Continued)

TABLE A.3 (SI) (*Continued*)

Properties of Superheated Steam

	P=0.20 MPa (120.23)				P=0.30 MPa (133.55)				P=0.40 MPa (143.63)			
T	*v*	*u*	*h*	*s*	*v*	*u*	*h*	*s*	*v*	*u*	*h*	*s*
Sat.	.8857	2529.5	2706.7	7.1272	0.6058	2543.6	2725.3	6.9919	0.4625	2553.6	2738.6	6.8959
150	.9596	2576.9	2768.8	7.2795	0.6339	2570.8	2761.0	7.0778	0.4708	2564.5	2752.8	6.9299
200	1.0803	2654.4	2870.5	7.5066	0.7163	2650.7	2865.6	7.3115	0.5342	2646.8	2860.5	7.1706
250	1.1988	2731.2	2971.0	7.7086	0.7964	2728.7	2967.6	7.5166	0.5951	2726.1	2964.2	7.3789
300	1.3162	2808.6	3071.8	7.8926	0.8753	2806.7	3069.3	7.7022	0.6548	2804.8	3066.8	7.5662
400	1.5493	2966.7	3276.6	8.2218	1.0315	2965.6	3275.0	8.0330	0.7726	2964.4	3273.4	7.8985
500	1.7814	3130.8	3487.1	8.5133	1.1867	3130.0	3486.0	8.3251	0.8893	3129.2	3484.9	8.1913
600	2.013	3301.4	3704.0	8.7770	1.3414	3300.8	3703.2	8.5892	1.0055	3300.2	3702.4	8.4558
700	2.244	3478.8	3927.6	9.0194	1.4957	3478.4	3927.1	8.8319	1.1215	3477.9	3926.5	8.6987
800	2.475	3663.1	4158.2	9.2449	1.6499	3662.9	4157.8	9.0576	1.2372	3662.4	4157.3	8.9244
900	2.706	3854.5	4395.8	9.4566	1.8041	3854.2	4395.4	9.2692	1.3529	3853.9	4395.1	9.1362
1000	2.937	4052.5	4640.0	9.6563	1.9581	4052.3	4639.7	9.4690	1.4685	4052.0	4639.4	9.3360
1100	3.168	4257.0	4890.7	9.8458	2.1121	4256.8	4890.4	9.6585	1.5840	4256.5	4890.2	9.5256
1200	3.399	4467.5	5147.3	10.0262	2.2661	4467.2	5147.1	9.8389	1.6996	4467.0	5146.8	9.7060
1300	3.630	4683.2	5409.3	10.1982	2.4201	4683.0	5409.0	10.0110	1.8151	4682.8	5408.8	9.8780

(Continued)

TABLE A.3 (SI) (*Continued*)

Properties of Superheated Steam

	$P=0.50$ MPa (151.86)				$P=0.60$ MPa (158.85)				$P=0.80$ MPa (170.43)			
Sat.	0.3749	2561.2	2748.7	6.8213	0.3157	2567.4	2756.8	6.7600	0.2404	2576.8	2769.1	6.6628
200	0.4249	2642.9	2855.4	7.0592	0.3520	2638.9	2850.1	6.9665	0.2608	2630.6	2839.3	6.8158
250	0.4744	2723.5	2960.7	7.2709	0.3938	2720.9	2957.2	7.1816	0.2931	2715.5	2950.0	7.0384
300	0.5226	2802.9	3064.2	7.4599	0.4344	2801.0	3061.6	7.3724	0.3241	2797.2	3056.5	7.2328
350	0.5701	2882.6	3167.7	7.6329	0.4742	2881.2	3165.7	7.5464	0.3544	2878.2	3161.7	7.4089
400	0.6173	2963.2	3271.9	7.7938	0.5137	2962.1	3270.3	7.7079	0.3843	2959.7	3267.1	7.5716
500	0.7109	3128.4	3483.9	8.0873	0.5920	3127.6	3482.8	8.0021	0.4433	3126.0	3480.6	7.8673
600	0.8041	3299.6	3701.7	8.3522	0.6697	3299.1	3700.9	8.2674	0.5018	3297.9	3699.4	8.1333
700	0.8969	3477.5	3925.9	8.5952	0.7472	3477.0	3925.3	8.5107	0.5601	3476.2	3924.2	8.3770
800	0.9896	3662.1	4156.9	8.8211	0.8245	3661.8	4156.5	8.7367	0.6181	3661.1	4155.6	8.6033
900	1.0822	3853.6	4394.7	9.0329	0.9017	3853.4	4394.4	8.9486	0.6761	3852.8	4393.7	8.8153
1000	1.1747	4051.8	4639.1	9.2328	0.9788	4051.5	4638.8	9.1485	0.7340	4051.0	4638.2	9.0153
1100	1.2672	4256.3	4889.9	9.4224	1.0559	4256.1	4889.6	9.3381	0.7919	4255.6	4889.1	9.2050
1200	1.3596	4466.8	5146.6	9.6029	1.1330	4466.5	5146.3	9.5185	0.8497	4466.1	5145.9	9.3855
1300	1.4521	4682.5	5408.6	9.7749	1.2101	4682.3	5408.3	9.6906	0.9076	4681.8	5407.9	9.5575

(*Continued*)

TABLE A.3 (SI) *(Continued)*

Properties of Superheated Steam

		P = 1.00 MPa (179.91)				P = 1.20 MPa (187.99)				P = 1.40 MPa (195.07)		
Sat.	0.1944	2583.6	2778.1	6.5865	0.163 33	2588.8	2784.8	6.5233	0.140 84	2592.8	2790.0	6.4693
200	0.2060	2621.9	2827.9	6.6940	0.169 30	2612.8	2815.9	6.5898	0.143 02	2603.1	2803.3	6.4975
250	0.2327	2709.9	2942.6	6.9247	0.192 34	2704.2	2935.0	6.8294	0.163 50	2698.3	2927.2	6.7467
300	0.2579	2793.2	3051.2	7.1229	0.2138	2789.2	3045.8	7.0317	0.182 28	2785.2	3040.4	6.9534
350	0.2825	2875.2	3157.7	7.3011	0.2345	2872.2	3153.6	7.2121	0.2003	2869.2	3149.5	7.1360
400	0.3066	2957.3	3263.9	7.4651	0.2548	2954.9	3260.7	7.3774	0.2178	2952.5	3257.5	7.3026
500	0.3541	3124.4	3478.5	7.7622	0.2946	3122.8	3476.3	7.6759	0.2521	3121.1	3474.1	7.6027
600	0.4011	3296.8	3697.9	8.0290	0.3339	3295.6	3696.3	7.9435	0.2860	3294.4	3694.8	7.8710
700	0.4478	3475.3	3923.1	8.2731	0.3729	3474.4	3922.0	8.1881	0.3195	3473.6	3920.8	8.1160
800	0.4943	3660.4	4154.7	8.4996	0.4118	3659.7	4153.8	8.4148	0.3528	3659.0	4153.0	8.3431
900	0.5407	3852.2	4392.9	8.7118	0.4505	3851.6	4392.2	8.6272	0.3861	3851.1	4391.5	8.5556
1000	0.5871	4050.5	4637.6	8.9119	0.4892	4050.0	4637.0	8.8274	0.4192	4049.5	4636.4	8.7559
1100	0.6335	4255.1	4888.6	9.1017	0.5278	4254.6	4888.0	9.0172	0.4524	4254.1	4887.5	8.9457
1200	0.6798	4465.6	5145.4	9.2822	0.5665	4465.1	5144.9	9.1977	0.4855	4464.7	5144.4	9.1262
1300	0.7261	4681.3	5407.4	9.4543	0.6051	4680.9	5407.0	9.3698	0.5186	4680.4	5406.5	9.2984

(Continued)

TABLE A.3 (SI) (*Continued*)

Properties of Superheated Steam

	$P=1.60$ MPa (201.41)				$P=1.80$ MPa (207.15)				$P=2.00$ MPa (212.42)			
T	v	u	h	s	v	u	h	s	v	u	h	s
Sat.	0.123 80	2596.0	2794.0	6.4218	0.11042	2598.4	2797.1	6.3794	0.099 63	2600.3	2799.5	6.3409
225	0.132 87	2644.7	2857.3	6.5518	0.116 73	2636.6	2846.7	6.4808	0.103 77	2628.3	2835.8	6.4147
250	0.14184	2692.3	2919.2	6.6732	0.124 97	2686.0	2911.0	6.6066	0.11144	2679.6	2902.5	6.5453
300	0.158 62	2781.1	3034.8	6.8844	0.140 21	2776.9	3029.2	6.8226	0.125 47	2772.6	3023.5	6.7664
350	0.174 56	2866.1	3145.4	7.0694	0.154 57	2863.0	3141.2	7.0100	0.138 57	2859.8	3137.0	6.9563
400	0.190 05	2950.1	3254.2	7.2374	0.168 47	2947.7	3250.9	7.1794	0.15120	2945.2	3247.6	7.1271
500	0.2203	3119.5	3472.0	7.5390	0.195 50	3117.9	3469.8	7.4825	0.175 68	31162	3467.6	7.4317
600	0.2500	3293.3	3693.2	7.8080	0.2220	3292.1	3691.7	7.7523	0.199 60	3290.9	3690.1	7.7024
700	0.2794	3472.7	3919.7	8.0535	0.2482	3471.8	3918.5	7.9983	0.2232	3470.9	3917.4	7.9487
800	0.3086	3658.3	4152.1	8.2808	0.2742	3657.6	4151.2	8.2258	0.2467	3657.0	4150.3	8.1765
900	0.3377	3850.5	4390.8	8.4935	0.3001	3849.9	4390.1	8.4386	0.2700	3849.3	4389.4	8.3895
1000	0.3668	4049.0	4635.8	8.6938	0.3260	4048.5	4635.2	8.6391	0.2933	4048.0	4634.6	8.5901
1100	0.3958	4253.7	4887.0	8.8837	0.3518	4253.2	4886.4	8.8290	0.3166	4252.7	4885.9	8.7800
1200	0.4248	4464.2	5143.9	9.0643	0.3776	4463.7	5143.4	9.0096	0.3398	4463.3	5142.9	8.9607
1300	0.4538	4679.9	5406.0	9.2364	0.4034	4679.5	5405.6	9.1818	0.3631	4679.0	5405.1	9.1329

(Continued)

TABLE A.3 (SI) (*Continued*)

Properties of Superheated Steam

		$P = 2.50$ MPa (223.99)				$P = 3.00$ MPa (233.90)				$P = 3.50$ MPa (242.60)		
Sat.	0.079 98	2603.1	2803.1	6.2575	0.066 68	2604.1	2804.2	6.1869	0.057 07	2603.7	2803.4	6.1253
250	0.087 00	2662.6	2880.1	6.4085	0.070 58	2644.0	2855.8	6.2872	0.058 72	2623.7	2829.2	6.1749
300	0.098 90	2761.6	3008.8	6.6438	0.08114	2750.1	2993.5	6.5390	0.068 42	2738.0	2977.5	6.4461
350	0.109 76	2851.9	3126.3	6.8403	0.090 53	2843.7	3115.3	6.7428	0.076 78	2835.3	3104.0	6.6579
400	0.120 10	2939.1	3239.3	7.0148	0.099 36	2932.8	3230.9	6.9212	0.084 53	2926.4	3222.3	6.8405
450	0.130 14	3025.5	3350.8	7.1746	0.107 87	3020.4	3344.0	7.0834	0.091 96	3015.3	3337.2	7.0052
500	0.139 98	3112.1	3462.1	7.3234	0.11619	3108.0	3456.5	7.2338	0.099 18	3103.0	3450.9	7.1572
600	0.159 30	3288.0	3686.3	7.5960	0.132 43	3285.0	3682.3	7.5085	0.113 24	3282.1	3678.4	7.4339
700	0.178 32	3468.7	3914.5	7.8435	0.148 38	3466.5	3911.7	7.7571	0.126 99	3464.3	3908.8	7.6837
800	0.19716	3655.3	4148.2	8.0720	0.16414	3653.5	4145.9	7.9862	0.140 56	3651.8	4143.7	7.9134
900	0.215 90	3847.9	4387.6	8.2853	0.179 80	3846.5	4385.9	8.1999	0.154 02	3845.0	4384.1	8.1276
1000	0.2346	4046.7	4633.1	8.4861	0.195 41	4045.4	4631.6	8.4009	0.16743	4044.1	4630.1	8.3288
1100	0.2532	4251.5	4884.6	8.6762	0.210 98	4250.3	4883.3	8.5912	0.180 80	4249.2	4881.9	8.5192
1200	0.2718	4462.1	5141.7	8.8569	0.226 52	4460.9	5140.5	8.7720	0.194 15	4459.8	5139.3	8.7000
1300	0.2905	4677.8	5404.0	9.0291	0.242 06	4676.6	5402.8	8.9442	0.20749	4675.5	5401.7	8.8723

(*Continued*)

TABLE A.3 (SI) (*Continued*)

Properties of Superheated Steam

	$P = 4.0$ MPa (250.40)				$P = 4.5$ MPa (257.49)				$P = 5.0$ MPa (263.99)			
Sat.	0.049 78	2602.3	2801.4	6.0701	0.044 06	2600.1	2798.3	6.0198	0.039 44	2597.1	2794.3	5.9734
275	0.054 57	2667.9	2886.2	6.2285	0.04730	2650.3	2863.2	6.1401	0.041 41	2631.3	2838.3	6.0544
300	0.058 84	2725.3	2960.7	6.3615	0.051 35	2712.0	2943.1	6.2828	0.045 32	2698.0	2924.5	6.2084
350	0.066 45	2826.7	3092.5	6.5821	0.058 40	2817.8	3080.6	6.5131	0.051 94	2808.7	3068.4	6.4493
400	0.073 41	2919.9	3213.6	6.7690	0.064 75	2913.3	3204.7	6.7047	0.057 81	2906.6	3195.7	6.6459
450	0.080 02	3010.2	3330.3	6.9363	0.070 74	3005.0	3323.3	6.8746	0.063 30	2999.7	3316.2	6.8186
500	0.086 43	3099.5	3445.3	7.0901	0.076 51	3095.3	3439.6	7.0301	0.068 57	3091.0	3433.8	6.9759
600	0.098 85	3279.1	3674.4	7.3688	0.087 65	3276.0	3670.5	7.3110	0.078 69	3273.0	3666.5	7.2589
700	0.11095	3462.1	3905.9	7.6198	0.098 47	3459.9	3903.0	7.5631	0.088 49	3457.6	3900.1	7.5122
800	0.122 87	3650.0	4141.5	7.8502	0.109 11	3648.3	4139.3	7.7942	0.098 11	3646.6	4137.1	7.7440
900	0.134 69	3843.6	4382.3	8.0647	0.119 65	3842.2	4380.6	8.0091	0.107 62	3840.7	4378.8	7.9593
1000	0.146 45	4042.9	4628.7	8.2662	0.130 13	4041.6	4627.2	8.2108	0.117 07	4040.4	4625.7	8.1612
1100	0.15817	4248.0	4880.6	8.4567	0.140 56	4246.8	4879.3	8.4015	0.126 48	4245.6	4878.0	8.3520
1200	0.169 87	4458.6	5138.1	8.6376	0.150 98	4457.5	5136.9	8.5825	0.135 87	4456.3	5135.7	8.5331
1300	0.181 56	4674.3	5400.5	8.8100	0.161 39	4673.1	5399.4	8.7549	0.145 26	4672.0	5398.2	8.7055

(*Continued*)

TABLE A.3 (SI) (*Continued*)

Properties of Superheated Steam

	$P = 6.0$ MPa (275.64)				$P = 7.0$ MPa (285.88)				$P = 8.0$ MPa (295.06)			
Sat.	0.032 44	2589.7	2784.3	5.8892	0.02737	2580.5	2772.1	5.8133	0.023 52	2569.8	2758.0	5.7432
300	0.036 16	2667.2	2884.2	6.0674	0.029 47	2632.2	2838.4	5.9305	0.024 26	2590.9	2785.0	5.7906
350	0.042 23	2789.6	3043.0	6.3335	0.035 24	2769.4	3016.0	6.2283	0.029 95	2747.7	2987.3	6.1301
400	0.04739	2892.9	3177.2	6.5408	0.039 93	2878.6	3158.1	6.4478	0.034 32	2863.8	3138.3	6.3634
450	0.052 14	2988.9	3301.8	6.7193	0.044 16	2978.0	3287.1	6.6327	0.038 17	2966.7	3272.0	6.5551
500	0.056 65	3082.2	3422.2	6.8803	0.048 14	3073.4	3410.3	6.7975	0.04175	3064.3	3398.3	6.7240
550	0.061 01	3174.6	3540.6	7.0288	0.051 95	3167.2	3530.9	6.9486	0.045 16	3159.8	3521.0	6.8778
600	0.065 25	3266.9	3658.4	7.1677	0.055 65	3260.7	3650.3	7.0894	0.048 45	3254.4	3642.0	7.0206
700	0.073 52	3453.1	3894.2	7.4234	0.062 83	3448.5	3888.3	7.3476	0.054 81	3443.9	3882.4	7.2812
800	0.081 60	3643.1	4132.7	7.6566	0.069 81	3639.5	4128.2	7.5822	0.060 97	3636.0	4123.8	7.5173
900	0.089 58	3837.8	4375.3	7.8727	0.076 69	3835.0	4371.8	7.7991	0.067 02	3832.1	4368.3	7.7351
1000	0.09749	4037.8	4622.7	8.0751	0.083 50	4035.3	4619.8	8.0020	0.073 01	4032.8	4616.9	7.9384
1100	0.105 36	4243.3	4875.4	8.2661	0.090 27	4240.9	4872.8	8.1933	0.078 96	4238.6	4870.3	8.1300
1200	0.113 21	4454.0	5133.3	8.4474	0.097 03	4451.7	5130.9	8.3747	0.084 89	4449.5	5128.5	8.3115
1300	0.121 06	4669.6	5396.0	8.6199	0.103 77	4667.3	5393.7	8.5473	0.090 80	4665.0	5391.5	8.4842

(*Continued*)

TABLE A.3 (SI) (*Continued*)

Properties of Superheated Steam

	$P=9.0$ MPa (303.40)				$P=10.0$ MPa (311.06)				$P=12.5$ MPa (327.89)			
T	v	u	h	s	v	u	h	s	v	u	h	s
Sat.	0.020 48	2557.8	2742.1	5.6772	0.018 026	2544.4	2724.7	5.6141	0.013 495	2505.1	2673.8	5.4624
325	0.023 27	2646.6	2856.0	5.8712	0.019 861	2610.4	2809.1	5.7568				
350	0.025 80	2724.4	2956.6	6.0361	0.022 42	2699.2	2923.4	5.9443	0.016 126	2624.6	2826.2	5.7118
400	0.029 93	2848.4	3117.8	6.2854	0.026 41	2832.4	3096.5	6.2120	0.020 00	2789.3	3039.3	6.0417
450	0.033 50	2955.2	3256.6	6.4844	0.029 75	2943.4	3240.9	6.4190	0.022 99	2912.5	3199.8	6.2719
500	0.036 77	3055.2	3386.1	6.6576	0.032 79	3045.8	3373.7	6.5966	0.025 60	3021.7	3341.8	6.4618
550	0.039 87	3152.2	3511.0	6.8142	0.035 64	3144.6	3500.9	6.7561	0.028 01	3125.0	3475.2	6.6290
600	0.042 85	3248.1	3633.7	6.9589	0.038 37	3241.7	3625.3	6.9029	0.030 29	3225.4	3604.0	6.7810
650	0.045 74	3343.6	3755.3	7.0943	0.041 01	3338.2	3748.2	7.0398	0.032 48	3324.4	3730.4	6.9218
700	0.048 57	3439.3	3876.5	7.2221	0.043 58	3434.7	3870.5	7.1687	0.034 60	3422.9	3855.3	7.0536
800	0.054 09	3632.5	4119.3	7.4596	0.048 59	3628.9	4114.8	7.4077	0.038 69	3620.0	4103.6	7.2965
900	0.059 50	3829.2	4364.8	7.6783	0.053 49	3826.3	4361.2	7.6272	0.042 67	3819.1	4352.5	7.5182
1000	0.064 85	4030.3	4614.0	7.8821	0.058 32	4027.8	4611.0	7.8315	0.046 58	4021.6	4603.8	7.7237
1100	0.070 16	4236.3	4867.7	8.0740	0.06312	4234.0	4865.1	8.0237	0.050 45	4228.2	4858.8	7.9165
1200	0.075 44	4447.2	5126.2	8.2556	0.067 89	4444.9	5123.8	8.2055	0.054 30	4439.3	5118.0	8.0987
1300	0.080 72	4662.7	5389.2	8.4284	0.072 65	4460.5	5387.0	8.3783	0.058 13	4654.8	5381.4	8.2717

(*Continued*)

TABLE A.3 (SI) (*Continued*)

Properties of Superheated Steam

	$P = 15.0$ MPa (342.24)				$P = 17.5$ MPa (354.75)				$P = 20.0$ MPa (365.81)			
Sat.	0.010 337	2455.5	2610.5	5.3098	0.007920	2390.2	2528.8	5.1419	0.005 834	2293.0	2409.7	4.9269
350	0.011 470	2520.4	2692.4	5.4421								
400	0.015 649	2740.7	2975.5	5.8811	0.012447	2685.0	2902.9	5.7213	0.009 942	2619.3	2818.1	5.5540
450	0.018 445	2879.5	3156.2	6.1404	0.015174	2844.2	3109.7	6.0184	0.012 695	2806.2	3060.1	5.9017
500	0.020 80	2996.6	3308.6	6.3443	0.017358	2970.3	3274.1	6.2383	0.014 768	2942.9	3238.2	6.1401
550	0.022 93	3104.7	3448.6	6.5199	0.019 288	3083.9	3421.4	6.4230	0.016 555	3062.4	3393.5	6.3348
600	0.024 91	3208.6	3582.3	6.6776	0.021 06	3191.5	3560.1	6.5866	0.018 178	3174.0	3537.6	6.5048
650	0.026 80	3310.3	3712.3	6.8224	0.022 74	3296.0	3693.9	6.7357	0.019 693	3281.4	3675.3	6.6582
700	0.028 61	3410.9	3840.1	6.9572	0.024 34	3398.7	3824.6	6.8736	0.021 13	3386.4	3809.0	6.7993
800	0.032 10	3610.9	4092.4	7.2040	0.02738	3601.8	4081.1	7.1244	0.023 85	3592.7	4069.7	7.0544
900	0.035 46	3811.9	4343.8	7.4279	0.030 31	3804.7	4335.1	7.3507	0.026 45	3797.5	4326.4	7.2830
1000	0.038 75	4015.4	4596.6	7.6348	0.033 16	4009.3	4589.5	7.5589	0.028 97	4003.1	4582.5	7.4925
1100	0.042 00	4222.6	4852.6	7.8283	0.035 97	4216.9	4846.4	7.7531	0.031 45	4211.3	4840.2	7.6874
1200	0.045 23	4433.8	5112.3	8.0108	0.038 76	4428.3	5106.6	7.9360	0.033 91	4422.8	5101.0	7.8707
1300	0.048 45	4649.1	5376.0	8.1840	0.041 54	4643.5	5370.5	8.1093	0.036 36	4638.0	5365.1	8.0442

(*Continued*)

TABLE A.3 (SI) (*Continued*)

Properties of Superheated Steam

	P = 25.0 MPa				P = 30.0 MPa				P = 35.0 MPa			
375	0.001 973 1	1798.7	1848.0	4.0320	0.001 789 2	1737.8	1791.5	3.9305	0.001 700 3	1702.9	1762.4	3.8722
400	0.006 004	2430.1	2580.2	5.1418	0.002 790	2067.4	2151.1	4.4728	0.002 100	1914.1	1987.6	4.2126
425	0.007 881	2609.2	2806.3	5.4723	0.005 303	2455.1	2614.2	5.1504	0.003 428	2253.4	2373.4	4.7747
450	0.009 162	2720.7	2949.7	5.6744	0.006 735	2619.3	2821.4	5.4424	0.004 961	2498.7	2672.4	5.1962
500	0.011123	2884.3	3162.4	5.9592	0.008 678	2820.7	3081.1	5.7905	0.006 927	2751.9	2994.4	5.6282
550	0.012 724	3017.5	3335.6	6.1765	0.010 168	2970.3	3275.4	6.0342	0.008 345	2921.0	3213.0	5.9026
600	0.014 137	3137.9	3491.4	6.3602	0.011 446	3100.5	3443.9	6.2331	0.009 527	3062.0	3395.5	6.1179
650	0.015 433	3251.6	3637.4	6.5229	0.012 596	3221.0	3598.9	6.4058	0.010 575	3189.8	3559.9	6.3010
700	0.016 646	3361.3	3777.5	6.6707	0.013 661	3335.8	3745.6	6.5606	0.011 533	3309.8	3713.5	6.4631
800	0.018 912	3574.3	4047.1	6.9345	0.015 623	3555.5	4024.2	6.8332	0.013 278	3536.7	4001.5	6.7450
900	0.021 045	3783.0	4309.1	7.1680	0.017448	3768.5	4291.9	7.0718	0.014 883	3754.0	4274.9	6.9886
1000	0.023 10	3990.9	4568.5	7.3802	0.019 196	3978.8	4554.7	7.2867	0.016 410	3966.7	4541.1	7.2064
1100	0.025 12	4200.2	4828.2	7.5765	0.020 903	4189.2	4816.3	7.4845	0.017 895	4178.3	4804.6	7.4057
1200	0.027 11	4412.0	5089.9	7.7605	0.022 589	4401.3	5079.0	7.6692	0.019 360	4390.7	5068.3	7.5910
1300	0.029 10	4626.9	5354.4	7.9342	0.024 266	4616.0	5344.0	7.8432	0.020 815	4605.1	5333.6	7.7653

(Continued)

TABLE A.3 (SI) (*Continued*)

Properties of Superheated Steam

	P = 40.0 *MPa*				*P* = 50.0 *MPa*				*P* = 60.0 *MPa*			
375	0.001 640 7	1677.1	1742.8	3.8290	0.001 559 4	1638.6	1716.6	3.7639	0.001 502 8	1609.4	1699.5	3.7141
400	0.001 907 7	1854.6	1930.9	4.1135	0.001 730 9	1788.1	1874.6	4.0031	0.001 633 5	1745.4	1843.4	3.9318
425	0.002 532	2096.9	2198.1	4.5029	0.002 007	1959.7	2060.0	4.2734	0.001 816 5	1892.7	2001.7	4.1626
450	0.003 693	2365.1	2512.8	4.9459	0.002 486	2159.6	2284.0	4.5884	0.002 085	2053.9	2179.0	4.4121
500	0.005 622	2678.4	2903.3	5.4700	0.003 892	2525.5	2720.1	5.1726	0.002 956	2390.6	2567.9	4.9321
550	0.006 984	2869.7	3149.1	5.7785	0.005 118	2763.6	3019.5	5.5485	0.003 956	2658.8	2896.2	5.3441
600	0.008 094	3022.6	3346.4	6.0114	0.006 112	2942.0	3247.6	5.8178	0.004 834	2861.1	3151.2	5.6452
650	0.009 063	3158.0	3520.6	6.2054	0.006 966	3093.5	3441.8	6.0342	0.005 595	3028.8	3364.5	5.8829
700	0.009 941	3283.6	3681.2	6.3750	0.007 727	3230.5	3616.8	6.2189	0.006 272	3177.2	3553.5	6.0824
800	0.011 523	3517.8	3978.7	6.6662	0.009 076	3479.8	3933.6	6.5290	0.007 459	3441.5	3889.1	6.4109
900	0.012 962	3739.4	4257.9	6.9150	0.010 283	3710.3	4224.4	6.7882	0.008 508	3681.0	4191.5	6.6805
1000	0.014 324	3954.6	4527.6	7.1356	0.011 411	3930.5	4501.1	7.0146	0.009 480	3906.4	4475.2	6.9127
1100	0.015 642	4167.4	4793.1	7.3364	0.012 496	4145.7	4770.5	7.2184	0.010 409	4124.1	4748.6	7.1195
1200	0.016 940	4380.1	5057.7	7.5224	0.013 561	4359.1	5037.2	7.4058	0.011 317	4338.2	5017.2	7.3083
1300	0.018 229	4594.3	5323.5	7.6969	0.014 616	4572.8	5303.6	7.5808	0.012 215	4551.4	5284.3	7.4837

TABLE A.4 (SI)

Properties of Compressed Liquid (Water)

	P = 5 MPa (263.99)				P = 10 MPa (311.06)				P = 15 MPa (342.24)			
T	v	u	h	s	v	u	h	s	v	u	h	s
Sat.	0.001 285 9	1147.8	1154.2	2.9202	0.001 452 4	1393.0	1407.6	3.3596	0.001 658 1	1585.6	1610.5	3.6848
0	0.000 997 7	0.04	5.04	0.0001	0.000 995 2	0.09	10.04	0.0002	0.000 992 8	0.15	15.05	0.0004
20	0.000 999 5	83.65	88.65	0.2956	0.000 9972	83.36	93.33	0.2945	0.000 995 0	83.06	97.99	0.2934
40	0.001 005 6	166.95	171.97	0.5705	0.001 003 4	166.35	176.38	0.5686	0.001 001 3	165.76	180.78	0.5666
60	0.001 014 9	250.23	255.30	0.8285	0.001 012 7	249.36	259.49	0.8258	0.001 010 5	248.51	263.67	0.8232
80	0.001 026 8	333.72	338.85	1.0720	0.001 024 5	332.59	342.83	1.0688	0.001 022 2	331.48	346.81	1.0656
100	0.001 041 0	417.52	422.72	1.3030	0.001 038 5	416.12	426.50	1.2992	0.001 036 1	414.74	430.28	1.2955
120	0.001 057 6	501.80	507.09	1.5233	0.001 054 9	500.08	510.64	1.5189	0.001 052 2	498.40	514.19	1.5145
140	0.001 076 8	586.76	592.15	1.7343	0.001 073 7	584.68	595.42	1.7292	0.001 070 7	582.66	598.72	1.7242
160	0.001 098 8	672.62	678.12	1.9375	0.001 095 3	670.13	681.08	1.9317	0.001 091 8	667.71	684.09	1.9260
180	0.001 124 0	759.63	765.25	2.1341	0.001 119 9	756.65	767.84	2.1275	0.001 115 9	753.76	770.50	2.1210
200	0.001 153 0	848.1	853.9	2.3255	0.001 148 0	844.5	856.0	2.3178	0.001 143 3	841.0	858.2	2.3104
220	0.001 186 6	938.4	944.4	2.5128	0.001 180 5	934.1	945.9	2.5039	0.001 1748	929.9	947.5	2.4953
240	0.001 226 4	1031.4	1037.5	2.6979	0.001 218 7	1026.0	1038.1	2.6872	0.001 211 4	1020.8	1039.0	2.6771
260	0.001 274 9	1127.9	1134.3	2.8830	0.001 264 5	1121.1	1133.7	2.8699	0.001 255 0	1114.6	1133.4	2.8576
280					0.001 321 6	1220.9	1234.1	3.0548	0.001 308 4	1212.5	1232.1	3.0393
300					0.001 3972	1328.4	1342.3	3.2469	0.001 377 0	1316.6	1337.3	3.2260
320									0.001 472 4	1431.1	1453.2	3.4247
340									0.001 631 1	1567.5	1591.9	3.6546

(Continued)

TABLE A.4 (SI) (*Continued*)

Properties of Compressed Liquid (Water)

	$P = 20$ MPa (365.81)				$P = 30$ MPa				$P = 50$ MPa			
T	v	u	h	s	v	u	h	s	v	u	h	s
Sat.	0.002 036	1785.6	1826.3	4.0139								
0	0.000 990 4	0.19	20.01	0.0004	0.000 985 6	0.25	29.82	0.0001	0.000 976 6	0.20	49.03	0.0014
20	0.000 992 8	82.77	102.62	0.2923	0.000 988 6	82.17	111.84	0.2899	0.000 980 4	81.00	130.02	0.2848
40	0.000 999 2	165.17	185.16	0.5646	0.000 995 1	164.04	193.89	0.5607	0.000 9872	161.86	211.21	0.5527
60	0.001 008 4	247.68	267.85	0.8206	0.001 004 2	246.06	276.19	0.8154	0.000 996 2	242.98	292.79	0.8052
80	0.001 019 9	330.40	350.80	1.0624	0.001 015 6	328.30	358.77	1.0561	0.001 0073	324.34	374.70	1.0440
100	0.001 033 7	413.39	434.06	1.2917	0.001 029 0	410.78	441.66	1.2844	0.001 020 1	405.88	456.89	1.2703
120	0.001 049 6	496.76	517.76	1.5102	0.001 044 5	493.59	524.93	1.5018	0.001 034 8	487.65	539.39	1.4857
140	0.001 067 8	580.69	602.04	1.7193	0.001 062 1	576.88	608.75	1.7098	0.001 051 5	569.77	622.35	1.6915
160	0.001 088 5	665.35	687.12	1.9204	0.001 082 1	660.82	693.28	1.9096	0.001 070 3	652.41	705.92	1.8891
180	0.001 112 0	750.95	773.20	2.1147	0.001 104 7	745.59	778.73	2.1024	0.001 091 2	735.69	790.25	2.0794
200	0.001 138 8	837.7	860.5	2.3031	0.001 130 2	831.4	865.3	2.2893	0.001 114 6	819.7	875.5	2.2634
220	0.001 169 3	925.9	949.3	2.4870	0.001 159 0	918.3	953.1	2.4711	0.001 140 8	904.7	961.7	2.4419
240	0.001 204 6	1016.0	1040.0	2.6674	0.001 192 0	1006.9	1042.6	2.6490	0.001 170 2	990.7	1049.2	2.6158
260	0.001 246 2	1108.6	1133.5	2.8459	0.001 230 3	1097.4	1134.3	2.8243	0.001 203 4	1078.1	1138.2	2.7860
280	0.001 296 5	1204.7	1230.6	3.0248	0.001 275 5	1190.7	1229.0	2.9986	0.001 241 5	1167.2	1229.3	2.9537
300	0.001 359 6	1306.1	1333.3	3.2071	0.001 330 4	1287.9	1327.8	3.1741	0.001 286 0	1258.7	1323.0	3.1200
320	0.001 443 7	1415.7	1444.6	3.3979	0.001 399 7	1390.7	1432.7	3.3539	0.001 338 8	1353.3	1420.2	3.2868
340	0.001 568 4	1539.7	1571.0	3.6075	0.001 492 0	1501.7	1546.5	3.5426	0.001 403 2	1452.0	1522.1	3.4557
360	0.001 822 6	1702.8	1739.3	3.8772	0.001 626 5	1626.6	1675.4	3.7494	0.001 483 8	1556.0	1630.2	3.6291
380					0.001 869 1	1781.4	1837.5	4.0012	0.001 588 4	1667.2	1746.6	3.8101

TABLE A.5

Gas Constant Values

Gas	Chemical Formula	Molecular Weight (kg/kmol)	English Units R $\frac{\text{ft. lb.}}{\text{lb.°R}}$	English Units c_p $\frac{\text{Btu}}{\text{lb.°R}}$	English Units c_p $\frac{\text{Btu}}{\text{lb.°R}}$	k=Cp/Cv c_v	SI Units R $\frac{\text{kJ}}{\text{kg·K}}$	SI Units c_p $\frac{\text{kJ}}{\text{kg·K}}$	SI Units c_p $\frac{\text{kJ}}{\text{kg·K}}$
Acetylene	C_2H_2	26.02	59.39	0.361	0.285	1.27	0.31955	1.5116	1.1933
Air		28.97	53.34	0.240	0.171	1.40	0.28700	1.0052	0.7180
Ammonia	NH_3	17.024	90.7	0.52	0.404	1.29	0.48802	2.1773	1.6916
Argon	Ar	39.94	38.68	0.124	0.075	1.667	0.20813	0.5207	0.3124
Butane	C_4H_{10}	58.12	26.58	0.415	0.381	1.09	0.14304	1.7164	1.5734
Carbon dioxide	CO_2	44.01	35.10	0.202	0.157	1.29	0.18892	0.8464	0.6573
Carbon monoxide	CO	28.01	55.16	0.249	0.178	1.40	0.29683	1.0411	0.7441
Ethane	C_2H_6	30.07	51.38	0.427	0.361	1.18	0.27650	1.7662	1.4897
Ethylene	C_2H_4	28.052	55.07	0.411	0.340	1.21	0.29637	1.5482	1.2518
Helium	He	4.003	386.0	1.25	0.753	1.667	2.07703	5.1926	3.1156
Hydrogen	H_2	2.016	766.4	3.420	2.434	1.405	4.12418	14.3193	10.1919
Methane	CH_4	16.04	96.35	0.532	0.403	1.32	0.51835	2.2537	1.7354
Nitrogen	N_2	28.016	55.15	0.248	0.177	1.40	0.29680	1.0404	0.7434
Octane	C_8H_{18}	114.14	13.53	0.409	0.392	1.044	0.07279	1.7113	1.6385
Oxygen	O_2	32.000	48.28	0.219	0.157	1.39	0.25983	0.9190	0.6590
Propane	C_3H_8	44.094	35.04	0.407	0.362	1.124	0.18855	1.6794	1.4909
Water vapor	H_2O	18.016	85.76	0.445	0.335	1.33	0.46152	1.8649	1.4031

TABLE A.6

Critical Constants

Substance	Formula	Molecular Weight	Temperature K	Temperature °R	Pressure atm	Pressure lbf/in.^{2}	Volume ft.^{3}/lb.mole
Ammonia	NH_3	17.03	405.5	729.8	111.3	1636	1.16
Argon	A	39.944	151	272	48.0	705	1.20
Bromine	Br_2	159.832	584	1052	102	1500	2.17
Carbon dioxide	CO_2	44.01	304.2	547.5	72.9	1071	1.51
Carbon monoxide	CO	28.01	133	240	34.5	507	1.49
Chlorine	Cl_2	70.914	417	751	76.1	1120	1.99
Deuterium (normal)	D_2	4.00	38.4	69.1	16.4	241	
Helium	He	4.003	5.3	9.5	2.26	33.2	0.926
Helium[3]	He	3.00	3.34	6.01	1.15	16.9	
Hydrogen (normal)	H_2	2.016	33.3	59.9	12.8	188.1	1.04
Krypton	Kr	83.7	209.4	376.9	54.3	798	1.48
Neon	Ne	20.183	44.5	80.1	26.9	395	0.668
Nitrogen	N_2	28.016	126.2	227.1	33.5	492	1.44
Nitrous oxide	N_2O	44.02	309.7	557.4	71.7	1054	1.54
Oxygen	O_2	32.00	154.8	278.6	50.1	736	1.25
Sulfur dioxide	SO_2	64.06	430.7	775.2	77.8	1143	1.95
Water	H_2O	18.016	647.4	1165.3	218.3	3208	0.90
Xenon	Xe	131.3	289.75	521.55	58.0	852	1.90
Benzene	C_6H_6	78.11	562	1012	48.6	714	4.17
n-Butane	C_4H_{10}	58.120	425.2	765.2	37.5	551	4.08
Carbon tetrachloride	CCl_4	153.84	556.4	1001.5	45.0	661	4.42
Chloroform	$CHCl_3$	119.39	536.6	965.8	54.0	794	3.85
Dichlorodifluoromethane	CCl_2F_2	120.92	384.7	692.4	39.6	582	3.49
Dichlorofluoromethane	$CHC1_2F$	102.93	451.7	813.0	51.0	749	3.16
Ethane	C_2H_5	30.068	305.5	549.8	48.2	708	2.37
Ethyl alcohol	C_2H_5OH	46.07	516.0	929.0	63.0	926	2.68
Ethylene	C_2H_4	28.052	282.4	508.3	50.5	742	1.99
n-Hexane	C_6H_{14}	86.172	507.9	914.2	29.9	439	5.89
Methane	CH_4	16.042	191.1	343.9	45.8	673	1.59
Methyl alcohol	CH_3OH	32.04	513.2	923.7	78.5	1154	1.89
Methyl chloride	CH_3Cl	50.49	416.3	749.3	65.9	968	2.29
Propane	C_3H_8	44.094	370.0	665.9	42.0	617	3.20
Propene	C_3H_6	42.078	365.0	656.9	45.6	670	2.90
Propyne	C_3H_4	40.062	401	722	52.8	776	
Trichlorofluoromethane	CCl_3F	137.38	471.2	848.1	43.2	635	3.97

TABLE A.7

Ideal Gas Properties of Air

T, °R	h, Btu/lb	p_r	u, Btu/lb	v_r	s° Btu/lb. °R
200	47.67	0.04320	33.96	1714.9	0.36303
220	52.46	0.06026	37.38	1352.5	0.38584
240	57.25	0.08165	40.80	1088.8	0.40666
260	62.03	0.10797	44.21	892.0	0.42582
280	66.82	0.13986	47.63	741.6	0.44356
300	71.61	0.17795	51.04	624.5	0.46007
320	76.40	0.22290	54.46	531.8	0.47550
340	81.18	0.27545	57.87	457.2	0.49002
360	85.97	0.3363	61.29	396.6	0.50369
380	90.75	0.4061	64.70	346.6	0.51663
400	95.53	0.4858	68.11	305.0	0.52890
420	100.32	0.5760	71.52	270.1	0.54058
440	105.11	0.6776	74.93	240.6	0.55172
460	109.90	0.7913	78.36	215.33	0.56235
480	114.69	0.9182	81.77	193.65	0.57255
500	119.48	1.0590	85.20	174.90	0.58233
520	124.27	1.2147	88.62	158.58	0.59173
540	129.06	1.3860	92.04	144.32	0.60078
560	133.86	1.5742	95.47	131.78	0.60950
580	138.66	1.7800	98.90	120.70	0.61793
600	143.47	2.005	102.34	110.88	0.62607
620	148.28	2.249	105.78	102.12	0.63395
640	153.09	2.514	109.21	94.30	0.64159
660	157.92	2.801	112.67	87.27	0.64902
680	162.73	3.111	116.12	80.96	0.65621
700	167.56	3.446	119.58	75.25	0.66321
720	172.39	3.806	123.04	70.07	0.67002
740	177.23	4.193	126.51	65.38	0.67665
760	182.08	4.607	129.99	61.10	0.68312
780	186.94	5.051	133.47	57.20	0.68942
800	191.81	5.526	136.97	53.63	0.69558
820	196.69	6.033	140.47	50.35	0.70160
840	201.56	6.573	143.98	47.34	0.70747
860	206.46	7.149	147.50	44.57	0.71323
880	211.35	7.761	151.02	42.01	0.71886
900	216.26	8.411	154.57	39.64	0.72438
920	221.18	9.102	158.12	37.44	0.72979
940	226.11	9.834	161.68	35.41	0.73509
960	231.06	10.610	165.26	33.52	0.74030
980	236.02	11.430	168.83	31.76	0.74540
1000	240.98	12.298	172.43	30.12	0.75042

(Continued)

TABLE A.7 (*Continued*)

Ideal Gas Properties of Air

T, °R	h, Btu/lb	p_r	u, Btu/lb	v_r	$s°$ Btu/lb. °R
1020	245.97	13.215	176.04	28.59	0.75536
1040	250.95	14.182	179.66	27.17	0.76019
1060	255.96	15.203	183.29	25.82	0.76496
1080	260.97	16.278	186.93	24.58	0.76964
1100	265.99	17.413	190.58	23.40	0.77426
1120	271.03	18.604	194.25	22.30	0.77880
1140	276.08	19.858	197.94	21.27	0.78326
1160	281.14	21.18	201.63	20.293	0.78767
1180	286.21	22.56	205.33	19.377	0.79201
1200	291.30	24.01	209.05	18.514	0.79628
1220	296.41	25.53	212.78	17.700	0.80050
1240	301.52	27.13	216.53	16.932	0.80466
1260	306.65	28.80	220.28	16.205	0.80876
1280	311.79	30.55	224.05	15.518	0.81280
1300	316.94	32.39	227.83	14.868	0.81680
1320	322.11	34.31	231.63	14.253	0.82075
1340	327.29	36.31	235.43	13.670	0.82464
1360	332.48	38.41	239.25	13.118	0.82848
1380	337.68	40.59	243.08	12.593	0.83229
1400	342.90	42.88	246.93	12.095	0.83604
1420	348.14	45.26	250.79	11.622	0.83975
1440	353.37	47.75	254.66	11.172	0.84341
1460	358.63	50.34	258.54	10.743	0.84704
1480	363.89	53.04	262.44	10.336	0.85062
1500	369.17	55.86	266.34	9.948	0.85416
1520	374.47	58.78	270.26	9.578	0.85767
1540	379.77	61.83	274.20	9.226	0.86113
1560	385.08	65.00	278.13	8.890	0.86456
1580	390.40	68.30	282.09	8.569	0.86794
1600	395.74	71.73	286.06	8.263	0.87130
1620	401.09	75.29	290.04	7.971	0.87462
1640	406.45	78.99	294.03	7.691	0.87791
1660	411.82	82.83	298.02	7.424	0.88116
1680	417.20	86.82	302.04	7.168	0.88439
1700	422.59	90.95	306.06	6.924	0.88758
1720	428.00	95.24	310.09	6.690	0.89074
1740	433.41	99.69	314.13	6.465	0.89387
1760	438.83	104.30	318.18	6.251	0.89697
1780	444.26	109.08	322.24	6.045	0.90003
1800	449.71	114.03	326.32	5.847	0.90308
1820	455.17	119.16	330.40	5.658	0.90609

(*Continued*)

TABLE A.7 (*Continued*)

Ideal Gas Properties of Air

T, °R	h, Btu/lb	p_r	u, Btu/lb	v_r	$s°$ Btu/lb. °R
1840	460.63	124.47	334.50	5.476	0.90908
1860	466.12	129.95	338.61	5.302	0.91203
1880	471.60	135.64	342.73	5.134	0.91497
1900	477.09	141.51	346.85	4.974	0.91788
1920	482.60	147.59	350.98	4.819	0.92076
1940	488.12	153.87	355.12	4.670	0.92362
1960	493.64	160.37	359.28	4.527	0.92645
1980	499.17	167.07	363.43	4.390	0.92926
2000	504.71	174.00	367.61	4.258	0.93205
2020	510.26	181.16	371.79	4.130	0.93481
2040	515.82	188.54	375.98	4.008	0.93756
2060	521.39	196.16	380.18	3.890	0.94026
2080	526.97	204.02	384.39	3.777	0.94296
2100	532.55	212.1	388.60	3.667	0.94564
2120	538.15	220.5	392.83	3.561	0.94829
2140	543.74	229.1	397.05	3.460	0.95092
2160	549.35	238.0	401.29	3.362	0.95352
2180	554.97	247.2	405.53	3.267	0.95611
2200	560.59	256.6	409.78	3.176	0.95868
2220	566.23	266.3	414.05	3.088	0.96123
2240	571.86	276.3	418.31	3.003	0.96376
2260	577.51	286.6	422.59	2.921	0.96626
2280	583.16	297.2	426.87	2.841	0.96876
2300	588.82	308.1	431.16	2.765	0.97123
2320	594.49	319.4	435.46	2.691	0.97369
2340	600.16	330.9	439.76	2.619	0.97611
2360	605.84	342.8	444.07	2.550	0.97853
2380	611.53	355.0	448.38	2.483	0.98092
2400	617.22	367.6	452.70	2.419	0.98331

Source: Abridged from Table 1 in Joseph H. Keenan and Joseph Kaye: *Gas Tables*. 1948.

TABLE A.7 (SI)

Ideal Gas Properties of Air

T	h	P_r	u	v_r	$s°$
		SI Units			
		T(K), h and u(kJ/kg), s°(kJ/kg·K)			
200	199.97	0.3363	142.56	1,707	1.29559
210	209.97	0.3987	149.69	1,512	1.34444
220	219.97	0.4690	156.82	1,346	1.39105
230	230.02	0.5477	164.00	1,205	1.43557
240	240.02	0.6355	171.13	1,084	1.47824
250	250.05	0.7329	178.28	979	1.51917
260	260.09	0.8405	185.45	887.8	1.55848
270	270.11	0.9590	192.60	808.0	1.59634
280	280.13	1.0889	199.75	738.0	1.63279
285	285.14	1.1584	203.33	706.1	1.65055
290	290.16	1.2311	206.91	676.1	1.66802
295	295.17	1.3068	210.49	647.9	1.68515
300	300.19	1.3860	214.07	621.2	1.70203
305	305.22	1.4686	217.67	596.0	1.71865
310	310.24	1.5546	221.25	572.3	1.73498
315	315.27	1.6442	224.85	549.8	1.75106
320	320.29	1.7375	228.42	528.6	1.76690
325	325.31	1.8345	232.02	508.4	1.78249
330	330.34	1.9352	235.61	489.4	1.79783
340	340.42	2.149	242.82	454.1	1.82790
350	350.49	2.379	250.02	422.2	1.85708
360	360.58	2.626	257.24	393.4	1.88543
370	370.67	2.892	264.46	367.2	1.91313
380	380.77	3.176	271.69	343.4	1.94001
390	390.88	3.481	278.93	321.5	1.96633
400	400.98	3.806	286.16	301.6	1.99194
410	411.12	4.153	293.43	283.3	2.01699
420	421.26	4.522	300.69	266.6	2.04142
430	431.43	4.915	307.99	251.1	2.06533
440	441.61	5.332	315.30	236.8	2.08870
450	451.80	5.775	322.62	223.6	2.11161
460	462.02	6.245	329.97	211.4	2.13407
470	472.24	6.742	337.32	200.1	2.15604
480	482.49	7.268	344.70	189.5	2.17760
490	492.74	7.824	352.08	179.7	2.19876
500	503.02	8.411	359.49	170.6	2.21952
510	513.32	9.031	366.92	162.1	2.23993
520	523.63	9.684	374.36	154.1	2.25997

(Continued)

TABLE A.7 (SI)
Ideal Gas Properties of Air

T	h	P_r	u	v_r	s°
		SI Units			
		T(K), h and u(kJ/kg), s°(kJ/kg·K)			
530	533.98	10.37	381.84	146.7	2.27967
540	544.35	11.10	389.34	139.7	2.29906
550	554.74	11.86	396.86	133.1	2.31809
560	565.17	12.66	404.42	127.0	2.33685
570	575.59	13.50	411.97	121.2	2.35531
580	586.04	14.38	419.55	115.7	2.37348
590	596.52	15.31	427.15	110.6	2.39140
600	607.02	16.28	434.78	105.8	2.40902
610	617.53	17.30	442.42	101.2	2.42644
620	628.07	18.36	450.09	96.92	2.44356
630	638.63	19.84	457.78	92.84	2.46048
640	649.22	20.64	465.50	88.99	2.47716
650	659.84	21.86	473.25	85.34	2.49364
660	670.47	23.13	481.01	81.89	2.50985
670	681.14	24.46	488.81	78.61	2.52589
680	691.82	25.85	496.62	75.50	2.54175
690	702.52	27.29	504.45	72.56	2.55731
700	713.27	28.80	512.33	69.76	2.57277
710	724.04	30.38	520.23	67.07	2.58810
720	734.82	32.02	528.14	64.53	2.60319
730	745.62	33.72	536.07	62.13	2.61803
740	756.44	35.50	544.02	59.82	2.63280
750	767.29	37.35	551.99	57.63	2.64737
760	778.18	39.27	560.01	55.54	2.66176
770	789.11	41.31	568.07	53.39	2.67595
780	800.03	43.35	576.12	51.64	2.69013
790	810.99	45.55	584.21	49.86	2.70400
800	821.95	47.75	592.30	48.08	2.71787
820	843.98	52.59	608.59	44.84	2.74504
840	866.08	57.60	624.95	41.85	2.77170
860	888.27	63.09	641.40	39.12	2.79783
880	910.56	68.98	657.95	36.61	2.82344
900	932.93	75.29	674.58	34.31	2.84856
920	955.38	82.05	691.28	32.18	2.87324
940	977.92	89.28	708.08	30.22	2.89748
960	1,000.55	97.00	725.02	28.40	2.92128
980	1,023.25	105.2	741.98	26.73	2.94468
1,000	1,046.04	114.0	758.94	25.17	2.96770

(Continued)

TABLE A.7 (SI)

Ideal Gas Properties of Air

T	h	P_r	u	v_r	$s°$
SI Units					
T(K), h and u(kJ/kg), $s°$(kJ/kg·K)					
1,020	1,068.89	123.4	776.10	23.72	2.99034
1,040	1,091.85	133.3	793.36	22.39	3.01260
1,060	1,114.86	143.9	810.62	21.14	3.03449
1,080	1,137.89	155.2	827.88	19.98	3.05608
1,100	1,161.07	167.1	845.33	18.896	3.07732
1,120	1,184.28	179.7	862.79	17.886	3.09825
1,140	1,207.57	193.1	880.35	16.946	3.11883
1,160	1,230.92	207.2	897.91	16.064	3.13916
1,180	1,254.34	222.2	915.57	15.241	3.15916
1,200	1,277.79	238.0	933.33	14.470	3.17888
1,220	1,301.31	254.7	951.09	13.747	3.19834
1,240	1,324.93	272.3	968.95	13.069	3.21751
1,260	1,348.55	290.8	986.90	12.435	3.23638
1,280	1,372.24	310.4	1,004.76	11.835	3.25510
1,300	1,395.97	330.9	1,022.82	11.275	3.27345
1,320	1,419.76	352.5	1,040.88	10.747	3.29160
1,340	1,443.60	375.3	1,058.94	10.247	3.30959
1,360	1,467.49	399.1	1,077.10	9.780	3.32724
1,380	1,491.44	424.2	1,095.26	9.337	3.34474
1,400	1,515.42	450.5	1,113.52	8.919	3.36200
1,420	1,539.44	478.0	1,131.77	8.526	3.37901
1,440	1,563.51	506.9	1,150.13	8.153	3.39586
1,460	1,587.63	537.1	1,168.49	7.801	3.41247
1,480	1,611.79	568.8	1,186.95	7.468	3.42892
1,500	1,635.97	601.9	1,205.41	7.152	3.44516
1,520	1,660.23	636.5	1,223.87	6.854	3.46120
1,540	1,684.51	672.8	1,242.43	6.569	3.47712
1,560	1,708.82	710.5	1,260.99	6.301	3.49276
1,580	1,733.17	750.0	1,279.65	6.046	3.50829
1,600	1,757.57	791.2	1,298.30	5.804	3.52364
1,620	1,782.00	834.1	1,316.96	5.574	3.53879
1,640	1,806.46	878.9	1,335.72	5.355	3.55381
1,660	1,830.96	925.6	1,354.48	5.147	3.56867
1,680	1,855.50	974.2	1,373.24	4.949	3.58335
1,700	1,880.1	1,025	1,392.7	4.761	3.5979

(Continued)

TABLE A.7 (SI)

Ideal Gas Properties of Air

T	h	P_r	u	v_r	$s°$
			SI Units		
			T(K), h and u(kJ/kg), s°(kJ/kg·K)		
1,750	1,941.6	1,161	1,439.8	4.328	3.6336
1,800	2,003.3	1,310	1,487.2	3.944	3.6684
1,850	2,065.3	1,475	1,534.9	3.601	3.7023
1,900	2,127.4	1,655	1,582.6	3.295	3.7354
1,950	2,189.7	1,852	1,630.6	3.022	3.7677
2,000	2,252.1	2,068	1,678.7	2.776	3.7994
2,050	2,314.6	2,303	1,726.8	2.555	3.8303
2,100	2,377.4	2,559	1,775.3	2.356	3.8605
2,150	2,440.3	2,837	1,823.8	2.175	3.8901
2,200	2,503.2	3,138	1,872.4	2.012	3.9191
2,250	2,566.4	3,464	1,921.3	1.864	3.9474

Source: Moran, M.J., and Shapiro, H.N., Fundamentals of Engineering Thermodynamics, 3rd. ed., Wiley, New York, 1995, as based on Keenan, J.H., and Kaye, J., Gas Tables, Wiley, New York, 1945.

Index

B

C

E

J

O

S

T

U

X

Y

Z

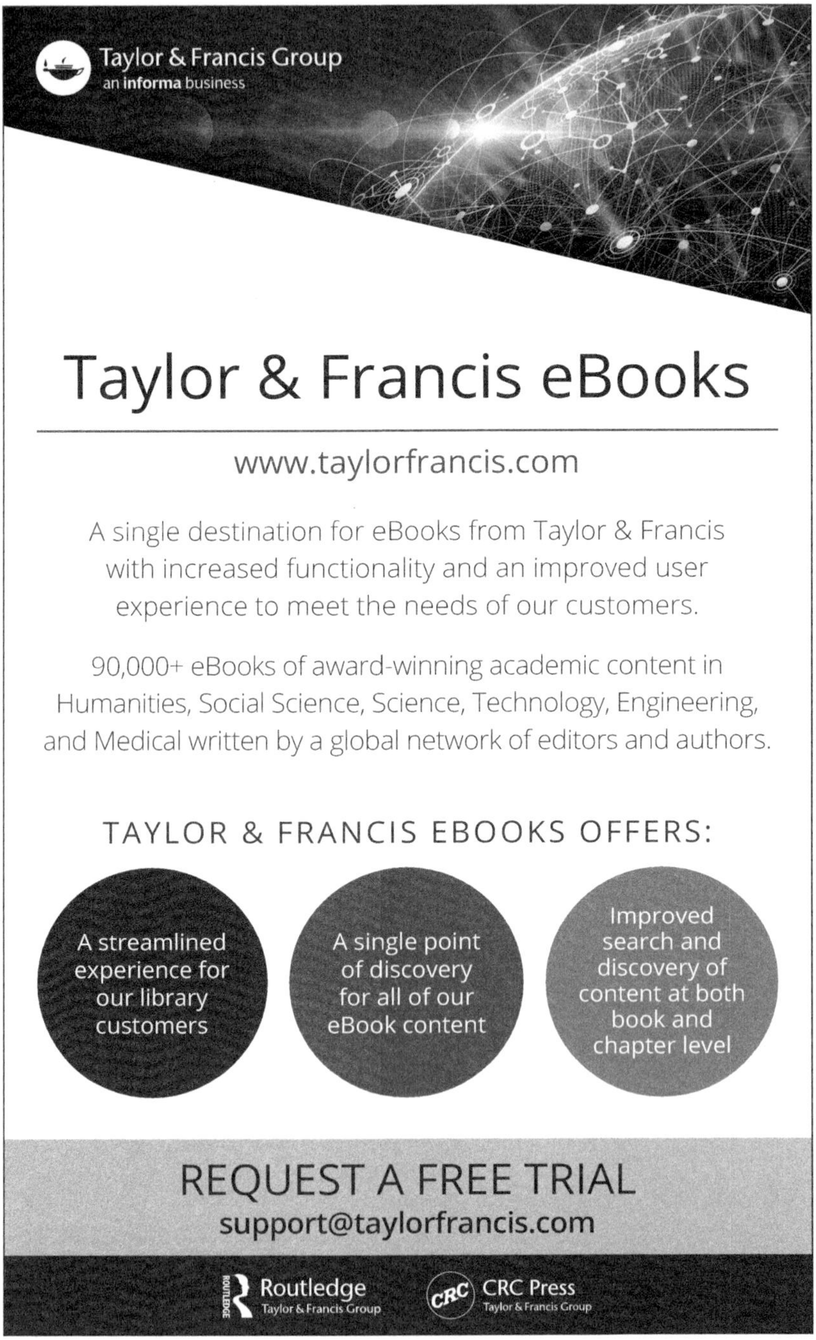

Taylor & Francis Group
an informa business
Taylor & Francis eBooks
www.taylorfrancis.com
A single destination for eBooks from Taylor & Francis with increased functionality and an improved user experience to meet the needs of our customers.
90,000+ eBooks of award-winning academic content in Humanities, Social Science, Science, Technology, Engineering, and Medical written by a global network of editors and authors.
TAYLOR & FRANCIS EBOOKS OFFERS:
A streamlined experience for our library customers
A single point of discovery for all of our eBook content
Improved search and discovery of content at both book and chapter level
REQUEST A FREE TRIAL
support@taylorfrancis.com
Routledge
Taylor & Francis Group
CRC Press
Taylor & Francis Group